Dec./1983.

Handbook of
GEOLOGY IN CIVIL ENGINEERING

McGraw-Hill Ryerson

takes pleasure in sending you this copy for review

Publication Date:

Published

Price:

$94.50

Please send a copy of your review to:

Barbara Meuleman
Trade Promotion
330 Progress Ave., Scarborough, Ont. M1P 2Z5
(416) 293-1911

Handbook of GEOLOGY IN CIVIL ENGINEERING

Nature, to be commanded, must be obeyed.
Francis Bacon

Robert F. Legget
Consultant
Ottawa, Canada

Paul F. Karrow
Professor, Department of Earth Sciences
University of Waterloo
Ontario, Canada

McGraw-Hill Book Company

New York   St. Louis   San Francisco   Auckland
Bogotá   Hamburg   Johannesburg   London   Madrid
Mexico   Montreal   New Delhi   Panama   Paris
São Paulo   Singapore   Sydney   Tokyo   Toronto

Endpapers: The map, reproduced here by permission, shows the surficial geology of part of the city of St. Paul, Minnesota, as depicted on one of a set of seven maps of the enginnering geology of the Twin Cities, published jointly by the U.S. and Minnesota Geological Surveys; details are given in reference 13.29

Frontispiece: This portrait of William Smith, the "Father of British Geology" and also an eminent civil engineer, is reproduced as a tribute to his memory and to his life's work, which after more than a century is still an inspiration to geologists and civil engineers in many countries of the world. (Portrait reproduced by permission of the Council of the Geological Society of London; signature by permission of Mr. T. Sheppard, Author of *William Smith: His Maps and Memoirs*.)

Library of Congress Cataloging in Publication Data
Legget, Robert Ferguson.
   Handbook of geology in civil engineering.

   Enl. ed. of: Geology and engineering. 2nd ed. 1962
   Includes bibliographies.
     1.Engineering geology. I.Karrow, Paul
Frederick, 1930–000 II.Title.
TA705.L4 1982   624.1'51   81-17218
ISBN 0-07-037061-3   AACR2

Copyright © 1983 by McGraw-Hill, Inc. All rights reserved. Printed in the United States of America. Except as permitted under the United States Copyright Act of 1976, no part of this publication may be reproduced or distributed in any form or by any means, or stored in a data base or retrieval system, without the prior written permission of the publisher.

1234567890   DOCDOC   89876543

Portions of this book were formerly published under the title *Geology and Engineering* by Robert F. Legget. Material for this book was also taken from *Cities and Geology* by the same author.

ISBN 0-07-037061-3

The editors for this book were Joan Zseleczky, Susan Thomas, and Ursula Smith; the designer was Mark E. Safran; and the production supervisor was Thomas G. Kowalczyk. It was set in Century Schoolbook by University Graphics, Inc.

Printed and bound by R. R. Donnelley & Sons Company.

CONTENTS

Preface ix
Acknowledgments xiii

PART 1 GEOLOGICAL BACKGROUND

1. The Civil Engineer and Geology 1-3
2. Some Outstanding Examples of Geology in Civil Engineering 2-1
3. Geology: An Introduction 3-1
4. Rocks: Their Nature and Structure 4-1
5. Soils: Their Origin and Deposition 5-1
6. Soil Mechanics and Geology 6-1
7. Rock Mechanics and Geology 7-1
8. Groundwater 8-1
9. Climate 9-1

PART 2 PRELIMINARY STUDIES

10. Site Investigation 10-3
11. Subsurface Investigation 11-1
12. Applied Geophysics 12-1
13. Geology beneath Cities 13-1
14. Geological Maps 14-1
15. Materials: Soil and Rock 15-1
16. Materials: Concrete 16-1
17. Materials: Manufactured Products 17-1

PART 3 CIVIL ENGINEERING WORKS

18. Drainage 18-3
19. Open Excavation 19-1
20. Tunnels 20-1
21. Underground Space 21-1
22. Building Foundations 22-1
23. Powerhouse Foundations 23-1
24. Bridge Foundations 24-1
25. Dam Foundations 25-1
26. Reservoirs and Catchment Areas 26-1
27. Grouting 27-1
28. Water Supply 28-1
29. Canals 29-1
30. Roads 30-1
31. Railways 31-1
32. Airfields 32-1
33. River-Training Works 33-1
34. Marine Works 34-1
35. Land Reclamation 35-1
36. Problems of Cold Regions 36-1

PART 4 SPECIAL PROBLEMS

37. Land Subsidence 37-3
38. Sinkholes 38-1
39. Landslides 39-1
40. Rockfalls 40-1
41. Erosion and Sedimentation 41-1
42. Problem Soils 42-1
43. Problem Rocks 43-1
44. Faults and Joints 44-1

45. Volcanoes and Earthquakes 45-1
46. Floods 46-1

PART 5 GEOLOGY AND THE ENVIRONMENT 47-1

47. Geology and Planning 47-3
48. Man-Made Geological Problems 48-1
49. Conserving the Environment 49-1
50. Geology and the Civil Engineer 50-1

APPENDIXES

A. Glossary of Geological Terms A-3
B. Geological Surveys of the World B-1
C. Geological Societies of the World C-1
D. Some Useful Journals D-1
E. Illustration Credits E-1

   Index of Proper Names N-1
   Index of Subjects S-1

PREFACE

This Handbook is intended to be of assistance to all civil engineers by demonstrating the vital importance of geology in all civil engineering work, by illustrating this importance through selected case histories from around the world, and by providing useful references wherein further and more detailed information may be obtained for use in the individual projects that make up the practice of civil engineering.

Every civil engineering project is unique. Superstructures may be identical for several jobs. Foundation strata—the geology of the building site—will never be. Small though the differences between the geology of any two sites may be, they are often significant. Geotechnical engineers are familiar with this infinite variety of site conditions. All civil engineers should be equally conscious of the ultimate dependence of their designs on the correlation of the actual geology of a site with the assumptions they have necessarily made about that geology.

Structural engineers, in particular, should have a vivid appreciation of the fundamental relation between geology and design. They will naturally leave detailed foundation design to foundation and geotechnical engineers. But a lively understanding of the variations possible in foundation-bed conditions, that is, in the geology of construction sites, will render their design assumptions more meaningful. It is therefore hoped that, like geotechnical and foundation engineers, structural engineers will find this Handbook of special use and of interest—even though it contains no mathematics.

The omission of mathematical symbolism is by deliberate decision of the authors, with the agreement of the publisher. Such symbolism could have been introduced in a number of places throughout the text. (One may today even study mathematical geology.) Any introduction of mathematics, however, although possibly helpful in one way, would have tended to obscure the fact that it is by the *observation* of natural features in situ against a background of accumulated knowledge, and so with steadily growing experience, that the geology of a building site must first be studied, and must constantly be kept under review until foundations are complete.

Quantitative studies are an essential element in the geotechnical part of site investigations. Theories of soil action, however, depend for their effective usefulness upon the assumptions made as to the properties of the soil or rock involved. The samples from which these properties are determined *must* be related to the natural context from which they have been removed. The testing of these samples of rock or soil to determine their properties can never be regarded as satisfactory or complete unless it is carried out with full knowledge of the geological context of the samples and of the geology of the location from which the samples come.

No computer is ever going to be able to replace human judgment in deciding upon the stability of rock or soil at a working face as excavation proceeds or upon the suitability of a foundation bed when excavation is complete. Advance calculations will assist, but, in

x   PREFACE

the final analysis, it is on-the-spot judgment that must lead to immediate decisions, judgment which comes from long experience based upon acute observation. For some, this experience will have been won in the field, their knowledge of geology being almost intuitive; for others, preliminary training in geology will have been sharpened and refined by experience in the field.

Geology, therefore, is the essential starting point for all site studies, as it must be for all geotechnical work. It is, correspondingly, the starting point for all foundation design, as it must be also for all overall site planning, whether the area is to be occupied by a small structure or by an entire town or industrial park. This Handbook has been designed to illustrate and illuminate this fundamental interrelation between geology and all civil engineering work.

Arrangement: To this end, the contents of the volume have been divided into five parts. The first (Chaps. 1 to 9) presents in summary form the essentials of geology, including a discussion of groundwater, with special emphasis on the geological aspects of geotechnique. Part Two (Chaps. 10 to 17) deals with the uses of geology in all work preliminary to the start of construction.

Part Three, the major part of the book, comprises Chaps. 18 to 36. Each chapter deals with one major branch of civil engineering, such as bridge foundations, with special reference to the importance of geology in associated design and construction work. Special problems, such as landslides, are considered in Part Four (Chaps. 37 to 46), and the text is brought to a conclusion by Part Five, the four short chapters (47 to 50) of which deal with environmental questions illustrating how civil engineers are indeed "stewards of the land." Four appendixes present a glossary of geological terms and short lists of geological surveys, societies, and journals.

Every effort has been made to ensure that each chapter is self-contained, to the extent that this is possible, even though this approach involves occasional repetition. For rapid reference, therefore, users may turn directly to the relevant chapter. The value of the Handbook will be enhanced, however, if Part One is read as background.

References: At the end of most chapters a list is provided suggesting further reading that may be useful for a fuller understanding of the chapter subject. Again, and as necessary, there is occasional repetition. Throughout each chapter, superscript numbers are used to indicate references to more detailed and complete descriptions of works and features discussed. Lists of these references are, again for convenience, placed at the end of each chapter.

To the maximum extent possible, these references are to publications and journals that should be available to readers who have access to adequate library facilities. Accordingly, a disproportionate number of references will be found to articles in three well-known journals—*Engineering News-Record, Civil Engineering,* and *New Civil Engineer.* It is unfortunate for reference purposes that those responsible for these fine journals have adopted pagination by issue rather than by volume (in postwar years in the case of the first two). Every effort has been made to list references in an easy-to-use manner.

Special note must be made of the changes in form and title of the most venerable of all civil engineering records in the English language, the *Proceedings of the Institution of Civil Engineers* (of England), founded in 1818. For those unfamiliar with this store of useful information, it may be noted that, after a few early volumes of *Transaction,* the Institution published for a century an annual volume of *Minutes of Proceedings.* In 1935 this became the *Journal of the Institution of Civil Engineers.* Since 1962 the *Proceedings* have been issued quarterly as *Part One* (Design and Construction) and *Part Two* (Research and Theory). The Institution's useful news journal, *New Civil Engineer,* does not carry volume numbers, each issue being numbered separately.

Case Histories: Examples from the practice of civil engineering around the world are used in this Handbook, especially in Part Three, to illustrate the varied ways in which geology can affect the structures or works under review. Official travels of the authors have permitted personal visits to many of the major examples cited outside North America. Works in almost 50 countries are used as such illustrative examples, although the majority of examples are, naturally, from North America. Some are of recent date. It was possible, for instance, to include a brief reference to the eruption of Mount St. Helens in May 1980, but many examples are from much earlier years, and a few are from the nineteenth century.

Unlike advances in those sciences where research is steadily rolling back the frontiers of knowledge, the significance of geology in relation to civil engineering remains the same. Designs become more complex, design methods more advanced; construction methods steadily improve in efficiency as construction equipment becomes more sophisticated and versatile. The geology of a building site, however, remains as it has been throughout recent geological time. Its influence upon work which may impinge upon it is unchanging, as important and significant now as it was when modern construction started. Any reader who may be puzzled by reference in a modern book to a dam constructed a century ago may be helped and encouraged by the wise old saying, "Newness is not a criterion for truth."

Geological Time: In all considerations of geology, even in its applications in civil engineering, the vast extent of the time span must always be kept in mind. A glance at Chap. 3, and especially at the diagram to be found therein illustrating the range of geological time (Fig. 3.2), will be helpful in making the necessary adjustment from thinking in terms of normal human time spans to thinking in terms of the more than four billion years now known to be the age of the earth. The diagram shows how small a fraction of this period has been taken up by the Pleistocene "incident" sometimes described as "the last million years." Yet most of the soils with which civil engineers have to deal have existed for only a part of even this period. Therefore, although in the long course of geological time nothing can be regarded as always insoluble, impermeable, or solid, any geological changes within human time spans are insignificant. The full range of recorded human experience reflects but a minute fraction of the total of geological events. So it is that observations from a century ago, in considerations of geology in civil engineering, are still valid and useful today.

Observations: The importance of *observation* has been mentioned; it runs like a thread throughout this Handbook. This word inplies far more than just the act of seeing. "You see," said Sherlock Holmes to his friend Dr. Watson, "but you do not observe." Many of the stories about Sherlock Holmes could usefully serve as supplementary reading to this Handbook. An earlier writer than Sir Arthur Conan Doyle, George Perkins Marsh of Vermont, said the same thing in a different way, stressing in magnificent Victorian prose the vital importance of the art of observation, an art that must be cultivated by all geologists and civil engineers. Its importance pertains especially when geology has to be considered in relation to engineering work, exposures often being available for observation for short periods only.

The observational method, a term that is now found in geotechnical literature, is a term used to stress the vital importance of accurate observations in the field—including observations of geology—as an essential complement of laboratory and theoretical studies. Dr. Terzaghi was preeminent in his meticulous field observations, and these always began with study of the relevant geology. It was far more than a passing jocular remark when this great leader said that one ounce of geology was always essential to leaven a pound of theory. And informed observation is the key to achieving this essential ounce.

The Pleasures of Geology: This Handbook deals with one of the most significant of the applications of the science of geology, its application "for the use and convenience of man." Brief note must be made, however, of the pleasure to be derived from a study of the science itself, without special reference to its application in engineering. It is hoped that one by-product of the use of this volume may be an increased awareness of what this pleasure can be.

Geology was once described as "the people's science," a pleasing reminder that it can be studied by all, without the aid of laboratories or of specialized training. It is, essentially, the study of the form of the surface of the earth on which we live and of the materials that make up this crust. Keen observation of natural features is the best of all starting points, observation which leads directly to a new appreciation of the beauties of the earth.

Some may be led to a more detailed study of minerals, some even to the collection of fossils. Sir William Van Horne, builder of the Canadian Pacific Railway, was an amateur fossil collector, one who even had a fossil named after him. The wide scope of the science gives range for many and varied individual interests. The list of books to be found at the close of Chap. 4 may be helpful; special note is made here of the little book by Bates and Kirkaldy as one of the best of all beginning guides known to the authors.

The Future: As all major cities of the world expand, as all the more desirable sites for dams, bridges, and other structures are steadily used up in developed countries, and as the younger nations of the world make strenuous efforts to provide good accommodation and services for their inhabitants—all in the face of a rapidly growing total world population—the need for the most efficient design and execution of civil engineering works is of mounting importance. The eminent desirability of utilizing underground space for appropriate purposes makes knowledge of local geology imperative. Such developmental work can be carried out only with the sure and certain knowledge of what site conditions will be found when construction has uncovered what was previously hidden from view. This knowledge comes only from the closest possible study of the geology of *all* building sites prior to design and construction.

An advertisement recruiting new engineering staff for Hong Kong said of that mushrooming city:

> Urban development is taking place very rapidly.... The natural terrain is extremely hilly, slope angles in excess of 30° being common, and the igneous rocks have generally weathered to great depths to form residual soils often overlain by colluvium. The climate is sub-tropical with an average rainfall of 2 m between May and October. These circumstances pose challenging geotechnical engineering problems, the most important of which is the prevention of landslides.

And this percipient statement, which appeared in an advertisement for staff, could be repeated with variations for a thousand cities around the world. It is a concise summary of what this Handbook is all about.

The proper use of geology as the starting point of all civil engineering achievement will become ever more important. Despite all the pressures that will come to bear on them through advancing technology and mounting demands, civil engineers should remember the words of Francis Bacon, written as the modern world began to emerge:

NATURE, TO BE COMMANDED, MUST BE OBEYED.

Robert F. Legget
Paul F. Karrow

ACKNOWLEDGMENTS

As explained in the Preface, geology does not change with the years; accordingly, much of the material for this Handbook is derived from an earlier book by R.F.L., *Geology and Engineering*, the first edition of which appeared in 1939, the second in 1962. Our first acknowledgment, therefore, is to all those who assisted with advice and information for *Geology and Engineering*.
Geology and Engineering.

Since the publisher has declared *Cities and Geology* (R.F.L., 1973) out of print, it has also seemed desirable to incorporate in this Handbook some of the information found there, especially the information concerned with geological problems in urban areas. Acknowledgment is therefore made to those who kindly assisted by providing information for *Cities and Geology*. This gratitude applies particularly to Professor Quido Zaruba of Prague, a revered and longtime friend of R.F.L., who during the early planning of this Handbook once again was generous in sharing his wise counsel and his lifetime of experience.

To all others who have kindly supplied information for this volume, we record our appreciation in this public way, supplementing the thanks we have expressed in private communications. We hope that the information so shared has been correctly used, but we naturally accept responsibility for any errors which may regrettably have escaped our most careful checking.

It would be impossible to list all the names of those who, down the years, have influenced us in our work and in the study of the vital application of geology in civil engineering. A general but sincere expression of our indebtedness must suffice, with particular mention of two individuals. Professor P. G. H. Boswwell of Imperial College, London, and formerly of the University of Liverpool, gave to R.F.L. the initial inspiration to work in this field and, eventually, to write about it, an inspiration now shared by his coauthor. Professor J. J. O'Neill of McGill University in Montreal proved to be another helpful guide. This Handbook is, in one way, a tribute to the memory of these men.

We are grateful to Mrs. Isabelle Noffke, who typed the final version of the complete text, always working with lively understanding and skill. Only authors know how much they are indebted to the editorial and production staffs of the publisher of their books. We are keenly aware and appreciative of the contributions of the staff of the McGraw-Hill Book Company to the final appearance of this Handbook. We would like to name our closest associates, but their names will be found on the reverse of the tital page; we invite all readers to associate our indebtedness with these names. Jeremy Robinson, editor in chief of architectural and engineering books in the Professional and Reference Books Division, orginally proposed this Handbook and has favored us with his full support.

Finally, we wish to record the sense of obligation we have to our wives, who have so graciously and helpfully lived with this Handbook while it was in preparation.

Robert F. Legget
Paul F. Karrow

DISCLAIMER

The authors know that there is a real place for women in the fields of geology and civil engineering. They are privileged to know women in both callings. Accordingly, they regret that the English language becomes stilted when the term *he/she* or *she/he* is frequently repeated. They therefore ask all women who use this Handbook to be good enough to accept their assurance that the pronoun *he*—and all allied words—is used throughout these pages in the generic sense.

part 1

GEOLOGICAL BACKGROUND

chapter 1

THE CIVIL ENGINEER AND GEOLOGY

1.1 The Science and the Art / 1-4
1.2 Training in Geology / 1-5
1.3 Practical Experience / 1-7
1.4 Employment of Specialists / 1-7
1.5 Geologists and Civil Engineering Work / 1-9
1.6 The Pattern of Civil Engineering / 1-10
1.7 Contract Plans and Specifications / 1-11
1.8 Earth and Rock Excavation / 1-13
1.9 Construction Operations / 1-15
1.10 Maintenance / 1-16
1.11 Conclusion / 1-17
1.12 References / 1-19

Every branch of civil engineering deals in some way with the surface of the earth, since the works designed by the civil engineer are supported by or located in some part of the earth's crust. The practice of civil engineering includes the design of these works and the control and direction of their construction. *Geology* is the name given to that wide sphere of scientific inquiry which studies the composition and arrangement of the earth's crust. This book is concerned with the application of the results of this scientific study to the art and practice of the civil engineer.

The relation of the science to the art is at once obvious and so intimate that general comment upon it might appear to be uncalled for; unfortunately, this is not the case. It may well be that the very intimacy of the relationship has been the main reason for its frequent neglect. Whatever the cause, the fact remains that, for a considerable period, civil engineering work was carried on in most countries with little conscious reference to geology or to geologists.

At the start of the nineteenth century, before engineering had become the highly specialized practice it is today, many civil engineers were also active geologists. William Smith is the outstanding example of these pioneers. Robert Stephenson combined geological study with his early work in railway construction, and other well-known figures in the annals of engineering history were also distinguished in geology.

1-4   GEOLOGICAL BACKGROUND

Today there is widespread recognition of the vital importance of the science to those who practice the art. Geology is commonly included as a basic subject in courses of training for civil engineers (it should be included in *every* civil engineering course); civil engineering papers contain frequent references to the geological features of the sites of works described; and soil mechanics, the generally accepted scientific approach to soil studies, provides a common meeting ground for civil engineer and geologist and a means of fostering their cooperation. The divorce of the science from the art persisted for so long, however, that their complete correlation may not be realized for many years. Some thought may therefore be given to general aspects of the contacts between the fields before the subject is considered in detail, so that users of this Handbook may have at the outset a clear idea of their fundamental relationship.

1.1 THE SCIENCE AND THE ART

Frequently in the past, geological considerations have featured prominently in the study and discussion of failures of civil engineering works; in fact, to some engineers geology may still be thought of as merely a scientific aid to the correct determination of the reasons for some of the major troubles that develop during or subsequent to construction operations. Valuable as is the assistance rendered by geologists and by the study of geological features in such "postmortem" considerations, the very fact that geological features may have had something to do with the failures suggests, with abundant clarity, that the best time to consult a geologist or to study geological features is *before* design and construction begin. In this way, the science can serve the art in a constructive rather than merely a pathological manner. It will later be seen, as applications of geology are considered in some detail, that not only can this constructive service of the science prevent possible future troubles but it can also suggest new solutions to engineering problems and often reveal information of utility and economic value even in preliminary work.

The more obvious effects of geological features on major civil engineering works may be seen in the underground railway services in London and New York. In London, because the city is built on a great basin of unconsolidated material, including the well-known London clay, tube railways, located far below ground level and built in clay that was easily and economically excavated, have provided an admirable solution to one part of the city's transportation problems. In New York, on the other hand, the surface of Manhattan Island on which the city is located is underlain to a considerable extent by Manhattan schist, and underground railways had to be constructed in carefully excavated rock cuttings just below surface level, as innocent visitors to that great city learn if they happen to stand on a ventilation grating when a train passes in the subway below.

Many similar instances of the profound effect of local geological characteristics upon major civil engineering works could be cited, but all would serve to emphasize the same point: how closely the science and the art are related and how dependent civil engineering work generally must be upon geology. The science of geology stands in relation to the art of the civil engineer in just the same way as do physics, chemistry, and mathematics. The importance of these sciences to the civil engineer is never questioned; they are always considered the necessary and inevitable background to civil engineering training. It would be inconceivable for any engineer worthy of the name to be unfamiliar with the chemistry of simple materials. Ignorance of the nature of the materials on which or in which the civil engineer is to construct his work should be equally inconceivable.

There is an important point of difference to be noted in the relation of geology and of parallel sciences to civil engineering. The latter sciences are utilized directly by the engineer, since mathematical and physical methods are important in many branches of design work. Geology, however, renders to civil engineering a service less obviously direct, in that

FIG. 1.1 Geology and Engineering: The Roman bridge at Lavertezzo, Switzerland.

the findings of the pure science are applied to the specific problems of the engineer. In the case of construction problems, for example, it is the task of the engineering geologist to state the probable difficulties, and the task of the engineer to overcome them; in the case of materials of construction, it is for the geologist to say where they may be found, and for the engineer to obtain them and put them to use.

This is a significant point of difference and it leads to another matter of great importance. The geologist, who has the whole of the earth's surface for a laboratory, encounters in every locality purely local problems that will not be duplicated elsewhere. Thus, every application of the science or its methods to engineering work will be in some respect unique. In this sense, too, there is a difference between the relationships of civil engineering to geology and to associated sciences. Although local characteristics will vary, the fundamental geological principles applying to them do not. And these guiding principles constitute the basis of geological study, study that can be seen to be an essential part of the training of every civil engineer.

There is today increasing concern over the conservation of the natural environment, especially in relation to construction projects. Environmental impact statements are now often mandatory before construction can be started. The preparation of these important documents must start with a consideration of geology since geological factors very largely determine the natural environment. Construction operations will inevitably disturb local geological formations, and so affect the environment. With sound design, environmental conditions can be restored when construction is complete; in some cases they are even improved from what they were before work started, yet this restoration must be assured before the necessary permits are granted. One of the greatest contributions that geology can make to the practice of civil engineering is in providing a firm base upon which valid environmental impact statements can be based.

1.2 TRAINING IN GEOLOGY

Most civil engineers who enter upon their professional careers do so by way of a university or technical college education, or an equivalent course of study, in preparation for examinations given by national professional societies. Thus consideration of the training of civil engineering students in geology may for convenience be confined to university courses. It will be noted that reference is made to training in geology and not to training in civil engineering geology or some such suggested course. There is no special branch of geology applicable to civil engineering. It would be possible and desirable, however, to have a special course of study concerning the application of fundamental geological principles and methods to civil engineering problems. Such a special course must be complementary to the study of geology as such, yet bear the emphasis implied by the title of this book. A study of the curricula of many universities and colleges will show that the value of a course of this nature is generally appreciated in university work.

The training in geology necessary for civil engineers must obviously be general; it must enable students to obtain a good grasp of the nature of the subject and of the character and interrelation of various branches of geology. Thereafter, in the usual college course, time does not permit detailed study of all branches of the subject, nor does such study seem desirable. Attention has to be concentrated on those branches of geology the applications of which are of special importance in civil engineering practice. These branches include physical geology, structural geology, and petrology. Study of geological maps and sections can be a valuable laboratory aid to the first two of these branches, and examination of thin sections of rocks in the petrological microscope is an essential complement to lectures on petrology. The actual making of rock sections is something that the student will never have to do in engineering practice, and it need only be touched upon in laboratory work. Similarly, it does not seem advisable that civil engineering students should spend too much time on the study of fossils, although an introduction to practical paleontological work will be a valuable stimulus to their interest in this branch of study.

Of fundamental importance in all training is experience in the field. It may be said with propriety that no course in geology for civil engineers can be regarded as in any sense complete without a reasonable period of time spent on geological survey work. Local conditions will dictate how fieldwork will be arranged, but a continuous period of one or two weeks spent in a suitable locality will usually be more effective than any number of shorter periods fitted into a regular schedule. Although local circumstances usually conspire to make it impossible to combine a geological field camp with the usual topographical survey camp, a combination of the two would be an ideal arrangement.

It has been suggested that the second part of geological training for civil engineers should be an introduction to the study of the applications of what has been learned of the science to the problems encountered in actual engineering practice. The word *introduction* is used advisedly, since this note refers specifically to college work. The average student will have had little experience on outside civil engineering work and will therefore have much to learn with regard to civil engineering construction. If, however, the student can start the study of construction in association with an introductory study of the applications of geology to such work, a double purpose will be served to great advantage. The usual college course does not let students devote much time to practical subjects, but courses on foundations and construction methods do present opportunities for instruction of the kind indicated.

Instruction in the classroom is but a cursory introduction to what is a lifelong study for the majority of civil engineers; considerations of the applications of geology to their work are either consciously or unconsciously an essential part of that basic experience which is the most prized possession of all members of the profession. How much more easily this experience can be gained if the engineer possesses at the outset a fundamental knowledge of geology will be evident from what has already been written. Without a correct attitude of mind toward geology, no civil engineer will benefit from instruction in the science, no matter how well it may be presented. Indeed, the development of this mental attitude should be the principal aim of those charged with instructing engineering students in geology. The task is no easy one, because the students will not have had any opportunity to learn to appreciate the vital importance of geological features in actual construction work.

Fortunately, the science is coming to be widely recognized as an important aid to the civil engineer. Perhaps the best indication of this recognition is to be found in the almost invariable mention of site geology in the papers on civil engineering work presented to professional societies. This situation is of real significance and is encouraging, since the presentation of any description of an engineering structure without some reference to foundation-bed conditions or to other corresponding geological data is equivalent to presenting a paper on a bridge without referring to the loading used in design or the materials used in construction.

1.3 PRACTICAL EXPERIENCE

An appreciation of geological features, and particularly of the characteristics of the materials that make up the earth's crust, is not acquired solely by training in geology. Many engineers, especially those who have spent much time on construction, possess this appreciation unconsciously as a product of their wide practical experience. In the past there have been many engineers—and there probably still are some—who have never troubled to find out what geology is, yet they know instinctively many of the matters herein discussed in solely geological terms. This point is emphasized since the dividing line between what may truly be called practical experience and a cultivated appreciation of geological features is quite indeterminate. For the sake of convenience, references throughout this book are confined to the trained attitude of mind toward geological features, but it is always to be understood that the practical experience of engineers unversed in geology is included, although not stated, in such references.

Knowledge derived from wide experience is not so common as some people suggest. It is almost intuitive in nature; certainly it is much more than merely factual knowledge derived from long observation. Some could not acquire such intuitive judgment even after a lifetime of outside experience; others acquire it easily and early in their lives. The suggestion may therefore be advanced that all so-called "practical experience" may not lead to sound intuitive judgment. Sometimes the "practical person" is the one who practices the theories of 30 years ago.

Whether the civil engineer gains an appreciation of geology by training or by intuition, that appreciation can be of good service in the field. Knowing, at least to some degree, what lies hidden beneath the surface of the ground, the engineer will be able to direct exploratory work more accurately. And in construction, every step taken in connection with excavation and foundation work will have a new and added significance for the resident engineer who has been awakened to all that geology can mean in the supervision of such work. He will not only be able to perceive the beauties of scenery more fully, but, in topographical work, will also have a better understanding of the origins of the many special features being surveyed.

1.4 EMPLOYMENT OF SPECIALISTS

One of the most important results of a desirable attitude toward geology should be that the civil engineer will know when to call for the services of an expert engineering geologist. As a general rule, engineers cannot hope to be more than amateur geologists, familiar with the science and its methods, appreciative of its value, but not fully qualified to carry out detailed investigations either in the field or in the laboratory. On small jobs and in routine civil engineering work, a general familiarity with the science will enable the engineer to tackle most of the problems that arise. On large jobs, however, and for unusual work, specialists must be consulted. The realization of the necessity for consultation can be taken as one of the hallmarks of a true engineer, rather than the reverse, as is sometimes erroneously believed.

On occasion, the engineer may meet with opposition when suggesting consultation with an engineering geologist. The leading argument against such a course is that similar works have been carried out successfully in the past without the special aid and additional expense of a geologist. So, it is asked, why not carry on the work without this extra assistance? Superficially, such an argument may appear difficult to refute unless thought is given to the parallel case of fire insurance premiums. There was a time when building owners never thought of paying out extra money for fire insurance, and owners then continued to live and prosper. But how many today would neglect to take advantage of fire coverage, despite the adequacy of modern fire-fighting equipment?

Thus the argument may be answered in economic terms. Of far more importance, however, particularly for the engineer's own satisfaction, is the impressive record of services already rendered to the art by geologists, known and unknown, a record to which the examples cited in this book pay some tribute. The number of disasters that have been prevented through the use of a geologist's advice can never be estimated, but the record of failures that have occurred when such advice has not been taken is at least an indication of what geology has contributed to civil engineering achievement. The important work of Dr. C. P. Berkey and his co-workers in connection with the Catskill Aqueduct for the water supply of the city of New York is a telling example of the constructive use of geology. The geological investigations associated with the tunnels that now exist under the Mersey River at Liverpool, England, and particularly the work of Prof. P. G. H. Boswell in connection with the great vehicular tunnel, offer similar evidence. The Hoover Dam on the Colorado River in the United States and many other successful dams testify silently to the value of the assistance that civil engineers have obtained from geological specialists. So also do bridges, large and small, around the world, the foundations of the great bridges of New York and San Francisco being especially notable testimonials.

Comments on these few examples of the geologist's contribution to engineering could be extended to fill many pages, but other examples of cooperative work quoted throughout this book will be even more impressive. Usually, engineering geologists are called upon to assist in engineering work in the capacity of consultants; the practice of having a geologist as a member of the official board of consultants, a practice so often followed in North American work, is a most satisfactory arrangement. During the years immediately prior to the Second World War, a few major engineering organizations added geological units to their permanent staffs. Experience during the war years, in both civilian and military engineering work, acted as a catalyst in this regard, as in many others, so that today almost all large civil engineering organizations have their own engineering geological staffs. Notable examples in North America, and pioneers in this practice, are the United States Corps of Engineers, the Tennessee Valley Authority, and the U.S. Bureau of Reclamation,* while the United States Geological Survey has its own engineering geology branch with headquarters in Denver, Colorado. This practice is widespread today, but the organizations noted here not only were early in the field, but have, through the work and

---

*In January of 1980 the Bureau of Reclamation, still part of the U.S. Department of the Interior, was renamed the Water and Power Resources Service. The Bureau's original name was restored just as this volume went to press.

the publications of their staffs, made significant contributions from which the entire engineering profession benefits. On smaller works, the individual consulting geologist can give similarly useful service. In most major cities, there is an increasing number of independent consultants who are serving in this way. Geological surveys in all countries are willing to assist when staff is available and conditions are appropriate. And the geological departments of universities are often willing and able to assist through the service of their specialist staff members.

FIG. 1.2   Geology and Engineering: The Copper River railroad bridge, Alaska.

1.5 GEOLOGISTS AND CIVIL ENGINEERING WORK

This book is not intended primarily for the use of geologists, but it may be useful for engineers to note that geologists welcome the opportunity to cooperate on civil engineering work; their only regret is that the opportunity does not occur more frequently. This suggestion is confirmed by the contributions made by geologists to the discussion of engineering papers, contributions such as the statement made many years ago by Dr. T. Robinson of the Geological Survey of Great Britain during discussion of a paper at the Institution of Civil Engineers, London.

> The records of the Geological Survey showed conclusively that closer cooperation between the geologist and the engineer would be greatly to the advantage of both, and it was a pity that there was no very direct way in which geologists could be kept informed of the progress of important excavations.[1.1]

Practically the only way geologists can learn of new exposures made by civil engineering operations is through the engineer in charge of the work. May it be urged, therefore, that, whenever possible, civil engineers advise the director of the appropriate geological survey and the head of the geological department of the nearest university of all excavation work of interest under their charge, so that geologists may at least have an opportunity to see the exposures before they are covered up. This courtesy demands little of the engineer, but it may lead to scientific information of great value. Examples, such as the tunnels under Boston, confirming this suggestion will be given in later pages.

When geologists are called in to advise in civil engineering work, they will have to act in conjunction with the engineers responsible for the work. Thus arises the need for cooperation between the civil engineer and the engineering geologist, the practical builder and the scientist. Their cooperation may lead to a valuable partnership, and it often is the source of considerable personal pleasure. Their partnership is, in some ways, a union of opposites, for the approach of the two to the same problem is psychologically different. The geologist analyzes conditions as they are found; the engineer considers how existing conditions can be changed so that they will suit a specific plan. The geologist draws on his analysis to cite problems that exist and suggest troubles that may arise; the engineer has to solve the problems and overcome the troubles. The final responsibility for the decisions involved must always rest with the engineer; but in coming to such decisions, the engineer will be guided by and will probably rely upon the factual information provided by the geologist. It is not without reason that it has been suggested that, although mining geologists have to be optimists, all engineering geologists should be confirmed pessimists.

This joint work calls for a fine degree of real cooperation. The engineering geologist has to remember that what the engineer wants is a clear picture, presented as concisely as possible, of the geological conditions related to the work, with a view to their practical utilization. On the other hand, the engineer must remember that the geologist is a geologist, not an engineer, and cannot be expected to deliver the kind of report that would come from another engineer. In many cases, the most effective results can be achieved if the engineer is able to give to the geologist, at the outset, a list of specific questions to be answered. The questions may relate to geological conditions, to the necessity for and the location of further exploratory works, such as boreholes and test pits, and to similar matters. The engineer should be willing to give the geologist the liberty of pursuing, within reasonable limits, any aspects of purely scientific interest that may develop in the course of the main task.

1.6 THE PATTERN OF CIVIL ENGINEERING

When a new project comes up for consideration, the civil engineer will initially require some firsthand knowledge of the locality in which the work is to be carried out. Preliminary investigations and studies will then be made, and when these are approved, complete contract drawings and specifications will be prepared. When financial arrangements are made, tenders (bids) will be called for, a contract awarded, and construction begun; when the project is complete, it must be periodically inspected and maintained in good order. Brief comment on the application of geology to each of these main divisions of civil engineering procedure may be helpful.

The civil engineer will generally obtain information about an area in which work is to be carried out by visiting the area, even if only for a hurried tour of inspection, and by studying descriptive literature on the district. If geological reports are included in the literature selected, and if the topography of the area is studied with due regard to the significance of the local geology, then the engineer will get a more vivid and more accurate picture of the district than if the geology were neglected. Since an elementary study of the geology of an area will demonstrate its relation to local scenery, a civil engineer trained in geology can visualize, after the necessary inspection and investigation, the general structure of the ground being considered and the origin of leading features of the local topography that will be of importance in the work.

As an example of broad concepts of local geology, the following extract may be quoted from a lecture on hydroelectric power development on the Rhine given by Dr. H. E. Gruner of Switzerland.

> In the stretch which concerns us, the Rhine cuts through the jurassic system, the tertiary system, the lower layers of trias, the gneiss massive of the Black Forest and some layers of the permian system. The power-plants are founded on each of the different layers, and the results have proved favourable in every case. Of still greater importance in their effect on the longitudinal profile of the Rhine, and thus upon its character as a source of energy, are the historical occurrences during its origin. Even in pre-glacial times the Rhine flowed through the same valley as it does today, but owing to the enormous quantities of gravel that the glaciers deposited during their retreat, the Rhine lost its old course and was partially forced into a new bed. Whilst scouring out its new bed it eroded the underlying rock, which it took more time to carry off than the gravel. This produced the different falls and rapids as we find them at Neuhausen, Reckingen, Schwaderloch, Laufenburg, Ryburg-Schworstadt, Rheinfelden and on the Kembsersill.... The soundings and geological studies that were needed for the different power-plants, enabled us to determine the old river-bed for almost its entire length. These pre-glacial river-beds of the Rhine are technically important in many ways; they always carry water and are thus well suited for water-supply works. During the construction of power-plants they may also become the source of disagreeable surprises if they are not thoroughly examined at first.[1.2]

This descriptive note gives a general picture of the geology of the sites of the plants described by Dr. Gruner; it could not have been written without a study of relevant geological reports. Although it is general, it indicates some of the geological problems that had to be faced during the construction of the Rhine water-power plants, and it suggests the kind of preliminary exploratory investigations that had to be undertaken.

Preliminary studies may be made in more detail if civil engineering work is to cover a large area. General reconnaissance surveys will probably have to be made, and general topographic maps either prepared or checked. Simultaneously with this work, geological reconnaissance can always be carried out with advantage; the local geology can be studied in more detail and correlated, although still in a general way with engineering requirements. It is not often that civil engineers will be called upon to undertake extensive work of this nature. When the need does arise, special organizations are usually recruited to undertake the work.

Having obtained a general idea of the district in which the work is to be done, the civil engineer will next proceed with preliminary plans and estimates. Gradually these will be evaluated and discussed, until finally an accepted scheme or design is evolved which can then be prepared in detail. All this work can properly be carried out only if the engineer possesses an adequate and detailed knowledge of the ground in which the work is to be located and of the natural materials available at or near the site. This essential information will be obtained by means of detailed geological fieldwork and by exploratory investigations such as boreholes and test pits. Preliminary exploratory work is so important that it is considered in detail in Part Two; it is mentioned here in order to show the logical association of geology with this leading phase of civil engineering work.

1.7 CONTRACT PLANS AND SPECIFICATIONS

The final design of the civil engineer will usually be incorporated in a set of contract plans and specifications, on the basis of which tenders for performing the work involved will be called for from contractors. In those projects which are carried out by direct administration, instead of by contract, a complete set of drawings equivalent to a set of contract plans will still be necessary, and the equivalent of a contract specification will be needed for the guidance of the engineers in actual charge of construction operations.

The preparation of these documents marks a definite change in the engineer's work and in his responsibilities. When issued to a successful bidder and made the basis of a formal contract, they become legal documents, entitling the contractor to certain rights and taking control of construction operations, to some extent, out of the hands of the owner and of the engineer as the owner's representative. If, therefore, the application of geology can in any way assist in making the preparation of contract documents more effective and less open to question, the science will be rendering particularly valuable service. It would appear certain, from considerations already advanced, that there are several ways in which assistance can thus be rendered.

It has already been mentioned that the chief aim of preliminary exploratory work and associated geological studies is to provide accurate information about subsurface conditions at the site of the proposed work and about the availability of suitable construction materials in the vicinity. The subsurface conditions will affect the design made by the engineer and also the construction methods adopted by the contractor; the availability of materials may have some bearing on the design adopted, especially from the economic angle, and will have an appreciable effect on construction planning. It is clear, therefore, that the engineer should include in the contract documents as much information regarding the site and the available materials as is possible.

On the contract plans, the engineer can do this by giving full details of the records of boreholes, test pits, and other subsurface explorations. These should be given not only in section but also in a general plan that shows their correlations with the location of the work to be constructed. Only with these records on hand can the contractor, if he is so minded, take advantage of all the information that the engineer has available relative to the work. In many cases, and when possible, it will be advisable to show the nature of the geological structure adjacent to foundations, instead of the useful pictorial representation of rock or unconsolidated material. In certain special cases, such as tunnel work, it will be desirable to show on the profile drawings the geological formations that are anticipated along the line of the work to be undertaken.

The same criterion should pertain in the preparation of civil engineering specifications. As a general rule, the opportunity to use geological information in specifications will arise in one or more of four ways:

1. In the provision for possible alterations in design due to variations encountered in subsurface conditions
2. In the provision of information relating to available materials of construction
3. In the clauses relating to methods of construction to be adopted
4. In reference to the measurement and payment for excavation

The first of these is closely related to preliminary exploratory work and so may conveniently be considered in Part Two. The second calls for the clearest and fullest explanation of all the facts known relative to the materials available; it will also be referred to in later chapters. The third touches upon a most important matter, but one that is generally and advisedly left to the selection of the contractor, usually with such a qualification clause as: "The contractors are to submit to the engineers a statement with drawings showing how they propose to carry out the works, but any approval of the engineers is not to relieve the contractors of any liability that devolves upon them under this contract."

This provision, as indeed the whole essence of a civil engineering contract, makes the engineer morally responsible for giving the contractor all available information which may

FIG. 1.3 Geology and Engineering: The Quintette Tunnels on the (former) Kettle Valley Railway, British Columbia.

assist in construction planning and methods; at the same time, it leaves the contractor free to apply his own accumulated experience and special skills to the most efficient prosecution of the work in view. Geological information included in the specifications, in addition to that shown on the contract drawings, will assist in fulfilling the engineer's obligation, and fullest use should be made of the opportunity thus provided.

1.8 EARTH AND ROCK EXCAVATION

The fourth instance, the opportunity to relate geological information to excavation, is of such importance that it will be considered in some detail. Contracts for civil engineering projects are designed in such a way that the work being carried out will be to the satisfaction of the owner, will fulfill the requirements of the engineer, and will provide due safeguards for the contractor. General conditions define the scope of the contract; specifications and contract drawings detail the design of the engineer; and, in the usual contract, quantities and unit prices define the extent of the contractor's operation and remuneration. Great care is always exercised in preparing these contract documents in order to avoid disputes, but success is not always achieved, as the record of court cases regarding engineering contracts makes clear. It is probably safe to say that no one feature of civil engineering contracts has been responsible for more disputes that the engineer's classification of material to be excavated as "earth" or "rock" and the consequent cost of this part of the work performed. As unit prices for rock excavation may be 10 or 12 times as much as the corresponding unit prices for earth excavation, the possibility of disputes arising from questionable classification of the material involved will be obvious.

The excavation of solid rock which has to be drilled and blasted is rarely questioned in this connection, nor is the removal of loose sand and gravel or soft clay. But in between these two extremes there may be materials that cannot easily be classified as either earth or rock unless a reference basis is adopted before contract operations begin. Such doubtful material is often termed *hardpan*. The use of this word should *always* be avoided by engineers, if they wish to obviate trouble. It is essentially a popular term and is sometimes applied to special local gravel deposits whose unusual hardness has been caused by partial cementing of the rock fragments. Hardpan has been the subject of many lawsuits, mainly because the best definition one can give to it is that it is material that proved harder to excavate than the contractor had anticipated. It is a name not generally included in geological nomenclature, and there are no satisfactory definitions of it in engineering literature. Glacial till (or boulder clay) is often described as hardpan, but in this case, as in practically all others, that material can be more accurately described as compact gravel, sand, and clay with boulders. If material of this kind has to be removed during construction operations, an indication of the methods necessary for its excavation should accompany its description in contract documents.

It might be thought that this is a situation in which a direct application of geological terminology would be of assistance. Unfortunately, this is not true, although an appreciation of the geological character of the materials involved will be of great assistance to the engineer. In geology, the term *rock* is used to describe all the solid constituent materials of the earth's crust (as explained in Chap. 4), and before going any further into geological terminology, one can see that more specific definitions are inapplicable. As a general rule in excavation work, the engineer is not interested in the type of rock that has to be excavated (and the contractor's interest is even less) but only in its character in relation to excavation methods.

Engineers have on occasion attempted to utilize geological information available to them in describing materials to be excavated, sometimes with unfortunate results. An example of this was the use of the term *cemented Triassic formation* in connection with an important contract for the construction of 1,830 m (6,000 ft) of the outfall of the Passaic Valley Sewer in New York Harbor. This term was said to be a good description of glacial till, but it was held by the courts to be misleading. In the course of construction, the contractor had encountered varieties of sand and clay that were not cemented, and construction troubles had resulted—troubles which finally became the subject of legal action.[1.3]

Thus, the use of geological terms alone in connection with excavation work may not be helpful, and the detailed divisions of the main geological rock types sometimes suggested in this connection should not be considered. What then is the proper scope of geology in excavation specifications? The following suggestions may serve as satisfactory guidelines.

1. Prior to the completion of contract plans, a careful study must be made, by means of boreholes, etc., of all material to be excavated.

2. If any material intermediate between loose soils and compact rocks is discovered, special tests should be conducted to discover its character insofar as excavation is concerned; this can be done through test boring by an experienced crew. Special care should always be taken to investigate the effect of water upon such material, since

frequently compact soil mixtures, which may be quite hard when dry, will disintegrate in the presence of water.
3. In the specifications, definitions of different classes of excavation should be rigidly stated in terms of the methods to be used to excavate the material.
4. These engineering definitions should, if possible, be correlated with the geological descriptions of the materials to be encountered; the geological descriptions should be written in the terminology which was used in connection with the test-boring results; no such indefinite and questionable terms as *hardpan* should ever be included.
5. The designations used for describing the material to be excavated should be as few and as simple as possible, e.g.:

Hard-rock excavation to designate excavation of granite and limestone which has to be drilled and blasted.

Loose-rock excavation to designate excavation of blocky limestone which does not require blasting but which cannot be removed by hand shovels or picks.

Soft-rock and earth excavation to designate excavation of disintegrated granite, clay, sand, and gravel material which can be handled with shovels and similar tools.

It may finally be noted that throughout the rest of this book, the term *soil* will refer to unconsolidated natural materials, and the term *rock* will refer to solid bedrock, either in place or excavated. Further comment on this important matter will be found in Chap. 19.

1.9 CONSTRUCTION OPERATIONS

Every cubic meter* that is removed during excavations, every unusual loading that is applied to a natural foundation bed, every pile that is driven into the ground—in fact, every operation in construction in which the existing condition of the earth's crust is affected—is associated with geological features of some kind. Preliminary investigations of the relevant geology should therefore be of considerable value not only to the resident engineer on construction work but also to the contractor who is undertaking the work. Throughout this book, most of the examples mentioned will confirm in one way or another the vital importance to contractors of advance geological information.

The geological information available at the beginning of a job can be fully effective, however, only if it is constantly compared with actual geological conditions as they are revealed during the progress of construction. It is essential, therefore, in all civil engineering construction work that a regular and constant watch be kept on geological formations as they are revealed and that in addition to the usual construction progress records, an adequate and complete record of leading geological features be kept. This can easily be done if at least one of the engineers on the job has been trained in geological fieldwork; in some cases, geological training has been made a prerequisite for appointment to a resident engineer's staff.

Geological information obtained as construction progresses has a threefold practical value. In the first place, such information acts as a check on the assumptions made with regard to geological conditions so that, in the preparation of the final design for the work

---

*The authors, with great respect, must dissociate themselves from this spelling; the International Organization for Standardization (ISO) has agreed on *metre,* and this spelling is now in general use except in the United States of America.

to be constructed, any variation from these assumed conditions can be incorporated into the design before it is too late. In the second place, the revelation of the actual geology of the working site enables the contractor to keep check on the suitability and efficacy of construction plans and plant. Finally, if the geological progress record is kept in a satisfactory manner, it may prove of inestimable value at some future time if further work has to be carried out at the same location.

In reference to the last point, brief mention may be made of a significant confirmation in connection with some cement grouting work carried out in parts of the famous Severn Tunnel in the southwest of England. This tunnel, which carries a main line of what used to be the Great Western Railway (now the Western Region of British Railways) under the river Severn, was started in 1873 and completed by the contractor, T. A. Walker, only after a fight against tremendous difficulties. Mr. Walker kept a complete geological record throughout the construction of the tunnel, which pierces the Trias and Coal Measures of England; the rocks encountered varied from marl and shale to limestone and sandstone.[1.4] The existence of this valuable and complete record assisted the engineering authorities of the railway company almost 60 years later when they wished to undertake the extensive cementation program.[1.5]

On the other hand, many examples could be given of the trouble and expense caused by the lack of comparable records for other works. A well-known tunnel in the United States, for example, had to be resurveyed, within seven years of its completion, in connection with the installation of a concrete lining, since no records were then available of the timber sets used, the final tunnel cross sections, or the geological strata encountered.

1.10 MAINTENANCE

Finally, all civil engineering works have to be regularly inspected and maintained in good condition by such measures as may be called for as a result of the inspection. This routine work must always be most carefully carried out. Not only must man-made structures be regularly inspected but also the ground adjacent to that on which they rest. Inspections of bridge piers, to check against scouring, and of dams, to check against erosion of foundation-bed material caused by leaks, are two of the more important aspects of maintenance work in which geological features may be of special significance, and reference will be made to them in later chapters. References will also be made to the regular checking and maintenance of other structures and works in which careful attention to geological features can add appreciably to the value of their inspection.

The regular inspection of civil engineering works, along with any maintenance which such inspection shows to be necessary, is often the first casualty when administrators have to reduce their budgets. This was demonstrated in vivid fashion in the late 1970s by the serious situation revealed when special inspections were made of highway bridges throughout much of North America. It is so easy to think that, since a structure looks safe, it is in satisfactory condition. Hidden from view, however, are the geological strata upon which the ultimate safety of all structures depends. Constant vigilance is as necessary with the static structures of the civil engineer as with the moving mechanisms of the mechanical engineer. Yet, while the need for adequate inspection and maintenance of moving mechanisms is rarely questioned, allocations for the regular inspection of foundations, especially if they are below water, are all too often the first victim of the budget cutter's axe. To help those who have to defend such vitally important expenditures, examples are cited in the appropriate chapters of this Handbook.

FIG 1.4 Geology and Engineering: The Royal Gorge suspension bridge over the Arkansas River, Colorado.

1.11 CONCLUSION

Geology has a vital part to play in all engineering operations concerned in any way with the ground, and thus the science is involved with the entire field of civil engineering. Most of the chapters in this volume are arranged to illustrate this interconnection in all the main aspects of civil engineering work—bridges, dams, water-supply and marine works to name a few. For ease of reference each chapter has been made as self-contained as possible. The chapters which follow in Part One provide basic background in the elements of geology, background essential for geological applications in civil engineering. Part Two is devoted to the importance of geology in all site investigations. Part Three takes up geological implications in specific engineering applications. Individual chapters in this part discuss drainage; excavation; foundation work; canals, roads, and railway construction; and other major engineering works. Special problems of importance to several branches of engineering work are reviewed in Part Four. The final four chapters, Part Five, deal with more general themes involving the environment. Appendixes supply guides to the geological surveys of the world and to sources of geological information; they also provide a small glossary of some of the more common geological terms that the civil engineer may encounter.

It will be noted, possibly with some surprise, that the treatment of the subject is generally descriptive; no mathematics is introduced into the discussion. This is a natural course to follow; it is one deliberately chosen. In their scientific work the authors are guided by Lord Kelvin's famous and oft-quoted dictum to the effect that "when you can measure what you are speaking about, and express it in numbers, you know something about it." But civil engineering is essentially an art, certainly in all its practical aspects, and its successful practice depends in large measure upon the exercise of sound judgment. Sound judgment comes from long experience based on acute observation; of nothing is this more true than in the study of the natural ground conditions upon which, or in which, civil engineering works have to be constructed. It is here that the judgment of the engineer, aided immeasurably by the skilled observations and studies of the engineering geologist, can make such great contributions. Such judgment is called for in solving difficult construction problems that must be solved in the field and without any delay; in selecting a final site from among various alternatives, with great economic issues dependent upon the correct decision; or in determining when excavation of a deep foundation may safely be stopped despite possible imperfections in the material to be used as a foundation bed for a great structure. Mathematics is of no direct assistance on occasions such as these, vital as its background role in relation to designs and computations may be. Here there is no substitute for good judgment and sound observation.

The ability to observe quickly and accurately is a most desirable attribute in civil engineers, all of whom could benefit from the training that Boy Scouts get with Kim's game. Those who know Rudyard Kipling's *Kim* will know the way in which Kim was trained in observation by being shown a variety of objects on a tray for a very short period, after which he was to remember all he could. The modern equivalent is far more than a pleasant parlour game; it is sound training in observation.

Field geology is based entirely upon such powers of observation, coupled with the ability to deduce three-dimensional structures from surface features. The civil engineer is trained to think in terms of three dimensions in relation to structural design; equally important, if not more so, is the corresponding ability to visualize the spatial character of underground conditions. The geologist visualizes these conditions automatically; the frequently used block diagrams in geological papers are an indication of this. It may not be easy, but it is possible for the civil engineer to gain such ability quickly. It is an ability that is of unusual service in field engineering work. Study of simple block diagrams is a first step. Books with colored block diagrams can be of even greater assistance. A particularly notable example, published some years ago in Great Britain, was a book profusely illustrated with colored-block cutaway geological diagrams that vividly illustrate quite complex geological structures.[1.6] But it is in the field, and under the tuition of a master teacher, that the art of developing this three-dimensional appreciation of ground structure can most effectively be gained.

Equally important is a full appreciation of the importance of water in geological study. Most engineers will have occasion to observe the remarkable effects of water on many solid materials and especially on soils. Dry soils cause little trouble; wet soils can cause havoc on a job. So is it in many geological formations. Therefore, the civil engineer must not only learn to visualize subsurface conditions in spatial terms but must always realize the added complications that the presence of varying amounts of groundwater may cause. That groundwater may vary in just the same way as surface water should become a commonplace in an engineer's thinking. And the relation of both to the weather must always be remembered, especially when field studies are carried out in fine dry weather. One of the basic tools of the engineer in the study of any construction site should be a simple summary of local weather conditions.

These concepts are basic to a full appreciation of the role that geology can and must play in the design and construction of civil engineering works. The concepts are relatively straightforward and simple, and their application calls for no unusual skills such as may be possessed by only the gifted few. The ability to "look below the surface," to realize that surface conditions may be vastly different from those only a few feet below ground level, to keep in mind the ever-present possibility of groundwater movement, to visualize the significance of unusual surface topographical features, and to be able to relate these to geological processes of the past—all these are characteristics of observation in which every civil engineer can be trained, even to a large extent self-trained. Such observational skills can be exercised not only on the construction job but in most travels; their application can add greatly to the pleasure of travel and to the enjoyment of scenery. In this book, however, they are shown specifically in their application to the successful prosecution of civil engineering work in all its variety and with all its continuing challenge.

The works to be described are carried out by human beings for the benefit of their fellows. It may well be remembered, therefore, that behind all impersonal discussions of scientific applications and engineering endeavors lie relationships between people and the cooperation of many of them working toward a common goal in rendering useful service to their community. This book would not be complete without at least a passing reference to this human background. It is in keeping with this thought that Chap. 1 has been titled "The Civil Engineer and Geology" and that the portrait that has been reproduced as a frontispiece is a portrait of a man who epitomized all that has been suggested as the attitude a civil engineer should entertain toward geology. For his geological work alone, William Smith will always be remembered with honor; but to civil engineers, his fame rests also on the way in which he applied his geological knowledge to his civil engineering work, work of no little importance a century ago. He indeed allowed his "faith" as a geologist to benefit his "works" as an engineer.

1.12 REFERENCES

1.1 R. T. McCallum, "The Opening-out of Cofton Tunnel, London, Midland and Scottish Railway," *Minutes of Proceedings of the Institution of Civil Engineers,* **231**:161 (1931).

1.2 H. E. Gruner, "Hydro-Electric Power Development on the Rhine," special lecture before Institution of Civil Engineers, London, 1935.

1.3 Anon: "Contractors Win Suit to Recover Losses due to Engineer's Misrepresentation," *Engineering News-Record,* **93**:21 (1924).

1.4 T. A. Walker, *The Severn Tunnel; Its Construction and Difficulties, 1872–1887,* Bentley, London, 1888.

1.5 R. Carpmael, "Cementation in the Severn Tunnel," *Minutes of Proceedings of the Institution of Civil Engineers,* **234**:277 (1933).

1.6 L. Dudley Stamp, *The Earth's Crust,* Harrap, London, 1951.

Suggestions for Further Reading

No better companion to this Handbook can be suggested than Professional Paper No. 950 of the United States Geological Survey. First issued in 1978 and almost immediately reprinted, this large (40.5- by 30.5-cm or 16- by 12-in) beautifully produced 98-page color folio is entitled "Nature to Be Commmanded . . . (Earth Science Maps Applied to Land and Water Management)." The editors are G. D. Robinson and A. M. Spieker. Following a helpful introduction about the use of maps, six case histories are described in detail, and the whole is illustrated with well over 150 photographs and diagrams, mainly in color. These cases should convince the most skeptical of the validity of the message of this Handbook.

There is today a growing number of smaller and more localized publications of the same general type illustrating the uses of geology in civil engineering, including planning. A good example is "Environmental Geology in Towne and Country" by W. C. Hayes and J. D. Vineyard, published by the Missouri Geological Survey (Rolla) in 1969 (43 pp.).

From the many now classical papers on the general subject of this Handbook, one must be mentioned: The Presidential Address to the Geological Society of America entitled "The Geologist in Public Works" by C. P. Berkey, to be found in the Bulletin of the Society, vol. 53, pp. 513–532 (1942).

For a broad review of the interrelations of geology and civil engineering based on wide and varied European experience, reference should be made to *Engineering Geology* (Elsevier), a 504-page volume by Quido Zaruba and Vojtech Mencl of Czechoslovakia, published in 1976. It is due to appear in a second edition in 1983.

As a first introduction to the science of geology for those who need it, the Golden Nature Guides are admirable, conveniently small in size (for pocket or handbag), well illustrated, and clearly written. The original *Rocks and Minerals* (1957) has now been joined by similar guides called *Fossils* (1962), *Landforms* (1971), and *Geology* (1972), all published by the Golden Press, New York (Western Publishing Co., Inc., Racine, Wisconsin).

chapter 2

SOME OUTSTANDING EXAMPLES OF GEOLOGY IN CIVIL ENGINEERING

2.1 The Snowy Mountains Scheme / 2-2
2.2 The Glen Shira Project / 2-6
2.3 The Kinzua Dam / 2-10
2.4 In Time of War / 2-12
2.5 Duisburg Harbor / 2-16
2.6 References / 2-19

It may be helpful to present at the outset of this volume some outstanding examples of the application of geology in civil engineering works, all of which projects have been visited by one of the authors. Later chapters will present examples chosen to illustrate the specific subjects under discussion. It is to be understood, however, that the importance of geology is not limited to such restricted applications, but is inherent in the overall planning for all civil engineering works. The first example, from Australia, demonstrates this well; here the basic geology of an area of almost 13,000 km² (5,000 mi²) had to be determined as the basis for the overall planning of an integrated water-power and water-diversion project. A second example is from Scotland, where a water-power development of two stages was planned and constructed in an area featuring some of the most complex geology to be found in the British Isles. The third example is of a project of a smaller scale, the application of detailed and long-term geological studies to the determination of a dam site in northern Pennsylvania. This case shows how the study of the geology underlying proposed dam sites, a study of features hidden from view and differing markedly from what surface topography would have suggested, prompted a change in location with a resulting monetary saving of at least $4 million. When savings of this magnitude can be shown to be due directly to the proper use of geology in civil engineering, even the most sceptical may be convinced.

The fourth example concerns the application of geology in military engineering, an application that has not been as well recognized as it might have been, other than by those who came into personal contact with some of the remarkable contributions that geology did make in both world wars. Military requirements of secrecy render it still impossible

2-2 GEOLOGICAL BACKGROUND

to describe one or two of the more significant examples in this field, but a summary account is presented here of some of the applications of geology to military engineering in both World War I and World War II.

Near the center of many wartime operations lies the great industrial district of the Ruhr, West Germany, served by the inland port of Duisburg on the Rhine, the busiest inland port of the world. Here, in great contrast to wartime operations, geology provided the basis for planning which ensured the future of the great port, despite steadily dropping water levels in the river Rhine. The lowering of Duisburg Harbor, a truly classic case of geology as applied in engineering, is given as the Chapter's last example. As such, it demonstrates clearly the main message of this Handbook: that the geology of sites on which engineering works are to be carried out must always be studied as the first step in planning, sometimes as a useful precaution, sometimes for delineation of geological problems that may affect design and construction, and sometimes—as at Duisburg—for providing the solution to a major problem that may have appeared, at one time, to be insoluble.

2.1 THE SNOWY MOUNTAINS SCHEME

About 160 km (100 mi) southwest of Canberra, national capital of Australia, and midway between Sydney and Melbourne, the Snowy Mountains stand as the highest part of the Great Dividing Range which lies on the border of the states of New South Wales and Victoria. A central belt, 24 to 32 km (15 to 20 mi) wide, trends north-northeast and is over 1,370 m (4,500 ft) above sea level; it includes Mount Kosciusco at 2,230 m (7,314 ft) the highest point in Australia. To the west, it is bounded by great fault escarpments that lead rapidly to the lowlands at an elevation of about 300 m (1,000 ft) above sea level; the headwaters of the Murray River and its tributaries, the Geehi and Tooma, and the Tumut River provide the main drainage. To the east, the high plateau descends more gradually to a tableland at 900 and 1,200 m (3,000 and 4,000 ft) above sea level, drained mainly by the Snowy River and its tributaries, the Eucumbene and Crackenback. The Snowy Mountains are dissected by many streams that have cut for themselves valleys as deep as 760 m (2,500 ft) below the plateau surface.

A masterly engineering plan, involving the construction of 16 large and several small dams, more than 145 km (90 mi) of large diameter tunnels, and more than 80 km (50 mi) of pipe aqueducts, has transformed the entire drainage pattern of this mountainous area. Water is now diverted through seven power stations (two of them underground) and a pumping station, with a total installed capacity of 3.74 million kW. After being used to generate power, the water is available, in the amount of 2.46 billion m^3 (2 million acre-feet) annually, for irrigation in the famous Murray and Murrumbidgee Valleys, and is diverted westward under the range. Australia is the driest continent in the world; its average annual rainfall is only 42 cm (16½ in)—half that of California—and its total runoff is about equal to that of the Columbia River, although the area of the country is about the same as that of the United States. Thus, the economic significance of the Snowy Mountains scheme is immediately apparent. Much of the limited rainfall falls on the Great Dividing Range, and the eastward-flowing waters run directly to the sea; by turning the water of the Snowy River westward, the Snowy Mountains Hydro-Electric Authority has put to good use the runoff from the part of the range which has the highest rainfall.

The geology of this region was first known in broad outline only as recently as 1948. Systematic mapping of the geology started in that year as a joint venture, and there was a little geological information available when the planning of the Snowy Mountains scheme started. As an early step, the Authority established an engineering geology branch which eventually included nine geologists and one geophysicist. The branch initiated detailed studies of sites under consideration for engineering works; along with the general mapping program that has continued, these studies obtained accurate information on the geology of this interesting area. Mount Kosciusko and about 1,000 km^2 (400 mi^2) of high land around it, extending down to 1,450 m (4,800 ft) above sea level, constitute the only part of Australia that was glaciated; three phases of glaciation have been recognized, and all the main features of glaciated country can be seen in this small area.

The greater part of the region consists of intrusive granites and granitic gneisses, hard metamorphic rocks that often show remnants of sedimentary bedding. The rest of the area is largely composed of highly folded sedimentary and metamorphosed sedimentary rocks generally of Ordovician age. These were originally muds, silts, sands, and some volcanic ash, subsequently compacted, cemented, and sometimes recrystallized. Belts of cavernous limestone have also been found. Sedimentary and volcanic rocks of Devonian age overlie Silurian rocks, mainly gently folded, along the middle section of the Tumut River where two of the major projects are located. Scattered areas of almost horizontal beds of basalt lava flows, of early Tertiary age, occur in one district and cap some of the highest ground. Limited deposits of alluvium along existing streams complete the geological picture of this unusually complex area. Because of the variety of rocks, the engineering of the Snowy Mountains scheme had to contend with most of the major types of geological problems encountered in other countries over much greater areas. Careful study of the geology was an essential part of the preliminary work for all the individual engineering projects. The Snowy Mountains scheme, which involved construction of an earth- and rock-fill dam 114 m (381 ft) high, was completed two years ahead of schedule. In the process, a world's record for tunneling progress—177 m (590 ft) driven from one heading in a 6-day week, and 209 km (697 ft) in 7 consecutive days' work—was set. Such remarkable progress is not unrelated to the accurate geological information available for both design and construction purposes.

One of the most serious geological problems encountered on this project was the extent and nature of weathered rock. Much of the area is covered with a mantle of residual soil of great thickness, a mantle derived from the weathering in place of the bedrock. It is unusual to find such soil in close proximity to glacial deposits, but this is merely one of the geological features that distinguished this enterprise. Determination of the extent of the rock weathering was naturally a problem to be faced in connection with every dam site, every tunnel route, and every potential powerhouse location. Investigations showed at an early stage that the type of rock most susceptible to weathering is, unfortunately, one of the most abundant in the area, a medium- to coarse-grained gray granite with a high content—(up to 20 percent)—of biotite mica. It was frequently found to be weathered to depths of 30 m (100 ft) below the surface, where it might then pass abruptly to relatively fresh rock. Even in completely weathered zones, large residual boulders of fresh or only slightly weathered granite surrounded by completely weathered material were often found; since they sometimes occurred in clusters, their detection by test drilling was unusually difficult. Weathering was naturally more prevalent in faulted zones than in solid rock; the Snowy Mountains contain many faults, though few are to be seen at the surface, and this further complicated subsurface exploration. Even the gneiss was extensively weathered in places; at the Tooma Dam, weathering extended to a depth of 42 m (140 ft) at the left abutment, and bands of moderately weathered rock were found in drill holes at a depth of 87 m (290 ft) below the surface.

Preliminary investigations for the project started with reconnaissance geological mapping to a scale of 1:16,000 (4 in to 1 mi) (see Fig. 2.1). This was followed by detailed geological mapping to a scale of 1:2,400 (1 in to 200 ft) for tunnel routes and of 1:600 (1 in to 50 ft) for dam sites. Diamond-core drilling then followed; all holes were located

2-4 GEOLOGICAL BACKGROUND

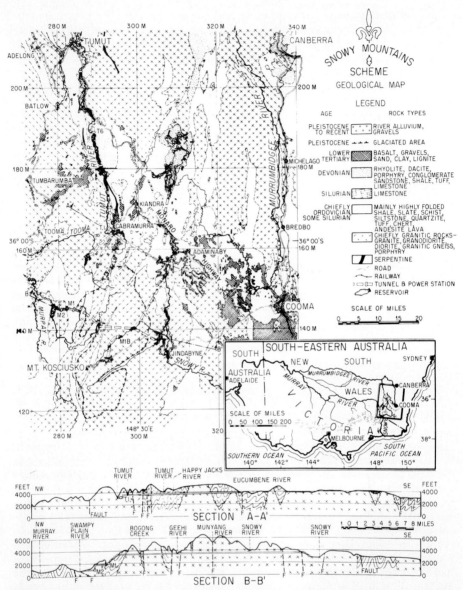

FIG. 2.1 Generalized geological map of the area utilized by the Snowy Mountains project in New South Wales and Victoria, Australia.

by the geologists on the basis of their mapping work. After the geological structure was revealed, most of the diamond drilled holes were subjected to high-pressure water tests, and thus were put to further use. With the aid of rubber packers, holes were tested in sections, and any losses of water were carefully measured and correlated with the structure revealed from the core. Finally, holes were logged electrically by means of a single-electrode electric logger. An instrument at the surface recorded, as the logger was lowered down the hole, the electrical resistivity and the changes in self-potential. Naturally, this logging could be carried out only in water-filled holes that were uncased, but it proved to be a valuable supplement and was easy to use; several hundred feet could be logged within the space of an hour. The resistivity of weathered granite, as well as that of clay seams or faulted zones, is considerably less than that of fresh granite; all these features showed up clearly on the recorded electrical logs. Geophysical methods of subsurface investigation, especially seismic refraction, were also used where appropriate. Study of the petrography of rock samples in the laboratory was a useful supplement to field investigations. Even though the area covered was almost 13,000 km^2 (5,000 mi^2), careful correlation of all preliminary investigation work gradually built up a general picture of the geology of the entire region, a picture that became increasingly valuable as the work progressed.

Specific examples will be cited later in this book, but to illustrate the application of the geological studies, the Tumut Pond Dam site may be mentioned, (See Fig. 2.2). This site was carefully studied in the manner described and good rock conditions were obtained, even though weathering extended to a depth of 18 m (60 ft) and limonite staining extended down to 51 m (170 ft) in test holes. Just upstream of the dam site, however, a fault zone had been detected by a careful study of aerial photographs and of the actual topography of the site. Upstream of the fault plane, a zone of crushed and sheared granite extended for about 90 m (300 ft); this and associated features of the fault severely limited the choice of the intake for the Tumut No. 1 pressure tunnel which was to convey water impounded by this dam. The machine hall of Tumut No. 1 power station, 23 m (77 ft) in maximum width, 31 m (104 ft) high, and 92 m (306 ft) long, is 330 m (1,100 ft) underground and located in closely jointed granitic gneiss and less-jointed granite. Following the preliminary geological studies, a site was selected which avoided the major faults of

FIG. 2.2 Tumut Pond Dam, a key unit of the Snowy Mountains project, Australia.

2-6   GEOLOGICAL BACKGROUND

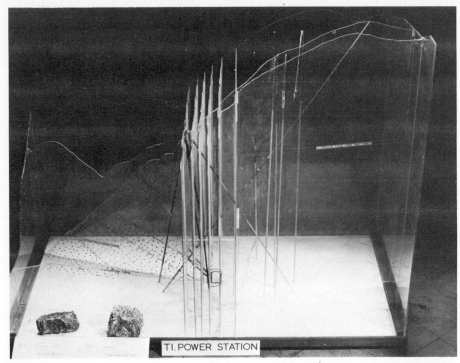

FIG. 2.3 Model of the location of Tumut No. 1 power station of the Snowy Mountains project, Australia (showing the utility of plastic for such purposes).

the region, and following the more detailed studies, the machine hall was sited chiefly in the less-jointed granite with a favorable orientation with respect to the joint pattern (See Fig. 2.3). Study of available rocks for use as concrete aggregate was another urgent preliminary; the biotite-rich granite gave reasonable concrete for the Guthega Dam and Tunnel (the first parts of the scheme to be completed), even though the biotite mica was found to be freed from its matrix by crushing in the fine aggregate. Other granites also proved to be satisfactory, but crushed diorite from a source convenient to the works was used as a main supply for Tumut Pond Dam and for the lining of its pressure tunnel. These are but typical examples of what geology contributed to this greatest of engineering projects of the Southern Hemisphere and, indeed, one of the outstanding multiple-purpose engineering projects of the world.[2.1]

2.2   THE GLEN SHIRA PROJECT

Different in scale but similar in significance is the Glen Shira hydroelectric project of the North of Scotland Hydro-Electric Board, located in the Highlands of Scotland close to the ancient town of Inveraray near the head of Loch Fyne. As a part of the progressive development of the water power available from even the relatively small rivers and streams of the Highlands, the flow of the river Shira and a number of adjacent streams has been controlled by dams and intakes; the resulting water is conducted by a pressure tunnel across a small divide in such a way as to take advantage of a sudden drop to sea level, giving a combined total head of about 330 m (1,100 ft). The main dam, a buttress structure of mass concrete 675 m (2,250 ft) long and 40 m (133 ft) in maximum height, impounds the flow of the river and local streams from a catchment of just over 34 km² (13 mi²). (An average annual rainfall of 2.55 m [105 in] makes this small catchment understandable.)

Local topography dictated a two-stage scheme. The consulting engineers took advantage of this to make the upper station, just below the main dam, a combined power-generating and pumping station; excess water is pumped back when available into the main reservoir from the lower reservoir which is formed by two dams separated by a large rock knoll. One of these dams is a concrete overflow section, 120 m (400 ft) long and 17.4 m (58 ft) high; the other is an earth-fill dam 180 m (600 ft) long and 16 m (53 ft) high. The pressure tunnel leads from the lower reservoir and is 6,315 m (21,050 ft) long; it has two main sections driven from four portals and one subsidiary entrance section; connections are made by the use of steel-pipe crossings over the streams that intersect the tunnel line with their valleys. The tunnel, 3 m (10 ft) in finished diameter, leads to an inclined shaft, steel-lined and with decreasing finished diameter (1.8 m or 6 ft minimum), that conveys the water to the small Clachan power station on Loch Fyne at sea level. The upper (Sron Mor) station generates 6,000 kW under 41.5 m (138 ft) head, the Clachan station 42,000 kW under 288 m (960 ft) head. (See Fig. 2.4).

FIG. 2.4 General view of the two impounding dams of the Glen Shira hydroelectric project, showing typical Scottish Highland topography.

The ingenious way in which topographical features were put to good use in planning the two-stage arrangement of the project, the use of the buttress type of dam, and the fact that the Clachan station was the first underground power station to be completed in Scotland distinguish the Glen Shira scheme as another example of first-class British civil engineering practice. In addition, the project is located in the Scottish Highlands, and all geologists know that the geology of the Highlands has presented some of the most baffling of all geological problems from the scientific point of view. Geology, therefore, might be expected to have played an important part in the successful prosecution of the Glen Shira works. This was certainly the case. The geology of every part of the works having been most carefully studied, geological difficulties were generally overcome. They revealed themselves, however, in one most unexpected way before the completion of the project.

2-8 GEOLOGICAL BACKGROUND

The area, part of the Dalradian metamorphic complex, is typical of the Grampians. Rocks consist in the main of phyllites, schists, and some epidiorite masses; there are thin intrusions of igneous rocks and a large zone of limestone north of the area covered by the works. The phyllites are soft and fissile and generally form poor foundations; the schists, on the other hand, are, on the whole, sound and massive. Foundations for the main dam, on a site dictated generally by topography, consisted of soft phyllites, some bands of quartzite, and an extensive zone of limestone. Loading tests were performed at the site because of the poor and varying bearing capacity of the foundation beds, and unit loads up to 111 kg/m^2 (25 tons/ft^2) gave acceptably small deflections. Pressure grouting was carried out beneath the dam; the average intake was 3.3 kg of cement per square meter of curtain (0.67 lb/ft^2). Under the concrete spillway part of the second dam, cement intake during grouting was 5.9 kg/m^2 (1.2 lb/ft^2); this showed the variability of the character of the rock, since the two dams were less than half a mile apart. A special stilling basin had to be incorporated in the spillway structure because of the nature of the bedrock and the necessity of avoiding its erosion.

Material conveniently available for the construction of the earth dam was from a glacial moraine. Study showed this to be essentially weathered mica schist, the particle size of which went down to rock flour. Its character was therefore extremely variable, and due allowance had to be made for this in the cross-sectional design. Owing to the high rainfall—an average of 0.38 m (15 in) of rain fell in each of four successive months during the construction period—construction of a rolled-fill dam was not easy; therefore a special installation of pore pressure gauges was included for continued observation of the behavior of the morainal material when in place and on rapid drawdown of water level in the reservoir in operation. Correspondingly, and because of the character of the earth fill, a thin flexible reinforced-concrete core wall was used, ingeniously hinged and sealed at the bedrock in the cutoff trench, from which pressure grouting was also carried out.

The pressure tunnel was carried through a variety of geological formations, the anticipated nature of which dictated the decision to make it circular in cross section. Only 5½ percent of the total length required steel-rib supports. Overbreak averaged 7 percent on excavation, but this increased somewhat before concrete work was completed because of weathering of the exposed rock. On the average 3.3 kg of explosives were used per cubic meter of rock excavated (5.5 lb/yd^3). The inclined shaft leading to the powerhouse was excavated through quartz schist with some igneous intrusions. The bedding planes were fortunately normal to the axis of the shaft; therefore, excavation was not a difficult operation, and dry working was obtained throughout. The steel liners were lowered into the shaft and concreted in when excavation was complete; finally, pressure grouting was carried out in the surrounding rock through plugs left in the lining.

As this grouting was proceeding, a bulge in the steel lining suddenly developed about halfway down the shaft and the steel plate was deflected inward by 178 mm (7 in). Since it was essential to get the station into operation, this part of the lining was temporarily stiffened, and after a careful study was made of the situation, the water was admitted. After about six months of successful operation, the tunnel was dewatered and the bulge was examined. The affected section was cut out and replaced. Observations at that time showed that possibly excessive pressure had been used for grouting, but a small trickle of water was also noticed. This was checked back and correlated with the geological section. It was then found that a fissured intrusion of epidiorite upstream of the point where the bulge occurred permitted a little groundwater to percolate as far as the quartz-schist zone downstream. Since this zone was more impervious, a buildup of hydrostatic pressure equivalent to the static head from ground surface occurred as groundwater accumulated. Figure 2.5 shows how this detail of geological structure could create the unexpected water pressure which had such an unfortunate effect. After repairs had been completed, a gauge

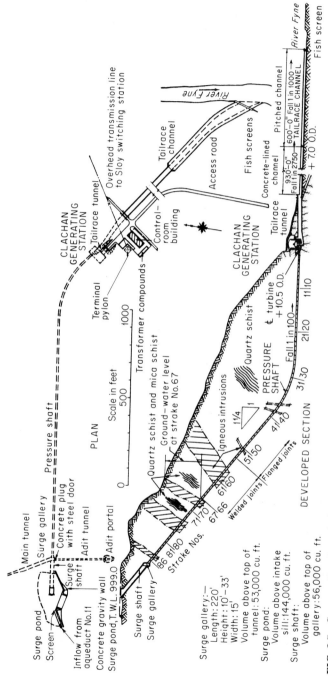

FIG. 2.5 Cross section through the surge shaft and pressure shaft of the Glen Shira hydroelectric project in Scotland.

that had been fitted to the lining at this point recorded water pressure up to 683 kPa (100 psi), a figure corresponding exactly with the static head.

The decision to locate underground the power station, a relatively small building containing only one generating unit, was made on the basis of careful economic studies linked with knowledge of the rock to be encountered. The decision was also in keeping with the stringent requirements for minimum interference with the local scenery, so justly famous. The rock is mica schist, including several crushed zones, and has a dip to the north-northwest of 30 to 40°. Although rock conditions necessitated a longer roof span than was originally contemplated, this was easily achieved in the design of the thin reinforced-concrete arch used as the roof, since its soffit was close to the surface. Provision of the necessary aggregate for the job also called for careful geological study. The nearest sand was 72 km (45 mi) away. Investigation of the properties of the various types of rock available on the job showed that epidiorite could best be used. Therefore, 500,000 tons of this rock, obtained from an outcrop close to the main dam, were crushed in a special plant that gave not only crushed rock for coarse aggregate but also appropriately graded rock sand. The project was completed in 1957 and has since supplied 80 million kWh per year to the Scottish electrical grid; it is being operated as a peak-load station to take full advantage of its water-storage possibilities.[2.2]

2.3 THE KINZUA DAM

Allegheny Reservoir is an important flood-control reservoir in the upper reaches of the Allegheny River in northwestern Pennsylvania and southwestern New York, extending about 48 km (30 mi) upriver from about 14.5 km (9 mi) east of Warren, Pennsylvania, to the downstream end of Salamanca, New York. It was created by the construction of the Kinzua Dam by the U.S. Corps of Engineers, the project being essentially complete by 1966. The reservoir when full covers 8,560 ha (21,180 acres), maximum flood storage level being 410 m (1,365 ft) above sea level. Flooding of the reservoir necessitated the relocation of 134 km (83 mi) of highways and 60 km (37 mi) of railways. The dam is a combined concrete gravity and earth-fill structure 584 m (1,879 ft) long, its crest 54 m (179 ft) above stream-bed level. Total investment in the project was $109 million, the dam structure accounting for $22.6 million. It is estimated that, excluding the major benefits contributed by the project to minimizing the damage from Hurricane Agnes in 1972, the direct benefits attributable to the Kinzua Dam amounted to 87 percent of the total cost of the project within its first ten years of operation (See Fig. 2.6).

In the vicinity of the dam, the Allegheny River flows generally southward from near Salamanca, New York, swinging to a westerly direction at Big Bend near the village of Kinzua. In this pleasant hill-and-valley country, there was an "obvious" location for the dam just downstream from the confluence of Kinzua Creek and the main river, at the upper end of an almost gorgelike section of the valley into which the river flows from a quiet stretch in the unglaciated Kinzua Valley. When the project was first given serious consideration in 1936, this site was the one naturally selected, three lines of NX (3 in) borings being put down on parallel axes. These revealed sound bedrock but at depths up to 22.5 m (75 ft) below stream bed near the center of the narrow valley. Estimates of cost, based on preliminary designs, were prepared on the assumption that the dam would be a solid concrete gravity structure.

Then occurred a 20-year delay in any further consideration of the project, detailed design studies being resumed in 1955. During this period, geologists of the Pittsburgh office of the U.S. Corps of Engineers, notably Shailer Philbrick to whose writings this summary is indebted, were able to study the complex geology of the Allegheny Valley

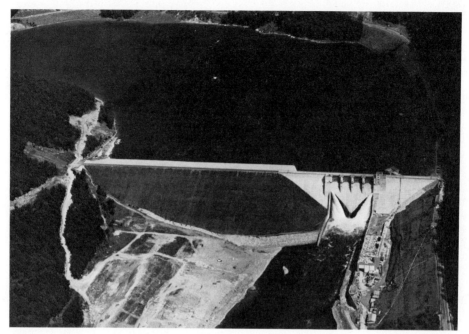

FIG. 2.6 The Kinzua Dam on the Allegheny River near Warren, Pennsylvania.

upstream and downstream of the location proposed for the dam at Big Bend. Present-day geomorphology is the final result of a complex succession of events associated with successive glaciations. Studies were not confined to the field. Every available earlier geological report on the area was examined, valuable information being found in several, notably in a report by J. F. Carll (1828–1904) of the Second Pennsylvania Geological Survey. In studies of gas and oil wells in this area Carll had noticed that the bedrock surface slopes downward to the north, the opposite of what would be expected from the topography and present direction of river flow. He deduced from this that the preglacial drainage of this area was to the north rather than to the south, as at present.

Philbrick and his fellow workers were able to confirm this deduction. Geophysical work and extensive test boring and sampling during 1955 and 1956 confirmed that in the vicinity of Big Bend there was a bedrock shelf 135 m (450 ft) wide at an elevation 11 m (37 ft) higher than the bedrock at the site which had seemed the obvious site in 1936. Wisconsin-age outwash materials which filled the deeper preglacial valley on the north side of the present river valley at the Big Bend location, would permit the use of an earth-fill dam for the major part of the river crossing. A concrete gravity structure founded on the bedrock shelf on the south side of the valley would be more than adequate for the necessary spillway structures. A combined structure was then designed on the basis of these geological findings, showing a savings in capital cost (as compared with the 1936 design) of over $4 million in 1957 dollars (See Fig. 2.7). Suitable material was available in the vicinity for the earth fill. A slurry-trench concrete wall 320 m (1,066 ft) long and 0.76 m (2½ ft) wide, was installed as a cutoff in the earth fill, embedded 2 ft into bedrock because of the pervious nature of the fill used and of the outwash material in the valley bed.[2.3]

It may be thought that it was the 20-year delay in the construction of the Kinzua Dam that enabled the geological studies to be so constructively utilized. The delay did permit

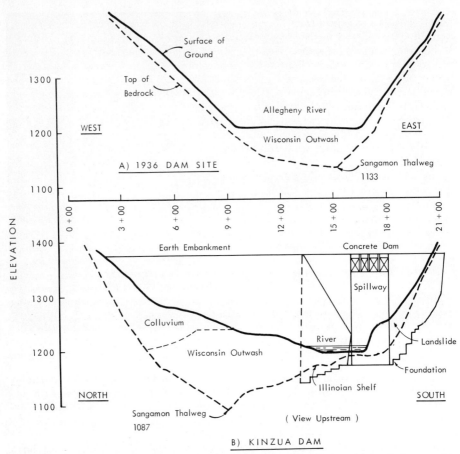

FIG. 2.7 Cross sections through the site proposed in 1936 and that later used for the construction of the Kinzua Dam on the Allegheny River, Pennsylvania.

investigations more thorough than usual to be made, but it is to be noted that the original deduction of reversed drainage in preglacial times was first propounded by Carll as early as 1880. In his paper describing this work, Philbrick lists four other earlier papers on the geology of the region which assisted in the elucidation of the geomorphology of the valley. It is also of significance that the test drilling in the vicinity of the site chosen in 1936, on the basis of perfectly acceptable topographic interpretation, although well done, could not of itself indicate that the original site could be improved upon if the underlying bedrock geology were more fully understood. Even the best of test-boring programs may prove to be less than fully effective unless carried out in full conformity with the geology of the region and not just of the site under study.

2.4 IN TIME OF WAR

It is easy to forget that civil engineering was originally an outgrowth of military engineering. Military engineering has always had its own special demands. As practiced in two world wars, it also gained from the application of geology, even under the imperative of front-line action, and further corroboration of the utility of geological applications in engineering work can be gained by a brief review of their use in time of war. The authors naturally share the worldwide hope that applications of geology in military engineering will in future be limited in character and extent. Applications in the two world wars were, however, so extensive that some reference to them seems desirable, if only to confirm the wide-scale importance of geology in all engineering works involving use of the ground.

Since there is a fairly general impression that geology was used as an aid to military engineering for the first time during the war of 1939-1945, attention may first be paid to a rather neglected chapter in the story of engineering geology, that is, its use during the First World War. The long campaign in Northwest Europe was probably the greatest ground battle that the world has ever known (and may it always keep this distinction). Desperate though it was, and high as was the toll of lives that it took, it is valuable as an illustration of the importance of geology. In 1914, no military geological establishment of any sort existed in the British Army; it is believed that the same was true of all the Allied armies and probably of those of the opposing powers. In April 1915, Lt. W. B. R. King, upon the nomination by the director of the Geological Survey of England, was appointed to work with the British War Office on geological problems. Later Captain King, he worked throughout the war on the staff of the chief engineer in France. Meanwhile, in far-off Australia, suggestions were advanced that this member of the British Commonwealth might contribute a special corps of geologists and miners, and an official offer was accepted at the end of 1915. In May 1916, the Australian Mining Corps, organized as a battalion for service in Gallipoli but too late for service there, arrived on the Western front. Major T. W. E. David, who served as adviser in matters of military mining and geology in positions of increasing importance as the war progressed, accompanied this battalion.

By the time of the armistice, five geologists were attached to the British General Headquarters; these men naturally received much assistance from French and Belgian geologists. In 1917, after the entry of the United States into the war, Lt. Col. A. H. Brooks, previously director of the Geological Survey of Alaska, was appointed geologist to the American Army in France; later an assistant geologist was also appointed. At the time of the armistice, plans had been made to increase this small American staff to a total of 17 geologists; most of them were actually appointed, though they were not called upon to serve.[2.4] In the record which he prepared of this wartime work, Lieutenant Colonel Brooks said: "The success of the British in gaining control of the underground situation must in large measure be credited to the refinement of the geological studies and their interpretations made by Lt. Col. T. Edgeworth David."[2.5] All who have studied the 1914 war will appreciate the significance of this tribute in relation to the military operations in France, and in a minor way in relation to the thesis of this book.

A necessarily impersonal treatment of geology and engineering may, therefore, be tempered for one paragraph in order to record something of this great pioneer in the field, a man still relatively little known outside his own Australia. Born in Wales in 1858, David went to Australia in 1882 and was appointed to the chair of geology at the University of Sydney in 1891. He retired in 1922 and died in 1934 as Professor Sir T. Edgeworth David, D.S.O.; he had received many other honors and was widely revered and respected. Some indication of his character is shown by the fact that he became a member of Shackleton's Antarctic expedition at the age of 40 and led the party of three that first reached the South Magnetic Pole. It was at the age of 57 (and when a grandfather) that he volunteered for service with the Australian forces in France. To read of his wartime activities, begun at an age now sometimes regarded as suitable for retirement, gives some slight indication of the inspiration that he must have been to his students and to all who worked with him.

The official records show how he applied his unusual geological skill to the needs of military engineering. Only in his private writing is one privileged to see that, even in the midst of the war, he never forgot his scientific calling. In the winter of 1917–1918, a fossil mammoth was discovered in an unusually deep dugout. The Germans captured this location in their offensive of March, 1918. In a letter written to his home at that time, David said, "The beastly Bosche has gone and captured my fossil mammoth, blast him! But the E. in C. promises to get it back again in a year." The site was actually recaptured four months later, but the Germans had removed the bones; David was most annoyed.[2.6]

An important part of the work of King and David was to assist with mining operations under the enemy's lines. One of the most important projects of this kind was that at Vimy Ridge. Figure 2.8 shows a section of the tunnel that was driven; the geological information was obtained from French maps and scientific papers and checked by shallow borings carried out under the most hazardous conditions. Since the tunnel was driven under the enemy's lines in the Louvil clay, noise was reduced to a safe minimum and the work was undetected. The attack on Vimy Ridge was launched on 9 April 1917, when the end of the tunnel was still 21 m (70 ft) away from the critical location. If the attack had been delayed one week, a much more successful military operation would probably have resulted.[2.7]

In similar tunneling at Messines Ridge, groundwater encountered at one point threatened the success of this immense mining project. David examined the working face at 21m (70 ft) below ground level and deduced that an old river bed was being pierced. He advised dropping a shaft some distance back from the face and then retunneling 6 m (20 ft) below; this was done and the water problem was solved. It was in this location that after 18 months' work by tunneling companies, work guided by geological advice, the greatest explosion ever used in ground fighting was set off. There is on record a wonderful letter from David explaining how he checked on some uncertain geological feature by operating a light test-boring rig with two privates in a small heading in the chalk at 34 m (114 ft) below the surface.[2.8]

Natural tunnels were also used for military purposes. Caves were discovered in the Arras district, for example, and as many as 100 were adapted and used at various times for sheltering Allied troops. New Zealand troops occupied 25 caves that gave warm and comfortable accommodation for 11,000 men. Geological examinations were always made in advance of using caves and prior to drilling the necessary ventilation shafts (for which 6-in boreholes were often used). It was found that the caves were really old quarries from which unweathered chalk had been removed in the seventeenth century for the rebuilding of the city of Arras.

Geology assisted in a different direction when in 1918, the British government complained to the Dutch government that a lot of Rhine gravel used in the construction of German pillboxes was being brought in by way of Dutch canals that were supposed to be neutral waters. This protest was passed on to the German government, which replied that such gravel was being used for peaceful purposes only. King and David then went to work, obtained samples of the aggregate from actual pillboxes, and found it to be a peculiar sort of basaltic rock which occurs only at Niedermendig on the Rhine above Cologne; this was later checked by more detailed studies in England.[2.9]

An even more remarkable detection was made during the war of 1939–1945 by a paleontologist, K. E. Lohman, on the staff of the American Military Geology Unit headed by W. H. Bradley. Lohman was able to determine the exact location of the launching sites of the small balloons sent up by the Japanese (so many of which reached the west coast of North America) by a study of the diatoms recovered from sand used as ballast in the balloons; the accuracy of the deduction was checked after the end of the war.[2.10]

The great mobility of military operations in the Second World War set a different pattern for the application of geology, recognized from the outset as an essential aid to mil-

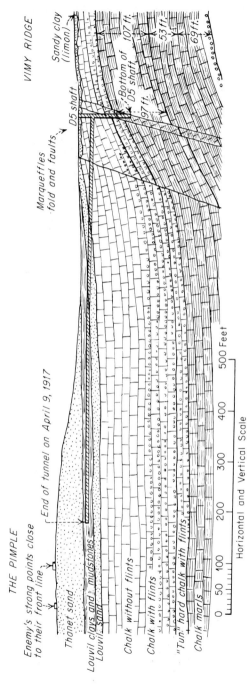

FIG. 2.8 Geological section showing the mining operations carried out during the First World War at Vimy Ridge; the tunneling was kept in clay as far as possible to ensure silence during the work.

itary intelligence and engineering. Trafficability maps assumed considerable importance and were produced in great numbers. Geologists studied the existing geological records of the countries involved and were aided by the results of aerial reconnaissance and, occasionally, by actual field investigations. Prior to the invasion of France, for example, geologists checked through French geological records and found that the beaches proposed for the landings appeared to be sandy but really had clay close to their sandy surface. This was first suggested, apparently, by one of Field Marshal Montgomery's scientific advisers who had spent a holiday on the Bay of the Seine before the war and had noticed the clay. Study of French records (including a paper in the *Bulletin de la Société Préhistorique Française*, noted by an official at the British Museum) confirmed these suspicions. The clay areas were then located on aerial photographs. By means of special commando raids, actual samples were brought back to England for study.

A beach with similar clay was located in Norfolk, England, and when the new heavy tanks were tried out there, they became bogged down. A new device was therefore developed which enabled a tank to lay its own path across a clay patch and to leave it in place for following tanks. In the actual invasion, the first tank out of each landing craft was fitted with this special device, and so the beaches were crossed.[2.11] Shortly afterward, however, geology appeared as an aid on the opposing side, for in the terrible fighting around Caen, any possibility of a surprise attack by the British was eliminated because the cavernous limestone of this area acted, in effect, as a sounding board for heavy vehicles. General Dietrich of the German Army said that he could observe the movement of British tanks merely by putting his ear to the ground, a trick he had learned in Russia.[2.12]

These incidents demonstrate the important role that geology can play, in one way or another, in actual combat. There remains to be mentioned another military by-product in the field of engineering geology, one of lasting value and of direct application to the needs of civilian life. Soon after the entry of the United States into the First World War in 1917, a subcommittee of the Council for National Defense suggested to the geological surveys of those states in which military cantonments were located that they might well prepare a bulletin on the topography and military geology of the areas around each cantonment; the subcommittee even gave an outline of what it thought such a bulletin might contain. So far as is known, only one major bulletin of this type was ever produced, although doubtless many interim studies and reports were prepared.

The published volume is Bulletin 23 of the Geological Survey of Ohio, a volume of 186 pages, issued in 1921.[2.13] It describes the geology of the Camp Sherman Quadrangle, in which the small city of Chillicothe is located; this historic town was once the capital of the Northwest Territories (of the United States, not Canada), and later still, an early capital of the state of Ohio. The volume presents a clearly written appreciation of the topography of the area under review, of its geology, and of the significance of topographical and geological features; the volume is directed toward the military needs of Camp Sherman but would be useful for any civil engineering work carried out in the area. It was an early example of a study in urban geology. Even today, this bulletin remains a little classic of its kind.

2.5 DUISBURG HARBOR

The river Rhine is one of the great inland waterways of the world, carrying a tremendous traffic from the North Sea to as far as Switzerland, with canal connections that enable it to serve large areas of France, West Germany, and the Low Countries. Its principal inland port is at the ancient city of Duisburg, at the mouth of the tributary Ruhr River, and the associated canal. The port facilities here include 45 km (24 mi) of masonry wharves enclosing a water area of 230 ha (570 acres), facilities capable of handling as much as 45 million tons of cargo a year. The harbor is used every year by over 50,000 shallow-draft vessels, its peak year having seen 27.4 million tons of cargo handled. Because of steady improvements for navigation on the Rhine starting at the end of last century, improvements involving much dredging, water levels in the river have been slowly dropping. The churning effect on bottom sediments caused by the propellors of heavily laden craft has increased this interference with natural conditions to such an extent that the water level in the harbor of Duisburg has dropped 2.4 m (8 ft) since the turn of the century. Hydraulic studies suggest that a further drop of 1.6 m (5 ft) is probable before the end of this century. The serious consequences of this for the operation of Duisburg Harbor can well be imagined. (See Fig. 2.9)

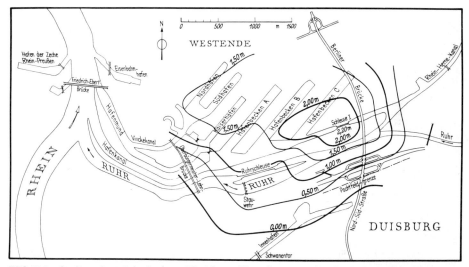

FIG. 2.9 Outline plan of the harbor of Duisburg, West Germany, showing estimated settlements.

The harbor authorities studied carefully all possible remedies. Enclosure of the harbor area by an entrance lock from the Rhine would have ensured constant water level, but locking in and out would have cut the efficiency of the harbor operations by possibly 50 percent. Rebuilding the wharf walls and dredging out the harbor bottom to give satisfactory clearance for vessels would have been possible, but estimates of the cost of this major construction operation were almost astronomical. Finally, thought was given to the geology underlying the city and harbor.

The entire harbor area is underlain by Upper Carboniferous beds of the Westphalian A series. In the lower Bochumer and upper Wittener beds occur four valuable coal seams. These are located approximately 100, 200, 400, and 600 m below ground level. The upper surface of the Carboniferous beds is about 65 m (213 ft) below ground level, being overlain by Tertiary and Quaternary deposits of gravel, fine sand, and silt—often described as marl—the silt being generally impermeable (See Fig. 2.10). The upper three seams of coal were known to contain something like 30 million tons of coal, but extraction of this had, naturally, been prohibited by law, since mining below such a critical area as the harbor could not be considered in view of the probable dangers of subsidence. This *Sicherheitspfeiler*, administered by the national mining authority, was essentially for the protection of the harbor and the surrounding city area. Since the geology was so well known, and

2-18 GEOLOGICAL BACKGROUND

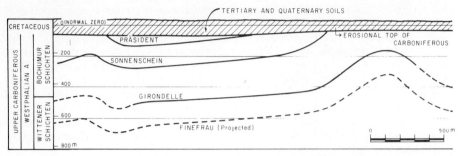

FIG. 2.10 Diagrammatic geological section beneath the harbor of Duisburg, West Germany.

since there had been a limited amount of mining beneath nearby canals, the idea was conceived of mining out the coal beneath the harbor, under the strictest controls and most expert supervision, so as to induce settlements of the amount necessary for the continued operation of the harbor, despite the dropping water levels. Negotiations were therefore initiated by the harbor authorities with the national mining and waterways authorities, in cooperation with a coal-mining company that already had a working shaft from which mining could be extended beneath the harbor area.

The harbor authorities prepared a plan of the harbor showing the subsidence that they would like to have achieved, varying downward from 2.0 m (6.5 ft) at the inner end of the harbor. The coal company, after detailed surveys by their expert staff, promised to achieve the subsidence desired with an accuracy of ± 5 percent. Permission for this bold scheme was eventually granted. More than two years were spent in preliminary planning and investigation, the underlying geology being mapped in greater detail than previously,

FIG. 2.11 Model of the geological structure beneath the harbor of Duisburg, West Germany, as used during the coal-mining operations.

FIG. 2.12 Typical masonry wharf wall in Duisburg Harbor, West Germany, showing very slight differential settlement.

all structures in the area to be affected being surveyed and photographed for checks on any possible future settlement. More than 3,000 reference points were established so that the progress of mining could be checked by the subsidence induced at the surface. Mining of the coal started in 1956 and continued, under the strictest controls, and at the hands of most experienced miners, until 31 July 1968 (See Fig. 2.11).

Work had then to be stopped, a little short of the ultimate goal, but with practically all the desired subsidence achieved, and with the accuracy guaranteed. Minor faults encountered underground had made further work undesirable, but the harbor was saved and its steady use can now be confidently anticipated. In addition to much complex unloading equipment along all main wharves, the area which was lowered contained coal tips, warehouses, several five- and six-storey industrial buildings, a coal-processing plant, several large tank farms, five shipyards, two steel-fabricating plants, and several miles of roads and railways. No damage was done to any structure, the entire operation being completed with only one or two minor effects at ground level (See Fig. 2.12). Some of those using these facilities did not even know that their installations had been lowered until told about it. And all was dependent upon accurate knowledge of the underlying geology.[2.14]

2.6 REFERENCES

2.1 D. G. Moye, "Engineering Geology for the Snowy Mountains Scheme," *Journal of the Institution of Engineers of Australia*, **30**:287–298 (1955); see also **32**:85–87 (1960) of same journal.

2.2 J. Paton, "The Glen Shira Hydro-Electric Project," *Proceedings of the Institution of Civil Engineers*, pt. I, **5**:593–632 (1956).

2.3 S. Philbrick, "Kinzua Dam and the Glacial Foreground," in D. R. Coates (ed.), *Geomorphology and Engineering*, Dowden, Hutchison and Ross, Stroudsburg, 1976 pp. 175–197.

2-20 GEOLOGICAL BACKGROUND

2.4 "The Work of the Royal Engineers in the European War 1914–1919; Geological Work on the Western Front," Institution of Royal Engineers, Chatham, England, 1922.

2.5 A. H. Brooks, *The Use of Geology on the Western Front* U.S. Geological Survey Professional Paper 128D, 1921.

2.6 M. E. David, *Professor David,* Arnold, London, 1937.

2.7 "The Work of the Royal Engineers . . ." op. cit., p. 28.

2.8 David, op. cit., p. 232.

2.9 Ibid., p. 256.

2.10 W. P. Woodring, Introductory Paper in The Bradley Volume, *American Journal of Science,* **258**A:3 (1960).

2.11 Chester Wilmot, *The Struggle for Europe,* Fontana Books, Collins, London, 1960 p. 219.

2.12 Ibid., p. 410.

2.13 J. E. Hyde, *Geology of the Camp Sherman Quadrangle,* Geological Survey of Ohio Bull 23, 4th ser., 1921.

2.14 R. F. Legget, "Duisburg Harbour Lowered by Controlled Coal Mining," *Canadian Geotechnical Journal,* **9**:374–383 (1972).

chapter 3

GEOLOGY: AN INTRODUCTION

3.1 Early History / 3-1
3.2 William Smith / 3-2
3.3 An Outline of the Science / 3-3
3.4 The Science Today / 3-4
3.5 The Earth's Crust / 3-5
3.6 Rocks and Minerals / 3-6
3.7 The Geological Cycle / 3-7
3.8 The Geological Succession / 3-8
3.9 Geological Maps / 3-13
3.10 Geological Fieldwork / 3-13
3.11 Conclusion / 3-14
3.12 References / 3-15

Geologists are concerned with every aspect of the composition and structure of the earth's crust. Their sphere of work is, therefore, worldwide; their main laboratory is the great out-of-doors wherein they examine rocks as those rocks actually occur in nature. Their considerations range from the beginning of time to far into the future. They study all that composes the crust of the earth and especially those materials of use to their fellow men. Yet geologists have sometimes been regarded as unduly academic persons. This paradox perhaps arises because of the extent of the field covered by the science of geology and because of its apparent remoteness from the things of every day.

3.1 EARLY HISTORY

This strange paradox—the contrast between the wide extent and importance of all that is now studied in the field of geology and the generally restricted appreciation of the significance of that study—is found to have existed from the earliest days of scientific endeavor. The idea of devoting a special branch of science to the study of the earth itself was not then contemplated, and even the references in early writings to "earth" as one of the four elements bear little relation to the modern concept of geology. Nicolaus Steno (1631–1686), at one time bishop of Hamburg and vicar apostolic of Denmark, is generally regarded as the founder of geology as an independent branch of science. Of special interest to engineers is the fact that Robert Hooke (1635–1703), who was professor of geometry at Gresham College when he was 30 and whose name is now associated with the law relating stress and strain, was the first man to suggest that fossils could be used to construct a chronology of the earth.

The name *geology* does not appear to have been used until the end of the eighteenth century. It is derived from the greek *geo*, the earth, and *logos,* a speech or discourse; the name is thus truly descriptive. As soon as geology was recognized as a branch of scientific inquiry, it developed rapidly, and before too many years had passed, the true extent of its scope was appreciated. The wide field thus made available for study encouraged, rather than discouraged, interested investigators so that geology, despite its late start, now ranks as one of the leading branches of natural science.

The first workers were mainly English and Italian, but other Europeans soon made their special contributions to the advance. J. G. Lehmann, a German who died in 1767, was the first, for example, to record the possibility of order in the arrangement of the rocks of the earth's crust. De Saussure (1740–1799), a Swiss who studied the Alps carefully, was the first to give the generic name to the science; previously it had been regarded as a part of mineralogy, a position that today is really reversed. James Hutton (1726–1797), a Scotsman of Edinburgh, was perhaps the first great name in the annals of the science; his *Theory of the Earth,* published in 1785, republished in 1795, and finally popularized by Playfair (1748–1819) in 1802, provided the basis for much of the great advance made during the nineteenth century. Hutton presented evidence to show that study of the present processes acting on the earth can be used to interpret the past. A. G. Werner (1749–1817), a German, was also a most potent influence in geology during these same years, drawing pupils from all over the world to his classes at Freiburg Mining Academy.

William Maclure (1763–1840) is known today as the "Father of American Geology," and Sir William Edmund Logan (1798–1875) will long be remembered for his pioneer work in Canada. Maclure's major contribution was the publication in 1809 of a booklet and geological map of the eastern United States; a more lengthy version appeared in 1817. Logan organized the Geological Survey of Canada, founded in 1842, and carried out many studies in eastern Canada related to mineral resources and stratigraphic successions. Louis Agassiz (1807–1873) is another name associated with the North American continent; he was a Swiss who emigrated to the United States of America when he was 42 years old, and there established his fame with his work on fossil fishes. He is generally regarded as the founder of the modern school of glacial geology. William Nicol, a lecturer at the University of Edinburgh, by his invention of the Nicol prism and of the method of preparing thin rock sections for examination through a microscope (a method later to be developed by Sorby), placed the whole science in his debt and laid the basis for much of the petrologic work of today. The names of Sir Charles Lyell (1797–1875) and Charles Darwin (1809–1882) cannot be omitted from this review, brief as it is; both were distinguished geologists, although Darwin is perhaps better known as a biologist. Their publications, notably *The Antiquity of Man* and *On the Origin of Species,* achieved fame far beyond the confines of the scientific world. Lyell is generally regarded as the founder of geology as it is known today. His masterly *Principles of Geology* remains one of the classics of the science, and is still well worth reading as background for more detailed studies.[3.1]

3.2 WILLIAM SMITH

One distinguished name was omitted from the foregoing list in order that it might be given special attention—that of the remarkable man whose portrait is the frontispiece of this volume. William Smith is now generally acknowledged as the "Father of British Geology." Born on 23 March 1769, of humble origin, he spent an appreciable part of his life of 70 years as a canal engineer in Somerset, England, and became later a land agent near Scarborough. The science was well founded when he first became interested in the structure of the earth on his journeys on horseback along the post roads of England. He was one of the first to introduce the concept of quantitative study and regular stratigraphical succession and so helped to establish the science on its modern foundations. The work for which he will ever be remembered is the preparation of the first real geological map of England and Wales. The "Map of the Strata," as it was first called, was constantly in his mind for a period of over 20 years, but it was not until the year 1813 that the work was definitely put in hand; William Cary was the engraver. The finished map, colored not unlike a modern geological map, was presented to the Society of Arts in April 1815 and was awarded a premium of £50, designated for the first such map to be produced.[3.2]

Throughout his life, William Smith practiced extensively as a consultant on geological matters, often with special reference to their application to engineering work. Before his death, he was honored by the Geological Society of London by being awarded in 1831 its first Wollaston Medal. The society had been founded in 1807, proof indeed that the comparatively new science was already active and virile. It is of interest to note that the society was thus founded some years before the Institution of Civil Engineers and almost 50 years before the American Society of Civil Engineers. Even more significant, however, is the fact that between the years 1799 and 1825, the period that saw the birth of geology as an organized branch of science, Laplace published his monumental work on astronomy, *Celestial Mechanics;* his earlier *Essay on the System of the World* (published in 1796) contains much that can be read with interest today. Geology had, therefore, much ground to make up, but the years between have seen the gap closed.

3.3 AN OUTLINE OF THE SCIENCE

Geology was originally a descriptive science. The geologist studied rocks in the field, recorded the results, assessed these for the region under investigation, and deduced from observation the structural arrangement and geological history of the rocks seen. Detailed study of the composition of rocks eventually led the geologist into the laboratory, first for microscopic investigations and later for chemical and mechanical analyses. Examination of fossil forms inevitably led to biological studies and, eventually, to that close affinity with biology that has resulted in the joint approaches found today in paleontology. Thus began the association of geology with other scientific disciplines that is so significant a feature of modern geological studies. There are today few branches of natural science that demand a working appreciation of so many sister sciences as does geology.

Of use to the engineer in applications of geology is a familiarity with the way in which that science has branched out into many subdivisions and become associated with other sciences in developing new branches of inquiry. Even though this can here be but a listing of names, the names will at least serve to guide the interested engineer to further study of any subject of special interest. Many of the names are self-descriptive even to those not acquainted with classical languages. *Mineralogy,* for example, is the study of minerals and has long been recognized as a scientific discipline in its own right, even though it is so closely associated with geology. *Crystallography* is, correspondingly, the more detailed study of crystal forms, a discipline not now limited only to geological studies but also widely used in chemistry and physics.

The prefix *geo* is a sure indication of study closely allied with geology. *Geochemistry,* a relative newcomer to the field, is clearly the study of the chemistry of the natural materials found in the earth's crust. *Geography* may not be regarded by geographers as a subdivision of geology (though the reverse is a far more questionable view), but there is a close link between the two disciplines. *Geotechnique* is an old term now gaining new usage as descriptive, generally, of the engineering study of the materials in the earth's crust; its close association with geology is discussed in Chaps. 6 and 7. *Geomorphology* is perhaps the most puzzling of this group of terms; derived first from German usage, it is now used by both geographers and geologists to describe the forms of the earth's surface and the processes acting upon it.

Other subdivisions are described by qualified areas of geology such as *physical geology*, the study of the present form of the earth's surface, its crustal structure, manner of origin, and the nature of the modifying processes at work upon and within it. *Historical geology* involves the synthesis of all those branches of inquiry which point the way to the evolution of the earth's crust as it is seen today; the detailed geological succession is one of the main results of this study. *Stratigraphy* is the closely allied investigation of the arrangement and correlation of layered rocks and the relation of these to earth forms of the past. *Structural geology* embraces the study of rock deformations under the action of stress in the earth's crust. *Marine geology*, a relatively new branch of the science, but one of increasing importance, is related to the seabeds of the globe; its development has been greatly aided by modern underwater techniques. *Pleistocene* (or *Quaternary*) *geology* is a special branch, related not to an area but to the most recent period in geological history.

Of the disciplines with somewhat more unusual names, names generally derived from Latin or Greek, *paleontology* has already been mentioned as the study of fossils and fossil life. *Petrology* is a companion study to mineralogy; it is the study of rocks in the widest sense, their mineral constitution, texture, and origin. As an indication of the extent to which detailed scientific study in the geological field has proceeded, one may note that even petrology now has its own subdivisions, such as *sedimentary petrology*. More frequently encountered in engineering work, however, will be the name *seismology*, the study of earthquakes, earthquake action, and associated vibrations, as well as of all vibrations in the crust of the earth from whatever cause.

There may be some who would expect to see seismology listed as a branch of *geophysics* rather than geology. Geophysics has been left for reference at the end of this summary because of its importance. It is, clearly, the study of the physics of the earth; in recent years, it has probably advanced at a faster rate, and upon a wider front, than any other scientific discipline outside the nuclear field. Among the various factors responsible for this phenomenal advance is the modern progress in scientific instrumentation. One cannot imagine geophysical studies except against a background of geology; correspondingly, one cannot think of geological study today without an appreciation of all that geophysical methods can do to supplement older forms of investigation. Here, then, are two complementary scientific disciplines, the joint progress of which is pushing back the frontiers of knowledge of the earth and of its constitution in a manner and at a rate undreamed of a few decades ago.

3.4 THE SCIENCE TODAY

It has been through the combined efforts of geologists and geophysicists that the modern concept of *plate tectonics* has been developed and has been so widely (although not yet universally) accepted as an explanation of the present arrangement of the continents. Many years ago it was noticed that the shape of the east coast of South America appeared to be similar to the west coast of the continent of Africa. Geological and biological similarities were also noticed and gradually codified. It required, however, the detailed study of the residual magnetism in rocks on both sides of the Atlantic and from the bed of this great ocean to demonstrate the certainty that the two continents were indeed at one time connected. *Drifting continents* has therefore become a term of real meaning. It is now believed that the land masses on the two sides of the Atlantic are still separating, at a current rate of about 6 cm a century. Although this exciting new development is of great scientific and academic interest, with wide implications, it does not affect the applications of geology to the problems of today, the applications with which this volume deals. It is, however, clear indication to engineers that geology as a science is going through a new age of discovery.

The same indication is reflected in the work that has been done in *extraterrestrial geology*. Detailed studies of the surface of the moon and of the other planets by means of the astronomer's tools gave interesting suggestions as to what the geology of the surface of these bodies would probably be. For the moon this has been triumphantly confirmed in detail by the findings of the astronauts who landed on its surface and by the samples they brought back. This detail of geological exploration attracted public attention to a surprising degree, even as it presented new problems to those who were privileged to study the "moon rocks."

It may be said that lunar geology has little to do with the subject matter of this book. It may also be noted, however, that one of the last group of three to make the journey to the moon was a geologist, Dr. Harrison W. Schmidt. In acknowledging a unique presentation made to him by the Geological Society of America, Dr. Schmidt said:

> The new challenge is in the application of earth science to the present. There are few of the answers to the major questions of the day that do not have their foundations in our science. Questions of energy, minerals, space exploration, environment, natural disasters, construction, expanded agriculture, and public education have real answers only in the framework of the Earth and its processes. The call each of us in the profession of geology must hear is a call of commitment, commitment to the rapid establishment of a framework by which rational answers to these questions can be found. We are starting late: thus, the race cannot help but be exciting.[3.3]

3.5 THE EARTH'S CRUST

Detailed knowledge of the nature and composition of the earth sphere is confined to an exceedingly thin crust, a crust much thinner than is generally realized. The diameter of the earth is slightly less than 13,000 km (9,070 mi). Geological investigation by means of boreholes has at present extended to about 7,600 m (25,000 ft) below the surface—no more than 0.05 percent of the total thickness. One may infer, utilizing stratigraphical relationships, that investigations may have extended in some places to a depth of 61,000 m (200,000 ft) but this is still a relatively small distance below the surface. Even the irregularities on the earth's surface—mountain ranges and ocean depths—are of small moment when compared with the whole globe. For the highest mountain, Mount Everest, is 8,848 m (29,028 ft) above sea level; the greatest depth to the ocean bed so far is in the Pacific Ocean, a depth that extends a distance of about 11,776 m (38,365 ft) below sea level. The total of these extremes amounts to only 20,624 m (67,663 ft) or 0.15 percent of the earth's total diameter; and if average land heights and ocean depths (420 and 3,600m, or 1,377 and 11,800 ft, respectively), are taken into account, the figure is only 0.03 percent. About three-quarters of the surface of the globe is covered with water; only one-quarter is exposed ground. The great land masses are made up of rocks of many different types and conditions, structurally arranged in an often surprising but always systematic manner.

3.6 ROCKS AND MINERALS

All rocks are composed of minerals. Minerals therefore are the unit constituents of the earth's crust, and as such they are of great interest and importance to the geologist. The mineral composition of rocks is also of importance to the engineer, for this composition enters to some degree into all applications of the science to engineering work. The study of minerals cannot therefore be neglected by the civil engineer.

Minerals are of many kinds; more than 2,000 separate and distinct kinds are known. Each has its own chemical composition and atomic structure and tends to form crystals whose shapes are determined by this structure. The detailed study and investigation of minerals in the laboratory is a fascinating occupation, but the civil engineer, unless engaged in some special task, will not become involved in such a refined area of geological work. Here the engineer will call on the specialist for assistance. It may, however, be mentioned that the laboratory work involves the use of sections of rocks and minerals, sections ground so thin that they can be mounted on glass slides and examined through a microscope. Using ordinary light and also polarized light, the geologist can determine the optical and physical properties of a specimen. Minerals have long been classified on the basis of chemical composition and crystal form. The larger divisions are determined by chemical similarities; the smaller groups and series within these divisions rest on similarities of crystal form. The more important chemical divisions are given in Table 3.1.

All *rocks* can be divided into three main groups: (1) igneous, (2) sedimentary, and (3) metamorphic. The third group is composed of derivatives from either of the other two. The names of these three great classes are descriptive. *Igneous* rocks (from the Latin *igneus,* of fire or fiery) were formed by the cooling of molten material, which is called, in general, *magma,* erupted from or trapped beneath the earth's crust. *Sedimentary* rocks were deposited in some geological age mechanically—through the agency of water, wind, or ice action—chemically, as in evaporitic salt deposits, or organically, as in the case of

TABLE 3.1 Some Chemical Divisions in Mineralogy*

Division	Typical example	Chemical composition
Elements	Graphite	Carbon
Sulfides	Pyrite	Iron Sulfide
Chlorides	Rock salt	Sodium chloride
Oxides	Quartz	Silicon dioxide
Sulfates	Gypsum	Hydrous calcium sulfate
Phosphates	Apatite	Calcium phosphate
Carbonates	Calcite	Calcium carbonate
Silicates	Feldspar:	
	Orthoclase	Potassium-aluminum silicate
	Plagioclase	Sodium-calcium-aluminum silicate
	Hornblende	
	Augite	Complex metallic silicates, containing sodium,
	Mica:	potassium, magnesium, iron, etc.
	Muscovite	
	Biotite	
	Olivine	Magnesium-iron silicate
	Chlorite	Complex hydrous magnesium-aluminum silicate
	Serpentine	Hydrous magnesium silicate

*Distinguishing features cannot usefully be summarized in tabular form; reference may be made to standard books on mineralogy.

coal. *Metamorphic* rocks (from the Greek *meta*, "between," "as denoting change," and *morphe*, "form," "shape") are rocks changed in some way from their original igneous or sedimentary form.

Since all sands, gravels, and clays are theoretically classed as sedimentary "rocks," one can readily appreciate the fact that such materials cover about three-quarters of the earth's land surface. One must not indiscriminately interpret this monopoly in terms of relative importance, nor is it a guide to the relative significance of the main rock groups for the geologist. For example, many of the most complex geological structures requiring investigation are those presented by igneous formations. For the civil engineer, however, the significance of the main rock groups is to some degree comparable with their relative occurrence, because of the general nature of civil engineering work. Construction operations are generally confined to the proximity of the surface of the earth, and consequently sedimentary rocks are more frequently encountered by the engineer than igneous and metamorphic rocks. The next two chapters present more detailed reviews of rocks and soils.

3.7 THE GEOLOGICAL CYCLE

The study of the processes that have led to the existing structural arrangement of the earth's crust may be approached by considering the similar processes at work in the world today. These are described generally as the *geological cycle* and may be presented as in Fig. 3.1. Denudation of existing rocks, or earth sculpture, is carried out by agencies such as severe changes of temperature (especially the freezing and thawing of ice in cracks),

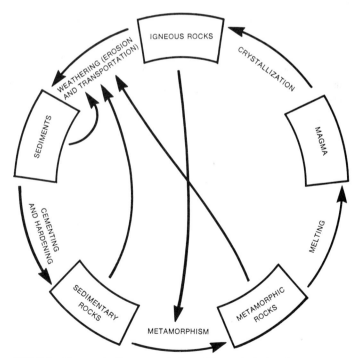

FIG. 3.1 Diagrammatic representation of the geological cycle.

the action of the wind (especially in desert regions), the action of rainwater on exposed soluble minerals and on rocks decomposed by water, the action of running water in eroding rock surfaces, the disintegrating and transporting power of the slow-moving ice of glaciers, and finally the erosive action of waves on all coastlines.

Sedimentation is due to glacial, wind, and water action and to gravity; the action of wind is generally confined to regions with dry climates. Deltas at river mouths and deposits that accumulate behind artificially constructed dams are obvious results of sedimentation. Other material is carried by rivers into the beds of oceans or lakes, although not all the material derived from earth sculpture reaches this resting place. But in this way a slow change is wrought in the configuration of the earth. Eventually, modifications lead to uneven pressures on the subcrustal portions of the earth; some movement is certain to result as these pressures disturb the equilibrium of the local crustal structure.

These two parts of the general cycle can be observed to a limited extent today. A true appreciation of the third part of the cycle requires more imagination and an excursion into the realm of geological time. It is true that there are some evidences of minor earth movement occurring now from time to time. One example which may be mentioned is the appreciable sinking of large areas of the Mississippi floodplain near New Madrid, Missouri, in 1811. Another is the subsidence of the shores of Lake Maracàibo in Venezuela and at the other locations, events which will be referred to later. Coal- and salt-mining subsidences are more familiar examples.

Within recent years, there have also been some violent manifestations of volcanic action, notably the complete disintegration of the Alaskan volcano, Katmai, last seen in 1912 and assumed to have been blown up by a severe volcanic explosion. Such effects are small when compared with the great earth movements of the past, that have given rise to the complicated geological structures known today. These latter-day events are significant, however, since they indicate that the solid earth is not a conglomeration of unusual structural materials but rather a body that is just as susceptible to stresses and their strain effects as any other structural material. The daily recording of definite load and relief obtained on seismographs that are situated near tidal water is another telling reminder of the elastic properties of the earth's crust. Even the activities of the civil engineer can lead to such earth movement, as will be shown later by examples such as the troubles encountered during major excavation work on the Panama Canal.

With the foregoing in mind, one may without too much difficulty imagine how great sections of the rock strata at surface level have been distorted, strained, and crushed when subjected to stresses which they could not resist. The movement can be divided roughly into the general raising and lowering of large sections of the earth's crust and the general bending and crumpling of smaller sections of the crust. After such disturbing action, natural physical conditions readjust themselves to the changed configuration of the earth's surface, and denudation continues. This denudation will affect the most exposed strata first, and those strata may be quickly worn away. An underlying stratum will then appear at the surface, and the cycle will continue. By imagining the results of several such cycles, one may appreciate the main reasons for the present complicated structure of the earth's crust.

3.8 THE GEOLOGICAL SUCCESSION

Despite all such earth movements, the original relation of the various rock strata and groups of strata, not only in one particular area but throughout the world, can usually be determined. Although the complete correlation is still far from finished, much has been done toward solving the many problems involved, and work toward this end is perhaps the supreme task of all geologists. In countries such as Great Britain, the general correlation can be regarded as complete, and the work now proceeding is generally on detailed local problems. In other parts of the world, however, no final results can yet be reported, and there is still much to do in linking the respective local systems, now being studied as parts of the whole. In the working out of all problems associated with stratigraphical succession, the study of special geological structures plays its part as one of the methods of the geologist. When the local succession of beds has been elucidated, it must be compared with similar arrangements of rock strata elsewhere in order to determine its place in the general geological timetable. This correlation is assisted by comparison of the detailed nature of the various rock types, their specific mineral composition, their general characteristics, and their fossil contents. As early an investigator as William Smith found that similar geological horizons have similar fossilized remains, a discovery that has withstood the test of time and forms one of the main foundations for the stratigraphical work of today.

Long years of intense study led gradually to the construction of a general classification of geological time called the *geological succession;* it may usefully be likened to a geological timetable. A summary of this table, which is generally applicable, is given in Table 3.2. Naturally, the main geological time spans are subdivided into many different intervals, each of which is capable of further local subdivisions; the table however, shows what

TABLE 3.2 The Geological Succession

Era	Age* (m.y.)	Period	Epoch	Character	Derivation of name
Cenozoic	2	Quaternary	Pleistocene	Ice age; age of Man	Early name for fourth division
	63	Tertiary	Pliocene Miocene Oligocene Eocene Paleocene	Age of mammals, flowering plants	Early name for third division of time and rocks
Mesozoic	145	Cretaceous		Age of reptiles and ammonites	Extensive chalk in England
	210	Jurassic			From Jura Mountains
	255	Triassic			Has three divisions in Europe
Paleozoic	280	Permian		Age of amphibians	From Perm, U.S.S.R.
	320	Pennsylvanian	Carboniferous		From Pennsylvania, USA
	306	Mississippian			From Mississippi, USA
	415	Devonian		Age of fishes	From Devonshire, England
	465	Silurian			From name of tribe in Wales
	520	Ordovician			From name of tribe in Wales
	580	Cambrian			From Roman Cambria (Wales)
Precambrian	To oldest rocks 3500			Beginning of life	Before the Cambrian

*Age data adapted from R. F. Flint, The Earth and Its History, Norton, New York, 1973.

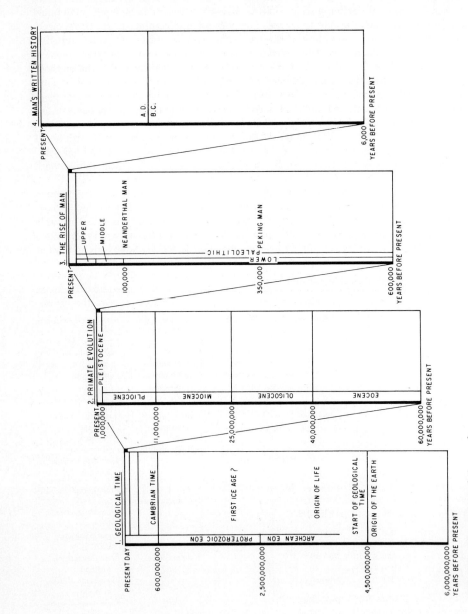

FIG. 3.2 The geological time scale.

are now regarded as the standard divisions. Figure 3.2 illustrates more graphically the long span of geological time.

The importance of paleontology, the study of fossils, will be evident from its vital contact with and use in stratigraphy. Even a slight knowledge of the nature of common fossil forms will demonstrate the variety of types of life now entombed in rock strata. It is readily seen that the biological aspects of fossil study constitute a further wide and complete field of investigation. Paleontologists themselves now concentrate in many cases on certain sections of their general subject. To these specialists, the civil engineer will always have recourse in connection with inquiries relating to fossils. Such inquiries are by no means so remote from normal civil engineering practice as might be supposed, although they may bear only indirectly on the immediate work in hand.

Civil engineers may on occasion use fossils to determine the exact nature of certain strata with special reference to other beds. More generally, however, the works of the engineer will reveal sections of fossiliferous strata that would otherwise remain forever hidden to the geologist. The skull of the now famous Rhodesian man may be mentioned in this connection. This was discovered, quite by chance, during extensive excavation work of the Broken Hill Mining Company in South Africa. In searching for lead, zinc, and vanadium, the engineer in charge of the work noticed the skull and saved it from destruction.[3,4] The discovery was of the greatest importance to all anthropological study. During excavation work for a new graving dock at Southampton, England, engineers uncovered a heavily fossiliferous section of the Eocene Lutetian beds. Apparently, the unusual fossils present were not noticed on the job; but through the cooperation of the engineers in charge of the work, geologists were able to examine the records of the strata, and scientific knowledge was thereby enriched. Other examples could be quoted, but all would have the same intent: to emphasize the scientific importance of the civil engineer's acquaintance with the nature and the value of fossiliferous remains so that any such evidence encountered in the course of engineering work will be brought to the attention of geological authorities.

Fossil remains may be of either animal or vegetable origin, and they may vary in size from minute organisms, which must be examined microscopically, to the great skeletons sometimes seen in museums. They consist of a mold, the petrified remains, or the chemically transformed body of a once-living creature—animal, insect, fish, bird, flower, or tree—preserved through the long annals of time in the surrounding rock stratum. The preservation processes can generally be understood, and the details involved need not be discussed here. Naturally, only certain types of life have been preserved. All rock strata are not fossiliferous; fossils are found only in some sedimentary and in some metamorphic rocks. Since the span of geological time is commensurate with the span of life on the earth, some demonstration of the evolution of living types might be expected in rocks of different ages, and this is found to be the case.

No attempt can be made here to describe even the leading typical fossils; one plate (Fig. 3.3) must suffice to demonstrate a few types as an introduction to fossil forms and possibly to serve as an incentive to further study in standard reference books. It may also be pointed out that the nature and type of fossils found in rock provide an accurate guide to both the climatic and the geographical conditions obtaining on the earth sphere at the time that the bodies now fossilized were alive. Arctic and tropical climates have alternated (broadly speaking) throughout certain stretches of geological time; such changes naturally affected the type of life extant. The fossil types are likewise a sure guide to the nature of the deposition of the rock in which they are found; coal beds (the most generally appreciated repository of fossil life), beds containing fossil footprints, and amber are three clear indications of the existence of some type of land surface. Other fossils are just as typical of lacustrine conditions, and still others just as surely denote marine conditions. Earth

FIG. 3.3 A few typical fossils: (*a*) trilobite of Cambrian age (550 million years), Utah (*b*) coral of Devonian age (375 million years), Ontario. (*c*) fern of Carboniferous age (300 million years), West Germany (*d*) ammonite of Jurassic age (175 million years), England (*e*) fish of Eocene age (40 million years), Wyoming (*f*) clam of Pleistocene age (200,000 years), California.

3-12

movements that have taken place since the original deposition during which the fossils were formed have resulted in wide and varied displacements of even marine deposits; for example, fishbeds have been found far inland and often at great heights above sea level.

A word may be added finally to explain, if not to defend, the international nomenclature adopted for the naming of fossils. Although so strange-looking to the newcomer, this nomenclature is nothing more or less than a simple type of scientific shorthand adapted to give a flexible system of classification, invoking the classical languages as an aid to this task. If this is remembered, *Strongylocentrotus drobachiensis* becomes a term of some meaning.

3.9 GEOLOGICAL MAPS

The main aim of much geological fieldwork is the determination of the present structure of that part of the earth's crust which is being examined and of its relation to neighboring and similar structures. Maps are used to record the results of these field studies. By a logical extension of the principles of topographic mapping, it is possible to indicate clearly and unmistakably on a map the clues that lead to the patient determination of the local geological structure. As Harrison has so rightly pointed out:

> Unlike topographic maps, which record data that can be gathered largely by mechanical, instrumental, or routine methods, the geological map, although in part objective and a record of (observed) facts, is also to a very large degree subjective, because it also presents the geologist's *interpretation* of these facts and his observations. A good geological map is the result of research of a high order.[3.5]

So important are geological maps to the civil engineer that Chap. 14 is devoted to a review of their preparation and use. In that chapter it will be seen that there are now special maps prepared for showing the significance of geological features in civil engineering ("engineering geology maps"). Publication of geological maps is naturally a complex and costly procedure, so it is usually the function of the national (or state, or provincial) geological survey. Most developed countries of the world now have a national geological survey as a well-established branch of their central government. Somewhat naturally, it is in highly developed small countries such as Belgium, Switzerland, France, and Great Britain that national geological mapping has advanced furthest. In all countries today, however, some geological maps will be available, general in nature though some of them may be. These will often prove to be of value in preliminary study of civil engineering sites, although necessarily only in a general way. All large-scale civil engineering works will require much more detailed geological field study. *Appendix B* of this book is a guide to geological surveys of the world; this guide should be of some assistance in leading those interested to the best sources of published geological maps.

3.10 GEOLOGICAL FIELDWORK

Geological studies in the field have naturally had frequent mention, even in this brief introduction to the science. It may usefully be stressed, therefore, that although a knowledge of the elements of geology can be gained in the lecture hall and even from private reading, in view of the many excellent introductory books now available, there is no substitute for the study of geology in the field. Both authors of this book can testify to the real insight into geology that each first gained as a student involved in geological fieldwork guided by an excellent teacher, fieldwork undertaken in quite different parts of the world. All courses of geology for engineers should include at least some well-planned field trips, with at least an introduction to geological mapping. An extended period in the field is even better. Once thus introduced to the joys of geological observation, anyone really interested can independently extend field experience, with the aid of one or another of the fine guides now available. Not only will the appreciation of geology be thus enhanced, but the enjoyment of the out-of-doors will take on new dimensions.

3.11 CONCLUSION

If independent geological investigations can be made without undue delay during fieldwork undertaken specifically to assist some construction operations, the civil engineer should appreciate its value and encourage the saving of effort thus achieved. Similarly, it behooves geologists engaged on engineering work to realize that what the civil engineer wants is not a complete geological survey of an entire area but sufficient information to be assured of the conditions to be expected in those parts of the earth's crust likely to be affected by the engineering operations. Mutual regard for the necessarily divergent viewpoints of the geologist and the engineer will lead to finer results from all joint effort. To this may be added the important suggestion that throughout many civil engineering construction operations, sections will be uncovered that may provide records of inestimable value to geologists, in some cases information that can be obtained only at that particular time. The courtesy of bringing these to the attention of geologists (the director of the local geological survey and the geological staff of the nearest university) cannot be too strongly commended.

So important can be this appreciation of the geological potentialities of excavation work, and correspondingly, so unfortunate can be its neglect, that this alone warrants the suggestion that every civil engineer should know at least the rudiments of geology. Unfortunately, the time has not yet come when all undergraduate civil engineering programs include a course in introductory geology. This gap in engineering education, slowly but steadily being closed, is one reason for the inclusion in this book of this introductory chapter. It is a chapter for civil engineers and not for geologists. Nobody realizes more fully than the writers the inadequacy of such a brief introduction to so great a science. It is intended to be no more than an introduction, but it may (and this is the authors' hope) act as an incentive to further study in some of the many fine books on all aspects of geology now available. For convenience, a selection of these as guides to further study follows. The attention of all engineer-readers of this volume is directed to the list; not only can these books assist in the field of engineering but they can also provide a background of geological knowledge that will add much to the pleasure of travel and to the appreciation of all that the beauty of nature can mean.

And if there should be any readers who consider it passing strange that engineers should be asked to consider scientific matters, even in connection with such natural features as rocks, while engaged upon their engineering work, they may reflect upon the fact, well recorded, that Alexander the Great (who died in the year 323 B.C.) attempted to survey his vast empire by organizing a special force whose task it was to maintain the condition of the major roads then existing. Alexander also made use of the services of these earliest of engineers for such scientific purposes as the collection of information regarding the "natural history" (including the geology) of the districts in which they were at work.[3.6] The association of geology with engineering has, therefore, a long tradition.

3.12 REFERENCES

3.1 For a vivid review of the history of geology see F. D. Adams, *The Birth and Development of the Geological Sciences,* New York, 1954.
3.2 T. Sheppard, *William Smith, His Maps and Memoirs,* Brown, Hull, England, 1920.
3.3 H. W. Schmidt, Response to Special Award, *Bulletin of the Geological Society of America,* **85**:1352 (1974).
3.4 Sir A. Keith, "An African Garden of Eden," *John O'London's Weekly,* **35**:890 (1934).
3.5 J. M. Harrison, "Nature and Significance of Geological Maps," *The Fabric of Geology,* C. A. Albritton Jr. (ed.), Addison-Wesley, Palo Alto, 1963, p. 225.
3.6 C. Singer, *A Short History of Scientific Ideas to 1900,* Clarendon, Oxford, England, 1959.

Suggestions for Further Reading

For those who wish to have a one-volume encyclopedic coverage of the science of geology, the authors have no hesitation in recommending *Principles of Physical Geology* by A. Holmes and D. L. Holmes, now in its third edition (1978), published originally in England by Thomas Nelson and Sons, and in the United States by the Halstead Press, New York. This 730-page volume, now regarded by many as a classic, gives a detailed, comprehensive treatment of all the main aspects of geology with which the civil engineer will normally be concerned.

The best of all introductions known to the authors is a small book of 215 pages, beautifully illustrated with 156 excellent color plates, entitled *Field Geology in Colour* by D. E. B. Bates and J. F. Kirkaldy, published (in 1976) by the Blandford Press of Poole, Dorset, England. Through its emphasis on field observations, so well illustrated by its plates, this inexpensive pocket-sized guide can be of real assistance to civil engineers who wish to become familiar with the science as background for its application in their work.

The *Glossary of Geology* of 749 pages, edited by R. L. Bates and J. A. Jackson, published in its second edition in 1980 by the American Geological Institute of Falls Church, Virginia, is far more than just a dictionary and can serve as a useful supplement to the volumes noted above. The *Cambridge Encyclopedia of Earth Sciences,* edited by D. C. Smith and published in 1982 by Cambridge University Press, gives similarly wide coverage.

There are today a large number of excellent general introductory volumes to the science of geology; one only will be mentioned since it was prepared especially for civil engineers. It is the *Manual of Applied Geology for Engineers,* a 378-page volume published by the Institution of Civil Engineers (London, England) and sponsored jointly by the Institution and the British Ministry of Defence so that it has appeared also as volume XV of *Military Engineering.* Despite the "Applied" in its title, it is an excellent introduction to the science.

A convenient introduction to the vast number of individual papers on all aspects of geology is provided by the series *Benchmark Papers in Geology,* now numbering over thirty, published by Dowden, Hutchinson & Ross Inc. and distributed by the Halstead Press. The series editor is Rhodes W. Fairbridge of Columbia University, each volume being under individual editorship.

The wonderful store of information available from the U.S. Geological Survey can best be tapped by consulting first the Survey's Circular No. 777 by Clarke, Hodgson, and North entitled "A Guide to Obtaining Information from the USGS 1979." Other surveys (listed in Appendix B) usually have similar guides.

Those who wish to know more about the history of geology will find interest in *Giants of Geology* by C. L. and M. A. Fenton, published in 1956 by Doubleday, a readable 333-page guide to the lives and work of leading geologists; and also in a standard reference in its field, *The Birth and Development of the Geological Sciences* by F. D. Adams, first published in 1938 and reissued as a paperback in 1954 by Dover Publications.

Geology by F. H. T. Rhodes (1972) and *Landforms* by G. F. Adams and J. Wyckoff (1971) are two of the excellent Golden Science Guides of the Western Publishing Company, which fit easily into the pocket or handbag and yet contain much essential information about the science, attractively presented.

A useful guide to the large volume of literature in the geological sciences now available is provided by *Books and Bibliographies in the Geosciences,* prepared and published by the American Geological Institute; copies may be obtained free from the Institute at 5205 Leesburg Pike, Falls Church, VA 22041.

chapter 4

ROCKS: THEIR NATURE AND STRUCTURE

4.1 Minerals / 4-3
4.2 Igneous Rocks / 4-3
4.3 Sedimentary Rocks / 4-5
4.4 Metamorphic Rocks / 4-6
4.5 Distinguishing Rock Types / 4-7
4.6 Special Structural Features / 4-12
4.7 Conclusion / 4-17
4.8 References / 4-18

Earth was the English word used in translation of the one Greek word used (as by Aristotle) to describe all the material making up the crust of the earth. The Romans differentiated between that part of the crust which is solid—*petra* rock—and that which was fragmented—*solum* (soil or ground). At the birth of modern scientific studies, clear distinction was made, as by John Evelyn in 1675, between rock and soil.[4.1] This distinction was in universal use in the English language, recognized by geologists and by engineers alike, until the early years of the twentieth century. (Topsoil was an additional term, used to denote the upper few inches of soil in which organic growth takes place.) William Smith, for example, used both words in his descriptions in his famous "Map of the Strata" as early as 1815.[4.2] At the end of the century, G. P. Merrill, in his famous *Rocks, Rock Weathering and Soils* published in 1897, uses the two important words in the accepted manner, both in his title and in his text.[4.3]

When the advance of scientific agriculture led to the start of modern studies of the topsoil and the immediately underlying soil horizons from which it is derived, early pedologists restricted their use of the word *soil* to describe only the material in which they were interested—the upper few feet, including the topsoil. It seems clear that this restricted use was due to a double translation of pioneer pedological papers in Russian, in which language there are two separate words to describe the earth material studied by pedologists on the one hand and, on the other, all the fragmented material above bedrock—including clays, silts, sands, and gravels—studied by geologists and engineers. This duplication of meaning for the same word has persisted, although there has long been mutual

4-2 GEOLOGICAL BACKGROUND

understanding among English-speaking pedologists, geologists, and engineers as to their difference in usage. The context in which the word is used generally gives clear indication as to which meaning is intended.

As the interrelation of pedology and geology has happily developed, some geologists have come to adopt the restrictive pedological use of the word *soil* so that there is today, most unfortunately, resulting confusion in geological literature. New terms have accordingly had to be selected to describe all the soil not included in pedological usage, terms such as *regolith, overburden,* and even *sediments* (despite the fact that many soils are certainly not the result of sedimentation). The limitations inevitable with these new terms are shown by such compound terms as *earthy regolith,* a lamentable expression which is to be found in a well-known geological textbook. Engineers, on the other hand, have continued to use the word *soil* in its original sense, while naturally making clear distinction between soil types such as clays, silts, sands, and gravels. Table 4.1 illustrates the variation in usage of this simple and most useful word. Throughout this Handbook, the word is used in its geological and engineering sense; correspondingly, the word *rock* is used to denote all bedrock.

TABLE 4.1 Variations in Usage of the Word *Soil*

CLASSICAL GEOLOGY	CIVIL ENGINEERING	HORTICULTURE	PEDOLOGY (Soil Profile)	GLACIAL GEOLOGY	Weathering Profile
Topsoil	Topsoil	Topsoil	A_1 Horizon — S	Zone 1	Humus
			A_2 Horizon — O L	Zone 2	Stable Silicates
Soil	Soil	Subsoil	B Horizon — U M	Zone 3	Clay and Oxides Enriched
			C Horizon	Zone 4	Carbonates leached
				Zone 5	Oxidised
				Zone 6 (unconsolidated rock)	Unaltered
				(consolidated rock)	Bedrock

The study of solid rocks dominates geological studies. Only a relatively few geologists are interested in detailed soil studies, but those few are just as dedicated to their specific branch of the science as the more numerous geologists are dedicated to their interest in solid rock. This ratio in scientific speciality is readily understood when it is remembered that most soils encountered in North America, for example, have been formed within the last two million years, whereas most of the other geological phenomena studied range back for a period of well over four billion years. In normal practice, the civil engineer will be concerned far more with soils than with bedrock. It is with soils that the most difficult problems arise. When problems do arise in work with rock, as in tunneling, the engineer will usually have recourse to the advice of specialists. Despite this, however, all civil engineers must be familiar with the principal rock types and with the elements of geological structure and stratigraphy, to which this short chapter is an introduction. The geology of soils will be considered separately in the chapter which follows.

TABLE 4.2 Some Common Minerals

Mineral	Specific gravity	Hardness	Crystal	Cleavage or fracture	Color
Quartz	2.65	7	Hexagonal	Curved like glass	Clear, milky, gray
Feldspar	2.5–2.8	6	Tabular	2-planar at 90°	White, gray, pink
Mica	2.9–3.1	2.5–3	Hexagonal	One plane	Black, brown, gray
Hornblende	3.1–3.5	5–6	Tabular	2-planar at 60°	Black
Pyroxene	3.2–3.7	5–6	Tabular	2-planar at 90°	Dark green
Calcite*	2.7	3	Rhombic	3-planar at 60°	White, pink, gray
Dolomite†	2.9	3.5–4	Rhombic	3-planar at 60°	White, gray, brown

*Reacts strongly with acid.
†Reacts weakly with acid.

4.1 MINERALS

Minerals are the basic building blocks of the earth's crust. Aggregations of minerals comprise rocks and soils. Minerals are classified mainly on chemical composition. The most important rock-forming mineral groups are the silicates, oxides, and carbonates. Each mineral species has physical and chemical characteristics which allow identification; the most common minerals are readily identified in a hand specimen. Crystal form (commonly not well developed), cleavage and fracture, hardness, specific gravity, reaction with dilute acid, and color are among the most significant properties easily observed in the field. Some familiarity with common minerals is useful in identifying various rock types. A few minerals and their characteristics are displayed in Table 4.2. Further information can be found in one of the widely available field guides to rock and mineral identification, and, of course, specimens may be seen in museums.

4.2 IGNEOUS ROCKS

Igneous rocks are of two main classes: *extrusive* (those poured out at the earth's surface) and *intrusive* (large rock masses which have not cooled in contact with the atmosphere). Initially, the rocks of both classes were molten magma, their present state resulting from the manner in which they solidified.

If a violent volcanic eruption takes place, some materials will be emitted, with gaseous extrusions, into the atmosphere, where they will cool quickly and eventually fall to the earth's surface as volcanic ash and dust. This type of action continues even today; dust on the dome of St. Paul's Cathedral in London is of volcanic and possibly cosmic origin. The main product of volcanic action is a lava flow, emitted from within the earth as a molten stream which flows over the surface of the existing ground until it solidifies. Extrusive rocks are generally distinguished by their usual fine-grained texture and by the "baking" of whatever rock stratum happens to underlie them.

Intrusive rocks, which cool and solidify under pressure and at great depths and which contain entrapped gases, are usually wholly crystalline in texture, since the conditions of cooling are conducive to crystal formation. Such rocks occur in masses of great extent, often going to unknown depths. Although originally formed deep underground, intrusive rocks are now widely exposed because of earth movement and erosion processes.

Hypabyssal rocks are intermediate in position between extrusive and major intrusive rocks. They occur in many forms, the main types of which are indicated in Fig. 4.1. *Dikes* are large wall-like fillings in the earth's crust cutting across normal bedding planes. *Sills*

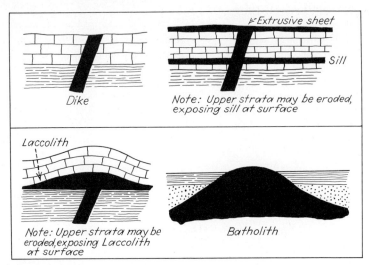

FIG. 4.1 Some forms taken by igneous rock.

are large comformable sheets intruded into other formations parallel to their structure. Common to all these occurrences is the baking of adjacent rock strata on *both* sides of the intrusion, as distinct from baking on the underside only as in the case of extrusive rocks.

Chemical analyses of igneous rocks show that they are essentially composed of nine elements: silicon, aluminum, iron, calcium, magnesium, sodium, potassium, hydrogen, and oxygen. These occur naturally in combination, generally as silicates, oxides, and hydroxides. Although the proportion of the oxides varies considerably, the chemical composition of rocks, considered as a whole, varies within quite narrow limits. Within these limits, despite the steady gradation from one composition to another and the varying and quite distinct mineral constitution of different rocks, certain groups can be distinguished. Chemical and mineral composition, in association with the mode of origin of the rock, has therefore been adopted as a basis for a general classification of crystalline igneous rocks. Silicon dioxide (silica), often crystallized as quartz, is one of the main mineral constituents of igneous rocks, and by the silica content a broad dividing line is fixed. Table 4.3 gives a list of the main types of igneous rocks and suggests broad lines of classification.

TABLE 4.3 Classification of Igneous Rocks

	Acidic (Quartz)	Intermediate (Little or no quartz)*		Basic (No quartz)
Most common minerals	Orthoclase	Orthoclase	Plagioclase	Plagioclase
	Oligoclase	Biotite		Olivine
	Mica	Hornblende	Biotite	Hypersthene
	Hornblende	Augite	Hornblende	Augite
	Augite		Augite	
Plutonic	Granite	Syenite	Diorite	Gabbro
Hypabyssal	Quartz porphyry	Orthoclase porphyry	Porphyrite	Dolerite
Volcanic	Rhyolite	Trachyte	Andesite	Basalt

*Reference should be made to the text for comment on the indefinite divisions noted.

4.3 SEDIMENTARY ROCKS

This great group of rocks may properly be regarded as secondary rocks, because they generally result from the weathering and disintegration of existing rock masses. These rocks are somewhat loosely named, since sedimentary is truly descriptive of but part of the group. Water is not a depositing agent in the case of all these secondary rocks, nor are they always found *stratified,* another title which has been suggested. In view of the common use of the word *sedimentary,* that term has been retained here and elsewhere in this book to denote this entire group of secondary rocks. Fossils are found almost exclusively in sedimentary rocks; nevertheless, sedimentary rocks are often lacking in visible fossil remains.

The distribution of sedimentary rocks over a wide area throughout the world is a result of the great land movements that have taken place in past geological eras. These movements are often vividly demonstrated by the existence of marine deposits, including fossil seashell remains, in places now elevated considerably above the nearest lake or seacoast. Marine deposits are found in the upper regions of the Himalayas and in many other parts of the world thousands of feet above sea level. The top of Mount Everest, the highest point of land in the world, is of limestone, a sedimentary rock.

Sedimentary rocks are generally found in quite definitely arranged beds, or strata, which were horizontal at one time, but are today sometimes displaced through angles up to 90°. This bedding, or stratification, is a direct result of the method of formation; the material was deposited evenly over a lake or sea bottom or in some tropical jungle swamp. Similarly, sedimentary rocks are being formed today by the mud washed down by rivers into lakes and seas and by such marine organisms as the coral of the tropical seas (Fig. 4.2). Measured by human standards of time, the building-up process is almost infinitesimally slow; geological time periods must be used as a basis for comparison.

FIG. 4.2 Well-stratified limestone, a typical sedimentary rock, near Port St. Mary, Isle of Man.

4-6   GEOLOGICAL BACKGROUND

Sedimentary rocks can conveniently be classified in three general groups. Briefly, the divisions include those rocks which are (1) mechanically formed, (2) chemically formed (evaporites), and (3) organically formed. The so-called "mechanical processes" leading to rock formation are the action of wind, frost, rain, snow, and daily temperature changes; all these can be classed as weathering influences, which lead to the formation of surface soil, rock talus (or scree breccia), and fine deposits of rainwashed material and blown dust such as brick earth and loess, as well as some special types of clay. Both physical and chemical weathering processes, acting together but with differing relative strength under differing circumstances, act to break down rock into fragments and solutions. The fragments form the clastic sediments—gravel, sand, silt, and clay—while the chemical solutions are carried away invisibly in surface or subsurface water. Water is the most effective agent for eroding, transporting, and depositing sediments. Finally, glacial action has been and still is a potent factor in rock formation, having given rise to extensive glacial deposits over wide areas of the world.

All these different types of deposits, when subjected to great pressure, and possibly to the action of chemicals and heat, will be transformed into rocks such as sandstone. Evaporites, their mode of origin indicated by their name, are chemically formed rocks; typical examples include rock salt, gypsum, anhydrite, and the various types of potash rocks found in some areas. Limestone is perhaps the best known of the organically formed rocks, being generally the accumulation of the remains of marine organisms; it occurs in many different forms. Coal and phosphate rock (coprolites and guano) are other rocks that are obviously of organic origin. Table 4.4 shows some of these interrelationships.

TABLE 4.4   Some Equivalent Forms of Sedimentary and Metamorphic Rocks

Sediment	Consolidated rock	Metamorphic product
Till	Tillite	
Gravel	Conglomerate	
Sand	Sandstone	Quartzite
Silt	Siltstone	Argillite or slate
Clay	Shale or mudstone	Argillite or slate
Lime mud	Limestone	Marble
Peat	Bituminous coal	Anthracite coal

4.4   METAMORPHIC ROCKS

Agencies that have helped to change sedimentary and igneous rocks into metamorphic rock types are many. The principal ones are the intense stresses and strains set up in rocks by severe earth movements and by excessive heat from the cooling of intrusive rocks or from permeating vapors and liquids. The action of permeating liquids appears to have been particularly important. The results of these actions are varied, and the metamorphosed rocks so produced may display features varying from complete and distinct foliation of a crystalline structure to a fine, fragmentary, partially crystalline state caused by direct compressive stress, including also the cementation of sediment particles by siliceous matter. Foliation is characteristic of the main group of metamorphic rocks; the word means that the minerals of which the rock is formed are arranged in felted fashion. Each layer is lenticular and composed of one or more minerals, but the various layers are not always readily separated from one another. It will be appreciated that these characteristics are different from the flow structure of lava and also from the deposition bedding which occurs in unaltered sedimentary rocks. *Schist* is the name commonly applied to

FIG. 4.3 An exposure of Pinyon conglomerate, nearly 600 m (2,000 ft) thick on the northwest margin of the Teton Range, near Survey Peak, Wyoming.

such a foliated rock, and the various types of schistose rock are among the best-known metamorphic rocks.

The nature of the original rocks from which metamorphic rocks were formed has been and still is a matter of keen discussion and the subject of much inquiry. Briefly, it may be said that some metamorphic rocks are definitely sedimentary in origin, some were originally igneous rocks, and some are of indeterminate origin. The presence of fossil remains in certain crystalline metamorphic rocks is proof enough of their sedimentary origin; and, on the other hand, uninterrupted gradation from granite and other igneous rock masses to well-defined schistose rocks is equal proof of the igneous origin of some metamorphic rock types. Of the former types, sedimentary in origin, marble (altered limestone) is a notable example; its appearance frequently demonstrates the organic remains of which it was originally formed. Schistose conglomerate is another well-defined sedimentary type. (Fig. 4.3). Quartzite can often be seen to have been formed from sand grains, and the great variety of slates are all obviously of clay or mudstone origin. The rocks classed generally as *schists* are of varying compositions; mica schist is a crystalline aggregate of mica and quartz and occasionally other minor minerals. *Gneiss* is a term somewhat loosely applied; it is generally used to distinguish a group of rocks similar to the schists but coarsely grained and with alternate bands of minerals of different composition.

4.5 DISTINGUISHING ROCK TYPES

The civil engineer has a natural interest in the study of rocks because they are used not only as materials of construction but also as foundation beds for many structures. The

4-8 GEOLOGICAL BACKGROUND

engineer should therefore be able to distinguish at least the main rock types seen in the field. For all the practical purposes of the civil engineer, field tests and simple microscopic examinations will suffice for identification of most of the common rock types. More detailed petrographical investigation will be called for only in exceptional cases, and then it will clearly be work for an expert. Valuable initial assistance can be gained by examination of well-marked hand specimens either in a university laboratory or in the display cases of a museum.

Simple equipment is all that is needed for the field investigation of rocks—a geologist's hammer, a tool that will accompany the civil engineer who is interested in geology on all out-of-doors excursions; a pocket magnifying lens for examination of the smaller crystal structures; a steel pocketknife for hardness tests; and a small dropper bottle of hydrochloric acid for the determination of mineral carbonates. A small magnet is often useful, since the mineral magnetite can be separated from other associated minerals (when crushed) by running a magnet through the mixture. The hardness test mentioned utilizes a hardness table, a selected scale based on the relative hardness of selected minerals. Of great use in preliminary testing, this scale is shown in Table 4.5

TABLE 4.5 Relative Hardness of Minerals

Hardness*	Mineral	Test characteristic
1	Talc	Can be scratched with a fingernail
2	Gypsum	
3	Calcite	Can be cut very easily with a penknife
4	Fluorspar	Can be scratched easily with a penknife
5	Apatite	
6	Feldspar	Can be scratched with a penknife but with difficulty
7	Quartz	
8	Topaz	Cannot be scratched with any ordinary implement; quartz will scratch glass; topaz will scratch quartz; corundum, topaz; and a diamond, corundum
9	Corundum	
10	Diamond	

*The numbers given are used as relative hardness numbers, relative only since the actual hardness value of talc is about 0.02, whereas that of a diamond runs into the thousands.

Investigation of rocks in the field necessitates close examination of a hand specimen which may be obtained from an outcrop by a sharp blow from the hammer. It must be emphasized that a clean, fresh rock surface must be used for all examinations. The word *fresh* is used to indicate that the rock surface must have been freshly broken, since most rocks weather on their exposed surfaces to some extent, and thus do not show their true character unless newly broken. From an examination of such a surface, one can usually see whether the rock is crystalline or not.

Of the noncrystalline rocks, *shales*, which are consolidated fine sediments, are usually hardened clay or mud and have a characteristic fracture. Generally dull in appearance, shale can be scratched with a fingernail. If it breaks into irregular laminae, the shale is argillaceous; if gritty, arenaceous; if black, it may be bituminous; if it effervesces on the application of acids, it is calcareous. *Slate* can easily be recognized by its characteristic fracture or cleavage and its fine, uniform grain; in color it may vary from black to purple or even green. All these rocks demonstrate their argillaceous nature by emitting a peculiar earthy smell when breathed upon.

Limestone is one of the most widely known sedimentary rocks; it can often be distinguished by its obvious organic origin as indicated by the presence of fossils, but a surer mark of distinction is that it effervesces briskly when dilute hydrochloric acid is applied to it. *Marble* is a crystalline (metamorphic) form of limestone, generally distinguished by its crystalline texture but always effervescent when treated with dilute acid. *Dolomitic limestone* is generally dark in color; it effervesces slowly when treated with cold hydrochloric acid, but more quickly when the acid is warm. *Flint* and *chert,* compact siliceous rocks of uncertain chemical or organic sedimentary origin, occur often as nodules in limestone beds.

Conglomerates and sandstones form another interesting group of sedimentary rocks; their granular structure suggests their sedimentary origin. *Conglomerates* are, as their name implies, masses of gravel and sand, waterborne, as denoted by rounded shapes, and cemented together in one of several ways into a hard and compact mass. *Sandstone* is the general term used to describe such sedimentary cementation of sand alone. *Quartzite* is a metamorphosed type of sandstone in which the grains of rock have been cemented together with silica so strongly that fracture takes place through the grains and not merely around them. *Grit* is a term sometimes used to denote a coarse-grained hard sandstone containing angular fragments. (See Fig. 4.4.)

The identification of igneous rocks and of some of the metamorphic rocks not yet mentioned is not quite so straightforward as the determinations so far described. These rocks are usually crystalline, but the crystals may vary in size from those of coarse-grained granite to those so minute that they must be examined under a microscope. Of the remaining metamorphic rocks, *serpentine,* a rock composed wholly of the mineral of the same name, is generally green to black, fairly soft, and greasy or talclike to the touch; the color may not be uniform. Serpentine is important to the civil engineer, since it is a potent cause of instability in rock excavation. *Gneiss* may be recognized by its rough cleavage and typical banded structure, which shows quartz, feldspar, and mica with a coarse structure. *Schists*

FIG. 4.4 Cross-bedded Wingate sandstone, Johnson Canyon, Kane County, Utah.

may be distinguished from gneiss by their essentially fissile character; in all schists, there is at least one mineral that crystallizes in platy forms (mica, talc, or chlorite) or in long oblong blades of fibers and gives the rock a cleavage parallel to the flat surface.

Granite is a typical example of igneous rocks; it is widely distributed and constitutes an important igneous rock type. It must however, be noted that the origin of some granites is still a matter of keen debate in geological circles. Granite is composed mainly of quartz (clear), orthoclase feldspar (white or pink), some mica, and possibly hornblende. All the crystals are about the same size, or even-granular, and the quartz (the last mineral to separate) occupies the angular spaces between the other crystals; this latter characteristic marks a *granitic structure*. (Fig. 4.5). Color may vary from pale gray to deep red, depending on the mineral content; an average composition, however, is 60 percent feldspar, 30 percent quartz, and 10 percent dark minor minerals.

The texture of igneous rocks varies from a coarse, even-granular structure to an aphanitic (without visible crystals) structure in which crystallization cannot be seen with the unaided eye. In a porphyritic structure, the constituent minerals occur as much larger crystals than the remainder (the large crystals are called *phenocrysts*); the other minerals may appear as a crystalline groundmass, or alternatively, they may be aphanitic. Finally, there are a few rare igneous rocks found in glasslike form which have not crystallized at all. As a general rule, the acid igneous rocks tend to be lighter in color than the basic

FIG. 4.5 Jointing in white granite, south of Tule Well, Pima County, Arizona.

rocks; granite, therefore, is pale because of the predominance of feldspar, a light-colored mineral. *Diorite* has a texture similar to that of granite, but it contains no free quartz. In *gabbro*—the corresponding basic rock—feldspar is subordinated but is still an important constituent; hornblende, pyroxene, and olivine are dark minerals which make the rock dark and give it a high specific gravity. *Dolerite* is a similar basic rock with a smaller grain size; *basalt* is the corresponding aphanitic rock, although occasionally phenocrysts will be found in it (see Fig. 4.6).

FIG. 4.6 Boulder of dolerite with phenocrysts, Grand Manan, New Brunswick.

Granite porphyry and *diorite porphyry* are similar to granite and diorite in composition, but they have a porphyritic structure, feldspar being the most usual phenocryst. The porphyries are a common group of rocks; they are hypabyssal and occur both in lava flows and as sills, dikes, and laccoliths. *Rhyolite,* an extrusive rock, corresponds in composition to granite, an intrusive rock. *Andesite* bears a similar relationship to diorite; it contains no quartz. The aphanitic type of both rhyolites and andesites is known as *felsite,* a name that includes most of the large group of light-colored aphanitic igneous rocks; the corresponding dark rocks are classed as *basalts.* Glasslike rocks are found only in the vicinity of cooled lava flows; *obsidian,* a lustrous, dark rock, is the most common variety. *Pumice* is simply a frothed type of glassy rock.

The foregoing list includes the main common varieties of igneous rock; it must again be emphasized that there is no hard-and-fast dividing line between a given variety and the adjacent one in scale. It cannot be too strongly urged that study of rock structures in situ is the only truly reliable method of becoming familiar with the rock types described. Indoor examination of hand specimens will help the beginner, but there is really no substitute for examination of fresh rock specimens in the field. Color, feel, surface appearance, field relationships, and the nature of fracture are all so important in distinguishing rock types that only through examination of actual specimens freshly obtained can one

4.6 SPECIAL STRUCTURAL FEATURES

It will be useful to describe briefly the main structural features encountered in geology, since they are regularly met with in normal civil engineering work. In their departure from what may be called *standard stratification,* these features warrant being described as special. It must be emphasized that the unusual features to be described are only the more important of special geological structural arrangements, each of the types described being capable of subdivision into many detailed classifications; other more unusual features may also be encountered.

Two widely used terms, *dip* and *strike,* are used to describe the present position of strata of rock with reference to the existing ground surface (Fig. 4.7). The strike of a rock

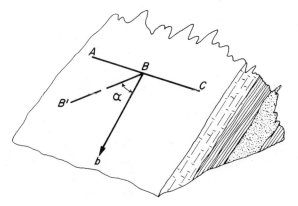

Strike is the compass direction of *ABC*
Angle of dip is the angle α
Lines *ABC* and *BB'* are in same horizontal plane
Lines *ABC* and *Bb* are on exposed rock surface

FIG. 4.7 Diagram illustrating the dip and strike of rocks.

layer is the compass direction of a line considered to be drawn along an exposed bedding plane of the rock so that it is horizontal; obviously, there will be only one such direction for any particular rock layer. The dip of the bed is the angle between a horizontal plane and the plane of the bedding, measured at right angles to the strike; it is thus a measure of the inclination of the bed to the horizontal plane. These two particulars, together with the usual topographical information, definitely locate a rock surface in space; as such, they are as invaluable to the civil engineer as they are to the geologist.

Sedimentary Rock Bedding The method of formation of sedimentary rocks leads in many cases to an uneven disposition of material and to an uneven distribution of pressure on deposits. Thus, a distinct variation in the physical qualities of a sedimentary bed at different levels, as well as changes in the thickness of a bed, may result. Sedimentary strata often thin out completely, which can cause confusion in geological mapping. Surface markings constitute another special feature of sedimentary rocks; one example is ripple marking similar to that seen on a flat, sandy beach after the recession of the tide. These variations are of minor importance, but another subsidiary feature is worthy of special note. It is what is called *cross*, or *false, bedding,* caused by a special process of deposition, i.e., by currents from varying directions.

Igneous Rock Features These have already been explained, insofar as the principal types of formation are concerned. They are mentioned here again to emphasize the fact that these several types of igneous formation may be encountered in structural geological investigations either in a perfect form (as shown in Fig. 4.1) or in an imperfect form caused by subsequent distortion or denudation.

Jointing If any typical rock mass is studied, it will soon be seen that in addition to bedding planes that may be visible (in a sedimentary rock), fractures also occur in other planes roughly at right angles to bedding planes. These fractures give rise to a blocklike structure, though the blocks may not be separated one from another. Such fractures are generally known as *joints,* or *joint planes;* they result from internal stresses either during the cooling of the rock or during structural displacement. They are not unconnected with the cleavage planes of the constituent minerals. Joints are sometimes filled with newer igneous rock that has penetrated into the cracks while still liquid or with mineral that has crystallized out from solution, e.g., quartz and calcite. Some remarkably intricate formations of this type may be found in which the filled joints are so small that they are barely visible. In sedimentary rocks, jointing is generally regular; in granite, it is often irregular; in basalt, it leads to the peculiar polygonal column formation that is a striking feature of such locations as the island of Staffa off the west coast of Scotland and the eastern Washington area where the Columbia plateau basalts occur. (See Fig. 4.8). Joints are of great importance to the civil engineer, and numerous references to them will be found in the main part of this book.

FIG. 4.8 Natural sculpture, the result of jointing in basalt, Grand Manan, New Brunswick.

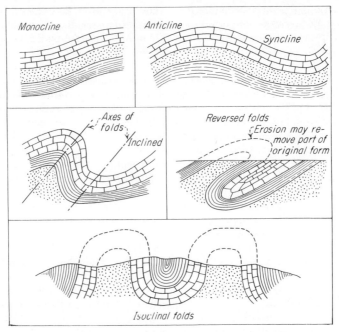

FIG. 4.9 Diagrammatic representation of some types of folds.

FIG. 4.10 Intense folding of layered sedimentary rocks near the Sullivan River, southern Rocky Mountains, British Columbia.

Folding When all the variable distortions of rock strata that can take place are considered, folding is perhaps the simplest structural feature, apart from a general raising or lowering of the whole of one part of the earth's crust. Basically, folding consists of the formation of simple regular folds, as shown in Fig. 4.9. The portions are termed *anticlines* and *synclines,* respectively, according to the type of bend. Anticlines are upfolds; synclines are downfolds, convex and concave upward respectively. This essentially simple type is not always found in practice. The first, and most general, variation is the inclination of the axis of the fold with the result shown. In certain special cases, the folding may not be confined to the one direction; if a rock mass is subjected to bending stresses all around, a domelike structure may occur. If the simplest basic type is extended, many variations can be obtained, such as *double folds, reversed folds,* and even the recumbent and fanlike structures indicated. At first sight, the latter might appear to be a purely hypothetical case, but several outstanding examples of this type of structure could be given, including the European mountain group of which Mont Blanc is a leading member. Photographs show something of the appearance of these folds in actual practice (Fig. 4.10). It will be realized that if the angle through which rock is folded is at all appreciable, unequal stresses will be set up in the rock mass. Normally, this is of no consequence, but to the civil engineer it can mean a great deal. If such distorted material is underground, civil engineering operations may interfere with this condition and bring about unexpected results when the previously restricted stresses are released.

Faulting When subjected to great pressure, the earth's crust may have to withstand shear forces in addition to direct compression. If the shear stresses so induced become excessive, failure will result; movement will take place along the plane of failure until the unbalanced forces are equalized, and a fault will be the result. Figure 4.11 shows in simple diagrammatic form the types of fault most generally encountered; the relative displacement of the various strata is clearly shown. In the simple fault, the terms most commonly used are indicated. The first fault shown is a normal fault in which the *hade* (or inclina-

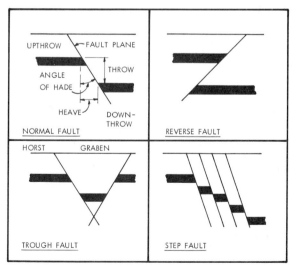

FIG. 4.11 Diagrammatic representation of some types of geological faults.

tion with the vertical) is always in the direction of the *downthrow*. The *throw* of the fault is the vertical displacement; the *heave,* the lateral displacement. In the diagram, the fault is shown as a plane surface; in practice, this is not always found. Frequently the rock on one or both sides of the fault is badly shattered into what is termed *fault breccia,* finely fractured fragments of rock, often retained in such close contact that they appear at first sight to be a solid mass. The direction of the main line of a fault is dependent on so many factors that it may bear no relation to the dip or strike of the surrounding rocks. Thus, for convenience, faults are characterized by their general direction as *dip faults* and *strike faults*. *Step faults* and *trough faults* are terms that will readily be understood by reference to the diagram. *Reverse faults* are the opposite of normal faults, having the hade away from the direction of the downthrow. As the diagrams show, faults are almost always inclined to the vertical, the upper face of inclined faults being known as the *hanging wall,* the lower face as the *footwall*.

A cursory consideration of these simple diagrams will suggest the complicated structures that can and do result from faulting, especially where the fault plane cuts across several rather thin strata. All these will be displaced, relative to one another, at the fault; after denudation has taken place, their surface outcrops will be confusing. In the study of such structural results, the basically simple nature of faults, coupled with the three-dimensional nature of the faulted section must be kept in mind (Fig. 4.12). The civil engineer's usually clear appreciation of three-dimensional drawings, based on solid geometry, will here be of great service. The study of solid geometry leads in particular to considerable simplification in the quantitative (as distinct from the qualitative) investigation of faulting, together with all allied calculations.

Faults may vary greatly in size. In length, they may range from a few feet to hundreds of miles; whereas in throw, they may include movements of a few inches or less (even pebbles have been found with perfect fault sections in them) to thousands of feet, as in the case of those in the plateau region of Arizona and Utah, some of which cross the Grand Canyon. Such dislocations may cause no unusual disturbance of topographical detail,

FIG. 4.12 A normal fault on beds of sandstone and fakes exposed at Mossend, Scotland.

although in some cases they may be the main factor affecting the physical character of large areas. In Canada, there are many notable conspicuous faults; for example, Lake Timiskaming, which separates Ontario from Quebec in the north, lies in a fault depression.

Denudation Denudation listed under special structural features will appear to be almost a contradiction in terms. Yet the direct effects of the erosive action of wind, water, and ice are so markedly a characteristic of almost all geological structures that the use of the term to include these several effects may properly be permitted. On all approximately horizontal beds, the effect of such action will not, as a rule, be unusual; it will, in fact, be generally uniform except where watercourses have cut their way somewhat deeply into the upper strata. Inclined strata, however, may be worn down unevenly, which will result in exposures of varying thicknesses. Erosive action will vary in the intensity of its results, according to the hardness of the strata encountered. The resulting differential action may give rise to unusual structural features such as those indicated in Fig. 4.1. Other notable results of erosive action are found in the history of river systems; this study is of special interest to civil engineers in view of the difficulty always caused by a buried river valley, a feature that will be mentioned repeatedly later in this book. The deposition of material so eroded need only be mentioned here, since delta formation and other sedimentary effects have already been considered. Apart from the erosive action thus generally described, glaciation produces special topographical effects which may almost be classed as unusual structural features. Because of their importance to civil engineers these features are considered in some detail in Chap. 5.

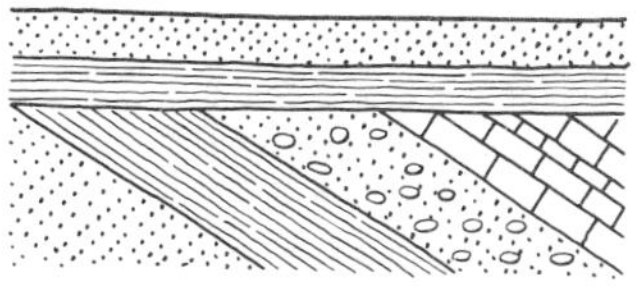

Unconformity This is an important effect of the deposition of sedimentary strata on previously deposited sedimentary beds, and is most clearly demonstrated by a diagram (Fig. 4.13). *Unconformity* is a term used to designate the unusual juxtaposition of several newer beds on older ones; it can be seen that an exposure of this contact would reveal no regular stratigraphical succession. Another term used in this connection is *disconformity*, which describes a juxtaposition of two series of beds with parallel bedding planes, the upper surface of the lower group having been eroded prior to the deposition of the upper group. Disconformity is a special case of unconformity.

FIG. 4.13 Unconformity of geological strata.

4.7 CONCLUSION

The small selection of photographs that has been chosen to illustrate this brief introduction to the geology of rocks will show the fascination of examining rock exposures in the field, even as it illustrates some of the complexities that will be faced by the civil engineer in the prosecution of work in bedrock. An interest in field geology also gives the engineer the opportunity for examining fossils in place and for starting a collection of fossil forms so studied. When one finds an exposure of varied rock strata now so displaced that they are almost vertical instead of horizontal (as they must have been originally), exhibiting some folding and even minor faulting, one can readily see why the unraveling of the geological history of such exposures can become a lifetime study, not only for the professional geologist but also, on a much more modest scale, for the amateur geologist which many engineers eventually become just because of the relevance of geology to their professional activity.

4.8 REFERENCES

4.1 John Evelyn, "Terra, a philosophical discourse on Earth," in *Silva, a discourse on forest trees*, 5th ed., Colborn, London, 1675.

4.2 T. Sheppard, *William Smith, His Maps and Memoirs*, Brown, Hull, England, 1920.

4.3 G. P. Merrill, *Rocks, Rock Weathering and Soils*, rev. ed., Macmillan, New York, 1911.

Suggestions for Further Reading

The Audubon Society Field Guide to North American Rocks and Minerals by C. W. Chesterman, published in 1978 by Alfred A. Knopf of New York, is a beautifully illustrated handbook of 850 pages, accurately described by its title. A corresponding pocket- or purse-sized guide is *Rocks and Minerals* by H. S. Zim and P. R. Shaffer, first published in 1957 by the Golden Press of New York as one of the Golden Nature Guides.

An Outline of Structural Geology by B. E. Hobbs, W. D. Means, and P. F. Williams, published by John Wiley & Sons of New York in 1976, is a useful 571-page introduction to this important aspect of geology. A corresponding British volume is *Elements of Structural Geology* by E. S. Hills, now in its second edition (1972) of 512 pages, published by Chapman and Hall (E. and F. N. Spon of London).

Atlas and Glossary of Sedimentary Structures is the title of a well-illustrated survey by F. J. Pettijohn and P. E. Potter. It covers the small structural features found in sediments (many of which affect the bulk properties of rock) and was published as a book of 370 pages in 1964 by Springer of New York.

Industrial minerals are often important to civil engineers. R. L. Bates is the author of a readable, 459-page coverage of these important materials, *Geology of the Industrial Minerals and Rocks*, published in 1969 by Dover of New York.

It is difficult to make a selection from the many books now available on geological structures. One volume may, however, be mentioned if only because of its widely acclaimed illustrations, which are truly excellent. It is now published as *Putnam's Geology;* the editors of the 789-page fourth edition are E. E. Lawson and P. W. Birkeland; the publisher is Oxford University Press of New York. The somewhat unusual title relates to the author of the first edition of this notable book.

chapter 5

SOILS: THEIR ORIGIN AND DEPOSITION

5.1 The Origin of Soils: Rock Weathering / 5-2
5.2 Pedology / 5-7
5.3 Residual Soils / 5-7
5.4 Transported Soils / 5-10
5.5 Glacial Action and Glacial Soils / 5-15
5.6 Soil Characteristics / 5-25
5.7 Organic Soils / 5-28
5.8 Permafrost / 5-29
5.9 Conclusion / 5-31
5.10 References / 5-32

The geology of soils is one branch of geological study that is of special importance to civil engineers. It is safe to say that the problems and difficulties encountered in engineering work, in both design and construction, are many times more frequent and more serious when structures have to be founded upon soil than upon bedrock. In spite of this fact, geologists as a whole have not devoted anything like the attention that has been given to "hard-rock geology" to the corresponding study of the geology of soils. There are many reasons for this relative neglect, but perhaps the main one is the economic unimportance of soils. Even this is a relative statement, since it is naturally based on the fact that so much officially conducted geological work has necessarily had to be concerned with mining; and in ordinary mining work, soils are at best regarded merely as a nuisance. The picture is changing. The increasing emphasis placed upon geological investigations in connection with groundwater surveys and inventories and with environmental impact studies has inevitably directed more attention to soils. The decreasing amount of sand and gravel in many developed areas is having a similar effect. The recognition that geology is steadily gaining as an essential aid in civil engineering is itself assisting in the same direction. It is now not uncommon for geological surveys to have sections devoted to Pleistocene geology. Corresponding groundwater sections are making their own contributions to a better understanding of the geology of soils. A review of geological literature leaves no doubt of the current general increase of interest in this somewhat neglected aspect of geology.

Soil mechanics is the universally accepted English-language term for the engineering study of such material. It is, admittedly, impossible to draw a hard-and-fast line of demarcation between soil and rock; some of the softer shales, for example, grade imperceptibly into the stiffer clays. On such a borderline of terminology, however, there is little need for argument. It is with soils, as herein defined, that the engineer has difficulties, meets problems, and undertakes vast earth-moving operations. (The word *earth* is sometimes a synonym for soil, but its use is generally confined to the construction job.) The engineer should therefore know something of the origin of soils and how transported soils have reached their present position. This knowledge will improve his understanding of soil characteristics, and it is an essential prerequisite to the full appreciation of the potential of soil mechanics in civil engineering practice.

5.1 THE ORIGIN OF SOILS: ROCK WEATHERING

Examination of almost any rock surface that has been exposed to the atmosphere for an appreciable period will show that it has been noticeably affected by this exposure; in many cases, disintegration of the surface layer will be apparent. The complex action caused by atmospheric factors results in the creation of soil and of the mechanically formed sedimentary rocks. The process of change resulting from exposure of rocks to the influence of the atmosphere is known as *weathering*. Dr. G. P. Merrill suggested that the word should be "applied only to those superficial changes in a rock mass brought about through atmospheric agencies, and resulting in a more or less complete disintegration of the rock as a geological body . . . it does not include those deeper-seated changes . . . during which the rock mass as a whole maintains its individuality and geological identity."[5.1] The term *alteration* is used to describe these internal rock changes which are often of the nature of hydration and which lead to the formation of new minerals in the rock mass.

So variable and uncertain are atmospheric influences that the process of rock weathering is always most complex; it naturally varies at different localities, at different elevations above the sea, at different times of the year, and with different rock types. It is possible, however, to classify generally the several agencies responsible for weathering, although the relative significance of each agent varies according to local circumstances. It is often difficult to separate weathering from *erosion,* the process by which the products of weathering are removed from their natural location, since the two are often simultaneous. For convenience, and in accord with the distinct identities of the two processes, they will be considered separately.

Agencies of Weathering The action of the atmosphere itself, i.e., of the mixture of gases that constitute the air, is practically negligible as a cause of weathering; only when water (or moisture) is present are the constituents of the air significant. Movement of the atmosphere, in the form of strong wind, is an important agency in arid parts of the world and is, in addition, a leading means of erosion. In dry areas, where winds can pick up and carry along in suspension solid particles such as sand grains, the grinding, or "sandblasting," effect of this solid matter on exposed rock faces can be a very effective mechanical agent of rock weathering. Figure 5.1 shows an extreme example of the type of scenery resulting from this action, examples of which on a smaller scale will be familiar. Variations in the temperature of the atmosphere are another cause of weathering attributable to the atmosphere itself; if the changes occur suddenly and cover an appreciable range of temperature, the internal stresses set up in an exposed rock mass may be sufficient to cause flaking off of surface layers. *Exfoliation* is a name frequently applied to this particular phenomenon, which is not at all uncommon (Fig. 5.2). Temperature changes are usually

FIG. 5.1 "Nature's architecture", illustrated in the unusual scenery on the Makran Coast, Iran, attributed to the action of sand-laden winds and torrential rains.

of more effect in the case of igneous rocks than in that of sedimentary rocks, especially with the coarser-grained igneous rocks, the several minerals in which may have coefficients of expansion that are sensibly different from one another. (See Fig. 5.3.)

Temperature plays an important part in one of the mechanical processes of weathering caused by water—the effect of freezing water. Rain falling on rock surfaces will naturally fill up any exposed cavities, open cracks, and joints in the rock mass and will tend to fill up the empty pore spaces of the rock. If freezing takes place, this entrapped water will exert a powerful disrupting force on the surrounding rock; the corresponding disintegration will often be considerable, even in a short period of time. Running water, in addition to being an important factor in erosion, will also act as a weathering agent if it is carrying solid matter in suspension and as a bed load. The removal of solid material from a river bed is primarily a process of weathering, although it becomes immediately a feature of transportation. The corresponding action of moving ice, as in a glacier, is equally important over large areas, and the quantity of soil formed in this way has been appreciable (Fig. 5.4).

Water plays another part in the general process of weathering. The impurities in water, which in nature is practically never found in its pure state, give it certain chemical properties which result in some of the most important weathering processes. Oxidation is one chemical reaction that may take place when rocks are in contact with rainwater; it is especially notable in rocks containing iron, as the brown stain of weathered surfaces often shows. The detailed reactions are complex; but hydrated iron oxides, carbonates, and sulfates are some of the products. All these reactions are accompanied by an increase in volume and a subsequent disintegration of the original rock mass. Finally, there is the action of rainwater as a solvent to be considered; in some localities this is the most potent of all aspects of weathering. The action is usually thought of in connection with limestones, but many other rock types are also affected to a lesser degree. Although pure water is a poor

FIG. 5.2 Exfoliating granite on lower Quarter Dome, Yosemite National Park, California.

FIG. 5.3 Exfoliation of granite boulder caused by a forest fire, northern Quebec.

FIG. 5.4 Yentna Glacier, Alaska.

solvent of the more common minerals, rainwater contains an appreciable amount of carbonic acid, which will in time decompose nearly all the rock-forming minerals. By "time," geological time is naturally meant.

There are a few minor weathering agencies. Possibly the most interesting of these are animals and plants. The action of bacteria is of real significance in connection with rock weathering, a matter now under active investigation. Ants and similar insects may assist the weathering process by carrying grains of soil and soil moisture underground; in this way the chemical action of water may begin to affect a rock mass. Finally, the effect that plant growth, especially the spreading of tree roots, can have in splitting up solid rock masses—and thus allowing other weathering agencies access to the interior of the rock—is a phenomenon frequently observed.

The possible complexity of the combined action of all or some of these agencies of weathering will now be clear. When the general process is considered in relation to the measure of geological time, the immensity of this great part of the natural cycle will be apparent. Speculations as to rates of weathering are interesting but of little practical significance; what *is* of value to the engineer is the evidence of weathering that is encountered in engineering work and the variety of soils that have resulted from weathering in the immediate geological past. Wind-cut topography may not be within the purview of every engineer, but evidence of the erosive power of running water will be familiar to all who work out-of-doors. Glacial action can be traced only in areas that have been ice-covered in the past; in these regions, a variety of typical topographical features may be noted. The chemical action of water is usually seen by more detailed observation; the weathered appearance of most exposed rock faces in temperate regions is due to this agency. Observations of rock outcrops in the field will show how weathering conforms to features of rock masses such as joints and bedding planes, how the extent of weathering may vary with the nature of the rock, how weathering may cause the original rock form to be retained until the decomposed rock is disturbed, and how in other cases the weathering may cause the decomposed or disrupted rock automatically to separate from the parent rock mass.

Rock Type and Weathering Observation will also show how the nature and rate of weathering are directly dependent on the type of rock exposed. Thus, to take extreme cases, a limestone may be partially dissolved away, leaving a cavity (such as many a famous cavern) and a residue of tough ferruginous clay, whereas a granite may weather only for a thin surface layer into "rotten rock," ultimately yielding sand and clay. Among the general rock types, many of the sedimentary rocks are themselves composed of weathered products (sandstone is a good example), so that it is to be expected that further weathering will be due merely to mechanical action. The other great group of sedimentary rocks, the limestones, is liable to be decomposed chemically. In the case of pure limestones, decomposition can be complete. With impure limestone, many records exist of a loss of weight of no more than 60 percent; the residue frequently is a tough clay.

The weathering of igneous rocks is not so simple; the precise process of weathering varies with the mineralogical content of the rock and, to some extent, with the grain size. The processes are complex, but fortunately, because of the recurrence of a few main mineral types in the more common igneous rocks, a broad outline of the processes of weathering can be obtained. Naturally, hard minerals will offer more resistance to mechanical weathering agencies than some of the soft ones. Similarly, the more stable the mineral (in a chemical sense) the greater will be its resistance to chemical change. In consequence, free quartz, being at once hard and chemically stable, is the most refractory of all common minerals to the action of weathering. Some silicates are but little affected, notably the harder types such as tourmaline and zircon. The weathering of mica depends on the form in which it occurs and the rock of which it is a constituent. Of the feldspars, the potash varieties are generally more refractory than the soda-lime types.

Products of Weathering Table 5.1 lists representative minerals and the products resulting from their weathering. From an examination of even such a simplified statement as this, one can readily appreciate that the chemical reactions causing the changes noted are involved and, in many cases, of long duration. Essentially, the combined effect on rock masses is a breaking down of the mineral contents into relatively stable products of decomposition. The weathering process does not stop at this point. Even the most stable minerals are in time affected by contact with rainwater; and if this water has leached through an upper layer of humus, carrying with it a small percentage of organic acids, its effect as a weathering agent will be increased. By the prolonged action of this moisture on silicate minerals, certain secondary products of weathering are gradually formed; these

TABLE 5.1 Some Leading Rock-forming Minerals

Mineral*	Main products of alteration
Quartz	Retains identity
Micas	Variable; clay minerals and hydrous silicates
Feldspars	Soluble matter and clay minerals
Hornblende } Augite }	Hydrous silicates
Olivine	Serpentine and iron oxide

*Arranged in descending order of resistance to weathering.

range down to colloidal material and consist of what may generally be termed the *clay minerals*. These secondary minerals may be generally classed into three main groups: (1) montmorillonite, (2) halloysite-kaolinite minerals, and (3) hydrated ferric oxide and alumina.

Many of the peculiar physical properties of clay are due to the properties of these minerals, especially those properties relative to moisture content. They are chemically reactive (particularly groups 1 and 3), being acidic in nature, and they possess the important absorptive property of ion exchange. This latter property can be of great importance in the treatment of clays when such treatment is necessary for engineering purposes. Amorphous material is also present in many clays, but the clay minerals, and their properties, are often of crucial importance when clay is used as an engineering material. Montmorillonite, chlorite, illite, and bentonite are, therefore, clay minerals which should be familiar in a general way to civil engineers.

Soils are thus seen to be the aggregation of the products of rock weathering; their formation is due to the simultaneous operation over long periods of time of one or more of several weathering agencies, the general nature of which is known. The coarser particles found in natural soil formations (above about 0.05 mm in diameter) consist generally of the more stable primary products of weathering—minerals that are relatively hard and chemically stable. The finer particles consist mainly of clay minerals, the secondary products of the weathering of silicate minerals. The properties of clay minerals are now under active investigation.

5.2 PEDOLOGY

Those who study soils for agricultural purposes (pedologists) refer to the weathered surface layers of soil as the *solum*. The effects of weathering naturally decrease with distance below ground surface, forming a sequence of layers, or *soil horizons*. Under the topsoil, or humic layer, is a highly weathered layer depleted in clay, oxides, and more soluble substances such as carbonates. This overlies a zone of enrichment in clay and oxides. Carbonates are retained below this in relatively unweathered *parent material*.

Pleistocene geologists recognize some weathering effects, particularly oxidation, extending some distance below this level. This may extend to the water table (see Chap. 8), but below this no weathering will usually be apparent. Figure 5.5 illustrates the major features of the various zones of weathering as viewed by pedologists and by many glacial geologists. The effects of weathering, both physical and chemical, which characterize the solum give rise to varying properties in the soil materials involved. Although most of the layers are so thin as to be of little concern in foundation engineering, they can be of importance in surface engineering operations such as the construction of highways and airfields. Much work has been done, notably by the Michigan State Highway Department, in correlating the soil properties as measured for agricultural purposes and those of importance in engineering.

5.3 RESIDUAL SOILS

The erosion and transportation of the products of rock weathering, phenomena which are so frequently associated with the weathering process, are not universal. Over many parts of the surface of the earth where the natural rock surface is approximately horizontal, weathering of the rock surface has proceeded without the soils so produced being moved from their original position. Soils of this nature are known as *residual soils*. The name is

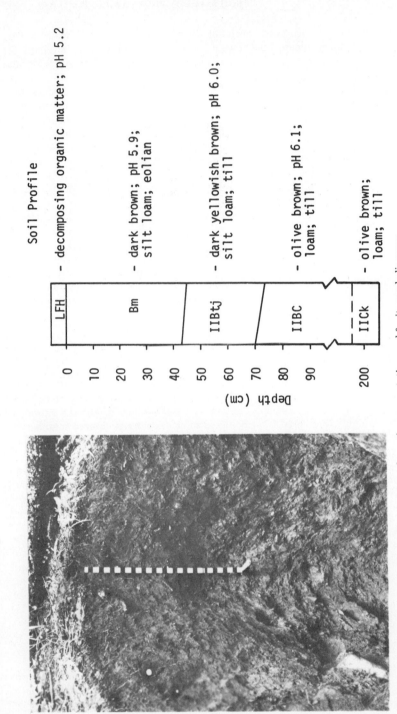

FIG. 5.5 Typical pedological soil profile (showing schematic representation used for its symbolic reproduction), Pend d'Oreille Valley, British Columbia.

accurately descriptive, since the soils are actual residues of the original rock; all soluble materials have been leached out by the long-continued seepage of groundwater. This chemical disintegration naturally becomes less active as depth increases; the alteration of the original rock gradually becomes less until finally the natural unaltered rock is reached. This gradual downward change from the soil to original rock is a distinctive feature of residual soils and one by which they can readily be distinguished. Another characteristic of residual soils is that they contain no minerals foreign to the locality; the entire mineral content is directly related to that of the underlying rock. Particles are sharp-edged, not rounded; fragments of the original rock may be found unaltered in the disintegrated stratum (Fig. 5.6).

In other respects, residual soils may vary. They may retain the structure and even the appearance of the original rock until disturbed in some way; on the other hand, some residual clays show no evidence at all of their origin. Some residual soils are sandy, but residual clays are more common than sands. Others possess peculiar characteristics and have been given special local or regional names. Only one of these—*laterite*—warrants mention in this discussion. This name is generally applied to soils of such tropical areas as India and parts of South America; because of the absence of organic material in the topsoil and the consequent lack of humic acid to act as a leaching agent, such soils have a high alumina and iron-oxide content. Clays of this nature are soft when naturally moist but harden on exposure to the atmosphere. The name laterite is sometimes applied in more temperate countries, such as the United States, to clays which have a high silica content, but its use in this way is largely governed by local practice.

Residual soils are found only in regions that have not been greatly disturbed during recent geological time. They are not generally found in areas that have been subject, for instance, to glaciation, but deep pockets are sometimes found in glaciated terrain. Some

FIG. 5.6 Unaltered Precambrian rock at left, but weathered granite in the form of residual soil in the center, Nový Knín, Czechoslovakia.

5-10   GEOLOGICAL BACKGROUND

of the more recent soil deposits which are, in a sense, residual soils are not subject to this restriction. They are the deposits of organic and inorganic material accumulated by vegetation; they usually occur in the presence of water, especially in peat bogs and marshes.

5.4 TRANSPORTED SOILS

Erosion has been repeatedly mentioned as one of the great modifying influences at work today on the surface of the earth. Through time, erosion has been responsible for the formation of most of the sedimentary rocks that cover three-quarters of the earth's surface; in the same way, it is today ceaselessly at work removing the products of rock weathering and depositing them elsewhere. Thus erosive processes are even now forming soils, some of which will in time become consolidated into the sedimentary rocks of the future. It has been estimated that very large volumes of the earth's crust have been eroded in this way; the resulting sediments give 80 percent shale, 15 percent sandstone, 5 percent limestone, and the corresponding unconsolidated materials.[5.2] Chapter 41 includes a discussion of present-day erosive processes and silting.

Aeolian Deposits   Although gravity is usually thought of as the agency ultimately responsible for the movement of disintegrated rock by erosive processes, the action of the wind is an additional agency of consequence. Deposits of soil formed by the wind are known as *aeolian deposits*. Sand dunes are an obvious and notable example of materials so deposited; the extent of some suggests their age. Their frequent existence near seacoasts shows that arid conditions are not imperative for the formation of aeolian deposits, although wind action is most potent in dry regions. Wind-deposited sand may show traces of stratification; its mineral grains will usually be fresh, since it has been formed by a mechanical weathering agency. The shape of the grains—well rounded and in some cases almost spherical—is the most notable characteristic of aeolian deposits, which may exhibit "frosted" surfaces.

A special type of aeolian deposit widely distributed throughout the world is that known as *loess*. It is very fine-grained and is often yellow (Fig. 5.7). It exhibits no stratification

FIG. 5.7   Loess as exposed in a road cut adjacent to the railway station, at Timaru, New Zealand.

but occurs in great massive beds which may be hundreds of feet thick. The material is composed largely of silt-sized particles which under microscopic examination can be seen to consist of sharp-edged fresh particles of minerals, such as quartz, feldspar, calcite, and mica; loess also has some clay content. It can fitly be described as an accumulation of wind-blown dust; its dustlike nature when in the air is evident in those areas in which it forms the surface of the ground—notably in China where ancient roads have cut themselves into the loess as deep trenches, where caves are excavated in it for human habitations, and where its characteristic color is used in names such as the Yellow River. Some extensive deposits of loess, in Europe and the midwestern United States were derived from the floodplains of the mighty rivers that drained from the great ice sheets of the last million years. A deposit somewhat similar to loess and found in the southwestern United States is known as *adobe;* it is used for making bricks and other molded articles by sun drying.

Gravitational Action The most obvious type of gravitational erosive action is that which leads to the accumulation of rock fragments at the foot of exposed faces of rock; the accumulated debris is generally classed as *talus.* In extreme cases, the erosive action may lead to the occurrence of rock avalanches. Here the force of gravity acts directly on the products of mechanical weathering (Fig. 5.8). These accumulate until disturbed in some way, often by erosion of the toe of the talus slope by running water. In extensive accumulations of talus, the mass of disintegrated rock may eventually reach a state of unstable equilibrium, and itself start to move. If the movement is sudden, an avalanche will result, but if it is exceedingly slow, it will be noticed only by long-term observations. The term *rock glacier* is sometimes applied to such phenomena, which are related to frost action.[5.3] Clearly, talus slopes are of importance in civil engineering work; if they are unrecognized and covered by a mantle of topsoil, they may be utilized as a foundation bed, with possibly disastrous results. Study of the local geology in association with test borings and drilling should disclose the true nature of the material. A gradation from dis-

FIG. 5.8 Rock-talus slopes on basalt cliffs at Dark Harbor on the west coast of Grand Manan, New Brunswick.

5-12   GEOLOGICAL BACKGROUND

integrated rock near the surface to fragments of fresh rock at greater depths will be a main distinguishing feature; all minerals encountered should be correlated with the adjacent natural rock formations.

Aqueous Deposits   Although aeolian deposits and talus are important types of deposits, the most significant and most widely distributed types of transported soil are those which have been moved by water and by the slow movement of ancient glaciers. *Glacial deposits,* as the latter are called, are so important and so closely related to topography that they will be treated separately in the next section. *Aqueous deposits* are also important and constitute probably the largest group of transported soils. Aqueous deposits may properly be called sediments, as they have been formed by deposition from either standing or moving water in a manner entirely similar to that in which sedimentation takes place at the present time. They are generally grouped into two main divisions: *marine* sediments and *continental* sediments. The former were deposited in the sea and the latter in fresh water, but in a variety of ways: some by river flow, some as deposits in fresh-water lakes, some as deltaic and detrital cone formations. All the processes of sedimentation can be observed today in some form or another, so that although there are many aspects of sedimentation as yet imperfectly understood, it is possible to classify the main ways in which material is deposited from moving water.

FIG. 5.9   The Pattullo Glacier, British Columbia, showing outwash soils, some in detrital fan formations.

Continental or terrestrial sediments may, in the first place, be transported by the headwaters of a running stream; disintegrated rock or eroded soil will be carried along by the streamflow, since the bed gradient will be steep. As the stream changes slope on entering a level area, its velocity will decrease, and a great deal of its load will be deposited as an *alluvial cone,* or *alluvial fan.* Material has been deposited in this way, probably, at all stages of the earth's history. Such deposits are to be distinguished in the vicinity of most important mountain ranges, and they are often of surprising depth and extent. The deposits show stratification, as would be expected from the nature of their formation; and because of the change in the volume of flow of all steeply graded streams, adjacent strata may vary greatly. Alluvial cone deposits usually show a major variation, from coarse sediment at the apex (possibly including boulders) to fine clays at the outer reaches. If the process of erosion continues without serious earth movement, the materials in an alluvial cone may themselves be eroded as the life history of the responsible stream develops (Fig. 5.9).

The final stage of river development is also a notable factor in sedimentation. After a stream has reached maturity, any increase in the volume of sediment brought down from its higher reaches will result in an *aggrading* of the river—a steady building up of the river bed. Several notable examples of this process may be seen today. Corresponding *natural levees* are built up at the sides of the normal river channel, and in times of flood the river water will extend beyond these where the terrain is flat, covering a large area on either side on which material is slowly deposited; this area is known as the *floodplain,* or *alluvial plain.* Material deposited in this way will also be stratified, but all of it will be fine-grained. Ancient deltaic deposits exhibit the same structural arrangement of beds; stratification is regular and grain size is uniform. River terraces, or *alluvial terraces,* often assist in the location of deposits of fine-grained sediments and sometimes of gravel (Fig. 5.10). They may be old floodplains below which the river has scoured out a deeper bed, possibly because of interference with the headwaters of the river. Alternatively, they may be evidence of earth movement that has led to a major alteration in the regime of the stream.

FIG. 5.10 River terraces at the junction of the Nicola River with the Thompson River, British Columbia.

5-14 GEOLOGICAL BACKGROUND

In desert regions, rivers may flow (often intermittently) into inland basins that are either dry or covered with only shallow water. The sedimentation that takes place in such regions when the rivers in flood bring down large quantities of material with them is obvious. If earth movement accompanies sedimentation, as has happened in the past, the depth of the deposits may become considerable. The material will be stratified, and as none can escape, it will be varied, often including layers of common salt and gypsum. Rivers flowing into lakes will tend to form deltaic deposits similar to those at outfalls into the sea or to those of marine deltas.

Marine Deposits Marine deposits are usually considered as having been formed in one of three regions: (1) the shore zone; (2) the shelf, or shallow-water, zone; and (3) the deep-sea zone, which extends beyond the edge of the continental shelf. It is helpful to recall that marine sediments are derived from the material transported by rivers and that eroded from seacoasts and also from organic remains of foraminifera and radiolaria. Except in the shore zone, therefore, fine-grained sediments are to be expected. The material in the shore zone, although often of considerable local importance, is really of a temporary nature, since it varies with changes in shore line and is itself constantly being changed. It is predominantly coarse and is usually classed as sand and gravel; its observed irregularity is characteristic.

Deposits in the shelf, or shallow-water, zone, often considered as extending to the 100-fathom (600-ft) line, are of great importance. The region is one of ceaseless change; the movement of the sea gradually sorts out the material deposited on the seabeds, and the finer material gradually works down into the deeper water. Interesting as are present-day deposits in this zone, the corresponding deposits of previous eras are of greater importance. Crustal movements involving parts of previous shallow-water zones have converted many of these sediments to land formations which cover surprisingly large areas and, because of repeated earth movement, are often of considerable depth. Correspondingly, parts of the continental shelf were above sea level in earlier times and so exhibit dry-land characteristics; for example, fossils found normally on shore can also be found in this zone.

Shallow-water deposits are usually flat and display stratification; they may show gradation from gravel to finer materials, but clays will sometimes be found close to original shore lines. Although usually inorganic, many shallow-water deposits contain lime-secreting organisms which have been responsible for the formation of limestones. Ripple marks are a frequent occurrence in these deposits; and when they are of clay, "mud cracks,"

FIG. 5.11 Bouldery till as exposed in road cut, St. Joseph Island, Ontario.

caused originally by the action of the sun on the wet clay, may still be seen. Shallow-water sediments sometimes display the tracks of land animals, and their fossil contents are often appreciable. The nature of the fossils can often serve as a basis for distinguishing between a marine shallow-water sediment and one laid down in fresh water; other than fossil content, the general characteristics of the two groups are markedly similar.

There remain for consideration the deposits of the ocean depths. Characteristic sediments are muds, organic oozes, and clays of varying color but remarkable uniformity of texture. Of great present-day interest, they are of little significance in the study of soils, since the existence of materials from the ocean depths now forming part of the surface of the earth has been traced in the case of a few islands only. Marine deposits when spoken of geologically are therefore generally to be regarded as shallow-water sediments (Fig. 5.11).

5.5 GLACIAL ACTION AND GLACIAL SOILS

Well over a century ago, Swiss naturalist Louis Agassiz reported the existence of boulders so far removed from their natural bedrock that he supported the view, already expressed by others, that they could have been transported only by glacial action. It was many years before this revolutionary idea was generally accepted; the persistence of the word *diluvial*, suggesting the great flood as the origin of deposits now known to be glacial, is an interesting reminder of this conservative attitude. Today, the importance of glacial action in the geological cycle is generally appreciated. The large erratic blocks so often preserved in public places when discovered in urban areas are significant indications of the power of this great natural process (Fig. 5.12). There are some blocks of this kind that weigh well over 1,000 tons, and some—even of this magnitude—are now at great heights *above,* as well as great distances away from, their original positions. Many readers will be familiar with boulders of this kind and so will appreciate the significance of glaciation as a factor in erosion. At one time this was probably the most influential of all erosional processes, since it is estimated that one-third of the land area of the world was once covered by ice. This Ice Age, as it is often called, is of great importance in connection with present-day soils and also with many civil engineering operations.

The existence of glaciers today in several parts of the world (notably in the European Alps and in North America) and of the ice sheets which cover Greenland and the Antarctic continent has made these phenomena generally familiar. Without such contemporary

FIG. 5.12 World's largest erratic blocks at Okotoks, Alberta, moved by ice action from Jasper 300 mi away; figure on right gives scale.

examples, the state of the world during glacial epochs would indeed be difficult to imagine. Even with these examples available, it is still far from easy to imagine practically the whole of Canada and a large part of the northern United States, most of northern Europe (including the British Isles), Asia, and part of South America covered with ice sheets varying in thickness up to several thousand meters. This is now known to have been the case during the recent Ice Age which occurred during the Pleistocene epoch. The proof of this has come from a fascinating accumulation of carefully observed facts linked to an outline of the effects of glaciation. It is known also that the ice receded several times; the interglacial periods were to some extent warmer than present-day climates in, for example, the area around Toronto, Canada, where some of the most important *interglacial deposits* yet to be examined have been found.

Why these incursions of ice took place is still a matter of conjecture. The engineer has to accept the accredited fact that glaciation did occur over a large part of the earth's surface and, recognizing this, must become familiar with all that a study of glacial geology can suggest of the possible effects of glacial features on engineering work. Figures 5.13

FIG. 5.13 Sketch map showing area of North America covered by Pleistocene ice.

and 5.14 show in broad outline the areas in North America and Europe that were covered with ice and in which, therefore, glacial geology is of practical importance.

There are two main types of glacier: the *mountain*, or *valley, glacier*, slowly moving down the valley that it fills; and the *continental glacier*, or *ice sheet*, of such extent and thickness as to override mountains and valleys. (The ice sheet over Greenland today, for example, is over 3,000 m, or almost 2 mi, thick.) Movement of the ice is similar in both cases, although rates of travel may vary considerably; some ice sheets remain stationary for long periods. Many interesting observations of the rate of travel of existing glaciers have been made.[5.4]

Glacial Erosion The action of glaciers in eroding and transporting material off the surface of the earth is well defined. As the mass of ice passes over the earth's surface, it may pick up loose and unconsolidated material directly, incorporating this into the lowest part of the ice mass. If some of the material is in the form of rock fragments and boulders and is pressed against solid bedrock as the ice passes over it, an abrasive action will result; the bedrock will be eroded, and thus more loose material—*rock flour*—will be picked up by the moving ice. It is known that thick ice sheets also exercise a "plucking," or "quarrying," action; the lower part of the ice formation adheres to large blocks of rock, tears them away from their natural position, and carries them along.

This erosive action has a definite effect on the topography of the land covered by the glacier. The effects of the actions of the two types of glaciers on topography are naturally different. Continental glaciers tend to level off the surface, eroding high points of land to

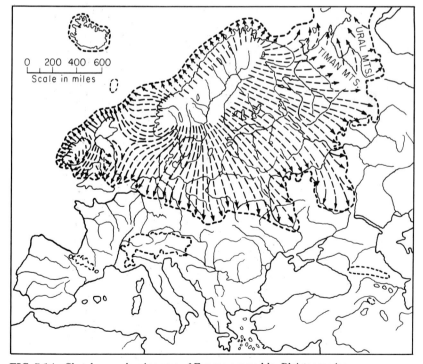

FIG. 5.14 Sketch map showing area of Europe covered by Pleistocene ice.

a greater extent than low-lying ground. The ice sheet will follow the contour of the ground that it covers and erode constantly throughout the whole course of its movement. On exposed rock surfaces, the characteristic polishing and markings caused by rocks embedded in the bottom of the moving ice sheet are to be seen (Fig. 5.15). The polishing is often remarkably smooth and uniform, broken only by the grooves and scratches—*striae*—made by projecting rock points; these markings are also found on some rock fragments encountered in glacial deposits. The polishing and markings provide good clues to glaciation, should this be in doubt. By studying these marks and by tracing the rocks found in glacial deposits back to their origins, geologists have discovered the centers from which the ice movements indicated in Figs. 5.13 and 5.14 radiated.

The main erosive effect of mountain glaciers is to smooth out the valleys through which they pass, often deepening them considerably and almost always giving to them a characteristic U-shaped cross section in distinction to the V shape of a young valley being eroded by a running stream (Fig. 5.16). A corresponding result is that valleys affected by glaciers are straightened; the projecting hillsides between the valleys of side streams are removed. Frequently, if the side-stream valleys do not have glaciers in them, they will remain at their original elevation while the main valley will be deepened. After the glacier has melted, this condition will give rise to *hanging valleys,* a topographical feature responsible for many waterfalls and offering potential water-power development sites. The deepening of valleys by glaciation has led to many beautiful fjords and similar inlets from the sea when a general depression of the land surface has followed the glacial period. Another topographical feature caused by erosion is the *cirque,* or steepsided, rounded end of a valley that has been glaciated; this differs considerably from the typical valley formation caused by running water. Cirques result from the initial stages of glaciation; snow and ice collect in the head of the valley and gradually accumulate debris eroded from the adjacent

FIG. 5.15 Effects of ice action over metamorphic rock of the Cross Lake group, Manitoba; striae, polishing, and gouging can clearly be seen.

SOILS: THEIR ORIGIN AND DEPOSITION 5-19

FIG. 5.16 Glacial valley in the Flint Creek Mountains near Anaconda, Montana, showing typical U shape.

FIG. 5.17 Roche moutonnée near Thessalon, Ontario; the ice moved from right to left.

valley slopes. *Roches moutonnées* are outcrops of rock that, having been passed over by a glacier, exhibit a striking and characteristic form, smoothed and rounded on the side opposed to the glacial movement and in a natural state on the opposite face (Fig. 5.17).

Types of Glacial Deposits The effect of glaciation on topography cannot be considered in relation to erosion alone; simultaneous deposition of the products of erosion has occurred and is frequently related to some of the erosive features already mentioned. Rock debris is carried in three ways by moving ice: at the bottom of the ice sheet; on its surface; and within the ice sheet, this material having worked its way in from the surface of the ground. During the process of melting, *glacial deposits* of great importance are laid down. Melting at the ends of glaciers can be observed today, but melting has not always been so confined and must be thought of as having been a much more extensive process during times of climatic changes of unusual degree.

Material deposited by glaciers is generally divided into two classes: glacial deposits proper and glaciofluvial deposits, which are laid down by water issuing from the glacier. Glaciolacustrine and glaciomarine deposits are those laid down in lakes and the seabed adjacent to glaciers. Among the deposits from the glacier itself, moraines will immediately come to mind, as they are often so striking a feature of modern glaciers (Fig. 5.18). An *end moraine* is the deposit formed by wastage of the end of a glacier; the moving glacier may have pushed forward the material deposited as its front face melted away, or the melting of the glacier may have just equaled in speed its forward movement. End moraines are therefore found as ridges across valleys; they are often of considerable height and are in plan usually convex down the valley. *Lateral moraines* and *medial moraines*, the deposits formed by the material carried on the surface of the glacier's sides and in its center, were often laid down in ridges parallel to its movement. They are frequently of considerable height, especially lateral moraines, some of which are over 300 m (1000 ft) high; in plan, they follow the direction of ice movement. Finally, the term *till* is applied to what is probably the most important of all types of glacial deposits—that deposited from the bottom of the glacier itself and consisting of material previously carried within the ice sheet at its lower surface. In thickness, this type of deposit may vary from 150 m (500 ft) to almost nothing; it produces a fairly level, gently undulating surface. Occasionally, the surface of the ground moraine will be interrupted by small, oval-shaped hills of unstratified sandy material which have their long axes in the direction of ice movement. They are called *drumlins* and are thought to be formed by actively flowing ice.

The melting of glaciers produces streams of water which, in the course of running away, carry with them at least some of the debris previously carried by the ice from which they flow (Fig. 5.19). Restricted in channel and swift-flowing at first, these streams gradually decrease in velocity and deposit their loads; this gives rise to stratified deposits of the debris from the glacier. In a manner comparable to that which causes alluvial fans at points of change in river flow, *valley trains* are formed by the outwash waters from mountain glaciers as they leave the cover of the ice and start their flow down the valley. *Outwash plains* are the corresponding formation of continental glaciers, although they are naturally on a larger scale (Fig. 5.20). Some glacial streams may flow under the ice sheet

FIG. 5.18 The Hugh Miller Glacier as it was in 1929, showing typical moraines.

FIG. 5.19 The first stages of soil formation; rock screes and morainal deposits at the foot of the Taweche Glacier in the Himalayas near Mount Everest.

FIG. 5.20 Excavation proceeding in glacial outwash deposits (close to dense glacial till), Lake St. John district, Quebec.

in tunnels, and the debris they carry with them will tend to be deposited in long, winding ridges along the paths of the hidden streams. As the ice melts, these ridges will be exposed; they are called *eskers,* and some of them are several kilometers long. Similar deposits sometimes formed at the mouths of such hidden glacial streams are called *kames;* they are roughly conical hills composed of poorly stratified glacial deposits. If a glacier had its end located in the waters of a lake, depositional features already described would be somewhat modified, and a deltaic formation would result; *glacial deltas* are a recognized, although infrequent, type of glaciofluvial deposit.

Since the different types of glacial deposit are associated with the results of the erosive action of the corresponding glaciers, the modification of topography made possible as a result of glaciation is surprising. Many major modifications have been detected and studied. Thus, it is now known that in North America, Lake Algonquin, a glacial lake which occupied the present basins of Lakes Huron, Michigan, and Superior, had outlets successively (or simultaneously) past the site of Chicago into the Mississippi, through the St. Clair River and Niagara, and through Georgian Bay into the Ottawa Valley. Similar examples from other regions can be found in publications dealing with local glacial features. Of far more importance to the civil engineer are the minor changes that glaciation effected, and these are particularly related to the deposition of glacial deposits in valleys that have been subjected to glacial erosion.

Buried river valleys or *bedrock valleys* will be frequently mentioned in later chapters. Although they may be encountered under any transported soils, they will probably cause most surprise in glaciated areas. Their origin is complex, including erosion by rivers in preglacial or interglacial times and by subglacial streams. (See Fig. 5.21.) After the melting of glaciers, a new drainage system has to evolve, and it is entirely possible that the

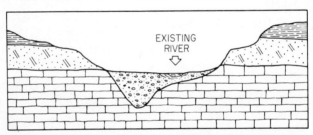

FIG. 5.21 Diagrammatic geological section illustrating the occurrence of buried valleys.

new course will have no relation to the buried valley. Buried valleys may be filled with permeable gravel which can be a potential problem in the planning of reservoirs but also a beneficial new source for water supply.

Terminal moraines often form natural barriers across the ends of eroded rock valleys (Fig. 5.22). If impervious, they may be the cause of lakes in the valleys and, similarly, they may sometimes be relied upon to form a part of the impermeable structure needed for the formation of a reservoir. In all such cases, because of the unstratified and variable nature of moraines, detailed subsurface examination must be made to obviate all possibility of leakage. In addition to covering up buried valleys, glacial drift will hide from sight all other irregularities of the bedrock surface; in some regions, these irregularities may be frequent and of appreciable magnitude, and they may account for some of the strange results obtained when test borings or well shafts are put down in glaciated areas. Well shafts are not frequently sunk in these areas, however, since the nature of the drift does

FIG. 5.22 Parallel moraines (the De Geer moraines) in the La Grande area near James Bay, northern Quebec.

not, as a rule, promote the accumulation of groundwater. Glaciofluvial deposits, on the other hand, particularly outwash plains and valley trains, are frequently good aquifers.

Nature of Glacial Deposits The nature of glacial deposits must finally be considered. The material transported by glaciers varies from the largest type of boulder to clays, including all gradations from one to the other; it is known generally as *till* but is often referred to as *drift* in British reports. Prof. W. O. Crosby, in a very early investigation, found that the drift around the Boston area in the United States, on the average and excluding all large stones, contained 25 percent gravel, 20 percent sand, 40 to 45 percent fine sand, and less than 12 percent clay. The material either was picked up by the glacier from the natural surface of the ground or was eroded by the glacial action and then transported. If this eroded material is that which is ground off the surface of bedrock, it will usually be rock flour—finely ground, fresh rock particles (its mechanical analysis generally places it in the class of silt) with no plastic properties. As deposited directly from the glacier, all gradations of transported material will be thoroughly mixed together; the resulting unstratified, unsorted mixture of unconsolidated material is that known as till. (See Fig. 5.23.) Because of its usual boulder content, the term *boulder clay* is sometimes used to describe this unsorted deposit, but the name is not strictly correct, since frequently the till contains no clay at all. A large percentage of the rock fragments will be angular in shape rather than rounded; many will exhibit scratches similar to the striae on polished bedrock surfaces; and the rock types encountered in the till will vary, usually carrying a portion of material from distant sources.

This same material constitutes the load carried away by outwash streams from the glacier and thus forms the glaciofluvial deposits already described. During transportation by the running water, and through the mechanical nature of its deposition, the material will be sorted according to size, at least to some degree. The coarsest material will be deposited first; the finest, last and farthest downstream. Glaciofluvial deposits are consequently stratified to some extent and consist of materials approximating the different soil groups. The exact nature of the deposits depends on the process of deposition. Thus

5-24   GEOLOGICAL BACKGROUND

FIG. 5.23   A close-up view of typical glacial till near Wroxeter, Ontario.

kames, because of their irregular mode of formation and although stratified, will usually consist of materials that vary so much between strata as to make the deposit useless as a source of one particular type of soil (such as sand or gravel) unless they are crushed. Eskers are not commonly encountered, but because of their formation from running water, they generally consist of coarser materials (sands and gravels). Valley trains, outwash plains, and glacial deltas all provide fairly regular deposits of gravels, sands, silts, and possibly clays; the type of soil will depend on the location, selected for trial relative to the formation of the deposit.

It will be seen that glacial deposits will not generally exhibit uniformity as a leading feature. This necessitates the exercise of the greatest possible care in investigating them before use. Attention must be devoted not only to the extent of the deposits but also, if one plans to use the soils, to their mineralogical nature. Because of the nature of their formation, glacial deposits will often consist of disintegrated rock that has weathered in a mechanical sense only; the constituent minerals will usually be fresh below the depth of recent weathering. If, therefore, any of these minerals are liable to weather chemically on exposure to the atmosphere, one must exercise discretion in deciding to use the deposit as, say, concrete aggregate. Similarly, with respect to the finer deposits, one must guard against mistaking rock flour for clay; but with modern methods of soil testing, this distinction is easy to make, even though particles in rock flour may be clay-sized.

Finally, brief mention must be made of boulders. Erratic blocks were mentioned in a general reference at the outset of this section; they typify the boulders that *may* be met in glacial drift. Examples from construction practice will be cited to show what trouble undetected boulders can cause. Such a boulder, encountered during the construction of the cutoff trench for the Silent Valley Dam, a water-supply project for Belfast, Northern Ireland, was described as being "as big as a cottage." Although this case is unusual, it illustrates vividly the dangerous possibilities of inadequate subsurface exploration in glacial drift.

If an engineer once recognizes the fact that he is working in drift-covered country, he should obtain diamond drill cores to a depth of at least 3 m (10 ft) when "rock" is struck. A careful examination of such cores should indicate whether the drilling is actually in bedrock or in a boulder (Fig. 5.24). Valuable information is gained if the engineer appreciates the leading features of the main rock types; if in doubt, he should consult an engineering geologist. This is one of the most important ways in which geology can shed light on an engineering problem.

FIG. 5.24 A "boulder pavement" at Sunnyside, Toronto, Ontario; the effect on construction if such glacial deposits are not penetrated or detected in subsurface investigations can well be imagined.

5.6 SOIL CHARACTERISTICS

In the study of soils, the geologist and the civil engineer take two extreme positions: the geologist is interested primarily in the origin of soils, and the civil engineer, having only indirect interest in origins, is vitally interested in types of soils and the properties peculiar to each type. Such extreme positions are tenable only in theory. In practice, the geologist has to differentiate among gravels, sands, and clays, at least in general terms; and the engineer, in order to have a proper appreciation of soil properties, must give some consideration to the origin of the particular soils being dealt with. Thus, a common meeting ground for the geologist and engineer is the consideration that both have to give to soil characteristics. Note may well be taken, therefore, of certain geological features affecting soil characteristics; this will be done without venturing into more detailed aspects of modern soil-testing work.

In general, residual soils and many sediments will vary in character from the surface downward. If the latter are stratified, identification of the respective strata may assist sampling operations. River deposits, especially those from swift-flowing streams, will probably be variable in content, as are also several types of glacial deposits. Marine deposits and floodplain sediments from mature streamflow tend to be more uniform. Whatever the origin of the soil, however, a general study of the local geology prior to taking soil samples will be of great value. When once the subsurface survey operations have been carried out, accurate records must be plotted to reasonable scales. In conformity with general engineering practice, one would quite naturally describe cross-sectional drawings of soil conditions as *soil profiles*. This is an expression to be avoided by engineers in their normal work, however, since it has been adopted by pedologists for a rather specific use. In order to avoid any further confusion in soil terminology, engineers should restrict their use of the term *soil profile* to the pedological use. Accordingly, all references to the graphical plots of cross sections of soil conditions, as derived from subsurface investigation, should be couched in terms such as *soil cross sections*. Careful study of such sectional records will be invaluable, particularly in showing where additional soil samples should be procured. Especially is this the case with river and glacial deposits. Only when this study of the record has been made can the sampling be regarded as complete.

Special note must be made of the terminology presently employed for soil classification. Each of the main group terms used—gravel, sand, silt, and clay—is based on the *size* of particles composing the soil, generally without respect to their mineralogical nature. This is in contrast to earlier practice in geological terminology. Occasionally, compound names, such as *quartz sand* or *limestone gravel*, are used to describe materials more accurately, but such usage is still infrequent. The more usual compound names have relation to the physical properties of the soil, e.g., *fine sand* and *tough clay*. It is probable that a more frequent use of geological compound names will develop as soil studies progress.

Terminology used in the field, as in the recording of test-boring results, provides another problem, since the selection of the names used to describe the soils encountered is still a matter of individual judgment. Experienced boring operators can usually be relied upon to differentiate correctly between different soil mixtures, but they use their own individual standards of comparison. Consequently, if a uniform and recognized standard is to be adhered to throughout the work, it is advisable, in all soil exploratory work, to correlate the soil names used by those responsible for taking field notes with the results of laboratory analysis and test (see Chap. 6).

In this respect, as in others, laboratory work on soils can be of assistance to the geologist as well as to the engineer. In the laboratory, for example, the shape of sand grains can be determined, and their mineralogical nature can be ascertained; this information can often be of real assistance to purely geological studies of sedimentary origins. In some cases, the information can be obtained by geological study in the field without the aid of laboratory facilities.

Gravels Gravels are accumulations of unconsolidated rock fragments which result from the natural disintegration of rock and which are at least 2 mm in diameter. From this minimum size, gravel may range up to the largest pieces of rock normally encountered. Names commonly given to larger fragments are *pebbles* (from 4 to 64 mm), *cobbles* (64 to 256 mm), and *boulders* (having a diameter greater than 256 mm, or about 10 in).

Gravels are widely distributed but are rarely found without some proportion of sand, and possibly silt, unless these finer constituents have been washed out subsequent to the deposition of the gravel. They are characteristic shallow-water or river deposits (including the river flows from glaciers); they may also be beach-formed, and some gravels are found in kames. The shape of the individual rock fragments and their relative mineralogical freshness will depend on the history of the gravel, but variations from rounded gravel to angular fragments may be met. Gravel that has been transported to such an extent that the individual fragments are rounded will consist of the more resistant rock types, often a mixed collection; but angular gravels, being comparatively new, may consist of fragments of even so relatively soft a rock as shale.

Sands Sand is the name applied to fine granular materials derived from either the natural weathering or the artificial crushing of rocks; sand ranges from 0.053 to 2.0 mm in diameter. Various subdivisions are used by different authorities to denote the gradations of sand-particle sizes; the most usual of these is to fix a diameter of 0.42 mm as the dividing line between fine and coarse sand. As it occurs naturally, a sandy soil mixture will frequently have some clay and silt mixed with the sand-sized grains. If the clay and silt content does not exceed 20 percent, the mixture is called a sand.

The origin and therefore the occurrence of sands are similar to those of gravels; the two are often found together in one deposit. Dune sands, which may today be found far inland as a result of earth movements, are the most uniform type of deposit; river sand often contains relatively large amounts of gravel, silt, and clay. Glacial deposits may contain good sand beds. Residual sands are sometimes found. As sands are often derived from the continued disintegration of gravel, it follows that they will generally be composed of the harder and more stable minerals. The popular conception that all sands consist of quartz particles is, however, far from true. Although some sands consist of almost pure quartz, most of them contain at least a small percentage of other minerals. Some consist primarily of minerals other than quartz; calcareous sands are sometimes encountered in limestone districts. As a rule, their mineral content is fairly stable, but glacial sands may contain fresh minerals liable to weathering on exposure.

The shape of sand particles may vary from completely rounded grains to angular fragments. The former are rare and are found generally in desert regions. The age of a sand deposit cannot be determined from the shape of the grains; shape depends principally on the rock from which the sand was formed and the history of the sand during transportation. Long travel in water will tend to round the grains, but this is a slow process, as is shown by the fact that sand at the mouth of the Mississippi is composed of angular grains. As a rule, river sands are more angular than those found in lake and marine deposits.

Silts Silt is the name now given to soils with particle sizes intermediate between those of sands and clays, i.e., from 0.053 to 0.002 mm in diameter. Most inorganic silts have little or no plasticity, and many glacial silts consist of such fresh minerals that they may properly be referred to as rock flour. Silts that exhibit some plasticity will contain flake-shaped particles, usually one or more of the clay minerals, possibly partially formed. Silts may readily be mistaken for clays; they frequently have the typical gray color of clay and the same apparent consistency when wet. Simple hand tests, such as shaking a pat of the wet soil in the hand, will usually distinguish the two; a sample of silt will lose water, and its surface will thus appear glossy. Dry silts will crumble much more easily than any clay; correspondingly, they will create dust easily on light rubbing. In a natural position, and especially when damp, silts will be almost impermeable. When disturbed, silt will flow easily if an excess of water is present. Local variations sometimes carry expressive names, such as *bull's liver,* a term often heard on construction jobs.

Clays Deposits of soil particles smaller than 0.002 mm in diameter and displaying plasticity when wet are known as a clay. The name is correspondingly applied to a soil mixture, although the clay and silt content combined must exceed 50 percent. Such a mixture will probably still display the characteristic properties of a pure clay. The essential qualification of plasticity shown by the soil mixture when wet is particularly to be noted; quartz may be so finely ground that it conforms to the size requirements of a clay, and yet it will have no characteristic clayey properties. It may also be noted that clay solids include particles of colloidal size; the colloidal content of clays is of great importance.

5-28   GEOLOGICAL BACKGROUND

FIG. 5.25   Typical section through varved clay at Steep Rock Lake, Ontario.

Clays are now known to be generally crystalline; the particle shapes of clay minerals are related to their peculiar physical properties (Fig. 5.25).

As will be noted from a study of soil origins and deposition, clays may be formed by all the main processes associated with rock weathering. They may be residual or transported. Transported clays may be floodplain river deposits of varying thickness and probably mixed with sand; they may be lake deposits now elevated as a result of land movement; or they may be estuarine or marine clays similarly raised from their original positions. The last are a particularly important group and occur in many parts of the world as large deposits. Finally, glacial clays—sometimes called boulder clays—constitute a widespread series of deposits. The name is correctly applied, however, to many glacial deposits, although the mixture of sand, gravel, and boulders usually present in such deposits often renders them useless as a commercial source of clay.

Unlike other soil types, clays are susceptible to pressure and so may vary from quite soft to extremely hard. It used to be thought that shales resulted when clays were subjected to large pressures for considerable periods, but this view is now questionable. The pressure may have been due to the presence of great depths of other deposits, subsequently eroded, or to that of an ice sheet possibly several thousand meters thick. Pressure of this magnitude cannot easily be duplicated in the ordinary laboratory; still less can the range of geological time be even remotely simulated in experimental work. Thus it is that when once a sample of such *consolidated* clay has been thoroughly disturbed (by remolding), it will lose its original properties, and these cannot be regained. Although overconsolidated clays can be detected in a laboratory by means of suitable tests, it is clearly of great assistance if their presence can be foretold from study of the local geology, and this is usually possible.

5.7 ORGANIC SOILS

Although organic soils are not frequently met with in the normal course of civil engineering work, they can cause real problems when they are encountered. They consist essentially of dead organic matter derived from former vegetation, and they often have a surface layer of living vegetal matter, deteriorated to differing degrees and holding amounts of water that may vary up to well over 1,000 percent by weight. The origin of these soils is obvious from their composition. They go by various names: *marsh, peat, bogs,* and *swamps* are the most general terms used in settled country, while *muskeg* is used in northern Canada. It is in the North that by far the most widespread occurrences of organic soil are found since, quite contrary to popular impression, the prevalence of organic soils decreases as tropical areas are approached. The rate of organic disintegration in tropical climates is so rapid that there is, to use an oversimplified expression, no time for organic soils to form. Organic soils are rarely of great depth; muskeg deeper than 10 m (33 ft), for example, is unusual in Canada. The character of the underlying soil is important in dealing with organic soils in engineering work. These soils occur frequently in depressions in the former land surface, a location that assists the accumulation of water. Their structure tends to make them spongelike, so that drainage of the water held by organic soils can be among the most troublesome of associated problems. (See Fig. 5.26.) Since these soils can

FIG. 5.26 Typical muskeg country. Road construction in northwest Ontario, showing drainage of muskeg; preliminary roadbed construction is on brushwood placed on the surface of the muskeg.

be regarded as assemblages of fossil vegetation, their age (if less than about 50,000 years) can be accurately determined by the carbon-14 method and their original status by means of the corresponding modern technique of pollen analysis. The interesting term *paleovegetography* has been coined to describe this relatively new branch of terrain study, which already has a system of classification in reasonably wide use for such types of organic soil as Canadian muskeg.[5.5]

5.8 PERMAFROST

Permafrost is met with even more rarely than organic soils in the practice of civil engineering, but it seems to have a degree of popular, and even of technical, interest quite incompatible with its limited distribution in the far northern and far southern portions of the globe. The top of Mount Washington in New Hampshire is a special example of permafrost in the mainland area of the United States, but most of the state of Alaska,

FIG. 5.27 Typical permafrost terrain; a view in the Mackenzie River delta, Northwest Territories, showing frost hummocks, water standing in depressions forming ground polygons (due to ice wedges), and in the background three pingoes or "ice volcanoes."

about one-half of the land area of Canada, and about one-third of the U.S.S.R. are underlain by ground that is perennially frozen. (See Fig. 5.27.) It is to describe this condition that the word *permafrost* is correctly used, even though Prof. Kirk Bryan made a valiant but abortive effort to have semantically accurate terminology adopted for this new branch of geological study when it came to the forefront following the Second World War.[5.6] Regrettably perhaps from the scientific point of view but fortunately for the sake of euphony, *cryopedology* has not come into general use to replace permafrost. Terminological inexactitude has, however, gone even further in this field, since the word *permafrost* is today all too frequently used to describe merely saturated soil that is perenially frozen. Strictly speaking, permafrost describes a condition rather than a material. It can be fully appreciated only in the light of a basic understanding of soil-temperature variation.

If the temperature of undisturbed soil be carefully measured throughout the year, a regular pattern of variation will be observed. Close to the surface, a daily cycle of temperature change corresponds to daily changes of air temperature. This diurnal variation is soon damped out, however, and the variation below about 0.5 m (1.5 ft) is annual only. At increasing depths, this annual variation is also damped out and retarded because of the laws of thermal transfer. In Ottawa, for example, a 6-month time lag exists at a depth of about 3.6 m (11.7 ft); the annual temperature variation is only 2 or 3°C at about 6 m (20 ft), since the temperature has reached the mean value about which the ground temperature above this depth is found to vary. It was originally thought that this mean ground temperature was always the same as the local mean annual air temperature, but this has been found not always to be true. Some carefully observed values suggest that the mean ground temperature is as much as 5°C greater.[5.7] The agreement is close, however, so that as one travels northward (in the Northern Hemisphere) one can expect the mean ground temperature to decrease, consistent with the decreasing mean annual air temperature. Eventually, the mean ground temperature near the ground surface will be 0°C and will go even below this as one proceeds still northward. Permafrost is used to describe the ground condition thus created, i.e., to describe all that part of the earth's crust which is at a perennial temperature of 0°C or lower. In general terms, the temperature of the ground, below the lowest level of influence by annual air-temperature variation, increases steadily at a rate of about 1°C for every 82 m (206.5 ft). When this increase is sufficient to offset the low temperature of the ground near the surface, the zone of permafrost has been passed. The depth of permafrost, therefore, varies from zero at its southern edge to well over 300 m (1,000 ft) in the Queen Elizabeth Islands of the Canadian Arctic archipelago.

Permafrost describes, therefore, this condition of perennially frozen ground. If the ground consists of solid rock—as so much of the Far North does—the fact that its temperature is below the freezing point of water is of little moment to the engineer, interesting though it may be to the scientist. If the ground consists of well-drained soils such as sands and gravels, normally in a dry condition, the temperature will again make little difference in the use of such ground for human activity. If, however, the ground consists of soils saturated with water, then one is immediately faced with a combination of ice and soil solid. And if, as is so frequently the case with northern soils, the frozen soil is a saturated silt or an unconsolidated clay, then the fact that it is frozen can lead to very serious consequences indeed if its natural state is interfered with in the course of engineering or building work. It is this aspect of permafrost which has attracted such wide attention, and the interest of the public has been drawn to pictures of drunken-looking buildings and of roads and airfields which have failed spectacularly. Today, however, as a result of research that is still being actively pursued by engineers and scientists in the Soviet Union, the United States, and Canada, means for satisfactory design in areas of permafrost have been developed; these means will be mentioned in later chapters. Many problems still remain to be solved, particularly in the scientific field, before an understanding of the thermal regime of the ground in northern regions is achieved; the exact nature of the protective action of muskeg upon permafrost is, for example, still imperfectly understood. From the simple standpoint of geological occurrence, however, permafrost presents no problem, once it is realized that a condition of the ground is being described and not some new and mysterious material.[5.8]

5.9 CONCLUSION

The condition of permafrost, as it is seen today, and the formation of organic soils are naturally relatively recent developments. The process of soil formation, however, is as old as most of the rocks now found in the crust of the earth. Sandstone is clearly derived from the products of the weathering of older rocks, and some sandstones are among the oldest rocks known. Slates provide similar evidence of early rock weathering, and shales are intermediate between the slates and the clays. Although most of these soil-formed rocks are now unquestionably solid rock, there are other examples of ancient soils that have not been greatly changed from their original form. These are usually clays, but sands and gravels of great geological age are also encountered. Typical are the Triassic clays of England, Cambrian clays near Leningrad, Ordovician clays in Estonia, and the clays usually associated with coal measures in many countries. An unusual combination of circumstances, however, was necessary to preserve these ancient soils; therefore, a relatively small percentage of the soil presently in the crust of the earth antedated the Pleistocene period. As Table 3.2 shows, this means that most soils are less than a million years old, quite juvenile by geological time standards.

It is for this reason that Pleistocene geology is of such great importance to civil engineers. To those engineers who have entered upon its study, Pleistocene geology usually and very quickly becomes a matter of genuine interest. It is a great common meeting ground for geologists and engineers, for the work of the Pleistocene geologist is often temporarily suspended until new soil exposures are revealed by the excavation work of the engineer. Many strange and perplexing variations in soil conditions can be explained, and sometimes foreseen, only by the engineering geologist who is well-versed in the Pleistocene history of the locality being investigated. Correspondingly, it is in the study of soils that the engineer will have an opportunity to cooperate with the pedologist. There is much to be gained, and nothing to be lost, from the closest possible collaboration in this field of mutual interest. Pedologists are well aware of the dependence of their detailed soil studies upon the geological origin of the soils with which they deal; it is implicit in most pedological literature. Geomorphologists, and all who are interested in the physical geology of the present surface of the earth, whatever title they may use to describe themselves, share this interest in the geology of soils, since so much of the earth's surface is now covered with a mantle of soil of recent geological origin.

The geology of soils is therefore a meeting place for several scientific disciplines. Long neglected, it is now gaining full recognition. This is reflected in the growing volume of geological literature dealing with soils, so remarkably summarized in Charlesworth's encyclopedic two-volume work, *The Quaternary Era*.[5.9] The use of the term *Quaternary*, and the alternative name *Pleistocene*, may possibly prevent engineers from realizing fully the wide interest that does now exist in the geology of soils—an interest in which they can share. Geological societies regularly hold meetings to discuss Pleistocene geology; although some discussions in this field may approach the academic, especially with regard to glacial chronology, there are few that will not have some interest for the engineer concerned with soils.[5.10] There is even an international organization which holds regular conferences to discuss the interrelation of national Pleistocene studies—the International Union for Quaternary Research (INQUA). In North America, besides the corresponding continental AMQUA organization, there are a number of local groups with the unusual title "Friends of the Pleistocene"; they conduct annual excursions for field study, the best type of soil study. In all this work, the soils engineer has a particular contribution to make; detailed study of the mechanical and physical properties of soils, a scientific discipline now called soil mechanics, has already contributed much to an understanding of soil characteristics and gives promise of an increasing potential in this field. To the interrelation of geology and soil and rock mechanics, attention may therefore next be directed.

5.10 REFERENCES

5.1 G. P. Merrill, *Rocks, Rock Weathering and Soils,* rev. ed. Macmillan, New York, 1911.

5.2 F. W. Clarke, *Nature,* **89**:334 (30 May 1912).

5.3 S. R. Capps, "Rock Glaciers in Alaska," *Journal of Geology,* **18**:359–375 (1910); see also same vol., pp. 549–553.

5.4 See *The Journal of Glaciology,* Cambridge, England, for papers on this and allied subjects.

5.5 N. W. Radforth, "A Suggested Classification of Muskeg for the Engineer," *Engineering Journal,* **35**:1199–1210 (1952), an early paper; see also I. C. MacFarlane, ed., *Muskeg Engineering Handbook,* University of Toronto Press, Toronto, 1969.

5.6 K. Bryan, "Cryopedology: The Study of Frozen Ground and Intensive Frost Action with Suggestions on Nomenclature," *American Journal of Science,* **244**:622–642 (1946).

5.7 C. B. Crawford and R. F. Legget, "Ground Temperature Investigations in Canada," *Engineering Journal,* **40**:263–269 (1957).

5.8 R. J. E. Brown (ed.), *Permafrost in Canada,* University of Toronto Press, Toronto, 1970.

5.9 J. K. Charlesworth, *The Quaternary Era,* 2 vols., E. Arnold, London, 1957.

5.10 R. F. Flint, *Glacial and Pleistocene Geology,* Wiley, New York, 1957.

Suggestions for Further Reading

One of the best of all reviews of residual soil known to the authors is by Don U. Deere and F. D. Patton and is to be found in a somewhat unusual publication. It is an 83-page paper entitled "Slope Stability in Residual Soils," to be found in *Proceedings of the 4th Pan American Conference on Soil Mechanics and Foundation Engineering* (held in San Juan, Puerto Rico), published in June 1971 by the American Society of Civil Engineers. Its value is enhanced by about 150 useful references.

Some indication of the widespread interest in the importance of residual soils is the fact that the American Society of Civil Engineers convened a geotechnical conference on this one subject in June 1982 in Honolulu. The Proceedings, *Engineering and Construction in Tropical and Residual Soils,* is a 735-page volume, published by the Society in 1982.

J. K. Charlesworth's magisterial two-volume review *The Quaternary Era,* published in 1957 by E. Arnold of London, is still one of the best of all guides to this recent period of geological time, when such a large part of the transported soils with which civil engineers have to deal was formed and deposited. An equally outstanding guide to glacial soils is the volume by R. F. Flint entitled *Glacial and Quaternary Geology,* published in 1971 by John Wiley & Sons of New York. From the many other volumes dealing with glacial soils and the (geologically) recent glacial periods, special note must be made of *The Last Million Years,* published by the University of Toronto Press in 1941. The author was A. P. Coleman of the University of Toronto and this was his last work, a popularly written summary of his major field of interest.

Geomorphology is a branch of geology concerned with soils. A classic introduction to this study is given in *Geomorphology* by A. K. Lobeck, published by McGraw-Hill of New York in 1939 as a 731-page book, with unusually good illustrations. A more recent review of the subject is to be found in *Process Geomorphology* by D. F. Ritter, published in 1978 by M. W. Brown of Dubuque, Iowa. And the Golden Guide entitled *Land Forms* by G. F. Adams and J. Wycoff (1971) provides an admirable introduction to the subject.

Clay Minerals are of great importance in the engineering study of soils. *Clay Mineralogy* by R. E. Grim, the second edition of which is a volume of 600 pages published in 1968 by McGraw-Hill of New York, is still the standard reference work. *Clay in Engineering Geology* by J. Gillott, published by Elsevier of New York in 1968, is a useful supplementary volume, especially on the engineering significance of the clay minerals.

That the literature on pedology, the study of soils from the agricultural point of view, is vast can be seen from the fact that the large *1957 Yearbook of the U.S. Department of Agriculture* is devoted entirely to soils. *Factors of Soil Formation* by Hans Jenny, a McGraw-Hill volume of 275 pages published in 1941, has been helpful to many engineers.

There are many excellent local guides to agricultural soils, some being publications of state or provincial agencies, some of even smaller governmental units. Special note may be made of *Highway Soil Engineering Data* by G. A. Miller, J. D. Highland, and G. L. Hallberg, an 109-page publication of the Iowa Geological Survey in 1979.

Part 19 of the 48 volumes that make up the 1982 *Annual Book of ASTM Standards* brings together, in a volume of 730 pages, all ASTM Standards on natural building stones, road and paving materials, and soil and rock. It is an invaluable reference. One of the standards (D2488) is entitled "Practice for the Description of Soils."

chapter 6

SOIL MECHANICS AND GEOLOGY

6.1 Historical Note / 6-1
6.2 Soil Mechanics Today / 6-3
6.3 Soil Testing / 6-4
6.4 Links with Geology / 6-6
6.5 Transported and Residual Soils / 6-8
6.6 Residual Soils / 6-12
6.7 Clay Minerals / 6-14
6.8 Some Contributions to Geology / 6-16
6.9 Conclusion / 6-18
6.10 References / 6-19

The unconsolidated materials found in the earth's crust, described as soil, constitute so large a part of the actual surface of the earth that few civil engineering operations, apart from rock tunneling, can be conducted without an encounter with soil of some type. Since foundations cannot always be carried to solid rock, the founding of structures on unconsolidated material is probably the most important part of foundation engineering. Troubles met during the construction of civil engineering works, and after their completion, are frequently due to failure of soils. Despite their significance in all phases of civil engineering work, soils were not studied and investigated by civil engineers in any general way until relatively recent years. The contrast between this neglect and the past century and a half of progress in the study and investigation of practically all other materials used by civil engineers is so marked as to be indeed a paradox. Nobody sells soils, however; they are generally a part of the natural order which the civil engineer has to accept as the basis of operations. It may be that the absence of commercial incentive is linked in some way with this past neglect; the development of testing and research laboratories under public control for public benefit was probably necessary in order to initiate the surprising progress in scientific soil study of the last five decades.

6.1 HISTORICAL NOTE

In the earliest days of modern civil engineering, the characteristics of soils appear to have received some attention, notably with reference to the pressure exerted on retaining walls.

6-2　GEOLOGICAL BACKGROUND

The theoretical work of Coulomb (1773) and of Rankine (1856) is fairly well known and appreciated. In France, remarkable pioneer experimental work was performed by Alexandre Collin as early as 1846 in connection with slides occurring during the construction of French canals. Not only did Collin investigate the curved sliding surface which he had noted in clay slips, but he experimented on the changes in the properties of clay caused by changes in moisture content.[6.1] Throughout the remainder of the nineteenth century, attention to the properties of soils appears to have been spasmodic. A paper by Sir Benjamin Baker, "The Actual Lateral Pressure of Earthwork" presented to the Institution of Civil Engineers in London on 5 April 1881, is a notable exception; the importance of geological structure in all such studies was stressed in the ensuing discussion.[6.2]

Not, apparently, until after the turn of the twentieth century were anything more than isolated soil investigations carried out, although notable individual contributions were made in the United States by Dr. Milton Whitney and Dr. George E. Ladd. Three major investigations undertaken in the early part of the present century directed the attention of civil engineers to soil problems and may rightly be regarded as the start of modern soil studies. These were the studies made in Sweden of the landslides that had occurred on the state railway system; the investigation of difficulties encountered during the construction of the Kiel Canal in Germany; and the intensive analysis made of the great landslides that interfered so seriously with the completion of the Panama Canal.

These investigations did not stand alone. In England a paper, "The Lateral Pressure and Resistance of Clay and the Supporting Power of Clay Foundations," was read by A. L. Bell at the Institution of Civil Engineers on 19 January 1915, a paper which, together with the long discussion that it provoked, contained much suggestive material on the mechanics of soils. The necessity of obtaining undisturbed samples—a practice that developed in subsequent years—was stressed by this author.[6.3] The paper was the precursor of a notable group of British publications on earth pressure and soil properties. In the United States, Dr. C. M. Strahan had begun his investigations of natural-soil roads, investigations which led to the presentation of his first paper in 1914. This paper may rightly be regarded as the start of modern stabilized soil-road practice, although some isolated papers on the subject had been published previously. In 1913, the American Society of Civil Engineers set up its first committee on the codification of bearing values for soils, and the work of this and of subsequent similar committees has been of great value.

These few references indicate the awakening of interest in the potentialities of scientific soil study that occurred during the early years of the present century and led eventually to the general recognition of soil mechanics as an important scientific aid to many branches of civil engineering. *Soil mechanics* was not used as a title to denote an entirely new branch of scientific study but rather as a convenient name for the recognition of the coordinated group of soil studies that were found to be necessary in the application of soil analyses and investigations to the several branches of civil engineering work. The acknowledged leader in this pioneer work was Dr. Karl Terzaghi. By his classic papers in *Engineering News-Record* and his work for the U.S. Bureau of Public Roads, at Massachusetts Institute of Technology, and particularly at Harvard University, Dr. Terzaghi gave an impetus to scientific soil and rock studies in North America that has had most fruitful results; his European publications and activities have been similarly effective for engineers of Europe. To the great joy of all who were privileged to know him, Dr. Terzaghi was able to see his own efforts develop fruitfully into a scientific discipline well recognized throughout the world before his death in 1963 at the age of 80.

There are now established courses of instruction in soil mechanics at leading universities and a steadily expanding volume of sound literature. The International Society of Soil Mechanics and Foundation Engineering, through its national committees, brings together workers in the field in almost 40 countries. Dr. Terzaghi was the president of the society from its formation in 1936 until 1957 and then its honorary president. The society has held conferences in Cambridge, Massachusetts (1936), Rotterdam (1948), Zurich (1953), London (1957), Paris (1961), Montreal (1965), Mexico City (1969), Moscow (1973), Tokyo (1977), and Stockholm (1981). The published proceedings of these international gatherings form a notable part of modern civil engineering literature. The writings have a strong geological overtone, as might be expected, for Dr. Terzaghi's studies were based upon a background of geology. In his own words: "I came to the United States and hoped to discover the philosopher's stone by accumulating and coordinating geological information. . . . It took me two years of strenuous work to discover that geological information must be supplemented by numerical data which can only be obtained by physical tests carried out in a laboratory."[6.4] The science of soil mechanics is now the medium through which this essential supplementary information is obtained, coordinated, and made available for use.

6.2 SOIL MECHANICS TODAY

One of the main features of modern soil mechanics is the impressive body of theory that has been developed dealing with all aspects of the states of stress in soils and with the deformation resulting from such stress conditions. Methods of calculation are now available for determining, theoretically, the stress which an imposed load will cause at any point below a foundation slab, the total pressure upon retaining walls and the pattern of its distribution down the wall, and the earth pressures to be expected upon buried structures such as large culverts. In a corresponding way, it is now possible to determine with accuracy (and with the aid of computers, if desired) the factor of safety against failure of a slope excavated in soil or constructed of soil. Possibly the greatest contribution to theoretical soil mechanics has been the theory of consolidation developed initially by Dr. Terzaghi. Using this theory, and given certain basic physical properties of the clay to be loaded, one can calculate the total settlement to be anticipated and the rate at which settlement will occur for any clay soil under load at its surface.

These examples are typical of the many theoretical approaches now possible to structural problems involving the use of soil. The qualification noted in the final example is, however, all-important. Excellent though the many theories are, they all depend upon certain *basic assumptions* with regard to the properties of the soil being used. The accuracy of the application of theoretical calculations therefore depends upon the accuracy of the assumptions made or upon the values for the soil properties provided. This leads to the second major branch of study in soil mechanics: the investigation of soils in the laboratory in order to determine experimentally the values that are required for use in calculations. Still in an active stage of development, the design and use of special equipment for soil testing have been rapidly advanced in the last three decades. In any modern soils laboratory there are *consolidometers,* rather simple devices for determining the consolidation characteristics of small soil samples under increasing increments of load; *permeameters* for the determination of soil permeability; and of greatest importance, two types of shear-testing machines.

For simple tests, and for shear determinations upon certain special types of soil, *direct shear boxes* are used; these are relatively simple machines in which the soil sample is placed in a split box so that it can be loaded vertically. While under load, the sample can be sheared along the break in the box by a horizontal force that can be measured. Of far more importance, however, are the *triaxial compression machines* that are now widely used. In these a cylindrical soil sample, sealed in a flexible membrane, is fitted into a larger transparent cylinder and held between top and bottom supports. The large cylinder

6-4 GEOLOGICAL BACKGROUND

is filled with an appropriate liquid to which pressure can be applied. With the restraining liquid under pressure and with loads applied vertically to the specimen through the top and bottom supports, the cylinder of soil can be subjected to a combination of three-dimensional stress, and loading can continue until the sample fails, usually on a definite shear plane.

Thus described in simple terms, soil testing might appear to be a simple business. So it often appears to be to the uninitiated. In practice, however, it requires much skill and delicacy of operation. The individual tests must be conducted meticulously, since the small size of the usual specimen and the relatively light loads call for refinements in technique not needed in most civil engineering laboratory techniques. And the soil samples must be handled expertly prior to the tests. Just as the accuracy of theoretical calculations depends on the accuracy of the soil properties assumed or determined, so will the accuracy of the results of the soil tests themselves depend upon the accuracy of soil sampling. Therefore, soil samples provided for laboratory experiments must, at the time of test, represent fairly the soil in situ from which they came and to which the results of theoretical calculations are to be applied.

Thus becomes obvious the third great division of soil mechanics: the procurement of good soil samples, their accurate description, and the necessary determination of general soil conditions at the site in order to demonstrate the validity of samples obtained. This involves the work of in situ soil testing. Here modern soil studies grade imperceptibly into the normal practice of civil engineering, for test boring and sampling (of a sort) were a regular feature of construction work long before soil mechanics had even been thought of as a separate discipline. The refinements that modern soil and rock studies have introduced into subsurface exploration, however, have transformed what was at best a rough-and-ready sort of procedure into a reliable, highly skilled, and thoroughly scientific operation. Most notable has been the absolute insistence upon so-called "undisturbed samples." This expression is another of those semantic inexactitudes of which civil engineers are so often guilty, since the removal of a soil sample from the ground naturally "disturbs" it. The word is used, however, to connote no disturbance whatsoever of the sample itself, and the retention in the sample of the exact moisture content that the soil has when in place. This last result is achieved by the immediate waxing of soil samples upon removal from a borehole or by the use of special sampling tools, or sleeves, in which samples can be left, waxed at the open ends, until they reach the laboratory.

6.3 SOIL TESTING

The actual techniques of soil sampling and of soil testing in the field are matters that do not require detailed treatment here; some notes by way of introduction to current methods will be found in Chap. 11. Suffice it to say that soil-sampling methods have been developed, even for cohesionless soils such as sands, that will give to the laboratory worker a specimen of soil that is as close to the natural condition of the soil in the ground as it is humanly possible to obtain. The first tests will usually be so-called "indicator tests"—those which determine generally the type of soil. These are used as a basis for consideration of the more elaborate mechanical tests and for purposes of accurate description. Usually the first such test made will be to determine the natural moisture content; this is done by drying a small sample in an oven under controlled and standard conditions. Moisture contents are always expressed as the ratio of the weight of the water contained to the weight of dry solid-soil matter, expressed as a percentage. Another introductory test will be made to determine the distribution of soil particles of different sizes, the mechanical analysis of the soil. This test is performed by sieving down as low as the standard No. 200

mesh, and determining smaller particles, with some degree of overlap, by means of a sedimentation method, naturally standardized, dependent upon an application of Stokes' law. The results are plotted to a semilogarithmic scale; Fig. 6.1 shows a few particle-size distribution curves for typical soils. It should be noted that, although it is especially useful for coarser soil mixtures such as glacial till, the record of mechanical analysis is not an infallible guide to soil properties for fine-grained soils. The subdivisions shown in Fig. 6.1 for particles of sand, silt, and clay size are those most generally used in North America.

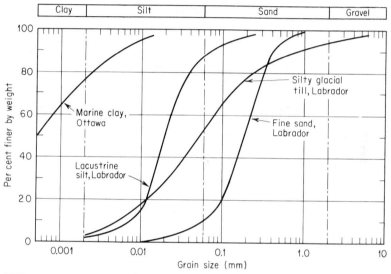

FIG. 6.1 Standard chart showing the mechanical analysis of particle-size distribution for four typical soils.

In normal laboratory procedure, soil samples will next be subjected to so-called "plasticity tests" in order to determine their Atterberg limits, limits of consistency named after the Swedish agricultural soil scientist who first suggested them. These limits are expressed as percentages of the weight of water, at the points shortly to be noted, compared with the weight of dry soil solid in the sample. If water is slowly added to a perfectly dry sample of fine-grained soil and uniformly mixed with it, the soil will gradually assume some cohesion, probably first forming lumps. It will eventually reach a plastic stage at which it can be rolled out in a long, unbroken thread upon a solid surface. A simple test has been standardized to indicate the limit of water content at which this plastic stage is reached; this limit is called the *lower plastic limit,* or more briefly the *plastic limit.* If more water is added and the mixing continued, the soil will gradually achieve the state of a viscous liquid. Another simple test, based on the flowage together of two parts of a pat of the soil-water mixture in a standard cup under standard tapping, has been accepted for general use to indicate this limit of consistency, the *(lower) liquid limit.* The difference between the liquid and the plastic limits is naturally a measure of the range of plasticity of the soil. It is, therefore, called the *plasticity index* and is expressed also as the usual percentage of moisture content. The third of these "limits" is a much lower value—that percentage of moisture content at which the soil sample stops shrinking as it dries. When this point is passed, the sample will usually grow lighter in color as the drying process continues. This is known as the *shrinkage limit,* but since its physical significance is not so great as that of the other two limits, it is not referred to so frequently.

The liquid and plastic limits are simple in concept and relatively easy to determine, but they are valuable indicators of soil characteristics. They are useful for comparative purposes, since soil samples from similar locations and horizons will have reasonably similar limit values. When plotted, as will be mentioned in a later section, they form easily recognizable patterns. In combination, not only do they give the plasticity index, but in association with the natural moisture content they give another factor called the *liquidity index.* This is obtained by dividing the difference between the natural moisture content and the plastic limit by the plasticity index. All these terms are self-descriptive. They appear almost universally in engineering reports of laboratory soil tests. As indicators of the general characteristics of a soil they are most useful. When considered in combination with the appearance of a soil and its mechanical analysis, they assist in the determination of an accurate description of the soils in terms now well accepted, terms such as "silty clay," "sandy silt," or "highly plastic clay."

Two comments must precede consideration of the place of geology in this overall picture of soil mechanics. First, it will be noted that no symbols or formulas have been used to describe the simple tests now standard in soil testing. This is in contrast with the practice of some writers on soil mechanics who seem to consider their papers incomplete unless they use mathematical symbols, preferably formulas, even to indicate the subtraction of one term from another. Symbols and formulas are, after all, merely a kind of scientific shorthand, with no intrinsic meaning. It seems desirable to use "longhand" here, since the concepts described are so basic and simple that their significance can actually be shrouded by the veil of mathematical symbolism. In the second place, it may be noted that all the simple tests described relate to the quantity of water present in a soil sample at different stages in its wetting or drying. This will not be the last time that the importance of water in relation to soils will be mentioned. It will gradually be seen that most of the subsurface problems that have to be faced by civil engineers in the conduct of their construction operations are caused by water and not by solid material as such, whether the material be soil or rock. The basic importance attached to water in soil mechanics, therefore, is symptomatic of the importance of water in all soil problems. Dr. Terzaghi once said that "On a planet without any water there would be no need for soil mechanics."[6.5]

6.4 LINKS WITH GEOLOGY

Soil mechanics has become a vital part of the scientific core of civil engineering. As soil mechanics deals with a major constituent of the earth's crust, so also does the practice of civil engineering; the close association of the two is therefore quite natural. At the same time, the scientific study of soils must have contact with geology; this, also, is natural and logical. From the outline given above, it can be seen that soil mechanics closely approaches a geological study when field investigations are involved. It can be said that soil mechanics gradually merges into geology at this one extreme and into structural engineering at the other. Therefore, to regard soil mechanics as a branch of geological study that has unfortunately strayed into civil engineering is just about as foolish a concept as to regard it as a subject of strictly engineering character with no necessary contact with geology.

Reverting to the general outline of soil mechanics given in the preceding section, one can appreciate that geology has few links with the theoretical side of engineering soil studies. Analysis of slope-stability problems may sometimes come up for consideration in geological studies, but the geologist usually has to accept the facts presented by nature without being concerned about whether a certain slope is potentially unstable or why some other slope has failed. At the same time, the mathematical concepts embodied in the theories of soil mechanics provide the geologist with a powerful analytical tool for those investigations of major phenomena that now supplement the older, but still invaluable, geological field studies. A masterly example of the direct application of these concepts to theoretical geological work is given by the two papers of Hubbert and Rubey on the mechanics of fluid-filled porous solids and the application of the hypothesis to overthrust faulting.[6.6] At the outset of their papers, these authors state that their attention was focused "upon the newly evolving science of soil mechanics with particular attention being paid to the treatises on this subject by Karl Terzaghi and associates, since it appeared that the phenomena of soil mechanics represented in many respects very good scale models of the larger diastrophic phenomena of geology."

In this fine example of the application of soil mechanics to a geological problem, the authors demonstrate how, given sufficiently high fluid pressures (pore-water pressures), very much larger blocks can be pushed over almost horizontal surfaces than would be possible under any other circumstances. The fact that the authors do not agree with all of Dr. Terzaghi's views and that, following discussion of their paper, they arranged for some large-scale shear tests in order to check some of their own views adds further interest to their work—work which constitutes a real challenge to all workers in the field of soil mechanics. To geologists, the work of Hubbert and Rubey provides an attractive explanation of such famous geological features as the great overthrust belt of western Wyoming and the adjacent states; their work also demonstrates how geology can be assisted by the use of techniques, both theoretical and experimental, borrowed from soil mechanics.

There are doubtless many other similar problems in geology the solution of which may one day be greatly assisted by contributions from soil mechanics. When studies of soil in the field are considered from the engineering point of view, however, one may well ask how they can possibly be conducted without due reference to the local geology. One would imagine that it would be impossible to neglect geology, even if not calling it by that name. All too often, however, soil studies have been conducted in the field without benefit from contact with geology in any form, recognized or unrecognized. Sometimes no harm has resulted, but this has been through good luck rather than through good management. Poor results have often been obtained through patent neglect of geological features; in all too many cases, money has been uselessly expended because subsurface explorations were not coordinated with the local geology; and occasionally the neglect of geology has had disastrous results. Some of these will be recorded later in this book, always with constructive intent, since the proper study of failures can assist so greatly in obviating similar troubles on future works performed under similar circumstances. It can therefore be stated without qualification that soil studies in the field, no matter how carefully conducted, are incomplete without some consideration of the appropriate local geology. Herein lies perhaps the chief contact between geology and soil mechanics.

The remarkable increase in the amount of attention that has been given to subsurface exploration on civil engineering projects during the last three decades has naturally had the fortunate by-product of providing the geologist with much useful information in the form of borehole logs. This is not unusual; in a similar way, borings for oil and gas exploration and for the development of new coal-mining projects are also constantly yielding new information. In the testing of soil samples in the laboratory, however, an even closer link is developing between geology and soil mechanics. Knowledge of the geological origin of a soil greatly assists the laboratory worker by suggesting the range of soil properties to be expected and by indicating special features that should be watched for during the progress of soil tests. And the detailed study of soil properties that is now possible with advanced soil-testing techniques is disclosing new information that is of particular value when considered in relation to the generally known pattern of the geological origin of the soil. This information sometimes fills in gaps in geological knowledge or refines generalized deductions drawn from field observations. Examples will be cited. In this way, perhaps more than in any other, the twin disciplines are being drawn closer together. It is not without significance that the name *Geotechnique* has been adopted as the title of one of the leading English-language journals in the field of soil mechanics.[6.7] The term *geotechnical studies* has been used to describe soil testing in its more general aspects, even though the testing is carried out for engineering purposes. Some well-known national soil-testing laboratories are happily known as *geotechnical institutes.* There is, therefore, steadily developing liaison between the scientific approach to the geology of soils and the engineering investigation of their properties. It is inevitable that this interrelation should progress to the mutual benefit of both geology and engineering and to the continued advance of human understanding of the most common of all solid materials.

In a paper published in 1955, Dr. Terzaghi emphasized the benefits of the relationship, noting: "The geological origin of a deposit determines the physical properties of its constituents.... Therefore, the knowledge of the relation between physical properties and geological history is of outstanding practical importance."[6.8] Dr. Terzaghi's statement might well be suitably exhibited in every soil mechanics laboratory.

6.5 TRANSPORTED AND RESIDUAL SOILS

In the preceding chapter, a major distinction was made between residual and transported soils. Aeolian soils, those transported by wind action, are easily recognizable when their geological origin is appreciated. The peculiar characteristics of some types of loess, owing to its open structure, will affect soil sampling and result in findings of unusual soil properties, especially with regard to the action of water on such soil. Here soil testing in situ is generally called for. A well-accepted practice today is to saturate loess in advance of placing loads upon it and then to consolidate the wetted material by external means. This increases its density, and when consolidation has proceeded far enough, shear strength will increase. The serious nature of this peculiar characteristic is shown by the fact that even the watering of a lawn has resulted in settlement of an adjacent house because of the unconsolidated underlying loess. Geological recognition of the presence of loess, therefore, can be a vital contribution to engineering soil studies.[6.9]

Alluvial soil, transported down rivers and streams, is usually readily recognizable if the watercourses are those of today. Soil of this type, however, may be found on what were the banks of rivers of ages past, rivers now superseded because of more recent geological events. Such conditions may not be at all obvious to the untrained eye, particularly if the area has been developed in any way for special uses. Knowledge of local geology will be a helpful guide in such cases, since the existence of buried watercourses can frequently be foretold from a general appreciation of the sequence of geological events in the area. Once their characteristics are recognized, then the extremely variable and probably unconsolidated alluvial soils can be anticipated in soil sampling and testing.

One of the most important contributions that geology can make to soil studies is at the same time one of the simplest—geology can give an indication of whether the site of a projected engineering work has been subjected to glacial action or not. If the site has been—and this will be the case for almost all the northern section of the Northern Hemisphere and for very small corresponding areas in the Southern Hemisphere—then the civil engineer must be on the alert for the great variation in glacial soils that may be found in relatively short distances, for the distinct possibility of encountering buried preglacial river valleys that were subsequently filled with glacial deposits that may themselves differ from the overlying soil, and for the possible effects of ice pressure upon soils in areas glaciated more than once, as is so frequently the case.

Glacial till is then a soil type to be expected. If a mixture of dense clay, sand, and gravel is found in boreholes, that mixture should *not* be characterized just as "hardpan" but should be most carefully studied, if necessary by means of block samples, in order to determine its characteristics under the conditions that will be imposed by the engineering use to which it will be put. If the direction of subsequent ice movement is known, for example, then high till densities and consequent toughness of the material, especially when dry, may be expected. If the material has to be used as a foundation bed, this feature will be desirable; if the material has to be excavated, this feature may be ruinous if not recognized before contract arrangements are made. If the till is of recent origin and if it was deposited in morainal form, then the reverse may be expected, and relatively low densities may be found with corresponding open structure (Fig. 6.2). Till of this kind was encountered in northern Quebec during railway construction; its perfect particle-size grading made it almost impermeable, but the addition of water filled its open voids, and after heavy rains it quickly acquired the consistency of the justly famous local *soupe aux pois* with results that can be imagined.[6.10] At least one case is known in which landslide movement shortly after till deposition had the effect of maintaining the till at an unusually high moisture content, a condition which was detected only by careful soil sampling and testing, and which, once recognized, called for special construction methods when the till was used as fill.

It might be thought that there could be no such variations with sands, but many engineers have discovered otherwise, sometimes to their cost. The density of sand can vary appreciably, depending on how the individual sand grains are packed together. It is a normal condition for sands to be close to their maximum density, but for reasons not yet fully understood but almost certainly dependent upon the way in which the material was

FIG. 6.2 Transportation troubles on loose glacial till in northern Quebec.

deposited, sands of low relative density are sometimes encountered. The relative density of sands, therefore, is an important soil characteristic that can best be measured indirectly in the field by means of penetration tests; it may also be tested by unusually careful sampling. Geology cannot here be a certain guide, but if sand of low relative density is encountered at a site, geological comparisons between its location and other local sand deposits can usually be helpful in further exploration. When sand of low relative density has been found, it can be consolidated to a greater and more desirable density by a variety of engineering methods, such as pile driving, the use of explosives, and the combined use of water and vibration.

Since glacial silts grade into glacial clays, they may be considered together. At one extreme, a coarse silt may consist of particles of fresh minerals that fully justify the name *rock flour,* essentially a very finely ground sand. When wet, such materials may appear at first glance to be identical with clays. Their true character may easily be detected when they are dry, but the use of simple soil tests will readily display the difference even in the wet condition. The Atterberg tests, for example, will give low values for the plasticity indices of silts. Once recognized, the rather difficult properties of glacial silts can be anticipated, and design and construction measures may be accordingly adapted. In some glaciated areas (along the north shore of Lake Superior, for example), it is possible to trace a gradual change in glacial deposits, all of which look alike, from silts (as described) to true glacial clays. The soil particles gradually change from fresh minerals to clay minerals, and the properties change correspondingly. Here is a case in which the simple tests of soil mechanics not only give the engineer the information needed but, at the same time, reveal information of considerable geological significance.

For all fine-grained glacial sediments, a check upon the natural moisture content and its relation to the lower liquid limit is a first requirement in soil testing. In many areas, especially on and near the Precambrian Shield in Canada, the natural moisture content may be appreciably higher than the lower liquid limit. When this phenomenon is first detected by those who have not encountered it previously, they immediately suspect that the test results are wrong.[6.11] But the results may be right. The unusual condition is again explained by the process of deposition. During the settling of the fine particles in the fresh water of glacial lakes, the particles came together in such a way that interparticle attraction led to what can best be described as a "honeycomb structure" of the solid particles. Excess water is therefore held between the particles and gives the resulting soil an artificially high moisture content. When the soil remains in its natural position, this unusual condition is of no moment and may be unsuspected. When such soils are disturbed, however, as they can be by engineering operations, the excess water may be released, thus quickly converting what had previously been a solid-looking material into a viscous liquid that will flow readily on low slopes until the excess moisture is lost and the soil "solidifies," having a new, lower moisture content. The manner in which soil of this type was successfully controlled in a vast mining operation is described in Chap. 19.

Geological information about the glacial origin of these clays can clearly be a most useful guide, pointing the way to the necessity of careful soil testing before the soil is moved at all. When such soil is detected, geological information can act as a constant reminder of the care that must be exercised in all engineering work, not only with soil to be moved but with natural slopes that may be disturbed. The phenomenon of *varving,* the deposition of such soils in thin, alternating light and dark layers probably representing annual sedimentation cycles, is of great geological interest, even though the varving will not usually have any significant effect upon engineering uses of the soil. (See Fig. 6.3.) Tests upon such soils, tests carried out for engineering purposes, have raised some interesting questions about the classical geological explanation of this widespread feature of glacial lake deposits.[6.12]

FIG. 6.3 Distortion in varved clays located in the center of a drained lake bed, Steep Rock Lake, Ontario.

Other glacial clays may have been deposited in seawater rather than in fresh water. Fortunately, the areas in which this has taken place are now fairly well recognized; the most extensive are in southern Scandinavia and in the valleys of the St. Lawrence and Ottawa rivers, with small extensions into connecting valleys. Clays so deposited are unusually sensitive; they, too, have abnormally high moisture contents. Accordingly, they are quite unstable when disturbed and have been the cause of many disastrous landslides in both the areas mentioned.[6.13]

Repeated references have been made to the indications that the Atterberg limits give of the character of fine-grained soils. This can be illustrated in a general way by a simple graphical record of the relationship of the lower liquid limit and the plasticity index for a group of typical glacial soils. This relationship is shown in Fig. 6.4. Admittedly, the chart is no more than an indicator of soil type, but it is a useful guide, especially when samples of unfamiliar soils are undergoing tests. Mention has already been made of another soil test that has considerable geological significance. This is the test upon enclosed soil samples, a test that relates their consolidation under load with time. The graphical record of this interrelation has a well-defined pattern which clearly reveals, for some soils, that the soil has been subjected to a previous long-term load that has already partially consolidated it. This value is known as the *preconsolidation load*. The load may have been caused by overlying soil that was removed for some reason long before the sample was obtained. Commonly, the load will be an indication of the weight of ice that has at some time in the past stood upon the soil in question. The preconsolidation load is an essential factor in the determination of probable settlements when the soil under test is actually loaded by a structure, but the geological significance of the test will be at once apparent.

Another result of ancient ice action must be briefly mentioned; although not often encountered, it is of unusual importance where it does exist. Just as the condition of permafrost (perennially frozen ground) exists in northern latitudes today, so is it almost certain that permafrost conditions must have existed at times of glacial advance in areas even to the south of the southern limits of glaciation. Many unusual results of ground

6-12 GEOLOGICAL BACKGROUND

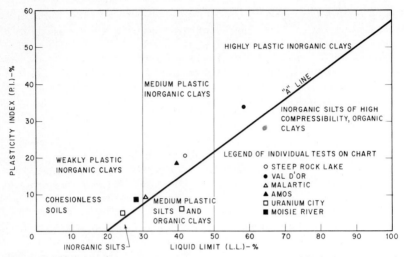

FIG. 6.4 Relation between the plasticity index and the liquid limit for a group of typical Canadian soils, showing also the Casagrande A line.

movement due to freezing may be observed in northern regions. These features are known generally as *periglacial phenomena*. The same types of ground movement must have taken place in glacial times in more southern regions when those areas, too, were perennially frozen. Naturally buried throughout all the intervening centuries, these distortions of the ground come to light only as a result of engineering excavation work. A puzzle until recently, these "fossil periglacial phenomena" have now been recognized and so are understood. In regions where they are known to occur, they may be anticipated; but they may usefully be kept in mind in all subsurface exploration, since when they are encountered, they can be the cause of otherwise quite inexplicable variations in subsurface conditions.[6.14] Subsurface exploration in areas of existing permafrost is difficult, and it is so specialized that it calls for brief mention here only as an indication that useful information is available for reference.

6.6 RESIDUAL SOILS

The origin of soils that have been formed by the direct in situ weathering of bedrock would appear to call for but little mention here, since the origin of such soils is usually obvious. There are, however, many areas in the warmer parts of the world where bedrock is far below the surface; the observable soils are of great age and their origin is not immediately recognized. Soil formation under tropical conditions is still only imperfectly understood; the origin of laterites, for example, is still a matter of debate. Laterites have accumulations of iron and aluminum oxides at the surface, with silica content generally leached out to a lower horizon. Black cotton soils constitute another major and important group in tropical areas, but they are well recognized, and their properties are being gradually codified.[6.15]

The physical and mechanical properties of residual soils naturally vary greatly, but they are determined in exactly the same way as those of glacial soils. The simple soil tests will sometimes give clear indication of varying origins for residual soils that may have the same appearance. Results of tests upon two soils encountered during field studies carried out for the Konkouré development in French Guiana are typical. Table 6.1 shows distinct differences between two soils found in the same location, soils which to all appearances were almost identical.[6.16]

Some residual soils, including the black cotton soils so common in India and the Far East generally, have rather unusual properties when considered for engineering use. Among the most difficult of all these soils to deal with are those derived from the weathering of volcanic rocks such as lava flows and volcanic ash. Material of this kind had to be used for the construction of the Sasamua Dam in Kenya, which was built to form a reservoir as part of the water-supply system for the city of Nairobi, located only 80 km (50 mi) south of the equator (Fig. 6.5). Among the unusual properties of the local clay used for this project are a somewhat high plastic limit, but a much lower plasticity index than would normally be expected; low density at optimum moisture content; considerable variation in the values obtained for the Atterberg limits, depending on the chemical used as a dispersing agent; and higher permeability but also higher shearing strength than would be expected for normal clays with similar liquid limits. This material was difficult to handle and caused many problems during construction. After the properties of the clay were fully appreciated, it was necessary to redesign several times the cross section of the earth dam, which has a maximum height of about 33 m (110 ft), and a length (curved in plan) of rather more than 300 m (1,000 ft).

TABLE 6.1 Comparison of Properties of Soils Derived from Dolerite and Schist[6.16]

Soil property	Doleritic material (% dry wt.)	Schistic material (% dry wt.)
Natural moisture content	37–43	13.5–37.5
Specific gravity of solids	2.80–2.98	2.73–2.85
Dry density, in situ	1.20–1.35	1.30–1.80
Liquid limit	56–74	30–67
Plastic limit	21.0–32.5	5.0–28.5

FIG. 6.5 Sasamua Dam, built to supply water for Nairobi, Kenya; the dam has since been raised.

6-14   GEOLOGICAL BACKGROUND

A fine record of the problems with the dam and its construction was prepared by Dr. Terzaghi, whose paper also showed the reason for the unusual properties of the Sasamua clay.[6.17] When the mineralogical content of the material was analyzed, it was found that the material consisted of almost 60 percent halloysite, about 16 percent goethite, and much smaller percentages of kaolinite, gibbsite, quartz, and mica. All the unusual properties of the clay could be explained by the assumption (applicable to a clay with a high halloysite content) that "most of the clay fraction of the soil occurs in the form of clusters or hard porous grains with rough surfaces, each of which consists of a great number of firmly interconnected clay mineral particles." Electronmicroscopic photographs confirmed this type of internal structure. Experiences with similar clays encountered at four other dams were compared with what was found at Sasamua; not only were the construction experiences similar, but analyses of the other clays also showed the presence of halloysite and goethite. Once the character of the unusual clay was understood, suitable test and construction methods were used, and the dam was completed and placed in service.

6.7 CLAY MINERALS

The last two examples cited, typical of many cases of the use of residual soils that could be mentioned, show that possibly the most important basic characteristic of fine-grained soils is the dominant type of mineral present. That all clays do consist of minerals—known generally as the clay minerals—is now well recognized, even though the minute size of the clay particles and the fact that they cannot be observed with the unaided eye make this designation seem strange to the uninitiated. Modern methods of mineralogical study, including the use of differential thermal analysis, the electron microscope, and the powerful tool provided by X-ray diffraction techniques, have made possible detailed analysis of individual clays and the study of different types of clay minerals (Fig. 6.6).

Prior to the introduction of X-ray diffraction studies in 1923, it was known that clays consisted of aluminum, silicon, water, and often iron. Such chemical analyses, although interesting, were of little assistance in understanding the properties of different kinds of

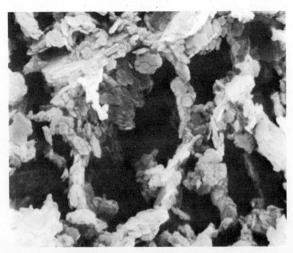

FIG. 6.6 Scanning electron micrograph, (scale: 6,000x) of freeze-dried flocculated Georgia kaolinite.

clay. It is now known that the clay minerals consist of silicates of aluminum and/or iron and magnesium. Some contain alkaline materials as essential components. Some argillaceous material may be amorphous, but this is not a significant component of normal clays. Most of the clay minerals have layered or sheetlike structures. The characteristic properties of these crystalline forms go far in determining the physical properties of a clay. The main clay minerals can be grouped together as kaolinite, halloysite, montmorillonite, illite, and chlorite; a few others are occasionally encountered, but those listed most commonly come to the attention of the worker in the field of soil mechanics. Of special significance is the fact that there is little bonding force between the successive layers in montmorillonite; therefore, water can readily enter between the individual sheets and cause swelling, a condition which can lead to trouble.[6.18]

The formation of clay minerals is the direct result of weathering, which was discussed briefly in Chap. 5. Weathering is an extremely complex process and is still not fully understood. Climate plays an important role: if a given kind of climate persists in one area for a very long time, the same products of weathering may result despite differences in the parent, or fresh, material. Vegetation may also have a profound effect: the pH value of the water that percolates down from the surface, thus acting as an important element in breaking down fresh minerals, will be affected by the character of vegetation in combination with local weather, especially rainfall. Interesting though the weathering process undoubtedly is, it is the results of weathering that are of concern to the student of soils. The fact that the clays have been derived from solid rocks, however, by a variety of processes prior to, during, and since transportation from their place of origin will naturally help to explain the otherwise strange variations in materials such as the Sasamua clay already mentioned. The variations in the "limits" of Sasamua clay, variations depending upon the treatment the material received before the Atterberg tests were carried out, are shown in Table 6.2.

TABLE 6.2 Atterberg Limits for Sasamua Clay[6.17]

Treatment before test	Liquid limit (% dry wt.)	Plastic limit (% dry wt.)
Natural state	87	54
Dried at 105°C and powdered in mortar	58	39
As above, but treated with tetrasodium pyrophosphate	47	37
Dried, powdered, and rehydrated 1 month	63	39

The sensitivity of clay with high halloysite content to the action of chemicals used for deflocculating in the standard sedimentation procedure for mechanical analysis will suggest that, although clays have long been generally regarded as "inert" materials by those engineers who were not chemically minded, they are not inert at all. Of special importance is the way in which the small electrical charges carried by soil colloids, and particularly by the particles of clay minerals, will react. The well-accepted limiting size for clay-sized particles, 2 microns or 2 μ, was not just a fortuitous choice, but a useful limiting point below which the surface properties of particles begin to dominate the chemistry of the material. Below this size, the electrical charges on individual particles tend to increase with decreasing size. It may seem that this discussion is getting too far away from soil mechanics and from geology, but the very important soil property of cation-exchange capacity can only be fully appreciated against this background. Under certain suitable conditions, some of the ions in clay will exchange; the most common exchangeable ions

6-16   GEOLOGICAL BACKGROUND

are calcium, magnesium, potassium, and sodium. As the ions change, properties may also change. If a change can be predicted and controlled, a useful method of altering clay soils is available.[6.19]

One of the earliest applications of this method is still of interest. At the San Francisco World's Fair in 1939, a 3-hectare (7-acre) lagoon was a prominent landscape feature. The inside of the lagoon was lined with a 25-cm (10-in) layer of calcium clay of loamy texture. Tests with fresh water showed high leakage losses almost immediately. Because of the intended purpose of the lagoon and the time schedule, this was a serious defect. Laboratory studies showed that base exchange would result from contact between the clay and seawater, with its high sodium content, and would increase the watertightness of the clay. The lagoon was therefore flooded with seawater for 45 days, after which fresh water was readmitted. Remarkably little leakage then took place, and the lagoon performed satisfactorily for the duration of the fair.[6.20]

A more practical example of the application of a complex surface chemical reaction, involving full appreciation of the character of clay minerals, could scarcely be imagined. It will serve well to emphasize the importance of clay minerals in even the most practical aspects of soil mechanics, in addition to their significance in connection with all laboratory tests upon fine-grained soils. It will now be clear that the geological origin of clays is of considerable, but indirect, significance in the engineering study of clay soils. The process of weathering is of equal importance, but again indirectly through the influence it has played in producing the clay minerals actually present in the samples being used. Differences in clay-mineral content will readily explain differences in soil properties, even those demonstrated so simply by the Atterberg limit tests. In most cases, such indirect evidence will naturally suffice for engineering purposes, as mineralogical analyses are necessary only in such unusual cases as with the clay at the Sasamua Dam. Knowledge of the significance that clay minerals may have, however, is invaluable to the student of soils, for if unusual characteristics are encountered, or suspected, a check on the clay minerals present—such as montmorillonite in relation to swelling clays—may prove of great utility.

6.8 SOME CONTRIBUTIONS TO GEOLOGY

It will have already become obvious that the interrelation of soil mechanics and geology is not a one-way street, even though this discussion has considered the matter generally from the point of view of the civil engineer. There is a steadily growing mutual respect and understanding between workers in the two fields, as each field contributes to the other. Throughout this book there will be found frequent references to the contributions that the civil engineer can make, and has made, to advance geological knowledge. The use that Hubbert and Rubey made of the theory and test procedures of soil mechanics has already been cited. It may be helpful, and of interest to both geologists and engineers, to mention briefly a few other cases in which soil studies that were carried out primarily for engineering purposes have led to significant geological results.

In a study of 163 undisturbed soil samples and 185 disturbed samples from test pits and boreholes in the part of the bed of glacial Lake Agassiz that lies south of the Canadian border in the Red River Valley, Rominger and Rutledge established five stratigraphic units within the lacustrine sediments.[6.21] Correlation of these units between three localities was aided by the application of statistical methods for study of the soil-test results. Profiles worked out on the basis of the results of consolidation tests indicated preconsolidation stresses as determined by laboratory tests and revealed a period of surface drying of the lake sediments that had not previously been detected; this was an important contribution to the knowledge of the geology of this greatest of the glacial lakes of North America. The authors stressed the valuable contribution that the geological study of the results of soil mechanics tests can give. Rominger continued the work described in the joint paper in a detailed study of the relationship of the plasticity and grain size of sediments from the bed of Lake Agassiz. He established a definite correlation between the Atterberg limits, as an expression of plasticity, and grain-size distribution, and he suggested that his results might be applicable on a much wider scale.[6.22] These two contributions present challenging suggestions regarding the valuable geological results which a large collection of simple soil tests from the same geological unit, in this case the bed of a well-defined glacial lake, can produce.

Study of glacial tills in southern Ontario has been greatly aided by the application of the simple techniques of soil mechanics. In a series of papers Dreimanis has recorded the way in which he has used soil-analysis methods common in soil mechanics to assist in his elucidation of the somewhat complex history of successive till deposits in this area.[6.23] Comparative study of the results of consolidation tests upon many samples of the Leda marine clay of the Champlain Sea area in the St. Lawrence and the Ottawa river valleys has enabled Crawford to show graphically the close interrelation between preconsolidation loading and elevation above sea level of the respective samples.[6.24] This is illustrated in Fig. 6.7. A glance at this record of the results of engineering soil tests will show clearly its geological significance. These two examples are quoted to show, on the one hand, that geologists can use the methods of soil mechanics directly in their own investigations, and on the other, that engineers can aid in the development of new geological information if they will remember the geological origins of the soils they are testing for their own utilitarian purposes.

Finally a different kind of example, but one of equal significance, may be mentioned. The delta of the Mississippi River is one of the most interesting recent geological phenomena in North America. Intense activity off the coast of Louisiana in the search for

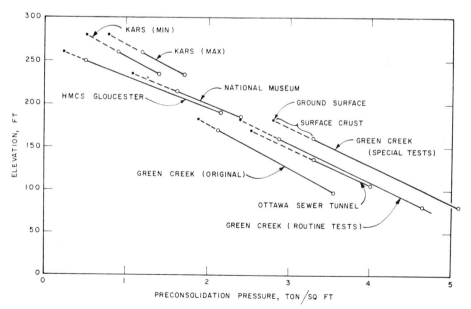

FIG. 6.7 Relation between preconsolidation pressure on Leda clay and its elevation above sea level at different locations in the valley of the Ottawa River, Canada.

petroleum has led to much offshore engineering. There are now over 2,000 wells in operation on the shelf off this coast. More than 100 test drilling rigs (more than 25 percent of all the offshore drilling rigs in the world) have been in operation at one time in this area. Supports for these very large drilling installations are in themselves engineering structures of some magnitude. The studies made for their foundation in the sea have included some deep borings and associated undisturbed sampling to great depths beneath the seabed. Fortunately, the geological significance of the results of these soil engineering investigations has been recognized. An unusually valuable paper, written jointly by one of the leaders in the geological study of the delta sediments and one of the engineers concerned with the drill-rig foundation studies, has been published. In summarizing some results, Fisk and McClelland show that soil tests upon the samples obtained for foundation investigation confirmed and extended important correlations between the shear strength of clay deposits and the type and abundance of clay minerals present.[6.25] At a depth of about 50 m (164 ft) below sea level, a sharp increase in strength was detected in one location, which clearly marked the top of the weathered surface of the late Pleistocene formation known to underlie the more recent deltaic deposits. Other changes in soil properties were noted at the same depth, leaving no doubt of the validity of the significant geological change thus detected so far from sight and normal observation.

6.9 CONCLUSION

The temptation to quote further from the paper by Fisk and McClelland is great, but this brief reference indicates clearly the great mutual benefit of such close cooperation between geologist and engineer in an area of unusual geological complexity. The full paper is worthy of close study. It is a challenging reminder of the value which advance geological information has for the engineer in planning subsurface exploration work and of the reciprocal value which a careful study of the results of the boring and the soil-testing work of

FIG. 6.8 Dr. Karl Terzaghi (center, flanked by F. E. Schmidt of New York and R. Tillman of Vienna) at the Quabbin Dike during the First International Conference on Soil Mechanics and Foundation Engineering, Cambridge, Massachusetts, 1936.

the engineer, when carefully correlated with known geological information, can have for the geologist. The paper vividly illustrates the dual character of the connection between geology and soil mechanics, a connection to which this chapter has been, necessarily, an introduction only. The interrelationship is still not fully realized by many geologists and is appreciated by possibly even fewer engineers.

There is, however, a steadily growing recognition of the benefit that can accrue from breaking down the barrier between the two disciplines. This is indicated by the growing number of papers that demonstrate the value of the links that have been sketched in general terms in this chapter. One outstanding paper of this kind—notable both in that it appeared in the fiftieth anniversary volume of *Economic Geology* and in that it was written by Dr. Terzaghi—is entitled "Influence of Geological Factors on the Engineering Properties of Sediments." It is a masterly treatment of this one major aspect of the general interrelationship that has been discussed. Dr. Terzaghi's concluding words, based as they are upon his lifetime of worldwide experience, may also serve to bring this chapter to a close. According to Dr. Terzaghi (pictured in Fig. 6.8), "The results of the detailed subsoil investigations performed in connection with engineering operations provide the geologist with a new source of significant information in the realm of physical geology."[6.26]

6.10 REFERENCES

6.1 A. Collin, *Landslides in Clay* (1846), W. R. Schriever (trans.), University of Toronto Press, Toronto, 1956.

6.2 Sir B. Baker, "The Actual Lateral Pressure of Earthworks," *Minutes of Proceedings of the Institution of Civil Engineers*, **65**:140–241 (1882).

6.3 A. L. L. Bell, "The Lateral Pressure and Resistance of Clay and the Supporting Power of Clay Foundations," *Minutes of Proceedings of the Institution of Civil Engineers*, **199**:233–336 (1916).

6.4 K. Terzaghi, In *Proceedings of the First International Conference on Soil Mechanics and Foundation Engineering*, **3**:13 (1936).

6.5 K. Terzaghi, "Soil Mechanics—A New Chapter in Engineering Science," *Journal of the Institution of Civil Engineers*, **12**:106 (1939).

6.6 M. K. Hubbert and W. W. Rubey, "Mechanics of Fluid-Filled Porous Solids and Its Application to Overthrust Faulting," *Bulletin of the Geological Society of America*, **70**:115–166 (1959); and in the same vol., W. W. Rubey and M. K. Hubbert, "Overthrust Belt in Geosynclinical Area of Western Wyoming in Light of Fluid Pressure Hypothesis," pp. 167–206.

6.7 R. F. Legget, "Geotechnique: New Word, Old Science," *Transactions of the Geological Association of Canada*, **12**:12–19 (1960).

6.8 K. Terzaghi, "Influence of Geological Factors on the Engineering Properties of Sediments," *Economic Geology*, 50th an. vol., 1955, pp. 557–618.

6.9 W. A. Clevenger, "Experiences with Loess as Foundation Material," *Proceedings of the American Society of Civil Engineers*, vol. **82** (SM3), paper 1025, 1956, pp. 1–26.

6.10 W. J. Eden, "Construction Difficulties with Loose Glacial Tills on Labrador Plateau," in R. F. Legget (ed.), *Glacial Till*, Royal Society of Canada Special Publication no. 12, 1976, pp. 391–400.

6.11 R. F. Legget and W. J. Eden, "Soil Problems in Mining on the Precambrian Shield," *Engineering Journal*, **43**:81–87 (1960).

6.12 W. J. Eden, "A Laboratory Study of Varved Clay from Steep Rock Lake," *American Journal of Science*, **62**:1223–1262 (1955).

6.13 E. Penner and K. N. Burn, "Review of Engineering Behaviour of Marine Clays of Eastern Canada," *Canadian Geotechnical Journal*, **15**:269–282 (1978).

GEOLOGICAL BACKGROUND

6.14 I. E. Higginbottom and P. G. Fookes, "Engineering Aspects of Periglacial Features in Britain," *Quarterly Journal of Engineering Geology,* **3**:85 (1970).

6.15 For a useful review and good bibliography see "Symposium on Airfield Construction on Overseas Soils," *Proceedings of the Institution of Civil Engineers,* **8**:211–292 (1957).

6.16 P. Simon and J. Vallee, "The Souapiti Fall; the Konkouré Development," *Travaux,* no. 296, 1958, pp. 193–202.

6.17 K. Terzaghi, "Design and Performance of the Sasamua Dam," *Proceedings of the Institution of Civil Engineers,* **9**:269–394 (1958).

6.18 For a full discussion see "Symposium on Physico-Chemical Properties of Soil: Clay Minerals," *Proceedings of the American Society of Civil Engineers,* **84** (SM2):1–102 (1959).

6.19 J. E. Gillott, *Clay in Engineering Geology,* Elsevier, Amsterdam, 1968.

6.20 C. H. Lee, "Sealing the Lagoon Lining at Treasure Island with Salt," *Transactions of the American Society of Civil Engineers,* **106**:577–607 (1941).

6.21 J. F. Rominger and P. C. Rutledge, "Use of Soil Mechanics Data in Correlating and Interpretation of Lake Agassiz Sediments," *Journal of Geology,* **60**:160–180 (1952).

6.22 J. F. Rominger, "Relationships of Plasticity and Grain Size in Lake Agassiz Sediments," *Journal of Geology,* **62**:537–572 (1954).

6.23 A. Dreimanis, "Tills, Their Origin and Properties," in R. F. Legget (ed.), *Glacial Till,* Royal Society of Canada Special Publication no. 12, 1976, pp. 11–49.

6.24 C. B. Crawford, "Engineering Studies of Leda Clay," in R. F. Legget (ed.), *Soils in Canada,* Royal Society of Canada Special Publication no. 3, 1961, pp. 200–217.

6.25 H. N. Fisk and B. McClelland, "Geology of Continental Shelf off Louisiana: Its Influence on Offshore Foundation Design," *Bulletin of the Geological Society of America,* **70**:1369–1394 (1959).

6.26 K. Terzaghi, "Influence of Geological Factors . . . ," loc. cit.

Suggestions for Further Reading

Fundamentals of Soil Behavior by James K. Mitchell of the University of California, published by John Wiley & Sons of New York as a 422-page quarto-sized volume, stands alone as a masterly review of all aspects of soil properties considered from the engineering point of view; it is warmly commended.

There are now, naturally, many useful volumes on soil mechanics in general. One of the early leaders in this field was *Soil Mechanics in Engineering Practice* by Karl Terzaghi and Ralph B. Peck, published by John Wiley & Sons of New York; it remains one of the outstanding volumes. *Introductory Soil Mechanics and Foundations* by George F. Sowers of Georgia Institute of Technology, now published as a 621-page fourth edition by the Macmillan Company of New York, has a useful introductory geological section.

The successive *Proceedings of the International Conferences on Soil Mechanics and Foundation Engineering,* listed on p. 6-3, now contain a wealth of valuable information on all aspects of the subject, including the links with geology. They are published by the host committees in the respective countries, sets being available for consultation in larger engineering libraries. Corresponding *Proceedings* for the Regional Conferences now held in several areas of the world between the International meetings, are similarly valuable collections of papers.

chapter 7

ROCK MECHANICS AND GEOLOGY

7.1 Rock Substance / 7.2
7.2 Mechanical Properties / 7.2
7.3 Discontinuities / 7.4
7.4 Groundwater / 7.6
7.5 Structural Geology / 7.7
7.6 Rock Characteristics / 7.8
7.7 Field Testing / 7.9
7.8 In Situ Stresses / 7.11
7.9 Field Observations / 7.12
7.10 Conclusion / 7.14
7.11 References / 7.15

The First Congress of the International Society of Rock Mechanics was held in Lisbon, Portugal, in 1966, almost exactly 30 years after the corresponding formal initiation of the International Society of Soil Mechanics. As with soil mechanics, there had been much individual activity in connection with the mechanics of rocks, as distinct from soils, prior to 1966, but the Lisbon meeting set the seal of international approval on this new discipline. Naturally and inevitably there is a good deal of overlap between the two associated disciplines. This is recognized in some countries by the existence of combined geotechnical groups, embracing both soil and rock mechanics. Elsewhere there is almost always close liaison between the pairs of national groups, as there is also at the international level. There are significant differences between the necessary approaches to the mechanics of solid rock and of soils. For convenience, therefore, two chapters in this Handbook deal separately with the interrelations of geology with the two disciplines.

Great advances have been made since 1966 in rock mechanics. No major work involving rock excavation, especially in mining, would be undertaken today without a thorough investigation of the mechanics involved. Correspondingly, in civil engineering rock mechanics is already playing an important role, the literature already reflecting sound and helpful advances. Just as with soils, so also with rock, an appreciation of geology is essential for the full benefits to be derived from applications of the laws of mechanics to rock stability.

7.1 ROCK SUBSTANCE

An early definition of rock mechanics, still valid, is that it is "the theoretical and applied science of the mechanical behavior of rocks; it is that branch of mechanics concerned with the response of the rock to the force fields of its physical environment." Some have criticized the emphasis upon mechanics in this definition, but if one studies the growing literature of the subject or speaks with its expert practitioners, it is clear that the subject is dominated, naturally and rightly, by mathematical analyses and syntheses of the forces acting upon rocks. If one picks up one of the better-known textbooks on rock mechanics, it will not be surprising to find the volume wholly given over to mechanics, with almost no reference to geology at all. (Even less understandably, there are books on soil mechanics with the same surprising omission.) Many of the mathematical solutions to problems in rock mechanics are elegant; they are naturally useful. In the equations developed for stability, however, there must always be a term relative to the properties of the rock with which the particular exercise is dealing.

In the case of studies of continuous, unbroken rock, *continuum mechanics* provides the necessary mathematical tools. Rock in place is not, however, always continuous and so a companion study is of rock in geometrically shaped blocks in contact with one another. Here the theoretical tools have come to be called *discontinuous*, or *clastic, mechanics*. Equally effective progress in developing helpful solutions for practical problems has here been made, but, again, equations must contain some term indicative of the characteristics of the rock being considered. Accordingly, it has been necessary to develop the concept of a perfect rock material. The term *intact rock* is sometimes used, the term indicating rock which is free from joints, bedding planes, breaks or shear planes, so intact that it can be tested, by means of samples, in a laboratory where its mechanical properties can be accurately determined. Some writers use the term *rock substance* to indicate this ideal material. Since these are terms which may not appeal to geologists, it is necessary to stress that some such theoretical, ideal material must be assumed if the theories now available through rock mechanics are to be applied. It is the task of the engineering geologist to show to what degree the rocks actually in place at the site being studied approach the ideal condition that has been assumed.

This procedure is a sound one, provided the limitations of the theoretical solutions are not forgotten. The elegance of the mathematical solutions to some slope-stability problems, especially if clastic mechanics has been used, assisted (as it now can be) by computer graphics, tends to obscure the fact that some geological feature of the actual rock face in the field may invalidate significant theoretical assumptions made initially. This is a common feature of engineering design, although the uncertainty of geological structure and of rock characteristics in the field makes the tempering of theory with practical considerations of rather more importance here than is usual. Accordingly, the remainder of this short chapter will be devoted to the practical considerations that must always be associated with the results of rock mechanics theory so that the latter may be used judiciously and with confidence. Full treatments of the mechanics will be found in the excellent texts on rock mechanics now available. A guide to them is provided at the end of this chapter, and full tribute must be accorded them prior to discussing some necessary qualifications.

7.2 MECHANICAL PROPERTIES

A good starting point is to consider the strength of rocks, since this is a factor which enters into all considerations of rock stability. It is something that can be studied, to a degree, by means of carefully procured samples tested in a laboratory. Sampling of rocks for this purpose is an easier matter than the sampling of soils, but care must be taken, naturally, to protect samples and to have them prepared for testing without damage to their integrity. Although shear stresses are those that are usually critical in rock masses, a first test will usually be of the compressive strength of the rock, a test that when properly carried out will also be the means of determining the value of Young's modulus.

Deere has been assiduous in assembling much useful information on this property of rocks.[7.1] He has prepared a valuable series of charts, relating compressive strength with modulus of elasticity. Peck has combined a number of these, as well as other useful concepts of Deere, in one chart which is reproduced as Fig. 7.1.[7.2] The value of this useful summation will be at once obvious. The general agreement in the strengths shown for such a wide variety of rocks, when related to the corresponding values for Young's moduli, is striking and significant. At the same time, however, the great range in the compressive strengths is equally impressive, especially when the use of the logarithmic scales is kept in mind. The groups of rocks shown are those most commonly encountered on engineering works and so the chart has a double value, not only illustrating the need for judicious use of the concept of "rock substance" in rock mechanics studies, but also as showing the wide range of qualities involved in the use of what is so loosely called "bedrock."

By an adaptation of the well-known triaxial testing machine used for triaxial tests on soil samples, an adaptation which allows for the higher pressures involved, laboratory values for the shear strength of rock specimens, and so the shear modulus for the rock, can also be obtained. Correspondingly, test methods are available for determining the porosity of rocks which may contain even small quantities of water. Density will be a normal first check on samples when they come to a laboratory for testing.

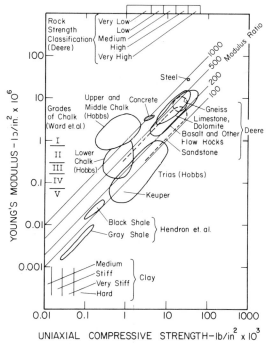

FIG. 7.1 Relation between the compressive strengths and Young's moduli for typical rocks and clays.

7-4 GEOLOGICAL BACKGROUND

The suite of mechanical properties thus obtained will be useful and will provide the necessary values for insertion in equations of stability. But it will have been obvious to the reader long before this that there are some major qualifications. If all rock substance were homogeneous and isotropic, all would be well, but it is far from this. Sedimentary rocks provide the most obvious and probably the most widespread exception, especially if bedding planes are well delineated, as they are in many limestones and shales. Even in sandstones the mode of deposition of the original sand is often obvious, though bedding planes may not be clearly evident. In all such rocks, the relation of the major axis of test loads with the bedding planes must be accurately determined, since there will usually be considerable difference between results from tests performed at right angles to bedding planes and those performed parallel to them. Differences of the same type may be found in testing some igneous rocks in different directions, each one to be related to the orientation of the sample in the field, since the fabric of mineral grains may be significant, especially in igneous rocks displaying flow structures and in foliated metamorphic rocks such as schists. The bonding of individual grains may similarly affect test results, since some bonding materials will lose strength when wet. All tests, therefore, should be carried out on dry and wet samples. Despite the cost, it is desirable to obtain a number of comparable test results on any major rock type involved in rock mechanics studies so that appropriate statistical correlations can be made.

All that has been said so far relates to rock that is reasonably uniform, if not homogeneous, typical of the sort of rock exposure that is found in temperate regions, where much of the work on rock mechanics has been carried out. But in warmer areas where rock weathering is a normal feature of the terrain, or in areas where geological processes have themselves created differentiation between varying horizons of the same rock, the determination of rock strengths that may be used in the application of rock mechanics theories becomes more complex. Figure 7.2 is a good illustration of the type of rock exposure that may be expected. It shows an exposure of andesite marked by severe weathering; four zones have been indicated that can readily be distinguished even by eye. Matula, in the useful paper from which this illustration comes, cites the case of a Triassic quartzite from the Carpathians, formed originally from beach sands, cemented during their diagenetic evolution by silica to form a solid, hard, and brittle rock.[7.3] The resulting sound rock has a strength of 2,840 kg/cm^2 (18 tons/in^2) and Young's modulus of 600,000 kg/cm^2 (3,800 tons/in^2). The region from which this rock comes was subjected to folding in Cretaceous times, with large scale faulting in some areas. Rock that has been subject to these disturbances has been shown to lose as much as two-thirds of its strength and modulus. Further Tertiary faulting and hydrothermal alteration have reduced the same rock, in places, to an almost cohesionless sand upon which tests could not even be conducted.

7.3 DISCONTINUITIES

The case just cited is, admittedly, extreme, but it serves to show how significant the geological history of rock may be, especially in relation to rock mechanics and the values to be used for unit rock strength. Geological history is even more important when the continuity of rock strata is considered. Only rarely will one encounter bedrock which is continuous within the area of study and which exhibits no discontinuities. Bedding planes in sedimentary rocks, and in many metamorphic rocks, and shear cleavages in many metamorphic rocks are other obvious discontinuities, while the joints to be seen by eye in the majority of rocks serve equally well to break the continuity that has to be assumed in

FIG. 7.2 An exposure of andesite in Czechoslovakia, showing different weathering zones determined by geologic observations.

theoretical considerations. Most serious of all is the complete rupture created by faulting, a phenomenon which may even lead to a change in rock type at the resulting fault.

As has already been made clear, rock mechanics takes discontinuities fully into consideration by its use of clastic mechanics. Even here, however, certain assumptions must be made as to the regularity of discontinuities and the uniformity of the resulting blocks. It is therefore essential that geological studies shall demonstrate the accuracy, or the inaccuracy, of the assumptions that must be made. Again, geological history is the first guide that must be utilized. The basic character of the rock (sedimentary, metamorphic, or igneous) will be the starting point, since that will indicate the origin of the rock and its major components. Knowledge of tectonic history since the rock was first formed will be the second guide. If there have been no significant earth movements, then jointing as a result of temperature changes will be the first possibility to be explored, once bedding planes (where they exist) have been determined.

If, however, there have been significant orogenic disturbances, then a careful geological survey of an appreciable area surrounding the site of the works in question will be essential. The overall geological structure for this area must be determined in order to see what folding has taken place since rock formation and whether there has been any faulting. This is a study that cannot be done merely by studying the geology at the site in question; a regional approach must be followed. The nature of folds was outlined in Chap. 4 (Fig. 4.9). From the simple diagrams then presented, it can be seen that whenever rock strata have been subjected to folding, those near fold axes, such as on the upper part of an anticline or the lower part of a syncline, will have been subject to tensile forces. Jointing is therefore to be expected, the joints being an expression of tensile failures. In the case of gentle folds, the joints may not be obvious to the eye, but they must be anticipated and special field investigations conducted in order to determine whether they are present or not.

The existence of faults will be indicated by a detailed geological survey of the region involved. They may not be evident at the surface if there is soil cover. Not only must the presence of faults be determined but so must be their location (as by inclined drill holes) and their throw. Stresses caused by major faulting may cause ruptures in rock some considerable distance from the fault itself, so that in all cases of faulted rock the most detailed reconnaissance must be carried out in advance of any decisions about rock excavation. Even then, it may not be possible to locate all failure surfaces, but their possible presence can be anticipated.

7.4 GROUNDWATER

Not only will discontinuities in bedrock affect the application of the results of theoretical rock mechanics studies but they will, in most cases, lead to problems with water. The discontinuities may be such that they provide tight contact between the fractured rock faces, but in most cases they will be open to some degree. They will thus permit the movement of groundwater and the entry of rainwater at the surface after heavy rains, assuming that light precipitation will be retained in the soil cover. The movement of groundwater through "solid rock" (as it may loosely be described), when discontinuities are available for its travel, can be remarkable. Examples are given later in this volume of troubles with water being encountered in tunnel works several hundred meters below the surface.

The resulting problems are related to practice and to design. When rock excavation is to be carried out in water-bearing rock, arrangements must be made for handling water during the course of excavation. Any estimate of quantity that can be made prior to the start of work will be of much assistance in planning construction methods and schedules. When excavation is complete, drainage facilities will have to be provided if the groundwater condition has been interfered with to the extent that flowing water enters the excavated area. It may even be necessary to calculate flow nets for the finished cross section in order to determine the exact nature of groundwater movement in the finished works. In all such investigations, the probable variation of groundwater throughout the 12 months of the year must never be forgotten.

In the application of clastic mechanics to the calculation of rock stability in the finished excavation, allowance can be made for the hydrostatic pressure that groundwater will induce in continuous-pore spaces. Equally significant will be the allowance for the pore pressures resulting from groundwater which will exist in the discontinuities, no matter how small. Such pressures will have a profound influence on the effective stresses in the rock mass, a concept now well understood and appreciated because of the pioneer work of Terzaghi in clarifying what had previously been a rather vague idea, even if sometimes applied correctly through intuitive thinking. The concept is one of great importance and must be rigorously examined for all cases in which groundwater may be present. If, for example, stability of rock slopes within a reservoir is being investigated, then the effective pressures that will exist after the water level in the reservoir has been raised must be carefully evaluated. These pressures may even have the effect of moving two sides of a steeply sloping reservoir inwards after water is stored, rather than outwards as might be thought, at first sight, to be the case. When groundwater is going to play a significant role in rock stability, it will be necessary, as a permanent precaution, to install measuring devices (piezometers) to record its position or pressure. Design work may be aided if these installations are made as soon as design work begins, so that assumptions made about groundwater based on overall preliminary studies may be checked and verified prior to their use in design calculations. The dynamic nature of groundwater (which will be stressed in the next chapter) as well as its annual variation (already mentioned) must always be kept in mind.

7.5 STRUCTURAL GEOLOGY

It will now be obvious that, if the significant contributions that rock mechanics studies can make to civil and mining engineering are to be fully effective, it is essential that their application be made against a full realization of the importance of structural geology. The assumptions as to the fabric of the rock being studied that must be made before theoretical studies can begin may be correct, but they also may not be. It is therefore necessary to have at least a general geological survey made of the area in which the works are situated, even when the nature of the rock or rocks to be considered is known. Only in this way can it be determined that the rock strata really are horizontal or, if not, at what dip they are inclined. The presence or absence of folding will readily be revealed by geological reconnaissance, as will the presence of faulting. If unconformities are present but are not immediately obvious by study of rock exposed at a site, then a careful geological investigation of the area around should reveal whether this further departure from a simple geological structure is present or not.

All these, and other minor features, are within the scope of structural geology, now well recognized as an important subdivision of the science. Structural geology has become a speciality in itself, as evidenced by the number of books which deal only with this one aspect of geology. Although civil engineers will not normally need to know more than the main types of geological structure, it is incumbent upon engineering geologists to be

7-8 GEOLOGICAL BACKGROUND

familiar with the main approaches taken in structural geology to the elucidation of complex structures. At the end of this chapter, therefore, a short list of useful references to structural geology is given.

7.6 ROCK CHARACTERISTICS

Every application of rock mechanics is unique, if only in so far as the geology of the site is concerned, since no two geological exposures are ever identical. Every specimen of rock that is tested gives results that can be applied to the location from which it came and to no other. As the volume of test information for different kinds of rocks has steadily accumulated, it has been only natural that similar information on the same rock type should be collected and assessed, this information in turn giving some general figures which suggest the range of values within which further test results for that particular rock may be expected to lie. The authors have indicated their appreciation of the worth of such "corporate values" by their inclusion of Fig. 7.1 and their reproduction of some of the values, developed by Deere, notably those for frequency of joints. These general characteristics of rock types seem to have a useful place, if employed with discretion. They have, however, been the subject of much discussion, there being some who dislike their use just because of the unique nature of each application of rock mechanics and the possibility of the misuse of such general figures.

This counter argument arises from the possibility that, if such values fall into the wrong hands, they may be used indiscriminately and thus eliminate the special investigation necessary for every site of an engineering work. It is easy to imagine the appeal that such a shortcut can have for a very parsimonious owner who has not had the benefit of sound professional advice. There is, therefore, an element of risk in the development, and especially in the publication, of such figures in an *Engineering Characteristics of Rocks,* as such compilations are commonly called. But the same risks are run in many other situations when uninformed entrepreneurs try to save money by using published general information, thinking that they can apply this to their own particular situation without having to spend money for professional services. In most cases, such unfortunate and dangerous practices are nipped in the bud as soon as an official application has to be filed in relation to some relevant public regulation.

The argument is, however, a serious one and therefore warrants mention even in so summary an introduction to rock mechanics as this. If all publications giving summary characteristics of rocks are clearly identified as being guides only, then surely public convenience will be served and minimum risk incurred. This seems to be confirmed by experience with *Engineering Characteristics of the Rocks of Pennsylvania,* a publication of the Pennsylvania Geological Survey.[7.4] It is a well-printed 200-page handbook which gives illustrated summaries of the main characteristics of the rock groups of the state and of each formation in the groups. These summaries are based upon "observations in road excavations, bridge abutments, railroad rights-of-way and quarries," as well as upon actual service records (Fig. 7.3). In two places, due warnings are given as to the need for detailed investigations for specific sites, and this same caution is made clear in the excellent introduction. A glossary and a useful list of references round out this admirable review of the main rock types of the state. One who has had experience with its uses has described the handbook in this way: "[It] has been well received and (with limited exceptions) used as intended. As with anything, even a highway, if misused, can be dangerous."[7.5] There is so much to be gained by describing the limits within which the properties of rocks may be expected to lie that the authors hope that many more such compilations

FIG. 7.3 An exposure of sandstone of the Chemung formation in Pennsylvania; this photograph is typical of the useful photographs (and summaries of test results) to be found in *Engineering Characteristics of the Rocks of Pennsylvania* (see Reference 7.4).

as that from Pennsylvania will be prepared, always giving due warning that they are to be used strictly as guides and not as substitutes for site investigation.

7.7 FIELD TESTING

The discipline of rock mechanics today embraces a sound body of theoretical knowledge, a comprehensive background of laboratory testing and analysis, and an increasing volume of case histories of "rock mechanics in action." But the discipline has also made a significant contribution to the development of testing in the field. When allied with necessary geological information, the results of careful field testing of rock can make important contributions to the solution of rock-stability problems.

Test drilling naturally takes pride of place in even a brief review of field-test methods; it is common to all site investigations. In specific studies of rock properties, the results of carefully supervised diamond drill work can give much valuable information. The state of the rock cores that are obtained becomes a matter of significance, the relative core recovery being one important factor. With good core recovery, some indication will be given of the amount of jointing present; some faults will naturally be revealed by drill-core records. Much good work has been done in photographing the sides of drill holes with specially designed borehole cameras and even with portable TV cameras which give useful images for immediate examination at the surface. By the use of ingenious "packers" in drill holes, sections of a hole can be isolated and then subjected to water pressure, so that the permeability of different horizons and probable losses of water through joint systems can be determined. Deere has proposed the term *rock-quality designation* (RQD) to denote, by reference to percentage core recovery, the state of rock in situ.[7.6]

Much useful information can, therefore, be obtained from a carefully laid-out program of test drilling, the choice of location of holes being assisted by an appreciation of the local geological structure. So useful (and often interesting) is this drill-hole information

7-10 GEOLOGICAL BACKGROUND

that it is all-too-easy to forget that it relates only to the rock in the immediate vicinity of the test hole, and so to only a minute proportion of the rock mass that is involved in any rock excavation or slope study. Geophysical methods, complementary to test drilling, can do something to extend knowledge of the properties of the rock mass. Methods are summarized in Chap. 12. The seismic method has generally been found to be most useful in investigations of the continuity and relative uniformity of rock masses. Especially valuable has been the use of suitably designed equipment for cross-borehole testing.

Strength tests in the field hold a special place in rock mechanics work; these tests give valuable information, but only at considerable cost. A worthwhile preliminary to any such testing is the use of the Schmidt hammer in a simple but effective type of "nondestructive testing." The standard hammer is allowed to drop onto a clean and suitable rock surface from a fixed height and its rebound is measured, giving figures which can usefully be applied in comparing rock types and in establishing the need for more elaborate testing (Fig. 7.4).[7.7]

Full-scale tests of rock in situ can be seen to be a desirable means of eliminating some of the uncertainties arising from the use of only theoretical and laboratory studies in major rock-stability investigations. These tests are, however, costly to conduct, even if they can be carried out while construction is proceeding and heavy equipment and skilled manpower are available. Only a few such large tests have been reported in the literature, although many more have doubtless been carried out. In 1950, the strength of bentonite seams in the Niobrara chalk on which the Fort Randall Dam is founded was in doubt. It was finally decided to carry out shear tests on blocks of the rocks in question, blocks left

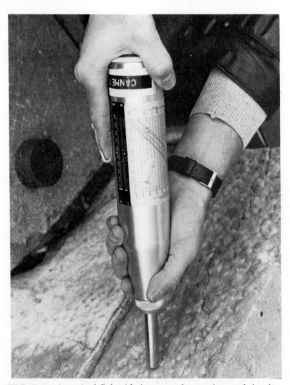

FIG. 7.4 A typical Schmidt hammer for testing rock in situ.

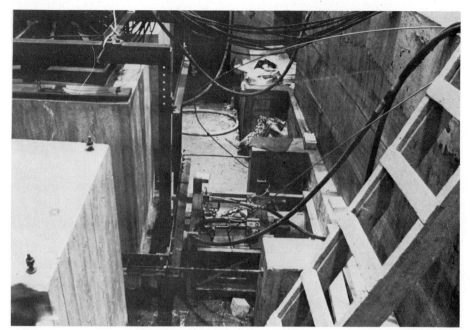

FIG. 7.5 Testing foundation rock in situ for the Carillon Dam, Ottawa River.

in place after rock around had been carefully excavated. The results were described by Thorfinnson.[7.8]

The Carillon Dam on the Ottawa River, midway between Montreal and Ottawa, is founded on horizontal or gently dipping strata of dolomitic and calcareous limestones with minor shale horizons, the strata being part of the Beekmantown formation of Paleozoic age. The strength of the shale strata was in question. Five test blocks, 0.76 m (30 in) square in plan, were prepared by the use of wire saws operating from 0.91-m (36-in)-diameter Calyx drill holes. When fully instrumented, the blocks were vertically loaded with concrete blocks and then subjected to shear forces by means of horizontal jacks. Consistent results were obtained from four of the five blocks, leading to confidence in design assumptions.[7.9] The fact that the duration of these tests was 15 weeks, and that they had to be conducted with a minimum of interference with the tight overall construction schedule will show that tests on this scale are not something to be undertaken lightly, useful as may be the results they give (Fig. 7.5).

7.8 IN SITU STRESSES

Field measurements of stresses in rock, and other factors involved in the behavior of rock masses, might be thought to be a subject of limited interest. In April 1977, however, there was held in Zurich, Switzerland, a full-scale international symposium devoted entirely to the subject of *Field Measurements in Rock Mechanics,* this being the title of the valuable two-volume proceedings that resulted from the meeting.[7.10] This is yet another way in which rock mechanics studies have so rapidly advanced.

There are available today, therefore, good instruments for the measurement of the stresses in rock in place. One of the pioneers in this field (one of the lone workers men-

7-12 GEOLOGICAL BACKGROUND

tioned earlier) was Hast in Sweden.[7.11] He observed, with the instruments he developed, that there were significant in situ stresses in the Precambrian rocks of his country. Since his time, and especially in quite recent years, there have been an increasing number of similar observations in other countries and areas, one of the most notable being that part of northeastern North America centering around Niagara Falls. Movements of "solid" rock in response to in situ stresses have been observed at Lockport (on the New York State Barge Canal) for more than 50 years and in wheel pits of one of the Canadian power stations at Niagara Falls since soon after the start of the century. "Pop-ups" in quarry floors are further evidence of these stresses (Fig. 7.6). In later chapters of this Handbook

FIG. 7.6 Pressure ridge that developed overnight in a limestone quarry in St. Louis County, Missouri.

more will be said about the serious problems caused in some engineering works by in situ stresses, problems which may be geologically significant. The matter is mentioned here to indicate yet another of the complications that can influence theoretical analyses of rock stability. If, however, the existence of in situ stresses is determined in advance of design studies, due allowance can be made for them.[7.12]

7.9 FIELD OBSERVATIONS

Since geology is so clearly a vital counterpart of rock mechanics, how best can the necessary geological information be obtained and presented? Normal geological maps, if available, will provide a sound starting point; if not, then geological survey work will have to be carried out and the geological map of the requisite area prepared. These regional maps, supplemented by geological sections, should give a good general picture of the geological structure of the area embracing the site under study and of the main rock types that will be encountered. These rock types must be described concisely in geological terms, i.e., their lithology must be stated. A typical description might be "granite, gray, coarse-grained, uniform," with additional comments such as "high strength" if test results are available.

In almost all cases, a more detailed geological map will be necessary, a map based upon intensive survey work immediately around the site in question. Again with the aid of good sections, and possibly of models, this will give to the designer a good three-dimensional picture of the geological structure to be dealt with. Faults would naturally be shown on such detailed maps, which would also give an indication of the frequency and nature of any joints that are present. The dip and strike of all strata would be clearly delineated as a result of this detailed study. If steep slopes are involved (as, for example, in the case of an open mine pit), aids such as plane-table surveying and photographic surveying may have to be invoked.

Joints are of such importance in rock-stability studies that special attention must be given to them and every effort be made to gain an accurate assessment of their frequency. This is tedious but important work. Deere has suggested a convenient subdivision of joint frequency.[7.13]

Very close	less than 5 cm apart
Close	from 5 to 30 cm apart
Moderately close	from 30 cm to 1 m apart
Wide	from 1 m to 3 m apart
Very wide	more than 3 m apart

The nature of joints must also be observed and recorded with care, i.e., whether they are open or tightly closed. In the case of faults that can be examined, the nature of the fracture surface and the character of the gouge filling must be determined. A number of con-

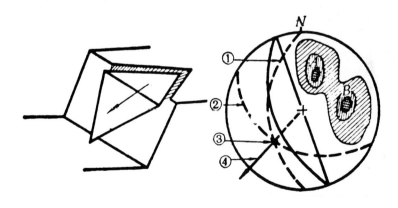

图 3

①代表对应优势 中心点 B 的 优势结构面； ②代表对应优势中心点 A 的优势结构面； ③组合 交点， ④组合交线及其滑动方向

FIG. 7.7 Polar diagram as used in rock-slope calculations, this diagram appearing in *Gongcheng Kancha* (no. 6, Fig. 3, p. 29 (1980)), journal of the Commission of Geotechnical Investigation of the Architectural Society of China, reprinted there from *Rock Slope Engineering* by E. Hoek and J. W. Bray (listed at the end of this chapter), an example of the internationalism of science.

7-14 GEOLOGICAL BACKGROUND

ventions and recording methods have been developed in rock mechanics work, some with a view to placing the information in a computer system. These will be found detailed in volumes dealing with methods in rock mechanics. Mention must be made, however, of the wide use of polar diagrams as convenient graphical aids. The basic Lambert projection is used, either in the form to give equal parts (since this has been found to be generally convenient) or in the stereo form. In each case a circular diagram is developed on one plane from the three-dimensional picture which study of the rock mass will give. These diagrams are capable of wide and varied use, as the typical example shown in Fig. 7.7 indicates.

7.10 CONCLUSION

This brief review will have shown what a powerful tool rock mechanics has now become in the study of rock stability and associated problems and how geological study is its essential counterpart. The great activity in the closing years of the twentieth century that can even now be foretold—activity in tunnel construction, in the extended use of under-

FIG. 7.8 Typical exposure in Hong Kong of residual soil, grading from solid granite bedrock up to surface soils, showing also blocks of unweathered granite within the soil.

ground space for a variety of purposes, and in general construction involving the use of many sites previously deemed too difficult to use—will require increasing applications of rock mechanics, with ever-growing confidence.

Such applications and such confidence will be possible only if geology is fully recognized as an essential part of the studies preliminary to the application of rock-mechanics theories. The structural geology of the region in which a site is being studied, the detailed geology of the area immediately around the site, the best possible prediction of joint conditions and of the possible presence of faults, and accurate lithologies of the several rock types that are going to be encountered—these are the vital contributions that geology has to offer to the steady advance of rock mechanics. Figure 7.8 is an illustration of a situation which exemplifies the necessity of preliminary geological studies in applications of rock-mechanics theories.

It is significant that the six main recommendations for needed research in a major report of the National Academy of Sciences in 1978 ("Limitations of Rock Mechanics in Energy-Resource Recovery and Development") all deal with essentially geological matters.[7.14] The first recommendation is for "research to determine and predict porosity, permeability and flow *in situ*"; another is for "research to improve the ability to map fracture patterns, particularly major fractures and faults, at depth." It was with good reason that, at the Lisbon Conference of 1966, President Muller gave as his first suggestion for future work, "geologists to the forefront."[7.15] This challenge from one of the pioneers of a new discipline still stands.

7.11 REFERENCES

7.1 D. U. Deere, "Geological Considerations in Rock Mechanics," in K. C. Stagg and O. C. Zienkiewicz (eds.), *Rock Mechanics in Engineering Practice,* Wiley, New York, 1968, pp. 1–19.

7.2 R. B. Peck, "Rock Foundations for Structures," *Rock Engineering for Foundations and Slopes,* 2 vols., American Society of Civil Engineers, New York, 1977; see pp. 1–20 of vol. 2.

7.3 M. Matula, "Engineering Geologic Investigations of Rock Heterogeneity," *Rock Mechanics—Theory and Practice,* Proceedings of 11th Symposium of American Institute of Mining, Metallurgical and Petroleum Engineers, 1970, pp. 25–42.

7.4 W. G. McGlade, A. R. Geyer, and J. P. Wilshusen, *Engineering Characteristics of the Rocks of Pennsylvania,* Pennsylvania Geological Survey Bulletin EG 1, 1972.

7.5 J. P. Wilsusen, Letter in *Newsletter of the Association of Engineering Geologists,* **19**(2):14 (April 1976).

7.6 Deere, op. cit., p. 15.

7.7 V. Hucka, "A Rapid Method of Determining the Strength of Rocks *in situ*," *International Journal of Rock Mechanics and Mining Science,* **2**:127 (1965).

7.8 S. T. Thorfinnson, "A Large Scale Field Shear Test on a Bentonite Seam," *Proceedings of the American Society of Civil Engineers,* **80** (1954).

7.9 C. H. Pigot and I. D. MacKenzie, "Carillon Foundation Studies," *Engineering Journal,* **44**:65–71 (October 1961).

7.10 K. Kovari (ed.), *Field Measurements in Rock Mechanics,* 2 vols., Balkema, Rotterdam, 1977.

7.11 N. Hast and T. Nillson, "Recent Rock Pressure Measurements and Their Implications for Dam Building," *Proceedings of the Eighth Congress on Large Dams,* 1964.

7.12 O. L. White, P. F. Karrow, and J. R. Macdonald, "Residual Stress Relief Phenomena in Southern Ontario," *Proceedings of the Ninth Canadian Rock Mechanics Symposium,* Ottawa, 1974, pp. 323–348.

7.13 Deere, op. cit., p. 14.

7.14 *Limitations of Rock Mechanics in Energy Resource Recovery and Development,* U.S. National Committee on Rock Mechanics, National Research Council NRC/AMPS/RM-78-1, Washington, p. 4.

7.15 L. Muller, Closing Address, *Proceedings of the First Congress of the International Society of Rock Mechanics,* **3**:91 (1967).

Suggestions for Further Reading

There does not yet appear to be any volume devoted to the geological aspects of rock mechanics, but some of the recently published volumes on the general subject of rock mechanics have introductory geological sections. The most useful source of relevant information is in the current literature now serving this important field, such as is to be found in the journals listed in Appendix D.

Rock Slope Engineering by E. Hoek and J. W. Bray, the third edition of which is a 1981 volume of 360 pages, is a publication of the Institution of Mining and Metallurgy (44 Portland Place, London, U.K.). It is one of several useful volumes on this subject, so important in mining engineering as well as in civil engineering. Another is *Pit Slope Manual,* a publication of the Canadian Department of Energy, Mines, and Resources (Ottawa) published in 1976. Yet a third is *Rock Slope Engineering Reference Manual,* a publication of the U.S. Department of Transportation, Washington, D.C.

On the more theoretical side are two notable publications of Chapman and Hall (E. and F. N. Spon of London): *Fundamentals of Rock Mechanics* by J. C. Jaeger and N. C. W. Cook (1976) and *Principles of Engineering Geology* by P. B. Attenwell and I. W. Farmer (1976).

A useful source of information, especially because of the excellent bibliography which it contains, is a Report by the U.S. National Committee on Rock Mechanics, published in 1981 by the National Academy of Sciences in Washington, D.C., under the title *Rock Mechanics Research Requirements for Resource Recovery, Construction, and Earthquake Hazard Reduction.*

The *Proceedings* of the regularly held U.S. and Canadian Rock Mechanics Conferences already contain a wealth of useful information. Published by a variety of agencies, they can readily be traced in the reference lists of university and similar libraries and through the reference service listed in Appendix D.

Although the subject of rock mechanics is still so relatively new, the ASTM *Annual Book of Standards* contains documents of value in Volume (Part) 19. ASTM also publishes a notable series of volumes containing papers known as STPs (Special Technical Papers) given at symposia which ASTM sponsors through its many active committees. STP 554 is of special relevance to this chapter; called *Field Testing and Instrumentation of Rock,* it is a volume of 188 pages published in 1974.

Structural Geology by M. P. Billings, now in its third edition (1972) published by Prentice-Hall, provides a most helpful introduction to this subject.

chapter 8

GROUNDWATER

8.1 Historical Note / 8.2
8.2 Characteristics of Groundwater / 8.3
8.3 Influence of the Nature of Rock / 8.6
8.4 Quality of Groundwater / 8.8
8.5 Influence of Geological Structure / 8.12
8.6 Springs / 8.15
8.7 Artesian Water / 8.16
8.8 Groundwater near the Sea / 8.19
8.9 Movement of Groundwater / 8.21
8.10 Groundwater Surveys / 8.22
8.11 Conclusion / 8.26
8.12 References / 8.26

Seventy-five percent of all American cities derive their public water supplies from groundwater. The water so obtained from beneath the surface of the earth represents about 20 percent of all the water used for all purposes in North America. Yet this large quantity is but a small fraction of the one million cubic miles of groundwater that, it is estimated, exists within half a mile of the ground surface, with probably an equal amount existing at greater depths. In the old world, although it is popularly assumed that the metropolis of Greater London obtains its water from the rivers Thames and Lea, yet it still gets one-sixth of its total supply from deep wells, continuing a practice that goes back many centuries. And in India at least 20 million acres (an area comparable with the total area of irrigated acres in North America) are irrigated with groundwater obtained from wells.[8.1]

These few figures demonstrate clearly how important a place groundwater occupies in the life of the world today. This importance will be appreciated even more when it is considered that the quantity of water within that part of the earth's crust relatively close to the surface is believed to be equal to one-third of the total volume of water in the sea. It is further estimated that water can exist to a depth of 10 km (6 mi) below surface level. This vast reservoir of water is therefore of great importance to those who have to work in the earth's crust, whatever their work may be. For the miner, groundwater may be a matter of life or death, and it may be all that swings a venture from success to disaster. Water below ground surface is also of vital importance to the civil engineer, not only as a source of water supply but also as the controlling factor in all drainage operations. Groundwater is a hazard to be countered in tunnel driving and other underground operations; it almost always adds to the complexity of foundation work—work in which it may occasionally be of unique significance.

8-2 GEOLOGICAL BACKGROUND

Despite this importance, there is probably no natural feature still so neglected in the practice of civil engineering as is groundwater. At one time this could also have been said about soils, but the phenomenal advance of soil mechanics has corrected that situation. Water beneath the surface of the ground, however, is all too often regarded merely as a nuisance in the prosecution of construction projects, and it is given due regard only when the troubles to which it can lead become serious. Important, therefore, as is the study of soils for the civil engineer, an appreciation of soils will not be fully effective unless there is an equal appreciation of the significance of groundwater and a basic understanding of its characteristics.

This chapter presents merely an introduction to the study of the water to be found in the earth's crust; the full study of the subject is a well-established scientific discipline of wide extent. To many, groundwater hydrology is of absorbing interest. The subject should at least have the recognition of every civil engineer and more than the passing acquaintance of all who are to be concerned with the execution of engineering works in the field. In particular, all who are responsible for carrying out subsurface exploration prior to the design and construction of any civil engineering project must have a lively appreciation of groundwater hydrology. Possibly the everyday familiarity with water has led to the widespread neglect of its significance when it is seen at the bottom of a hole in the ground. Examples shortly to be cited should remove any such casual response for all time from the mind of the interested reader.

8.1 HISTORICAL NOTE

The existence of groundwater has probably been realized from the dawn of history. The coyote and possibly other animals will dig down to water if it is close enough to the surface to be reached by the animal's scratchings. One of humanity's earliest documents, the twenty-sixth chapter of the Book of Genesis, discloses that biblical man had a thorough familiarity with groundwater conditions; other references can be traced throughout the Bible. It is well known that the Romans were familiar with the use of wells; in England they had one well 57 m (188 ft) deep, and they supplemented their well supplies by means of adits driven in the chalk, a method which some consider a modern development.[8.2]

Throughout the Middle Ages, groundwater continued to be widely used, although its origins were not understood. Almost until the end of the seventeenth century, people generally conceded that springwater, as observed on hillsides, could not possibly be derived from rain. Many explanations were advanced to explain its origin; one of the most interesting was that, owing to the curvature of the earth, the water in the middle of the ocean was actually at a higher altitude than the springs, and thus furnished the necessary head. The idea that springs had their origin in the sea was often based on passages from the Bible, such as Ecclesiastes 1:7—"All rivers run into the sea, yet the sea is not full; unto the place from whence the rivers come thither they return again."

Not until the sixteenth century were these views questioned. Bernard Palissy (1510–1590) played an important role in developing the theory of the infiltration of groundwater. Pierre Perrault (1608–1680), one of the first to put hydrology on a quantitative basis, related the total rainfall and runoff for the basin of the river Seine in France. This work was roughly done, but it may be regarded as the starting point of modern hydrology, especially as it was just about this time that Edmund Halley (1656–1742) conducted the first known quantitative experiments on evaporation and thus proved definitely the origin of rainfall. Thereafter, hydrological work developed steadily into the scientific study of today, although it was hampered to some extent by strange ideas that persisted with regard to water witching, or divining.

A scientific approach to groundwater problems was thus established when modern engineering began in the early years of the last century. This approach was adopted and developed by several early civil engineers, at least two of whom made notable contributions to the subject. One of them was William Smith; the second was Robert Stephenson, whose name is well known in connection with the early development of British and other railways. William Smith was one of the first to relate the study of groundwater to the study of geology; going further than this, however, he applied the results of such combined studies to the solution of engineering problems. The works that he carried out show clearly that he had a very vivid conception of the main principles of groundwater location and movement. Two examples which may be mentioned are the water-supply system that he developed for Scarborough and the drainage of ground at Combegrove, near Bath, in the south of England.[8.3]

8.2 CHARACTERISTICS OF GROUNDWATER

That part of the rainfall which is absorbed into the ground is important in many ways. In some areas this absorption accounts for almost the entire rainfall, as in the drier sections of Australia where there are many rivers that disappear and others that flow only for certain limited portions of the year. Even more striking is the elevated area of two million acres to the north of Casterton and Coleraine in the state of Victoria. This area acts as the "intake" for the Great Murray River Artesian Basin; it has an average rainfall of 62.5 cm (25 in) and no runoff.[8.4] This is an extreme case. If an average case is considered, that part of the rainfall which does seep into the ground may be subdivided as follows:

1. That absorbed directly by plants
2. That drawn by capillary action to the surface and there evaporated
3. That which unites with molecules of mineral substances and so becomes chemically fixed
4. That which flows directly into the sea (in coastal districts) through springs and underground channels
5. That which escapes at the surface of the earth through springs or by feeding rivers
6. That which is retained in the earth

The first two divisions are important in agriculture, the last two in engineering, both civil and mining; the fourth is sometimes important in engineering work.

Civil engineering considerations of groundwater refer almost wholly to that part of the water which is retained in the ground; Fig. 8.1 is here included in order to make clear the general relationship of all subsurface waters. The term *vadose water* has been used in more than one way, but the diagram indicates its correct use; it describes that water which is still in the zone of aeration, i.e., in the part of the crust which is not saturated with water. General vadose water affects the drainage of shallow excavations and is naturally related to surface drainage and irrigation. Below the zone of saturation, what is termed *internal water* is indicated; it is located in that deep part of the earth's crust (estimated by some as a depth of 10 km or about 6 mi) where rock pressure is so great that no interstices occur in which free water can exist. This water and that associated with rock magma are primarily of geological interest. They are mentioned since, in connection with certain artesian basins and the drainage of some mines, water from plutonic sources, i.e., water that has not previously existed as atmospheric or surface water, is undoubtedly encountered. It is known as *juvenile*, or *plutonic, water* and may be of two kinds: (1) that which has been imprisoned in the earth's interior since its formation or (2) that which has been created by chemical combination of primitive hydrogen with oxygen of external origin.

One of the most striking examples of plutonic water was met during the driving of the Simplon Tunnel. Water was encountered that was so hot that workmen were scalded to

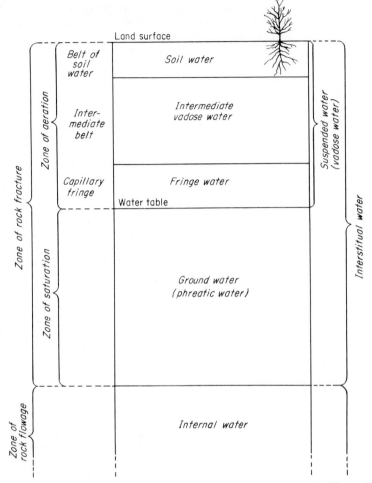

FIG. 8.1 Diagram showing the main subdivisions of groundwater (after Meinzer).

death. It was found to contain no chlorine, whereas all surface waters contain at least a small percentage of this element.[8.5] During the construction of the foundations for the Harlem Hospital on the upper east side of New York City, a daily flow of several million gallons of groundwater was encountered. Its temperature was a steady 20°C. It was effectively sealed off, but since there was no obvious source of the heat it contained, it may have been of plutonic origin.[8.6]

There are many notable hot springs, such as those near Kivu in central Africa, at Karlovy Vary in Czechoslovakia, and at Bath in England, which are thought to be of internal origin. Similarly, there are the *fumaroles suffaroli* of northern Italy, the steam of which must have a similar origin. Some students of the matter have suggested that probably the sea is fed by plutonic water emerging in the seabed. At Bendigo, Australia, there are mines that have a zone of workings in which meteoric (surface) water occurs; below this zone is a perfectly dry one; below that, at a depth of 1,370 m (4,500 ft) from the ground, springs appear.[8.7] Controversy regarding plutonic water has centered chiefly on the Great Australian Artesian Basin. In this basin, there are wells from which water is obtained so hot that it can be used for making tea or cooking. The origin of this water is an interesting topic and is mentioned here in order to emphasize the fact that in the construction of very long tunnels this unusual factor of considerable importance may be encountered.

Connate, or *fossil, water* is water trapped in sediments which have have been raised and distorted at some time in the past. Such trapped water may also be encountered in subsurface engineering work. As can be seen in Fig. 8.1, the zone of saturation is topped by a water table. The concept of a water table is a most useful one, apart from the fact that the use of the word *table* suggests a level surface, and a water table is rarely if ever level. One can readily imagine that the distance to a water table varies; it may be great or small, depending on local geological conditions.

If the earth's crust were of uniform composition, all that is essential for a summary of groundwater characteristics would have now been said; but underground conditions are far from uniform, and so the disposition of groundwater is one of the most complicated problems in geology. Essentially, the disposition of water below the saturation level depends on two main factors: (1) the texture of the local rocks, their composition and relative porosity; and (2) their structural arrangement and relationship to neighboring rocks. The variations possible in both these factors will make clear the complexity of the distribution of groundwater.

Although the conditions underground are variable, the water retained there is still governed by the natural laws governing surface water. Its movement is in accordance with the law of gravity, apart only from the minor motion due to capillary action. It may be well to emphasize the fact that groundwater does move. The vertical oscillations of the upper surface of saturation, or the water table, will be readily appreciated since so many direct results of this movement are observed, such as the alteration of well-water levels and the intermittent flow of springs. That groundwater is often moving in a horizontal direction corresponding to the slope of the water table is perhaps not quite so clear. A vivid explanation of this underground travel has been given by Dr. Herbert Lapworth.

> It was well recognized that underground water could travel for vast distances. For instance, it was known that in the artesian basins of Australia underground water traveled for hundreds of miles; and there were cases in the United States where it traveled 200 miles, with pressures which ran up to 500 feet head at the surface and something like 2,000 feet head at the bottom. In Algerian Sahara it was known, from borings, that water came a distance of 300 miles from the Atlas Mountains, passing along definite channels, some of which had actually been traced for a length of 70 miles.[8.8]

The cases quoted by Dr. Lapworth are admittedly extreme, but they indicate clearly how extensive underground travel can be. A more utilitarian example is the water supply of the city of Leipzig, Germany, obtained from an underground source that is almost a "stream," as it is 3.2 km (2 mi) wide and 12 m (40 ft) deep. Another example is that of a 550-m (1,798-ft) well at Grenelle, Paris, which draws water from Champagne, 160 km (100 mi) away.

The usual means of directly gauging the position of groundwater, relative to the ground surface, is to measure the water level in wells or boreholes; this measurement is indicative of the local water table. Care must always be exercised after the sinking of a borehole to allow sufficient time for equilibrium to be restored in the local groundwater, so that the observed level will not be a transitory value. In most areas, groundwater levels vary throughout the year in response to local rainfall and water use. This annual variation is of great importance in excavation work, both surface and subsurface, and also in foundation design, road performance, and frost action in soils.

In deep boreholes, other factors may influence water levels. These factors include the effects of tides, atmospheric pressure, winds, and earthquakes and, where the groundwater is held in a compressible aquifer, even such transient loads as passing trains.[8.9] The effect of tides upon groundwater levels near the sea is regularly observed as a rhythmic variation in many wells. Other effects are of relatively minor extent, although they can be appreciable in some areas, such as southern Florida. In a well at Miami, for example, a variation in water level of 1.4 m (4.5 ft) was once observed; it was caused by an earthquake in the Dominican Republic 1,200 km (750 mi) away.[8.10] Even though variations in water level such as these will not normally be encountered in civil engineering work, they are worthy of this brief record, as they indicate clearly the dynamic character of groundwater and its conformity with the laws of physics.

8.3 INFLUENCE OF THE NATURE OF ROCK

The variation between the distribution of water in ground consisting in one case of uniform coarse sand and in another of solid igneous rock is an obvious illustration of the effect of the composition and texture of rocks on groundwater distribution. Between these extreme cases occurs a wide range of conditions of varying complexity. The influence that the nature of rocks can have on groundwater will now be considered, but first, some definitions are necessary. *Perviousness* and *permeability* are words commonly used in this connection; they may be defined as the capacity of a rock to allow water to pass through it. A pervious rock has communicating interstices of capillary or supracapillary size. Its degree of permeability cannot be correlated with any standard scale but is usually expressed as a volume of flow per unit of cross-sectional area per unit of time (such as liters per square meter per day). *Porosity,* on the other hand, is a measure of the interstices contained in any particular volume of the rock; it is generally expressed as a percentage and indicates the aggregate volume of interstices to total volume.

Strange as it may appear at first sight, a high degree of porosity is no assurance of perviousness. Clay, for example, has a high porosity; examples have been found of newly deposited Mississippi clay with a porosity of between 80 and 90 percent. When it is saturated, however, it becomes impervious; the water it contains is held firmly by molecular attraction. Despite the fact that porosity and permeability are not always synonymous, the convenience of the property of porosity as an index to the water-bearing capacity of solid rocks and as one that can easily be investigated in the laboratory has resulted in much attention being devoted to it. Table 8.1 presents some typical values for well-known types of rock.

TABLE 8.1 The Porosity of Rocks

Type of rock	Maximum porosity* (percent)
Soil and loam	Up to 60
Chalk	Up to 50
Sand and gravel	25 to 30
Sandstone	10 to 15
Oölitic limestone	10
Limestone and marble	5
Slate and shale	4
Granite	1.5
Crystalline rocks generally	Up to 0.5

*These figures are presented as illustrative only of the general trend of porosity in relation to various rock types. Tests on actual samples should be performed if a value for porosity has to be used.

It must be emphasized that these figures are approximate only. For example, the porosity of the chalk found in southern England varies from 26 to 43 percent; that of the chalk found in Yorkshire is about 18 percent; and that of the chalk of Antrim, Ireland, is sometimes less than 10 percent. Typical of possible variations in porosity are the following figures: seven sample cores of Bunter sandstone (usually permeable and without joints and fissures) taken from depths varying from 90 to 225 m (300 to 750 ft) gave porosity values from 13.2 to 30.2 percent and proved to have a permeability ranging from 0.02 to 7.3 l/m^2 (0.06 to 20.9 gal/ft^2)* per 24 hours.[8.11] These figures serve to demonstrate the wide variation possible in a fairly regular type of rock. The variation is more strongly emphasized by samples taken from a 12-m (40-ft) bed of this same rock at approximately 30-m (100-ft) centers; the porosity even under these circumstances varied from 3.58 to 20.3 percent. Despite this appreciable porosity, the permeability in the direction of bedding, i.e., perpendicular to the cores, was very small indeed, and checks showed that leakage through this rock was practically all taking place along joints and fissures. That this should occur in such a relatively good "water-bearing rock" as sandstone is indeed surprising; the fact that the same feature has been demonstrated to be the explanation of the water-bearing properties of some chalk deposits and even some blocky and fissured clays will thus more easily be appreciated.

It is sometimes observed that, in addition to flowing along joints and fissures, water traverses chalk along the bands of flints and rubble that frequently distinguish chalk strata. Another feature to be noted in connection with both sandstones and limestones is the significance of the material that acts as a natural cement joining the individual grains to form the solid rock. An unusual but striking example of this is given by Monk's Park and Ketton stones. Both are British oölitic limestones in which the oölitic grains are microporous and similar in size and structure. In the Monk's Park stone, however, the intergranular spaces are filled with a nonporous matrix of crystalline calcite, whereas in the Ketton stone, they form interconnected pores. The Ketton stone is therefore pervious and yields water freely, whereas the Monk's Park stone is impervious. It is sometimes thought that porosity decreases with increase of depth as a result of the pressure of the superincumbent rock. It is suggested that the Chalk of England, usually regarded as the most porous British rock, gives a very small yield because of the pressure of overlying rock on this relatively soft material; a good example is found under Crystal Palace Hill in South London.

Confirmation of the importance of fissures and joints in connection with the movement and storage of groundwater is available from many sources. In all but pervious rocks such as sandstone and some limestones, fissures are usually the leading factor in determining the water-bearing capacity of a rock. The distribution of joints is a matter of some complexity. It seems clear that beyond a certain depth the distribution and the size of joints diminish appreciably. The critical depth may be about 100 m (328 ft). Below this, a regular diminution of supply from joints may be expected, and therefore an economic pumping limit will soon be reached. In any case, pumping is not normally economic below depths of about 250 m (820 ft). Although water may be normally available below these levels, it cannot usually be regarded as "commercially" available unless under artesian pressure, a special feature to be considered later.

Distribution rather than the mere location of groundwater is therefore the prime question in underground water surveys; for this reason, some general notes on the water-bearing properties of the more common rock groups will next be presented.

Sands and Gravels Since these materials are both porous and pervious, they may be classed as almost ideal water-bearing strata. Their wide use as artificial filtering and

*Gallons are given throughout in U.S liquid standards.

Clays and Shales As a general rule, these will be useless as sources of groundwater. Although clays are often wet, the water present is not readily available. Hard shales may yield water at joints.

Sandstones Sandstones are variable in texture and composition; some may be almost impermeable and others may be so pervious that water actually squirts out from them when under pressure, as from some of the Bunter sandstone of England. Pervious sandstones form an admirable source of groundwater, since in addition to giving a high yield, they constitute an effective filtering medium. For example, water from deep wells in New Red sandstone is almost invariably clear, sparkling, and palatable.

Limestones This extensive group of rocks is second only to sandstones as a source of groundwater. Its importance is well indicated by the existence of underground solution channels, associated caverns, and other familiar features, as well as by the wide dependence on chalk strata as underground reservoirs. The water-bearing characteristics of chalk have already been noted. It may be further observed that since water in contact with limestones dissolves a small quantity of the rock, the use of limestones as sources of water tends always to increase the yield obtained from them. It is calculated that with every million gallons of water pumped out of the chalk of the south of England, about 1,500 kg (3,300 lb) of chalk is also removed; thus pore spaces and any existing underground channels are correspondingly enlarged.

Crystalline Rocks In general, these rocks are not classed as water-bearing, although few are absolutely dry when encountered in excavation. When decomposed, as the result of weathering or some other cause, and also when fractured and fissured, they may yield appreciable quantities of good potable water. Fresh crystalline rocks in massive formation will not generally yield any useful quantity. Granites, when decomposed, may have a relatively fair yield; quartzites, slates, and marbles may yield a useful supply because of jointing; gneisses and schists, unless they have decomposed badly, cannot be relied upon for any appreciable quantity. In Cornwall, England, several important towns obtain their water from wells in granite; it is probable that the water collects in the blanket of disintegrated rock at the surface. In Maine, there are many wells in other igneous rocks, but their yield is usually small unless jointing is very marked.

8.4 QUALITY OF GROUNDWATER

The quality of groundwater is a matter of vital importance, whether the water is to be used for industrial or for domestic purposes. In general, groundwater is free from bacteria, since the passage of the water through the ground strata constitutes a natural filtering process. This does not remove the vital necessity of the bacteriological examination of all groundwater that is to be used for domestic purposes, especially when the arrangement and nature of the strata would permit contamination of the water stored in them from surface sources.

Groundwater is usually found at a fairly constant temperature, close to the average annual air temperature for the locality. Since the temperature is usually moderate, the water is cool in summer and warm in winter—one feature that commends it as a source of domestic water. The even temperature of groundwater has taken on an added significance in connection with its use as cooling water. The development of air-conditioning systems in many large buildings in cities calls for steady supplies of cooling water, supplies not large in themselves but which become appreciable when many buildings are dependent on the same source of supply. Consequently, in New York, for example, several large theaters were permitted to use well water for cooling air-conditioning systems, provided that (1) a closed circuit was used for the water to obviate any possibility of contamination and (2) all water used was returned to the ground by means of wells. The rise in temperature of the cycled water averaged about 6°C. In some American cities, the extensive use of groundwater for this purpose resulted in serious problems, but the practice is now generally discontinued.[8.12]

There was at least one location at which groundwater was used for domestic supply in which the reverse effect pertained. Riverside, California, obtained a supply of 7,600 lpm (2,000 gpm) from an artesian well 294 m (965 ft) deep-installed in 1931. The temperature of the water was 43°C and in consequence a special and elaborate cooling system had to be installed by the water authority in 1936. Two adjacent wells, each 330 m (1,100 ft) deep, provided cool water (Fig. 8.2). Clearly, some peculiar underground structural arrangement

FIG. 8.2 Cooling plant for groundwater supply to the city of Riverside, California; a view taken before this unique plant was destroyed in March 1938 by severe floods on the Santa Ana River, which changed its course and now flows over the site of the plant.

must be held responsible for this unusual condition.[8.13] Unfortunately, this unique plant was destroyed in March 1938 by severe floods on the Santa Ana River, which changed its course and now flows over the site of the plant.

Groundwater will almost certainly contain dissolved solids and gases. Of the gaseous impurities, methane may occasionally be encountered in water that has not traveled far from surface deposits or that has been in contact with deep-buried strata of organic origin. The gas is usually caused by the decomposition of organic matter in the absence of free oxygen. It is dangerous, and if it is present in such quantities that it is liberated from the water on reaching atmospheric pressure, it can be a source of trouble. Hydrogen bisulfide is occasionally met and is easily detectable by odor if present in any appreciable quantity; its origin may vary, but a possible source is the interaction of organic acids from surface deposits with underground sulfates. The gas is easily removed by aeration; its presence is often the chief feature of medicinal springwaters. Carbon dioxide is the most important gaseous impurity of groundwater; its origin is generally the atmosphere. It gives water a sparkle which is not unpleasant, and it is not therefore objectionable. But in water, the gas makes a weak solution of hydrocarbonic acid, which acts as a solvent for several different rock constituents, and for this reason the presence of the gas is often significant.

Free carbon dioxide gas is occasionally encountered in civil engineering work in sufficient quantities to be troublesome. On the aqueduct that conveys water from Owens Valley to the Los Angeles area in California, severe corrosion was discovered in the steel pipelines; tubercles formed very quickly, and constant cleaning and painting were necessary for adequate maintenance. Careful tests were made of the water at the entrance to the aqueduct, but no peculiarities were found. Further checks were made, and it was eventually found that between the two ends of a 2,100-m (7,000-ft) tunnel the carbon dioxide content increased from a negligible amount to 4.2 ppm. The tunnel was therefore dewatered and examined; the gas could actually be heard hissing as it escaped through cracks in the concrete lining. The tunnel penetrates what is locally known as Soda Hill, a granitic rock formation containing many badly fractured fault zones; hot limestone beds are thought to exist nearby. The consequent acidity of the water was clearly the cause of the corrosion troubles. The remedy adopted was to lead the gas through special chases into a pipe beneath the tunnel invert which, in turn, was connected with a new adit, 600 m (2,000 ft) long, through which the accumulated gas could escape.[8.14]

Pure water will dissolve only 20 ppm of calcium carbonate and 28 ppm of magnesium carbonate, but water containing carbon dioxide will dissolve many hundreds of parts per million of the solids. These and similar solid impurities in groundwater may give rise to the main purification problems of the waterworks engineer. It follows that water obtained from limestone strata is always suspect, even without test. The dissolved carbonates give the water what is termed *temporary hardness,* since the dissolved solids can be removed by simple chemical processes. If the dissolved solids are the corresponding sulfates, the resultant hardness of the water is termed *permanent,* since it cannot be removed by simple processes used for eliminating the carbonates. Chemical analyses are necessary to determine the degree of hardness of the water; if this exceeds about 200 ppm of calcium carbonate the water requires *softening.* Clark's lime process, a chemical method, is often used to remove temporary hardness. For permanent or mineral hardness, various processes exist, but zeolites are generally used in all of them. They have the property of exchanging nonhardening constituents for the hardening constituents present in water; when exhausted, zeolites can be regenerated by treatment with a brine solution. Artificial zeolites formed of complex silicates are now available.

Iron salts are another impurity in groundwater that can be of serious consequence in industrial work, e.g., dyeing, and they are a nuisance in connection with domestic supplies. Soft water is likely to contain less iron in solution than hard water; the maximum iron content for hard water is generally about 1 ppm, but it depends on the other salts present. Manganese salts may also be present; they are often associated with iron salts, though always in small quantities. They affect the taste of water and give black stains, whereas iron salts produce red or brown stains. Deferrization of such water is a simple matter in principle but difficult in actual practice. It is essentially a chemical process, aimed at the production of ferrous hydroxide as a coagulant which can be precipitated or removed by filtration.

One of the most unusual construction accidents caused by groundwater ever to be recorded was related to iron deposits. During the construction of a tunnel in Milwaukee, five workmen were killed in the access shaft to the tunnel; it was found that their deaths were due to carbon-dioxide poisoning. Pumping operations had reduced the groundwater level so that the excavated rock was exposed to air for the first time. This caused a change in its ferric content to the ferrous state, and this reaction liberated carbon dioxide. When compressed air was applied to the tunnel, it forced some of the liberated gas into the shaft; quite unsuspected, the gas soon proved fatal to the five men who were working in the shaft.[8.15]

Trace elements, a term now familiar to the informed reader, describes those minute quantities of the rare elements which are essential for healthful living; the absence of the necessary minute quantity of fluorine in water, for example, causes dental decay. Correspondingly, there are other elements, minute traces of which can be fatal. They are not unrelated to groundwater, as the records can demonstrate. There were, for example, several cases of arsenic poisoning some years ago north of Kingston, Ontario; poisoning of this kind, unless it is due to obvious sources of arsenic, is a most unusual medical occurrence. The presence of arsenic was finally detected in the farm well water used by the victims. This, in turn, was traced to minute quantities of ferrous arsenate existing naturally in the limestone stratum through which the well had been excavated.[8.16]

In Los Angeles, just before the Second World War, it was found that citrus leaves were turned yellow by the presence of minute quantities of boron in the water used for irrigation. This water came from the Owens Valley Aqueduct supplying Los Angeles and had been used for 15 years previously with no ill effects. Careful investigation showed that the quantity of boron detected was not injurious to human beings but was dangerous to some of the crops grown in the Los Angeles area. Years were spent in tracking down the source of the sudden occurrence of boron, and the investigation is an example of "engineering detective work" at its best. It was finally determined that 75 percent of the troublesome boron content was being contributed by the Hot Lake area in Long Valley. The boron concentration was found to increase at points where hot gases and hot water bubbled up through lakes and springs which fed into the aqueduct. By taking the aqueduct out of service for its usual maintenance period at times when boron concentrations were highest, much of the boron could be deposited in Owens Lake (a temporary reservoir). Later, as a further expedient, water containing boron was spread on land formerly irrigated but since abandoned, then allowed to seep to discharge points at lower elevations, whence it could be pumped back into the aqueduct; this natural filtration was sufficient to remove almost all traces of the troublesome element.[8.17]

The presence of high concentrations of calcium and magnesium sulfates in groundwater can cause very serious trouble with concrete work in contact with the ground. Occurring chiefly in heavy clay soils, high-sulfate groundwater is common in western Canada. It was recognized as early as 1908, when problems were encountered with deterioration of concrete in Winnipeg. By 1918 these problems had become so serious that the Engineering Institute of Canada was requested to appoint a committee to look into the matter. In 1921, Dr. C. J. Mackenzie, then dean of engineering at the University of Saskatchewan, was appointed chairman of such a committee. The services of Dr. T. Thorvaldson were obtained, a major research was undertaken, and the satisfactory results are recorded in Chap. 16 where concrete is considered as a material. Groundwaters in western Canada still have high sulfate contents. But once the unusual character of the relevant groundwater has been determined, the problem can be solved by use of sulfate-resisting cement.[8.18]

The same problem is so widespread in Great Britain that in the years immediately preceding the Second World War the Institution of Civil Engineers began an investigation of the effect of sulfates in soil upon cement products. A summary of the initial findings of this research project has been published by the Institution. It was found that the sulfates of calcium, magnesium, and the alkali metals occur widely in the Mesozoic and Tertiary clays of Great Britain, including such widely known formations as the Keuper marl, the Lias, and the Oxford, Kimmeridge, Weald, Gault, and London clays. Superficial deposits, generally glacial, were substantially free of sulfates; thus, it is only where the geological structure is such that the older formations come to the surface that trouble with sulfates is encountered on any appreciable scale.[8.19]

Sodium chloride remains for mention as the final important mineral impurity liable to be encountered in underground water. It may have its origin in the sea, as explained later

8-12 GEOLOGICAL BACKGROUND

in this chapter, but may also be of mineral origin. Rock-salt deposits are an obvious source of supply; some are "mined" by means of brine solutions. Although this system of mining is of great interest from the point of view of utilizing groundwater for mining purposes, it can have unusually serious consequences if the workings are so close to the surface that settlement can result. This will be a matter for discussion in Chap. 37.

The quality of groundwater has so far been treated by considering the nature of possible impurities; it may be useful to add a brief summary of what may be expected from typical water-bearing strata. The only type of clay from which water supplies are regularly obtained is boulder clay, which may sometimes have a pervious structure. Water thus obtained may be quite hard; traces of hydrogen sulfide are often found. Sands usually give a fairly pure water, but occasionally they yield water with a small ferruginous content. The same may be said of water from sandstones; if the cementing material is calcium sulfate or calcite, the water can be quite hard. Water from limestone, chalk, or similar rocks will almost certainly be hard, but occasionally other impurities will also be present. Water from some of the East Anglian chalk in England, for example, contains so much ferrous carbonate that it must be treated.

To deal with the possible impurities in groundwater in this brief space necessarily means that only a sketch of the subject can be presented. This is done with one main purpose in view—to demonstrate the relation of such impurities to the geological conditions of the ground from which the water is being obtained. This correlation is useful for two reasons. First, preliminary study of the local geology will at least suggest the possible impurities to be expected in any prospective groundwater supply; and, second, accurate knowledge of the relevant geology will demonstrate the course of any known impurities and indicate the probability of any increase or decrease of the amount of contamination. All civil engineers should be able to gain at least a general idea of the geology of the source area of the water supply of a town they may be visiting merely by noting the relative hardness of the local water when they first wash their hands. If the water is hard, as from limestone, lathering with soap will be slow.

8.5 INFLUENCE OF GEOLOGICAL STRUCTURE

Underground conditions affecting groundwater differ from the ideal case not only because of the wide variation of materials in contact with the water but also because of the way in which the various rock strata are arranged, in general and in relation to one another. The first variation from the ideal that may be considered is the alternation of pervious and impervious beds. If the impervious stratum constitutes the surface layer, no water will normally penetrate to the potential underground reservoir provided by the pervious bed. Possibly a more usual case is that in which the pervious bed is at the surface; rain will soak through until it reaches the main body of groundwater retained above the impervious stratum from which supplies can be drawn by means of boreholes or wells. The arrangement is capable of duplication, and a "sandwich" structure can be imagined. Such structures actually occur; the Middle Lias formation in Northampton, England, only 28 m (93 ft) thick, contains no less than five water-bearing horizons separated by impervious clays, any four of which may be present at one location. The condition illustrated in sketch A of Fig. 8.3 will be interfered with only if the impervious stratum is pierced. This would seem hard to imagine, but cases have occurred in which wells have been drilled too deep. At Kessingland, near Lowestoft in England, the bottom of an old well penetrating 15 m (50 ft) of gravel was pierced, and the water tapped by the well ran down to the lower stratum. A similar instance happened in the drilling of a 21-m (70-ft) well at Nipigon in Ontario.

A further variation will be obtained when the strata are inclined instead of level, as indicated in sketch B. In this case, the inclined impervious stratum will constitute a bar-

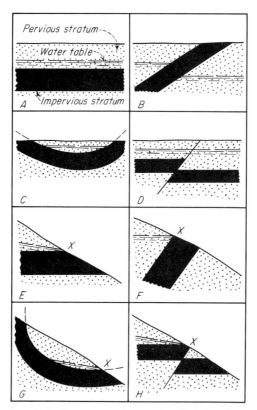

FIG. 8.3 Diagrams illustrating the effect of geological structures on groundwater distribution.

rier between the two pervious beds, so that the elevations of the respective water tables need not and probably will not be the same. A similar effect is noted when alternating layers constitute part of a fold; many arrangements are naturally possible, but only one general case is illustrated. Sketch C shows how water will collect in such a distorted stratum. It may be noted that the geological structure underlying the city of London is of this type. Finally, sketch D demonstrates the effect that a fault may have on the distribution of groundwater in alternating strata. The variations possible in this simple case are dependent on the relative thickness of the strata, the nature of the fault, and the throw of the fault. In the case illustrated, the barrier provided by the impervious stratum remains, but a slight increase in the throw of the fault would leave a gap in the barrier, with consequent alteration of the groundwater conditions. The shatter zone generally found in rock bordering a fault plane may also alter the water-bearing character of a rock by providing water passage along the planes of fracture.

The four simple cases already considered are illustrated in sketches E, F, G, and H, with the surface of the ground inclined; the altered groundwater conditions are indicated diagrammatically and need no elaboration. It will be seen that two pecularities are immediately introduced. At the points marked X, bodies of groundwater will come into contact with the atmosphere. This leads directly to the second feature, viz., that the surface of the groundwater will not be level but will be inclined at the hydraulic gradient necessary for flow to take place through the material of the stratum. This is a general condition, as

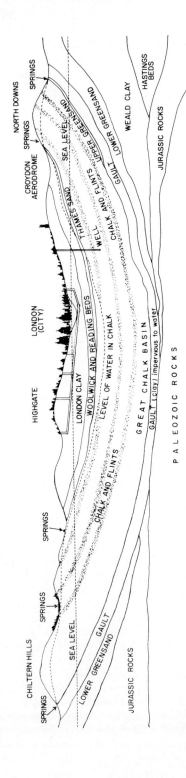

FIG. 8.4 Simplified geological section through the London Basin of southern England.

has already been explained. An important consequence is that, as a rule, the "water table follows the ground surface"—a broad statement, but useful as a guide.

Of particular significance is the variation in the yield of water obtained from the same rock at different positions along the range of a fold in which the rock has been distorted. If the folding has been anticlinal and is of relatively short radius, there will be a distinct tendency for fissures to develop. Similarly, if the folding has been synclinal, there will be a tendency for the rock to "tighten up" and have its usual perviousness reduced; this tendency is most marked in the softer rocks. This has been demonstrated in connection with the water supply of the municipality of Richmond, London, where the construction of a well and over 3,000 m (10,000 ft) of adits in the chalk failed to develop an appreciable supply because the location was on a syncline in the chalk formation (Fig. 8.4).

8.6 SPRINGS

Springs sometimes provide a useful and pure water supply, but often an engineer will have to investigate their capacity to ensure a continuous supply. Alternatively, an engineer may be called upon to study the source of springs in connection with possible water-lowering operations, such as pumping for water adjacent to springs already in use. Again, as springs often accompany instability of ground and especially of steep natural slopes, their course may have to be traced in drainage operations. Sketch E (Fig. 8.3) represents in simple fashion the geological arrangement necessary for the existence of steadily flowing springs; if the principle therein presented is applied to local conditions, one can readily see the nature of many springs. The flow may vary; but because of the relatively slow movement of groundwater, a spring will not dry up except in very dry spells. This slow movement when related to rainfall explains the lag between major movements of the water table and heavy rainfall, a lag that may amount to as much as six months in temperate climates and on which depends the real efficiency of groundwater supplies.

Although springs that flow intermittently are not numerous, they have an interest all their own (Fig. 8.5). Sketch F (Fig. 8.3) shows the basis of one type of intermittent spring.

FIG. 8.5 Springs coming out of limestone bedrock in the valley of the Saugeen River, Ontario.

If the surface level of the groundwater drops below X, as it might through depletion elsewhere during dry weather, the spring previously existing will disappear, returning only when the groundwater has been so replenished that the level again raises above X. The principle illustrated may again be applied to local geological conditions and thus explain natural phenomena such as the "breaking out of the bournes" in England after the early spring rains (*bourne* is a name applied in southern England to this type of intermittent spring). Bournes are a feature of the chalk uplands and also of the oölitic limestones of the Cotswold hills in England; they occur frequently in northern France. Names of villages in England betray the existence of such springs; Winterbourne Zelston and Waterbourne Whitechurch are two delightful examples from the county of Dorset. Many springs discharge into the sea because of local geological conditions; this could well happen in the situation given in sketch G (Fig. 8.3). In the Persian Gulf, swimmers dive to the sea bottom with leather bags which they fill with drinking water from spring discharges.

8.7 ARTESIAN WATER

Water is said to be *artesian* when the groundwater rises either up to or above ground level as soon as the water-bearing bed is pierced; local geological structure, in which the water is stored in the bed under hydrostatic pressure, accounts for this phenomenon. Its name is one of the few in connection with subsurface conditions that is not self-descriptive; it originated from that of the province of Artois, France. The term *subartesian* is used to denote wells in which the groundwater is under hydrostatic pressure but not to such an extent that it is forced to ground level; as soon as it is freed it comes to rest at some point below ground level but above the top level of the water-bearing bed. The usual cause of artesian pressure is the hydrostatic head of that part of the body of groundwater which is confined by impervious strata above the level at which it is tapped. This condition may arise in a synclinal structure of alternating pervious and impervious beds, as illustrated in Fig. 8.6; two alternative structural arrangements are illustrated in the same figure, and other similar conditions can be imagined.

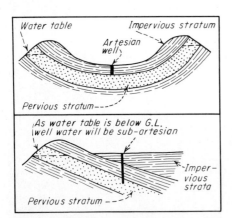

FIG. 8.6 Simplified geological sections illustrating artesian groundwater conditions.

Artesian water may quite possibly exist beneath impermeable strata which form a river bed, with results such as the rather unusual occurrence shown in Fig. 8.7. This photograph was taken after a test borehole at the site of the Shand Dam in Ontario had penetrated glacial till and entered the underlying Guelph dolomite beneath the waters of the Grand River.

There are many examples of artesian waters, but naturally the water encountered in Artois may be said to be the leading example. The groundwater in the well-known chalk deposits there does not find its natural level until the overlying Tertiary clay is pierced. Despite the general adoption of the name, the use of artesian water was revived rather than initiated at Artois, since it had been extensively used in much earlier periods, particularly in northern Africa. At Thebes, for example, shafts were sunk, and excavation was continued as bores 15 to 20 cm (6 to 8 in) in diameter to water-bearing sands, in many cases 120 m (400 ft) from the surface.

FIG. 8.7 Groundwater under artesian pressure flowing out of a drill-hole casing pipe in the bed of the Grand River, Ontario, at the site of the Shand Dam.

Artesian conditions may exist in the form of artesian slopes or artesian basins. In both, but especially in the former, artesian water may be present at depths to which it is uneconomical to drill. Local conditions and demands will determine the relative economy of using such supplies, and the inevitable variation of conditions will explain the great variation in the depths to which artesian wells have been sunk. In Berlin, Germany, and in St. Louis, Missouri, it has been necessary to drill to depths of 1,000 m (3,280 ft); in England, a well 476 m (1,585 ft) deep at Ottershaw appears to be about the maximum depth utilized in that country; in Australia, a well at Boronga, New South Wales, is 1,302 m (4,338 ft) deep, whereas along the Atlantic seaboard of the United States, artesian supplies at depths of a few hundred meters are relatively common. One famous well on this coast is the 30-cm (12-in) well at St. Augustine, Florida, which originally supplied 38 million liters (10 million gal) a day from a depth of 420 m (1,400 ft).

Although the areas in which artesian water can be tapped are often relatively small, there are, on the other hand, some extensive areas under which groundwater is stored under pressure. One of the best known of these is the Great Dakota Artesian Basin in North and South Dakota, which has an area of about 39,000 km^2 (15,000 mi^2). It is widely used as a source of water supply; the water-bearing bed is Dakota sandstone, and it has been the subject of detailed groundwater investigation. Particular attention has been paid to the possibility that some if not all of the artesian pressure may be due to the weight of the superincumbent rock strata. Pressure due to the weight of rock is still a subject of debate in geological circles; it is of interest to note that the subject was discussed by Thales about 650 B.C. and later by Pliny and that it was suggested that this pressure might be the agent responsible for elevating seawater to the level of springs.

Australia is a continent in which artesian water is of major importance; the Great Australian Artesian Basin is almost worthy of classification as one of the wonders of the world (Fig. 8.8). Having an area of more than 1.5 million km² (590,000 mi²), it is located in the states of South Australia, Queensland, and New South Wales. Over 3,000 artesian wells draw on the supply that it provides—a daily flow of over 2,250 million liters (600 million gal). In form, the basin approximates the ideal complete artesian basin; the eastern rim is tilted up to a greater altitude than the western. The main aquifer extends as a practically continuous body under the whole of the basin and consists, in general, of Jurassic sands occurring as very soft sandstones. Artesian conditions proper extend over the greater part of the basin; but in the western section, the head is sufficient to give rise only to subartesian conditions. The water obtained is of good quality, but its temperature is high, extremely so in the parts which reach a depth of 1,500 to 1,800 m (5,000 to 6,000 ft). Experiments in New South Wales have shown that the temperature varies almost directly with depth; records range from a temperature of 30°C at 210 m (700 ft) to 50°C at 900 m (3,000 ft). There are many notable records associated with this great basin, the drilling of an experimental well at Careunga, near Moree, New South Wales, to a depth of 1,200 m (4,014 ft) is worthy of mention. Within this depth, three separate artesian flows were

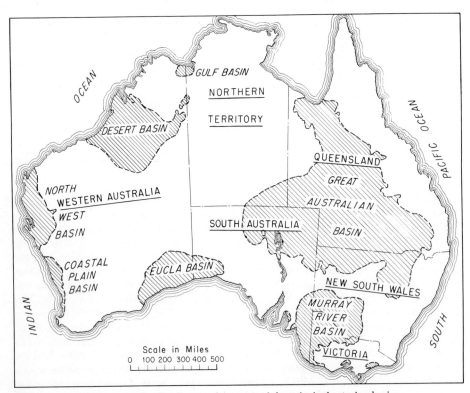

FIG. 8.8 Sketch map of Australia showing the extent of the principal artesian basins.

encountered at different pressures, with different compositions, and with different emissions of methane gas; records of the three flows were kept separately.

There are several other large and important artesian basins in Australia; the most notable one is the Murray River Basin, lying in the states of South Australia, New South Wales, and Victoria; it has an area of 278,000 km² (107,250 mi²). It is of special interest to engineers to note that the conservation of these great natural reservoirs of groundwater has been recognized in Australia as overriding ordinary state boundaries. In consequence, since 1912 a series of regular interstate conferences has been held, attended by geology and engineering delegates from most of the states concerned; records have been exchanged, experiences discussed, and general conclusions arrived at with regard to the many problems involved. The printed records of the conferences, together with the relevant state geological bulletins, constitute a most valuable part of groundwater literature.[8.20]

One might expect that civil engineers, with their basic training in hydraulics, would have no difficulty in appreciating the fundamentals of artesian pressure and that they would be ever mindful of its potential. Unfortunately, the records prove otherwise, for there are many cases in which the movement of structures has been attributed to artesian pressure beneath foundations, particularly when construction pumping has temporarily lowered the water table and thus temporarily concealed the final head of water that will be effective. The matter can be illustrated by brief reference to experience during the construction of the Conchas Dam in New Mexico. The main section of this dam is a concrete gravity structure, placed in separate monoliths which reach a height of 60 m (200 ft) and which are founded upon a porous sandstone where groundwater is under artesian pressure. This sandstone extends 12 m (40 ft) below the base of the dam and is underlain by 9 m (30 ft) of a dense shale, which overlies another stratum of porous sandstone where groundwater is also under artesian pressure. Pumping operations during construction lowered the effective pressure; high pressure grouting was carried out under the monoliths. When pumping was stopped, movement of some of the monoliths was noted—up to a maximum lift of 2.5 cm (1 in), despite their immense size. The matter was most carefully studied and it was determined by laboratory tests that the elastic properties of the sandstone alone could not account for the rise observed; this led to the conclusion that the movement was associated with the swelling characteristics of the shale stratum.[8.21]

8.8 GROUNDWATER NEAR THE SEA

The presence of sodium chloride in groundwater has already been mentioned during discussion of the impurities that develop through contact of groundwater with rocks having some soluble content. There is, however, another possible source of sodium chloride and allied salts in groundwater, encroaching seawater in coastal areas. Although originally a phenomenon of almost academic interest, the matter has now assumed most serious proportions in several parts of the world where increasing seawater encroachment is interfering with the purity of groundwater used for domestic supply. Basically, the matter is one of simple hydrostatics. Since the specific gravity of seawater is slightly greater than that of fresh water, a state of equilibrium will exist when the two are in contact in a porous medium, as illustrated in Fig. 8.9. This superimposing of a lens of fresh water over seawater is called the *Ghyben-Herzberg effect* after two of those who have investigated the matter in theory. The whole of the Florida peninsula and such islands as Oahu, on which Honolulu is situated, are most certainly completely underlain, at depth, by seawater. The state of equilibrium illustrated in the simple diagram is, however, a delicate one. It can very easily be disturbed, and once disturbed, it cannot readily be restored. Overpumping

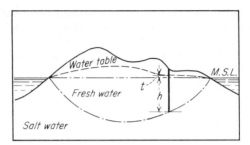

FIG. 8.9 Diagrammatic section illustrating the elements of the Ghyben-Herzberg effect when fresh water overlies salt water.

the fresh water will permit an advance of the boundary between fresh water and salt and will be one sure way of interfering with the natural state of balance.

This balance has been upset in many parts of the world. Wells along the shores of the Thames Estuary and in the Liverpool district of England, wells along much of the California coastline, wells tapping the remarkable groundwater reserves under Long Island, New York, all have been made brackish when salt water became mixed to varying degrees with fresh water because of excessive drawdown of the latter. The matter is one of almost worldwide importance, having been investigated intensively in England, Holland, Japan, and North America. Some of these investigations will be mentioned in Chap. 48, but the matter is directly related to civil engineering operations in coastal areas and is so clearly illustrative of the geological implications of groundwater movement that some further reference to it here seems warranted.

The presence of brackish water in wells is indicative of the existence of water-bearing, i.e., pervious, strata. If the wells are adjacent to the sea, it will normally follow that at some point the fresh water will be in contact with the infiltrated seawater. The zone of diffusion between the two is relatively small, and thus any appreciable change from the normal position of the groundwater (such as can be caused by excessive pumping) will result in a corresponding change in the position of the salt water. As the average salt content of seawater is about 35,000 ppm, and the normal maximum salt content for water for domestic use is about 3,000 ppm, it will be clear that it takes but a small contamination of fresh water by seawater to make the former unfit for human use. Rainfall onto the exposed land surface restores balance by replacing fresh water lost to evaporation and transpiration.

Herzberg's original investigation of the interrelation of seawater and fresh water was carried out in Germany about 1900. His results and those of other investigators have been fully confirmed by field studies in the coastal lands of Holland and in other areas. The findings apply generally to all geological conditions in coastal areas and so should be kept in mind not only with regard to pumping groundwater in coastal areas but also in connection with any civil engineering works which may interfere with natural groundwater conditions.

The case of Angaur Island, one of the Palau Islands of the Pacific, about 1,280 km (800 mi) southwest from Guam, is an illustration of the serious results that simple excavation can have under such circumstances. Angaur Island, roughly triangular in shape, has an area of only 8.3 km^2 (3.2 mi^2). Valuable phosphate deposits on the island have been mined steadily since 1908, although the commercially available deposits are now exhausted. In 1938, the Japanese introduced power-operated equipment in order to speed up the mining operations. With the availability of such improved methods, large excavations were soon completed. Some of these large "holes" went appreciably below sea level. Water from the fresh-water lens overlying the main body of seawater formed pools in the excavations. Because of the increased volume of water required to fill the "holes," as compared with that previously required to fill merely the voids in the coral, movement of the groundwater took place, flow of seawater was induced, and the previously fresh and usable groundwater supply on the island was contaminated. Remedial measures carried out by U.S. forces after the war included sealing off artificial lakes and reducing their size, using coral rock as rubble, with as much fine material included as possible.[8.22] Admittedly this is a small and rather isolated case. It is, however, a vivid reminder of the delicate equilibrium that exists wherever seawater and fresh water are in contact beneath the ground, a situation that should never be forgotten when civil engineering works, especially pumping operations, have to be carried out near the seacoast.

8.9 MOVEMENT OF GROUNDWATER

Some of the examples given have illustrated the fact that groundwater does move. Surprising though it may seem, this concept is not so widely appreciated as might be imagined. In general, the movement follows D'Arcy's law, amended to allow for the pervious medium through which flow takes place. Much work has been done in developing the necessary theory, and correspondingly, much experimental work in both laboratory and field has been carried out in checking upon theory. Well established as is the study of groundwater hydrology, it can never be forgotten that the theory assumes certain specified subsurface conditions and that the practical application of the theory demands accurate information about subsurface conditions—in other words, an appreciation of local geology. Despite the fact, therefore, that this volume is concerned with geology and not with hydrology, brief reference to the movement of groundwater is desirable in order that readers may have a well-rounded picture of this vital aspect of subsurface conditions.

Reference may usefully be made to the movement that will take place when pumping is started in an open well that has been successfully completed to below the water table. The water level in the well will naturally fall when pumping begins and will continue to do so until the hydraulic gradient in the material surrounding the well is such that flow into the well will occur at the rate equal to that at which water is being removed by pumping. It can be shown theoretically that, for uniform material, the curve that the hydraulic gradient will assume between the water level in the well, under equilibrium conditions, and the unchanged natural level of the water table will be a parabola. As this condition will obtain all around the well, a *cone of depression* will result; the apex of the inverted conical form of the groundwater surface will be the water level in the well. The intersection of the cone of depression with the natural water table will, theoretically, be a circle which will mark the limit of the range of interference of the pumping operation. Any other well located within this circle will be interfered with to some degree; its effective yield of water will be reduced. Such interference can be a critical matter in built-up areas where it is desired to have wells close together so that groundwater resources will be used to maximum extent. It is surprising to note how wide these ranges of influence may be in very pervious strata. In Liverpool, England, for example, the effect of pumping has been noticed as far as 3.2 km (2 mi) away from a pumped well; in this instance, the intervening ground was, in general, a pervious red sandstone. When drainage works are considered, in Chap. 18, even greater distances at which groundwater levels were affected by pumping will be recorded.

The basic concept of the cone of depression, illustrated in Fig. 8.10, is simple, although detailed groundwater hydrological studies can lead to complex mathematics. Simple as

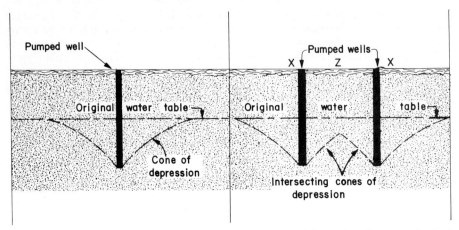

FIG. 8.10 Simplified section illustrating the form of the cone of depression when groundwater is pumped.

the concept is, it is surprising how frequently it is neglected, or forgotten, when problems with groundwater are encountered in civil engineering. The dynamic character of groundwater, therefore, is something that all civil engineers should keep in mind, since it can have such profound influence upon their construction operations.

8.10 GROUNDWATER SURVEYS

Not only for the purpose of discovering whether groundwater will provide a satisfactory supply of potable water at a new site, but also for discovering whether groundwater will be encountered in underground construction operations, determination of groundwater conditions becomes a vital part of preliminary site investigations. When water supply is involved, the quality of water must also be known with some certainty in advance of its actual use. Obtaining this information is the objective of groundwater surveys. These cannot be considered directly comparable to topographical surveys, since they are most complex and generally call for trained and expert hydrological advice. Local geological conditions are always involved, and so a working knowledge of local geology is a prerequisite for any groundwater investigation. In addition, familiarity with the hydrological characteristics of the rocks met with is essential before even general principles can be applied to the particular problem. Thus it is that the services of an expert are always desirable and often essential in a complete groundwater survey. The following note is only an outline of what is involved in this survey work; it is presented so that engineers engaged in groundwater problems may know how to regard them and may appreciate the advisability of expert assistance when the necessity arises.

The problem is not merely that of finding water, as is often and popularly imagined, for there are probably few parts of the earth's surface under which water cannot be found at some depth. The engineer's problem usually is to find water at such a depth, in such quantities, and of such quality that it can be utilized economically in the public service. Despite this economic limitation, finding water is an important part of groundwater investigations. It seems probable that plants and other vegetation were the first indicators of groundwater to be used regularly, and vegetation still constitutes a fairly reliable guide.

In the first century B.C., Vitruvius listed plants known to him to indicate the presence of groundwater near the surface of the ground and even endeavored to suggest when they were reliable guides.

The subject has received little concerted attention until comparatively recent years. Scattered references to water-indicating plants, shrubs, and trees occur in geological and botanical literature, but it has been left to the United States Geological Survey to publish a general review of existing knowledge. This survey was prepared by Dr. O. E. Meinzer, who suggested the term *phreatophytes* to denote those plants that habitually grow in arid regions only when they can send their roots down to the water table.[8.23] This publication shows that the vegetation encountered will be some indication of the depth to the water table, and it shows that there is at least some basis for the idea that plants indicate the quality of the groundwater. In other parts of the world, information with regard to plants that may serve as good guides can generally be obtained from observant local inhabitants.

The search for water can be likened to prospecting for minerals; water is, in a sense, a mineral, although of a special kind. The two branches of investigation have sometimes been closely related, as when Vogt in his water investigations discovered one of the world's richest potassium salt deposits at Mulhouse, Alsace. They have also been associated throughout a long period of time, since both have been the object of attention by "diviners," those gifted individuals who are said to be influenced by a divining rod when they carry it over a deposit of minerals or groundwater. Divining rods have been used not only for the alleged locating of water and minerals but also for attempting to find hidden treasure and even lost children, criminals, and corpses. Divining rods appear to have been used by the Scythians, the Medes, and the Persians; Marco Polo found them in use in the Orient; and there are people who believe that the rod of Moses was a divining rod. Originally, incantations were used in conjunction with the rods, and it is said that in Cornwall, England, the belief still exists that the rod is guided to the deposit by pixies. For about a century (1669–1760), religious authorities debated whether or not water divining was an invention of the devil.[8.24]

Despite these mystical associations, the use of the *virgula divina,* as it was known to the ancients, has been treated more realistically for some considerable time. Probably the first scientific account was that in *De Re Metallica* by Georgius Agricola (1490–1555) (Fig. 8.11). Since then, a constant stream of literature has emerged from those who feel strongly about the matter; almost 600 relevant works were published between 1532 and 1916. There have been a special journal and a society in Germany devoted to the study and encouragement of the *Wünschelrute* (wishing rod). In 1946, however, the official use of dowsers in West Germany was prohibited by order of the inspector general for water and energy. There is a British Society of Dowsers, and international conferences on *rhabdomancy,* the modern name for the ancient belief, have been held. One delegate at the first such conference in 1932 is reported to have suggested that divining could aid in climatology and in "hydromineral therapeutics." This conference agreed that the principles of rhabdomancy are established but that there is some charlatanism; it proposed that a union be formed in order to keep out the charlatans.[8.25]

Whether groundwater is located by a diviner or in some other way, its presence is due to the local geological structure in association with surface-water conditions and the character of the local strata, all of which may be determined in a scientific manner. For the person aware of these basic facts, the attraction of rhabdomancy is reduced. This attraction is mentioned quite seriously; even today, one may meet senior engineers who, caring little for all that geology can do for them, will employ water diviners. It cannot be denied that some people are influenced physically by the presence of water. The writers, as also probably many readers, have good friends who are dowsers but who, fortunately, have such respect for geology that they do not presume to place dowsing in the same category.

A—Twig.   B—Trench.

FIG. 8.11 An early illustration of the divining rod in use; a woodcut from *De Re Metallica* by Georgius Agricola, first published in 1556.

There is not, however, any agreement as to how divination works. The subject is discussed in an illuminating manner by the great scientist, Sir J. J. Thomson, who relates in his autobiography that in the early years of the present century he examined many samples of water from wells and found that "all contained the radio-active emanation from radium." This element only retains its activity for four days so that if the water is stagnant it soon loses its radioactivity.[8.26] This suggests a possible explanation for dowsing. The distinguished author goes on to relate that the most active water he tested came from a brewery, a fact he was forbidden to publicize at the time. (An American friend suggested to him that, if such a thing had happened in the United States, the whole country would have been quickly plastered with advertisements for "Dick Smith's Radio Beer.") Thomson relates also that he once met a dowser who could tell whether a bottle of medicine would do him any good merely by passing his dowsing rod over it. Following the lead of Sir J. J. Thomson, one may readily admit that the observed movement of divining rods or forked sticks in the hands of some individuals is not charlatanry, but neither is it a guarantee of finding water; the existence of groundwater is completely dependent upon local geological and hydrological conditions and can be determined scientifically. As a distinguished and sympathetic French writer on the subject of dowsing has said, "It is no good looking for water where geology tells us there is none."[8.27]

Contour maps provide one means of estimating the quantity of groundwater available; this quantitative work is the other main branch of groundwater survey work. Various methods may be used; their relative suitability is dependent on local conditions and the data available. One method is to attempt to prepare a water balance sheet for the area considered, measuring the total runoff and rain, estimating the losses due to evaporation and absorption by plants, and determining the remaining balance, which may normally be assumed to be the addition to the groundwater supply. Measurements may also be made with regard to actual discharge of groundwater into watercourses from areas in which the water table is known to be reasonably steady. A time lag will always be found between the falling of rain and an increase in groundwater flow; but with this determined, estimates can be made of the percentage of the rainfall available as groundwater. A good example of a survey somewhat on these lines is available in connection with the valley of the Pomperaug River, Connecticut; the valley has a catchment area of 230 km^2 (89 mi^2).[8.28] The most reliable method and that most widely employed is to determine and use the *specific-yield factor* for the rock in question; the one serious obstacle to the application of this method is the possible variation in the nature of the local rocks, since such variation may invalidate the calculations to some extent.

Specific yield means the "interstitial space that is emptied when the water table declines, expressed as a percentage of the total volume of material that is unwatered." For the measurement of specific yield, various methods have been suggested, some of which can be utilized in a laboratory on samples of the materials encountered in the area under investigation; other methods necessitate field observations. Laboratory methods are convenient and interesting, but the possible variations, not only of the materials tested but also of their structural arrangement in the area being studied, make these methods of doubtful value unless unusual precautions are taken. On the other hand, they can often provide a useful guide to probable yield of groundwater before major operations are undertaken in the field, since the most effective field methods for estimating specific yield depend on the recording of water levels underground during pumping or recharging. These observations can be made only after boreholes have been sunk to required levels. Records of groundwater levels will always be the most useful guide to groundwater supplies, since inevitably they reflect all the subsurface variations that might be disregarded in the application of any experimental test results. Even in the use of recorded water levels, however, observations from adjoining boreholes and wells should be carefully correlated with one another and with what is known of geological conditions in the vicinity; in this way any underground irregularity may be detected from the data available, if such detection is possible.

To select any one example to illustrate the wide field covered by this branch of groundwater study involves a difficult choice. But a record of one of William Smith's many engineering activities in England may be cited in view of its historic interest and its intrinsic illustrative value. In a letter dated 24 February 1808, from John Farey to Sir Joseph Banks, the following descriptive note has been found:

> We met the Reverend Mr. Le Mesurer, Rector of Newton-Longville near Fenny-Stratford, who related his having undertook to sink a well, at his parsonage house, within a mile or two of which no good and plentiful springs of water were known, but finding clay only at the depth of more than 100 feet, was about to abandon the design: Mr. Smith, on looking into his map of the strata, pointed out to us, that Newton-Longville stood upon some part of the clunch clay strata, and that the Bedford limestone appeared in the Ouse river below Buckingham, distant about eight miles in a northwest direction, and he assured Mr. L. that if he would but

perservere, to which no serious obstacles would present themselves, because all his sinkings would be in dry clay, he would certainly reach this limestone, and have plenty of good water, rising very near to the surface; Mr. L. accordingly did persevere in sinking and bricking his well, and at 235 feet beneath the surface (the first 80 feet of which were in alluvial clay with chalk and flints, etc. similar exactly to what I have uniformly found on your estate at Revesby, and in the bottoms of many of your fen drains) the upper limestone rock (8 feet thick) was reached and found to be so closely enveloped in strong blue clay as to produce not more than 9 feet of water in the well in the course of a night; from hence, an augur hole was bored in blue clay, for some distance, to the second limestone rock which produced a plentiful jet of water, which filled and has ever since maintained the water, I believe almost up to the surface of the ground.[8.29]

8.11 CONCLUSION

Among the subsurface geological conditions with which the civil engineer may have to deal, groundwater must be considered of primary importance.

In their own experience the writers have become increasingly impressed with the overall importance of groundwater in civil engineering work, with the relative simplicity of the principles of groundwater (at least in theory), and with the widespread comparative neglect of the subject, very often just because of its inherent simplicity and the fact that water is so "common" a substance. Groundwater must be taken into consideration whether the works be large or small. On the largest project with which one of the authors was associated, control of groundwater was the key to the solution of a major problem, and on one of the very smallest jobs with which he assisted, exactly the same condition prevailed. In both cases, and in many others, it could be pointed out that if there had been no groundwater present, there would have been no problem. Once it became evident that water, in association with soils that would have been quite stable if dry, was the cause of the trouble, the engineers knew that if the water could be removed, or controlled, the troubles would be quickly minimized.

The fact that this simple truth proved to be the critical factor will perhaps serve to explain why this chapter has been placed so early in the Handbook, in a position intended to emphasize the importance of groundwater in civil engineering work. Groundwater characteristics are considered even before necessary preliminary site investigations are discussed, since adequate subsurface investigation *must* include as much study as is possible of local groundwater conditions. Such study must be in addition to the study of local geology which, in all too many cases, is thought of as the geology of rocks alone.

8.12 REFERENCES

8.1 L. B. Leopold, *Water, A Primer,* Freeman, San Francisco, 1974.

8.2 O. E. Meinzer, "The History and Development of Groundwater Hydrology," *Journal of the Washington Academy of Sciences,* **24**:6 (1934).

8.3 T. Sheppard, *William Smith, His Maps and Memoirs,* Brown, Hull, England, 1920.

8.4 R. Lockhart Jack, *The Geology of the County of Jervois ... with Special Reference to Underground Water Supplies,* Geological Survey of South Australia Bulletin no. 3, Adelaide, 1914, p.29.

8.5 F. Fox, "The Simplon Tunnel," *Minutes of Proceedings of the Institution of Civil Engineers,* **168**:61 (1907).

8.6 "Pumps and Weight Keep Hospital out of Hot Water," *Engineering News-Record,* **158**:46–48, (20 June 1957).

8.7   J. W. Gregory, "The Origin and Distribution of Underground Waters," *Water and Water Engineering*, **29**:321 (1927).

8.8   Discussion of E. O. Forster Brown, "Underground Waters in the Kent Coal Field and Their Incidence in Mining Development," *Minutes of Proceedings of the Institution of Civil Engineers* **215**:77 (1923).

8.9   C. G. Parker and V. T. Springfield, "Effects of Earthquakes, Trains, Tides, Winds and Atmospheric Pressure Changes on Water in Geological Formations of Southern Florida," *Economic Geology*, **45**:441–460 (1950).

8.10 Ibid.

8.11 Report of discussion at a meeting in Aberdeen of the British Association for the Advancement of Science, *Water and Water Engineering*, **36**:623 (1934).

8.12 J. F. Hoffman, "Planning to Use Water for Air Conditioning Design," *Heating, Piping and Air Conditioning*, **29**:96 (1957).

8.13 "Hot Well Water Cooled for Use in City Distribution System," *Engineering News-Record*, **118**:130 (1937).

8.14 "Corrosiveness of Aqueduct Reduced by Gas Drainage," *Engineering News-Record*, **117**:790 (1936); see also, same journal, "Drains Remove CO_2 from Tunnel," **120**:885 (1938).

8.15 T. C. Hatton, "Ferrous Strata Develops CO_2 in Tunnel Shafts," *Engineering News-Record*, **89**:706 (1922).

8.16 J. Wyllie, "An Investigation of the Source of Arsenic in a Well," *Canadian Public Health Journal*, **28**:128–135 (1937).

8.17 "Boron Creates Special Problems in Los Angeles' Water Supply," *Engineering News-Record*, **125**:113–114 (1940).

8.18 E. G. Swenson (ed.), *Performance of Concrete*, University of Toronto Press, Toronto, 1968 (several papers).

8.19 G. E. Bessey and F. M. Lea, "The Distribution of Sulphates in Clay Soils and Groundwater," *Proceedings of the Institution of Civil Engineers*, **2**:159–181 (1953).

8.20 Reports of the interstate conferences on artesian waters, starting with that in Sydney (1912); regular proceedings published by the respective state government printers.

8.21 "Artesian Water Pressure Lifts Dam," *Engineering News-Record*, **123**:485 (1939).

8.22 C. K. Wentworth, A. C. Mason, and D. A. Davis, "Salt-water Encroachment as Induced by Sea-level Excavation on Angaur Island," *Economic Geology*, **50**:669–680 (1955).

8.23 O. E. Meinzer, *Plants as Indicators of Groundwater*, U.S. Geological Survey Water Supply Paper no. 577, Washington, 1927.

8.24 A. J. Ellis, *The Divining Rod; A History of Water Witching*, U.S. Geological Survey Water Supply Paper no. 416, Washington, 1934.

8.25 "Divination by Rods," *Water and Water Engineering*, **34**:147 (1932).

8.26 Sir J. J. Thomson, *Recollections and Reflections*, G. Bell, London, 1936, p. 160.

8.27 Vicomte Henri de France, *The Modern Dowser*, G. Bell, London, 1936.

8.28 O. E. Meinzer and N. D. Stearns, *A Study of Groundwater in the Pomperaug Basin, Connecticut, With Special Reference to Intake and Discharge*, U.S. Geological Survey Water Supply Paper no. 597(b), Washington, 1928.

8.29 T. Sheppard, op. cit., p.245.

Suggestions for Further Reading

Water was the subject of the usual comprehensive *Year Book of the U.S. Department of Agriculture* for 1955 (Washington); it remains an admirable reference book on groundwater and all other major aspects of water. Far more extensive is the wealth of information to be found in the many *Water Supply Papers* of the U.S. Geological Survey. One title which illustrates the wide coverage of this fine series is *Urban Growth and the Water Regime* by J. Savini and J. C. Kammerer (W.S. Paper No. 1591-A of 1961). The current listing of these valuable publications should always be consulted before the start of any major groundwater investigation.

Only infrequently does one find a major technical publication emanating from a commercial manufacturing company. A good example of this quite limited group of books is *Ground Water and Wells,* a 440-page volume in its second edition (1972) published by and available from the Johnson (Screen) Division of Universal Oil Products Co., St. Paul, MN 55165.

The National Water Well Association (500 W. Wilson Bridge Road, Worthington, OH 43085) has developed strong technical standing in all aspects of groundwater, well reflected in the pages of its valuable journal *Groundwater,* of which twenty volumes had been produced by 1982. Two examples of the interest shown by larger societies are *Groundwater Management,* Manual No. 40 of the American Society of Civil Engineers issued in 1972 and reprinted in 1982 (216 pages) and *Manual on Water* (C. E. Hamilton, editor) of the American Society for Testing and Materials, a 471-page volume published in 1978 as STP No. 442A.

Some of the broader aspects of groundwater are treated in *Groundwater, Soil and Terrain Management* (D. R. Coates, editor), a useful 140-page volume published in 1973 as Contribution No. 2 of *Publications in Geomorphology* of the State University of New York at Binghamton. Typical of the many excellent regional guides now available is *Ground Water in Pennsylvania* by A. E. Becker, a 42-page bulletin of the Topographic and Geologic Survey of Pennsylvania, a fifth printing being dated 1973.

In the last two decades a number of fine textbooks on various aspects of groundwater have appeared; the following is a mere sampling of useful titles. *Hydrogeology* by S. N. Davis and R. J. M. de Weist is a useful general review published in 1966 by John Wiley & Sons of New York. *Groundwater* is a comprehensive treatment in a volume of 604 pages by R. A. Freeze and J. A. Cherry, published in 1979 by Prentice Hall, Inc. *Groundwater Hydrology* by D. K. Todd is another Wiley publication which appeared in its second edition in 1980. And a volume of 269 pages dealing only with *Groundwater Resources: Investigation and Management* by S. Mandell and Z. L. Shifton was published by Academic Press, Inc., of New York in 1981. Mention should also be made of one of the pioneer volumes in this field, *Ground Water* by C. F. Tolman, published by McGraw-Hill in 1937.

chapter 9

CLIMATE

9.1 The Hydrological Cycle / 9.2
9.2 Rainfall / 9.3
9.3 Temperature and Wind / 9.5
9.4 Climatic Records and Hythergraphs / 9.7
9.5 Conclusion / 9.11
9.6 References / 9.11

The majestic flow of mighty rivers to the sea, the burbling of streams, the trickles of springs—all these are reminders of the dynamics of the water system that seems to distinguish this earth as a planet. The slow movement of groundwater is yet another reminder that, as was said so long ago, "all the rivers run into the sea; yet the sea is not full." Once the dynamics of water are appreciated, then the necessity for a cyclical movement is at once clear. The *hydrological cycle* is the term used to denote the constant movement of water—by evaporation from open surfaces, by rainfall and snowfall as precipitation, by infiltration and direct runoff into the ground and watercourses, and so back to lakes or sea for the continuation of the cycle (Fig. 9.1). Further thought makes clear that this is but a part of the vastly greater dynamic system of air and water around the entire globe, a system influenced by the earth's rotation and to a large extent controlled by movements of the "heavy air" in the lower 5½ km (3½ mi) of the atmosphere. The dynamics of this great system create what is known everywhere as the weather, the "weather machine" being a phenomenon of Nature vast in extent, fascinating in its complexity.

Climate is defined as the weather at a particular site over a long period of time, or—in effect—the generalized weather for any location. It is this localized result of the general weather cycle that is of such importance to both engineers and geologists. The civil engineer must design all his structures in relation to all relevant climatic factors—snow loads on roofs, wind loads on all structures above ground, rainfall and its disposal, and the possiblity of changing levels of the water table. The operations of the construction engineer are so obviously determined by climatic factors that it would be tedious to describe them; but all with experience in construction know that any major variation from average climatic conditions at a building site can interfere with progress and change an anticipated profit into a serious loss. The geologist is interested in the long-term effects of climate since it is the weather that is principally responsible for those current changes in the form

9-2 GEOLOGICAL BACKGROUND

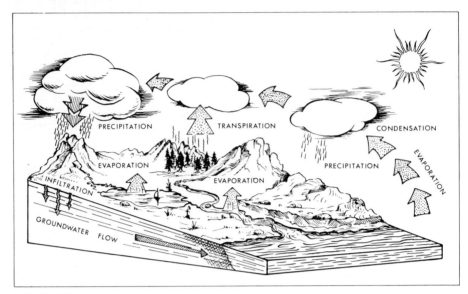

FIG. 9.1 The hydrological cycle.

of the earth's surface that are studied under the general title of geomorphology. The weathering of rocks and soils is clearly dependent upon the relevant climate, while floods and their devastating effects are perhaps the most dramatic evidence of all of the importance of climate in human affairs.

It is, therefore, somewhat surprising to find such widespread neglect of climate as one of the natural features that must be studied in advance of design for all civil engineering works. There are, naturally, many civil engineers and geologists who have become so interested in the weather, as one of the most complex of all natural phenomena, that they have made its study their hobby, some essaying amateur forecasting. But they are the exception. So it is that this separate short chapter has been included in this Handbook specifically to direct attention to the importance of climate (and so of weather) in the interrelation of geology and civil engineering.

9.1 THE HYDROLOGICAL CYCLE

The hydrological cycle provides a good starting point since it is at once so obvious and so important to civil engineers. The driving force of the system, it is well to recall, is the solar radiation that falls on the earth's surface; 47 percent of this radiation is transformed into heat. The resulting changes in temperature around the globe are the basic cause of air movements, and so of the movements of water. Seventy percent of the surface of the globe is covered with water, land surface being only 30 percent, so that evaporation from the sea (in particular) is on a scale that is difficult to visualize. Equally is it beyond normal thought to realize that there are, it is conservatively estimated, 1.25 billion km^3 of water on the globe; of this 97 percent is the salt water which forms the oceans. Of the remaining 37.5 million km^3 of fresh water, about 87 percent is locked up, occurring as ice in polar regions and in glaciers, all but 10 percent of it in Antarctica. This leaves a mere 4.7 million km^3 as the fresh water with which all are familiar. Of this, about 90 percent constitutes groundwater. It will be seen that the remaining surface water, on which life depends—that held in lakes, rivers, and streams—amounts to a mere 0.01 percent of the overall total.

These figures provide a useful background against which to consider the dynamics of the whole system. Solar radiation is basically responsible for the phenomenon of evaporation of water into the atmosphere from all surface waters. A small quantity of water will enter the atmosphere by transpiration from living organic matter, such as trees. A small part of the atmospheric water will be seen evidenced as clouds which, under appropriate conditions, will precipitate the water they hold as rain, hail, or snow. That portion of precipitation which falls onto open water, by far the largest part, will repeat this first part of the cycle directly. It is that part which falls on the land that is of special significance in human affairs. Some will run off the surface directly, especially if it is bare rock or soil well protected by a natural cover of thick grass, and so find its way into watercourses and eventually back to the sea. That part which does not run off will percolate into the ground, either through cracks in the surface or directly into soil if it is at all pervious and if it is not already saturated. Figure 9.1 illustrates the cycle graphically.

The types of "soil water" were illustrated in Fig. 8.1, a diagram that represents an average condition. If, however, a site is subjected to continuous and heavy rainfall, then the water table will rise to the maximum elevation it can attain, as determined by local topography, and the soil above it will become, in the course of time, saturated. When this happens, runoff will be the movement of any further precipitation, instead of percolation into the ground. If not anticipated, this situation can have serious results. It is accordingly essential that the long-term records of local rainfall be always considered in all site studies so that the "groundwater depletion" (as it has been called by Thornthwaite) can be estimated and the possible results of future heavy rains determined.[9.1]

Appreciation of the hydrological cycle should therefore be an integral part of the background to all site investigations. Groundwater problems will be the better dealt with when the dynamic aspect of subsurface flow is recalled. Rainfall can be of critical importance for some construction sites; its local significance can be determined only from a study of all available records from the nearest recording station. Winds are not usually of crucial moment on construction, even though they must be known before designs can be completed. Local temperature variations are always important in design, in construction, and so in geological studies of any site. To these three aspects of climate, more detailed attention will now be given.

9.2 RAINFALL

It is common knowledge that rainfall is influenced by topography, and topography is always an expression of the local geology. There is, therefore, an intimate relationship between rainfall intensity and local geology. The highest rainfall experienced in Canada, for example, is to be found in British Columbia on the west coast of Vancouver Island where average annual precipitation exceeds 7 m (21 ft) on the seaward (west) side of the mountainous core of the island. Some of the driest areas in Canada are also to be found in British Columbia, but in the interior, protected from the west by the mountains of the Coast Range. Here annual rainfalls of less than 18 cm (7 in) are experienced in parts of the Fraser River valley only 350 km (220 mi) from the west coast of Vancouver Island.

This mountainous province of Canada provides vivid examples, also, of local variations in rainfall of the type that can be of extreme importance in large-scale civil engineering work. The city of Vancouver occupies a beautiful setting at the mouth of the Fraser River, its southern sections built upon some of the flat-lying land of the river's delta. Its northern

9-4 GEOLOGICAL BACKGROUND

boundary reaches into the foothills of the Coast Range, this great coastal mountain barrier being responsible for the uplifting of saturated air blown onshore from the Pacific. Despite this setting, it is still surprising to find that annual rainfall varies in Vancouver from 0.76 m (30 in) in its southern parts to 3 m (120 in) in the northern suburbs, a ratio of 4 to 1, and this within the same municipality.

Study of local rainfall records, therefore, must be an early part of all site investigations. Where no local recording stations exist, expert advice should be taken as to how best the nearest available records can be extrapolated for the site being studied (Fig. 9.2). And in

FIG. 9.2 Typical instruments for a simple weather-recording station: Stevenson screen for thermometers on left, rain and snow gauges on right.

such cases, early installation of a recording rain gauge should be seriously considered, if the nature of the project warrants this expenditure. The closest attention must be paid to the relation between the rainfall during the period of site study and the long-term average, since if any major deviation from the normal is occurring, then suitable allowance for the effect of this on natural features must be made. In particular, any unstable slopes should be carefully investigated, since excessive rainfall will often be found to be responsible for many movements of potentially unstable ground. The reverse naturally holds true; slopes that appear to be quite stable in very dry weather may display instability when continuous heavy rain falls on the site. The Handlová landslide in Czechoslovakia, to be described in Chap. 39, is a telling reminder of the vital necessity of continuous vigilance and regular study of current rainfall records wherever unstable ground is known to exist.

Rainfall is a major contributor to flood flows in rivers and all smaller watercourses. If a site under investigation includes any watercourse, however small, every effort must be made to determine the extent of the floodplain and to correlate this with relevant rainfall records so that the necessary allowance can be made for possible future flooding. Floodplain mapping is now well developed in many urban and developing areas, but the danger of building on known floodplains remains. And the possibility of heavy rains causing such dangers can usually be determined by careful study of local rainfall records.

Since climatic records in many regions are so geologically short, the engineer would do well to develop a sensitive eye to reading the record in the landscape. No better example can be cited than the desert regions. These dry regions, with low annual rainfall *on the*

FIG. 9.3 Construction site—wet; what a difference the rain makes.

average, seemingly contradict the weather records; the sensitive eye can see that the surface has been etched and eroded mainly by the effects of running water. While long intervals, perhaps years, may pass between storms, the rare intense storms provide spectacular evidence of the power of flowing water.

Drainage facilities will be dependent upon maximum runoff flows, and so upon intense rainfall. This is well recognized in civil engineering design procedures, but it is desirable to see for one's self what such runoff means, in addition to using the standard figures for design calculations being worked out in a comfortable dry office. It is a quite serious suggestion that no major design should be completed and, correspondingly, no site investigation should be regarded as complete, until the site in question has been seen during the heaviest possible rainfall (Fig. 9.3). Such site visits may be very uncomfortable but they are almost essential. Once such a visit has been made, the significance of climate in site studies, as well as in design procedures, will need no elaboration. During construction, heavy rainfalls will be experienced on most sites. It is, accordingly, standard practice on larger jobs to install a small weather-recording station so that there will be available an accurate record of climate, and especially of rainfall, throughout the duration of the project.

9.3 TEMPERATURE AND WIND

Temperature variations throughout the year constitute a natural feature usually taken for granted and seldom thought about in any depth except by those with special interest in the weather. These variations are, however, of great importance in civil engineering works and they influence geological phenomena such as weathering. The average annual cycle of temperature should therefore be studied as a part of all site investigations. Maximum temperatures may affect concreting operations. Minimum temperatures will dictate the necessity of special precautions for winter operations, if below-freezing weather should be experienced. Alternations between freezing and thawing—freeze-thaw cycles—can be of importance in relation to the durability of exposed materials, i.e., rocks, either in place or as used in construction. The duration of cold throughout a winter, when considered in association with the nature of the local soil, will affect the penetration of frost into the ground.

Temperatures are therefore an important factor in the local climate and are recorded at all weather stations, usually as maximum and minimum for each day. When these are plotted on a chart which already contains a record of the long-term average temperatures, then daily variations from the normal can readily be seen, and judgments made accordingly. The duration of winter cold is measured as the total of *degree-days* for the season, a degree-day being the average daily temperature subtracted from 18°C. Not only is this concept essential for calculating the freezing index for a site, but it is also a vital element in the design of buildings and of the necessary heating equipment for them, as well as for insulation requirements. Figure 9.4 illustrates the change in annual summations of the freezing index (one use of degree-days) as one goes northward in North America. It should be noted, however, that these isolines indicate *duration* of winter cold and not necessarily very low temperatures. Low temperatures on the Arctic coast, for example, at times are no lower than those in the Midwest and on the Canadian prairies.

The duration of the winter has another effect of much importance in site studies. As the use of the underground, for appropriate purposes, steadily increases (as outlined in Chap. 21), a knowledge of local ground temperatures becomes correspondingly important. Figure 9.5 illustrates a typical annual chart of ground temperature variation. In summer, the temperature of the ground close to the surface will be somewhat less than the local air temperature at midday, but still warm. Correspondingly, in winter the temperature close to ground surface will be just slightly higher than that of the adjacent air, unless there is a local cover of snow, in which case the difference will be more marked. Due to the thermal properties of the ground (whether soil or rock), the subsurface temperature will change rapidly with depth, decreasing in summer and increasing in winter. It is found that there is a hysteresis effect at a modest depth below the ground (3 m or 10 ft near the Canadian-U.S. border), due also to the insulating effect of the ground, and a corresponding time lag of up to six months. (This explains why, in northern regions, trouble with frozen pipes is experienced at its worst in the spring, not in the depth of winter).

Eventually, at a depth of not more than 10 m (33 ft) in temperate parts of the world, the annual variation of ground temperature becomes imperceptible. Below this, the temperature will increase slowly due to heat coming from the earth's core, a typical rise being 1°C for every 90 m. This general pattern is almost universal for temperate regions. A little thought will suggest that the mean ground temperature must be the mean annual air temperature for the locality, or very close to it. Where snow covers the ground for a part of each winter, there will always be a difference between the two figures of one or two degrees. Appreciation of the general pattern shown in Fig. 9.6 will assist with many problems involving underground works, but the phenomenon of ground temperature variation has still further significance.

Since the mean ground temperature at any location is close to the average annual air temperature for that location, it follows that as the latter climatic factor decreases, so also will mean ground temperature. From Fig. 9.4 it can be deduced that the average air temperature will steadily decrease the further one goes north. There will be some point, on each meridian, at which the average annual air temperature will be at freezing point, and beyond this will fall increasingly below 0°C—so also will the mean ground temperature, with a slight lag. In all areas where the average ground temperature is below freezing, the ground is in the condition of *permafrost,* this word (although semantically inaccurate) being now universally used in the English language to denote this *condition* of the ground; it does not denote any special material.

Local air temperatures, therefore, are indeed significant in site studies, quite apart from their relevance to human comfort. Not many site studies will have to be carried out in permafrost areas, but it is worth noting that one such study in northern Canada did not take ground temperatures into consideration. The location was in the discontinuous permafrost zone, a fact undetected until construction was well advanced; necessary remedial work can best be left to the imagination. In all regions where frost is regularly experienced, and that includes a large part of North America and of Europe, troubles may be experienced during construction or with the operation and maintenance of structures; soil temperatures are not something of significance only in far northern parts.

It is in the northern parts of the globe that *winds* become so important, since the combination of high winds with low temperatures can give equivalent wind-chill temperatures of unusual severity for human activity. These are beyond the scope of this Handbook, but wind as a climatic factor cannot be completely overlooked. In dry regions, it is a weathering factor of importance, since wind erosion, especially with driven sand, is a potent geological agent. It has been responsible for the widespread deposits of loess. Drifting of sand dunes provides current evidence of the importance of the wind.

9.4 CLIMATIC RECORDS AND HYTHERGRAPHS

Every major country today has a national weather service, responsible for recording all weather features throughout the country and for publishing these in convenient form. Typical are the series of fine bulletins entitled *Climates of the States* prepared by the National Oceanic and Atmospheric Service and available, for each state of the Union, from the superintendent of documents in Washington, D.C. Information given in a typical bulletin includes "freeze data," long-term average air temperatures, and precipitation by month for locations throughout the state, as well as the extreme values so far recorded. Helpful maps with isolines for air temperatures and mean annual precipitation are also included. With every passing year, the duration of the available records increases and so all such publications are regularly issued in revised and updated versions. In Canada, the Atmospheric Environmental Service of the Canadian Department of the Environment provides similar climatic information, as well as useful bibliographies and advice as to the availability of further and more detailed information.

On first looking at one of these climatic summary publications, the tables in which most of the records are presented appear rather forbidding. Those accustomed to reading tabular material will be able to select the information for the location in which they are interested without difficulty and to appreciate the general pattern that it presents. It takes an unusual grasp of figures, however, to visualize the "picture" of the annual variation in climate for one particular location. An elegant solution to this difficulty was proposed by Griffith Taylor in his geographical writings, but Taylor's solution, the *hythergraph,* has not yet achieved widespread use.[9.2] It is the view of the writers that it is a device that should be featured in every site report and in many other geological and engineering publications. It is illustrated in Fig. 9.6, which presents several contrasting *hythergraphs.* By this simple system of plotting monthly average air temperatures and average precipi-

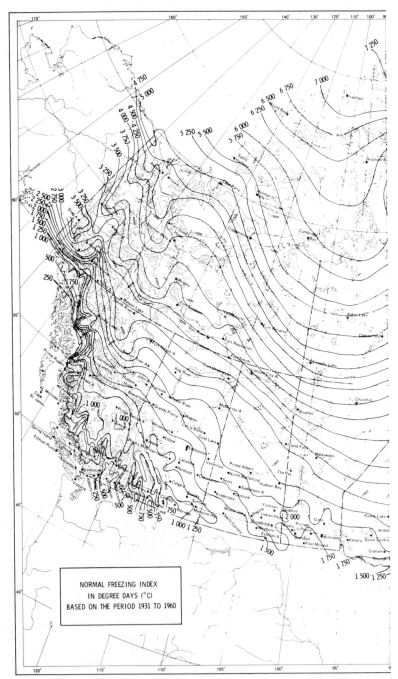

FIG. 9.4 Isolines for the freezing index, similar to diagrams for other climatic factors such as degree-days, increasing to the north.

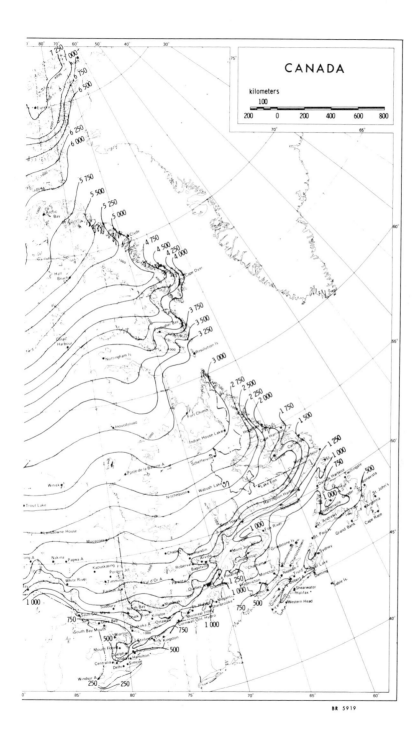

9-10   GEOLOGICAL BACKGROUND

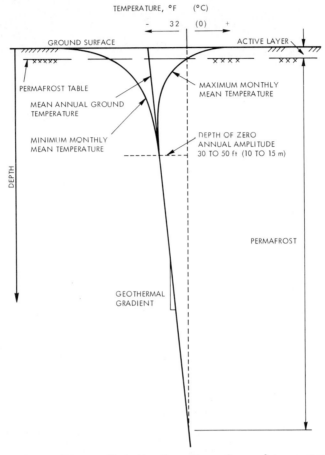

FIG. 9.5 Diagram illustrating the pattern of ground temperature variation, illustrating also the occurrence of permafrost.

tation, a 12-sided polygon, which is unique to each location, is obtained. As can be seen, the hythergraph provides a ready means of comparing, at a glance, the climate of any one location with that of another.

In individual hythergraphs it is possible to include also a record of prevailing wind. Starting at each monthly temperature-precipitation point, a line is drawn to a length representing, to some scale, the intensity of wind for that month, and at an angle to the vertical corresponding to the wind compass direction; an arrow completes each vector. These have not been used in Fig. 9.6 since the necessarily small scale would have made the resulting diagram too complex; but it can be done, thus giving all main climatic factors for one location in one simple diagram. So useful are hythergraphs as an aid to site study that the writers wish that they could use them freely in later chapters of this book, when describing cases in which climate was a decisive factor. This would be impractical, but the authors do commend the use of hythergraphs in all geotechnical reports upon site investigations.

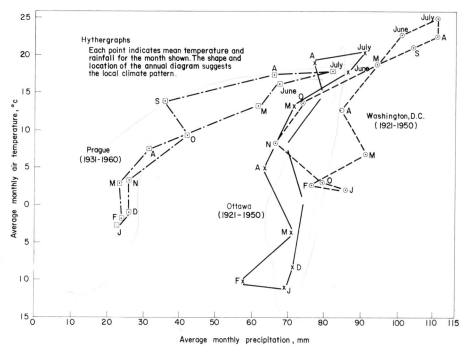

FIG. 9.6 Typical hythergraphs, a convenient means of comparing climates of different locations.

9.5 CONCLUSION

There is an old saying attributed to Mark Twain that "everyone talks about the weather but nobody does anything about it." Although powerless to change the weather, the engineer has a responsibility to be aware of what it holds in store, both from the standpoint of structural design and problems during construction. "Forewarned is forearmed" is as appropriate for climatic factors as it is for geological factors. Both are essential parts of the environmental context or framework in which the engineer must work. The most careful assessment of the properties of geological materials under certain "normal" weather conditions may be defeated by major changes in those properties under "abnormal" conditions. Climate, therefore, must always be a vital part of all site investigations.

9.6 REFERENCES

9.1 C. W. Thornthwaite, "An Approach to a Rational Classification of Climate," *Geographical Review,* **38**:55–94 (1948).

9.2 Griffith Taylor, *Canada,* Methuen, (London), 1947, p. 119.

Suggestions for Further Reading

The weather is a matter of such general interest that it is not surprising to find an extensive relevant literature widely available. Consultation with almost any local library will disclose the availability of one or more of numerous popular books. These provide useful starting points for more detailed studies of climate by civil engineers, once initial interest has been aroused. The well-known Golden Guides provide a most helpful and convenient pocket-sized introduction through their volume entitled *Weather,* by P. E. Lehr, R. W. Burnet, and H. S. Zim.

In all developed countries there are relevant national societies such as the American Meteorological Society, the Royal Meteorological Society (of the U.K.) and the Canadian Meteorological and Oceanographic Society. These bodies publish transactions or proceedings which are valuable collections of information, as are the journals which they all publish, ranging from strictly scientific periodicals to more popular magazines such as *Weather* and *Weatherwise.*

In addition to the regular publication of climatic information by national agencies, there are now available nongovernmental books giving overall climatic summaries. Elsevier Publishing Company of New York has a complete series on world surveys of climate, a typical title being *Climates of North America* by R. A. Bryson and E. K. Hare (420 pages, published in 1974).

On a broader scale, *Climate, Past, Present and Future* by H. H. Lamb is a two-volume work of over 1,500 pages, published in 1972 and 1977 by Chapman and Hall (of London), who also gave us *Climatology* by A. A. Miller, the 9th edition of which was published in 1961 as a book of 332 pages. *Climatic Change* is the title of two separate volumes, one edited by J. Gribbin and published by Cambridge University Press in 1978, the other edited by H. Shapley and published by Harvard University Press in 1953.

Two admirable recent general books are *The Atmosphere* by F. Lutgens and E. J. Tarbuck (the 440-page second edition of which was published by Prentice-Hall, Inc., in 1982) and *The Weather Book* by R. Hardy, P. Wright, J. Gribbin, and J. Kingston, first published in 1982 (224 pages) by Michael Joseph of London.

Statistics play a vital role in all climatic studies. Peter Lumb has illustrated this well in a paper called "Statistics of Natural Disasters in Hong Kong 1884-1976" to be found in *Proceedings of the 3rd International Congress on the Application of Statistics and Probability in Soil and Structural Engineering,* held in Sydney, Australia, in 1979. The holding of this congress confirms the importance of statistical studies to many of the topics dealt with in this volume.

part 2

PRELIMINARY STUDIES

chapter 10

SITE INVESTIGATIONS

10.1 Economics of Preliminary Work / 10-4
10.2 The Pattern of Preliminary Work / 10-7
10.3 Regional Geology / 10-10
10.4 Aerial-Photo Interpretation / 10-13
10.5 Geological Surveying / 10-18
10.6 Detailed Site Studies / 10-20
10.7 When Construction Starts / 10-21
10.8 As-Constructed Drawings / 10-21
10.9 Conclusion / 10-22
10.10 References / 10-23

Before undertaking any design work for a project, a civil engineer must have full information about the material on which the structure is to be founded or in which the construction is to be carried out. This will necessitate an examination of the site of the work before designing is started and a thorough investigation of the site before detailed designs are prepared. Truisms? Admittedly, but they are truisms that have often been neglected in the past, sometimes with disastrous results and almost always with consequent monetary loss. This last statement might be characterized as too sweeping were it not possible to confirm it indubitably by many examples from engineering practice. To the civil engineer with even a slight appreciation of geology, neglect of what is the obviously essential course of making thorough and complete preliminary investigations before embarking on construction will appear to be utter folly. An engineer who failed to make such investigations might with propriety be compared to a surgeon who operated without making a diagnosis or to a lawyer who pleaded without having any prior discussion with a client.

To the engineer who has not studied geology, these analogies might appear farfetched; in many cases the nature of the surface of the ground and experiences on similar adjacent works might seem enough evidence to warrant making certain assumptions about subsurface conditions, assumptions on which designs can be based. It may be that such reasoning has led to the neglect of preliminary explorations in the past. Similarly, it is probable that unquestioning reliance on the information given by test borings alone, considered without relation to the local geology, may have sometimes resulted in unforseen difficulties in con-

10-4 PRELIMINARY STUDIES

struction and may have led engineers to distrust the value of such preliminary work. All too often, however, it has been the difficulty of obtaining the necessary funds before the start of construction that has prevented engineers from carrying out the explorations that they know to be necessary. This results from a shortsighted policy on the part of some owners which engineers may find hard to understand.

Discussion of this attitude, and of other factors that may have furthered the neglect of subsurface explorations on civil engineering works in the past, will not be profitable. Most civil engineers today recognize the supreme importance of this preliminary work. Readers who may entertain doubts about its value will not have to read far beyond this chapter before they find evidence that should set all their doubts at rest. But all engineers may have to deal with owners who have not had an opportunity to visualize the necessity of exploratory work. The following section therefore presents arguments that may be of use in demonstrating the economic value of such preliminary expenditure.

10.1 ECONOMICS OF PRELIMINARY WORK

Economics may be too technical a word to apply to the following brief considerations of preliminary work. It will serve, however, to describe a discussion of the cost of such work relative to the total cost of a project and of the savings that may be effected and the difficulties that may be avoided by such a precautionary measure. Preliminary work, as here intended, is held to include geological surveys, geophysical surveys, test pits, borings, drillings, and any further exploratory measures that may be called for in special cases. Examples of the total expenditure on such work are not easy to obtain for complete projects, because the totals are usually so small that they cannot suitably be featured in the summaries of cost that form most valuable parts of published descriptions of engineering works. It has been possible, however, to assemble some figures for interesting examples of tunnel, dam, bridge, and building construction; these are given in Table 10.1.

Records from works of such differing character and in such varied parts of the world show clearly that the cost of preliminary work relative to the total cost of a civil engineering project is small indeed. It is believed that the figures are generally typical for normal engineering work. If the costs thus presented are compared with other percentages

TABLE 10.1 Proportional Cost of Subsurface and Site Investigations for Leading Types of Civil Engineering Projects

Type of work	Range of cost of site investigation (as % of total cost)
Tunnels	0.3 to 2.0
Dams	0.3 to 1.6
Bridges	0.3 to 1.8
Roads	0.2 to 1.5
Buildings	0.2 to 0.5
Mean values	0.26 to 1.48
In round figures	0.25 to 1.50

Source A summary prepared from many detailed costs in the records of the authors.

which are usually included in civil engineering estimates, their significance will be appreciated. Special thought might perhaps be given to the comparison of this usual 1 or 2 percent for the total cost of site investigations with the item of "10 percent for contingencies," which is so frequently included in civil engineering estimates. For some types of work, especially marine work, contingencies must be allowed for in any estimates of cost; but, in general, this broad item is often used to allow for just those uncertainties which thorough subsurface exploration would assist in eliminating. Viewed in this light, the cost of preliminary work may not appear to be so high as may at first be thought by an owner who is faced with the necessity of spending some money in order to find out the feasibility of the project being planned.

Further argument may be needed, however, to persuade a dubious owner. If cost records are not sufficient, actual examples from engineering practice may be drawn upon to complete the picture that the engineer wishes to present. This book contains many examples that might thus usefully be quoted, but for convenience a few may be summarized here.

Consideration may first be given to cases in which adequate preliminary work was *not* carried out. For example, a contract for one tunnel included a large bonus payment for every day gained in completion before a certain fixed date. The engineers based their estimate of construction time on a calculated rate for excavating the tunnel through the quartzite exposed at the two tunnel portals. The tunnel line pierced an anticline, however, and it was found that the quartzite lay over easily excavated shale. The tunnel was driven very quickly and the contractor's bonus was so large that it led to a legal battle.

An even more serious case was that of a dam for water supply which was to be built in a valley as an earth-fill structure with a concrete core wall carried to rock at an assumed maximum depth of about 18 m (60 ft) below the valley floor. This figure was derived from test borings which were stopped at what was believed to be solid rock but which later proved to be boulders of a glacial deposit. The core wall finally had to be carried to a depth of 59 m (196 ft), and the cost of the dam was increased correspondingly by a very large amount (Fig 10.1).[10.1] An example from bridge engineering practice is one in which a bridge was built upon a surface of "hardpan" in the river bottom, but no examination of this surface was made because of the difficulty of taking borings in a river current of 8 km/h (5 mph). When constructed, the bridge pier sank out of sight, causing the loss of the two adjacent spans of the bridge and a number of human lives. An examination was made later, and it was found that the "hardpan" was only a thin stratum with boulders overlying a deep layer of soft clay (Fig. 10.2).[10.2]

These examples show that trouble can be encountered through neglect of preliminary investigations. What of the constructive contributions that such work can make to civil engineering practice? Many examples could be quoted, although the actual monetary savings effected are usually hard to assess. The successful completion of many notable tunnels, such as those under the river Mersey in England or those constituting the Catskill and Delaware aqueducts for the water supply of New York, which will be referred to in more detail in Chap. 28, provide striking evidence of the value of preliminary geological studies and underground investigation. Evidence supplied by the Hoover Dam and a vast number of smaller dam structures founded on strata previously investigated is also striking. From the records of bridge construction, there may be mentioned, to similar effect, the San Francisco-Oakland Bay Bridge, the foundations for which are referred to in some detail in Chap. 24; despite their unusual size and the record depths to which they were carried, their construction agreed closely with what was anticipated as a result of the preliminary exploration, on which the sum of $135,000 was expended.[10.3]

A somewhat unusual example may be given to illustrate the value of underground exploration for building foundations. The Lochaber water power scheme in Scotland has

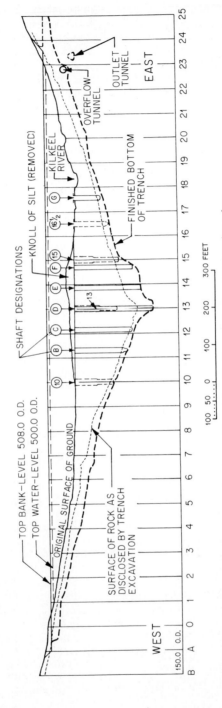

FIG. 10.1 The Silent Valley Dam in the Mourne Mountains, Northern Ireland; a cross section taken soon after the excavation was finally completed.

FIG. 10.2 Boulders from Wisconsin till, typical of those which can cause trouble if not detected, here grouped as a part of the landscaping of the area in which they were uncovered at relatively shallow depths.

its powerhouse located at the end of a steel pressure pipeline 987 m (3,240 ft) long which descends from the slopes of Ben Nevis. The discharge from the Pelton wheels enters a tailrace about 900 m (3,000 ft) long excavated partially in rock tunnel, partially in open rock cut, and partially in open cut through drift deposits. As originally planned, the powerhouse was to have been much nearer the river Lochy; the tailrace was to have been only 30 m (100 ft) long, and the length of pipeline 1,860 m (6,200 ft). Further investigations were conducted, however, through an extensive program of shallow test borings all over the available sites for the powerhouse and along corresponding routes for the tailrace. In all, 312 m (1,040 ft) of borings was carried out at a total cost of £1,340 (about $7,000). Results so obtained were carefully plotted, mainly to locate contours on the rock surface underlying the surface deposits; all borings were carried at least 1 m (3 ft) into the rock to make sure that boulders had not been encountered. On the basis of this information, 16 complete schemes for the powerhouse location were prepared and their overall costs estimated; the final decision was for the arrangement already described, for which the combined cost of tailrace, pipeline, and powerhouse excavation, on the average 12 m (40 ft) deep, was kept to a minimum (Fig. 10.3). A substantial saving, which amounted to many times the cost of the exploratory work, was thus effected over the original plans.[10.4]

10.2 THE PATTERN OF PRELIMINARY WORK

When a civil engineering or building project is first contemplated, the site will almost always be examined in a preliminary way, and such topographical maps as are available will be obtained for office study. As plans develop, more detailed and possibly more extensive surveys will be required, and more accurate maps may be made; upon the basis of these, economic studies can proceed and the feasibility of the project can be evaluated. Finally, if the work is to proceed, detailed designs are prepared, the necessary contract documents and drawings are assembled, tenders are called, a contract is awarded, and the job gets under way. This is a familiar pattern, followed with only minor variations for the vast majority of construction jobs, large and small. What is meant, then, by preliminary work supplementary to these well-recognized steps? As has already been indicated, an absolutely essential complement of these preliminary studies of the topography of the construction site is an equally careful study of the geological strata to be encountered in excavation and to be utilized as foundation beds and construction materials. The proce-

10-8   PRELIMINARY STUDIES

FIG. 10.3 The Lochaber Works of the British Aluminium Company at Fort William, Scotland, the tailrace running (at first in tunnel) from the powerhouse to Loch Linnhe, off left.

dure for this part of the preliminary investigation is equally straightforward, equally simple in its essentials, equally easy to follow.

Even a general reconnaissance of the site can be used to good advantage by a cursory examination of those features that betray something of the local geology. General examination of the physiography of the country around will give at least some idea of the geological history, e.g., whether it is glaciated or not. But more accurate knowledge of the regional geology will usually be available (for all but remote sites in unmapped country) through the medium of published geological reports. Real study of the site, therefore, may well begin in a library. The records of the local geological survey will naturally be the starting point; for this reason, a concise list of geological surveys around the world is given in Appendix B. Officers of geological surveys will always be helpful in directing inquirers to sources of geological information most useful for the site in question. These sources may include information privately available at other locations, such as universities. Geological reports will usually include geological maps and sections, the careful study of which is an essential preliminary to work in the field; such maps are described in Chap. 14. The extent of this procedure will vary from cases of wild, inaccessible country for which only the most general information is available to city building sites for which complete geological information is readily available in printed form.

Similar study of the records of the geology encountered in adjacent works is a corresponding step. For urban sites, this may be rewarding; for isolated sites, there may be no such records. Assembly of all available information from test borings put down in the area of the proposed work, if not already publicly available (as it now is in some cities, as noted in Chap. 13), can often be most rewarding. Examination of well records in the area of the work will often yield information of value.

Further records to be sought out and studied, especially for civil engineering works (such as highways and airports) that do not involve deep excavation, are those of agricultural soil studies, frequently summarized conveniently in descriptive memoirs on the soils in counties or similar local regional units. Naturally, these valuable records are available only for the more developed parts of the world, but they are always helpful when they are available. Often associated with such soil studies are descriptive works dealing with the physiography of regions. Although general in character, such volumes can often provide a useful background to more detailed site investigation. An outstanding example is a volume by Chapman and Putnam describing the physiography of southern Ontario, a well-written and interesting book accompanied by a most useful map of the area; it is mentioned as typical of the sources of information available to civil engineers that would not normally come to their notice in ordinary practice[10.5]

When all such information has been obtained and studied, examination of the site itself is the next essential. In times past, this meant going to the site, but today a first approach may be made through the medium of aerial photographs. The potential of aerial-photo interpretation is outlined later, so important a part of modern site investigation has it become. When all possible information has been obtained from aerial photographs, with alternative sites possibly greatly reduced in number and limited in area by this ingenious method, site study itself must be undertaken. A geological survey is the first site investigation called for on all but the most restricted sites and those in urban areas. The geological survey can often be done in remote locations concurrently with topographical survey work but by different personnel. Although naturally the work of geologists, geological survey work should at least be understood in broad outline by the engineers who are to use the results it gives; it is therefore also outlined in a later section.

Geological survey work may be supplemented by geophysical investigations, depending on the area involved and the local geological features revealed by the earlier studies. Chapter 12 is devoted to a review of geophysical methods of exploration, now so vital a part of many major civil engineering operations. When all possible progress has been made with such methods, and by careful geological surveying, final confirmation of deduced subsurface conditions will always be necessary through actual determination by means of test holes, either as test pits or test borings. Frequently, these operations are all that a civil engineer thinks of when considering preliminary site studies, but it can now be seen that they form only a part, though a vital part, of a much broader operation. Test boring and test drilling, and the use of the information they give, are given further treatment in the next chapter.

In this fieldwork, not only will rock and soil be studied but also groundwater—its location, its quality, and the impurities it contains. In areas suspected of having high sulfate content in their soils, analysis of groundwater is a particularly important part of the field study. Soil and rock samples obtained during the course of fieldwork will frequently be subjected to laboratory tests. On large jobs, field testing of soil and rock may also be necessary, either by tests on existing ground surfaces or in specially excavated test shafts, tunnels, or pits. And in very unusual cases, extraordinary test procedures, such as trial excavations, the placement of trial sections of fill, pumping tests to determine permeability in situ, or trials with grouting methods, may be necessary.

Actual practice will naturally vary from one job to the next, and the extent of preliminary work will depend on the size and location of the job. The overall pattern, however, remains the same—first, a search for all available written information on the regional geology; second, a study of the site, first through aerial photogrpahs (if available) and then on the site itself, a study in which all the techniques of geological surveying and geophysical prospecting may be used; third, the development of a test-drilling program based on what has been determined of the local geology and checked continually, as it progresses, against the gradually unfolding picture of the local geological structure so that maximum advantage can be obtained from every test hole sunk; and finally, the prosecution of such special tests and field investigations as the work so far carried out suggests are necessary in relation to the structure to be built.

10.3 REGIONAL GEOLOGY

A general appreciation of the regional geology of the area in which a civil engineering project is to be carried out can be of such direct assistance to site investigation, and yet is so often overlooked, that a brief explanation of its potential may be helpful. General geological reports on specific areas, such as quadrangles in the map framework of the United States, will be found to be a regular form of publication by national and state geological surveys. Published as descriptive memoirs, usually accompanied by a geological map, they will have been prepared for purposes other than assisting with engineering site investigation. At the same time, the overview which they give of the geology of the area under review can be most helpful.

Geological formations encountered as bedrock will be a central feature of many such reports; knowledge of the bedrock to be encountered at a project site is a good starting point for all site studies. If bedrock is limestone, for example, even the most general report may be expected to give some indication as to whether solution of the limestone is encountered as a local feature; the dangers inherent in building in such karstic areas will be described later. The character of soils over the area will be explained, i.e., whether they are residual or transported, as will also the glaciation of the area if this has taken place. If multiple glaciations have taken place, this will be recorded. This information can be of great value in indicating the possibility of two or three strata of till, since buried tills are often of unusual toughness due to the action of overriding ice. The direction of ice movements across the area will also be significant in indicating some features of the glacial deposits that may be encountered.

The origin of clays in the region should also be outlined in general geological reports. In southern Scandinavia and in the valley of the St. Lawrence River and its tributaries (including the Richelieu River and Lake Champlain), large areas are found in which the local clay has been deposited in salt water, with consequent unusual characteristics. About 20,000 mi^2 of the St. Lawrence Valley is covered with a marine clay, popularly called the Leda clay (because of a common fossil found in it) but more accurately described as the Champlain clay. The clay has an open microstructure because of its mode of deposition. This results, in many cases, in natural moisture contents that are higher than the corresponding liquid limits, indicating clearly that such soils are highly sensitive. Accordingly, they can lose strength rapidly if disturbed and, because of the breakdown of the internal structure, will not regain that strength through thixotropic processes. Further, these soils can lead to large surface settlements if, for example, their natural moisture contents are reduced, as by the action of tree roots in dry weather. In recent years, Champlain clays have been the subject of much useful research, a helpful review being that of Penner and Burn.[10.6] Figure 10.4 shows the area in which these clays will almost certainly be encountered in site studies. Prior knowledge of the regional geology will, accordingly, assist greatly in planning the site investigation. Corresponding areas in southern Scandinavia are now equally well known.

Entirely different are the properties of glacial deposits found in and around New York City. Wisconsin glaciation extended just beyond the tip of Manhattan, a terminal moraine being a distinctive feature across Staten Island and Long Island. This moraine interfered with drainage of the melting ice sheet, the result being the formation of the glacial lakes Flushing, Hudson, and Hackensack. Soil was deposited in each of them, but the resulting sediments are far from identical. Lake Flushing occupied an area covering parts of the boroughs of Manhattan, Brooklyn, and Queens. Although it is popularly believed that New York is featured by bedrock almost at the surface, these glacial deposits cover an extensive part of the great city. Those from Lake Flushing are of considerable importance; they have been investigated in detail as a result of foundation studies for many of the large buildings built over them.

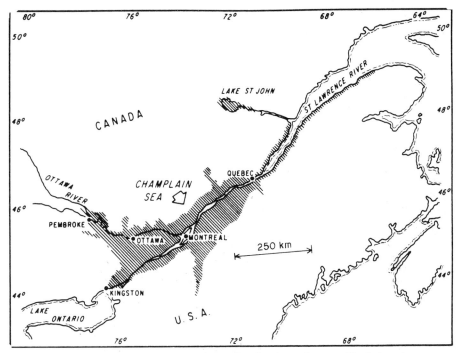

FIG. 10.4 Map showing the extent of the Champlain Sea in northeastern North America.

In earlier days, these buildings were founded on piles to bedrock, but more recently the glacial clay-silt-sand deposits have been used as bearing surfaces, as will be explained later in Chap. 42. The corresponding studies of the "bull's liver," as the deposits are colloquially known, have been extensive, showing a variable succession of varved silty fine sand, silt, and clay. Many laboratory tests on these materials have shown them to be preconsolidated, showing that they must have been overridden by ice near the close of the glacial period, even though this is a conclusion contrary to some geological deductions. Most fortunately, a fine summary of this experience with these soils has been written up by Parsons, his paper being a prime reference for all who are engaged in engineering work in the area once covered by Lake Flushing.[10.7] Figure 10.5 shows the regional geology deduced from this extensive subsurface exploratory work, one of the innumerable reciprocal contributions of engineering to geology, a matter discussed further in Chap. 50.

Different again, and from yet another vast depository of useful regional geological information—the publications of national geological surveys—is the summary report by Scott and Brooker on the great area featured by swelling clay-shales of the Midwest. The Bearpaw formation is part of the widespread Upper Cretaceous rocks found in this region. First named in Montana, the formation reaches its widest extent in southern Saskatchewan, being a problem rock that is therefore truly international in occurrence. Tests conducted on the material in geotechnical laboratories show that the shale is overconsolidated and that it is composed generally of silt- and clay-sized particles, dominated by clay minerals of the montmorillonite type. The overconsolidation is greater than that due to the present overburden, thus giving a clue to the geological history of the area. The properties of the Bearpaw formation naturally lead to many problems in engineering practice, some of which are directly dependent upon the local geology, since deposits vary throughout the great area of its occurrence.[10.8] Once again, therefore, there is a good general guide

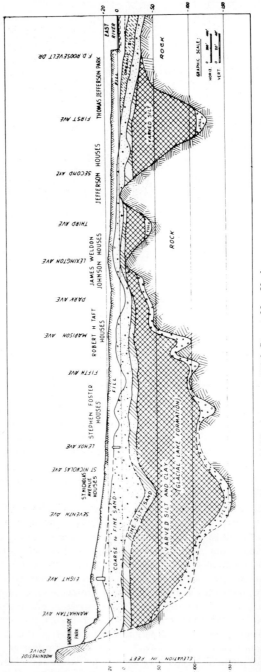

FIG. 10.5 Geological section across upper mid-Manhattan at 113th Street, New York.

FIG. 10.6 Graben structure in landslide area on south side of the Red Deer River at Dorothy, Alberta, typical of topography underlain by Bearpaw shale.

available for use by all who have occasion to contemplate engineering works in the region featured by the Bearpaw formation, another illustration of the fundamental importance of an initial study of local geological records in all site investigations (Fig. 10.6).

British experience has provided useful examples of possible variations in subsurface conditions even though the bedrock at two adjacent sites may be identical. Oldham in Lancashire and Stoke-on-Trent in Staffordshire are less than 40 miles apart, similar in topography, in history, and in development, and both are susceptible to problems due to coal-mining subsidence. Superficial judgment would conclude that the geological conditions underlying the two cities were similar, especially as both are underlain by Carboniferous strata. In fact, subsurface conditions are quite different because of fundamental differences in geological deposition. The Oldham coal beds are much more variable in vertical sequence and in horizontal continuity than those at Stoke-on-Trent. Depth to the bedrock at Oldham varies from nothing to 30 m (100 ft), and the bedrock is overlain by varying depths of till, whereas at Stoke-on-Trent a uniform glacial clay, always between 2 and 3 m (6 and 10 ft) thick, overlies the relative horizontal surface of the bedrock. Once these differences in the regional geology are appreciated, then it can be seen that site investigations in Oldham will be more difficult and more expensive than will corresponding studies at the other city.[10.9]

Before any work is done in the field for a new program of site investigation, therefore, every effort must be made to check on all available regional geological reports so that the general character of the local geology may be grasped before the planning of actual work on the site is started (Fig. 10.7, next page). Such "library research" (so often neglected) can be supplemented through the medium of aerial-photo interpretation for projects that are to cover a large area.

10.4 AERIAL-PHOTO INTERPRETATION

All engineers who have had the opportunity of viewing some geologically interesting country from the air must have been impressed by the value of such bird's-eye views (Fig. 10.8). It is not, therefore, surprising to find that concurrently with the remarkable development of aviation has gone a similarly significant development in the use of aerial photographs for mapping and survey purposes, as well as for generalized geological studies.

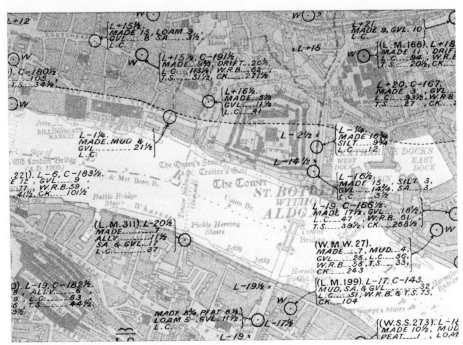

FIG. 10.7 Small part of one of the "6-inch maps" for London, England, showing the wealth of information that may be available for urban areas.

It was only in 1919 that a real start was made in taking photographs from airplanes. Early photographs were made by adapting the rough-and-ready experience gained by a few intrepid flyers in the First World War. The first maps to be produced from aerial photographs were issued in 1923. In 1931, the first use of aerial photographs for highway location appears to have been made. Not until 1935 does there seem to have been any widescale use of aerial photography for geological study; a significant application was made by the Dutch in that year in a reconnaissance study of New Guinea.[10.10]

Great progress has been made since that time. Photography from the air is now used in all states as an aid to route location; it is also used in the provinces of Canada and by the government of Canada for road location in the Far North. Despite the fact that Canada is the third largest country in the world, and that its Arctic archipelago is wild and inhospitable in the extreme, the entire country has now been photographed from the air. So the record can be continued. A new tool of great significance is now available for the civil engineer and geologist; it can be of invaluable assistance in general regional geological studies. With air photos now available, for some locations for several decades, it is possible, by comparing views of the same feature taken at intervals, to get a vivid impression and even quantitative information about geomorphological processes such as shoreline erosion.

The analysis of the features to be seen on aerial photographs and their interpretation in terms of local geology have now been developed as well-accepted techniques to which there are some excellent printed guides.[10.11] As with so many other modern technical developments, the services of the expert are advisable in order to derive maximum benefit from aerial photographs of areas to be studied. An increasing number of civil engineers and geologists are now getting training in this field, and aerial-photo interpretation is ceasing to be the novelty and thing of mystery that it appeared to be in its early devel-

FIG. 10.8 "Geology from the air"—a vivid example of geological structure as revealed from the air; west side of the Butler Range, British Columbia.

opment. Aerial photographs are especially useful in betraying old earth movements such as landslides; some major slips, undetectable on the ground, have been located in this way. The use of stereophotographs is particularly helpful in work of this sort. It is now possible to recognize many of the surface manifestations of permafrost from aerial photographs, and this is especially useful in the region of discontinuous permafrost (i.e., on the southern boundary of perennially frozen ground).[10.12] Muskeg provides its own peculiar patterns when viewed from the air; since these patterns are most revealing when seen in color, the use of strip-color aerial photographs has become a powerful technique in route location through terrain made difficult by extensive muskeg deposits (Fig. 10.9).[10.13]

FIG. 10.9 Muskeg country in Ontario as seen from low-flying aircraft. Scale is given by automobile on the right.

10-16 PRELIMINARY STUDIES

Location of sand and gravel deposits, always a matter of concern in major civil engineering work, is greatly assisted by aerial-photo interpretation. These deposits may be located by careful study of physiographic features and local microrelief, of the forms of gullies as indicating the results of erosive processes, and of the examination of soil tones and vegetal pattern in relation to microrelief (Fig. 10.10). Mollard and Dishaw have recorded work in so locating most of the available sand and gravel deposits in the plains of western Canada, a task extending over more than a decade.[10.14] Over 2,000 prospects were mapped in an aera of 84,000 km² (32,000 mi²). In one season's work, 340 prospects were mapped and good material was found at 305 of them. In 227 of these locations, there were no pits excavated at the time of their discovery on aerial photographs. Experiences of this kind have suggested an accuracy in such work of at least 75 percent, which readily suggests the economy possible with the judicious use of aerial photographs.

A similar approach is used for assessing the general geology of an area, including study of photographic tone, ground texture, the pattern of physiographic features, especially of drainage, the shape of unusual features, and the interrelation of all these aspects with existing knowledge of typical geology for the area under review. The information thus derived from the photographs can be indicated on maps that may themselves be prepared from aerial photographs; this is but one interlocking of photogrammetry and photo interpretation. The result can be a useful guide to further, more detailed site investigation. It will be clear that the larger the area to be investigated, the more useful will be the use of aerial photographs; as the area being studied gets smaller, so will the utility of aerial photographs decrease until they cease to be a necessary or useful aid at all, as in the case of urban building sites.

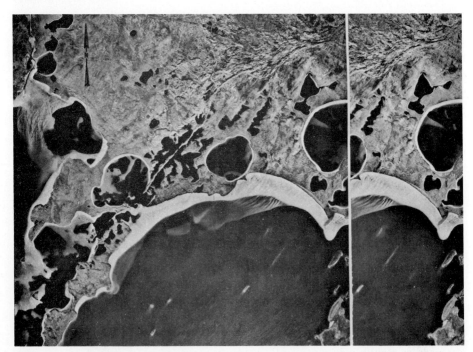

FIG. 10.10 Identifying characteristics of extensive sand and gravel deposits are well shown in the upper part of the aerial view east of Lower Post, British Columbia; kettle holes and channel scars reveal sand and gravel of outwash deposits.

This point directs attention to the overriding fact that, even with optimum use of the best possible photographs and the most expert interpretation, aerial-photo interpretation is not a substitute for investigations on the ground, but only a most useful preliminary. The efficient use of aerial photographs can save considerable time and effort in eliminating obviously unsuitable sites or routes; their proper interpretation can point clearly to features calling for special study on the ground. But in the final analysis, accurate knowledge of the subsurface conditions at the site of engineering works must be obtained by careful subsurface investigation. And this must, in turn, be based upon at least a general knowledge of the geology of the site; thus, even with the aid of aerial-photo interpretation, study of the geology of the site on the ground is almost always a further preliminary to actual subsurface investigation. Even geological mapping is today often aided by air-photo interpretation.

Ground truth is the term now generally used to indicate the essential checks upon the ground that must be made in association with even the best of aerial-photo interpretation. It is a term, and a procedure, used and welcomed by the most enthusiastic advocates of this remarkable example of applied photography. The term will be found, for example, in the first paragraph of a paper describing one of the most extensive and urgent uses of photo interpretation in recent years, the survey of the proposed route for the Trans-Alaska Pipeline System.[10.15] The route of this facility traverses 1,270 km (789 mi) of varying terrain between the Arctic and the southern coast of Alaska, including 950 km (590 mi) of permafrost. Before the pipeline could be designed, the most accurate information possible in the time available on all ground materials, permafrost conditions, and the environment had to be obtained.

This work started in the manner just recounted, i.e., by a careful study of all available published information. A preliminary soil boring program was then initiated, with borings up to 16 km (10 mi) apart. The generalized soil conditions along the line thus determined were gradually refined by a continuing boring and soil test program, involving over 3,500 borings by January 1974. Concurrently, however, a photomosaic was prepared, from aerial photographs, for a strip 2 mi wide, to a scale of 1:12,000 on which all test borings were indicated (Fig. 10.11). Study of the mosaic enabled a generalized map to be prepared of the physiographic provinces through which the pipeline would be build. Use of results from the test borings in association with the mosaic—determination of ground truth—permitted the tight schedule for construction to be met without the usual closely spaced borings that would have been necessary had ordinary standards been applied. It was estimated that something on the order of 20,000 borings would have been necessary had such ordinary standards been followed. The physiographic provinces were further subdivided into 55 typical *landforms*, and on this basis the line was successfully designed and constructed.[10.15]

For the proposed oil pipeline from the Canadian Arctic up the full length of the valley of the Mackenzie River, a pipeline not yet built but actively considered during the early 1970s, an even more extensive program of air-photo terrain study was carried out. An area approximately 3,600 km (2,000 mi) in length, and from 5 to 50 km (3 to 30 mi) wide, was studied in detail by means of air-photo interpretation, aided by test drilling and sampling on the ground, to determine ground truth. The extent of the information thus obtained, on the terrain through which the several possible routes for the pipeline might run, is indicated by the fact that the final report on the work ran to four volumes. Fortunately a concise and convenient summary is available as a first guide for those who have to meet similar challenges. As with the Alaska study, a preliminary study of available literature was made, as background for work in the field, no less than 272 references in the literature being cited in the complete report.[10.16] This number is given to indicate the amount of information that is already available, even for work in such an isolated part of North

10-18　PRELIMINARY STUDIES

FIG. 10.11 An aerial view of part of the route of the Alaska pipeline; here the pipeline (center) is elevated on steel piles to obviate ground thawing; Middle Fork of the Koyukuk River on left, Yukon-to-Prudhoe Bay highway on right, location in the central Brooks Range (lat. 67°35′ N).

America as the Mackenzie Valley. It will help to emphasize what has already been said about the essential requirement of a "literature search" as the starting point for all site investigations.

A development of space-age technology, satellite photos have joined traditional air photos as part of the data base for many modern geological studies. Taken at higher elevations, of larger areas, they have provided a broader view of the earth's surface and have led to the discovery of many new faults in its crust. They are of particular interest in mineral exploration (including petroleum) and in the siting of nuclear power plants. One might assume that such small-scale photography is of little use in engineering geology, but this is not entirely true. The U.S. Geological Survey displayed a mosaic of such photography covering the continental United States at a geological meeting in 1974; resolution of detail was such that the land survey sections were clearly visible.

10.5　GEOLOGICAL SURVEYING

It should not be difficult for an engineer to appraise the general nature of the local geology at a site, provided he entertains an appreciation of the importance of the science to the work. He should have little difficulty in obtaining assistance when necessary with regard to geological problems from the respective geological survey, the staff of a local university, or a consulting geologist who has specialized in engineering work. The engineer will be able to assist, either personally or through his staff, with the necessary geological survey work, which is the job of a specialist. His cooperation will certainly be needed in placing before the geologist a succinct account of the work to be carried out and the specific questions relating to subsurface conditions which are of importance.

The initial simplicity of geological field observations is almost dangerous, certainly to a beginner. Although so simple, involving little more than accurate observations with the unaided eye and simple measurements with hand instruments, they call for an unusual concentration of attention and involve a background of experience. After making a general reconnaissance of the area to be surveyed, the geologist must carefully examine all exposures of solid rock, noting and recording the nature of each exposure and its dip and strike. These direct observations are, however, but the start of investigation. Although in topographic mapping, record is made only of what can be seen, in geological mapping the contacts of adjacent beds are the main goal of the survey, and these are quite often hidden from sight by surface deposits of drift material. Thus, the direct examination of rock exposures must be followed by a detailed examination of all topographical features that may give a clue to the hidden geological structure beneath.

Escarpments, indicating a relatively hard stratum; unusual deviations of watercourses and waterfalls, also often indicating a hard-rock layer; the elevation and location of springs; these and many other topographical features come within the critical review of the geologist in the field. The natural slope of the ground must always be remembered, especially when consideration is being given to such surface features as are demonstrated by the soil; here the upper limit of all special effects is clearly the only one that can be used with certainty. Naturally, a contact of two adjacent beds can only rarely be traced, or even seen, except in isolated places; its general location must be obtained by inference from the accumulated evidence of other observations. In a similar way, the various structural features noted in Chap. 4 will not often be found in anything but minor and infrequent locations. Occasionally complete sections of rock folds are exposed; but generally, folding will be revealed only after the study of numerous dip observations. Faults are sometimes shown a little more clearly, especially if they are of comparatively recent origin and their effect on topographic detail is still unobscured. Glacial structures and effects are sometimes clearly indicated.

It cannot be too strongly emphasized that these notes are no more than an introduction to geological field methods; mention is made only of the principal points. Experience in the field under expert guidance is the only sure training, although much help can be obtained from the writings of experienced geologists. One excellent guide is listed at the end of this chapter. The impression may have been given that geological surveying can be carried out only in country exhibiting noticeable topographical variations and frequent outcrops of solid rock. This is not the case, although it must be admitted that mapping the structure of exposed solid sections of the earth's crust is comparatively easy when compared with mapping a drift-covered area. In addition to geological maps showing bedrock, however, a second complete set of geological maps may exist, maps which show the nature and structural arrangement of the superficial deposits above the solid rock formation. The superficial deposits are sometimes characterized as drift (although they include all recent deposits and not only glacial till), and the maps are sometimes described as *drift maps*. The importance of this branch of geological surveying to the civil engineer will immediately be obvious. Attention was first given to "maps of drift" by the Geological Survey of England in 1863, and since that time they have gradually come to be regarded on an equal footing with solid maps, especially in countries where the surface has been glaciated and therefore has a soil mantle of glacial deposits.

Mapping methods for soil are similar to those for solid-rock strata, but there is added emphasis on the close examination of the soil constituents. The surveyor may be unwittingly aided by the excavation work of burrowing animals, not only because of the spoil dumps which they accumulate outside their burrows but also because of the well-known habits of the main species, e.g., the habit which rabbits have of choosing dry sand deposits. The hand auger provides a simple and ready means of obtaining actual sections through deposits with the expenditure of little energy or cost—in marked contrast to the use of test drilling in rock strata. Drift maps are naturally complementary to solid maps, covering only those areas of ground on which solid rock is not exposed. The actual structure of any particular area covered by both solid and drift maps may therefore be visualized by imagining the latter superimposed on the former. There are areas in which it is difficult, if not impossible, to map the underlying rock strata; if these strata are situated at a great depth below surface, the drift map will constitute the only geological map for the area. Chapter 14 deals with geological maps in some detail.

Because soils are derived from parent material whose location can usually be readily determined from the overall geological picture of a region, detailed study of soil constituents is now being developed as an indirect method of "prospecting." *Geochemical prospecting* is already a term of real meaning. Vogt, a pioneer in this work in Norway, noticed many years ago that the health of grazing cattle was affected by changes in the chemical composition of the soils on which they grazed. These changes were due to variations in the underlying bedrock, which Vogt mapped. Study of vegetation is therefore also developing as a special aid to geological surveying, and *geobotanical prospecting* is another term that may be encounterd and with which the civil engineer should be familiar.

Naturally, such studies of soils are closely linked with the soil-survey practice of the agricultural soil specialist; soil surveying is an important part of all pedological work. The field methods employed are identical with those used in geology, but they are usually carried out in more detail, since the pedologist is concerned with even minor changes in soil type. Correlation between overall geology and the general pattern of agricultural soil distribution will usually be noted in the records of pedological soil surveys. Correspondingly, there is a growing volume of information with regard to the interrelation of pedological soil subdivisions and the engineering properties of the soils so classified, especially in highway and airport engineering.[10.17]

10.6 DETAILED SITE STUDIES

Knowledge of the geology of a site is but the beginning of detailed site investigation. Deductions as to subsurface conditions must be checked by actual examination. This may be done with test pits, hand borings, machine borings, and drilling, with experimental shafts and tunnels excavated when necessary in special cases. In this actual probing of the underground, samples of materials to be encountered can be procured for testing in appropriate laboratories. Observations of groundwater will be made in all such test borings and excavations, special observation wells being often necessary for full understanding of groundwater movement. All this work is so important that it is dealt with in some detail in the next chapter of this Handbook.

It is summarized here in order to sketch in outline the full scope of modern site investigation. Against the geological background so far outlined, it can be appreciated that all actual probing of the underground, and all special trial excavations, must be carried out in close liaison with what has been deduced from the surface about the geology underlying the site. This is so obvious a requirement that, as is too often the case with procedures that are obvious, it is sometimes neglected, especially by those who wish to see contracts for test boring and drilling work established with firm estimates of total cost. All such work must be arranged for on a flexible basis, as will later be explained, so that full advantage can be taken of the guidance so often provided by correlation of drilling results with anticipated geology.

There is available one further technique of great value for subsurface study, especially for preliminary work, to be used in advance of test-boring work. This is the application of geophysical methods to the study of ground conditions. In the hands of experts, this has become a powerful tool. It is especially valuable for route surveys and for studies in areas where test boring cannot be carried out (such as in urban areas), at least until the start of construction has been assured. On the other hand, very simple devices, such as magnetometers, can be used for small, straightforward investigations. All geophysical studies, however, are again indirect—as are geological deductions—and so must be checked by putting down test borings before the results they give can be used with certainty. Fortunately, as will be seen, there is today a useful accumulation of information demonstrating the accuracy of many geophysical surveys; despite this, caution must always be exercised when this invaluable aid is used for important work, and ground truth must always be established. Chapter 12 provides an introduction to the use of geophysical methods for site studies.

10.7 WHEN CONSTRUCTION STARTS

There is a widespread but mistaken belief that the work of the geotechnical engineer, charged with the investigation of the subsurface at a projected building site, is finished when the report summarizing such studies has been completed and submitted to the responsible authority. That is really only the "halfway mark," since the geotechnical engineer should have a continuing responsibility for studying and recording the subsurface conditions as they actually *are,* when exposed by excavation. It is naturally always hoped, and expected, that the conditions will be just as predicted from the site investigation, but about this there can be no certainty. Throughout this Handbook examples will be found where conditions as revealed when excavation was carried out differed markedly from what had been anticipated. Sometimes these variations can have serious consequences for the progress of the job and even—in extreme cases—necessitate changes in design. Excavation may also reveal unsuspected features of the geology which, while not affecting construction adversely, are of such significant, academic interest that they should be recorded before being hidden again from human sight.

Site investigation, therefore, is not finished until excavation has been completed and work is about to start, for example, on the placing of the foundation structures or on the lining of a tunnel. It may be difficult to persuade owners that such an extension of site investigation work is imperative, but imperative it is, as all-too-many examples from practice can testify. From the professional point of view, such checking on actual conditions is also of vital importance. It is only in this way that defects in investigational procedure can be detected, and methods and tools improved. When conditions are found to be just as expected there will naturally be a sense of satisfaction, but when variations are found, variations previously unsuspected, there is an equally important challenge to see where the methods used for preliminary work went wrong. And before any excavation is covered up by permanent structures, there is one further vital operation to be performed.

10.8 AS-CONSTRUCTED DRAWINGS

When a structure is built, one of the most important requirements during construction is the careful recording of all changes, even the most minute, from what is shown on the contract drawings. All with construction experience know well that, of all the tasks on the job, the most tedious and difficult is the recording of such changes on contract drawings which, when so amended, become the *as-constructed drawings*. With construction proceeding all around, there is always so much to be done immediately that the temptation to skimp on recording accurately the exact dimensions and details of the structure as built is very great. But as-constructed drawings are *absolutely essential,* not only as a record of what has been built for payment and other current purposes, but also to have as a permanent record in case the day should come, as it does so often, when changes to the structure have to be made or an extension has to be constructed. At such times, the absence of as-constructed drawings can be costly and even hazardous.

Important as are such drawings for what is built, they are even more important—if such an expression can be allowed—for the ground conditions upon which the structure is to be erected. Preparation of as-constructed drawings for what is revealed by excavation is, therefore, the final responsibility of the geotechnical engineer and the proper conclusion of all site investigation work. Not only is a record of the geology actually encountered in excavation essential for current purposes but, just as with the structure, such records are essential and invaluable if ever work has to be carried out adjacent to the building site, as for an extension. Anyone who has had to carry out such building extension work, where as-constructed drawings of the original excavation were not available, will know how frustrating, costly, and even dangerous is the absence of such simple records. These records, so easy to prepare when excavation is complete, give information that will be impossible to obtain again without new excavation—but they must be made right then and there.

Some qualification, perhaps, should be made to the use of that word "easy." Recording the conditions revealed when excavation is complete is, in itself, an easy operation. But it has to be done at a critical time in the overall construction procedure. With excavation complete and the foundation bed approved by the responsbile engineer, there is almost always a keen desire to get the concrete in place, to get started on the "real job" (as so many think of it). Records of the foundation bed often have to be made, therefore, amid an extremely busy construction scene, with the first concrete forms ready to be placed and supervisors anxious to see the first concrete in place. Despite all the pressures that such timing may create, the recording of the foundation beds and other features of excavation must be well and accurately done, if the record is to be fully useful. One of the writers can testify from personal experience how difficult it is, on occasion, to explain to a good construction foreman why such "messing about with soil" is so very important. But the reasons are valid and vital. Careful explanation will usually convince even the most impatient man on the job that the necessary recording is essential to the well-being of the job.

Finally, there is perhaps no one operation that is as vivid a reminder that no two foundation beds are ever identical as is recording what excavation reveals. The fascination of geology lies partly in its infinite variety, and site investigation is yet another way in which the wide scope of geology is brought to light. Whatever are the geological conditions revealed, the record of them will add a small quota to the steadily accumulating body of information on the geology of the area in which the building site is located. As will be seen later in this Handbook, there have been occasions when geological information revealed by excavation, or even by samples from test boring, has provided unique facts that aided in interpretation of the local geological structure. One never knows when this may happen, this possibility being yet another aspect of site investigation that makes every study a challenge and every excavation virgin territory for geologists to explore.

10.9 CONCLUSION

Site investigation is one of the most crucial parts of civil engineering practice. It is, correspondingly, the supreme challenge to the geotechnical engineer. The geology underlying the site being investigated provides the basis for all the work that must be done in exploring underground conditions in order to determine whether the site can safely be used for the purpose intended.

No two sites are ever the same. Two sites may be similar, but even the slight differences that will always exist may prove to be unusually significant. Every investigation of a building site is, therefore, a voyage into the unknown. As buildings get larger and more complex, the investigation of building sites increases in importance and extent. As the more obvious and so the more convenient sites for larger structures, such as bridges and dams, are progressively used up, site investigations for these larger projects will become more complex and will often expose conditions that will tax the ingenuity of even the most competent designer.

It has been said, in the reply to the discussion of papers describing the design and construction of one of the world's greatest bridges of recent years, that "most site investigations prove to be inadequate."[10.18] This is a hard saying. A glance at most of the cases described in the main sections of this Handbook might appear to give grounds for supposing that the statement is correct. It is to be remembered, however, that the cases selected for summary description here include, by their very nature, some geological feature that caused difficulties. For most of these cases, site investigation probably was inadequate. There are, however, throughout the world very many times the number of cases herein noted which were designed, constructed, and are now operating with complete satisfaction, without any of the troubles that will have to be described in later pages. For this vast majority of civil engineering structures, site investigation was entirely adequate. In some cases, it may have been adequate because local geological conditions were so favorable that they compensated for any deficiency in investigaiton that may have existed. In general, however, and to an increasing degree, site investigations in civil engineering practice have been satisfactory. It is the whole burden of this Handbook, and especially of the philosophy propounded in this chapter, to assist in ensuring that all site investigations of the future will be entirely adequate for the successful design and construction of all civil engineering works needed, as they will be throughout the world, for the steady improvement of physical standards of living.

10.10 REFERENCES

10.1 G. McIldowie, "The Construction of the Silent Valley Reservoir, Belfast Water Supply," *Minutes of Proceedings of the Institution of Civil Engineers*, **239**:465 (1936).

10.2 H. S. Jacoby and R. F. Davis, *Foundations of Bridges and Buildings*, 2d ed., McGraw-Hill, New York, 1925, p. 385.

10.3 C. S. Proctor, "The Foundations of the San Francisco-Oakland Bay Bridge," *Proceedings First International Conference on Soil Mechanics and Foundation Engineering*, Cambridge, Mass. 1936, p. 183.

10.4 Personal communication from Sir William Halcrow; for a general description of these works, see W. T. Halcrow, "The Lochaber Water Power," *Minutes of Proceedings of the Institution of Civil Engineers*, **231**:31 (1931).

10.5 L. J. Chapman and D. F. Putnam, *The Physiography of Southern Ontario*, 2d ed., University of Toronto Press, Toronto, 1966.

10.6 E. Penner and K. N. Burn, "Review of Engineering Behaviour of Marine Clays in Eastern Canada," *Canadian Geotechnical Journal*, **15**:269–282 (1978).

10.7 J. D. Parsons, "Glacial Lake Formation of Varved Silt and Clay," *Proceedings of the American Society of Civil Engineers*, **102** (GT6): 605–638 (1976).

10.8 J. S. Scott and E. W. Brooker, "Geological and Engineering Aspects of Upper Cretaceous Shales in Western Canada," *Geological Survey of Canada Paper no. 66–37*, 1968.

10-24 PRELIMINARY STUDIES

10.9 E. R. Hassall and P. T. Rankilor, "A Comparison of the Geological and Mining Aspects of the Urban Redevelopment of Oldham, Lancashire with Stoke-on-Trent, Staffordshire," in *Engineering Geology of Reclamation and Redevelopment,* Geological Society of London, Engineering Group, University of Durham, England, 1973, p. 69; and personal communication from P. T. Rankilor.

10.10 R. G. Ray and W. A. Fischer, "Geology from the Air," *Science* **126**:725–735 (1957).

10.11 R. N. Colwell (ed.), *Manual of Photographic Interpretation,* American Society of Photogrammetry, Washington, 1960.

10.12 R. J. E. Brown, *Permafrost in Canada,* University of Toronto Press, Toronto, 1970, pp. 33–34.

10.13 I. C. McFarlane (ed.), *Muskeg Engineering Handbook,* University of Toronto Press, Toronto, 1969; see chap. 3 on air-photo interpretation of muskeg.

10.14 J. D. Mollard and H. E. Dishaw, "Locating and Mapping Granular Construction Materials from Aerial Photographs," *Bulletin of the Highway Research Board* no. 180, Washington, 1958, pp. 20–32.

10.15 R. A. Kreig and R. D. Reger, "Preconstruction Terrain Evaluation for the Trans-Alaska Pipeline Project," in D. R. Coates (ed), *Geomorphology and Engineering,* Dowden Hutchinson and Ross, Stroudsburg, Pa., 1976, pp. 55–76; see also R. A. Kreig, "Terrain Analysis for the Trans-Alaska Pipeline," *Civil Engineering,* **47**:61–65 (July 1977).

10.16 J. D. Mollard, "Airphoto Terrain Classification and Mapping for Northern Feasibility Studies," in R. F. Legget and I. C. McFarlane (eds.), *Proceedings of the Canadian Northern Pipeline Research Conference,* National Research Council, (Ottawa) 1972, pp. 105–127.

10.17 *Field Manual of Soil Engineering,* 5th ed., Michigan Department of Transportation, Lansing, 1970.

10.18 W. C. Brown and M. F. Parsons, "Bosporus Bridge," *Proceedings of the Institution of Civil Engineers, pt 1,* **58**:506 (1975); for discussion, see *Proceedings,* **60**:503 (1975).

Suggestions for Further Reading

Attention must first be directed to *Field Geology in Colour* by D. E. B. Bates and J. F. Kirkaldy, published in 1976 by the Blanfford Press of Poole, Dorset, U.K.; its helpful text and splendid illustrations, together with its convenient pocket size, making it an ideal companion for all engaged on any major study of a new site for civil engineering works. As the best of all introductions to the use of aerial photos in geological fieldwork known to the authors, Professional Paper No. 373 of the U.S. Geological Survey can be recommended unreservedly; its title is *Aerial Photographs in Geologic Interpretation and Mapping* by R. C. Ray, a 230-page manual published in 1960 and reprinted in 1977.

Another U.S.G.S. Professional Paper well illustrates the message of this chapter: No. 551 of 1967, a 93-page account by D. J. Varnes and G. R. Scott called *General and Engineering Geology of the United States Air Force Academy Site, Colorado.* And the study of the site for the Warragamba Dam in Australia (referenced again at the end of the next chapter) is another admirable example of all that a site study should be.

It may be helpful to note here that the International Association of Engineering Geology now has a Working Commission on Site Investigations that is bringing together experience in this important field from around the world which, when assessed and eventually published, will be of worldwide benefit.

chapter 11

SUBSURFACE INVESTIGATION

11.1 Correlation with Geology / 11-3
11.2 Test Boring and Sampling / 11-7
11.3 Test Pits, Shafts, and Tunnels / 11-10
11.4 Test Drilling / 11-13
11.5 Large-Diameter Test Holes / 11-16
11.6 Field Tests / 11-18
11.7 Supervising Exploratory Work / 11-20
11.8 Utilizing the Results / 11-22
11.9 Claims for Extras / 11-24
11.10 Conclusion / 11-27
11.11 References / 11-29

Subsurface predictions from geological study and from air-photo interpretation, good as they may be, and sufficient for very general reconnaissance purposes, are yet indirect. For use in every branch of civil engineering work, such predictions must be checked by actual penetration of the subsurface, either by means of exploratory boring and sampling of the soil and rock encountered or by actual excavation of small shafts or adits, which will also yield samples of the materials encountered for later testing in a laboratory. Exploratory excavations are naturally limited in scope, by cost, in the first place, and by the problems of deep penetration if more than shallow depths are to be studied. Test boring and drilling with modern tools and techniques, however, can economically reach any depth likely to be needed in site investigations for civil engineering works. These techniques have changed but little in essence over the years, although being refined steadily with improved equipment. This is well shown by the fact that the masterly post-second-war report on subsurface exploration by Hvorslev was reprinted in its entirety in 1968, being still regarded as probably the best guide to the subject in the English language.[11.1]

Drilling for other purposes has made spectacular advances. Worldwide searches for oil and gas, as well as for precious minerals, have led to quite remarkable progress. Drilling in search of gold ore in South Africa, for example, has involved well over 100 holes between 2,500 and 3,500 m (8,250 and 11,500 ft) in depth, one hole having been taken to a depth of 4,350 m (14,300 ft). But the cost of this deep drilling is high, as much as $100,000 or more for a 3,000 m (10,000 ft) hole. Even more extensive has been the drilling

11-2 PRELIMINARY STUDIES

carried out in the search for oil and gas. One hole in Texas has been drilled to a depth of 7,724 m (25,000 ft), while in Europe a drilling at Münsterland, West Germany, was taken to a depth of about 5,900 m (19,500 ft). Planning was carried out, and some preliminary work was started before the project was cancelled, for drilling through the earth's crust to penetrate the Mohorovicic discontinuity; for obvious reasons this came to be called the Mohole. The project would have involved drilling beneath 5,600 m (18,000 ft) of seawater to a total depth of 10,500 m (34,500 ft). The fact that such a project could even be contemplated is clear indication of the progress made in deep drilling and of what is possible when the necessary funds become available.[11.2]

Even more remarkable has been the work done in deep drilling into the seabed, both for oil and gas discovery and for scientific purposes. Some of the massive drilling platforms now in place on the seabed in a number of locations around the world themselves cost several million dollars, giving some indication of the cost of deep drilling on this scale. Equally remarkable is the work being done throughout the oceans of the world by the scientific research vessel, the *Glomar Challenger* (Fig. 11.1).

Much had been learned about the ocean floor by inference from depth-sounding, which determined the relatively detailed topography of the ocean floor in the 1950s. Surveys of the magnetic properties of the ocean floor led to the discovery of intriguing bands of normal and reversed magnetization. Following the demise of the Mohole project, a group of oceanographic institutions pooled their resources (JOIDES program) to investigate, by deep-sea drilling in a systematic manner, the nature of the sequences of sediments and rocks under the ocean floors. Enormous quantities of new information have resulted, and knowledge of the ocean floors has been increased tremendously. It was found, for example, that ocean floors are generally much younger than the continents, and the changing dynamic relationships between continents and oceans have been greatly clarified. Major

FIG. 11.1 The *Glomar Challenger*, justly famous deep-sea drilling vessel.

engineering problems had to be solved to achieve the desired results of drilling hundreds of meters into the sea floor under hundreds or even thousands of meters of often turbulent seawater.[11.3]

All these major operations are, however, on a scale far removed from the more modest requirements of site investigation for civil engineering works, even though they do illustrate the potential that now exists for subsurface exploration far beneath the surface of the earth. One of the main problems in advance explorations of building sites is obtaining any money at all for this essential preliminary work. It must, therefore, be carried out efficiently and economically. This means that, if it is to be fully effective, it must be carried out in close association with geological studies of the site and, as work proceeds, in the closest correlation with geological predictions. This, in turn, makes it quite essential that contractual arrangements for the conduct of test boring and drilling must always be flexible, so that the program can be changed, extended, or reduced, depending on the way in which the gradual accumulation of facts from the test bores confirms or disproves the underground conditions anticipated on the basis of surface geological observations. It is probably in the conduct of subsurface exploration that the engineering geologist and the civil engineer come closest together, since the proper conduct of the work requires the closest attention of both. The actual work of excavation, test boring, or drilling is an engineering operation, the details of which lie outside the scope of this Handbook. An outline of the methods will, however, be presented as a background for proper appreciation of the basic philosophy of subsurface exploration as an essential part, but only a part, of site investigation, which is, in turn, so necessry for the proper execution of exploratory work.

11.1 CORRELATION WITH GEOLOGY

It cannot be too strongly emphasized that underground exploratory work must always be considered supplementary to and conditioned by previous studies of the local geological structure. Brief consideration of the possibilities of complex structures existing beneath relatively simple-looking ground surfaces will confirm the validity of this suggestion. It is unusually important in country that has been subjected to glacial action and where glacial drift now covers an original rock surface which may be entirely unrelated to present-day topography. Even without the existence of glacial conditions, however, disastrous results have been known to occur when reliance was placed on the results of test boreholes that had not been correlated with the local geology.

Consider as an example (not taken from actual practice but typical of many a river valley) the conditions shown in Fig. 11.2. Cursory surface examination of exposed rock in

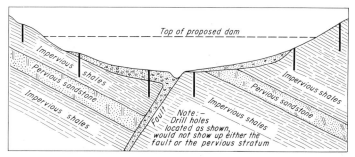

FIG. 11.2 A geological section showing how a fault may be undetected by test drilling.

the immediate vicinity of the proposed dam would disclose outcrops of shale only. Test borings might be put down to confirm what these rock outcrops appear to suggest; and if they were located as shown—as well they might be if the local geology were not considered—they would give an entirely false picture of the subsurface conditions across the valley, possibly with disastrous results. How could geology assist in such a case? Any geological survey made of this dam site would include at least a general reconnaissance of the neighborhood, and it is almost certain that, either by observation of the outcrops along the river bed or through some peculiar features of local topography, the fault would be detected. Even if this were not done directly, a detailed examination of the outcrops of shale on the two sides of the valley would almost certainly reveal differences between the two deposits, possibly minute, possibly even of their fossil contents, but enough to show that they were not the same formation and thus to demonstrate some change in structure between the two sides of the valley.

Figure 11.3 shows a case in which casual surface examinations and the use of the information given by the boreholes shown would be misleading because of the existence of folds in the strata, the outcrops of which are drift covered. Figure 11.4 illustrates the necessity, where the strata involved dip steeply across the site being investigated, of correlating the strata encountered in one drill hole with those pierced by the adjacent hole. If this were not done, and hole 3, for example, had been put down only as far as the point marked A, the existence of the fault would have been undetected unless it were discovered through surface geological investigations.

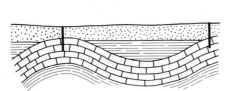
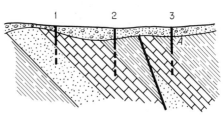

FIG. 11.3 A geological section showing how a fold may be undetected by test drilling.

FIG. 11.4 A geological section illustrating the desirability of correlating steeply inclined strata in test drilling.

A geological survey, even of a preliminary nature, will ususally reveal whether or not an area has been subjected to glacial action. This information will be particularly valuable in the prosecution of underground exploratory work, since it will at once suggest some of the unusual subsurface features that may be encountered. The study of underground conditions in glaciated country is often complicated by boulders and buried river valleys. Those in the valley of the Rhine have already been mentioned. Many of the examples cited later in this Handbook, such as the Mersey Tunnels in England, will illustrate the great importance of buried valleys and the necessity of their detection before construction starts.

One of the most remarkable examples of extensive buried valleys, and the consequent necessity for careful correlation of test drilling with site geology, is in the St. Maurice Valley on the north shore of the St. Lawrence River in Quebec. Before any of the major water-power plants that now distinguish this large river were constructed, the valley had been topographically surveyed, and its geology studied on the ground and from the air (in a very early example of aerial geological reconnaissance, 1928–1929). Carried out under great difficulty in view of the rough country, the study showed that "the St. Maurice is not now flowing in its pre-last-glacial course but that it occupies parts of several old valleys ... being made up [in its present course] of parts of several earlier river systems."[11.4] These old courses were delineated so successfully that the later test boring and drilling at the respective power sites, with this prior warning, located exactly the old buried valleys with consequent benefit to design and construction.

The examples illustrated in the three sections are not given to show that test borings and similar investigations are often faulty. Such a conclusion would, in fact, be incorrect, because in all the cases cited the results from the boreholes may have been correct, but their *interpretation* might not have been, if the results were not correlated with the local geology. From study of the cases described and many other similar instances, the following general guides for underground exploratory work may be suggested. (For convenience, the word *boreholes* will be used to describe all such subsurface work.)

1. No boreholes should be put down before at least a general geological survey of the area has been made.
2. Boreholes should always be located in relation to the local geological structure.
3. Boreholes should, whenever necessary, be carried to such depths that they will definitely correlate the strata encountered in adjacent holes by an "overlap" into at least one bed.
4. In exploring superficial deposits, one borehole at least should, whenever possible, be carried to rock.
5. In all cases of superficial deposits extending to no great depth, test borings should be taken into the rock for some specified distance, never less than 2 m (6 ft), but more than this if the nature of the work warrants the extra cost or if large boulders are liable to be encountered.
6. Unusual care must be exercised in putting down test borings in areas known to have been subjected to glacial action, especially with regard to checking all rock encountered during drilling (as in 5) and for the possible existence of buried river valleys.
7. The three-dimensional nature of the work must always be remembered; for example, three boreholes properly located will define exactly the thickness, dip, and strike of any continuous buried stratum having a uniform dip.
8. Planning and contractual arrangements for the conduct of all test boring and drilling must be kept flexible, so that changes can be made immediately at the site, as the picture of the underground conditions gradually unfolds.
9. Special attention must be paid to groundwater encountered in test holes, careful observations being made of levels before work starts each day in holes that are in the process of being put down.
10. At least one test hole on every site should be cased and fitted with the necessary screen and filter arrangements, so that it may serve as an observation well for groundwater for the longest possible duration.

Methods adopted for penetrating unconsolidated materials and solid rock are naturally different, even though they may often have to be employed in the same hole. It is convenient to describe the former as test borings and the latter as test drillings; test pits and other excavations are sometimes used as an additional precaution in superficial deposits. Exploratory work will therefore be considered under these three headings. Table 11.1 summarizes the various methods currently used in subsurface exploration.

TABLE 11.1 Summary of Main Methods of Direct Subsurface Investigation

Method	Materials in which used	Method of work	Method of sampling	Value in foundation-bed investigations
Hand-auger boring	Cohesive soils and some granular soils above water table	Augers rotated and withdrawn carefully for removal of soil	Samples from augers always very disturbed	Satisfactory only for shallow investigations for roads, airports, and small buildings
Test pits	All soils above the water table	Hand excavation, necessary bracing, and due safety precautions	"Undisturbed" samples can be cut out as required	Most valuable; soil conditions can be studied in situ
Wash borings	All soils except the most compact	Washing inside a driven casing and retaining "sediment"	Material so obtained a rough guide only	Almost valueless; dangerous if limitations of method not fully appreciated
Dry-sample boring	All soils except the most compact	Washing inside a driven casing; sampling tool used at bottom of hole	Relatively "undisturbed" samples usually obtained	Most reliable of simple methods of boring; groundwater information lacking
Undisturbed sampling	All soils except the most compact	Forcing sampling tool into cased hole to get continuous core of soil	Best possible "undisturbed" samples	Best available method of soil sampling, with a wide variety of excellent sampling tools available
Core drilling	All solid rock and very compact soil	Rotating power-driven coring tools with diamond, shot, or steel cutters	Cores cut out and recovered from holes	Best method of studying bedrock and boulders
Caissons, trenching, and tunnelling	All types of ground	Regular construction methods used, with due safety precautions	"Undisturbed" samples obtained from working faces	Best of all methods, but expensive; essential for major works

11.2 TEST BORING AND SAMPLING

The purpose of underground investigations must be clearly appreciated at the outset. Only in special cases is the purpose merely to find out where the hidden surface of rock lies. As a rule, it is also necessary to identify the material penetrated before rock is reached, not only for purposes of construction but also for the preparation of designs, especially when foundations are to be constructed in the unconsolidated material. For important foundations, and when rock has to be excavated, the nature of the rock penetrated by drilling must also be investigated. Usually, therefore, the aim of preliminary work is not only to provide information about underground boundaries between essentially different materials but also to procure samples of the materials for laboratory analysis and testing. In view of the attention now paid to the testing of all soils, it is almost always of the utmost importance to obtain relatively undisturbed samples.

The simplest type of exploration in unconsolidated material can be done by probing with a steel rod; in shallow ground, depths to rock may sometimes be explored in this way, but the method is extremely limited in its application. The next stage involves the use of a soil auger, a boring tool described by its name, which may be obtained equipped with many special devices for penetrating different kinds of material and for procuring samples that, although not undisturbed, are often satisfactory for preliminary testing in a laboratory. The auger bores the hole, and samples of cohesive material are obtained by withdrawing the tool and cleaning off the material adhering to it. Holes through sand have to be lined with pipes, but special drills and devices with which samples may be obtained are available for this type of work. A soil-auger set, complete with the usual extra fittings and capable of boring to 9 m (30 ft), weighs less than 45 kg (100 lb) when packed for carrying and is relatively inexpensive. It is strange that more use is not made of this equipment, at least for small-scale work.

The soil auger could be very usefully substituted for a type of boring, generally called *wash boring,* which was widely adopted by engineers in the past. In wash boring, a hole is excavated by washing inside a pipe casing which is forced into the ground as the hole is made; material excavated is trapped in a washtub, where it is examined and identified (if possible). Wash borings may be useful in some cases to show where a hidden rock surface is located; but as a means of exploring unconsolidated material, they are worse than useless, since they can give very misleading information. They reveal the material penetrated only in a thoroughly disturbed state, generally with much of the finer material washed away; they give no clue to underground water conditions and no sample of any use at all for testing. A simple modification of wash boring provides a much more satisfactory method. A cored hole is sunk by washing, but the washing is stopped at intervals (determined by the nature of the ground) so that a pipe or special sampling device may be lowered into the hole, forced into the undisturbed material at the bottom of the hole, and withdrawn with a sample enclosed. Washing out then continues until another sample is required. This method also has the disadvantage of not immediately disclosing underground water conditions.

Limited in scope and efficiency though they be, auger boring and wash boring should not be written off completely. They are simple to conduct, and the equipment needed is to be found on even the smallest construction job. Many a borrowed carpenter's auger has been blunted irremediably by being used (with permission!) as a soil auger. If the limitations be kept in mind, especially of the character of the material washed up from a wash boring, then the two methods have a place in subsurface investigations—in isolated locations for preliminary site studies or for quick checks on conditions underground when no other equipment is available. Reliance upon results obtained only from wash borings has, in the past, led to many sad experiences and even to some disasters, the natural character of soils penetrated (especially till) being often completely destroyed by the action of washing. Wash borings, therefore, are to be avoided whenever possible and used only with the greatest discretion when no other course is open.

Dry coring, either by rotary drilling, by percussion sinking of holes, or by continuous sampling with well-designed sampling tools, is eminently desirable for all modern site studies. Soil-sampling tools can then be used to obtain undisturbed samples at every desired depth in the holes formed by drilling or driving, when continuous soil sampling is not required. The design of samplers for this purpose has progressed considerably since the need for undisturbed samples was reaffirmed by the advance of studies in soil mechanics. (The word "reaffirmed" is used, since Alexandre Collin stressed as long ago as 1846 the need for samples of clay that were in their "natural state.")[11.5] A variety of excellent tools is now available for obtaining good samples from all the main types of soils. Piston samplers are a common type, but for friable soils the so-called "Dennison sampler" is useful, because loss of the sample is minimized by spring guards. Freezing, chemical injection, and impregnation with asphalt have all been used in developing methods for procuring undisturbed samples of cohesionless soils such as sands. The necessity of getting undisturbed samples arises from the vital importance that the actual density of such soils in place may have; if undisturbed soil is in a loose state, subsequent compaction, as by vibration, may cause trouble.

A parallel development has been the design of improved methods for obtaining what are now commonly called "Shelby-tube" samples, taken by forcing thin-walled steel tubing into the undisturbed ground, without the aid of wash boring; the steel tubing is so arranged that the lower section can be removed for sample retrieval. Soil sampling in this way has now become an accepted practice in soil mechanics. Measurement of the force required to penetrate successive soil strata, if carefully recorded, can itself, in association with sampling, be used as a guide to soil properties.

Emphasis must again be placed upon the importance, for the proper study of subsurface soil conditions, of obtaining good undisturbed soil samples. Pioneer work in this direction was carried out in the exploratory work that preceded the construction of the San Francisco-Oakland Bay Bridge. Undisturbed samples were obtained along the line of the bridge from depths up to a maximum of 82 m (273 ft) below low-water level; thus, the unusually deep foundation for this great structure could be designed with accuracy and confidence.[11.6]

On the east coast of the United States, another pioneering test-boring program was carried out in 1935 and 1936 in connection with the Passamaquoddy tidal-power scheme then projected by the United States government. This work was all in exposed coastal waters and was hampered by severe winter climatic conditons, tidal currents in excess of 8 knots, and a tidal range of 8.4 m (28 ft)—a combination of adverse conditions that could hardly be equalled. Despite this, however, test sampling was conducted and core drilling into the rock over a large area of the sea near the coast was carried out in conjunction with geological investigations; thus foundation-bed conditions for the structures then proposed were accurately investigated. The deepest hole put down during these interesting operations was in 36 m (120 ft) of water and through 50 m (164 ft) of unconsolidated material above the rock, which was then core-drilled.[11.7]

More than 20 years later, the Passamaquoddy scheme being still under investigation, further exploration was made of the sea bottom from a fully equipped barge that had to be towed 4,830 km (3,000 mi) to the site. The barge was moored with long anchor lines, and a 30-cm (12-in) casing pipe was lowered through its center, decreasing in size to a 15-cm (6-in) pipe, ingeniously guyed under water to the ends of the barge to give requisite strength against bending. With this arrangement, undisturbed samples of marine clay 12.5 cm (5 in) in diameter were satisfactorily obtained (with a fixed-piston sampler) from depths as great as 97.8 m (326 ft) below the surface, despite the high tidal range and swift currents already mentioned. Core drilling, through a 10-cm (4-in) casing in the seabed, to a penetration of 4.6 m (15 ft), was also carried out (Fig. 11.5).[11.8]

It is not only in the sea that deep water has to be penetrated in procuring soil samples. For the formation of the Tappan Zee Bridge across the Hudson River, test boring was carried out to depths of 88.5 m (295 ft), and undisturbed soil samples were obtained for laboratory testing. It is, however, somewhat natural that the most difficult test-boring operations are those conducted in the open sea. A significant record of successful work of this kind is gradually being built up; one of the most hazardous examples yet undertaken is the investigation of the bed of the English Channel in connection with revived hopes for a channel tunnel. Test borings were made to depths of 67.5 m (225 ft), and samples were obtained; the work was done in the open sea in depths of water up to 48 m (160 ft), with added hazard caused by the strong currents that sweep through the Straits of Dover on every tide. This work naturally involved the use of specially equipped ships (Fig. 11.6).[11.9]

If there is any possibility of using fixed stagings, however, even in the open sea, they are to be preferred; the cost of their construction is usually more than repaid by the added security and continuity of the work. In exploring possible sites for oil-well derricks in the

FIG. 11.5 Working platform and rigging for deep drilling at the site of the proposed Passamaquoddy tidal-power project, Maine, an early (1935) seabed investigation in 36 m (120 ft) of water.

FIG. 11.6 Converted landing craft *Ian Salvor III* in the English Channel, taking test borings in the seabed in 1959–1960 as part of the investigations for the proposed channel tunnel.

deep waters of Lake Maracaibo in Venezuela, test boring was carried on up to 16 km (10 mi) offshore from welded tubular-steel platforms that were floated out to location and then secured in place by means of steel spuds 36 m (120 ft) long. From such platforms, soil samples were obtained from depths up to 75 m (250 ft).

These were all big jobs; they are mentioned to show that there is no reason why subsurface exploration cannot be carried out anywhere that civil engineering works may be constructed. On small jobs, however, just as much as on the largest jobs, the same attention must be paid to test-boring work in order to ensure accurate records of all soil strata penetrated and the best possible undisturbed soil samples for testing in the laboratory. In this way, preliminary geological studies can be corrected, confirmed, or extended, with a view to determining accurately the subsurface conditions that will be encountered when construction commences.

11.3 TEST PITS, SHAFTS, AND TUNNELS

Another means of investigating superficial deposits is the test pit, an excavation large enough for a person to dig in comfortably. In practically all cases, test pits must be lined with timber, the design and placing of which must always be carefully checked. Clearly, there is a limiting depth to which test pits may reasonably be carried; hence their main use is for the investigation of relatively shallow depths of unconsolidated material and particularly for the study of surface deposits of gravel, sand, or clay for use in construction. Test pits have the advantage that the deposits penetrated can be easily examined in place, and undisturbed samples can be obtained, when necessary, from the bottom of the pit. They will disclose underground water conditions in the material penetrated, but clearly they can be used only on dry land. For extensive exploratory work, test pits are usually more expensive than test borings; but on small jobs for which a drilling outfit would have to be specially procured, they may sometimes prove economical.

When test pits are used to investigate surface deposits of construction material, it will be necessary to take regular samples of the materials encountered. Sampling is a matter that does not always receive the attention that it deserves in engineering work; it is not just the haphazard collection of small quantities of material that one might imagine it to be. Two objectives are important in sampling: (1) obtaining samples typical of average conditions for all material to be investigated and (2) obtaining samples representative of maximum and minimum characteristics of the material.

When investigating material to be used in construction, both objectives must be remembered, although the former will usually be the controlling one unless the material is extremely variable in composition. Samples must always be taken at regular intervals in materials of the same appearance; from a large number of these, thoroughly mixed at the site, one or more average samples should be taken. These may then be used for testing, but they must always be accompanied by at least one unmixed sample, selected as typical for the stratum of material being studied, in order to check on the correctness of the average samples obtained. (Naturally, this procedure applies only to materials that can easily be mixed and not, for example, to stiff clay, from which a number of individual samples must be selected.) All samples should be most carefully marked with full information about the location from which they have been taken; it is most exasperating to have a good set of samples which cannot be identified.

Simple windlasses can be used in test pits for removal of excavated material and samples, but there is naturally a maximum depth to which pits can be sunk with such simple means. When handpower for such haulage has to be abandoned in favor of mechanical equipment, the exploratory operation becomes an engineering task and must be subject to the usual preliminary economic and design studies. This will be the exception, however, rather than the rule, the greatest use of test pits being for relatively shallow exploration. They are so valuable for this that it is surprising that they are not more widely used than they are. Test pits have the advantage that they can be started with the simplest of tools and, with experienced supervision, can be carried down to depths of several meters safely, conveniently, and economically.

It is occasionally found on major projects that subsurface conditions are so critical, and so uncertain, that nothing less than a major test shaft will suffice. One of the early projects of the Tennessee Valley Authority, the design of the Gilbertsville Dam on the Tennessee River, provided such a case. The ground consisted of limestone bedrock overlain by about 15 m (50 ft) of water-bearing soil. It was decided that a shaft would be the best means of determining soil properties, the character of the rock surface, and the way in which steel sheet piling behaved when driven through the local overburden; in addition, a shaft would provide a means for inclined drilling into the bedrock. A ring of steel sheet piling was therefore driven to rock, and excavation was carried down in it in the open, with bracing installed as required. A bad blow occurred, however, 4.2 m (14 ft) from bedrock, and it was then decided to freeze the soil around the shaft. This was done with equipment that the TVA had on hand, using a brine-circulating system; thus, the shaft was unwatered and excavation was completed. Bedrock was found to be a dense, black siliceous limestone into which 25 of the 50 steel piles had penetrated, as deep as 12.5 cm (5 in); six of the interlocks between piles had, however, failed. Test drilling was then carried out from the bottom of the shaft, as shown in Fig. 11.7. It is to be noted that the drill holes from the base of the shaft were steeply inclined. It was in this way possible to explore a comparatively large area—or more properly, a large volume—of ground from the single shaft.[11.10]

The sinking of this shaft, as can be seen, was a major construction operation. So also are the construction of special exploratory tunnels in advance of the excavation of major tunnels or for unusual explorations such as those into the lower parts of massive landslides. Indicative of the extent of this kind of major subsurface exploration work was the award of a contract valued at $1.3 million for a 2,500-m(8,200-ft)-long pilot tunnel beneath Loveland Pass, about 100 km (60 mi) west of Denver, prior to the construction of a major highway tunnel, about which more will be said when tunnels are considered later in this Handbook.[11.11] Equally remarkable has been the construction of a full-size 20-m (65.6-ft)-long section of subway tunnel in Atlanta, Georgia, purely for experimental and

11-12 PRELIMINARY STUDIES

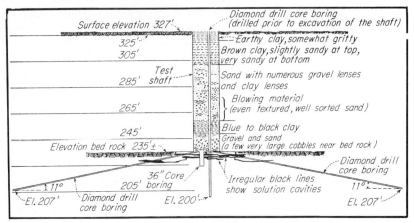

FIG. 11.7 Cross section through special test shaft at Gilbertsville Dam (TVA) showing use of inclined test holes for subsurface investigation.

exploration purposes, prior to the building of Atlanta's new subway system.[11.12] Prior to the construction of a new water-power plant on the Columbia River at Revelstoke, British Columbia, special study had to be given to an old major landslide, the foot of which would be inundated when the power dam was complete and closed. The slide has a volume of about 1.5 billion m³ (1.95 billion yd³). Its stability had to be assured; it is not surprising to find that one of the measures taken to study its condition in advance of final construction was the driving of an adit 250 m (820 ft) long into the slide near its base (Fig. 11.8).[11.13] Exploratory tunnels were used to similarly good effect in advance of final designs for parts of the Snowy Mountains project in Australia.

FIG. 11.8 View inside the test adit constructed to explore the Downie slide near Revelstoke, British Columbia.

FIG. 11.9 Test drilling into the bed of the Vltava River, adjacent to the fourteenth century Charles Bridge, with Hradčany Castle and the Cathedral of St. Vitus in the background.

11.4 TEST DRILLING

When rock surface is reached, either in a test pit or in test boring, a change in method is necessary if the rock has to be penetrated. Usually the rock is penetrated, if only to determine its nature and to make sure that it is solid rock and not a boulder. Various methods are available, but reference need here be made only to core drilling, in which a cylindrical hole is drilled around a central core which is periodically broken off from the bedrock at its lower end and removed from the hole for examination. The rotary drilling machine used for this purpose is fitted with a special bit equipped with black diamonds (diamond drilling), chilled-steel shot (shot drilling), or removable steel cutting teeth. The choice of equipment will depend to some extent on the nature of the rock; the size of hole will depend on the drilling tool available and on the anticipated depth of the hole. Core drilling is a highly specialized operation, necessitating skilled workmanship and experienced supervision. Percussion drilling is another means of penetrating rock, but it is not widely used for exploratory work.

The idea of diamond drilling originated in Switzerland in 1863. The first machine was hand-operated, but a steam-driven machine was used as early as 1864 in the Mont Cenis Tunnel between Italy and France. The bit speed was only 30 rpm, and the penetration was about 30 cm (12 in) per hour. The first United States patent for a steam-driven dia-

11-14 PRELIMINARY STUDIES

mond drill was issued in 1867; by 1870 a 250-m (750-ft) hole had been drilled with the first American machine in the search for coal near Pottsville, Pennsylvania.[11.14] Steam continued to provide the motive power for diamond drilling until the First World War, after which gasoline engines took over. With higher speeds possible and with greater flexibility because of the use of lightweight materials, diamond drilling has steadily improved, so that today it is an efficient and reliable process in the hands of expert operators.(Fig. 11.9).

In some types of soft or disintegrated rock, it will sometimes prove difficult to obtain complete sections of core by drilling, and estimates will then have to be made, based on the operator's observations and experience, with regard to the exact nature of the material. The amount of core recovered will vary with the type of rock penetrated; typical average values are quartzite, 90 percent; granite, 85 percent; sandstone, 70 percent; limestone, 60 percent; shale, 50 percent; slate, 40 percent. These figures are naturally only a guide to what may be expected.

The action of a core drill will not, in general, affect the bedrock through which the hole is drilled; diamond bits, especially, bore a clean, smooth hole. Consequently, it is possible to utilize cored holes for other purposes. When the rock is to be subjected to water pressure (as in a dam foundation), drill holes can be capped and filled with water which is then kept under observation while pressure is applied to it; note can be made of whether the hole "holds water" or not (Fig. 11.10). An interesting investigation can sometimes be

FIG. 11.10 Diamond drilling at the Lower Notch, Montreal River, Ontario, many years before a dam was constructed at this site.

carried out in drill holes of relatively large diameter (about 10 cm or 4 in and over) by the use of a periscope device equipped with an electric light below the inclined mirror. When this is lowered into the hole, the appearance of the rock walls can be examined with a fair degree of certainty. This simple idea has been greatly extended by the development of ingenious "borehole cameras," specially designed instruments that fit into even small-diameter boreholes and secure a photographic record at any required depth and through the 360° exposure around the hole. One of the earliest of such instruments was the NX camera developed by geologists of the U.S. Corps of Engineers; this camera was used in the investigation of a fault condition in foundation rock encountered during the construction of the Folsom Dam.[11.15] The instrument is designed as a slim cylinder that fits into a 7.5-cm (3-in)-diameter borehole. Through a rotating conical mirror it can photograph throughout the 360° of the borehole onto a flat 35 mm film; the process is reversed when the developed film is viewed by those responsible for the investigation. At the Folsom Dam, not only did a camera of this type reveal details of the critical fault zone, but it showed up delicate changes in rock coloring and fractures as small as 0.25 mm (0.01 in) in thickness, as well as the surface of the groundwater table.

Steady improvement has been made to instruments for borehole photography since they were first introduced (Fig. 11.11). There are now commercially available stereoscopic cameras with which stereo photos have been taken at depths up to 3,000 m (10,000 ft); others can carry down a borehole enough photographic film to take 600 pairs of stereo

FIG. 11.11 Borehole camera suspended from lowering equipment.

photos without being withdrawn to the surface. By photographing a compass suspended beneath the camera, the exact orientation of faults, seams, and other geological features can be recorded.

Further improvements have been made in this field in Europe; an electronic borehole camera has been developed which, in association with the appropriate circuits, will project an image directly onto a television screen aboveground. This instrument will also operate in 7.5-cm (3-in)-diameter boreholes, giving a reasonable image on a 56- by 18-cm (21- by 7-in) screen. Since the idea of combining television and geology may seem strange to some, it may be appropriate to note here that underwater closed-circuit television cameras have been successfully employed for unusually difficult riverbed exploration. A notable example was the study of the bed of the Columbia River before designs and plans could be made for the closure of the Dalles Dam. Preliminary surveys by divers proved to be unsatisfactory because of high velocities, and so television was used for the final investigations. The camera, suspended by a steel cable, was operated from a strongly moored barge. By this means, the character of the riverbed, i.e., whether it was bare rock or covered with gravel, could be distinguished. On the basis of this information, plans for the critical closure operation were successfully developed, and the gap was finally closed after over 2.3 million m^3 (3 million yd^3) of fill was placed.

Reference has been made only to vertical boreholes, since the great majority of holes are vertical, especially unusual ones such as those necessary for undersea investigations. Diamond drilling had to be carried out, for example, in the Strait of Canso in eastern Canada in order to prove up the bottom of the strait prior to the placing of a massive rock causeway which joins the island of Cape Breton with the mainland of Nova Scotia. Using a tubular steel frame for supporting casing and drill rods through 58 m (186 ft) of water, the contractor successfully carried out regular diamond drilling after passing through and casing 15 m (50 ft) of soil. This was a relatively small job, when compared with some of the test drilling carried out in the open sea in advance of the placement of large oil-drilling platforms. It is mentioned, however, as a reminder that in civil engineering practice, such innovations were developed long before current major platform oil-drilling projects made them a necessity.

All drill holes need not be vertical. It is, in fact, essential that civil engineers in particular remember that holes can be drilled at any desired angle, since for some investigations inclined holes, horizontal holes, or even holes inclined upward may be necessary. A good example of the use of inclined holes is found in the investigation of rock conditions in the bed of the Hudson River, New York, outlined in Chap. 20. Horizontal holes were used extensively in the investigation of rock conditions in the abandoned tunnels utilized for the Pennsylvania Turnpike. In this case, working faces were available, and knowledge of the rock to be penetrated as the tunnels were completed was essential. Accordingly, about 1,050 m (3,500 ft) of horizontal holes were drilled to supplement the 3,000 m (10,000 ft) of vertical holes in the thorough study made of the geology of the turnpike tunnels. One of the horizontal holes was successfully drilled to the almost unprecedented length of 435 m (1,450 ft) from the east heading of the Tuscarora Tunnel.[11.16] And in the necessary exploration of Ripple Rock, British Columbia, before its demolition, appreciable overhead drilling was carried out when the access tunnel extended out under the two rock mounds that had to be removed by blasting.[11.17]

11.5 LARGE-DIAMETER TEST HOLES

A natural extension of the techniques of drilling small-diameter holes in rock was to develop drilling rigs capable of drilling holes of such large diameters that a person could be lowered into them. Such holes been drilled in the past in other branches of work, and their value in civil engineering practice was quickly recognized. They are drilled with machines of the "Calyx" type, with diameters varying normally up to 1.1 m (42 in), although holes up to 1.8 m (72 in) in diameter have been drilled. The method of drilling and removal of the cores is similar to that followed for smaller holes, although breaking off the core from the bedrock is sometimes difficult and necessitates the use of special wedging devices or of blasting. The holes are of avail only if leakage of groundwater into them can be taken care of by a small pump; special precautions must always be taken to keep a supply of fresh air at the bottom of the holes, particularly if blasting has been used for core removal.

When the holes are drilled to the requisite depth and cleaned out, the geologists and engineers in charge may be lowered down in suitable cradles; with the aid of portable lights, they can carefully inspect the surrounding rock exactly as it occurs in place; they can investigate boundaries between beds, study fissures, and make a thorough and complete exploration of the rock with certainty and convenience. If the holes are drilled after grouting of the foundation beds has been carried out, the efficacy of the grouting operation can thus be checked; this is a most valuable feature in view of the inevitable uncertainty regarding the penetration of grout. The cost of holes of this size may be considerable, but it is commensurate with the great advantages that they present for underground investigation (Fig. 11.12).

The use of these large-diameter holes in civil engineering work appears to have developed initially in the United States, although they have now been used successfully in several other countries. Early applications were mostly in connection with dam-foundation work, as for the Grand Coulee and Norris dams. A particularly significant early application was at the site of the Prettyboy Dam, constructed to impound water for supply to the city of Baltimore, Maryland. The site is physically a good one; the valley is quite deep; but geological conditions required unusual precautions in the excavation of the cutoff trench below the base of the main structure, a procedure which called for placing about 145,000 m^3 (190,000 yd^3) of concrete. The rock formation beneath the dam is mainly mica schist with some limestone, gneiss, and intruded quartz, and it has been twice subjected

FIG. 11.12 Typical Calyx drill core (diameter 36 in) laid out for inspection of core losses.

to earth movement. Faulting was therefore to be expected. The exposed rock had undergone weathering to a considerable degree. It was realized that extreme care would have to be taken in blasting for the excavation of the foundation bed. As a preliminary operation, it was decided to use Calyx core drills to sink several deep shafts, 0.9 m (36 in) in diameter, so that the consulting geologist could be lowered into them to study the formation of the schist in place. By this means, the geologists were able to prepare accurate sections showing the position of all faults and large seams, as well as the direction of their strike and dip. Disintegration of the schist on the hanging-wall side of a major fault, traced in this way, considerably increased the volume of excavation. Because of the use of wide-diameter holes, it was possible to obtain truly representative samples of the schist, the study of which constituted an interesting investigation in pure geology.[11.18]

Large-diameter drill holes of this type were widely used during the great program of dam building that distinguished the early years of the Tennessee Valley Authority. One particularly interesting application of large-diameter drill holes was developed by the engineers of the TVA in exploring the unusually complex underground structure on which the Watts Bar Dam is founded. At various locations across the river at the dam site selected, large holes were drilled by means of the cofferdam arrangement sketched in Fig. 11.13. This worked admirably; the rock was penetrated to depths up to 21 m (68 ft), which permitted complete inspection of the sections thus revealed.[11.19] The potential of this drilling technique was well demonstrated in the investigations for the cross-Florida barge canal. A Calyx drill was successfully used for sinking a 1.1-m (42-in) casing through 15 m (50 ft) of soft Ocala limestone, which underlay sand at the surface, in order to reach underlying dolomite, in which a 0.9-m (36-in)-diameter Calyx hole was then drilled in the normal way.[11.20] One of the most extensive jobs ever undertaken with Calyx drills was, strangely enough, not for subsurface investigation but for the forming of holes in rock for penstocks. Five 1.8-m (6-ft)-diameter Calyx holes were successfully sunk 437 m (1,435 ft) through granite on the Cañon del Pato hydroelectric development on the Rio Santa in the Peruvian Andes. The penstocks thus formed connect a 9-km (5½-m) pressure tunnel with an underground powerhouse, which houses five 25,000-kW generators.[11.21]

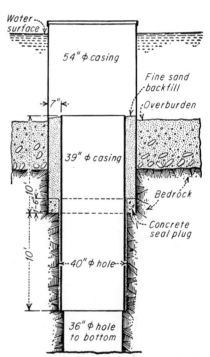

FIG. 11.13 Telescoping casing and cofferdam as used by TVA for carrying out 0.9-m (36-in)-diameter test hole drilling at the site of the Watts Bar Dam on the Tennessee River.

11.6 FIELD TESTS

These more costly and elaborate means of subsurface exploration—the use of large-diameter holes, the sinking of large shafts, and the driving of test tunnels—must not obscure the fact that, for the great majority of civil engineering projects, relatively simple exploration techniques are all that will be necessary. That the use of small-diameter borings and drill holes should be made as useful as possible is obvious. There has gradually developed, in consequence, a combination of test boring and drilling, with the carrying out of simple tests of soil and rock in situ, using the holes initially put down for exploration purposes. One obvious measurement that can be made without much difficulty is that of resistance to penetration of casings or sampling tools as they are forced into the ground. Thus has developed the *standard penetration test* in which a specially designed device is forced into the soil stratum exposed at the bottom of a test hole, while rate of penetration and force are recorded. Correlation of such results with the penetration of larger units such as samplers is possible.

Another type of device that can be lowered into a borehole is one equipped with small vertical vanes. If this is forced into the undisturbed soil at the bottom of a test hole and then turned with a torque-measuring device, values of the shear resistance of the soil can be obtained and, in due course, correlated with more accurate values obtained on samples in a laboratory. Such vane shear tests are not confined to small devices that can be lowered into test holes. Convenient portable instruments have been developed for use at or near the ground surface or where handwork is possible, such as at the bottom of a test pit. Once the idea of utilizing boreholes for carrying out field tests is accepted, the wide scope of this supplementary aspect of subsurface exploration can readily be appreciated. In holes both in soil and rock, permeability tests can readily be conducted in the field. Details of these and other tests are to be found in some of the guides listed at the end of this chapter.

It seems desirable to stress the relative simplicity of both the tools needed and the procedures used for that type of relatively shallow ground exploration that is so necessary for the usual type of civil engineering project. Such sophisticated equipment, both tools and instruments, is now available that it is easy to forget that, when far removed from the availability of elaborate drilling and boring equipment, sound and useful exploration can still be carried out with the aid of very simple tools. If the right approach is taken, assuming nothing other than what the initial study of local geology has suggested, useful information can be obtained. Modern methods and equipment should, naturally, be used whenever possible and economical. The main requirement, however, is a correct approach to the task, and a critical assessment of all information gained by actual sampling of the subsurface.

Almost all of the examples from practice given throughout this volume reflect the practice of the last few decades, most of them from relatively recent years. A few examples from the early days of civil engineering still have sound lessons to suggest and are therefore presented in appropriate places. One such can be used with regard to site exploration. In the 1860s Canada was building its first main-line railway, the Intercolonial Railway from Montreal to Halifax. The chief engineer was Sandford Fleming, one of the towering figures in early engineering in North America. He had a number of major bridges to design, two of the most important being across the two branches of the Miramichi River in northern New Brunswick (Fig. 11.14). Located near the junction of the two branches, and so only a quarter of a mile apart, the two sites were assumed to be identical; preliminary simple test boring seemed to confirm this. Actual excavation showed that foundation beds were different. Fleming immediately had better augers prepared and carried out more accurate test boring into the riverbed; naturally, no "modern" equipment was available.[11.22] Not only did he obtain in this way a better appreciation of the material on which the bridge piers had to be founded, but he got his blacksmith to make small plates attached to boring rods, had these lowered to the bottom of some of the test holes, loaded them, recorded the settlement, and so deduced the character of the soil from penetration tests similar to those of today, but made over a century ago.[11.23]

FIG. 11.14 Load test on the first pier of what is now the Canadian National Railways bridge over the northwest arm of the Miramichi River, New Brunswick, 1870.

11.7 SUPERVISING EXPLORATORY WORK

It will be obvious, even from this brief discussion, not only that the accuracy of underground exploratory work is of supreme importance but also that the conduct of the work often requires exceptional skill and wide experience on the part of the person in charge. In engaging workers to undertake subsurface exploration, therefore, employers should give particular attention to their experience and reliability. It is often advantageous to carry out all exploratory work by direct administration, since the ultimate extent of the work is never known when the work is begun. Direct supervision and employment of experienced workers constitute probably the most satisfactory procedure. Many large engineering organizations, such as highway and public works administrations, maintain special test-boring divisions for conducting all their regular exploratory work.

On civil engineering work, however, carried out by consulting engineers or by engineering offices that are not able to maintain regular boring crews, it will be necessary to engage an outside test-boring contractor for this work. The necessary contract documents must be prepared with special care. Because of variations that may be revealed as the work proceeds, it is most important to ensure a wide degree of flexibility, not only in the extent of the work but also in the location of test holes. It will be clear, therefore, that under no circumstances whatsoever should a test-boring contract be awarded on a "lump-sum" basis. Unit prices must be secured, probably with a guaranteed minimum number of holes and total depth of drilling or boring, with modified prices for operations in excess of certain specified limits. Although the usual practice of calling for bids will probably have to be followed, the award of the contract should not be made on the basis of price alone; due regard should be paid to the experience and reliability of the respective contractors who bid and to the test-boring and drilling equipment they have available, descriptions of which should be required with the tenders.

Exploratory work is of value only if a complete and accurate record of the results is obtained for the use of the engineer in charge and all advisers. This is so obvious that there would seem to be little need to emphasize the necessity of obtaining accurate records. Not infrequently, however, recording is left in charge of the test-boring foreman, supervised by occasional visits of an engineer; this is often done even when a considerable amount of money is being spent on the exploratory work. Even an experienced and conscientious foreman should never be left in charge of the records, not only because he will probably not be well versed in record work but also because it is important that the results of the exploratory work be checked by an independent observer.

As a general rule, therefore, a member of the engineer's staff, either a qualified engineer or engineering geologist, should always be present throughout exploratory work in order to watch its progress and keep the necessary records. If a geologist can undertake this task, so much the better, but this will be possible only when a large geological field staff is available. If being consulted about the work for which the testing is being done, a geologist should occasionally visit the site of the exploratory work while the work is in progress so that the results may be seen firsthand and discussed with the people in charge. When it is realized that the success or failure of an entire construction project may depend on the accuracy of records of exploratory work, this insistence on accuracy will be appreciated.

It is clear that a precise record of the nature of all strata must be kept in order to achieve the second objective of subsurface investigations—study of the materials encountered. This record will be supplemented by samples of each stratum and by the cores obtained in core drilling. The importance of preserving soil samples in their undisturbed state has already been stressed; it cannot be overemphasized. Arrangements must be made for the proper and immediate sealing (with paraffin wax or other suitable sealant) of all soil-sample tubes right at the drilling site, so that the samples, once they are removed from the ground, will lose no moisture. If soil is needed for visual inspection, additional samples should be procured for this purpose, since the samples to be used for laboratory testing must not be handled at all on the job after they are sealed. If this point ever has to be emphasized to field engineering staff, a simple demonstration of the rate at which water will evaporate from a soil sample, carried out by exposing a small sample of moist soil on a delicate chemical balance in an ordinary dry atmosphere, will usually be more effective than any argument.

Rock cores must also be carefully stored; the most convenient way of doing this is to use special core boxes—flat wooden boxes divided into narrow compartments, each wide enough to hold one core, and, for convenience, of standard length (Fig. 11.15). The cores are placed in these compartments as they are obtained, care being taken to place them in compartments corresponding exactly in position to the location of the core in the holes; thus, gaps will have to be left periodically to allow for the inevitable core losses. Obviously, large-diameter rock cores cannot be kept in storage boxes, but arrangements are usually made to have the various sections of core, as they are secured, laid out in order and in line, adjacent to the hole, so that they may readily be examined by those interested.

Finally, a prime requirement of all records of exploratory work is that materials encountered should be accurately described. If it is possible to examine samples of all the materials after the hole has been made, it will not be imperative to keep accurate descriptive notes as the work proceeds, provided all samples and cores are properly correlated with the progress records. It will be a distinct advantage, however, if accurate terminology is used in the day-to-day records. The engineer who is keeping records should therefore be familiar at least with main rock groups and the distinctions between the various grades of unconsolidated material. The latter should always be described by the use of the appropriate geological term—gravel, sand, silt, or clay, or a combination of two or more of these—together with an accurate notation of the physical conditon of the material, e.g., whether it is hard packed or very loose. Terms that are essentially popular, such as "hard-

FIG. 11.15 Drill-core box (job-made) for ready storage and examination of cores from diamond drilling.

pan," or local in application, such as "pug," should be avoided. If an exact measure of the state of the material is desired, this must be obtained by using a penetration device of some type. Naturally, the more uniform such field descriptions are, the more useful they will prove to be. There is no widespread agreement, even in the English-speaking world, on terms to be used for the field description of soils, but there are available a number of printed guides.

11.8 UTILIZING THE RESULTS

Underground explorations such as have been described in the preceding sections can be fully effective only if they are correlated, without delay, with the results of geological survey work at the site being investigated. This requirement provides another convincing argument in favor of having an engineer or engineering geologist in constant attendance at all exploratory test work, since the correlation can best be carried out while the work is in progress. The obvious, and most useful, means of combining the results of the two methods of investigation is to draw tentative geological sections, based on survey work, along lines on which test holes are to be put down; the records of these holes can then be plotted to scale on the section as work proceeds. It may be necessary to use a distorted scale for this plotting; but provided this fact is not lost sight of and that a definite scale is used to relate the boreholes and sections with one another, the resulting record can be easily interpreted. By means of this simple device, it will be possible to keep constant check on the progress of holes, to stop them when they have gone far enough, to locate new holes in order to clear up doubtful points revealed by the section, and generally to see that no effort is wasted and no necessary information left unobtained.

The problem, however, is a three-dimensional one. Although a general appreciation of the three-dimensional aspects of many drawings will usually enable the engineer to follow the progress of subsurface exploratory work merely by a study of drawings, those who are not accustomed to "three-dimensional thinking" will be greatly aided by an actual model of the subsurface information as it is gradually revealed by a test-boring and drilling program. The use of models for this purpose, on all but the very smallest jobs, is now well-established practice. In their simplest forms, such models can consist merely of vertical sticks, colored to correspond with the different strata encountered and located on a plan of the works so as to be in correct relation one with another. The steady development of the underground picture, even through such a simple and inexpensive medium, is always revealing. For special jobs, and for complex underground conditions, more elaborate arrangements can be made.

One of the most extensive models ever developed for study of subsurface conditions was that used by engineers and geologists of the TVA to illustrate the exploration program carried out at the site of the Kentucky Dam, near Gilbertsville, Kentucky. The bedrock is a siliceous and chert limestone. The dam is a concrete and earth-fill structure 2,595 m (8,650 ft) long. The exploratory program extended over four years because of the complexity of underground conditions encountered; in all, 817 borings were put down, aggregating, 30,189 m (100,631 ft); this constituted one of the most extensive of such investigations ever carried out in North America. Progress of the work was followed by means of a peg model constructed of 5-mm (³⁄₁₆-in) steel pins, suitably colored and arranged in plan to a scale of 1:100. The resulting model was over 30.5 m (100 ft) long, but it proved invaluable not only to those engaged upon the work but also to the consultants during their regular visits of inspection to the dam site (Fig. 11.16).[11.24]

The basic idea of such models cannot be improved upon, but the development of plastics has permitted some variation in the type of model. It is now possible to form transparent models, embedding pegs to illustrate test borings; thus a vivid impression of underground conditions is given by means of a solid model that can be handled and viewed from a variety of angles. And in the case of extensive underground excavation, the process can be carried still further and a model of the excavation itself prepared, marked or colored to correspond with the geological strata that are to be encountered, as deter-

FIG. 11.16 Model of the Kentucky Dam site near the mouth of the Tennessee River, Kentucky; the total length of the model is about 30 m (100 ft), every drill hole being represented by a painted rod.

11-24 PRELIMINARY STUDIES

FIG. 11.17 Part of the transparent plastic model of foundation beds at the site of the proposed Auburn Dam, California, showing some of the extensive subsurface investigations.

mined from test borings, test shafts, and tunnels. (See Fig. 11.17) With the aid of such models, contractors can readily visualize the character of excavation work and can plan their drilling and blasting operations long in advance of actually encountering the changes in strata with which they have to contend.

Just as the completion of the initial geological survey of a site will usually enable the civil engineer to prepare preliminary plans for a project, plans on which the program of exploratory work will depend, so will the completion of the exploration program permit preparation of the final designs in detail. The combination of the results of the exploratory work and the geological survey, in all but exceptional cases, will enable the engineer to obtain a complete picture of the underground structure at the site insofar as it will affect his plans. He will know to what limitations his design is subject, will be able to calculate with a fair degree of accuracy the amount of material that will have to be excavated and will know what natural construction materials are available within easy reach of the site. Thus will he utilize the findings of his preliminary work, embodying the results in his contract plans and specifications.

11.9 CLAIMS FOR EXTRAS

In many specifications and on many contract plans, a clause similar to the following used to be found:

> Drawing X contains details of borings that have been made at the site of the work, but their accuracy is not guaranteed; and intending contractors are required to take, before they tender, the borings that they may deem necessary to satisfy themselves as to the accuracy of the information regarding local conditions conveyed by the plans and specifications.

If this qualification is considered at all seriously, it will be seen to evidence a surprising paradox: the entire design has presumably been based upon the information that the engineer now suggests the contractor not use in case it is wrong. The clause is probably a carry-over from the days when preliminary investigations were not always comprehensive and when the methods available did not permit even the engineer to place great reliance upon the results obtained. Today, however, the clause is an anachronism. If there is any need for each of the contractors who intend to bid on the work to take individual sets of trial borings, then the engineer's design cannot be assumed to be free from doubt, and the expenditure of all the money necessary to take the borings may be doubly wasted. On the other hand, if the engineer's design is based on accurate exploratory results, there is usually no need to include the clause. A final necessary comment on this strange feature of contract documents is that the courts do not always support the intention of the clause, viz., to put the onus of anticipating satisfactory foundation conditions on the contractor. In this connection, it was stated editorially in *Engineering News-Record* some years ago (and the statement is still valid):

> The established view of the courts on the matter roughly appears to be this: The owner is responsible for unforeseen costs to the contractor when the engineer's borings are found to contain inaccuracies or fraudulent misrepresentation. The owner also is responsible when the engineer does not reveal to the contractor his complete record of preliminary investigations even when he has reason to doubt their accuracy.
>
> On the other hand, the owner is not responsible for the fact that incomplete borings do not reveal hidden ledges of rock in an earth bank, buried cribs along old waterfronts, and the like. ... Lower courts have penalized owners when unforeseen difficulties have caused the contractor to sue, but only under exceptional circumstances have higher courts ruled that the risk of such discovery is the owner's rather than the contractor's, provided always that there has been no concealment by the owner.[11.25]

These words were written in 1938. In the years since then, there has been such marked improvement in all aspects of site investigation that a more liberal view of this difficult matter has been current during recent years. The essential requirements—that the owner does not conceal from the contractor any information available that pertains to subsurface conditions, and, naturally, that there is to be no fraudulent misrepresentation—remain. Based, however, on the assumption that site investigation has been well and carefully done, as is now very general practice, relevant contract clauses today usually allow the contractor to claim extra payment for extra work or for changes necessitated by unforeseen subsurface conditions, provided the owner is notified as soon as these are encountered. Typical of good modern approaches to the problem is the following contract clause that has appeared in documents covering works carried out for the government of the United States.

> Changed Conditions. The Contractor shall promptly, and before such conditions are disturbed, notify the Contracting Officer in writing of: (1) subsurface or latent physical conditions at the site differing materially from those indicated in this contract, or (2) unknown physical conditions at the site, of an unusual nature, differing materially from those ordinarily encountered and generally recognized as inhering in work of the character provided for in this contract. The Contracting Officer shall promptly investigate the conditions, and if he finds that such conditions do so materially differ and cause an increase or decrease in the cost of, or the time required for, performance of this contract, an equitable adjustment shall be made and the contract modified in writing accordingly. Any claim of the Contractor for adjustment hereunder shall not be allowed unless he has given notice as above required; provided that the Contracting Officer may, if he determines the facts so justify, consider and adjust any such claim asserted before the date of final settlement of the contract.[11.26]

11-26 PRELIMINARY STUDIES

In one of the many discussions of this matter that have appeared in civil engineering literature, even this clause was subjected to some criticism, notably in relation to the arbitration of disputed claims. All commonly used clauses were reviewed in a study by R. F. Borg, his paper including a most useful comparative table.[11.27] As a result of the discussion that his paper generated, Borg suggested a new general clause, the first part of which reads as follows:

> The work to be performed below the surface is based upon the available data. The subsurface conditions, quantities, dimensions, classes of work and the borings, such cores or samples as are available, all as shown or described in the contract documents are agreed upon by the parties as embodying the assumptions on which the contract price was determined. The unexpected shall be deemed to have occurred if actual subsurface conditions, quantities, dimensions, or classes of work differ materially from those which were shown or indicated. . . .

The remainder of this interesting suggested contractual requirement deals with advice about changed conditions, estimating their cost, and the provision of independent arbitration if agreement on extra payments cannot be achieved.[11.28]

In the background of all these discussions, and contract provisions, is the fact that there can never be any certainty about subsurface conditions until excavation is complete. This has been said before; it is a statement that cannot be too often repeated, since there are still some who think, for example, that contract drawings should show not only the records of all available test borings but that the engineer should indicate the ground conditions between borings by direct interpolation and should so indicate by marking, for example, the soil-rock interface by solid lines between adjacent boring records. This should *never* be done. The borings alone should be shown as factual information. If any linking up is done, it must be done in broken lines with clear indication that the broken lines show assumed, but not assured, ground conditions—and even this is best omitted.

Throughout this Handbook will be found examples of works in which subsurface conditions were found *not* to be as inferred from test borings, borings sometimes well but more often poorly carried out. To illustrate the danger of assuming uniform conditions between adjacent holes, the reader is asked to examine Fig. 11.18. This is an actual case from Warsaw in Poland.[11.29] Test borings were carefully put down in advance of the construction of a new expressway in this fine, modern city (rebuilt after its complete destruction in World War II). The upper drawing shows the geological profile as deduced from the records of the boreholes. The lower drawing shows the ground conditions as they were found to be when excavation was carried out. Details of the various strata encountered are here of no special importance, the message being rather that even careful interpolation between adjacent test holes failed to indicate anything like the real structure of the subsurface. Admittedly this is an extreme case, but extreme cases can happen.

Another solution to this very valid difficulty is that adopted by certain public engineering organizations. For example, in the case of dam construction, the site of the dam may be cleared and the foundation bed may be prepared by direct administration and then a contract awarded for the construction of the dam structure alone, about which there should be few serious disputes. This method, however, is applicable only to certain projects and is by no means free of disadvantages. It may be suggested with propriety, therefore, that the best of all methods for avoiding contractual disputes regarding foundation-bed conditions is to have the preliminary and exploratory work so well carried out that there remains, when construction begins, small chance of encountering serious variations from assumed subsurface conditions. (See Fig. 11.19.)

SUBSURFACE INVESTIGATION   11-27

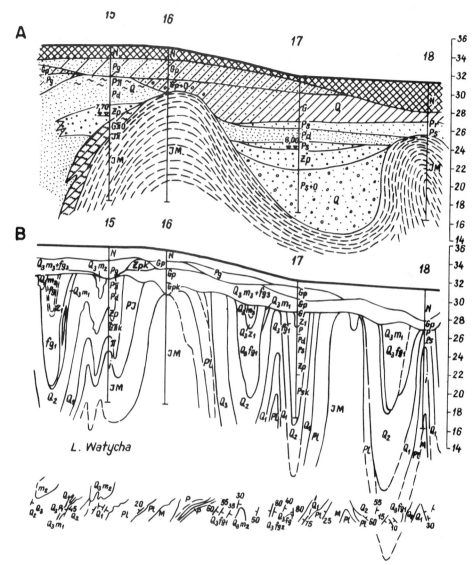

FIG. 11.18 Subsurface conditions along part of a new expressway in Warsaw, Poland, (a) as deduced from a study of test-boring records, and (b) as actually revealed by excavation.

11.10 CONCLUSION

The local geology must be the starting point for all well-conducted site investigations. Test borings, test drilling, and field tests of soil and rock can only be fully effective when considered against the background provided by knowledge of the geology underlying the site being studied. Although only rarely appreciated, it is an instinctive reliance upon geological continuity that permits engineers to design their structures on the basis of a

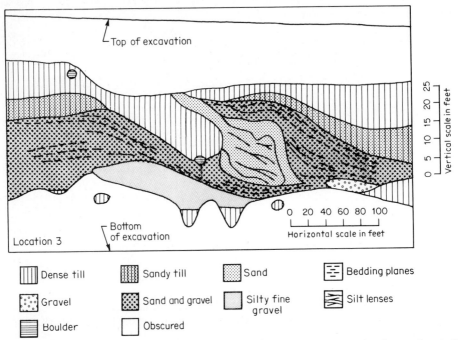

FIG. 11.19 Stratification of till, as exposed in one section of the excavation for closure of part of the main Cornwall Dike of the St. Lawrence Seaway and Power Project.

relatively few carefully selected test bores. Consider, for example, the design of a simple bridge pier, measuring 20 by 3 m (65 by 10 ft) in plan. Sinking two 7.5-cm (3-in)-diameter test borings at the site of the pier would be a reasonable approach to the investigation of this relatively small site. If the boreholes were carried to a depth of 15 m (50 ft) and soil samples obtained, and if it were assumed (for simplicity) that the prism of ground beneath the pier that would be stressed when the pier was loaded had its sides at 45° to the vertical, a simple calculation would show that the volume of ground tested and sampled—ground upon which the design was to be based—was only 0.0007 percent of the volume of ground that would be under stress.[11.30] This is limited sampling indeed, and yet it would be regarded as "good practice," thanks to the unacknowledged contribution of geology. Fortunately, this uniformity of ground conditions is of widespread occurrence; study of the local geology will usually, but not necessarily always, suggest when variations are to be expected.

There is yet one further aid to site investigation to be mentioned. It involves the interaction of physics and geology. Civil engineering work here benefits from the development work done in applying geophysics to ground exploration for other purposes, generally on very much larger scale than is ever required in building site investigation. Accordingly, the use of applied geophysics is considered in a separate chapter, but it must here be stressed that, while it can be a most useful tool in civil engineering work, it can only be applied after an initial study of the local geology; and its results must be correlated with what is known of the geology beneath the area surveyd. For all detailed studies, the results obtained from geophysical work must be checked by test borings at selected locations, ground truth being again an essential counterpart of this valuable, but indirect, method of site investigation.

11.11 REFERENCES

11.1 M. J. Hvorslev, *Subsurface Exploration and Sampling of Soils for Civil Engineering Purposes*, Waterways Experiment Station, U.S. Army, Vicksburg, Miss., 1949; reprinted by American Society of Civil Engineers, New York, 1968.

11.2 D. C. Findlay and C. H. Smith, *Drilling for Scientific Purposes*, Department of Mines and Technical Surveys G.S.C. Report 66–13, Ottawa, 1966.

11.3 R. Revelle, "The Past and Future of Ocean Drilling," paper given at 10th anniversay symposium, 30 October 1978, Joint Oceanic Institutions, Washington.

11.4 I. B. Crosby, "Drainage Changes and Their Causes in the St. Maurice Valley in Quebec," *Journal of Geology,* **40**:140–153 (1932).

11.5 A. Collin, *Landslides in Clay, (1846),* W. R. Schriever (trans.), University of Toronto Press, Toronto, 1956.

11.6 "Sampler for Hard-to-Hold Soils," *Engineering News-Record,* **137**:366–367 (1946).

11.7 A. L. Dow, "Foundation Exploration in Deep Water," *Engineering News-Record,* **119**:635–640 (1937).

11.8 R. L. Perkins, "Floating Rig Takes Core Samples in Deep Swift Water," *Engineering News-Record,* **159**:60–62 (1957).

11.9 J. M. Bruckshaw, J. Goguel, and H. J. B. Harding, "The Work of the Channel Tunnel Group," *Proceedings of the Institution of Civil Engineers* **18**:149–178 (1961).

11.10 H. B. Gough, "A Wet Shaft Frozen Tight," *Engineering News-Record,* **122**:666–668 (1939).

11.11 "Pilot Bore is Laboratory for Twin Road Tunnels," *Engineering News-Record* **173**:38–39 (13 August 1964).

11.12 D. Rose, "Research Tunnel in Atlanta, Georgia," Association of Engineering Geologists abstract vol., Hershey, Penna,. 1978, p. 16.

11.13 F. Patton and A. Imrie, "Southern British Columbia," *Guidebook to Field Trips,* Association of Engineering Geologists, Seattle, 1977, pp. 14–39; see also R. L. Brown and J. F. Psutka, "Structural and Stratigraphic Setting of the Downie Slide, Columbia River Valley, British Columbia," *Canadian Journal of Earth Sciences,* **17**:698–709 (1980).

11.14 W. L. Fornwald, "Recent Developments in Soil Sampling and Core Drilling," *Proceedings of Ninth Annual Conference on Geology as Applied in Highway Engineering,* Charlottesville, 1958, pp. 31–38.

11.15 "The Bore-hole Camera," *Engineering News-Record* **150**:39 (1953).

11.16 "Diamond Drills Explore Rock Strata in Turnpike Tunnels," *Construction Methods,* **58**:98 (1943).

11.17 V. E. Dolmage, E. E. Mason, and J. W. Stewart, "Demolition of Ripple Rock," *Transactions of the Canadian Institute of Mining and Metallurgy,* **61**:382–395 (1958).

11.18 "Redesign and Construction of Prettyboy Dam," *Engineering News-Record,* **111**:63–67 (1931); see also illustrated articles in *Construction Methods,* April and July 1932.

11.19 H. R. Johnston, "Underground Exploration with Calyx Drills," *Engineering News-Record* **120**:436–438 (1938).

11.20 H. A. Scott, "Calyx Drill Aids Soft-Ground Examination," *Engineering News-Record* **137**:102–104 (1938).

11.21 "Big Calyx Drill Sinks 1,400-ft. Holes," *Construction Methods, August 1947, pp. 78–81.*

11.22 R. F. Legget and F. L. Peckover, "Foundation Performance of a 100-Year-Old Bridge," *Canadian Geotechnical Journal,* **10**:504–519 (1973).

11.23 F. L. Peckover and R. F. Legget, "Canadian Soil Penetration Tests of 1872," *Canadian Geotechnical Journal,* **10**:528–531 (1973).

11.24 A. V. Lynn and R. Rhoades, "Large Peg Model Reproduces Damsite Borings," *Engineering News-Record,* **124**:372 (1940).

11-30 PRELIMINARY STUDIES

11.25   "Look to Your Borings," *Engineering News-Record*, **121**:11 (1938).

11.26   U.S. Standard Form 23A, art. 4, as revised in March 1953.

11.27   R. F. Borg, "Changed Conditions Clause in Construction Contracts," *Proceedings of the American Society of Civil Enginers*, **90**:(CO 2):37–48 (1964) and **92**(CO 1):78 (1966).

11.28   G. F. Sowers, "Changed Soil and Rock Conditions in Construction," *Proceedings of the American Society of Civil Engineers*, Paper no. 8509, **97**(CO 4): 257 (1971).

11.29   J. Bazynski, "Studies of the Landslide Areas and Slope Stability Forecast," *Bulletin of the International Association of Engineering Geology* no. 16, Krefeld, W. Ger., 1976, p. 77.

11.30   The authors are indebeted to A. C. Stermac for this suggestion.

Suggestions for Further Reading

Despite many advances in the detailed design of boring and drilling equipment, the principles of subsurface investigation by means of drilled or bored holes remain the same. Accordingly, the fine report of Hvorslev, *Subsurface Exploration and Sampling of Soils for Civil Engineering Purposes*, although first published in 1948, remains unchallenged as the major reference work on the subject in English, a status well shown by its reprinting in 1976 by the American Society of Civil Engineers.

Correspondingly, so well established are the basic methods now employed in subsurface investigations that soils engineers, working together through ASTM Committee D-18 on Soil and Rock for Engineering Purposes, have now produced a useful collection of ASTM Standards in the field. This is no indication of complacency about methods; all Standards are kept regularly under review and must be officially reviewed and updated at stated intervals of years.

Part (volume) 19 of the *1982 Annual Book of ASTM Standards* therefore again calls for mention. It should be available for ready reference by all users of this Handbook. The majority of the 120 Standards which it contains deal with some aspect of soil or rock; others deal with natural building stones and road and paving materials. The address of the Society is 1916 Race Street, Philadelphia, PA 19103.

So many of the Special Technical Publications of this same society deal with various aspects of site investigation that the Society's current publication list is well worthy of consultation. Typical is STP 750 of 1981, a 218-page volume dealing only with *Acoustical Emissions in Geotechnical Engineering Practice*, the editors being V. P. Drnevich and R. E. Gray.

Useful papers on various aspects of subsurface investigation will be found in the sets of volumes of the proceedings of the successive International Conferences on Soil Mechanics and Foundation Engineering, and of the corresponding regional meetings. *Recommendations on Site Investigation Techniques* has been published by the International Society on Rock Mechanics as the final report of one of its committees, dated 1975.

Professional civil engineering Societies have been similarly active in this field by convening conferences and publishing records and also specially sponsored reports of their proceedings. The American Society of Civil Engineers, for example, published in 1974 a 464-page report of a conference sponsored by the Engineering Foundation called *Subsoil Investigation for Underground Excavation and Heavy Construction;* the Society's own Manual No. 56 (61 pages, 1976) covers *Subsurface Investigations for the Design and Construction of Foundations for Buildings*. The Institution of Civil Engineers (U.K.) published in 1970 (and reprinted in 1972 and again in 1978) a most useful 328-page report called *In Situ Investigations in Soils and Rocks*.

In modern subsurface investigations, geophysical methods are often employed in association with test boring and drilling. In advance of the following chapter, therefore, and as a reminder of the close association of the two approaches, mention should be made of *Applied Geophysics for Geologists and Engineers* by D. H. Griffith, revised edition, 242 pages, published in 1981 by Pergamon Press. South Africa has been prominent in the application of geophysics in civil engineering work. Nine papers from a South African symposium on this subject will be found in the winter 1978 issue (vol. 15) of *Bulletin of the Association of Engineering Geologists*. Spread through volumes 2 and 3 of *Transactions of the South African Institution of Civil Engineers* will be found useful papers from an earlier symposium held in 1952.

chapter 12

APPLIED GEOPHYSICS

12.1 Magnetic Methods / 12-3
12.2 Seismic Methods / 12-4
12.3 Gravitational Methods / 12-7
12.4 Electrical Methods / 12-9
12.5 Applications in Civil Engineering / 12-11
12.6 Investigation of Shallow Deposits / 12-16
12.7 Geophysics and Groundwater / 12-18
12.8 Conclusion / 12-20
12.9 References / 12-22

Geophysical methods of subsurface exploration constitute one further means of determining underground conditions. The methods can be a useful and economical supplement to a test-boring program; they are briefly described here with some illustrative examples from civil engineering practice. Fully to appreciate the value of the methods, however, demands an awareness of the contributions which physics is now making to the development of geological knowledge. As an introduction, therefore, to a summary account of geophysical methods of subsurface exploration, a note on the broad scope of these contributions may be helpful.

Classical geology was essentially a descriptive science, based on careful observation in the field and in the laboratory, aided by careful reasoning in deducing general principles from individual observations. With the turn of the twentieth century, however, increasing attention was paid in geological studies to the contributions that could be made to geological problems by application of the principles of other sciences. Chemistry proved of such assistance that geochemistry is a branch of study now recognized in its own right. Botany has made its contributions also, mainly through paleobotanical studies; the development of pollen analysis has placed a most useful tool in the hands of the Pleistocene geologist. Statistics, similarly, is receiving steadily increasing attention from geologists; statistical analysis is already an accepted technique in sedimentary petrology.

It is by the application of physics to geology that the greatest transformation in geological study has taken place. *Geophysics* has been recognized now for almost a century as a scientific discipline of major status, linking the two older and more traditional branches of human inquiry. A witty writer has gone so far as to describe geophysics as "geology once removed." Like many a witticism, this happy saying carries with it more than a germ of truth. When one hears geophysics defined as "the science concerned with the constitution, age, and history of the earth, and the movements of the earth's crust," it is difficult to distinguish geophysics from geology, as defined by some workers. There is, however, a real distinction between the two branches of study, since geophysics, as now generally understood, starts with the solar system and considers the earth in relation to this system; its consideration of earth problems is predicated upon an appreciation of the globe as a whole. In a way, it can be thought of as providing a framework within which classical geological studies can helpfully be viewed; the general methods of geophysics when applied to detailed geological problems broaden the limits of human understanding of the world in a rewarding and challenging manner.

Geophysical studies led to the estimate of $4,500 \times 10^6$ years as the most probable age of the earth. The size of the earth and its distance from the sun are such that in combination they lead to a mean temperature at the earth's surface that, under the influence of the thin protective atmosphere surrounding it, permits the existence of water as a major constituent of the surface covering. (Water is extremely rare elsewhere in the universe.) Although spoken of as a sphere, the earth is really an oblate spheroid; its mean radius varies from 6,378 km at the equator to 6,356 km at the poles. The relative significance of these figures can be illustrated by imagining the earth reduced to the size of a 50-cm-diameter globe. The maximum and minimum depths and heights known anywhere on earth could then be represented by no more than 1.6 mm. The highest mountains would be represented by no more than 0.3 mm, and the average elevation of the continents by the thickness of a mere coat of varnish. Since human observations have to be confined to this extremely thin crust, exploration of the interior of the earth has had to be made by indirect measurements and by careful deductions based upon them. Geophysical studies have thus shown that the average density of the earth is 5.5 g/cm^3. The significance and interest of this figure become apparent when it is realized that the average density of the rocks that form the crust of the earth is no more than 2.8 g/cm^3. Explanation of the difference leads to detailed study of the constitution of the earth, a matter of great scientific interest but somewhat outside normal engineering thinking.

The composition of the crust itself, however, is a starting point for engineering considerations of the earth. It appears almost certain, for example, that the oceans are underlain by a relatively thin crust of basaltic rocks (5 to 6 km or 3 to 4 mi thick) and there seems to be abundant evidence to support the idea that this crust is continuous beneath the continents, even though the explored crust is composed, in general, of lighter and more siliceous rocks. It is a logical development of such studies to explore the temperature conditions within the earth. One estimate suggests that the temperature at the earth's center is about 5000°C, the corresponding pressure being 3.5×10^6 atmospheres. It is possible to measure directly the heat balance at the surface of the earth, although this is not easy. Such measurements suggest that of the observed heat flow at the surface not more than 20 percent can come from the original heat of the earth and that most of it is due to other sources, mainly radioactive. Because of the slow rate at which heat is being lost from the central part of the earth, this source of heat is of little direct significance with regard to the temperature of the ground near the surface. This is a matter of great importance in engineering in relation to such matters as the heat losses from structures, the performance of cold-storage buildings, and the widespread extent of permafrost in northern regions. Here is an important meeting point for engineering and geophysics, with climatology as a singularly important factor in determining the pattern of ground-temperature variation. Long-term studies of climate verge into geophysics and are naturally linked closely with geology.

The enduring rocks remain, despite all environmental changes such as worldwide climatic changes and the recession of glaciers, affected by these changes only superficially and only through eons of time. To the physical properties of this great mass of material which makes up the thin crust of the earth, geophysicists have devoted continuing attention from the time of the earliest linking of physics with geology. The residual magnetism of rock masses, the speed with which shock waves move through the crust, the effect of variations in the density of adjacent rock masses, and even the electrical properties of the crust have excited the curiosity of scientists for many decades. As methods were developed for measuring these and associated properties—by the further development of the simple dip needle, the invention of seismic recorders, the precise study of gravity, and the application of electrical techniques to the study of the ground—those who were farsighted saw the practical potential of such methods for subsurface exploration in the search for minerals and oil as well as in the determination of underground conditions for the special purposes of the civil engineer. So arose the widespread activity of today known as "geophysical prospecting" or "geophysical exploration."

It may be useful to emphasize that geophysical methods constitute only another exploratory aid to geological surveying and that, as applied to civil engineering, they must never be regarded as anything more than aids. They will not disclose more than will a good set of boreholes and drill holes—and usually not so much—and they can never be used without specific and constant correlation with geological information. In fact, a preliminary geological survey is essential before geophysical methods can be applied with any certainty of success, since they necessitate knowledge of certain general conditions of local geology if their interpretation is to be at all effective. The most favorable condition occurs when rock underlies a shallow superficial deposit and the physical characteristics of the two are markedly different. This condition, in connection with civil engineering work, occurs most frequently as a deposit of glacial drift overlying a solid rock body, and it is to the study of the geology of such deposits, especially as regards the depth to bedrock, that geophysical methods have generally been applied in civil engineering practice.

12.1 MAGNETIC METHODS

The phenomenon of terrestrial magnetism has been known for a long period, and it is not surprising that magnetic methods of investigation are the oldest of all geophysical methods for studying underground conditions. On the earth's surface, there is a magnetic field which is relatively constant in direction and strength and which can be measured by a magnetic needle. When the needle is freely suspended on a vertical axis, it will come to rest in a direction known as the *magnetic meridian;* when it is suspended on a horizontal axis, it will come to rest at an angle with the horizontal called the *magnetic dip,* if the axis is perpendicular to the magnetic meridian. The earth's magnetic field is not absolutely constant; it varies slightly at any locality in a regular manner and has a slow seasonal change and a rapid daily variation; it may also be affected irregularly by magnetic storms. In addition, the field may vary in an unusual manner between one location and an adjacent point, owing to the presence underground of some material possessing the property of a permanent magnet. Minerals of iron most notably possess this property. Magnetite, or lodestone, an oxide of iron, and pyrrhotite, a sulfide of iron, are the minerals most notable for this property. Some other minerals and rocks, however, also have weak magnetic effects; by employing suitably sensitive instruments it is possible to map the distribution of certain types of concealed rock and mineral formations by detecting these effects.

12-4   PRELIMINARY STUDIES

De Castro, in the seventeenth century, appears to have been the first to realize that hidden masses of magnetic material might cause local magnetic anomalies and that these masses might be detected by determining the magnetic changes that they cause. Since that time, instruments, based essentially on the simple magnet, have generally been developed for measuring these anomalies; the more usual types of such instruments are known as *magnetic variometers,* and they may be of either the vertical or the horizontal type. In place of a magnet, a rotating coil may be used to measure the vertical or horizontal magnetic force; the induced current is measured by suitable electrical devices.

The modern magnetic variometer is a highly complex instrument, but despite this, magnetic methods of investigation are capable of relatively wide use in their restricted field. One significant development has been the use of specially designed airborne *magnetometers* for mineral exploration over wide areas from low-flying aircraft. Some notable discoveries have been made in Canada by this means. Aeromagnetic maps are now available for large parts of the earth's surface. This geophysical method contrasts markedly with the earliest simple magnets, such as those used by William Wales, who was sent to Fort Churchill on Hudson Bay by the Royal Society in 1768 to conduct various scientific experiments, including a series with the dip needle.[12.1] This isolated location is but 322 km (200 mi) from the scene shown in Fig. 12.1.

FIG. 12.1   Airborne electromagnetic surveying instrument in use in northern Manitoba, an historic location as noted in text.

12.2   SEISMIC METHODS

Earthquakes have been studied by scientists for a long period, but scientific methods have been applied to their investigation only in the last century and a half. As early in the development of this branch of study as 1846, Robert Mallet suggested that artificial earthquakes (and thus earthquake waves) could usefully be created for experimental purposes by exploding gunpowder on land or on the sea floor. This idea is the basis of geophysical seismic methods; since the vibrations set up by earthquakes, either real or artificial, do not travel at the same speed in different media, the existence of a change of medium may be detected. Although the idea was suggested at so early a date, and some experiments were later conducted along the lines suggested, it was not until the early part of the twentieth century that satisfactory instruments were devised for measuring and recording the vibrations reaching the points of observation. Since then, the original instruments have been improved and new ones have been developed, so that seismic methods of underground investigation are today widely used.

Artificial earthquakes are produced by blasting with powerful explosives. High-strength gelatins and dynamites are most suitable, and the charges should be buried in order to obtain best results. The two types of waves so set up in the ground—elastic earth waves, as they are sometimes known—are due, respectively, to longitudinal and to transverse vibrations. The latter may be principally in a vertical direction or, alternatively, in a horizontal one. Other minor types of waves may be generated, but they are subsidiary to the two main types. Of these, the longitudinal waves travel faster than the transverse waves and so will be the first to reach the point of observation. Both types of waves travel through different kinds of rock with different velocities, and they will be refracted as they pass from one medium to another. This fact is the basis of geophysical seismic methods of investigation; records of observed vibrations are taken at different distances from the location of the explosion, and the results are correlated with known facts about wave travel in different media.

If two rock strata are to be investigated, the upper one allowing waves to pass through it at low velocity and the lower one at high velocity, a charge of explosive will be detonated at some convenient location in the upper stratum. Waves will then be dissipated in all directions through this stratum; some, by traveling directly through the low-velocity medium, will reach the observation points that had previously been set up at different distances from the explosion point. Some waves will reach the surface of the high-velocity lower bed. It has been found that some of this energy will travel along the boundary surface at velocity equal to that of the waves in the lower stratum. The energy is being continually diffracted back to the ground surface through the upper layer, and eventually some will reach the observation points already set up for the purposes of the investigation. The relation of the time of travel by this route to that of the waves which proceed directly through the upper stratum will clearly depend on the relative velocities in the two media, the depth of the upper layer, and the relative positions of the observation points. It will be clear that a point can be reached at which the second path of the wave will become the quicker; and if simultaneous readings are taken at all observation points, it should be possible to find the position of this point (or, actually, this circle) within reasonable limits. Determination of this point is the object of this type of seismic investigation (Fig. 12.2).

A reflection method has also been developed in which observations are taken at a convenient distance from the point of explosion of the time required for waves to travel downward to the surface of the lower stratum and then to be reflected therefrom to the point of observation. The practical difficulties of eliminating the recording of the waves reaching the observation point directly, so that the very much weaker reflected waves may be noted, have been overcome, and the method is credited with some remarkable results. It is being widely used, especially in exploration for oil. A difficulty sometimes encountered, especially in glacial soils, must be mentioned. If a dense material, such as a till, overlies a less-dense material, such as sand, a "velocity inversion" may take place and give misleading results. When used over a buried valley, this reflection method may give unsatisfactory results for the materials in the valley, since reflections may be off the sides of the valley instead of from the bottom. In such cases, resistivity or gravity methods may be more satisfactory.

The instruments used for recording the vibrations reaching the points of observation are specially designed field-model seismographs. They are of two main types, the mechanical and the electrical. In the former type (now largely superseded by the latter), the necessary magnification of very small vibrations is achieved by mechanical or optical methods; in the latter, this is done by electrical means. The essential part of both types of

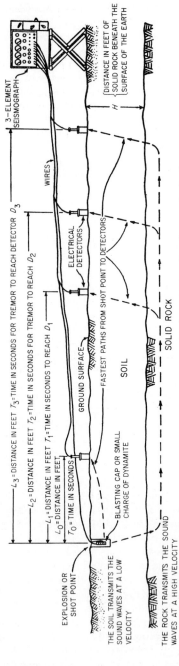

FIG. 12.2 Diagrammatic section illustrating the principle of the seismic-refraction method of subsurface exploration.

seismograph is a heavy mass suitably mounted by a nonrigid suspension system in a box or cover. The inertia of the mass will tend to keep it in a state of rest, whereas the cover will tend to move in accordance with vibrations reaching it through the ground; the instrument thus measures the movement of the cover relative to that of the mass. Usually, a clockwork mechanism that produces a continuous photographic record of the vibrations is included; time intervals and the instant of firing the explosive charge are also marked on the same record strip. Normally, several seismographs are used, as already indicated; in the electrical instruments, the detector parts can be used at each observation point, and the records can be transmitted electrically to a control instrument on which all the results are photographed simultaneously.

In recent years, steady improvement in instrumentation has been achieved, but the scientific principles involved have naturally remained unchanged. A simplification in technique has been introduced by the commercial production of an easily operated recorder, actuated by the waves created by the fall of a heavy sledge hammer upon a steel plate in contact with the ground. The method is simple to use and presents interesting possibilities. It has already been used successfully on engineering work (Fig. 12.3).

FIG. 12.3 Simple seismic prospecting with sledge hammer and steel plate.

12.3 GRAVITATIONAL METHODS

The fundamental law of gravitation was announced by Isaac Newton in 1687; it stated that the force of attraction between any two bodies is directly proportional to the product of their masses and inversely proportional to the square of the distance between them. Originally, the law was associated only with large bodies such as those of the solar system. Scientists soon turned their attention, however, to the attraction exerted by large mountain masses. Bouguer appears to have been the first to experiment in this direction; he carried out some experiments at Mount Chimborazo (in South America) between 1737 and 1740 in which he used a pendulum at a height of 2,700 m (9,000 ft) and again at sea level. Similar work was done in Scotland by Maskelyne in 1775 and in the following century by various experimenters. The results of these early experiments were not as expected, but they were not explained until many years later, after a further series of important observations had been made in India. From the study of these, there was finally formulated the principle of isostasy, which is "the basis of all accurate work on the variation of gravity at the earth's surface. [It] ... implies that above some depth below the surface, neighboring vertical columns of the crust contain the same mass."[12.2]

Determinations of the value of gravity at different locations are still made by means of pendulum-type instruments, although the instruments of today are specially designed and constructed so that they may be used in the field as well as in the laboratory. Many interesting gravity determinations have been carried out from submarines by Dr. Vening Meinesz. They have also been carried out on shipboard by means of long-period instruments that are not affected by wave motion. With such instruments, an accuracy of about one part in a million can be attained, but this is not sufficient to measure the variation of gravity between closely adjacent stations. That variations do exist is known from the different values obtained by gravity determinations carried out at different points. The changes that can thus be measured by pendulum instruments are due to extensive buried masses of relatively dense material, some of which have been located in this way.

This method of detecting underground structure was greatly extended by the use of an instrument generally known as the eötvös torsion balance, which would record changes in the horizontal component in gravity with an accuracy of one million-millionth part of the total gravitational attraction. Today, however, gravity measurements are more generally made with *gravimeters,* or gravity meters, the basis for the operation of which is shown in a simple sketch in Fig. 12.4. It is thought that the principle was first suggested by Sir John Herschel; it depends upon the extension of a balanced spring as the instrument is moved from one station to another. Modern models are very compact devices, little larger than a large vacuum flask; it is possible to survey up to 100 stations a day with such modern instruments.

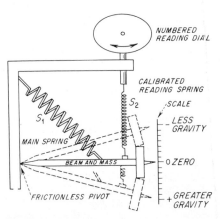

FIG. 12.4 Diagram showing the essentials of a gravity meter.

Like other geophysical methods, a torsion balance can be used to detect the existence of two adjacent rock strata of different densities; clearly, the best results will be obtained if the change in density occurs near the ground surface. The instruments will be affected not only by changes in the rock density but also by any change in level of the adjacent ground and by the presence near the point of observation of obstructions such as buildings or trees. In addition, special care has to be taken in areas covered by glacial drift, since any large boulders in the drift may affect the balance almost as much as the structural feature being investigated. The use of the instruments is consequently a most delicate matter, but it is possible, by taking a series of observations, to eliminate the effect of any feature other than that which has to be investigated.

12.4 ELECTRICAL METHODS

The various materials that constitute the earth's crust possess electrical properties of wide variation; the two most commonly used in geophysical work are conductivity and its reciprocal resistivity. The differences in conductivity of different rock types are so large that the range of variation is much greater than that of any of the other properties so far discussed; this may be a partial explanation of the wide use to which electrical subsurface prospecting methods have been put. Some sulfide minerals, as a result of chemical changes occasioned possibly by groundwater, have within them small electrical currents which circulate in the adjoining ground. A good deal of experimental work has been done in investigating such currents, but the method is not generally applicable because of its specialized nature. The more usual methods of investigation depend on passing a current through a section of the earth's crust between two (current) electrodes, placed at a fixed distance apart, and exploring the nature of the ground adjacent to or between them by means of two or more (potential) electrodes inserted into the ground at specially selected points (Fig. 12.5).

The basic electrical theory underlying the several developments of this system is not new, but it was not until the beginning of the present century that successful field methods were evolved. Daft and Williams, working in England, appear to have devised the first satisfactory equipotential method, which has since been improved and extended by other investigators.[12.3] By means of suitable equipment, lines of equal potential are traced out on the ground, and the shapes so obtained are compared with the perfectly symmetrical distribution that would be obtained in homogeneous ground; the differences between the shapes are an indication of the presence of buried material having electrical properties different from those of the material at the surface. Alternatively, the resistivity of sections of the ground may be determined in a similar way. Practical difficulties at first prevented the general use of this method, but these were gradually overcome; notable advance was made in 1912 and 1925 by Schlumberger and by Gish and Rooney, following the work of Wenner, working independently in different countries.

A third group of methods may be classed generally as electromagnetic methods; some of these are related to the reception of hertzian waves, measured at different points; the variation in intensity is an indication of geological changes in the adjacent ground. Other electromagnetic methods are based on the induction of currents in the ground under the action of alternating magnetic fields and the measurement of the distortions from results that should be obtained in homegeneous material, an operation that can now be carried out even from low-flying aircraft. It will be seen that these methods cover a wide field of electrical theory with which the civil engineer will be generally unfamiliar. He will therefore have to allow specialists in this class of work to apply the methods for him.

The methods most usually applied in civil engineering practice, however, are related to resistivity measurement, and so a word or two may be added with regard to them. It must again be noted that if the methods are to be fully effective, a general conception of the local geology must be available before they are applied. Particularly is this true with regard to the possible presence of groundwater, since the resistivity of certain types of rock varies appreciably with their water content and also with the salinity of that water. A general mode of operation is to measure the average resistivity of a volume of earth by gradually increasing the distance between the potential electrodes and plotting the results so obtained in the form of a curve relating resistivity and electrode spacing. A curve having an upward trend will suggest that resistivity is increasing with depth, whereas an abrupt change in curvature will indicate, if no other complicating features are present, a change in the nature of the material underground at a depth approximately equal to the distance between the electrodes at which the change occurs. (See Fig. 12.6.)

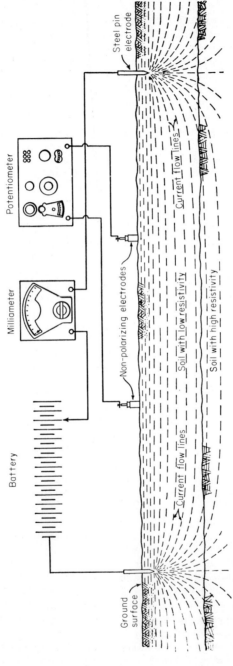

FIG. 12.5 Diagrammatic section illustrating the electrical-resistivity method of subsurface exploration.

FIG. 12.6 Electrical-resistivity geophysical apparatus used for the survey of the site for the South African National Building Research Institute.

It is of some interest to note that the instrument known as the "megger," which will be familiar to some civil engineers as that used for testing the resistance of buried grounds in power-station construction, can be used, by means of a special adaptation, for measuring ground resistivity. Results obtained with it compare favorably with those obtained with other methods. Another development of interest and of great utility in special circumstances is the use of resistivity methods in wells or drill holes, known usually as "electrologging." Three insulated conductors are lowered down the hole, with their ends at different depths; the deepest will be the current electrode, and the other two the potential electrodes. The recording instruments are read at the surface; by means of special operating devices, great speed can be achieved—up to 300 m (1,000 ft) of hole can be examined in an hour. This application is becoming of increasing importance, its use in glacial stratigraphy having already yielded most worthwhile results.

12.5 APPLICATIONS IN CIVIL ENGINEERING

It will be seen that the main methods of geophysical exploration, in their practical form, are developments of the twentieth century; some are of relatively recent invention. Despite this, geophysical methods applied to civil engineering have already achieved notable results. An early example of significance was in British Columbia. The Bridge River Tunnel was completed in 1930 for the British Columbia Electric Railway Company, Ltd. The 3,960-m (13,200-ft) tunnel passes through Mission Mountain at a depth of 720 m (2,400 ft) below the summit; it was constructed in connection with hydroelectric development and is located 175 km (110 mi) northeast of Vancouver. A geological survey of the surrounding area was made by Dr. V. Dolmage of Vancouver before bids were called; the

12-12 PRELIMINARY STUDIES

structure disclosed is shown in Fig. 12.7. Driving the tunnel started at the two ends, and work at the northern end proceeded without trouble. About 360 m (1,200 ft) in from the southern portal, however, bad conditions were encountered as a very pronounced fault zone was met. The greenstone was found to be finely crushed and extensively altered to talc and serpentine, and the whole zone was highly water-bearing; so much pressure was exerted on the timber supports that a short length of the timbers was completely demolished.

The probability of encountering such material had been foretold by Dolmage and confirmed by test borings, but it was necessary to find out how wide the zone was. This was done by the application of the resistivity method, then used by the Schlumberger Electrical Prospecting Methods. It was found that the high water content of the fault zone had a resistivity much different from that of the surrounding hard greenstone, and it therefore proved possible to determine the width of the zone on three trial locations from the tunnel center line. The second of these was adopted, and the tunnel was diverted by 45 m (150 ft) from its original location through this fault zone. As a result of the geophysical survey, it was estimated that the troublesome section of tunnel would be 111 m (370 ft) wide. Excavation proved it to be about 81 m (270 ft) wide, with another short section of water-bearing rock some 18 m (60 ft) farther on. The value of this determination to the engineers in charge of the tunnel work should be obvious, since construction methods could be devised for driving through this relatively short stretch of water-bearing material with the knowledge that a return of satisfactory conditions could soon be expected.[12.4]

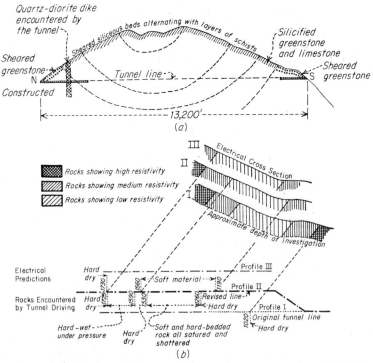

FIG. 12.7 Diagrammatic section showing essentials of the geology at the Bridge River Tunnel, British Columbia.

Geophysical methods applied to dam-foundation problems have been principally concerned with determining the depth to the surface of solid rock through superficial deposits. What is believed to be the first application of geophysical methods to civil engineering work in North America was made in 1928 at the site of the Fifteen Mile Falls water-power station of the New England Power Association on the Connecticut River near Littleton, New Hampshire. The local bedrock of Precambrian schist and occasional intrusions of granite was overlain by glacial deposits of boulder clay, gravel, and sand. The presence of boulders made test-boring operations expensive, and so an electrical-resistivity method was employed for investigating the two available sites. A buried valley was known to exist beneath the glacial deposits, and its presence was detected under an overburden of 90 m (300 ft). Other positive results were obtained. Test boring and drilling were carried out as a check; and for those holes which were related to the previous electrical prospecting, the depths predicted varied generally between 69 and 118 percent of the true depths.[12.5] Similar work was done in some of the early studies on the international section of the St. Lawrence River; drilling in this case revealed an average error of only 6.6 percent.[12.6]

In Europe, some early pioneer work was done in applying geophysical methods to the detailed study of rock strata used as the foundation beds for dams. Messrs. Lugeon and Schlumberger carried out early experiments at the site of the Sarrans Dam, located on a tributary of the river Lot, La Truyère, in the department of Aveyron, France. The dam is of the gravity type, slightly arched, 103 m (345 ft) high, and contains 480,000 m^3 (600,000 yd^3) of concrete; it is founded throughout on granite. Drill holes and exploratory tunnels were used extensively in the investigation of foundation-bed conditions throughout a period of several months; this work disclosed several zones of weathered granite, transformed into a granular mass by the alteration of the feldspar and mica. A geophysical electrical survey of the ground around the dam site was later carried out, and some interesting results were obtained, notably in a correlation of a decreased value for the resistivity with a known zone of moist crushed granite; from this finding it was concluded that "the resistivities studied from the surface actually bring into evidence the direction of geological features which otherwise could be discovered only by drilling exploration or excavation work."[12.7]

Throughout the world, geophysical methods are now frequently used for preliminary investigations at potential dam sites and at the proposed locations of other types of civil engineering structure; these investigations are almost always carried out in association with a test-drilling program, and the combination of the two methods has usually proved to be an economical and speedy means of subsurface investigation. Typical of such work was the study of the White River Dam site in South Africa in an area underlain by continuous granite, considerably weathered at the surface. Moisture content of this decomposed material was low (not more than 5 percent) because of seepage and other losses, including transpiration of vegetation. Its resistivity was consequently high (40,000 to 200,000 ohm cm). Resistivity of the solid granite was, correspondingly, at least 10^5 ohm cm, so that it was possible to distinguish between fresh and decomposed granite fairly readily by resistivity methods. A dike of diabase which follows the riverbed was traced by the application of magnetic methods. Table 12.1 gives actual comparisons of depth to rock as determined by test holes and the corresponding deduced depths from the geophysical work. Resistivity measurements in this investigation were considerably influenced, for some distance below the surface, by an increase in the moisture content of the surface soil and underlying decomposed granite following very heavy rainstorms, a variation indicative of the potential accuracy of the method.[12.8]

Australian engineering practice has also presented interesting and instructive examples of geophysical subsurface investigations, one being the use of this procedure for some of the dam sites in the great Snowy Mountains Hydro-Electric scheme. One of these is

TABLE 12.1 Example of Correlation of Test Boring and Geophysical Results

Borehole no.	Depth to bedrock (ft)		Percentage difference
	Geophysical	From borings	
36A	16	15.1	6
37A	18	16.4	10
38A	25	21.0	19
39A	10	8.5	18
40A	8	6.9	16
41A	16	13.5	18
42A	6.5	6.5	0
43A	10	8.5	18
44A	26	31.8	−18
45A	25	20.7	21
46A	40	30.8	30
47A	34	31.2	9
48A	78	81	−4

Source From J. F. Enslin, "Geophysics as an Aid to Foundation Engineering," *Transactions of the South African Institute of Civil Engineers*, **3**:53–54 (1953).

located near the highest point in Australia, Mount Kosciusko, 2,194 m (7,313 ft) above sea level, in the vicinity of the only glaciated region in Australia (three stages of Pleistocene glaciation have been recognized). The dam site lies on Spencer's Creek, a tributary of the upper Snowy River, at an elevation of about 1,698 m (5,660 ft); the valley is flat-bottomed because of glacial action and is underlain generally by slightly weathered granite. A potential dam site was presented by the break through a large terminal moraine created by the natural stream; unusually careful subsurface investigation was called for, since it was planned to use the moraine itself, if possible, as the left abutment for the dam. Geological mapping, test drilling, test pits, and seismic surveys were all employed by the engineering geologists of the Authority in their thorough study of the site before it was finally adopted for use. Much difficulty was experienced because of the presence of granite boulders, usually considerably weathered; in particular their presence complicated permeability tests in the moraine. The work was successfully completed, and seismic results checked closely with the depths to rock established by test boring. The proposed dam has now been constructed and, together with the associated water-power plant, is performing satisfactorily. When seen by one of the writers, the dam, in its rugged surroundings and under a cover of snow, was more reminiscent of northern Canada than of the popular, and generally correct, impression of the Australian scene.[12.9] (See Fig. 12.8.)

Probably the most extensive use of geophysical methods in the practice of civil engineering has been in the United States; all the larger federal engineering organizations, and notably the Bureau of Reclamation, the Bureau of Public Roads, and the Corps of Engineers, have used geophysical methods regularly in their preliminary exploration work. Typical of these uses was the application of the seismic-refraction method for determining the bedrock profile at the site of the Englewood Dam on the Cimarron River in Oklahoma, a project of the Reclamation Bureau. Four test boreholes were available as controls when the first geophysical work was done; 15 seismic depth points were then determined with good agreement with predictions from the test-hole records. At a later date, 12 additional drill holes were completed on the projected axis of the dam. Eight were

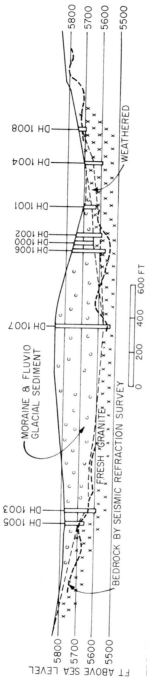

FIG. 12.8 Section through the site of the Kosciusko Dam on Spencer's Creek, part of the Snowy Mountains project, Australia, showing results of the seismic survey.

in locations that gave further checks upon the geophysical work and showed that the seismic depths to bedrock were accurate within an average error of 0.9 m (3 ft); the average depth to bedrock was 9.9 m (33 ft). The seismic method appeared to give equally reliable results in sand and gravel and in clay overburdens. Direct comparison of the costs of the respective methods was possible, and this suggested a ratio of about 7 to 1 for the cost of test drilling as compared with that for the geophysical work.[12.10] It must be remembered, however, that the two methods are really complementary to each other.

12.6 INVESTIGATION OF SHALLOW DEPOSITS

One of the most effective applications of geophysical methods, both seismic and electrical, is in the investigation of shallow surface deposits, as in exploration work for sand and gravel deposits and in the classification of highway excavation before the material is moved. Several states of the United States now include this work as a part of the regular operations of their highway departments. The state of Minnesota was a pioneer in the use of electrical prospecting methods for locating road material (sand and gravel); it is reported that over 100 prospective sites were tested for this purpose in a little more than a year and that 98 percent of the holes subsequently dug have verified the predictions of the electrical results.[12.11]

Although the progress made in the application of geophysical methods to highway engineering in the United States has not yet been so extensive as earlier experiences suggested, there is now available abundant evidence to demonstrate the efficiency and efficacy of the methods in this branch of civil engineering when terrain conditions are suitable. In the state of Massachusetts, for example, the U.S. Geological Survey some years ago had a steady program of seismic investigation in connection with highway construction; the glacial, and therefore varied, character of local surface deposits made the method particularly suitable. Well over 200 separate studies were completed; where it has been possible to check on results by means of test borings, good agreement has been found. Five traverses, with six shot points were usually achieved in one working day.[12.11]

For those sections of the 200-km (123-mi) Massachusetts Turnpike (running westward from the vicinity of Boston to the border of New York State) in which an appreciable depth of cut was necessary with no adjacent exposure of bedrock, seismic investigation was used in the selection of the final location. Again, it was the glacial character of the surface deposits and the corresponding uneven surface of bedrock that dictated the suitability of geophysical work as a supplement to a test-boring program. Seismic instruments were mounted on four-wheel-drive trucks for ease of movement; three parallel lines were generally run along selected sections of the route, one on the proposed center line, the others parallel to it and between 15.0 and 19.5 m (50 and 65 ft) to right and left. In a period of nine months, approximately 4,000 shots were fired, resulting in 45,000 instrument readings; the information thus obtained was used for the preparation of more than 1,400 cross sections which assisted in final route selection. In general, results from the seismic work checked with depths to bedrock obtained by borings within 60 cm (2 ft). Some difficulty was experienced with the layer of seasonal frost encountered during winter months; the frozen layer constituted, in effect, a separate geological stratum, but despite the difficulty thus introduced, the work was continued successfully throughout the winter. The sites for two major bridges on the turnpike were also investigated in part by seismic methods.[12.12] Geophysical methods were also used with success in the planning and design of the Ohio Turnpike to which further reference is made later in this book.

Sweden is another country in which geophysical methods, and the seismic method in particular, have long been used in relation to civil engineering work; a useful summary account of Swedish investigations is available in English.[12.13] Stimulated by some early German experiments, Swedish geophysicists carried out their first seismic-refraction tests in 1922, a year before corresponding work was first done in North America. Thereafter applications developed slowly but steadily, with the commonly experienced rapid increase following the Second World War. Many thousand investigations have now been carried out with seismic methods in Sweden, generally, but not exclusively, at water-power sites, and mainly in connection with relatively shallow overburden deposits.

Experience has shown that one seismic crew, consisting of an observer and six helpers, can profile 150 m (500 ft) per day. This involves up to 900 m (3,000 ft) of travel, with 25 to 30 shots for either five or six seismograph setups; a maximum rate of 400 m (1,300 ft) in one day has been reached, although greater speeds than this are usual in North American practice. The method is now regarded by Swedish workers as reasonably foolproof for suitable terrain. Repeated checks on the accuracy of seismic work suggest that an error of no more than 10 percent can now be almost guaranteed. Swedish workers and those of other countries have also applied the same methods to determinations of the depth of ice in glaciers; this application is greatly extending the knowledge of glacier formation and travel.

South Africa has provided an unusual example of the application of two seismic methods to the study of the subsurface conditions at a building site—that for the new buildings of the Council for Scientific and Industrial Research, including, appropriately, the headquarters of the National Building Research Institute known to one of the authors.[12.14] Geologically the site was somewhat unusual; it was underlain in part by quartzite up to the line of a fault which was found to cross the site roughly from east to west; north of the fault there was considerable depth of decomposed diabase covered by sand of varying composition. A dike intersected the quartzite. A test-drilling program, coupled with an electrical-resistivity survey, had given a reasonably accurate picture of the subsurface conditions and had revealed the existence of a large block of quartzite overlying the diabase and located exactly where the National Building Research Institute was to be built. This part of the site was therefore further studied by means of a new seismic method developed by D. I. Gough; work proceeded at night because of the high ambient noise level in the daytime. This further study delineated the quartzite block and also revealed the presence of a smaller block at a depth of about 12 m (40 ft) under the same building site. Test borings gave a rather puzzling correlation, but this was explained later when excavation for the building showed that four boreholes had all struck smaller fragments of quartzite; the results of the seismic survey were then found to be substantially correct.[12.14]

Far removed from this example from South Africa in character and location, was the use of seismic-refraction survey, in close cooperation with a test-drilling program, for the study of a proposed new outlet channel from Lake Winnipeg, a channel which was part of a regulation project for this large Canadian lake. It was essential to select a route in which the necessary dredging would be in soil, since the local bedrock was Precambrian granite. Exploration of the two-mile-long route by drilling alone was estimated to cost $3 million and to take two years to complete. Geophysical exploration provided the only possible alternative (Fig. 12.9). All that was needed was a continuous record of the depth to rock, but an accuracy of 1.5 m (5 ft) was desirable. The work was carried out in winter and was complicated by the fact that frozen shallow lakes covered part of the area to be surveyed, the ground consisting of unfrozen peat (muskeg) and glacial clays with some patches of till.[12.15]

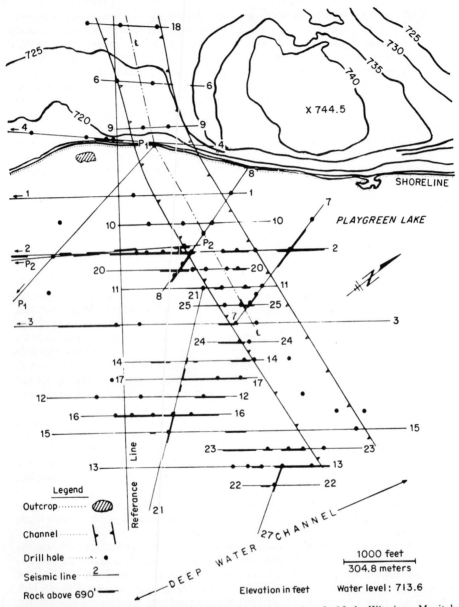

FIG. 12.9 Plan of boreholes and geophysical lines at the north end of Lake Winnipeg, Manitoba, showing shallow rock locations (i.e., above el. 690).

12.7 GEOPHYSICS AND GROUNDWATER

Can geophysical methods be used successfully to determine the existence and position of groundwater? The question arises since it relates to what is often the one matter that cannot be foretold in preliminary work from geological investigations alone; test boreholes are often necessary to check only on groundwater conditions. It is reasonably clear that geophysical methods by themselves will not "discover" groundwater. Still, they can be of considerable assistance in determining subsurface conditions that are favorable to the occurrence of groundwater, although a drill hole must always be put down to determine the actual presence of water. It has been established, for example, that the resistivity of a rock depends mainly on two factors, the porosity of the rock and the salinity of the solution held in the pores of the rock, which, in open-textured rocks, may be groundwater. Thus gravel, even when water-bearing, sometimes has a very high resistivity; this is explained by the possibility that the water present may be exceptionally pure and free from electrolytes. In general, resistivities are high for dense, impervious rocks and low for porous, water-bearing rocks. Bruckshaw and Dixey give the following typical values for resistivity (i.e., the electrical resistance of a cm^3 of material when the current flows parallel to one edge).[12.16]

Fresh crystalline rocks	20,000 to 1,000,000 ohm cm
Consolidated sedimentary rocks	1,000 to 50,000 ohm cm
Recent unconsolidated formations	50 to 5,000 ohm cm

This variation in resistivity provides a clue to the possibility of the direct determination of groundwater by electrical-prospecting methods. As waterlogging will affect the resistivity of any one material, the upper surface of groundwater occurring in a uniform medium will, in effect, be the boundary between two media with different resistivities, the wet and the dry material, respectively. This simple ideal case will rarely occur in practice; but by a development of the idea that it represents and its extension to three- and four-layer conditions, it presents possibilities for direct groundwater determination. On some trials, satisfactory results have been achieved by a careful use of this method correlated with detailed knowledge of the local geology, but on other trials, results have been disappointing.

Expert opinion is still divided on the ultimate possibilities of direct determination of groundwater by geophysical methods, but in view of the disappointments experienced in some field trials, it should be noted that there are available records of a number of quite successful determinations. Despite the uncertainty that still surrounds direct determination of the presence of groundwater by geophysical methods, there is an indirect approach that can be used, and has been used, with confidence. This method reveals, at a cost generally lower than that of comparable test holes, the local geological structure, from which the probability of the presence of groundwater may be determined. An obvious example is the revelation, by geophysical methods, of a depression or buried valley in the rock surface under porous superficial deposits, a depression in which groundwater might be expected to collect. The area that can be covered in a given time by geophysical methods, once the equipment is available in the field, as compared with the test holes which may be bored or the test pits dug, is so great that the resulting economy in water-finding operations will be obvious. This work is no different from the usual geophysical-prospecting work in connection with civil engineering projects; but because of its association with water supply, in which the average person is directly interested, there is a tendency to consider such work to be of a special nature.

It is the interpretation of results so obtained which is, in a sense, a specialty and which necessitates a fine appreciation of the geology of the entire neighborhood and of hydrological principles governing the local groundwater distribution. Notable work has been done with the aid of geophysical methods in finding groundwater in several states of the United States, in South Africa, in the Middle East, and in the U.S.S.R. In South Africa, savings of 66 percent in the cost of drilling successful water boreholes have been reported, and doubtless, similar economies have been experienced elsewhere.[12.17] Further advances may confidently be anticipated in this branch of applied geophysics, all of which will be of benefit either directly or indirectly to civil engineers in many branches of their own work.

12.8 CONCLUSION

It is possible to quote almost astronomical figures to illustrate the amount of money and the degree of effort that are today being expended on geophysical investigations, particularly in the United States and Canada, in the search for oil. Excellent publicity, coupled with the undoubted fascination of such "scientific" methods used to explore deep beneath the surface of the ground, has made the public well aware of the current extent of geophysical exploration.

There is a real danger that this popular enthusiasm for geophysical work will be translated into injudicious appreciation of its place in civil engineering. It is useful therefore, to recall that the requirements of underground investigation in civil engineering and in oil prospecting (for example) are essentially different; the accuracy and certainty demanded in the former are incidental in the latter to the discovery of oil. With this qualification must always be associated the thought that, in civil engineering, geophysical methods are no more than a special means of exploring subsurface conditions, always in association with trial drill holes and boreholes and always supplementary to preliminary geological surveying which, indeed, must be carried out before the methods can be usefully applied. All the methods are subject to definite geological restrictions—rocks of essentially different physical character must be in contact, and the strata encountered must be fairly uniform in respect to their physical character. Finally, all information obtained as a result of geophysical prospecting must be studied and utilized only when properly correlated by a geologist with the maximum available information regarding local geological conditions. Despite these qualifications and necessary restrictions, geophysical methods are a powerful and useful tool available to the civil engineer in preliminary and exploratory work. They must naturally always be used by experts; it is for this reason that no details of the methods or instrumentation have been given in this summary account.

It should, perhaps, be emphasized that the use of geophysical methods in civil engineering must represent but a very small percentage of the general uses of these techniques. Accordingly, there is continuous progress being made in the development of new methods and improved instrumentation for these more general uses that can sometimes be applied to civil engineering problems. Side-scanning radar (SLAR), for example, although still available only on a restricted basis in view of the classified nature of much of the information about it, has been used with success for the "remote sensing" (to use the common term) of areas in which civil engineering works were to be carried out. SLAR views of major faults, such as the San Andreas fault in California, show vividly what potential this method has for demonstrating unusual geological features. Radar has also been applied in other ways, notably in an impulse radar system mounted in a mobile antenna from which repetitive electro-magnetic pulses are radiated into the ground, coupled electromagnetically to the surface. Under suitable conditions this system will provide continuous profiles of soil-rock interfaces with a high degree of accuracy.

The use of microwave radiometers for determining the thickness of dust on the surface of the moon was investigated in 1965. This work led to the idea that this same technique might be used for subsurface investigation on a wider scale. An experimental survey was therefore conducted in an area near Cool, California, to see if the existence of hidden caverns in the local Calaveras limestone could be detected from the ground surface in this way. The necessary equipment was housed in a mobile laboratory, on a 1½-ton truck, all work being carried out adjacent to an existing highway. Soil cover varied from 3.6 to 18 m (11.8 to 59 ft). No locations of any cavities were known when the survey started, but the microwave radiometer results, when checked against boring results, were found to have detected caverns in almost every case along the route surveyed. The further development of methods that can be used in this way, and between adjacent boreholes to check on the ground in between them, is an urgent need, and one that is being prosecuted with vigor by a number of commercial and governmental agencies. Despite the degree of sophistication that will one day be achieved, the relevance of geology to all such applications will still remain.

Beyond the practical utility of geophysical methods in civil engineering practice, there is the further aspect of their use which was noted at the outset of this chapter as warrant for this special attention to just one means of subsurface exploration. The appreciation by civil engineers that basic physical laws are as relevant to the materials constituting the earth's crust as they are to materials handled in the laboratory is an extremely important "by-product" of the use of geophysical methods for exploration. This awareness, when carried into the field and especially on major civil engineering works, can prove to be invaluable. Engineers may well note the significance of the studies of ground temperatures that have enabled geologists to detect ore anomalies. They can take careful note of the extension of echo sounding (another geophysical method) to the determination of the character of underwater sediments. On a more restricted scale, they can consider the similarity of the vibrations due to pile driving to the effect of shocks induced for purposes of seismic investigation, realizing that the two are identical in character though different in degree. They can understand why the dynamic design of machinery foundations is not a matter of simple rule-of-thumb (such as the classic advice to use 1 ft^3 of concrete in a foundation block for every indicated horsepower of the power input to air compressors), but a matter that depends for its complete solution upon a detailed study of the physical laws governing the transmission of vibrations in the materials supporting the foundation.

Finally, an appreciation of geophysics may sometimes suggest a solution to the unusual sort of simple problem that so frequently characterizes construction operations. An example may usefully be cited to support this generalization and to bring this chapter to a close in an appropriate manner. During the construction of a power plant in northern Ontario, an unavoidable accident in a cofferdam caused a clay bank to fail. The clay happened to be one of the sensitive glacial clays found in the vicinity of the Precambrian shield of Canada, with the result that the bank failure led to an extensive clay slide, or more literally, clay flow slide. The accident occurred at night, and the flow slide completely engulfed a large mobile crane that was in the cofferdam area. Workers viewing the site on the following morning could see no sign of the crane, but the magnitude of the slide suggested that it must have been moved from the position in which it was last seen. Test holes dug at random and test drilling over an area of 90 by 21 m (300 by 70 ft) failed to reveal any trace of the crane. It was then that a geophysicist, Dr. Lachlan Gilchrist of the University of Toronto, was called in by the engineers responsible. Since the object being sought was made of steel, the magnetic method provided a possible solution. Using a sensitive magnetometer, Dr. Gilchrist made a careful survey of the slide area along lines parallel to and at right angles to the magnetic meridian, taking observations at intervals of 1.5 m (5 ft). Two anomalies were immediately evident when the results were plotted. Digging was started immediately at the first of these, and the crane was located 3 m (10 ft) below the surface of the slide material. The second anomaly proved to be caused by large bundles of reinforcing steel that had also been buried by the slide. *The Case of the Miss-*

12-22 PRELIMINARY STUDIES

ing Steam Crane might be a very pedestrian title for a thriller, but it may possibly provide a suitable mnemonic for users of this Handbook. It may serve to remind them of the remarkable possibilities presented to them in their practice of the art of civil engineering by the judicious combination of physics and geology.

12.9 REFERENCES

12.1 J. Dymond and W. Wales, "Observations on the State of Air, Winds, Weather Made at Prince of Wales Fort on the Northwest Coast of Hudson Bay in the Years 1768–1769," *Philosophical Transactions of the Royal Society* **60**:137 (1770).

12.2 O. T. Jones, "Geophysics," (James Forrest Lecture), *Minutes of Proceedings of the Institution of Civil Engineers*, **240**:699 (1936).

12.3 H. Shaw, *Applied Geophysics*, Science Museum publ. London, 1936, p. 43.

12.4 I. B. Crosby and S. F. Kelly, "Electrical Subsoil Exploration and the Civil Engineer," *Engineering News-Record*, **102**:270 (1929).

12.5 E. G. Leonardon, "Electrical Exploration Applied to Geological Problems in Civil Engineering" in *Geophysical Transactions, Transactions of the American Institute of Mining Engineers* **97**:99 (1932).

12.6 Crosby and Kelly, loc. cit.

12.7 M. Lugeon and C. Schlumberger, "The Electrical Study of Dam Foundations," *Water and Water Engineering* **35**:609 (1933).

12.8 J. F. Enslin, "Geophysics as an Aid to Foundation Engineering," *Transactions of the South African Institute of Civil Engineers* **3**:53–54 (1953).

12.9 D. G. Moye, "Engineering Geology for the Snowy Mountain Scheme," *Journal of the Institution of Engineers of Australia*, **30**:287–298 (1955).

12.10 D. Wantland, "The Application of Geophysical Methods to Problems in Civil Engineering," *Bulletin of the Canadian Institute of Mining and Metallurgy*, **46**:288–296 (1953).

12.11 C. R. Tuttle, "Application of Seismology to Highway Engineering Problems," *Ninth Annual Conference on Geology as Applied to Highway Engineering*, Charlottesville, 1958, pp. 49–59.

12.12 V. J. Murphy, "Seismic Profiles Speed Quantity Estimates for Massachusetts Turnpike," *Civil Engineering,* **26**:374–375 (1956).

12.13 H. Hedstrom and R. Kollert, "Seismic Sounding of Shallow Depths," *Telus,* **1**:24–36 (1949).

12.14 D. I. Gough, "The Investigation of Foundations by the Seismic Method," *Transactions of the South African Society of Civil Engineers,* **3**:61–70 (1953); see also J. F. Enslin, op. cit., pp. 55–58.

12.15 J. W. Prior and W. K. Mann, "Location of a By-Pass Channel Using a Combined Seismic-Drilling Program," *Proceedings of the 24th International Geological Congress*, sec. 13, Montreal 1972, pp. 217–224.

12.16 J. M. Bruckshaw and F. Dixey, "Ground Water Investigation by Geophysical Methods," *Water and Water Engineering* **36**:261, 368 (1934).

12.17 "Geophysical Prospecting for Water," *South African Mining Engineering Journal,* **47**:44 (1936).

Suggestions for Further Reading

Because space is limited, reference must be confined to two classic volumes: the second edition of *Physics and Geology* by J. A. Jacobs, R. D. Russell, and J. T. Wilson, McGraw-Hill, 1972; and the fourth edition of *Applied Geophysics in the Search for Minerals* by A. S. Eve and D. A. Keys, Cambridge University Press (U.K.), 1957.

chapter 13

GEOLOGY BENEATH CITIES

13.1 Early Urban Studies / 13-2
13.2 Urban Geology / 13-3
13.3 Archival Records / 13-12
13.4 Utility Records / 13-15
13.5 Methods of Investigation / 13-16
13.6 Filled Ground / 13-17
13.7 What Some Cities Have Done / 13-19
13.8 What Every City Should Do / 13-21
13.9 References / 13-22

Throughout the world, a large proportion of civil engineering work is carried out in urban areas, necessarily so in view of the many services that are now required for urban life. Most large buildings are located within city boundaries, and the provision of their foundations presents to the civil engineer some of the most challenging problems for restricted sites. The investigation of sub-surface conditions for works within urban areas is, therefore, just as important as elsewhere, and all the disciplines involved (with the possible exception of aerial-photo interpretation) must be brought into play. There is, however, a complication in that all too often the surrounds of the site to be investigated will be covered up by the roadways and sidewalks of city streets. Nothing about the underlying geology can be determined, in all but exceptional cases, from surface observations.

It is surprising how this simple fact appears to have influenced so much thinking and writing about urban planning. The fact that geology does not stop at city boundaries—absurd though such an idea may seem to all who use this volume—is all too often completely neglected, as though urban areas were in their way quite special places, all their problems being at or above the surface. If this seems to be an extreme statement, consider the contents of the large, modern urban atlas of 20 large American (i.e., U.S.) cities, published by one of the most prestigious of university presses under the distinguished sponsorship of five eminent organizations.[13.1] Five maps are usually included for each of the 20 cities, maps presenting a vast amount of useful information, but geology is not mentioned once! Incredible though this omission is, it is sadly typical of modern lack of attention to geology in urban development.

13-2 PRELIMINARY STUDIES

Civil engineers, charged with the design and construction of urban public works, are well aware that the geology beneath city streets will be just as complex, as interesting, and as vital to their work as is geology of sites far beyond city limits. Because of the usually restricted size of urban building sites, and the complications caused by the existence of underground works throughout all urban areas, they know also that the details of the subsurface for all sites within municipalities must be known with an unusual degree of accuracy. The presence of so many existing structures in every urban area would suggest that somewhere, most probably in the city hall or its equivalent, there will be a convenient collection of subsurface information, assembled from these earlier construction operations and ready for use as a guide for site investigations for new works. Would that this were so. In all but exceptional cases, such information banks do not exist. Fortunately, however, there are some additional aids to help with urban site investigation, aids supplementary to those already considered for the general case. It is for this reason, and because of the unusual importance of all site studies within cities, that this separate chapter is included. It will then be seen that the geology beneath city streets is a subject of remarkable importance in civil engineering.

13.1 EARLY URBAN STUDIES

The widespread neglect, in recent years, of urban geology is in distinct contrast with the situation a century ago, even though modern geology was then just starting its exciting development. In 1862, for example, Professor Eduard Suess published in Vienna a 300-page treatment of the geology, as it was then known, underlying his lovely city.[13.2] In 1885 there was published another fine volume of 138 pages, entitled *Rocks of Philadelphia*, the author being Angelo Heilprin of the Academy of Natural Sciences of that city. In its pages, he takes his readers on walks along roads and trails around the city, describing in an interesting way the geology to be seen.[13.3] In England, the Reverend Charles Kingsley (the well-known author of *Westward Ho!* and other books) gave a series of lectures to young men in 1872 under the auspices of the Chester Natural History Society. These lectures were later published (in 1877) in a small volume with the title *Town Geology*.[13.4] Although nontechnical, the lectures present a vivid appreciation of the importance of geology in urban planning. Typical is this brief extract:

> It does seem to me strange, to use the mildest word, that people whose destiny it is to live, even for a few short years, on this planet we call the earth, and who do not at all intend to live on it as hermits...should in general be so careless about the constitution of this same planet, and of the laws and facts on which depend, not merely their comfort and their wealth, but their health and their very lives, and the health and the lives of their children and descendants.

Make due allowance for the rolling Victorian prose of Charles Kingsley, and these words set the scene for this chapter.

In 1900 a 50-page paper having the title "On the Geology of the Principal Cities in Eastern Canada" was presented to, and published by, the Royal Society of Canada. The author was a distinguished member of the staff of the Geological Survey of Canada, H. M. Ami, who had seen from his own geological work that: "What the drill has to penetrate in any one of our larger centres of activity in Canada, before reaching the old Archaen or original crust of the earth in this portion of the North American continent covered by the areas under discussion, is a question not only of interest but also of economic value."[13.5]

These are but a sampling of the evidence of a healthy interest in urban geology in the latter part of the last century. (See Fig. 13.1.) Little has been found that indicates com-

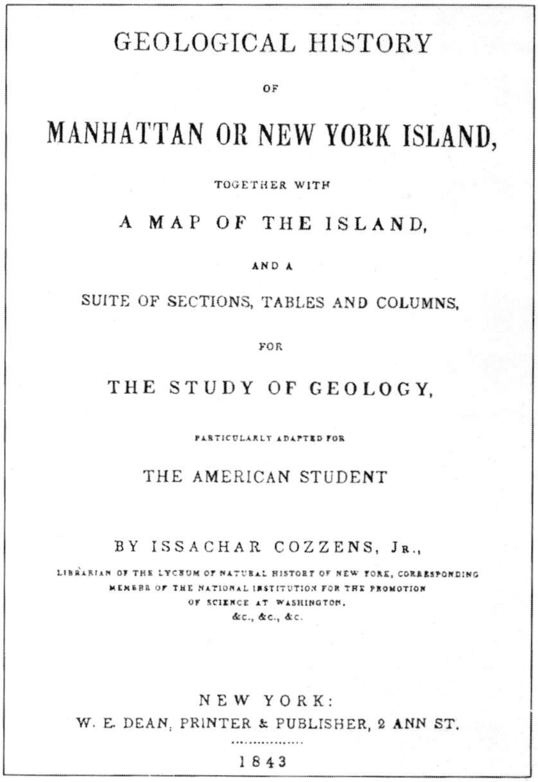

> GEOLOGICAL HISTORY
>
> OF
>
> MANHATTAN OR NEW YORK ISLAND,
>
> TOGETHER WITH
>
> A MAP OF THE ISLAND,
>
> AND A
>
> SUITE OF SECTIONS, TABLES AND COLUMNS,
>
> FOR
>
> THE STUDY OF GEOLOGY,
>
> PARTICULARLY ADAPTED FOR
>
> THE AMERICAN STUDENT
>
> BY ISSACHAR COZZENS, Jr.,
>
> LIBRARIAN OF THE LYCEUM OF NATURAL HISTORY OF NEW YORK, CORRESPONDING
> MEMBER OF THE NATIONAL INSTITUTION FOR THE PROMOTION
> OF SCIENCE AT WASHINGTON,
> &c., &c., &c.
>
> NEW YORK:
> W. E. DEAN, PRINTER & PUBLISHER, 2 ANN ST.
>
> 1843

FIG. 13.1 A reminder of earlier interest in geology beneath cities; title page of an 1843 volume describing the geology of Manhattan.

parable interest in the first third of the present century. One of the first indications of renewed interest in the subject was the publication in 1936 of a set of four sheets which, together, showed the engineering geology underlying Warsaw, the capital city of Poland.[13.6] These sheets were the work of Professors Sujkowski and Rozycki. The geology they showed, including the depths through the overburden to bedrock, was based upon a detailed study of the records of several thousand test borings and test pits. Even with this example (and a few others) available, the general lack of interest in the urban subsurface continued until the years following the Second World War. Many cities are now tackling this matter with vigor. It will be possible to mention only a selection, but, as an indication of the material that may be available for the reader's own city and as suggestive of what every city should be doing, the examples which follow may be helpful.

13.2 URBAN GEOLOGY

In place of searching for published records of regional geology, as for larger site investigations, here search must be made for published compilations of urban geology. These may be available from a variety of sources. The respective national or state (or provincial) geological survey will naturally be the best starting point. Some surveys have published

13-4 PRELIMINARY STUDIES

this useful information in the form of maps, others as smaller scale maps supplemented by descriptive memoirs. If there is a local engineering society with its own publications, it may prove to have information of value. Local natural history organizations, even though normally thought of as "amateur" bodies, will sometimes be found to have issued useful geological guides. Correspondingly, some of the more active local museums have interested themselves in the geology underlying their own cities, with resulting publications of use. And, quite naturally, local geological societies, or local sections of national societies, have likewise sometimes provided such guides for public information and enjoyment.

Boston, Massachusetts Boston must be given pride of place in any listing, since in the first published copy of the *Journal of the Boston Society of Civil Engineers* (1914) there appeared a paper by J. R. Worcester, entitled "Boston Foundations," in which an assembly of boring records was featured. As a subsequent development of the interest aroused by this paper, the society set up a Committee on Boston Subsoils, the purposes of which were stated as follows: "The purpose of this Committee is to gather data regarding the character of the subsoils in Boston and adjacent areas, and to present it to the Society in such form as to add to the general knowledge and to make it available for reference by any who may wish to get a clear idea of the geological construction under this City."[13.7]

This committee made an extensive report in 1931 and a second one in 1934. With the cooperation of those making borings in the city area, the committee assembled records of over 9,000 such tests, locations of which were shown on 16 sectional maps. The borings are so well located that it was possible to prepare six geological sections through the city at many points, to make maps of the rock and boulder-clay contours, and to locate old shorelines of Boston and Cambridge (Fig. 13.2). The committee gave special consideration

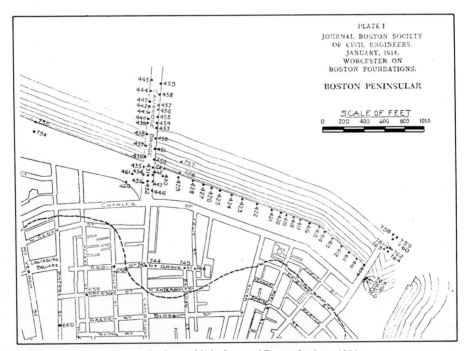

FIG. 13.2 Small portion of the first published map of Boston borings, 1914.

to the complex groundwater situation under the city area; it made tests of local soils and compiled a classification of these. The rock and boulder-clay maps were made by the geologist member of the committee who had started a study of the area as early as 1923.

This important work continued through the years, still under the aegis of the Boston Society of Civil Engineers, at one time in association with the Emergency Planning and Research Bureau, Inc. Starting in 1949, the Society's Foundations Committee published a series of papers giving in concise form, and accompanied by key maps, full details of the boring logs they have available for various parts of the Boston area.

These reports were issued in 1949, 1950, and 1951. The committee resumed its work in 1967 and published another set of records in 1969. This invaluable collection of records for one city is now being used by the U.S. Geological Survey for the preparation of geological maps of the Boston area, thus combining voluntary and official efforts. It has been well said that "Boston has probably been more completely probed, drilled, cored and investigated than any other American city."

New York, New York New York City can, however, give Boston a "good run for its money." As early as 1902, the U.S. Geological Survey published the New York City Folio as no. 83 of its *Atlas*; this was followed in 1905 by the publication of Bulletin no. 270, *The Configuration of the Rock Floor of Greater New York*, based on a study of over 1,400 borings. The municipal engineers of the City of New York were active in this field since as early as 1915 and published several notable papers in their journal. Planning the West Side Highway in 1933 provided a spur to further collection of boring records; these were assembled in the office of the borough president of Manhattan as the start of what has proved to be probably the most extensive collection of subsurface information available for any city. In the thirties a Works Progress Administration (WPA) project led to the preparation of "The Rock Map of Manhattan," enlarging still further the scope of earlier work. More than 17,000 borings were plotted on this map when it was first prepared; the number of records available has greatly increased in the years since then. A similar project was started at the same time for that part of New York outside the borough of Manhattan; for this mapping of earth and rock borings project, 27,000 boring records were used. Through the years, these fine record maps and compilations of surface information have been extended and improved. All records today are available for public consultation at the various borough offices.[13.8]

New Orleans, Louisiana Another of the few early approaches to doing something about recording urban geology was taken in New Orleans. Action was initiated in 1934 at the request of the Louisiana Engineering Society, and a survey of foundation records was made as a project of the WPA of Louisiana. The results of the survey were published in a notable volume in 1937 *(Some Data in Regard to Foundations in New Orleans and Vicinity)*. Through the cooperation of local engineers, architects, and officials, over 200 boring records and details of over 100 examples of foundation design in the New Orleans area were assembled. As the city is located near the mouth of the Mississippi River, the local geological formation is typically deltaic, consisting of sand and clay strata extending to depths greater than 107 m (350 ft). Several interesting features of these beds were noted: a layer of cypress tree stumps, whose trunks had been cut off, lies a meter or so below the surface, and a second similar layer lies about 8 m (25 ft) below mean gulf level; a human skull was found in undisturbed clay at a depth of 11 m (37 ft) below ground surface; and layers of oyster shells occur about 15 m (50 ft) down.[13.9]

Winnipeg, Manitoba Yet another early approach was that taken in Winnipeg. At the instigation of the Winnipeg branch of the Engineering Institute of Canada, a Committee on Foundations in Winnipeg was set up in May 1937. The function of the committee was to investigate and report on (1) why difficulties are experienced with foundations, (2) how best to repair faulty construction, and (3) proper design for new foundations. In August 1937, the committee submitted an interim report, discussing the formation of Winnipeg soils and subsoils, the foundation of new buildings, and the cause of settlement of existing buildings. Although described as an interim report, this document was unusually complete and is still used; it has remained the only such report, since the project did not advance further after its good start. More work is in prospect, however, and subsurface records obtained since this report was published may be made available. In general, limestone lies between 18 and 21 m (60 and 70 ft) below ground level, and the intervening strata are a succession of clay and silt beds superimposed on a layer of glacial till, a common type of subsurface pattern for urban areas in the glaciated areas of central and eastern Canada and the northeastern United States.[13.10]

San Francisco and Oakland, California Until after the Second World War, however, only a few such isolated examples of recording the geology beneath city streets as have so far been described are known to have existed. One of the most significant of the early postwar efforts in this direction was the work of the U.S. Geological Survey in producing two fine maps for the San Francisco Bay Area. In 1957, there was published a map showing the engineering geology of the Oakland West Quadrangle, California, followed in 1958 by a companion map of the geology of the San Francisco North Quadrangle, California. The ground surface is shown by contours at 25-ft intervals above an altitude of 25 ft, and at 5-ft intervals below that; bedrock contours are at 100- and 25-ft intervals, respectively. All who know San Francisco will appreciate how valuable such a map can be, not only because of the rugged character of the rock surface, which can be observed at the surface, but also because of the concentration of large buildings where the geology might be expected to be erratic. The published map confirms this inference; it shows, for example, that near the Ferry Building, at the foot of Market and Mission streets, bedrock lies nearly 90 m (300 ft) below the present surface, whereas it is exposed at the surface just a few blocks away to the northwest and southeast. The bedrock contours show clearly that this strange variation is due to an old buried river valley, a geological feature not uncommon but one that can play havoc with foundation designs if unsuspected and undetected (Fig. 13.3). The local section of the American Society of Civil Engineers has also been active in assembling engineering information on subsoils, some of which has been summarized in a useful general paper.[13.11]

Saskatoon, Saskatchewan These few examples have been selected to illustrate some of the varied sources that should be consulted in any search for information on local urban geology. The city of Saskatoon, in the center of the Canadian prairies, may usefully be mentioned therefore, since here one of the interdisciplinary geotechnical groups of Canada (associated with the Canadian Geotechnical Society) heard a talk about urban geology at one of their regular meetings and decided to do something about it. A committee was established under Dr. Earl Christiansen and, in a remarkable cooperative voluntary effort, material was assembled on all aspects of bedrock and surficial geology, pedology, climate, and geotechnology—with subsections on groundwater, slope stability, engineering properties of local soils, land use potential, etc. The information covered an area with a radius of 32 km from the center of Saskatoon. With financial assistance from the National Research Council and the Saskatchewan Research Council, the assembled material was published in 1970 in a 68-page folio which includes 54 maps and diagrams, most of them color-printed; the maps are drawn to a scale of 4 miles to the inch, which necessitated a page size of 31 by 37.5 cm. The *Saskatoon Folio* (as it has become known) is well known across Canada, the Canadian Geological Foundation having distributed copies to the

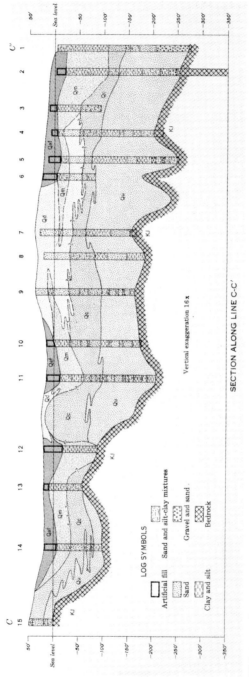

FIG. 13.3 Section of the geology beneath San Francisco, from 20th Street at Mission to the Ferry Terminal (from the map mentioned in text).

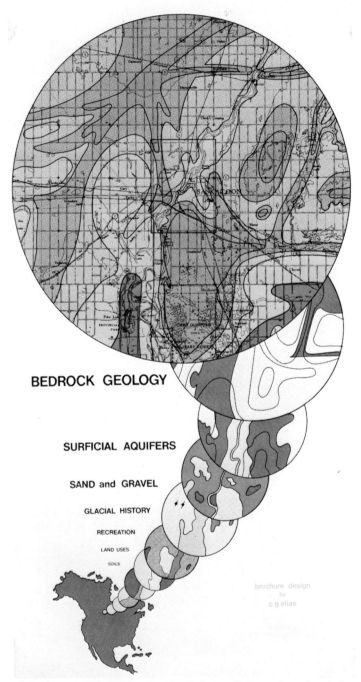

FIG. 13.4 Diagrams illustrating the *Saskatoon Folio*, a volume descriptive of the physical features of Saskatoon, Saskatchewan.

offices of most of the city engineers of Canada; it is hoped that this effort will be copied and improved upon in other cities where the information is not already available in some other form.[13.12] (See Fig. 13.4.)

Montreal, Quebec One Canadian city has produced much useful information about its subsurface directly, with supplementary material from both the provincial government of Quebec and the government of Canada. In 1966, the Montreal City Planning Department (Service d'Urbanisme) published as their Bulletin no. 4, a color-printed brochure of convenient size (by the use of folding plates), illustrating very clearly the general bedrock and soil geology of the island of Montreal and the immediately adjacent areas (Fig. 13.5). Relief and drainage characteristics are also shown, as are land slopes and local forest coverage—all the basic material for the initial phases of planning. The city's Public Works Department, jointly with the Geological Survey of Canada, has produced a useful brochure on the local soils and their engineering properties. Finally, an emeritus professor of McGill University (T. H. Clark) assembled all available information on the bedrock of the Montreal area, showing his synthesis on a fine geological map which accompanies his 244-page memoir, published as Geological Report no. 152 of the Geological Exploration Service of the Quebec Ministry of Natural Resources.[13.13]

Pittsburgh, Pennsylvania On a somewhat different scale is a publication of the Topographic and Geologic Survey of Pennsylvania dealing with the geology of the Pittsburgh area. This was prepared in cooperation with the Pittsburgh Geological Society, another example of that fine association of volunteer and official effort that happily dis-

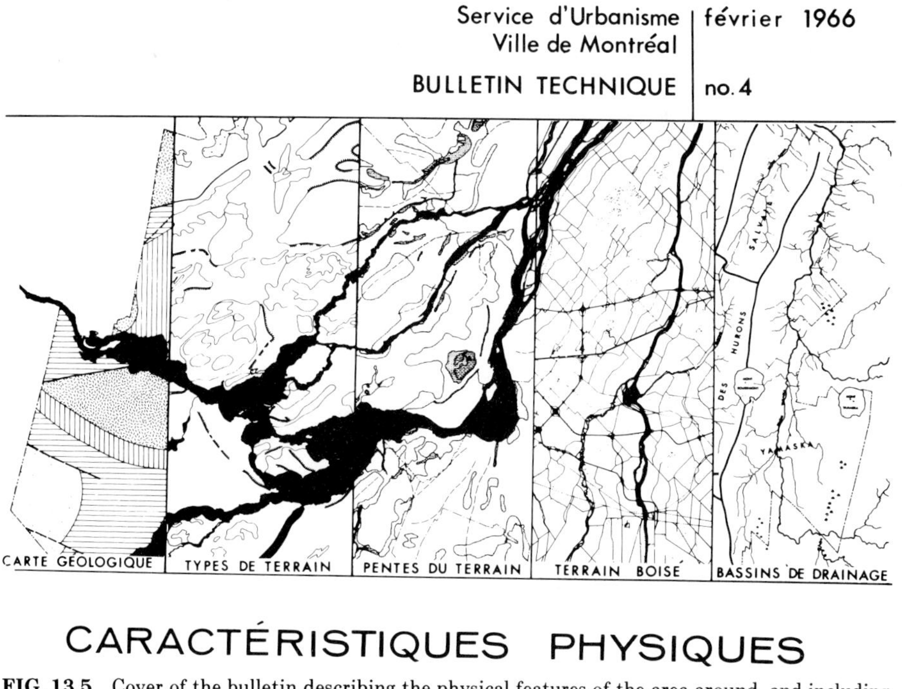

FIG. 13.5 Cover of the bulletin describing the physical features of the area around, and including, the island of Montreal, Quebec.

FIG. 13.6 The city of Pittsburgh, at the junction of the Allegheny and Monongahela Rivers.

tinguishes so much geological work carried out with the public interest in mind. This publication, for example, was prepared "mainly for the interested adult and the secondary school earth science teacher," but it provides, at the same time, an excellent beginning for any site investigation in this great city (Fig. 13.6). The four pages of references to publications relating to the geology of the Pittsburgh area alone make the use of this publication worthwhile as a starting point for the more detailed studies required for urban civil engineering work.[13,14]

Prague, Czechoslovakia Some reference must be made to corresponding work on urban geology outside North America. The capital city of Czechoslovakia provides a fine example since it is located in an area of complex geological conditions which have influenced the development of the city ever since the Middle Ages. The bedrock of the urban area consists of folded shales and quartzites of the Ordovician. The Paleozoic rocks are overlain by Cretaceous sandstones and marls, the nearly horizontal beds of which form the flat-topped hills at the outskirts of the city. The greatest part of the area is covered by Quaternary surface deposits—terrace gravels and loess sheets. Apart from this intricate geological structure, human activity has contributed to the complexity of local foun-

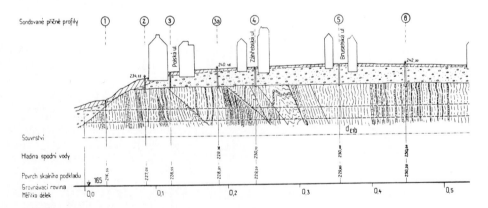

FIG. 13.7 Diagrammatic cross section through the city of Prague, Czechoslovakia, showing the underlying geology.

dation conditions. Prague, having been inhabited continuously for at least 1,000 years, has had its surface conditions changed by excavations and man-made fills to such an extent that, in many cases, the original relief is difficult to ascertain.

Former experience and records have been published in a series of papers dealing with the geological conditions of the Prague district (Fig. 13.7). The inner city area was described in detail in a paper by Q. Zaruba, *Geological Features and Foundation Conditions of the City of Prague* (Czech with English summary). A map on the scale of 1:12,500 accompanies the paper. The work includes the results of researches and borings carried out for the design of the subway, hydrogeological conditions of the area, and a critical evaluation of earlier literature.[13.15]

The geological and foundation conditions of building sites have been investigated by various bodies; at present this work is mainly done by the Czechoslovakian Institute of Engineering Geology. The results of this research work are recorded in boring records and geological reports, maintained in the so-called "Geofond" kept by the Central Geological Institute in Prague. It is compulsory to deliver to the Geofond profiles of boreholes and duplicates of reports of all geological explorations carried out in the city area. The locations of borings are plotted on a 1:5,000 map and indicated by the numbers under which they are registered. These archives are available for the use of interested inquirers.

Since World War II, a new map of foundation soils of the urban area to the scale of 1:5,000 has been developed under the sponsorship of the Board of Town Planning. It serves as the basis for new regulation plans and for the design of new districts. It will be described in some detail in the next chapter of this Handbook.

It must again be emphasized that geological maps and reports on local geology for urban areas are but the starting point for site investigations within cities for civil engineering works. They are invaluable for this purpose since, without some general knowledge of the underlying geology, no detailed site studies (such as test boring and drilling) can properly be planned. From the few examples that have been cited, it will be evident that local geology can have a profound influence on the physical development of a city. A complete volume on this one subject (*Cities and Geology*) has been published.[13.16] The concern of this chapter, however, is to show how the best possible picture of the subsurface conditions beneath urban building sites, or along the line of projected tunnels, can be obtained in advance of design and construction. The local geology is the starting point. There are two other possible sources of useful information that should be used, when pos-

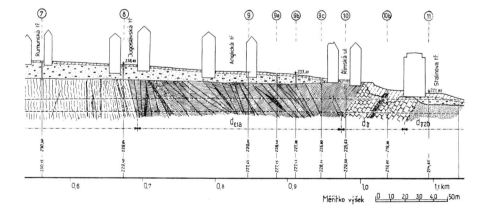

13-12 PRELIMINARY STUDIES

sible, before actual probing of the site is started—archival records and the records of buried utility services.

13.3 ARCHIVAL RECORDS

There are available for all urban areas some records of their past history. Some may be very sketchy and so not of much use in site studies, but others may reveal information of great value. In earlier times, the local public library was the usual repository for old city maps, old prints showing the location before modern growth began, and similar historical records. Today, many cities have their own specialist archival staffs, and in the case of some larger cities, their own archives. Search should therefore always be made for any such local archival material, and especially old city maps, starting with the repository of such records in the municipality itself. Now that archival material is being properly garnered, search is also warranted in the archives of the respective state (or province), since they may be found to have the older and so more valuable local records. And in special cases even the national archives should be consulted, their map rooms often containing unsuspected treasures.

The twin towers of the World Trade Center at the foot of Manhattan Island in New York City are known around the world. Many engineers must have wondered about their foundations, which rest on the well-known Manhattan schist. The necessary excavation was 21.3 m (70 ft) deep, and covered an area of six city blocks (Fig. 13.8). As part of the site investigation, archival material was consulted, in the form of maps going back to the year 1783. It was found in this way that, in those early days, the shoreline of the Hudson River extended as far as the building site and was at that time built up with rock-filled timber cribs. With this, and similar archival information available, detailed studies of the site were made with useful foreknowledge, and the excavation contractors, knowing what would be encountered, were greatly aided in their work.[13.17] Similar experiences have assisted site investigations, and so foundation design and construction, in other cities in the United States, notably San Francisco and Pittsburgh.

Despite the relative youth of most Canadian cities, the same approach has proved of value there. Records in the 70-year-old city of Edmonton, Alberta, enabled a local consultant to prepare a complete atlas of the old coal workings that still exist under the central part of the city, to the lasting benefit of all concerned with subsurface investigations in this area.[13.18] A new garage building in the center of Canada's capital city of Ottawa had to be located on a site which, the responsible engineer recalled, might have been an old quarry used by the Royal Engineers and their contractors in the building of the Rideau Canal between 1826 and 1832. Check was made in the Public Archives of Canada. Not only was a plan of the workshop area of the Royal Engineers discovered but also one of the earliest photographs ever taken in Ottawa, both showing the old quarry (Fig. 13.9). With this archival material available, test borings were located appropriately, their interpretation greatly assisted by the old records. In a corresponding manner, old maps were found to be useful aids to modern investigations for two building projects in Nova Scotia, one found in the Provincial Archives, and the other in a local city hall.[13.19]

In a record that has been published of these and similar cases, reference is made to the proclamation of President Ulysses S. Grant in 1876 calling on all county administrations then existing in the United States to prepare county histories as their centenary projects. Over 200 such county histories were prepared. Although geological and groundwater information is incidental to the social histories in these historical documents, there is some, and this information provides another source of possible use for site investigations.

FIG. 13.8 Excavation for the foundations of the World Trade Center, New York City, the scale given by the figures in the group of visitors and the subway tunnel crossing the site; tieback anchors seen on the right.

Neglect of old records, or a failure to search thoroughly for relevant archival material, can naturally have unfortunate results. If a contractor, for example, starts on a foundation-excavation job expecting to encounter nothing but soil and encounters unsuspected rock-filled timber cribs (a very common early form of construction in North America), the work will be interfered with and there will be the inevitable, and justified, claim for extra payment. Old unsuspected works even beneath the lower limit of excavation, or pile driving, may cause trouble to a completed building. In July 1974, serious settlement took place under one of the new buildings of the University of Kent in Canterbury, England. Investigation showed that this was possibly due to the partial collapse of an old railway tunnel, completed in 1830 for the Canterbury and Whitstable Railway (the great George Stevenson being the engineer). This tunnel ran immediately beneath the affected building, the pile foundations of which would have had the effect of transferring the relatively light building loads closer to the tunnel than shallow foundations would have done. The tunnel had to be backfilled as a part of the remedial work.[13.20]

Once the interrelation of archival material and site investigation is appreciated, the information exchange may sometimes work in reverse. Test borings in the Boston area, for example, revealed the existence of " . . . [old] culverts to drain swamps and springs,

13-14   PRELIMINARY STUDIES

FIG. 13.9   Garage of the Chateau Laurier (hotel), Ottawa; the limestone face indicates where the old quarry was located before designs were completed.

and wharf-like structures at the old shore line [which have been] unearthed and recorded in the course of geologic mapping [adding] . . . yet another chapter to the already rich historical records of Boston."[13.21]

Even more remarkable was an experience in Zürich, Switzerland. A new multistorey commercial building was to be erected on a choice building site to the east of the city's center and about 300 m (1,000 ft) from the edge of the Lake of Zürich. The city is of ancient foundation and it is known that so-called "prehistoric" lake villages existed along the original shores of the lake, the present built-up shoreline being formed by fine retaining walls founded on reclaimed land. Dr. Armin von Moos was charged with the site investigation for the new building. The city's archaeological staff indicated to him that the site might possibly be over part of the original shoreline, requesting that watch be kept for any artifacts in the cores of test borings. The young engineer charged with the supervision of the test-boring work noticed a "piece of stone" in one core; it proved to be a flint arrowhead. Thus alerted, archaeologists of the city were able to arrange for the entire excavation over the building site to be carried out by archaeologists, under scientific control, while the building superstructure was constructed and completed, supported on the surrounding basement walls and specially constructed central supports. Remains from five different civilizations were thus discovered, the oldest dating back to 4,000 B.C., and over 10,000 piles were uncovered (remains of an extensive lake village), to the enrichment of historical and archaeological records.[13.22] It is improbable that any such discovery could ever be made in North America, but the experience is worth recording here briefly as a reminder of the value of archival material in all site investigations, when it is available, and also of the vital importance of professional supervision of test-boring work.

13.4 UTILITY RECORDS

When search is made for useful information to aid in urban site investigations, as for example by inquiry at the local city hall, it will usually be found that the municipality will have good records of the utilities buried beneath its streets, even though little if anything is known about the underlying geology. Records of buried utilities are clearly essential, as may so readily be appreciated when one looks at the maze of pipes and cables uncovered when excavation has to be made in a street near any civic center (Fig. 13.10). A common pattern is to have a coordinating committee responsible for the maintenance of all such records, their accurate recording on large-scale maps and, in many cases, with absolute control over any disturbance of streets before check has been made on the records and formal approval issued. This is the pattern followed in the city of Philadelphia, for example, and in Toronto, now Canada's largest city.

Local utility records will not show geology beneath streets, but they do have value to the civil engineer in two ways. They must be consulted before any test drilling or boring is done in urban areas and the necessary permission accordingly obtained. Beyond this, they show how such invaluable records can be prepared by cooperative effort, and at little direct expense, once the will to maintain them is there and the necessary facility for record keeping is available. Exactly the same sort of arrangement for the recording of geological conditions beneath city streets could, and should, be found in every municipal adminis-

FIG. 13.10 Typical view of the utilities buried beneath most streets of most cities, this being at the intersection of Queen and Yonge streets, Toronto, at the start of excavation for Toronto's first subway.

tration. Such information is as valuable as are utility records, and it could just as readily be codified for public benefit.

The first of these points may be illustrated by a personal experience of one of the authors. In 1944 when Toronto started the planning for the first section of its subway system, it was necessary (in accordance with what has already been said) to supplement the information about the geology underlying the downtown area by putting down test holes to determine accurately soil and bedrock conditions. The route of the first part of the subway lies directly under Yonge Street, the main south-north artery of the city. Utility plans were consulted (and the necessary permission obtained) before any test holes were finally located. One of these had to be close to the main downtown intersection (Yonge and Queen streets, for those who know Toronto). The roadway could not be used because of traffic congestion so the hole had to be located on one of the four corner sidewalk intersections. Only under one of these was enough space between the pipes and cables shown on the utility plans to permit the sinking of a test hole. Even there, an "empty" square area 60 cm (24 in) on the side was all that was available—and it was bounded by a high-voltage electrical cable, a multi-circuit telephone cable, and high-pressure gas and water lines. Based on the relevant utility drawing, the hole was located accurately and drilling commenced. One of the authors stood with the drillers while the hole was sunk. Nothing was struck and the hole was satisfactorily completed. When, later, excavation uncovered this busy street intersection, it was possible to check the drill hole location and it was found to be accurate, in relation to the surrounding utilities, within less than an inch.[13.23] Utility records are, therefore, a vital part of urban site investigations.

13.5 METHODS OF INVESTIGATION

Despite all logistical difficulties, such as those indicated in this last example, investigation of the actual subsurface conditions at urban building sites is always imperative. The same methods as have already been generally described will be used—the sinking of test borings in soil, test drilling in rock, the carrying out of special field tests in boreholes or in special excavations. Because of the inevitable complications, such work will be more costly than usual and must be conducted so as to cause as little inconvenience as possible in the use of the streets where investigation has to be carried out. It might be thought that geophysical investigations have no place in city work, but experience has shown otherwise.

Investigations in the city of Detroit have shown what can be achieved with relatively simple equipment and a minimum of interference with municipal services. By means of a six-trace seismograph with a suitable control unit, giving a continuous record on sensitized paper, it was found that in some cases a hard blow from a sledge hammer was sufficient to give useful records for shallow depths (Fig 13.11). Inserting part of a stick of 40 percent dynamite in a hole 1.2 m (4 ft) deep was effective for rock depths up to 30 m (100 ft). Proper precautions naturally have to be taken in carrying out such work in city areas, but with the accurate plans of underground utilities which have just been mentioned, the placing of small charges can usually be safely arranged. In one test case in Detroit, rock was located at a depth between 37 and 43 m (123 and 142 ft) with an error of only 3 percent. At a midtown site in Detroit, depths of rock up to 30 m (100 ft) were successfully determined for the purpose of establishing pile lengths by work carried out on a Sunday; the results were available for use on Monday morning. Writing of this work, R. H. Wesley states that geophysical exploration "is not a universal tool, but intelligently used in conjunction with drilling and soil testing [it] can clear up many doubts as to what will be encountered when construction begins."[13.24]

FIG. 13.11 Simplicity of equipment needed for hammer seismic profiling.

Seismic investigations were carried out in a somewhat similar manner for the start of the Prudential Center in the Back Bay district of Boston. A large area, much of it occupied by railroad tracks, was ultimately cleared for this great development, of which the 55-storey Prudential Tower was the first unit. Test borings had shown that the geology underlying the site consisted of up to 3 m (10 ft) of fill overlying sediments of the former tidal flats; sand with some gravel formed the base of the tidal deposits below which was a thick stratum of Boston blue clay. A thin layer of compact till overlaid the bedrock of Cambridge argillite. The Tower was to be founded on caissons carried to bedrock. The use of seismic detection enabled the deductions from boreholes to be confirmed for the area to be occupied by the tower. The only unusual feature of the work was the necessity for placing the half-pound charges of dynamite in holes specially drilled into the upper part of the clay stratum, after charges placed in shallow holes in the fill material had proved to be unsatisfactory. Good accuracy was obtained as was shown from caisson records during construction.[13.25]

13.6 FILLED GROUND

Many areas within cities are today at a higher level than the original ground level because material has been deposited on the original ground, with the result that ground level of today is the upper level of such fill material. In the older cities of Europe, this is a very common condition, the fill being the debris from older buildings which have been either demolished or destroyed by fire (as was all too often the case), the resulting rubble being left in place to form, eventually, a new building area.

In North American cities, on the other hand, filled ground is commonly a depression in the original natural surface that has been deliberately filled in—to bring it up to an acceptable elevation—by the dumping upon it of surplus material. Today, such dumping would usually be carefully controlled, and essential compaction of the soil used for fill

Fig. 13.12 What may lie beneath the surface of filled ground, a normal way of disposing of chunks of demolished reinforced concrete.

ensured, so that the site could be safely used for building. But such cases will be the exception to those encountered in site investigation work in urban areas. Accordingly, the greatest care must be taken in test boring on city sites to ensure that filled ground is not mistaken for anything else. The first few feet of test boring are, therefore, often the most critical. An accompanying photograph (Fig. 13.12) shows the sort of "fill" that might be encountered where control has not been exercised over fill placement. It is a good rule to expect the worst where filled ground is detected, so that every effort necessary may be made to determine what will be encountered when excavation starts.

FIG. 13.13 Once buried beneath the streets of the city of Exeter, England; the mediaeval bridge, now happily restored (as can be seen) as a part of redevelopment of the central part of the city.

Here is where the study of old records can prove of unusual value, if carried out before test boring is started. The location of abandoned quarries which have been filled in poses special problems and dangers if not detected in advance. And although quarries would not be expected within city areas today, in earlier days the founding settlement, when small, might well have permitted quarrying on its outskirts, now absorbed into the modern municipality. The possibility of filled ground, therefore, presents a special hazard in all urban site investigations, but, if anticipated, its presence is readily detected in test drilling and boring, with consequent benefit to design and construction.

13.7 WHAT SOME CITIES HAVE DONE

Every site investigation within an urban area, when completed, will have provided one more item of information about the geology beneath the streets of the city. Naturally, its first use for the owner of the building project will be to assist with final foundation design, with the preparation of contract documents, and finally with the supervision of actual construction. When the foundation structure is complete, the site information will have served its purpose. The natural sequel is for it to be stored with the basic "papers" relating to the project, in effect to be buried in office files. If, however, there is some way by which this information can be associated with similar records from adjacent sites, a picture of the geology underlying far more than just the building site will begin to evolve. With the assembly and coordination of enough records, a picture of the geology beneath the whole city can be obtained. In the past, there has sometimes been objection to the sharing of such "private" information, despite the fact that it has fully served its intended purpose for the owner who paid for it. When, however, the city itself, or some well-recognized agency, makes itself responsible for this further use of site studies for the public good, then such objections can be overcome to public benefit and the refinement of all future site studies. Boston is a good example of what can be done in this direction by a local engineering society. There are now other examples that can be cited. Brief reference may be helpful in encouraging similar efforts elsewhere.

Johannesburg, South Africa Starting about 1960, the city engineer of this modern city of the Rand started to assemble subsurface information, following an appeal to consulting engineers from the city council. This and other material was used by a graduate student at the University of Witwatersrand in the preparation of a master's thesis, presentation being in the form of a colored geological map of the main city area. This was printed by order of the city council and widely used. Interest so aroused led to the setting up of a special committee by the Johannesburg branch of the South African Institution of Civil Engineers jointly with the city (the deputy city engineer being chairman). This committee has developed a complete subsurface-information-retrieval system into which all new site records are fed. This information bank is available for consultation by all interested in the city's underground, clearly to public advantage.[13.26]

Tokyo, Japan In the early 1950s a staff member of the Building Research Institute of Japan started to collect such subsurface records as he could locate for sites in the city of Tokyo, now one of the largest cities of the world. When over 3,000 such records had been assembled, with the full support of the Director of the Institute, a subsoil map of Tokyo was prepared. A beautifully printed volume containing all the boring logs, and the map divided up into small double-page sections, was also published (Fig. 13.14). This work attracted much attention and was drawn to the notice of the Japanese Research Council which, in turn, recommended to the government of Japan the extension of this sort of

MINISTRY OF CONSTRUCTION
BUILDING RESEARCH INSTITUTE
建設省建築研究所 監修
東京都建築局
TOKYO METROPOLITAN BOARD OF CONSTRUCTION

東京地盤図

GEOTECHNICAL PROPERTIES OF TOKYO SUBSOILS

東京地盤調査研究会
RESEARCH GROUP ON TOKYO SUBSOILS

工学博士 北澤五郎 Dr. of Eng. GORO KITAZAWA
工学博士 竹山謙三郎
理学博士 鈴木好一
大河原春雄
工学博士 大崎順彦 Dr. of Eng. OHSAKI YORIHIKO
編 著

Gihô-dô
技 報 堂

JAPAN. BUILDING RESEARCH INSTITUTE

FIG. 13.14 Title page of the volume containing records of the geology beneath Tokyo.

service to all the leading cities of Japan. The idea found favor, with the result that a special committee was established to further the project. Financing on a 50-50 basis was arranged with major cities. Instructions were prepared and published so that the work in different cities would be in reasonable agreement. Today, therefore, all major cities of Japan have an accurate knowledge of the geology beneath their streets, the detailed pictures being refined with the submission of each additional record.[13.27]

France Yet another pattern is to be found in France. Here the national geological service—le Bureau de Recherches Géologiques et Minières (BRGM) with headquarters at

Orléans, 112 km south of Paris—has as one of its prime responsibilities the collection and publication of information on the geology underlying all the urban areas of France. The headquarters staff at Orléans (about 350) is supplemented by about 250 workers in 11 regional offices; one of the chief functions of each regional office is to develop this urban geological service in close association with authorities of the municipalities in the region. Here, therefore, is a national system of urban subsurface information available for general use.[13.28] A similar system is now in force, but on a voluntary basis, in Switzerland, the work in this highly developed country having begun with postwar studies started by an engineering geologist, of the geology underlying Zürich, the records now maintained at the Swiss Federal Institute of Technology (ETH of Zürich).

Twin Cities, Minnesota As this volume is going to press, a major urban geological project is being published by the state geologist of Minnesota, in association with the U.S. Geological Survey. When search was made, no less than 186,000 logs of test borings in the Twin Cities area (St. Paul-Minneapolis) were found. Over 4,000 "key" borings were selected and the records they provided coded into a computerized data base; the borings are located in an area of 175 km^2, including the downtown business districts of both cities. With this computerized information, maps were prepared showing the bedrock geology and topography, the thickness of overburden, and the Quaternary geology. The maps have already been used for the planning of underground conduits for a Twin Cities district heating system. It is confidently anticipated that this is but the start of the wide use of this subsurface information, information which will naturally aid all future detailed site studies within the area covered by the maps.[13.29] Part of one of these significant maps appears on the endpaper of this Handbook.

13.8 WHAT EVERY CITY SHOULD DO

These are examples of what some countries and cities have done by way of gathering up, codifying, and correlating information about the geology beneath their streets; it may not be amiss to conclude this chapter with a plea for further action in this direction. Knowledge of the local urban geology is, it must be stressed, only the starting point for the necessary detailed site studies for civil engineering works, but such knowledge can be of invaluable aid in the planning and conduct of localized investigations. Correspondingly, every site study can add its quota to the overall picture, to subsequent general benefit.

In the North American context, national and state geological surveys have already done much and may be expected to do even more to encourage by example, and stimulate with advice, the preparation of local maps of urban geology. Ultimately, however, this should be a municipal responsibility; the public surveys cannot be expected to provide the complete service for all urban centers, as is done in smaller European countries such as France and Switzerland. It is essentially a cooperative matter. Some cost is involved, and so municipal councils will have to be persuaded that this is a vital public service, its minimal cost being but a fraction of the savings possible (even with their own public works) when such a system exists. Nothing is better than a demonstration and this calls for voluntary effort. It will have been noticed that local engineering societies and local sections of national engineering societies have been prominently associated with these cooperative ventures. In the field of civil engineering, there seem to be few more useful tasks that such local groups of engineers can perform in association with local geologists— to their mutual benefit. The functions of the local engineering group are often in question. Here is an essentially local task which requires concerted action, which takes years to complete, and which results in lasting benefit to the community. It is one that might usefully be adopted by many more engineering groups than appear yet to have entered this field; it would greatly benefit the profession by steadily removing uncertainty about subsurface conditions, and it would enrich scientific records. One of the happiest of all meeting grounds for civil engineers and geologists is in this joint study of what may be so truly called *urban geology*.

13.9 REFERENCES

13.1 J. R. Passonneau and R. S. Wardman, *Urban Atlas—20 American Cities*, M.I.T. Press, Cambridge, Mass., 1966.

13.2 E. Suess, *Der Boden Der Stadt Wien*, Bramuller, Vienna, Austria, 1862.

13.3 A. Heilprin, *Town Geology: The Lesson of the Philadelphia Rocks*, Academy of Natural Sciences, Philadelphia, 1885.

13.4 C. Kingsley, *Town Geology*, Daldy, Isbister, London, 1877.

13.5 H. M. Ami, "On the Geology of the Principal Cities of Canada," *Transactions of the Royal Society of Canada*, 2d ser. VI, Sec. IV, Ottawa, 1900, pp. 125–173.

13.6 Z. Sujkowski and S. Z. Rozycki, *Mapa Geologiczna Warszawy*, Warsaw, 1936.

13.7 J. R. Worcester, "Boston Foundations", *Journal of the Boston Society of Civil Engineers*, **1**:1 (1914); (see also same journal, **18**:243 (1931) and **56**:131 (1969).

13.8 M. J. O'Reilly, "Subsurface Work in the Department of Public Works," *Municipal Engineers Journal* **40** (1954).

13.9 *Some Data in Regard to Foundations in New Orleans and Vicinity*, Works Progress Administration and Board of State Engineers, Louisiana, 1937.

13.10 Report of the Committee on Foundations in Winnipeg, Manitoba, *Engineering Journal*, **20**:827–829 (1937).

13.11 J. Schlocker, M. G. Bonilla, and D. H. Radbruch, *Geology of the San Francisco North Quadrangle, California*, U.S. Geological Survey Map I–272, 1958.

13.12 E. A. Christiansen (ed.), *Physical Environment of Saskatoon*, Saskatchewan Research Council and National Research Council of Canada, Ottawa, 1970.

13.13 *Physical Characteristics of the Region*, Bulletin Technique no. 4, Service d'Urbanisme, Montreal, 1966. See also: J. Hode-Keyser, "Géologie de Montréal," *Proceedings of the Sixth International Conference on Soil Mechanics and Foundation Engineering*, **3**:114–133 (1965).

13.14 W. R. Wagner et al., *Geology of the Pittsburgh Area*, Topographic and Geologic Survey of Pennsylvania, General Geology Report G 59, 1970.

13.15 Q. Zaruba, *Geologicky Podklad a Zakladove Pomery Cnitrni Prahy* (*Geologic Features and Foundation Conditions of the City of Prague*), Svazek 5, Geotechnica, Prague, 1948.

13.16 R. F. Legget, *Cities and Geology*, McGraw-Hill, New York, 1973.

13.17 M. S. Kapp, "Slurry-Trench Construction for Basement Wall of World Trade Center," *Civil Engineering* **39**:36 (April 1969).

13.18 R. Spence Taylor, *Atlas: Coal Mine Workings of the Edmonton Area*, privately published by the author, Edmonton, 1971.

13.19 R. F. Legget and K. N. Burn, "Archival Material and Site Exploration," in press.

13.20 "Tunnel Strengthening Too Late at Canterbury," *New Civil Engineer*, no. 101, 18 July 1974; see also same journal, no. 103, 1 August 1974.

13.21 "Boring Data from Greater Boston," *Journal of the Boston Society of Civil Engineers*, **56**:131 (1969).

13.22 A. von Moos, in private communication.

13.23 R. F. Legget and W. R. Schriever, "Site Investigations for Canada's First Underground Railway," *Civil Engineering and Public Works Review*, **55**:73–79 (1960).

13.24 R. H. Wesley "Geophysical Exploration in Michigan," *Economic Geology,* **47**:57–63 (1952).

13.25 W. J. Murphy and D. Linehan, "Engineering Seismology Applications in Metropolitan Areas," *Geophysics* **27**:213–220 (1962).

13.26 Professor J. E. Jennings and Dr. A. B. A. Brink and associates, Johannesburg, in private communcation.

13.27 Since this volume is printed entirely in Japanese, it cannot be referenced in the usual way, but it can be identified as *The Subsurface of Tokyo.*

13.28 Dr. J. S. Scott (Ottawa) and Dr. J. Rigour, BRGM (Orléans, France), in private communication.

13.29 R. F. Norvitch and M. S. Walton (eds.), *Geologic and Hydrologic Aspects of Tunneling in the Twin Cities Area, Minnesota,* U.S. Geological Survey Map I–1157, 1979.

Suggestions for Further Reading

The geology beneath cities has been so consistently neglected up to very recent years that there is no useful reading in English that can be suggested. The Geological Society of America will be publishing, almost concurrently with this volume, the fifth volume in their series of Reviews in Engineering Geology. This will deal with *Geology under Cities* and it will contain papers describing the geology beneath Washington, D.C.; Boston; Chicago; Edmonton, Ontario; Kansas City; New Orleans; New York; Toronto; and the Twin Cities of St. Paul and Minneapolis. The editor is R. F. Legget. The *Bulletin of the Association of Engineering Geologists* in its post-1982 issues will feature papers describing the geology under individual cities, this having been adopted as an editorial policy by the Association. It is greatly to be hoped that in these and other ways this significant gap in the literature of geology (and also of applied geology) will gradually be closed.

Geology in the Urban Environment, edited by R. O. Utgard, G. D. McKenzie, and D. Foley (Burgess Publishing Company, Minneapolis, 1978) is a useful 355-page collection of papers that emphasizes the message of this chapter.

chapter 14

GEOLOGICAL MAPS

14.1 Geological Maps and Their Use / 14-1
14.2 Geological Map Making / 14-3
14.3 The Endpaper Map / 14-6
14.4 Maps for Engineering Geology / 14-6
14.5 Engineering-Geology Maps / 14-11
14.6 International Cooperation / 14-14
14.7 Conclusion / 14-15
14.8 References / 14-16

If geology is to be utilized to full advantage in site investigations, it is necessary for the results of geological studies to be communicated to those responsible for these vital preliminary parts of civil engineering projects. This can be done through the medium of the printed word in reports; photographs can provide useful supplementary information. The complete picture of the local geology can, however, best be presented by means of a general geological map. Superficially these may look little different from the colored topographical maps with which all civil engineers will be familiar. In fact they differ considerably, containing a great deal more information than is apparent at first glance. The basic principles of geological mapping should, therefore, be familiar to civil engineers.

This chapter gives an outline of the main features of geological maps and their preparation. A specimen geological map, which is reproduced in color as the endpaper of this volume, will be explained. The special adaptations of geological maps for engineering uses will then be explored and some examples given of the useful approaches now being made to the preparation of geological maps for planning purposes; such preparation now occurs on an international basis. Color is an essential feature of the best geological maps. The publisher has assisted by permitting the use of color for the endpaper maps. Other illustrations will have to be in black and white, but the reader is asked to keep in mind the added value that the use of color makes in map interpretation.

14.1 GEOLOGICAL MAPS AND THEIR USE

Standard topographical maps form the base for all geological maps. Exposures of different strata of bedrock will be marked by shading (black and white) or in different colors,

14-2 PRELIMINARY STUDIES

according to some accepted standard. Contacts between adjacent strata will be marked, where possible, with indications of the dip and other geological features such as dikes, also clearly marked. A completed geological map for country of normal structure will therefore be a fairly simple-looking topographical map, specially colored and marked. It represents, however, far more than the usual topographical map ever indicates and therefore deserves careful and detailed study. A number of publications deal solely with the quantitative interpretation of such maps, a matter that can only be touched upon here. The basis of all interpretative study is a realization of the fact that the topographic signs and, in particular, the contour lines represent the surface of the ground only, whereas the geological markings indicate subsurface structure as well.

The relation of the two leads to certain definite results which will gradually be realized after some study of geological maps. An elementary observation is that the projected distance between contacts shown on the map is no indication whatsoever of the thickness of the stratum in between, since this depends on the inclination of the beds and on the difference in elevation of the ground surface between the two contacts. An appreciation of this is fundamental. Certain characteristics will become evident after study of different types of maps; the most common of these characteristics can readily be memorized as four general rules:

1. Lines on the map representing boundaries of strata that are horizontal will be parallel with contour lines.
2. Lines on the map representing boundaries of strata that dip into a hillside will wind less than do the corresponding contours.
3. Lines on the map representing boundaries of strata that dip away from a hillside will wind more than the contours of the hill, except when the dip exceeds the average inclination of the hill.
4. In valleys, strata will show up on maps with a V shape leading upstream in all cases except those in which the dip is downstream at a greater angle than the inclination of the valley floor.

The reasons for these rules will become clear from a study of geological maps of different and differing regions; then the interpretation of new maps will be much easier, and some idea of the structural relationship of the various rocks in an area can be obtained from a study of the geological map alone without the aid of drawn-out sections. Since quantitative study is also possible and is often desirable, a small subsidiary branch of specialized study has arisen which may almost be called "geological geometry" or "geological graphics." By measuring boundary locations and elevations on maps, the geologist can make accurate calculations of the actual (not projected) thicknesses of the beds being studied.

Remembering that geological maps are a two-dimensional representation of an essentially three-dimensional structure, the reader can appreciate that, prior to the completion of many geological maps, it is necessary to plot a cross section along some special line across the area under investigation; this must be done in order to study the inferred structure from another angle. Similarly, when a geological map has been completed, one or more sections will almost always be advisable in order to present a more vivid idea of the structure studied than can ever be obtained from a study of the surface map alone. It is not unusual to find one or more such sections printed in the borders of officially published geological maps. Naturally, the vertical scale used is distorted, but to engineers this feature of geological sections will not be strange. The preparation of these sections from the base map obtained is a rather straightforward matter. The dip of the rock strata at the several contact planes plotted will indicate the general direction of the beds; and when these are drawn out, the fundamental structure of the rocks will often become clear immediately. An important point to note when using such sections is that, if the section taken does not align exactly with the direction of the dip of the beds, a projected dip will show in the section. The relation of the direction of the section and the dip of the main strata should always be stated.

In order to get as much information onto geological maps as possible, in a style convenient for use, systems of "shorthand" have been evolved for indicating concisely features of significance. Figure 14.1 gives some typical symbols. All well-printed maps will include a list of the symbols used, so that there need not be any confusion in using maps from different agencies which may not have identical symbols, although the most common signs are naturally in almost universal use.

Almost all parts of the world are now covered by geological maps, if only to a very small scale for more remote areas. Most developed countries will have geological maps to a scale of $1:500,000$ available, but these will usually be of little use in engineering work. As national geological survey work proceeds, an increasing number of $1:50,000$ scale geological maps are becoming available. Prepared and published by national, state, or provincial geological surveys, these maps form an unrivaled source of general geological information, a source to be supplemented naturally by more detailed mapping for individual sites. Appendix B lists all geological surveys throughout the world and so will be a guide to the availability of geological maps and associated information.

14.2 GEOLOGICAL MAP MAKING

Note may be made of the interest to be derived from the fieldwork involved in the preparation of geological maps. As in many other outdoor activities, nothing can quite take the place of actual experience in the field. If geology can be thus studied under expert guidance, more can be learned in the field in a short time than is possible through many months of book study. Not only will geology become "real" and the sense of the three-dimensional structure of the earth's crust become second nature in the mind of the interested engineer, but a new feeling for scenery and the beauty of the land will be gained. The attention of readers who are attracted to these possibilities may be directed to the activities of local natural history clubs and geological clubs (in the larger centers). Contact with a local university may lead to participation in university geological fieldwork arranged for students. It was in such fieldwork that both the authors of this Handbook gained the most lasting of their early experiences of geology.

The main information to be recorded on field maps is noted carefully in the field on the topographic map in use; for this purpose it is naturally advantageous to adopt a special shorthand such as is shown in Fig. 14.2. To avoid possibility of error it is essential that each observer adhere to a rigid system in all fieldwork. Different types of deposits may be indicated either by lettering or by characteristic shading or coloring for each exposure. The limits of rock outcrops must also be indicated clearly; one method used is to draw a line around the boundary of the outcrop, coloring or shading the side of the line corresponding to the start of the drift coverage of the solid rock.

At the conclusion of each day's work, all observations must be clearly and accurately inked in, either on the map used in the field or on a special copy of the base map retained for office use only, but preferably on both. As such notes gradually accumulate on a map, geological structure will slowly reveal itself. The limits of the surface exposure will first appear in a general way; later the boundary-contact lines will become traceable with some degree of accuracy; structural features such as faults will appear in correlation with the rock strata; and finally the cause of certain topographical features may become plain. Nat-

DESCRIPTION	SYMBOL
Bedrock outcrop	
Limit of outcrop	
Contact	
Contact, showing dip	45 90
Overturned contact, showing dip	65
Approximate contact	
Strike and dip of beds	50
Fault, showing relative horizontal movement	
Fault, showing bearing and plunge of apparently downthrown block	D 65 (Normal) D 43 (Reverse)
Normal fault, (hachures on apparently downthrown side)	
Anticline, showing crestline	
Anticline, showing crestline and direction of plunge	
Anticline, showing crestline and plunge	20
Dome	
Syncline, showing troughline	
Syncline, showing troughline and direction of plunge	
Syncline, showing troughline and plunge	15
Glacial striae	
Line of stratigraphic section	
Vertical shaft	
Inclined shaft	
Portal or adit	
Portal and open cut	
Trench	
Prospect pit or open cut	
Mine dump	

FIG. 14.1 A selection of symbols used on geological maps by the U.S. Geological Survey.

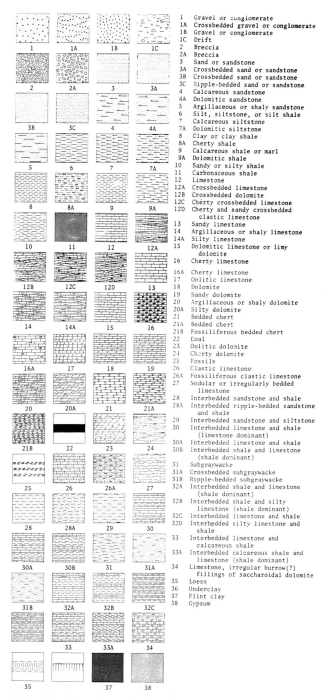

FIG. 14.2 Typical symbols for indicating different rocks types in geological sections, as used by the U.S. Geological Survey.

urally, the progress of this elucidation of geological structure will vary from one locality to another, but usually the surveyor will reach a point at which the map can be finished with some degree of confidence. (See Fig. 14.3)

In finished form, the field map will have the various strata exposed at the surface differentiated by several wash colors or by special shading, preferably the former. Different surveys have different standard color forms; and whenever possible, the local standard arrangement should be followed. Observed contacts can be marked as full black lines; inferred contacts as broken black lines. Faults are sometimes marked as heavy (full or broken) white lines superimposed on the base color but more often as "wiggly" lines. Other details noted on the original map will be retained, indicated in a general way by pen and ink markings. Such finished maps should always be accompanied by an index table or legend of the colors used and possibly of the special signs incorporated in the map. (See Fig. 14.4.)

14.3 THE ENDPAPER MAPS

In order to give at least an indication of the appearance of a colored geological map, the endpapers of this volume have been used to illustrate a small portion of one of a set of seven maps which, together, constitute one of the very best examples known to the authors of geological maps prepared for engineering purposes. The full title of the set is *Geologic and Hydrologic Aspects of Tunneling in the Twin Cities Area, Minnesota*, the editors being Ralph F. Norvitch and Matt S. Walton. Issued as Map I–1157 of the U.S. Geological survey in 1979, the set is the result of a cooperative effort of the national survey with the Minnesota Geological Survey, both in cooperation with the U.S. Department of Transportation. The title tends to obscure the great value of this remarkable publication, not only as a guide for use locally in the Twin Cities area but also as a model that can well be followed in other urban areas throughout the world. If users of this Handbook wish to have at hand an example of how geological information can be made readily available for engineering use, they should have a copy of this publication.

With the ready assistance of agencies and individuals in the Twin Cities area, records of about 4,000 test borings and 90 water-well records were assembled, plotted, and studied, in association with what was already known of the geology of this important area. The maps cover the central city areas of St. Paul and Minneapolis to a scale of 1 : 24,000. The attempt was made to use records from 25 test borings for each 10 mi^2, their location being shown on Plate 1 (Introduction and Data Base). Plates 2 and 3 show, respectively, the surface geology and the underlying bedrock geology, Plate 4 giving the corresponding depth of drift over the entire area, i.e., the depth to bedrock. Plate 5 presents useful geological sections which make vividly clear the geological structure underlying the Twin Cities; these sections are accompanied by useful information on engineering properties of the main soils. Plate 6 presents in graphic form the hydrogeologic setting of the area, Plate 7 elaborating on this and suggesting constraints on tunneling. This was the main purpose of the maps, but, as will be seen, they will prove useful for all civil engineering purposes in the Twin Cities.

14.4 MAPS FOR ENGINEERING GEOLOGY

Useful as general geological maps are for several purposes, one of them being in preliminary studies of sites for civil engineering works, their use in connection with civil engineering is restricted. One reason for this, as will have been noted, is that most general

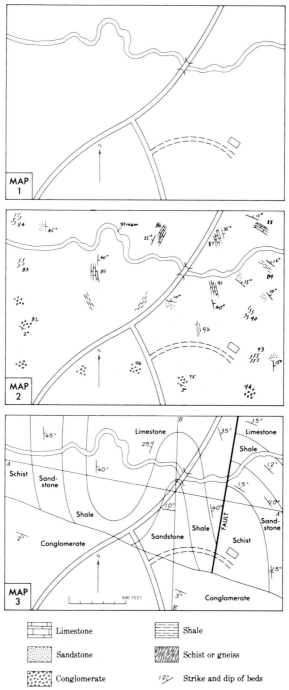

FIG. 14.3 Evolution of a geological map. Map 1 is the base map showing only topographical features; map 2 is the same map as marked up by a geologist from field studies, dip angles indicated at the dip-and-strike symbols, other numbers being index numbers for field notes; map 3 shows the boundary lines between beds, as determined by the geologist.

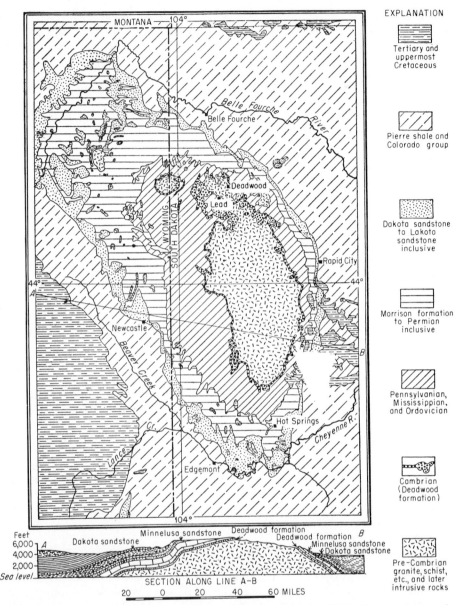

FIG. 14.4 A typical geological map and section, although without the coloring that assists so much in their use; this map shows part of the Black Hills in South Dakota and Wyoming, including a geological dome structure.

maps show bedrock only, whereas most civil engineering work is carried out in soil overlying bedrock. Some surveys have a long tradition of publishing maps which do show soils. Notable are the "drift maps" of the Geological Survey of Great Britain (now the Institute of Geological Sciences), but similar maps are to be found in the lists of other national and state surveys. The small scale of many general geological maps is another difficulty; this limitation is slowly but steadily being improved as national surveys increase their detailed coverage and publish more maps to larger scales, commonly 1:50,000. A third and final limitation, even with large-scale maps, is that they show only the geological conditions for the area covered and do not attempt to interpret these in terms that will be useful to civil engineers.

As will already be obvious, the information that the civil engineer needs is incorporated in general geological maps, but it has to be "dug out" and made available in suitable form. Such interpretations call for three qualifications, as Eckel pointed out in an early paper on the use of geological maps for engineers.[14.1] The engineer or geologist who is to interpret the map must have a sound knowledge of the fundamentals of geology, as well as a clear understanding of the facts needed by the engineer and how these facts will be used; the interpreter must also have a "genuine desire to put geology to practical uses."

Fortunately, there is available for civil engineers a most unusual but a singularly valuable publication of the U.S. Geological Survey (through its engineering geology and groundwater branches) that well illustrates the great utility of geological maps in engineering work. This is a folio of maps entitled *Interpreting Geologic Maps for Engineering Purposes*.[14.2] It includes six maps of the same area, the Hollidaysburg Quadrangle in south-central Pennsylvania, all printed to the well-known high standards of the USGS. First comes a standard topographical map of the area, followed by the standard geological map. Since the area contains mountainous country, the geological map is a vivid example of the utility of color in representing geological formations. There follow four interpretative maps, prepared after office and field study by the staffs of the two branches. These show how the local geology affects, respectively, foundation and excavation conditions, the availability of construction materials, water supply (both surface and underground), and the selection of sites for engineering works. This list alone again illustrates the vital place that geology can and should occupy in the planning of civil engineering operations. The folio should convince even the most skeptical of the great value of preliminary site study through the medium of good geological maps. It should be a valued item in all civil engineering libraries. Although first issued in 1953, it still provides an excellent starting point for an appreciation of the great value of general geological maps in civil engineering work.

Much more extensive studies of the same sort, all starting with general geological maps, have been carried out in more recent years. As examples there may be mentioned the several series of maps, to a scale of 1:24,000, that have been prepared as part of a major study of the Front Range Urban Corridor, the rapidly developing area centered around Denver, along the foothills of the Rocky Mountains. Following a dispute about land use, the Denver Regional Council of Governments entered into an agreement with the U.S. Geological Survey in 1967, this being the first such cooperative arrangement with a regional planning body made by the Survey. The objects of the work were three: to interpret the geology of the area so that it would be of maximum use to engineers and others engaged in planning work; to develop, from the use of the first maps produced, evaluations of these maps so that future work could be improved; and to stimulate among local government agencies and their officials an appreciation of the value of geology in all their work.

As a typical early example, there may be outlined the "Preliminary Engineering Geologic Map of the Golden Quadrangle, Jefferson County, Colorado" issued in 1971.[14.3] It was issued in six sheets, in black and white printing only. Sheet no. 1 showed an outline of the geology considered from the engineering point of view, different strata being indicated by symbols and letters. The signs so used were all explained on the second sheet, and typical cross sections were featured on the third. The other three sheets were all interpretive in nature, presented in the form of similar large tables for the superficial materials, the sedimentary bedrock, and the igneous and metamorphic bedrock, respectively. The columns presented such information as the physical character of each material encountered; the corresponding topographic expression and susceptibility to weathering; surface drainage; groundwater characteristics; foundation and slope stability; and known, reported, and probable uses of those materials that could be used when excavated. All this material was derived from a study of the basic geology, a study made by those who fully understood the purposes for which the maps were being prepared; special fieldwork supplemented the study of the basic geology. Later maps in this notable project have been issued (with color printing) by the U.S. Geological Survey in association with the Colorado Geological Survey; a typical map shows gravel sources in the Greater Denver area.[14.4]

These folios, as they are known, have done much to show to nongeologists the wealth of information that can be gleaned from general geological maps when such maps are interpreted by those who know their geology and appreciate the information needed in civil engineering work and in all phases of land planning. As a further example, there may be mentioned the folios of Knox County, Tennessee, issued by the U.S. Geological Survey in 1972–1974 in the form of 13 separate maps, all based on the general geology of this important area, an area which has the city of Knoxville as its principal urban center.[14.5] The maps show:

A. Land slopes and urban utilization
B. General geology
C. Distribution of sedimentary rocks
D. Structural geology
E. Groundwater yield potential
F. Areas with abundant sinkholes
G. Basins drained by sinkholes
H. Soil associations
I. Physical characteristics of soils
J. Overburden related to type of bedrock, etc.
K. Engineering characteristics of overburden
L. Categories of relative feasibility for septic-tank filter fields
M. Areas of possible flooding
N. Mineral resources

(The capital letters refer to the main index number for the folio—USGS Map 11-767.) All maps were published in 1972 except for the last listed—mineral resources—which was published in 1974.

The availability of such useful folios of interpretations of geological maps, prepared specifically for the use of engineers even though often denoted by reference merely to planning, is still so little appreciated in connection with site studies for civil engineering works that one further example will be cited. It indicates yet another source of such useful information, since it is a publication of a state geological survey. Circular 497 of the Illinois State Geological Survey, published in 1976, might not attract attention by its designation as a "circular."[14.6] It is, in fact, a brochure of 76 pages, accompanied by seven most useful interpretative maps, all serving to illustrate the use suggested by the brochure's title—*Geology for Planning in the Springfield-Decatur Region of Illinois*. The brochure is an admirably clear description of the geology of the area covered, with attention mainly devoted to the soils, the region being a famous one in the study of North American glacial deposits. Good references are given as a guide to further information and a listing of more than 800 wells, the records of which give useful information on local groundwater conditions, as well as depths to bedrock in many cases. The maps, all in black and white but easy to use, present information on:

1. Surficial materials (i.e., soils)
2. Water resources (with well locations marked)
3. Thickness of unconsolidated deposits (soils)
4. Coal and oil resources
5. Industrial minerals
6. Geology related to general construction
7. Geologic conditions related to waste disposal

The mere listing of the subjects presented by these maps indicates clearly how much useful information can be derived from general geological maps, when supplemented (as necessary) by some extra fieldwork. Geological maps, therefore, are not merely exercises in academic geology but can be tools of great value when put to use in connection with civil engineering works. It is desirable, therefore, that all civil engineers should be familiar with such maps. It is for this reason that the endpaper maps were included in this Handbook. At the end of this chapter some references are given to useful publications on geological maps and their use, so that those who wish to study them in some detail may be guided to readily available sources of information.

14.5 ENGINEERING-GEOLOGY MAPS

The steadily increasing number of interpretative geological maps prepared for civil engineering use has naturally directed attention to the possibility of preparing geological maps specifically for such uses. This comes as a development of interpreting general-type geological maps, a development actively pursued in European countries, although now well recognized elsewhere, as will be explained in the next section of this chapter. Various approaches have been taken in this work; some of the maps are little different from the interpretative maps prepared on the basis of an ordinary geological map but marked with reference to one specific aspect of ground use by civil engineers. Landslide probability maps from both European and North American practice provide one such example. As a guide to the possibility of landslides in an area under study, maps of this kind can be of service in preliminary assessments and as guides to necessary field examinations. All civil engineers will recognize, however, that there can be no substitute for actual examination of sloping ground on foot.

It must be admitted that, with the enthusiasm that such practical applications of geology have generated, there have been published some engineering-geology maps that tend to stray a little too far from the rightful field of the geologist and into the realm of engineering. Some maps have been published, for example, using the three well-known colors—red, yellow and green—to indicate, from geological considerations, whether an area is dangerous, borderline, or safe for various engineering functions. Bearing capacity for foundations and ease of excavation are two subjects that may be noted as being capable of such representation. Unless, however, it is made crystal clear that such maps provide a guide only, they present the possibility of misuse, in which case they do a disservice rather than making a useful contribution. And even though the necessary warning may be clearly printed on the maps, experience shows that, as always, "there are none so blind as those who will not see," and the warnings may be disregarded. On the other hand, engi-

14-12 PRELIMINARY STUDIES

neering-geology maps prepared by those who fully appreciate the limitations of all such aids can be of inestimable value; and so they have proved to be in many parts of the world.

With particular reference to work in cities, where engineering-geology maps have their greatest use, the civil engineer engaged in studying a site for some proposed structure needs to know the depth to bedrock, the general nature of the soils overlying the bedrock,

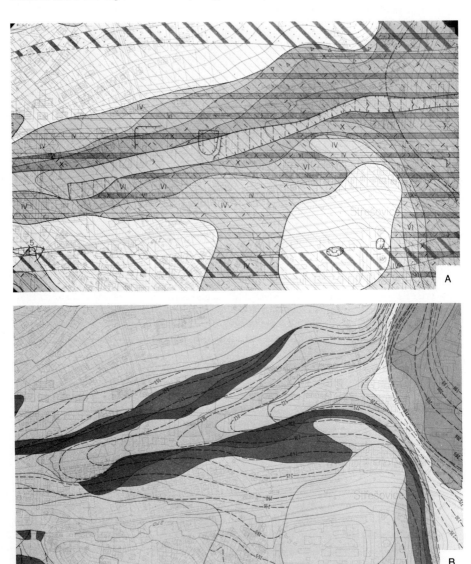

FIG. 14.5 Small sections of one of the quadrangle maps now available for the city of Prague, Czechoslovakia, showing: map A—geological conditions 2 m below the surface; map B—the thickness of the overburden overlying bedrock; map C—the hydrogeological map, showing maximum level of the water table and the direction of groundwater flow; and map D—the documentation map, showing boreholes, etc., from which information has been obtained.

groundwater conditions, and the probable accuracy of this information for the site in question. With accurate information on these three major geological features, the necessary detailed site investigation can be planned with reasonable confidence. So logical is this approach that it is really no surprise to find that a quite general approach to the preparation of engineering-geological maps has been developed. This approach calls for four sheets for each area, one showing the bedrock geology and depths to bedrock, one the nature of the soils above bedrock, one the hydrogeological conditions, and one the points at which the data providing the basis for the information given has been obtained.

The writers know of no better example of this fine approach to the preparation of engineering-geological maps than in the city of Prague, Czechoslovakia.[14.7] By permission of the state geological survey, Fig. 14.5 is presented to show one small section of one of

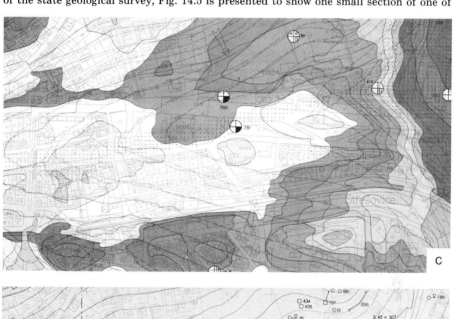

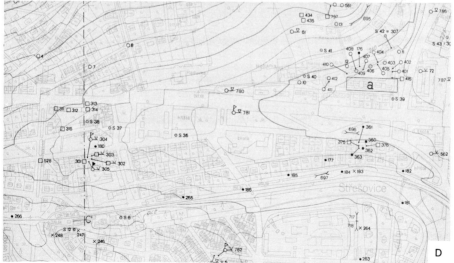

14-14 PRELIMINARY STUDIES

about 70 quadrangle maps, the entire city area having been divided into such quadrangles for mapping purposes. The four maps are shown here in black and white, although the originals are color-printed and so (in the case of the first three) much more vivid in presentation. The work was initiated by Professor Quido Zaruba, its most recent refinement being under the direction of Dr. R. Simek of the Institute for the Design of Communication and Engineering Construction (PUDIS) of Czechoslovakia.

Map A shows the general geological conditions for the quadrangle (of which only a small portion is reproduced in Fig. 14.5). By means of shading and the use of color, the geological conditions at a depth of 2 m below ground surface are shown. Surficial materials between this elevation and the ground surface are not shown on this map. Bedrock is shown by hatched lines (brownish-grey in the originals), and the soils above by colored areas for the first stratum, and by colored horizontal stripes if there is a second. Depths to bedrock from ground surface are marked, and the dip of bedrock is indicated in the usual manner. The map also clearly indicates such features as gravel pits, cuttings, and quarries, when these extend below the 2-m depth.

Map B shows the thickness of the overburden, the differing thicknesses being indicated by a color scheme, each color representing a 2-m interval (i.e., 0–2 m, 2–4 m, etc.). Bedrock contours are indicated by the usual type of contour lines, full lines being used when depths are certain, broken lines when they are inferred. Map C is the hydrogeological map, showing groundwater conditions. Depths to the water table are shown, again by a color scale, colors changing with each 2-m depth. Arrows indicate the direction of groundwater movement. Symbols give an indication of the chemical properties of the groundwater, when this information is available, so that any potential danger to materials that are to be placed in the ground can be foretold in a general way, and the need for special care, and further water testing, will thus be indicated.

Map D is, rather obviously, the "documentation map," a record of the points from which subsurface information, incorporated in the other maps, has been procured. As many as 1,000 points may be marked on any one quadrangle map. These will include locations of test borings, test pits, trenches, outcrops of rock, wells, exploration drifts (if any), and springs. An ingenious system of symbols, which may be seen even on this small portion of a quadrangle map, has been evolved to indicate such further information as whether groundwater was encounted in a test hole, whether a sample of groundwater was taken from a hole or well; whether soil or rock samples were taken, with a special notation if field tests were carried out at the location in question.

From such maps it is possible at a glance to gauge the accuracy of the information shown on maps A, B, and C for the site under review. Of even greater use is the fact that this map will show what, if any, test borings or other explorations have been made on or in the vicinity of any particular site. In this way, much time in searching can be saved and the planning of the detailed site investigation greatly facilitated. As more information becomes available, it is added to the master maps so that knowledge of the geological conditions beneath the whole of the area of Prague is not only already on record but is constantly being refined and improved. It will not be difficult for readers to assess for themselves the value of such a set of maps—did it exist—for city areas with which they are acquainted. The monetary savings alone that such maps make immediately possible warrants this somewhat extended reference to what one city has already done.

14.6 INTERNATIONAL COOPERATION

Some indication of the rapidly increasing attention being given to the preparation of special geological maps for civil engineering purposes was the publication in 1976 of a guide to their preparation by UNESCO. This international agency had supported the efforts of the Working Group on Engineering Mapping established by the International Association of Engineering Geology (IAEG), the publication being the first report of this group.[14.8] Membership represents seven countries, the chairman being from Czechoslovakia, the secretary from the United States, and the editor from Great Britain, countries that are mentioned to indicate the truly international character of this useful publication. The guide sets out the principles to be followed in producing engineering-geological maps and the objectives that must be kept in view in the light of the necessary interrelation of civil engineering and geology in this branch of applied geology. Techniques that may be used in preparing such maps are given. More than one-half of the 79-page brochure is taken up with examples of different types of engineering-geological maps, maps selected from those of seven countries.

One of the examples illustrated in the publication is the set of maps prepared by the U.S. Geological Survey at the request of the U.S. Department of Transportation for a study of the "Northeast Corridor," between Washington, D.C., and Boston, Massachusetts. In contrast to examples already mentioned, these maps were small scale (1:250,000), designed to give the overall picture of the physical conditions along this densely populated and highly industrialized part of the United States. Illustrated in the UNESCO guide are small sections of one sheet of each of the sets of five which show the bedrock along the Corridor and the nature of the surficial deposits, respectively; a third set of maps show earthquake epicenters, geothermal gradients, and major excavations and borings along the route. Each set is accompanied by tables of useful information on the physical characteristics of the rocks and the soils shown on the maps, some parts of which are also reproduced in the guide.

Members of the responsible Working Party "realize that [their] guidebook is not an exhaustive treatment of the subject, [but] hope that more complete versions may be prepared in the future . . . ," and they solicit comments and suggestions. The publication is given such extended notice in this chapter to indicate to users of this Handbook the activity currently taking place on an international basis. Further developments in these useful aids to engineering work, therefore, may confidently be anticipated. With the international cooperation already so happily demonstrated by this UNESCO publication, some degree of uniformity in the presentation of geological information on maps prepared for engineering use may be looked for, as agreement is reached on basic international standards. With civil engineering activity already so international in its coverage, this advance will be one of inestimable value.

14.7 CONCLUSION

If readers will examine map D in Fig. 14.5, they cannot but be impressed by the vast amount of subsurface information available for this small portion of the city of Prague, the amount of information indicated by the large number of records which were used in compiling the associated maps. Similar amounts of subsurface information are available for almost all cities, if only such information can be traced and coordinated as it has been in Prague and a few other major cities. Every new site investigation in urban (and rural) areas provides one more contribution to the general fund of knowledge of the underground of the area in question. Readers are, accordingly, asked to keep this in mind when conducting their own site studies. If these studies are assisted by information obtained from some type of information bank, it is small return to provide—in due course—a summary of the subsurface conditions revealed by the new site study. If no such information bank is available, and so no engineering-geological maps to assist with initial planning of the site study, then an equally valuable contribution to the advance of local civil engineering, and geology, will be to promote the idea of developing such a central store of subsurface information, preferably under city auspices. As such cooperative efforts spread, the availability of engineering-geological maps will likewise increase, to the lasting benefit of subsurface engineering and the further development of knowledge of the local geology, covered though it now may be by the pavements and sidewalks of city streets.

14.8 REFERENCES

14.1 E. B. Eckel, *Interpreting Geologic Maps for Engineers*, American Society for Testing and Materials Special Technical Publ. no. 122, Philadelphia, 1951, pp. 5–15.

14.2 *Interpreting Geologic Maps for Engineering Purposes* (folio of six maps; scale 1 : 62,500), U.S. Geological Survey, Engineering Geology and Ground Water branches, 1953.

14.3 M. E. Gardner, H. E. Simpson, and S. S. Hart, "Preliminary Engineering Geologic Map of the Golden Quadrangle, Jefferson County, Colorado," no. 2 in *Front Range Urban Corridor, Colorado: Environmental Geologic and Hydrologic Studies*, U.S. Geological Survey and Colorado Geological Survey, Folio MF–308, 1971.

14.4 D. E. Trimble and H. R. Fitch, "Map Showing Potential Sources of Gravel and Crushed-Rock Aggregate in the Greater Denver Area," no. 9 in *Front Range Urban Corridor* . . . , U. S. Geological Survey and Colorado Geological Survey, Map I–856–A, 1974.

14.5 I. D. Harris, W. M. McMaster, J. M. Kellberg, and R. A. Laurence, Map I–767 A to N, Folio of Knox County, Tennessee, U.S. Geological Survey, 1972, 1973, and 1974.

14.6 R. E. Bergstrom, K. Piskin, and I. R. Follmer, *Geology for Planning in the Springfield-Decatur Region, Illinois*, Illinois Geological Survey Circular 497, Urbana, 1976.

14.7 Z. Krelova, *Engineering-Geological Maps of the City of Prague*, Institute for the Design of Communications and Civil Constructions, Prague, Czechoslovakia, 1973 (An English summary of explanatory pamphlet in Czech).

14.8 W. R. Dearman (ed.), *Engineering Geological Maps: A Guide to Their Preparation*, Earth Science Publ. no. 15, UNESCO, Paris, 1976.

Suggestions for Further Reading

The paper by Eckel, referenced above, is still a useful introduction to geological maps for civil engineers; the principles involved have not changed since it was published in 1951. Geological maps are mentioned and usually illustrated in the general books on geology that have already been listed, but there are a few books devoted solely to the preparation and use of these end products of geological field work. Two examples are *Geological Maps and Their Interpretation* by F. C. Blyth, published by St. Martin's Press of New York in 1965, and *Geological Maps* by B. Simpson, published by Pergamon of New York in 1968.

J. L. Thomas is the author of an *Introduction to Geological Maps* published by George Allen & Unwin, Ltd. (London and Winchester, Mass.). *Basic Geological Mapping* is the title of a more recent volume by J. W. Barnes, an 128-page book published by Halstead Press of New York, one of the texts developed for use in the Open University of Great Britain.

Another Professional Paper of the U.S. Geological Survey (No. 837 of 1974, 48 pages) calls for mention since it is a stimulating, almost philosophical discussion of *The Logic of Geological Maps with Reference to Their Interpretation and Use for Engineering Purposes*, the author being David J. Varnes.

chapter 15

MATERIALS: SOIL AND ROCK

15.1 Soil for Earth Dams / 15-2
15.2 Bentonite / 15-10
15.3 Puddle Clay / 15-10
15.4 Soils for Filters / 15-12
15.5 Broken Rock or Stone / 15-12
15.6 Building Stone / 15-17
15.7 Conclusion / 15-23
15.8 References / 15-25

Most of the materials used by civil engineers in the construction of the projects they design are obtained directly or indirectly from constituents of the earth's crust. Probably little thought is given to this fact in connection with the use of steel and other metals, for example, but the discovery and working of metallic ores are branches of economic geological work of great importance to civil engineering. With such indirect contacts the civil engineer is rarely concerned, but there are other materials that are frequently used, the geology of which is of direct significance. The construction of an earth dam, in which local deposits of unconsolidated materials are used, provides an obvious instance. The use of clay for the manufacture of bricks, tiles, and similar products is clearly related to the geology of the clay thus utilized. The peculiar properties of the natural materials used for the production of limes and plasters are, to a large extent, determined by geological features. Finally, the extensive use of stone for building construction and the steadily growing use of "artificial building stone"—if concrete may be so described—are intimately connected with the geological history and mineralogical nature of the rocks thus employed.

Civil engineers accustomed to work within the limits of North America may be inclined to "write off" the uses of such natural materials as "old fashioned." All who have seen a modern high earth dam, however, will know differently. Far more important is the fact that the use of indigenous materials is a prime requirement of much building construction in developing countries. One has only to think of the small brick plant to be found outside almost every village in India (and there are something like half a million villages) and the still-common use of cut stone for so much building in the Middle East to realize that the use of Nature's own materials is still a vital part of worldwide civil engineering work, a

15-2 PRELIMINARY STUDIES

part that can be aided greatly by an appreciation of local geology. Examples, in what follows, will be selected generally from North America, but the special needs of other parts of the world should be kept in mind.

Despite the fact that all the materials mentioned may be supplied for use after having been inspected and approved by independent testing organizations, the civil engineer should be at least generally familiar with their origin and nature, particularly as he may one day be faced with the problem of selecting and approving such materials without aid when engaged on pioneer work. No treatment of geology as applied in civil engineering would therefore be complete without at least an introduction to the geology of these natural materials of construction. Even a brief summary of this vast subject will serve to direct attention to the very real importance of geology in relation to materials that can, all too easily, be taken for granted. For convenience, some sources of more detailed information are noted at the end of this chapter. It will be seen that the specialized literature available is of wide scope; in general, this is an indication of the attention that has been devoted to this contact of geology with civil engineering.

The economic importance of natural building materials has, quite naturally, led geological surveys, bureaus of mines, and similar public agencies to take an active interest in the location of deposits of what are frequently called "industrial minerals." There are, therefore, more published papers available for consultation in this part of the application of geology to civil engineering than in any other. A particularly good example, but typical of many such publications, is a bulletin of the U.S. Geological Survey, *Geology and Construction Material Resources of Morris County, Kansas.*[15.1] It was prepared in cooperation with the State Highway Department of Kansas as a part of the program of the U.S. Department of the Interior for the development of the Missouri Basin; some of the engineering features of this program will be referred to later. Of particular value for engineers not versed in geological terminology is a table in the bulletin listing the outcropping stratigraphic units and showing what construction materials may be obtained from each. The table starts with the system of the strata, then gives the group. It next gives general geological sections, with the thicknesses noted and the symbols used on the map; then the names of formations and members are listed, and finally, an indication of the construction materials that may be available from each member is given. The report discusses the availability of aggregate for concrete, mineral fillers, riprap, structural stone, road metal (the old English word is used), and finally subgrade and embankment materials. This bulletin is an excellent example of what such a regional survey should be and can show.

The natural materials discussed in this, and the two succeeding chapters, are generally classified as *industrial minerals.* Their economic importance is well illustrated by the fact that in Canada—so often thought of as a supplier of products of the mines such as nickel, gold, uranium, and potash—the annual value of industrial minerals is now about 12 percent of the total production of the mineral industry, recent annual totals being valued at over $500 million. Natural building materials are, therefore, an important part of the materials used in civil engineering work.

15.1 SOIL FOR EARTH DAMS

Despite the usually restricted meaning of the word "earth," its use to describe dams constructed of natural, unconsolidated materials has now been generally accepted. Earth dams are essentially simple structures. Frequently, they are the most economical of all types possible. They are probably the most common type of small dam; but in recent years, many earth dams have been constructed that may truly be classed as major engineering structures; some are almost 300 m (1,000 ft) high. Their simplicity resulted in their use in primitive communities, as exemplified by some of the notable ancient dams of India. Their economy is often due to the fact that they can frequently be constructed of material found adjacent to the dam site, a feature well demonstrated by some of the great natural earth dams.

The preliminary investigations at the site of an earth dam must naturally include due attention to the local geology. Studies will not be restricted to the dam site, however, if an earth dam can be considered as a possibility for the final design; subsurface studies must be continued around the site until uniform beds of suitable soils have been located, beds that are sufficiently extensive to serve as sources of material for the dam. This work will ordinarily be conducted along standard lines; test borings, test pits, and possibly geophysical methods will be used to supplement and confirm the results of surface observations. Samples of material in sufficient quantity to permit detailed study will be required; they need not be undisturbed, since the soils sampled will have to be moved if they are to be used for dam construction. If preliminary tests of grain size, permeability, and porosity show that the material is suitable, more detailed investigations of the sources of supply will be necessary. These investigations may be carried out by means of more frequent test borings and sampling; in some cases, test cuts with an excavator are made. Close study of the local geology, especially that relating to the origin of the soils, can obviously be of great service in this work.

Up to relatively recent years, the design of earth dams, involving selection of the material, proportioning of the cross section, and a decision as to the necessity of a concrete or clay core wall, was based to a large extent on previous experience and on general engineering judgment. The utilization of scientific design methods initiated a notable change of procedure, one which the use of the results of soil mechanics studies has developed still further. Today, none but the very smallest earth dams should be designed on any basis other than a complete and detailed analysis of all the factors involved; modern soil-testing methods, scientific stability calculations, and accurate field control should be utilized. There are few types of graded soil mixture containing a reasonable amount of fine material that cannot successfully be used for earth-dam construction, so that once a suitable source of material is located, its geology ceases to have any special significance in relation to design. Construction methods will depend to some extent on the type of soil being used. Dams built in rolled layers are frequently composed of material high in sand and gravel content, at least for the downstream side of the core wall, if one is included in the design. The ease with which material can be spread and rolled is a determining factor which automatically eliminates many soils with an appreciable clay content.

Three examples will be briefly noted to illustrate the influence that geology may have upon the design and construction of earth dams. The first major rolled-earth-fill dam in eastern Canada was built in 1939 across the Grand River in southwestern Ontario as a pioneer water-conservation structure in this part of the Dominion (Fig. 15.1). Having a maximum height of 22.5 m (75 ft), the dam has a central mass concrete spillway section, typical of the kind of dam commonly used in such locations and for water-control purposes. Study of an alternative design, in which glacial till obtainable on the site of the dam itself and from nearby locations would be used, demonstrated the undoubted economy of this type of structure. Flanking the central concrete section, therefore, are two massive embankments containing about 380,000 m^3 (500,000 yd^3) of compacted glacial till, which was obtained from borrow pits at the dam site and required little, even in the way of additional moisture, to render it suitable for placing. Compacted densities up to 216 kg/m^3 (135 lb/ft^3) were readily obtained. The other physical properties of the compacted till were such that it was possible to design the cross section of the dam without the use

15-4 PRELIMINARY STUDIES

FIG. 15.1 The Shand Dam on the Grand River, Ontario, a dam built of glacial till with no core wall.

of any core wall (as was necessary in earlier practice), properly placed filters ensuring its stability. The use of this widely distributed material directed attention to the potential of glacial till as a most valuable material of construction.[15.2]

Very different was the use of soil for the construction of a multiple-purpose dam on the Neusa River high in the Andes Mountains of Columbia, 56 km (35 mi) north of Bogotá and almost 3,000 m (10,000 ft) above sea level. With a height of 45 m (149 ft) above riverbed and a total length of 345 m (1,150 ft), this dam contains just over 760,000 m³ (1 million yd³) of rolled fill. The original design contemplated a rolled core of compacted clay between two shells of compacted Guadalupe sandstone. Study of the weathered clay from the right abutment, thought to be suitable for the impervious fill, showed it to consist very largely of the mineral halloysite, which possesses low strength when saturated. The design, therefore, had to be changed. It was found that an impervious fill with adequate strength properties could be obtained from the thin-bedded soft shales found also in the Guadalupe formation on the left bank at the dam site. The sandstone could be rolled with heavy rollers in order to break it down effectively and could apparently be used to form the two pervious shells. When work started, inspection of the borrow areas showed that it would be difficult, if not impossible, to separate properly the shales from the sandstone. Therefore, the idea of separate pervious shells was abandoned, and the entire dam was made of a compacted rolled mixture of the two materials. A satisfactory structure resulted.[15.3]

Another high dam, high in location and in design, is the main dam of the Southern California Edison Company's Mammoth Pool development in the High Sierras. Located in a narrow gorge on the San Joaquin River, the dam is 99 m (330 ft) high and 249 m (830 ft) long; it retains water for passage through a pressure tunnel, 12,000 m (40,000 ft) long and 6 m (20 ft) in diameter, which leads to a penstock supplying a powerhouse in which there are two 75,000 kva units. Figure 15.2 shows the precipitous sides of the gorge in which the dam was built, a location seemingly ideal for a concrete dam, either arched or of gravity section. Economic studies, however, showed clearly that a rolled-earth structure would be most economical, by a margin of at least $1 million. It was found that 80 percent of the required fill was available in borrow pits only 3.2 km (2 mi) away from the dam site, and the road from the borrow pits to the dam was on a 6 percent downgrade, ideal for hauling. The remaining material came from a variety of local sources; filter material was dredged from the riverbed below the dam site. In all, about 3.8 million m³ (5 million yd³)

FIG. 15.2 The Mammoth Pool earth-fill dam under construction in the narrow gorge of the south fork of the San Joaquin River, California.

of fill was required; a large proportion of it had to be well watered in the borrow pit before it was moved in order to give the necessary moisture content when rolled in place.[15.4]

These three examples, so different from one another, well illustrate the profound influence that geology can have on the selection of materials for the construction of earth dams. Most cases, somewhat naturally, will exhibit no such special features, but it is, as always, the unexpected that must ever be kept in mind. One further example may be cited to illustrate how a geological search for suitable materials may provide unusual benefits. The Yale Dam of the Pacific Power and Light Company is 97 m (323 ft) high, damming the Lewis River in the Pacific Northwest to provide a working head of 75 m (250 ft) in an adjacent powerhouse (Fig. 15.3). It has a total volume of about 3 million m^3 (4 million yd^3). Early plans called for a rock-fill structure, but the only rock available at the site broke up too finely, making its handling uneconomical. Plans were therefore changed to an earth-fill dam. Close at hand, a large deposit was located of silty clay of a grade suitable for earth-dam construction. To ensure enough material during all types of weather, and since winter placement was contemplated, the general contractors went so far as to cover 10 ha (25 acres) of the deposit with an asphalt blanket.

Further search revealed an alternative supply of what was locally called a "torrential conglomerate," an almost perfectly graded mixture of soil sizes—a mixture looking like gravel—which, when compacted, was almost impervious. It develops from sedimentary deposits made without segregation on upland terraces. It was at first thought that only a limited amount of this ideal material was available, but further site investigation proved that the natural deposits contained more than enough for the 450,000-m^3 (600,000-yd^3) core of the structure. Since placement of fill during wet weather was unavoidable, in view of the local climate and the urgency of getting the structure completed at the earliest possible time, it was originally planned to place the silt-clay during dry weather, using the conglomerate soil mixture only during wet weather. So well did the latter material behave

15-6 PRELIMINARY STUDIES

FIG. 15.3 The Yale Dam on the Lewis River in the Pacific Northwest under construction, looking downstream.

under wet conditions, however, that when the amount available was found to be sufficient for the entire core, plans were changed and the conglomerate material was used throughout. Compaction was by rubber-tired rollers, and it proved possible to place the material and compact it in all but the wettest weather. Some water had to be added during dry spells to give the necessary optimum moisture content (Fig. 15.4).[15.5]

The Yale Dam illustrates well the beneficial result of intensive geological study of the area around a dam site when searching for suitable fill material. Usually this search is carried out on foot, in ways already described, but much can be done, at least in a preliminary fashion, by careful study of large-scale aerial photos of the area around the dam location. For one notable Canadian dam, built on the Bersimis River on the North Shore of the Gulf of St. Lawrence and involving 3.5 million m³ (4.5 million yd³) of fill, available air photos were studied carefully by interpretative experts in their own offices, without any visit to the remote site of the works. They were able to detect, from their knowledge of ground erosion patterns, where there was a nearby deposit of a clay-type soil (and in a similar way, they located deposits of gravel for concrete and road construction). Subsequent test drilling on the sites so selected proved the accuracy of the photo interpretation, and all the soils required for the project were obtained as had been foretold by these somewhat unusual geological studies.[15.6]

This close attention to the geology of the sites for earth dams and of the areas surrounding them, combined with advances in soil mechanics, has resulted in a real advance in civil engineering practice. When site studies can assure a plentiful supply of suitable soil, and tests in soil mechanics laboratories show that the materials so found have the necessary physical properties, then earth structures can safely be used to heights previ-

FIG. 15.4 The Yale Dam on the Lewis River as completed, looking upstream with Mount St. Helens in the far background, before its recent eruption.

ously regarded as unattainable. In Chap. 25 reference will be made to some of the great earth dams of the world, some now exceeding 300 m (1,000 ft) in height. Here, attention may be directed to more mundane structures such as the dikes that must often be constructed as extensions at both ends of a major dam to retain reservoir waters when land values are such that flooded areas must be restricted or to accomplish other purposes.

The International Power House at Cornwall-Massena, spanning the River St. Lawrence as a part of the St. Lawrence Seaway and Power development, is a case in point. All the early plans for this great civil engineering project, planned and constructed by the United States and Canada jointly, were based on a two-stage development. Eighteen miles of dikes were needed for the single-stage project actually built, dikes extending from the ends of the powerhouse dam, especially on the Canadian side, in order to limit the flooding of valuable land (Fig. 15.5). In a similar way, many miles of dikes were necessary to form the new canal along the south shore of the St. Lawrence River facing the Island of Montreal, in the entrance section of the Seaway. In both cases, most detailed geological surveys were carried out, followed by an intensive program of subsurface investigation. Glacial till was found in considerable quantities, sufficient for all the earth fill required. Dikes consisting of compacted glacial till were therefore designed and built, with heights up to 22.5 m (75 ft) and with broken-stone facings serving as protection and as pervious outer blankets. Till for the canal dikes was, in general, excavated from the canal prism, that for the powerhouse dikes from specially located borrow pits in the area to be flooded.[15.7]

15-8 PRELIMINARY STUDIES

FIG. 15.5 Canadian end of the joint Cornwall-Massena powerhouse on the St. Lawrence River, showing dikes constructed of glacial till and extending from the powerhouse structure.

Another project in which this combination of good geological reconnaissance and modern soil testing, design, and construction methods was critical was the Grand Rapids water-power project of Manitoba Hydro on the western shore of Lake Winnipeg. Here the Saskatchewan River, draining much of the Canadian prairies, discharges into the lake down what were so rightly called "the grand rapids." The very name suggests a good site for the development of water power, and so it was, a powerhouse near the lake being able to utilize the full fall in the rapids. The area above the rapids, however, was so flat that many miles of dikes would have been necessary to retain water in the reservoir to the level which engineering economics could determine was necessary for storage. Until the advent of modern soil studies, this proved an insuperable obstacle.[15.8]

In summary, when an earth dam is a possibility for any river project, geology can be a powerful aid in locating suitable supplies of material. The areas to be surveyed in any search for suitable soil around the site for a dam will normally be so extensive that, without preliminary knowledge of the local geology, any hunt for suitable material will be a "hit-or-miss" procedure. Knowledge of the geology can point the way to the best areas in which detailed subsurface investigations should be pursued. Once suitable material has been located, and proved by the necessary laboratory tests, engineering design can proceed, leading to the economic studies upon which selection of the design to be adopted will be based. This is all in the province of the civil engineer, but successful completion of the design process will always be aided by initial geological considerations. Figure 15.6 indicates the wide scope of such investigations.

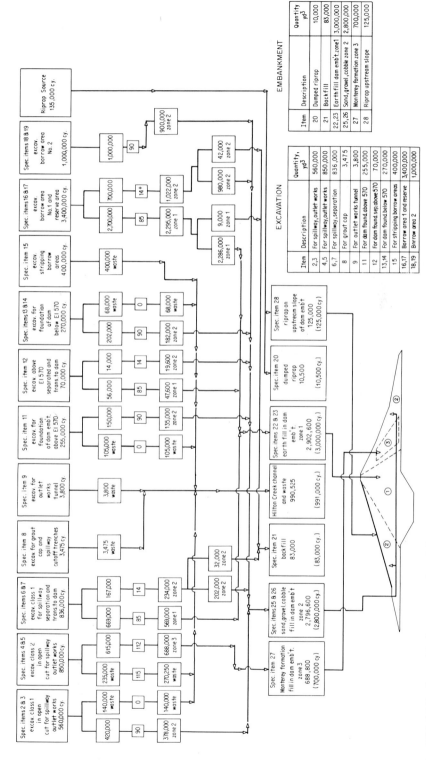

FIG. 15.6 A graphical representation of the many uses of soil on a major dam project.

15.2 BENTONITE

Bentonite is a special soil type that has achieved prominence in civil engineering because of its use in the "slurry-trench" method of constructing cutoff walls. This will come up for mention in later sections of this Handbook. Bentonite is a clay derived from the alteration of volcanic dust and ash deposits and composed mainly of montmorillonite. Owing to the capacity of this mineral to absorb water within its crystal lattice, bentonite will swell considerably upon the addition of water. This property led to its early use as a constituent of "drilling mud" in oil-well drilling. Thence it became known in civil engineering circles and is now widely used not only, as indicated, in slurry-trench excavation but also as a sealant in open excavations and as grouting. Extensive commercial deposits are found in the Black Hills of South Dakota. A 4% mixture of bentonite and water has been found suitable for grouting up leaking joints in a sewer. Such a mixture has also been used for filling expansion joints, held in place by metal strips once installed. Bentonite is a soil, therefore, worth remembering when difficult problems with water are encountered.[15.9]

15.3 PUDDLE CLAY

It was common practice until the 1930s to have in the center of many earth dams a core wall constructed of *puddle clay,* natural clay mixed with water to form a plastic mass which can easily be handled and which can be "worked" into position to form a continuous wall of material. The puddling was usually done in a pug mill of special design. Puddle clay has also frequently been used to seal leaks in canal banks and similar water-retaining structures. The term is one that is somewhat loosely applied, since many types of clay can successfully be used for the purposes indicated. The three main requirements of the puddle clay used in dam construction are tenacity, imperviousness to water, and the property of drying without seriously cracking or shrinking. These properties have been judged in the past by means of simple field tests. Although these tests have been superseded by the more scientific tests evolved in soil-mechanics laboratory practice, they may be given at least this passing attention, since they have been successfully applied to the puddle-clay requirements of many notable earth dams. Tenacity was often tested by making up a roll of the puddle clay about 2.5 cm (1 in) in diameter and 30 cm (12 in) long; this roll had to remain as one piece when suspended by one end. Permeability was tested by making a bowl of clay, filling it with several gallons of water and comparing the loss therefrom with that from a similarly filled impervious basin. Rough and ready as they may now seem to be, tests such as these have proved of great service in the past.

Advances in soil mechanics as applied to the design of earth dams and similar water-retaining structures have naturally relegated puddle-clay tests and usage to "practices of the past." But there may still be cases, especially in isolated locations far removed from laboratory facilities and with no modern construction equipment available, where this "old-fashioned" method of using soil may still prove useful. And it still has its occasional uses in modern structures and in late-century engineering practice where normal methods cannot be used. A good example of its modern use was for forming a cutoff wall for stopping seawater intrusion into the subsidence basin of the Wilmington and Long Beach harbor areas in southern California. In all, almost 3,600 m (12,000 ft) of puddle-clay core wall, 0.8 m (2.6 ft) wide and 7.5 m (25 ft) deep, was constructed; a novel aspect of the project was the use of bentonite slurry to maintain the open excavated trench until the clay was placed, an idea more recently applied to other construction operations.[15.10] Since information about material used to form puddle clay is sadly lacking, Table 15.1 is included as a reference that may be useful to any who have to neglect the scientific advances of soil

TABLE 15.1 Properties of Puddle Clays[25,19]

Name of reservoir	Proportion of clay, percent	Proportion of sand, percent	Proportion of water when pugged to max. plasticity, percent	Contraction on drying in bar 100 mm long, percent	Percolation of water through a layer of clay 2 in thick under a head of 7 ft, cc			Tenacity of molded briquettes 1 in^2 in section when dry, lb
					After 24 hr	Total in 432 hr	During last 24 hr	
1. Pontian Ketchil (Singapore Municipality)	94.6	5.4	29.8	7.2	Nil	Nil	Nil	65
2. Blaen-y-cwm (Ebbw Vale Steel Company)	79.0	21.0	21.0	5.0	0.8	11.6	0.6	67
3. Carno (Ebbw Vale U.D.C.)	71.5	28.5	20.0	8.0	1.4	11.4	0.5	73
4. Taf Fechan (Merthyr Tydfil Corporation)	71.5	28.5	17.4	2.9	0.2	2.2	0.1	
5. Silent Valley (Belfast Water Commissioners)	60.0	40.0	20.9	4.0	Nil	Nil	Nil	156
6. Burnhope (Durham County Water Board)	57.3	42.7	20.1	5.9	Nil	Nil	Nil	107
7. Alwen (Birkenhead Corporation)	50.6	49.4	15.0	2.5	2.3	17.6	0.6	30
Average	69.2	30.8	20.6	5.1	0.7	6.1	0.2	83

mechanics and utilize again this old and well-tested method of using one of Nature's own materials.

15.4 SOILS FOR FILTERS

If only as a reminder of the widespread utility of soils for special purposes, this note is included to remind readers that sands and gravels are still widely used for a variety of filtering purposes. Probably the main such use is for providing the graded filter beds of water-supply projects. Long experience, and more recent laboratory investigations, have shown that filter sands must possess certain specific characteristics if they are to be satisfactory for this special use. The effective size of grains should lie between 0.25 and 0.50 mm (0.01 and 0.02 in) with 10 percent being finer and 90 percent coarser, by weight. Of more geological significance, however, is the usual requirement limiting the amount of calcium and magnesium carbonates in the sand. Knowledge of the probably geological origin of the sand will assist in preliminary assessments of this property. Silica sand is naturally desirable, but specifications will usually permit minute quantities of other materials. Ordinary sand, therefore, can prove to be a material of significance. When its use as a basic constituent of glass is recalled, it may not be regarded, as it is so often, as a nonengineering material.

15.5 BROKEN ROCK OR STONE

Not only are soils used directly as engineering materials in modern civil engineering work, but so also is rock—in finished form as building stone, in fragmented form as crushed stone, and as roughly quarried broken rock used for a variety of purposes. The first use will be considered later in this chapter, the second in the following chapter. Here brief consideration will be given to the use of broken rock or stone for the construction of rock-fill dams, as riprap for the protection of exposed faces of dams and embankments, and for the building of simple embankments. In all these cases, it is not just "rock" that is being used, but a particular form of rock with its own special geological characteristics, qualities that may be found to be reasonably constant for the same type of rock if geological location or deposition are similar.

Although it is not necessary (usually) to qualify broken rock by its geological name in contract documents, it is certainly essential that its geological origin and composition be considered before any rock is accepted for use. This simple statement might be thought to be uncalled for in this book, but, as another example of neglect of the obvious, Eckel has recorded that:

> In one large midwestern city . . . the same nearly uniform limestone formation has been tested several hundred times, at a cost to the city of around $100 for each set of tests. Yet is has apparently never occurred to anyone that all of the test results are nearly identical and that use of a geological map showing this formation might make most of the testing unnecessary.[15.11]

So, also, with every use of broken rock; its geology is a first requirement of investigation.

Rock-fill dams have been increasingly used with changing economics; refinements have steadily been made in their design; notably, sloping impervious cores, or blankets, are a feature that has greatly widened the potential use of the rock-fill dam. As in the case of rolled-fill earth dams, selection of a rock-fill dam will depend jointly on the availability of a supply of suitable material and a study of the economics of alternative designs. Experience with the San Gabriel Flood Control Dam 1 is a useful reminder of the care that must be taken in the selection of durable rock for rock-fill dams. With standard tests now accepted and tried by long experience, the suitability of rock proposed for a dam structure can be readily determined well in advance of design.

Typical of recent practice is another structure in the High Sierras, the Courtright Dam on the Helms Creek, a tributary of the Kings River, California; this project of the Pacific Gas and Electric Company contains 1.15 million m^3 (1.5 million yd^3) of rock and rises 93 m (310 ft) high above stream bed (Fig. 15.7). Located also in a narrow canyon that might

FIG. 15.7 The Courtright Dam on Helms Creek, California, during the placing of rock fill.

appear to be more suited to a concrete structure, the Courtright rock-fill dam was shown by detailed studies to be the most economical of all available solutions, since it was possible to blast granite down from the canyon walls to form the dam. As no impervious fill was economically available, a reinforced-concrete upstream membrane was constructed to provide the necessary cutoff; the rock-fill structure provided the necessary support. Rock was dumped or blasted directly into place, and high-pressure water sluicing finished the operation; blocks of rock, put into position by cranes, formed a relatively smooth face on which the upstream concrete membrane was placed.[15.12]

The use of such igneous rock conveniently adjacent to the dam site often presents the most economical solution to the problem of dam design. Even much weaker rocks, however, can be used if necessary, after requisite site studies and laboratory testing have been carried out. When the U.S. Corps of Engineers was charged with constructing a dam across the Missouri River at Gavins Point, near Yankton, South Dakota, suitable fill materials in the quantity needed for the dam were not available. But excavation for the adjacent spillway cuts and powerhouse was going to involve the removal of 8.3 million m^3 (11 million yd^3) of Niobrara Chalk, 380,000 m^3 (500,000 yd^3) of Carlile shale, and 1.5 million m^3 (2 million yd^3) of glacial deposits as the overburden. The Niobrara Chalk is widely distributed east of the Rocky Mountains, from Canada to Texas, in a formation varying in thickness from 150 m (500 ft) in southwestern Nebraska to 54 m (180 ft) at Gavins Point. The volume to be excavated was close to that required for the main dam embankment. A good deal of information on use of the chalk was available from experiments con-

15-14   PRELIMINARY STUDIES

ducted earlier at the Fort Randall project, where a large test embankment had been constructed. The chalk rock is so weak that it can readily be broken down with suitable construction equipment (which at the Gavins Point project included a compacting roller weighing 27,000 kg or 60,000 lb when filled), and so it was decided to use the chalk for the dam at Gavins Point, which was to have an impervious core.

Excavated material was spread in 12-in lifts, broken up with a spike-tooth roller; additional moisture was provided for the material, which was then compacted in place with a tamper-type roller. Naturally, an extensive series of tests was conducted before designs were completed and construction was started. The satisfactory experience gained showed that, even with such a weak rock, adequate investigation and field testing permitted its use for an important dam structure.[15.13]

Riprap for the protection of erodible surfaces, such as the slopes of earth-fill dams and embankments, is another widespread use of broken rock. The long standard "one-man stone" still holds its own as one of the most effective types of slope protection when durable rock is selected for use and the riprap is carefully installed. There is at least one case on record where a carefully designed and placed porous concrete protection apron failed, had to be removed, and was then replaced by "old-fashioned" rock riprap at a cost of over $3 million.[15.14]

In the selection of rock for riprap, the greatest care must be taken in studying the character, durability, and quarrying characteristics of any rock available; here, too, geology is of prime importance. In parts of the world where hand labor can still be employed economically for the placing of embankment protection, granite "setts" may still be seen in use, hand-placed on sand-clay beds, with asphalt joint-filling. Perhaps the most unusual bank protection, geologically, is to be found along the banks of the upper reaches of the river Rhine where hand-placed blocks of diorite (basalt) may be seen, carefully interlocked as a result of their hexagonal shape (Fig. 15.8). The hexagonal shapes, which form ideal protection from the wash caused by the heavy river traffic, are the result of stresses developed during cooling of the molten magma from which the rock was formed. This columnar jointing is what lends great interest to well-known locations such as the Giant's Causeway in northern Ireland; it is especially interesting to see it put to such profitable use, as one can when sailing on the Rhine.

FIG. 15.8   Minor repairs to the protective revetment on the banks of the river Rhine, for which natural hexagonal blocks are used in places.

Among the tests necessary before any rock can be adapted for use as riprap are those which will indicate its susceptibility to deterioration as a result of alternate wetting and drying. Behind all such practical tests must come an appreciation of the character of the rock *in situ*. Much work has been done on systems of classification, based on observed features. In Finland, for example, a system has been developed for describing the natural "brokenness" of bedrock. Use is made of four types of natural jointing (cube, slab, wedge, and irregular); frequency of jointing; four degrees of weathering; and the quality of joints, whether tight, open, or filled.[15.15] With such a description available, a description based entirely on geological observations, the suitability of the bedrock in question for civil engineering uses can be the better appreciated. Methods that must be used for excavation naturally loom large when bedrock has to be selected for use as fill. One of the best-documented cases of field studies of rock excavation was for the design of a major embankment for one of Great Britain's modern highways, the location of the embankment being such that the completed fill was used also as a retaining dam for the local water supply.

Near the city of Huddersfield, some major road construction was carried out under the supervision of the engineer and surveyor of the county council of West Riding in Yorkshire. The new Lancashire-Yorkshire Motorway, for the construction of which the West Riding County Council was the agent for the British Ministry of Transport, connects the Great North Road up the east coast of England with the Stretford-Eccles Motorway (Road M62) in the west. Passing through the Pennines involved some very heavy engineering work, including rock cuts as much as 46 m (150 ft) deep and an embankment as high as 65 m (213 ft). This very high embankment was adjacent to the city of Huddersfield, and so advantage of the construction was taken to use the embankment as a rockfill dam to impound water for the Huddersfield Water Undertaking. This most economical dual project was therefore under two engineering jurisdictions, that of the county for the excavation and all road work and that of consulting engineers for the impounding dam. For the latter use, broken or crushed rock had to have certain very specific properties in the way of density, durability, strength, and grading. On the other hand, excavation of the great quantity of rock from the roadway cutting would be desirably carried out using the heaviest equipment possible, if it could be done without blasting, and with strict limitation if blasting were necessary. Accordingly, very extensive field trials of both excavation (using different methods) and fill placement were carried out in 1964 at the site of the works in order to ensure that the rock could be successfully used and to obtain information that would permit the preparation of an adequate specification for the main contract.

The trials were carried out at the western end of one of the deepest of the proposed cuttings, a cutting that would have a maximum depth of 46 m (150 ft). The trial embankment was constructed 1.2 km (0.75 mi) farther to the west. The trial drilling and excavation was confined to the top 18 m (60 ft) of the cutting. The strata to be traversed by the new road form part of the Millstone Grit series of rocks, one of the great divisions of the carboniferous rocks of Great Britain. The trials were carried out in a sandstone known as the Midgley Grit, which is generally massive in character but with many open joints. The work was carried out on a day-labor basis by an experienced general contractor under unusually close supervision. Most careful records were taken of all aspects of the trials, even including cinephotography, from the results of which a sound film was made for displaying to contractors when they came to tender on the main contract. The jointing in the rock permitted the use of massive rippers hauled by 385-hp tractors, but the resulting grading of excavated rock was not satisfactory. Multiple-row and single-row drilling and blasting were tried using a variety of explosives. Corresponding variations were used in the deposition of the broken rock in the test embankment and for its compaction in place (Fig. 15.9).

15-16   PRELIMINARY STUDIES

FIG. 15.9 Specimens of broken rock laid out for inspection during preliminary full-scale trials prior to the building of Motorway M62 near Huddersfield, Yorkshire, England.

Laboratory analyses using standard and special tests were run on samples of the excavated rock. These showed that the rock was not suitable for coating with tar or bitumen, that it could be used as aggregate for concrete of specified low-compressive strength only, and that the rock, although not acceptable for use as a subbase in road construction, could be used to within 15 cm (6 in) of the surface of the subbase. The results of the major field trials cannot usefully be summarized, but they are clearly set out in the excellent paper that was published soon after these trials were carried out. This reference can be recommended for careful study to all who have to face similar problems in the use of large quantities of excavated rock for embankment construction or, indeed, for any purpose other than dumping to waste.[15.16]

It should be added, to avoid any misunderstanding, that it is only rarely in the conduct of civil engineering work in North America that full-scale trials of this kind have to be carried out. The process of design based upon the accumulated experience of the profession and the steady development of sound theoretical methods is so far advanced that, when materials are under the control of the design engineer, a project can be designed completely in the design office with certainty that when constructed it will perform as intended. When dealing with the ground, however, the materials are not under the control of the designer, who has to accept what is found and "make the best of it." This is why foundation engineering will always remain an *art*, depending on experience and the ability to assess, right on the job, the best way of dealing with soil and rock conditions as revealed by preliminary investigation. Correspondingly, when large quantities of material have to be excavated, most careful assessment of the results of preliminary investigations is essential, and, when necessary, as in this case from Huddersfield, actual full-scale trials must be carried out to remove all uncertainty. This is of special importance when working in isolated areas or in countries without the advantage of extensive test laboratory facilities. But in all cases, no matter what the location, careful study of the geology of the bedrock to be excavated for use is an essential first step.

15.6 BUILDING STONE

"The style of a national architecture," wrote John Ruskin, "may evidently depend, in a great measure, upon the nature of the rocks of the country." Although modern methods of construction have in many instances reduced building stone to a mere facing, this forthright statement remains a telling reminder of the close connection between geology and the successful use of *building stone,* as specially cut and trimmed rock is generally called. The granite of Egypt and the marble and limestone of many European countries bear this out, as also such typically popular descriptions as "The Limestone City" (Kingston, Ontario) and "The Granite City" (Aberdeen, Scotland). The relationship provides one of the few points of contact between civil engineering and geology that appear to have been generally and continuously recognized, at least to some degree, throughout the past century.

The study of the geology of building stones may, indeed, have sometimes induced too great a regard for geological features and the neglect of other physical characteristics or even of quarrying methods. This was well shown at a very early date in connection with the stone used for the construction of the new Houses of Parliament in Westminster, England, about the year 1840. In 1838, a commission was appointed to select a suitable building stone for the work; William Smith and Thomas de la Beche, the first director of the Geological Survey of Great Britain, were members. After an exhaustive examination of many supplies of stone, a dolomitic limestone from Anston, near Mansfield, was selected as the most suitable. In the year 1861, just about 20 years later, another government commission had to be appointed to consider the "Decay of the Stone of the New Palace of Westminster," so serious had the disintegration of the selected stone become.

The fault appears not to have been in the geological selection but in the fact that there was no inspection of stone at the quarry before it was shipped. It has been stated that if only 15 percent of the original stone had been rejected at the quarry (as showing signs of weathering or of being traversed by vents), trouble would have been obviated. Confirmation of this was given by the prior use of the same stone, but quarried and placed under strict supervision, for the construction of the Jermyn Street Museum in London within about 1.5 km (1 mi) of the Parliament buildings. After almost a century's exposure to the London atmosphere, the stone still presented a splendid appearance when the building was officially closed in 1917. (See Fig. 15.10.) It is, perhaps, of more than passing interest to note that the establishment in this building of the British Museum of Practical Geology (predecessor of the well-known and justly famous South Kensington Museum of Geology), which was for so long the headquarters of the British Geological Survey, was largely due to this investigation. The Jermyn Street building was provided to accommodate the many specimens of British building stones assembled by the Commission of Enquiry during the search for an appropriate stone.[15.17]

Geology is only one of the factors on which the suitability of a building stone depends, but it is one of prime importance. Unless the geological nature and mineralogical content of a natural rock are satisfactory, other tests of its suitability as a building stone are useless. There is, fortunately, an extensive literature on the use of natural building stone (or *dimension stone* as it is sometimes called) to which the geologist and civil engineer may turn with profit when specific studies have to be made. Well-recognized standard tests are available. In view of the decrease in the use of natural stone in building and engineering work in the Western World, however, a civil engineer rarely has to consider opening up a new quarry for dimension stone; most of that used today comes from one or another of the large, well-established stone suppliers. In the less advanced parts of the world, however, new stone supplies will still have to be developed, if only because of variations in national economies. In India, for example, it is still more economical to build gravity dams

FIG. 15.10 Doorway of the (old) Jermyn Street Museum, London, for many years the headquarters of the Geological Survey of Great Britain, showing the excellent preservation of the stonework referred to in the text.

of hand-placed masonry than of mass concrete, even with the aid of the most efficient of modern construction machinery.

Availability of Suitable Rock The availability of suitable rock, an obvious prerequisite for the use of cut stone as a structural material, can be determined only by means of a geological survey. In open country, this may be a simple matter; but in drift-covered areas, the services of a trained geologist may be essential.

Relative Cost of Cut Stone Cut stone will not ordinarily be used on civil engineering works, except as decoration, unless it proves to be an economical alternative to concrete. Delivered cost of the stone will depend on ease of quarrying, extent of handling, length of haul, and cost of dressing (including wastage). Quarrying and dressing costs will be determined by the nature of the rock to be worked; massive formations of hard igneous rocks are obviously more expensive to work than those of blocky limestone, which can be quarried without the aid of explosives and can be readily dressed. Geological investigations will assist, therefore, in cost determination, particularly in securing an estimate of the essential homogeneity of the rock mass.

Appearance of Dressed Stone The final appearance of dressed stone in a structure will depend on its texture, its mineralogical composition, and the weathering to which it is exposed. In civil engineering work, a beautiful appearance is not often the controlling factor in the selection of a building stone, although a satisfactory appearance will always be desirable for any stone that is to be exposed to public view. Uniformity of color and texture will usually be desirable; this can generally be checked by visual examination of the rock mass before quarrying, provided one has an appreciation of its geological nature.

Strength and Weight of Building Stone These physical properties are of obvious importance in connection with the use of stone in civil engineering work. They depend on the geological nature of the rock, as a glance at any table of weights and strengths of building stones will make clear. As one indication of the influence of geological nature on these properties, the fact that fine-grained rocks are usually stronger than coarse-grained rocks may be noted. The strength of porous rocks (particularly sandstones) depends on the quantity of water present in them; tests should therefore always be made on both wet and dry samples. The variation in strength may amount to 50 percent of the test results for the dry stone; thus, it is advisable that ample safety factors be applied if dry stone is to be used.

Durability of Building Stones Durability is a prime requirement of a building stone that is to be used in civil engineering work. Expressed in simple terms, the durability of a stone is its capacity to retain its original size, strength, and appearance throughout a long period of time, while performing the work intended for it in the structure concerned. Essentially, durability is related directly to the normal weathering of rocks. The first and most obvious test of a building stone is therefore to be made by visiting the quarry from which it came and there examining the evidences of weathering on the naturally exposed surfaces of the untouched rock (Fig. 15.11). If this were always possible and were an invariable practice, some troubles with building stones would be avoided. Conditions of exposure at a quarry, however, may not be comparable to those to which the cut stone is to be subjected. This is especially true if the stone is to be used in city areas where the atmosphere may be charged with soot and the waste gases of industry, contaminants which may tend to form weak acids in combination with atmospheric moisture. Therefore, a second test of a very general nature is to examine other stone obtained from the same working and exposed to the same type of atmospheric conditions as those contemplated. Such observations can then be evaluated in light of the length of exposure.

The Use of Building Stone The large number of variable factors influencing the durability of building stone suggests that there may be restrictions to be observed in connection with its use, restrictions that will assist in its preservation. Some of these may be noted. The advisability of always setting sedimentary building stones parallel to their bedding planes is most important. If dressed stone is to be used, this direction should be determined before the stone is dressed. In the case of porous stones, every possible precaution should be taken to facilitate drainage of the stone, particularly as some stones become more pervious the wetter they get. No ledges on which water can collect should be left unprotected, particularly if the exposure is in a city area. The greatest care must be taken in the selection and use of mortar for jointing. Portland-cement mortar should not be used with sandstones or with some limestones. Pitting of high-quality limestone (and sandstone) has been observed in buildings in Liverpool, England. Although the exact reason for this was uncertain, the accompanying efflorescence suggested the portland cement backing to the stone as the cause.[15.18] All mortar for stone should be kept as low in

FIG. 15.11 The Rock of Ages granite quarry, at Barre, Vermont, the scale of operations well shown by the size of the men on the staging just above the center.

lime and cement content as possible. Some authorities suggest that pointing should be left rough, and not troweled off, in order to be sure that the joint is not sealed tight. Limestones and sandstones should never be used together in alternating positions.

Stone has been a leading building material for a very long time; in all probability, it was first used in ancient Egypt. When the methods developed by the Egyptians for quarrying stone are reviewed, they give a new perspective to modern stonecutting. For quarrying the granite from which so many of their notable monuments were made, they got rid of the surface layers by alternately burning papyrus on them and drenching them with cold water; if no natural fissure could be found, a trench was formed by pounding with balls of dolerite. Wooden wedges were inserted in fissures or in formed trenches, driven tight when dry, and then wetted. Levers, sledges, dragropes, and unlimited slave labor were used to move hewn stone blocks, the handling of which would challenge even an expert rigger on a modern construction job. The Great Pyramid was built of limestone in this way about 4700 B.C. Recent studies of the way in which it has weathered since its outer facing of hard, white limestone was removed from the ninth century on (for providing stone for mosques and other buildings in Giza and Cairo) suggest that it will last at least another 100,000 years.[15.19]

Even today, cut stone is still usually required for those parts of civil engineering structures where resistance to wear is highly important, as for such items as dock sills, lock-gate quoins, and large valve seats. And the real beauty of a carefully designed and well-constructed building stone exterior is still frequently called for, especially in the Old World, when the finished structure of a civil engineering project must blend as satisfactorily as possible with natural beauty that has had to be temporarily disturbed. The permanence of such buildings is well shown when a stone building has to be demolished to make way for one of more modern design, the permanence of which may perhaps be in keeping with current trends in aesthetic design.

Sources of Building Stone Quarries will naturally be the usual source of rock that is to be worked into dimension stone. Even to this there are occasional exceptions, as observant visitors to the campus of the University of Saskatchewan in Saskatoon, in the middle of the Canadian prairies, may appreciate. All visitors are attracted by the appearance of the well-proportioned older buildings of the university, all faced with a buff-colored limestone. The use of this building stone for all but a few of the more recent buildings gives to this campus an architectural unity that is unusual and pleasing. (See Fig. 15.12.) Those with any geological interests will naturally ask about the source of this building stone, especially if they know that the outcrops of bedrock nearest to the city are about 320 km (200 mi) away, in the northeast corner of the province of Saskatchewan, just south of the southern edge of the Precambrian Shield. The limestone, both dolomitic and calcareous, comes generally from the Ordovician formation known as the Red River formation, some from two allied Ordovician formations, and some from the Interlake formation, which is

FIG. 15.12 Buildings of the University of Saskatchewan, Saskatoon, all constructed of local field stone.

Silurian in age. These Paleozoic rocks do outcrop in the location mentioned, but the building stone was not obtained from the outcrop. Its origin was in what is called a *boulder train*, a collection of glacial erratics, scoured out from the underlying bedrock by glaciers and deposited from the glacial ice in a fanlike formation, the apex of which is approximately 40 km (25 mi) to the northeast of the city of Saskatoon.[15.20]

When early farmers started to clear the land around Saskatoon and to the northeast of the city, they found many of these boulders, which they cleared in the usual manner and piled up to the extent that early farm equipment made possible. Those responsible for the first buildings at the university saw in these boulders an economical source of building stone. The boulders were hauled to the building site and there worked by masons into the desired shapes for building purposes. Admittedly, this is an unusual case, but the boulder train near Saskatoon is far from unique, and there are other unusual sources of good building stone occasionally encountered.

Development of Standards for Cut Stone There was a time when some thought that the days of building stone were numbered, but its enduring qualities and the technological advances made by building-stone producers have combined to ensure its continued importance as a building material. Massive masonry is still regularly used for monumental buildings and for the more ornamental parts of modern frame buildings, such as entrances. The main use of stone today, however, is in the form of cut stone or masoned stone, thin slabs of stone cut to close tolerances and used as a facing material for exterior walls or for special interior finishes. The most modern types of cutting machinery are now used for the shaping of cut stone, with elaborate controls in the largest plants, but the selection of the stones to be cut still depends upon the art of the quarryman. Since, in developed countries, the stone quarries still in use are the larger and older ones (small quarries developed for special jobs having usually become obsolete), the principal building stones in use today are the stones that are "tried and true," with information gathered from many years of experience in their use now available (Fig. 15.13).

The significant place still occupied by these materials in modern building is indicated by the fact that there are now available standard specifications for use in the procurement of cut stone, either in the form of dimensioned slabs of stone or even in the form of "sandwich slabs" of stone combined, through the medium of special adhesives, with insulating material. Typical is Standard No. C568–79 of the American Society for Testing and Mate-

FIG. 15.13 Exposure test wall of the National Bureau of Standards, now at Gaithersburg, Maryland, containing samples of building stone exposed to natural weathering.

FIG. 15.14 Slabs of sand-blasted texture-vein marble and reinforced concrete being set in place in the American Savings Association building in Wichita, Kansas.

rials, a standard developed by a truly representative committee including representatives not only of stone producers but of stone users.[15.21] It is entirely appropriate that this attention should be devoted to this long-standing building material, allied so closely as it is with the geology of the bedrock from which it comes. (See Fig. 15.14) Well-established producers of building stone have expert technical staffs ready to supply to those who need it the best available information on the properties and the proper use of the stones they sell. In newer countries, such a service is not normally available, but there are available most helpful guides to the use of dimension stone, guides especially prepared for use in developing countries. A notable example is a comprehensive treatment of the subject published by UNESCO, one of the many valuable contributions in this field from Dr. Asher Shadmon.[15.22]

15.7 CONCLUSION

There are few interfaces of geology and engineering of more general interest than the use of natural materials, especially as cut stone, for engineering and building purposes. To walk round an ancient city, such as Prague, in the company of an expert engineering geologist, such as Dr. Quido Zaruba, is to have a minor course in geology as the sources and characteristics of the many different building stones used in that lovely city are pointed out. Not all can have the privilege of such guided tours but many leading cities now have available printed guides to their geology, including building stones, or even to building stones alone. The U.S. Geological Survey has issued two such publications, a brilliantly illustrated guide, *Building Stones of Our Nation's Capital*,[15.23] and a fascinating treatment of geology in relation to the history of Boston, *The Geology and Early History of the Boston Area of Massachusetts: A Bicentennial Approach* (which includes a section on building stones).[15.24] North of the border, the capital city of Canada, Ottawa, has an equivalent publication, *Guide to the Geology and Scenery of the National Capital Area*, which also includes a treatment of the stones used for the main buildings of the city's center.[15.25] Guides such as these can serve as admirable introductions to the study of geology.

It is not only in the larger centers that these interfaces of geology and building are to be seen. There can be few, if any, urban areas anywhere that do not exhibit, to the observant visitor, some indication of the importance of geology in providing natural building materials. In the Marble House at Newburyport, Massachusetts, built in 1892 by W. K. Vanderbilt, there are, for example, 21 different kinds of marble on view. At the other extreme, as one wanders round the village of Worth Matravers in Dorsetshire, in the south of England, it does not take long before one notices that every home is built of stone, and of the same stone, this being the Purbeck stone, justly famous for centuries.

FIG. 15.15 Orton Hall on the campus of Ohio State University. Opened in 1893 as the home of the University's Department of Geology, it now houses the Geological Museum and the Orton Library of Geology. The building is unique in that it is constructed of 40 different kinds of Ohio building stone, all arranged in stratigraphic sequence. The main and second floors, for example, are of stone of Mississippian age. The main entrance steps are of limestone of Devonian age, and the basement is of dolomite and limestone of Silurian age. A 30-ton boulder in front of the building (not shown) was brought from Canada by glacial action.

Anyone who travels as far as Peru and sees the ancient stonework of the Incas, featuring massive stones fitted so tightly together that one can not insert a knifeblade in the unfilled joints, will realize that despite all the advances of modern technology, there are ancient works still unrivaled today. (It is believed that the Incas discovered the use of soil organic matter as a natural chelating material.[15.26]) These stones are many centuries old and yet exhibit no signs of weathering. Their isolated environment is one reason for this; another is that the granites and porphyries used are resistant to weathering. So also are many other natural building stones, as well exemplified by Cleopatra's Needle now in New York City. A careful study of this famous obelisk suggests that its pre-New York history is responsible for what had previously been taken to be weathering caused by the atmosphere of its present location.[15.27]

A much wider-ranging study of weathering within urban areas was recorded in a classic paper of a century ago. The famous Scottish geologist, James Geikie, used his eyes as he walked about the streets of his own city of Edinburgh. He saw the value to be attached to the weathering of tombstones, the dates upon them being automatic guides to the duration of exposure, the finish of the stones assisting in the identification of the rock from which they had been cut. "Rock Weathering as Illustrated in Edinburgh Churchyards" was the title of the paper in which he recorded his observations, a useful reminder of how acute observations of very ordinary things can sometimes yield scientific information of value.[15.28]

Finally, should there be any readers of this chapter who are still dubious about such attention being given to natural building materials, their doubts can be answered by even a cursory glance at those chapters dealing with natural stones to be found in the magisterial review by the U.S. Geological Survey, *United States Mineral Resources*, published in 1973.[15.29]

15.8 REFERENCES

15.1 M. R. Mudge, C. W. Matthews, and J. D. Wells, *Geology and Construction Material Resources of Morris County, Kansas*, U.S. Geological Survey Bulletin 1060–A, Washington, 1958.

15.2 R. F. Legget, "An Engineering Study of Glacial Drift for an Earth Dam, near Fergus, Ontario," *Economic Geology*, **37**:531–556 (1942).

15.3 A. E. Cummings and R. B. Peck, "Local Materials in High Andes Prove Suitable for Rolled-Fill Dam," *Civil Engineering*, **23**:293–296 (1953).

15.4 "High Earth Dam Plugs Narrow Canyon for Power," *Engineering News-Record*, **164**:44–49 (4 April 1960).

15.5 "Earthfill in Wet Weather," *Engineering News-Record*, **149**:49–50 (7 August 1952).

15.6 "Air Photos Spot Fill for Dam in Wilderness," *Engineering News-Record*, **151**:47 (29 October 1953).

15.7 F. L. Peckover and T. G. Tustin, "The St. Lawrence Seaway: Soil and Foundation Problems," *Engineering Journal*, **41**:69–76 (September 1958).

15.8 R. H. Grice, "Hydrogeology of the Jointed Dolomites, Grand Rapids Hydroelectric Power Station, Manitoba, Canada," *Engineering Geology Case Histories*, no. 6, Geological Society of America, Boulder, Colo., 1968, pp. 33–48.

15.9 E. C. Hallock, "Bentonite Stops Sewer Leakage," *Engineering News-Record*, **124**:39 (4 January 1940).

15.10 A. D. Rhodes, "Puddled-Clay Cutoff Walls Stop Sea-Water Infiltration," *Civil Engineering*, **21**:71–73 (1951).

15.11 E. B. Eckel, "Research Needed in Engineering Geology," Presidential Address to Colorado Scientific Society, Denver, 1951, p. 8.

15.12 "Builders Save Millions on Two Faced-Rockfill Dams," *Engineering News-Record*, **159**:38–44 (29 August 1957).

15.13 H. W. Sikso, "Chalk Used as Construction Material at Gavins Point Dam," *Civil Engineering*, **27**:612–614 (1957).

15.14 H. G. Jewell, "Rock Riprap Replaces Porous Concrete Slope Protection at Santee-Cooper Project," *Civil Engineering* **18**:2–6 (1948).

15.15 H. Niini, "Engineering-Geological Classification and Measurement of the Brokenness of Bedrock in Finland," *Second International Congress of the International Association of Engineering Geology*, sect. IV, São Paulo, Brazil, 1974, pp. 6.1–6.7.

15.16 H. Williams and J. N. Stothard, "Rock Excavation and Specification Trials for the Lancashire-Yorkshire Motorway, Yorkshire (West Riding) Section," *Proceedings of the Institution of Civil Engineers*, **36**:607–631 (March 1967).

15.17 Sir John S. Flett, *The First Hundred Years of the Geological Survey of Great Britain*, H. M. Stationery Office, London, 1937, p. 33.

15.18 S. H. Ellis, Alfred Holt and Co., Liverpool, Eng., by personal communication.

15.19 K. O. Emery, "Weathering of the Great Pyramid," *Journal of Sedimentary Petrology*, **30**:140–143 (1960).

15.20 A. F. Byers, University of Saskatchewan, Saskatoon, by personal communication.

15.21 See ASTM Standard C568–67 for Dimension Limestone and ASTM Standard C170–50, Standard Method of Test for Compressive Strength of Natural Building Stone, both in ASTM Book of Standards, pt. 12, Philadelphia, 1969.

15.22 A. Shadmon, *The Development Potential of Dimension Stone*, U.N. Department of Economic and Social Affairs, New York, 1976; see also A. Shadmon, "Quarry Site Surveys in Relation to Country Planning," *23rd. International Geological Congress* **12**:125–132 (1968).

15.23 *Building Stones of Our Nation's Capital* (based on information from C. F. Withington), Inf–74–35, U.S. Geological Survey, Washington, D. C., 1975.

15.24 C. A. Kaye, *The Geology and Early History of the Boston Area of Massachusetts: A Bicentennial Approach*, U.S. Geological Survey Bulletin 1476, 1976.

15.25 D. M. Baird, *Guide to the Geology and Scenery of the National Capital Area*, Misc. Report no. 15, Geological Survey of Canada, Ottawa, 1968.

15.26 A. Schatz and V. Schatz, "Soil Organic Matter as a Natural Chelating Material," *Compost Science*, Sept./Oct. 1970, pp. 28–31.

15.27 E. M. Winkler, "Weathering Rates as Exemplified by Cleopatra's Needle in New York City, *Journal of Geological Education*, **13**:50–52 (1965).

15.28 J. Geikie, "Rock Weathering as Illustrated in Edinburgh Church Yards," *Proceedings of the Royal Society of Edinburgh*, **10**:518–532 (1879–1880).

15.29 D. A. Brobst and W. P. Pratt (eds.), *United States Mineral Resources*, U.S. Geological Survey Professional Paper no. 820, Washington, 1973.

Suggestions for Further Reading

A good modern treatment of one of the most venerable of building materials is *Stone: Properties, Durability in Man's Environment* by E. M. Winkler, a fine volume published in 1974 by Springer-Verlag of New York. The most comprehensive review of the cut-stone industry on a worldwide basis known to the authors is *The National Dimension Stone Industry*, a 129-page report to the South African Department of Planning (Pretoria) made publicly available in 1971. The test wall at the National Bureau of Standards illustrated in this report is described in the Bureau's Building Materials and Structures Report No. 125 of 1951 by D. W. Kessler and R. E. Anderson.

chapter 16

MATERIALS: CONCRETE

16.1 Cement / 16-2
16.2 Sand for Aggregate / 16-4
16.3 Crushed Stone as Aggregate / 16-5
16.4 Aggregate: General / 16-6
16.5 Aggregate: Sources of Supply / 16-9
16.6 Aggregate: Unusual Materials / 16-13
16.7 Concrete: Sulfate Attack / 16-14
16.8 Concrete: Alkali-Aggregate Reaction / 16-16
16.9 Conclusion / 16-18
16.10 References / 16-20

Concrete is one of the three main materials of construction utilized in modern civil engineering work. It is, perhaps, that which should be most closely studied by civil engineers. The quality of a given piece of timber cannot be varied, and steel also has definite physical properties when supplied for use. But the quality and strength of concrete are under the control of the engineer in charge of construction. This fact provides at once an opportunity and a responsibility which are met by continued study of and instruction in concrete technique.

The use of concrete in modern practice probably dates from about the year 1850; but as in the case of cement, this was a rediscovery rather than an innovation. The fame of the Romans as the master builders of olden days is due in no small measure to their use of a kind of concrete, *structura caementicia*, which they made from pozzolan cement mixed with broken fragments of stone and tile. They placed the material into timber forms, the marks of which can still be seen in places, and they even used pumice as aggregate when lightweight concrete was needed. The great dome of the Pantheon and other vaulted roofs testify to their skill in the use of this early concrete, lacking only reinforcing steel to make it the counterpart of that used today.[16.1]

Despite the number of factors that determine the final character of concrete, the aggregate is always of prime importance. When it is recalled that, in an average concrete, 70 percent by volume and possibly 80 percent by weight consists of the coarse aggregate used, this importance will be more readily appreciated. The attention usually devoted to the testing of coarse aggregate is not always in accord with its importance in the finished product. Compared with the care exercised in the testing of cement, the checks made on the suitability of aggregates are sometimes insignificant. Although the old saying that "the strength of a concrete is the strength of its aggregate" cannot be upheld as always accurate, it does give a vivid indication of the general importance of the aggregate in a concrete mix. It must be added that a good aggregate will not ensure good concrete; the possible variations in grading, mixing, placing, and curing (quite apart from climatic conditions during setting) are so great that they may far outweigh the initial advantages presented by an unusually sound and strong aggregate. On the other hand, the use of a poor aggregate is one of the easiest ways of obtaining concrete that will be a source of constant trouble.

In most modern urban areas concrete can be procured today from central mixing plants and delivered to the construction job in accordance with any stated specification, all phases of production being under strict control and inspection. This efficient service naturally tends to obscure the fact that concrete is a manufactured product, made from materials the geology of which determines their characteristics and indeed their suitability for use in concrete. Since on more isolated construction projects and on works in developing countries, the contractor may not have the advantage of ordering concrete for delivery from a central plant, attention to the constituents and manufacture of concrete is still important in civil engineering. Accordingly, this chapter presents a summary account of the materials used in concrete, with some warnings of the problems that may be encountered if geology is not recognized as an essential background in material selection.

Portland cement, or one of the special cements developed from it, is the key constituent. Although now available in the familiar bagged form or in bulk supply in almost all locations where construction is carried out, it is still derived from the calcining of appropriate natural materials. Sand and gravel, as found in nature, will often provide a suitably graded mixture such as can be used directly as the aggregate, coarse and fine, for mixing with cement, provided the source materials do not contain any injurious ingredients. Crushed stone, usually mixed with sand or rock fines to provide the necessary well-graded mixture, must be used where sand and gravel are not available for aggregate. The use of crushed stone will be of increasing importance as economically available supplies of gravel are used up around the larger urban centers.

This Handbook is not a manual of civil engineering practice but only a guide to the importance of geological factors in all civil engineering work. Despite this, however, it should here be said that, if due care is used in the selection of the aggregate for a concrete mix and if the appropriate cement is used, then with proper mixing, handling to the final location, and placing (*not* "pouring") into well-constructed forms and with careful curing, a material can be obtained that should be virtually permanent, of the strength required, and impervious to water. When looking at a bad example of modern concrete that has disintegrated because of poor selection of materials, inadequate processing, or unusually severe exposure, one has only to think of Roman "concrete," still in place and sound after 2,000 years of service, to be reminded of what good concrete should be. Attention to the geology of materials used in concrete will not, of itself, ensure a perfect material—but it will help.

16.1 CEMENT

Portland cement was patented in 1824 by an Englishman, Joseph Aspdin, and so named because he obtained his limestone from the famous Portland quarries. Since he had no good ball mills for grinding his limestone prior to its calcining, he gathered powdered stone from the main road at the quarry; this led to his being arrested and charged with theft of public property. It was Aspdin who discovered that by fusing together a suitable mixture of limestone and clay (or shale), one obtained a clinker which, when finely ground, would yield a cement that would set admirably in the presence of water. A small quantity of iron oxide is also incorporated in modern cements, the manufacture of which is now a highly refined and scientifically controlled process. The limestones used can have a wide variation in hardness, texture, and chemical composition; but magnesia, free particles of silica, and sulfur are undesirable constituents. Clays used in cement manufacture are often impure, but they cannot contain gravel or any free solid particles. Relative proportions of the several materials can be determined by calculation when the nature of the constituents is known; this is one of the several phases of cement manufacture that are the work of experts. Portland cement is distinguished by its relatively high strength and its slow and even setting when mixed with water. Its exact chemical composition is complex, but by means of modern methods of analysis, accurate knowledge of its makeup has now been obtained.

In addition to standard portland cement, special cements are available for a variety of unusual purposes. Thus, white cement for decoration is obtained by the use of clays that burn to a final white color. Aluminous cement, having an unusually high alumina content, is produced by fusing together a mixture of limestone, bauxite, and coke; it is unusually quick setting and is said to be more resistant to the action of seawater than ordinary portland cement. Tests have shown clearly that it is more resistant to the action of acidic moorland waters.[16.2] In Canada, a special high-silica cement has been developed that is extremely resistant to the peculiar alkaline soil conditions found in some parts of the west of the Dominion.[16.3]

Although manufactured cements are now in general use, they represent an ingenious adaptation of natural materials to give an imitation of a natural product. If a limestone itself has a very high clay content and certain other impurities, it may give a natural cement when calcined at a higher temperature than that used for the production of lime. Although not common, deposits of this kind are used—sometimes with small admixtures—for the manufacture of cement. One of the very early master builders of North America, Lt. Col. John By of the (British) Royal Engineers, found an exposure of argillaceous limestone in 1827 from which he made his own "cement" (described as better than any obtainable from England) for his masonry work in constructing the Rideau Canal, a military defense work linking the fortress of Kingston on Lake Ontario with what is now the site of Ottawa, Canada's capital city.[16.4] A large, modern cement manufacturing works is now located just a short distance from the little quarry from which John By obtained his limestone (Fig. 16.1).

Natural cements usually set more rapidly than portland cement and have a somewhat lower strength; they are naturally rather variable in composition. Natural cement is sometimes called "Roman cement;" it appears to have been discovered by a Mr. Parker of London, England, in 1796, who took out a patent in that year for the manufacture of what he called "Roman cement from the septario of clayey limestone found in the London clay formation of the island of Sheppey, England." It was found later that other deposits would yield similar natural cements.

Mention of the name Roman in connection with cement is a reminder of one of the most fascinating aspects of early Roman engineering—the discovery and use of *pozzolan*. This is a volcanic ash found in great beds near Naples at Puteoli from which was derived the ancient name *pulvis Puteolanus*. Puteoli is now called Pozzuoli, from which the modern name has been derived. The Romans found that, if this ash was mixed with lime instead of the sand ordinarily used, a watertight "cement" was obtained which would set under water. This mixture has been found in the lining of the water channel in the Aqua Marcia, dating back as early as 144 B.C. It was subsequently the principal constituent of

FIG. 16.1 A modern cement rock quarry and works (the Montreal East plant, Quebec, of Canada Lafarge Cement Company).

the "concrete" so boldly used in later Roman structures. Modern Italian builders still use the same material; in other countries, similar deposits of volcanic ash are used to make pozzolan cement. Smeaton used pozzolan in the construction of the Eddystone Lighthouse (1757), one of the pioneer structures of modern engineering. The name is often loosely used; a slag cement made up of a mixture of finely ground blast-furnace slag and lime is sometimes thus described. It would seem that the name should rightly be restricted to cement containing natural volcanic ash.[16.5]

16.2 SAND FOR AGGREGATE

Sand is used as a material of construction not only as filling and as a porous foundation blanket (as for roads) but also, to a wide extent, as a filtering medium and as a constituent of mortars and concrete. In view of these important applications, it must be pointed out that the term *sand* is used to describe naturally granular material of a certain grain size, irrespective of the shape of the grains, their uniformity, and their mineral composition. The formation of sands during the process of rock weathering and the way in which they are transported before reaching their final position affect all three of these leading characteristics of sand.

The mineralogical nature of sand is important, especially in view of the popular belief that all sands are composed of quartz particles. Although quartz is frequently a major constituent, pure silica sand is the exception rather than the rule. Depending on the nature of the rock from which it was formed and on the erosive action to which it has been subjected, many other minerals may be found as constituents. Mica will sometimes be present; feldspar is more common; even particles of shale occasionally occur. The presence of such materials in sand makes a detailed examination of many sands imperative before their use as concrete aggregate. Feldspar is not a thoroughly stable mineral, and its presence in some sands seems to be responsible for "hair cracking" in concretes. Shale particles and soluble salts may be equally injurious.

Specifications for sand for concrete usually exclude all sands containing deleterious substances and restrict the clay and silt content of sand to a small amount, often 5 percent or less. Usually they also exclude sand that is not "clean," a term used to signify freedom from organic matter such as tannic acid derived from surface vegetation. Geology can often assist here by suggesting the mode of origin of sands, the probable extent of deposits, and the probable origin of organic matter; geology thus presents a guide to the condemnation of either a whole deposit or just an upper layer.

Uniformity of grain size is important in connection with special sands used for standard concrete tests and for filtering media. Sands that have been subject to marine action are often fine-grained and fairly uniform. River sand may have similar properties, but it is likely to be more variable. Glacial sands frequently are mixed with gravel and are not generally uniform. These several types of sand have fairly characteristic grain shapes, none of which, incidentally, is rounded as a general rule. Rounded grains are generally an indication of a wind-deposited sand.

16.3 CRUSHED STONE AS AGGREGATE

Crushed rock is used for a variety of purposes in civil engineering; it has the advantage over natural gravel of having angular particle shape, and its limiting size and grading can be controlled as desired. It is widely used as railway ballast and as a base course for roads. It is most useful as a special type of fill material, especially for drainage work. It is also very widely used in coarse aggregate for concrete making; since the resulting concrete is in effect an artificial conglomerate of the aggregate used, the geological significance of this application will be discussed in the next section.

A common practice on construction jobs is to utilize any rock that has to be excavated as a source, when crushed, of concrete aggregate. If the rock is suitable for concrete work, a qualification discussed later, this practice may lead to great economy in construction. It comes within the experience of many civil engineers, and it might therefore be expected that the correlation of the nature of leading rock types with the operation of rock-crushing plants would have received wide attention in civil engineering literature. This does not appear to be the case. Even though the rock-crushing plant can often prove to be the key unit in the construction plant layout for a major undertaking, it is all too often the least appreciated and the least understood, since rock crushing is very generally regarded as a rather crude operation. It is far from that; it calls for just as much engineering attention as the most complex operation on the job, and it requires an appreciation of the geological significance of the types of rock to be crushed.

The effect of different rock types on crushing machinery is well shown by a record of crusher jaws lasting only ten days when crushing granite, whereas a pair of exactly similar jaws lasted seven years when used for crushing limestone. This surprising difference is best explained by variations in the toughness and hardness of rocks. Hardness is generally measured by testing the abrasive quality of rock in a machine in which a sample is pressed on a revolving steel disk, and toughness is usually measured by impact tests on rock samples. The strength of rock specimens tested in compression can be related to their behavior when being crushed into rock fragments. The shape of the fragments obtained from a crusher is largely influenced by natural planes of weakness in the rock, but in general, particle shape may be influenced also by the nature and direction of the blow struck by the crusher mechanism, i.e., by the type of crusher being used. Grain size is also important; coarse-grained rocks are poorer in quality than corresponding fine-grained rocks. Geological age, in the case of sedimentary rocks, is also significant.

The variable factors are so numerous that no very specific statements can be made with regard to the crushing qualities of individual rock types. The following general statements are a guide. Granites possess variable crushing qualities, although their behavior when crushed is related to their resistance to impact. Porphyries are, in general, very tough and hard. Basalts naturally show a wide variety of results, but crushing strength is found to be related to porosity, possibly produced by partial decomposition. Quartzites are moderately tough, but gneisses are much less so because of their foliated texture. The nature of the cementing material governs the performance of many sedimentary rocks during crushing.

A careful geological examination of samples of any rock to be crushed should therefore be made before plans for rock crushing are completed. If this examination can be supplemented by physical tests on samples and also by petrological examinations, a good general idea of the probable behavior of the rock during crushing will be obtained. Special attention should always be paid to the natural planes of cleavage. The natural tendency to break along these planes can be offset to some degree by the use of a suitable type of crusher. Variation in particle shape can have an appreciable effect on the workability of concrete and therefore on the amount of cement necessary. A change to a correct type of crusher may sometimes result in a savings of one bag of cement per cubic yard; the monetary saving that this can mean on a large construction job can readily be imagined.

The economic importance of crushed rock is well shown by the fact that in 1970 it represented an expenditure of over $1,475 million in the United States. This figure includes the value of construction aggregate, railway ballast, riprap, and rock for highway subgrades. As further evidence of the scale of crushed-rock operations there may be cited the construction of the Fontana Dam of the TVA which necessitated the quarrying and crushing to size of more than 6 million tonnes of rock for concrete aggregate alone. The rock used was the Fontana quartzite which was exposed in a special quarry giving good access to the deposit between the underlying and overlying beds of slate, 270 m (900 ft) apart. The high silica content of the quartzite made drilling slow and expensive; an average of only 1.5 m (5 ft) of hole was obtained with the 23-cm (9-in)-diameter drill bits before resharpening was necessary. An average of 1.4 kg (3.2 lb) of explosives was used for each ton of rock excavated.[16.6]

16.4 AGGREGATE: GENERAL

A satisfactory aggregate must meet five main requirements: it must be clean; it must be durable; it must be correctly graded in size; it must be uniform, both in grading and in quality; and it must not possess any chemical characteristic that will react with portland cement after mixing and placing. Grading can be controlled in the crushing and screening plant in the case of a crushed rock. Uniformity can be similarly controlled by close inspection of the supplies of aggregate. Cleanliness can be obtained if suitable measures are taken, and durability can be determined before an aggregate is selected for use. Both these properties are dependent on the geological history and composition of the rock to be used.[16.7]

The cleanliness required of coarse aggregate is similar to that required of fine aggregate; organic impurities are the most objectionable "dirt." These impurities are usually in the form of weak organic acids or their derivatives, which may seriously impair the setting of the cement. One of the earliest occasions on which the serious effects of organic matter on concrete was noted was during the construction of the Quinze and Kipawa lake dams in the upper reaches of the Ottawa River, Quebec, by the Department of Public Works, Canada, in 1910 and 1911. The organic matter could not be detected by eye, since it had formed transparent coatings around the particles of the siliceous aggregate used. Eventually it was found that the impurity could be removed by washing the aggregate in a 3% solution of commercial caustic soda. This was therefore done in a special plant. The concrete made with aggregate not so treated disintegrated rapidly.[16.8]

A colorimetric test, involving the use of a similar solution of caustic soda, is now almost universally employed as an indicator of the presence of organic matter in aggregate. The origin of organic matter may be surprising. In one case, for example, sand which appeared of admirable quality was found to be contaminated with tannic acid derived from oak trees which had grown above it. It is of special interest to note that organic acids will not be encountered in calcareous sands or in limestone or similar calcareous rocks, since the acids can react with the calcium carbonate in these materials. This simple geological fact will often prove a useful guide in locating good aggregate. In "limestone districts," clean aggregates will usually be widespread; this is the case in Ontario, west of the Niagara escarpment, apart from a small area near Windsor.

Aggregate must often be washed with water not only for removing organic impurities but also for washing out surface dirt and dust. Washing is also imperative with some types of crushed rock in order to remove the rock dust attached to the rock fragments as they leave the crushing plant, since the dust may be a potential source of trouble. Minute particles of feldspar formed during the crushing of some granites affect the quality of concrete, and other injurious minerals present in rock may have similar serious effects. The detection of these injurious minerals is related to the study of the durability of a proposed aggregate in a way similar to the way in which petrology is used in building-stone investigations. Mineral content can therefore be investigated by physical tests on porosity, by freezing and thawing, and by soaking in salt solutions, always coordinated with petrological study of the rock under the microscope. Injurious minerals in building stones will be injurious minerals in aggregate and for the same reasons. Possibly the most serious defect is the presence of any material that will swell in contact with water. This last property is of unusual importance in connection with concrete, since many types of rock that would not be considered as building stones will often be used as concrete aggregate.

The temptation to use rock that has to be excavated on a construction project as a cheap concrete aggregate has already been mentioned. It has probably been responsible for much of the poor-quality concrete seen today. There are not many jobs so fortunately placed as was the construction of the Denver Hilton Hotel in Denver, Colorado (Fig. 16.2). On the site of the hotel a light-pink pea gravel was found, a material which had to be excavated and which was then used as concrete aggregate. The gravel was so suitable for the unusual architectural treatment of the hotel building that acid was used to etch the concrete of the precast frames so that the aggregate would be exposed to view.[16.9] In many cases, engineers may have been forced to use unsuitable rock in this way against their better judgment. Limestone with any clay content, or so-called "argillaceous limestone," looks sound and durable when freshly quarried, but it is unsuitable for use as aggregate. Shales and slates are similarly undesirable; Fig. 16.3 shows an extreme case of disintegration arising from the use of shale as aggregate; the rock is shown in place in the foreground. Other types of shale, if relatively strong, can sometimes be used, although the typically laminated structure is always a source of weakness. Clay in any form is therefore to be avoided in concrete aggregate; petrological examination will assist in its detection.

Porosity of rock is just as important a factor in aggregate as in building stones. In addition to the usual penetration of rainwater into exposed aggregate, it must be remembered that when the concrete is placed in position it has a high water content so that the

16-8 PRELIMINARY STUDIES

FIG. 16.2 The Denver Hilton Hotel, showing the facing of precast concrete panels described in the text.

FIG. 16.3 Disintegration of concrete pedestal supporting a log flume, due (at least in part) to the use of local shale as aggregate. The shale, in situ, can be seen in the foreground. Lower part of the pedestal has been patched, but disintegration has continued.

aggregate will be in a saturated condition. This may lead to strange results, especially if the concrete is subjected to frost before the outer part has had time to set permanently. The freezing of water thus trapped in the aggregate may sometimes lead to the disruption of otherwise sound concrete.

This is a reminder that the aggregate used is only one of the variables involved in the preparation of good concrete. Possibly for this reason it is difficult to specify exactly the desirable characteristics of a high-grade concrete aggregate. The prohibition of such generally recognized materials as gypsum, anhydrite, and chert is frequently inserted. Some specifications also prohibit any clay content or limit it to a very small percentage. Metamorphic rocks of doubtful value can also be excluded by reference to fragments having a foliated structure. There remains a number of possible potentially dangerous mineral constituents, but it is impossible to list them all in the usual specification. To overcome this difficulty, a general clause in the contract specifications restricting the use of aggregates to those obtained from approved sources of supply may sometimes be imposed without hardship to contractors; alternatively, the engineer may himself arrange for the supply of satisfactory aggregate on the job. These are precautions by which close check can be kept on this important constituent of concrete.

Fortunately, there are now available well-proven standard tests to which proposed aggregates can be subjected. They are not infallible, but they provide a useful guide. They must always be supplemented by petrographic tests of the mineral content of the material under test since in this way deleterious materials can usually be detected, even though their presence may not be demonstrated by standard tests. The American Society for Testing and Materials has long had a very active committee charged with the development of standard tests for concrete. A guide to some of the standards developed by this committee is given at the end of this chapter. As with all ASTM test methods, these are constantly under review, not only within the private discussions of the committee and its subgroups, but also from time to time in public meetings and through special publications. One such publication of value—*Significance of Tests and Properties of Concrete and Concrete-Making*—was issued in 1978. It is noteworthy that one of the early chapters of this 882-page volume is a treatment of petrographic examination.[16.10]

16.5 AGGREGATE: SOURCES OF SUPPLY

That there are almost unlimited sources of supply of material for concrete aggregate is the superficial impression of those who have had no direct contact with the making of concrete. The actuality is that in some parts of North America, even in 1982, sources of economically available aggregate supplies were reaching a critical stage—headlines such as "Manitoba Running Short of Gravel" being not unusual, and this even for areas which have been glaciated. The economics of handling material for aggregate is one determinant, since the large quantities involved inevitably mean that beyond a critical distance from every concrete-using location (changing prices make it useless to cite a specific figure), the cost of hauling aggregate, be it sand and gravel or crushed rock, will be more than the cost of procuring the material. It is for this reason that there is always such a temptation to use material excavated on a construction job as the aggregate for the necessary concrete. Sometimes, this can safely be done, but Fig. 16.4 shows what can happen if the material excavated is not suitable for concrete-making.

Search for suitable aggregate for large construction projects which cannot be served from central plants is, therefore, an extremely important preliminary to construction. A typical procedure is study of the area around the site by means of air photos (an area up to 4,000 mi^2 being not unusual), selection of preferred sites which look suitable, study of these on the ground and detailed subsurface investigation of promising areas, sampling the materials so revealed, and then the thorough testing of all such samples to ensure that the materials are suitable for the concrete that is needed. Economic studies follow. Regional geological reports will assist in this work. Some areas have geological reports on potential aggregate supplies, a good example being *Sand and Gravel in Southern Ontario*, published in 1963.[16.11] This example is cited since it was issued before the recent remarkable increase in the use of concrete in this part of Canada—a fivefold increase in the per capita use which, with a doubling of population, has resulted in a tenfold increase in total

16-10 PRELIMINARY STUDIES

FIG. 16.4 Disintegration of concrete in the structure of a university stadium, due (at least in part) to the use of a local argillaceous limestone as aggregate.

quantity of aggregate within two decades.[16.12] A similar situation will be found in many other parts of North America where there are high concentrations of population.

Alternative sources of supply are therefore being widely studied. Crushed rock is an obvious first substitute for the more common sand and gravel, thus making petrographic examination still important. Kansas City has shown what can be done by mining suitable aggregate from beneath the surface and then using the mined-out space; this will be described in Chap. 21. Another possibility is the use of aggregate unsuitable by itself but which can be processed so as to remove objectionable constituents.

Siltstone, shale, and chert are among the most common deleterious materials sometimes found in gravel. At first sight it might be thought impossible to separate out small fragments of these materials from the large volumes that have to be handled from any ordinary pit. The problem is a common one in the mining industry. Ingenious use has been made of the regular method of separation used with ores for the benefication of gravels. First tried by engineers of the Royal Canadian Air Force in construction of a new runway at Rivers, Manitoba, *heavy media separation* (HMS, as the process is known) has now come to be used widely, one of the first installations in the United States being a floating plant on the Ohio River (1952). In the work at Rivers in 1947, fragments of shale occurred in the local gravel, with deleterious effect upon concrete. The nearest alternative source of gravel was 265 km (166 mi) away; the use of this would have added $3/m^2 of runway. After laboratory tests had proved that the use of HMS was practicable, a full-scale installation was developed and operated successfully for less than half the cost of bringing in the alternative gravel. The method used is based on the varying specific gravity of the gravel particles. By placing them in a liquid of a specific gravity intermediate between that of the impurities to be removed and that of the principal constituent of the gravel, those that are lighter will float, and so separation becomes easy.[16.13]

Sands and gravels, even if they do contain impurities, can, therefore, readily be processed into materials entirely suitable for the diverse purposes to which they may be put. (See Fig. 16.5.) As already indicated, they may be found in a number of the physical features left on a landscape following glacial action. They are also to be found in all beaches,

FIG. 16.5 A typical modern plant for crushing, washing, and screening sand and gravel at Stouffville, Ontario.

those of the present day and those of earlier days that may now be at high elevations because of uplift of the land. They come also from the beds of rivers and lakes into which they have been transported from the upland in one of the ordinary processes of physical geology. In some parts of the world, this underwater source is of the greatest importance. Probably nowhere is it of more importance than along the river Rhine, in the upper reaches of which is one of the greatest collections of "underwater gravel pits" in the world, as all travelers on this great river can see so clearly. It has been said jocularly that "half the sand and gravel for Europe comes from the Rhine." Although this must be an exaggeration, the constant stream of 1,000-tonne barges filled with clean sand and gravel that one passes when sailing up the Rhine is a wonderfully impressive sight (Fig. 16.6). The fact that well over 1 million tonnes have been handled in one year through the Ruhr port of Duisburg alone is some indication of the extent of this particular example of sand and gravel production. Dredging from the river channel itself amounted to over 1 billion tonnes in 1967; much of this sand and gravel was sold for industrial uses.[16.14]

The same source of sand and gravel is available on most of the major rivers of Europe and North America, although to nothing like the same extent. Visitors to the rebuilt city of Warsaw in Poland can see large dredgers at work on the Vistula River right in the center of the city. One of the most unusual examples, and yet one that is of special interest geologically, is the provision of good material for fill purposes at the eastern entrance to the harbor of Toronto. This remarkable inland harbor is completely protected by the Toronto Islands, which are actually (in geological language) a "compound recurved spit," the entire peninsula—as it was originally—having been formed and now being nourished by soil that is eroded by wave action from the Scarborough Bluffs some kilometers to the east. The *littoral drift* that features this part of the northern shore of Lake Ontario is from east to west. As the cliffs are eroded, the material so washed down is water-sorted, the coarser material moving in a regular way along the shore until it meets the outflow of the river Don, near to what is now the center of Toronto. This originally deflected the drift out from shore, with the resulting formation of the islands. The drift still continues and is well appreciated by local engineers and contractors. It has been estimated that

FIG. 16.6 Typical sand and gravel washing and reclaiming plant on the bank of the river Rhine, West Germany, the material obtained by dredging.

about 300,000 m³ (415,000 yd³) are eroded each year from the cliffs, about 10 percent of which is the coarse sand that has gradually built up the islands. The remainder is swept along the bed of Lake Ontario. Dredging this material has regularly been carried out off the eastern entrance, over 10 million tonnes of sand having been obtained in this way within a 22-year period. As with other underwater sources of sand and gravel, this supply is self-replenishing, and it is washed clean by its movement in the lake.[16.15]

Indicative of the attention being given to all possible alternative sources of concrete aggregate are the studies that have been, and are being, made into the use of sand and gravel dredged from the sea. As but one example, in 1964 the U.S. Corps of Engineers conducted a sand-resources survey along the New Jersey coast between the 4.5 and 30-m (14.8- and 98.4-ft) water depths on what is known as the Continental Shelf. Here 2,660 km (1,650 mi) of seismic refraction survey was undertaken, and 198 cores of the sediments encountered in the survey were obtained. Between Sandy Hook and Cape May a quantity of 2.15 billion m³ (2.8 billion yd³) of recoverable sand and gravel was located in this way. The survey was concentrated upon five segments—Sandy Hook, Manasquan, Barnegat, Egg Harbor, and Cape May, all names that will be familiar to those who have occasion to visit Atlantic City either for business or pleasure. Typical of the results was the finding of over 1.8 billion tonnes of sand and gravel in an area of 88,000 ha (217,000 acres) off Cape May. It must be noted that this was just a survey; much study would be necessary before any large quantities could be safely dredged so close to a coastline, but the significance of the survey will be obvious.[16.16]

In Europe, dredging sand and gravel from the sea is already being carried out on a steadily increasing scale. The Netherlands, having almost no exposed bedrock, has to rely upon sand and gravel for all concrete aggregate; about 12 million tonnes are used each year for this purpose. Up to 10 percent of the total is imported from Germany, most of the rest is dredged from large Dutch rivers, and now something over 300,000 tonnes is being obtained from the sea. These figures were given in one of the papers presented at a conference on sea-dredged aggregate for concrete arranged by the Sand and Gravel Association of Great Britain in December 1968. The proceedings of this gathering constitute a most useful guide to current thinking with regard to the problems associated with the use of material dredged from the sea. They also indicate clearly the importance that this source of sand and gravel has assumed in Great Britain, providing already about 10 percent of the total British consumption.[16.17]

16.6 AGGREGATE: UNUSUAL MATERIALS

A modern requirement is the specially dense concrete needed in nuclear plants for shielding purposes. Among the special aggregates used to meet this requirement are limonite and barite for modest increases in density; magnetite and ilmenite for medium increases; and ferrophosphorus or even steel punchings for extra-heavy concrete. Various mixtures of these several materials are naturally possible. Some difficulties have been experienced with the use of ferrophosphorus, but ilmenite has proved highly satisfactory and is probably the most widely used special aggregate for nuclear installations. Densities of 394 kg/m^3 (246 lb/ft^3) have been obtained with ilmenite from St. Paul in the province of Quebec.[16.18] Densities up to 250 kg/m^3 (157 lb/ft^3) have been obtained for "ordinary"concrete, using basalt as aggregate.

One of the most widespread building materials in the Pacific area is one that was generally regarded as suitable only for specialized work until the years of the Second World War. *Coral* very quickly came into its own, and before the end of the war it became well known as the "Pacific lifesaver." There are many types of this organically formed rock, but that most useful as aggregate occurs in reefs and ledges and has characteristics similar to those of soft limestone. At least 90 percent of coral is calcium carbonate; common densities range from 112 to 176 kg/m^3 (70 to 110 lb/ft^3) in loose volume. It fractures into large lumps and so does not give much of the fine material also required for concrete aggregate. So-called "finger-coral" is much lighter, with densities from 960 to 1,280 kg/m^3 (60 to 80 lb/ft^3), loose volume; it is soft, fragile, and porous and does not make good concrete.

The best type of coral for building purposes is generally found on the ocean side of islands rather than on the shores of lagoons and, correspondingly, on the windward sides; the best of all is found on shores perpendicular to prevailing winds, where there is a strong current. Clearly, the reasons for these variations are biological rather than geological, but knowledge of the variations has been used with good effect in the selection of coral as concrete aggregate, not only during the war years, but since then for the construction of massive concrete structures in connection with nuclear test work. Coral has also been widely used as rock fill and has been found to have good self-binding properties, so that when mixed with small quantities of cement it will yield a satisfactory mixed-in-place, stabilized blend for economical wearing surfaces. It is worth noting that coral concrete has been made with seawater in the absence of fresh water, and the results have, apparently, been successful.[16.19]

Selection of aggregate for concrete is, therefore, far from the simple procedure that might at first be expected. It may even include the unexpected. One of the most unusual "aggregate experiences" was encountered during the construction of the Friant Dam in California for the U.S. Bureau of Reclamation. Sand and gravel obtained from gravel bars in the San Joaquin River were used as aggregate. Placer gold was detected in some of the preliminary tests, and arrangements were made for the installation of a $20,000 piece of extra equipment so that the small quantities of gold could be extracted before the sand and gravel became concrete. By the end of the job, gold to the value of $176,000 (net) had been recovered; this bonus was shared equally between the Bureau and the general contractors for the dam.[16.20] There have been a few similar experiences at other dams in California. Even in so mundane a matter as processing aggregate for concrete, therefore, the unexpected may prove of interest and of value.

16.7 CONCRETE: SULFATE ATTACK

There are some areas of the world where groundwater will be found to have unusually high contents of sulfates, of calcium, magnesium, and the alkali metals. These compounds are derived from the surrounding soils, some clays having high soluble constituents. Some clays of England sometimes have this characteristic; clays underlying the Canadian prairies "contaminate" in this way the groundwater they contain. These are but two areas of many, and checking groundwater for its sulfate content is a necessity for all locations where concrete is to be placed below ground level. An arbitrary classification has been suggested, based on the sulfur trioxide content of groundwater. Where this is below 30 parts per 10,000, there is little cause for concern; from 30 to 100 parts is an indication that further study is necessary; when over 100 parts, then precautionary methods will be essential.[16.21] The sulfates will react with portland cement in concrete and cause deterioration of the concrete which, if undetected, can lead to structural difficulties (Fig. 16.7).

FIG. 16.7 Damage done to the concrete roof of a tunnel structure in western Canada by sulfate-bearing groundwater, ordinary portland cement having been used.

In western Canada, for example, a Winnipeg sewer made of portland cement concrete collapsed in 1910. Study of this failure pointed to interaction of groundwater and the cement in the concrete. This initiated a long-term basic research program, headed by Dr. T. Thorvaldson of the University of Saskatchewan, which led to recognition of the chemical reactions responsible and, eventually, to the development of sulfate-resistant cement now available (under various designations) in most countries. Prior to this, however, the start of sulfate attack had been detected in a 154-km (96-mi)-long concrete aqueduct of the Greater Winnipeg Water District, an aqueduct made with high-quality, dense portland-cement concrete. That section of the conduit exposed to this sulfate attack was opened up, the offending clay removed, gravel used for new backfill, and all groundwater in the vicinity of the line drained to dry wells.[16.22]

MATERIALS: CONCRETE   16-15

The economic effects of sulfate attack on concrete can, therefore, be serious indeed. The St. Helier Hospital at Carshalton in Surrey, England, was built in 1938; it has a capacity of 750 beds (Fig. 16.8). The main accommodation was provided in four multistorey ward blocks, with central services contained in one main service block, subways connecting the different parts of the complex. The foundations were of concrete, all placed on a brown clay stratum that overlies the well-known London clay. In March 1959 some of the foundation concrete was exposed in the course of routine maintenance work and was found to be seriously deteriorated. This led to a more general study which showed that the foundations were deteriorating steadily.

> Samples of the substrata taken [from] boreholes showed that while the ground was of adequate load-bearing strength, the sulphate content was very high. It was evident that much of the existing concrete had suffered from chemical attack and the cement, which was of a high aluminous type, had weakened through these conditions and the warm damp atmosphere.[16.23]

In a major underpinning operation, the entire building complex had to be provided with new foundations, at what trouble and expense can best be imagined (Fig. 16.9).

Groundwater quality, therefore, is a vital element in the satisfactory use of concrete below ground level. Even civil engineering operations can affect the quality of groundwater, as Lea has recorded in a case from Bexley, in Kent, to the south of London. A tunnel was being constructed with bolted precast concrete segments as lining, the concrete being made with sulfate-resisting cement. Compressed air had to be used. When the air pressure was released and the groundwater returned to the periphery of the tunnel, small leaks were noticed at the connecting bolts. Study revealed that the sand adjacent to the tunnel contained pyrite, typical for the Woolwich beds through which the tunnel was

FIG. 16.8 St. Helier Hospital, Carshalton, Surrey, England, the foundations of which had to be replaced because of concrete deterioration.

FIG. 16.9 Typical of the cramped working spaces in which the St. Helier Hospital foundations had to be replaced (Carshalton, Surrey, England).

driven. Compressed air from the tunnel caused the pyrite to be oxidized to ferrous sulfate and sulfuric acid. Even the minute quantities thus formed were enough to produce a pH value of 1.8 in groundwater. Proof of this explanation was given by a rise at one point in the pH of the groundwater from 1.8 to 3.5 over a period of a few weeks, following the end of compressed-air work, and at another location from 3.0 to 4.5 in three years. A secondary lining was installed and was found to be satisfactory.[16.24] The example well illustrates the importance in civil engineering work of the chemistry of groundwater.

16.8 CONCRETE: ALKALI-AGGREGATE REACTION

Portland cement, because of its chemical nature, produces a highly caustic solution when mixed with water for the manufacture of concrete. The alkali content of cement is usually denoted as the equivalent content of Na_2O. "High-alkali cement" is that which has an alkali (Na_2O) content of more than 0.6 percent. It has long been known that all rocks and gravels will react to some extent with the caustic solution caused by cement when all are mixed into concrete. If limited in extent, this reaction can be favorable to the quality of concrete. When, however, it is excessive, expansion and cracking of the concrete in question may result.

Excessive expansion of concrete due to alkali-aggregate reaction was first reported from California in 1940.[16.25] Once recognized, the same phenomenon was found in Idaho,

Arizona, Nebraska, and Kansas, as well as in some other states. These early cases were all due to reaction between the alkalis in the portland cement and certain reactive silica constituents in the aggregates used—mainly cherts, opals, chalcedonies, and other cryptocrystalline quartz, as well as rhyolites and certain other volcanic rocks. Publicity about this serious problem led to its recognition also in Australia, Great Britain, and especially in Denmark, where widespread trouble with concrete was found to be due to the use of a flint stone as aggregate.

Much research was carried out on the problem and eventually standard tests were developed by ASTM, the use of which enabled this type of alkali-aggregate reaction to be identified. Work in Canada in the 1950s on unexpected concrete problems in the Kingston area of Ontario showed that these standard tests were not sufficient to determine accurately the possibility of trouble with certain dolomitic limestones, another reaction clearly being involved.[16.26] Work in Nova Scotia, in the 1960s, extended the range of problems still further, since some sedimentary rocks in the Appalachians and the Precambrian Shield were also found to be susceptible to reaction with high-alkali portland cement.[16.27] So important is this matter that there is today an extensive literature covering all aspects of the alkali-aggregate problem. In summary, however, it may be said that three main groups of rocks regularly used as concrete aggregate may react with high-alkali portland cement:

1. Rocks in which the microforms of silica, the amorphous and microcrystalline phases, are most frequently found; generally glassy to cryptocrystalline rhyolites, dacites, latites, and andesites
2. Some types of dolomitic limestone, typically gray, very fine-grained, dense, and of close texture
3. Phyllites, graywackes, and argillites which contain fine-grained quartz and phyllosilicates[16.28]

To go further into detail would involve discussion of the complex reactions that have now been determined and which explain the slow, but serious, expansion that takes place when high-alkali cement interacts with some of these rock types. In some locations, the expansion has taken place so slowly, and has been so limited, that it has not caused any serious problems. In consequence, it may not have been recognized, but the presence of hairline cracking in exposed concrete surfaces or the occasional "crazing" of concrete pavements may indicate that alkali-aggregate reaction has been taking place (Fig. 16.10). As the supplies of long-used, reliable aggregates are steadily consumed, necessitating the use of new and possibly untried sources, the problem becomes the more serious.

Accordingly, the selection today of any rock for use as a new source of aggregate in making portland-cement concrete involves most careful testing, in advance, of the possibility of an alkali-aggregate reaction occurring. Standard (ASTM) tests are now available for evaluating specimen batches of concrete using the proposed aggregate, but expert petrological examination is now also essential. Information is available which should make it possible to detect any unsuitable aggregate in advance.[16.29] Under some circumstances, it may still be economically necessary to use aggregate that has been shown to react adversely with high-alkali cement. This can be done by using low-alkali cement. The difference in the alkali content of cements is due to differences in the materials used for manufacture in association with manufacturing methods. Alkali contents will usually be found to be reasonably steady from any one cement plant, some Canadian plants, for example, producing relatively high-alkali cements. To obtain an alternative supply from a more distant plant may involve extra expenditure, but it is easy to see that this may be fully justified if sound concrete is to be obtained.

FIG. 16.10 Crazing in a pavement at Kingston, Ontario, typical of damage done by alkali-aggregate reaction.

16.9 CONCLUSION

Concrete is so versatile a material, and is now in such wide use, that the lack of attention to the geology of its constituents is, perhaps, one of the most serious aspects of the neglect of geology by civil engineers. The situation is happily changing. Recognition of the alkali-aggregate problem in some areas has answered long-standing questions as to why some concrete was better than concrete elsewhere. There is today no excuse for any deterioration of concrete due to sulfate action from groundwater. With sound methods of mix design, good mixing and placing, and proper curing in adequate forms, concrete can be produced today that will be all that the designer requires when the structure is complete and have an indefinitely long life as an admirable "artificial rock." It should offend all civil engineers to see any concrete that does not conform to these high standards.

Enough has been said to show that selection of aggregate is today not just a matter of good judgment (although this is still essential); it is a choice that must be based on examination of the geological history of the material proposed for use, the results of standard tests, and expert petrographic examination. In many places throughout North America, the economics of aggregate supply have already become critical. This is a situation that is not going to improve as the quantity of concrete used daily continues to increase. The search for aggregate, therefore, has become in itself an interesting exercise; sources previously neglected are now worthy of reconsideration, and unusual possibilities for a supply of aggregate, a supply coming sometimes as almost a by-product, are being given careful attention. Two examples may be cited in conclusion.

Urban zoning regulations are frequently an impediment to the use of natural materials as aggregate when the removal of such materials would contravene some limitation upon land use or cause a social nuisance, even though the materials are in every respect suitable for use in concrete. Many cities similarly must now regret the fact that lack of planning in earlier years allowed buildings and other structures to be erected over sand and gravel deposits, for example, the removal of which would have benefitted planning and yielded good aggregate at the same time. One exception to this disquieting picture was a case in Montclair, New Jersey, when a stone company was permitted to quarry more than 2,700,000 tonnes of high-quality traprock that formed an escarpment on state land adjacent to the campus of a state college. When quarrying was finished and the area cleaned up, the college gained 80,000 m² (96,000 yd²) of good level ground immediately adjacent to the campus, while the company obtained a fine supply of excellent crushed rock for aggregate (Fig. 16.11).[16.30]

FIG. 16.11 Traprock being "mined out" for use as aggregate from the campus of the state college, Montclair, New Jersey.

The city of Dayton, Ohio, has for many years obtained its main water supply from an elaborate installation of wells, said to be one of the most extensive in the world. The city is underlain by a far-ranging deposit of sand and gravel, forming an ideal aquifer. Groundwater removed by pumping has been replenished by seepage from the Mad River which flows in the city over a gravel bed. This supply was augmented in earlier years by allowing river water, when clear, to flow into specially excavated trenches connected with the river. In the 1950s the American Aggregates Company approached the Dayton Water Department and offered to mine gravel, under the city's direction, in order to form open areas of water while obtaining the gravel it required, gravel for which it was willing to pay the city a royalty for every ton removed. An agreement was entered into, and Dayton now possesses four pleasant lakes with an area of 140 ha (340 acres), which it obtained for nothing, with a bonus from the royalties. Its water supply has been protected and enhanced by percolation from the lakes. And the citizens of Dayton now have the advantage of boating on these lakes or fishing in them, the wisdom of the city administration well shown by the fact that fishing is allowed on alternate days to boating.[16.31]

16.10 REFERENCES

16.1 A. P. Gest, *Engineering (Our Debt to Greece and Rome)*, Longmans, New York, 1930, pp. 43–47.

16.2 W. T. Halcrow, G. B. Brook, and R. Preston, "The Corrosive Attack of Moorland Water on Concrete," *Transactions of the Institution of Water Engineers*, **33**:187 (1929).

16.3 A. G. Flemming, "The Development of Special Portland Cements in Canada," *Engineering Journal*, **16**:215–223 (1933).

16.4 R. F. Legget, *Rideau Waterway*, University of Toronto Press, Toronto, 1972, pp. 49, 193.

16.5 A. P. Gest, loc. cit.

16.6 J. B. McKamey, "Aggregate Production at Fontana Dam," *Engineering News-Record*, **135**:232–238 (1945).

16.7 E. G. Swenson and V. Chaly, "Basis for Classifying Deleterious Characteristics of Concrete Aggregate Materials," *Journal of the American Concrete Institute*, **27**:987–1002 (1956).

16.8 K. M. Cameron, "Public Works," *Engineering Journal* **20**:297–311 (1937); and information from E. Viens when chief of Testing Laboratories, Department of Public Works, Canada.

16.9 "Precast Window Frames Cover Hotel," *Engineering News-Record*, **163**:38–40 (22 October 1959).

16.10 *Significance of Tests and Properties of Concrete and Concrete-Making*, American Society for Testing and Materials Special Technical Publ. no. 169B, Philadelphia, 1978.

16.11 D. F. Hewitt and P. F. Karrow, *Sand and Gravel in Southern Ontario*, Industrial Mineral Report no. 11, Ontario Department of Mines, Toronto, 1963.

16.12 *Towards the Year 2,000: A Study of Mineral Aggregates in Central Ontario*, Special Report of Ontario Ministry of Natural Resources, Toronto, 1974 (with maps).

16.13 F. E. Hanes and R. A. Wyman, "The Application of Heavy Media Separation to Concrete Aggregate," *Bulletin of the Canadian Institute of Mining and Metallurgy*, **55**(603):489–496 (1962).

16.14 Information kindly supplied by Shipping and World Traffic Publications, Ltd., Basel, Switzerland.

16.15 D. F. Hewitt and S. E. Yundt, *Mineral Resources of the Toronto-Centred Region*, Industrial Mineral Report no. 38, Ontario Department of Mines, Toronto, 1971.

16.16 D. B. Duane, "Sand and Gravel Deposits in the Nearshore Continental Shelf, Sandy Hook to Cape May, New Jersey," abstract vol. for Atlantic City meeting, Geological Society of America, 1969, p. 53.

16.17 A. F. G. Lewis (ed.), *Sea-Dredged Aggregates for Concrete;* Proceedings of a symposium, Sand and Gravel Association of Great Britain, London, 1969.

16.18 H. S. Davis, "High Density Concrete for Reactor Construction," *Civil Engineering*, **26**:376–380 (1956); and "Heavy Concrete Mixed in Transit," *Engineering News-Record* **163**:43–50 (1959).

16.19 J. R. Perry, "Coral: a Good Aggregate in Concrete," *Engineering News-Record*, **135**:174–180 (1945); see also D. L. Narver, "Good Concrete made with Coral and Sea Water," *Civil Engineering*, **24**:654–658 (1954).

16.20 "Gold Recovered from Friant Dam Aggregate," *Engineering News-Record*, **130**:397 (March 1943).

16.21 "Concrete in Sulphate-Bearing Clays and Groundwaters," *Water and Water Engineering,* **43**:574–577 (1939).

16.22 W. D. Hurst, "Experience in the Winnipeg Area with Sulphate-Resisting Cement Concrete," in E. G. Swenson (ed.), *Performance of Concrete,* University of Toronto Press, Toronto, 1968, pp. 125–134.

16.23 "Foundation Reinstatement Without Interference to Services," *The Engineer,* **220**:791 (12 November 1965).

16.24 F. M. Lea, "Some Studies on the Performance of Concrete Structures in Sulphate-Bearing Environments," in E. G. Swenson (ed.), *Performance of Concrete,* University of Toronto Press, Toronto, 1968, p. 58.

16.25 T. E. Stanton, "Expansion of Concrete through Reaction between Cement and Aggregate," *Proceedings of the American Society of Civil Engineers,* **66**:1781–1811 (1940).

16.26 E. G. Swenson and R. F. Legget, "Kingston Study of Cement-Aggregate Reaction," *Canadian Consulting Engineer,* **2**:38–46 (1960); see also J. E. Gillott, "Petrology of Dolomitic Limestones, Kingston, Ontario, Canada," *Bulletin of the Geological Society of America,* **74**:759–778 (1963).

16.27 M. A. G. Duncan and E. G. Swenson, "Investigation of Concrete Expansion in Nova Scotia," *The Professional Engineer in Nova Scotia,* **10**:16–19 (1969); see also E. G. Swenson, "Interaction of Concrete Aggregates and Portland Cement—Situation in Canada," *Engineering Journal,* **55**:34–39 (1972).

16.28 J. E. Gillott, "Alkali-Aggregate Reactions in Concrete," *Engineering Geology,* **9**:303 (1975).

16.29 J. E. Gillott, "Properties of Aggregates Affecting Concrete in North America," paper to Engineering Group of the Geological Society of London, February 1975.

16.30 R. L. Bates, "Mineral Resources for a New Town," *Geoforum,* **6**:169–176 (1975).

16.31 K. A. Godfrey, "Sand and Gravel—Don't Take Them for Granted," *Civil Engineering* **47**:55–57 (March 1977).

Suggestions for Further Reading

Concrete has become such a versatile and important material of construction that a very large specialized literature about it is now available. Despite this, the attention given to the geological aspects of aggregate for concrete has been comparatively neglected in written works. In one of the main indexes to North American publications on concrete, the word *geology* does not once appear!

An early publication about aggregate in which geological aspects are touched upon was a Special Technical Publication of the American Society for Testing and Materials, STP No. 83 of 1948, which deals with mineral aggregates in 240 pages. A later STP, last revised in 1978 as No. 169B, is an 882-page compendium called *Significance of Tests and Properties of Concrete and Concrete Making Materials.* And STP 597 of 1976 (113 pages) has the intriguing title *Living with Marginal Aggregates.*

This attention by ASTM to concrete and its constituent materials is naturally reflected in the existence of a considerable number of Standards, some of which describe tests for the detection of troublesome aggregate (as mentioned in the text). Part (volume) 14 of the *1982 Annual Book of ASTM Standards* is a volume of no less than 882 pages containing just *Standards on Concrete and Its Constituent Materials.* The Society also publishes a semiannual journal called *Cement, Concrete and Aggregates.* Publications of the American Concrete Institute and especially its *Journal* are naturally of great value.

Problems with concrete aggregate are now engaging the attention of engineering geologists. A notable contribution to discussion of these problems is a group of seven papers, "The Stability of Concrete Aggregates" (edited by W. J. French), which appeared in the *Quarterly Journal of Engineering Geology,* vol. 13, pp. 205–316, in 1980. Accompanying bibliographies are especially useful.

If any readers should still be skeptical about the serious and widespread problems that geologically unsuitable aggregates can present, they will be interested in the account of what appears to be the first recorded case of alkali-aggregate reaction in a dam in Great Britain. It appeared in the July 1980 issue of *New Civil Engineering International* (no. 21), the case being that of a dam in the island of Jersey. Correspondence in the following issue contains reference to similar experience in New Zealand.

chapter 17

MATERIALS: MANUFACTURED PRODUCTS

17.1 Bricks / 17-1
17.2 Tile and Other Structural-Clay Products / 17-5
17.3 Lime and Plaster / 17-6
17.4 Lightweight Concrete Aggregates / 17-7
17.5 Volcanic Ash / 17-8
17.6 Fly Ash / 17-9
17.7 Mining Wastes / 17-9
17.8 Steel Slag / 17-10
17.9 Conclusion / 17-10
17.10 References / 17-11

The factory-made materials used in such large quantities in civil engineering construction may not seem to have any direct connection with geology. Brief consideration, however, will show that all are ultimately derived from natural products, predominantly geological. The mining, selection, and processing of such raw materials will not normally enter the province of the civil engineer. There are, however, a few building products whose manufacture from basic materials is a relatively direct process. Bricks are perhaps the most widely used and important of such simple products. Although bricks and similar materials can be purchased in all developed countries to strict specifications from large suppliers, they are still manufactured in small local plants in many parts of the Third World. Since much construction will be carried out in developing countries in the next few decades and since the use of indigenous materials in all such work is eminently desirable, this short chapter has been included mainly as a reminder of the importance of geology in the selection of materials for the manufacture of bricks and allied products.

17.1 BRICKS

Civil Engineers are rarely required to become brickmakers; but as in the case of other materials of construction, the source of bricks and the leading features of their manufacture should be familiar to those who have to use them frequently; civil engineers are numbered in this general company. Bricks are of three main kinds: concrete bricks, sand-lime bricks, and fired-clay bricks. All have some relation to geology. The first two types are

17-2 PRELIMINARY STUDIES

manufactured by specialized processes, in one case from mixtures of portland cement and aggregate and in the other from sand and lime. Chapter 16 is an introduction to the importance of geology in concrete-making.

The peculiar properties of many clays that make them so useful have been known from the very early years of human history, as recent archaeological work has demonstrated. Working rules based on long-established practice had to suffice until comparatively recent years for the guidance of brickmakers. Today a vast amount of scientific information is available, obtained not only in connection with brickmaking but also in connection with the ceramics industry generally, and this has made brickmaking a specialized and highly technical industry.

The two properties that have made clays so important a source of useful articles are their plasticity, by reason of which they can be molded into desired shapes when wet, and the way in which they will harden when subjected to the action of heat, thus establishing the molded shape permanently. The term *plasticity* is here used in relation to molding processes; if too plastic, a clay may be difficult and sticky to mix; and if deficient in plasticity, it may break up as it is being molded. Mixtures of clay and sand are therefore sometimes used in order to obtain a satisfactory material for the molding process. Shrinkage of clays is another important property in connection with brickmaking; uniformity of shrinkage is desirable. The reactions that take place when clays are heated are complex. Generally, however, two recognizable stages are of importance when bricks are made: that at which the clay vitrifies, or becomes hard by a fluxing together of the whole mass, which still retains its shape; and that at which the whole mass becomes soft or viscous, as a result of excessive fluxing, and loses its shape. Clearly, in the manufacture of bricks in large kilns, an appreciable temperature interval between the two stages is desirable (Fig. 17.1).

A combination of these properties is what determines the suitability of a clay for brickmaking, provided the clay is uniform, clean, and free from gravel and boulder intrusions. Thus it will be apparent that although most clays can be baked to give some sort of bricks, by no means will all give a brick that can satisfactorily be used in construction. The presence of gravel in clays is an objectionable, although not an insuperable, difficulty; but if pebbles of limestone are present, or if the clay is calcareous, it will be advisable to continue the search for a suitable brick clay elsewhere. It is in the location of possible sources of clay supplies that the contact between geology and brickmaking is made, since when a uniform clay supply has been located, its suitability for brickmaking depends primarily upon its physical and chemical properties, which have to be determined by suitable laboratory tests.

Adobe clays, which contain sand, are often used for making sun-dried bricks in hot climates. Common bricks are made from a wide variety of suitable clays. Special "engineering bricks," or high-quality pressed bricks, are made from especially pure, high-quality clays or shales and sometimes from inferior fireclays. The latter are clays that will stand a high degree of heat without fusing; their use for making firebricks is well known. They are frequently found in association with coal deposits, in adjacent strata from 0.3 to 1.5 m (1 to 5 ft) thick, but not all clays found with coal are fire-clays. Paving bricks are made from clays that can be accurately machine-molded and that vitrify at a moderate temperature.

It is sometimes thought that the color of bricks is an indication of their quality. This is so in the case of the famous Staffordshire blue bricks of England, made from clays and marls of the county of Stafford, which contain from 7 to 10% iron oxide; the bricks are burned at unusually high temperatures. But this is a rare exception, for, in general, the conception is erroneous. Pure clay would be white, but the usual color found in most clay deposits is due to oxides and carbonaceous matter. When clays are burned, those which are white usually remain so; but colored bricks result from colored clays, although the tint

FIG. 17.1 Red-burning shale being excavated at Canton, Ohio, for the Belden Brick Company.

may change. The effect of iron oxides is generally to give a red or buff color; if lime is present, its effect may counteract that of the iron oxide and produce a cream-colored or yellow brick. Terms such as "red burning" and "buff burning" are used to describe the final color of brick clays after they have been vitrified.

In the course of the long-established business of brickmaking there has gradually developed, especially in Europe, a definite correlation between clay deposits of well-defined geological position and their suitability for brickmaking. Information on this correlation and the areal distribution of the clays so described is usually obtainable from local geological surveys, as is information about suitable shales for brickmaking.

Finally the fact may be mentioned that in certain parts of the United States in which there are no outcrops of solid rock for use as railway-track ballast, burned-clay ballast has been extensively and successfully used. The same material has been used even for such a purpose as the riprap protection for the slopes of earth dams. Material of this type was specially developed for use on the Keystone (Kingsley) Dam of the Central Nebraska Public Power and Irrigation District, a rolled-earth structure that required 420,000 m² (500,000 yd²) of riprap for protection. The nearest quarry rock suitable for this purpose was 300 km (186 mi) away from the dam site, and the long rail haul rendered it unusually expensive. A type of random ceramic block was developed by fusing together ordinary bricks, giving a weight of over 1,200 kg/m³ (2,000 lb/yd³) and all the requisite properties for riprap in the way of soundness, absorption, toughness, and wear.[17.1] The ceramic riprap has now been in use for 40 years and has performed well. It is an unusual illustration of the wide use to which clay products are put in the practice of civil engineering.

Bricks occupy an important place in the building activity of many developing countries, probably nowhere more so than in India. The use of indigenous building materials is of such vital importance to the developing world that attention may usefully be drawn to Indian practice in brickmaking. Figure 17.2 shows a typical simple brickmaking plant such as will be found in the outskirts of most Indian villages. The principal source of the clays used for brickmaking are the surface alluvial deposits from all the main rivers of India, the same soils which support the extensive agricultural practice of the subcontinent. They are naturally of recent origin and can be readily excavated by hand to depths generally not exceeding 2 m (6 ft). In some areas, silt-laden river waters at the monsoon seasons are collected in large tanks in which the silt is allowed to settle and then dry out, providing a further source of material suitable for brick making.

The actual making of bricks is a relatively simple hand operation, firing being carried out in simple kilns, the chimneys of which often provide an interesting feature of village landscapes. The work is traditional and has changed but little over the years. One result is that there are no publications that can be referenced for use in brickmaking in other developing countries. Research work is now in hand, however, notably at the Central Building Research Institute at Roorkee, while there are useful papers on the clays of India in general available from the Indian Council of Agriculture and the National Institute of Sciences. The Geological Survey of India also has publications that can be of assistance.[17.2]

FIG. 17.2 Typical Indian village brickyard.

17.2 TILE AND OTHER STRUCTURAL-CLAY PRODUCTS

An extension of the use of clay for the provision of building materials was the development of structural-clay products, as they are commonly called, such as hollow tile and architectural terra-cotta. These products are widely used in connection with buildings constructed with steel or reinforced-concrete frames. Some idea of their economic importance is given by the fact that in the year 1969 42.3 million tons of clay was processed in the United States for manufacture into brick and tile products.

Common to the manufacture of all these articles are processes similar to those used in brickmaking. Fine-grade clays are used, and because of the finished shapes, structural-clay products are processed under more rigid control than common brick (Fig. 17.3). All

FIG. 17.3 Starring Potteries plant at Littleborough, Lincolnshire, England; the fireclay used is mined from a 1.2-m (4-ft) stratum underlying an extensive coal seam.

the clays are regularly submitted to detailed testing and analysis in order to guard against an excess of impurities, such as iron, magnesia, or lime. Stoneware is usually made from refractory clays which burn to a dense mass at relatively low temperatures; clay mixtures, sometimes with sand added, are frequently used. Sewer pipes have to be glazed during manufacture, and this is done with salt or with a special glazing material. It is said that a high iron content is of assistance, but the main requirement of clays for pipe making is that they be high in fluxes so that they vitrify readily. The materials most commonly used for structural tiles are fireclay of first and second grades, high-grade brick clays, and suitable clay shales. Some of the latter have to be blasted; regular quarrying operations may therefore have to be planned and directed in connection with tile manufacture, in addition to the mining operations that are necessary for the supply of some fireclays.

Manufacturing methods for these varied products are interesting but generally outside the scope of this book. Two details may perhaps be mentioned. In many plants, it is usual to store the clay or shale out-of-doors for an appreciable period before use. Broken up and exposed to the atmosphere in this way, the clay or shale will disintegrate and soften. This result, although often so unfortunate in civil engineering construction, is here an advantage, since the weathered material can be easily worked. Color is important in connection with some structural-clay products, especially hollow tiles to be used for partition walls. With uniform clays, color can be controlled during manufacture; but it is possible—and the practice is often followed—to introduce iron oxide (and other substances) during manufacture in order to ensure a uniform shade of the desired color.

17.3 LIME AND PLASTER

Although the importance of lime and plaster in civil engineering work has declined with the steady advance in the use of concrete and concrete products, limes are still used in connection with building work and masonry construction and so are of significance for the engineer. Lime and quicklime are calcium oxide and are never found in nature; impure forms are prepared by calcining limestones, the calcium carbonate of which loses its carbon dioxide and yields calcium oxide mixed with whatever impurities were present in the original limestone. The distinguishing property of lime is the "slaking" that takes place when water is put into contact with it; calcium hydroxide results from the strong chemical combination of the two. The slaking manifests itself by a general effervescence and disintegration of the usual lumps in which quicklime is obtained from a lime kiln. The slaked lime has the property of setting when exposed to the atmosphere, and for this reason it is used in building mortars.

The impurities in limestone may vary in quantity from a slight trace to a large percentage of the total volume of rock. They determine in large measure the nature of the lime that will be obtained when the limestone is calcined. The impurities cannot often be distinguished by eye, but they can be roughly determined by solution of the calcium carbonate in hydrochloric acid. When an engineer has to test a limestone as a possible source of lime, the easiest course will usually be to calcine a small quantity of the natural rock and then examine the properties of the resulting lime. The more usual impurities are siliceous and clay materials; they have the effect of slowing up the slaking of the lime and developing its "hydraulic" property, i.e., its ability to set when immersed in water. "Hydraulic limes," as they are called, are commonly yellowish, in contrast to the white of pure quicklime; they set slowly and have little strength unless mixed with sand.

The impurities that cause these variations in the properties of limes naturally depend on the formation of the limestones from which the limes are made. The impurities may therefore vary considerably, even between the top and bottom of a bed of rock. In the quarrying of limestone for lime making, care must therefore be exercised with regard to the exact character of the rock being obtained. For example, in the well-known Lias beds of England, the quantity of aluminum silicates present may vary from 8 to 64 percent; the corresponding quantity of calcium carbonate may vary from 90 to 34 percent.[17.3] Limestones that contain fossils may produce a lime with variable and uncertain slaking properties because of the peculiar calcining of some fossil material. Small particles that will not slake may occur; they subsequently take up water and tend to disintegrate the mass of material surrounding them.

Manufacture of lime is carried out in coal-, gas-, or wood-heated kilns: calcining takes place at about 900°C, or somewhat less if steam is present in the kiln. Higher temperatures will cause "clinkering," or partial fusion of the rock, if clay impurities are present. The resulting quicklime will be obtained in blocks corresponding to the original blocks of rock; it is usually slaked in special boxes before being used. Various names, such as "fat lime," "rich lime," and "poor lime," are used to denote different grades; the rate and nature of slaking are the usual criteria. These descriptive terms should be accepted and used with caution, as they are essentially local in significance. If the lime has an appreciable percentage of magnesium oxide in it, derived from the magnesium compounds present in the original rock, it will be designated as a "magnesian" or "dolomitic" lime.

The plasters that play so important a part in finishing processes in building construction are derived mainly from deposits of gypsum, the hydrous sulfate of calcium. These occur as natural deposits, usually as a solid rock mass but in some places as unconsolidated sandy material. Gypsum is fairly widely distributed; one of the most notable deposits of high-grade gypsum is in the extreme north of Nova Scotia on Cape Breton Island (Fig. 17.4). It is often quite pure, having probably originated by the evaporation of oceanic

FIG. 17.4 Working face in gypsum quarry, Dingwall, Cape Breton Island, Nova Scotia.

waters and the consequent concentration of the dissolved salts. It is often found in association with common salt. Gypsum will frequently change to anhydrite, the anhydrous calcium sulfate, some of which will be found in most gypsum deposits. Unlike gypsum, anhydrite is of limited commercial value and so has to be carefully distinguished and discarded. When pure gypsum is heated to a temperature between 120 and 200°C, it loses most of its combined water and gives an amorphous substance commercially known as "plaster of paris." When this is mixed with water, it takes up by chemical combination as much as was previously lost and sets as a hard and compact mass. Special plasters may be obtained by variation of the calcining process and by the admixture of other materials with plaster of paris.

17.4 LIGHTWEIGHT CONCRETE AGGREGATES

There are a number of materials that are widely used as lightweight aggregates for concrete. These may conveniently be grouped into four categories: (1) those occurring naturally, (2) those produced as by-products in manufacture, (3) those manufactured from natural materials, and (4) those which result in unusually lightweight concrete. Pumice, lava, and tuff—all volcanic rocks—are materials in the first group; they have long been used as aggregate. Fairly widely distributed, these materials require little processing beyond washing and screening before use. Cinders provide a well-known member of the second group, but they are not generally regarded as a geological material, even though they are, ultimately, derived from the earth.

The slag that is produced in the manufacture of steel has at least the appearance of a naturally occurring rock. It is used as aggregate both in its normal condition and after being processed by one of the several methods now available for producing a lightweight product; "foamed slag" and "expanded slag" are two terms in common use to describe this specially prepared by-product. Manufactured aggregates are largely derived from certain types of clay that possess the property of expanding rapidly, or "bloating," when heated to the point of incipient fusion. Bloated clays have been used for the making of concrete for several decades; most of the material of this sort is produced by one or another of the specialized methods developed for the actual process of bloating. In areas where suitable rock for aggregate is scarce, the possible use of clay that can thus be processed should always be kept in mind. Other aggregate manufacturing methods under development, and in use to a limited extent, depend upon the well-known process of "sintering" commonly followed in metallurgical work. Finally, when very lightweight concrete is required, it may be produced by the use of an expanded vermiculite, a special variety of mica, or of expanded perlite, a type of volcanic glass. In each case, the material has the property of expanding, or "popping," when heated to the point of incipient fusion, and this property is the basis of the preparation of the material for use in concrete (and even plaster aggregate).[17.4]

17.5 VOLCANIC ASH

In areas where there has been volcanic action, the possible presence of volcanic ash must be kept in mind, since this can be a troublesome material if encountered on construction. Its name describes its character; it is the fine-grained, loose deposit formed after violent volcanic eruptions and often carried by the wind for surprising distances from its point of origin. It will readily retain its "fresh" character for long periods, especially when protected by vegetal or other cover. During the construction of Alaskan military bases at the start of the Second World War, volcanic ash was encountered that had originated in the explosion of Mount Katmai in 1912. In one project under development for military purposes, a 46-cm (18-in) layer of ash was found all over the area. Its abrasive quality greatly increased wear on construction equipment and even affected wearing apparel and leather.

During building of a wartime road on the island of New Britain, the Seabees excavated a sharply sloped hill that was found to consist of volcanic slag, or compacted volcanic ash. Trials showed that, when naturally fragmented to the size of pea gravel and rolled into place, the volcanic material formed what was practically a monolithic slab, an admirable wearing surface for heavy traffic. Its use spread and it was put to good service in paving landing fields and service areas. Demand for the material became so great that difficulty was experienced with normal methods of excavation, since drilling the necessary holes for explosive charges was slow. With the ingenuity so often demonstrated in times of war emergency, those responsible soon solved this problem by bringing up a Sherman tank and firing 75-mm shells directly into the face of the pit. Armor-piercing shells were found to be most suitable, giving 25-cm (10-in)-diameter holes 3 m (10 ft) deep and doing in half an hour three times as much work as had been possible for a normal crew working a 12-hour shift.[17.5]

17.6 FLY ASH

Another type of ash that is achieving importance in civil engineering practice, perhaps because of the increasing amount of it that is becoming available, is the ash collected in the flues of powerhouses in which boilers are fired with pulverized solid fuel. Admittedly it is not a naturally occurring geological product, but it is of such interest, and of such increasing importance, that it calls for at least brief mention. It is usually a fine gray powder, having silica, alumina, and iron oxide as its main constituents. It is relatively inert but has the valuable property of producing a pozzolanic effect when used in the making of concrete with portland cement. For such use, fly ash should have a silica content of not less that 35 percent and a loss on ignition of not more than 12 percent. Other more detailed requirements can be specified. The material has a bulk density of between 800 and 1,100 kg/m^3 (50 and 70 lb/ft^3). Because of its pozzolanic property, it gives increased resistance against chemical attack to any concrete in which it is used, a reduction in the heat generated during setting, a reduction in the rate at which the concrete will gain strength, and a corresponding extension of the period during which this effect takes place. Fly ash is, therefore, a valuable "man-made geological product." One of its main disadvantages is that, since it is man-made, its properties are extremely variable—even in ash from the same power station—so that unusual control has to be exercised over its properties in order to ensure its satisfactory use as an addition to ordinary concrete.[17.6]

The quantities becoming available (now over 30 million tons a year in the United States) make the disposition of fly ash an increasingly important industrial problem. Much research and experimentation is being carried out into possible uses of this "waste" material, and into its proper handling when it has to be used as fill. In some areas, bricks, blocks, and tiles are being manufactured with an appreciable fly ash content. When no such uses are possible and the ash has to be deposited as fill, end dumping or other simple modes of disposition are common. For small fills this may not be too serious, but for larger fills, and for any fills that are to be used for the support of structures or which are adjacent to locations where laborers may be at work, loose fly ash dumping should not be permitted. The material can be compacted, and thus made the more stable, in the same way as soil, with optimum water content of between 20 and 30 percent.[17.7] Even though utilization of this waste material is fraught with possible dangers, it can be usefully disposed of, but only under strict control and provided its physical properties are known (Fig. 17.5).

17.7 MINING WASTES

In addition to the vast quantities of fly ash now available, other waste materials may have their uses. There may be mentioned studies that have been undertaken into the possible use of gold-mine wastes. Typical are the ample quantities of waste material that had to be disposed of in large-scale gold mining when the gold content of river gravels was minute. Hydraulic mining of such gravels was once widespread in parts of California, but in 1884 the U.S. Circuit Court of Appeals prohibited uncontrolled deposition of hydraulic mining wastes. As a result, many gold-dredging operations had to close down; some were able to continue when means of controlling the waste were available. Such was the case on the Bear River where the Combie Dam, near Auburn, now "impounds" 2.5 million tons of mine debris. This material is ideally suited for the manufacture of calcium silicate products, and so research has been done on using this waste material in the manufacture of calcium silicate bricks and lightweight concrete blocks.[17.8] As more normal supplies of the basic materials for brickmaking and lightweight-concrete manufacture decrease, this is an alternative source that may one day prove to be of value.

17-10 PRELIMINARY STUDIES

FIG. 17.5 Wimbleball Dam, Devonshire, England, the concrete of which contains fly ash as an admixture.

17.8 STEEL SLAG

Another common waste material from industry is the slag that has to be disposed of from steelmaking. It has become well and favorably used for a variety of purposes, the simplest of which is as fill in place of crushed rock. Many years of satisfactory experience with this artificial-geological material have led to its almost unquestioned use. An experience in Canada, however, sounded a warning note even here. A small single-storey building at the Canadian end of the international bridge across the St. Mary's River at Sault Ste. Marie was found to be subject to uneven settlement, despite the light loads imposed by the building on the fill material beneath. Careful study revealed that the building was founded on a well-compacted fill composed of slag from the local steel works. The slag was that produced in the open-hearth furnaces of modernized industry and had a basic flux not found in the slag produced in the customary blast furnaces. Once this was known, open-hearth slag was studied in the laboratory and found to be capable of providing vertical expansions of more than 10 percent by volume. When a record of this experience was published, discussion revealed quite a number of similar experiences.[17.9] Even steel slag, therefore, must be studied before use, and if found to be from open-hearth furnaces, its use must be restricted to those locations where expansion with time is of no consequence.

17.9 CONCLUSION

These brief notes provide but an introduction to an application of geology that is of singular interest and economic importance. To those expert in the fields touched upon, the brevity of the treatment may appear to be almost foolhardy. The intention, however, has been not to attempt detailed consideration of the geological aspects of any of the building materials mentioned but, by this general coverage, to illustrate the geological origin of many of the materials most commonly used on construction. With aggregate delivered by rail or truck from some source off the job; with brick and tile inspected and tested against standard specifications and delivered from a modern manufacturing plant; with cement, lime, and plaster delivered in convenient bags direct from the plants; and with even cut stone sawed to finished dimensions at the quarry and delivered on the job in packages as a "finished product," it is all too easy for the engineer on the construction job to forget that these and the other materials mentioned are derived from the earth, have a geological origin, consist of minerals that can be predetermined, and so are subject to study and analysis in just the same way as are other materials that are truly the products of the earth.

Especially is this true of natural stone, whether in crushed or finished form. And with just a little care taken in advance, notably by carrying out a few check petrographical analyses, possible trouble can be avoided in those cases, fortunately few in number, where deleterious minerals are present. It is, perhaps, not without significance that one of the very few petrographic papers in the literature of civil engineering known to the writers is one to be found in the *Journal of the Institution of Civil Engineers*. This paper describes a "postmortem" examination of bad staining in granite, staining found to have been caused by the weathering of an unstable mineral, in this case a type of dark mica.[17.10] The great activity in concrete-research circles in connection with alkali-aggregate reaction will probably ultimately have much beneficial effect, despite the difficult problem with which such research is concerned. Because of the studies being carried on in this field, more civil engineers than ever will realize that "rock" is not just an inert material to be accepted for use merely on the basis of mechanical tests, but is rather a complex assemblage of minerals, whose study is a procedure that may be of the greatest possible benefit to their own work.

17.10 REFERENCES

17.1 D. A. Buzzell, "Fused Ceramic Material for Riprap on Keystone Dam," *Engineering News-Record,* **120**:787–788 (1938).

17.2 Dinesh Mohan, Director of the Central Building Research Institute, Roorkee, India, by personal communication.

17.3 R. F. Sorsbie, *Geology for Engineers,* Griffin, London, 1911, p. 331.

17.4 T. Whittaker, *Lightweight Concrete in America,* National Building Studies Special Report no. 13, H. M. Stationery Office, London, 1953.

17.5 C. A. Cunney, "Blast Holes into Volcanic Slag," *Civil Engineering,* **16**:71 (1946).

17.6 P. G. K. Knight, "Pulverized Fuel Ash as a Construction Material," *Proceedings of the Institution of Civil Engineers,* **16**:419–432 (1960).

17.7 A. M. DiGioia and W. L. Nuzzo, *Fly Ash as a Structural Fill,* Reprint JPG–70–9, American Society of Civil Engineers, New York, 1970.

17.8 T. C. Hansen, C. W. Richards, and S. Mindess, "Sand-Lime Bricks and Aerated Lightweight Concrete from Gold Mine Waste," *Materials Research and Standards* (ASTM), **10**:21–24 (August 1970).

17.9 C. B. Crawford and K. N. Burn, "Building Damage from Expansive Steel Slag Backfill," *Proceedings of the American Society of Civil Engineers,* **96**(SM4):1325–1334 (1969), with discussion in four subsequent issues.

17.10 B. H. Knight and R. H. Knight, "Rapid Staining in Granites Used in Civil Engineering Work," *Journal of the Institution of Civil Engineers,* **9**:545 (1938).

Suggestions for Further Reading

This brief chapter touches upon such a variety of materials that only a general reference to sources of further information is warranted. Geological surveys (national, state, or provincial) frequently have useful information about the basic materials from which these manufactured products are produced. Since the materials are manufactured in commercial plants, unlike other materials dealt with in this Handbook, there are usually excellent trade associations that link together the producing companies. These include in their work the provision of useful technical information about their products. The several national brick associations, for example, have a number of publications designed to aid the proper use of their products, a typical title being *The Geology and Mineralogy of Brick Clay*. For overall production and similar figures, reference should be made to national statistical agencies or to such publications as Professional Paper No. 820 of the U.S. Geological Survey, to which several references are made in this Handbook.

part 3

CIVIL ENGINEERING WORKS

chapter 18

DRAINAGE

18.1 Stephenson and the Cone of Depression / 18-5
18.2 Natural Drainage / 18-7
18.3 Gravity Drainage / 18-9
18.4 Drainage of Open Excavations / 18-13
18.5 Well Points / 18-17
18.6 Effects of Drainage / 18-22
18.7 Acid Mine Drainage / 18-26
18.8 Some Unusual Cases / 18-27
18.9 Conclusion / 18-31
18.10 References / 18-31

Drainage works are here placed first among the works of the civil engineer in the prosecution of which geology has great influence since, as all with experience in construction know well, the control of water is always important and sometimes vital to the successful completion of construction work. The subject may seem to be mundane to the uninitiated. One can still hear it described by some as "messing about with water," although anyone who has had to deal with difficult water problems on the job will not so describe it. Control of groundwater is a prerequisite for successful construction practice. And control of groundwater by drainage, when necessary, depends absolutely upon the hydrogeology of the site, which must therefore be investigated thoroughly before any plans for water handling are made.

Drainage is probably the oldest of all types of civil engineering activity, going back even into the realm of mythology. The draining of the swamp of Camarina against the orders of Apollo will be found mentioned by Erasmus in his famous book, *In Praise of Folly.* Some of the earliest tunnels were drainage tunnels, through which water flowed by gravity from areas that had to be drained. The Romans executed some mighty works of this kind. One justly famous Roman drainage work was the tunnel from Lake Fucino. This tunnel was driven through limestone for a distance of 5.6 km (3.5 mi) and is 1.8 by 3.0 m (6 by 10 ft) in theoretical section. It is said that 30,000 men were employed for 11 years on its construction.

There are many other examples from the past, but only one will be mentioned. Lake Copais is now an area of about 25,900 ha (100 mi²) of dry land, fertile and well cultivated, located near the east coast of Greece. It lies generally about 90 m (300 ft) above the level of the sea from which it is separated by a steep range of hills. The lake was drained in 1886 by a French company. After drainage had lowered the water level, the engineers found that the ancient inhabitants (of the prehistoric Minoan period, 2000 B.C.?) had drained it previously. The works discovered by the French included some ancient drainage canals which led to some of the *Katavothras,* natural tunnels in the form of large fissures through the limestone rock. Unfortunately, although these interesting natural tunnels were so effectively used in the past, the largest of them is now blocked up, and so a modern drainage tunnel had to be constructed. North of the drained area in the pass to the sea at Laryma, there are 16 shafts which appear to have been put down by the prehistoric workers mentioned with a view to tunnel construction, although the work had not proceeded far. The whole region is of great interest to the geologist and the civil engineer.[18.1]

In the Middle Ages, long before the start of modern civil engineering, the Dutch became expert in drainage works in connection with the reclamation of low-lying land around their coasts and in the Zuider Zee. One of their foremost drainage engineers, Cornelius Vermuyden, was invited to England about the year 1621 to assist with a breach in the embankment of the river Thames near Dagenham in Essex. He stayed on and did notable work in reclaiming wetlands on the borders of Yorkshire and Lincolnshire. Together with other Dutch engineers, Vermuyden was also engaged on the drainage of large areas of the Fenlands in Norfolk and Cambridgeshire. The Fenmen objected strongly, thinking that their traditional occupations of fishing and wildfowling would be affected adversely (early environmentalists!). Oliver Cromwell's championship of the Fenmen was a noteworthy episode in his early career.[18.2]

Toward the end of the eighteenth century, civil engineering as it is known today got slowly started. Drainage of bogs and wetlands came before the building of most of the early canals; but both drainage and canal building needed an intuitive sense of geology on the part of the designing engineer, especially an appreciation of the movement of groundwater through porous media. One of the greatest of the early British engineers was William Jessop. Despite his eminence and the important early works which he carried out—in drainage, canal construction, and harbor development—the first worthy biography of this fine man appeared only in 1979. It is replete with examples supporting what has just been said about drainage and canal works. Between 1760 and 1830 many thousands of acres of fenland in Britain were drained and reclaimed for use as a part of the improvement of agriculture during that period. Eleven projects carried out by Jessop are described by Hadfield and Skempton, one or two of the projects being in direct succession to the earlier works of Vermuyden.[18.3]

One of Jessop's canal projects well illustrates his approach to water problems on construction. He was the engineer for the Grand Junction Canal between the Midlands and London, a canal still to be seen today. Near Blisworth, a tunnel was necessary to link two major sections of the canal, a temporary railway having been constructed around Blisworth Hill. It was to be a "soft-ground" tunnel (i.e., through soil), almost 2,700 m (9,000 ft) long, 5.4 m (18 ft) high, and 5.0 m (16 ft 6 in) wide, a sizable tunnel even by modern standards. An experimental heading was driven which showed that the groundwater could be handled by direct drainage. In consequence, a small drainage tunnel was excavated parallel with the line of the main tunnel, and for almost its full length, with 360 m (1,200 ft) of cross headings. With groundwater drained in this way, the tunnel was excavated from 19 vertical shafts and completed by March 1805. It is believed to be the first tunnel, in Britain at least, to have been driven after predraining of the ground to be penetrated.

A contemporary of William Jessop was William Smith, the civil engineer and geologist whose portrait is the frontispiece to this volume. Much of Smith's engineering practice was also in the carrying out of drainage works and in canal construction. Unlike most of the other early engineers, William Smith did write descriptions of some of his works. In 1806, for example, he published *Treatise on the Construction and Management of Water-*

meadows. In this he describes how he drained Prisley Bog for the Duke of Bedford, a notable piece of work for which he received a medal from the Society of Arts.[18.4] In his description of this project, Smith has this to say: "Hence it appears necessary for the designer of plans of irrigation to be fully master of *the art of draining, which cannot be well understood but by a knowledge of the strata*" [italics added].[18.5] (The phrase shown in italics is typical of the way in which he usually referred to his geological studies, a habit which led to the nickname of "Strata Smith.")

In North America also there was sound appreciation of the fundamentals of drainage and their dependence upon geology. Thomas Roy was a civil engineer of note who lived in Toronto from 1834 to 1842, serving as one of the first engineers to the little settlement. In 1841 he, too, published a small book, a book entitled *Remarks on the Principles and Practice of Road-Making as Applicable to Canada*. He says in this:

> Drainage is an affair of primary importance in road-making and requires much skill to execute it in a proper manner. The ditch ought to be upon that side from whence the flood water flows towards the road. . . . *The strata of the soil should be carefully studied*, and means used to convey all water from springs, however small, into the ditch. It may even be necessary to carry the process of draining far beyond the area of the road, but no general rules will apply to such a case [italics added].[18.6]

It is not surprising to find that Roy stresses the importance of good drainage in preventing frost damage to road surfaces. This was in 1841, yet the lessons have still to be learned 140 years later, as examples in Chap. 30 will show.

18.1 STEPHENSON AND THE CONE OF DEPRESSION

The examples so far given might be attributed to the innate common sense of these early engineers, even though it is sense that has been far from common in engineering circles down through the years. There is another early example, however, that shows such a remarkable grasp of groundwater hydrology that it is worthy of detailed mention. This example is taken from the construction of the Kilsby Tunnel on the main railway line connecting London, Rugby, and Birmingham. It became a part of the main line of the London and North Western Railway from London to Scotland, being now one of the more important tunnels on the "West Coast" route of British Railways. It is passed through (now at high speed on BR's modern electric trains) a few kilometers south of Rugby station. Its completion was a major problem when the original line was built.

It was completed in 1838 at a total cost of some £300,000, as against an estimated cost of £90,000, and only at the expense of some lives. Unfortunately, preliminary exploratory work missed striking a pocket of water-laden sand and gravel of the inferior oölites in the Lias shale; when the tunnel works encountered it, extraordinary difficulties were met because the deposit proved to be a veritable "quicksand"; at one time, construction work had to be stopped. But Robert Stephenson was the engineer, and that remarkable man solved the problem in a characteristically satisfactory manner. Shafts were sunk, and steam pumps were erected on the line of the tunnel (Fig. 18.1). The following extract from his report is of interest:

> As the pumping progressed the most careful measurements were taken of the level at which the water stood in the various shafts and boreholes; and I was soon much surprised to find how slightly the depression of the water-level in the one shaft influenced that of the other notwithstanding a free communication existed between them through the medium of the sand, which was very coarse and open. It then occurred to me that the resistance which the water

18-6 CIVIL ENGINEERING WORKS

FIG. 18.1 Pumping machinery at the Kilsby Tunnel, England, from a drawing by J. C. Bourne, 1837.

encountered in its passage through the sand to the pumps would be accurately measured by the angle or inclination which the surface of the water assumed towards the pumps, and that it would be unnecessary to draw the whole of the water off from the quicksand, but to persevere in pumping only in the precise level of the tunnel, allowing the surface of the water flowing through the sand to assume that inclination which was due to its resistance....

The simple result, therefore, of all the pumping was merely to establish and maintain a channel of comparatively dry sand in the immediate line of the intended tunnel, leaving the water heaped up on each side by the resistance which the sand offered to its descent to that line on which the pumps and shafts were situated.[18.7]

Here, in the direct and simple words of this great engineer, is an accurate description of what today would be called the *cone of depression,* effected on both sides of the tunnel heading by the pumping at tunnel level. It was Robert Stephenson's own thinking since, on this project, he even disregarded the advice given to him by his distinguished father, George Stephenson. The quoted account was made in a report in 1841 to the London and Westminster Water Company, where it was buried and forgotten for half a century. It was quoted by Professor Boyd Dawkins in the James Forrest Lecture which he gave to the Institution of Civil Engineers in 1898. This was published the same year in the *Minutes of Proceedings of the Institution,* at that time one of the main sources of information on civil engineering in the English-speaking world.

The principles of drainage were well known, therefore, at the turn of the century, as were also the importance of the geological study of ground to be drained and the principles of groundwater movement. But the neglect of the information then, and still, available may be further illustrated by a little-known incident in the life of that very great inventor, Thomas Edison. He had learned of the discovery of nickel at Sudbury in northern Ontario. Needing nickel for his experiments on storage batteries, he was moved to go up to Sudbury, then at the start of its remarkable development. Using a magnetometer (being one of the first to do so), he located an anomaly in the drift-covered eastern part of what is now called the Basin. A team of men was employed to excavate a shaft; this was sunk to a depth of 10 m (35 ft), but work was abandoned in 1903 because the workers struck "quicksand" at the bottom of the shaft.

Brilliant as he was, Edison had not apparently read in the literature of civil engineering and so did not appreciate that the sand encountered was "quick" only because of the presence of groundwater, which could have been drained by suitable pumping. (Most geotechnical laboratories today have a "quicksand device" to demonstrate this simple fact.) In 1916 a new operator successfully sank a shaft just 6 m (20 ft) away from the Edison shaft, and nickel ore was found 4.5 m (15 ft) below the level of the bottom of the earlier shaft. This is now the site of the development of the Falconbridge Nickel Mines, Ltd., one of the great producers of the Basin. It is interesting to speculate how the history of Sudbury might have been changed if Edison had read Stephenson.[18.8]

18.2 NATURAL DRAINAGE

"Vanishing rivers" are a feature of many areas underlain by cavernous limestone. Irregular erosion of limestone beneath ground surface, by the long-term action of groundwater, will come up for frequent mention in this Handbook. While often detrimental to the use of such areas for engineering works, this natural feature can be of great utility in providing natural drainage, but only when the local geology is known with certainty, so that such underground disposal of wastewater will have no harmful effects elsewhere.

The use of such natural cavities provided the arrangement used for taking care of the surface drainage from a new bypass road built around the ancient city of Winchester, England, just before the Second World War. Most of the new road lies in the valley of the river Itchen, one of the famous trout-fishing streams of England, whose waters are inviolate from such possible pollution as road drainage. Fortunately it was possible to collect all surface runoff from the road into ditches and to discharge these ditches through "soakways" into the local chalk. Although actual cavities can rarely be used for the disposal of surface drainage, exactly the same procedure is being followed when water is allowed to penetrate from the surface of the ground into underground permeable strata; the "cavities" are the voids in the structure of the permeable material. This has long been a well-accepted method of drainage. When the pervious stratum is at the surface, drainage into it is an obvious procedure. An interesting modern example is shown in Fig. 18.2. When,

FIG. 18.2 Soakway drains into limestone on the A64 Malton bypass road in North Yorkshire, England; fill material is wrapped in Terran fabric to minimize the washing in of fine material.

however, the pervious material is some distance below the surface and is overlain by less permeable material, then drainage has to be conducted to it down specially excavated bores.

These have long been called *dumb wells,* a singularly lucid name since the bores do act as wells in reverse. Throughout the world, there are many well-known examples of this type of drainage. In South Australia, the famous Dismal Swamp was drained, with partial success, by boring dumb wells from the surface into the underlying limestones. Dumb wells have long been used in Lincolnshire, England, for draining the upper estuarine clays into the Lincolnshire limestone. During the construction of the London tube railways through the London clay, at least one contractor saved himself a lot of money by boring a dumb well from his excavation, deep as it was below the surface, into the underlying Chalk, through which he drained his workings, having previously determined the exact position of the Chalk and the fact that the normal water table was below the level of the tunnel he was excavating. The use of dumb wells for both surface and subsurface drainage has been described; it will be appreciated that only rarely are circumstances such that these wells can be used for the drainage of underground works. More usually, the only recourse in these instances is to collect drainage water into sumps for pumping to the surface and disposal in natural watercourses.

One of the most unusual cases involving the use of natural underground drainage channels is of such a nature that it may well be termed a "cautionary tale." The town of Bellevue, Ohio, is located over cavernous limestone. This had long been known so that, following earlier practice, disposal of the town's sewage in recent years was economically effected by discharging it into the underground. An estimated 1,500 sinkholes were used at one time for this purpose. With the growth of the town and the development of modern standards of sanitary safety, the Ohio Water Resources Board, in 1946, ruled that the use of any well for disposal of sewage "or other material considered deleterious to the potable underground water" was to be prohibited throughout Ohio. The state health department considered the Bellevue procedure a potential menace. Plans were therefore prepared for a modern sewerage system and sewage treatment plant. Septic tanks were used for treatment as a temporary measure, but the town is now well served by a fine treatment plant.

It must be added, although not in any critical way, that prior to receiving the edict of the state health department, the town had had another kind of warning, a warning which illustrates one of the fundamentals of groundwater hydrology. On 26 June 1937, rain so heavy as to constitute a cloudburst fell in the area to the south of the town, reaching its greatest intensity close to the town itself. It was estimated that between 17.5 and 18.0 cm (7 and 8 in) of rain fell on the town in nine hours. Ordinary surface drainage facilities were naturally taxed beyond their capacity, and surface flooding occurred. The underground caverns into which the town had been discharging its surface drainage were very quickly filled, the normal underground flow from the caverns being susceptible to exactly the same laws of hydraulics as are surface waters. The underground system therefore overflowed, and geysers 4.5 to 6.0 m (15 to 20 ft) high broke through into the streets in the lower parts of the town. Since the overflow was not all rainwater, but included sewage that had previously been discharged in the usual way, the distress of the town can well be imagined. Time took care of the emergency, and Bellevue drainage soon returned to normal.[18.9]

There is one other system of natural drainage that is coming to be increasingly used, especially in road and airport construction when building has to take place over soft clay, i.e., over clays with high moisture contents. As is sometimes the case, modern practice is a reversion to a very old French technique, but now applied with full knowledge of the processes involved. This is the use of what are called *sand drains.* These consist essentially of vertical, uncased "piles" of sand or other porous material, placed at predetermined spacings in order to provide an adequate drainage system from the stratum through which they pass, either down to a pervious stratum beneath or up to the surface. The pressures exerted by fill placed over an installation of such vertical drains will assist the consolidation of the wet material and thus hasten the stabilization desired for road construction over the fill.

Many notable sand-drain installations are found in North America. The most extensive sand-drain project one of the authors has seen was in Finland, where sand drains were installed in preparation for the construction of a new highway from Helsinki to Turku. The highway required more than 500,000 m (310 mi) of vertical drains, carried to a maximum depth of 22 m (72 ft) in an unconsolidated marine clay, and installed over a portion of road 1 km (0.6 mi) long. Estimated maximum settlement of the drained clay under the fill designed to carry the road was 4 m (15 ft). At the time of the writer's visit, 1.15-m (3.8-ft) settlement had taken place in the six weeks since the drains had been installed. The new road was then completed, and the fill section over the sand drains has since performed quite successfully.[18.10]

The design of sand-drain systems naturally involves the application of soil mechanics principles, once the clay to be drained and consolidated in this way has been sampled and tested. It will be obvious, however, that unless the geology of the clay stratum is known, the drains may not be as effective as hoped. It is significant, therefore, that one of the most important publications dealing with sand drains is a classic paper by P. W. Rowe entitled "The Influence of Geological Features of Clay Deposits on the Design and Performance of Sand Drains," published by the Institution of Civil Engineers. The author gives some striking examples from British and United States practice to support his thesis that "the real drainage behaviour of a deposit as a whole depends on the geological details of its formation. Quite small layers, veins of silt along fissures, or organic inclusions can transform the permeability of the mass compared with that of small samples."[18.11]

18.3 GRAVITY DRAINAGE

Before considering drainage by pumping, note must be made of the extensive drainage projects that can be, and have been, carried out utilizing gravity flow. This usually involves the construction of drainage tunnels, and so, for major projects, a careful survey of the topography and geology of the area to be drained is an essential preliminary. If the geology is not suitable, then real troubles can be experienced. As one of the following examples shows, even when the local geology makes tunnel driving difficult because of the presence of water, tunnel construction may still be economical if the drainage accomplished makes valuable mine workings accessible.

An early example is provided by the lead and zinc mines of northeastern Wales, bordering on the estuary of the river Dee; these mines have been worked for many years, always with difficulty because of the presence of excessive water in the surrounding strata. This is a "block and checkerboard" arrangement of limestone, with many cross veins that provide for a free flow of groundwater. As early as the year 1818, the mines were drained by a tunnel known as the Halkyn ("deep-level") Tunnel, which started in a hillside at Nant Flint at an elevation of 60 m (200 ft) above Ordnance Datum; it was driven thence in a southerly direction, eventually for a distance of 8 km (5 mi). After most of the accessible ore had been removed from the space thus drained, the Milwr, or Holywell-Halkyn, Tunnel was started in 1896 from the banks of the Dee, 57 m (190 ft) below the level of the older tunnel. In a similar way, it drained a large volume of limestone and thus permitted a great extension of mining operations.[18.12] Figure 18.3 illustrates this extensive drainage operation diagrammatically.

18-10   CIVIL ENGINEERING WORKS

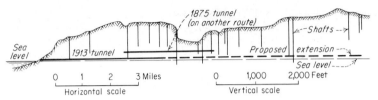

FIG. 18.3   A generalized section through the drainage tunnels from the Halkyn mines, North Wales.

The removal of the water from the mines raised no difficulties, since the water drained directly to the river Dee. It has to be remembered, however, that underground drainage is usually a reciprocal operation, for the water drained from one area must be disposed of elsewhere. If the water is contaminated, this may give rise to serious problems. It may create drainage problems elsewhere. In some cases, water drained from the underground for another purpose can be put to use as a water supply. Thus, what is believed to be meteoric water, appearing as flows from coal seams in the Rishton colliery, Lancashire, England, is pumped out of the mine and, by means of a suitable connecting system, augments the public water supply of the town of Accrington.[18.13] Less constructive, perhaps, but undoubtedly effective was the work of the Royal Engineers in the First World War in directing a supply of water into an abandoned mine gallery under Messines Ridge, whence it percolated through to one of the lower German counter headings from which it had to be laboriously pumped out.

Similar major gravity-drainage projects have been carried out in North America. One of the most remarkable and extensive was in the Leadville area of Colorado. Valuable zinc, lead, and manganese ores had been mined for half a century, but water problems in the mines were always present and grew increasingly more serious. Because of the geological structure of the mining area, no one operator could drain his workings without in turn draining adjacent properties. Finally, the U.S. government, through its Bureau of Mines, decided to undertake the construction of the necessary main drainage tunnel. Known, naturally, as the Leadville Tunnel, it is 5,190 m (17,300 ft) long and 2.7 by 3.2 m (9 ft by 10 ft 6 in) in cross section. The great expenditure its construction involved was justified by the fact that it was estimated at the time that 3 million tons of lead and zinc ores and 1 million tons of manganese ore were then under water and could not be taken without the main drainage scheme.

The geology of the region is complex, consisting of badly faulted granite, quartzite, limestone, and porphyry, covered with glacial soils that are water-bearing, groundwater being thus available for seeping into the many open faults. Two earlier and smaller drainage tunnels (the Yak and the Canterbury) had already been driven, but they could not serve the main mining area. It was known in advance, therefore, that tunnel driving would be difficult. It was, successful completion of the tunnel being one of the outstanding achievements of United States tunneling history. So much water was encountered in one section that a bypass tunnel had to be developed. Test holes at the surface showed a water table 54 m (180 ft) above the tunnel level; hence, compressed air could not be used. An old device of mixing oats with cement grout proved to be the only way of controlling water at one point; much ordinary grouting was also necessary. (See Fig. 18.4.) It was not surprising that the experienced project manager described it as the "toughest ground I ever saw," but the job was finished, the mines were drained, and the ore was recovered. Mining has been completed but drainage still flows from this famous tunnel (Fig. 18.5).[18.14]

A similar tunnel project for the drainage of another lead-zinc mine with copper and silver was carried out in Peru, again under extremely difficult conditions due to water

FIG. 18.4 One of the many leaks into the Leadville, Colorado, drainage tunnel, while under construction.

problems in the headings of the tunnels. The plural is used since the drainage of the Casapalca mines was effected by driving twin tunnels, each 11 km (7 mi) long and of horseshoe cross section, 3 by 3 m (10 by 10 ft), local topography being such that the great length was essential. The two tunnels were driven on a 0.3 percent grade, one being 2 m (6 ft 6 in) above the other, cross headings connecting the two at intervals. The upper tunnel was (and is) used for ventilation, the lower one as the drainage tunnel, flows of water of such volume being encountered that without the added facility of the second tunnel, the project might not have been completed. Driving was in limestone, the tunnel being planned to finish 387 m (1,290 ft) below the bottom of the deepest shaft in the mine.[18.15]

It is to be expected that drainage of water is often a feature of tunnel construction. Some of the examples cited in the next chapter illustrate this well. The Tecolote Tunnel was one of the most difficult ever to be carried out in North America, but other countries have experienced similar tunneling problems. One of the most difficult tunneling projects yet on record was the Rokko Tunnel built for the State Railways of Japan, as one link of the new San-yo line between Osaka and Kōbe. Excavated in badly jointed granite through Rokko Mountain, the tunnel made slow progress in some sections, because joints allowed groundwater access to the workings and badly disintegrated rock was encountered. To advance one of the eleven headings used for the tunnel excavation just 400 m (0.25 mi), the contractor had to develop 1.2 km (0.75 mi) of branch and bypass tunnels and to drill

18-12 CIVIL ENGINEERING WORKS

FIG. 18.5 Exit from the Leadville drainage tunnel today, the tunnel still acting as a drain even though mining has been discontinued.

8 km (5 mi) of 5-in-diameter holes, fanning out from the tunnel line, in attempts to divert the groundwater from the working faces. The tunnel is 16 km (10 mi) long, 10 m (33 ft) wide, and 8 m (27 ft) high. Despite all difficulties, it was completed in 1979 (Fig. 18.6).[18.16]

These tunnel jobs were major construction operations, but gravity drainage is, naturally, not confined to projects on this scale. It is often utilized for small jobs when topography is suitable, such applications being so straightforward that no more than this brief mention is necessary. Sometimes, however, unusual combinations of local geology and topography present opportunities for unusual gravity-drainage facilities. Such a case was the vertical shaft installed on the campus of the University of British Columbia at Vancouver. This unusually beautiful campus is located on Point Grey, a headland with a general land elevation about 90 m (300 ft) above sea level. Even local residents have to admit that Vancouver does experience heavy rainfall. This created a problem of disposal of runoff from the university area. The headland consists of fine sands through which it is easy to excavate. A vertical shaft was therefore sunk from a specially built intake structure, below ground level and lined with concrete. Because of the velocity which the falling water would have, a helical spiral groove was built into the concrete through the use of special wooden form work. From the bottom of the shaft, the wastewater flows by gravity into the sea.[18.17] Although the exact circumstances of this installation will be rare, the idea behind

FIG. 18.6 Leakage into the pilot heading for the Rokko Tunnel, Japan, during construction.

it will serve as a reminder that gravity is still the most economical method of moving water and so should always be kept in mind when drainage problems have to be faced.

18.4 DRAINAGE OF OPEN EXCAVATIONS

It might be thought that applications of the theory of groundwater flow in the development of drainage schemes would be confined to water deep below ground level. This is not the case at all since, if the nature of such movements is not also fully appreciated for drainage of water at or near the surface of the ground, and especially in open excavations, efficiency may be sacrificed when drainage starts. Pumping will usually be necessary, but only after local geology has been studied and the most suitable form of pumping arrangement decided upon.

A useful example is on record from Arizona. In the Salt River valley of that state, 82,000 ha (203,000 acres) of improved farmland was threatened with serious waterlogging due to excessive irrigation. The surface materials consisted of sandy clay, loam, and loess, underlain by clay and caliche, which sloped gently toward a river channel. Drainage by means of open channels proved to be ineffective because of the nature of the surface materials, and the water table rose 0.4 m (1.4 ft) per year after irrigation started. Soon 26,000 ha (64,000 acres) had groundwater within 3 m (10 ft) of the surface. After the local conditions were studied, especially the local geological formation, pumping was decided upon; this was an economic possibility in view of available, cheap, off-peak power. Pumping was started in 1923, and the waterlogged area was quickly reduced.[18.18]

In all such cases, the location of pumps and the depth to pump intakes will depend upon movement of groundwater in the soil to be drained, so that the concept of the cone of depression is relevant and important even in such simple cases. Exactly the same idea is used in a modified form in the application of drainage tile for shallow surface drainage.

18-14 CIVIL ENGINEERING WORKS

Long used in European farming, tile underdrainage was introduced into North America in 1835 by a Scottish farmer, John Johnston, with an installation near Geneva, New York, that is still in use. His pioneer contribution to good farming is memorialized at his farm by a bronze plaque sponsored by the American Society of Agricultural Engineers (Fig. 18.7). Civil engineers might well have shared in this recognition, for the use of tile drains around building foundations, at the rear of retaining walls, etc., is in the same tradition; its effect is to induce gravitational flow of groundwater away from soil in which water is not desired.

Brief mention must be made of that part of groundwater which is close to the surface but which cannot be drained gravitationally. Long neglected by civil engineers, although of obvious importance to the agriculturist, "soil water" is now appreciated as being important in many branches of engineering work, particularly in connection with the stability of foundations on clay soils. Again, the relevance of geology will be clear, since if it is known that buildings are to be founded upon a clay soil susceptible to shrinkage when drying, then special study must be given to local rainfall records in order to determine the potentiality of trouble. In such investigations, once the local geology is known, Thornthwaite's evaluation of local soil-water balance through consideration of rainfall and evapotranspiration will be of great value.

In earlier days, drainage of large excavations by means of pumping was limited by the capacity and types of pumps available. This situation was changed, gradually but dramatically, as submersible pumps became available after about 1925. There is now a well-

FIG. 18.7 Bronze plaque on the memorial cairn to John Johnston near Geneva, New York.

established practice of using such pumps in gravel-packed wells when the size of the excavation and the local geological conditions make this the economical solution to a drainage problem. The gravel packing, sometimes of two or even three sizes of gravel, is needed to provide the equivalent of a filter around the intake to the pump in order to obviate undue erosion of the surrounding soil. In some cases, submersible pumps are left in place after the completion of construction in order to provide a permanent drainage facility, if movement of the local groundwater is such that this may be necessary when the final structure is in operation. A number of major construction projects around the world have been assisted by the use of deep wells and submersible pumps, a notable example being the building of the Albert Canal in Belgium. Two other examples may be briefly described.

The construction of graving (or "dry") docks is naturally almost always an unusually difficult operation in civil engineering. The areas involved are always large; the very nature of the structures necessitates their being immediately adjacent to deep water. Unwatering problems have, therefore, accompanied the construction of most of these notable structures, which are not numerous, even on a worldwide scale, so that the record of each case is well worth careful study.

In the building of the King George V Graving Dock in Southampton, England, ten 35-cm (14-in) pipes, gravel packed in 60-cm (24-in) holes, formed the drainage wells for dealing with the artesian water that was found by test borings under part of the area now occupied by this great dock.[18.20] Permanent groundwater relief pipes were installed in the completed structure.

A second example comes from the record of outstanding construction projects carried out during the Second World War. The great west coast dock of the U.S. Navy at San Diego (the location was not mentioned in the published description of the work, dated 1944), with an entrance sill 13.5 m (45 ft) below mean low water, was designed to accommodate the largest naval vessels of the time. In view of the general character of the site, the original plans were to build the dock by underwater methods. Careful soil studies over the site, however, suggested that it might be possible to build the dock in the dry, and a full-scale field test by the U.S. Bureau of Yards and Docks confirmed this suggestion.

The site was therefore dredged to grade and then protected by a rock-fill seawall, behind which sand fill and an impervious silty-loam blanket were placed. Thirty-six 9,000-lpm (2,000-gpm) deep-well pumps were then installed, as shown in Fig. 18.8, together with three rows of well points which circled the construction site and stabilized the soil slopes. In all, 1,600 well points were used in this way, yielding up to 9,000 lpm (2,000 gpm). Twenty-five days after all 36 main pumps were in operation, the entire site had been dewatered to an elevation of 18.6 m (62 ft) below sea level; the average pumping rate was 165,000 lpm (36,000 gpm). A continuous pumping rate of from 90,000 to 113,000 lpm (20,000 to 25,000 gpm) was necessary to keep the water level down to the required level. Pumping was continued for eight months, by which time the structure was complete. The deep-well pumps and motors were later used as the permanent pumping installation for the dock. Some settlement of adjacent ground was caused by the dewatering, but careful soil studies contributed to the solution of the difficulties that might have resulted had it not been determined that the settlement was to be of limited extent.[18.21]

Very different was the use of deep-well pumps in the late 1960s for a major ground dewatering problem during the construction of the 13-km (8-mi) Welland Bypass of the Welland Canal in southwestern Ontario, close to Niagara Falls. Limestone bedrock lay about 30 m (100 ft) below ground level and was overlain by varied deposits of till, sand, and gravel, with an upper stratum of heavy clay. Water under subartesian pressure was present in the bedrock. Many wells had been drilled to secure this water for domestic purposes, so that any interference with the groundwater level, when groundwater had

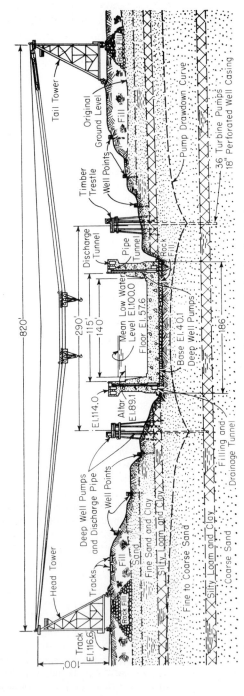

FIG. 18.8 Cross section through the site of the San Diego naval graving dock, showing drainage arrangements.

been released by penetrating to the bedrock, would cause much local inconvenience. A major test-pumping program using two deep wells was therefore carried out prior to the start of construction. Careful observations of groundwater were taken over an area of 650 km² (250 mi²) around the route of the bypass canal. These showed a drop of level of 30 cm (12 in) even 19 km (12 mi) away from the pump location, a vivid reminder of the flow of groundwater through permeable ground (Fig. 18.9).

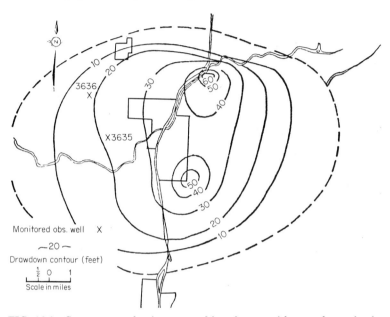

FIG. 18.9 Contour map showing range of interference with groundwater levels during test pumping at the Welland Canal, 30 June 1969.

Results of the tests permitted the development of a deep-pumping program for the construction of the two tunnels under the by-pass canal. These were constructed "in the dry," with consequent great economy, by a major application of the same principle as that used by Robert Stephenson at the Kilsby Tunnel, although here carried out much more easily with modern deep-well pumps. Temporary arrangements were made for supplying water to all homes dependent upon the well supplies which were interfered with during construction. Once construction was complete, limited pumping had to be continued from a permanent pumping installation in the vicinity of the two tunnels.[18.22]

18.5 WELL POINTS

Most drainage problems on civil engineering works are not on the scale of the projects just described. Fortunately, there is available another means of handling groundwater on smaller jobs. This method involves the use of *well points*, which are simple to install (when done by experts), effective, and economical. Various proprietary systems incorporating well points (a self-descriptive term) are available for this type of work. Basically, the system consists of a special pump and a number of well points for lowering the underground water table below the lowest excavation level. In this way the material to be excavated is predrained and so converted from the wet state, in which its behavior may be treacherous, to the dry; thus excavation progress is facilitated and water troubles are obviated. The lowering of the groundwater level is achieved by means of special well points with riser pipes about 5 cm (2 in) in diameter. The well points are fitted with special jetting nozzles, and immediately above these is wrapped a triple layer of bronze screen mesh; a standard area is 2,260 cm^2 (350 in^2).

The well points are jetted into the ground to a depth 1.5 m (5 ft) or more below bottom grade level and are located close to the area to be excavated in such a way that they will not be disturbed when excavation is complete. They are spaced at such distances apart that the cones of depression around each well point intersect; thus, the water level midway between the points is lowered below any level at which it might cause trouble. All the points are connected with a header pipe leading to the pump, which is set in operation when the complete system is ready for working. Successive cones of depression therefore completely surround the area of excavation. If the pumps are properly installed and of sufficient capacity, they will prevent the rise of groundwater into the area, and excavation can proceed within the area in the dry. The system can be extended and modified in many ways, notably in the excavation of deep trenches; successive surfaces of depression are as indicated in Fig. 18.10. Bold and simple in conception, the system can naturally be fully effective only under skilled direction and with a sure and certain knowledge of the geological conditions obtaining at the site, since the success of its operation depends on the travel of the groundwater in the strata encountered.

An example familiar to one of the writers may be cited; this was on the construction of the Victoria Park pumping station, a part of the duplicate water-supply system of the city of Toronto, Ontario. The buildings of this station cover an area 26 by 89 m (87 by 297 ft); during their construction, excavation had to be taken to a depth about 10 m (35 ft) below the level of Lake Ontario, the shore of which was only about 30 m (100 ft) away from one of the longer sides of the excavated area. Excavation proceeded through stiff clay for

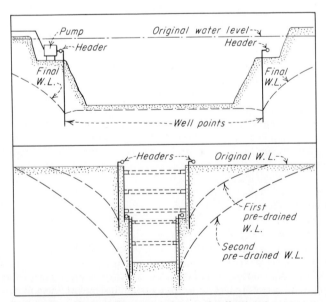

FIG. 18.10 Diagram illustrating the use of well points in open excavation and for trench excavation in water-bearing ground.

about 9 m (30 ft) of the total depth of 14 m (46 ft); small water boils then made their appearance. Auger borings were taken all over the site to check those originally taken, and a few thin water-bearing sand seams were encountered. A well-point system was therefore put in, and well points were jetted down to full depth all around the area to be excavated. The excavation was then completed without encountering water. It is of special interest to note that methane gas as well as water was "pumped out" by the Moretrench pump; the gas, which was lit and burned, was believed to have been the result of decaying organic matter entrapped with the sand deposits.[18.23]

A similar installation had been made some years before during the construction of the North Toronto sewage treatment plant in a 51 by 75 m (170 by 250 ft) area consisting of swamp mud, clay, sand, and gravel; the groundwater table of this area was lowered by 4 m (13 ft) during the progress of excavation. In other cases the facility of working in the dry—made possible by the use of the well-point system—has permitted the simplification of foundation design. The large Holland Plaza Building in New York City, for example, was designed to be supported by a concrete-pile foundation placed in a clayey sand. A well-point system dried up the excavation, and a heavy reinforced-concrete mat foundation was used instead.[18.24]

At first sight, it would appear that the well-point system would operate only in porous materials such as sands and gravels. It will, however, give equally effective results even in some types of clay. The quantity of water pumped out of clay is naturally very much smaller than that pumped out of more permeable material. Careful study has suggested that in such cases the suction caused by the pumping action results in a slight vacuum at the base of each point; this leads to an unbalanced pressure at the nearest exposed face, and atmospheric pressure outside prevents water from leaving the clay and, indeed, tends to hold the material at a slope far in excess of that at which it would naturally stand. At the other extreme, well points have been successfully used for dewatering softer types of rock. If the points can be driven into place, or placed in suitably drilled holes, they can frequently change a wet and difficult job into one that is as "dry as a bone," provided always that careful study of the strata to be drained has been made in advance.

Use of well points in sand strata is an obvious application; one of the most extensive of early installations was for the construction of an outfall sewer into Jamaica Bay, New York City, where excavation was carried out in an artificial peninsula of sand fill that was made of dredged sand deposited over some of the original muddy, marginal bay bottom (Fig. 18.11). So extensive was the installation, with three levels of well points in operation at once, that over 2,100 m (7,000 ft) of 20-cm (8-in) header pipe was in use at one time; the pumps handled as much as 60,000 lpm (800,000 gph).[18.25] During the construction of the Denison Dam by the U.S. Corps of Engineers on the Red River between Texas and Oklahoma, an extensive well-point installation was used to dry up not only an area in which the closure section of the dam was to be constructed but also an area in which slumping had occurred so that excavation had to be carried on under rigid control.[18.26] Water-bearing coarse gravel was successfully dried out by well points over an area of 76.2 by 60.9 m (250 by 200 ft) during the construction of a power plant of the Pennsylvania Electric Company at Warren, Pennsylvania, immediately adjacent to the Allegheny River; on this job four *permanent* bronze well points were installed before construction was complete. It was hoped that they would provide a continuing pure-water supply for boiler feed and general purposes, but water quality was not quite good enough when tested.[18.27] Well points can frequently demonstrate their double utility in this way.

Deep excavations in the Key West district of Florida were not regarded with favor by contractors for a long time because of the groundwater difficulties encountered in the local oölite, a soft white rock composed of calcium carbonate, organically formed. In 1954, an attempt was made to drain this unusual material with the aid of well points; the initial

FIG. 18.11 Excavation in swamp mud, predrained by a well-point installation, seen on the right, for the placing of 2.1-m (84-in)-diameter precast concrete pipe as an outfall from the Jamaica sewage works, New York.

installation at Key West was made for the excavation for a sewage lift-pump station, 12 m (40 ft) square and 7 m (22 ft) below the surface, with groundwater standing 60 cm (2 ft) below the surface. Open pumping reduced the water level by 2.1 m (7 ft) and well points, placed in specially drilled holes, were successfull in pulling the water down the remaining distance; special attention was paid to the spacing of the points in order to ensure proper approach velocities so that screen clogging would be avoided (Fig. 18.12).[18.28]

Figure 18.13 illustrates graphically an unusual geological condition that was encountered during the building of the west anchorage for a cable-stiffened suspension bridge over the Lempa River in El Salvador. Well points here had three functions: to dry up the area in which excavation was to proceed; to reduce hydrostatic pressure in the underlying soil strata in order to eliminate any possibility of blowing in the bottom of the excavation; and to stabilize the soils adjacent to the large area excavated. Some trouble was encountered with boils caused by artesian water coming up unplugged test boreholes that had penetrated the underlying strata; this is a reminder of the necessity of plugging all such holes with clay, or preferably bentonite, when preliminary investigation is complete, lest the same holes cause unsuspected trouble when encountered again after construction begins.[18.29]

So widespread has the use of well points for dewatering excavations become in construction that it is only right to add that there have been applications where problems were encountered. In all such cases known to the writers, however, the problem arose not from any defect in the well-point system but because the geology of the ground being dewatered proved to be different from what had been expected, usually despite careful site investigation. Geology was, therefore, the culprit. Two such cases have been faithfully

FIG. 18.12 A well-point installation keeping dry an excavation in Miami oölite at Key West, Florida.

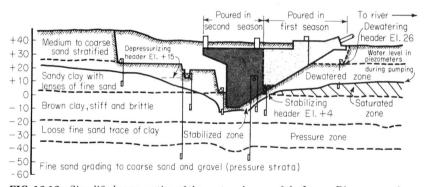

FIG. 18.13 Simplified cross section of the west anchorage of the Lempa River suspension bridge, El Salvador, showing the soil conditions which necessitated the three-stage well-point installation.

recorded for the benefit of others working in similar areas; both involved the use of well points on construction projects in Florida in which local limestones were encountered.[18.30] As has been so rightly stated, "The greatest mistake in the Miami area is to take one boring (only) or to evaluate the subsurface on the basis of a previous block several blocks away."[18.31] This is wise advice for all sites which are to be dewatered by means of well points, not only for those in the Miami area.

18.6 EFFECTS OF DRAINAGE

Examples have already been cited which show the great distances over which groundwater may be affected by pumping for drainage at one spot. This follows directly from the laws governing the flow of water in porous media such as soils. Despite the obvious inevitability of such widespread effects of pumping, they have all too often been forgotten in the past, sometimes with unfortunate results. One notable example occurred 14.5 km (9 mi) east of Harrisburg, Pennsylvania, in the Hershey Valley, a pleasant area underlain generally by an Ordovician limestone of the Beekmantown formation. For many years, the Annville Stone Company had been mining the high-calcium Annville limestone at a location 2.4 km (1.5 mi) northeast of the factory of the Hershey Chocolate Company. The stone company had pumped 16,000 lpm (3,500 gpm) out of its workings without serious effect. In May 1949, the company suddenly increased the pumping rate in its lower workings to 30,000 lpm (6,500 gpm). Within a short time, groundwater levels were affected to varying degrees throughout an area of 2,600 ha (10 mi^2). Over 100 sink holes developed, some with serious effect. Since the Hershey Company was affected, it started recharging groundwater from the surface in December 1949, in order to restore groundwater levels at its plant. It began with 16,000 lpm (3,500 gpm), but soon increased this to 45,400 lpm (10,000 gpm). Groundwater levels immediately started to rise again, but the pumping rate of the stone company also increased from 29,500 to 36,500 lpm (6,500 to 8,000 gpm). After attempting to have the recharging operation stopped by the courts, without success, the stone company adopted a program of grouting around their operations. With the recharging continuing, groundwater conditions in the area around the mine were thus effectively restored by May 1950. The field investigations carried out in connection with this case constitute a splendid example of a good groundwater survey.[18.32]

More serious, since human lives were involved, is an example from South Africa. The city of Johannesburg, center of the world-famous gold-mining operations of the Rand, has seen the location of active mining change from within city limits to new areas to the west of the city. As mining has followed the gold-bearing reefs, operations have had to go deeper. Since the ground is water-bearing, pumping has had to be extended downwards. In 1960 a major dewatering program was initiated in the Far West Rand mining district some 65 km (40 mi) west of the city. A thick bed of Transvaal dolomite and dolomitic limestone (up to 1,000 m or 3,280 ft thick) overlies the Witwatersrand gold-bearing beds. Thick vertical dikes cut across the dolomite and divide the great bed into compartments, one of which, the Oberholzer, is at the center of the new mining area. Large springs flowed in the vicinity of these dikes prior to 1960, but the big pumping program soon dried these up and then started to lower the water table in the compartment.

Between 1962 and 1966 eight sinkholes larger than 50 m (165 ft) in diameter and deeper than 30 m (100 ft) had appeared, together with many smaller ones (Fig. 18.14). In December 1962 a large sinkhole suddenly developed under the crushing plant adjacent to one of the mining shafts. The entire plant disappeared and 29 lives were lost. In August 1964 a similar occurrence took the lives of five people as their home dropped 30 m (100 ft) into another suddenly developed sinkhole. The mechanism by which these catastrophic subsidences develop is not yet fully understood, but it appears to be related to the prior existence of caverns in the dolomite, the roofs of which are in such unstable equilibrium that the lowering of the groundwater changes ground conditions sufficiently to reduce the necessary support for the roofs, which then collapse with such dire results as are here summarized.[18.33]

Serious effects may even follow the development of simple gravity drainage if the local geology has not been carefully studied in advance. A classical case comes from the history of one of the most remarkable railways in North America, the White Pass and Yukon

FIG. 18.14 Sinkholes which developed suddenly in the West Rand area near Johannesburg, South Africa, after deep pumping had started for mine drainage.

Railway, linking the sea at Skagway, Alaska, with Whitehorse, capital of the Yukon Territory, 176 km (110 mi) away but 900 m (3,000 ft) higher. The line was built, mostly by manpower, in the closing years of the last century in response to the excitement generated by the Klondike gold discovery. The White Pass is crossed at an elevation of 870 m (2,900 ft) above sea level, only 33 km (21 mi) from the coast, yet there is only one small tunnel on the line, a fact indicative of the way the railway follows the dictates of local geology.

The line passes what used to be Lewes Lake, 130 km (81 mi) from Skagway. The originally selected location here ran along the shore of this lake but was so broken by coves and bays that the engineers responsible decided to lower the water level of the lake by 4 m (14 ft) in order to use a shallow underwater bench for their line. A channel was therefore excavated through the ridge that clearly controlled the water level of the lake. It would be too much to have expected those responsible, working as they did under the imperative of the Klondike gold fever and with the Soapy Smith gang continually interfering with construction operations, to have conducted careful preliminary geological studies in that wild and inhospitable area. They could not have known, therefore, that they were excavating through a glacial moraine which, as soon as the natural stream bed through it was disturbed, would be so easily eroded by the increased velocity of the lake discharge that an immense gully would soon be formed, draining the lake 21 m (70 ft) instead of the intended 4 m (13 ft) and leaving about a dozen small pools in place of the original lovely lake. These pools can be seen from the train today; in some, lake trout still exist (Fig. 18.15).[18.34] Although this is, admittedly, an unusual case, it remains a cautionary tale and a reminder of what *should* be done in the way of preliminary geological investigation—when time and bandits permit.

These examples come from large projects; exactly the same influences of pumping, or even of uncontrolled or undetected drainage, can happen on the smallest job. In urban areas this can be very serious, especially if timber piles have been used to support struc-

18-24 CIVIL ENGINEERING WORKS

FIG. 18.15 A small lake in southern Yukon Territory, Canada, which was drained, by mistake, when a soil barrier to the lake was disturbed during the building of the White Pass and Yukon Railway.

tures on the assumption that the water table would remain sensibly constant so that no rotting of the wood, due to alternate wetting and drying, need be considered. Groundwater control during urban construction operations is, therefore, of vital importance; it can be effected only on the basis of detailed knowledge of the local subsurface, the detailed urban geology.

FIG. 18.16 Upper sections of timber piles under the public library building, Boston, Massachusetts, showing deterioration attributed to varying water levels.

An interesting and historic example is that of the great tower of Strasbourg Cathedral, designed in 1439; its stone footings were originally supported on timber piles. The installation of a new drainage system in 1750 lowered the water table appreciably and in time caused the tops of the piles to decay. Serious settlement of the tower resulted. Some years ago, the tower was jacked up after its columns had been encased in concrete, and new reinforced-concrete foundations were installed.[18.35] Several other European buildings of historic interest have been similarly affected; but as a second example, there may be mentioned a famous building of the New World, the Boston (Massachusetts) Public Library (Fig. 18.16).

About the year 1929, cracks were noticed in the building; and after investigation by the city authorities and their consulting engineers, it was found that the tops of many of the timber piles supporting the structure had decayed, some having rotted completely. About 40 percent of the building had to be underpinned; affected pile heads were cut off and replaced with concrete. The foundation strata in this part of Boston consist of boulder clay, blue clay, and silty sands overlying bedrock; pile foundations are therefore usual in this city area. The trustees of Trinity Church, which is adjacent to the library building, were alarmed at the possibility of similar trouble and had an investigation made in their behalf by X. Henry Goodenough, Inc., consulting engineers of their city. The engineers' detection of the cause of low groundwater in this vicinity is an interesting example of groundwater exploration carried out at relatively low cost.

The consulting firm finally revealed, in association with city authorities, that groundwater was leaking away into a low-level sewer constructed about 1912 through the area being investigated. The sewer was partially blocked off by a suitable dam, and groundwater levels in the area rose to some extent. The test dam was therefore left as a permanent feature.[18.36] Figure 18.17 shows groundwater contours before the leakage was checked; as a telling piece of evidence showing the possible danger of groundwater leakage in pervious soil strata, it would be hard to equal. Although the construction of the dam in the sewer was to some extent effective, further remedial measures were taken in 1955. At the northeast corner of Copley Square, perforated metal pipes designed to serve as rechargers

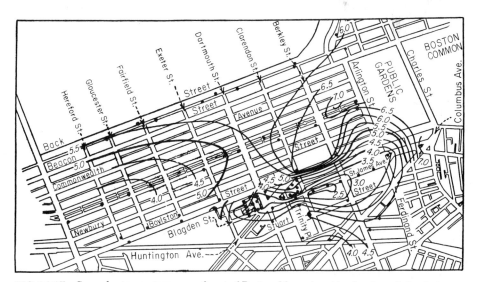

FIG. 18.17 Groundwater contour map of part of Boston, Massachusetts, showing relation between areas of low groundwater and the low-level sewer under St. James Avenue.

of the groundwater in the affected area were connected to the sewer to take water from the flow through the sewer which is high even under dry-weather conditions. It is thought that the Boylston Street Tunnel of the Boston subway also had an influence upon this groundwater situation, which is under continued careful observation.[18.37]

The Boston incident represents a case of unexpected and quite uncontrolled drainage, but it is so striking an example of what *can* happen when engineering works are in direct contact with groundwater that it carries a most useful message. Not only timber-pile foundations are affected by lowering of the local water table. The bearing power of soils, particularly of clays, may be appreciably changed by an alteration in their moisture content. Allied troubles include the shrinkage of clays as they dry out and the possibility that the finer particles will be washed out of waterlogged sand as groundwater recedes or is drawn from it. All these conditions have been responsible for notable foundation troubles. To take another Old World cathedral as an example, the troubles experienced in connection with the foundations of St. Paul's Cathedral, London, are related to local groundwater conditions. The troubles apparently began with the start of excavation for a deep sewer close to the cathedral site by the Corporation of the City of London in 1831. "Hundreds of tons" of quicksand or silt were removed by steam pumps set to work at a construction shaft; but after protests by many eminent engineers and architects (including Thomas Telford and John Rennie), the work was abandoned, and the shaft was filled up; damage, however, had already been done. Following careful investigations, all foundation work in the neighborhood of the cathedral is now subject to severe restrictions.[18.38]

Before any drainage project is initiated, therefore, the most careful study must be made, on the basis of the local geology and the known groundwater conditions, to ensure that the effects of the cone of depression around all pumping units will not have objectionable impact on neighboring property or structures. Legal records contain all-too-many cases of claims for damages due to neglect of this simple guide, which is of special importance in all urban areas.

18.7 ACID MINE DRAINAGE

There is one potential danger with the drainage from mines, and especially from coal mines, that calls for this separate reference, as a matter of emphasis. The mineral pyrite (iron sulfide) is often found in association with coal deposits. In the presence of air and water, this mineral will be subject to a chemical reaction which eventually yields sulfuric acid and ferrous sulfate. In abandoned coal mines, therefore, sulfuric acid may be expected as a pollutant of any groundwater present in the mine. If this is drained into flowing streams, the streams will in turn be polluted. That this is no theoretical picture is well shown by the fact that in 1966 it was estimated that no less than 22,500 km (14,000 mi) of streams in the United States were thus polluted by mine drainage discharges which contained 3.5 million tons of sulfuric acid, most of this pollution occurring east of the Mississippi River. The fact that there were then more than 20,000 abandoned mines in the country shows how complex this problem is.[18.39]

The cause of the pollution of such drainage is known. The problem is one of control, a topic peripheral to the subject matter of this book. Since, however, mine drainage may be encountered from mines anywhere in the world, it is perhaps worth recording that the first approaches to control of this aspect of drainage in the United States appear to have been made in 1933, when the federal government first started sealing up abandoned mines under the provisions of the Works Progress Administration. More recent efforts have seen significant progress made in the control of this danger. Many states now have the necessary enabling legislation, Pennsylvania having been in the lead in prohibiting the dis-

FIG. 18.18 Drainage from a disused mine in Pennsylvania; color (bright yellow) is needed to indicate the nature of the wastewater, which has since been controlled.

charge of polluted drainage from all active coal mines. (See Fig. 18.18) Methods of chemical treatment have been developed that result in pure effluents and a chemical sludge which can then be transported to suitable disposal areas; pollution of running streams being thus eliminated.[18.40] The possibility of pollution is one special aspect of drainage which, although limited in occurrence, should never be overlooked in designing drainage works.

18.8 SOME UNUSUAL CASES

It is a truism to observe (yet again) that no two civil engineering projects are ever the same. Accordingly, the long history of civil engineering contains many examples of projects that presented drainage problems unique to their sites. Although they are unlikely to be duplicated elsewhere, all such cases have something to contribute to the general appreciation of the close interrelation between drainage and the local geology. Three examples will therefore be briefly noted.

The first relates to the changing pattern of water supply in the borough of Brooklyn, New York, a matter which will come up for mention in Chap. 28 when water supply is

18-28 CIVIL ENGINEERING WORKS

considered. The New York Transit Authority has had to install a system of deep-well pumps and underground pumping stations around the Newkirk Avenue station on their Nostrand Avenue line, as the only solution to a groundwater problem that may well be unique. When this line was built in 1915, the water table was 5 m (16 ft) below the base of rail. At that time, Brooklyn and adjacent areas were obtaining their public water supply through an extensive pumping program, drawing water from the pervious water-bearing beds beneath Long Island.

As pumping steadily increased and more land was covered up, thus diverting normal infiltration, groundwater levels dropped steadily. Seawater intrusion resulted. So serious did the situation become that pumping eventually had to be stopped; Brooklyn and much of the southwest end of Long Island now get their public water supply from the surface sources of the New York Board of Water Supply. Pumping in the vicinity of the Newkirk Avenue station (by the New York Water Service Corporation) was stopped in 1947. By 1952 groundwater levels had risen to 82 cm (2 ft 6 in) above base of rail level in the station, with results, such as leakage, that can be imagined. Various attempts have been made to correct the situation. The final solution is the permanent pumping installation that has been constructed (Fig. 18.19).[18.41] This situation could not have been foreseen when the subway line was built. The costly solution will, however, serve as a vivid example, to all wbo know of it, of the long-term significance of groundwater movements beneath civil engineering structures.

Another case of rising water levels necessitating the construction of special drainage facilities is on record from Mexico. Lake Tequesquitengo (an oval lake 6.4 by 3.2 km or 4 by 2 mi) is a weekend resort area, an hour's drive from Mexico City. Legend has it that there is an abandoned village somewhere in its depths but such "old wives' tales" were given no heed when, between 1942 and 1951, developers built a fine resort hotel and a number of attractive homes around its shores. The water level of the lake dropped, start-

FIG. 18.19 Newkirk Avenue subway station, on the Flatbush Avenue line in Brooklyn, N.Y., showing controlled drainage in which seepage from the rise in the water table is conducted to special sump pumps installed at the station.

DRAINAGE   18-29

ing in 1951, and so development encroached still farther towards the water. In 1957, however, the water level started rising and by 1958 was 0.9 m (3 ft) higher than its old level, and by the next rainy season 0.9 m (3 ft) above that. Springs and a stream provided water supply into the lake; underground drainage through erosion channels in the underlying limestone, together with evaporation, apparently balanced the inflow and normally kept the lake level in equilibrium. Geologists blamed the 1957 earthquake in Mexico for the changes in subsurface conditions which must have reduced the flow through natural underlake drainage channels. The damage to lakeshore properties due to the high water levels can best be imagined; renovation of the hotel alone was estimated to cost $50,000 (in 1960). Siphoning and pumping were ruled out as solutions, the final decision being to drive 2.4-m (8-ft) -diameter drainage tunnel, 810 m (2,700 ft) long, through an appropriate section of the surrounding hills and to construct a control dam and an open cut for leading excess water from the lake to gravity drainage through the tunnel, all at a cost of $400,-000.[18.42] The drastic solution worked well.

It is pleasing to feature as the third example of unusual drainage facilities a project distinguished by superb engineering in which all the admonitions ventured in this chapter were fully reflected—detailed study of the local geology around the site, special study of the groundwater conditions, and careful investigations of the possible effects of the construction and operation of the project upon the surrounding area, its inhabitants, and their work. La Compagnie Nationale du Rhône (of France) is responsible for developing the series of low-head hydroelectric plants which, eventually, will generate power from almost the entire fall of this lovely river, from its source in the Lake of Geneva to the Mediterranean Sea at Port St. Louis. In the ultimate development of the river, the fourth plant coming up from the sea will be that known as Donzère-Mondragon (located between the two towns with these names). An entirely new bypass canal, 28 km (17.5 mi) long, was constructed on the east bank of the river, the drop in level between the two ends being 20.7 m (68 ft). With masterly planning, this drop in head is now transformed into power (330,000 kW) in a power station about two-thirds of the way down the bypass. The powerhouse was, therefore, built "in the dry" together with the adjacent ship lock, water transportation on the Rhône being economically important. Interesting as they are, details of the plant and lock must be passed over and attention directed to the upstream and downstream bypass canals (Fig. 18.20).

It will be clear that the level of the water in the canal immediately upstream of the powerhouse will be at almost the same level as that which it had when entering the bypass from the river. Accordingly, it is here considerably above the ground level of the surrounding terrain, which is most valuable agricultural land dependent upon a reasonably constant groundwater level. Correspondingly, the water level in the bypass immediately below the powerhouse (in what would normally be called the tailrace) will be approximately the same as that of the river where the bypass canal joins it again; this canal is therefore a deep cut excavated into the natural terrain. Dikes had to be constructed to maintain the necessary water level in the upstream part of the canal, material excavated from the bed of the new canal being used for this purpose. Tested fully before construction commenced, the material was so proportioned in the dike cross sections that some seepage would take place, draining into ditches constructed along the outer toes of both dikes to such a depth that the water level in the ditches (once construction was complete) maintained the previous level of the water table. Agriculture continued just as previously, therefore, once groundwater conditions had been restored to equilibrium after construction was finished (Fig. 18.21).

Exactly the reverse problem had to be faced downstream since, if no special provisions had been made, the groundwater would have found its own cone of depression down to the water level in the tailrace. Accordingly, special canals were constructed at ground level

FIG. 18.20 An aerial view of the Donzere-Mondragon hydroelectric plant on the River Rhone, France. This view looking upstream shows the headrace canal and, in the foreground, the tailrace canal.

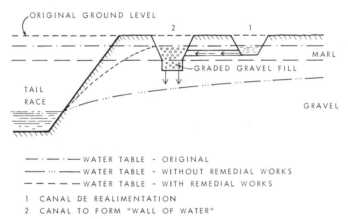

FIG. 18.21 Diagrammatic section showing control of groundwater along the Donzère-Mondragon tailrace.

on both sides of the deep bypass-tailrace canal, with ingenious arrangements for keeping them filled to the correct level. Unlike all ordinary canals, however, these two *canaux de ré-alimentation* were designed to leak, enough water seeping out of their inverts (by design) so that the water table on the land sides of the canals would be maintained at its existing elevation, leakage on the water side of these truly named "feeder" canals naturally seeping down into the water flowing down the tailrace. Completed in 1952, this great plant has been in steady operation ever since; visitors to the lovely farmland around, unless unusually percipient, have no idea of the ingenious combination of engineering and geology that maintains groundwater levels beneath the prosperous farmlands, despite the two canals.[18.43]

18.9 CONCLUSION

Of necessity this review of the relevance of geology to drainage in civil engineering practice is merely a brief survey. Despite this, it may be noted that reference has been made to (1) agricultural engineering, (2) foundations for bridges and buildings, (3) dam construction, (4) the building of canals, (5) roads and railways, (6) dock and harbor engineering, (7) irrigation, (8) land reclamation, (9) quarrying, (10) sanitary engineering, (11) subway and tunnel construction, and (12) mining engineering. Most of these special branches of work are the subjects of following chapters in this Handbook. All the branches of civil engineering thus treated separately could well have been included in this review of drainage, did space permit. It should be clear, therefore, why this chapter is the first of those that deal directly with engineering practice. The handling of water, be it surface water or groundwater, whenever it occurs on civil engineering projects, is of the utmost importance and is always dependent on geology.

Without an appreciation of subsurface conditions—both the character and arrangement of strata rock and soil and the presence, character, and potential movements of groundwater—the planning of the best and most economical drainage facilities may be a difficult achievement. Accordingly, the carrying out of site investigations for civil engineering works should always be so conducted that the investigation leads to the clearest possible picture of subsurface conditions, especially of any possible unusual interrelation between groundwater and the strata in which it occurs, always mindful of the fact that annual variations in the level of the water table are probable.

It is strange that water, the most common of materials while at the same time one that is essential for life, should so often be neglected, and not only in construction operations. Those who are involved with research for the construction industry (building research, as it is usually called) know well that most of the problems coming to their attention involve the presence of water, often in the wrong place. So also in the laboratory study of soils, it was not until the vital importance of the exact moisture content of soils in situ was fully appreciated that real advance in understanding the mechanics of soils was possible. One of Dr. Karl Terzaghi's percipient throw-away lines was: "On a planet without any water, there would be no need for soil mechanics."

There is water on this planet, however. Controlling the water that encroaches on his works is one of the greatest challenges to the civil engineer; in meeting that challenge he will find that geology can render invaluable aid.

18.10 REFERENCES

18.1 A. J. Dean, "The Lake Copais, Boeotia, Greece: Its Drainage and Development," *Journal of the Institution of Civil Engineers,* **5**:287–316 (1937).

18.2 J. W. Gough, *Sir Hugh Myddleton,* Clarendon, Oxford, England, 1964, p. 88.

18.3 C. Hadfield and A. W. Skempton, *William Jessop, Engineer,* David and Charles, Newton Abbot, England, 1979.

18.4 J. Phillips, *Memoirs of William Smith LLD,* J. Murray, London, 1844, p. 49.

18.5 T. Sheppard, *William Smith: His Maps and Memoirs,* Brown, Hull, England, 1920, p. 116.

18-32 CIVIL ENGINEERING WORKS

18.6 T. Roy, *Remarks on the Principles and Practice of Road-Making as Applicable to Canada*, Rowell, Toronto, 1841, p. 14.

18.7 Boyd Dawkins, "On the Relation of Geology to Civil Engineering," (James Forrest Lecture, 1898), *Minutes of Proceedings of the Institution of Civil Engineers*, **134**:254–277 (1898).

18.8 R. F. Legget, "Famed Inventor Pioneers Canadian Prospecting Technique," *Canadian Consulting Engineer*, **15**:58, 59 (Sept. 1973).

18.9 "Vertical Sewer System May Be Abolished," *Engineering News-Record*, **136**:823 (1946); see also "Unusual Flood in Ohio Town," *Engineering News-Record*, **119**:40 (1937).

18.10 U. Soveri, Helsinki, Finland, by personal communication.

18.11 P. W. Rowe, "The Influence of Geological Features of Clay Deposits on the Design and Performance of Sand Drains," Paper 7058S in *Supplementary Volume to the Proceedings of the Institution of Civil Engineers*, ICE, London, 1968.

18.12 J. L. Francis and J. C. Allan, "Driving a Mines Drainage Tunnel in North Wales," *Transactions of the Institution of Mining and Metallurgy*, **41**:236 (1932).

18.13 "Public Water Supply from Colliery," *Water and Water Engineering*, **38**:294 (1936).

18.14 "Tough Tunneling at Leadville," *Engineering News-Record*, **134**:313–318 (1945).

18.15 "Twin Tunnels Will Drain and Ventilate Flooded Peruvian Mine," *Engineering News-Record*, **173**:24–25, (3 September 1964).

18.16 "Miners Drain Mountain for Rail Tunnel," *Engineering News-Record*, **184**:22–23, (2 April 1970).

18.17 Neville Smith, University of British Columbia, Vancouver, by personal communication.

18.18 J. C. Marr, *Drainage by Means of Pumping from Wells in Salt River Valley, Arizona*, U.S. Department of Agriculture Bulletin 1456, 1926.

18.19 "Agricultural Engineers Dedicate Monument to Tile Drainage Pioneer," *Agricultural Engineering*, **16**:454–455 (1935).

18.20 M. G. McHaffie, "Southampton Docks Extension," *Journal of the Institution of Civil Engineers*, **9**:184–236 (1938).

18.21 D. R. Warren, "Novel Construction Plan for Graving Dock Suggested by Soil Studies," *Civil Engineering*, **14**:323–328 (1944).

18.22 R. N. Farvolden and J. P. Nunan, "Hydrogeologic Aspects of Dewatering at Welland," *Canadian Geotechnical Journal* **7**:194–204 (1970); see also R. J. Conlon, R. G. Tanner, and K. L. Coldwell, "The Geotechnical Design of the Townline Road-Rail Tunnel," *Canadian Geotechnical Journal*, **8**:299–314 (1971).

18.23 R. F. Legget, "The Wellpoint System: Application to an Excavation in Water-Logged Ground in Canada," *Civil Engineering and Public Works Review*, **31**:229 (1936).

18.24 Thomas F. Moore, The Moretrench Corporation, Rockaway, N.J., by personal communication.

18.25 "Wellpoints in a Pumped-Sand Fill Facilitate Outfall Sewer Construction," *Engineering News-Record*, **137**:238–240 (1946).

18.26 T. C. Gill, "Wellpoints Dewater Denison Dam Closure Area," *Civil Engineering*, **16**: 108–110 (1946).

18.27 "Draining Coarse Gravel with Wellpoints," *Engineering News-Record*, **137**:638–639 (1946).

18.28 E. D. Wood, "Wellpoints Master Oölite," *Engineering News-Record*, **152**:35–37 (6 May 1954).

18.29 B. J. Prugh, "Anchorage Excavation Tests Versatility of Wellpoints," *Civil Engineering*, **24**:580–582 (1954).

18.30 J. H. Schmertmann, "Dewatering Case History in Florida, *Proceedings of the American Society of Civil Engineers*, (CO3), **99**:377–393 (1974); see also B. J. Prugh, "Pressure Relief System Tames Florida Boil," *Civil Engineering*, **28**:582–583 (1958).

18.31 B. J. Prugh and E. P. Simcic, "Use of Wellpoints at Miami," *Engineering-News Record,* **154**:10 (26 May 1955).

18.32 R. M. Foose, "Ground-water Behaviour in the Hershey Valley," *Bulletin of the Geological Society of America,* **64**:623-646 (1953).

18.33 J. E. Jennings, "Building on Dolomites in the Transvaal," *The Civil Engineer in South Africa,* **8**:41 (1966); and additional information from Professor Jennings.

18.34 W. D. MacBride, "The White Pass Route," *The Beaver,* Outfit **285**:18-23 (1954).

18.35 "Modern Engineering to Save Mediaeval Tower," *Engineering News-Record,* **154**:505 (1923).

18.36 B. F. Snow, "Tracing Loss of Groundwater," *Engineering News-Record,* **117**:1-6 (1936).

18.37 "Drainage in Reverse at Copley Square," *Engineering News-Record,* **154**:47 (28 April 1955).

18.38 F. Fox, *Sixty Three Years of Engineering,* J. Murray, London, 1924, p. 196.

18.39 "Mine Acid: A Growing Pollution Problem," *Engineering News-Record,* **177**:26-28 (8 December 1966).

18.40 W. N. Heine and E. F. Giovannatti, "Treatment of Mine Drainage by Industry in Pennsylvania," *Proceedings of the American Society of Civil Engineers,* **96**(SA3):743-755 (1970).

18.41 A personal communication from Commissioner J. J. O'Neill.

18.42 "Flooded Owners Tunnel a Drain," *Engineering News-Record,* **164**:112-114 (23 June 1960).

18.43 Henry, Bonnier, and Mathian, "Nappe de la plaine alluviale de la rive gauche du Rhône entre Donzère et Mornas," *Compte Rendu des Troisièmes Journées de l'Hydraulique,* Alger, 12-14 avril 1954, pp. 147-162.

Suggestions for Further Reading

Not surprisingly, it is difficult to suggest any supplementary reading specifically related to drainage. The subject is so diffuse, despite its supreme importance, that beyond the well-known books on hydrogeology and the hydrology of groundwater for general guidance, there are few volumes available that can be of help. *Construction Dewatering* by J. P. Powers, a 404-page volume published in 1981 by Wiley Interscience, deals with a special aspect of drainage, but one more closely associated with construction practice than with geology.

The references given at the end of the chapter show the variety of publications in which articles on current drainage practice are to be found. One reference not there listed is a printed symposium called "Vertical Drains," which occupies 100 pages of *Geotechnique* vol. 31 (1980), a contribution of such value that it was reprinted as a separate publication in 1982 by the Institution of Civil Engineers (U.K.).

The manufacturers of commercial products used in drainage, notably metal culverts of various types, have published a number of most useful brochures and volumes on the correct use of their products. These are fine contributions to better drainage practice, but they are so numerous and varied that to mention one or two would involve making invidious distinctions. This general reference will, it is hoped, encourage readers to obtain copies of such publications when needed.

chapter 19

OPEN EXCAVATION

19.1 A Major Example of Excavation / 19-3
19.2 Problems with Faults / 19-8
19.3 Economics of Open Excavation / 19-9
19.4 Open Excavation in Soil / 19-11
19.5 Support for Excavation / 19-17
19.6 Sinking of Shafts / 19-23
19.7 Control of Groundwater / 19-26
19.8 Fill for Embankments / 19-30
19.9 Excavation of Rock / 19-31
19.10 Quarrying in Civil Engineering / 19-37
19.11 Conclusion / 19-38
19.12 References / 19-40

The digging of a hole in the ground—for open excavation work is frequently so regarded—appears superficially to be a relatively simple matter. All experienced engineers know otherwise. The problems introduced are sometimes intensified by their very simplicity. The economics of engineering designs involving open excavation may be initially more indefinite and in consequence more involved than in many other branches of design work. Excavation methods and rates of progress are clearly dependent on the material encountered and its geological structure, the two factors that alike determine also the finished cross section of the excavation. All too often, these vital factors do not receive the study they deserve. Unforeseen problems are not the same hazard in embankment construction as in excavation work, but the history of fill placement bears testimony to the fact that problems of serious moment are not unknown. These are often associated more with the effect of filling on existing strata than with the nature of the fill material itself.

Open excavation consists essentially in removing certain naturally formed material, within specified limits and to certain definite levels, in the most expeditious and economical manner possible. This necessitates knowing with some degree of accuracy the nature of the materials that have thus to be handled, their relative structural arrangement, their behavior when removed from their existing position, the possibility of meeting water during excavation, and the possible effect of the excavation operations on adjacent ground and structures. When these preliminary requirements are not fully answered before excavation begins, trouble may be encountered and money lost. This neglect of preliminary investigation work in open excavation is to some extent understandable: such work looks easy, and the unit prices generally charged are so low that it is often thought that a few cubic meters more or less will make little difference. The few can easily become dangerously many.

Preliminary investigations and constant check on excavation work as it proceeds are therefore clearly necessary. The preliminary work will follow the lines already suggested, with the addition of such special investigation as is found to be needed. The resulting geological maps should show rock contour lines where possible or the contour lines for any stratum so different from the surface material as to affect excavation methods and progress. The geological sections prepared will assist the engineer in visualizing the structure of the mass of material to be excavated. The usual care must be taken to investigate the possibility of encountering groundwater. The sections prepared across and along the site to be excavated will show generally the relation of the hole to be made to adjacent strata and structures, and possible damage may in this way be foreseen. Complete and detailed records of all geological features encountered are just as necessary in excavation work as in other types of civil engineering work, since such records will act as a constant check on the validity of information deduced from preliminary exploration and surveys and will often serve as a warning of possible unforeseen future troubles. Similarly, such records will be of interest to geologists and may reveal structural details of great value. A close watch should be kept for fossiliferous strata, and whenever possible, interested geologists should be granted permission to study fossil beds.

As an illustration from practice of the various facets of open excavation there may be cited the first rail subway in Canada, that of the Toronto Transit Commission. The first section of this railway is 7.5 km (4.6 mi) long; it extends in a northerly direction from near the shore of Lake Ontario in downtown Toronto and follows Yonge Street, the main traffic artery of the city. Prior to the completion of designs, every boring record that could be found for the central city area was studied. A test-boring program was then laid out on the basis of this study, and holes were put down at about 120-m (400-ft) intervals right through the heart of the city. Soil samples were taken from the holes for soil testing, and groundwater levels were observed in some of them until construction commenced. It was found that the lower 0.8 km (0.5 mi) of the subway box structure would be founded on shale, overlying which, and so constituting the bulk of all excavation for the "cut-and-cover" methods to be followed in construction, was a mixture of glacial till and glacial sands, gravels, silts, and clay. It was known that the geologically famous interglacial beds of the Toronto area would be encountered in the excavation. A local advisory geological committee was therefore formed in advance of the start of the work in order to provide liaison between interested geologists and the Transit Commission.

On the basis of the information obtained through these preliminary studies, designs and contract documents were prepared. Those tendering on the work were given access to all the information obtained on soil and groundwater conditions, with suitable qualifications as to its limitations; in the result, all bids received were remarkably close. Excavation in the cut-and-cover operation was generally by large diesel shovels; spoil was trucked up ramps to street level. Complete soil profiles were recorded in step with excavation; soil samples were taken at intervals of 15 m (50 ft) and later lodged with the Royal Ontario Museum for safekeeping, where they are available for study by interested geologists. Much useful geological information was gained that has contributed to further understanding of the interglacial beds. The work was "unexciting" from the engineering point of view, because soil conditions proved to be almost exactly as predicted by the preliminary study. No unusual soil problems were encountered, even in some of the difficult underpinning jobs that had to be done adjacent to the subway. Some minor claims were received by the commission after the work was completed (naturally), but all were settled satisfactorily and without recourse to law. As an indication of public interest, the Royal Ontario Museum arranged a special exhibit on the geology and construction of the subway, an exhibit which attracted much attention (Fig. 19.1).[19.1]

The preparation of limited areas of excavation for special foundations, as for buildings, bridges, and dams, will be considered in the chapters devoted especially to these subjects. In general excavation work, the dip of the strata to be encountered is a factor of major importance. If the strata are horizontal, or approximately so, excavation work will be relatively straightforward, and side slopes can be determined with some degree of certainty. If the bedding is appreciably inclined, excavation methods will be affected, and hazards possibly increased. The side slopes must be so selected as to be in accord with the natural slopes given by the bedding. Faults that cross the area over which excavation is to be carried out may cause serious trouble; in soft ground, it may be difficult, if not impossible, to detect them before digging starts, and hazards will thus be increased. Finally, all excavation work is influenced primarily by the nature of the material to be excavated. Broadly speaking, this can be classified as unconsolidated material (soil) and rock. It will be convenient if the general problems of excavation, problems already indicated, are considered under these two headings. Note may again be made of the vital importance of adequate definitions in contract specifications of materials that have to be excavated, since the difficulties frequently encountered with claims most often occur in connection with open excavation. The suggestions already advanced for overcoming these difficulties are therefore of particular importance for this work.

19.1 A MAJOR EXAMPLE OF EXCAVATION

Despite the apparent simplicity of excavation work, and the relative ease of defining what is meant by soil and rock, it is probably fair to say that in civil engineering work as a whole more claims for extras have been received for alleged incorrect classification of material to be excavated than for any other cause. For quite natural reasons, these claims are not often referred to publicly. So serious did they become, however, in connection with the building of the St. Lawrence Seaway and Power Project (on both sides of the border) that some reference to this particular instance is warranted.

The overall concept of the project will be familiar—the provision of an 8-m (27-ft) waterway, capable of handling oceangoing vessels, from the head of the navigable River St. Lawrence at Montreal to Lake Ontario and thence into the Upper Lakes. This involved new locks and a canal from Montreal to Lake St. Louis; locks at Beauharnois; two lift locks and a guard lock in the international section of the St. Lawrence, designed integrally with the international powerhouse which utilizes almost the full fall from Lake Ontario to Lake St. Francis to generate over 1.5 million kW; the Welland Ship Canal (opened in 1932) providing, with some deepening, the link between lakes Ontario and Erie; all combined with much dredging of channels and the construction of a large number of ancillary works, some of which (such as the Long Sault Dam) are major structures in their own right by any normal standard.[19.2]

By international agreement, the two lift locks in the international section of the river were located on the United States side. All the earlier, smaller locks had been built by Canada and were located on the Canadian side, where geological conditions are favorable. The Eisenhower and Snell locks are notable structures, with the same lock dimensions as other Seaway locks—24 m (80 ft) wide, 258 m (860 ft) from pintle to pintle, and a minimum of 9 m (30 ft) of water over sills. They are located near the eastern end of the Wiley-Dondero Canal, a straight channel almost 16 km (10 mi) long, all but the western end

FIG. 19.1 Display model of soils encountered during the first stage of construction of Toronto's subway, as exhibited in the Royal Ontario Museum, Toronto.

being a major new cutting excavated through dry land, with a bottom width of 133 m (442 ft). (See Fig. 19.2.) This involved over 15 million m³ (20 million yd³) of excavation, in advance of which extensive test boring, soil sampling, and soil testing were carried out and a demonstration test pit was dug. On the contract drawings for the canal, *till* is clearly indicated in the borehole records thereon plotted, although two tills are not differentiated. In the associated specification it is clearly stated that "the materials to be encountered . . . will consist predominantly of compact to very compact glacial till." On the contract drawings (and specifications) for the powerhouse works, the terms *sand, gravel,* and *clay* are used in various combinations. Many of the borings are indicated as "wash" borings, taken some years before construction started; the only mention of till is made in reference to a few borings taken along the line of the old Cornwall Canal in Canada, some distance removed from the location of the powerhouse works (Fig. 19.3).

It is doubtful whether the use or nonuse of the term *till* made much difference to the bids received. Bidding was keen from contractors with wide experience in earth moving who clearly wanted the jobs, to some extent, it has been suggested, for prestige purposes. Early unit prices for excavation of soil varied from $0.31/yd³ to $0.58/yd³; $1.26 was a later figure. Of the contractors involved with the canal, one went bankrupt, another defaulted, a third entered a claim for extra payment of $5.5 million on a contract awarded for a total price of $6.5 million. Closely associated with the channel excavation was work on the United States half of the power development; this also involved large quantities of exca-

FIG. 19.2 The St. Lawrence Seaway and Power Project; excavation in progress for the Wiley-Dondero Canal (in the U.S.).

19-6 CIVIL ENGINEERING WORKS

FIG. 19.3 Exposure of typical glacial till in excavation for the Eisenhower Lock of the St. Lawrence Seaway and Power Project.

vation. Here, on contracts with a total value of $89.1 million, claims for extra payment amounted to $27.6 million (or 31 percent of the contract price); settlement was for $4.8 million (or only 17 percent of the amount claimed). The chief complaints were that the till—for that is what the "sand and gravel" was—was "cemented," some of the material requiring ripping or even blasting before being moved by scrapers, and that the marine clay was very difficult to handle with normal equipment because of its stickiness.[19.3]

As is usual, contractors were supposed to have satisfied themselves as to the nature of the work on which they had tendered, an understanding clearly reflected in the small amounts given in settlement of claims. Quite apart from what might have been observed in the vicinity as to the character of the materials to be moved, the glacial drift (till) had been under renewed detailed geological study since 1952, early maps resulting from this work being on "open file" at the New York State Geological Survey in Albany.[19.4] In the comprehensive paper summarizing this geological work, a report which includes observations made on the Seaway excavation, MacClintock and Stewart list about 150 papers dating back to 1843 dealing with the surficial geology of the St. Lawrence Lowland, at the heart of which lies the Seaway.[19.5] As early as 1910, Fairchild had proposed two periods of glaciation over the area and thus suggested two tills. All papers after that date clearly indicate that till is the predominant soil beneath the upper stratum of marine clay, the well-known Leda clay of the St. Lawrence Valley. The upper (Fort Covington) till was well recognized; the lower (Malone) buried till had been suggested, directly or indirectly, by several geologists. Since the lower till had been subjected to submersion and then to the pressure of the ice of the second glaciation, its compact nature was a certainty (Fig. 19.4).

All this geological information was available before the bids for the Seaway and Power

FIG. 19.4 Excavation starting in the Wiley-Dondero Canal where a dragline could be used for removal of upper soil.

excavation were submitted, but it does not appear to have been fully utilized. Some of those tendering might have preferred to rely upon the experience of other contractors, rather than upon the "academic" opinions of geologists. Immediately adjacent to the Wiley-Dondero Canal lies the Massena Power Canal, excavated at the turn of the century to lead water from the St. Lawrence to the new power station then constructed for the Aluminum Company of America. There were still alive (in 1957) men who had worked on this job and who could therefore have testified that the original contractor for this excavation also went bankrupt, apparently because of exactly the same difficulties with the same soils as were encountered on the Seaway and Power excavation and no more than a mile away.

About 80 km (50 mi) downstream is the Beauharnois Power Canal on the St. Lawrence, now a part of the main Seaway channel. It had been excavated between 1929 and 1932 in exactly the same type of marine clay as was encountered in 1956 and 1957. And when the Welland Ship Canal was built in the 1920s, the same type of problem in excavating tough till was encountered, leading to vast claims for extra payments and, indirectly, to the failure of a leading Canadian contractor. In a major account of this work, published in 1929, this statement appears:

> . . . the rock surface was overlaid by glacial till. Over the northern one-third the rock dropped much lower, and was overlaid by very hard boulder clay of the first glaciation, over this again being the ordinary till of the second glaciation.[19.6]

Admittedly, the Welland Canal is 400 km (250 mi) upstream, but the two areas had almost identical histories during and since the last glaciation.

This brief account has had to be based upon news reports of the time, personal recollections, discussions with some of those involved, and study of the few documents that are cited. Only one relevant United States paper has been traced, and this concludes with a reference to "unsound and inadequate exploration data" used in presenting, from the point of view of contractors involved, a review of physical properties of the "basal till."[19.7] This experience on the St. Lawrence Seaway and Power Project could provide invaluable information for the guidance of all who have to deal with till, especially its excavation. It is to be hoped that one day a full account will be prepared and published.

19.2 PROBLEMS WITH FAULTS

Naturally, there are construction projects on which the open-excavation work must be through material that actually varies from rock to soft, unconsolidated deposits. Some unusual excavation work of this kind was encountered during the opening out of the Cofton Tunnel on the Birmingham-to-Gloucester section of the (then) London, Midland and Scottish Railway (now British Railways) in England in 1925. The tunnel was constructed between 1838 and 1841 and formed a part of the route taken by this line of railway in crossing a fairly high ridge of land immediately south of Birmingham. Because of modern traffic requirements, it had to be opened up and reconstructed as a cutting. The strata shown up in the cuttings on either side of the tunnel were soft, false-bedded sandstones with thin beds of marl (north) and thick beds of hard sandstone and beds of tough marl (south); for the civil engineer, therefore, they can be classified generally as both rock and soil. No records of the construction of the tunnel and the strata passed through were available, but the geological survey map of the district showed a major fault traversing the tunnel.

Troubles due to earth movements were therefore anticipated, and some of a serious nature were encountered very early in the work (Fig. 19.5). Before the final 73 m (80 yd) of tunnel was demolished, a careful geological survey of the surrounding country was made in order to attempt a deduction of the hidden structure yet to be encountered. The survey was considered in conjunction with the records of dip and details of faults made from the start of the work. This investigation was carried out in the midst of construction, and although incomplete, it was of great service. The evidence tended to show that a tunnel, instead of an open cutting, had originally been built because of a realization of the troubles to be encountered. Examination of the geological survey map showed that if the original line had been located either 180 m (200 yd) west or 275 m (300 yd) east of its actual position, little or no trouble would have been experienced from the Longbridge fault. The opening out was finally completed satisfactorily, but only after great difficulties had been surmounted. At least one member of the staff of the Geological Survey of Great Britain was in constant attendance at the work.[19.8] The paper describing the work is one

FIG. 19.5 Opening up of Cofton Tunnel, England, showing a hanging-wall slip in the west slope on a fault plane.

of those rare civil engineering publications which deal with the surmounting of unusual difficulties in construction, and as such it is worthy of special study.

19.3 ECONOMICS OF OPEN EXCAVATION

The end of a tunnel is often the beginning of an open cut; the location of the change of section is always a matter demanding careful consideration by the designing engineer. The basic criterion, but not the only one, is the relation of the respective costs of construction of tunnel and open cut; geological structure is intimately associated with a study to determine such costs. For an open cut, the volume to be removed per foot of length is naturally much greater than for an equivalent tunnel, but the unit cost of excavation will be lower, since the work can be done in the open without underground hazards and with excavating machinery that cannot be used underground. If the side slopes can be trimmed off to a satisfactory angle, no equivalent to the lining of a tunnel will normally be necessary; but if the side slopes have to be restricted, retaining walls at the foot of the slopes may involve a greater cost than tunnel lining. Based on the figures thus obtained, comparative estimates of cost for tunnel and cut can be calculated; when these are applied to locations with increasing depth to grade, a section can be selected as that at which the change in type of construction can be made most economically.

Often this result will be the final one; but true economy has not been studied in investigations of this type, unless some consideration is given to geological structure. Geological sections along the center line of the proposed cut and at right angles to it (at several stations)—together with a detailed material study—are essential. The proposed cut can then be seen in relation to underground drainage, the dip of the strata, and possible unusual underground structural relationships; in consequence, important modifications of design may have to be made.

Consider such a section as that sketched in Fig. 19.6. In the tunnel, drainage of underground water need not be considered, as no water will reach the tunnel; but in the cut, drainage will certainly present serious problems. Water will be continually seeping out at contact A; and if the quantity of water is sufficient, it may tend to wash some of the sand out onto the face of the slope. In order to escape, the water must run down the face of the shale, thus tending to weaken it. There might even be some travel of water along bedding planes of the shale, lubricating them and giving rise to dangerously unstable conditions which might lead to serious slips. Alternatively, through this unrestricted drainage of the underground porous bed, unstable conditions might be induced in the clay stratum above, with the consequent possibility of serious earth movement.

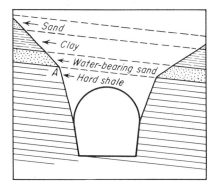

FIG. 19.6 Simplified diagram showing a possible relation between tunnel and open excavation dependent upon geology.

In all such cases, and they are by no means rare, standard calculations of relative economy may not therefore present a true picture. In the particular case illustrated, it will be advisable, even at the expense of increased first cost, to extend the tunnel construction beyond the theoretical economical limit in order to avoid the possibility of the difficulties indicated. Instances will occur in which no such simple geological structure as is sketched will be encountered. Faulted ground, as has just been seen, is a menace in both solid rock and soft material; and although it causes difficulties in tunnel work, it may cause greater difficulties in open excavation, owing mainly to the greater freedom which open excavation gives for material on fault planes to move in large masses, movement that increases unnecessary excavation. (Fig. 19.7)

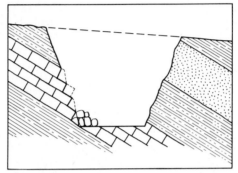

FIG. 19.7 Simplified geological section showing how the dip of rock strata may affect open excavation.

In open excavation work not associated with tunnel construction, overall dimensions and depths to finished grade will normally be determined by considerations other than excavation costs, so that no question of relative economy will normally arise. Thorough preliminary investigations are necessary all the same, and these may reveal special conditions which do affect the economic aspect of the work undertaken. For example, subsurface explorations may show that a slightly different location will avoid troublesome material. Finally, geological conditions will have a bearing on the necessity of finishing off side slopes, either with or without retaining walls, when there is no restriction on how an excavated area is to be bounded. The local geology will affect generally the economic cross section of the cut desired in this case, since the structural relationships of the strata to be exposed will affect the nature of side slopes, quite apart from the nature of the material. A point to be especially watched in this connection is the question of drainage; the adoption of simple and relatively inexpensive drainage systems can readily lead to appreciable modifications in the side slopes adopted.

Each of the two great modern water-power projects at Niagara Falls provides an illustration of one aspect of excavation economics. For the Sir Adam Beck No. 2 Project of the Hydro Electric Power Commission of Ontario, the two large pressure tunnels lead to a final 3.5 km (2.25 mi) of open canal, designed for a flow of 1,130 cms (40,000 cfs), excavated through the upper strata shown on Fig. 19.8. Careful correlation of the cross section of the canal necessary for hydraulic reasons with the exact location of the rock strata, which are inclined slightly to the horizontal, showed that if the invert of the canal was kept in the Grimsby sandstone, drilling and mucking would be facilitated and a better canal section would be obtained from the structural point of view. The canal section was therefore gradually widened, in pace with the gradual rise of its invert and of the level of the surface of the Grimsby formation, in order to provide the requisite hydraulic cross section. This project thus provides one of the most vivid examples imaginable of the interrelation of geological conditions and civil engineering design.[19.9] The canal excavation was notable further in that it was depicted on a very large illustrated sign erected for the information of visitors to the great construction job while it was in progress; this sign is shown in Fig. 19.8. The rising invert level can clearly be seen.

FIG. 19.8 Sign displayed by Ontario Hydro during the construction of the Sir Adam Beck power development at Queenston, near Niagara Falls.

Immediately across the Niagara River is the equally impressive power project of the Power Authority of the State of New York, constructed a few years after the Ontario Hydro project, but similar to it in overall conception. The great powerhouses face each other diagonally across the Niagara River below the rapids that distinguish the Niagara Gorge. The United States station is somewhat upstream from the Canadian; with the interconnecting high-tension lines clearly in view, they typify the joint development of that portion of the international waters that can be diverted from the world-famous falls without destroying their natural beauty. Since the Niagara project is the equivalent of the old and the new United States stations, the capacity of its main water conduits is 83,000 cfs; water is abstracted from a point above the falls about 3.2 km (2 mi) to the west of the Canadian intake. From the intake, the American water is conveyed to its powerhouse in twin concrete conduits, each equivalent in size to six double-track railroad tunnels. These are not pressure conduits, as on the Canadian side, but are reinforced-concrete arched structures within cuts excavated in the upper formations shown on Fig. 19.8. Because of the local topography and the route followed by the conduits, a route generally to the east of the city of Niagara Falls, New York, the depth of the necessary twin rock cuts was as much as 45 m (150 ft) below the original ground surface, and maximum depth in solid rock was 43.5 m (145 ft). Figure 19.9 is a typical view of this impressive example of rock excavation; the total volume of excavation for the complete job of 30 million m³ (39 million yd³) was in keeping with the majestic size of the twin conduit excavation.[19.10] The contrast between the designs of the main conduit for the two adjacent projects is striking to the interested observer; other factors, in addition to strict economics, were probably involved. Reference to the rock heave in the invert will be found in Chap. 43.

19.4 OPEN EXCAVATION IN SOIL

Excavation in soil may vary from work in good, clean, sharp gravel or glacial till to digging in soft, cohesive clay. Between the two extremes lie the many and varied substances classed generally as soils or earth or by some other such generic name, in nearly all of

FIG. 19.9 Rock cut 42 m (140 ft) deep near the outlet of the east conduit of the Niagara power project of the Power Authority of the State of New York, Niagara Falls.

which cohesion is a property of some significance. It can be seen that in considering the mechanics of these materials, as distinct from solid rocks, conditions other than those studied in the ordinary strength-of-materials laboratory must be investigated. It is here that soil mechanics has made such a great contribution to this field of civil engineering; the determination of slope stability, based on laboratory investigations of the shearing strength of the soils to be encountered, is a notable part of soil mechanics theory. All that soil mechanics can do, however, is of little avail without accurate knowledge of the geology of the site to be excavated. The exact interrelation of the various soil strata, and particularly their relation to local groundwater conditions, must be known with certainty if excavation is to proceed as planned and without trouble. There is probably no branch of "soil work" in which soil mechanics and geology must be so closely associated as in the determination of slopes to excavation. That this is no new observation is shown by the comments made by Alexandre Collin more than a century ago.[19.11]

Preliminary Considerations in Open Excavation In preliminary considerations of open excavation work, the bottom grade level will be determined either by factors depending on other parts of the works involved or by general economic considerations. In any event, the effect of the side slopes on the total quantity of material to be excavated, and therefore on the total cost, will be considerable. In ordinary railway cuttings, for example, the volume of material to be excavated to form flat side slopes, nonproductive excavation as it might be called, can be far larger than the volume of material to be moved from that part vertically above the level base of the cut—the material that must productively be moved. It is clear, therefore, that side slopes must be as steep as possible, consistent with safety, in order to minimize the volume of material to be excavated.

In the preliminary as well as in the final determination of such side slopes, geological factors should be given the fullest consideration. In the past, this has not always been done; side slopes used in design have often been based on the past experiences of those concerned with the work. On the basis of practical experience, tables of angles of repose for materials of various kinds have been built up and are in general use today. Although these tables undoubtedly have their use, unquestioning reliance on them is not advisable. Since it is almost impossible to correlate the broad material classification in these tables with the detailed results of test borings, the procedure is without a logical basis and so is the more surprising in view of normal engineering method. It is true that the side slopes of many cuttings stand up quite successfully; it is also true that trouble has been encountered with the sides of many such cuttings and that large quantities of material have probably been unnecessarily removed from the side slopes of many cuttings.

The inevitable use of assumed angles of repose in very preliminary designs may be permitted; but it is suggested that before the completion of any final designs, a full study should be made of the geological structure of the complete prism of material to be removed, not only to discover the geological arrangement of the material but also to investigate the effect of removal of the prism on adjoining geological structure and to determine what soil mechanics investigations are advisable. Sands and gravels are usually unaffected by exposure to the atmosphere. Clays are different, and special attention must always be paid to the effect of exposure to the atmosphere on any clays that are to be uncovered during excavation. Moisture content of clay is a critical factor in such determinations; it can profitably be studied in the laboratory if undisturbed samples are obtained for testing. The old construction "dodge" of covering up exposed steep faces of clay was evolved in order to maintain unchanged the water content of clay; it is a device that will often result in considerable saving in excavation. Because of the mechanical properties of some clays, even the progress of excavation may be influenced by their presence. Experience on the St. Lawrence Seaway and Power Project is a telling reminder of the problems that may be experienced with till.

Methods Used in Open Excavation The foregoing comments on open excavation work in soil deal with points that are so obvious to all who have had any experience in this class of work that they are almost banal. Those, however, who have had experience will appreciate that, although the suggestions made are so obvious, they are often neglected or not even considered. Knowledge of all the strata to be penetrated in excavation work is absolutely essential before work begins, and this is true irrespective of the methods to be used for excavation. On large jobs, the use of automotive scrapers and tractors is now almost universal. So well have these construction tools been developed by their manufacturers that average unit prices for excavation on large North American jobs have changed but little during the last 30 years, despite a threefold increase in the general price index. But as the experience on the St. Lawrence power project demonstrated in no uncertain manner, even the best of equipment will not operate efficiently in material for which it was not designed. Geological study, therefore, is a prerequisite for getting the best out of earth-moving equipment.

On more confined jobs, it is often necessary to limit the extent of excavation, or its effect on neighboring property, by the use of retaining walls of steel sheet piling, an aid to civil engineering construction the use of which has grown steadily during the last three decades. Here, too, knowledge of the exact nature of the materials through which the piling is to be driven is an essential preliminary. The severe vibration set up by the driving of the piles, although usually unimportant in sand and gravel, may produce strange results in clay, especially if the driving is under water. One of the authors has encountered a small pier construction in which steel piles were driven into overconsolidated marine clay which naturally showed up in the preliminary wash boring as "very hard." When disturbed by the pile driving and consequent churning action, the clay changed character completely; it became almost fluid in the neighborhood of the piles and gave a greatly reduced passive pressure as resistance to movement. While the pile-driving operations were in progress, a large dredge was engaged in excavating a channel in the clay. When some distance from the pier, the steel cutterhead of the dredge had to be used to break up the clay sufficiently for it to be pumped; but starting at about 120 m (400 ft) from the site of the pile driving, a change in the nature of the clay became apparent. It became progressively softer as the neighborhood of the piles was approached until finally the steel cutter was stopped and the clay was removed by direct pumping.

The presence of boulders can be a serious impediment to the driving of steel piling; in the interpretation of the results of the preliminary test boring which is always necessary in this branch of work, an appreciation of the local geology can be of great assistance; knowledge of the glacial nature of strata encountered, for example, suggests the possible presence of boulders even if the test borings do not show any. Furthermore, if steel piling is to be driven through the full depth of unconsolidated material at a particular site in order to achieve toehold in the rock below, a study of the local geology will sometimes be the only means whereby a reasonable estimate can be made of the nature of the rock surface and whether or not it will be weathered to such an extent that the piles can penetrate it at least for a little way. In this branch of work, therefore, as in others, not only will an appreciation of the significance of geological features be of service to the civil engineer, but in many instances a direct application of geological information will assist in both the design and the construction of his works.

Examples of Open Excavation Work Instead of illustrating the foregoing remarks with what might be called a "conventional" example, attention will be invited to one phase of an excavation job known to the writers which involved the moving of more soil than was moved in the building of the Panama Canal. Over 230 million m^3 (300 million yd^3) of soil was moved from the bed of Steep Rock Lake in northwestern Ontario before all the valuable iron ore beneath the lake bed was accessible for open-pit mining. In 1943 and 1944 the lake, with an area of about 26 km^2 (10 mi^2), was drained by a major river-diversion project. The natural bed of the lake consisted of varved clays with high sensitivity and a liquidity index always greater than 1; this soil index property was sometimes as high as 1.5. By a vast dredging operation the soil covering the ore bodies was removed easily, but the surrounding slopes had to be finished off in a stable condition. This involved careful cutting with high-pressure water jets to a finished outline that was based on most careful geological studies, studies carried out with simple equipment made on the job and closely associated with continuing laboratory studies of the properties of the soils involved. Basically, the method used was to trim the soil to a slope of 1 in 3 and to limit the vertical height of each slope to 6 m (20 ft). A horizontal berm, its width depending on the shear strength of the soil, was then excavated and graded gently away from the face of the slope to obviate drainage onto the erodible face of the excavated soil. In this way, all the slopes shown in Fig. 19.10 were successfully finished off; vegetation finally completed the stabilizing process. The most striking example of what could be achieved with this very unstable soil, which, due to its high moisture content, ran like pea soup when disturbed, was the forming of a natural dam from the undisturbed material, a dam 48 m

FIG. 19.10 General view of the south end of Steep Rock Lake, Ontario, after the draining of the lake and the start of excavation. The old shoreline may be recognized near the crest of the surrounding hills; all the soil seen was potentially unstable.

(160 ft) high, to retain an undrained part of the lake bed while on the other side excavation proceeded down to one of the ore bodies 60 m (200 ft) below the crest level of this unusual barrier.[19.12]

Excavation at Steep Rock Lake was, admittedly, an unusual case, but it well illustrates what success can be achieved with relatively simple methods, even when dealing with very sensitive soils. Almost at the other extreme was the excavation for one of the largest automobile parking garages in the world. This garage, which lies under Pershing Square in the heart of Los Angeles, is generally unsuspected by visitors to this part of this vast city as they look at the gardens and trees that make the square such a pleasant urban oasis. Beneath the garden, however, is a three-level garage capable of holding 2,000 automobiles, the lowest level about 9.7 m (32 ft) below street level. The garage occupies an area of about 2 ha (5 acres). It was opened in 1952 after a fast construction schedule. Careful preliminary investigation of the subsurface and the necessary soil testing had shown that all excavation would be in clean, fine brown sand, sand and gravel mixed with a little clay, fine gray sand and silt, and a local siltstone commonly known as Puente shale. The area would generally be dry, but some groundwater at a depth of about 6 m (20 ft) was expected in the northern part of the square. Excavation was carried out over the entire site by 2-cubic-yard shovels that were even able to excavate the shale, although with difficulty. (Incidentally, hauling away the 200,000 m^3 (260,000 yd^3) of excavated material was permitted only between the hours of 6 P.M. and 2 A.M.) The relatively limited quantity of groundwater was handled easily by intermittent pumping from a deep sump well at the north end of the site (Fig. 19.11).

Because of the generally dry site and its great area, an ingenious system was developed for foundation construction, a system based on the accurate knowledge of subsurface conditions that had been obtained. Excavation was carried out in two 4.8-m (16-ft) lifts. Con-

FIG. 19.11 Excavation in Pershing Square, Los Angeles, in 1951, for the construction of the parking garage now hidden beneath the central garden.

currently, a large drill rig surrounded the entire site with 1.5-m-diameter holes, spaced at 2.7 m (9 ft) centers, going below final grade level, and belled out at their bottoms for the placing by hand of concrete footings. In these holes, precast concrete columns were then placed, each weighing 9 tonnes. Soil was left in place at its angle of repose up to the tops of these columns while the rest of the excavation was completed and the foundations were placed for the whole garage except for the one bay around its entire perimeter. As the massive concrete structure rose to its finished level, 75 cm (30 in) below finished garden level, it was used to support struts that were then fixed to the individual columns all around the site. With these supports in place, the remaining excavation of the sloping bank could be completed, and the final perimeter bay of the garage constructed, incorporating the precast columns placed when the job began.[19.13]

The size of the two excavation projects just described is far removed from the normal small excavation encountered on ordinary civil engineering construction. The necessity for due attention to geology remains the same, irrespective of the size of the enterprise. Unfortunately, however, although problems with small excavation projects sometimes do finish up in the courts, there are very few examples described in the literature, and the writers must refrain from using too many examples from their own personal experiences. One useful example, however, has been recorded and it is, unfortunately, typical. The boiler house basement for a large building in Massachusetts involved excavation to a depth of 7 m (23 ft). Only one test hole was put down in the area of the boiler house, and only to a depth of 4.3 m (14 ft). Instead of the glacial overburden encountered over the main part of the site, the contractor encountered bedrock at a depth of 4.5 m (15 ft) in this deeper cut. The bedrock was hard schistose sandstone of Pennsylvanian age (the Rhode Island formation). Blasting could not be used, but, fortunately, the contractor was able to remove the rock with a large tractor-hauler ripper, although at greatly increased cost. His claim for extra payment led to the "usual" argument. Although payment was eventually agreed to, all the trouble would probably have been avoided if only site investigation had been more thorough.[19.14]

19.5 SUPPORT FOR EXCAVATION

Reference has already been made to the necessity for supporting the sides of some excavations when side slopes cannot be employed around the edges of the areas excavated. The most extreme cases are naturally the excavations carried out in narrow trenches, such as for sewer and water main installations. Details of this kind of work lie in the field of engineering practice, but it can at least be said that exactly the same attention to subsurface conditions, and to the geology of the strata encountered, is as necessary here as in the largest size of project. Workers' lives are involved and so the safety of such excavations is of paramount importance. Because human safety is involved, it is in some ways even more necessary than usual to have detailed knowledge of the subsurface before laborers are allowed to work in the confined spaces in narrow trenches. Experienced construction men know better than to allow any workers to proceed into unsupported cuts excavated in clay, even though the sides may stand vertically when first excavated. Similarly, in advance of the installation of pumping arrangements, those with experience will know the old but well-tried construction "dodge" of dumping gravel into what might be called "quicksand," a solution the efficacy of which depends on the very nature of the "quick" condition in water-bearing sand.

On the larger scale, however, the necessary support of the sides of excavations—especially in urban areas—must be planned and designed well in advance of the start of construction, based upon the best possible information about subsurface conditions. The ingenious solution used for the Pershing Square excavation in Los Angeles has been mentioned briefly. Another major urban excavation was that for the Carlton Center in Johannesburg, South Africa. Seldom if ever before have four complete blocks in the center of a large, modern city been closed off for a great building complex and excavated to almost 30 m (100 ft) below street level, as was the case in the middle of Johannesburg in the late sixties. A 202.7-m (665-ft)-high office tower and an associated office building are the main features of the 2.4-ha (6-acre) Carlton Center, located in the heart of this South African metropolis. Necessary excavation resulted in a 120.9- by 152.4-m (400- by 500-ft) hole in the city's center, excavation in one corner being planned to go to 29 m (95 ft). Subsurface conditions consist of a complex system of igneous and metamorphic rocks—diabase, dolerite, quartzite, and shale—weathered to depths varying up to about 30.5 m (100 ft) and overlain by a thin veneer of transported soil. The boundary between transported and residual soils can always be distinguished by a "pebble layer" that is an unvarying indicator in the vicinity. Since groundwater was present in the weathered rock, pumping was essential. Water-level gauges were consequently installed in all the streets around the site. Regular movement monitoring was carried out to ensure that no damage accrued to neighboring buildings. All these had been carefully surveyed and observations recorded before excavations started. Support of the excavation—involving a ringbeam of prestressed concrete wide enough to be used as a roadway—makes the engineering features of this project of great interest. Figure 19.12 shows the surrounding reinforced concrete wall and the supported street. Accurate knowledge of all subsurface conditions made possible the safe and efficient carrying out of this immense excavation in so critical a location.[19.15]

The usual method of supporting the sides of excavations, large and small, in soil is to drive around the site, in advance of excavation, either steel sheet piling (the necessary

FIG. 19.12 Excavation for the multiblock Carlton Center, Johannesburg, South Africa, the scale given by the size of the men in right background; the large strut and surrounding beam are of reinforced concrete.

supports being placed as the work progresses) or "king piles" (usually steel H sections) at regular intervals, suitable lagging being placed between adjacent piles as the hole deepens. Such temporary walls can be supported either by horizontal systems of bracing, such as those used at Johannesburg, or by inclined struts founded either on rock or on the central part of the concrete foundation slab, such as those used on the Pershing Square job in Los Angeles. Naturally, contractors have developed variations of these methods but, until relatively recent years, temporary supports for excavations were usually of this general type.

Even with such straightforward systems, troubles could develop because of unexpected ground conditions. An account has been published of one such case, an account which starts out with these words: "When a contractor digs in New York City, he expects to find almost anything!" The job that prompted this remark was the foundation for an extension to the Mount Sinai Hospital in Manhattan, involving excavation over an area of 30 by 60 m (100 by 200 ft). Test borings put down prior to the start of work showed rock but gave no indication of broken rock. When a shovel started excavation in advance of the driving of steel H piles, it encountered rock broken into large chunks. This was then found to be the condition all over the site, the broken rock having apparently come from earlier subway excavation, dumped at this location since the original ground level formed a depression in the vicinity of the hospital. Plans had to be changed and drilled caissons put down. Cable-operated drills were able to penetrate the boulders and chunks of broken rock, whereas the original plan of driving the steel sections would have been impossible. Steel H sections were embedded in the holes sunk by the drills to depths 1.2 m (4 ft) below the surface of bedrock. Excavation then proceeded, but with great difficulty because of the variation from broken rock to ledge rock, steel walers being welded in place as depth increased.[19.16]

Slurry-Trench Method of Support The practice of civil engineering construction has long been distinguished by ingenuity. Two excellent examples are two major new methods of supporting the sides of excavations in soil. Both depend for their success upon an unusually accurate knowledge of subsurface conditions. The first is known as the "slurry-trench" method, or the "Milan method," since it was first developed by an Italian firm in Milan in 1950. Its first major use in North America appears to have been on the second section of the Toronto subway at a location where working space was at a premium and speed of construction was a necessity (Fig. 19.13). Two 60-cm (2-ft) -wide trenches were excavated along the alignments of the two sides of the finished subway box structure. The upper 1.5 m (5 ft) were provided with concrete sides to serve as guides for the special buckets (slightly less than 60 cm or 2 ft wide), which then proceeded to excavate the trenches to a final depth of from 7 to 16 m (24 to 54 ft). The special feature of the method is the use of a bentonite slurry in the trenches, the density of the slurry being such that it will support the exposed soil safely as the excavation proceeds. "Tremie" concrete is then placed in the completed trench, after suitable cages of reinforcing steel have been lowered into place and firmly secured in proper position. As the concrete fills the trench, the bentonite slurry is forced out into specially prepared containers for reuse, or into the

FIG. 19.13 An early North American example of the slurry-trench method for assisting excavation, used in Toronto during the first extension of the Toronto subway; cage of reinforcing steel ready for lowering.

next section of trench ahead. When the concrete has set, a normal reinforced-concrete wall is in place, and excavation can proceed when suitable arrangements have been made for the support of the wall.

Use of bentonite slurry is long-standing practice in oil-drilling work so there was some experience available as a guide, but its application to civil engineering practice naturally called for special skills and new design features. For the Toronto subway job, the two concrete walls were connected by suitable bracing just below ground level as excavation proceeded to the level of the top of the subway box section. After special excavation for this, the concrete roof was placed and connected suitably to the two walls; excavation continued and eventually the concrete invert section was similarly placed, with the central wall, the previously formed trench-concrete walls now serving as the two outer walls of the box section.[19.17]

This is, naturally, an abbreviated description of a complicated construction schedule, but the essentials of the system will be clear. Equally obvious will be the absolute necessity for consistent soil conditions along the slurry trenches so that there can be no possibility of the slurry's escaping. If, for example, a pocket of gravel or an unsuspected lens of gravel is encountered, the slurry can be lost, with serious damage to the trenches. Accurate subsurface investigation is, therefore, a prerequisite and an essential part of the slurry-trench method. (See Fig. 19.14.)

FIG. 19.14 Detail of the slurry-trench method as used in Toronto (Fig. 19.13), showing the trench, slurry filled, and cage of reinforcement.

Methods akin to the bentonite slurry-trench method had been used previously, as by the use of clay slurry in the excavations for some pier foundations in Chicago. It is, however, the density attainable with the bentonite mixture (up to 1,280 kg/m^3 or 80 pcf) that makes the "Milan method" so effective. Not only have concrete walls placed in slurry trenches come into use for unusually difficult foundation excavations, but they have also been used for other purposes. Cutoff walls had to be constructed, for example, below 1,620 m (5,400 ft) of earth embankment during the construction of the Wanapum Dam on the Columbia River, the embankment serving first as a cofferdam and then being incorporated into the main dam structure. Careful advance tests were made before the slurry trenches were constructed, some as deep as 24 m (80 ft) and from 2 to 3 m (7 to 10 ft) wide, the entire project requiring about 500,000 barrels of slurry. A similar use of the system has been made on the Kinzua Dam as will be described in Chap. 25.[19.18]

Tieback Method of Support The second relatively new method consists of providing support for the walls temporarily surrounding an excavation by tying them back into the ground behind that which is actually being retained by the walls. Even this capsule description shows immediately that, once again, accurate knowledge of the subsurface conditions around the site being excavated is absolutely essential for the safe use of the method. The idea is believed to have originated with Andre Coyne, a distinguished French civil engineer; it was pioneered in the United States by Spencer, White and Prentis, foundation contractors of New York who patented the idea of prestressing the tieback anchors. As a comparison of accompanying illustrations will show, the tieback system gives the great advantage of a working site "uncluttered" by the usual bracing; in confined sites this can be a great aid to construction. Safety, however, is dependent upon the performance of the tieback anchors. Figure 19.15 shows a usual arrangement, the inclined ties being well anchored into the adjacent bedrock, the exact location and strength of which must naturally be known. As with other special methods in construction, the basic idea is simple, the application a matter of considerable skill.[19.19]

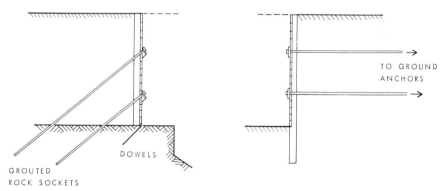

FIG. 19.15 Sketch illustrating the basic idea of the tieback method of anchoring excavation retaining walls.

Local geological conditions may sometimes dictate the use of the tieback method. One such case was the foundation for a new nine-storey building in Syracuse, New York, that had to be built immediately adjacent to a nurses' residence of the New York State Teaching Hospital. Geology of the site was most complex, nine different rock types (dolomites, siltstones, clay-shales and clay) being distinguished in the test borings, some weathered, only the ninth suitable for the foundation. Driving normal supports into this complicated subsurface was impracticable. Drilled-in caissons were therefore used, being supported by tiebacks anchored into the sound bedrock.[19.20] A similar system was used for the excavation for the 4,000-car garage for office employees of the House of Representatives in Washington, covering an area of 81 by 114 m (270 by 380 ft), with an average depth of 12 m (40 ft). Tiebacks were here secured in soil by the use of well-designed earth anchors, the tie rods being from 7 to 16 m (24 to 55 ft) long, inclined at 25 to 35° to the horizontal. Subsurface conditions were known accurately in advance of design.[19.21] (Fig. 19.16)

Unusually difficult geological conditions were encountered during the construction of the foundation for the new Hamilton Plaza building in White Plains, New York, adjacent to a large building on one side with busy streets on the other three sides. The local gneiss was of such poor quality that the upper 7.5 to 9.0 m (25 to 30 ft) could be excavated without blasting. A sharp fold in the gneiss near the center of the site resulted in smooth and micaceous planes dipping into one wall of the excavation at angles varying from 38 to 70° to the horizontal. This really tough construction problem was solved by the use of steel cable tiebacks, each placed at such an angle to the horizontal as to compensate for the local dip of the rock. No less than 1,780 ties were thus used, requiring approximately 96 km (60 mi) of steel cable.[19.22] With the weak rock secured in this way, 160,000 m^3 (210,000 yd^3) of rock was excavated (as well as 38,000 m^3 or 50,000 yd^3 of soil) to depths of from 19 to 22 m (65 to 75 ft). Even these depths were exceeded in the excavation for the foundations of three modern buildings in Los Angeles which went to depths of 34 m (112 ft) in the local soft, dark gray siltstone. All the foundations were successfully completed with the use of tiebacks, 2,700 tiebacks being used for one job alone.[19.23] (Fig. 19.17)

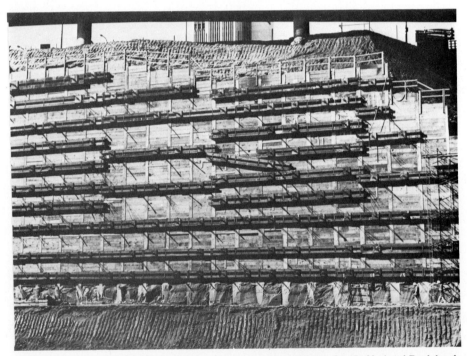

FIG. 19.16 Tiebacks in use for foundation excavation of the Security Pacific National Bank headquarters building, Los Angeles, California, in places over 100 ft deep.

FIG. 19.17 Tiebacks in use in sand for the foundation excavation of the Justice Center, Cleveland, Ohio.

An Example of Methods Combined The unusual subsurface conditions at the site of the World Trade Center in New York, discovered through study of archival records, were described in Chap. 13. They posed a problem of the greatest complexity in view of the size of the area that had to be excavated, the depth to be attained (21.3 m or 70 ft below street level) and the obstructions that would be encountered. The solution was found in the use of a combination of both the new methods that have just been described. A slurry-trench wall was constructed and then supported by tiebacks anchored into the local schist. A further reason for the use of the slurry-concrete walls was the necessity for not interfering with the local groundwater conditions around the site. The continuous, impervious wall provided by the slurry-trench method provided the watertight enclosure that was so essential. All the material excavated from this vast hole in the heart of Manhattan was trucked to an adjacent site in the Hudson River that had been surrounded by a sheet-pile retaining structure. Dumped into the enclosed water, the fill has given a new section of land in this highly valued area, a section estimated to be worth $90 million to the City of New York.

Since the space excavated contained two circular tunnels in which subway trains were operating, stretching from one side to the other and supported during excavation on a steel truss bridge, the entire project was one of extraordinary fascination to all who are interested in unusual foundation construction. But almost all who visit the two great towers today will probably be so fascinated by what they see above ground that they will give no thought to the rock below, nor to all that went into the design and construction of the works below ground upon which the great building now rests so securely (Fig. 19.18).[19.24]

19.6 SINKING OF SHAFTS

One of the most specialized parts of open excavation is the sinking of shafts, often for access to tunnel workings. There is a wealth of experience to draw upon from the mining

FIG. 19.18 Excavation from the foundations of the World Trade Center, New York City, being used to create new land on the waterfront of the Hudson River; see also Fig. 13.8.

industry; indeed, shaft sinking is a construction operation shared by civil and mining engineers, many a major mine shaft being the work of civil engineering contractors, many a mining engineer being consulted about problems with shafts on civil works. There are few enterprises in civil engineering that look so easy and yet are so often fraught with difficulties. One has only to think of the experience of Thomas Edison at Sudbury to be reminded yet again of this fact. Wide employment of shaft sinking is the prerogative of the few, but to watch an experienced shaft-man at work in the planning of his operations is fascinating. The suggestions which follow are second nature to him; his skill is shown in the manner in which he deals with problems and with the equipment which he will assemble for his operations.

The first requirement is, as always, an accurate knowledge of the strata to be penetrated and the location (and possible variations) of the groundwater table. For shallow shafts, well points may be used, but the depth to which this method will be effective is limited. Pumping from any great depth will usually be impracticable, not only because of the amount of water to be pumped, in view of the size of the cone of depression, but also because of the possible effects of the pumping operation upon surrounding property, or at least upon local groundwater conditions. Accordingly, other means have been adopted for the sinking of shafts in water-bearing strata; one of the most common in recent years has been long used in mining operations, the freezing of the ground surrounding the shaft. Suitable conditions for such an application must be definitely known before the project can be considered; that is, there must be regular strata of water-bearing material in which water is present to such an extent that difficulty will be experienced in opening up the excavation in any other way. Once these conditions are determined, suitable-sized pipes are sunk (generally by jetting) all around the area and connected to a refrigerating system which, when operating, gradually freezes solid the ring of material around the area of excavation; this ring acts as a temporary shell for the excavation work.

Freezing has been used as an aid in civil engineering work for many years. There are early records of Russian construction men forming cofferdams in rivers by working in winter periods and allowing ice to form in relatively simple enclosures as the water level was gradually lowered. The first patent for the application of artificial refrigeration in engineering work appears to have been taken out in 1883 in Germany, and the first application of that process to shaft sinking was carried out in that same year for a mine in Schneidlingen, Germany. The first application in North America seems to have been around a shaft for the Chapin Mining Company of Iron Mountain, Michigan. Since those early days, methods have been refined, theories of freezing action in soils have been well developed (notably in the U.S.S.R.), and the number of applications has steadily, but slowly, increased. The use of the freezing process is costly; its success is wholly dependent upon the certainty and uniformity of ground conditions all down the shaft; and since it is temperature-dependent, there is always an element of risk attached to it.

Some of the most remarkable shafts of recent years have been those sunk in the province of Saskatchewan for reaching large deposits of potash. Depths up to 900 m (3,000 ft) have been penetrated in order to reach the potash bed, an unusual problem being the solubility of the potash in water. This solubility makes leakage into any shaft a matter of unusual seriousness in any case in which the potash bed has been reached. Because of the water-bearing Blairmore formation, at least seven of the major shafts in this area were sunk with the aid of freezing. Further work with grouting was necessary in some cases even after a permanent lining had been put in place. These were, however, somewhat specialized operations in what is, itself, a specialized field of work (Fig. 19.19).[19.25]

A good example from the more normal practice of civil engineering was the application of this technique in the excavation of the large shafts used during the construction of a vehicular tunnel under the river Scheldt at Antwerp, Belgium. The shafts were 21 m (70 ft) in diameter and 26 m (87 ft) deep; and because of the construction method adopted, they were entirely free from bracing. They penetrated water-bearing sand with an interbedded turf layer. For the freezing process, 116 holes were bored, one set on a circle of 25-m (86-ft) diameter, and one set on a circle of 23-m (78-ft) diameter; the spacing between the holes was 1.3 m (4.5 ft), and all were sunk to a clay stratum 27 m (90 ft) below ground level. A 15-cm (6-in) pipe was sunk and sealed in each hole, and a 5-cm (2-in) pipe, open at the bottom, was inserted in each of the larger pipes to form the necessary circulating system for the brine solution which was used to lower the temperature of ground and water. It took four months to complete each of the requisite cylinders of ice and frozen ground; they were then maintained in a frozen state until the shaft construction was complete.[19.26]

On a smaller scale, and in an urban environment, was the use of the freezing process in connection with the access shaft for a new sewer tunnel under the East River in New York City. The New York City Public Works Department would not permit interference with the local groundwater condition at the site of the access shaft in Manhattan because of possible settlement of adjoining buildings. The use of compressed air and freezing were the alternatives available. The contractor adopted freezing for the 4.3 m- (14.5-ft) -diameter shaft that he had to sink to a depth of 37 m (123 ft) in order to reach solid rock; the frozen shaft was then lined with concrete, and the freezing was stopped as work continued down for another 60 m (200 ft) in solid rock.[19.27]

Under exceptional circumstances, other more specialized methods can be used in order to "solidify" soil that is giving trouble. Several chemical-injection methods are now available, some of a proprietary character. They have been successfully applied to shaft work but, like all other methods, they depend for their success upon an accurate foreknowledge

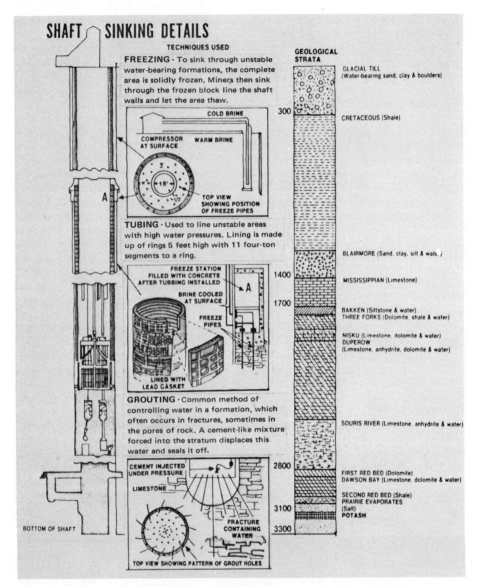

FIG. 19.19 Diagram illustrating methods of shaft sinking at potash mines in Saskatchewan, meaningful photographs of this operation being impossible to obtain.

of the strata to be penetrated. Since these methods require a specialist's knowledge for their successful application, their availability may be noted without further comment here, but reference to them will be found in Chap. 26.

19.7 CONTROL OF GROUNDWATER

The freezing process has also been used to control movements of soil due to groundwater on a few major cases of open excavation. One of the most remarkable, and most successful, of these isolated examples was the control of earth movement, which might have been serious had it developed further, during the construction of the Grand Coulee Dam for the U.S. Bureau of Reclamation. The dam is located on the Columbia River, Washington, and is one of the major dam structures in the world. At the dam site, silt had originally filled the valley of the river to a depth of about 150 m (500 ft), but the river had subsequently worn this down to a depth of between 12 and 15 m (40 and 50 ft). During this erosive process, the river channel swung from side to side of the valley, with the result that slides of the silt were frequent. Excavation in the dried-out river bed encountered the toes of many of these old slides and in this and other ways so disturbed the stability of the remaining silt that many slides occurred during construction.

The silt is a very fine rock flour, containing from 20 to 25 percent colloidal material; when undisturbed, the material stood up well, but as soon as it was moved, it proved unstable on any slope steeper than 1:4, even when relatively dry. Drainage by wells and tunnels assisted in controlling slides, as did also slope correction; but near the center of the east excavation area, the bedrock was intercepted by a narrow gorge, 36 m (120 ft) deeper than the average bedrock elevation, in which silt created an unusually serious slide. Since a 5-cubic-yard shovel could make no headway against the slide, the engineers finally decided to freeze an arch of the material in such a way as to form a solid blockade of the gorge behind which excavation could proceed. (See Fig. 19.20.) This was done with the aid of 377 special freezing points and a circulating ammonia-brine-solution refrigerating system which had a capacity of 72,000 kg (160,000 lb) of ice per day. The frozen dam was about 12 m (40 ft) high, and was located on top of a concrete and timber crib structure 10.5 m (35 ft) high; it was 6 m (20 ft) thick and 30 m (100 ft) long, and cost $30,000. Although freezing started when the slide was moving into the excavation at a rate of 60 cm (2 ft) per hr, the dam was successfully completed and was estimated to have saved its own cost in the excavation which did not have to be done; in addition, several weeks of

FIG. 19.20 Flow of silt in the east excavation area of Grand Coulee Dam on the Columbia River, Washington, stopped by an arched dam of frozen material.

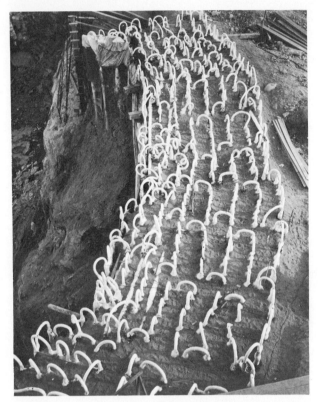

FIG. 19.21 A close-up view of the freezing arrangement used to stop the flow of silt at Grand Coulee Dam (see Fig. 19.20).

very valuable time had been saved in connection with work in the cofferdammed riverbed (Fig. 19.21).[19.28]

Another application of the freezing process was in providing a cutoff for groundwater which would otherwise have reached the open excavation for the Gorge Dam on the Skagit River in the state of Washington, but such applications are so rare that attention must be directed rather to yet another specialized method for the control of water in soil, when the soil is suitable. This is by an application of electrical direct current with suitable electrodes for causing the phenomenon of electroosmosis in clays and silts. First used in Europe, the method was introduced to North American practice by Professor Leo Casagrande. As early as 1807 Reuss had discovered that if an electrical potential is applied to a porous diaphragm, the water will move through capillaries towards the cathode as long as current flows. Helmholtz explained the phenomenon in 1879. Much experimentation had to be carried out before its practical application to soils was possible but this was achieved, one wartime application in Norway giving notable results. Well points are usually used as the negative electrodes and steel rods may be driven into the ground to be drained midway between electrodes.

An early application in North America was in the excavation of the foundation for a power plant at Bay City, Michigan. Excavation was in wet silt, and although a well-point system operated at high vacuum was used successfully for the first stage of excavation, generally in medium sand, excavation for the second stage, generally in sandy silt with seams of sand, proved difficult with the standard well-point installation. Electroosmosis was therefore tried, the well points already installed being used as negative electrodes, the deeper parts of the excavation standing up well.[19.29]

A more extensive and unusual application was in the stabilization of a part of an almost completed earth-fill dam in Ohio. A 24 m (80 ft) high embankment on the West Branch of the Mahoning River is about 3,000 m (10,000 ft) long. With 2,700 m (9,000 ft) completed, movement of the central part of the dam was observed just before the fill was finished. Careful studies by the U.S. Corps of Engineers revealed that high pore pressures in the clay on which the embankment was founded were responsible. In the course of time, they would have dropped, but remedial measures had to be applied to permit the dam's completion. This was achieved by an installation of about 1,000 electrodes, roughly one-third of them cathodes (special well points); old rails served as the anodes. Pore pressures in the clay were reduced and the project completed, although at considerable additional cost for the electroosmosis installation (Fig. 19.22).[19.30]

Both the freezing method and electroosmosis are elaborate to install and relatively costly, so that their use is naturally limited to exceptional cases. When normal methods of water control (detailed in the foregoing chapter) fail, then it may prove economical to use one or other of these special methods, if it is known that soil conditions are suitable

FIG. 19.22 Electrode for the application of electroosmosis being set in place at the West Branch Dam on the Mahoning River, Ohio, for reducing pore water pressures during construction.

19-30 CIVIL ENGINEERING WORKS

for their use and that no unusual geological features will be encountered in the subsurface to interfere with their successful operation. Specialists must naturally be engaged for their successful application, but all civil engineers should be familiar with their potential in case they should ever encounter one of the special problems to which these methods can usefully be applied.

19.8 FILL FOR EMBANKMENTS

Some of the most notable advances in soil mechanics have been in the way in which soil, once removed from its natural location, can be used as an "engineered material" in construction. By laboratory study of adequate samples, the compaction characteristics of soils can be determined, maximum densities at optimum moisture content now being a regular requirement for the placing of all fill material other than the smallest amounts. (Here it must be noted, with regret, that another of the semantic differences between the practice in geology and in civil engineering is in the use of the words *compaction* and *consolidation*. They are here used in the engineering sense, i.e., *compaction* is the act of compacting soil with mechanical devices at a specified moisture content in order to achieve maximum density; correspondingly, *consolidation* is the process by which a fine-grained soil will compress under any applied load.)

Investigation of soil to be used as fill will follow standard procedures as described for site investigation; sampling will, correspondingly, be carried out in accordance with standard methods, varied as may be necessary according to the geology of the soils. The old problem of bulking has virtually disappeared in the use of soil as fill, although it is still of significance when rock has to be used, fragmentation of the naturally occurring compact rock mass leading to an increase in volume when it is placed as fill. Correspondingly, laboratory tests will indicate the strength properties of the remolded soil so that accurate designs can be prepared for the dimensions of all fill structures. Settlement of completed embankments is also becoming a thing of the past; if the fundamentals of soil compaction are properly applied, based upon adequate soil mechanics tests on the soils to be handled, there is today no excuse for any appreciable settlement. The state of California, for example, without fear of trouble, regularly places finished pavement on top of highway fills up to 27 m (90 ft) in height, shortly after completion of the fill. Soil mechanics techniques are a regular part of its engineering control, as they now are in all highway organizations (Fig. 19.23).

FIG. 19.23 Fill being compacted during construction of a modern highway near Windsor, Ontario.

Unfortunately, the same cannot be said with regard to embankment foundations. Here soil mechanics techniques are not yet widely employed, even though embankment foundations are also susceptible to analysis, and so to control, on the basis of proper preliminary investigations, including thorough test drilling over all the area to be affected by the load from the embankment. If the weight of material contained in even an average-sized embankment is calculated and reduced to the form of a load per unit area of original ground surface, the result will be found to be surprisingly high. Although it decreases toward the sides of the fill, this loading is directly comparable to that induced by concentrated structures, and the stress distribution in the foundation strata must be investigated accordingly. Progressive settlement will inevitably take place while the foundation beds take up their new loading, unless the fill stands on solid rock. If the strata are not uniformly strong, failure of one bed can occur, and the fill may collapse. The investigation of fill stability thus becomes a part of the study of soil mechanics; both the nature of the fill and foundation materials and the mechanics of the combined structure have considerable bearing upon the success of the fill construction.

Stability investigations can be made only when accurate information on the local strata is available; thus, the need for preliminary exploratory work is again made clear. So many fill failures, great or small, will be known to most engineers that space will not be taken to mention even one here; instances such as those in which a new fill completely disappears below the original ground tell their own story to the engineer whose appreciation of geology is a constant reminder of what may lie below a seemingly solid surface. It may be worth noting, however, that one of the many soil problems encountered in the Pittsburgh area is the stability of the dump piles of excavated soil from new highway construction and similar work. Excavated material often has to be dumped in one of the many ravines that make this part of Pennsylvania scenically so attractive. This course has to be followed because of the limited availability of the level ground normally used for this purpose. The slides that sometimes occur in these "fills of excavation" are attributable to the steep slopes developed and to the apparent stability of the natural slopes of the ravines that can obscure, on occasion, the necessity for giving the same attention to slope stability here as on level ground. Unfortunate consequences of placing fill have also been experienced on the shore of Lake Ontario to the east of Toronto. Here the majestic Scarborough Bluffs, composed of till, are being slowly eroded, but erosion was found to be accelerated when fill material was dumped at their foot in a valiant, but vain, attempt to restrict erosion!

19.9 EXCAVATION OF ROCK

The problems that may be encountered in the excavation of solid rock necessitate preliminary determination of the structural arrangements of rock strata, the nature of rocks to be encountered, the possible presence of water, and the resistance of exposed rock to weathering influences.

The structural arrangement of rock strata will affect considerably the side slopes to be adopted in design, since if the general dip is toward one side of the cutting, there will be a tendency for shatter rock to fall off the ends of successive layers unless it is trimmed well back; a reverse effect will naturally be found on the opposite side of the cutting. Structural arrangement will also affect underground water problems; the arrangement of rock strata will determine the general lines of water movement, and in connection with the porosity or impermeability of beds, the levels at which water may be encountered. Angles of dip and other special structural features such as fault planes will obviously have a considerable influence on excavation methods.

Of great importance is the *nature of the rock* to be excavated. By "nature" is meant not merely the general classification of the rock but a vivid conception of the actual properties of the particular type of rock to be encountered all over the area to be excavated. The value of general classification is by no means unimportant—indeed, this is a first essential—but each class of rock can vary so considerably that, for civil engineering purposes, it must be further described by some indication of its physical properties. Thus, sandstone can vary from a hard and compact rock to material that is little better than well-compacted sand; granite can prove to be one of the hardest of all rocks, and yet it may be found in such a state of decomposition that exposure to the atmosphere will lead to its immediate disintegration. All preliminary geological information must therefore be checked thoroughly, and the general description of rock types, obtained from borings, must be considered in relation to the physical state of the rock samples so found. Similarly, it is essential to know with some degree of accuracy, before work begins, how the rock to be encountered will stand up after long exposure; this is necessary in order to obviate possible future accidents and to keep maintenance charges to a minimum.

If due allowance is made for exceptional variations in physical properties, the following notes will serve as a general guide. The stability and permanence of igneous and metamorphic rocks when exposed to the atmosphere can generally be relied upon unless the rocks are badly weathered. If they are badly fissured, jointed, or shattered by local faulting, the greatest caution must be observed in finishing off slopes. Caution is especially required for work done during severe winter weather, as in Canada, where frost action may temporarily slow local disintegration. Sedimentary rocks must be considered with great caution and judgment in such work. The presence of clay in any form, even as a shale which may be exceedingly hard when first exposed, must be regarded with suspicion, particularly if the bedding planes are inclined to the horizontal at any appreciable degree. The action of the atmosphere may soon reduce this material to an unstable state, with consequent extra trouble and expense. When clay is known to be present, ample allowance should be given in the design of side slopes, and the economics of protecting the exposed faces with light retaining walls should be investigated. Sandstone and limestone, if in a firm and solid state, will stand with vertical or almost vertical faces; but since both types of rock vary from a sound, solid state to material that can be crumbled in the hand, every case must be considered on its own merits.

Adequate *drainage facilities* are naturally an important feature of all rock-excavation work. They will be provided by one of the standard methods available in civil engineering practice, but geological considerations may affect them to some degree. For example, in the construction of a new railway in El Salvador and Guatemala (by International Railways of South America), many cuttings were made through a conglomerate called *talpetate*. The side slopes in all these cuttings had to be left as steep as possible in order to carry off the torrential tropical rains before they led to erosion of the conglomerate.[19.31] Similarly, drainage in all types of shale must be arranged to remove surface water as quickly and surely as possible in order to minimize its actions on the rock. Limestone may also be affected by excessive contact with water.

The usual method of drilling and blasting may not always be suitable when the exact *condition of the rock* to be moved is known, as instanced by work carried out in preparing the foundation bed for the Prettyboy Dam in Maryland. A similar case occurred during the construction of the main spillway for the Fort Peck Dam in Montana, a dam which is founded on local Bearpaw shale. This is a relatively soft rock, which can be excavated without the aid of blasting. To excavate the 150,000 m³ (200,000 yd³) taken out, three different types of cutting machine were used: an auger which drilled holes 1.5 m (5 ft) in diameter and two special adaptations of electrically operated coal-cutting machines, one used for cutting vertical slots between holes and the other for horizontal undercutting (Fig. 19.24). The Bearpaw shale disintegrates when exposed to the atmosphere, and for

FIG. 19.24 An early application of machinery to rock excavation; coal-cutting saw making horizontal cut in shale in the excavation for training wall at spillway for the Fort Peck Dam, Montana.

this reason a 1.2-m (4-ft) cover layer was left in place over the excavated area at the start of excavation; the finally finished surface was sprayed with a bituminous paint against which the concrete was immediately placed.[19.32] Not only does this rock disintegrate when exposed, but it exhibits a slow rebound after excavation, owing to the release of internal stresses previously resisted by the superincumbent rock now removed. This singularly unfortunate feature will be dealt with in connection with the founding of dams on shale such as that of the Bearpaw formation.

Another example of interest was the excavation for the powerhouse substructure at the Wheeler Dam on the Tennessee River in the United States. Extensive preliminary exploration work disclosed that the rock at the site consists of an "alternating series of nearly horizontal layers of pure limestone and a cherty siliceous rock, the latter occurring in much thicker strata than the former." More than 380,000 m³ (500,000 yd³) of rock had to be removed to depths varying to 17 m (57 ft) below the existing stream bed, and owing to the nature of the rock to be encountered, it was deemed advisable to avoid any blasting close to the faces of the deep cuts in which the power units were to be placed. A system of close-line drilling was therefore developed and carried out with wagon drills all around the site; the rock so enclosed was drilled and blasted by the usual methods, except that no blasting was done within 9 m (30 ft) of the line-drilled faces, since here the rock broke away cleanly (Fig. 19.25). The undisturbed face of the rock was of great importance in the structural design of the powerhouse, and the elimination of overbreak made the method economical.[19.33] Many other examples could be given, but the purpose of all would be but to confirm the importance of geological features in rock excavation.

Methods of Rock Excavation In view of the basically elementary character of rock excavation as a construction operation, it is not surprising that no startling advances in technique have been made in the last few decades. On the other hand, meticulous attention to detail—coupled with steady advances in the equipment used for drilling and mucking, in the procedures used in drilling, and in the nature of explosives used in blasting—have greatly improved the efficiency of rock removal. In Sweden, rock-excavation

19-34 CIVIL ENGINEERING WORKS

FIG. 19.25 Line-drilled excavation for turbine pits in the powerhouse area of the foundations for the Wheeler Dam, Tennessee River, Tennessee.

practice (as well as tunneling) has been outstanding, especially in the adoption of small, mobile, one-man drilling rigs. Correspondingly, the wide adoption of removable bits and the improvement in the wearing properties of the bits have led to great advances in the handling of drill steel and therefore in the operation of drilling in general. Swedish rock experts have not hesitated to experiment with innovations in blasting techniques when conditions warranted it; one unusual example is the blasting of overburden and rock at the same time.

A new sea-level canal was constructed to cut off a sharp bend in the Motala River between Bråviken Fjord and the port of Norrkoping, some 145 km (90 mi) southwest of Stockholm. The Lindo Canal has a length of 6.5 km (4 mi); the first section is 3.2 km (2 mi) long. Its depth is 9 m (31 ft) and its bottom width 54 m (180 ft); thus it is wider than some of the major ship canals of the world; 5,000-ton vessels can pass each other while under way in the channel. The total project involved the removal of about 3.8 million m^3 (5 million yd^3) of material, most of it by dredging; 20,000 m^3 (26,000 yd^3) was rock. At the entrance to the canal, glacial till with overlying clay occurred on top of the local bedrock to depths of 9 m (30 ft); the clay was saturated and thus awkward to remove, although removal would normally be the procedure before drilling the rock to be blasted. In order to obviate this difficult job, a technique was developed for drilling quickly through the overburden.

With this technique, special drilling rigs with 16-m (52-ft) feeds were adapted for driving a casing pipe simultaneously with sinking drill steel through the overburden, to be ready for drilling as soon as rock was encountered. Drill steel was withdrawn when the hole had been drilled to depth, and the casing pipe was left in place. Plastic pipe was then set in place, with the aid of a suitable collar, over the drilled hole, and the casing pipe removed. Loading of the holes was carried out by compressed air through the plastic tubing; as many as 40 to 50 cartridges were loaded in one hole in this way and tamped into place. One major blast carried out in this way used 12,000 detonators and 55,000 kg (123,000 lb) of dynamite in 4,750 holes; the blast moved 27,500 m³ (36,000 yd³) of rock and 76,000 m³ (96,000 yd³) of overlying clay and till (Fig. 19.26). The high ratio of dynamite to rock moved was due not only to the extra effort required to move the overburden but also to the necessity of fragmenting the rock to a finer degree than usual to facilitate its removal by dredging.[19.34]

Details of the blasting technique have been mentioned in this example because of its unusual character; since these details are normally in the province of the expert, the geologist or civil engineer will not usually be directly concerned with them. But as in so many other branches of work herein described, even the best of blasting techniques will be of only partial avail if the geology of the work site has not previously been studied so that it is known with certainty. Especially is this so when blasting has to be carried out in proximity to existing structures. In the past, the difficult operation has generally been left to the expert. Many civil engineers must have had the joy of watching a really expert "powder man" carefully set his charges for some hazardous excavation work, such as the removal of rock to provide room for a new turbine installation inside a powerhouse containing operating machinery. But the innate art of such experts should be capable of translation into measured terms; this is gradually being achieved.

Knowledge in this field was extended by some notable tests conducted in an area to be flooded by the St. Lawrence Seaway and Power Project. In this series of comparative tests, observations were made of the effects of blasts set off at varying distances from abandoned houses founded on different materials.[19.35] As experience such as this is accumulated, vibrations due to blasting will gradually cease to be the hazard they can now sometimes

FIG. 19.26. "Double blasting" of rock and overburden on the Motala River canal, Sweden; this explosion was the result of detonating 551 tons of dynamite in 4,750 drill holes.

19-36 CIVIL ENGINEERING WORKS

prove to be when adequate precautions have not been taken before blasting. Such precautions should include noting all preexisting damage to buildings. Cracking, especially, should be noted; if carefully recorded, it will often be found to coincide exactly with the damage reported after blasting has taken place. One of the authors knows of a large rock-excavation job which had to be carried out in the vicinity of an important hotel. At the end of the job, the "usual" claim for damages caused by blasting was received; the claim, however, was not pressed when it was divulged that the contractor had rented a room in the hotel where he had maintained a vibration record throughout blasting operations; this record showed that the vibrations caused by the blasting were not so serious as those caused by heavy traffic.

The records of modern civil engineering contain so many fine examples of rock excavation illustrating how accurate knowledge of the geology involved facilitated the work that it has been difficult to select the few examples that space permits. Brief mention must be made, however, of one of the most unusual structures (if so it may be called) recently erected, the radar and radio telescope built at Arecibo, Puerto Rico (Fig. 19.27). A reflector of 261-m (870-ft) radius had to be constructed to a high degree of accuracy, and a movable line feed suspended above it from three concrete towers, one 105 m (350 ft) high. A geological study of possible sites led to the selection of a site in coral limestone country, characterized by large sinkholes. Weathering is responsible for unusually large sinkholes in this particular type of limestone, so that the amount of excavation was minimized by utilization of a natural depression. Despite the advantages of this natural site, the contractor still had to remove 210,000 m³ (270,000 yd³) of rock, generally by blasting except for final trimming, and to place 150,000 m³ (200,000 yd³) of compacted fill in order to shape up the final surface to conform as closely as possible with the spherical reflector.[19.36]

In great contrast, but showing what can be achieved with modern methods of rock excavation even in such tough rock as granitic gneiss interlaced with mica, feldspar, and

FIG. 19.27 The Arecibo Observatory, Puerto Rico, part of the National Astronomy and Ionosphere Center, operated by Cornell University under contract with the National Science Foundation, the big "dish" of the telescope requiring special excavation in a natural depression.

FIG. 19.28 The Beaucatcher Tunnel, North Carolina, showing its proximity to the deep cut here required for Interstate Highway 240, which is being completed in stages.

quartz, was the excavation of 1.6 million m^3 (2.1 million yd^3) for Interstate Highway 240 through Beaucatcher Mountain, North Carolina (Fig. 19.28). There were 499 structures within 600 m (2,000 ft) of the cut area, including a 65-year-old water reservoir, while Beaucatcher Tunnel (on U.S. Route 70) lies only 48 to 69 m (160 to 230 ft) from the big cut. On the basis of test shots, consultants controlled the blasting strategy. Slopes were presplit with 7.5-cm (3-in) holes on 75-cm (30-in) centers; 11.2-cm (4.5-in) holes were used for production blasting, sunk 7.5 to 9.0 m (25 to 30 ft) on a 2.4 by 2.4-m (8 by 8-ft) pattern. The job was successfully completed.[19.37]

19.10 QUARRYING IN CIVIL ENGINEERING

Building stone used in civil engineering will usually be obtained from an established quarry, and so it is not often that large-scale quarrying operations will be necessary in the course of normal civil engineering practice. The two most common types of work requiring special supplies of quarried rock are the construction of rock-fill dams and the use of rock for the construction of rock embankments or special structures such as mound breakwaters. If the quantity of rock required for projects of this or any other kind is appreciable, the civil engineer will be well-advised to obtain the services of a quarry expert. On some projects, however, civil engineers may be called upon to open up small quarries that do not warrant the employment of an expert; an appreciation of geology can then be of good service.

19-38   CIVIL ENGINEERING WORKS

After the site of the quarry has been stripped, the exposed rock surfaces should be studied carefully; special attention should be paid to the dip and strike of the strata, to the presence of any unusual features such as folds, and, particularly, to the jointing of the rock. On the strength of the information so obtained, the engineer will be able to open up a working face in the most advisable way, taking advantage of the direction of dip, strike, and jointing to facilitate both the blasting and the removal of rock; jointing is of particular importance. Methods of operation will depend on the amount of rock to be moved, the rate at which it is required, and the type of rock to be quarried. If a large output is not required, it may be desirable to utilize for drilling the regular job hand drills and jackhammers, in which case a series of low faces and narrow benches will probably prove most suitable. If the job is a large one, however, consideration should be given to the possibility of using larger drilling equipment, high faces, and deep snake holes which can be sprung before being finally shot and thus lead to economy in operation.

The location and depth of drill holes and the necessary kind and amount of explosive to be used are matters that will have to be determined accurately for every new quarry face opened up. The problems involved are not dissimilar from those involved in the normal excavation of rock in grading work. As the U.S. Bureau of Public Roads has carried out field studies in connection with highway grading, some of the conclusions reached may usefully be cited, even though they were obtained some years ago. A study of 71 different excavation projects showed that over half were poorly blasted and that only 15 were classed as "good." The consequent average increase in operating time for shovels was 42 percent; and as the volume of rock handled was less than the dipper capacity, the average overall efficiency was only 50 percent.[19.38] It is particularly to be noted that as rock drilling is an expensive operation (1 ft of hole is roughly equivalent to about 1 lb of dynamite in North America), economic practice will tend to minimize drilling.

Only infrequently do civil engineering projects call for quarrying on a major scale, apart from the construction of rock-fill dams. One job that did need a lot of rock was the construction of the Hiwassee Dam of the TVA. All the aggregate, including even the sand, necessary for this great project had to be obtained from a specially opened quarry, as the rock from excavation was unsuitable. Good graywacke rock was found at a suitable location; in all, 1.3 million m³ (1.7 million yd³) of rock was quarried. Interesting experiments were carried out in order to check the use of 22.5-cm (9-in) drill holes as compared with 15-cm (6-in) holes. The rate of drilling the two sizes was found to be about the same, although the 22.5-cm (9-in) holes filled with explosive produced twice as much fragmented rock per foot of hole as did the 15-cm (6-in) holes.[19.39] Far more extensive was the rock-quarrying operation called for by the construction of the new rock-fill embankment across Great Salt Lake to replace the old trestle of the Southern Pacific Railroad. What was described at the time as "the biggest controlled explosion in history" involved the use of 800,000 kg (1.79 million lb) of explosive in one blast; this resulted in over 1.5 million m³ (2 million yd³) of suitably fragmented rock (Fig. 19.29). Location was Promontory Point on a peninsula extending into the lake from the north. Total quantity of rock required for the complete job was 10 million m³ (13 million yd³), all of which was a hard quartzite.[19.40]

19.11 CONCLUSION

Some open-excavation work is called for on almost every civil engineering project; it is small wonder therefore that practice in this field is so diverse and so interesting despite the overall simplicity of the basic operation. Improvement of well-accepted techniques may be expected to continue; new features will be added in matters of detail. There are

FIG. 19.29 Blasting 1.5 million m³ (2 million yd³) of quartzite in the quarry supplying fill for the rock-fill embankment of the Southern Pacific Railroad across Great Salt Lake, Utah.

signs that some unusual methods may be available that can successfully be applied to excavation work. In rock excavation, for example, experiments in the fusion cutting of rock (and concrete) have been carried out. First tried in mining practice for the sinking of 15.2-cm (6-in) -diameter blast holes in tough taconite ore, the method has been commercially applied to the cutting of holes through concrete, always a delicate operation in existing structures. Obviously costly at the present time, the method presents possibilities for special rock-excavation work that may one day prove worthy of exploitation.

Finally, there is already looming on the horizon of civil engineering practice the possible use of atomic explosions as an aid to major excavation. Experiments conducted by the U.S. Atomic Energy Commission and carefully observed and recorded by engineering geologists of the U.S. Geological Survey at the test site adjacent to Yucca Flats, Nevada, have provided vivid evidence of the potentialities of this peaceful use of atomic energy. The bedrock in which the Nevada tests were conducted is a volcanic tuff of three degrees of toughness. In the later blasts, exploded in underground chambers, "solid" rock 330 m (1,100 ft) away from the blast was moved as much as 30 cm (1 ft). After the largest explosion (up to 1959), a 1.5-m (5-ft) fault movement of wide extent was observed at the surface. In many cases the flow of groundwater was stopped by the blasts, which disrupted the strata and sealed joints in the tuff. Rock surrounding the blasts was brecciated all around, in one case for a distance of 24 m (80 ft) from the center of the blast. And one year after the explosions had taken place, it was found that one-half the heat generated was still in the ground around the blast center.

These few facts, selected from the mass of information now available, at once point to the possibilities for the application of this extraordinary explosive power in civil work but serve also as a reminder that any consideration of the peaceful uses of atomic energy raises questions that go far beyond the normal concerns of civil engineering economics. Far more significant is the fact that, despite the steady worldwide rise of prices in practically every field of activity, the cost of excavation on civil engineering projects is still the "best bargain in construction." (Fig. 19.30)

FIG. 19.30 The pool of Gibeon, near El-Jîb, Jordan, part of the waterworks carried out by the Gibeonites in the eighth century B.C., the shaft being 25 m (82 ft) deep, a salutary reminder of the long tradition of open excavation work.

19.12 REFERENCES

19.1 R. F. Legget and W. R. Schriever, "Site Investigation for Canada's First Underground Railway," *Civil Engineering and Public Works Review*, **55**:73–80 (1960).

19.2 R. F. Legget, *Canals of Canada*, Douglas, David and Charles, Vancouver, 1976.

19.3 "St. Lawrence Seaway and Power Project," (News Reports), *Engineering News-Record*, **155**:31 (13 Oct. 1955), **158**:32 (18 April 1957), **158**:31 (13 June 1957), **160**:25 (15 May 1958).

19.4 P. MacClintock, *Glacial Geology of the St. Lawrence Seaway and Power Project*, New York State Museum and Science Service, Albany, 1958.

19.5 P. MacClintock and D. P. Stewart, *Pleistocene Geology of The St. Lawrence Lowland*, New York State Museum and Science Service Bulletin no. 394, Albany, 1965.

19.6 "Welland Ship Canal IV," *Engineering*, **128**:192 (1929).

19.7 A. B. Cleaves, "Engineering Geology Characteristics of Basal Till: St. Lawrence Seaway Project," *Case History Series*, no. 4, Geological Society of America, Boulder, Colo., 1963, p. 51.

19.8 R. T. McCallum, "The Opening-out of Cofton Tunnel, London, Midland and Scottish Railway," *Minutes of Proceedings of the Institution of Civil Engineers*, **231**:161 (1931).

19.9 "On Schedule at Niagara," *Engineering News-Record*, **151**:33–38 (10 Sept. 1953).

19.10 H. P. Cerutti, "Twin Conduits for Niagara," *Civil Engineering*, **30**:50–53 (July 1960).

19.11 A. Collin, *Landslides in Clay (1846)*, W. R. Schriever (trans.), University of Toronto Press, Toronto, 1956.

19.12 R. F. Legget, "Soil Engineering at Steep Rock Iron Mines, Ontario, Canada," *Proceedings of the Institution of Civil Engineers*, **11**:169–188 (1958).

19.13 C. A. McMahon, "Los Angeles Constructs 2,000 Car Underground Garage," *Civil Engineering*, **21**:689 (1951).

19.14 O. C. Farquhar, "The Rhode Island Formation as 'Rock Excavation,'" in O. C. Farquhar (ed.), *Economic Geology in Massachusetts*, University of Massachusetts, Amherst, 1967, p. 11.

19.15 R. A. Heydenrych and B. Isaacs, "The Excavation and Stabilising of the Carlton Centre Basement," unpublished paper, 1967; see also "Record Slip Formed Core Rises from Tricky Excavation," *Engineering News-Record*, **185**:22 (3 December 1970).

19.16 A. DiGiacinto, "Rx for a Rock-Choked Hospital Excavation," *Engineering News-Record*, **165**:46–50 (6 October 1960).

19.17 "Toronto's Cover-then-Cut Subway," *Engineering News-Record*, **164**:37–39 (3 March 1960).

19.18 "Bentonite Slurry Stabilizes Trench, Keeps Groundwater Out," *Engineering News-Record*, **164**:42–46 (11 February 1960).

19.19 "Tiebacks Remove Clutter in Excavation," *Engineering News-Record*, **166**:34–35 (8 June 1961).

19.20 R. E. White, "Pretest Tiebacks and Drilled-in Caissons," *Civil Engineering*, **33**:36–38 (April 1963).

19.21 W. S. Booth, "Tie-Backs in Soil for Unobstructed Deep Excavation," *Civil Engineering*, **36**:46–49 (September 1966).

19.22 "Deep Excavation in Bad Geology Secured by 60 Miles of Tiebacks," *Engineering News-Record*, **191**:31 (11 January 1973).

19.23 J. C. Nelson, "Earth Tiebacks Support Excavation 112-ft Deep," *Civil Engineering*, **43**:40–44 (November 1973).

19.24 M. S. Kapp, "Slurry-Trench Construction for Basement Wall of World Trade Center," *Civil Engineering*, **39**:36–40 (April 1969).

19.25 W. J. S. Ostrowski, "Design Aspects of Ground Consolidation by the Freezing Methods for Shaft Sinking in Saskatchewan," *Bulletin of the Canadian Institute of Mining and Metallurgy*, **60**:1145–1153 (October 1967).

19.26 S. A. Thoresen, "Shield-Driven Tunnels near Completion under the Schelde at Antwerp," *Engineering News-Record*, **110**:827–832 (1933).

19.27 "Deep Freeze to Keep Shaft in the Dry," *Engineering News-Record*, **163**:25 (3 September 1959).

19.28 G. Gordon, "Arch Dam of Ice Stops Slide," *Engineering News-Record*, **118**:211 (1937).

19.29 "Electricity Dries Out Wet Silt," *Engineering News-Record*, **150**:36–39 (30 April 1953).

19.30 "Electro-Osmosis Stabilizes Earth Dam's Tricky Foundation Clay," *Engineering News-Record*, **176**:36–45 (23 June 1966).

19.31 "A New Interoceanic Railway in Central America," *Engineering News-Record*, **98**:474–476 (1927).

19.32 "Tough Shale Bored and Sawed for Fort Peck Spillway," *Engineering News-Record*, **116**:37–39 (1936).

19.33 "Wheeler Dam Construction Enters Final Year," *Engineering News-Record*, **115**:259–261 (1935).

19.34 S. Brannfors, "Blasting Without Removing the Overburden," *Civil Engineering*, **29**:780–781 (1959).

19.35 T. D. Northwood and A. T. Edwards, "Experimental Blasting Studies on Structures," *The Engineer,* **210**:538–546 (1960).

19.36 T. C. Kavanagh, "1,000-Ft. Telescope Takes Shape in Mountain Hollow," *Engineering News-Record,* **170**:22–26 (10 January 1963).

19.37 "Artful Blasting in New Cut Protects Old Tunnel Nearby," *Engineering News-Record,* **201**:24 (28 September 1978).

19.38 A. P. Anderson, "Some Studies of Drilling and Blasting in Highway Grading," *Public Roads,* **12**:293–302 (1932).

19.39 C. E. Blee, "Drill Hole Size Tests at Hiwassee Dam," *Engineering News-Record,* **123**:643 (1939).

19.40 "Big Controlled Blast Makes Molehill out of Mountain," *Engineering News-Record,* **159**:28–29 (15 August 1957).

Suggestions for Further Reading

Excavation in soil and rock is so widespread an operation in construction and one with such a long history, that it is not surprising to find that most relevant publications deal with specialized methods adopted to deal with unusual problems. The *Excavation Handbook* by H. K. Church is an exception; this 976-page McGraw-Hill volume of 1980 is a useful review of all major aspects of the vast subject. *Blasting Operations* by C. Hemphill is another McGraw-Hill book which in 272 pages deals with one aspect of excavation in rock.

Problems of material classification continue to plague contractual settlements, and so attention may be directed to studies of this difficult matter by David L. Royster of the Tennessee Department of Transportation. One report on this work was made to the 55th annual meeting of the Transportation Research Board (Washington, D.C.) in January 1976; a later paper was published in vol. 15 of *Bulletin of the Association of Engineering Geologists* for 1978 (pp. 341–354) entitled "Excavation Characteristics: Designations for Materials Identified in Field Investigations."

The freezing process in excavation has now been the subject of an international conference, held in Bochum in 1978. The *Proceedings* were published in 1979 by Elsevier of New York as a 558-page volume edited by H. L. Jessberger. Regular meetings are held to discuss the slurry trench method, resulting in useful publications; the subject is dealt with at length in *Slurry Walls* by P. P. Xanthakos, a 704-page McGraw-Hill volume of 1979. And *Tiebacks in Foundation Engineering and Construction* are dealt with by H. Schnabel in a 176-page McGraw-Hill book of 1982.

chapter 20

TUNNELS

20.1 Historical Note / 20-2
20.2 A General Note / 20-2
20.3 Preliminary Work / 20-4
20.4 Tunnels for New York / 20-7
20.5 Underwater Tunnels / 20-14
20.6 Pressure Tunnels / 20-18
20.7 Tunnel Shapes and Linings / 20-22
20.8 Overbreak / 20-24
20.9 Construction Methods / 20-26
20.10 Grouting / 20-31
20.11 Special Geological Problems / 20-32
20.12 Construction Records / 20-39
20.13 Tunnels under Boston / 20-44
20.14 Conclusion / 20-47
20.15 References / 20-50

Tunneling must have been one of the earliest construction activities of man. It may be that natural tunnels and other results of water action such as "swallow holes," frequently found in limestone and similar formations, first suggested to early man the idea of an artificial passage through rock. The rock or cliff dweller may have been an example to other human beings in distant ages; but however came the inspiration, excavated underground dwellings and temples, the first tunnels, are to be found in many of the ancient civilizations that have been investigated in recent years. Excavated passages beneath the earth were early utilized for purposes other than residence. One of the first of the more general utilitarian purposes for which tunnels were specially constructed was for drainage, often in connection with mining and quarrying work. Water supply and road construction also necessitated the digging of tunnels at an early date in human history.

Today tunnels are used for the same purposes. They facilitate transportation and are often used in the generation of water power as well as in the provision of water supply. Exceptional uses are to be found, but tunnels generally can still be classed as they were 2,000 years ago—either as aqueducts or as viaducts. It is indeed strange to reflect that this ancient branch of construction has changed but little in achievement, despite great advances in method throughout the centuries. A tunnel is and can be only a tunnel, having a certain length and a certain cross section. The long Alpine tunnels are indeed magnificent achievements, but so also are many of the long tunnels of the pre-Christian era, especially when the construction methods then in use are considered.

20.1 HISTORICAL NOTE

Brief mention of some ancient feats of tunneling may therefore rightly lay claim to a place in even the most cursory review of the art. Although geology as a separate branch of science was not recognized in these early years, brief reference here to a few of these tunnels is instructive. Examples of tunnels of some sort are to be found in almost all ancient civilizations, notably in Egypt where some of the rock-cut galleries of the early Egyptian kings are over 225 m (750 ft) long. In Malta one may see underground temples and gathering places, at least 5,000 years old, hewn out of solid sandstone with flints which must have been brought to the island from the mainland. These early tunnels were generally built through the softer types of solid rock, but some were excavated through unconsolidated materials and so required immediate lining for stability. Another notable example was the tunnel under the river Euphrates, probably the first submarine tunnel of which any record exists. It was 3.6 m (12 ft) wide and 4.5 m (15 ft) high and was built in the dry, since the river was temporarily diverted.

The Romans were preeminent in early tunnel construction because of their achievements and because of the improvements in method that they effected. They appear to have introduced the use of fire into tunnel construction, utilizing the principle (known to others earlier) that a heated rock, if suddenly cooled, will crack to some extent and so make excavation easier. They also most probably employed vinegar instead of water as a cooling agent when working in limestone and similar rocks, utilizing the acid nature of the vinegar to disintegrate the rock chemically as well as physically. The sufferings of the slaves who had to apply such methods can hardly be imagined. It is known also that the Romans utilized intermediate vertical shafts and even inclined adits in the construction of their longer tunnels, notably of the tunnel built for the drainage of Lake Fucino. Volcanic tufa was another of the softer rocks pierced by these intrepid builders; one notable tunnel through this material, that which gave the road between Naples and Pozzuoli passage through the Ponlipio Hills, was 900 m (3,000 ft) long and 7.5 m (25 ft) wide.[20.1]

The Middle Ages saw no advance in the technique of tunnel building, and even the introduction of gunpowder (first used in tunnel work during the period from 1679 to 1681 at Malpas, France) had little immediate effect. It was the extension of canals and the introduction of railways that finally initiated the great advance of the last 200 years, during which most of the tunnels now in use were constructed. Although so rudely built and simple in conception, ancient tunnel works were inevitably dependent upon geological considerations—not only in design but also in construction; this is evidenced by the possibility of the use of vinegar and the specially hard nature of the flints and corundum utilized for cutting tools. Simultaneously with advance in tunneling technique has gone the development of geology, and so today the two are intimately associated.

20.2 A GENERAL NOTE

Of all the activities of the civil engineer, tunneling is without question that to which the study of geology can most fitly and usefully be applied; geological problems alone affect design and construction methods once the general location and basic dimensions of a tunnel are determined. Except in the case of special tunnels, such as those partially driven through an artificial deposit of clay (e.g., the Sixtieth Street and the Fourteenth Street tunnels in New York) and those isolated examples built as tunnel linings prior to having fill placed over them (e.g., the Golden Circle Railroad in the Cripple Creek mining district of Colorado, where the line was carried in this way across what was later to be a refuse dumping ground), all tunnels are driven through a part of the earth's crust. Accurate location and construction methods depend therefore on the rock through which the tunnel is to be driven; the necessity of lining must usually be determined by the behavior of the rock when exposed to the air and possibly to water. A thorough geological investigation before construction begins is, therefore, of paramount importance in all tunnel work.

Basically, accurate geological sections along all possible routes available for the tunnel are the first requirement; knowledge of the characteristics of the rock to be encountered is the second and is of little less importance. With this information at hand, the engineer can decide whether or not the construction of the proposed tunnel is a practical and economical possibility. If designing a water tunnel in which the water is to be conducted under a pressure head (hereinafter called a *pressure tunnel*), the engineer can determine with some semblance of accuracy what cover of rock must be allowed between the tunnel line and the ground surface. The necessity for, and the design of, an artificial lining for the tunnel can be generally determined, and the likelihood of having to grout the rock adjacent to the tunnel can usually be foretold from information gained in a geological investigation. Finally, such information will give to the engineer at least some indication (often quite accurate) of the percentage of overbreak that is likely to be encountered when construction is under way and for which allowance must be made in estimates. To the contractor for tunnel work, geological information is equally vital; the entire construction program and construction methods depend on the material to be encountered, and the main hazards, such as underground water, are determined solely by geological conditions. On at least one major tunneling project, preliminary geological studies suggested the possibility of finding valuable mineral ores within the tunnel, the sale of which might have paid part of the construction cost. Unfortunately, not a trace of ore was found.

These several aspects of the application of the results of geological investigation to tunnel work will be considered in some detail and illustrated by examples from practice. There are some general observations which first demand attention. Clearly, the ideal condition in tunneling is to encounter one easily excavated material only, a material which contains no water-bearing fissures and which is unaffected when exposed to air. Rarely is anything approaching such an ideal material found; the London clay which most of the London tube railway tunnels penetrate in part is perhaps as close to the ideal as is obtained anywhere. Normally, tunnel locations can be changed only for economic considerations, and so no search for suitable tunneling material is ordinarily feasible; rock conditions at a particular location have to be accepted by the engineer and explored thoroughly in order to render design and construction as certain as possible.

In general, Archean or Precambrian rocks, the oldest types geologically, are difficult to excavate; construction in such formations is consequently relatively expensive. Paleozoic rocks, on the other hand (younger, geologically), are usually the most simple to excavate, and, consequently, they afford more economical construction. Formations of recent origin increase construction difficulties. Tunneling becomes generally more difficult as the formation becomes younger, and recent sand and gravel deposits are particularly awkward to drive through. Novel methods have been adopted in tunnel work for penetrating such materials; the practice of freezing the water in water-bearing strata is perhaps one of the most ingenious.

As in every other type of civil engineering work, economic considerations generally predominate in tunnel design and construction. Here, too, geology is of avail, since not only will the materials to be penetrated affect the actual cost of construction, but they will definitely affect the speed at which progress in excavation may be anticipated. From estimates, detailed comparisons of the costs of various alternative routes for the work proposed can be prepared. In the Kinlochleven hydroelectric scheme in Scotland (25,000 kW with a head of 300 m or 1,000 ft), such a study led to the adoption of a plan calling

20-4   CIVIL ENGINEERING WORKS

for an open reinforced-concrete conduit, following the contour of the hillside, as the main aqueduct. In the neighboring Lochaber scheme (90,000 kW with a head of 240 m or 800 ft), the main aqueduct is a tunnel; the selection of this route was based on a similar and extensive economic study utilizing preliminary geological information.

20.3 PRELIMINARY WORK

Normally, no deviation from general survey methods will be necessary in obtaining the requisite geological sections along possible tunnel routes. By their nature, however, tunnels will usually be located in such a way that accurate preliminary correlation of the nature of the strata to be expected in the tunnel with surface conditions will be difficult. Subaqueous tunnels, for example, can be readily checked only from indications on adjacent dry land; tunnels under cities will pierce strata covered by built-up city areas; and tunnels through high mountain ranges will be at great depths below the surface in some places—over 2,100 m (7,000 ft) for the Simplon Tunnel in Switzerland. In many such cases, it will be impracticable, if not impossible, to put down the usual exploratory drill holes, which generally serve to confirm the deductions of geological survey work. It is for this reason that tunnel-construction records (to be mentioned later) are of vital importance.

Clearly, the first task of the geological adviser on a tunnel project is to indicate whether the anticipated geological conditions are favorable for tunnel construction. The civil engineer must then decide whether or not the tunnel can be constructed economically or if at all. It is of interest to note that, in connection with the Simplon Tunnel, it has been stated:

> Had the geologists been quite accurate in their preliminary investigation and reports, and had they truly and correctly anticipated the dangers and obstacles that were eventually met with in soft rock, the "Great Spring" or river of cold water, the high temperatures and hot springs, and the "creep" or lifting of the floor, no one would have dared to undertake the contract, and the tunnel would never have been constructed.[20.2]

Perhaps it is as well for human welfare that in this instance preliminary geological advice was not perfect. It should be added that the geological structures met in the European Alps are some of the most complex known in the science. (See Fig. 20.1.)

As has been noted, this tunnel is in places over 2,100 m (7,000 ft) below the surface of the ground, piercing high ranges of the Alps, so that it is not surprising that preliminary predictions were not accurate; they did, however, give a reasonable conception of what rock would be encountered (gneiss, mica schist, and some limestone, or "sugar marble"). An even more serious result of incomplete preliminary information occurred during the construction of the Lötschberg Tunnel, from Kandersteg to Goppenstein, between October 1906, and September 1911. The tunnel is 14.5 km (9.04 mi) long, and it was believed that it would penetrate solid rock (granite) throughout. Unfortunately, about 3.2 km (2 mi) in from one portal, the drilling broke into an ancient glacial gorge, now filled with detritus and followed by the Kander River, and in an instant 6,100 m^3 (8,000 yd^3) of material rushed in and 25 men lost their lives. The heading had to be bulkheaded off and the course changed; only in this way was the tunnel eventually finished, finally being 0.8 km (0.5 mi) longer than had been anticipated.[20.3] Had geophysical methods of exploration then been available, trouble might have been avoided.

As can be seen from these brief references, and the information presented in Table 20.1, experiences in driving tunnels through the Alps have given to the engineering profession varied and invaluable experience. The geology of the Alps is so complex and the

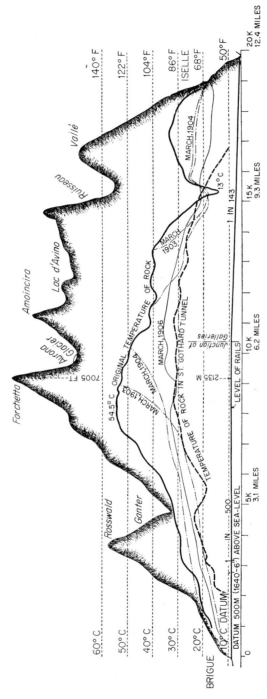

FIG. 20.1 Section through the Alps along the center line of the Simplon Tunnel, showing temperatures encountered here and also in the St. Gotthard Tunnel.

depth of the tunnels is so great beneath the peaks that, even with the most meticulous preconstruction investigations, nothing can be certain about driving Alpine tunnels other than that the work will be difficult. Together the tunnels in this region constitute probably the world's greatest concentration of major tunnels. It is difficult to resist discussing them at greater length. Suffice to say that the driving of the 11.5-km (7.2-mi) -long Mont Blanc highway tunnel from Hameau des Pélerins in France to Entrèves in Italy not only provided European motorists with a welcome cutoff (saving almost a day's driving on north-south journeys) but provided a fascinating comparison of different methods of tunnel driving between the Italian and French ends. Differences in rock types—schist and granite at the French end and calcareous schist, quartzite and granite at the Italian end—and especially the severe problems encountered in the southern heading with inflows of groundwater make any direct comparison of the differing methods invalid. U.S. drilling equipment and mucking in rail cars featured the French approach, Swedish equipment and mucking in diesel trucks the Italian. In both headings much loose and rotten rock was encountered, necessitating the extensive use of rock bolting and steel supports. The tunnel was started in 1958 and opened for use in 1964.[20.4]

Even the Mont Blanc Tunnel has now been surpassed in length by the St. Gotthard vehicular tunnel, roughly paralleling the 60-year-old railway tunnel (actually passing beneath it at the northern end). The tunnel runs a total length of 15.4 km (10.2 mi), all in Switzerland, and its construction was also featured by varying rock conditions and much trouble with inflows of groundwater. This tunnel was necessitated by a steadily increasing volume of north-south highway traffic, which had already saturated the special facilities for transporting automobiles through the railway tunnel. Started in 1970, the St. Gotthard Tunnel was opened for use in 1980.

Tunnels of the magnitude of these Alpine tunnels have to be constructed only infrequently. Problems can be experienced, however, even in quite short tunnels, if preliminary subsurface investigation is inadequate. One example will illustrate what has already been said about the absolute necessity for relating geological studies to the engineering problems to be faced—i.e., the necessity of applying engineering geology and not just academic (or "pure") geology.

The Broadway Tunnel through the Grizzly Peak Hills on California Highway No. 24 in the outskirts of Oakland was designed to provide improved highway connection to the east. It was completed in the midthirties, but for reasons that will shortly be obvious the problems associated with its excavation were not publicly described until 1950. There were two parallel bores, about 900 m (3,000 ft) long and 11 m (36 ft) wide. A special highway district was formed to build the facility. The district engaged a prominent geologist; he made a favorable report which was mentioned in the contract documents (and made available to contractors) but not officially endorsed or quoted. A combination of six experienced contractors was awarded the job, and it now seems clear that they accepted this preliminary report and made no geological study of their own. Very little in the way of supports or lining was anticipated, but the driving of a pilot drift showed how erroneous this impression was. None of the ground penetrated could be left unsupported prior to lining. Two bad cave-ins took place, one taking three lives. The contractors withdrew from their contract, which was completed by other constructors; the original contractors sued the highway district, unsuccessfully, for over $3 million. Very few who use the tunnel today know of this tangled history, but there are few more telling examples of how inadequate geological advice, unrelated to proper geotechnical investigations, can be so misleading (Fig. 20.2).[20.5]

The example of the Lötschberg and Broadway tunnels serves to demonstrate that, in the case of deep tunnels, even expert geological investigation will not always reveal difficulties to be encountered. Another example of serious trouble in tunnel work occurred in the construction of the Las Raices Tunnel, Chile, for the Transandine Railway. It is 4,470

FIG. 20.2 Portal of the Broadway tunnels, Oakland, California, the tunnel mentioned in the text being that on the right.

m (14,885 ft) long and pierces a section of the Andes. Surface indications (and there were few rock exposures above the tunnel) suggested that it would be driven solely through porphyrite; thus, no trouble with water was anticipated. Driving disclosed that the porphyrite rock was generally firm, in some places cracked and fissured with bands of clay, and that there was some infiltration of water. Practically without warning, a break-in occurred at about 525 m (1,750 ft) from one face, at a junction of a full section of the tunnel and a smaller leading section; a large amount of mud poured into the tunnel through quite a small hole and trapped 42 men. Fortunately, they were eventually rescued by means of a special rescue tunnel (the construction of which was a remarkable piece of work). The significance of the failure geologically, however, was that the break-in of 2,000 m³ (2,616 yd³) of material caused a hole to form in the ground above the tunnel. This earth movement greatly facilitated the investigations of the failure. It was found, mainly from the presence of a boulder of granodiorite, that the tunnel had tapped and drained an ancient glacial bed which had been filled in with river deposits. The official report stated, "The accident could not have been foreseen for a rivulet which joins the River Agrio, 150 metres (450 ft) upstream of the tunnel mouth, forms a waterfall, the lip of which is 30 metres (100 ft) higher than the roof of the tunnel, this seeming to show that the thickness of rock over the tunnel was ample." Again, it will be seen what trouble can result from underground conditions caused by glacial action of the past.[20.6]

Diamond drilling along the line of a proposed tunnel is made a supplement to preliminary geological investigation whenever possible. The results of this exploration are positive so far as they go; but their application to tunnel work must be considered in conjunction with the geological survey work also available. Cooperation of this kind has prefaced the construction of many important tunnels, among which are several of those that serve the great metropolitan area of New York. So important are these tunnels that a brief description of some of them will be useful.

20.4 TUNNELS FOR NEW YORK

At the outset, perhaps, it should be said that some of the early experiences in tunneling in the New York area were not very fortunate. The first Hudson River tunnel was started

20-8 CIVIL ENGINEERING WORKS

in 1874, but apparently no exploratory work had been done in advance. The work met with repeated misfortunes; it was interrupted several times and was completed only in 1908. Another example of early work, in which test soundings were utilized but apparently not with geological correlation, is that of the East River Gas Tunnel, constructed between 1891 and 1894 for the purpose of taking three gas mains across the East River to the city of New York. About 10 pipe soundings were made in the two river channels under which the tunnel was to pass, and these indicated a solid rock bottom throughout, on which assumption a contract was let and work started. Trouble with water was soon encountered, however, and eventually the tunnel ran into a vein of soft decomposed rock. Construction methods had to be radically changed; compressed air and a shield had to be used before the work was finally completed, somewhat naturally not under the original contract.[20.1]

Just after the turn of the century, the Board of Water Supply of the City of New York initiated one of the greatest water-supply undertakings that the world had seen, the Catskill water-supply project, which even in this day of great projects is still a monumental piece of engineering (Fig. 20.3). It consisted essentially in the construction of two large impounding reservoirs in the Catskill Mountains, to the north of New York, and of an aqueduct 176 km (110 mi) long, to convey the water so obtained to the city area. The project involved other associated works, some of which are themselves of considerable magnitude. Two large pressure tunnels were also constructed beneath the city for the distribution of the Catskill water; City Tunnel No. 1 was completed in 1917, and City Tunnel No. 2 in 1936. The successive chief engineers to the Board of Water Supply have been aided by many distinguished consultants, only three of whom can be mentioned here—Professors W. O. Crosby, J. F. Kemp, and C. P. Berkey, all famous geologists, who were appointed in 1905 and 1906 to advise on geological matters affecting the works. From the very start of the work in 1905, the prosecution of this great enterprise was characterized by the closest possible cooperation between geologists and engineers (Fig. 20.4). In the words of Dr. Berkey:

> In this project virtually nothing was taken for granted. Every new step was the subject of special investigation with the avowed purpose of determining the conditions to be met; and, when these were determined, the plan of construction and design of the structure were brought into conformity with them. In this manner specifications could be drawn with sufficient accuracy to avoid most of the dangers, mistakes, and special claims commonly attending such work. Very few features or conditions were discovered in construction that were not indicated by the exploratory investigations, and such as were found proved to be of minor significance and were cared for at moderate expense.

When it is considered that the country traversed by the aqueduct (as is shown so clearly by the geological section along the whole route shown in Fig. 20.3) "exhibits so great a variety of natural features and physical conditions that virtually every individual section of this aqueduct presented special geological problems" and that the dam sites were generally marked by great deposits of glacial drift with ancient river valleys thus covered from view, this remarkable result will be more fully appreciated.

Fortunately for the engineering profession, the results of this cooperative work were made generally available by those in charge of the work through technical books, papers, and articles contributed to engineering journals.[20.7] The paper by Dr. Berkey and J. F. Sanborn, presented to the American Society of Civil Engineers in 1923, is already an engineering classic; it is safe to say that the practice that it describes is responsible for some of the change in attitude of the civil engineering profession toward geology which has taken place since the turn of the century, certainly in North America. For those interested in a brief account of the geological features affecting engineering work in and around

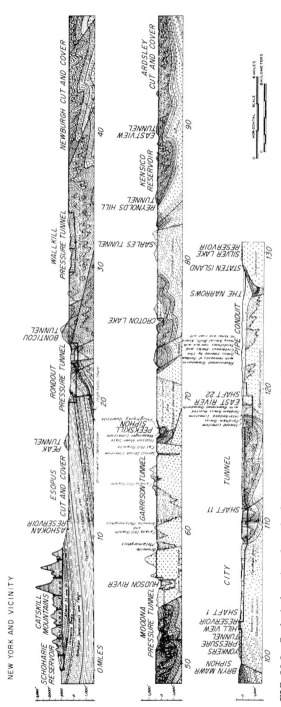

FIG. 20.3 Geological section along the Catskill Aqueduct supplying New York City.

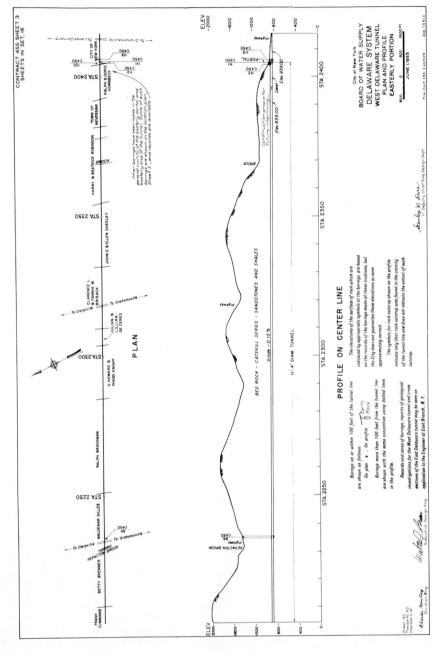

FIG. 20.4 A typical contract drawing for a New York water supply tunnel.

the city of New York, reference may be made to *Guidebook 9,* prepared for the Sixteenth International Geological Congress held in 1933. Two of Dr. Berkey's statements from this interesting handbook have already been quoted; the following quotation is a description of one of the most unusual parts of the Catskill Aqueduct, the Hudson River crossing, the general location of which can be seen in the geological section.

> On the basis of preliminary field studies, therefore, the Storm King crossing was selected because it appeared, from field evidence, that at this place a tunnel could be driven in a single formation of essentially the quality of a granite without encountering a fault beneath the gorge. . . .
>
> It was commonly believed, however, that the Hudson River followed a great fault throughout much of its course and that this was the chief reason for its straight course and apparent independence of other structural control. In the preliminary studies therefore, three important questions had to be determined before suitable specifications could be drawn—first, the depth of the channel, so that the grade of the tunnel could be fixed; second, the kind of rock to be penetrated, so that cost could be estimated; third, whether any special weaknesses or dangers were likely to be encountered, so that suitable provision for them could be made.
>
> When explorations were undertaken at this particular site great difficulty was encountered in determining the depth of channel, and very much greater depth was found than was anticipated from previous geologic knowledge. . . .
>
> The difficulty was increased because the Hudson carries a heavy river traffic. The boring equipment had to be anchored in midstream in the main traffic channel, where there was always danger of interference. Furthermore, the gorge is filled with mixed material in which large boulders are numerous. It was necessary to begin a boring with very large casing (18-inch) so that reductions could be made in passing through these obstructions.
>
> After more than a year, borings to bedrock were successfully made on both sides of the river, finding granite on both sides. The gorge was found to be at least 500 feet [152 meters] deep for a width of 3,000 feet [914 meters]. How much deeper it might be out in the center of the gorge no one could tell. A boring placed in mid-channel finally succeeded in getting to greater depth, reaching 765 feet [233 meters] without touching the rock floor. . . .
>
> This fact, together with the slow rate of progress being made by the borings, finally led to the adoption of another plan for testing the ground beneath the river. The final working shaft on each side of the river was put down to a depth of about 200 feet [61 meters], and at that level a room was cut in the side wall, where a diamond drill was placed. This was set up to drill at an angle so as to penetrate the ground beneath the gorge out under the river. The first two borings, one on each side, were set to reach a depth in the center of 1,400 feet [427 meters] below sea level. These were successfully run in sound rock, and it was then decided to drill two others at a smaller angle, cutting the central ground at a maximum depth of 950 feet [290 meters] below sea level. In these borings sound rock, of granite type, was found continuously across the gorge. By this time the boring in the river had reached its maximum depth of 765 feet [233 meters] without touching bottom. With these data in hand, it was considered unnecessary to carry explorations further. . . .
>
> When construction was finished the conditions uncovered in the Hudson River pressure tunnel were essentially as indicated by the exploratory and other investigations. The Storm King granite is continuous across the gorge. There are no great faults in the gorge at this point. The rock is sound and presented no special construction problem. [See Fig. 20.5.]

This extensive program of exploratory work is typical of the work done by the Board of Water Supply. On the 29-km (18-mi) City Tunnel No. 1, for example, 13,800 m (46,000 ft) of exploratory borings were put down, and more than 15,000 m (50,000 ft) were drilled for City Tunnel No. 2; all were put down in conformity with geological advice. The same pattern was continued and the experience gained on the earlier tunnels was applied when New York undertook the much greater Delaware project, which involved the most extensive tunneling work ever carried out on any project. From the Rondout Reservoir, west of Poughkeepsie, the main aqueduct runs as a deep pressure tunnel for 136 km (85 mi) to

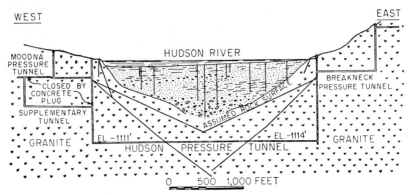

FIG. 20.5 Geological section across the Hudson River, New York, at Storm King, based on exploratory borings for the Catskill Aqueduct crossing (indicated in the section, Fig. 20.3).

the Hill View Reservoir of the Catskill scheme; together with the connecting City Tunnel No. 2, the total length of this one tunnel is almost 170 km (105 mi). The East Delaware Tunnel is 40 km (25 mi) long and connects the Pepacton and Rondout reservoirs, and the West Delaware Tunnel (70 km or 44 mi long) connects the Cannonsville and Rondout reservoirs; the latter was holed through in January 1960, and marked the beginning of the last stage of this vast water-supply project, designed to satisfy the water demands of New York City until the end of the century.

A succession of papers by T. W. Fluhr has kept the profession informed of the geological conditions experienced through all this hard-rock tunnel work and has presented to geologists details of geological structure that would otherwise have been unknown.[20.8] To summarize the geology of the Delaware tunnels is clearly impossible; suffice it to say that the rock formations encountered were similar to, and an extension of, those already described in connection with the Catskill tunnels. A number of major faults were encountered, but as a result of careful preliminary work in test drilling from the surface, and in some cases drilling ahead from working faces, soft-ground conditions were always predicted and successfully planned for in the tunnel driving. One of the most critical sections was where the main tunnel passes under the Kensico Reservoir of the Catskill scheme; poor ground was detected in this area. The grade of the tunnel was therefore dropped to an elevation of -195, or more than 300 m (1,000 ft) beneath the water level in the reservoir; after unusually careful test drilling, special precautions were taken in driving, and the tunnel was driven and then lined without trouble. This experience showed that faulting and rock decay, under the conditions here encountered, diminish rapidly with depth. The methods developed on the Delaware tunnels for driving through major faults have provided a wealth of valuable information, now fortunately available on record for the information of all who encounter similar difficulties in tunnel work elsewhere.

The vehicular tunnels which so greatly aid traffic to and from Manhattan Island were also marked by unusual geological conditions. Possibly the most difficult of all was the Queens-Midtown Tunnel, holed through successfully in 1939. Figure 20.6 shows the geological character of the ground penetrated by this tunnel. Even in this summary form, it can be seen to be unusually complex, involving tunneling with a shield in soft ground, artificial ground (formed by the clay blanket), and solid rock; in some sections the lower parts of the shields were in rock and the upper in loose soil. Compressed air naturally had to be used; a maximum pressure of 104 kg/cm² (37½ psi) proved to be necessary. Illustrative of the conditions that had to be faced is the fact that, at one time, one of the working

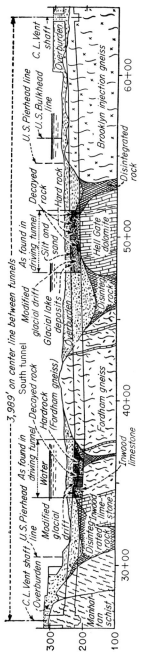

FIG. 20.6 Diagrammatic section across the East River, New York, showing geological conditions encountered during the driving of the East River vehicular tunnel.

faces was in 2.4 m (8 ft) of lumps of coal (deposited long before in the riverbed) of egg size and larger at the top; then 2.7 m (9 ft) of sand, clay, gravel, and boulders; and finally 2.4 m (8 ft) of sound rock in the invert. Under such conditions air was lost at surprising rates, even breaking up and eroding the clay blanket in the riverbed, but the face was never lost and the tunnel driving was completed without a single fatality due to bends. It is difficult to imagine a more hazardous tunnel job; its successful completion is a tribute alike to the contractor's skill and to the very detailed foreknowledge of the subsurface conditions that was made available by geological studies.[20.9]

There may finally be mentioned the Wards Island sewer tunnel which was constructed during the period from 1935 to 1937. Preliminary borings, although carefully taken, did not disclose the extremely soft, almost fluid, consistency of some decayed and disintegrated rock deep underground, material which finally necessitated a complete change of grade in the river section. The tunnel was constructed to deliver sewage from interceptor sewers on Manhattan Island to the new treatment plant on Wards Island in the East River, and the original plans were made for a depth of 89 m (297 ft). This proved too shallow, and the tunnel was finally put through at a depth of 153 m (510 ft) below water level. The conditions encountered are clearly seen in Fig. 20.7, which also shows the location of the drill holes originally put down and the additional holes drilled after the work was temporarily closed in December 1936, following the discovery that the seam of soft chloritic mass (of the consistency of fine mud) made further progress at that level practically impossible. One of the original borings had passed through this material; but as the hole had been drilled from the river, it had not been possible to secure such reliable samples as usual, and the wholly unsubstantial quality of the material was underestimated. The cross section illustrates how the secondary drill holes disclosed satisfactory conditions at a lower grade; work was resumed in March 1937, and the headings were holed through two months later.[20.10]

20.5 UNDERWATER TUNNELS

Visitors to London, England, who have occasion to use Wapping Station on the East London line of the London Underground may notice a simple plaque bearing these words, "Thames Tunnel at Wapping designed by Sir Marc Isambard Brunel (1769–1845) and completed in 1843. Isambard K. Brunel (1806–1859) was Engineer-in-Charge 1825–1828." Thus was the centenary of the death of the younger Brunel marked at the site of the first underwater tunnel ever to be completed; the plaque was erected in 1959 by London Transport. The tunnel is still in regular use; its completion marked a victory of unusual ingenuity and heroic courage over what seemed at times to be insuperable obstacles. The ground conditions encountered were almost as varied as those met in the Queens-Midtown Tunnel, New York, although there was no solid rock present. The contrast between the two jobs, one done with crude and simple equipment, the other with all the aids of modern construction techniques, is a vivid reminder of the progress made during the last century in construction methods.

It is said that Brunel got the idea of the shield with which he built the tunnel while examining a specimen of *Teredo navalis* in a piece of oak; he noted how the little mollusc bored its way into the wood, working under the protection of its own shell. He presented the idea to the Institution of Civil Engineers in 1824; a company was formed; Brunel was named engineer. Test borings were taken across the river near the ferry that the tunnel was designed to supersede. The opinions of "eminent geologists" were obtained, to the effect that the tunnel should not go lower beneath the riverbed than was absolutely necessary. This is one case in which geological advice was less than helpful. Trouble was

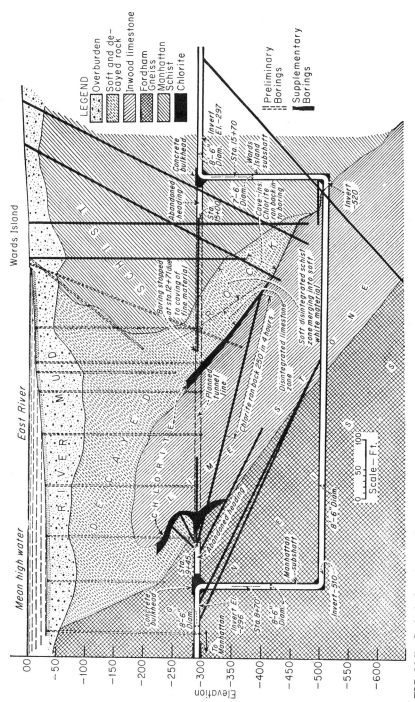

FIG. 20.7 Geological section on the center line of Wards Island sewer tunnel, New York, as revealed by test drilling, showing the necessary relocation of the tunnel because of chlorite seams.

20-15

encountered soon after construction commenced, owing to inadequate cover, and on 18 May 1827, a bad blow occurred and the tunnel was flooded. This was but the first of many worries. The works were shut down for six years because of lack of funds, but Brunel's personal enthusiasm eventually led to the resumption of operations in 1836, and the tunnel was successfully completed in 1843. His great granddaughter has told of Brunel's practice of personally examining samples of soil from the working face every two hours, day and night. Presumably on the insistence of his wife, Brunel had a basket arrangement rigged up outside his bedroom window so that the samples could be pulled up in the night for him to examine. Following his examination, he wrote his instructions, and these were sent down in the basket and back to the job. This went on for four years. It is not surprising to read that Brunel and his wife found themselves waking regularly every two hours throughout the night for many years after the tunnel was completed. In the face of such devotion to duty, it is sad to record that Brunel never did get paid all of his promised professional fee as engineer to this first of underwater tunnels.[20.11]

Better geological advice was obtained for the second Thames tunnel, a small foot tunnel between London Bridge and Limehouse which is still in existence but not in use. P. H. Barlow was the engineer; he also used a shield, although of a new design. Since he was advised by geologists of the thickness of the London clay in this area, he located his tunnel, which is 2.1 m (7 ft) in diameter, well below the riverbed and in London clay throughout; as a result, the job was completed between 26 April and 8 October 1869. In great contrast is the modern Dartford-Purfleet Tunnel, 1,410 m (4,700 ft) long and 9.1 m (30.5 ft) in external diameter, which crosses under the Thames at a location that necessitated driving through solid chalk and a variety of soils including Thames gravel. The use of compressed air was essential; it was at the time said to be the largest tunnel yet constructed under air. Even with air, however, some of the ground to be penetrated had to be pretreated by grouting with a specialized process; grout tubes were inserted from the surface of the river in advance of tunneling. A typical working face was 3.9 m (13 ft) of clay, 2.1 m (7 ft) of grouted sand and gravel, and 3.3 m (11 ft) of solid chalk. Excavated chalk was pulverized, mixed with water, and pumped as a slurry to the surface.[20.12]

The first underwater tunnel in North America to be completed appears to be one that is little known because the usual account of its construction was not published in any of the leading civil engineering publications of the time. This was the tunnel conveying the Grand Trunk Railway (now Canadian National Railways) beneath the St. Clair River from Sarnia, Ontario, to Port Huron, Michigan. Joseph Hobson was the chief engineer; the general location of the tunnel had been selected before his appointment. He made the final detailed location in 1884 and then proceeded to have test borings put down at 15 m (50 ft) intervals throughout the length (1,800 m or 6,000 ft) of the tunnel. Even today, this would be regarded as examplary site investigation; in 1884, it was probably unique. The tunnel was located in the gray clay-till that here overlies the bedrock, minimum clearance between crown of tunnel and the riverbed being 4.8 m (16 ft). The tunnel now carries a single track; it is 6.0 m (19 ft 10 in) in diameter and is lined with cast iron segments (another innovation). Started in 1889, it was holed through on 30 August 1890. Even though compressed air had to be used for the 690 m (2,301 ft) under water, the work was completed with no difficulties, the worst accident on the job being a broken arm. The tunnel is still in active use, a superb example of early engineering and a constant reminder of the value of adequate subsurface investigation.[20.13]

Since underwater tunnels always involve hazards, one or two more examples may usefully be given. During the construction of the Almendares vehicular tunnel, connecting two of the most popular residential districts of Havana, in Cuba, unusual geological conditions were revealed by the intensive subsurface investigations carried out before completion of designs (Fig. 20.8). Two approaches, totaling about 300 m (1,000 ft), join an

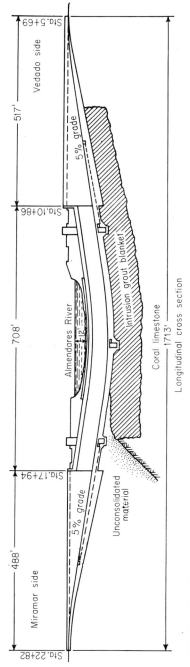

FIG. 20.8 Simplified geological section along the center line of the Almendares Tunnel, Havana, Cuba.

underwater tunnel section 212 m (708 ft) long. On the Miramar side, ground conditions consisted of several layers of peat, overlying clay, and silt with sand and gravel lenses. Approximately halfway across the Almendares River, outcropping coral limestone was encountered and continued as an available foundation bed for most of the remainder of the structure. Extensive testing revealed that water could be controlled in the soil deposits by the use of well points, and the concrete approach structure was thus constructed in the dry. The coral rock was so porous that pumping it dry would have been impossible. A 7.5 m (25 ft) depth of the coral was therefore grouted in advance of construction by means of the intrusion method. Thus, 51,000 m^3 (67,000 yd^3) of rock was "solidified," permitting the driving of steel sheet piling to form a cofferdam in which excavation proceeded in the dry to a depth of 11.4 m (38 ft) below river level with only minor seepage. All expenditure on preliminary boring and testing was many times repaid by the ease with which construction was carried out after the precautionary grouting program had been completed.[20.14]

Japanese engineers and contractors have been responsible for notable underwater tunnels linking the islands which form that most interesting country. Two tunnels link the islands of Kyūshū with the main island of Honshu, a railway tunnel completed in 1944 and the Kammon vehicular tunnel started in 1939 but completed only in 1957 because of the cessation of work during war years. Of the underwater tunnels so far completed, only the Mersey Tunnel, to be mentioned shortly, is larger. The Kammon Tunnel replaced a busy ferryboat service and roughly parallels the railroad tunnel. It is 3.45 km (2.16 mi) long and provides two roadways on two levels with footpaths for cyclists and pedestrians. Penetrating diorite and porphyrite, it is a hard-rock tunnel throughout, but many faults had to be passed, since the rock had been badly affected by earthquake shocks that occur in this part of the world. Cement grouting on an extensive scale was necessitated in the underwater section of the tunnel because of the decomposed rock in the vicinity of the faults; the work was completed without trouble, despite the fact that its construction was delayed by the war.[20.15]

These important tunnels pale almost to insignificance when compared with the double-track railway tunnel being constructed between Honshu and the northern island of Hokkaido. The Seikan Tunnel, constructed for Japanese National Railways, goes under Tsugaru Strait, the underwater section being 21 km (13.5 mi) long and 99 m (330 ft) below the seabed. Long approach tunnels beneath high ground at both approaches give a total length of 52 km (32.3 mi), making this the longest railway tunnel in the world. Access shafts were sunk to 219 m (730 ft) below the seabed, and from their bottoms pilot tunnels were bored, with service tunnels above them, aligned close to the main tunnels. A main purpose of the pilot tunnels was to probe the geology (Fig. 20.9). No summary can do justice to the Seikan Tunnel, which approaches completion as this volume goes to press. Suffice to say that it is in the main a hard-rock tunnel, about one-third of its total length being in andesite; another long section is through volcanic ash; faults caused much trouble not only of themselves but by creating major inflows of water to the tunnel workings. The cross section of the main tunnel is 11.2 m (37.6 ft) wide and 8.5 m (28.4 ft) high; it is being excavated by drilling and blasting, although tunnel-boring machines have been used for the pilot and service tunnels. The Seikan Tunnel will long remain one of the great tunnels of the world.[20.16]

20.6 PRESSURE TUNNELS

The design of pressure tunnels presents to the civil engineer some of the most serious difficulties in tunnel work. Three separate problems always have to be faced. In some way, the engineer must make sure that the material surrounding the water will be impervious.

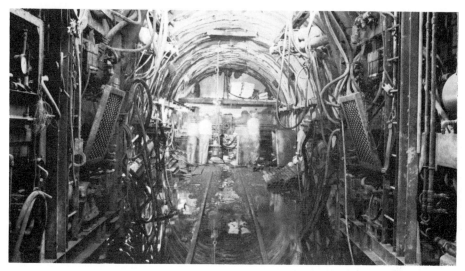

FIG. 20.9 The pilot tunnel for the Seikan Tunnel, Japan, being driven in advance of the main bore.

A means must be secured for withstanding the pressures set up by the unbalanced head on the water passing through the tunnel. Finally, as a lining is almost invariably used, both to satisfy these requirements and to reduce frictional resistance to the flow of water, the design must take into consideration the possibility that the tunnel will be empty and that groundwater will tend to exert a considerable pressure on the outside of the lining. The engineer could naturally design a pressure conduit (of steel or reinforced concrete) that would meet all the requirements called for by the foregoing conditions and neglect entirely the existence of the surrounding rock. Such a course would be far from economical. True engineering is the attainment of the economic solution to the problems faced; and so in the design of pressure tunnels, the civil engineer seeks the cooperation of the geologist so that the best advantage can be taken of the rocks to be encountered in reducing the lining used to its economic minimum.

The three distinct problems must be emphasized; they each have a separate bearing on the final solution, although they are sometimes confused in discussion. Almost all important pressure tunnels are lined today, and such linings are usually of concrete. Mass concrete has almost negligible tensile strength, and when under tension, even when reinforced, it will open up in minute cracks. If it is to be impervious to the water it is retaining, as it must be in the majority of cases (since the surrounding rock will not always be completely impervious to water), the lining must clearly be made from the finest quality concrete, placed carefully in position, and well compacted in the forms. This is essentially an engineering matter, but it is stressed here, since the rock excavated from the tunnel will probably be used, if possible, as aggregate, and a geological opinion of the rock's suitability for this purpose will have to be given before construction starts. The effect of the prolonged contact of portland-cement concrete with the rock to be encountered must also be considered. A prediction of how the rock will break will aid in the preparation of estimates of lining-placing costs.

The structural design of the lining is governed by the anticipated maximum water pressure, from which can be calculated the total bursting pressure that will be exerted on the tunnel lining. This pressure has to be resisted by the lining and the surrounding rock. At tunnel portals and where the cover of rock (the minimum distance from the tunnel

20-20 **CIVIL ENGINEERING WORKS**

perimeter to ground level) is small, the surrounding rock cannot be reckoned on at all, and the lining must be designed on standard lines as a reinforced-concrete conduit, neglecting entirely the effect of the surrounding rock. When the rock cover is appreciable, especially when it can be selected to suit design requirements, the geological nature of the surrounding rock is naturally of importance. Various methods have been suggested for arriving at the allowance to be made in design for the strength of the superimposed rock; the minimum assumption is that the actual weight of the column of rock vertically above the tunnel is all that can be assumed to resist movement due to the water pressure. Admittedly, any selection of the allowance to be made will be empirical, but a knowledge of the state and nature of the rock strata to be reckoned on in such calculations will tend to rationalize the decisions finally made.

Designing the lining for a pressure tunnel is an exercise in engineering judgment; although application of the laws of mechanics can suggest certain dimensions for the component parts, all calculations depend ultimately upon the accuracy of the assumptions made regarding the rock to be penetrated. Here again the absolute necessity of accurate knowledge of the rocks to be encountered during tunnel driving, as well as the necessity of taking samples so that the mechanical properties of the rock may be assessed for design purposes in advance of construction, cannot be too strongly emphasized. Study of rock mechanics is rendering assistance in such work, but again, accurate knowledge of the rocks to be encountered is the only basis upon which rock-mechanics results can safely be used.

This essential approach was well illustrated by the carefully conducted studies which were a part of the design of a major Canadian hydroelectric project, the Sir Adam Beck No. 2 development at Niagara Falls, completed in 1956. Duplicating the existing Queenston development, with the new 1.2-million-hp powerhouse adjacent to, but upstream from, the old one, the intake and main conduit arrangements for the new project differed from those previously used. Instead of using an open canal, skirting the city of Niagara Falls, Ontario, the Hydro Electric Power Commission of Ontario decided, after exhaustive engineering and geological studies, to use twin 13.5-m (45-ft)-diameter, lined concrete tunnels to convey the water from the intake along 8.8 km (5.5 mi) of its journey to the powerhouse. Figure 20.10 illustrates the well-known succession of geological strata at Niagara Falls and the relation of the twin tunnels to it. Although some of the strata shown are not so competent as could be desired theoretically (some of the shales, for example, disintegrate upon exposure to the atmosphere), the rock in which the tunnels were driven was entirely adequate; the Irondequoit limestone which forms the crown of the tunnel is massive and competent, excellent in all respects for this purpose. Core samples were taken well in advance of design as part of the thorough exploration program; when tested, these gave a range of rock properties, so that a design could be prepared with confidence. From the start of construction, a careful program of measurement of possible rock movements was carried out.

The first significant results confirmed design expectations. By extrapolation, a maximum inward movement of the walls of the first tunnel, on the major horizontal diameter, of 2.7 cm (1.08 in) 30 years after placement of the lining is probable; this is based on the movement of almost 2.5 cm (1 in) recorded by the time that the lining was placed, about 180 days following excavation. Of special interest is the fact that the corresponding inward movement for the second tunnel, excavated after the first, was only about one-quarter to one-third of that in the first, as measured at several points; this is an indication of the stress relief in the rock due to the adjacent excavation. The instrumentation used and the records of rock movement so far obtained on this great project should be of special value to all concerned with pressure tunnels of comparable size and depth.[20.17]

There have been some failures of pressure tunnels; since constructive study of failures is always valuable, two may be mentioned, not in any critical sense but as indicating the

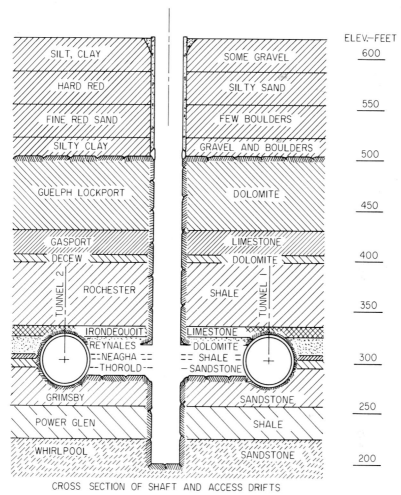

FIG. 20.10 Cross section of rock strata at Niagara Falls, Ontario, showing the location of the twin tunnels for the Sir Adam Beck power development near construction shaft no. 1.

hazards associated with such tunnels. Leakage from a partially lined pressure tunnel was blamed as the cause of a massive landslide which severely damaged the Whatsham power plant building and transformer station in British Columbia in 1953. Water was conveyed through two 3.6-m (12-ft) -diameter tunnels, 3.2 km (2 mi) long, to the small power station in which 25,000 kW was generated. Only the ends of the tunnels had been provided with concrete as an economical feature of design.[20.18]

Rather more serious was a case from Australia. During the years from 1923 to 1928, the city of Sydney, New South Wales, constructed a pressure tunnel under a major part of the city in connection with its water supply. The tunnel is 16 km (10 mi) long and was designed initially with an internal diameter of 3 m (10 ft); it is constructed throughout in Hawkesbury sandstone, of Triassic age, of varying character, fissured and jointed, and marked by current bedding. Before the completed tunnel was put into service, sections of it were tested. During the test, the lining ruptured, to a varying extent, in all but one of the sections tested, although the pressures used were not unduly high. In general, the failures occurred (1) where bad rock and fissures had been encountered and (2) where the cover of solid rock above the tunnel was theoretically inadequate. The test water was found issuing at the surface of the ground, and from the accurate test readings taken it was seen that part of the lining and the surrounding rock had moved laterally 1.3 to 1.9 cm (0.50 to 0.75 in) in the direction of the axis of the tunnel and at an inclination conforming to that of the bedding plane of the sandstone. As a remedial measure, the whole tunnel had to be lined with a continuous steel pipe. The published description of the work is a fascinating engineering document and repays much study. Although the failure may have been due to faulty structural design, it was possibly associated with a lack of appreciation of the true nature of the sandstone rock when in place and also when used as aggregate for the concrete of which the lining was formed.[20.19]

20.7 TUNNEL SHAPES AND LININGS

The lining of tunnels can naturally be considered under two general classifications, corresponding to the two general uses to which tunnels are adapted. In road and rail tunnels, a lining may be necessary to support the pressure exerted by the material in which the tunnel is excavated; a lining may likewise be necessary to cover up the exposed material on the tunnel perimeter to protect the material from atmospheric influences (and possibly the action of locomotive or automobile exhaust gases and vapors). Water tunnels, on the other hand, must often be lined to present a smooth surface to the flow of the water through the tunnel and thus to reduce friction losses to an economic minimum; it may also be necessary to provide an impervious lining to prevent leakage of water into the surrounding rock, if this is at all fissured or pervious. Pressure exerted by surrounding material and, in the case of pressure tunnels, by the water inside the tunnel must be considered. Finally, the material penetrated must be protected, if necessary, from the action of continued exposure to water and, possibly, air. These distinctive functions of linings are emphasized here, since sometimes they appear to be confused. Brief consideration will show that water tunnels are generally lined irrespective of the material through which they are driven, while traffic tunnels may or may not need such additional attention before completion. The few tunnels constructed for the passage of canals through high ground are governed by the characteristics of both types of lining.

The determination of the final shape of a tunnel cross section will also be affected by the material penetrated. In water tunnels, a circular section is the most economical from the hydraulic point of view, but considerations of construction methods may result in a horseshoe section with an inverted arch bottom. In soft ground and in poor rock, structural considerations of the design of lining will generally necessitate the use of a full circular section. Tunnel roofs are usually designed as semicircular arches even in quite sound and solid rock; the remainder of the section (in traffic tunnels) is proportioned on the basis of economic considerations. An interesting variation from this latter criterion is given by the Roman road tunnel, already mentioned, through the Ponlipio Hills; this tunnel has its cross section in the form of a pointed arch 7.5 m (25 ft) wide, 6.6 m (22 ft) high at the center of the tunnel, but 22.5 m (75 ft) high at the ends; the increase was intended to improve the illumination of the tunnel. As the total cost of a tunnel varies almost directly with its cross-sectional area, the importance of the correct determination of this area will be clear. Economic considerations control this, but as has been indicated, in some cases the nature of the material to be penetrated will override basic financial factors.

An interesting sidelight on the early application of geology to civil engineering problems is given by the difficulty encountered by the younger Brunel (Isambard Kingdom) in directing the construction of the Box Tunnel on the Great Western Railway about 1835. Built on a gradient of 1 in 100, the Box Tunnel is today a leading feature on the main line of the Western Region of British Railways, and it is still, to a large extent, unlined. At the time of its construction, however, in addition to violent opposition from local landowners, Brunel had to put up with adverse comments from eminent geologists. The most severe critic was the famous Dr. William Buckland of Oxford, who maintained, even before the Institution of Civil Engineers, that the unlined portions of the tunnel (driven mainly through limestone) were dangerous; the "concussion of the atmosphere and vibration caused by the train," he said, would make the rock fall away. Brunel was unimpressed by this so-called "academic advice" and went ahead with the tunnel as he had planned it.[20.20] An equally famous railway tunnel in North America, the Hoosac Tunnel in northwestern Massachusetts, completed in 1873, then and still the longest railroad tunnel east of the Mississippi and the first tunnel in which nitroglycerine was used for blasting, was not lined and is still in excellent condition after continuous use since its opening.[20.21]

Methods of lining are not within the scope of this book. It may, however, be mentioned that in soft ground segmental cast-iron or pressed-steel plates or precast reinforced-concrete block work is now in almost universal use as the lining for resisting ground pressure (having succeeded the brickwork once extensively used). If necessary, such linings can have a concrete surface coating applied to them for the purpose of presenting a smooth water surface. In solid-rock tunnels, timber framework used in construction may be left in place as a semipermanent lining; in bad ground, it may be replaced by masonry or surrounded by a solid concrete lining. Mass or reinforced-concrete linings are now used almost universally for covering up exposed surfaces liable to disintegrate. Concrete-placing methods follow standard practices, but there is at least one case in which geology may be said to have contributed significantly even to the placing of concrete lining. For the Ontario Hydro tunnels at Niagara Falls just mentioned, centrally mixed concrete was used for the linings which gave a finished diameter of 13.5 m (45 ft). The invert concrete was placed by means of a self-propelled traveler working on rails, into which concrete was dumped from a "boot traveler" located beneath a number of specially drilled boreholes through which concrete was dropped from the surface. These holes were 30 cm (12 in) in diameter, sunk by churn drills, and lined with steel plates that were grouted into place. The very regular geology along the tunnel route permitted the confident adoption of this unusual means of concrete transport.[20.22]

As the cost of lining a tunnel may amount to as much as one-quarter of its total cost, the importance of this feature of design will be obvious. Preliminary geological considerations will therefore be of great value in determining the design of the tunnel cross section. In igneous and metamorphic rock known to be solid, a lining will probably be unnecessary for a traffic tunnel. Should there be disintegrated or shattered rock, most treacherous conditions may have to be countered, conditions that not only require permanent linings but that may give trouble during the temporary support of the exposed material during construction. In the Simplon Tunnel, for example, much difficulty was experienced at a distance of about 4.4 km (2.75 mi) from the Italian face because of pressures exerted by decomposed calcareous mica schist; even 40-cm (16-in) rolled-steel beams used as temporary strutting buckled. Quick-setting cement provided the final solution to the problem then presented, although at a cost of £1,000 per yard of tunnel. In other places, rock in the tunnel floor was forced up; but in sections driven through solid rock, even at the maximum depth below the surface, no deformation occurred; the lightest type of masonry lining proved adequate.[20.2]

For a variety of reasons, sedimentary rocks produce most tunnel-lining problems. As a result of their mode of origin, sedimentary rocks are likely to change in character within relatively short distances; their original depositional characteristics often make them susceptible to changes in pressure; and the combination of these two factors frequently renders them of an unstable composition, easily and quickly affected by exposure to air and water. Furthermore, the bedding planes of sedimentary formations naturally affect the stability of any tunnel section bored through them; the angle made by the line of the tunnel with the dip and strike of the beds is a most important feature. The underground water carried in fissures and along bedding planes will also affect the design of a lining. All these factors can be estimated, to some degree, from preliminary geological considerations.

20.8 OVERBREAK

Unless the specification covering the construction of a tunnel under contract is most carefully drawn up, with due regard for all the geological conditions anticipated, the payment for overbreak is almost certain to be one of the matters that will cause the most trouble at the conclusion of the contractor's operations. The term *overbreak* is one now generally used in civil engineering practice to denote the quantity of rock that is actually excavated beyond the perimeter previously fixed by the engineer as the finished *excavated* tunnel outline. It will be realized that in the case of a lined tunnel, not only may the contractor receive no payment for rock excavated beyond this line, but also additional concrete filling will be necessary to fill up the resulting cavity. For engineer and contractor, therefore, the question of overbreak is an important one.

A close study of records and descriptions of early tunnels suggests that overbreak was not seriously considered in the early days of modern construction, possibly because many tunnels were then paid for on the basis of a unit length of completed tunnel. In modern work, payment is more usually made per unit volume of excavation and per unit volume of concrete or other lining. Under such circumstances, it will be readily seen that overbreak must be kept to a minimum; and as the quantity will depend primarily on the nature of the rock penetrated, given equally good driving methods, preliminary geological investigations are of great importance in estimating what overbreak will be.

In connection with soft-ground tunnels, the term is rarely met, since such works can generally be taken out exactly to the neat line of excavation, and construction equipment (shields, lining plates, etc.) can be designed to fit to this line. Close-grained igneous rock will usually break closer to the theoretical section than sedimentary or metamorphic rock, although certain types of chalk and compact sandstone will also give good sections. As overbreak will depend on the detailed nature and structural arrangement of the rock strata encountered, it cannot often be determined with great accuracy in advance. When, however, the nature of the rocks has been determined by preliminary geological investigations, the tunnel engineers, because of their previous practical experience, and the geologist, because of his intimate knowledge of the characteristics of the rock types, will together usually be able to make a close estimate of the way in which the rocks will break. It will then be possible to prepare the relevant parts of the specification so that they will conform as closely as possible with the situations that develop as the work proceeds. The following represents a typical way of dealing with this matter; the items suggested refer to that part of a specification dealing with payment for excavation.

1. No points of solid rock must project beyond a line, called the *neat line* of excavation, fixed at a distance from the inside perimeter of the finished tunnel section equal to the minimum thickness of lining required.

2. No flat areas of rock of more than, say, 3 ft² must occur within a specified minimum distance from the finished perimeter, usually a few inches more than the minimum lining thickness.
3. All excavation and concrete lining will be paid for up to, but not beyond, another tunnel perimeter called the *pay line,* although all cavities beyond this line must be carefully filled, generally with concrete of an inferior mix.

The fixing of this pay line is usually the most difficult part of the design of the tunnel cross section. The location of the pay line will depend upon the nature of the rocks to be met; the way in which they will tend to break; the necessity of temporary timbering and whether this will have to be incorporated in the finished lining or not; and last but not least, the anticipated dip and strike of the rocks to be met in relation to the axis of the tunnel. The overbreak is the volume of excavation that actually has to be taken out beyond this line. Clearly, the engineer will want to fix the line so that a contractor will not be faced with the possibility of much overbreak, a feature that will be reflected in the unit prices tendered for excavation by those contractors who appreciate the significance of geology and in the disputes over claims made at the end of the job by those contractors who do not.

Two photographs are reproduced showing finished cross sections of the Lochaber water power tunnel in Scotland, as excavated. The first (Fig. 20.11) shows the shape of section obtained when the drilling was almost perpendicular to the strike of the mica schist; the marks of the drill steel are clearly visible; the overbreak here is very small. The second view (Fig. 20.12) shows the shape of section obtained when drilling was approximately along the strike; the section is almost rectangular owing to the way in which loosened blocks of the schist have fallen away. These views illustrate more vividly than can any

FIG. 20.11 Main tunnel of the Lochaber water-power project, Scotland; a view near heading 3E showing the effect of drilling at right angles to the strike of the mica schist.

20-26 CIVIL ENGINEERING WORKS

FIG. 20.12 Main tunnel of the Lochaber water-power project, Scotland; a view near heading 5W showing the effect of drilling along the strike of the mica schist.

words the dependence of overbreak upon structural geological conditions as well as upon the nature of the rock that is penetrated.

This emphasis on what might seem to be a detail of tunneling procedure may seem to those without experience on tunnel work to be misplaced. It must be remembered, however, that not only has the extra rock represented by overbreak to be removed from the tunnel but, in the case of all lined tunnels, the space occupied by this extra excavation must be filled with concrete. Excessive overbreak can, therefore, lead to great increases in cost over initial estimates. Overbreak can be experienced with almost all types of rock and is even a factor when bores are made using tunneling machines, if sections of loose rock are encountered. Typical was the case of a small drainage tunnel excavated in Toronto through Ordovician (Dundas) shale for which blasting was necessary. The concrete lining, instead of being 0.3 m (12 in) thick as designed, ranged actually in thickness from 0.6 to 1.2 m (24 to 47 in) in thickness, at what extra cost can be imagined—and this was in easily excavated shale.[20.23]

20.9 CONSTRUCTION METHODS

To attempt to deal briefly with tunnel construction methods in such a chapter as this is to be forced onto a course between the Scylla of a statement of somewhat obvious facts and the Charybdis of a detailed exposition of drilling, blasting, excavation, and tunnel boring, treatment of which belongs more properly to a volume devoted solely to tunnel construction. Yet the attempt must be made, since, as in other features of tunnel work, construction methods depend primarily on the geological strata encountered. Broad classification of the material will determine generally the type of construction to be adopted. The exact nature of the material, together with local practice, will determine the detailed variations of the main method in use—for example, the use of top, bottom, or central headings in rock tunnels and the use of the English, Belgian, German, Austrian, or other schedule of excavation in soft-ground tunnels. These considerations all depend on the geological nature of the materials, although methods will be determined by experienced construction personnel rather than by geologists. Special construction problems will almost invariably necessitate specialized investigation, and attention will therefore be directed to a few of the more important unusual conditions that may be encountered.

In very deep tunnels, underground temperatures may sometimes cause inconvenience and adversely affect the progress of tunnel excavation. The study of such temperature variation can proceed only on the basis of actual experience, and it is therefore not possible to predict accurately the conditions to be met. The temperature inside the Lochaber Tunnel in Scotland (during construction) rarely varied appreciably from 12°C (54°F). In the Simplon Tunnel, on the other hand, a considerable range of temperatures was experienced, as is shown in Table 20.1. Mine workings have given some useful data on this aspect of tunnel work. Exhaustive tests have been carried out on this matter at the Lake Shore gold mine in northern Ontario, tests which show that the rock temperature there rises 1°C (1°F) with every 98 m (163.4 ft) of depth below surface. In Great Britain, a corresponding figure has been found to be 1°C (1°F) for every 32 m (60 ft) of depth. It has been during the construction of the long tunnels that pierce the European Alps that the most serious difficulties due to abnormal temperatures have been encountered. Table 20.1 summarizes some of the relevant information.

Other construction problems arise frequently on short tunnels and are often encountered in the usual course of civil engineering practice. They may arise from the proximity of other underground works, the safety of which must be guarded. For example, an unusual problem had to be faced during the construction of the Breakneck Tunnel on the

TABLE 20.1 Rock Temperatures in Some European Tunnels

Tunnel	Mt. Cenis	St. Gotthard	Arlberg	Simplon	Lötschberg
Length, miles	8.0	9.3	6.4	12.3	9.0
Elevation of portals, ft above sea level	4,162 / 3,765	3,756 / 3,638	3,995 / 4,271	2,080 / 2,253	4,002 / 3,936
Elevation of "crown"	4,248	3,788	4,297	2,312	4,067
Maximum elevation of mountain range in profile	9,673	9,384	6,658	9,315	9,568
Maximum cover	5,425	5,596	2,361	7,003	5,501
Maximum rock temperature, °F	85.1	87.4	65.3	132.8	93.2
Tunnel ventilation, cu ft air/sec, up to	247	35–71	106	1,236	388–883
Date constructed	1857–1872	1872–1882	1880–1883	1898–1906	1906–1911
Rocks encountered	Limestone, calcareous schist, gneiss, schistose sandstone	Gneiss, mica schist, serpentine, and hornblende (fissured)		Slate, gneiss, and calcareous mica	

Source Reproduced, with some additions, from I. Schoklitsch, *Hydraulic Structures*, vol. 1, p. 15, by permission of the publishers, the American Society of Mechanical Engineers, and the J. R. Freeman Trust Estate, New York, 1937.

New York Central Railroad in the state of New York now some years ago. An existing double-track tunnel, extensively used and only partly lined, had to be enlarged, and a new double-track tunnel had to be constructed parallel to the existing bore, with a 9-m (30-ft) rock wall in between them. The material penetrated was a hard granite gneiss. The northern portals are close to one of the siphon shafts carrying the Catskill water supply from the deep-water tunnel under the Hudson River. Clearly, the excavation work had to be done in such a way that the siphon shaft would not be affected by the excavation methods used. Because of the compact nature of the gneiss, vibrations from explosions, if used, would be transmitted without much damping over considerable distances. The excavation of the rock was therefore all carried out by the drilling of 60-cm (2⅜-in) holes, 1.8 m (6 ft) deep, all around the final outline of the tunnel section, spaced at 3-in centers. The 16-cm (⅝-in) gaps between each hole were then broached, after which the isolated core of rock was broken up and removed by standard methods. The operation was successfully completed, and the finished appearance of the rock surface is shown in the accompanying photograph (Fig. 20.13).[20.24]

As indicated when linings were discussed, material may be encountered that is affected by exposure to air. Certain types of clay and shale are particularly susceptible to the influence of the atmosphere, and a common practice (as followed, for example, in Toronto tunnel work) is to spray the shale with a thin coat of cement as soon as it is exposed. Sometimes more elaborate methods have to be utilized, as in the case of the construction of the P. L. M. Railroad tunnels on the line between Nice, France, and San Dalmazzo, Italy. Rock consisting of almost pure anhydrite was encountered in the Col de Braus Tunnel for a length of about 1 km (1,000 yd) and in the Caranca Tunnel for a few hundred meters. Water reaching the anhydrite in the former tunnel from adjacent Jurassic limestone caused the exposed rock to increase about 30 percent in volume. The trouble became more serious during the time when work was halted from 1914 to 1919. When excavation

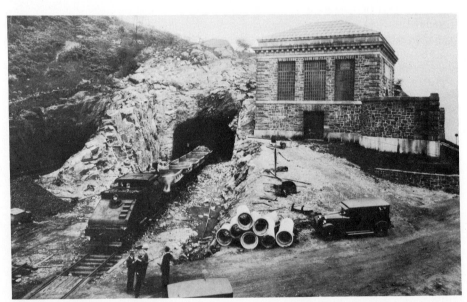

FIG. 20.13 Breakneck Tunnel of the old New York Central Railroad, New York, showing the portal of the new tunnel on right; note the proximity of the shaft house for the Catskill water-supply tunnel to the right of the new tunnel.

and lining were finally completed, the water which had not been completely drained continued its action, and a portion of the masonry lining was badly ruptured. Aluminous cement was used for its reconstruction, and an elaborate drainage system was installed; it appears that a satisfactory result has been obtained. This experience was utilized in the later construction of the Caranca Tunnel. As soon as mucking was completed, the rock was covered with coal tar; the lining was built in with the least possible delay, using aluminous cement; and after its completion, coal tar was injected under pressure behind the lining. The tunnel was put into use in 1921 and has given satisfactory service since.[20.25]

Methods of excavation have naturally advanced in recent years in keeping with general progress in mechanization and in equipment controls. Interesting though they are, these are engineering matters, reference to which will be found in some of the suggested readings at the end of this chapter. Brief mention must be made, however, of the remarkable developments in tunnel-boring machines, now so widely used that the acronym TBM is common in tunneling circles. Even though Colonel Francis Beaumont in 1883 used a simple type of boring machine for the pioneer exploration tunnel for the long-discussed Channel Tunnel, it was not until after the Second World War that the modern TBM started its phenomenal development. An interesting review of progress has been published, based on the excavation of diversion tunnels for dams on the Missouri River. Tunnels for the Fort Peck Dam (1934) saw trials of two coal saws adapted for tunnel work, but the experiment was not successful and was abandoned. For the Fort Randall Dam (1949) trials were resumed using a jumbo mounted circular-operating ring saw and some success was achieved. At the Oahe Dam (1954), four different types of specially designed TBMs were used with real success; rates of advance in faulted shale twice as fast as by conventional methods were achieved. A geologist of the U.S. Corps of Engineers was in constant attendance throughout this work at Oahe Dam, mapping every advance; predictions based on his detailed geological mapping proved of value to the tunnel contractor. And the machines so well tested on this major job were, in large part, the forerunners of the elaborate TBMs of today (Fig. 20.14).[20.26]

Postwar trials of excavating machines, at first based on coal-working machines, were tried on other jobs. As early as 1953, a coal digger was used, successfully, on a 1,200-m (4,000-ft) sewer tunnel in Euclid, Ohio, as but one example. Thereafter, advance was rapid. One of the latest developments has been the invention in England (later refined in Japan) of a "Slurry Mole," a TBM using bentonite slurry at the working face in poor ground, by an ingenious application of the same principle as that used for slurry-trench construction. High-pressure water jets have been incorporated in some shields, and elaborate controls have reduced the manpower necessary for the smooth operation of TBMs, when the going is good. The advance that has been made is indicated by the fact that no less than 382 TBMs were at work in Japan in 1979, this country being said to be carrying out more tunneling work than the rest of the world put together. A variety of types of TBMs are now available. They are costly machines and so are economical only for tunnels beyond a critical size. TBMs weighing 850 tons, with cutter heads operated by 9,850-kW motors, have been supplied for excavating 10.5-m (35-ft)-diameter tunnels in Chicago (Fig. 20.15).[20.27]

Tunnel-boring machines are indeed so remarkable a development that it is all too easy to allow their vaunted efficiency to obscure the fact that their use makes even more essential than usual the most accurate knowledge possible of the geology (including groundwater) along the route to be followed by the machine. There are already on record cases in which machines have been trapped underground by some unsuspected bad ground condition. All that has so far been said, therefore, about tunnel geology and the imperative of the best possible preliminary investigations must be reemphasized for all cases in which the use of TBMs is contemplated. One failure can more than wipe out all the savings that

FIG. 20.14 An early tunnel-boring machine used for a 9-ft bore through hard sandstone and limestone under Toronto, Ontario.

are possible with uninterrupted machine operation. As but one example, and one reportedly caused by the unsuspected presence of a "buried valley" between two boreholes, there may be mentioned the collapse of ground around a mole of the most modern type when it had completed 80 percent of an 8,475-m (27,800-ft) tunnel, 3.9 m (12 ft 11 in) in diameter, for a major new sewer on the island of Montreal. It took more than a year for the necessary remedial works to be completed (using compressed air) and for the mole to be rehabilitated.[20.28] Accurate knowledge of tunnel geology is, therefore, the basis of all successful TBM operations.

Compared with the use of modern boring machines, other aids to tunnel construction are apparently "unexciting," but they are nonetheless important. Freezing, as described for open excavation, may occasionally be the only solution when tunneling in very poor ground. In driving a double-track mass transit tunnel beneath a riverbed in Tokyo, Japan, a Japanese contractor froze 37,000 m³ (1.3 million ft³) of soil and then had to use a coal-machine excavator to dig out the frozen earth; the work was done successfully.[20.29] Rock bolts have ceased to be the temporary expedients of earlier years and are now a valuable tool for the stabilizing of weak rock faces, an operation which calls for careful design. Prereinforcement of rock, by inserting reinforcing bolts in holes drilled from a working face inclined away from the tunnel axis and ahead of the face, is a method of great promise.[20.30] And grouting is, as in so many construction operations, a vital aid to tunneling.

FIG. 20.15 A large, modern tunnel-boring machine breaking through one of the major Chicago drainage tunnels; the scale can be gauged from timbers in the foreground.

20.10 GROUTING

Grouting has become such a powerful aid in most construction operations that Chap. 27 has been devoted to this one subject. So important is it in tunnel work that brief reference here seems advisable, especially as grouting is usually employed for reducing the flow of underground water into tunnels. Its application and the necessity of its application, if anticipated before construction, therefore depend on the geological nature of the strata to be encountered. The methods adopted for grouting in tunnel work are similar to those generally used; a cement mixture is usual, although sometimes special chemical additions are injected first to ensure cementation of fine cracks.

Fissures or joints must clearly exist in the rock if it is to be grouted satisfactorily. Fissures or general porosity of the rock are presupposed in the necessity of grouting. Excessive inflow of water, structurally weak rock unable to support itself across the tunnel arch, and material liable to disintegrate when further exposed to the atmosphere are conditions which result from such rock characteristics. In all these cases, if the anticipated trouble is determined beforehand, as is generally possible through careful geological investigation, grouting can prove an effective remedy and prevent serious trouble. The construction of tunnels through water-bearing strata, such as coarse gravel, can be materially simplified in the same way.

20-32   CIVIL ENGINEERING WORKS

Leakage of river water into the famous Severn Tunnel of the (old) Great Western Railway in England was practically stopped by the application of a specialized cementation process many years after the tunnel's construction. The construction of the new Mersey vehicular tunnel was greatly facilitated by the same process. Similar precementation in water-bearing gravel facilitated the construction of some of the inclined escalator shafts for the London tube railways. In a section of one of the P. L. M. railroad tunnels in France, a tar injection was successfully adopted to coat voids behind the tunnel lining and to prevent further disintegration.

The work carried out in the Severn Tunnel may be mentioned in more detail. The tunnel was started in 1873, but progress was slow; in 1881, the river water broke into the tunnel workings, passing through open marl beds. These were successfully plugged up by depositing clay from schooners sailing in the river above the breaks. Operating conditions in the tunnel have always been troublesome, and the Great Western Railway Company (now part of British Railways) in the years 1924 and 1929 had to follow the earlier practice and dump clay above obvious danger spots. After prolonged investigation, aided materially by the geological record kept by the original contractor, it was decided to grout the natural strata behind the brick lining throughout a critical section of the tunnel. This work was carried out in two contracts; about 3 million kg (6.5 million lb) of cement was used in 2,100 holes totaling 2,850 m (9,500 ft) in one contract; and 5.5 million kg (12 million lb) was used in 2,400 holes totaling 3,150 m (10,500 ft) in the second. The existence of large voids between the brick lining and the surrounding strata, even beneath the invert of the tunnel, was proved by the travel of the cement. The riverbed above the tunnel was patrolled at low tide during the grouting operations; but in only one instance did cement actually travel to the surface; there it formed a protective covering over the mouth of a fault in the marl.[20.31]

Chemical grouting has also played a part in assisting with the control of groundwater in tunnels; one interesting application was in the tunnels of the Pennsylvania Turnpike. More than 50 years after contractors quit work on the ill-fated South Penn Railway, the seven tunnels then partially driven were adopted for the use on the great highway that now utilizes the old right-of-way. The completed tunnels vary in length from 1,053 to 2,035 m (3,511 to 6,782 ft). About half the length of the turnpike tunnels represents enlarged headings of the old railroad tunnels; the remainder were new bores. Detailed geological studies were made under the direction of A. B. Cleaves; the local geology was unusually complex, involving 19 formations of shales and sandstones. Careful studies of water seeping into the tunnels revealed pH values from 3.4 (highly acidic) to 8.3 (highly alkaline); the variation was in itself a reflection of the complex geology. Special provisions were made for drawing off the highly acidic waters to prevent damage to the concrete lining; vitrified-clay tile drains were widely used for this purpose. Gunited cover was provided for exposed rock in cavities above the finished lining. After a decade, seepage of groundwater through the lining in some of the tunnels was proving to be a nuisance, and a program for sealing them was initiated. A specialized chemical grouting system was used in which an expanded shale lightweight aggregate served as backfill. Considerable success in sealing the tunnels was achieved.[20.32]

20.11 SPECIAL GEOLOGICAL PROBLEMS

Almost all the tunneling problems so far mentioned can properly be described as geological; this would be true no matter how many examples were cited. The presence of unwanted groundwater is by far the most common, but there are other problems in tunnels, problems also dependent upon geology, that may be no less serious. This section provides a quick overview of these special aspects of tunnel work, the relative space devoted to each problem being proportional (roughly) to its incidence in practice.

Groundwater The presence of groundwater is often the main source of trouble in tunnel construction. It introduces at once the necessity of drainage facilities from all headings; and when the tunnel grades cannot be chosen to facilitate such drainage, the additional trouble and expense of pumping is necessitated. If water is present in any appreciable quantity, it will impede construction work of any kind; if the ground is soft and liable to be affected seriously by the drainage of water through or from it, the use of compressed air for all tunneling operations may become essential. It is of the greatest importance, therefore, before construction starts, to have as accurate information as possible about the groundwater conditions likely to be encountered. It may safely be said that no major civil engineering operations that are to be carried out below surface level should be started before something is known about the level and flow of groundwater at the site.

It is equally important to keep careful check on all water that is met with during construction. This will not only permit those in charge to determine the source of the water and therefore to seal the water off in some way, but it will also permit them to check so far as possible the course that the water follows and to make sure that no serious undermining or cavitation is being caused. In deep-tunnel work, *plutonic water* (water that has not come from the atmosphere or the surface of the earth) may be encountered; in this case, no such precautions can be taken. During the construction of the Simplon Tunnel, for example, alleged plutonic water was encountered at one point; a hot spring and a cold spring of water, differing by many degrees in temperature, issued into the tunnel within 1 m (3.3 ft) of each other. In contrast to this, in a test made during the construction of the Moffat Tunnel in Colorado, calcium chloride was dumped in a lake 420 m (1,400 ft) above the tunnel line, and traces of the chemical were found in the water entering the tunnel no more than two hours later.[20.33]

The more general factors affecting the presence of water in underground rock structures are considered in some detail in Chap. 8. These factors hold true for general tunnel work, although they are often complicated and sometimes present special features. The problem presented by groundwater in tunnel work is the reverse of that usually to be solved, since the water is not wanted in the tunnel, and its flow must therefore be stopped if this is at all possible. Only rarely can an underground water flow be sealed off at its source, even if the source is known, although this was done in the Severn Tunnel in England. Grouting is therefore generally the first recourse of the engineer, provided the quantity of water encountered is such as to render economic the expenditure of this additional money and provided also that it can be definitely stated, from knowledge of the geological formation concerned, that the grouting will be effective.

There may usefully be mentioned the experiences gained in connection with driving the two tunnels under the river Mersey at Liverpool, England, to which brief references have already been made. The Mersey is slightly less than 1.6 km (1 mi) wide at its mouth, where it separates the cities of Liverpool and Birkenhead; its banks are largely taken up with the great system of tidal docks necessitated by its 9-m (30-ft) range of tide. So great was the traffic across the river at this point that a tunnel crossing was discussed as early as about the year 1800. In 1885, the Mersey Railway Tunnel, which still provides for an important electric railway service, was opened. And in 1934, the Queensway Vehicular Tunnel, at that time the largest subaqueous road tunnel in the world, was opened. The local geological formation is the Bunter sandstone, a red sandstone of the Triassic system; the tunnels lie almost wholly in the middle of three Bunter beds. The sandstone is hard, porous, massively bedded, and jointed; the strike is approximately north-south, and the dip is between 2 and 5°. Faulting is a characteristic of this geological structure, and a Liverpool geologist, G. H. Morton, suggested in 1861 that a fault would be found below the bed of the Mersey. Throughout the district the bedrock is covered with glacial drift, boulder clay, sands, and gravels, and these deposits extend across the riverbed. In 1873, T. Mellard Reade, a distinguished Liverpool civil engineer and geologist, predicted that a buried river valley belonging to preglacial topography would be found under the present bed of the Mersey.

Construction of the Mersey Railway Tunnel started in 1879. Boreholes were sunk, and Sir Francis Fox, the engineer for the work, talked to Mr. Reade. The latter stated in a paper that he maintained the correctness of his forecast of the buried valley.

> ... notwithstanding that I was given a section showing a series of borings taken by the first promoters of the tunnel, which was supposed to prove a rocky bed all through. Fortunately the engineers who carried out the work, forewarned, took further borings, with the result that the *level* of the tunnel was lowered. The trial headings found rock all the way through, but in the actual construction of the tunnel the bottom of the buried channel was cut through by the roof of the tunnel for about 300 feet, and I had the satisfaction of seeing the verification of my prediction without disaster to the undertaking. If the levels had not been lowered the result might have been disastrous failure.

The contractors encountered the buried channel suddenly and withdrew all their men from the workings, but the clay cover held, and work was eventually resumed and successfully completed. Mr. Reade's concise and modest statement illustrates vividly the value of this application of his studies of the glacial geology of the Liverpool district.

The fault predicted by Mr. Morton was also found, and both features were again encountered in building the first modern vehicular tunnel. When this tunnel was started in 1925, Professor P. G. H. Boswell undertook the extensive geological investigations that were necessitated by the increased size of the new tunnel (13.2-m or 44-ft internal diameter), its location seaward from the Mersey Railway Tunnel, and the natural desire for economy in location and construction. Construction of the earlier tunnel assisted in these investigations, but a special study had to be made of the anticipated bottom level of the buried valley (Fig. 20.16). This hinged on the elucidation of the fact that the old drainage system flowed in the direction opposite to that of the present river. Thus, the level of the new tunnel could be kept reasonably high; the actual minimum clearance between the tunnel excavation and the bottom of the valley was only 90 cm (3 ft) at one spot. The old valley, although filled with water-bearing sand, was covered with impervious boulder clay, and so safety could be assured; because of the existence of this clay, which there came down to the rock surface, the railway tunnel could be completed, even though it did pierce through the rock surface. (See Fig. 20.17.) It was not practicable to put down boreholes in the river (owing to the range of tide and to the amount of shipping), and so all exploratory work was done by drilling ahead of the working face, up to 45 m (150 ft) away. Many of these holes were used later as grout holes. The total cost of this extensive preliminary and exploratory work amounted to less than 0.2 percent of the total cost; it was a remarkably good investment.[20.34]

A singular difference between the two tunnels is related to groundwater. The lining of the railway tunnel was not made watertight, so that even today about 22,600 lpm (5,000 gpm) have to be pumped out of its drainage sumps; the water has the composition of the Mersey and comes through the fissures and joints in the sandstone. The flow of water is *not* related to the size of the tide, but it does vary sympathetically with the rainfall. Pumping had to be carried out during the construction of the vehicular tunnel, and this permitted many interesting observations with regard to groundwater in the vicinity. The lining of the first vehicular tunnel is watertight, and the surrounding rock was grouted so

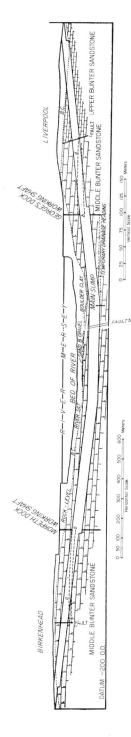

FIG. 20.16 Geological section along the center line of the underwater section of the Queensway Vehicular Tunnel under the river Mersey, Liverpool, England; *a-a* represents probable water level after construction pumping stopped, *b-b* represents water level under conditions of maximum pumping.

FIG. 20.17 Queensway Vehicular Tunnel under the river Mersey, Liverpool, England; a view showing final phase of rock excavation with permanent lining in place; scale may be judged by the two men in center background.

that, after its completion, groundwater conditions reverted to their previous state (Fig. 20.18).

In soft-ground tunnels, grouting may be equally effective but only for material, such as gravel, which the grout can penetrate sufficiently. Freezing has already been mentioned as another method (although of limited application) for sealing off troublesome underground water. Still another method is to use the special well-point installations utilized in open excavation work, with the well points jetted ahead of the working face, which is thus kept dry by the usual pumping process through the screens in the points. Finally, there is the more drastic procedure of actually lowering the underground water table to such an extent that the ground through which the tunnel has to be driven is above this level and so in the dry.

No record of groundwater problems in tunneling would be complete without a final reference to what has been called the "hottest and wettest tunnel ever driven," the Tecolote Tunnel of the Cachuma project of the U.S. Bureau of Reclamation. The tunnel is 10 km (6.4 mi) long and 2.1 m (7 ft) in diameter; it conveys water from the Cachuma Reservoir on the Santa Ynez River to the coastal area of California near Santa Barbara. It therefore penetrates the Cretaceous sedimentary rocks of the Santa Ynez Mountains, and it was known in advance that a major fault would have to be crossed by the tunnel. Trouble in driving was expected, based on experience gained with two roughly parallel tunnels driven many years before (the Mission and Doulton tunnels).

As one of the special precautions taken, therefore, the U.S. Geological Survey arranged to survey all springs and wells within 16 km (10 mi) of the new tunnel in case they dried up, since water flows, gas, and heavy ground had been encountered in the two earlier bores. Methane and hydrogen sulfide were found in small quantities and caused some trouble; deformation of temporary shoring also occurred. But it was a steady increase in

FIG. 20.18 West portal of the second Mersey vehicular tunnel (The Kingsway) under the river Mersey, Liverpool, England, with one tunnel in use, its twin then still under construction.

the quantity and temperature of water entering the tunnel that was the real hazard and led to a temporary shutdown of the work in 1953. Work was resumed the next year, and the tunnel was successfully holed through early in 1955, but only after further and almost incredible difficulty. The flow of water reached a maximum of 34,000 lpm (9,000 gpm) and the temperature reached 47°C (117°F); thus, working conditions at the face were almost unbearable.

Although hot water had been expected during the driving of the two earlier tunnels, since there is a well-known hot spring 3.2 km (2 mi) away, none had been encountered. A special combination of unusual geological conditions, at a rather local level, was responsible for the heat and water encountered in the Tecolote Tunnel; the heat was thought to be residual heat in the ground from geologically recent faulting. At first, attempts were made to grout around the tunnel in order to slow up the flow of water, but eventually the tunnel was completed without the aid of grouting; special pumping equipment was installed that, fortunately, was able to deal with the 34,000-lpm (9,000-gpm) flow. Men going to work at the face had to travel through the "hot" section of the tunnel, immersed in water up to their necks; they rode in dump cars specially provided for this job. It is small wonder that an Indian visitor described the job graphically as what he thought "the Christian hell" might be like. The experience is a salutary reminder that, even with the most expert geological advice, as was here available, minor local variations in geological structure in extremely distorted strata can give quite unexpected results; it points to the necessity of continual attention to every detail of the geology encountered in tunneling until the job is complete. This case also provides an illustration of how adversity can be turned to some good effect; the excessive flow from the tunnel was piped to the Santa Barbara area to relieve a serious water shortage; thus, the trouble in the tunnel proved to be the salvation of the local water authority (Fig. 20.19).[20.35]

Presence of Gases Occasionally, and more especially in soft-ground tunnels, methane gas may be encountered. If it is not detected, fire may result or, worse still, explosion. Methane is also lethal if inhaled. If, therefore, study of the local geology gives any suggestion that methane may be encountered, the most stringent precautions must be taken, such as the use of spark-proof equipment and the presence of gas detectors of a rugged type near the working face. The local shale rock in the Toronto area sometimes contains methane which can be released in excavations, as was shown at the site of Victoria Park pumping plant. When, therefore, the first section of Toronto's subway was constructed, the contractors were warned of this possibility and took all necessary precautions; fortunately no methane was encountered, but the possible danger had been recognized. There have been other recent tunneling jobs, however, in which methane was encountered unexpectedly, sometimes with disastrous results, including the loss of lives of some tunnel workers. In shales, in clays, and in any soils in which any organic matter has been detected, no chances can be taken, and the possibility of methane being present must be anticipated.

FIG. 20.19 Tunneling crew at the Tecolote Tunnel returning to their working heading, riding in water-filled mucking cars since the temperature at this location was 97°F (36°C) when the photo was taken.

Carbon monoxide has been encountered in tunnel work. Two men were killed when going to inspect the working face of a flood-relief sewer in the English Midlands, after a three-day suspension of work. It was reported that carbon monoxide and carbon dioxide, escaping from a former waste dump under which the tunnel was being driven, were responsible for their deaths.[20.36] Even more remarkable was the presence of nitrogen in shaft sinking for a new London tube railway station. Compressed air was applied to control groundwater conditions in a nearby shaft; it was thought that the extra pressure forced the "dead air" in the voids of intervening gravel (the oxygen having long since been dissipated) into the open working, with this potentially dangerous result.[20.37] Adequate ventilation arrangements provide one safeguard, but the necessary constant vigilance can be assisted by an appreciation of every aspect of the local subsurface conditions.

Weathered Rock Weathering of bedrock in situ is another problem calling for repeated mention in this volume; such weathering can have serious consequences if encountered in tunnel work. In areas featured by residual soils, bedrock may be altered to saprolite and so appear superficially still to be rock, even structural features being unaltered until the material is disturbed, when it reveals itself as soil. This was apparently the cause of much delay and trouble on the Wilson Tunnel in Honolulu, an 832-m (2,775-ft) vehicular tunnel with straight alignment. The first 570 m (1,900 ft) of the tunnel was driven without difficulty, but caving occurred, with loss of life, when the local lava rock changed to saprolite.[20.38] In preliminary studies for sinking a shaft for access to a new cable tunnel to be constructed under the river Thames between Tilbury and Gravesend, borings had suggested that solid chalk would be encountered, but when bedrock was reached, it proved to be badly disintegrated to a depth of 6 m (19.7 ft). The fracturing had left the bedrock in place, with the superficial appearance of solid rock; the fracturing was explained as probably being the result of exposure to earlier permafrost conditions.[20.39] Similar fractured siltstones are encountered in Czechoslovakia.

20.12 CONSTRUCTION RECORDS

In all tunnel construction work, it is of the greatest importance that accurate, complete, and up-to-date geological records be maintained from the start of construction. This can be done most conveniently by preparing a geological map of the tunnel route and a geological section along the line of the tunnel. The exact nature of the rock excavated must be observed after every round has been fired, and the direction of strike and dip of the rock must be recorded regularly. All this information should be carefully and continually compared with the geological section along the tunnel, predicted from preliminary investigations before construction started, and with similar records obtained at other headings in the tunnel. In addition, it must be compared with surface topography and geology above the tunnel and in its vicinity in order that as complete a picture as possible may constantly be available for study. In this way, a close check can be kept on the relation of observed to anticipated formation; any unusual departure from the latter can be checked as soon as it is discovered, and its implications can be investigated. Alternatively, if difficulties are encountered underground because of fissures or faulting, surface conditions may serve as a guide to future similar troublesome spots, once a correlation has been established.

The geological record so obtained is of great interest and importance to the geologist in another direction entirely; it gives information obtainable in no other way and, if the tunnel is to be lined or used as an aqueduct, at no other time. Such records can therefore be of supreme value to the community as a whole, and all civil engineers in charge of tunnel work should therefore invite the attention of the director of the local geological survey to any section available for inspection and should furnish to the survey a copy of the final section obtained. The courtesy, involving no expense and but little trouble, should likewise be extended when possible to the head of any university geological department in the immediate neighborhood. The discretion of scientists is such that no engineer need ever hesitate to take the step here suggested.

There are today, most fortunately, many examples of tunnels that have yielded invaluable geological information, thanks to such cooperation as is here suggested. Even radioactivity in rocks has been recorded in some Swiss tunnels. Quite the most important instance known to the writers is recorded in the moving autobiography of the great German geologist, Hans Cloos. Entitled *Conversation with the Earth* and available in English translation, it may be recommended without reservation to all engineers interested in geology; it is a most human and stimulating document, penned (as it was) shortly before the death of Dr. Cloos as a "valve and a means of salvation from the confinement, brute force, and untruth of the times." Dr. Cloos was greatly interested in the Rhine Graben, a geological feature of western Germany about which controversy had raged for many years.

He tells of the interest to him, as possibly providing a clue to this geological riddle, of the proposal to build a new approach to the main railway station of the lovely city of Freiburg in the Black Forest near the Swiss border. Although the original contract for the work was signed in 1910, the years of the First World War and postwar conditions led to such a delay that it was not until August 1928, that work was actively resumed. The tunnel had to be 509 m (1,696 ft) long and was to pierce the Lorettoberg, below the line on the Gunterstal. Its western portal would be cut into sandstone and its eastern into gneiss, so that it might reveal the exact interrelation of the young sediments of the Rhine Valley (limestone, sandstone, marl, and clay) and the ancient gneiss of the Black Forest and the Vosges. Dr. Cloos visited the work regularly and in April 1929, he was able to see the excavation completed and to receive the answer that he had so long been anticipating.

> There it was right before my eyes ... a furrow ran up the wall into the ceiling and down the other side, as neatly as if it had been done with a knife ... dipping 55° away from the mountain and downward below the plain.... I took pencil and paper and drew as much of the three-dimensional picture as I could reduce to lines and planes. For the wet spongy stone would not be there very long. It was to be sheathed in, and only a little window would be left open through which posterity may catch a glimpse of the extraordinary phenomenon that I have been privileged to see.[20.40]

Unfortunately, it has proved impossible to get a photograph of this unusual "window."

In addition to their vital importance to pure geology, construction records often have practical application and general utility. Should any trouble develop in the future operation of the tunnel, the cause may often be traced to some unsatisfactory condition of the surrounding ground, if this information can be obtained from the records available. When tunnels, especially pressure tunnels, are being examined after continued periods of use, a geological section will always help to conserve time and increase the value of the survey by indicating the sections that need special attention. If a railway tunnel is to be increased in cross section, an accurate geological record of the original work will be of inestimable value. And if there should unfortunately be litigation with regard to the construction of a tunnel, as is sometimes caused by disputes over overbreak, a geological record of the tunnel line will often be a deciding factor in a court of law.

As an example of tunnel construction in which close attention to geology during construction was an important feature, the main tunnel of the Lochaber Water Power scheme

(Scotland) may be cited. The tunnel extends from the main reservoir for the project, Loch Treig, to the side slope of Ben Nevis (Great Britain's highest peak) overlooking Fort William, on Loch Linnhe; it is slightly more than 24 km (15 mi) long and has an effective average diameter of about 4.5 m (15 ft). The tunnel passes through rock strata which constitute one of the most complicated parts of Great Britain's geology. Many types of rock were encountered, from shattered mica schist to granite and baked schist of excessive hardness (some requiring three sets of drill steel in each hole drilled into them). Prior to construction, a survey of the route selected for topographical reasons was made by E. B. Bailey (later director of the Geological Survey of Great Britain), who for many years had made a close study of the rocks to be encountered. The preliminary information then obtained was confirmed to a remarkable degree by the geological section revealed by the excavation.

During the construction of the tunnel, Dr. Bailey made a complete survey of it and collected many rock samples; in this way the detailed mapping of the surface above has been completed. Since several interesting facts not shown by surface outcrops were discovered, the public records were thereby enriched. Most of the staff of the resident engineer (one of the writers was a member for all-too-short a period) had a working knowledge of geology and so were able to observe and record the varying geological features encountered in the course of their regular engineering work. From among the many interesting features of the work, one may be mentioned. At the Loch Treig end of the tunnel, excavation proceeded in soft mica schist, much disturbed by faulting and lines of movement. There were seams of clay in this section, and the rock adjacent to them was soft; special timbering and cast-iron lining had to be used. After two of these disturbed areas had been passed, it was noticed that they corresponded to the lines of cliffs on the hill above the tunnel; accordingly, succeeding lines of cliff were surveyed, and the presence of bad rock in the tunnel was predicted thereafter to within a few feet of where it actually occurred. The value of such work to the contractors needs no elaboration.[20.41]

Two cases may be cited in conclusion in which the absence of construction records led to serious and unfortunate consequences. The first long railway tunnel ever to be constructed was the Woodhead Tunnel on the line connecting Manchester and Sheffield in the Pennine district of northern England. Twin single-line tunnels, one completed in 1845 and the other in 1852, are 4,772 m (15,906 ft) long, making them the fourth longest tunnel in the United Kingdom. Disintegration of the brick lining in the 1940s made maintenance difficult, and daily traffic of over 80 trains on each line left little time available for the necessary repairs. In 1948, when the electrification of this important rail link was pending, it was decided that a duplicate double-track tunnel, parallel to the existing twin tunnels, would be constructed. Owing to a necessary slight realignment, the length of the new tunnel is 4,811 m (16,037 ft), making it now the third longest. In cross section, it was excavated as 9.2 m (30 ft 6 in) wide and 7.9 m (26 ft 4 in) high in the center; a mass concrete lining reduced the effective width to 8.1 m (27 ft). With the existing tunnels alongside, it might have been thought that construction of the new tunnel would be a fairly straightforward operation, but this did not prove to be the case. Since no technical account of the planning and execution of the old tunnels could be traced, no firsthand record of the behavior of the shale and sandstone rock that had to be penetrated was available. On the other hand, there was discovered a large and detailed longitudinal section along the original tunnel and through the five construction shafts. This was the section made in 1845 by the original resident engineer and deposited in the museum of the Geological Survey; it showed the strata encountered, the fossils found, and the limits of all geological features. In the report of the government's inspecting engineer (of 1852), reference was made to "bulging of the sidewalls" for a distance of about 15 m (50 ft), a condition that was corrected before the tunnel was approved for public use.

20-42   CIVIL ENGINEERING WORKS

The new tunnel was completed toward the end of 1953; there had been a great deal of trouble with rockfalls and even with the method originally planned for excavation, owing mainly to the character of the black argillaceous shale that had to be penetrated for much of the total length. At about 270 m (900 ft) from the Woodhead end of the tunnel, just as the transition from slabby sandstone into the shale had been made in the enlargement to final section of the pilot tunnel first excavated, a very large rockfall took place, involving a length of 30 m (100 ft) of tunnel; the cavity thus formed extended to a height of 21 m (70 ft) above the invert. The fall crushed the steel-rib supports that had been erected, as shown in Fig. 20.20. The accident delayed work at this location by six months, but eventually the difficult section was passed, excavation completed, and the tunnel lined for final use. Construction records from the earlier tunnels would almost certainly have disclosed similar trouble, even though the bores were smaller in cross section. The case is a useful warning of the troubles that may be encountered with shale; the record that is now fortunately available of this tunnel job should be consulted by all who have major tunnel work to carry out in shale that is in any way of dubious character.[20.42]

The second case also comes from England; it involves the railway tunnel at Clifton Hall on a short branch line used for mineral traffic between Patricroft and Molyneux

FIG. 20.20   Roof collapse between 270 and 291 m (900 and 972 ft) from the portal of the new Woodhead Tunnel, built for British Railways.

Junction, not very far from the Woodhead tunnels. The tunnel was only about 1,170 m (3,900 ft) long; it was straight in plan, lined with brick, of horseshoe shape, 7.4 m (24 ft 9 in) wide and 6.7 m (22 ft 3 in) high. It was constructed in 1850 and used until the time of the Second World War. At that time it was closed, but it was reopened for limited freight traffic in October 1947. It was regularly and carefully inspected. On the morning of 13 April 1953, the "ganger" responsible for this stretch of line noticed a small amount of brick rubble on the track; he immediately reported this, and swift measures for detailed inspection were taken by the railway engineering staff concerned. It was clear that the brick lining at the location noted was under stress; arrangements were made to have steel ribs made up so that the tunnel arch could be reinforced. Before these could be installed, early on the morning of 28 April, the tunnel crown failed, precipitating a mass of wet sand and rubble into the tunnel. In itself this would have been serious, but the debris and sand came from an old construction shaft that had been located at this point; the contents of the shaft were precipitated into the tunnel and a crater formed suddenly at the surface. Three houses had been built over the top of the unknown and completely hidden shaft; all collapsed into the shaft and the five residents of two of the houses were killed.

The official inquiry brought out the fact that the district engineer responsible had no knowledge at all of the existence of the old shaft, since tunnel records had been destroyed when the district engineer's office was badly damaged in an air raid in 1940, and the records that were left were lost in a fire in 1952. Records were found later that did disclose the fact that eight shafts had been used during construction and that the work had been difficult because of the presence of water-bearing sand. Considerable timbering had been necessary in the shafts. It appeared that some of the old timbers had been built into the brick lining and had decayed in the course of time; ultimate failure of the rotted timber had the tragic result described. No blame could be attached to any of those responsible for the tunnel, but Brigadier Langley, the inspecting officer for the Ministry of Transport and Civil Aviation, included this warning in his official report; it is a statement that well warrants quotation.

> The loss of the tunnel records contributed materially to this accident, and the events leading up to it have shown only too clearly the danger that arises when vital knowledge is not readily available. The maintenance staff should know of the existence of old shafts and other features which may cause weaknesses but in many cases the only records are the original construction drawings which with the growth in the number of documents to be preserved in engineers' offices, may possibly be overlooked.... I recommend, therefore, that all tunnel records be reviewed and that any special features be brought to the notice of the maintenance and examining staff.[20.43]

This procedure was immediately adopted by the civil engineer of the London Midland Region of British Railways; it is one which could well be widely followed. The Clifton Hall Tunnel has since been filled in completely, and its traffic has been diverted to alternative routes.

Neglect of essential tunnel records is not confined to Great Britain. There is, for example, one well-known tunnel in North America in which some improvements had to be effected; it was then found that there were no records at all of the tunnel as excavated! Attention must rather be directed to the many cases in which accurate records of tunnels and their geology have been made. From the many cases that could be cited, the records of the tunnels under the city of Boston constitute such an excellent example that they may be described in some detail, since they illustrate so well a number of the important facets of tunnel geology discussed in this chapter.

20.13 TUNNELS UNDER BOSTON

Metropolitan Boston is located around the deeply indented bay that forms the famous Boston harbor, with the Charles River coming in from the west, past Cambridge, and the Neponset River from the south. The local bedrock geology is most complex, but it has naturally been intensively studied so that its main features are well known. The main feature is what is locally known as Boston basin; this is marked on Fig. 20.21. The Boston beds that occupy this basin are generally divided into the Roxburgh conglomerate and the Cambridge argillite. To the north, the latter is in contact with the Lynn volcanic rocks. Over all these folded and jointed rocks lie the surface soil deposits with an interest all their own, as will be explained elsewhere in this volume. From Fig. 20.21 it will be seen that all the tunnels marked lie within the Boston basin, except for the Malden Tunnel that crosses its northern boundary. All the tunnels other than the Dorchester Bay Tunnel (built in the 1880s) were projects of the Metropolitan District Commission of the Commonwealth of Massachusetts, through its construction division. Geologists on the staff of the commission and special consultants were able to carry out the necessary preliminary investigations, including test borings into rock as necessary, so that all the recent tunnels were completed without undue difficulty, all by experienced contractors.

The Main Drainage Tunnel is 11.4 km (7.1 mi) long; for a large part of its length it runs under the waters of the harbor. It was constructed between 1954 and 1959. It consists of two main headings connecting three shafts, two of which were used as main construction shafts. Between shaft A and shaft B the tunnel is 3 m (9.8 ft) in diameter; between shaft B and shaft C, 3.45 m (11.3 ft) in finished diameter. The full tunnel is lined with a concrete lining up to 60 cm (24 in) in thickness so that the tunnels as excavated were slightly more than 3.6 m (11.8 ft) and 4.05 m (13.3 ft) in diameter, respectively. The rocks penetrated were the Cambridge argillite, which proved to be the most competent, requiring steel supports for only 11 percent of its exposure, and the Dorchester shale of the Roxburgh formation, the least competent, requiring supports for 87 percent of its exposed length. A short stretch of conglomerate in the Dorchester formation was sound, the shales and argillites being the rocks that required almost continuous support. Much jointing and many minor faults were encountered, but these did not seriously influence the need for support.[20.44]

The City Tunnel extension constructed between 1951 and 1956 is almost the same length as the Main Drainage Tunnel (11.4 km or 7.1 mi). It is also through solid bedrock for its entire length, its invert being between 30 and 117 m (98 and 384 ft) below mean low sea level, with ground level above the tunnel rising to 60 m (196 ft) above this datum. The tunnel has a finished diameter of 3 m (9.8 ft) but an excavated diameter of about 4.05 m (13.3 ft). It conveys water from the Chestnut Hill Reservoir in western Boston to the southwestern part of Malden. The tunnel, located within the Boston basin, penetrated the same type of rocks as did the Main Drainage Tunnel, but they proved to be much better for tunneling, only 5.6 percent of the total length requiring steel supports; one-third of this was in weak shales. Another third was related to dikes, the remainder to shear zones at faults in the rocks, joints, and fractures. The tunnel cuts diagonally across one of the major folds within the Boston basin and so revealed geological information of much interest. One hundred and six faults were mapped, most of them with apparent displacements of only a few meters, but in 17 cases the displacement amounted to more than the height of the tunnel (Fig. 20.22).[20.45]

The Malden Tunnel was the last to have been built, its construction taking place in 1957–1958. It is slightly less than 1.6 km (1 mi) long and forms part of the Spot Pond Brook Flood Control Project. Figure 20.21 will show why it was known in advance that the tunnel would pass through the northern boundary of the Boston basin. Since this was

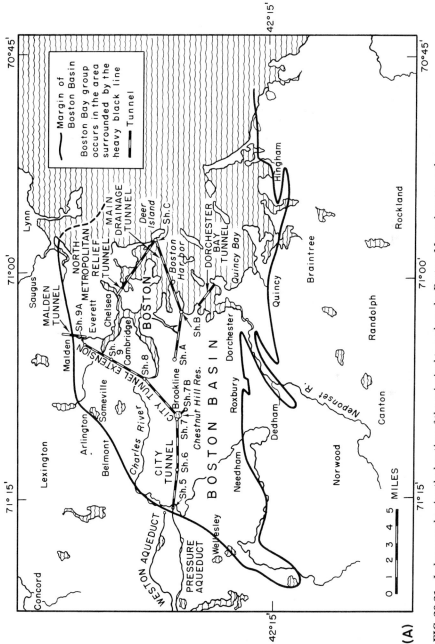

FIG. 20.21 Index map showing the location of the principal tunnels under Boston, Massachusetts, and the outline of the Boston basin.

FIG. 20.22 Construction view in the City Tunnel of Boston, Massachusetts, showing the effect of the dip of the bedrock on the tunneling operation.

believed to be a major fault, bad driving was to be expected. The location of the contact was determined with reasonable accuracy by test borings and drilling. Excavation proved the accuracy of the predictions, driving in the vicinity of the major fault—for the existence of the fault was clearly demonstrated when excavation got to the contact—being difficult. Fifty-two percent of the total length of tunnel required steel supports, all in the vicinity of the fault. The actual contact of the Lynn felsite to the north and the Cambridge argillite to the south was tight on the west side of the tunnel but had a gouge-filled separation of about 2.5 cm (1 in) on the east side. The effect of the faulting, however, had been to shear the rocks on both sides very seriously, with the result that a great deal of water was encountered, all of which could be attributed to the fractured condition of the rock. Pumping rates varied but reached an average of 5.5 million liters (1.45 million gallons) per day for February 1958, in contrast to the relatively dry state of the two other tunnels that have been described. Invert of the tunnel is 82.5 m (270 ft) below the local datum, land surface varying between 1.8 and 17.1 m (6 and 56 ft) above. The tunnel has a finished internal diameter of 3.75 m (12.3 ft) and was mainly driven from the southern of the two shafts at its ends.[20.46]

The structure of the Boston basin has been intensively studied not only by members of the staff of the U.S. Geological Survey (one of whom, Dr. L. LaForge, wrote a masterly report upon it) but also by Professor Marland P. Billings of the department of geology at Harvard University. Dr. Billings has himself studied in detail the geology of the tunnels just described and has encouraged his students to do the same. The result has been a notable series of papers presented to the Boston Society of Civil Engineers, the oldest engineering society of the United States, and published in their journal. Although the geology of these tunnels was studied as excavation progressed, the most detailed mapping that Dr. Billings directed (with the cooperation of the staff of the Metropolitan District Commission) was done after the tunnel walls had been washed down just prior to the placing of the concrete lining. This involved intensive work during a short period, but methods of mapping were developed that made possible the recording of the complete tunnel sections to a scale of 1 to 240. Copies of the resulting folios were placed in the custody of the construction division of the commission and with the department of geological sciences of Harvard. The permanent value of these folios can well be imagined; they can indeed be studied by all, since, with the aid of ingenious graphical symbols, small-scale sections through the tunnels were included in the BSCE papers in which they were described.

More than description was involved, however, for the study of the geology has revealed answers to long-standing questions about the Boston basin. In the case of the City Tunnel extension, the tunnel revealed "for the first time that the Roxburgh and Cambridge formations are actually facies of one another ... [diagrams showing] that the Cambridge type of lithology inter-fingers with the Roxburgh type of lithology...." The explanation is that the Boston Bay group of sediments that became the present bedrock was deposited in a basin, the movements of water across which, in combination with a large deltaic formation, resulted in the variety of rock types in the two formations. More remarkable was the geology as revealed in the Malden Tunnel, for here the major fault, long suspected, between the Cambridge formation and the Lynn volcanic rocks to the north was clearly to be seen. It was naturally examined with unusual care and recorded in detail, confirming the deductions made from study of the geological structure of the basin. These Boston tunnel papers can be commended without reservation as splendid examples of the sort of record that should be prepared and published for all major tunnels and for all those exhibiting unusual geological features.[20.47]

20.14 CONCLUSION

Many more examples could be quoted from the records of tunnel practice around the world, notably from Japan. The availability of this material is probably due to the fact that every tunnel job is unique at least in some one respect and often in several. Descriptions of tunnel construction consequently figure prominently in civil engineering literature, a fortunate feature in connection with any study of the application of geology to civil engineering work in view of the dependence of all tunnel work on geological conditions. This dependence will probably be clear from the examples that have already been quoted; the constructive contribution that the science has made to the art will be obvious. Notable among available records is the compilation by the California Department of Water Resources of complete records for 99 tunnels of various size, shape, character, and geology. Charts prepared from this study provide one of the most comprehensive estimating guides of this kind available in civil engineering literature.[20.48]

As a final note, reference may fitly be made to cases in which even the nature of the material encountered has permitted the use of special methods during construction. In connection with tunneling in soft ground, for example, the character of the clay encountered in London, Chicago, and Winnipeg permits its removal in large slabs "cut off" the working face by means of specially shaped power-operated knives. In great contrast to this work of peace may be mentioned the unusual procedure developed during the excavation of mine tunnels in the chalk of northern France during the First World War. In order to soften the chalk before excavation and to reduce the noise of the underground working, auger holes were bored into the working face and filled with vinegar, a strange reversion to the practice of the Romans.[20.49]

Many great tunnels will still be built; the story of man's tunneling activities is an ever-widening one. Even as this volume goes to press, renewed consideration is being given yet again to one of the most widely discussed of all tunnel projects not yet undertaken, a project which will probably one day build the tunnel under the English Channel to link England and France (Fig. 20.23). As early as 1800, a proposal to construct a tunnel was

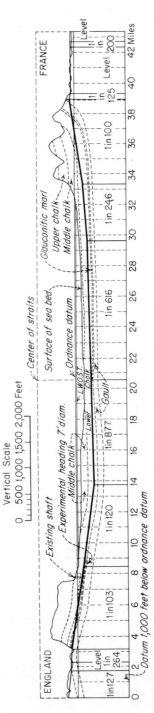

FIG. 20.23 Generalized geological section along the line of the proposed Channel Tunnel beneath the Straits of Dover between England and France.

made; in 1867, the first definite plans appear to have been advanced; and thereafter public interest in the project has continued. A great deal of geological investigation was carried out; 7,000 soundings across the channel were taken in 1875, for example, and almost 4,000 samples were obtained from the seabed. Exploratory shafts were sunk, and there exist today several trial headings which extend for some distance under the sea. The consensus of general geological opinion is that the proposed tunnel (63 km or 39 mi long) should be constructed throughout in the Lower Chalk Measures, although the continuous existence of this formation across the channel can only be surmised. M. Fougerolles has worked out in detail a scheme for disposing of the excavated chalk by grinding it into a slurry and pumping it upward into the sea. A lengthy report on the project was made to the British government in 1930; it was favorable, but the government subsequently decided against the project, not on technical grounds but as a matter of policy. Further and even more exhaustive studies were made jointly by France and Great Britain in postwar years; some consideration was even given to the construction of a bridge as an alternative. Again, the project was set aside, but as this book goes to press there are signs of a revival of official interest.[20.50]

For the benefit of those readers who may think that the foregoing reference to pumping out tunnel excavations is facetious, it may be mentioned that during the construction of sewer tunnels for the twin cities of Minneapolis and St. Paul, Minnesota, some tunneling work was carried out in the local white sandstone formation. Although this rock has a good appearance, it can be cut with a high-pressure water jet. This means of cutting was adopted to break up the working faces; a 5-mm ($\frac{3}{16}$-in) jet at 21 kg/cm² (300 psi) pressure was used. The rock quickly disintegrated to the state of fine sand, and the resulting slurry was pumped, by means of centrifugal pumps, along the tunnel and up the working shafts to settling bins at street level (Fig. 20.24).[20.51] And with this reference to solid rock being pumped through a 10-cm (4-in) pipeline, perhaps this chapter should be brought to a close. (Fig. 20.25)

FIG. 20.24 Intercepting sewer tunnel in Minneapolis-St. Paul, showing jetting of the St. Peter sandstone and the pumping of the resulting sand slurry, as an excavation process.

FIG. 20.25 Frozen working face in morainal material (silt, sand, gravel and boulders) of the Milchbuck Tunnel, for traffic improvement in Zurich, Switzerland.

20.15 REFERENCES

20.1 For some of these historical notes the authors are indebted to C. Prelini, *Tunneling*, 6th ed., Van Nostrand, Princeton, N.J., 1912.

20.2 Francis Fox, "The Simplon Tunnel," *Minutes of Proceedings of the Institution of Civil Engineers*, **168**:61 (1907).

20.3 "The Lötschberg-Simplon Tunnel and Its Construction," *The Engineer*, **112**:633 (1911).

20.4 "Mont Blanc Tunnel will be the World's 'Longest' Shortcut," *Engineering News-Record*, **162**:30 (26 March 1959); see also same journal, **164**:45 (28 April 1960) and **168**:56 (31 May 1962).

20.5 B. N. Page, "Geology of the Broadway Tunnel, Berkeley Hills, California," *Economic Geology*, **45**:142 (1950).

20.6 "The Las Raices Tunnel, Chile," *Civil Engineering and Public Works Review*, **28**:335 (1935).

20.7 C. P. Berkey and J. F. Sanborn, "Engineering Geology of the Catskill Water Supply," *Transactions of the American Society of Civil Engineers*, **86**:1 (1923); see also C. P. Berkey, *Geology of the New York (Catskill) Aqueduct*, New York State Museum Bulletin 146, 1911.

20.8 T. W. Fluhr, "Engineering Geology of the Delaware Aqueduct," *The Municipal Engineers Journal*, **27**:91 (1941).

20.9 "Driving the Queens Midtown Tunnel," *Engineering News-Record*, **124**:29 (1940).

20.10 "Tunnel Looped Under a Fault," *Engineering News-Record,* **119**:220 (1937).

20.11 Celia B. Noble, *The Brunels, Father and Son,* Cobden-Sanderson, London, 1938, p. 80.

20.12 J. Kell, "Pre-treatment of Gravel for Compressed Air Tunneling under the River Thames at Dartford," *The Chartered Civil Engineer,* March 1957, p. 3.

20.13 R. F. Legget, "CNR Tunnel under St. Clair River Still in Use After Almost 90 Years," *Canadian Consulting Engineer,* **21**:56 (September 1979).

20.14 H. C. Boschen, "Havana Traffic Tunnel Built in the Dry Under Almendares River," *Civil Engineering,* **23**:447 (1953).

20.15 R. D. Coren, "Kammon Highway Tunnel, Japan," *The Military Engineer,* **49**:346 (1957).

20.16 "Japanese Tackle Water to Drive Record Tunnel," *Engineering News-Record,* **192**:16 (4 April 1974).

20.17 A. D. Hogg, "Some Engineering Studies of Rock Movement in the Niagara Area," in *Engineering Geology Case Histories,* vol. 3, Geological Society of America, Boulder, Colo. 1959, p. 1.

20.18 J. V. Clyne, *Report on Inquiry . . . into the Circumstances of Landslides at Whatsham,* Government of British Columbia, Victoria, B.C., 4 March 1954.

20.19 G. Haskins, "The Construction, Testing and Strengthening of a Pressure Tunnel for the Water Supply of Sydney, N.S.W.," *Minutes of Proceedings of the Institution of Civil Engineers,* **234**:25 (1932).

20.20 "Isambard Kingdom Brunel: 9 April 1806–15 Sept. 1859," *Proceedings of the Institution of Civil Engineers,* **14**:9 (1959).

20.21 F. Walker, (ed.), *Daylight through the Mountain,* Engineering Institute of Canada, Montreal, 1957.

20.22 J. R. Glaeser, "Lining Concrete Pumped into Place," *Civil Engineering,* **23**:753 (1953).

20.23 K. Y. Lo, M. Devata, and C. M. K. Yuen, "Performance of a Shallow Tunnel in a Shaly Rock," *Proceedings of the International Symposium on Tunneling,* London, 1979.

20.24 Chief engineer, New York Central Railroad Company, by personal communication.

20.25 Le chef du Service de la Voie et des Batiments de Sud-Est, S.N.C.F., Paris, France, by personal communication.

20.26 L. B. Underwood, "Machine Tunneling on Missouri River Dams," *Proceedings of the American Society of Civil Engineers,* **91**, CO 1, Paper no. 4314, 1965.

20.27 For a useful general review, see "Boring Machines Will Drive the World's Tunnels," *Engineering News-Record,* **187**:22, (22 July 1971); information also from G. W. Murphy Industries, Houston.

20.28 N. Hancock, "Contractor Successfully Rescues Mole Trapped by Cave-In," *Engineering and Contract Record,* **90**:54 (October 1977).

20.29 "Soil on Ice: Cooling Pipes Toughen Weak Riverbed Silt," *Engineering News-Record,* **200**:98 (18 May 1978).

20.30 C. E. Korgin and T. L. Brekke, "Field Study of Tunnel Pre-Reinforcement," *Proceedings of the American Society of Civil Engineers,* **104**(GT 8):1091–1108 (1978).

20.31 R. Carpmael, "Cementation in the Severn Tunnel," *Minutes of Proceedings of the Institution of Civil Engineers,* **234**:277 (1933).

20.32 C. W. Stickler and A. Allan, "Chemical Sealing Stops Leakage in Tunnels of Pennsylvania Turnpike," *Civil Engineering,* **24**:722 (1954).

20.33 R. H. Keays, "Construction Methods on the Moffat Tunnel," *Transactions of the American Society of Civil Engineers,* **92**:69 (1928).

20.34 D. Anderson, "The Construction of the Mersey Tunnel," *Journal of the Institution of Civil Engineers,* **2**:473 (1936); see also P. G. H. Boswell, "The Geology of the New Mersey Tunnel," *Proceedings of the Liverpool Geological Society,* **17**:160 (1937).

20-52 PRELIMINARY STUDIES

20.35 E. R. Crocker, "Hottest, Wettest Tunnel Holed Through," *Civil Engineering,* **25**:142 (1955); see also *Technical Record of Design and Construction of the Tecolote Tunnel,* U.S. Bureau of Reclamation, Denver, September 1959.

20.36 "Two Gassed in Mini-Tunnel," *New Civil Engineer,* **139**:6 (17 April 1975); see also same journal, "CO_2 Death Verdict on Mini-Tunnellers," **149**:4 (26 June 1975).

20.37 H. D. Morgan and J. V. Bartlett, "The Victoria Line, The Project," *Proceedings of the Institution of Civil Engineers,* suppl. vol. 1969, Paper 7270S, 1969, p. 377.

20.38 "Honolulu Plugs Wilson Tunnel," *Engineering News-Record,* **154**:22 (24 February 1955); see also same journal, "Hawaiian Tunnel Will Get New Start," **156**:25 (9 February 1956).

20.39 C. K. Haswell, "Thames Cable Tunnel," *Proceedings of the Institution of Civil Engineers,* **44**:323 (1969).

20.40 H. Cloos, *Conversation with the Earth,* E. B. Garside (trans.), Routledge, London, 1954.

20.41 B. N. Peach, "The Lochaber Water Power Scheme and Its Geological Aspects," *Water and Water Engineering,* **32**:71 (1930).

20.42 P. A. Scott and J. L. Campbell, "Woodhead New Tunnel: Construction of a Three-Mile Main Double-Line Railway Tunnel," *Proceedings of the Institution of Civil Engineers,* pt. I, **3**:506 (1954).

20.43 "Report of the Collapse of the Clifton Hall Tunnel," Ministry of Transport and Civil Aviation, H. M. Stationery Office, London, 1954.

20.44 D. A. Rahm, "Geology of the Main Drainage Tunnel, Boston, Massachusetts," *Journal of the Boston Society of Civil Engineers,* **49**:319 (1962).

20.45 M. P. Billings and F. L. Tierney, "Geology of the City Tunnel Extension," *Journal of the Boston Society of Civil Engineers,* **51**:111 (1964).

20.46 M. P. Billings and D. A. Rahm, "Geology of the Malden Tunnel," *Journal of the Boston Society of Civil Engineers,* **53**:116 (1966).

20.47 M. P. Billings, "Geology of the Boston Basin," in *Studies in New England Geology,* Memoir no. 146, Geological Society of America, Boulder, Colo., 1976, p. 5.

20.48 "Procedure for Estimating Costs of Tunnel Construction," *Investigation of Alternative Aqueduct Systems to Serve Southern California,* app. C, California Department of Water Resources Bulletin no. 78, 1959. See also "Tunnel Estimating Improved: Tied to Geology," *Engineering News-Record,* **163**:64 (17 December 1959).

20.49 W. G. Grieve and B. Newman, *Tunnellers,* Jenkins, London, 1936, p. 136.

20.50 For a good overview, see T. Whiteside, "The Tunnel in the Chalk," *The New Yorker,* **42**:105 (18 November 1961) and **42**:104 (25 November 1961).

20.51 "Hydraulic Mining on Twin Cities Sewer Tunnel," *Engineering News-Record,* **115**:627 (1935).

Suggestions for Further Reading

Tunneling in Rock by E. E. Wahlstrom is a 250-page volume published by Elsevier of New York in 1973 that presents a useful review of its subject. *Tunneling Technology* by T. N. Nasiatka is the title of a 1968 Bulletin of the U.S. Bureau of Mines that features a helpful bibliography. It is in the proceedings of the numerous conferences on tunneling that are now being held regularly that much current useful information can be found, additional to that to be found in the pages of the journals listed in Appendix D. Proceedings of the annual conferences of the International Tunneling Association (the fifth of which was held in Atlanta in 1980), for example, are published by Pergamon Press. The Institution of Mining and Metallurgy (44 Portland Place, London WIN 4BR) publishes useful reports of British tunneling meetings, *Tunnelling '82* being the most recent as this volume goes to press. *Tunnels and Shafts in Rock* is a 417-page publication of the U.S. Corps of Engineers of 1978.

chapter 21

UNDERGROUND SPACE

21.1 Historical Note / 21-2
21.2 Use of Old Mines and Quarries / 21-6
21.3 Underground Powerhouses / 21-10
21.4 Underground Fuel Storage / 21-14
21.5 Underground Spaces for Human Use / 21-18
21.6 Some Problems / 21-25
21.7 Conclusion / 21-26
21.8 References / 21-26

The above title might suggest that this chapter is a duplication of the preceding treatment of the geology of tunnels. This is not the case, the term *underground space* being used here to indicate excavated (or natural) space beneath the ground surface in which human activities can proceed or which can be used for storage purposes. Methods of excavation are similar to those used in tunnels, but, to a steadily increasing degree, the dimensions of spaces used for underground activities are considerably larger than even the largest tunnel cross sections. Underground power stations are becoming increasingly well known although they have been in use throughout the twentieth century. All bulk oil storage in Sweden is now being located underground for safety and economy.

The use of underground space for conducting manufacturing and other human operations, and even for residential purposes, is now also increasing. Again, it is a reversion to a practice rooted in the depths of antiquity. Modern uses have the great advantage of preserving the visual environment, a matter of widespread social concern, but they have the further advantage of conserving the use of energy. With the world's energy situation steadily growing more critical as the end of the twentieth century approaches, all aspects of energy conservation are becoming increasingly important. It is for this reason that this chapter is devoted to a review of appropriate uses of underground space and the ultimate dependence of this desirable development upon geology.

So important is the conservation of energy, and yet so little appreciated is the potential that the use of the underground presents for such conservation, that further discussion of this potential may well preface the treatment of actual underground installations. The normal pattern of ground temperature variation (explained in Chap. 9) is the starting point. As depth below ground surface increases, the amplitude of ground temperature variation will decrease. There will be diurnal variations close to the surface and annual variations as depth increases, until at a depth of about 10 m (33 ft) in temperate climates the ground temperature becomes sensibly constant. This constant ground temperature is close to, but usually not exactly equal to, the local annual average air temperature. In northern parts of the United States and in southern Canada, this temperature will be found to be not far from 10°C (50°F). If this temperature is compared with the comfort temperature required in buildings (say, 22°C or 72°F), it will be clear at once that, to bring underground space up to a state comfortable for human use, the heating load will be constant throughout the year and will be appreciably less than that for aboveground structures. This will lead to considerable economy in the capital cost of heating equipment; no air conditioning will be required in summer; but, of greatest importance, the energy consumption will be much reduced from that required for normal aboveground buildings. When it is recalled that between one-quarter and one-third of all the energy used in North America is for the heating and cooling of buildings, the significance of the use of underground space for appropriate purposes becomes crystal clear.

One especially appropriate use is for what is commonly called "cold storage." In aboveground installations, unusual provisions for insulation have to be made, and when perishable goods are to be stored, duplicate refrigeration equipment is essential in case there is a failure of the prime cooling plant. If, for example, there is a power failure at a cold storage plant, and no standby power is available, the chilled space will warm up by about 1°F per hour, with what serious results can well be imagined. Below ground, cold storage space does not require anything like the same degree of insulation; initial capital costs for refrigerating plants will be much reduced; and if power supply should fail, the chilled space will warm up at a rate of only about 1°F per day. Beyond this, and of even more importance, is the much reduced power requirement for the normal operation of such plants.

Fortunately, there are available accurate figures from modern installations to support these suggestions. Bligh and Hamburger have reported that in Kansas City, where 10 percent of all cold storage in the United States is now located (as will shortly be explained), installation costs expressed as dollars per square foot are $30 above ground, and $8 to $18 below ground for comparable cold storage space; operating costs, on the same basis, are $0.12 aboveground but only $0.01 for underground installations, the reduction in energy consumption being comparable.[21.1] Estimates of operating cost for the Brunson manufacturing plant in Kansas City are $3,200 per year for heating units (using 75,000 Btu/hour) in the underground plant as constructed but $50,000 to $70,000 per year for heating units in an equivalent aboveground plant, which would require 2 million Btu/hour. These figures represent valid savings in energy. Although many underground installations have been made for reasons other than the conservation of energy (such as underground power stations located below ground for convenience, safety, and economy in construction cost), each one of them illustrates what can be done once the underground is accepted as a part of the physical world which, with good design, can be used for many purposes, always provided the local geology is satisfactory. Before modern examples are described, it will be useful to take a brief glance at historical examples, if only to show again something of the great achievements of builders of the past.

21.1 HISTORICAL NOTE

Located in the desert part of Jordan about 480 km (300 mi) due east of Cairo, Petra is a phenomenal place even today, many centuries after it ceased to be an active city (Fig. 21.1). It was very largely an underground city, its temples and halls hewn out of the local

FIG. 21.1 Ancient rock-cave dwellings at Petra, the scale of which can be seen from the human figures in the foreground.

red sandstone that makes it such a colorful place. Geologically it is located in a great rock basin on the eastern side of the Wadi el-Araba, the great Rift Valley that runs from the Jordan depression to the Gulf of Aqaba. The country around had been traditionally the home of the Horites, who were cave dwellers, predecessors of the Edomites of the Old Testament. It is uncertain whether any specific reference to Petra can be found in the Bible, but the city was certainly in use in the sixth century B.C. It developed into a major trading settlement, reaching the peak of its activity around A.D. 100, and lasting until the great Mohammedan conquest of 629–632. For over a thousand years, therefore, this underground city served the diverse purposes of its desert residents, becoming an important religious center early in the Christian era. The appearance today of its temples and its carvings is a salutary reminder that modern man was not the first to think of gaining security by "going underground."[21.2]

Petra was not alone. The neighboring valley of the river Nile has long been regarded as one of the cradles of civilization. There, too, are to be found temples and other spaces for human use carved laboriously out of the solid rock. Probably the best known today are the temples of Abu Simbel, if only because of the international effort that went into their preservation. Located 280 km (175 mi) upstream of the site of the (new) Aswan High Dam, the great carvings in the immense cliff of the local pink sandstone are so remarkable that they would probably have been long regarded as one of the wonders of the ancient world had they not been lost to sight until 1813, when they were rediscovered, almost completely buried in sand, by the Swiss explorer Burckhardt. Not only did the discoveries include the frequently illustrated four immense figures of Rameses II (20 m or 66 ft high), which date the carvings as being probably older than 1200 B.C., but they included as well the great temple, an excavation in the cliffside 16.2 by 17.4 m (53 by 57 ft), with a corridor connecting it to an inner sanctuary that extends 60 m (200 ft) into the solid rock. The raising of this masterpiece to obviate its being flooded by the rising waters of the Nile on completion of the High Dam is one of the epics of modern foundation engineering, but the temples are mentioned here only to lend further confirmation to the early use of underground shelter.[21.3]

Even more remarkable are the vast and numerous caves still to be seen today in the triangle formed by Kayseri, Nevsehir, and Nigde in Cappadocia, Turkey. In the early years of the Christian era, there was an underground city here (at Derinkyu) with a population of 50,000, connected by an 8-km (5-mi) tunnel with another large underground city. But perhaps attention should be directed to the caves of northern France which have been known and used since the earliest days of human settlement. Those bordering the river Seine are notable in this respect. Near Vétheuil, hills of chalk are almost honeycombed with caves, many of which have been used for the shelter of men and beasts for hundreds of years. In more recent years, the caves have proved to be admirable (and natural) bomb shelters. One of the greatest of Norman chiefs made his home in a group of caves known as Le Grand Comombier (the great dovecote), which includes one cave forming a room over 100 m (328 ft) long. Of more interest to some readers will be the great caves at Pommeroy Park. These are known to have been used by the Romans as a convenient quarry for building stone, but their fame today arises from the fact that they provide almost perfect storage for the maturing of champagne.[21.4] Millions of bottles can be seen by the visitor to the caves, all naturally under careful guard and constant attention (Fig. 21.2 and Fig. 21.3). In the south of France, close to the valley of the Rhône, is

FIG. 21.2 Wine storage in the Cave de Vouvray, typical of the way in which caves in limestone are used in France for this significant purpose.

FIG. 21.3 A meeting of the Grand Council of Les Chevaliers de la Chantepleure in a limestone cave used for maturing wine in the valley of the Loire, France.

the greatest of all these underground wonders, the Bramabian, with its underground river and 9.6 km (6 mi) of galleries. The underground river of Labouiche is said to be the longest in the world, more than 3.2 km (2 mi). Visitors can sail more than 1.6 km (1 mi) by boat. The underground torrent of the Cigalière, with its waterfalls 18 m (60 ft) high, is in the same area. Then there are chasms such as the *gouffre* Martel which goes down 420 m (1,400 ft), 270 m (900 ft) of which have been explored.[21.5]

Possibly even better known are the extensive caves in the Pyrenees between France and Spain. Here some of the world's most extensive underground networks of interconnecting caves and tunnels have made the area a mecca for speleologists (*speleology* is recognized as the name for the rather unusual outdoor science of cave exploration). Norbert Casteret, one of the most intrepid underground explorers, has written an exciting book about this area and his travels in it, or rather under it. Of special interest is an incident he relates regarding a famous hole, Le Trou du Toro, high among the peaks of the Pyrenees, down which disappear the rushing waters of one of the mountain streams. There had long been speculation about where the water reappeared, and the general belief was that it came out lower down the same valley, which is on the Spanish side of the border. When a hydroelectric development was being planned for this valley, Spanish engineers had to find out exactly what did happen. Speculation had to give way to fact, especially since there was also a rumor that the water penetrated beneath the mountain ridge through an underground passage, which the speleologists had not been able to explore, and reappeared in France. Under the leadership of Casteret, therefore, engineers introduced a suitable quantity of dye into Le Trou du Toro. Older inhabitants of the lovely upper Garonne countryside still talk of their amazement when, on the morning of 20 July 1931, they saw their beloved river running a vivid green. Only the Spanish engineers failed to share in the general excitement, since they now had no claim to the water they had planned to use.[21.6] Use of the underground for human activity has a history almost as long as that of the human race. But modern civilizations have modern reasons to revert to this ancient practice.

21-6 CIVIL ENGINEERING WORKS

21.2 USE OF OLD MINES AND QUARRIES

Wherever there are unused mines or underground quarries conveniently located adjacent to or even in city areas, their use for purposes other than for mining can often prove to be economical. They should not be overlooked, therefore, in planning. In time of war there has always been a demand for such "safe" space for storage.[21.7] The Second World War saw great activity in this direction. The salt mines at Hungen, north of Frankfurt, were used by the Germans to store their choicest art treasures during the years of war, including the world-famous bust of Nefertiti, the beautiful queen of Egypt. One hundred tonnes of treasures from the British Museum were stored, during the same period, in a disused tunnel near Aberystwyth in Wales, while the famous Elgin Marbles were stored in one of the London underground railway tunnels. In Poland the old salt mines at Wieliczka near Cracow are used permanently as a picture gallery. It is known that the mines from which the famous Bath Stone of England is obtained were put to good use as underground workshops. This was a wartime expedient, but in times of peace, the same mines have been used for growing mushrooms. The even temperature, good ventilation, steady relative humidity, and absence of sunlight combine to make the location ideal for mushroom culture. At one time, 5.2 ha (13 acres) underground were used for this purpose, producing between 113,000 and 136,000 kg (250,000 to 300,000 lb) of mushrooms per year.

The even temperature of underground mines is of significance for much more sophisticated uses. Mention of mushroom-growing, even though this is an important business enterprise, should not give the idea that the use of underground space is confined to such unusual operations. This same area of Wiltshire, where excavations started in Roman times, housed many secret operations of the British government during the Second World War, including a complete aircraft factory. Even today one large area adjacent to the underground mines from which Bath Stone is still quarried, is in use for a large classified

FIG. 21.4 A view inside one of the disused limestone mining areas in Kansas City, Missouri, showing the naturally level floor and competent rock roof.

governmental operation. Quite the most remarkable use of "mined-out" space, however, is to be found in and around the great American city of Kansas City.

This important Midwest urban area is in the middle of a 225-km (140-mi) wide belt of exposed Pennsylvanian beds. These are divided into a number of subgroups, one of which, the Kansas City group, is widely exposed in the vicinity of Kansas City. The group consists of alternating limestones and shales. Many rock escarpments in the area, up to 15 m (50 ft) high, expose different members of the group, quite the most important of which is the Bethany Falls limestone, an excellent building stone. From the closing years of the nineteenth century, quarrying, and eventually mining, of this and associated beds of limestone has steadily increased. The value of such nonmetallic mining is now estimated to be at least $25 million per year. The limestone has been used not only as building stone, but also for making lime and portland cement, for concrete aggregate, and for road construction. Although mined from the surface in early days, the Bethany Falls limestone in particular has now been mined for many years by underground quarrying, many of its exposures permitting easy access from roads in the area, with a minimum of preparatory work for approach to good faces of the rock.

The general method of mining has been to leave standing large pillars of rock to serve as roof supports. The generally level character of the beds results in level floors in the mined-out areas. (See Fig. 21.4.) Groundwater gives little trouble, being encountered only occasionally, so that these large underground areas are generally dry, with the usual constant temperature and steady relative humidity. The total thickness of the Bethany Falls limestone, of from 6 to 7.5 m (19.5 to 24.5 ft), results in mined-out space that is almost ideal for underground operations. There are still about 15 active underground limestone-mining operations in the vicinity of Kansas City, excavating about 4 to 5 ha (10 to 12.5 acres) each year. But it has been estimateed that there are already 1,160 ha (2,870 acres) of mined space beneath the area around the city, with 4.7 million m^2 (5.6 million yd^2) of finished space available for warehousing and other purposes in 21 different mined areas (Fig. 21.5).

As far back as 1938, Mr. Amber N. Brunson, President, Brunson Instrument Company, in Kansas City, started plans for an underground facility to house his very delicate operations of manufacturing precision instruments.... He purchased [a] site and started mining out the space to his rigid requirements, selling the quarried limestone to help defray expenses on the overall development. Today his company is operating in an attractive, efficient plant, a marvel of ingenuity....[21.8]

Today, most if not all of the limestone mining in Missouri is being carried out with eventual use of the mined-out space in mind. The pillars, instead of being left in place to no definite plan, are now geometrically spaced as to give efficient cleared spaces (Fig. 21.6). Floors are properly graded, and entrances are left with sufficient roof cover to give portals that are safe. So valuable is this space becoming that it has been quite seriously suggested "it might sometimes pay to quarry the rock and throw it in the river, if necessary, to make the mine safe for the future, because the value of the rock mined is but a small portion of the extended value of the space." This is some indication of how rapidly the economics of this mining operation have changed. But limestone is still needed in increasingly large quantities, and so underground space in Missouri generally will probably still come from suitable mined-out limestone quarries for some time yet to come.

Almost 60 ha (150 acres) have now been put to use for industrial purposes in the Kansas City area alone. The J. G. Nicholls underground industrial park, 16 km (10 mi) from the city center, has an area of 4.8 ha (12 acres) converted into office, manufacturing, and storage space for the individual uses of five tenants. The Epic Manufacturing Company finds underground conditions most suitable for its toolmaking business. Woodward,

FIG. 21.5 One of the entrances to the old limestone mine workings in Kansas City, Missouri, showing the rock formations involved and one of the railway tracks that leads to areas now used for storage.

Clyde, Sherard and Associates, consulting soil and foundation engineers, appropriately find the constant temperature and relative humidity conducive to accurate soil testing. Ventilation must naturally be provided, and, with this, any necessary adjustment to air temperature and humidity. The plant necessary for this, however, involves but a fraction of both the capital and operating costs required in a normal building. Rock bolting to secure safe roof rock is occasionally necessary, but in the areas that have been well mined, little extra work is usually found to be necessary prior to the finishing of the space for use. Railroad spurs from conveniently located tracks of the Missouri Pacific and other railways have been constructed into the inside of some of the areas used for warehousing. The Pixley Company of Independence, Missouri, has a total area of 413,000 m² (500,000 yd²) of serviced underground space, including railroad facilities, now available for use, all excavated by removal of Bethany Falls limestone. Even larger—over 600,000 m²—is the underground property of the Centropolis Crusher Company, similarly serviced and located on the Manchester Trafficway, two blocks from U.S. Highway 40. One of the other large developments has even been said to contain a clubhouse for underground trout fishermen.[21.9]

This may be apocryphal but the reality of the small underground city, in which over 2,000 people are gainfully and comfortably employed, is the best of all arguments to show

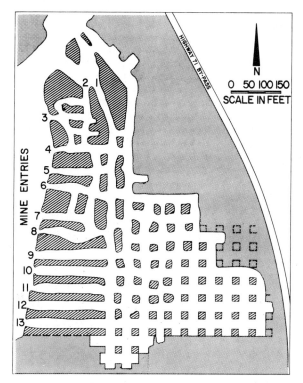

FIG. 21.6 Plan of old limestone workings in Kansas City, Missouri, showing the random spacing of pillar supports in older workings but the regularly spaced pillars in areas mined out with future use envisaged.

just what can be done by an enlightened use of existing underground space. There are other uses to which old mines adjacent to cities have been put. Naturally not within city limits, but in some cases not too far from urban areas, abandoned salt mines have been used for the underground storage of petroleum products, just as natural gas fields when depleted can be used for the same purpose. This is a highly specialized exercise, always carried out by large petroleum producers in association with public authorities and with all due safety precautions; it does not warrant more than this brief recognition here. It is, however, further indication of the use to which old mines can be put. Even more noteworthy is the use in the Soviet Union of an abandoned salt mine near the township of Solotvino in Carpathia as a hospital for bronchial asthma sufferers. Patients spend between 7 and 12 hours a day in the converted mine at a depth of 202 m (610 ft) below ground. No medicines are used, but doctors are said to believe that the climate within the mine, coupled with remedial exercises, renders medicine unnecessary.[21.10]

In all urban and regional planning, therefore, close attention should always be paid to any unused space beneath the ground surface of the area being planned. It may provide economical and satisfactory solutions to problems of space; its use will be energy-conserving and the visual environment will benefit. There are problems to be faced in the utilization of such space, as outlined later in this chapter, but they are capable of engineering solutions if the local geology is suitable, the inevitable concomitant.

21.3 UNDERGROUND POWERHOUSES

The concept of placing power stations (almost always water power or pumped storage stations) underground is now so well known that it has ceased to be regarded as a novelty of design. It is essentially a development of the years since the Second World War, although prior to the end of the nineteenth century, one of the first Canadian power stations at Niagara Falls used the underground to a degree. Water turbines were located at the bottoms of deep circular excavations but were connected by steel shafts to generators at surface level, and so the installation (still in operation) cannot truly be classed as underground. The first completely underground station of which any record has yet been traced is the Snoqualmie Falls station of the Puget Sound Light and Power Company, located about 40 km (25 mi) east of Seattle. Built in 1899 and still in operation with much of the original equipment, it was truly a pioneer civil engineering project. The powerhouse excavation, all in competent basalt, is only 60 m (200 ft) long, 12 m (40 ft) wide and 9 m (30 ft) high, but it is located 81 m (270 ft) below surface level and leads into a tailrace tunnel 135 m (450 ft) long.[21.11] These figures are small indeed when compared with some of those shortly to be quoted, but the station has a special place in the annals of North American civil engineering.

In Europe, the first such plant was that built in 1911 for the Porjus project in Sweden. The first Italian underground station was that at Coghinas in Sardinia; it was built in 1926. Today Italy is one of the leaders in this phase of hydroelectric design, having over 100 underground power stations already in use. Despite the fact that the local geology must give solid rock of excellent quality for such large excavations, there are even underground power stations in the Highlands of Scotland, famed in geological circles for some of the most complex hard-rock geology to be found anywhere. And in the Ceannacroc station, the local schist is proudly exposed to view, as shown in Fig. 21.7. Together with the associated Glenmoriston station, it was constructed in altered sedimentary and

FIG. 21.7 Interior of the Ceannacroc underground water power station in the Highlands of Scotland, showing the exposed mica schist.

igneous rocks, mainly schists and granulites with close graining imposed during metamorphism. Rock bolting was used to reinforce the exposed roof of the station; much grouting was done since there were water inflows during construction; reinforced concrete barrel-vault roofs were used; but, as can be seen, much wall rock was left exposed to view.[21.12]

This brief reference focusses attention on all the main problems associated with this specialized use of the underground (and with other uses also)—namely, the necessity for an accurate foreknowledge of bedrock conditions, rigid field control and inspection as excavation proceeds, ready means for dealing with water inflows, and readiness to use rock bolts, either directly or with prestressing and other means of rock support, when this is necessary. Reinforced concrete may be used, but more frequently recourse is had to the use of bolt-supported wire mesh with sprayed cement or concrete cover. Only rarely has the adequacy of preliminary information been in question, although not infrequently actual rock conditions as revealed by excavation have not been exactly as had been hoped for from the best of preliminary study. One major case (the Kariba powerhouse in Africa) is *sub judice* as this volume goes to press. When the record of this important case is publicly available, it will provide information of great benefit to all future underground work.[21.13]

Accordingly, it would be repetitious to go into detail of further examples, but, in order to show in what varied rock types underground powerhouses have been constructed, a few examples from around the world may be noted. In Sweden more than one-half of all water power stations are underground. Two early examples were the Kilforsen power plant (285,000 kW) and the Stornorrfors plant which, with an installed capacity of 520,000 kW, was the largest Swedish power station when completed. Sweden is fortunate in being underlain by competent Precambrian igneous rock, most suitable for excavation, both these stations being excavated in sound granite and gneiss. Over 2 million yd³ had to be excavated for the Stornorrfors tailrace tunnel alone, but this is 4 km (2.5 mi) long with a cross section 26.5 by 16.0 m (87 by 52 ft) wide. Many innovations in rock drilling and excavation and in shaft sinking were featured on this project, which was a milestone along the steady development of underground construction (Fig. 21.8).[21.14] Similar use of the

FIG. 21.8 Excavation for the tailrace of the underground Stornorrfors water power station on the Ume River in Sweden.

underground was being made at about the same time (the mid-50s) in France, notably in the Montpezat powerhouse; this station utilizes water obtained in a complex scheme involving the use of an old volcanic crater as the storage reservoir for the flow from three tributaries of the upper Loire. Under a head of 900 m (3,000 ft), 105,000 kW can be generated.[21.15]

In the United States, an important underground station was built by the Pacific Gas and Electric Company, this being the Haas station on the Kings River project, California (Fig. 21.9). Excellent massive granite was the local bedrock, and this aided in the economy which the underground station presented since, by going underground with certainty, it was possible to substitute 600 m (2,000 ft) of unlined tailrace tunnel for steel penstocks above ground.[21.16] More recently, and with practice improved on the basis of experience over the last three decades, the Hongrin underground pumped storage plant was successfully completed in Switzerland, on the shores of Lake Geneva at Veytaux. Here the local rock was Dogger limestone and limestone schist; unusually careful procedures were followed for excavation, and extensive use was made of special prestressed rock anchors.[21.17] Over 1,000 prestressed rock anchors and almost 4,000 rock bolts were used on a German project. It is another underground pumped storage plant, near Hemfurth, West Germany (the Waldeck II plant), which was built in mixed strata of slate and graywacke, a coarse, grey sandstone. So vital was the reinforcing of the rock, after excavation, that a special

FIG. 21.9 Interior of the underground Haas water power station of Pacific Gas and Electric Company, California.

TV camera (7.5 cm, or 3 in, in diameter and 1.5 m, or 5 ft long) was used by job geologists to survey about 10 percent of the holes drilled for the anchors.[21.18]

On the other side of the world, both the Tumut power stations of the Australian Snowy Mountains project were constructed under ground. The Tumut 2 station was built entirely in granite or granitic gneiss, which was most thoroughly explored before designs were completed. Figure 2.3 is merely an indication of the exhaustive preliminary studies made for this major project, studies which resulted in satisfactory construction progress and the revelation of geological features almost identical with those predicted from exploration, with no major changes necessitated during the progress of the work.[21.19] In great contrast, due entirely to geology and not to the preliminary engineering and geological studies, was the experience in Belgium at the Coo-Trois-Ponts pumped storage plant in the Ardennes near Liège. Topographically, the site was ideal, but the local geology consisted of almost horizontal thin strata of quartzites and phyllites of unusual brittleness. A major test boring program was carried out before construction started, and an inspection tunnel 295 m (984 ft) long was driven, but when excavation started, the faulting of the rock was worse than expected. Coupled with the character of the rock, this led to unusual difficulties in construction. Eventually the difficulties were successfully overcome with the aid of all the devices already mentioned. Some of the rock bolts used to give assured strength were as much as 21 m (70 ft) long, some supporting as little 10.8 m² (116 ft²) of exposed rock.[21.20]

Canada has yet to be mentioned. Suffice it to say that, starting with the Chutes des Passes station for Aluminium, Ltd., on the Péribonka River in the Lake St. John district of Quebec, in which 750,000 kW are generated, Canada has consistently had some of the largest underground power stations of the world. There need be mentioned only the Outardes III station, 21 m (70 ft) wide, 42 m (140 ft) high, and 124 m (415 ft) long, built in late sixties; the Portage Mountain station in British Columbia, built at about the same time, with its main hall 25 m (85 ft) wide, 267 m (890 ft) long, and 30 m (100 ft) high; the

FIG. 21.10 The main hall of the Kemano underground water power station of the Aluminum Company of Canada, British Columbia, during final stages of excavation.

21-14   CIVIL ENGINEERING WORKS

FIG. 21.11 Construction of the Churchill Falls underground waterpower station in Newfoundland-Labrador in progress; the man in the foreground shows the scale; 7 million hp are now generated in this station of the Churchill Falls (Labrador) Corporation.

Kemano power station (another Aluminum Company project) in British Columbia, 342 m (1120 ft) long, 24.6 m (80 ft) wide, and 41.7 (137 ft) m high (Fig. 21.10), which was eclipsed by the Churchill Falls station in Labrador (Fig. 21.11); and this now by the LG-2 station, the first of four vast projects which form the great James Bay scheme of Hydro Quebec. The LG-2 station is 26.5 m (87 ft) wide, 47.3 m (155 ft) high, and 483.4 m (1,580 ft) long, excavated in Precambrian granitic gneiss cut by numerous pegmatite and diabase dikes (Fig. 21.12).[21.21]

21.4 UNDERGROUND FUEL STORAGE

The demands of the Second World War led to developments in the use of specially excavated vaults for a variety of storage purposes, initially for defense but in postwar years for products for civilian use also. The engineering and geological features of such works are not different from those discussed in connection with tunnels; the shapes and sizes of the excavations alone are different. Because of the unusual character of some of these storage arrangements and because they depend for their success so completely upon geological factors, this summary review appears to be advisable.

FIG. 21.12 Excavation in progress for La Grande underground water power station of le Societé d'Energie de la Baie James in northern Quebec, east of James (Hudson) Bay.

In view of all that Pearl Harbor meant to the Unites States, it is not surprising to find that one of the first large underground storage installations was constructed during the late war years near that port in the Hawaiian Islands for the storage of naval diesel and fuel-oil supplies. Twenty cylindrical vaults, each 100 ft in diameter and 250 ft high, with domed inverts and crowns, were constructed as steel containers and set in a concrete lining within rock excavations of appropriate shape. Excavation was generally in a thick series of volcanic lava flows with the Hawaiian names of *aa* and *pahoehoe*. Grouting was carried out when the tanks were completed in order to fill any interstices in the rock adjacent to the lining and also to prestress the concrete lining in compression before the tanks were filled.[21.22] When the war closed, it was found that vast underground storage facilities had been constructed in Germany, Japan, Italy, Czechoslovakia, and Sweden. In some cases, these large excavations were used not only for storage but also for manufacturing and allied activities. About 5 million m³ (6½ million yd³) of rock was excavated from one part alone of one of the largest German installations, an indication of the extent of such defense facilities.

With the imperative of war removed, such unusual activities could not normally be considered for civilian use, but other ideas for underground storage were developed. One of the first of these was the result of studies by Harald Edholm of the Swedish State Power Board who wondered if the immiscibility of oil and water could not be utilized in underground storage facilities for oil. Using an abandoned feldspar mine on the east coast of Sweden, Edholm developed a system which worked exactly as he had predicted. The mine was cleaned out, and boreholes were installed to ensure a constant supply of water at the bottom of the vast storage cavern; the oil floated on top of this, and suitable arrangements were made for getting the oil into the "mine" and out again when wanted. Oil is now stored to a depth of 78 m (260 ft); the extent of the storage area is reflected in the fact that a special rowing boat has to be kept for inspecting the subterranean oil lake. It was estimated that a comparable "standard" oil-storage depot would have cost about five times what the full operation of the abandoned mine actually costs.[21.23]

The use of abandoned oil and gas wells for the storage of refined petroleum products is a regular practice in the oil industry; doubtless, this practice was responsible for the development of the idea of using artificially made underground cavities for the same purpose. Notable among such modern installations is that of the Imperial Oil Company of Canada in the "backyard" of its large refinery at Sarnia, Ontario. A vast salt bed lies 660 m (2,200 ft) below the plant and extends under much of this part of Canada and the adjacent United States. By means of rotary drilling methods, a borehole was sunk into the salt bed; two sets of pipes were installed in the hole, and solution of the salt began. In three months 1.45 million L (384,000 gal) of water was used, and 10.5 million kg (23 million lb) of salt was dissolved; this created a natural storage cavern with a capacity of 30,000 bbl, which was later expanded to 50,000 bbl; an adjacent cavity was similarly formed with a 40,000-bbl capacity. Butane and propane were then pumped into the caverns and so stored; brine is pumped back as the lighter petroleum products are withdrawn from storage. The economy of this method of storage will be obvious; even the saving in the cost of the ground necessary for "normal" storage tanks is itself enough to pay for much of the subterranean work.[21.24]

More than a million bbl of fuel was being stored in North America in this way by 1958, with probably more than 25 million bbl in abandoned oil and gas wells. These figures illustrate the remarkable scale of the vast underground storage project of the Standard Oil Company of New Jersey at Linden, New Jersey. Here there was no salt stratum to be mined easily, but only the red shale that characterizes so much of the geology of New Jersey. Through a 90 m (300-ft) shaft only 105 cm (42 in) in diameter, 107,000 m³ (140,000 yd³) of rock was removed after being mined out to form a vast honeycomb pattern of underground tunnels; these tunnels gave a total storage capacity of 675,000 bbl, all to be used for the storing of liquified petroleum gas.[21.25]

One more North American example may be cited. The Standard Oil Company of New Jersey, in a search for economic oil storage, found two adjacent abandoned slate quarries at Wind Gap, Pennsylvania. (Abandonment of the quarries is in itself a commentary upon changing building technology.) The company interconnected the two quarries and carried out the necessary rehabilitation work to make them serviceable for their intended purpose and to give a storage capacity of 42 million bbl. At the time of its completion, this was the largest single oil-storage facility in the world. Because of the clean lines to which the slate had been excavated, it was not too difficult to fit the necessary pontoon floating roof; the tightness of the slate formation was naturally an essential part of the plan for the utilization of the quarries. As oil is pumped out of one quarry, water is pumped into it from the other in order to keep the top level approximately the same at all times. This project again involved the principle of the immiscibility of oil and water.[21.26]

These were just a few of the storage installations developed in the decade following the end of the Second World War. They have been completely eclipsed, in scope and size, by developments in the succeeding two decades. In the United States, the Sun Oil Company now has a million-bbl oil-storage installation, excavated 127 m (425 ft) below ground level at a major tank farm 32 km (20 mi) southwest of Philadelphia, for storing liquified propane gas. Excavation was in solid granite and consisted of 2,100 m (7,000 ft) of storage tunnels 7.5 m (25 ft) wide and 10.5 m (35 ft) high on a rectangular layout, 19.8 m (66 ft) square pillars being left in place for support. All access to the workings was through a single 1.5 m (5-ft) -diameter shaft. Rock temperature is 16°C (61°F) and the gas will be stored at this temperature, its corresponding pressure being 6.3 kg/cm² (90 psi). Total cost was $22 million.[21.27]

Even more remarkable is the latest British gas-storage facility, located beneath the sea off the coast of Yorkshire. Based on Canadian and French experience with gas storage in old salt workings, and after exhaustive exploration of the geology of the seabed off the village of Atwick, British Gas decided to make its own storage cavities by leaching out the salt from a bed 1,650 m (5,400 ft) below sea level. This necessitated preliminary "temporary" works costing $2.5 million, consisting of an access shaft and elaborate launching arrangements for the seawater mains through which the leaching water was pumped, the resulting solution discharging into the sea. It was planned to leach out 28 million m³ (1 billion ft³) for each of six wells, in pairs, total construction period being estimated at six years; pumping started in the summer of 1974. Total cost was estimated at $15 million as compared with $22.5 million for an equivalent surface installation.[21.28]

The safety, economy, and visual effectiveness of underground storage for bulk fuels somewhat naturally led to the concept of forming suitable underground unlined cavities in soil, where no bedrock was available, by the freezing method already mentioned. Some such storage units were in successful operation for a time, but trouble has been experienced, notably in New Jersey and in England. Two tanks constructed in this way at Canvey Island on the river Thames in the late 1960s were abandoned in 1975 after high running costs rendered surface storage more economical. Despite the fact that the tanks were used to store liquified natural gas at $-162°C$ ($-248°F$), which might have been expected

FIG. 21.13 A typical rock cavern used for the underground storage of oil in Sweden, the scale indicated by the human figures.

to maintain the frozen cylinder of soil forming the tank, fissures developed, resulting in leakage and subsequent frost heaving at ground level.[21.29] More will be found about the properties of frozen soils in Chapter 36.

This relative failure in underground storage is noted in order to give a true overall picture since, when good bedrock is available, underground storage of both gas and oil is now well-accepted practice. Today, all new bulk oil storage in Sweden is placed underground, in caverns specially excavated in the competent Precambrian bedrock which underlies so much of this country (Fig. 21.13). Because of advances in rock excavation methods and equipment, the cost of such storage is about two-thirds the cost of equivalent surface storage. So extensive is underground storage now in Sweden, as elsewhere in Scandinavia, that Swedish engineers convened in 1977 the first international conference on underground storage. With such a restricted subject, one might have expected a relatively small gathering. Over 1,000 people attended, 57 countries and every continent being represented. Reports were presented indicating major underground storage facilities for oil, gas, liquified gas, and even of molasses in 24 countries. The three volumes of proceedings from this gathering provide an up-to-date review of world practice in this developing field.[21.30]

Those attending (one of the writers was there) were permitted to inspect such installations as that of the Swedish State Liquor Board where 50 percent of the wine and spirits consumed in Sweden are stored underground, with no energy needed for heating or cooling since the ground temperature is ideal for this sensitive storage. Typical of oil storage facilities is that at Nynäshamn, 50 km (31 mi) south of Stockholm on the Baltic Sea. It has a capacity of 900,000 m³ (31.7 million ft³) of oil. Groundwater is present in the Precambrian bedrock, but because of the differing specific gravities of oil and water, the groundwater has the effect of confining the oil in its caverns; seepage has to be pumped from the bottom of the caverns at such a rate that the delicate oil-water balance is maintained. Figure 21.13 shows better than any words the appearance of a typical Swedish oil-storage cavern before being filled with oil, the pipes being those required for the pumping of water and oil. Underground storage will assuredly come into wider use, when local geological conditions are favorable.

21.5 UNDERGROUND SPACES FOR HUMAN USE

The fact that the remarkable use of underground space in Kansas City utilizes worked-out mining space might suggest that only when such abandoned mines are available can use of the underground be considered for purposes other than bulk storage. This is not the case at all, since all around the world there are now underground spaces that were specially excavated for regular human use. As with other uses of the underground, initial use of space below ground surface for human activity arose out of wartime and postwar requirements, notably those for civil defense, and notably (again) in Sweden (Fig. 21.14).

Under the revised Swedish Civil Defense Law of 1 July 1960, all communities of over 5,000 people must be equipped with properly designed "normal shelters." These are usually provided in basements and similar cellars; the legal requirement shows clearly the attention given to civil defense in this strategically located Scandinavian country. The 14 largest cities and certain important communes are exempted from this provision, but new buildings in the communities around these areas must have normal shelters designed to take care of an overload of people. It is planned to move out 90 percent of the populations from these larger centers in case of emergency. The 10 percent of these urban populations required for controlling all emergency operations are to be accommodated in deep rock shelters. More than 100 of these unusual facilities have already been constructed at a cost running into tens of millions of dollars. These shelters must have a rock cover of at least 15 m (50 ft), but some have as much as 30 m (100 ft) of rock above them. Quite a number of these excavations have been designed so that they can be put to use in peacetime; hence, the investment in them will not be wasted. Typical is the pumping installation for the water-supply system of Göteborg, on Sweden's southwestern coast.[21.31]

Much of the design work in connection with these excavations is related to planned resistance to the effect of bomb explosions; one general feature is the construction of free-standing concrete structures inside the rock excavations to serve as private spaces for people using the shelters. All design, however, is completely dependent upon the existence of satisfactory rock conditions at the site selected for strategic reasons. Not the least of the associated problems is the effect that major rock excavation work will have on the groundwater conditions in the rock around and above the resulting opening. Serious consideration was given to this and to similar problems in the proceedings of an international symposium on large permanent underground openings held in Oslo, Norway, in September 1969. The fact that such an international meeting was held will confirm the growing importance of this increasingly frequent interference with the geology beneath cities.[21.32]

Of comparable underground installations in North America, one of the most notable

FIG. 21.14 Swedish destroyers in a completely underground marine dock, the scale given by the size of the man near the bow of the nearest vessel.

21-20 **CIVIL ENGINEERING WORKS**

of those that have been publicly described is undoubtedly the command combat operations center of the North American Air Defense Command (NORAD) near Colorado Springs, Colorado. Completed in 1962, this fully equipped underground installation is capable of accommodating a total staff of 700, with all services necessary for their living, such as water supply, power supply, and sewage disposal. Excavated almost completely in the local Pikes Peak granite, the center is located in Cheyenne Mountain, reached from the main road running south from Colorado Springs. The access road takes off to the west at Fort Carson, climbing steadily into the Front Range Mountains to the portals of the two access tunnels that are 360 m (1,180 ft) above the start of this 4-km (2.5-mile) road. Even the building of these roads was a construction operation of note; naturally they had to be completed before excavation could start. Preliminary test borings had indicated that the rock was sound enough to permit the excavation of the large chambers necessary for the complex, but since excavation was to be in granite, it was known that close examination of the rock would be necessary as excavation progressed. The two approach tunnels are 789 and 366 m (2,580 and 1,200 ft) long, respectively, 8.7 m (28.5 ft) wide, and 6.75 m (22.1 ft) high. Where the tunnels meet, the great complex of underground chambers starts, a typical excavation being 39 m (128 ft) long, 13.5 m (44 ft) wide, and 18 m (62 ft) high. Rock separation between chambers is as much as 30 m (100 ft). Within the excavations,

FIG. 21.15 Construction of the NORAD underground command center at Colorado Springs, Colorado, in progress, showing framework and reinforcing steel for the reinforced-concrete dome used at the main tunnel intersection.

three-storey steel "buildings," spring-mounted because of possible danger from blast shocks, were erected, and it is in these that the work of the center is done.

A geologist of the U.S. Corps of Engineers mapped the granite exposed as the access tunnels were excavated. This mapping revealed two major joint systems, and all indications were that they would persist into the large chambers. One of the joint systems was oriented in a direction parallel to the length of the largest chambers, the size of which prevented the use of any of the conventional means of roof support, should this be found to be necessary. Accordingly, the direction of the chambers was reoriented to a more favorable direction, and excavation was completed as planned. Correspondingly, close geological observation of the excavation at one of the most vital intersections of two main chambers revealed in advance of its completion poor rock conditions. Highly weathered rock was encountered, as was close jointing extending in several directions, and this despite the 420 m (1,400 ft) of rock cover above. Various methods of dealing with this problem were considered, the final solution being the construction of a reinforced-concrete dome at the base of the intersection chamber and its jacking up into place below the imperfect rock roof; an integrated reinforced-concrete lining to the chamber was also provided (Fig. 21.15). These details of structural design are mentioned since they were developed during the course of the work and were based on the acute observation of the geology as exposed by excavation, a procedure that calls for repeated mention in this volume. Few examples could be so clearly illustrative of the need for constant vigilance in connection with subsurface conditions until excavation is complete.[21.33]

Canada has a corresponding underground defense installation for the companion NORAD station, which is also the headquarters of the Canadian Air Defence Command, located at North Bay, Ontario, beneath the airport (Fig. 21.16). It was excavated in competent granitic gneiss which caused no unusual problems during construction. Rock bolts and wire mesh guards were used as necessary, but maintenance has been relatively simple. Over 230,000 m³ (300,000 yd³) of rock was excavated to form two long entrance tunnels

FIG. 21.16 Entrance to the NORAD underground command center at North Bay, Ontario, the access roadway indicative of the general size of the installation.

21-22   CIVIL ENGINEERING WORKS

and the main halls. Two of these are 120 m (400 ft) long, 18 m (60 ft) high, and 13.5 m (45 ft) wide, and in each stands a three-storey, steel-framed building. A third cavern, almost as large, contains a large diesel power plant, all three being connected by cross tunnels. The station has been in steady use since 1963.[21.34]

Around the world, there are numerous other defense installations in specially excavated space below ground, many of them still on classified lists. Notable amongst those which have been described are the great caverns that now honeycomb the Rock of Gibraltar, one of the largest of which is used for a water-supply reservoir for the town below. Much of the limestone excavated was used as fill, in the sea, for providing the base for an extension to the main runway of the strategic airport which serves Gibraltar.[21.35] Peacetime uses of underground space are, however, of more relevance today. Major installations have been excavated to provide space for sewage plants, archives (an ideal use), libraries, and garages, to mention only the more obvious uses to which underground space can so productively be put.

Archival material must be protected from direct sunlight and kept at constant temperature and humidity; underground storage provides, therefore, ideal conditions. State archives in Norway and Sweden are so located, again in competent Precambrian rock. The Church of Jesus Christ of the Latter-Day Saints (the Mormon Church) has shown in

FIG. 21.17 Entrance doorways to the underground storage facilities, in the granite of the Wasatch Front, Utah, of the genealogical department of the Church of Jesus Christ of the Latter-Day Saints.

North America how the underground can be utilized for this same purpose. The Church's archives are now located in a complete underground installation, excavated in the granite of the noteworthy Wasatch Front. Six storage caverns, located 210 m (700 ft) below ground surface, provide 6,000 m² (65,000 ft²) of storage, at a constant temperature of about 15°C (59°F) and relative humidity of 40 to 50%, with the added advantage that the atmosphere is dust-free. Entrance to this unique facility is through three adits driven into the Front from road level (Fig. 21.17).[21.36]

Libraries can also be located below ground with advantage. This has been demonstrated at the University of Illinois in Urbana, where a new undergraduate library was constructed entirely below ground level, excavation being in glacial till.[21.37] A 7,000-m² (75,000-ft²) extension to the justly famous Radcliffe Science Library of Oxford University was constructed entirely below ground level, one objective being to preserve the visual delight of the University Museum building. Excavation was from ground level in the usual manner, tiebacks being used to support the vertical sides of the finished excavation. (See Fig. 21.18.) Two test holes had shown that the ground consisted of about 60 cm (2 ft) of top soil overlying a light-brown, coarse-to-medium sand containing some gravel. The sand stratum extended to about 5.1 m (17 ft) below ground level, being underlain by a stiff, grey fissured clay with silty zones and a thin, weathered crust. Both test holes were cased for observation of the groundwater. (So quiet has the new library proved to be that consideration had to be given to providing "background noise" in order that those using the library should not be too disturbed by the unaccustomed absence of noise!)[21.38]

Garages are now a public necessity. Because of their nature, it is difficult, if not impossible, to make them visually attractive, and they always represent an unproductive use of land. Reference has already been made to the major downtown public garage in Los Angeles that was built below ground level. There are a growing number of smaller garages now following the same pattern. Congestion in the historic center of Quebec City in Can-

FIG. 21.18 Excavation for the underground extension of the Radcliffe Science Library of the University of Oxford, England.

ada was diminished by construction of a three-storey underground garage adjacent to the city hall, all excavated in solid rock. The noble hill on which the upper part of the city stands, so well known from pictorial views, is founded on the Quebec City formation of middle Ordovician age. This consists of hard, fine-grained limestone, very dark shale, and at times, beds of limestone conglomerate.

Queen's University at Kingston, Ontario, together with the adjoining Kingston General Hospital, faced a serious parking problem. A proposed extension to the hospital made action imperative. Careful planning and subsurface investigation led to an ingenious solution. Possession of a treasured playing field between the university and the hospital was obtained for a period of 12 months. Top soil was excavated and preserved; glacial deposits were excavated down to the underlying Trenton limestone; this was then excavated to the depth requisite for a two-storey underground garage, and all the rock excavated dumped into Lake Ontario, a few hundred yards away, under the direction of the City of Kingston which thus gained a new waterside park.[21.39] The only complaint about the new facility is that it should have been made three storeys instead of two! All that visitors to this lovely small campus see today are two pedestrian entrances, in the form of concrete staircase shelters, on the university side and the main automobile entrance in the sloping face of the edge of the playing field, facing the General Hospital. The playing field itself looks as it always did, so that it is not surprising that many visitors do not know that there is a large parking garage beneath it (Fig. 21.19). Within the garage, the simplicity of the necessary structure is obvious, exposed limestone forming all the walls, a reinforced-concrete framework supporting the upper floor and roof.

Sewage treatment plants are, likewise, not an aesthetic addition to the urban landscape, despite the best efforts of designers. Four of the five plants which serve Sweden's capital city of Stockholm are now located deep underground. The first was so located for defense reasons, the other three after careful economic studies showed that they would be preferable to plants at the surface (with the added advantage of preserving the visual environment).[21.40] From what has already been said, it will be appreciated that the most favorable geological conditions underlying Stockholm facilitated this unusual solution to the common problem of sewage plant location (Fig. 21.20). But even less favorable geology can, with careful planning based on meticulous preliminary subsurface investigation, be

FIG. 21.19 Vehicle and pedestrian entrance to the underground garage serving Queen's University, Kingston, Ontario, and the adjacent Kingston General Hospital, showing clearly the famous playing field, the integrity of which had to be preserved.

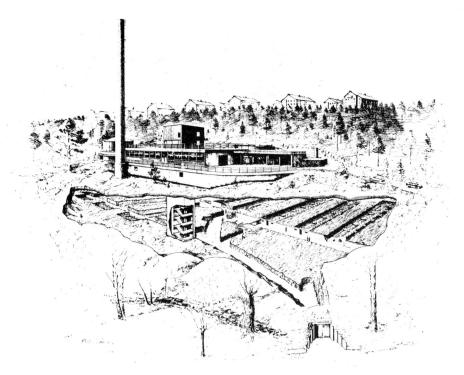

FIG. 21.20 An artist's conception of the Eölshall plant, one of the four underground sewage treatment plants serving the city of Stockholm, Sweden.

adapted through sound design to the safe housing of these and other facilities, with all the benefits already noted.

21.6 SOME PROBLEMS

As in all branches of civil engineering work, there are special problems in the use of underground space that must be given most careful consideration in design. Power supply for lighting and ventilation must be guaranteed for continuous service; an alternative supply must, therefore, be included in design in case of partial failure of the main public supply. Exit requirements are of paramount importance, in case of fire, flooding, or accident; dual exits, clearly marked, must be available for all who are to work underground. Fire, as always, is an ever-present danger, but with suitable design of ventilating facilities, the serious problem of smoke can be dealt with rather more easily than in buildings above ground. These matters are normal for consideration in good civil engineering design. Two more problems are essentially geological in nature.

Groundwater conditions in the area in which any underground installation is to be located must be known with certainty, and for at least a full year's cycle, before designs can be completed. As already noted, the presence of groundwater around unlined oil-storage caverns is an advantage, since the difference in the relative specific gravities of oil and water prevent the loss of oil into surrounding rock fissures by confining it to the caverns. Space that is to be used by people, however, must be dry, and there must be no possibility

21-26　CIVIL ENGINEERING WORKS

of any change in groundwater conditions that might lead even to minor flooding of any excavated space. Once recognized, the potential of any trouble from this cause can be examined with certainty in advance, and design decisions made accordingly.

A second potential source of trouble is the possibility that the rock excavated to provide the underground space, while quite satisfactory in its natural location, may be affected by exposure to air, if it is to be so exposed by the final outlines of the space planned for use. Accordingly, another preliminary study in underground space investigations involves determining whether or not the rock types to be encountered are likely to deteriorate if exposed to the air. The only real problem that has yet had to be faced in the extensive use of underground space in Kansas City, for example, is due to the exposure, in some areas of the Hushpuckney shale. When exposed to air, and the moisture that air normally contains, this material undergoes slow chemical change. Expansion results and there have been, in consequence, slight problems with floor movement. The process is a complex one, but it is under active review in research laboratories.[21.41] It does not detract from the remarkable achievement in the use of the underground at Kansas City, but it does serve as a very valuable reminder of the absolute dependence upon geology of this use of space beneath the surface of the ground.

21.7 CONCLUSION

As the energy situation throughout the world becomes steadily more serious, geology will play an increasingly significant role in assuring future energy supplies. Not only will it be involved to an increasing degree in the worldwide search for new sources of natural petroleum products, but in all efforts directed to tapping the heat of the earth, geology will be the determinant. If, as seems probable, increasing dependence is placed upon the use of solid fossil fuels, geology will again be of unusual importance in the discovery of new deposits and the working of those already known. And for the design and construction of all the power plants needed to supply the increasing demand for centrally generated power, engineering geology will have an important part to play.

It is hoped that this chapter will have made clear that, beyond all these important applications, geology can assist also in the essential conservation of energy by assisting with the development of underground space for appropriate purposes. When local ground conditions are suitable, such development can bring great energy saving in many cases. Lest it be thought that this unusual view, still infrequently encountered, is peculiar to the authors, it may be mentioned that there is now an active American Underground Space Association, with its own professional journal, *Underground Space,* its headquarters (as this book goes to press) being at the University of Minnesota. The State of Minnesota is taking an active interest in the use of underground space, a delegation from the state legislature attending the Rockstore Conference in Stockholm. And as indicating the interest of geologists in this subject, it will be found that the Geological Society of America entitled one of the sessions at its 1978 annual meeting (in Toronto) "Geology beneath Cities," a title later applied to a GSA volume containing the papers given at that meeting and some other papers.[21.42] The geology beneath modern cities is, indeed, of vital importance.

21.8 REFERENCES

21.1 T. P. Bligh and P. Hamburger, "Conservation of Energy by Use of Underground Space," *Legal, Economic and Energy Considerations in the Use of Underground Space,* National Academy of Sciences, Washington, 1974, p. 103.

21.2 M. A. Murray, *Petra, the Rock City of Eden,* Blackie, London, 1939; see also A. Kennedy, *Petra, Its History and Monuments,* Country Life, London, 1925.

21.3 E. E. White, "Saving the Temples of Abu Simbel," *Civil Engineering,* **32**:34 (August 1962).

21.4 R. Gibbings, *Coming down the Seine,* Dent, London, 1953, p. 88.

21.5 *The Observer,* (London), 27 August 1939.

21.6 N. Casteret, *Ten Years under the Earth,* B. Mussey (trans.), Dent, London, 1939.

21.7 P. Laurie, *Beneath City Streets,* Allen Lane, London, 1970.

21.8 *A New Concept in Space,* Missouri Resources and Development Commission, Jefferson City, 1961; see also *Guidebook to the Geology and Utilization of Underground Space in the Kansas City Area, Missouri,* Association of Engineering Geologists, 1971.

21.9 "Building Underground. Factories and Offices in a Cave," *Engineering News-Record,* **166**:58 (18 May 1961).

21.10 "Russia Puts Old Salt Mine to New Use as a Hospital," *Toronto Globe and Mail,* Reuters, 16 November 1970.

21.11 R. F. Legget, "The World's First Underground Water Power Station," *Underground Space,* **4**:91 (1979).

21.12 "Underground Powerhouse in Poor Rock," *Engineering News-Record,* **160**:63 (6 February 1958).

21.13 "The Story of Borehole 4, Kariba North," *New Civil Engineer,* **27**:10 (8 February 1973).

21.14 T. Goransson, "Stornorrfors: Milestone in Rock Excavation History," *Engineering News-Record,* **160**:39–44 (30 January 1958).

21.15 W. G. Bowman, "Exciting Is the Word for French Hydro," *Engineering News-Record,* **154**:36 (30 June 1955).

21.16 "PG & E Goes Underground for Power," *Engineering News-Record,* **159**:51 (19 September 1957).

21.17 M. Buro, "Prestressed Rock Anchors and Shotcrete for Large Underground Powerhouse," *Civil Engineering,* **40**:60 (May 1970).

21.18 "Sea of Anchors and Bolts Keep Powerhouse Cavern Shipshape," *Engineering News-Record,* **189**:39 (20 July 1972).

21.19 I. L. Pinkerton and E. J. Gibson, "Tumut 2 Underground Power Plant," *Proceedings of the American Society of Civil Engineers,* Vol. 90, PO I, Paper no. 3835, New York, March 1964, p. 33.

21.20 "Pumped Storage Project Is a Bad Rock Nightmare," *Engineering News-Record,* **184**:28 (1 January 1970).

21.21 L. Hamel and D. Nixon, "Excavation of World's Largest Underground Powerhouse," *Proceedings of the American Society of Civil Engineers,* vol. 104, CO3, Paper no. 13995, New York, 1978, p. 333.

21.22 "Huge Underground Vaults Built Oiltight," *Engineering News-Record,* **135**:311 (1952).

21.23 "Oil Storage in Virgin Rock," *Civil Engineering and Public Works Review,* **47**:311 (1952).

21.24 "Subterranean Storehouses," *Imperial Oil Review,* June 1955, p. 11.

21.25 "Carving Out a Cavern Through a 'Needle's Eye'," *Engineering News-Record,* **160**:36 (23 January 1958).

21.26 "Roofed-in Slate Quarry Now Stores Oil," *Engineering News-Record,* **153**:24 (16 September 1954).

21.27 "Largest Mined Gas Storage Cavern Carved from Granite," *Engineering News-Record,* **196**:24 (1 January 1976).

21.28 "Yorkshire Salt Seam To Be Leached Out to Form Huge Gas Storage Reservoir," *New Civil Engineer,* **96**:26 (13 June 1974).

21.29 "Gas Men Abandon Canvey Earth-Pit Fight," *New Civil Engineer,* **169**:20 (20 November 1975).

21.30 M. Bergman (ed.), *Storage in Excavated Rock Caverns,* 3 vols., Pergamon, Oxford, 1978.

21.31 O. Albert, "Shelters in Sweden," *Civil Engineering,* **31**:63 (November 1961).

21.32 C. O. Morfeldt, "Significance of Groundwater at Rock Constructions of Different Types," *Proceedings of the International Symposium on Large Permanent Underground Openings,* Oslo, 1969, p. 311.

21.33 T. O. Blaschke, "Underground Command Center," *Civil Engineering,* **34**:36 (May 1964); see also "Builders Blast Out Underground Fortress," *Engineering News-Record,* **167**:38 (16 November 1961).

21.34 A. D. Margison, "Canadian Underground NORAD Economically Achieved," *Underground Space,* **2**:9 (1977).

21.35 J. C. Cotton, "The Tunnels in Gibraltar," *The Civil Engineer in War,* **3**:229 (1948).

21.36 *Records Protection in an Uncertain World,* Genealogical Society of the Church of Jesus Christ of the Latter-Day Saints, Salt Lake City, 1975.

21.37 C. Fairhurst, "Going Under to Stay on Top," *Underground Services,* vol. 2 no. 3, Brentwood, Eng. 1974.

21.38 "Radcliffe Science Library," *Building,* **231**:47 (27 August 1976).

21.39 G. McCahill, "Underground Parking at Queen's University," *Underground Space,* **1**:347 (1977).

21.40 S. Edlund and G. E. Sandstrom, "Stockholm Puts Sewage Plants Underground," *Civil Engineering,* **43**:78 (September 1973).

21.41 R. M. Coveney and E. J. Parizek, "Deformation of Mine Floors by Sulfide Alteration (Kansas City)," *Bulletin of the Association of Engineering Geologists,* **14**:313 (1977).

21.42 R. F. Legget (ed.), *Geology Beneath Cities,* in press.

Suggestions for Further Reading

The American Underground Space Association (AUSA), with its office at 221 Church Street, University of Minnesota, Minneapolis, MN 55455, is a relatively new organization dedicated to the promotion of the use of economical underground space for appropriate purposes. It already has a number of general publications available, because the geological influences on the use of such space are generally recognized. Its bimonthly journal *Underground Space* (published by Pergamon Press) is now well established; some papers which it contains are referenced in this Handbook. All interested in the use of underground space may make contact with the Association at the given address.

One useful record published by AUSA is *The Potential of Earth Shelters and Underground Space* (T. L. Holthusen, editor), which is the 500-page volume of proceedings of a conference held in Kansas City in 1981. Records of similar Scandinavian conferences, such as are listed in the preceding references, contain much information of value. Since the economics of the use of underground space are of prime importance, note may be made also of the 161-page volume of proceedings of another conference held at the University of Nevada in 1980 called *Economical Underground Space with the Use of Modern Mining Practice,* a publication of the Multipurpose Excavation Group of Minneapolis.

Textbooks on the subject have yet to come, but *Underground Excavation in Rock* by E. Hoek and E. T. Brown is a notable volume of 532 pages published by the Institution of Mining and Metallurgical Engineers (44 Portland Place, London WIN 4BR).

chapter 22

BUILDING FOUNDATIONS

22.1 Influence of Geological Conditions on Design / 22-3
22.2 Foundations on Bedrock / 22-4
22.3 Foundations Carried to Bedrock / 22-6
22.4 Foundations on Soil / 22-10
22.5 Groundwater / 22-13
22.6 Effect of Trees / 22-18
22.7 Piled Foundations / 22-20
22.8 "Floating" Foundations / 22-23
22.9 Caisson Foundations / 22-24
22.10 Preloading of Foundation Beds / 22-27
22.11 Buildings on Fill / 22-29
22.12 Settlement of Buildings / 22-30
22.13 Prevention of Excessive Settlement / 22-36
22.14 Buildings over Coal Workings / 22-39
22.15 Underpinning / 22-43
22.16 Precautions on Sloping Ground / 22-44
22.17 Garages / 22-46
22.18 Small Buildings / 22-49
22.19 Conclusion / 22-50
22.20 References / 22-54

All engineering structures must be supported in some way by the materials that form the upper part of the earth's crust. There thus exists an inevitable connection between geological conditions and foundation design and construction. Dams, bridges, and structures associated with transportation routes are usually of considerable size and so achieve in the popular mind an importance not strictly in accord with fact, for the major part of engineering construction is made up of a large number of smaller projects. It is to this great group of miscellaneous structures that many civil engineers have to devote their attention. They know that small works may cause trials and difficulties no less than large projects; among these difficulties, foundation problems are far from unimportant. To describe the supporting elements of this great miscellaneous group of structures as "building foundations" is to use a general term in a singularly wide way. But so varied are the types of work involved that generalization is inevitable. On the other hand, the main types

22-2 CIVIL ENGINEERING WORKS

of foundation design are so well recognized that a general discussion of them can safely be made without fear of oversimplification. In addition, the founding of building structures represents a relatively large part of this miscellaneous foundation work, and many of the associated problems are common to general foundation practice.

Building foundation work was, for many years, largely a matter of rule of thumb, restricted only by such building regulations as were relevant. All too often, construction methods, which are of vital importance in many foundation operations, were left entirely to building contractors, and sometimes even the final details of foundation design were left to their discretion. In the hands of a competent construction company, this practice may not have been too objectionable; but in other cases, it led almost certainly to unsatisfactory foundation work. Today, the foundation of structures is coming to be adequately recognized as an important part of building design. On many projects, large and sometimes small, architects are cooperating with engineers in this branch of building work which is essentially in the domain of the engineer. Preliminary investigations are now more thoroughly made, and building plans are frequently prepared in association with a definite scheme of construction; the contractor is thus protected against the uncertainties that he might otherwise have to face. Simultaneously, the science of foundation design has been developing. Theoretical design methods are being correlated with the results of their application in practice; and by the improvement of field and laboratory soil-testing

FIG. 22.1 Two similar apartment buildings in Toronto, that on the right founded on bedrock through concrete caissons, but that on the left founded on a concrete raft bearing on to glacial soil overlying the bedrock, illustrating the effect that local differences in underlying geology can have on foundation design.

methods, better advantage is being taken of the possibilities that many building sites present.

The design of building foundations, and of all other foundations, consists of three essential operations: (1) determining exactly the nature of the foundation beds that are to act as a support, (2) calculating the loads to be transmitted by the foundation structure to the strata supporting it, and (3) designing a foundation structure to fit the conditions ascertained as the result of operations (1) and (2). Building loads may be affected in a general way by the local geology; but for any particular region they will, as a rule, be related only to structural design. Design of the foundation, however, is completely dependent on the nature of the ground underlying the building site. The determination of ground conditions is essentially a geological problem. When the foundation-bed conditions are known, selection of the type of foundation to be used is a matter of engineering judgment. In the actual preparation of design features, due account must be taken of all geological details liable to affect both the construction and the successful performance of the foundation. Soil mechanics studies make a major contribution to design work of this kind, but even the accurate design made possible in this way may be invalidated if a full study of the geology of the underlying strata has not also been made (Fig. 22.1).

This broad outline provides the necessary background, so that consideration may now be given to leading aspects of the application of geology to building-foundation work. The fact must be emphasized that geological conditions alone are to be discussed—a warning necessary in view of the incomplete picture of foundation engineering that the following pages display. Adequate determination of geological conditions at a building site is but one of the three essential components of complete foundation design, and its isolation from the other two parts, and particularly from the final detailed correlation of loading and ground conditions, necessarily distorts its significance in foundation work. Foundation failures, however, are rarely due to faulty structural design; it is safe to say that, in general, they are related in some way to failure of the foundation beds to carry the loads put upon them. It is, therefore, an extremely important part of foundation work that is here discussed and one that, either directly or indirectly, is associated with the design and construction of every civil engineering structure.

22.1 INFLUENCE OF GEOLOGICAL CONDITIONS ON DESIGN

Ground conditions at a building site may be classified as one of three general types according to foundation possibilities.

1. Solid rock may exist either at ground surface or so close to the surface that the building may be founded directly upon it (Fig. 22.2).
2. Bedrock may exist below ground surface but at a distance that may economically be reached by a practical form of foundation unit so that the building load can be transmitted to it.
3. The nearest rock stratum may be so far below the surface that the structure will have to be founded upon the unconsolidated material overlying the rock.

The influence of geology is evidenced directly by this broad classification, since the three types of ground condition are the result of geological processes of the past. The city of New York, in certain parts, typifies the first classification; some of the buildings of Manhattan are founded directly on the Manhattan schist which used to outcrop over part of that famous island. Much of the area covered by the city of Montreal is of the second

FIG. 22.2 Crown Center development on Signboard Hill, Kansas City, Missouri, under construction, foundation beds of alternating strata of shale and limestone being exposed, the latter used for all major foundations.

type; in this city Paleozoic bedrock is overlain by unconsolidated materials of the Pleistocene and Recent periods which vary in thickness, but which in some places reach more than 30 m (100 ft). The city of London may be taken as an example of the third type; the great London Basin with its deposits of London clay, sands, and gravel overlies the chalk, rendering impracticable the foundation of buildings on rock. It would be of interest to study the influence that ground conditions of these broad types have had on architecture, but engineers will readily appreciate the significance of the groups without the aid of this comparison.

22.2 FOUNDATIONS ON BEDROCK

Determination of site conditions in the case of the first of the three main types of foundation i.e., when bedrock is at or near the surface, will be a relatively simple matter. Conditions of this type are rare in urban areas, possibly because cities are usually located on riverbanks where unconsolidated deposits are the rule rather than the exception. Nevertheless, the first detailed service of geology will be in connection with the estimation of the soundness of the rock, with a determination of the significance of any structural features revealed, and possibly with a determination of the allowance to be made for seismic disturbances. Rarely will there be a question of whether the bearing power of rock is sufficient to withstand building loads; but if compression tests of the rock are made, great care should be taken to see that they are made with specimens loaded in a direction corresponding to that at the site. The difference in the bearing power of sedimentary rocks when tested parallel to the stratification and at right angles to it may be as much as 50 percent. As sedimentary rocks include those rock types which have low bearing capacities, this point is sometimes of importance. Typical figures for slate are 500 kg/cm² (7,220 psi) at right angles to bedding and 360 kg/cm² (5,180 psi) in the direction of bedding. These are average figures obtained in connection with the foundation of the Boston Parcel Post Building in the United States.[22.1] Values for the usual bearing stress utilized for various rocks types will be found in engineering handbooks and treatises on foundation design; independent tests for all but the smallest projects are usually advisable. Field tests of rock in situ (as described in Chap. 7) will be necessary in the case of unusual loadings or unusually weak rock.

Where bedrock is covered with a thin mantle of soil which is to be removed when construction starts, the exact nature of the rock will be seen only when the surface has been cleaned off. When this was done at the site of the electromagnetic plant at the Clinton Engineer Works in Clinton, Tennessee, following a test boring program, two differing rock conditions were found. On one side of a small valley, the formation was the Conasauga shale, which had weathered in the upper few feet to a soft brown clay. Weathered shale, with some clay seams, started at about 1.5 m (5 ft) below the surface. Only at depths of about 3.6 m (12 ft) was unweathered shale encountered, but it was sound and quite adequate for the foundation loads in view. On the opposite side of the small valley in which the site was located (a site covering a large area for the 175 structures involved), quite different conditions were encountered. Here the formation was dolomitic limestone of the Knox formation, inclined away from the center of the valley with a dip of 55°. Erosion had taken place in many of the exposures of the upended beds, the crevices so formed being generally filled with soft residual clays, with many boulders.

Since the location of indiviudal buildings was determined by operating needs, these site conditions had to be accepted. After cleaning of the rock, grouting through original cored test holes was carried out, and concrete footings were built over this grouted area. For the fourth of the process buildings, further complications were found, since here the "rock" encountered by the test borings proved to be large, flat-topped, irregular boulders, some 6 m (20 ft) below the surface, so closely packed that normal excavation methods for removal of the soil around could not be used. The whole area was washed down and then entirely covered with a heavy, solid concrete mat which had the effect of safely spreading foundation loads on to the embedded boulders and the intervening limestone[22.2]

Another sort of problem with bedrock near the surface was encountered in the building of the St. Vincent Hospital in Portland, Oregon. A comprehensive site investigation program was carried out in advance of design, but when excavation got going, the engineers were surprised to find two narrow, buried, rubble-filled channels crossing the site. The channels had been missed in both the test boring and seismic survey studies, which had shown the site to be underlain by sound basalt. Construction was under way when this discovery was made. More detailed studies were immediately put in hand, since it was clearly impossible to change the site. It was decided to cover over the two rubble-filled channels with concrete plugs, to cover the entire site with a 1.5-m (5-ft)-thick blanket of compacted rockfill, and then to construct a concrete mat foundation for supporting the main central tower structure of the hospital. Suitable adjustments were made to the foundation designs for the low-rise structures adjacent to the main tower, where such adjustments were necessary because of the channels.[22.3]

Figure 22.3 shows how suddenly such buried valleys can affect building sites. The building, well known to the writers, is founded throughout on limestone bedrock. A 23-m (75-ft)-long extension was needed to the rear of the building (facing the camera). Fortunately, the presence of a buried valley, eroded in the underlying limestone in the vicinity, was suspected. A test boring put down 22.5 m (74 ft) from the corner marked with an arrow went down 27 m (88 ft) without encountering rock. At this point boring was stopped and a decision was made to use another site where bedrock was near the surface.

FIG. 22.3 Rear view of the Building Research Center, Montreal Road Laboratories of the National Research Council, Ottawa; all the buildings are founded on limestone bedrock, but the arrow indicates location of the test boring mentioned in the text.

22.3 FOUNDATIONS CARRIED TO BEDROCK

Where rock does not outcrop at the surface in the vicinity of a building site, underground exploratory work will be essential. If this investigation discloses the existence of bedrock within a *reasonable* depth below the surface (thus suggesting the existence of the second general type of ground condition), test boring, drilling, and, if necessary, geophysical surveying should be carried out as extensively as possible all over the site in order to determine accurately the contours of this rock surface. Simultaneously, groundwater observations should be made. The word *reasonable*, although frequently so objectionable, is here used to cover the many variable local conditions that may affect the depth to which it proves economical to carry foundations to rock. The maximum depth so far utilized appears to be 75 m (250 ft); this was for the foundations of the Cleveland Union Terminal Tower, Cleveland, Ohio.[22.4]

A variety of engineering methods is available for transferring the building load down through the overburden to the bedrock. Choice of the method will depend upon a study of the economics and feasibility of the alternatives; such a study can be carried out with full satisfaction only against a background of complete information regarding subsurface conditions. If the depth is not too great and if the soil is free of boulders, the use of end-bearing piles is a method frequently followed. If steel piles are to be used, the possibility of corrosion must be studied; methods are now available for checking this possibility in advance of construction. If concrete piles are to be used, the possibility of high sulfate content in the groundwater must be investigated, so that any necessary precautionary measures may be taken. If cast-in-place concrete piles are to be used, the greatest care must be taken in studying the character of all soils to be penetrated to make sure that they are appropriate for the use of this type of foundation unit. Some cases of serious building settlement due to failure of this type of pile have occurred. If wooden piles are contemplated, similar care must be taken to study all possible variations in groundwater level, since some wood deteriorates under alternating wet and dry conditions.

For greater depths of overburden, for ground that contains boulders, and for carrying very heavy building loads, the use of some type of caisson or cylinder of concrete, either cased or uncased, will often provide an appropriate solution. For soft blue clay as is found in the Chicago area, for example, a simple type of open caisson (the "Chicago well") has proved satisfactory for depths well over 30 m (100 ft). Some of the most difficult foundation jobs yet undertaken have had recourse to this type of foundation unit. If groundwater is present, air pressure may have to be used; in every case, however, accurate subsurface information is the first requirement.

Experience in the Detroit area shows how important it is to know also the full character of the underlying bedrock. Detroit, like Chicago, is located in the glaciated area of North America; the foundation-bed conditions there, although unusual, are not unique and may be found in other cities similarly located, e.g., Winnipeg. The bedrock is fissured limestone, overlain by clay strata of varying types, with a layer of very hard glacial till ("boulder clay" or "hardpan" being local names) immediately above the rock. Depths to this impervious layer vary up to 36 m (120 ft). If it is not pierced, foundation design and construction are not unusual; shallow foundations may bear on to clay strata, and deep foundations on to the impervious layer through the medium of piles or concrete piers. If the glacial till is pierced at all, or if a crack is encountered in it, subartesian water may rise as high as 30 m (100 ft) above the rock. During the construction of the Greater Penobscot Building, 169.5 m (565 ft) high, having a total foundation load of 91 million kg (200 million lb), these conditions were encountered. Compressed air was applied to bell-mouth, open concrete caissons, which were to have been founded on the impervious layer, and they were instead carried right to rock. The water has an unusually high sulfur content, which made it injurious to concrete. Efforts to counteract this additional trouble included grouting of the bedrock around the base area of concrete caissons before the impervious layer was pierced.[22.5]

So much construction is carried out in the cities around the Great Lakes, in the United States and in Canada, that brief mention must be made of some problems that may be encountered even with this apparently straightforward method of founding buildings. In Buffalo, for example, a typical profile shows up to 4.5 m (15 ft) of miscellaneous fill material, then 6.0 m (20 ft) of clay and silt, with compact sand, silt, and gravel extending to bedrock at depths usually over 18 m (60 ft) below the surface. Excavation in this combination of material is not as easy as in Chicago, and steel end-bearing piles, driven to rock, are the usual medium used for transferring foundation loads to safe bearing. More than 50,000 tons of steel piles have been used in this way in this one city area. A further complication is the presence of boulders, which frequently interfere with pile driving.[22.6]

Even when excavation is such that "Chicago wells" can be used, there are dangers in sinking unlined shafts in advance of filling them with concrete. In the 1960s there were several cases of slight failures in such concrete columns because unsuspected inclusions of soft clay caused them to fail under load. It requires only one deviation from the anticipated soil profile, or one lapse in inspection procedures, to create a defect in what has been designed as a continuous column of concrete or, more usually, reinforced concrete. This was vividly shown by a case in Montreal where a new three-storey, reinforced-concrete building had to be demolished because of the failure of cast-in-place concrete piles. An unsuspected stratum of weak organic material, not strong enough to withstand the pressures exerted upon it while piles were being driven, was the cause of the failure.[22.7]

Not only must the character of the bedrock be known with certainty, but the contours of its surface must also be known. In many areas, a relatively flat bedrock surface can be almost guaranteed, but even so, one or two test borings strategically located will always be a good insurance policy. If the uniform surface of the bedrock is in any doubt at all, then adequate test boring must be carried out to make sure that the nature of the rock surface over the whole of the building site is known. One glance at Fig. 22.4 should be enough to establish this point in the mind of the most skeptical of readers. This is a model illustrating, to scale, the foundation of a modern office building in the center of Oslo, Norway, on Fridtjof Nansen's Plass, close to the famous town hall of the lovely Norwegian capital city. One corner of the building is carried on concrete piers resting directly on rock; the other is supported on steel piles 50 m (164 ft) long, driven to rock through the soft, sensitive clays that distinguish this part of Scandinavia (as they do the St. Lawrence Valley in Canada). Despite the almost phenomenal foundation-bed conditions, the foundation shown has proved to be quite successful; maximum settlement (despite troubles during construction) has amounted to no more than 3.8 cm (1.5 in)[22.8]

In mining areas, special attention must be paid to the location of mine workings in relation to buildings that are to be located above them. The Pittsburgh area presents many examples of foundation difficulties due to the neglect of, or the insufficient attention given to, this feature. On the other hand, it presents also some notable examples of successful solutions to problems of this sort. An outstanding example is provided by the foundations for the Veterans Administration general medical hospital, a 10-storey building with two basements, located over an abandoned coal mine 30 m (100 ft) below the surface. No details of this mine were initially available, apart only from the evidence given by some surface cavings. Subsurface exploration, not so much of the rock as of the voids from which coal had been abstracted, was conducted by means of 7.5 cm (3-in) (NX) core borings, supplemented by three 75-cm (30-in)-diameter Calyx holes. Comparative estimates of the cost of the alternatives

FIG. 22.4 Cutaway view of model of modern office building on Fridtjof Nansen's Plass, Oslo, Norway, showing variation in bedrock levels.

presented by the subsurface conditions led to the decision to grout up the old mine workings, since these were partially backfilled and experimental grouting had demonstrated satisfactory results. All the old workings were therefore consolidated in this way at a cost of $524,151, and on this artificially "reconstructed" rock foundation this great building stands today, a landmark of greater Pittsburgh.[22.9] In Zanesville, Ohio, on the other hand, a 3.5-million-L (800,000-gal) elevated steel water tank, that had to be constructed over old mine workings for which no adequate plans were available, was supported on one main central and 12 circumferential concrete columns, all carried about 15 m (50 ft) below ground surface and so about 10.5 m (35 ft) into rock, to a level below the mine workings. Of special note is the fact that the columns were reinforced only for their upper portions, where the supporting sandstone rock was relatively weak; the lower parts in hard shale were plain concrete.[22.10] Since problems caused by coal mining under building sites are not peculiar to the Pittsburgh area, further reference to this matter will be found in Chap. 37.

Finally, as an indication of what can be encountered in such an area as the island of Manhattan, Fig. 22.5 is reproduced; it shows a section through the foundation excavation for the Chase Manhattan Bank's headquarters in downtown New York. Water-bearing silt and the presence of boulders in the so-called "hardpan" made the final stages of excavation most difficult; the Joosten chemical solidification system was applied at the east end of the building site and at other trouble spots, and the excavation was completed

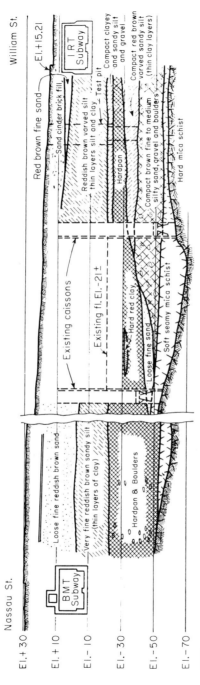

FIG. 22.5 Simplified cross section through the foundations of the Chase Manhattan Bank, lower Manhattan, New York City.

satisfactorily without any loss of soil, which, as can be seen from the section, could have had serious effects upon adjacent structures.[22.11]

It is of importance to note that if the surface of the bedrock slopes at an appreciable angle, special precautions may have to be taken in designing and constructing a foundation. This restriction is obviously important in the construction of buildings founded directly on rock. Such a restriction is also of some significance in buildings constructed on the second type of ground condition, where solid rock is within a moderate distance of the ground surface, although it is not actually used as the foundation bed. The action of gravity naturally affects all loose material lying above a solid-rock surface; and if the latter is lubricated (by groundwater, for example), the whole mass may slip down the slope of the rock.

This was the explanation generally recognized by engineers familiar with the serious movement of the Cahuenga Pass multiple-arch retaining wall constructed in Los Angeles in 1925. This wall is 135 m (450 ft) long and its height varies up to 18 m (60 ft); the centers of buttresses are 9 m (30 ft) apart. The wall was founded on bedrock at its ends but not in the central portion. Rock was here 12 m (40 ft) below ground level; spread footings were therefore used and carried to a depth of 6 m (20 ft). Movement of this central section was observed before all the fill had been placed; it continued even after the loading on the wall had been reduced, and in July 1927, it amounted at one place to 43 cm (17 in) outward and 38 cm (15 in) downward. The load on the spread footings was limited to 39,000 kg/m^2 (8,000 psf), but the underlying rock sloped steeply away from the wall, so the conclusion was that soil strata and wall were moving down on the rock surface as a unit.[22.12]

An allied problem is to have stratified bedrock available as a foundation stratum but dipping at such an angle to the horizontal that the stability of the upper layers is doubtful. In parts of New York City, the Manhattan schist is so inclined; it is a feature that is encountered elsewhere not infrequently. Local details will determine the best solution of the problems thus presented, but a usual method is to drill the surface rock and anchor it by dowels to layers that are so far below the surface as to be beyond the range of possible movement.

22.4 FOUNDATIONS ON SOIL

The third general type of ground condition, where soil is used as the foundation bed, is probably the most common of all, particularly if due consideration is given to the vast number of small structures that could be founded on rock (by methods already described) but that need not be, because of their relatively small weight. Adequate subsurface investigation is again imperative if the foundation design is to be satisfactory. There was a time when it was generally thought that to carry test borings down as far as the length of bearing piles was enough; today, this limitation of test boring would often properly be regarded as tantamount to waste of the cost of what exploratory work was done. Even before the type of foundation to be used has been selected, accurate and extensive knowledge of subsurface conditions is essential. As a general rule, exploratory work should give definite results for a depth *at least equal to twice the width of the structure* and greater than this if possible, especially if the presence of relatively soft strata is suspected. The extra cost of penetrating a few feet more in test borings, when once the boring equipment is set up, is negligible compared to the value of the results so obtained.

The vital necessity of obtaining subsurface information to such depths can be illustrated by a brief glance at one of the fundamental concepts of soil action. It may first be observed that foundation loads may be transmitted to unconsolidated strata in two general ways: (1) by constructing a continuous raft, or spread footing, which rests directly on the surface stratum, and (2) by driving piles into the soil or by constructing piers therein, their tops structurally connected in a suitable way to the bearing columns of the structure. A variation of the first type is to have the load from a single structure distributed through a number of isolated footings. The bearing strength of the second method may depend for its efficiency on piles or cylinders whose ends rest on a reasonably hard stratum, or it may be the result of skin friction. Other minor variations might be listed, but these methods include generally all leading classes of foundation structure for unconsolidated strata.

How do these transmit the load from the building to the soil in which they rest? Some discussions would suggest that loads simply disappear into the ground in some mysterious manner. Figure 22.6 is a graphical illustration of the way in which loads are actually dissipated by transfer of stress to steadily increasing volumes of soil as the distance from the foundation structure increases. The lines of equal pressure indicated in Fig. 22.6 are colloquially termed *bulbs of pressure,* a convenient figure of speech, even though terminologically inexact. The proportional reductions of stress shown are typical values; exact figures for any set of assumptions made can be calculated for any given set of conditions; this is one of the great contributions of soil mechanics to foundation design. Simple though the diagram is, it is at the same time one of the most significant in the study of foundations; it shows at a glance the vital necessity of having accurate knowledge of subsurface conditions to a depth of at least twice the width of the structure to be supported.

When this precaution is not taken and a building is erected without accurate subsurface information, trouble may develop if, for example, there is a buried stratum of weak soil beneath the site. There are on record all-too-many cases in which serious settlement has occurred from this cause. The following account is typical. A large building was carefully planned and designed; test borings were taken and the results carefully studied before construction began. Prior to its completion, this building was discovered to have

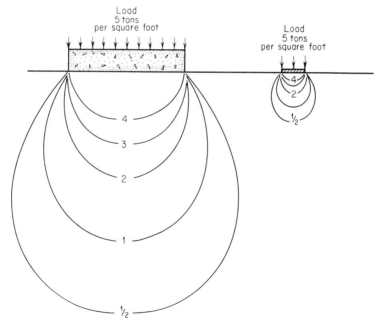

FIG. 22.6 Simple "bulb of pressure" diagram illustrating the effect of the size of a loaded area on the stress distribution beneath it.

moved out 10 cm (4 in) and settled 10 cm (4 in) at one place in its front wall. New borings were put down to a depth of 26 m (87 ft); the original borings had stopped 13 and 9 m (42 and 31 ft), respectively. In the new tests, it was found that a stratum of "very soft black clay and fine sand" existed at a depth of 16.5 m (55 ft) below the surface; the upper strata were clay, with some sand of varying consistency. The front of the building was about 27 m (90 ft) wide. The necessary underpinning and pile driving to correct the unsatisfactory condition thus discovered cost about $18,000—"probably four times what the cost would have been had the piles been driven before the wall was built."[22.13]

This aspect of foundation design in relation to soil strata is so important that a further example from ordinary practice will be cited, again not an exceptional case but one having general significance. The Westinghouse Electric and Manufacturing Company Building in Philadelphia, Pennsylvania, was constructed on a site, boring records for which were supplied by the previous owners. These showed 2.4 to 3.0 m (8 to 10 ft) of loose fill and then uniform clay and sand to rock level, 13.5 m (45 ft) below the surface. Bearing-pile foundations were therefore decided upon for the 10-storey building. The concrete pile contractors drove test rods to determine the length of piles, which was fixed at between 7.5 to 9.0 m (25 to 30 ft), a penetration shown by the test rods to be satisfactory. When eight of the ten storeys had been constructed, the building was found to be settling noticeably. It had already sunk 10 cm (4 in) on one side. New core borings and an open test caisson were therefore put down to rock, a mica schist which was found 17 m (57 ft) down; the strata passed through were 8 m (27 ft) of fill, 1.2 m (4 ft) of peat, 30 cm (1 ft) of gravel, 4.5 m (15 ft) of silted peat, 2.4 m (8 ft) of silt, and 60 cm (2 ft) of coarse sand and gravel, instead of the fill, clay, and sand expected. The weakness that permitted the excessive settlement was at once obvious; an elaborate and ingenious underpinning operation had to be undertaken immediately, and the footings had to be supported on 37.5- and 40.0-cm (15- and 16-in) pipe piles, concrete filled and carried to rock.[22.14]

Accurate knowledge of subsurface conditions at a building site naturally includes full information regarding the properties of the soils to be encountered. This information is obtained by careful laboratory testing of good (undisturbed) soil samples. With the knowledge of basic soil properties and the application of theories of consolidation, it is now possible to calculate in advance of construction the probable total settlement of the building and also the rate of settlement. This is the settlement of the building to be expected because of the increase of stress in the soils supporting it; the fact that all buildings founded upon soil settle to some degree is now generally recognized. Correspondingly, safe bearing capacities for soil can be calculated, safe against excessive settlement and against overstressing of the soil immediately under the foundations. These are matters in the province of the engineer; they are mentioned here in order to emphasize the absolute necessity of accurate subsurface information at building sites, a matter with which the geologist can frequently assist.

With the progress being made in the development and application of soil mechanics to foundation engineering, foundation failures should steadily decrease in number and magnitude. Study of failures of the past, carried out with constructive intent, can, however, be of real assistance to modern design. An extreme example of the difficulties caused by ignorance of soil behavior is provided by the National Museum Building in Ottawa, a massive red sandstone structure built in 1910 and now somewhat unusual in appearance, since the elaborate upper part of the entrance-porch tower had to be removed shortly after the building was opened (Fig. 22.7). Serious settlement was noted before the building was completed; before the tower was removed, the main porch had separated from the rest of the building by 35 cm (14 in) at roof level. Modern study has shown that the supporting soil was just at the point of actual shear failure when the load was so fortunately reduced. The inside of the building until 1975 exhibited to those who knew where to look,

FIG. 22.7 The National Museum Building, Ottawa; the inset view shows the entrance porch as originally constructed and before its tower was removed as a safety precaution.

evidence of differential settlement that almost had to be seen to be believed; at least 30 cm (12 in) of settlement was visible in one restricted area. Curious suggestions about the cause of the settlement were made at the time of the trouble; one eminent geologist stated that there must be a fault beneath the site. The real explanation is that the building was founded on over 30 m (100 ft) of the sensitive marine clay already mentioned, a clay which was not strong enough to carry the heavy stress concentrations induced by this splendid example of Victorian "wedding-cake" architecture.[22.15]

22.5 GROUNDWATER

The necessity for accurate knowledge of the groundwater conditions at every building site must again be mentioned, so often are these conditions forgotten, or considered only at the time of making the site investigation, with no thought being given the probable variation in the elevation of the water table throughout the year. Examples were given in Chap. 18 of building foundations with which trouble was experienced due to changes in the water table elevation, one notable case being in Boston. The problem of falling groundwater levels is not peculiar to Boston; it is, indeed, one of the most widespread problems of civil engineering works. In San Francisco, a 14-storey building near Sansome and Bush streets had to have its entire load transferred from 1,100 wooden piles to 210 concrete piers as early as 1930 because of water table variations.[22.16] In Milwaukee, Wisconsin, as in all-too-many cities, the same problem has, unfortunately, had to be faced.

How can such foundation troubles be prevented and remedied? One method, followed in some building regulations, is to make sure that all piles are either cut off or embedded in concrete at a level well below the minimum elevation possible for the local groundwater. An alternative method is to use "composite" piles in areas of fluctuating groundwater level, i.e., to use protected sections within the range of water-level variation and timber below this level. Another practice was followed in connection with the foundation of the Northwestern Mutual Life Insurance Building in Milwaukee, Wisconsin (Fig. 22.8). Foundation beds consist generally of glacial strata, with bedrock 75 m (250 ft) below ground surface; the water table at the time of construction (1930) was 5 m (17 ft) below the surface. Timber bearing piles were used, and lengths of 10-cm (4-in) pipe, fitted with screw

FIG. 22.8 Head office building of the Northwestern Mutual Life Insurance Company in Milwaukee, Wisconsin, the more recent section of the building at the rear.

caps and strainer bottoms, were built into the foundation slabs. Readings of water levels in these observation pipes have been taken every three or four months since the building was completed. It has proved necessary to add water to about one-fifth of the pipes in order to maintain the groundwater level at the required elevation, but this has been done without difficulty and the foundations have performed satisfactorily. Naturally, a complete record of the levels of the groundwater observed in the pipes is continuously maintained.[22.17]

One of Britain's most noble buildings is the Cathedral Church of Winchester, ancient capital of England. Generally known for its historic interest and beauty, the great building is famous also in the annals of foundation engineering. Groundwater comes close to the surface in the vicinity of the church; this has been handled in a manner well worth reporting here. An extensive scheme of reconstruction was prepared in the early 1900s, a scheme which involved shoring the outside of the building; supporting the great arch vaulting inside to prevent collapse; using steel tie rods when necessary, in association with an extensive use of high-pressure grouting in all the damaged masonry; and then underpinning all of the exterior walls down to a bed of gravel. This last operation had to be carried out in the complete darkness created by the black groundwater since, with the equipment available at that time, pumping would probably have removed some of the silt and possibly disturbed the foundation beds in other ways. The work involved excavating from beneath the walls, piers, etc., in succesive pits, down to the gravel, and then placing concrete by hand up to the underside of the masonry of the walls. The entire job was done in a period of 5½ years by one man, an expert diver named William Walker, his work being regularly inspected by Sir Francis Fox, who also wore a diver's suit. All was successfully completed, and the walls of the cathedral have shown no signs of any structural trouble since then. The building was reopened, with the reconstruction complete, at a great service held on 14 July 1912.[22.18] A statue of William Walker was later placed in the cathedral in grateful recognition of his notable work, possibly a unique tribute to a real builder.

Fifty years after this work, a new hotel had to be built within little more than 51 m (167 ft) of the cathedral. The record of the earlier work was available, and the same groundwater condition had to be dealt with, in such a way, naturally, as to cause no damage to the cathedral. Well points, all connected into a larger pipe "header," were used in closely spaced rows to surround the area that had to be pumped dry (Fig. 22.9). This advance in construction methods permitted the drying out of the excavation for the hotel with no difficulty. Naturally, the most careful records were maintained, including groundwater levels in test holes around the site; the most important of these test holes was actually inside the cathedral. With excavation complete, and the building of the superstructure proceeding, the groundwater was allowed to regain its normal position, the foundation of the hotel having been designed accordingly. Of special interest is the fact that about one-third of the excavation for the building had been completed before the general contractor had even been appointed, much less started work—and all by archaeologists!

FIG. 22.9 Excavation for the foundations of a new hotel at Winchester, England, showing the header for the well-point dewatering system (a horizontal pipe left of center) and the great cathedral in the background.

They found and recorded an archaeological "treasure trove," including remains of a Roman villa and of a Roman road.[22.19] In a similar way, Walker's laborious hand excavation at the cathedral site had revealed unexpected items from the past; these included a rats' nest containing a carpenter's rule between 300 and 400 years old and pieces of parchment that, when deciphered, proved to be from a service of 700 years ago.

Groundwater may cause troubles with building foundations by being at such a level that a foundation will tend to "float," to move upward instead of downward, as is usually the case. In the construction of the new general care building of New York's Harlem Hospital (mentioned in Chap. 8 in connection with the unusually high temperature of the water encountered in the excavation), the volume of water handled by the well-point system installed by the general contractor was so great that the system was kept in operation until 12 storeys of structural steel and the first three concrete floors were in place; this was done in order to give sufficient deadweight to counteract the calculated hydrostatic pressure. Despite the intensive study given to this water, no agreement about its origin was reached among the experts consulted.[22.20]

When similar circumstances are encountered with structures that do not have sufficient intrinsic weight to offset anticipated hydrostatic pressure, other measures have to be taken. When the sewage-treatment tanks of the Ley Creek plant at Syracuse, New York, have to be emptied for maintenance or repair (about once every three months), the superintendent must first set in operation a permanent well-point installation that lowers the groundwater level around the tanks to a sufficient depth to avoid floating the empty tanks. Soil conditions in this area include a thick stratum of what is locally known as "black sand," a medium-to-medium-coarse sand, high in calcium and iron salts, underlying a loam and clay blanket at the surface.[22.21]

More unusual was the problem faced during the construction by the New York City Housing Authority of five 14-storey apartment houses adjacent to the boardwalk at Coney Island. Figure 22.10 shows the local soil conditions and the proximity of two of the adjacent buildings, which could naturally not be disturbed during construction of the new blocks. Careful laboratory study of samples of the organic silt from both beds showed that this material had been overconsolidated in the past by an overriding sand dune. Hence, the foundations of the large apartment blocks would have to be so located as to give some degree of buoyancy, in order to reduce the additional pressure they would exert on these silt beds, deep below the surface though they are. This meant excavation below the level of existing groundwater, which, in turn, meant unusually careful control of groundwater-lowering operations to ensure the safety of the existing adjacent buildings. Detailed pumping tests were undertaken, and a large number of observation wells, through which the changes in groundwater level could be continuously observed, were installed. The job was completed successfully by the use of a judicious combination of a well-point system pumping water *out* of the excavation for the apartment blocks and a diffusion system, operating simultaneously, pumping water back *into* the ground in the vicinity of the existing buildings. The full system was in operation between the months of June and September, during which period the foundations for all five apartment blocks were completed. When the system was closed down, careful checks showed no evidence of any settlement of the existing structures or of any damage because of the manipulation of the groundwater conditions in the area.[22.22]

It is believed that this was the first installation of such a diffusion or recharging system in association with a well-point system; conditions were unusual, but the success of the project depended upon the most complete study of subsurface conditions made as a preliminary to planning and design. In great contrast are the problems met almost every day in areas where the local subsoil is a shrinkable clay, as is so frequently the case. Here,

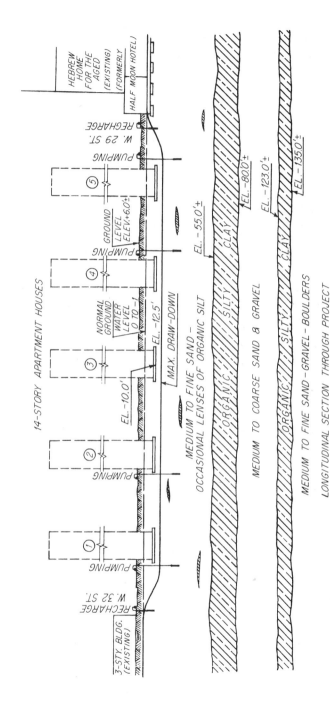

FIG. 22.10 Cross section through foundation strata for buildings in Coney Island, New York, in relation to groundwater conditions.

22-17

22-18   CIVIL ENGINEERING WORKS

even small houses may be affected by the drying out of the clay. An unusually dry period or even the action of tree roots may cause a serious groundwater depletion.

22.6 EFFECT OF TREES

This is one further aspect of water and foundations that the writers mention with regret, since they are tree lovers and want to see the cities of the future given all the charm and serenity that beautiful trees can bestow. Trees can, however, ruin buildings. They are living things and require water, much water, for their growth. Young leaves can contain up to 90 percent of their weight in water; even tree trunks can contain 50 percent. Although the formation of 100 g (3.5 oz) of cellulose (the main content of wood) requires 55 g (2.0 oz) of water, a tree will lose correspondingly by transpiration almost 100,000 g (220 lb) of water, or 1,000 times as much as its own gain in weight. The tree obtains this water through its root system from the soil in which it grows. The root system, formed mainly of fine roots feeding into the few main thick roots to be seen when a tree is uprooted, may account for 10 percent of the total weight of a tree. The root system of an oak tree, for example, can total several hundred kilometers in length. When the Suez Canal was excavated, roots of a tamarisk tree were found at a depth of 30 m (100 ft) below ground level. Correspondingly, tree roots have been traced for a horizontal distance of 30 m (100 ft) from the tree itself. This is the intricate system of feeder channels through which a tree obtains its essential water supply. If growing in sandy or gravely soils and the water supply fails, as by a sudden lowering of the water table, trees will quickly die. If in clay soils, especially those with high natural-moisture contents, trees will draw water from the minute pores within the clay. As the clay dries out, the root system will spread into the clay that still contains water. Some trees are greater users of water than others, and so more of a hazard when growing in clays that have the unfortunate property of shrinkage with decreasing water content. The broadleaf deciduous trees—poplar, alder, aspen, willow, and elm, in that order, followed by maple, birch, ash, beech, and oak—are the worst "offenders."[22.23]

Normally trees will draw their water supply from the upper part of the soil in which they grow, which is itself supplied by percolation of rain. In times of normal rainfall, this will be a system in equilibrium, even for the heavy water users. When, however, a period of low rainfall is experienced, the deep-seated root system will go into action and draw water from previously unaffected clay. And if this is a shrinkable clay, shrinkage will

FIG. 22.11 The sidewalk of Metcalfe Street, Ottawa, showing the effect of tree roots in causing differential soil settlements.

inevitably follow, with unfortunate results for any buildings, streets, or other structures that may be supported by the stratum. Experiments in Ottawa have shown surprising movements in the local Leda clay in a dry summer (Fig. 22.11). A vertical settlement of 13 mm (0.5 in) at a distance of 6 m (20 ft) from a row of elm trees, and at a depth of 3.9 m (12.8 ft), was typical and shows clearly what damage can be wrought in this way in periods of very dry weather on buildings founded on shrinkable clay.[22.24]

Older residential buildings in the center of the city of Ottawa provide all-too-good examples of the differential settlements (and, in a few cases, actual failures) caused by the elm trees that so grace the roads by the side of which they stand, following a number of dry summers. Figure 22.11 shows the effect even on sidewalks. Ottawa is very far from unique in this respect, however. Examples are on record from Africa, Australia, Burma, China, India, Palestine, the Sudan, Belgium, and Texas. There are probably very few areas of the world where some evidence of cracking in buildings founded on clay soils, cracking caused by trees, could not be found. Some remarkable examples have been recorded from Great Britain. Typical was the case of two-storey brick houses so badly cracked that their corners had to be shored up, and all because of Lombardy poplars within 6 m (20 ft) of the houses. A large theater, built of brick, in Stamford Hill, London, was seriously cracked because of the root action from a row of Lombardy poplars that had been planted, presumably, to act as a screen for the large expanse of brick wall—the wall that cracked. The crack was 44 mm (1.7 in) long at the top of the wall at the rear of the

FIG. 22.12 Cracks in a major building in London, England; damage was due to the poplar trees that may be seen on the right and necessitated temporary supports for the wall.

theater building (Fig. 22.12). For special local reasons, the remedy in this case involved deep underpinning of the wall in question.[22.25]

The solution of this unusual problem involves, as always, accurate foreknowledge of the subsurface conditions. If it is found, through tests on soil samples in the laboratory, that buildings have to be founded upon shrinkable clay, then the necessary precautions must be taken in design. Footings should be taken down as deep as practicable, but, of much greater importance, landscaping near the structures must be strictly controlled, particularly with regard to the siting of trees. A good working rule is to ensure that no tree is placed nearer to buildings on such foundation beds than a distance equal to the total height to which the tree may be expected to grow. Landscape architects may object to such a requirement, but they should be convinced if shown, in detail, some of the examples that have been summarized in the foregoing paragraphs.

If trouble develops after a building has been erected, then the easiest solution is to remove the offending trees, regrettable though this procedure may be. If for any reason such a course is not possible, then the only solution is to underpin that part of the foundation that has been or may be affected. This is always a costly procedure, and so the economics of the situation call for special study. Cracks that have developed can be covered up, but a more desirable procedure is to get Nature to reverse itself, by restoring the natural moisture content of the clay after the trouble-causing trees have been removed. To suggest that cracks in a building can be cured by the simple procedure of cutting down a tree and leaving a garden hose running continuously for some days will cause surprise to the uninformed, but it is a simple solution that works. It is also a vivid reminder of the importance of knowing the exact state of the unseen materials upon which buildings are founded.

22.7 PILED FOUNDATIONS

The four main methods of transmitting light loads to soil strata can be outlined as: (1) the use of individual footings under columns, (2) the use of a continuous raft, or spread footing, under either a complete building or one section of it, (3) the use of bearing piles driven to such depths that they will safely carry the loads transmitted to them from footings, and (4) in extreme cases, the use of excavated cylindrical shafts in which concrete piers (suitably reinforced) can be formed to transmit the loads from footings to a safe bearing stratum.

In earlier days, it was common practice to "drive piles" if ever there was any question of the competency of surface soil strata to carry the building loads in prospect. Bearing piles, when circumstances require their use, can serve as excellent foundation units; buildings all over the world testify to this. But certainly in some cases, foundations just as satisfactory could have been obtained without the use of piles, while in some exceptional cases, the use of piles has actually weakened the foundation bed intended for use. If, for example, site investigations show that the building area is underlain with sensitive clay, such as has so often been mentioned in these pages, then pile driving must be viewed with extreme caution, since in many of these clays the vibrations set up by pile driving are enough to reduce their effective bearing capacity by partial "liquefaction." There is in the province of Quebec a large church that had to be completely underpinned after having been founded on piles in complete disregard of the extremely sensitive clay on which it was founded; piles had proved satisfactory for supporting a church in a neighboring village which happened to have quite different subsurface conditions.

The fundamentals of pile action in soil must be fully appreciated before piles are even considered as possible foundation units. This is illustrated in Fig. 22.13 where it will be seen that the concept of the "bulb of pressure" (using the common term) naturally applies to piles. If the length of piles is great in comparison with the width of the foundation (as is often the case), then the soil will be stressed to a considerable depth beneath the bottom tips of the piles. The neglect of this fact has led to serious settlements of buildings founded on piles, below the bottom tips of which existed soil strata of insufficient strength to resist, without undue compression, the loads thus imposed so far below ground surface.

For those who do not fully appreciate the necessity of knowing with certainty geological conditions to a depth of at least twice the width of every structure, and more than this if piles are to be contemplated, the problem is compounded in that the best of load tests carried out on single piles will fail to indicate the danger of using a fully piled foundation where there is a weak stratum below. Reference to Fig. 22.13 will clearly show why. From the many examples of neglect of this sort that can, most unfortunately, be cited, there may be mentioned the case of a 20-storey hospital founded on a recent alluvial deposit 30 m (100 ft) thick. The deposit was mainly a soft clay, but it contained appreciable strata of firm-to-dense sand. It was decided to found the building on piles driven to about 14 m (46 ft) below ground surface to a 2-m (6-ft)-thick sand stratum. More than 10,000 piles were so driven; within a year of completion, differential settlement had begun; within 20 years the settlement at the center of a dish-shaped area was nearly 1 m (3 ft), all due to consolidation of clay *beneath* the sand stratum to which the bearing piles had been driven.[22.26]

Bearing piles, both end-bearing and friction-bearing, are excellent foundation units when properly used in appropriate ground conditions, and there are many piled foundations performing quite satisfactorily around the world. It is, therefore, in one way unfortunate that so many "horror stories" of the misuse of piles are to be found in the annals of civil engineering. But this is unavoidable, and if the failures are described with constructive intent, some advantage is gained, since the cases can then be cautionary tales. Some failures of piled foundations are directly due to lack of appreciation of pile action

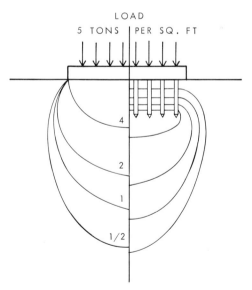

FIG. 22.13 Simple "bulb of pressure" diagram illustrating the redistribution of stresses in foundation strata with the use of bearing piles.

22-22 CIVIL ENGINEERING WORKS

in relation to the geological conditions at the site; others are caused by mistakes in pile design or driving, matters with which this Handbook is not concerned.

One further illustration of a pile failure due to geology may usefully be given, if only to show how serious can be the lack of proper attention to subsurface conditions. The case took place half a century ago and so can now be referred to quite impersonally. A large grain elevator was to be built at Durban, South Africa on a site underlain by a surface stratum of sand overlying a thick stratum of clay. The properties of the clay were not fully appreciated, because an inadequate sampling was taken from the ten boreholes that were put down in advance of design and construction. As a result, the length of the bearing piles for supporting the structure was reduced from that originally intended. The clay was a normally consolidated deposit with low shear strength and high sensitivity, and it was therefore susceptible to considerable loss of strength by such disturbance as the driving of piles.

Three hundred and eight piles for the annex were driven first, and the concrete foundation structure with tunnels was completed about two months after pile driving. Settlement, even without further loading, started as soon as the foundation structure was complete and accelerated after the building of the superstructure, a settlement of almost 65 cm (25 in) taking place in less than one year. Similar settlement took place under the working house. Work had to be stopped and an attempt made to underpin the structures

FIG. 22.14 General view of the mill and grain silo complex at Randfontein, Transvaal, South Africa.

to rock, but even this proved difficult because of the character of the clay. A new foundation was designed, carried to rock, and the elevator was eventually completed but at considerable loss of time and great extra expense. The published records of this case are still helpful reading and vivid reminders of the vital importance of a full understanding of subsurface conditions before piles are included in any foundation design.[22.27]

By way of contrast (even though it is a structure not founded on piles), another South African grain elevator may be mentioned briefly, since its excellent foundation design illustrates so well the message of this Handbook. Located at Randfontein, it is the tallest such structure in South Africa; it is shown in Fig. 22.14. Initial auger test borings came to refusal at depths up to 10 m (33 ft) on Black Reef quartzite. Just before construction was about to start, investigations at another site one kilometer away revealed a stratum of *wad* (an unusual type of residual soil) 2 m thick and up to 12 m (40 ft) below the surface—therefore, beneath the quartzite. Foundation design was changed; the entire site was excavated to a depth of 13 m (43 ft). This disclosed complex rock conditions, including four step faults (shown in Fig. 44.10). Two exploratory shafts were sunk to a depth of 24.5 m (81 ft) in order to be certain of rock conditions. A new design was prepared for the foundation, including a massive concrete mat and belled-out short caissons, and the great structure was satisfactorily completed.[22.27]

22.8 "FLOATING" FOUNDATIONS

Some structures have to be constructed on sites below which weak and sensitive soils exist, soils which must be accepted as the foundation beds to be used. One solution is to use what are colloquially called "floating" foundations, an accurate description even though the first impression it gives is slightly misleading. Even weak soils manage to sustain their own weight. At any depth below the surface, the soil below that level is carrying the weight of the superincumbent soil, without any settlement. If, therefore, a depth of soil can be removed the weight of which is equal to the weight of the structure to be built on the site, then the weight of the structure can replace the weight of the soil, and no settlement of the resulting sunken structure should result. This is sound in theory, but the actual execution of such designs involves much ingenuity in construction methods. Figure 22.15 shows one ingenious means of achieving the ideal result. More conventional construction methods were used for the foundation of the head office building of the New England Mutual Life Insurance Company in Boston's Back Bay district; this building is 10-storeys high and has an 87-m (290 ft) tower. Ground conditions, from the surface down, were found to be 6 m (20 ft) of man-made fill, 6 m (20 ft) of alluvial silt, 3 m (10 ft) of oxidized, hard yellow clay (on which the building was founded), 21 m (70 ft) of glacial clay, and finally 9 m (30 ft) of compact sand and gravel above bedrock. Groundwater is only 3 m (10 ft) below the surface. The final design consisted essentially of a 12-m (40-ft)-deep concrete box, the excavation for which just about balanced in weight the total building load of 130,000 tons, thus inducing no basic change of stress beneath the final foundation.[22.28]

A corresponding example from Great Britain is to be found at the Grangemouth oil refinery of Scottish Oils, Ltd., on the Firth of Forth. Here the subsurface conditions consist of a very soft (and weak) clay for a depth up to 9 m (30 ft), underlain by a 2.7-m (9-ft) layer of stony clay, below which gray silty (estuarine) clay extends more than 30 m (100 ft). Previous structures, with bearing pressures as low as 5,468 kg/m² (1,000 lb/ft²), had settled at the site, and a 33-m (110-ft) chimney had settled 60 cm (2 ft) out of plumb. Even with cast-in-place, bulb-ended concrete piles, the new structures also settled. For the latest structures, therefore, and after careful soil testing, a total load of 37,000 tons

22-24 CIVIL ENGINEERING WORKS

FIG. 22.15 "Floating box" foundation structure being sunk in position near Georgetown, Guyana.

was carried on 15 separate cellular caissons, which were sunk to depths up to 7.5 m (24½ ft), forming, in effect, a set of "floating" foundations for the refinery. The largest of the caissons was 51 by 51 m (170 by 170 ft). They were made with thin, reinforced-concrete walls, constructed in the first instance at ground level, resting on reinforced-concrete pads which were removed when sinking had to start. Soil within the cells was excavated by grab; a concrete plug sealed the bottom of each cell when the correct depth to give the required buoyancy was reached.[22.29] Looking at the new refinery today, with its complex equipment and highly instrumented operations, one finds it hard to credit that all the structures are supported on such weak soil, so successful has this foundation design proved to be, based as it was on a very thorough survey of subsurface conditions. And there is little need to emphasise that any use of "floating" foundations can only be decided upon on the basis of the most accurate possible knowledge of the geology underlying the building site.

22.9 CAISSON FOUNDATIONS

Japanese engineers have used a not dissimilar method, which depends to an unusual degree upon complete and accurate knowledge of subsurface conditions, for the construction of over 20 buildings in the Tokyo district. When the subsurface conditions are known with certainty, and are suitable, a cutting edge is built on the surface of the ground and upon this the basement of the building is constructed as a "superstructure" above ground. When the resulting structure has the requisite weight, sinking is started by excavating beneath it; control is exercised by the progress of excavation under the cutting edge, and sinking is induced by the weight of the structure. (See Fig. 22.16.) This idea has long been used for the construction of bridge piers. Construction procedure for this daring method is naturally an engineering matter of unusual interest, but it will be appreciated that its success depends entirely upon exact knowledge of subsurface conditions. Boulder-free soils are obviously desirable; the most appropriate conditions are those given by alluvial or lake deposits of uniform soft or medium clay.

The largest building yet to be constructed in this way is the Nikkatsu International Building in Tokyo, which has a four-storey basement, nine storeys above ground, and a three-storey penthouse; it measures 97.5 by 66.0 m (325 by 220 ft) along the streets on which it faces at right angles and occupies a full, triangular city block. Six borings and five wells were sunk to depths of 45 m (150 ft) in determining subsurface conditions. Beneath the subsoil, a blue-gray clay extends for 15 m (50 ft), underlain by gravel and sand beds. Groundwater was encountered at a depth of 12 m (40 ft), but not in any quantity until a gravel bed was reached. These are almost ideal conditions for the sinking method. Three basement floors were built above the ground before sinking started; the caisson then weighed 25,000 tons. The record of the successful and speedy completion of this $5 million project is an outstanding example of fine engineering based upon adequate preliminary geological information (Fig. 22.17).[22.30]

The idea has now been applied also in Europe, where a 530-car parking garage, with seven levels below grade, has been constructed in the same manner in Geneva, Switzerland. Here subsurface conditions consisted of 7.5 m (25 ft) of water-bearing sand and gravel overlying 21.3 m (70 ft) of soft clay that grades into a good, firm clay on which the caisson was founded. The versatile mineral bentonite was put to good use again, as a lubricant this time, injected in an annular ring around the caisson formed by the slightly wider diameter of the cutting shoe as compared with the structure itself (Fig. 22.18). Again, suitable soil conditions and the corresponding absence of boulders in the soil were an absolute prerequisite for employment of this method.[22.31]

FIG. 22.16 Commercial building, to have five storeys below ground and nine above, being constructed by the Takanaka caisson method, the structure seen being the complete concrete basement complex of the building in process of being sunk to final depth by controlled excavation.

22-26 CIVIL ENGINEERING WORKS

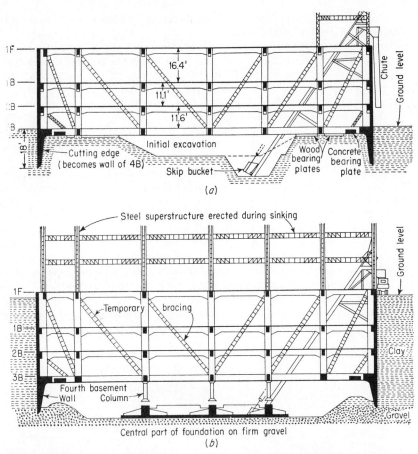

FIG. 22.17 Sections through the foundation caisson for the Nikkatsu International Building, Tokyo, Japan, showing the way in which the caisson was transformed into the permanent foundation structure.

A different type of caisson foundation structure had to be used, as the only economical solution, for a new eight-storey-high storage warehouse in the port area of Stockholm. Expansion of the port facilities made the new building essential, but the only site available was one previously considered as wasteland and used for the dumping of spoil from excavations, including large blocks of rock up to 2 m³ (2.6 yd³) in size. Below the original seabed was a stratum of glacial till, providing a geological combination of unparalleled difficulty. Heavy loads from the building had to be transmitted to rock. It was therefore decided to build the complete lower basement in "dry dock," even though the basement measured 47 by 54 m (158 by 180 ft) in plan, float this out to the site, and there sink it on to specially leveled sand pads. The second "storey" was then constructed, holes being left in the structure through which heavy steel pipes, 95 cm (37 in) in diameter, could be lowered, being rotated by special equipment, serrated cutting edges penetrating the miscellaneous material encountered before rock was reached (Fig. 22.19). Special arrangements were made at the foot of each shaft for transmitting the heavy loads that each would have to carry to the underlying bedrock. Construction details of this unusual approach to

FIG. 22.18 The basement complex for the Rive-Centre garage in Geneva, Switzerland, being sunk in place by controlled excavation.

founding a major warehouse were complex; all were necessitated by the unusual subsurface conditions, thoroughly explored before designs were completed.[22.32]

22.10 PRELOADING OF FOUNDATION BEDS

For some buildings, deep basements are not required, and yet potential settlement of weak soil may have to be faced if a site must be used at which the only material available to an economical depth consists of unconsolidated soils of low strength. Here again adequate preliminary information and full knowledge of the geological history of the soil can assist with the necessary engineering solution. If laboratory tests upon good soil samples show consistent values for the consolidation characteristics of the soil, it is possible to calculate how to induce a degree of consolidation of the soil in place by preloading it with a load that can be removed when necessary settlement has taken place. If the permanent structure can then be built, excessive settlements can be eliminated; the actual amount of settlement will be controllable to a reasonable degree of accuracy. The method is an old one, frequently practiced in the past (on the basis of "horse sense") for approach fills to bridges and similar earthworks. Use of preloading by Sandford Fleming in 1870 has been mentioned, but long before this, in 1819, the same idea was used by Thomas Telford during the building of the Caledonian Canal in Scotland.

The application of the techniques of soil mechanics, coupled with accurate subsurface exploration in advance of construction, has converted this very practical approach into a refined and predictable procedure. Again, the details of its application lie in the field of engineering; again, the success of its application depends upon accurate subsurface information. Of the many examples now available from modern practice, there may be men-

FIG. 22.19 Precast box foundation for warehouse in Stockholm Harbour, Sweden, approaching completion in its dry dock, before being towed to the building site and sunk into place.

tioned the construction of the Cathedral of Mary Our Queen, in Baltimore, a monumental structure 112.5 m (375 ft) long and up to 28.2 m (94 ft) wide, with two stone towers 39.9 m (133 ft) high. Here, preloading in the form of a pile of earth 9 m (30 ft) high, spread over an area of 36 by 21 m (120 by 70 ft), was carefully designed to obviate the differential settlement which the extra load imposed by the two towers would have caused in the underlying soil formed of weathered micaceous rock (Fig. 22.20).[22.33]

An unusual example was provided by the rehabilitation of a large wartime building at Port Newark, New Jersey, by the Port Authority of New York and New Jersey. Built in

FIG. 22.20 Preconsolidating the building site for the Cathedral of Mary Our Queen, Baltimore, Maryland, by overloading with soil.

FIG. 22.21 Site of the Iona Island sewage treatment plant, Vancouver, British Columbia, preloaded with fill which, when removed, will have induced settlements of up to 1.2 m (4 ft) in the underlying soft silt and clay.

1941 by the U.S. Navy, the building (measuring 165 by 76 m or 510 by 256 ft) was essentially a structural steel frame supported on 25.5-m (84-ft) wooden piles. The Port Authority acquired the building in 1963 and wished to use it as a warehouse, but studies showed that the support of the floor was not adequate for this. Newark Bay, on which the building is located, is underlain by red shale bedrock over which there is a fairly uniform stratum of red clay 21.3 m (70 ft) thick. Over this is about 7.8 m (26 ft) of black organic silt, and on this the usual heterogeneous collection of miscellaneous fill material so often found around the shores of long-established harbors. At the site of the building, the fill consisted of cinders. In keeping with the Port Authority's experiences in normal preloading of outside areas, it was decided to apply the same procedure to the floor inside this building. The entire building was therefore filled to a depth of 3.6 m (11.8 ft) with soil brought in by trucks, giving an additional floor load of 6,350 kg/m^2 (1,300 lb/ft^2). Left in place for 14 months, the fill was all removed by scrapers in January 1966, when the floor was found to have settled 0.43 m (17 in). A new floor was then built that, careful calculations suggested, might settle an additional 38 cm (15 in) in the ensuing five years, a small amount that could readily be arranged for in the final floor design.[22.34] Figure 22.21 illustrates another example of the use of preloading in foundation engineering.

22.11 BUILDINGS ON FILL

Brief note must be made of one further construction procedure occasionally necessary with building foundations. It is sometimes necessary to construct a building on a site that

22-30   CIVIL ENGINEERING WORKS

is not naturally level, requiring either fill or excavation for the preparation of a level area which can then be used as the foundation bed. Correspondingly, it is sometimes necessary to make additions to existing buildings, some preparation of the adjacent areas being first necessary for this purpose. In times past, both these situations have all too often led to differential settlements (readily explicable from what is now known about soil action). Today no such undesirable effect need be contemplated. Soil studies, and especially the understanding of soil compaction at optimum moisture contents, can now result in the placement of accurately compacted soil as fill in any area and to any normal depth, with the certainty of minimum settlement.

It is of interest to find that this fact was recognized as early as the immediate postwar years, when the uneven site for the great Clinton Engineer Works in Tennessee had to be prepared for a large U-shaped building with 750-m by 120-m (2,500 by 400-ft) legs and a connection 120 m (400 ft) across. It was suggested, as one of three alternative approaches to the necessary leveling, that soil compaction be tried. This method was adopted; optimum moisture contents of all fill were determined; and soil at these moisture contents placed in 15-cm (6-in) layers by compacting with sheepsfoot rollers.[22.35] At the time this was a revolutionary solution which gave completely satisfactory results. Today, such use of compacted soil under strict control is standard practice in civil engineering, its successful use dependent always upon known and acceptable subsurface conditions.

22.12   SETTLEMENT OF BUILDINGS

All buildings settle, as does every other structure erected upon the surface of the ground. It is strange how this simple fact puzzles the layman, but every civil engineer and geologist knows well that foundation beds, be they of rock or soil, are just as susceptible to the imposition of loads as are all other solid materials. The stress induced by building loads will be reflected in the strain in the supporting materials. With properly designed foundations, the amount of compression, or consolidation in clay soils, will be so small as to be unseen by eye. When "settlement of buildings" is discussed, therefore, it is almost always excessive settlement that is under review or, in some cases, unacceptable differential settlement between parts of a building which causes structural problems. Differential settlements have been exhaustively reviewed in a classic paper.[22.36] Although ultimately dependent upon some inadequacy of the foundation beds, the problem of settlement falls mainly within the domain of the structural engineer and so will be mentioned only incidentally in what follows. Excessive settlements of buildings have occurred all too often in the past, one or two cases having achieved worldwide recognition. Only very rarely has settlement gone beyond the point of excessive vertical movement, soil or rock being stressed to the point of actual failure. There is, however, one famous case where this did happen; the lessons to be learned from it, and the ingenuity displayed in the correction of the failure, make it an essential example for citation in this chapter.

Construction of a large grain elevator at Transcona, Manitoba, was started in 1911. The site chosen was adjacent to one of the world's largest railway yards, just outside Winnipeg, to facilitate the rapid handling of grain from the Canadian prairies. The drying house was 18 m (60 ft) high and measured 5.6 by 9.0 m (18 by 30 ft); the adjoining workhouse was 54 m (180 ft) high and 21.0 by 28.8 m (70 by 96 ft); and the bin structure was 30.6 m (102 ft) high and 23.1 by 58.5 m (77 by 195 ft). The 65 circular bins, arranged in five rows, were capable of storing 36,000 m³ (1 million bu) of grain. A reinforced-concrete raft foundation 60-cm (2-ft) thick, supported the bin structure, which weighed 20,000 tons when empty. Storage of grain started in September 1913. On 18 October, when 31,500 m³ (875,000 bu) were in the bins, a vertical settlement of 30 cm (1 ft) occurred within an hour

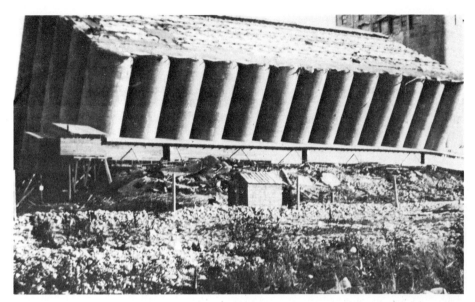

FIG. 22.22 An unusual photograph of the failure of the Transcona grain elevator, Winnipeg, showing the upheaval of clay caused by the soil shear failure.

after movement was first detected. The structure began to tilt to the west and within 24 hours was resting at an angle of 26°53′ to the vertical; its west side was 7.2 m (24 ft) below its original position, and the east side had risen 1.5 m (5 ft) (Fig. 22.22).

Fortunately, because of the monolithic nature of the structure and its sound construction, no serious damage was done to the bins, apart from their displacement. Accordingly, and through an outstanding underpinning operation, the bins were forced back into a vertical position, supported on concrete piers carried to rock, and the entire elevator rehabilitated. It has been in steady use ever since and can be seen today by travelers to Winnipeg on the Canadian Pacific Railway. The site is underlain by Ordovician limestone upon which rest deposits of glacial till, sand, and gravel. Then come 12 m (40 ft) of glacial lake clays in two distinct layers of equal thickness, clays deposited in glacial Lake Agassiz. Three meters (10 ft) of recent alluvial deposits and outwash complete the subsurface profile. How much was known of the subsurface conditions when the elevator was constructed is not now clear, but recent investigations of the failure, investigations using the techniques of soil mechanics. leave no doubt that the failure was caused by overloading of the glacial lake clay. Ultimate bearing capacity, determined on the basis of laboratory tests on samples, was found to be 31,300 kg/m^2 (6,420 psf), as compared with a calculated bearing pressure at failure of 30,300 kg/m^2 (6,200 psf), reasonably close agreement that confirms the value of studies in soil mechanics for foundation design when associated with accurate knowledge of subsurface geological conditions.[22.37]

Admittedly, the failure of the Transcona elevator was (fortunately) an extreme case. Closer to the normal practice of civil engineering was the slight tilting of a reinforced-concrete water tower at Skegness in England. The 33-m (110-ft) tower was founded on a reinforced-concrete mat; it tilted 60 cm (2 ft) out of plumb when first filled with water. Test boring revealed that the subsoil consisted of about 9 m (30 ft) of water-bearing sand with some interbedded peat, naturally irregular, all underlain by glacial till which, if it had been investigated before construction commenced, could have been used as the foundation bed without difficulty. As it was, the tower base had to be surrounded with steel sheet piling, driven to varying depths, deep on the "low" side and shallow on the "high" side. Open, large-diameter boreholes were then drilled on the high side with their bottom elevation well below the bottom level of the steel piling. This side of the tower base was then loaded still further with sandbags, until the excess pressure started to force some of the subsoil from beneath the footing up into the holes. Control of settlement was effected by removing varying quantities of the extruded soil from the open boreholes. In a period of six weeks, the tower had righted itself; when the tower was level, the steel piling was all driven down to the glacial till, thus enclosing the remaining subsoil beneath the tower which, under this constraint, gave all the bearing capacity required.[22.38]

All records of building subsidence are reminders of the fact that the earth's crust is not the solid immovable mass so popularly imagined. When solid rock outcrops at the surface, it provides a foundation stratum of material that is as susceptible to stress and strain as any other solid matter. If unconsolidated material constitutes the surface layer, it provides a foundation bed even more liable to evidence movement under load, and long-time settlement may be anticipated if some clays are loaded beyond a certain limit. Modern foundation design aims not at eliminating settlements (the impossible) but at so controlling them that the structure supported will exhibit no undesirable effects. Should there be any reader of these words to whom the idea that all buildings settle seems strange, let that reader reflect on the quite typical fact that a part of the Tower of London—that symbol of stability—is now known to rise and fall as the tide in the river Thames ebbs and flows.[22.39]

It may be helpful to mention briefly a few other famous examples, not in any recriminatory sense—for all the buildings in question were erected long before foundation engineering had even been recognized as an important aspect of construction—but to illustrate still further the vital dependence of even the most beautiful of buildings upon the geological conditions beneath the ground on which they stand. The Leaning Tower of Pisa naturally comes first to mind in this connection (Fig. 22.23). Located in northwestern Italy, the famous tower was started in 1174 but not completed until 1350. It has continued to tilt since then; the present displacement is over 4.8 m (16 ft) in its total height of 53.7 m (179 ft). Its foundation consists of a circular slab 19.2 m (64 ft) in diameter, with a central hole 4.5 m (15 ft) in diameter. Whether or not there are piles under the slab is still uncertain. Foundation strata consist of a bed of clayey sand 3.9 m (13 ft) thick, underlain by 6.3 m (21 ft) of sand resting on a bed of brackish clay which is of unknown thickness. Prewar investigations and laboratory tests suggested that the tilting is attributable to the clay layer 8.4 m (28 ft) below the foundation slab. A careful study was conducted, and remedial work was begun. The masonry of the foundation slab and ring wall was strengthened with cement grout under low pressure, and, as a final measure, the sand stratum surrounding the foundation was consolidated by the use of a chemical method employing a single gel medium injected at a low viscosity. There is some doubt whether this treatment was really effective, since movement of the tower continues, as recorded in a special modern observation room in the tower about 9.0 m (30 ft) above ground level; the movement now is at the rate of 1 mm (0.04 in) per year. It is estimated that it will take 200 years, at this rate of movement, before the tower is in real danger. Many suggestions have been advanced for rectification of the trouble, so it is to be hoped that this estimate will not be put to actual test. Since the bearing pressure under the base on the strata noted above has been estimated to be 87,500 kg/m^2 (8 tons/ft^2), the wonder is that the tower has not settled more than it has.[22.40]

The Taj Mahal in Agra, India, is widely regarded as one of the most beautiful buildings ever to be erected. Fortunately, it is in no such state as the quite beautiful Tower of Pisa, but to the observant eye, the Taj Mahal betrays the fact that, despite its ethereal appear-

FIG. 22.23 The leaning tower in Pisa, Italy.

ance in moonlight, it is not supported in any mysterious way but rests, as do all buildings, upon the ground. The north wall is not often seen by visitors; it abuts on the river Yamuna, and its foundations are therefore subject to varying water levels, especially in time of the monsoon floods. Cracks in the cellar immediately beneath the plinth terrace on the north side and in the eastern superstructure of the mausoleum can be attributed to trouble with the foundations of the north wall many years ago, troubles which were so satisfactorily dealt with at the time that no evidence of any recent movement has been observed. Examination of the foundations of the north wall some years ago revealed the condition shown in Fig. 22.24, a prosaic representation indeed of this lovely monument to a beloved wife, but one which, at least to readers of this volume, may add still further to appreciation of this masterpiece of Indian building and architecture.[22.41]

Even so famous a North American symbol as the Washington Monument in the heart of the capital city of the United States has had settlement troubles. This great masonry shaft, 167 m (555 ft 5 in) high above its foundation, weighing 81,120 tons, settled about 14.5 cm (5.75 in) in the 50 years after its completion to full height in 1880. It is believed that further settlement is improbable, provided the ground around the monument is not disturbed, a restriction that has already interfered with plans for landscaping in the vicinity. Started in 1848 by public subscription, the monument had risen to a height of 45 m (150 ft) by 1854 when funds ran out. In 1876, before Congress appropriated funds for completion of the shaft, a committee of engineers was appointed to study what had already been done. It was then found that the shaft was 4.5 cm (1.75 in) out of plumb and that the area and depth of the footing should be increased before the tower was completed. These recommendations were followed, and the incomplete shaft was brought

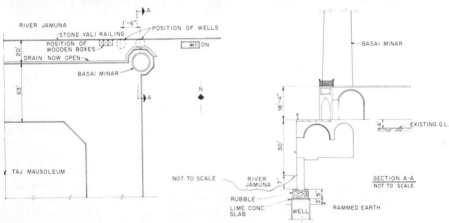

FIG. 22.24 The Taj Mahal, Agra, India, and sketches showing details of a part of its foundations.

back into plumb by an ingenious system of loading in connection with the underpinning, a job which was personally supervised by Gen. T. L. Casey. It is significant to note that, under General Casey's supervision, test borings were sunk, but only to a depth of 5.4 m (18 ft); thus they were still in the water-bearing sand and gravel stratum upon which the monument rests. The fact that these borings were not carried deeper is clear indication of the state of the art of foundation engineering at that time. When deeper borings were put down in 1931, a complex pattern of subsurface conditions was revealed; a layer of soft blue clay overlies bedrock, which was reached about 27 m (90 ft) below the present ground level at the foot of the monument. Since the thickness of the clay stratum varies considerably beneath the footing, the initial differential settlement is not surprising.[22.42]

This same blue clay underlies the Lincoln Memorial, not far away from the Washington Monument, and caused many troubles with the construction of that most noble memorial. Much better soil conditions were fortunately found under Washington's most famous building, the White House, when it was reconstructed in 1951, but even here a layer of clay, 13.5 m (45 ft) beneath the surface, gave cause for some concern in view of the loadings to which the reconstructed mansion would subject this soil. The techniques of soil mechanics were applied to the foundation design. In contrast with the sad record of other famous buildings noted, it is satisfying to record that actual settlements here agreed very closely with those calculated before reconstruction began and were minimal.[22.43]

One of the most elaborate buildings of the New World, erected to serve the arts, has achieved fame also because of its settlement; the Palace of Fine Arts in Mexico has already settled 3 m (10 ft), and its elevation is still falling in relation to streets outside the influence of the building. The building was begun in 1904 and completed 30 years later. (See Fig. 22.25.) It consists essentially of a structural-steel frame with its exterior covered with Italian Carrara marble; interior partitions are brick. The building measures roughly 80 by 117 m (267 by 390 ft); its total weight of 58,500 tons is carried on a concrete-mat foundation, which weighs 46,000 tons. Subsoil conditions at the site of this building consist of about 50 m (165 ft) of so-called "clay," with some interbedded layers of sand and sandy clay; the high moisture content of the main stratum is generally responsible for the remarkable settlement. The building was designed by an Italian-American architect on the basis of an architectural competition. Although the architect spurned engineering advice (a recorded saying of his was that "if the structure is pleasant to my eye it is struc-

FIG. 22.25 The Palace of Fine Arts and the adjacent Tower Latino Americana in Mexico City, the differing foundations of which are described in the text.

22-36 CIVIL ENGINEERING WORKS

turally sound"), it has been stated that he did engage an engineering consultant to design the foundations, although no records appear to exist of the design assumptions. The architect was advised to tear down the beginning of the building when serious settlements were discovered at the start of construction, but he refused to take this course, and the results are all too well known today.[22.44]

22.13 PREVENTION OF EXCESSIVE SETTLEMENT

The record of excessive settlements of older buildings could continue, but enough has been said to show what can happen if the fundamentals of good design are not followed. The other side of the picture is given by the innumerable buildings of today, all over the world, which are performing quite satisfactorily in accordance with their designs and with no excessive or differential settlements. Two examples may be cited, one relatively new and in direct contrast with the sad story of the Palace of Fine Arts in Mexico City; the other, an older building, illustrates what sound engineering intuition could do before the development of modern foundation engineering. This second case has been selected also since there are available for it no less than 60 years of continuous leveling records, the longest such record yet to appear in the literature.

Immediately across the Place la Alamada in Mexico City, and so facing the Palace of Fine Arts, is the modern Tower Latino Americana completed in 1951 and performing with complete success, exactly as was anticipated in design. A most extensive and careful investigation of the subsurface was carried out in advance of design. A series of test borings was put down to a depth of 70 m (240 ft) below ground level, undisturbed samples of soil being taken at every change in soil type. In a typical boring, over 30 distinctly different strata of soil types were sampled and tested. They included volcanic ash, pumice sand, and even some gravel. The results of laboratory tests on the soil samples and study of the local groundwater situation enabled one of the most ingenious foundation designs yet applied to a large building to be successfully prepared and eventually constructed.

This foundation achievement was essentially a matter of excellent engineering, but it was so closely allied with detailed knowledge of every aspect of the underlying geology that a summary description of the construction may not be inappropriate. Once the properties of the various foundation beds were known, it was possible to design a foundation structure consisting of a concrete slab supported on end-bearing piles driven to good bearing on a stratum of sand 33.5 m (110 ft) below ground surface. These piles went right through the first and thickest stratum of volcanic clay, the cause of most of the famous settlements. The piles were driven from the bottom of a relatively shallow excavation, 2.5 m (8 ft 2 in) deep, the material excavated being debris with sand and clay humus in which a quantity of Aztec pottery remains were found. The major problem was in excavating for the two basements required and for the foundation slab structure, the necessary depth being 13 m (42 ft 6 in), and this with the groundwater close to the surface. The site was therefore surrounded with a solid wall of wooden sheet piling that, swelling when in contact with water, gave a good tight enclosure. Excavation then proceeded, but only partially down to the full depth. The gridiron of concrete beams needed for the foundation was placed in specially prepared trenches excavated to the full depth of 13 m (42 ft 6 in). The panels between intersecting beams were then excavated one by one, the foundation slab concreted between them and, as soon as possible, loaded with sand and gravel, to give a load on the underlying ground equivalent to a good part of the total load to be expected from the building. This "false loading" was completed when the foundation was finished and then gradually removed; restoration of the groundwater to its original position was slowly permitted as the steel superstructure and its cladding were steadily erected (Fig. 22.25).

Anticipated settlements—settlements caused by the transfer of the total weight of the building (only 24,000 tonnes as compared with the 58,500 tonnes of the Palace of Fine Arts) to the piles, from the piles to the sand bearing stratum, and through it to the strata below—were carefully calculated. The building has naturally been under close observation since completion, and settlements have been exactly as calculated over 30 years ago. Since the ground in Mexico City may settle quite apart from any settlement of well-founded buildings, the first floor of this great building was designed to be movable so that it could be adjusted, if necessary, to give access always at street level. It has not been necessary to make any adjustment to the floor, tribute not only to the design of the foundation but to the way in which interference with groundwater conditions in Mexico City is now under control.[22.45]

Far to the north of this interesting city is Victoria, the capital of the Canadian province of British Columbia. One of its dominant buildings is the Empress Hotel, opened in 1908 "when the West was young." "The Empress" (as it is known to most Canadians) has become an institution, as gracious hotels can do, afternoon tea there being a ceremony that is an invariable feature of many visits to Victoria. The hotel was built and is still owned by the Canadian Pacific Railway Company, whose president in the early years of the century was Sir Thomas Shaughnessy, a legendary figure in railroad history. He decided on the location of the hotel, at the head of an inlet that was being developed as the inner harbor of Victoria and immediately adjacent to the new legislative buildings of the province. When chosen, the site was a muddy bay (James Bay) completely flooded at high tide. The engineering staff that would be responsible protested as to the unsuitability of the site for a large building, but Sir Thomas, being the man he was, had his way, and there the hotel was built (Fig. 22.26).

FIG. 22.26 The Empress Hotel, Victoria, British Columbia, built on land reclaimed from the sea.

The city of Victoria ceded the land and built a seawall across the bay to enclose the desired site, pumping from the harbor an average of 3.9 m (19 ft) of fill over the entire area to bring it up to the level required for building. Borings were naturally taken, in 1904, and again in 1913 when the hotel was extended; shafts were also excavated, presumably for visual inspection of the soils encountered. The profile thus revealed can be fully appreciated only against the recent geological history of the Victoria area. When the last glacier retreated from this area about 13,000 years ago, the land had been so depressed by the great load of ice that the sea level was 83.8 m (280 ft) above its present level, relative to the land. Rebound of the land occurred by about 11,000 years ago, but in this period a blanket of silty clay (the Victoria clay) was deposited over the entire area to depths up to 30 m (100 ft). During the next 2,000 years, the level of the sea dropped to 9 or 12 m (30 or 40 ft) below its present level, thus exposing the newly deposited clay. The clay therefore dried out near its surface, giving the Victoria clay its characteristic weathered crust that, at the hotel site, is about 4.6 m (15 ft 6 in) thick. After that, the sea level rose to its present elevation, and a new layer of sediment was deposited over the weathered surface of the Victoria clay. These geological details are given to illustrate how geology can explain a soil profile that would otherwise be puzzling.

A log of a typical boring at this interesting site shows 9.7 m (32 ft 6 in) of dredged material overlying the original surface, then 3.3 m (11 ft 5 in) of "soft mud," 2.7 m (8 ft 10 in) of "hard yellow clay," and 13.5 m (44 ft) of "blue clay on gravel and sand." The "blue clay" is the Victoria clay and the "hard yellow clay," the weathered crust of this deposit. With such soil conditions, the original foundation engineers naturally decided to use piles. Timber piles were therefore driven to support all footings. If these had been driven to the underlying sand and gravel, they would have provided good bearing, but for reasons that cannot now be determined, and although the piles on the western side of the hotel were driven to this good bearing material, the piles over most of the site were driven only into the blue clay. The loads from the heavy masonry building were appreciable, and so settlements were soon observed, most acutely when additions were made to the hotel in 1913. Newly appointed foundation engineers clearly appreciated, even at that early date, what had happened and attempted to undo some of the damage by arranging for large quantities of soil from under the south end of the building to be excavated; these voids were made permanent by concrete construction and are there today. Metallic studs were also installed in an ornamental stone course that ran all round the hotel building, and settlement observations were then started. In 1970, therefore, almost 60 years of settlement records were available and, kindly provided by the CPR, were the subject of a detailed research study aided by modern soil mechanics techniques. This enabled the history of the foundation to be traced back to its start, and assurance was given that the future settlement would be very small indeed. It was thus confirmed that the building was in good condition for long-continued use.[22.46]

The surface of the hotel site must have started to settle as soon as the pumping of the fill upon it was completed, partly because of the natural compaction of the pumped material but also because of the consolidation of the underlying Victoria clay owing to the weight of this fill material. (Six meters or 20 ft of fill would be a load of about 11 tonnes/m^2 or 13.2 tons/yd^2 on the original surface.) This initial settlement of about 60 cm (2 ft) affected the entire site. When the piles were driven after the necessary excavation for the basement of the hotel and the footings were loaded with the weight of the superstructure, settlement continued due to this additional loading, except where the piles rested on the firm sand and gravel. Accordingly, the northeast corner of the building remained at the elevation it had when constructed, but the southeast corner (where the Victoria clay was thickest and where the piles did not rest on the sand and gravel) had settled 0.48 m (19 in) by the time of the construction work in 1913, and it has settled about 0.24 m (9 in) since then. There is today, therefore, a 0.72-m (28-in) difference in level between these two extreme corners of the original hotel building (since substantially enlarged with no foundation problems like this). The building is so large and was so very well constructed that it shows no evidence of this unusual condition. No visitors to this famous hostelry would ever suspect that the polished floors were not exactly level. Only those who have read the notable paper describing the recent studies would be tempted to see if they could find evidence of the differential settlement. Rolling billiard balls on the corridor floors would be about the only way to confirm inside the building what the leveling surveys have shown outside the building, but this would be a pastime inappropriate for the Empress Hotel of Victoria. It remains a splendid building and a classic example of the influence of geological site conditions upon long-term building performance.

22.14 BUILDINGS OVER COAL WORKINGS

One special subsurface condition, which may not be revealed by the usual program of site investigation, is the presence of old coal workings beneath urban areas now being used for building purposes. Study of old civic records will usually disclose the existence of this hazard, if it exists. Few cities are so fortunate as Edmonton, Alberta, for which an *Atlas of Coal Workings* was produced by a local geotechnical consultant, a publication based on an exhaustive study of all available early records of this relatively new city of the Canadian West.[22.47] In older cities, and especially in older countries, the fact that coal workings do exist in building areas will usually be known, so that arrangements can be made at the design stage to take care of any unexpected settlements that may develop. Some of these arrangements are ingenious.

An unusual example is provided by the special supports and foundations for four circular, reinforced-concrete water tanks, each 10.8 m (36 ft) in diameter with its top about 9.6 m (32 ft) above ground level, designed by H. C. Ritchie for the Urban District Council of Heanor, Derbyshire, England (Fig. 22.27). The tanks were built to replace a reservoir that had failed because of subsidence of supporting ground, which had colliery workings beneath. The same site had to be used; and although the foundations could be carried to

FIG. 22.27 Reinforced-concrete water storage tanks, at Heanor, Derbyshire, England, showing the unusual three-point bearing design.

a stratum of hard shale, future subsidence of the substrata had to be anticipated. Each tank is therefore carried on three points of support (instead of the usual four or more), in order to obviate indeterminate stresses at any stage of irregular settlement. The feet are not structurally continuous with the supporting foundations but rest freely on them, and the load is transmitted through special steel bearing plates and shoes which will permit the jacking up of the structure at each individual support, in this way the tanks may be raised and releveled as subsidence takes place.[22.48]

An alternative method has been adopted by the Nottingham County Council in England for the design of new school buildings that are known to be located over coal mines and so liable to damage from ground subsidence. Modern in design, the school buildings have light, structural-steel frameworks. These were constructed on flexible, concrete-slab foundations, reinforced to take up variations that might develop in ground level, and an adjustable joint connection was incorporated in the cross bracing adjacent to column bases; thus the design was given sufficient flexibility to take care of all "normal" settlements due to mining subsidence. Details of the steel design are shown in Fig. 22.28. The design has been widely adopted in other areas where settlement due to mining subsidence is a problem.[22.49]

Problems with old coal workings are by no means confined to Great Britain. Two

FIG. 22.28 Structural-steel framework developed for school buildings in Nottinghamshire, England, showing adjustable joint to allow for differential settlement.

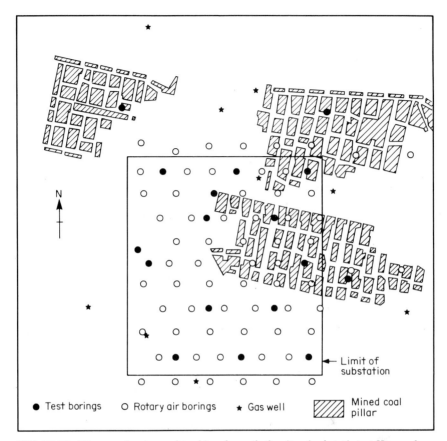

FIG. 22.29 Diagram showing coal workings beneath the site of substation at Haywood, West Virginia.

examples from North America have already been briefly described. A more recent example was the foundation for an important electrical substation at Haywood, West Virginia, where it was necessary to use a site which was known in advance to be underlain by old coal workings. This was evidenced through sinkholes on the site, and ground instability even caused some slides, one of 76,450 m³ (100,000 yd³). The initial part of the substation was located so as to avoid this unstable ground, but detailed site investigation showed it to be located immediately over old coal workings, as shown in Fig. 22.29. The ground profile was in general soil, from 3 to 18 m (10 to 60 ft) in depth, overlying rock (limestone, claystone, shale, coal, siltstone, and silty shale), with depths of rock over the coal workings varying from 3 to 6 m (10 to 20 ft). Site investigation was aided by the use of old mine maps, which were correlated with boring records (a slight error in location of the mapped coal workings being thus located). The result of 82 borings showed, as the condition of the coal workings beneath the site: coal pillars 48%, crushed coal pillars 5%, caved material 35%, and large voids 12%. A number of alternatives for the design of the foundation units were considered. Eventually, it was decided to sink 75- and 122-cm (30- and 48-in) -diameter, steel-cased caissons to rock; these would be founded on firm bedrock beneath the coal stratum. Even with this solution, an additional problem had to be faced; since the

22-42   CIVIL ENGINEERING WORKS

groundwater was, somewhat naturally, acidic, ordinary steel could not be used. A special proprietary corrosion-resistant steel pipe was selected for use. After installation, all pipes were concrete filled and the reinforced-concrete grade-beam framework was then constructed over them.[22.50]

Very different are some of the problems with buildings in relation to coal in the countries of central Europe. Typical is the situation in Czechoslovakia, in an area known to one of the writers. To the east and northeast of Prague in the lovely land of central Bohemia are some of the most important and valuable brown-coal deposits of Europe. Only in relatively recent years have these vast beds of coal been worked extensively. They lie close to the surface and so can readily be worked by surface mining. In one pit, coal is brought right out of the pit by belt conveyor to a screening plant and thence directly into the bunkers of a large steam-power plant located on the edge of the mine. The deposits are, therefore, being worked efficiently, but overall planning has been hampered by the fact that small villages and some fairly large towns were sited, many years (if not centuries) ago, right on top of these valuable coal beds. Some of the larger towns had to be moved. Buildings that could profitably be replaced by more modern buildings were destroyed when their replacements were ready, but historical buildings were preserved to the extent that was possible. This movement of complete towns and displacement of rural settlements in the interests of winning coal is, as can be imagined, a fascinating chapter in the history of European mining, with the complex local geology the root cause of all the disruptions.

Only one case can be mentioned from the many interesting by-products of this accident of geology, as it may well be called. In the largest of the brown-coal areas, about 70 km (43.5 mi) to the northeast of Prague, lies the thirteenth-century town of Most. Covering an area of 390 ha (965 acres), with a population of 54,000, Most sits right on top of a lignite bed, 27.3 m (90 ft) thick, that contains 90 million tonnes of high-quality brown coal only 1.5 to 15.2 m (5 to 50 ft) beneath the narrow streets of the old town. Moving the town was clearly essential, and modern apartment blocks in the new Most, about 3.2 km (2 mi) away, are now providing more up-to-date accommodations (even though not so picturesque) for the town's inhabitants. A central point of the old town had been the six-

FIG. 22.30   The church at Most, Czechoslovakia, while being moved from its original location as part of the relocation of the town to permit further coal mining.

teenth-century gothic Church of the Ascension. Its singularly beautiful interior has three naves; it is 59.3 m (195 ft) long, 26.4 m (86 ft) wide, and 31.3 m (102 ft) high. Its total weight is estimated to be more than 12,000 tonnes. The Czechoslovakian government decided that this beautiful church must be saved, even though the masonry of the adjoining bell tower was too badly fractured to make moving it practicable. The church itself, however, was carefully "caged," or reinforced with a steel framework, before moving and after it had been underpinned on to new supports. When secure in this temporary frame, it was moved 870 m (2,850 ft) and downhill by 10.5 m (34 ft) to a newly prepared site (Fig. 22.30). Even the move was complicated by old mine workings, a 152-m (500-ft) stretch of the route being over old mine shafts that had to be specially backfilled and strengthened.[22.51]

22.15 UNDERPINNING

Some of the most difficult of all foundation problems are encountered when, for one reason or another, existing buildings (and especially valuable old buildings) have to be "underpinned" to a firm foundation bed, in effect, being given a new foundation and so a new lease of service. Although it is a highly specialized part of foundation engineering, it goes without saying that underpinning can be carried out only when subsurface conditions are known with absolute certainty and accuracy. The risks involved in interfering with existing foundations are far too great to permit of any doubt whatsoever that the proposed underpinning operation cannot be carried out exactly as planned. What is involved is the "picking up" of loads on existing columns by some form of temporary expedient and their transfer to new and reliable foundations. The decay of wooden piling beneath buildings is one problem with foundations that can necessitate underpinning, but so also can progressive settlement that finally becomes so serious that it must be corrected. Underpinning is, naturally, a civil engineering operation of unusual complexity, always the work of experts. Equally is it always dependent upon accurate knowledge of every detail of the subsurface. Examples are chiefly of note because of the underpinning methods employed, but one unusual case may be briefly cited, for reasons that will be obvious.

All visitors to Stockholm will know the outside appearance of the great Royal Palace located in what is known as the Old Town, immediately opposite the Grand Hotel on the inner harbor (Fig. 22.31). The palace, built at different periods in the last 2½ centuries,

FIG. 22.31 The Royal Palace, Stockholm, Sweden, the underpinning of part of which is described in the text.

22-44 CIVIL ENGINEERING WORKS

with its wings founded on the site of an older palace that burned in 1697, is said to have 700 rooms. It has long been the residence of the Swedish royal family, but its two wings have suffered from serious settlement, as much as 0.6 m (24 in) in the last half century. Some remedial work was done in the 1920s, but trouble continued, due possibly to the increased load of the concrete then placed, the heavier and increased traffic on roads adjacent to the palace, and the slowly dropping groundwater levels. (This drop in groundwater levels is possibly associated in part with the known rise in the level of all of Scandinavia due to rebound following the last glacial period.)

The two palace wings rest partially on the remains of the former building that burned. A timber raft near the main building and timber piles were used as foundation units, the mat resting on 3 to 6 m (10 to 20 ft) of miscellaneous fill. Beneath the fill is a thin stratum of clay with some sand, then a bed of compact gravel from 9 to 23 m (30 to 75 ft) thick overlying bedrock. Using the concrete mat placed beneath the building in the 1920s as a working platform, the specialist contractor for the underpinning operation sank 29 cylinders through all these strata to bedrock. These were cleaned out, filled with concrete, and used as the new foundations for the two wings. This was difficult work, carried out in the confined quarters of the old stone-arch basement, but it was successfully completed. In the excavation of the old fill, quite a number of archaeological items were discovered, such as a shoe from the sixteenth century and ceramic ovens, of unusual interest to the resident of the palace, King Gustav Adolf VI, himself a distinguished archaeologist; all were passed to the Stockholm City Museum, where they may now be seen.[22.52]

22.16 PRECAUTIONS ON SLOPING GROUND

There is still another geological problem with building foundations which can be mentioned here only briefly; it is one of more local incidence, but when it does arise it can be serious indeed. It is the possible instability of sloping ground used for the foundations of buildings. Since large buildings necessarily need level areas for their siting, the sloping-ground problem usually affects residential construction. Sloping sites would not normally be chosen for building, but the mounting value of real estate in developed city areas has increased the use of such sites, especially where beautiful vistas further enhance the value of the land. Slope-instability problems in relation to building foundations, therefore, are found mainly in city areas, especially in those cities having hilly terrain (Fig. 22.32). The west coast of North America immediately comes to mind, but other locations have had similar troubles. In the area of greater St. Louis, Missouri, where the local loess is a soil most susceptible to water, many houses built too near the tops of slopes have shown evidence of damage after heavy rains or when drainage was interfered with. In the beautifully named suburban municipality of Bellefontaine Neighbors, for example, the slide of a steep slope into a borrow pit in 1957 damaged 10 houses on St. Cyr Drive; three of them were, unfortunately, a complete loss.[22.53]

It is in the Los Angeles area, however, that this type of trouble has been most widespread and serious. In 1952, when heavy rains occurred in Los Angeles following seven very dry years, extensive damage was caused to many hillside properties; estimates of the value of the damaged property ran into millions of dollars. Many lawsuits resulted, since in areas of developed property, damage on one site may be affected by, or may itself affect, other adjacent properties. Before the following winter, the city of Los Angeles passed a grading ordinance which amended the city building code and placed some 60 percent of the city's area under special regulations. Rains in more recent years have caused further damage. Local studies have continued, and public interest has been aroused. In 1957, for example, a Geological Hazards Committee of the City of Los Angeles was formed volun-

FIG. 22.32 Small landslide in Portland, Oregon, caused by heavy rains on a steeply sloping hillside, typical of slope-instability problems in urban areas.

FIG. 22.33 One way of protecting steeply sloping hillsides—sheets of plastic used by householders in Portland, Oregon, to protect their properties; "plastic raincoats" put to good use in preventing unwanted water from reaching soil, in order to prevent further slides such as the small one to be seen on the left.

tarily by local geologists and engineers to assist civic authorities in dealing with this problem. Most dramatic and personally tragic are the losses of homes due to movements of soil and rock on slopes; such movements are usually due to the action of uncontrolled water. But masses of soil and rock are moved irrespective of houses, destroying useful land and causing all kinds of interference with normal municipal services.[22.54]

The causes of such trouble are exactly the same as those outlined in Chap. 39 and so need not be given here; the problem of building foundations on sloping ground is only a special example of ground instability. Along the coast roads in the Los Angeles region, there is the added hazard of undercutting by the sea, but even here it is the effect of uncontrolled surface water and groundwater that causes most of the trouble. The application of geological principles to the problems encountered in constructing buildings on sloping sites is probably the most perfect example in engineering work of the old tag that "a stitch in time saves nine." For the expenditure of a very small sum, an examination of any site for a proposed small building can be made by a competent professional adviser. If good advice so received is followed, damage that might run into thousands of dollars can often be avoided. A few simple hand borings, examination of the local geology, study of groundwater conditions in the vicinity and of surface drainage arrangements, and above all, a close scrutiny of the records of local weather—such simple steps when followed by an expert can lead to sound advice on how a sloping site can be used or, alternatively, whether it should be avoided. (See Fig.22.33.)

22.17 GARAGES

In order to illustrate the wide variety of responses by designing engineers to unusual geological conditions beneath the sites of the buildings they design, a variety of specialized groups of buildings could be cited. One special group alone must suffice, garages having been chosen for brief mention because of the unusual solutions sometimes developed for their design, as in the two examples which follow. One involves the construction of a building over filled ground, settlement of which was anticipated and allowed for in design. The second case comes from Europe, where the lack of any suitable available urban land, coupled with the increased use of private automobiles, led the city of Geneva to build a major garage underwater rather than underground.

At Hiroshima, Japan, a major automobile producer needed a large storage building in which up to 10,000 passenger cars could be stored prior to export. The open-floor building designed to meet this unusual requirement is 314 by 71 m (1,046 by 238 ft) in plan and four storeys high and is provided with the necessary ramps for access (Fig. 22.34). The site available was underlain by 60 m (200 ft) of silt with some overlying fill, materials which were known to be susceptible to consolidation once the building load was applied, with an eventual settlement of 1.5 m (5 ft) being anticipated over the next three years. Time did not permit the adoption of any methods (such as preloading) for obviating the settlement. The design was therefore prepared to allow for the major settlement that was predicted. The foundation structure consists of a network of reinforced-concrete beams, making 9-m (30-ft) squares. At each intersection, concrete stub columns extend 55 cm (22 in) above the tops of the beams, each stub having carefully set into it four 38-mm (1.5-in)-diameter, steel anchor rods, each well anchored into the reinforced-concrete base structure. The 351 columns which support the upper floor are open-corner, steel hollow boxes, each 50 cm (20 in) square, with a 48-mm (1.9-in)-thick steel shoe plate, through which the anchor rods extend. Extensions to these anchor rods will be fitted as necessary. As the building settles, the anchor rod nuts are loosened, and four 50-ton-capacity jacks, one to each anchor rod, are used at each column to raise the building. With the jacks still in

FIG. 22.34 (a) Reinforced-concrete storage building for automobiles at Hiroshima, Japan, with (b) a detail of the adjustable (patented) column bases, by means of which the entire building can be leveled up following settlements which were fully anticipated in the design.

place, steel shim plates are placed and the jacks are then removed.[22.55] As an acceptance of an unusual geological condition, this design takes some beating.

Far removed from Hiroshima is the lovely lakeside city of Geneva, Switzerland. It has the usual urban automobile problem, with parking an urgent necessity. The Canton of Geneva owns the beds of both the Rhône River and Lake Geneva from which the river flows. Once the idea of using the bed of the river as the location for a major garage had been accepted, the Canton granted a local business group a rent-free concession to build and operate an underwater 1,450-car garage. The location chosen was between the Mont

22-48 CIVIL ENGINEERING WORKS

FIG. 22.35 The site of the Mont Blanc parking garage at Geneva, Switzerland, before it had been pumped out, the location beneath the exit from the lake clearly delineated.

FIG. 22.36 The Mont Blanc underwater garage in Geneva, Switzerland, under construction. All that can now be seen at this location are beautifully landscaped entrance ramps and escalators for pedestrians in the waterside park.

Blanc Bridge (the structure that links the two banks of the lake at the outflow of the Rhône) and the smaller, downstream Bergues bridges. The riverbed here consists of 22 to 29 m (72 to 98 ft) of compact clay, with strength characteristics well able to sustain the load of the large reinforced-concrete structure of the garage while in the dry, the whole building being constructed within a large, cofferdammed area (Fig. 22.35). Access and exit ramps lead to a major six-lane city street which parallels the river's left bank. Construction scheduling was in itself a remarkable building feat, but once completed and the cofferdam removed, the entire 138- by 67-m (460- by 222-ft) reinforced-concrete structure became, in effect, a submerged floating box, with a depth of water over it varying from 2.5 to 3.9 m (8 ft 6 in to 13 ft), the garage sloping on a 2% grade to fit with the original grade of the riverbed (Fig. 22.36). There are now many successful underground garages, but the Geneva garage is believed to be the only major underwater repository for automobiles, its conception the result of accurate knowledge of the riverbed in which it was constructed.[22.56]

22.18 SMALL BUILDINGS

The examples cited in this chapter are almost all large structures, inevitably so since they create the heaviest loads to be imposed upon the ground, any failure to perform as designed involving considerable financial loss. Small buildings, however, can be subject to all the problems that affect larger structures and so must at least be mentioned if this chapter is to be a reasonably complete guide to the significance of geology in the founding of buildings. Reference again to Fig. 22.6 will be a useful reminder that the loads from all buildings, large and small, stress the ground beneath them; hence, this must be strong enough to sustain the increase of stress caused by the building loads. The loads from small buildings, such as residences, will usually be relatively so small that almost all normal soils will sustain them.

Sometimes, however, small buildings have to be erected on soil that is clearly not strong enough for the normal type of foundation structure. Typical was a new two-storey firehouse in San Francisco. It had to be located on the basis of fire-fighting logistics, rather than on choice foundation-bed material. The result was that it had to be built on a site underlain by 6 m (19 ft) of fill material overlying bay mud extending about 30 m (100 ft) to bedrock. Any normal design would have imposed loads on this material which would have resulted in major settlements. It was therefore decided to float even this small structure. A 1.2-m (4-ft) -deep, ribbed, reinforced-concrete foundation mat was therefore designed; lightweight concrete was used for the superstructure and other features of design included to keep the dead load to the minimum possible. The dead load amounted to 1.22 million kg (2.7 million lb), but this was almost exactly the same as the weight of the fill removed. Only the smaller live load (122,500 kg or 270,000 lb) will therefore increase the stress in the underlying soil. This is estimated to cause a settlement of no more than 30 cm (12 in) in the next 50 years.[22.57]

This example shows well how devices generally used for large structures can, when necessary, be adapted for much smaller buildings. Groundwater problems will be the same but can usually be guarded against by having open-tile drains surrounding the lowest part of the foundation structure (house basements, for example). Building on floodplains is to be avoided at all costs. Any chance of slides from adjoining sloping ground must be thoroughly investigated. The possibility of an "innocent-looking" site being actually located over filled ground, and possibly over an earth-covered site used earlier as a refuse dump, must never be forgotten. So the list can continue, the investigation of the building site being just as necessary as that required for major structures although on a much more modest scale.

22-50   CIVIL ENGINEERING WORKS

Fortunately there are available many guides to site selection for the use of the small builder and prospective homeowner. One may be mentioned, with appreciation; it is selected because it was prepared as a service to their community by the members of the Pittsburgh Geological Society. Entitled *"Lots" of Danger,* it is a simple, quarto-sized document written in nontechnical terms and amusingly illustrated so as to be easy reading for Everyman.[22.58] Its subtitle is *Property Buyers' Guide to Land Hazards in Southwestern Pennsylvania;* its contents are just that. The title, and the reference to hazards, were probably deliberately selected in order to bring home to the man on the street that even small buildings may be subject to unsuspected dangers. In a similar way, a careful study has been made by geologists in Texas of the real estate brokers' and builders' knowledge of geology and other physical site determinants.[22.59] It was found that, although the "brokers and builders are aware of and interested in the geologist's contribution ... they indicated that the contribution must be both informative and intelligible." This is useful advice for all who have to deal with the problems of foundations for small buildings.

22.19 CONCLUSION

Users of this Handbook may notice, and be surprised by the fact that this chapter is almost the longest in the book, although it deals with such an apparently simple subject as the foundations for buildings. Even at that, the treatment of this vast subject has necessarily had to be in summary form, restricted—as was said at the outset—merely to the geological factors affecting the founding of that large group of structures generally designated as "buildings." The very familiarity of building work, and the apparent simplicity of most building foundation projects, has led, and unfortunately still leads, to a lack of attention to subsurface conditions quite out of line with that which is now generally accepted as necessary for such major structures as dams and bridges.

Lest it be thought that this is but a subjective view of the authors, it must be recorded that in 1977 the U.S. General Services Administration (GSA) admitted that in the previous ten years they had spent more than $10 million to repair faulty foundations beneath public buildings for the construction of which they had been responsible. One of two studies of this situation carried out for the General Accounting Office (GAO) showed that of 28 buildings studied, difficulties with foundations and "unanticipated soil conditions" had been experienced in 15, or more than 50 percent. These were all buildings costing more than $2 million each and under construction on 30 June 1963. In connection with the two studies made, it was said that inadequate geotechnical information was an industrywide problem and one not peculiar to GSA.[22.60] If this assertion is correct, then this chapter could well be a great deal longer than it is.

The writers do not think that the general situation is quite as serious as this suggestion indicates. The two reports for GAO have certainly assisted GSA in their more recent work. Examples cited in this chapter show what excellent work in building foundation is now being done. There is still progress to be made, however, before failures due to inadequate subsurface information for building foundations are eliminated from the construction scene. With the wealth of information about this subject that is now available, there is no excuse whatsoever for the neglect of that vital part of the building design process—full investigation of the subsurface conditions beneath the building site and for some distance around, with special emphasis on groundwater conditions. In earlier days, such neglect could sometimes be explained, despite the problems it caused. It may be helpful to show, by a summary of one more famous example, what trouble can result even when inadequate site investigation is entirely explicable and so understandable. The example is the foundation of one of the great buildings that now dominate the famous Raffles Quay in the city of Singapore.

The Asia Insurance Building is a steel-framed, 18-storey structure, founded on a site reclaimed from the beach during the last century (Fig. 22.37). The beach formation beneath the site varies from 3 to 4 m (10 to 14 ft) in thickness; it was thought to be underlain by fill and decomposed shale and sandstone. Four borings were sunk on the site in 1949 and came to refusal from 9 to 13 m (30 to 42 ft) below the surface. A strong chisel was used to obtain chips from the sandstone encountered at the bottom of the holes, since there was no diamond drilling equipment available in Singapore for penetrating deeply into the rock. A cylindrical type of foundation was then designed, with bearing intended for the hard layer encountered, and construction was started by a local contractor. It was soon found that the presumed bedrock was really boulders of weathered shale and sandstone of Triassic formation, common to Singapore. Granite boulders were also encountered, but these had originated in ballast deposited on the beach from ships using the port many years before. Lack of control over the cylinder sinking added to the difficulties by permitting soil boiling and consequent loss of ground in the area around.

Work had to be stopped, therefore, and since proper drilling equipment was by then available, further subsurface investigation was undertaken. Diamond drilling operations did not readily distinguish between the hard clay that was encountered and the sandstone boulders, but clay was found to depths exceeding 30 m (100 ft), and samples were obtained. Cylinders already placed were therefore grouted up and then underpinned. This was a most critical construction operation, but it was successfully completed; the bottoms

FIG. 22.37 The Asia Insurance Building on Raffles Quay, Singapore.

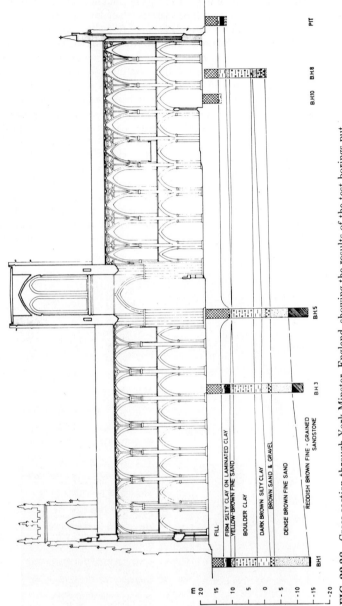

FIG. 22.38 Cross section through York Minster, England, showing the results of the test borings put down prior to the reconstruction in 1970.

of the new cylinders were belled out at appropriate depths to give the calculated bearing values determined safe for the clay on the basis of careful laboratory tests of the samples obtained during the diamond drilling operations. The bearing pressure had to be reduced from the original value of 109,360 to 38,300 kg/m² (10 to 3.5 tons/ft²).[22.61]

Much of the interest of this job lies in the engineering methods followed in dealing with the revision of the foundation design, but as a cautionary tale, the case is of great value. It vividly demonstrates the absolutely vital role that accurate subsurface information plays in the foundation design of buildings—a large building in this case, but size is not a determinant. Special note must be made of the fact that lack of proper test-drilling equipment in Singapore in the immediate postwar years was in part responsible for the trouble that developed. Similar neglect has been known, however, in other locations with no such limitations; only the fact that foundation-bed conditions have been so much better than those below Raffles Quay in Singapore has saved many another building from similar difficulties.

Today, no such untoward development should even take place at the start of the construction of any building, large or small. For minimal expenditure, the equivalent of a small insurance policy, a study of the urban geology at the building site can be made, archival records that may have any bearing on the site can be studied, foundation records of neighboring building can be examined, and the necessary check test borings (and possibly drilling) can be carried out. As a part of this survey, the groundwater situation at the site can be determined and, if time permits, arrangements made for its observation throughout the annual cycle, or as much of this as possible. Geotechnical investigations can follow, all coordinated so as to give the most accurate picture possible of the subsurface beneath the site. Uncertainties will remain, as always, but study of the local geology will give a reasonable indication of what these may be, so that they can be provided for in contract documents. And when the record of subsurface conditions as actually revealed by excavation is carefully recorded, and filed with civic authorities, the general picture of the urban subsurface will be enhanced and made the more valuable for building projects

FIG. 22.39 " . . . and we can save 700 lira by not taking soil tests." (© *Engineers Testing Laboratories*)

of the future. Admittedly, this is the ideal situation, but it is an ideal that has already been realized in some cities, as Chap. 13 makes clear. It is an ideal that should be the goal of all concerned with building foundations in every city.

Once the importance, occasional complexity, and invariable interest of building foundations are appreciated, the study of older buildings takes on new significance and value. Guests of the Cumberland Hotel in London who are so interested, can discover, for example, that beneath this vast, city-block structure, a tributary of the famous Tyburn Ditch still flows.[22.62] All who admire the restored majestic beauty of York Minster in England will know that by far the most important part of the reconstruction of this great church was the entire rebuilding of the foundations of the central tower, now open for inspection by an imaginative arrangement of the lower crypt (Fig. 22.38).[22.63] All observers of the White House in Washington will know that, following its reconstruction, it is as stable and solid as its fine appearance suggests.[22.43] Practically all major old buildings have a story attached to their foundations, if it can be found. Even the building often described as the most beautiful in the world, Chartres Cathedral in France, has its story. The justly famous West Front, with its exquisite carved figures, can be seen at a glance to be unusually placed, almost flush with the front of the two great towers instead of set back somewhat as would be expected. And the explanation is that there was a foundation-bed failure in its original, inset, position, back in the fourteenth century.[22.64] Building foundations have indeed a long and interesting history.

22.20 REFERENCES

22.1 H. E. Sawtell, "Foundations of the Boston, Mass., Parcel Post Building," *Journal of the Boston Society of Civil Engineers,* **22**:29 (1935).

22.2 L. Kerr and P. Brown, "Process Building over Faulted Rock," *Engineering News-Record,* **135**:795 (1945).

22.3 L. R. Squier, "Mat Foundation Spans Rubble Channels," *Civil Engineering,* **40**:61 (August 1970).

22.4 E. A. Prentis and L. White, *Underpinning,* rev. ed., Columbia University Press, New York, 1931.

22.5 C. H. Dickinson, "Foundation Work Hampered by Soft Ground and Seamy Rock," *Engineering News-Record,* **99**:424 (1927).

22.6 Mason Hill, Buffalo, by personal communication.

22.7 J. Feld, "Foundation Failure," *Civil Engineering,* **43**:89 (June 1973).

22.8 L. Bjerrum, Oslo, by personal communication.

22.9 S. S. Philbrick, "Cyclic Sediments and Engineering Geology," *Report of the 21st International Geological Congress,* pt. 20, Copenhagen, 1960, pp. 49–63.

22.10 "Tank Piers Carried 50 Ft to Rock Through Old Coal Mine," *Engineering News-Record,* **110**:190 (1933).

22.11 C. W. Campbell, "Chemicals Seal Foundation for New York Building," *Civil Engineering,* **27**:693 (1957).

22.12 F. A. Noetzli, "Multiple Arch Retaining Wall Damaged by Slip," *Engineering News-Record,* **98**:146 (1927); see also in the same journal, "Cahuenga Pass Arch Retaining Wall Continues Slipping," **99**:681 (1927).

22.13 H. A. Mohr, *Exploration of Soil Conditions and Sampling Operations,* Harvard University Graduate School of Engineering Bulletin no. 208, Cambridge, 1937.

22.14 "Undiscovered Substratum of Peat Complicates Foundation Job," *Engineering News-Record,* **91**:192 (1923).

22.15 C. B. Crawford, "Settlement Studies on the National Museum Building, Ottawa, Canada," *Proceedings of the 3rd International Conference on Soil Mechanics and Foundation Engineering,* **1**:338 (1953).

22.16 "Foundation of Fourteen-Story Building Replaced under Basement Floor," *Engineering News-Record,* **105**:496 (1930).

22.17 "Artificial Groundwater for Wood Piles," *Engineering News-Record,* **107**:70 (1931).

22.18 F. Fox, *Sixty-Three Years of Engineering,* J. Murray, London, 1924, p. 125.

22.19 E. E. Green, "An Account of the Problems Caused by Ground Water, Encountered During the Construction of the New Wessex Hotel at Winchester, for Trust Houses Limited," *Proceedings of the Institution of Civil Engineers,* **28**:171 (1964).

22.20 "Pumps and Weight Keep Hospital out of Hot Water," *Engineering News-Record,* **158**:46 (20 June 1957).

22.21 "Permanent Wellpoints Keep Sewage Tanks from Floating," *Engineering News-Record,* **150**:47 (26 March 1953).

22.22 J. D. Parsons, "Foundation Installation Requiring Recharging of Ground Water," *Proceedings of the American Society of Civil Engineers,* vol. 85, CO 2, Paper no. 2141, New York, 1959.

22.23 R. F. Legget and C. B. Crawford, "Trees and Buildings," *Canadian Building Digest,* no. 62, Ottawa, 1965.

22.24 M. Bozozuk and K. N. Burn, "Vertical Ground Movement near Elm Trees," *Geotechnique,* **10**:19 (1960).

22.25 A. W. Skempton, "A Foundation Failure Due to Clay Shrinkage Caused by Poplar Trees," *Proceedings of the Institution of Civil Engineers,* pt. 1, vol. 3, London, 1954, p.66; see also M. J. Hammer and U. B. Thomson, "Foundation Clay Shrinkage Caused by Large Trees," *Proceedings of the American Society of Civil Engineers,* vol. 92, SM 2, Paper 4956, New York, 1966.

22.26 K. Terzaghi and R. B. Peck, *Soil Mechanics in Engineering Practice,* Wiley, New York, 1948, p. 476.

22.27 L. E. Collins, "Some Foundation Experiences in the Durban Area," *Transactions of the South African Institution of Civil Engineers,* **4**:1 (1955); see also: A. B. A. Brink, *Engineering Geology of Southern Africa,* vol. 1, Building Publications, Pretoria, 1979, p. 225.

22.28 "Building on Soft Clay," *Engineering News-Record,* **123**:692 (1939).

22.29 C. W. Pike and B. F. Saurin, "Buoyant Foundations in Soft Clay for Oil Refinery Structures at Grangemouth," *Proceedings of the Institution of Civil Engineers,* pt. III, vol. 1 London, 1952, p. 301; see also J. B. Sissons, "Geomorphology and Foundation Conditions Around Grangemouth," *Quarterly Journal of Engineering Geology,* **3**:183–191 (1971).

22.30 A. C. Mason, "Open-Caisson Method Used to Erect Tokyo Office Building," *Civil Engineering,* **22**:944 (1952).

22.31 "Caissons Dig out a Seven-Story Basement," *Engineering News-Record,* **166**:42 (6 July 1961); see also M. A. Brugger, "Garage Riv-Centre à Genève," *L'Entreprise,* no. 15–16, Zürich, 1961, p. 1.

22.32 H. W. Hunt, "Building Foundation Built as a Caisson for Port Storage in Stockholm," *Civil Engineering,* **35**:60 (January 1965).

22.33 "Earth Compacts Earth for Cathedral Base," *Engineering News-Record,* **154**:41 (28 April 1955).

22.34 "Surcharging a Big Warehouse Floor Saves $1 Million," *Engineering News-Record,* **176**:22 (3 March 1966).

22-56 CIVIL ENGINEERING WORKS

22.35 J. D. Watson and O. R. Bradley, "Compacted Fill Equals Natural Ground," *Engineering News-Record,* **135**:810 (1945).

22.36 A. W. Skempton and D. H. MacDonald, "The Allowable Settlements of Buildings," *Proceedings of the Institution of Civil Engineers* pt. III, vol. 5, London, 1956, p. 727.

22.37 A. Baracos, "The Foundation Failure of the Transcona Elevator," *Engineering Journal,* **40**:973 (1957).

22.38 "Righting a Tilted Water Tower," *Engineering News-Record,* **105**:300 (1930).

22.39 T. E. Stanton, "Engineering Research" (The James Forrest Lecture), *Minutes of Proceedings of the Institution of Civil Engineers,* **232**:400 (1923).

22.40 For a useful review, see J. T. Mitchell, V. Vivatrat, and T. W. Lambe, "Foundation Performance of the Tower of Pisa," *Proceedings of the American Society of Civil Engineers,* vol. 103, GT 3, Paper no. 12814, 1977, p. 227.

22.41 Sir Harold Williams, India, by personal communication.

22.42 D. H. Gillette, "Washington Monument Facts Brought Up to Date," *Engineering News-Record,* **109**:501 (1933).

22.43 R. E. Doherty, "The White House Made Safe," *Civil Engineering,* **22**:482 (1952); see also D. M. Burmister, "Foundation Studies for the White House," *Columbia Engineering Quarterly,* New York, March 1952, p. 1.

22.44 J. H. Thornley, C. B. Spencer, and P. Albin, "Mexico's Palace of Fine Arts Settles 10 Ft.: Can It Be Stopped?" *Civil Engineering,* **25**:357 (1955).

22.45 L. Zeevaert, "Foundation Design and Behaviour of Tower Latino Americana in Mexico City," *Geotechnique,* **7**:115 (1957).

22.46 C. B. Crawford and J. G. Sutherland, "The Empress Hotel, Victoria, British Columbia: Sixty-Five Years of Foundation Settlements," *Canadian Geotechnical Journal,* **8**:77 (1971).

22.47 R. Spence Taylor, *Atlas of Coal Workings under Edmonton, Alberta,* privately publ., Edmonton, 1971.

22.48 P. H. Ogden, "Adjustable Water Tanks at Heanor," *Civil Engineering and Public Works Review,* **33**:131 (1938).

22.49 "Building over Coal Mine Yields as Ground Settles," *Engineering News-Record,* **164**:39 (7 January 1960).

22.50 J. A. Collins, "Substation over Old Mine gets Firm Footing," *Civil Engineering,* **43**:41 (1973).

22.51 J. Skopek, "Geotechnical Problems Encountered in Moving the Church at Most," *Canadian Geotechnical Journal,* **16**:473 (1979).

22.52 "Sweden's Royal Palace Gets Underpinning to Stop Settling," *Engineering News-Record,* **175**:48 (23 September 1965).

22.53 A. L. Baum, St. Louis, by personal communication.

22.54 "A Geological Challenge," *Geotimes,* **2**:8 (November 1957).

22.55 "Automobile Warehouse, Built on Fill, Expected to Sink," *Engineering News-Record,* **188**:16 (16 March 1972).

22.56 "Swiss Water Down Parking Problem," *Engineering News-Record,* **184**:16 (30 April 1970).

22.57 "Firehouse, Built on Mud, Is Designed to Float," *Engineering News-Record,* **190**:7 (31 May 1973).

22.58 Jane L. Freedman (ed.), *"Lots" of Danger,* Pittsburgh Geological Society, Pittsburgh, 1977.

22.59 C. C. Mathewson and D. W. Ruckman, "Geological Needs and Knowledge of Real Estate Brokers and Builders," *Geology,* **2**:539 (1974).

22.60 "GSA's Faulty Foundations Are a Multi-Million-Dollar Problem," *Engineering News-Record,* **199**:37 (6 October 1977); see also same journal, "Foundation Woes a Chronic GSA Problem," **199**:10 (27 October 1977).

22.61 W. J. R. Nowson, "The History and Construction of the Foundations of the Asia Insurance Building, Singapore," *Proceedings of the Institution of Civil Engineers,* pt. I, **3**:407 (1954).

22.62 "The Cumberland Hotel, London," *Building,* **140**:6 (January 1933).

22.63 D. J. Dowrick and P. Beckman, "York Minster Structural Restoration," *Proceedings of the Institution of Civil Engineers,* Supplementary Paper no. 7415S, London, 1971, p. 93.

22.64 K. Clark, *Civilisation,* B.B.C. and J. Murray, London, 1969, p. 50.

Suggestions for Further Reading

Further reading that is essential in connection with the design and construction of foundations for buildings is that contained in the local building regulations for the municipality in which the building is to be erected. These have the force of law and so must be followed. They should reflect the accumulated experience gained in dealing with the foundation beds normally encountered in the area in question, along with the local urban geology.

An increasing number of municipalities are now legally adopting for their own use one or other of the national or regional building codes, prepared by expert representative committees. The complexity of modern building and the rapid advance of building technology usually make this procedure an economical solution to one of the most important civic responsibilities in connection with urban physical development, especially for smaller cities.

Several states now have excellent building codes, kept up to date with the best available technical assistance. Some voluntary associations likewise produce and keep up to date building codes that are widely used. Many of these efforts are now being coordinated so that building codes have long ceased to be the "impediments to good building" they were sometimes considered to be in earlier years.

Although the provisions in such documents for the design and construction of foundations can be usefully codified, suitable allowance must always be made for possible variations due to any unique aspects of foundation beds in each locality, i.e., of the local geology. In some cases, building codes (the legal-type documents) have supporting documents of a technical nature associated with them. For example, the *National Building Code of Canada* (prepared by a representative national committee under the aegis of the National Research Council of Canada) has associated with it a special Supplement. The *Canadian Foundation Manual,* now in its second edition, for which the Canadian Geotechnical Society is responsible, is a guide to good foundation practice. Some latitude is necessary in the use of such documents to allow for variations dependent upon local geology.

Building codes by themselves, therefore, are not the complete guides to design and construction of foundations which the unthinking might assume. They set out, as they must, the minimum requirements that must be followed in design in the interests of public safety. These must be applied, in connection with foundations, only after close study of all available information about the local geology, "further reading" of a special kind.

Helpful information of a more general type is to be obtained not only from current periodical literature, as indicated by the references above, but also from special publications such as the records of conferences where experience is shared. Typical is STP No. 670 of ASTM which records such experiences in *Behavior of Deep Foundations,* edited by R. Lundgren (616 pages, 1979).

The importance of last-minute inspection of the foundation beds on which a building is to be supported (when this is possible) has been stressed. Recognition of the importance of inspection has been given by the appearance in 1982 of a volume with the title *Systematic Construction Inspection* by R. W. Liebing, a book of 119 pages published by John Wiley & Sons of New York.

chapter 23

POWERHOUSE FOUNDATIONS

23.1 Some General Considerations / 23-2
23.2 Geothermal Power / 23-5
23.3 Tidal Power Plants / 23-8
23.4 Water Power Plants / 23-10
23.5 Thermal Plants / 23-20
23.6 Nuclear Plants / 23-26
23.7 References / 23-31

The current energy situation presents the world with some of its most serious problems of physical development. It is not a crisis that will pass, but a situation of growing complexity and urgency. It has been estimated that the peak in world crude-oil production will occur around the year 2000, within two decades of the publication of this volume. The peak of world coal production is similarly thought to be probable around the year 2150, not a long period ahead in terms of human history. Naturally, there are differing views as to these estimates of fossil fuel production, but there can be no argument about the eventual depletion of fossil fuels; they cannot provide the ultimate answer to the world's need for controlled energy. Geothermal energy provides an alternative source of power, but one of limited extent and, again, a source of dubious continuity. Even though tidal power and water power plants can provide power almost indefinitely, the amount of tidal power that can be developed seems to be restricted and available water power can even now be estimated as a strictly limited amount. Nuclear power appears to many to be the answer to the world's increasing energy demands. It probably must be used as a major source of power in the immediate future, but there is a growing consensus that it cannot be contemplated as a permanent solution to the energy demand, if only because of the problems associated with it. Solar energy is the one certain major source, but the problems of economically developing it on a major scale, either directly or indirectly, still await solution.

Against this general background, the requirements of the immediate future can the better be appreciated. While research into the control of solar energy, and on a lesser scale, of geothermal energy, are being aggressively pursued, the steadily increasing energy demands of all countries must be met by the best possible use of fossil fuels, the maximum use of water power, and the development of tidal power, all in keeping with the rapidly changing economic scene. So vital has an assured supply of centrally generated power become (for all cities in particular), and so little reserve is usually available in view of mounting power demands, that safety in the design of power stations and expedition in their construction are paramount. Even though the problems associated with their design and construction are similar to those encountered in other civil engineering work, it seems advisable to include in this Handbook this short chapter to show how some of these problems were dealt with in the construction of power stations now in successful operation, to illustrate some unusual problems that have been met with in power station work, and to emphasize the fundamental importance in all such work of full and proper attention to geology in all aspects of site investigation.

Geology is naturally a vital part of the discovery and exploitation of all fossil fuels, as it is also in the exploration for and use of geothermal power sources; it holds the key to the discovery of uranium ores. In this chapter, however, consideration is limited to the study of geological conditions at the sites at which power is to be developed for public use. Almost pedestrian in apparent interest, as compared with discoveries of fossil fuels or uranium, this particular application of geology is one of the most important branches of engineering geology today, as is well shown, for example, by the attention now being given to the geology underlying all sites proposed for nuclear power stations. Before examining some instructive case histories, there are some general considerations that may usefully be outlined.

23.1 SOME GENERAL CONSIDERATIONS

The severe limitations upon site selection for power stations impose great restraints, sometimes to the actual elimination of sites that may be suitable from purely engineering and economic points of view. All thermal stations, no matter what fuel is used, must be close to a large body of water for cooling purposes. Geothermal stations must be located where ground temperatures have been found to be suitable. Tidal power stations must be built in the sea at locations where the tidal range makes such developments possible, and water power stations must naturally be sited in accordance with the topography that makes the fall of sufficient water an economical source of power. When, to these restrictions, one adds the problems of design presented by the unusually heavy and dynamic loads to be found in power stations, then the development of adequately safe and economical power station designs can be seen to be an engineering challenge of unusual complexity.

The necessary location of thermal stations adjacent to bodies of water such as rivers and lakes usually involves the use of riverside or lakeside sites, locations which almost automatically present geological conditions with some undesirable features. The dangers inherent in the use of floodplains in river valleys is a recurring theme throughout this Handbook; such locations may have to be used for thermal power stations, calling for unusual ingenuity in design. Only very rarely will any source other than lakes or rivers be available, apart only from occasional use of seawater for cooling purposes for stations located on seacoasts. One station for which it had been hoped to use groundwater for boiler feed and for other purposes requiring pure water in the plant is the thermal station of the Pennsylvania Electric Company at Warren, Pennsylvania, located close to the Allegheny River (Fig. 23.1). Well points had to be used during construction for the dewatering of excavation. Before construction was complete, four *permanent* bronze well points were installed adjacent to the 75- by 60-m (250- by 200-ft) area that had been dewatered in the water-bearing coarse gravel that here formed the foundation bed.[23.1] It was found, unfortunately, that the quality of groundwater was not good enough for feed

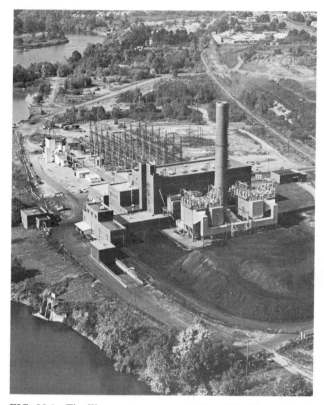

FIG. 23.1 The Warren steam power station of the Pennsylvania Electric Company, in the construction of which well points played an important part.

water, but the possibility is one worthy of being considered whenever well points have to be used and hydrogeological studies show that the level and character of the groundwater are suitable and liable to remain reasonably constant.

Not only is the supply of water an essential for all thermal plants but so also is the supply of the necessary fuel—oil, gas, coal, or uranium. Fuel has almost always to be brought to the plant from afar, but, in Europe in particular, an increasing number of thermal power plants have been built immediately adjacent to coal deposits, such as the open pit mines for the brown coal of Czechoslovakia.

The situation on some of the rivers of Pennsylvania, those which run through areas distinguished by outcrops of the famous Pennsylvania coal seams, is probably unique. In the early days of coal mining, only desirable coarse coal was sold; the finer "river coal" was dumped directly into adjacent streams or was eroded from spoil dumps in which it was piled. The wet process of preparing anthracite is another source of fine coal particles washed into waterways; the ordinary coal-washing process has had similar effect. It has been estimated that collieries in the Schuylkill Valley have been responsible for as much as one million tons of silt and coal washed into the river in one year. As a part of the operations of the Interstate Commission on the Delaware River Basin, a $35 million dredging program was initiated in 1944 to remove the accumulated silt and fine coal ("culm") that had been steadily spoiling the Schuylkill River; much of the project was

23-4   CIVIL ENGINEERING WORKS

self-liquidating because some of the coal content of the dredged material is recovered in special plants.[23.2]

The Pennsylvania Power and Light Company built a special plant for the reclamation of 500,000 tons of coal annually from the bed of the Susquehanna River. This followed the successful operation since 1938 of the 30,000-kW steam plant at Holtwood using only reclaimed river coal as fuel. Over 3 million tons of coal were used in this plant in its first 20 years of operation. The Holtwood plant was increased by the installation of a 66,000-kW steam generator for which the new coal-reclaiming plant was started in 1951; studies showed that the known reserve of submerged coal would last approximately 35 years. The Safe Harbor Dam, built to create the necessary head for the associated water power plant, had resulted in the accumulation of vast quantities of the river coal in the bed of the reservoir so formed. The new coal plant was therefore located just above the dam (Fig. 23.2). Coal and silt are dredged out by a 35-cm (14-in) hydraulic rotary cutter dredge, transported by barge to an unloading wharf, and then pumped through a 30-cm (12-in) line to the coal-washing plant, located on a convenient site 82.5 m (275 ft) above river level. The finished product is transported 13 km (8 mi) by rail to the Holtwood steam plant and there stored until required for use. The entire plant was designed in full conformity with the "clean-streams" program of the Commonwealth of Pennsylvania. The fact that an investment of $6 million for this unusual plant was warranted is further indication of the economic significance of the process of erosion by river flow, even though in this case the entry is on the right side of the balance sheet.[23.3]

A reverse problem of thermal stations is the disposal of fly ash, when the fuels being used result in large accumulations of this "waste" material. Efforts are being made to find

FIG 23.2 Dredge No. 117 of the Pennsylvania Power and Light Company at work dredging coal tailings in the Susquehanna River, Pennsylvania, as an integral unit of the Manor plant.

uses for the large quantities of this material produced by some power stations, but in an increasing number of cases, some means of disposal in bulk must be found. British engineers have suggested the use of "littoral drift in reverse," the utilization of the constant movement of seabed material along most coasts under the action of along-shore currents. The idea was conceived that, if fly ash was deposited in the sea, it would be moved naturally away from the point of deposition, all finer material eventually being deposited in ocean deeps.

One station for which this idea has been suggested is that at Blyth in northern England where 500,000 tons of fly ash might be disposed of in this way. The station is located close to the coast, adjacent to Blyth Harbor. Study of the character of the fly ash showed that 80 percent of it consisted of particles finer than the No. 100 sieve. Engineers studying the disposal problem found that material finer than No. 100 sieve was never a constituent of the material on a number of British beaches; the same thing has been found in a study of 400 American beaches. It was therefore concluded that 80 percent of the Blyth fly ash would be removed by the normal action of the sea if discharged with the wastewater from the plant. Study showed that one other power station—that at Casablanca—used this system, but there the discharge was only 50 tons per day. Colliery wastes have been dumped into the sea, as at the Cambois colliery, without any of the material staying on the adjacent beach. This unusual proposal has yet to be tried on the scale contemplated, but it presents interesting possibilities that should become economically more valid with every passing year.[23.4] In every such case, the most careful preliminary investigations will be necessary in order to ensure that the material will not be deposited at some other location along the coast in question and that there is no possibility of any interaction between the waste material and seawater, or sea life. If assurance can be gained on these points, then the method is one of unusual interest.

Since no two water power stations are ever the same, there are no further useful generalizations that can be made about the relationship of geology to the design and construction of such stations. At the same time, this inevitable variety gives to many water power installations unique geological aspects that will always be of interest, and often of use in relation to other plants. This was well illustrated in Chap. 2, where the Snowy Mountains project of Australia and the Glen Shira scheme in Scotland were selected from the innumerable cases available to show in a general way the dependence of all civil engineering works upon site geology. Descriptions of a few more cases will follow in this chapter, therefore, each illustrating but one of the many problems that may have to be faced when controlling the great force of falling water for human use and convenience.

23.2 GEOTHERMAL POWER

In view of current activity, in both the United States and Canada, in investigating all locations at which geothermal energy may be tapped, a brief note on geothermal energy is called for. Hot springs have been used at least since the time of the Romans, and probably long before that, the hot water naturally evidencing the high ground temperatures of bedrock at the location of all such springs. The use of this thermal energy for commercial purposes appears to have started at Larderello in Italy in 1777, when borax was recovered from natural steam which comes out of the ground at this location, its condensation leading to the descriptive name *fumaroles suffaroli*. It was at this same location in Italy that power was first commercially developed from natural steam in 1905. In the years following the Second World War more than half of all the power generated from geothermal energy (389,000 kW) came from the 14 Larderello installations situated not far from Pisa (Fig. 23.3).[23.5]

FIG. 23.3 Larderello steam power plant in northern Italy, operated with natural steam from the *fumaroles suffaroli*.

New Zealand was then second only to Italy in the use of this natural source of power, with 192,000 kW produced in its Wairakei plant. Development of this plant followed an intensive study of geothermal power in general and in relation to the active volcanic region of New Zealand, in which heat loss from the ground surface is from 20 to 100 times greater than the normal rate. The report summarizing these studies is still a useful document today. Smaller plants are now operating in Iceland, Japan, Mexico, and the Soviet Union, with studies anticipating such plants being developed in a number of other countries such as France, Turkey, and Chile.

In the United States, an area known as The Geysers, 145 km (90 mi) north of San Francisco, has long been a natural attraction. This is one of the few areas in the world where dry steam escapes from the ground (Larderello being another). The unusual character of the area was discovered by prospectors in 1847, but it was not until 1922 that any attempt was made to utilize the steam. In that year a 20-kW generator was driven by the use of natural steam. In 1955 the Magma Power Company leased extensive areas, and by 1957 six wells had been successfully sunk to depths of several hundred feet. Pacific Gas and Electric Company now has ten installations at The Geysers, generating 421,000 kW, with further expansions in view, so that it is now the world's leader in the use of geothermal power (Fig. 23.4). The natural steam is run through centrifugal separators at the well head, to eliminate dirt and small stones, and then piped directly to steam turbines which drive electrical generators in exactly the same way as in a normal thermal station.[23.6] Although the heat thus used clearly comes from the heat of the earth's magma, the exact mechanism by which it is transferred to water is still only imperfectly understood. Further developments of this source of power are expected, in the United States and elsewhere. It would be thought that the associated freedom from pollution would make such developments welcome, but California projects have been seriously interfered with by problems of ownership rights and by environmentalists. Geological problems are, therefore, subsidiary to legal difficulties. Since volcanic rocks are an almost invariable part of geothermal manifestations, foundation and other site problems are usually minimal.

Hot springs are much more common throughout the world than steamflows from the ground, as evidenced by the number of health spas which utilize such warm or hot water for medicinal purposes. Flows are usually limited in quantity, but there are a number of

FIG. 23.4 General view of The Geysers, California, showing a steam plant of Pacific Gas and Electric Company and steam lines providing the plant with natural steam.

locations where naturally hot water is utilized for commercial purposes. Quite the most extensive of such uses is in the small northern country of Iceland where hot groundwater has been put to the most extensive and remarkable use. Nearly one-half of the total population of the country (just under 200,000) live in houses that are warmed in this way by natural heat. Practically the entire population of the pleasant capital city of Reykjavík are now served by heat from the ground, with the result that the city is probably the most smoke-free of all capital cities of the world (Fig. 23.5). Hot springs have been known in Iceland since time immemorial, as have active volcanoes, but it was not until about 1925 that modern utilization of the hot water evidenced in the country's many hot springs and geysers was really started.

Today, 72,000 of the 80,000 inhabitants of the capital city live in naturally heated houses. The heat comes from groundwater obtained from three areas—Reykip, with 70 free-flowing wells, 300 to 600 m (1,000 to 1,950 ft) deep, giving 300 L (60 gal) per second at 86°C (187°F); Laugarnes, in the center of the city, with 11 boreholes, 650 to 2,200 m (2,130 to 7,200 ft) deep, giving about 300 L (60 gal) per second at 130°C (266°F), and Ellidaar, in the eastern part of the city, discovered only in 1967, where 5 boreholes, 860 to 1,600 m (2,800 to 5,250 ft) deep, give 165 L (33 gal) per second at 100°C (212°F). Over 100,000 m^2 (120,000 yd^2) of land is now covered by glass and warmed by natural heat, most of the enclosed area being used for cultivation. All the many swimming pools throughout the country are similarly warmed to convenient temperatures. The unusually comfortable living conditions enjoyed by Icelanders are provided, therefore, by geology—suitably aided by excellent modern engineering practice in controlling for public convenience the heat transferred by groundwater from the interior of the earth.[23.7]

23-8 CIVIL ENGINEERING WORKS

FIG. 23.5 The city of Reykjavík, Iceland, showing some of the storage tanks for the naturally hot water used to heat buildings in this capital city.

23.3 TIDAL POWER PLANTS

Of all the sources of power in nature, the tides probably appeal to the imagination more than any other. In locations such as the Bay of Fundy in eastern Canada, where tides range up to 13.5 m (45 ft), it is tantalizing to see huge volumes of water being moved twice a day in each direction, the energy involved being obvious even to the uninformed observer. In the early years of the industrial era, some small tidal mills were built and successfully used, notably in the east of England, but they were small indeed and their construction did not involve the problems that the building of larger modern plants will involve. Serious as are these problems, it has been the economic aspects of tidal power development that have delayed its use on any wide scale for so long. The variation in the times of high and low water each day, the variation in tidal range throughout the lunar month, and the variation in ranges throughout the year all combine to render essential the coupling of any tidal power project either with a major pumped storage installation or, more desirably, with a power network which will enable other stations (such as thermal plants) to supply the extra power necessary in order to match power production with power demand. Despite this, as the cost of other sources of power increases, the economic viability of tidal power will become more assured, and some developments are to be expected before the twentieth century closes. Brief mention here of tidal power is therefore desirable.

The problems of design are considerable, but they involve no more than a development of designs already successfully used in water power stations. Problems of construction will be the more serious, in view of the unavoidable location of tidal power plants, i.e., in the sea, with water levels changing continually every day. Much ingenuity in construction methods will be called for and will doubtless be forthcoming. First, however, must come unusually careful site investigation, again in the open sea, again with constantly varying water levels. That such investigations have been successfully carried out was shown in Chap. 11. Similar site studies, on an extensive scale, have been carried out in Canada in the headwaters of the Bay of Fundy, in an assessment of a number of different potential sites as part of an overall study by the Fundy Tidal Power Corporation.

In both France and the Soviet Union, modern tidal power developments have been constructed. The French plant is located at the mouth of the Rance River, on the north-

FIG. 23.6 Le Rance tidal power station of Electricité de France, on the northern coast of France near St. Malo.

west coast of France, close to the mediaeval city of St. Malo (Fig. 23.6). A project of the governmental power agency, Electricité de France, the station commenced operation in 1965 and continues to give good service. The power it produces is considerably more costly than that produced in the other stations of the system, but this was known in advance, the installation being in the nature of a major experiment and one that has proved to be successful. The powerhouse itself is the necessary impounding dam containing twenty-four 10,000-kW units. It is flanked at one end by a rock-fill embankment and a concrete spillway structure, with a ship lock at its other end, the total length of the combined structures being about 720 m (2,400 ft). Site studies in advance of design showed that competent gneiss was continuous across the site, at depths up to 12 m (40 ft) below low tide level. Fortunately, the bedrock surface was generally exposed, with only occasional deposits of sand and gravel which had to be removed. A compressed-air caisson was used for giving access to workmen to the seabed where they cleaned the rock surface, light excavation preparing it for the placing of concrete slabs, on to which caissons for forming the

FIG. 23.7 The Kislogubsk tidal power plant of the U.S.S.R., on the northern coast of the Soviet Union, west of Murmansk, facing the Barents Sea; the complete power station was constructed in a temporary dock near Murmansk and then towed to its final location.

two necessary cofferdams were floated into position. It was estimated that the cost of the cofferdams represented about one-third of the total cost of the plant.[23.8]

Profiting by this experience, although for a much smaller plant, engineers of the U.S.S.R. designed their first tidal power plant so that the entire concrete structure could be prefabricated in a sheltered location and then floated into place, thus eliminating the need for cofferdams. The plant is located on the Arctic coast at Kislogubsk, near the mouth of the Ura River not far from Murmansk. High tides are experienced all along this coast, so that further major installations are planned. This first plant, equipped with one 400-kW unit, was placed in commission in 1970 (Fig. 23.7). Again, site conditions were favorable, bedrock being exposed where the small concrete structure had to be located. Some underwater blasting was necessary to give a level bed, loose rock being removed with a clamshell bucket, sand and gravel providing the necessary bearing surface.[23.9]

These two installations, so different in size and in execution, illustrate well the unusual significance of site conditions for tidal power plants. For such plants of the future, site investigations are bound to be challenging indeed.

23.4 WATER POWER PLANTS

Since no two water power plants are ever the same, every such plant exhibiting some feature of special interest, it is difficult to present examples that will be helpful in a general way. The difficulty is compounded by the fact that the powerhouse building itself, unlike the case with thermal and nuclear plants, may be a small part of the entire project. At least one dam will almost always be necessary and often tunnels, with pipelines needed for high-head plants, the location and installation of which sometimes give rise to geological problems of great interest. Construction of water power plants therefore involves many different types of work, frequently posing problems which may be critical until construction is complete. And when water power plants are in operation, the change in hydrogeological conditions which they cause may sometimes create problems even after construction has been completed.

It is impossible to consider even these major geological aspects of water power plant design and construction within the confines of this short chapter. The authors have therefore selected half-a-dozen cases known to them, each of which well illustrates at least one major problem such as may be encountered in water power development. They are located in the United States, Canada, the United Kingdom, and New Zealand. The example from the southern hemisphere, although the problem developed over half a century ago, yet remains the most remarkable example of the interrelation of geology and civil engineering work known to the authors. It will therefore, be described first.

The Arapuni Development The Arapuni water-power scheme is located on the North Island of New Zealand, about 190 km (120 mi) south of the city of Auckland which it supplies with power (Fig. 23.8). It was constructed between 1925 and 1932 under the direction of, and in the later portions by, the Public Works Department, since all public power supply in New Zealand is under government control. The scheme consists essentially of a main dam of the curved gravity type, 58 m (192 ft) high, with a crest length of 91.5 m (305 ft) on a 75-m (250-ft) radius, which diverts the water of the river Waikato into an open headrace canal 1.2 km (¾ mi) long, finishing with a spillway dam and penstock intake. Steel penstocks, built into rock tunnels, lead to the powerhouse, the capacity of which is 150,000 kW operating under a head of 52.5 m (175 ft). The geology of the Waikato Valley is complicated; the valley has been the scene of repeated volcanic activity, and as a result the course of the river and its gradient have varied widely and often. Old

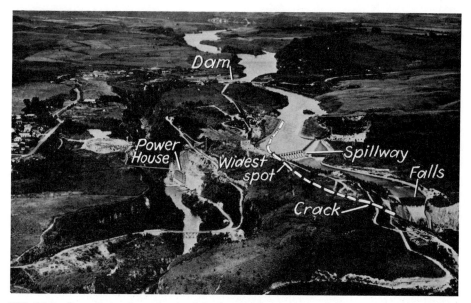

FIG. 23.8 General view from the air of the Arapuni water power project, New Zealand, showing location of the crack the occurrence of which is described in the text.

river courses have been filled with erupted matter, and consequent denudation has not always followed the old courses. The result has been that several old river channels now occur at varying heights above the existing riverbed in the vicinity of the Arapuni works; one of these was utilized as the main part of the headrace canal.

The rocks encountered were chiefly volcanic tuff and breccia. The vitric tuff is supposed to have been ejected as incandescent dust from vents in the ground and then to have fused together as it cooled. At the bottom of river valleys, however, where the probable presence of water caused quicker cooling, the rock, though of the same chemical and mineralogical composition as the solid tuff, is completely unconsolidated. The solid tuff may contain up to 30 percent water which may be held in ultramicroscopic pore spaces, as the rock is durable and moderately hard. Below the tuff occur varying beds of tuff and breccia, including a pumaceous breccia on which the main dam is founded. Below this last stratum, softer tuffaceous material extends to great depths. These volcanic deposits are uneven; they also proved to be so elastic that "the absence or presence of 10 feet of water in the gorge as the diversion tunnel was opened or shut, caused decided and opposite tilts to be registered on the seismograph in the powerhouse." On 7 June 1930, while the dam was retaining water and the powerhouse was under load, a crack occurred in the local country roughly parallel to the flow of the river (Fig. 23.9). It was widest (5 cm, or 2 in, across) where the spillway joined the penstock intakes, and it extended for about 600 m (2,000 ft). Observations then made showed that "the whole mass of country, about 600 m (2,000 ft) long, 45 m (150 ft) thick, and 120 to 240 m (400 to 800 ft) wide, was bent over towards the gorge." The power station was shut down, and the lake drained; "as the lake fell, the country recovered its position, the cracks closing except where jammed by drawn-in debris, and the powerhouse, suspension bridge, etc., regaining their original positions."

This most unusual occurrence was naturally investigated and studied closely by expert geologists and engineers. The consensus was that the movement was due to leakage of water from the headrace canal, which affected the adjacent volcanic rock to such an extent

23-12 CIVIL ENGINEERING WORKS

that movement took place. The remedial measures undertaken, therefore, centered on the provision of a waterproof lining to the headrace. After this was installed, apart from some minor troubles, the works functioned satisfactorily, and they do so today. While the remedial works were in progress, many unusual features developed. Thus, as the water level behind the dam was being lowered, gas consisting of 96 percent nitrogen was found to escape from the rock in the headrace. The gas was unusual in that it contained no oxygen, and its volume may have reached the surprising figure of 1,400 m^3 (50,000 ft^3) instead of the 700 m^3 (25,000 ft^3) mentioned in the published description. Much grouting was carried out; and while the necessary holes were being drilled, water was frequently lost; a fire hose was turned into one hole, finishing at a depth well below river-water level, without any trace of escaping water being found in the vicinity.

In other holes, great difficulty was experienced because the drills encountered vegetation, which had been growing on ground formed of early volcanic deposits when covered by later volcanic material. Timber was secured from holes and appeared to be quite sound; although in the opinion of geologists it had been there for at least 10,000 years, the bark was still intact. One of the writers was given a small cask assumed by many to be made of walnut, but which was actually made from a wood sample obtained from the deepest of the three lava-buried forests by F. W. Furkert, engineer-in-chief for all the work at Ara-

FIG. 23.9 Arapuni water power project, New Zealand; the only detailed photograph obtained of the crack, the view taken adjacent to the main intake structure.

puni. (It is now displayed in the building of the Geological Society of America in Boulder, Colorado). In a notable paper, Mr. Furkert described fully the troubles that developed and the remedial measures undertaken. It is a fascinating engineering document, as will be clear from even this brief summary, and it is an inspiring record of the successful surmounting of seemingly insuperable difficulties. The evidence thus presented by the rock movement at Arapuni shows clearly how the works of the civil engineer do affect the earth's crust. Although in most other cases the stresses set up in natural formations do not cause actual rupture of bedrock, the experience at Arapuni is a timely reminder of the strains that inevitably accompany such stresses, in rock as in other solid materials.[23.10]

The Bonneville Development Quite different were the problems that had to be faced in the development and subsequent enlargement of the Bonneville power project on the Columbia River, one of the major engineering works of the northwestern part of the United States. A postdepression undertaking, it was much criticized when first built (1934), since there seemed to be no obvious outlet for the power it generated (Fig. 23.10). As this book goes to press, Bonneville is being more than doubled in size, eventually to have a capacity of 1,076 MW (megawatts), the increase in size being possible because of the implementation of the Columbia River control works, mainly in Canada, giving greatly increased minimum flow down this international waterway.

The site of the dam and powerhouse is interesting, complicated as it is by the existence of the Cascade Slide, a great landslide which occurred about 700 years ago, after the river had cut a deep gorge close to one bank, a gorge estimated to have been at least 60 m (200 ft) deeper than the modern channel. The slide deflected the river from its course and altered the topography all around the site selected for the dam; preliminary investigations, therefore, had to be carried out with unusual care. A railroad and a highway on each side of the river required relocation to bring them above reservoir level. The unusual part

FIG. 23.10 Bonneville Dam and powerhouse, as originally constructed, on the Columbia River, Oregon, a view taken just as impounding of water had started; the large area to the left of the dam is the ancient Cascade Slide.

of this work was to obtain a stable location for the railroad on the Oregon shore some distance above the dam site, where a smaller and more recent landslide had given the railroad company much trouble even before the dam construction started. The slide area is about 2.4 km (1½ mi) long and 900 m (3,000 ft) wide; its slow movement is attributed to the "lubrication" by groundwater of an underlying stratum of shale. Solution of the problem included extensive drainage works and a major retaining wall.[23.11]

The accompanying aerial view of the original installation, taken just after a start had been made at impounding water behind the dam, shows how the ship lock, powerhouse and spillway dam utilize the full width of the river. The only possible location for the still-larger second powerhouse was on the northern shore, where the ancient landslide had occurred. It was therefore known that subsurface conditions would be complex and that even with the most extensive and careful subsurface investigation, conditions as revealed by excavation would be uncertain. As one connected with the extension work has said: "There are 100-foot rocks next to sand." Below this heterogeneous old slide material there still remains the original alluvium, which has been the cause of seepage around the end of the spillway dam. Beneath this is poor-quality bedrock which loses strength when exposed to air. An area of 16 ha (40 acres) had to be excavated, mainly in this material, to provide the site for the new 558-MW (megawatts) powerhouse. It was decided to surround the entire area with a slurry-trench cutoff wall; this was constructed by two large contracting groups for a cost of about $20 million. Four and a half million m^3 (6 million yd^3) of excavation was then removed from this area, down to bedrock surface, in some places 30 m (100 ft) below mean water level in the river above the existing powerhouse.[23.12]

The railway running along this bank of the river had naturally to be relocated before other work could start. For this purpose, a single-track, concrete-lined tunnel was constructed through the old slide material, involving some of the most difficult tunneling anywhere in recent years. The tunnel is 420 m (1,400 ft) long and cost about $12 million. A pilot bore was driven through sections that were known, from preliminary investigations, to be in unusually difficult material. The main cross section (7.5 by 10.5 m, or 25 by 35 ft, horseshoe shaped) was then driven, fortunately with recourse to very little blasting. A 7.5-cm (3-in) layer of concrete was sprayed on exposed surfaces with little delay, but steel ribs, spaced as close as 75-cm (2.5-ft) centers, had to be used in much of the tunnel. The ribs were also covered with sprayed concrete prior to the construction of the continuous 53-cm (21-in) thick, reinforced-concrete lining. Every phase of the second part of the Bonneville development faced problems created by the old slide material, problems which persisted until all structures were completed. The full record of this remarkable construction operation will be an engineering classic. (Fig. 23.11.)

The Cheakamus Development An ancient landslide of a very different character was a complicating factor in the construction of the Cheakamus water power development in British Columbia, almost due north of the Bonneville plant, across the international border. The powerhouse, with a capacity of 140,000 kW, is located on the Squamish River, at an elevation but little above sea level. Water reaches the turbines through steel pipes under a head of 337.5 m (1,125 ft), flowing to the pipes through a 10.4-km (6.5-mi) long tunnel. All these works were notable but involved little that was unusual. In order for the water retained in Stillwater Lake to flow through the tunnel, however, the lake level had to be raised by 19.5 m (65 ft), and this required the construction of a dam, 26.4 m (88 ft) high, across the Cheakamus River at a suitable location. Preliminary studies showed that the bed of this river at the only possible locations for such a dam consisted of rock debris, clearly the result of an ancient slide. Geological studies, notably by W. H. Mathews, had shown that about a century ago a major rock slide, probably involving 15.2 million m^3 (20 million yd^3) of rock debris, had flowed down a small creek emerging from Lake Garibaldi

FIG. 23.11 Bonneville Dam and powerhouses as now completed, looking east; the new powerhouse is on the left.

in the Coast Range of western Canada, filling the adjacent valley of the Cheakamus River for a distance of 4 km (2.5 mi) up and downstream to a depth of up to 45 m (150 ft), burying a forest previously there and, when natural conditions had readjusted themselves, forming Stillwater Lake.

Dr. Karl Terzaghi was consulted by the owner, the British Columbia Hydro and Power Authority. He studied the problems of the site in his own inimitable way. On the basis of his observations, he posited that a dam could be built at the site selected for topographical reasons, a matter about which there had been grave doubt. As always with difficult jobs, he wrote an account of the problems, an account which became the last publication of his lifetime. Based on a detailed study of the site and aided by earlier geological studies, without which subsurface conditions would have been inexplicable, Terzaghi decided that the dam could be built which would raise the water level of the lake sufficiently to take advantage of the potential power development. Completed between 1955 and 1957 as a rock-fill structure, using rockfall debris as the main material, the dam is about 450 m (1,500 ft) long and 26.4 m (88 ft) high. Dr. Terzaghi visited the site no less than 18 times during the course of construction, adjusting the design as excavation revealed new features of the complex geology of the heterogeneous rockslide material. The dam stands today, performing quite satisfactorily, one of the supreme examples of adapting most unfavorable geological conditions to the service of humanity (Fig. 23.12).[23.13]

The Kelsey Development Far to the north of the pleasant sites of these two North American dams, 684 km (425 mi) due north of Winnipeg, Manitoba, on the banks of the Nelson River which flows into Hudson Bay, stands the Kelsey power development. Built

23-16   CIVIL ENGINEERING WORKS

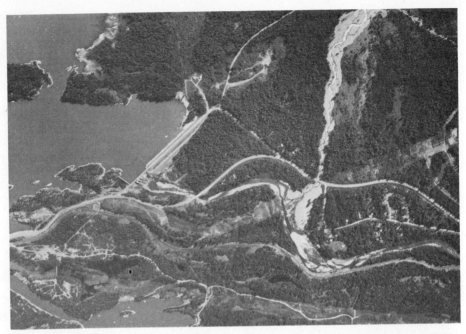

FIG. 23.12 The Cheakamus Dam of British Columbia Hydro and Power Authority on the Cheakamus River, built of rockslide debris on top of the old slide.

to supply power to the nickel plant at the new town of Thompson, Manitoba, the powerhouse is equipped with five 31,000-kW turbines, operating under a head of from 15.0 to 15.5 m (50 to 55 ft). The station went into operation in 1960. Local bedrock is a medium-to-dark-gray paragneiss, thought to be a granitized sedimentary complex. Local soils are glacial deposit so there was ample, good material available for the main concrete powerhouse and spillway structure, the associated rock-fill dam, and the dozen dikes (low dams across depressions in the perimeter of the reservoir area) needed for completion of the project. In summary, it was a fine example of a modest-sized, modern water power plant, constructed with little difficulty, apart from problems of cold-weather work (with which Canadian contractors are familiar). The logistics were greatly aided by a branch line to the site from the Hudson Bay Railway, a notable part of Canadian National Railways.

There was, however, one complication. Two of the dikes (nos. 2, east and west) were located across low-lying swampy ground instead of across the more usual depression in the exposed bedrock surface. The location of the plant is north of the southern boundary of permafrost in Canada. Preliminary investigations confirmed that the ground beneath both of these two dikes was in a frozen condition. When, therefore, the reservoir was filled, the latent heat available from the water stored in it would gradually change the temperature of the ground forming the bottom of the reservoir and, in the course of time, would also warm up the ground beneath the dikes, thawing the frozen soil; this would then result in subsidence of the ground surface, and so of the dike structures. Fortunately, by the time these dikes had to be designed, enough was known about the performance of frozen ground to enable advance calculations to be made of the rate of warming up of the ground and of the magnitude of settlement. This was done and the dikes were designed accordingly, with the foreknowledge that over a matter of years they would gradually settle. They were carefully designed as compacted, sand-fill embankments, with sand drains installed in the ground beneath each dike in advance of dike building, their cross sections being such that, when necessary, additional sand fill could be added along the crests to retain the crest level necessary for maintenance of the reservoir (Fig. 23.13). The National Research Council of Canada, through its Division of Building Research, installed sensitive ground temperature measuring and other instrumentation beneath the reservoir and beneath one of the dikes. It is satisfying to record that the two dikes have performed exactly as anticipated, maximum settlement after seven years of operation being about 2 m (7 ft), all of which was made up by the regular placement of new fill in accordance with the original design.[23.14]

The Foyers Development In Scotland, on almost the same latitude as that of the Kelsey station, but with greatly different climatic conditions because of its island location and the moderating influence of the Gulf Stream, stands the Foyers pumped storage project of the North of Scotland Hydro-Electric Board. Located in the Scottish Highlands, it discharges into Loch Ness, one of the chain of narrow lakes marking the line of the Great Glen, and the Caledonian Canal. The complexity of the bedrock geology of the Scottish Highlands is well known. Unusual care was therefore taken with preliminary investigations for this project. As always, topography dictated the general arrangement of the works. Loch Mhor, previously used as head pond for a small water power plant at the same general location, became available as the main storage reservoir for the pumped storage when the old station was abandoned. Existing dams on Loch Mhor, after rehabilitation, provided the necessary storage. A tunnel about 3 km (1.8 mi) long conveys water, generally under low pressure until close to the powerhouse, where its invert drops through a combined high-pressure surge shaft. This leads into steel pipes, forming steel lining to the last sections of tunnel, the water then flowing directly into the turbines. These have a capacity of 300 MW.

Fourteen boreholes were put down along the line of the tunnel and seismic surveys were also conducted, resulting in "a remarkably accurate interpretation of (geological) conditions by Dr. D. I. Smith of the Institute of Geological Sciences, Edinburgh." The assumptions made for purposes of design were confirmed in the field, and so it may be wondered why the Foyers installation is included in this brief list. A representative of the general contractor for the works had this to say:

FIG. 23.13 One of the saddle dikes for the Kelsey water power project of Manitoba Hydro, built in northern Manitoba over perennially frozen ground; evidence of some subsidence (due to thawing of frozen ground, as calculated in advance of construction) can be seen.

From the Contractor's viewpoint the appraisal of ground conditions was unusual in that greater reliance was placed on the trial shaft by Loch Ness and on the geological survey than on conventional site investigation tests and boreholes. More borehole information in the vicinity of both upper and lower control works would have been valuable, particularly in relation to the nature and depth of the overburden, but it is unlikely that additional boreholes elsewhere would have justified the extra cost. Rock conditions within the zone of the Great Glen fault adjacent to Loch Ness were too variable to permit detailed prediction, but elsewhere the geological survey provided an accurate forecast for construction purposes.[23.15]

This is helpful and constructive comment which usefully directs attention to the fact that civil engineering works for power developments (as for all other purposes) have not only to be designed but also constructed. The excellent engineering design of this important project was well served by the preliminary site investigation, but the general contractor ran into real difficulty in constructing his temporary works for the reason noted in the quotation.

In order to construct the upper control works (a reinforced-concrete gatehouse leading into a drop shaft which connected with the low-pressure tunnel), a circular cofferdam 29 m (95 ft) in diameter had to be constructed. Steel piling was selected for forming this temporary structure, the limited subsurface information available suggesting that there would be little difficulty in driving the piles to the necessary depth. It proved, in practice,

FIG. 23.14 Tunnel intake structure of the Foyers pumped storage water power project of the North of Scotland Hydro-Electric Board, showing glacial till which caused difficulties during construction.

to be impossible to drive them to depths greater than 7 m (23 ft) because of the toughness of the till and the presence in it of cobbles and boulders (Fig. 23.14). The steel piling had to be withdrawn and a ring of excavation formed by drilling holes (which were cased) with heavy auger rigs. Three months of double-shift working had to be expended on this work. When the ring was completed, the augered holes having been filled with gravel as the casings were withdrawn, the steel piles were then used again, being driven into the artificially formed gravel ring. Inevitably some delay in the completion of the works within the cofferdam resulted from the difficulties thus experienced. It is to be noted that it was, again, the properties of the glacial till in situ which caused the difficulty.[23.15] It cannot be emphasized too strongly that when excavation (or pile driving) has to be undertaken in glacial till the very greatest care must be taken with preliminary subsurface investigation in order to determine the actual properties of the till. Knowledge of the geology of the site will always help, especially in suggesting whether any buried tills are present, since these will probably be unusually compact. Similar cases will be found in all the main chapters of this Handbook, problems with till being possible hazards wherever this widespread glacial deposit may be encountered.

The Kootenay Development Exactly the reverse type of problem had to be faced in the construction of the $39 million powerhouse of the Kootenay Canal project in British Columbia. Following the construction of the Libby Dam in the United States, on the Kootenay River, a much increased minimum flow in the river was available for power generation. Instead of increasing the capacity of five existing plants, it was decided by the British Columbia Hydro and Power Authority to build a new plant fed by a new 4.8-km (3-mi) canal with a capacity of 850 cms (30,000 cfs). The best site for the powerhouse involved 1.9 million m³ (2.5 million yd³) of excavation into a sloping hillside, the soil being predominantly a glacial silt. This was water-bearing and so the danger of liquifaction due

FIG. 23.15 Electrodes for the application of electro-osmosis for slope stablization on the Kootenay power canal of the British Columbia Hydro and Power Authority.

to blasting and the passage of heavy equipment had to be faced. It was eventually decided to utilize electroosmosis to stabilize the soil while excavation proceeded (Fig. 23.15). The work was carried out in six stages, the same procedure being repeated as excavation worked up into the hillside. Thirty-cm (12-in) -diameter, open-ended steel pipes were driven at the head of each slope, on 6-m (20-ft) centers, down to bedrock, which in places was over 30 m (100 ft) deep. Pipes were cleaned out by water under high pressure and then backfilled with sand. In each such sand column a 32-mm (1.25-in) well point was inserted, encased in a 50-mm (2-in) pipe. Another parallel set of 50-mm (2-in) pipes was driven to the same depth, 3 m (10 ft) away, to act as the anodes. Direct current at 150 volts was then applied, the well points activated, and excess water removed, enabling excavation to proceed with safety. It was estimated that groundwater flow increased in the ratio of 1,000 to 1, as compared with what well points alone could do, up to 150 lpm (40 gpm) being removed. The dewatering contract was for about $1 million and it proved to be a wise insurance policy, since the work was completed without difficulty.[23.16]

23.5 THERMAL PLANTS

Electroosmotic stabilization of silt greatly assisted also in the construction of the Weadock steam plant at Essexville, Michigan. One section of excavation had to be carried to depths of 11 m (37 ft) below the level of the adjacent river. Steel sheet piling was used for surrounding the site. The first cut of about 6 m (20 ft) in depth (to El. 565 ft) was in medium sand and was completed without difficulty, ordinary well points being used for dewatering. Excavation below this level, down to El. 546 ft, was in a silty sand, and here some movement of the steel sheet piling started. This had to be prevented because of close proximity of underground ducts and other parts of the station. Electroosmosis was therefore applied, with a voltage of 90 and electrode spacing of 4.5 m (15 ft). With the system in operation, excavation was satisfactorily completed with no further movement of the sheeting.[23.17]

Problems with Uplift Pressure Problems of construction of steam power plants due to the proximity of the river or lake from which water is to be drawn have led to many other and varied solutions. In the case of the Shawnee steam plant of the TVA, on the Ohio River downstream from Paducah, Kentucky, the station is founded on a permeable sand and gravel base, which is covered naturally with an impervious blanket of loess. The dewatering necessary for construction was similar to the drainage control that would be necessary for the completed plant, in order to keep uplift pressures at varying stages of the river under rigid control. Using information about earlier installations made available through the U.S. Corps of Engineers, the TVA designers decided to install a system of 84 deep wells around the perimeter and under the powerhouse building. The typical well is a 20-cm (8-in), slotted, wood-stave pipe surrounded by a graded gravel filter. All the wells are connected to a common collection header, offset 1.5 m (5 ft) from the center lines of the several strings of wells to allow for the pumping of any one well without interfering with the others. Each well has a 60-cm (24-in), reinforced-concrete riser for the top 60 cm (2 ft), the last 60 cm (2 ft) of the wood-stave pipe being unslotted. Seven of the wells were designated as permanent pumping points, the remainder having their risers capped and fitted with cast-iron manhole frames and covers.

Permanent pumps for uplift control have capacities of 9,500 lpm (2,500 gpm) under the normal 7-m (26-ft) total head, operation of the pumps being automatically controlled by a float switch (Fig. 23.16). As can well be imagined, the whole system is somewhat complex, the foregoing being but a summary of its salient points. It is, however, a fine

FIG. 23.16 The Shawnee power plant of the Tennessee Valley Authority, showing permanent pumping arrangements being installed.

example of the control of unusually difficult site conditions by an ingenious application of well-tried engineering methods, all dependent upon the accurate knowledge of subsurface conditions which was obtained before the first construction work started.[23.18]

An alternative method of controlling uplift pressures on the base of a steam power plant is featured in another of the TVA stations, that at Johnsonville. Here the bedrock on which the station is founded is chert, but it is water-bearing. The plant is adjacent to the shore of the Kentucky Reservoir with the possibility, therefore, of a head of water acting on the lowest foundations of up to 15 m (50 ft). This could have been balanced by the addition to the foundation structure of a deadweight, in the form of additional concrete, but 50,000 tons of extra concrete would have been necessary for this purpose. As an economical alternative, resulting in an overall saving of some $400,000, the designers arranged to tie down the foundation by means of reinforced-concrete "cells," drilled into the chert bedrock. Two hundred and twenty holes, generally 75 cm (30 in) in diameter and penetrating 5.7 m (19 ft) into the rock, were drilled; there was some variation in detail but the dimensions given were those for the majority. Drilling was not easy, in view of the character of the chert, but churn drills using special rotary star bits successfully completed the job, holes being cleaned out by a suction baler. A cage of reinforcing steel was then lowered into each hole, prior to the placing of high-quality concrete. Reinforcing bars were splayed out above the top of the hole, to be bonded into the foundation slabs but in such a way that the cells would not carry vertical loads. Sides of the drilled holes were fortunately rough, thus increasing the pull-out resistance of each unit.[23.19]

Problems with Weak Foundation Strata Problems of a quite different nature were faced in the design and construction of the South Bay power plant of the San Diego Gas

FIG. 23.17 The South Bay steam power plant of the San Diego Gas and Electric Company, features of the construction of which are described in the text.

and Electric Company at Chula Vista, California. The site for the plant was a low, flat, marshy area on San Diego Bay, previously occupied by saltwater evaporation ponds (Fig. 23.17). Initial installation was two units, each of 142,000-kW capacity, so foundation loads were considerable. The site had to be prepared also for the installation of additional units. Site investigation showed that the soft surface soils extended down to 6 m (20 ft) below the surface; older and firmer soils then followed, but bedrock was not reached by a boring put down to a depth of 63 m (210 ft). End-bearing piles or caissons would have provided a desirable foundation design but they were not possible. Friction piles were considered after preliminary studies had shown that a concrete mat foundation would be subject to too great a settlement. Advances in the engineering uses of soils made it possible to consider another solution, that which was adopted—removal of the soft surface soil down to an elevation 3 m (10 ft) below mean sea level and its replacement with carefully compacted soil of a character suitable for supporting the powerhouse loads. Preliminary estimates of cost suggested a total of $350,000 for a piled foundation and only $100,000 for a compacted-soil foundation design, an estimate that proved to be slightly more than the actual cost. The soil finally selected for the compacted fill was a beach-deposited, wind-blown dune sand obtained from the Coronado side of San Diego Harbor. It was not what would be called "select" fill material, but the cost of using it was estimated at only 70 percent of the cost of using a more desirable coarse sand from the Tia Juana River valley.

Excavation over the site of the soft surface soils was a somewhat delicate construction operation. Initial excavation was with scrapers; while this was in progress drainage ditches, backfilled with clean crushed rock, were formed around the site. Perforated-concrete, 105-cm (42-in) -diameter standpipes were installed at each corner to serve as pump sumps, groundwater being thus pumped into the bay from the area to be excavated. Lighter excavating equipment had to be used as the original groundwater elevation was approached, the final cut to the underlying clay-marl being carried out with a backhoe, the operation of which did not disturb the soil surface on to which the compacted sand fill was to be placed. An initial layer of crushed rock, 30 cm (12 in) thick, was placed first and bulldozed into position. This was followed by a 5-cm (2-in) layer of well-graded pea gravel to obviate movement of the fine sand fill into the crushed rock base. A 15-cm (6-in) layer of fine-to-coarse sand was next deposited, and then the main sand backfill was placed. Bottom dumping from 15-cubic-meter (20-cubic-yard) dump trucks was followed by spreading of the sand into 30-cm (12-in) layers. The sand was saturated before compaction, always under careful field control. Compaction was effected by three passes of a suitable vibratory roller, an average compacted dry density of 1,600 Kg/m^3 (100.2 lb/ft^3) being obtained, this representing 98 percent optimum compaction.[23.20]

Continuous and accurate field control of all backfill operations was naturally essential for the success of this unusual foundation but this was achieved and the plant stands today, its capacity now 706 MW, a tribute to the contribution that soil mechanics can make to such a difficult situation. There are at least three other power plants in California alone successfully founded on compacted fill over weak foundation strata. The method can be used only on the basis of the most careful site investigation, with full appreciation of the geological character of the soil strata involved. The same is true when, as yet another possibility if weak soils underlie a power station site, the foundation soils are preloaded in order to minimize the settlement that would otherwise occur when foundation loads are applied. Preloading is common practice with smaller structures such as bridge piers, as will be seen in Chap. 24, but the same principles can also be applied to the large areas required for power stations.

This was well shown in the design and construction of the 250-MW steam power plant of the Cajun Electric Power Cooperative, Inc., located 35 km (22 mi) upstream of Baton Rouge, Louisiana, in the west alluvial floodplain of the Mississippi River. This site is underlain by 1.8 to 2.4 m (6 to 8 ft) of medium clay, overlying 22.5 to 30.0 m (75 to 100 ft) of soft clay and loose silt, with occasional pockets of organic clay, typical deposits for floodplains of major rivers. Dense sand suitable for end-bearing piles exists only at or below 30 m (100 ft) beneath the surface. Predicted settlements, if no treatment were given to the site, were so high as to be unacceptable. Friction piles therefore presented what might be called the conventional solution, but preloading of the upper soil strata in order to reduce significantly the settlement to be expected was a possibility. Modern soil studies permitted this alternative to be studied in advance of final design, with such satisfactory results that it was adopted as the approved design, the savings in cost of $164,000 over a piled foundation being naturally a major consideration. Soil conditions over the site were somewhat variable, necessitating detailed field testing and control. Under the cooling towers, large structures with critical loads, 1.2 m (4 ft) of the surface clay was removed, the area being backfilled with sand in order to ensure homogeneous support for the reinforced-concrete "boxes" on which the cooling towers were to be erected.

The contractor was allowed to use as surcharge the topsoil that had to be removed from the entire site. A surcharge load of twice the uniform design load was decided upon; this would give twice the settlement to be expected from the actual loads, thus reducing the period during which the surcharge had to remain in place. (It is this necessary time interval that sometimes restricts the adoption of this method of site preparation for heavy foundation loads). A final surcharge 3.6-m (12-ft) thick was used over most of the foundation area, some variations being necessary as initial settlements were carefully observed. These showed up variations in the underlying soil properties, information unobtainable by test boring except at prohibitive cost. Surcharge was left in place for a period of nine months, during which a settlement of about 30 cm (12 in) took place under one of the cooling tower areas. Smaller areas for special structures were also preloaded in different ways. For tank foundations, for example, the tanks were erected and then filled with water (heavier than the oil ultimately to be stored). In view of the many power plants that must be located in major floodplains, this case is of unusual significance.[23.21]

A useful example of preliminary site investigation in a very different locale was that for the Battersea steam power station of the Central Electricity Board of England. This station is located on the south bank of the river Thames in London and dominates the local skyline, having become a London landmark. Preliminary study of available records of borings in the vicinity suggested that the London clay might be reached at a depth of 9.0 to 10.5 m (30 to 35 ft). Some local records indicated that the clay was at a much greater depth, and one showed that a well at the Nine Elms Brewery had failed to encounter the clay at all. From a study of these records and of a map prepared by the Geological Survey, it was anticipated that, with the London clay at about the depths indicated, considerable depressions in its surface might be encountered. This proved to be the case.

Figure 23.18 is a key plan of the site; the irregular spacing of the boreholes represents the efforts made to trace the depth and extent of the depressions in the clay. In all, 73 test holes were bored, and eight trial pits were sunk. These showed the alluvial gravels of the Thames floodplain overlying the London clay. Since water accumulated in the pits, it was decided to surround the entire building site with a retaining wall in order to exclude the water; it was further decided to found the station on the London clay. The Geological Survey was given full information from the test borings and was consulted about the results. The two main depressions detected from the borings were later confirmed as excavation proceeded and were found to be of the "pipe" type and not a continuous channel. They may have been caused by subsidence of the clay, following solution of the underlying chalk.[23.22] The cost of the necessary preliminary work is often a difficult matter to estimate in any undertaking, and building construction is no exception to this. It will be useful to point out, therefore, that this thorough investigation for the Battersea power station cost only 0.2 percent of the total cost of the station structure—a percentage that is not unusual and that illustrates vividly the economy of adequate subsurface exploration.

Problems with Subsidence The examples which have been cited show well the almost infinite variety of problems that have to be faced in the design and construction of foundations for thermal power plants. One more example must be given in order to illustrate a problem which, although not unique, is fortunately rare. The Sam Bertron steam power plant of the Houston Lighting and Power Company is one of five plants of this company located on the Houston Ship Channel. As will have to be explained in Chap. 37, the area around and including the city of Houston has been subsiding for some years, and even when remedial measures have been implemented, the subsidence that has already taken place will remain. When built, the Sam Bertron plant was at an elevation of 6 m (20 ft) above mean sea level. By 1972, ground elevation at the plant had been lowered to 3.8 m (12.8 ft) above mean sea level. The plant is located about 6.4 km (4 mi) above the Baytown Tunnel where, in 1961, a hurricane tidal surge of 5.0 m (16.8 ft) was recorded. The combination of local ground settlement and the possibility of future tidal surges placed the power plant in some danger. Some remedial measures were essential. Soil conditions at the site consisted of a bottom channel deposit of recent soft, organic gray clay (usefully described as "mucky," since its shear strength was in places as low as 488 to 976 kg/m^2 (100 to 200 psf), overlying reasonably strong Pleistocene clay (with shear strengths of about 4,882 kg/m^2 or 1,000 psf), the contact being generally at about 6 m (20 ft) below sea level. This suggested that a levee in front of the plant could be designed, but further bor-

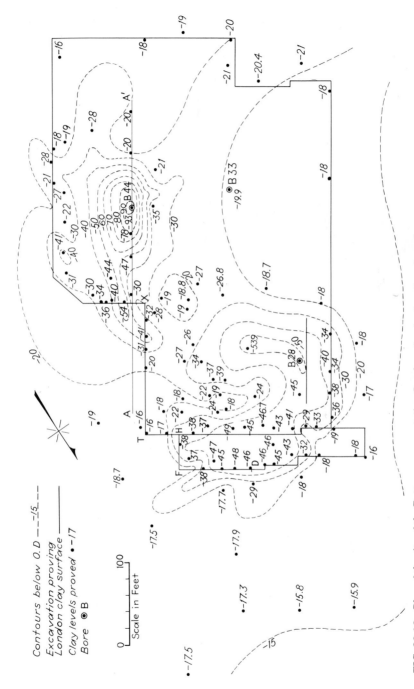

FIG. 23.18 Plan of the site of the Battersea power station, London, showing location of test bores and the contours of the surface of the London clay.

23-26 CIVIL ENGINEERING WORKS

FIG. 23.19 The Sam Bertron steam power plant of the Houston Lighting & Power Company, showing the special cofferdam, the purpose of which is described in the text.

ings revealed that the channel had been overdredged, leading to depths of up to 7.5 m (25 ft) of the "muck" at the bottom of the channel, elevation of the Pleistocene clay in the vicinity of the station's intake structures being 12 m (41 ft) below sea level. These changed conditions rendered impossible initial designs. The intake structures naturally had to have clear access to the water of the channel except at times of emergency.

An alternative design was therefore prepared and constructed, consisting of a "permanent cofferdam" of circular cells constructed of steel sheet piling driven into the channel bed, openings left between pairs of cells being closed by specially designed concrete-and-steel gate structures (Fig. 23.19). These were founded on limestone fill which replaced the organic surface clay, steel sluice gates installed in each structure being normally in the open position but being capable of rapid closure if ever dangerous high water levels are experienced. Embankments of soil serve as levees to protect the plant on land.[23.23] The engineering design of this unusual installation is most interesting, but it is to the need for the construction of such a major structure that attention is directed. Ground subsidence is not an academic rarity. It can and does sometimes have a profound effect on engineering structures and so is a matter that can never be forgotten or neglected.

23.6 NUCLEAR PLANTS

Nuclear power plants are thermal generating stations in essence, but with special features because of the use of enriched uranium as fuel and the unusual precautions that have to be taken to ensure the safety of all operations. In view of the number of such plants that must be built in the remaining part of this century and of the stringent regulations covering the design and construction of nuclear plants already in force in almost all countries, this section is but a brief introduction to this latest division of power plant engineering. The geology underlying all nuclear plants can be seen, at a mere glance, to be of fundamental importance.

One of the earliest major nuclear plants was the 580-MW Sizewell plant of the Central Electricity Generating Board of Great Britain. Started in April 1961 and commissioned in 1965, it was the largest such plant in the world when completed. It is located just over 160 km (100 mi) northeast of London, on the Norfolk coast. One reason for the selection of its location was the satisfactory site geology. Test borings encountered nothing but well-graded, compact sand down to depths of 48 m (160 ft), beyond which test boring was not continued. With the two reactors enclosed in one building and a common equipment and control center, foundation loads were high. So satisfactory was the sand found to be that the reactors are founded on a 2.4-m (8-ft) -thick concrete slab. No piles were necessary under this, although piles were used in a few places for highly concentrated loads.

The plant occupies an area of 9.6 ha (24 acres), so initial clearing was an operation of some magnitude. More than 300,000 m^3 (400,000 yd^3) of material had to be moved, including some large, reinforced-concrete antitank barriers erected during the Second World War, a hazard that is fortunately not too common. An extensive program of sand testing was carried out, in-place density being a property of special significance. Although the sand in its natural location was fairly dense, it was found that with the modern compaction methods used on the job even higher densities could be obtained in sand placed as fill. In a typical test pit, wet densities were as follows: fill at the surface, 1,517 to 1,580 kg/m^3 (95 to 99 pcf); fill 60 cm (2 ft) below the surface, 2,000 kg/m^3 (125–126 pcf); natural sand at 1.5, 2.0, and 2.4 m (5, 6, and 8 ft) below the surface, 1,725, 1,780, and 1,860 kg/m^3 (112, 115, and 120 pcf), respectively. The California bearing ratio was found to be the best control test. Vibrating rollers weighing up to 3.75 tons were used for compaction. Eight to twelve passes of vibrating rollers were used, passing over 45 cm (18-in) thick layers, for compacting the sand at all crucial locations, such as the main foundations. Densities as high as 2,050 kg/m^3 (128 pcf) were obtained in this way.

Many of the engineering features of this plant are of interest but the only one involving the geology of the site directly was the construction of the two 3.3-m (11-ft) diameter tunnels for the cooling water intakes. These had to be driven in sand, under the sea, to intake caissons 540 m (1,800 ft) offshore. Compressed air was therefore necessary. When the site of the intakes was reached, vertical shafts were raised from each tunnel with special covers onto which the intake structures were lowered, having been floated out as completed precast units (equipped with false bottoms) from a casting yard on shore (Fig. 23.20).[23,24]

FIG. 23.20 A diver preparing to fit a 3-ton cap onto one of the underwater shafts of the circulating water tunnels (9 m or 30 ft below the surface) at the Sizewell nuclear power station, Suffolk, England.

The first nuclear generating plant of the TVA was that at Brown's Ferry; it achieved some notoriety later for reasons not connected with its site. The site finally selected was chosen after all aspects of its intended use and of its probable impact on the area adjacent to the site had been considered. Meteorological and hydrological features were studied as was the seismic history of the region. Cooling water supply was ensured since the site selected is on the north shore of Wheeler Lake, one of the TVA-controlled reservoirs. This location gave good access by barge, connection to the nearest railway (12.8 km or 8 mi away) being found unnecessary. The geology of the site was intensively studied, first in a regional sense and then in detail, with the aid of 34 test holes in the area of the most critical structures and 46 elsewhere on the site. In only 11 of the 80 holes was any evidence of cavities found, and then only a total of 5.1 m (17 ft); 2.7 m (9 ft) was found in the 34 holes beneath the main areas to be loaded. Drilling and examination of the cores gave no indication of any deep or extensive development of solution cavities in the underlying limestone. It was therefore concluded that the local bedrock would serve well as the foundation bed for the plant, even with the high loads that would be involved.[23.25]

Different again were the conditions at the Salem nuclear generating station of the Public Service Electric and Gas Company of Newark, New Jersey. This plant is located at the southern end of a man-made plot of land in the Delaware River, called Artificial Island, about 63 km (39 mi) south-southwest of Philadelphia (Fig. 23.21). The island was formed of dredged material deposited here around a natural sandbar in the river at the end of the last century by the U.S. Corps of Engineers. The station has two identical units, each rated at 1095 MW, their size being an interesting reflection of the advances made in nuclear power generation in the few years immediately following the building of the Sizewell station. Thirty-five borings, 10 cm (4 in) in diameter, were put down on the site, all carried to a depth of 60 m (200 ft) below grade, which was about 2.7 m (9 ft) above mean sea level. Subsurface conditions were found to be reasonably uniform under the site. Hydraulic fill and alluvium were generally found immediately below the surface to a depth of 7.5 to 9.0 m (25 to 30 ft). Then followed 1.5 to 3 m (5 to 10 ft) of coarser sands and gravel, below which the local Kirkwood formation extended to about 21 m (70 ft) below grade. The formation consists of gray silts and clays with some basal sands; it was deemed to be unsuitable for supporting the loads from the plant. From the 21-m (70-ft) level to about 42 to 45 m (140 to 150 ft) below grade is the Vincentown formation, a silty sand stratum that is in places well cemented. It is dense and was considered suitable for the powerhouse loads. Beneath it are dense sand formations down to 540 m (1,800 ft) below the surface where igneous and metamorphic bedrock is located.

Since seismic forces had to be allowed for in the station design, special tests were made of the Vincentown formation to ensure its safety with respect to possible "liquefaction," should seismic shocks be experienced. Soil densities were found to be greater than those in areas where no "liquefaction" damage occurred at Niigata, Japan, where an earthquake on 16 June 1964 resulted in much damage from this cause. Possible shear forces resulting from seismic shocks eliminated the possibility of using piles down to the Vincentown formation for the major loads. It was therefore decided to excavate down to the top of this formation for the area required for Class I (most sensitive) loads. No suitable fill could be found in the vicinity, and so it was finally decided to use a lean concrete as fill from the Vincentown formation up to the base level for foundation structures. The lean concrete was made using a minimum of 3 bags of cement and 178 kg/m^3 (300 lb/yd^3) of fly ash, with fine and coarse aggregates such that a minimum strength of 105 kg/cm^2 (1,500 psi) at 28 days would be obtained.

The excavation of this area alone was an engineering project of magnitude, the area being 1.2 ha (3 acres), depth of excavation being 21 m (70 ft) below grade. The site was surrounded with circular steel sheet-pile cell units with interconnecting arcs, dewatering being achieved through 300 wells, each 27 m (90 ft) deep, so arranged that they could maintain the groundwater table at 22.5 m (75 ft) below grade. Other structures at this notable plant were supported on steel, concrete-filled pipe piles driven into the Vincentown formation. The plant has special interest, therefore, in that two different means were used to transfer loads to the same formation, 21 m (70 ft) below original grade. The vital importance of accurate knowledge of all the subsurface conditions for the successful completion of this unusual foundation job will be clear.[23.26]

Nuclear power plants in steadily increasing numbers have been built in many countries of the world since the early pioneer plants of the 1960s. In every country using nuclear power there has been a corresponding growth in the number and rigor of the regulations which must now be satisfied before a license to build a new plant can be issued. Little is to be gained, therefore, by any further descriptions of geologically interesting installations. Attention will rather be directed to regulations current in the United States, one of the world leaders in nuclear plant technology. Even this treatment is difficult since, under mounting public pressure, regulations are being repeatedly revised and changed.

Two general comments must first be made about all regulations for the siting of

FIG. 23.21 The completed Salem nuclear power plant of the Public Service and Gas Company, New Jersey, on an island in the Delaware River; the special foundations for the plant are described in the text; the boat seen is used by the company for educational purposes.

nuclear power plants. A first reading of many of these regulations, and of some of the presentations that are made by environmentalists at the public hearings now commonly held in relation to nuclear regulations, suggests rather definitely that some of the officials responsible for these regulations and presentations have no conception of the site studies made for every major civil engineering work, studies which are the central theme of this Handbook. There is an implied suggestion that the careful steps that are enumerated for site selection for nuclear plants are something new, whereas they have been common practice in civil engineering for decades, as so many of the examples cited in this volume show clearly. The need for extra certainty at nuclear plants, especially with regard to seismic risks and the presence of faults at building sites, can naturally be understood, although even these matters are regularly considered for many civil engineering projects, such as major dams. In the second place, the necessity for preparing regulations in codified form necessarily means a subdivision of the disciplines involved in site selection, whereas in practice, as all civil engineers and engineering geologists know well, the approach must be an interdisciplinary one with the closest cooperation between all those involved. In current U.S. regulations, for example, there are separate sections on geology, hydrology (much of it actually hydrogeology), meteorology, and even geography. While regulations must naturally be followed meticulously, watch must be kept against any rigid subdivision in the application of the several disciplines involved in site studies in order to ensure the absolutely essential coordinated approach to final site selection that must be made.

In 1979, regulations in the United States with regard to reactor safety were the responsibility of the Nuclear Regulatory Commission (NRC). The report of the President's Commission on the Accident at Three Mile Island was, however, so critical of NRC that this situation might change. The accident at the Three Mile Island plant was not related to the siting of the plant but rather to its operation. Accordingly, it may still be helpful to summarize the 1979 situation with regard to the siting of nuclear plants. The following notes are based on the most useful current summary known to the authors, Paper no. 14615 of the American Society of Civil Engineers. This paper is a report of the society's task committee on nuclear facilities siting, a report which is commended to all who wish for a more complete summary of the situation in late 1979.[23.27]

A building permit issued by NRC is necessary before construction of a nuclear power plant can commence. Issuance is based upon consideration by NRC of the applicant's *Preliminary Safety Analysis Report* (PSAR). Full description of the site proposed must be given in this report, including meteorological, hydrological, geological, and seismological characteristics. An *Environmental Report* (ER) must be submitted with the PSAR. Both submissions are independently reviewed for NRC. The major geological considerations that have to be fully reviewed in an application include the possible presence of faults and an assessment as to the probability of movement taking place on them and the possibility of the occurrence of such hazards at the site as landslides, land subsidence, uplift, harmful groundwater movements, and volcanic effects. Construction methods and materials planned for use in construction must be reviewed. Static and dynamic stability of the structures proposed, with special emphasis on seismic probability, must be considered. Detailed seismic investigations, geological and geophysical, are mandatory. Possible flooding of the site is a matter that must be investigated under the heading of hydrology, whereas "terrain analysis" comes under the heading of geography and includes such geomorphic features as terrain ruggedness.

All these matters have already been mentioned elsewhere in this volume, but, naturally, they take on increased significance when considered in relation to the possibly disastrous consequences of any accident to the structures containing a nuclear power facility, the seriousness of which cannot be minimized. The first conclusion of the ASCE task committee is worth quotation in this context:

Insofar as technical aspects of siting are concerned, there is a wealth of experience indicating the general adequacy of technology in this area. Indeed, the application of this technology to the siting of other industrial facilities of equal comparable risk could probably provide an overall benefit. This is not to indicate that no unknowns or uncertainties remain, or that no room for improvement exists. Despite the recognized difficulties, the Committee shares the view that definition of geologic and seismic characteristics of potential sites remains the area in greatest need of additional effort to reduce current conservatism.

Many more nuclear power plants will almost certainly be built before safer and more desirable sources of public power become economically available. The engineering geologist, working in concert with civil engineers, has an important role to play in ensuring that the sites for all nuclear plants are safe beyond all possibility of doubt. Time-consuming as they may prove to be, the detailed requirements of official regulations must be followed exactly. These requirements should be taken as a challenge. They can certainly act as a useful checklist for the engineering geologist to use in ensuring that all the provisions, modified as necessary, form a part of the normal procedure in site studies. In this way, the special precautions that have to be taken in the siting of nuclear power plants can have widespread benefit.

23.7 REFERENCES

23.1 "Draining Coarse Gravel with Wellpoints," *Engineering News-Record*, **137**:638 (1946).

23.2 J. H. Allen, "Coal Fines to Offset Cost of Culm Elimination," *Civil Engineering*, **16**:395 (1946).

23.3 P. Levin and D. I. Smith, "Six-Million-Dollar Plant Recovers Fine Coal from Susquehanna River," *Civil Engineering*, **24**:435 (1954).

23.4 F. L. Harwood and K. C. Wilson, "An Investigation into a Proposal to Dispose of Power Station Ash by Discharging It into the Sea at Low Water," *Proceedings of the Institution of Civil Engineers*, **8**:53 (1957).

23.5 "Harnessing Hot Springs Around the World," *Nature*, **230**:209 (1971): see also "Geothermal Resources Gather a Head of Steam," *Engineering News-Record*, **186**:30 (6 May 1971).

23.6 A. W. Bruce and B. C. Albritton, "Power from Geothermal Steam at the Geysers Power Plant," *Proceedings of the American Society of Civil Engineers*, vol. 85, PO 6, Paper no. 2287, 1959, p. 23; see also "Boil, Boil, Toil and Trouble: Earth's Cauldron 'Brews' Energy," *Financial Post* **68**:E7 (2 November 1974); see also "Automatic Steam Gathering Reduces Emissions at 'The Geysers,'" *Civil Engineering*, **50**:78 (June 1980).

23.7 J. Zoega and G. Kristinsson, "The Reykjavík District Heating System," *Proceedings of the United Nations Conference on Resources*, **2**:21 (1961).

23.8 W. G. Bowman, "French Promise Tidal Power by 1965," *Engineering News-Record*, **170**:32 (18 April 1963).

23.9 L. B. Bernstein, "Russian Tidal Power Station Is Precast Offsite, Floated Into Place," *Civil Engineering*, **44**:46 (April 1974).

23.10 F. W. Furkert, New Zealand, by personal communication; see also F. W. Furkert, "Remedial Measures on the Arapuni Hydro-Electric Scheme of Power Development on the Waikato River, New Zealand," *Minutes of Proceedings of the Institution of Civil Engineers*, **240**:411 (1935).

23.11 C. P. Berkey, "Foundation Conditions for Grand Coulee and Bonneville Projects," *Civil Engineering*, **5**:67–71 (1935).

23.12 "Bonneville Dam Squeezes in New Powerhouse," *Engineering News-Record*, **199**:64–65 (22 December 1977).

23-32 CIVIL ENGINEERING WORKS

23.13 K. Terzahgi, "Storage Dam Founded on Landslide Debris," *Journal of the Boston Society of Civil Engineers,* **47**:64 (1960).

23.14 D. H. MacDonald, R. A. Pillman, and H. R. Hopper, "Kelsey Generating Station Dam and Dykes," *The Engineering Journal,* **43**:87 (October 1960); see also G. H. Johnston, "Dykes on Permafrost, Kelsey Generating Station, Manitoba," Canadian Geotechnical Journal, **6**:139, 1969.

23.15 J. H. Lander, F. G. Johnson, J. R. Crichton, and M. W. Baldwin, "Foyers Pumped Storage Project: Planning and Design," *Proceedings of the Institution of Civil Engineers,* vol. 64, pt. I, London 1978; pp. 103–117; and D. D. Land and D. C. Hitchings, "Construction," in the same vol., pp. 119–136.

23.16 "Electro-Osmosis Stabilizes Powerhouse Site," *Engineering News-Record,* **192**:22 (7 February 1974).

23.17 "Electro-Osmotic Process Stabilizes Power Plant Foundation," *Civil Engineering,* **23**:424 (1953).

23.18 "Site Dewatering Wells Stay Put to Control Uplift," *Engineering News-Record,* **152**:36 (1 April 1954).

23.19 "Anchor Cells Hold Down Foundation (and Cost)," *Engineering News-Record,* **145**:40 (10 August 1950).

23.20 V. A. Smoots and P. H. Benton, "Compacted Earth Fill for a Power-Plant Foundation," *Civil Engineering,* **31**:54 (August 1961).

23.21 A. B. Williamson and D. B. Patin, "Foundation Soil Preload Saves $164,000," *Civil Engineering,* **48**:61 (March 1978).

23.22 C. S. Berry and A. C. Dean, "The Constructional Works of the Battersea Power Station of the London Power Company, Ltd.," *Minutes of Proceedings of the Institution of Civil Engineers* **240**:37 (1937); see also F. H. Edmunds, "Some Gravel Filled Pipes in the London Clay at Battersea," *Summary of Progress for 1930,* pt. II, Geological Survey of Great Britain, 1931.

23.23 R. R. Farid and E. J. Davis, "Cofferdams—Fast Flood Protection," *Civil Engineering,* **48**:67 (September 1978).

23.24 H. T. Mead, "Nuclear Power Station is Built on Sand," *World Construction,* **15**:26 (June 1962).

23.25 J. E. Gilleland, "Siting of Brown's Ferry Nuclear Plant," *Proceedings of the American Society of Civil Engineers,* vol. 95, PO 2, Paper 6827, 1969, p. 195.

23.26 M. Kehnemuyl, "Foundations for Salem Nuclear Generating Station," *Civil Engineering,* **40**:52 (May 1970).

23.27 "Nuclear Facilities Siting," *Proceedings of the American Society of Civil Engineers,* vol. 105, EE 3, Paper no. 14615, 1979, p. 443.

Suggestions for Further Reading

The conferences and publications of the International Association for Bridge and Structural Engineering (with its office c/o École Polytechnique, Zürich, Switzerland) are a most useful complement to the regular meetings and publications of such bodies as the Structural Division of the American Society of Civil Engineers, for information on bridge design and construction.

The dangers associated with scouring out of riverbeds around bridge piers are so significant that attention is directed to the following three further references. "Scour Problems at Railway Bridges on the Thompson River, B.C.," a paper by C. R. Neill and L. R. Morris, will be found in the *Canadian Journal of Civil Engineering* (vol. 7, pp. 357–372, 1980), published by the National Research Council of Canada, Ottawa. In *Civil Engineering* (New York) vol. 33 for May 1963, two companion papers are to be found on pp. 46–49; the first (by L. Stabilini) is "Causes and Effects of Scour at Bridge Piers"; the second (by C. J. Posey) is "Protection of Threatened Piers." A much earlier paper, "Safe Foundation Depths for Bridges to Protect from Scour" by R. W. Stewart is in volume 9 of the same journal (1939), pp. 336–337.

chapter 24

BRIDGE FOUNDATIONS

24.1 Importance of Bridge Foundations / 24-2
24.2 Special Preliminary Work / 24-2
24.3 Design of Bridge Piers / 24-9
24.4 Design of Bridge Abutments / 24-15
24.5 Precautions Against Settlement / 24-20
24.6 Earthquakes and Bridge Design / 24-22
24.7 Scouring Around Bridge Piers / 24-23
24.8 Some Construction Requirements / 24-24
24.9 Cofferdam Construction / 24-27
24.10 Some Unusual Cases / 24-31
24.11 Inspection and Maintenance / 24-36
24.12 Conclusion / 24-41
24.13 References / 24-42

The founding of bridge piers and abutments is at once so special and so important a part of civil engineering work that of itself it calls for detailed consideration. Bridge construction is, moreover, such a widespread operation, falling at some time to the lot of most practicing civil engineers, that it merits more detailed study than some other specialized types of work. Memories (not perhaps too pleasant) of schoolbook illustrations of the piled foundation of a Roman bridge across the Rhine must not be allowed to deflect attention from a subject even the history of which is full of interest. Of the many famous early bridges, London Bridge, over the Thames in England, is one of the best known. According to a third-century Roman writer, there was a bridge across the river just above its mouth as early as A.D. 43, and records indicate the continuous existence thereafter of a river crossing of some type at this location. A reference of particular interest is that which tells of King Olaf (St. Olaf) coming in A.D. 1014 to aid King Ethelred of England against the Danes who held London. His fleet "rowed quite up under the bridge and then rowed off with all the ships as hard as they could down stream [having secured ropes to the piles supporting the bridge]. The piles were then shaken at the bottom and were loosened under the bridge" which gave way, throwing all the defenders ranged upon it into the river.[24.1]

Although such a barbaric use of bridge foundations fortunately has few modern applications, the further history of London Bridge is of special significance in this study. The

24-2   CIVIL ENGINEERING WORKS

structure known so well through illustrations in history books appears to have been completed in the early part of the thirteenth century. Fynes Morrison, writing in 1671, states that it was founded on "pakkes of wool most durable against the force of water," but this is probably a garbled reference to the tax on wool which enabled the king to pay his share of the cost. Since the waterway of the river was reduced by this multiarched structure to a width of 58.5 m (195 ft), such swift rapids developed between the arches that many persons lost their lives in passing through; the old saying is that "London Bridge was made for wise men to go over and fools to go under." An act of Parliament passed in 1756 ordered all the buildings on the bridge to be removed and the two central arches rebuilt into one arch; this work inevitably diverted the main flow through the opening and set up serious scouring, which eventually led to the necessity of demolishing the bridge altogether. A modern structure now occupies a site parallel to and adjacent to this famous crossing.[24.2]

This is but one of the ancient bridges the piers of which have given rise to trouble. Records are extremely scarce, unfortunately, but it can safely be said that scouring out of the foundation beds adjacent to bridge piers has been, in the past, a major cause of trouble. The piers of ancient bridges rarely failed because of excessive loading on the foundation beds, if only because of the limitation of span length imposed by the structural materials available. The two defects mentioned can be regarded as the two main possibilities of failure to be investigated in the design of bridge piers. Both are essentially geological in character.

24.1   IMPORTANCE OF BRIDGE FOUNDATIONS

As a necessary preliminary, the fundamental importance of geology in the founding of bridges must be stressed. However scientifically a bridge pier may be designed, the whole weight of the bridge itself and of the loads that it supports must be carried ultimately by the underlying foundation bed. The piers and abutments of a bridge are relatively "uninteresting" to the keen structural engineer. Herein should lie, however, more than passing interest, for all too often the design of the foundations is left to the structural engineer responsible for the design of the superstructure. Many structural engineers appreciate the difference between the two types of design work and their relative importance. Some, however, may not—perhaps naturally, since the careful consideration of the materials available as foundation beds and the forecasting of the forces that may in time affect them are different concerns from the determinate mathematical calculations relating to the arrangement of steel, reinforced concrete, or timber to be used for the superstructure.

It may be that in some cases an assumption regarding the relatively small cost of foundations compared with the total cost of a bridge may militate against giving sufficient attention to their design. Actual cost records, however, show that the cost of foundations (piers and abutments) often almost equals the cost of superstructure, even on large bridges, as the typical figures presented in Table 24.1 clearly demonstrate. Whatever may have been the cause, it cannot be denied that the importance of bridge-foundation design has not always been fully recognized, thus betraying on occasion the basic assumption of the superstructure designer—that the pier- and abutment-bearing surfaces will provide fixed and solid pedestals on which can be supported the structure the engineer conceives and that there need be no fear of any serious movement in the future.

24.2   SPECIAL PRELIMINARY WORK

The first considerations in bridge location are generally those of convenience and economy; foundation-bed conditions usually take a subsidiary place, for the prime require-

TABLE 24.1 Some Typical Costs of Bridge Projects, Showing The Relative Costs of Superstructures as Compared with Substructures

Bridge	Type	Cost of Superstructure (a)	Cost of Substructure (b)	Ratio a : b	Reference
Cherry St. Bridge, Toledo, Ohio	7 concrete arches + bascule	$545,482	$531,103	1.02 : 1	Am. Soc. Civil Eng. Trans., **80**:789 (1916)
Hell Gate Bridge, New York	Steel arch, 977.5-ft span	$2,000,000	$1,700,000	1.18 : 1	Am. Soc. Civil Eng. Trans., **82**:882 (1918)
Sydney Bridge, Australia	Steel arch, 1,650-ft span	£3,202,000	£1,046,000	3.06 : 1	Inst. Civil Eng. Min. Proc., **238**:193 (1935)
Iskandar Bridge, Fed. Malay States	7 steel arches, 160- to 148-ft span	£51,217	£76,883	0.67 : 1	Inst. Civil Eng. Min. Proc., **240**:598 (1937)
Tees (Newport) Bridge, Middlesbrough, England	Steel lift bridge, 259-ft 6 in span	£201,302 (excluding approaches)	£65,688 (2 piers)	3.06 : 1	Inst. Civil Eng. Min. Proc., **240**:598 (1937)
Chelsea Bridge, London	Self-anchored suspension bridge	£149,000	£75,000	1.98 : 1	Jour. Inst. Civil Eng., **7**:420 (1938)
Thousand Islands Bridge, Canada-United States	Group of structures, including 2 suspension spans, steel arch, etc.	An extreme case, in view of unusually favorable foundation-bed conditions:→		5 : 1	Civil Eng., **8**:408 (1938)
Bridge across the Tigris, Amara, Iraq	Steel girder and cantilevers	$303,454	$232,879	1.3 : 1	Proc. Inst. Civil Eng., **16**:46 (1960)
Hindiya Bridge, India	10 46-ft reinforced-concrete spans and 1 102-ft steel arch	$147,000	$173,000	0.85 : 1	Proc. Inst. Civil Eng., **8**:15 (1957)
Clifton Bridge, Nottingham, England	See p. 24-20	£224,175	£87,652	2.56 : 1	Proc. Inst. Civil Eng., **14**:465 (1957)

ment of a transportation route is that it connect its terminal points by the shortest convenient route consistent with topographical configuration. When crossings of any natural defile have to be made, considerations of cost usually limit the choice of site to that calling for the shortest possible structure. The bridge engineer is therefore not usually given much choice of location; in consequence, often the foundation-bed conditions at a site so determined have to be accepted, and, if it is possible to do so, a suitable foundation structure has to be designed for those conditions.

This limitation of site necessitates the most complete information possible with regard to the geological conditions at the site. A still more potent reason for obtaining full geo-

24-4 CIVIL ENGINEERING WORKS

logical information is that once the construction of bridge piers is started, their respective locations cannot be changed except in most unusual circumstances. More than the usual degree of certainty must therefore be attached to the design and anticipated performance of bridge piers and abutments. There is yet a further reason for this special care in preliminary investigations. Bridges, as a rule, are constructed to cross river or other valleys—depressions below the normal level of the ground which by their existence suggest some departure from normal geological structure. In districts that have been subjected to glacial action, an older riverbed or other depression now completely hidden by subsequent glacial or river deposits may be found well below the existing riverbed. Such a condition, even if known in advance, can have a serious effect on design. If the existence of a buried valley is not discovered before construction begins, it can lead to untold difficulties.

Again, riverbeds will probably contain many types of deposits, including boulders; and if preliminary work is not carefully done and correlated with geological considerations, an extensive boulder deposit can easily be mistaken for solid rock. For example, a bridge near Cornwall, Ontario, failed in 1898, with the loss of 15 lives, because a pier was founded on a boulder occurring in a layer of "hardpan," which had not previously been explored by borings and which proved to be only about 60 cm (2 ft) thick. Scouring in the vicinity of one of the bridge piers disclosed the clay beneath; the pier eventually tipped over and dropped two of the bridge spans into the St. Lawrence, which it crossed (Fig. 24.1).[24.3] The wreckage of the original superstructure was salvaged in 1958 in connection with the building of the great St. Lawrence Seaway and Power Project. (It was the same glacial till which led to this bridge failure that caused such trouble in excavation work for the power project.) Another example is given by the construction of the Georges River Bridge, New South Wales, Australia. Georges River flows into Botany Bay, 19 km (12 mi) south of the city of Sydney. Work toward the construction of a toll highway bridge to replace existing

FIG. 24.1 Remains of the railway bridge across the St. Lawrence River near Cornwall-Massena, shortly after the central pier had failed on 6 September 1898, dropping two spans into the river with serious loss of life.

vehicular ferries was begun in 1923. Borings were put down at three possible sites for the bridge by an experienced foreman; the borings at the site finally selected showed solid rock at depths below bed level varying between 10.5 and 14.1 m (35 and 47 ft) at regular intervals across a river section about 450 m (1,500 ft) wide, the rock at the sides of which was known to dip steeply. On the basis of this information, a through-truss bridge of six main spans, supported on cylinder piers, was designed, and a lump-sum contract was awarded. During construction, it was found that rock actually existed at only two of the seven main piers; additional borings taken to depths up to 39 m (130 ft) failed to disclose any solid rock at all at the other pier sites, and what is even more strange, they disclosed no stratum harder than "indurated sand" (Fig. 24.2). Construction had to be stopped and designs changed; in consequence, the bridge took five years to build instead of two and cost 27.6 percent more than the contract price.

Discussion of the paper in which this work was reported to the Institution of Civil Engineers naturally emphasized the rigid necessity of having borings most carefully watched by a trained observer. The absence of geological references in both paper and discussion suggests that neglect of geological features may have been a contributory cause of the trouble that was experienced.[24.4] Although this is an unusual and possibly exceptional example, the construction of the Georges River Bridge is a most telling reminder of the supreme importance of preliminary geological information in bridge design and of the vital necessity for professional supervision of test boring work.

Another reason for devoting unusual care to the study of the geological conditions at bridge sites in all cases of river crossings is the fact that so much of the ground surface involved is hidden below water. Dependence has thus to be placed on the results of the underwater borings, correlated with geological information secured on the adjacent shores. Where solid rock is encountered, this calls for no unusual attention, provided the exposed surfaces of the rock show no signs of disintegration or weathering; but if any part of the foundation bed consists either wholly or partially of clay, then it is desirable—in most cases imperative—to obtain samples of the clay in as undisturbed a condition as possible. Suitable test-boring devices have been developed with which undisturbed samples of clay and other unconsolidated materials can be obtained, even through great depths of water. One such piece of equipment was developed especially for investigating underwater soil conditions at a bridge site—that of the San Francisco-Oakland Bay Bridge in California. This project is one of the largest yet undertaken; it will long remain one of the outstanding bridges of the world. Since the design and construction of all piers were determined only after the most thorough preliminary investigations, the subsurface conditions will be briefly described.

San Francisco is separated from the district of Oakland by the entrance to San Francisco Harbor. Yerba Buena Island stands in the center of the harbor and divides it into the East Bay and the West Bay. For many years, transportation across the harbor was maintained by ferryboats, but a long-projected bridge scheme finally reached the construction stage in 1933, and the great bridge was officially opened on 12 November 1936. Early in the planning of the scheme, the engineers in charge decided upon a program of borings and soil tests to enable them (1) to determine the nature of the subsurface materials, (2) to ascertain the most desirable location for the center line of the bridge, (3) to determine the best location for individual piers, and (4) to select a logical basis for the design of the piers. Preliminary jet borings enabled the engineers to prepare a contour map of the underlying solid-rock surface of the harbor, including the West Bay bridge area. With the aid of additional wash-pipe borings and diamond-drill core borings into the rock, they prepared a final design for the West Bay crossing. Piers were located and designed; all were founded on solid rock and constructed by means of caissons, the behavior of which could be accurately foretold.

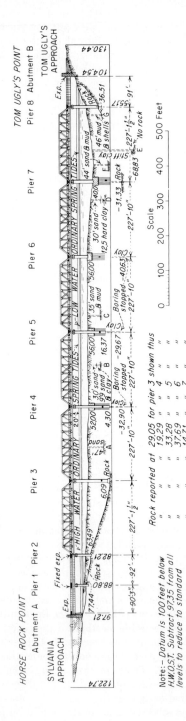

FIG. 24.2 Cross section along the Georges River Bridge, New South Wales, Australia, showing the results of test borings and the foundation strata as revealed during construction.

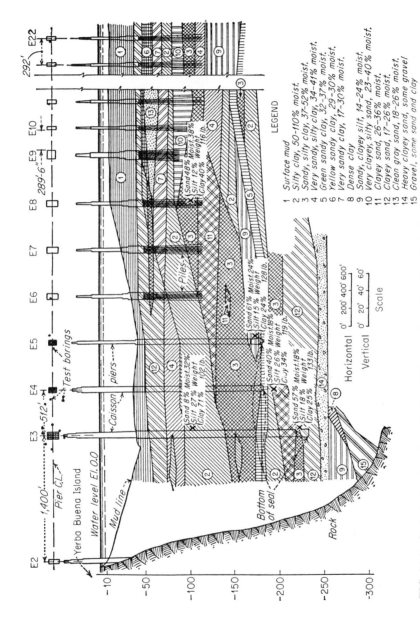

FIG. 24.3 Geological section along the line of the east channel section of the San Francisco-Oakland Bay Bridge, showing results of subsurface investigation and the positions of the piers.

24-8 CIVIL ENGINEERING WORKS

The East Bay crossing presented quite distinct problems, since rock was not found by borings at practicable depths, and it became necessary to determine the nature of the overlying unconsolidated material. Cores were obtained and hermetically sealed without removal from the sampling tube on reaching the deck of the drill barge; they were tested later at the University of California. When the containers were opened, perfect cores were generally found, although in some cases a slight swelling was noticed, possibly due to the change in internal pressure in the sample as it came up to the surface. Material was obtained in this way from depths of 82 m (273 ft) below water level. The section on the final center line of the bridge as finally disclosed by the subsequent tests is shown in Fig. 24.3, from which the completeness of the subsurface survey will be evident; the use of such information in the design of pier foundations will be to some degree self-evident.[24.5]

The example just described is in some ways unique, although many features of the preliminary investigations have general application. Adequate test borings, not only along the line of the selected bridge site but on either side of it; careful study of core specimens so obtained; and the correlation of this information with the geological structure of the adjoining dry ground should present a reasonably accurate structural picture of the foundation beds which will enable the designing engineer accurately to locate and to design

FIG. 24.4 Central anchorage pier of the San Francisco-Oakland Bay Bridge, anchoring two suspension spans; a pictorial representation showing subsurface conditions and construction details.

the bridge abutments and piers. (Fig. 24.4.) Finally, the necessity of taking all test borings deep enough below the surface of solid material (and especially of unconsolidated material) must be stressed. Loadings from bridge supports are always relatively concentrated and often inclined to the vertical. It is therefore doubly necessary to be sure that no underlying stratum may fail to support the loads transmitted to it, even indirectly, by the strata above.

An interesting example of trouble due to this cause was furnished by the failure of a highway bridge over the La Salle River at St. Norbert, Manitoba. The bridge was a single reinforced-concrete arch, with a clear span of 30 m (100 ft); the spandrels were earth-filled. The roadway was about 9 m (30 ft) above the bottom of the river, and the height of the fill placed in each approach was about 6 m (20 ft). The bridge abutments were founded on piles driven into the stiff blue clay exposed at the site and thought to overlie limestone bedrock, as the results of preliminary auger borings and the record of an adjacent well seemed to indicate. Failure occurred by excessive settlement; the north abutment dropped 1.2 m (4 ft), and bearing piles were bent and broken. Subsequent investigations disclosed the existence of a stratum of "slippery white mud" (actually bentonite) about 7.5 m (25 ft) below the original surface; failure of this material to carry the superimposed load resulted in the bridge collapse. Local soils are sediments from an ancient glacial lake and usually overlie compacted glacial till, under which is limestone carrying subartesian water. The existence of this water complicated the underpinning of the bridge foundation, but the work was successfully completed, and the bridge was restored to use. The existence of the bentonite was previously unknown in the vicinity; it illustrates the uncertainty of glacial deposition. The occurrence has a special interest for engineers: although the bearing piles were driven into the compacted till ("hardpan"), settlement of the abutment occurred as the result of failure of soft material underlying this.[24.6]

24.3 DESIGN OF BRIDGE PIERS

Although this chapter is not meant to be an introduction to bridge-pier design, some attention to design requirements is necessary in order to obtain an adequate appreciation of the relation of geological conditions to this design work. Generally speaking, there are four types of loading, one or more of which may have to be provided for in design: (1) vertical loads, possibly of varying intensity, from truss or girder spans or suspension-bridge towers; (2) inclined loads, again of varying intensity and possibly varying direction, from arched spans; (3) inclined tensions, from the cables of suspension bridges; and (4) horizontal thrusts due to the pressure of ice or possible debris, the flow of water impinging on the piers, and the wind acting on the bridge superstructure and piers. In earthquake regions, allowance must also be made for seismic forces that may act upon the piers. Combinations of these several loads will give rise to certain maximum and minimum unit pressures to be taken on foundation beds; from considerations of these results and of the nature of the strata to be encountered, the type of foundation can be determined.

The process of estimating the load that the foundation strata at the site of a bridge pier will safely support is generally similar to the same operation for ordinary foundation work. The usual precautions must be observed with regard to strata below the surface that might fail under load. There are two unusual features to be noted, and both are reductions which may be applied, under certain circumstances, to the calculated net load on the base area. The first is the allowance that can be made for the natural material excavated and for the displacement by the pier of water; and the second is the reduction that can be allowed for the skin friction on the sides of the pier because of the usually large surface area exposed as compared with the base area. These two factors are obvi-

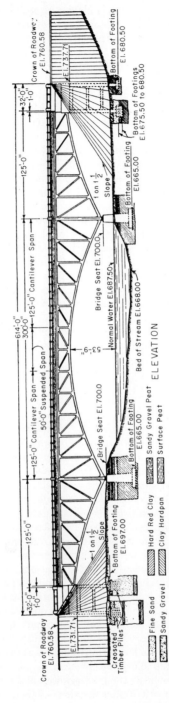

FIG. 24.5 Elevation of the Mortimer E. Cooley Bridge over the Manistee River, Michigan, showing also foundation strata and the special abutment structures.

ously dependent on the nature of the strata penetrated. The estimation of the first is straightforward, but that of the second is generally a matter of past experience or of experiment during pier sinking, coupled with use of the results of careful laboratory tests upon soil samples from the bridge site.

Throughout all determinations of safe foundation loads, the nature of the local strata does play an important part; it may even dictate the use of hollow piers to reduce unit loads or of such unusual structures as the open reinforced-concrete framework abutment supports adopted for the Mortimer E. Cooley Bridge across the Manistee River in Michigan. This singularly beautiful bridge, consisting of two 37.5-m (125-ft) deck truss steel cantilever arms supporting a 15-m (50-ft) suspended span and balanced by two 37.5-m (125-ft) anchor arms, has its deck level about 18 m (60 ft) above the level of the ground on either side of the river; the ground consists of varying strata of unconsolidated materials which were accurately explored. In order to keep foundation loading on this material to a minimum, the open framework design shown in Fig. 24.5 was adopted and successfully used.[24.7]

When preliminary investigations have shown that the foundation-bed material will have poor bearing capacity, consideration may be given to the use of artificial methods of consolidating such material to improve its bearing capacity. That this is no new expedient is shown by some of the older records of bridge building, e.g., the account given by Leland (antiquary to King Henry VIII), who wrote in 1538 concerning the Wade Bridge in England, that "the foundations of certain of th'arches was first sette on so quick sandy ground that Lovebone (Vicar of Wadebridge) almost despaired to performe the bridge ontyl such tyme as he layed pakkes of wolle for foundation." Although this use of wool has been disputed, the record is interesting as demonstrating that some artificial means was used to improve bearing capacity. Modern methods are described elsewhere in this book. Grouting is naturally the most common method, but chemical consolidation has been suggested for some bridge works. Another approach to the problem is to consider leaving the steel piling of the pier cofferdam in place in order to confine the foundation-bed material and thus prevent lateral displacement; in this way bearing capacity will be increased to some extent.

The foundations for the Tappan Zee Bridge that carries the New York Thruway across the Hudson River for a distance of 4.5 km (2.8 mi) between Nyack and Tarrytown provide an even more unusual approach to the problem of minimizing loads on strata that are deficient in bearing capacity. Here was a case in which the site of the bridge, selected on the basis of most careful location studies, had to be accepted even though the bedrock that is so dominant a feature of the local landscape drops off under the bridge to depths as great as 420 m (1,400 ft) below water level; rock elevation throughout the whole length of the bridge is too low to be reached by normal bearing piles. The approach spans are therefore carried on friction-bearing piles, driven into the silt and sand and gravel that form the riverbed. For the large piers carrying the 363.3-m (1,212-ft) cantilever main-channel span, however, and for the immediately flanking piers, other methods had to be considered (Fig. 24.6).

The design finally adopted used buoyant, reinforced-concrete box foundations to take about two-thirds of the dead load of the superstructure; the remaining part of the dead load, and the live load, are taken by 75-cm (30-in) concrete-filled pipe piles for the four main piers and by 35-cm (14-in) steel H piles for the four other buoyant boxes; in each case piles and boxes are ingeniously connected together. The steel piles had to be driven to depths up to 52.5 m (175 ft), but the concrete pipe piles went as deep as 102.0-m (340 ft) below water level, being driven through clay and then gravelly clay after the sand and gravel had been penetrated. The hollow piles were mucked out to full depth by waterjet and airlift techniques and then concreted by a special intrusion method in which grout

FIG. 24.6 The Tappan Zee Bridge over the Hudson River, New York, looking east, showing main spans which are supported by the special piers described in the text.

was forced into preplaced aggregate. The grouting technique also had the effect of consolidating the seamy gneiss and decomposed sandstone, which constitute the underlying bedrock, so as to give the necessary bearing capacity. Boulders added to the complications, but all foundations were successfully completed as planned and on schedule. Details of the piers are admittedly an engineering matter, but it was the geology of the site that dictated such a bold design.[24.8]

Among the special problems to be faced in bridge-pier design in the solution of which geological information can usefully be applied is that of predicting possible settlement of piers when loaded; unequal settlement can have a serious effect on certain types of bridge structures, notably continuous-truss spans, vertical-lift bridges, and fixed arches. An accurate knowledge of foundation-bed conditions and of the probable behavior under load of the material in these strata will usually permit accurate predictions of settlement. What happens when uneven settlement does take place is well illustrated by the failure of piers 4 and 8 of Waterloo Bridge, London; the whole bridge had to be taken down and a new structure erected. Described by Canova as "the noblest bridge in the world worth a visit from the remotest corner of the earth," Waterloo Bridge was constructed from 1811 to 1817 to the designs of John Rennie, who took especial care to protect all his pier foundations—timber rafts on timber piles bearing on gravel—against scour. Progressive settlement of the two piers mentioned became serious in 1923; the total settlement of pier 4 exceeded 75 cm (2½ ft) and naturally caused an arching action between piers 3 and 5.[24.9]

This case is cited mainly because of its general interest; as an example it can hardly be called typical, since the conditions encountered were relatively unusual. In a general way, settlement may occur from one or another of the following causes: (1) displacement of part of the bed by scour; (2) lateral displacement of the foundation bed due to lack of restraint; (3) consolidation of the underlying material; or (4) failure of an underlying stratum. Only the settlement caused by condition 3 can be called controlled settlement; the other three types are of a nature that may cause serious trouble to the structure. All types can be foretold, in the majority of cases, provided adequate preliminary information is obtained. Dangerous possibilities can therefore be avoided, and the controlled settlement accurately predicted, as in the case of the San Francisco-Oakland Bay Bridge foundations described above.

Provision against unequal settlement of piers in the case of relatively small span bridges has assumed considerable importance in recent years owing to the development of the rigid-frame type of structure, an essential feature of which is that the abutments are unyielding; this means that any settlement of piers must be almost uniform. For rigid-frame structures founded on clay, some interesting expedients have been adopted. The difficulties of clay as a foundation bed will later be fully stressed. One such expedient (employed at a Canadian National Railways bridge at Vaudreuil in Quebec) is to isolate the bridge foundation from the bearing piles and to transmit the load through a specially tamped layer of crushed rock.[24.10] This is an interesting return to an old practice (although for a new reason). The foundations of the present London Bridge over the Thames in England bear on clay through a similar layer of crushed rock and timber piles; these foundations were installed almost a century ago.

Stresses set up in foundation beds during the bridge construction, especially those which may be set up during unbalanced loading caused by irregular construction scheduling, must also be carefully considered in design. During the construction of the Broadway Bridge, Saskatoon, Saskatchewan, in 1932, a five-span, reinforced-concrete arch structure, the concreting of the arches proceeded in varying stages; as a result the piers tilted when carrying the dead load of only one adjacent arch rib (even though this was supported on centering). Careful records were taken, and a maximum deflection of 15 mm (0.6 in) was recorded (Fig. 24.7). In this instance, the result was anticipated.[24.11]

The inclined tensions, mentioned above as the third type of loading possible, are generally transmitted to anchorages in solid rock; in this case design resolves itself into providing for shearing resistance in the rock, which, together with allowance for the dead weight of the anchorage, will be sufficient to balance the tensile forces in the bridge cables. Sometimes the inclined tension in bridge cables has to be taken up wholly by concrete piers. The Ile d'Orléans (suspension) Bridge, in Quebec, is one such example. The sus-

FIG. 24.7 The Broadway Bridge over the South Saskatchewan River at Saskatoon, Saskatchewan.

24-14   CIVIL ENGINEERING WORKS

pension span of this bridge is flanked by long approaches from the shores; the main cables are secured in anchor piers, one of which is founded on rock, but the other on sand. Naturally, the stability calculations for these piers had to take into consideration all stabilizing factors, while keeping the unit toe and heel pressures within the limit of the foundation-bed material. Friction between concrete and rock and concrete and sand, therefore, had to be considered as well as the shearing value of both strata. As a result of these studies, inclined H-beam piles were driven into the sand underlying one of the anchor piers and were left so as to project into the concrete of the finished pier in order to give the necessary increase in stability.[24.12]

Another unusual example is provided by the double suspension spans of the West Bay crossing of the San Francisco-Oakland Bay Bridge; the two 663-m (2,210-ft) suspension spans join and are anchored into a central anchorage pier 150 m (500 ft) high above rock level. The engineering design of this structure is of unusual interest; its peculiar relevance to this discussion is that, under one of the most severe combinations of loading considered, a unit pressure on the rock of 865,000 kg/m^2 (79 tons/ft^2) may occur, a figure that could be contemplated with safety only after completion of the very thorough preliminary studies made for this bridge project and already described (Fig. 24.4).

As a final example, there may be mentioned the building of the Burford Bridge as part of a bypass road across the river Mole in Surrey, England, which is one of the most unusual contacts of special geological features with the design of bridges known to the writers. The bridge is a single, reinforced-concrete arch span of 24 m (80 ft), 30 m (100 ft) wide between parapets, with specially selected brick facing. Many who know the Mole's lovely valley, located some 40 km (25 mi) to the south of London, will know also that it is as geologically interesting as it is scenically beautiful. The river flows over the local chalk formation which has been so eroded in places by the action of the dissolved carbon dioxide in the river water as to have underground cavities large enough to receive the whole normal flow of the stream. Trial borings were put down at the projected site of the new bridge and carefully checked to see if any such "swallow holes" in the chalk were revealed. Two soft spots were located, and these proved to be solution channels of this kind, having almost vertical sides and filled up with alluvial matter. It was at first intended to secure adequate bearing for the road and abutments over the holes by driving reinforced-concrete bearing piles into the material found in the holes, but further studies led to a more

FIG. 24.8 Foundations for the Burford Bridge on the Mickleham bypass road, Surrey, England, reinforced-concrete domes covering the "swallow holes."

unusual solution which could be adopted with more certainty. Concrete domes were constructed over each of the holes, domes founded on circular ledges cut in the chalk around the tops of the holes; the largest dome is 17.4 m (58 ft) in diameter with a rise of 2.4 m (8 ft). The holes were filled up to the undersides of the domes, and the filling was then covered with waterproof paper and used as the lower form for concreting the domes (Fig. 24.8). Each dome was furnished with an access shaft connecting to a manhole at road level by means of which the engineers may inspect the swallow holes from time to time to see that no dangerous undercutting or further erosion of the chalk is taking place.[24.13]

24.4 DESIGN OF BRIDGE ABUTMENTS

In addition to having to support, at least partially, loads that the piers of a bridge have to carry, the abutments of a bridge may and often do have to resist another load, that of earth pressure against the face and wing walls of the abutment structure. Design may therefore be a complicated matter, combining the difficulties of both bridge-pier and retaining-wall construction. In the case of abutments retaining artificially formed fills such as those of approach embankments, difficulties of design and construction can generally be foreseen and so guarded against. Careful placing of fill material, working away from the abutments; adequate provision for drainage by the installation of cross drains along the lower part of the inner faces of abutment structures; connection of drains to suitable weep holes which cannot become plugged up; and supervision of all fill settlement are perhaps the leading matters for attention in this type of work.

Abutments located on sloping ground, as so many of them have to be, may present different problems; for in addition to having to retain the pressure of earth backing, often with considerable surcharge, they may be subjected to forces set up by the instability of the whole hillside slope; or, alternatively, their foundations may be made insecure by earth movement on slopes below. Again, the forces acting on them are far from symmetrical, so that balancing (especially during construction) is a matter which often calls for great ingenuity in design. A few examples will illustrate the type of problem presented and the dependence of abutment design on foundation-bed conditions.

Bridges for which solid rock provides abutment support will be familiar to almost all engineers, since this is an ideal condition and one which is essential for certain types of arch designs. A famous English bridge illustrates in an interesting way a slight variation from the ideal case; one of the abutments of the Grosvenor Bridge across the river Dee at Chester is founded partially on rock and partially on sand. The bridge, built in the year 1833, has a clear span of 60 m (200 ft) and a rise of 12 m (40 ft). Figure 24.9 illustrates the construction that had to be adopted at the north abutment, where the outcrop of solid rock suddenly terminated and was succeeded by a deep stratum of loose sand to which the arch thrust was transmitted be means of timber piles.[24.14]

It is naturally with arch bridges that abutments are such critical parts of bridge design. Figure 24.10 shows an unusual example, demonstrating the possibilities in design when rock conditions are excellent. The view shows one of the supporting structures for the Ivy Lea International Bridge that crosses the St. Lawrence River in the lovely Thousand Islands region between Ontario and New York State. Here gneiss bedrock was at the surface; the Thousand Islands are exposures of the Frontenac Axis, the projection of the Precambrian Shield extending into New York and forming the Adirondack Mountains. With such sound rock available the engineers were able to combine in this one structure a main abutment for a long steel arch, an anchorage for cables from one of the main suspension spans, and the necessary support for vertical columns under that part of the bridge joining the arch and the suspended span.

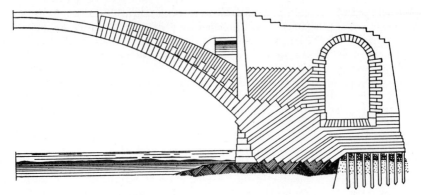

FIG. 24.9 Cross section through the abutment of the Grosvenor Bridge, Chester, England, showing variation of structure with changes in foundation strata.

FIG. 24.10 Combined pier, anchorage, and abutment structure for viaduct (overhead), suspension span, and steel arch—all component parts of the Thousand Islands International Bridge over the St. Lawrence River at Ivy Lea, Canada.

Figure 24.11 shows a graceful, steel-arch bridge which spans the Volta River in Ghana; it is a two-hinged structure with a clear span of 259.5 m (865 ft), making it the seventh longest such bridge in the world when built. Eight months were spent on preliminary geological studies; rocks at the site are metamorphosed Paleozoic sediments (indurated shales and quartzites). The quartzites form most of the bedrock in the area; they are highly folded. Even with this advance knowledge, however, more difficulty was experienced than had been anticipated when the rock was exposed for the abutment on the west bank. Interbedded shale and quartzites were found to a degree unexpected; the saw-tooth design prepared for the rock excavation had to be modified to take into account the actual dip and strike of the rock, the level of which was not quite as expected, since some of the few auger borings used had suggested bedrock where only boulders were found. With modification of the design, however, the west abutment was successfully completed, and this fine bridge is now serving to link two major parts of this virile country.[24.15]

When no rock is available, abutment designs must follow methods suggested in the case of pier design; subsoil conditions must be thoroughly investigated, and the foundation-structure design determined after a full consideration of test results. In many cases of abutment designs on unconsolidated material of low bearing capacity, bearing piles will have to be employed to provide for the anticipated load. Such piles, unless driven at a batter, will offer small resistance to lateral movement; for this reason, quite a few abutments have been moved inward from their original positions by the excess of earth pressure behind them over the stabilizing forces. Earth pressures, and, similarly, the frictional resistance between foundation and foundation bed must receive the most careful attention in all such designs. In some cases, the only solution possible is to brace one abutment against the other by means of underwater struts. One example of this type was the Summit Street bascule bridge across Swan Creek in Toledo, Ohio, in which a series of braced concrete struts supported on piles, each 1.2 m (4 ft) thick and 2.7 m (9 ft) wide, placed in 4.8 m (16 ft) of water, had to be installed after abutment movements had seriously

FIG. 24.11 The Volta River Bridge, Ghana, West Africa, looking toward the west abutment.

24-18 CIVIL ENGINEERING WORKS

affected the bridge operation.[24.16] Many similar cases could be cited; another well-known example is Bridge No. 16 over the Welland Canal in Canada.

One would not imagine that bridge abutments would ever move *inwards* against approach embankment fill, but this did happen with seven overpass bridges built at the extreme eastern end of the Macdonald-Cartier Freeway in Ontario (connecting Toronto with the Quebec highway system and so with Montreal). All were excellently designed and constructed bridges supported on end-bearing piles driven to rock, but within three years of completion slight movements of abutments away from the road had occurred. The explanation appears to be that the approach embankments were founded on sensitive Leda clay; settlements of this marine deposit were anticipated in design; settlement would be at its maximum under the full depth of fill but small at the edges of the fill slopes. The "dishing" configuration of the ground beneath the fill would initiate slight movements of the fill around toward the point of greatest settlement; the mechanics of the resulting "system" account for the corresponding movements of the abutments.[24.17]

Not only the design of abutments but even that of bridge superstructures may be affected by the geological structure of the abutment site. Nowhere else has this been better demonstrated, perhaps, than in the case of the Kohala Bridge carrying the Rawalpindi-Kashmir Road across the Jhelum River on the rugged mountainous boundary of the Punjab and Kashmir. Originally constructed as a three-span girder bridge having a 39-m (130-ft) center span and two 27-m (90-ft) approach spans, the unstable condition of the hillside at the Punjab abutment caused movement of the girders toward the Kashmir side; the Punjab approach span was seriously damaged in 1929 and finally wrecked in 1931 through serious landslides occasioned by heavy rains. W. T. Everall, then deputy chief engineer, North Western Railway of India, discovered a large fissure in the hillside about 60 m (200 ft) above the level of the bridge; this fissure occurred in a slope of loose soil and rock detritus and was indicative of future movement. In consequence, after the failure of the Punjab approach span, Mr. Everall had a special design for reconstruction accepted and carried into effect, cantilevering a 21-m (70-ft) span back from the first main river pier and connecting it by means of a light suspended span to the hillside road level. Suitable drainage work and stone pitching were also carried out. Although future earth movement has been well guarded against, the intent is that, should a landslide occur, it will

FIG. 24.12 The Kohala Bridge across the Jhelum River on the Rawalpindi-Kashmir road, Pakistan; the "floating span" may be seen on the right, leading from the abutment on the sloping river bank which caused so much trouble.

displace the suspended span which can safely ride up and over the end of the cantilever span. Thus little damage will be done to the main bridge structure, and the span can be easily replaced after movement has stopped. The accompanying photograph shows this feature clearly (Fig. 24.12). Engineering ingenuity has here accepted geological forebodings and anticipated trouble in a particularly imaginative manner.[24.18]

An unusual and extreme case of settlement may next be mentioned. The Pemberton Loop line of the old London, Midland and Scottish Railway (now British Rail) crosses the Leeds and Liverpool Canal by Bridge No. 14 near Wigan, England. The ground adjacent to the bridge site consists of alluvial and glacial deposits overlying the Middle Coal Measures of the carboniferous formation; the ground surface was originally a meter above water level in the canal. Owing to extraction of coal from the strata immediately under the bridge site and the adjacent area, continual subsidence took place, and the fields around gradually became an extensive lake. Correspondingly, the plate web girders of the bridge had to be jacked up and the abutments built up under them to compensate for the settlement. By the beginning of 1935, the headstones on which the girders had originally rested had sunk 3.9 m (13 ft) and were barely visible above the canal water level. Subsidence was continuing and its final extent could not be ascertained. The engineers therefore decided to construct timber trestles to replace the abutments (although they were to

FIG. 24.13 Abutment of Bridge No. 14 on the Pemberton Loop of British Railways near Wigan, England, showing timber supports above steel pile foundation; successive capstones in the masonry show the progressive settlement.

24-20 CIVIL ENGINEERING WORKS

be left in place); the trestles were supported on two series of special boxes constructed of steel sheet piling, all piles being driven as deeply as possible. Jacking up could then easily be done from the trestles, which were intended to serve until subsidence has ceased, when the brick abutments can be permanently rebuilt (Fig. 24.13). The building of the steel pile boxes was carried out without interference to traffic and constituted a most interesting piece of construction.[24.19]

24.5 PRECAUTIONS AGAINST SETTLEMENT

If settlements due to future mining operations are known to be a possibility at a bridge site, superstructure designs can be prepared accordingly. A notable example of this type of "controlled" design is that of the Clifton Bridge at Nottingham, England, across the river Trent. Coal mining is an important local activity; preliminary studies showed that some valuable coal seams, as yet unworked, ran directly under the bridge location. The structure was therefore designed to take account of possible settlement, should mining take place. The resulting structure is a prestressed-concrete main span of 52.5 m (275 ft), with two end spans of 37.5 m (125 ft) and three 27-m (90-ft) end viaducts, all built on a skew of 24°. Structural engineers will appreciate the complication that this feature introduced to the design; the aesthetic appeal of the resulting 240-m (800-ft) river crossing can, however, be generally appreciated. The main span had to be designed as two cantilevers with a central suspended span, all statically determinate because of the anticipated settlement; arrangements were included in the design for the spans to be jacked up off the tops of the main piers whenever this became necessary. The piers are founded on the Keuper marl, which extends to depths of over 150 m (500 ft) at the bridge site. The marl was badly weathered to a depth of 8 m (27 ft), but below this level it was hard and durable, containing bands of siltstone and very stiff clay; the formation is one of the many soil

FIG. 24.14 Clifton Bridge, over the river Trent, Nottingham, England, during construction.

horizons in Britain that are not of geologically recent formation. Consolidation tests showed that the marl, so well known to geologists, was really a highly overconsolidated clay, so that it would swell if exposed and would deteriorate if unprotected from water (Fig. 24.14).

It remains to be added only that, just as the complex design for the bridge was complete, it was found that a geological fault between the colliery and the bridge site made profitable working of the coal seams improbable. The local city council therefore bought out the rights for the working of the coal seams beneath the bridge for a nominal sum and thus eliminated all possibility of settlement from this cause. It was too late to change the design, so the bridge was built as planned and as here described, except for minor alterations such as the permanent covering in of the jacking pockets left in the piers.[24.20]

Even gold mining has to be considered in relation to ground subsidence in some areas. Johannesburg has already been mentioned in connection with the recent change to deep mining for its famous gold ore. In earlier days, the ore was found almost at the surface and within the limits of what is now the great modern city. The old workings have long since been unused, but they are still there. They necessitate severe restriction of modern building in a great wide strip of land (the old reef) running through the city from east to west immediately to the south of the main city center. Ground movements still take place, and so this area has limited capability of development. It was, however, ideally located for road construction when a great system of freeways was planned in the early sixties. Land for the east-west freeway, most conveniently located (as can be imagined) to serve

FIG. 24.15 Construction of a major bypass highway through the city of Johannesburg, South Africa, showing the fixed viaduct section and the start of construction of adjustable piers to allow for possible future settlements.

24-22 CIVIL ENGINEERING WORKS

central Johannesburg, was acquired at relatively low cost. From the planning point of view this was all to the good, but the design and construction of this modern highway presented most unusual problems. Ground movements over the old workings had to be included as definite possibilities, and yet a wide, overhead, modern highway had to be constructed with certainty of performance. Adding to the complications was the fact that some gold still remains in some of the old shallow workings. Although then uneconomic to remove, it had to be left available for removal in case the price of gold should go up. The engineers responsible were able to lay down limits of movements to be allowed for in design, given strict control over any future mining. These restrictions are considerable: a 5-cm (2-in) change in grade in 30 m (100 ft), longitudinal movements of up to 95 mm (about 3 ft) in 30 m (100 ft) in any individual span, and a maximum cumulative movement of 25 cm (10 in) in a group of spans. The engineering solution is one of the greatest interest—the supporting piers, which weigh as much as 600 tonnes, are so mounted on bearings that, if and when necessary, they can be jacked either vertically or horizontally to compensate for any ground movement that may displace them (Fig. 24.15). Few of the motorists who now use this fine highway can ever realize upon what an unusual structure they are riding.[24.21]

24.6 EARTHQUAKES AND BRIDGE DESIGN

In regions susceptible to earthquake shocks, seismic forces have to be carefully considered in bridge design; when bridge piers are high, there must accordingly be full allowance made in their design for seismic forces. Among the highest bridge piers yet built were those for the Pitt River Bridge that carries the Southern Pacific Railroad and U.S. Highway 99 over a part of the reservoir formed by the Shasta Dam. The piers have a maximum height of 108 m (360 ft), of which over 90 m (300 ft) is now submerged beneath the waters of the Shasta Reservoir. Because of the location, the possibility of earthquakes had to be considered. A detailed study of this unusual case was therefore carried out by the U.S. Bureau of Reclamation, as a result of which it was shown that, under earthquake shocks and when submerged, these piers would no longer act as elastic structures fixed at their bases but would act as rigid structures, rotating at their bases. This conclusion led to design criteria which can be summarized in the requirement to design the piers for all ordinary loads without considering earthquake effects, and then to design them for anticipated earthquake forces, so that the resultant force would be just at the edge of the pier base.[24.22]

The Pajaro River Bridge, which is 148 km (92 mi) south from San Francisco, and carries the Southern Pacific coast line to Los Angeles, once provided vivid evidence of what an earthquake shock can do to a bridge. A new structure at the same location now demonstrates the best in modern bridge design in taking into account the fact that the bridge spans not only the Pajaro River but also the well-known San Andreas fault, which here follows the river channel for some distance. At the time of the earthquake of 18 April 1906, the bridge consisted of five simple deck-girder spans with a total length of 138 m (460 ft). All five spans were moved by the earthquake vertically and all but one horizontally. In some cases the motion was enough to throw the spans off their supports; one span was left hanging precariously over the edge of one of its piers.

When the bridge was reconstructed in 1944, advantage was taken of the opportunity to redesign it as a three-span, continuous-deck girder which, with the addition of one 25-m (86-ft) side span, gave a total length of 135 m (450 ft). In this case, the continuity over the central supports in the riverbed was deemed to be advantageous in case of any serious movement of the piers. Clearance has been left at the free end of the continuous girder for a reasonable amount of longitudinal movement. Instead of the usual type of support,

FIG. 24.16 The Southern Pacific Coast Daylight Express crossing the Pajaro River Bridge, showing in the lower center of the photo the special bearing support at the top of the concrete pier, mentioned in the text.

special rockers were designed, permitting more movement than usual. Concrete projections above the rocker shoes will uphold the girder, should it be moved off its supports (Fig. 24.16). A heavy pendulum has been mounted on a special fitting on one of the piers, so arranged that movement of the center of the bridge in excess of 19 mm (¾ in) will cause movement of the pendulum to open a circuit, de-energize a relay, and cause the automatic block signals controlling access to the bridge to be set at danger. Abutments and the ends of the girders are similarly connected; they will set the signals at danger for any movement in excess of 25 mm (1 in). A recording device attached to the pendulum has recorded movements due to seismic shocks up to 16 mm (⅝ in), but as yet no tremor has been so great as to actuate the tripping of the approach signals.[24.23]

Seismic forces were naturally considered in the design of the majestic Golden Gate suspension bridge (which must be mentioned, if only briefly, even in such a cursory review as this). The northern pier and anchorage are founded on diabase and basalt and on Marin sandstone, respectively, the southern pier and anchorage on serpentine. In the design of the main piers, a horizontal force of 0.10 g was allowed for, and for the anchorages and superstructure, there was a corresponding allowance of 0.075 g.[24.24] With its main span of 1,260 m (4,200 ft), this great structure still excites all who see it for the first time.

24.7 SCOURING AROUND BRIDGE PIERS

The consequences of bridge construction across a waterway involve three main effects, which may generally be foreseen: (1) the construction of piers and abutments will generally decrease the effective cross-sectional area of the stream and thus inevitably increase the velocity and raise the water level (usually very slightly) upstream from the bridge; (2) the existence of piers as obstructions to the streamflow will set up eddies around the piers and may possibly institute crosscurrents in the stream, tending to change it from its normal course below the bridge site; and (3) the combined effect of increased average velocity and eddies may disturb the equilibrium of the bed material between piers and so lead to scouring. All these results represent interference with natural conditions and therefore call for the application of geological information in the engineering solution of the many problems they bring up. An example of the serious effect of a change in permanent water level, almost classical in nature, is the scouring around the pier foundations of the old Westminster and Vauxhall bridges over the Thames, London, in the early years of the nineteenth century following the removal of old London Bridge and the consequent lowering of the water level in the stretch of the river immediately upstream. This opening of the river channel restored normal flow conditions at the site of the bridges mentioned; since the foundations had not been designed for these conditions, both structures in course of time had to be completely rebuilt.[24.25]

Geological factors may affect runoff calculations appreciably, and it is from such calculations that estimates can be prepared showing what hydraulic conditions a bridge will have to withstand. On many tropical rivers with low-lying banks, bridges have to be designed on the assumption that they will be completely submerged in flood periods; naturally, therefore, the geological stability of the foundation bed for piers, abutments, and approaches must be assured before a submergible design can be entertained. An interesting and unusual example has been described; this is the bridge across the Nerbudda River near Jubbulpore in India. It is 366.6 m (1,222 ft) long, consisting of six 29.4-m (98-ft) and eight 13.8 m (46-ft) spans, all reinforced-concrete arches; it is founded throughout on basalt. At flood periods, the bridge is completely submerged.[24.26]

River-training works frequently have to be carried out in connection with bridge construction projects on rivers with relatively unstable beds. If the river approaches to such bridges have not been suitably "trained," the volumes of water passing between the several piers may not be the same; the flow between pairs of piers has actually been reversed in some cases. It is not only on the great rivers flowing through alluvial plains that river-training and associated problems occur. Scouring of a riverbed may happen on even the smallest stream if its normal state is interfered with, as by the construction of bridge piers in midstream. It is safe to say that scouring of bed materials from around the foundation structures supporting bridge spans has been responsible for more bridge failures in the past than has any other cause.

In olden days, there was more tendency to set up scouring action under bridge spans because of the relatively small spans and consequent relatively large cross-sectional pier area as compared with water area; a contributory factor was that construction methods then possible did not permit foundations other than timber piling much below low-water level. Construction methods today suffer from no such limitations, and the general use of longer spans does not cause serious constriction of waterway areas. The problem still remains, however, and even in the records of modern bridge building, there are cases of serious scouring of foundation-bed material; two of these are mentioned toward the end of section 24.11. Despite the relative simplicity of the problem, scouring of riverbeds around bridge piers still takes far too great a toll of property and, occasionally, of lives. It has been estimated that in one ten-year period $15 million worth of damage was done to highway bridges in the United States by this one cause.[24.27] In a British study of bridge failures, 66 of the 143 cases reported upon were due to scour around the piers.[24.28]

24.8 SOME CONSTRUCTION REQUIREMENTS

The design of bridge-foundation structures can rarely be considered without specific relation to the problems of constructing the piers so designed. Limitations in the restriction of a watercourse during construction, requirements of navigation, requirements of traffic across the bridge in the case of reconstruction work, depths of water and tidal range, and the depth below water to foundation-bed level are all features that must be considered in relation not only to structural stability but also to construction methods. Here, too, the geological nature of the underlying strata must be correlated with other design factors.

Construction of piers will generally be carried out by one of three main methods: the use of open cofferdams (working either in the dry or in water), the use of open dredging caissons, or the use of compressed-air caissons. There are, of course, other special methods available. The general arrangement of foundation strata will be a potent factor in determining which of these construction methods shall be followed. In the case of cofferdam work, for example, subsurface conditions will indicate fairly accurately the length of the piles required and will at least suggest what resistance to penetration will have to be overcome in driving. Should varying strata be encountered, possibly a combination of two of the methods, especially of the second and third, will be advisable, but such a possibility can be considered only if the details of the strata are accurately known.

The development in recent years of modern construction techniques has opened up new possibilities for bridge-pier design that, in cases of poor foundation conditions, can sometimes prove much more economical than the older and more conventional methods. Bridge piers can readily be constructed by the use of long steel H piles driven to rock, the piles possibly encased between bed and water level with sheet-pile cofferdams in which aggregate can easily be placed and converted into solid concrete by specialized grouting techniques. Such piers give all necessary support above water level, provide requisite protection down to riverbed level, and derive their bearing capacity from the column action of deeply buried steel piles transferring loads to bedrock. There are now quite a number of bridges in Canada successfully supported by long steel piles; one of the most notable is that which carries the line of the Canadian National Railways from Noranda to Senneterre in northwestern Quebec over the Kinojevis River. Steel piles up to 52.5 m (175 ft) long were used to form bridge piers in material that could best be described as "soup."[24.29]

When subsurface conditions at bridge sites are bad, therefore, there are now available to the designing engineer a variety of approaches to pier design. By way of contrast, there may be mentioned the dilemma facing Isambard Kingdom Brunel, another of the noted early British railway engineers, when he came to design the Royal Albert Bridge at Saltash over the river Tamar on the Great Western line to Penzance. Divers first revealed the poor foundation conditions in 1847; Brunel decided that he must know more about the riverbed. A wrought-iron cylinder, 1.8 m (6 ft) in diameter and 25.5 m (85 ft) long, was therefore fabricated, towed out to the site of the necessary central pier between two barges, and sunk through the water and mud to bedrock. From this cylinder 175 borings were made into the rock, and its profile was plotted; a column of masonry was even built on the rock and carried up to the level of the riverbed. Only then was Brunel satisfied; but he knew that unusual construction methods would be necessary to give him the pier he required. Caissons had been used as early as 1720 by Gabriel, but the caisson designed by Brunel for the Saltash Bridge must surely be one of the pioneer structures of construction practice. It measured 11.1 m (37 ft) in diameter, and 27.0 m (90 ft) in height and weighed 300 tons; it was floated out into place and fitted with an inner cylinder to give access to the "diving-bell" compartment that eventually extended to rock level. Dense beds of oyster shells in the river were only part of the troubles that had to be overcome before the pier was well founded, but overcome they were (Fig. 24.17). They were recalled appreciatively by British engineers in 1959 on the occasion of the centenary of the opening of the bridge.[24.30]

If caissons are used, one of the first problems to be faced is the skin friction which must be overcome in sinking the caisson through the foundation-bed material. Prelimi-

FIG. 24.17 Brunel's historic Royal Albert Bridge at Saltash, Cornwall, England, carrying a single line of British Railways.

nary estimates of the skin friction to be overcome may be guided to some extent by knowledge of the strata which will be passed through, although the determining factors are so variable that only actual experience can be relied upon to give an accurate estimate of the resistance to movement. Overcoming the resistance can sometimes be facilitated by the use of jetting pipes around the cutting edge of the caisson, to discharge water under pressure, but this course can be followed only when it is known with certainty that it will not seriously disturb the surrounding foundation-bed material.

Trouble with caissons is often experienced because of uneven settlement. This may be due to purely mechanical causes, but often it is due to variations in the underlying strata. In a general way, these variations may be foretold if a full soil survey has previously been carried out. So many factors are involved that it must be added that information on soil properties can be only a contributory source of information relative to possible movements and that such information must be considered in conjunction with other relevant facts. Study of the tilting of bridge-pier caissons (several cases of which are on record) shows that such occurrences are generally due to a combination of causes; ground conditions are usually the most important cause, either because of uneven settlement due to varying strata or because the soil material does not "break down" under the cutting edges of caissons in the way anticipated. The tilting (through 42°) of the 19,000-ton, reinforced-concrete open caisson for the east pier of the Mid-Hudson Bridge at Poughkeepsie, New York, seemed to be due to this latter cause; a stiff stratum of clay and sand, in association with the design of the dredging pockets, resulted in excavation extending more than 3 m (10 ft) below the cutting edge. Sudden collapse of the unsupported wall of the bed material seems to have been the cause of the serious tilting that occurred, the righting of which constituted a construction operation of unusual interest and ingenuity.[24.31]

Knowledge of subsurface conditions in advance of the start of actual work will always be of assistance in the selection of construction methods and may even suggest new methods. This seems to have been the case with what is now known as the "sand-island method," which was apparently originally evolved by M. R. Hornibrook, Ltd., for their contract for the construction of the Grey Street Bridge, Brisbane, Australia, in November 1927. The main part of the structure consists of three reinforced-concrete arch spans, each 71.4 m (238 ft) from center to center of piers. The river pier had to be founded in depths of water up to 15 m (50 ft), with rock varying from 24.9 to 32.1 m (83 to 107 ft) below high-tide level; reinforced-concrete cylinders were to be used for the two central piers, and two caissons for the main pier on the south bank. Preliminary information revealed mud as the surface stratum, and this, together with other information, led to the construction of artificial "islands" at each pier site, islands formed of cylinders of steel sheet piling filled up to above water level with sand. Thus, the cutting edges could be set up in the dry, and the sinking of the cylinders and caissons could be carried out under constant control. The steel piling was salvaged after the foundation structure was well founded in the riverbed.[24.32] A similar method has been successfully applied to other leading bridge contracts in various parts of the world, notably the Suisun Bay bridge of the Southern Pacific Railroad in California and the Mississippi River bridge at New Orleans, Louisiana. An outstanding example is the Philadelphia-Gloucester suspension bridge.

A further construction requirement of design is that some means must usually be provided for thoroughly inspecting and, if necessary, cleaning the surface of the foundation bed on which the bridge pier or abutment is to be founded. In the case of bridges, which will use low unit-bearing pressures and will be founded on unconsolidated materials, such inspection is not so important as it is in cases where solid-rock foundation beds will be used. Personal inspection of rock surfaces is always a necessary supplement to even the best preliminary core drillings; the engineer must check on the structure of the rock stratum and possible surface disintegration. The rock surface must also be properly cleaned of all unconsolidated material so that good bond may be obtained between concrete and rock. There is on record at least one case, that of the Union Pacific Railroad's bridge over the Colorado River in the United States, in which the piers had to be rebuilt only 15 years after the date of their original construction (1925) because the rock-bearing surface had not been properly cleaned before concreting was carried out.[24.33]

24.9 COFFERDAM CONSTRUCTION

Consideration of the essentially geological aspects of cofferdam construction has been left for discussion in this separate section as a matter of convenience because, in addition to providing facilities for the construction of bridge piers and abutments, cofferdams are used for much general foundation work. They are now so common a feature of construction work that they are generally accepted without much thought being given to their development. It is sometimes believed that they are a relatively modern innovation; frequent reference is made to their "early" use on the river Thames in England for the construction of Waterloo Bridge (from 1809 to 1817), to which extended reference has already been made. The following extract from Vitruvius, dated probably about 20 B.C., is therefore of special interest.

> Then, in the place previously determined, a cofferdam, with its sides formed of oaken stakes with ties between them, is to be driven down into the water and firmly propped there; then, the lower surface, inside, under the water, must be levelled off and dredged, working from beams laid across; and finally, concrete from the mortar trough—the stuff having been mixed as prescribed above—must be heaped up until the empty space which was within the cofferdam is filled up by the wall.... A cofferdam with double sides, composed of charred stakes fastened together with ties, should be constructed in the appointed place, and clay in wicker baskets made of swamp rushes should be packed in among the props. After this has been well packed down and filled in as closely as possible, set up your water screws, wheels, and drums, and let the space now bounded by the enclosure be emptied and dried. Then dig out the bottom within the enclosure.[24.34]

Cofferdams designed to be pumped dry (as distinct from those used merely to provide a still-water area) perform one main function, that of retaining the surrounding water and unconsolidated materials and thus providing an exposed dry area of river, lake, or seabed on which construction operations can be carried out. The retention of sand, clay, or other foundation-bed material, e.g., after excavation has proceeded inside a cofferdam lower than the bed level outside, is not a difficult matter; it is a special case of the retaining structures already considered. The retention of water above bed level is also a relatively simple matter provided the piling used to form the cofferdam has watertight joints; calculations for the necessary bracing to support the water pressures are straightforward. One main problem remains, that of preventing the ingress of water around the lower edge of the piling—in other words, that of preventing "blows," or piping.

The usual cofferdam structure consists of a single continuous wall of interlocking piling, driven into the foundation strata for a distance sometimes determined by the resistance offered to pile driving, sometimes merely on the basis of previous work, but almost always in an empirical manner. Experience must ever be a guide in such matters, and the actual evidence presented by the driving of the sheet piling into the strata to be met with will be a telling indication of the underground conditions encountered. Cofferdams generally have to be designed prior to construction; materials have to be ordered before they can be used. Estimates of piling penetration necessary are therefore an essential preliminary; and in this estimating work, accurate information on the strata to be encountered will be of value.

Permeability is naturally a soil characteristic of leading importance in such considerations, but it cannot be considered alone. Sands and gravels, when revealed by test borings, will naturally suggest the necessity of deep penetration, and if sufficiently permeable, they may even dictate the use of a clay blanket all around the outer face of a cofferdam as a means of sealing the direct path that would be taken by water when pumping inside began. Clays, on the other hand, will generally provide an impermeable barrier to the seepage of water, provided the piling is driven below any influence from surface disturbance. In cofferdams constructed in tidal water, or in any watercourse liable to fluctuation of water level, the greatest care must be exercised with all cofferdam structures in clay. Under certain maximum water-level conditions, the clay will be compressed by one face of the piling and may possibly deform permanently, leaving a gap between piling and clay when the water level changes to the other extreme. This type of water passage is a far easier course for the flow of water to follow than even the pores in sand and gravel; and if this condition is allowed to develop unobserved, serious damage may result, and a bad blow occur.

These notes indicate the importance that subsurface conditions occupy even in this detail of construction; they emphasize, above all else, the necessity of knowing as accurately as possible the nature of the material to be penetrated. Again, careful tests are necessary to distinguish various soil types. The writers know of at least one case in which a material that to the untrained eye appeared to be clay actually proved to be compacted limestone flour (with practically no clay content at al). It was assumed to be clay in cofferdam design, and yet it proved in practice to be as porous as a sieve, leading to most serious trouble with "blows."

Another way in which geological advice can be of appreciable assistance in cofferdam work is in foretelling, with some degree of accuracy, the presence of boulders in the strata through which piles have to be driven. The presence of even one boulder can on occasion cause as much trouble in constructing and pumping out a cofferdam as all the rest of the work put together, as all engineers who have had such an experience will know. Test borings may possibly show the existence of boulders; but unless the boulder formation is closely packed, it is quite probable that the few test borings put down on the actual site of a bridge pier may miss any boulders that may be present. On the other hand, knowledge of the nature of the geological strata to be encountered in foundation work will at least suggest whether or not boulders are to be expected, e.g., as in glacial deposits, in which case extra care must be exercised in the preliminary test-boring work. Foundation records for neighboring works have often been found to be of great value in an application of geological considerations; the value of such records has already been generally stressed but may be specially emphasized for this case.

Some cofferdam construction which encountered unusual difficulties with boulders was an important part of the building of the three new bridges across the Cape Cod Canal in Massachusetts in 1934 and 1935. Wash borings had disclosed the existence of glacial deposits of sand and granitic gravel containing numerous granite boulders. Specifications were drawn up calling for foundations to be constructed inside steel sheet-pile cofferdams driven to predetermined depths. What was not known and was revealed only by the foundation operations were the variable amount of boulders, their relatively large size, and their distribution. The contractor for the work estimated that 10 percent of the piling might strike boulders that would have to be removed; actually, as much as 40 percent of the piling in one cofferdam was so obstructed; and at all the main cofferdams for the three bridges, serious difficulties were encountered (Fig. 24.18). It was found, for example, that the disturbance caused by pile driving and the operations for removing the boulders so disturbed the sand that it went "quick" and thus resulted in large and frequent boils. Ordinary pumping methods to facilitate the removal of the boulders failed to keep pace with the flow of water into the cofferdams, and eventually an extensive well-point installation had to be used both inside and outside the sheeting. The presence of boulders had been revealed during the original construction of the canal 20 years before, but whether or not this earlier construction indicated the presence of boulders in the amount encountered at the bridge piers was a matter of controversy.[24.35]

Cavernous limestone comes up again for mention when attention is directed to an unusual bridge construction job of the Kentucky State Highway Department across the Barren River at Bowling Green; the structure was a four-span, continuous-plate girder,

FIG. 24.18 Cofferdam for the Bourne south channel pier, Cape Cod highway bridges, showing irregular driving of sheet piling due to the presence of boulders, some of which may be seen between tracks and cofferdam.

with each span 33 m (110 ft) long. The bedrock across the riverbed is limestone, extensively honeycombed below its surface for several meters and filled with mud pockets. Up to 3 m (10 ft) of the cavernous limestone had to be excavated over the sites of the two river piers, which were constructed within sheet-pile cofferdams. The steel piling could not be driven into the limestone, so pile driving and excavation had to be carried out concurrently. The necessary blasting disturbed the natural sealing of the rock, and strong flows of water into the cofferdams commenced. Various expedients were tried in attempts to cut off the water, but eventually very extensive grouting programs had to be carried out; as many as 9,000 bags of cement were used for this purpose around the two relatively small cofferdams.[24.36]

An entirely different sort of groundwater problem was encountered by the Illinois Division of Highways during the construction of a 1,424-m (4,745-ft) bridge over the Illinois River at Peoria. The Peoria Water Works Co., a private utility that supplies water in the city, obtains its supplies from wells extending into a sand and gravel stratum, separated from the riverbed by layers of clay and shale of varying thickness. Some of the H bearing piles for the main piers supporting the three-span, continuous-through bridge over the main channel were designed to pass through this water-bearing stratum; the water company was naturally concerned over the possibility of pollution of its supply. Extensive studies were made, and an unusual schedule was worked out for the control of water level in the cofferdams from which the steel piles were to be driven; the water level was held down to riverbed level while pile driving was in progress and allowed to rise only when pile driving was complete and thick concrete seals had been placed.[24.37]

Geology assisted with one of the most ingenious solutions to the "cofferdam problem" known to the writers. Two 8.4- by 18.0-m (28- by 60-ft) open cofferdams were needed for construction of the main piers of the Deer Island suspension bridge in Maine, on the Atlantic coast, due south of Bangor. Test borings showed that the bedrock surfaces on which the piers were to be founded were steeply sloping away from the land, a common feature of this rocky coastline. The substructure contractor therefore decided to clean off the areas of the piers to bedrock, place dowels in the rock for the locating of the cofferdams, and then frame the steel sheet-pile cofferdams, complete with bracing, on shore, the bottoms of the piles carefully trimmed to follow the exact contours of the rock surface. The contractor had available a 250-ton-capacity floating derrick, so the handling of the "tailor-made" cofferdams (as they were naturally called) was not too difficult a job. The

FIG. 24.19 The "tailor-made" cofferdam for the Deer Island suspension bridge, Maine, showing how it was transported to the site.

cofferdam boxes were fabricated in New York, barged to the bridge site, and then lowered into place (Fig. 24.19). Bags of dry-batch concrete were placed by divers around the edges of the cofferdams, concreting then proceeded normally, and the 614-m (2,048-ft)-span bridge was soon completed.[24.38]

24.10 SOME UNUSUAL CASES

As with all other types of civil engineering projects, no two bridge piers are ever the same. There is something to be learned from every pier that is ever constructed. The field of practice is so large, however, that only a few especially useful examples can be cited in this review of the significance of geology in the founding of every bridge pier. There are a few such cases that do not fit into the sections into which this chapter has been divided for convenience; they are therefore grouped here. No treatment of bridge foundations however brief, for example, would be complete without some reference to the original Forth Bridge, that majestic structure carrying the main railway line to the north of Scotland that spans the Firth of Forth near Queensferry, just to the north of Edinburgh (Fig. 24.20). Even today, when so many other great bridges have been constructed in all parts of the world, it still excites imagination, especially when it is recalled that it was opened for use in 1890.

It was, indeed, the geology of the site that made the great triple cantilever design possible. The north main pier is founded on basalt, the bedrock underlying the Fifeshire shoreline. The south main pier at Queensferry is founded on the sandstone which underlies the thick deposit of till characteristic of this shoreline. The till extends into the Firth where it fills the lower part of a 180-m (600-ft)-deep preglacial gorge that exists under the center of the channel, the course of an ancient river Forth. The existence of this gorge might have eliminated any possibility of such a design as the masterpiece of Sir John Fowler and Sir Benjamin Baker. Geology, however, had created Inchgarvie Island as a pinnacle of the Fifeshire basalt protruding above water level, so located that it provided an ideal location for the large central foundation, and so dimensioned that the individual piers could be separated rather farther apart than those for the other two foundations, giving the central span the extra stability the designers wished to have. And, as a bonus, the basalt (whinstone in Scotland) excavated from the island for the founding of the piers

FIG. 24.20 The railway bridge across the Firth of Forth, Scotland.

24-32 CIVIL ENGINEERING WORKS

proved satisfactory for the concrete which was needed in addition to the granite facing of all the piers, the granite coming from Aberdeen. Because of splendid maintenance down the years, the Forth Bridge is as sound today as when it was dedicated on 4 March 1890 by H.R.H. The Prince of Wales. This 2.4-km (1.5-mi)-long structure will always be an inspiration to civil engineers, while geologists can take satisfaction from the part that geology played in its conception, design, and construction.[24.39]

Fifteen years before the opening of the Forth Bridge, another notable railway bridge was opened for use in the wilds of Canada; it, too, is still in daily use, although the superstructure has (naturally) been once renewed. The bridge is one of two six-span bridges which now carry the main Halifax-to-Montreal line of Canadian National Railways over the northwest and southwest branches of the Miramichi River of New Brunswick. The two branches of this famous salmon river join 3.2 km (2 mi) upstream of the town of Newcastle. Both rise in the forests of New Brunswick and to all appearances are identical in character—or so it was assumed when simple borings were taken in advance of the building of the two bridges, under the direction of Sandford Fleming, chief engineer of the Intercolonial Railway (as the line was originally called). The borings suggested that bedrock would be reached at depths between 13.5 and 15.0 m (45 and 50 ft) below high-water on spring tides (HWOST). Foundations were therefore designed as timber caissons, to be floated into place, filled with tremie concrete, and topped by masonry piers above water level. When work started, it was soon found that the borings were inaccurate. Fleming immediately ordered new and better borings, this in 1870. Instead of bedrock, sand and gravel were encountered, overlying "silt" in the northwest branch and "clay" in the southwest branch, bedrock being reached at a depth of 33.6 m (112 ft) in the former and 27 m (90 ft) in the latter.

Fleming was a great engineer, and gifted with unusual intuition (soil mechanics would not be a recognized field for another 65 years!). He deduced that the piers as designed for the southwest bridge would be safe; work proceeded on this basis with most satisfactory results. Dubious about the northwest bridge, he had special "penetration tests" made on small plates at the bottom of boreholes (almost three-quarters of a century before such tests were "discovered" in modern soil studies). Using the results so obtained, he enlarged the caissons and, against the advice of experts consulted by the government of Canada but with his own chairman's approval, carried on. Knowing from his penetration tests that settlements would take place, he hit upon the idea of preloading the completed piers with

FIG. 24.21 Canadian National Railways bridge over the northwest arm of the Miramichi River, near Newcastle, New Brunswick; the pier nearest to the camera is that shown in Fig. 11.15.

loads slightly greater than the total loads to which they would be subjected, thus accelerating settlement. The piers were finished, the superstructure erected, and the bridge has performed satisfactorily to this day (Fig. 24.21). Level records taken by Fleming have been correlated with modern studies to give a 100-year record of performance.[24.40] The interesting and significant difference between the soils in the two branches of the Miramichi River is wholly geological in origin, the northwest branch flowing through country underlain generally by Precambrian rock while the southwest branch runs over younger (Mississippian) rock.

Preloading of bridge piers (and of other large foundations) is now a recognized technique in civil engineering. A modern example has special interest in that the necessity for prestressing arose, just as at the Miramichi bridges, because of incorrect results from test borings. A steel arch bridge with a skew span of 241.5 m (805 ft) was the desirable solution to the problem of getting the Trans Canada Highway across a gorgelike section of the Fraser River in British Columbia (Fig. 24.22). Test borings having shown bedrock to be at a depth of 9 m (30 ft) at the locations of both abutments, a fixed steel arch with a rise of 34.6 m (115 ft 6 in) was designed. When the ground was opened up, large boulders were found at the supposed bedrock elevation (*again* must be added). Actual depths to bedrock were 60 m (200 ft) on the west bank and 45 m (150 ft) on the east. Soils above rock level consist generally of sand and gravel with one stratum of clay at a depth of about 30 m (100 ft). The site could not readily, or economically, be changed. The arch could be (and

FIG. 24.22 Steel arch bridge on the Trans Canada Highway over the Fraser River, British Columbia, the foundations of which were preloaded.

was) changed to a two-hinged arch with some increase in the quantity of steel, so that all question of fixing moments would be removed, although settlements had still to be obviated. Five alternative approaches were considered, preloading with the full load of, and thrust from, the completed bridge being finally decided upon. Massive concrete abutments were designed and were anchored to bedrock by steel cables which were then stressed, with jacks, to the full bridge load. The anchored abutments were used as reaction blocks for the horizontal loads which were then applied by more jacks to the soil "behind" the abutments, again to the ultimate maximum value. Accurate knowledge of the soil strata thus stressed permitted assurance to be placed on their future behavior, assurance which was fully justified.[24.41]

Approaching Vancouver, the Trans Canada Highway crosses the Fraser River again on another graceful steel arch with long approach viaducts, as it crosses the floodplain of the river just upstream of the head of its great delta (Fig. 24.23). Even more difficult foundation conditions were experienced here, not because of any failure of the preliminary test-boring program but because of what this extensive program had revealed. Borings taken to depths of over 60 m (200 ft) revealed a general pattern of flood plain deposits of sand and silt (in which all the north approach spans were founded) overlying tough till-like deposits. In the vicinity of the river channel, and especially on the north bank, 4.5 m (15 ft) of organic silt was first encountered; this was found to overlie soft organic silts and clay silts to about 12 m (40 ft); then a stratum of compact peat under which a clay stratum extending in places to 15 m (50 ft) below the surface was revealed; sand strata of varying density with some silt occurred next in the subsurface profile—to a depth of about 33 m (110 ft)—and then sand and gravel for the next 3 m (10 ft); finally, to depths of about 57 m (190 ft) the borings revealed soft-to-firm sensitive clays with some silt. To found the main piers for a 360-m (1,200-ft) span steel arch on such material called for the greatest ingenuity in design.

A number of alternative designs were studied, the solution adopted being novel indeed,

FIG. 24.23 The Port Mann Bridge carrying the Trans Canada Highway over the Fraser River near its mouth, adjacent to Vancouver, British Columbia.

probably unique in the history of bridge engineering. Holes 80 cm (32 in) in diameter were drilled from the surface to the top of the till. Into these holes were lowered 60-cm (24-in)-diameter, steel pipe piles with heavy steel tips; the lower sections of these piles were filled with concrete to a height of 4.5 m (15 ft). This material served as a cushion for the drop hammer which then drove the piles by being actuated *inside* the piles. In this way, disturbance of the clay strata was avoided and sound bearing obtained. So weak was the clay that the annular ring (10 cm or 4 in wide), which was filled with mud during the pile driving, was finally refilled with cement-clay grout which displaced the mud as it was placed. The four main piers for the Port Mann Bridge were founded in this way, the engineering details of the design being a landmark in the history of geotechnical engineering.[24.42]

One of the longest and greatest of North American bridges also had its main piers founded almost 60 m (200 ft) below water level, this being the great suspension bridge across the Straits of Mackinac in northern Michigan, having, with its approaches, a total length of 7,800 m (26,000 ft). The geology of the site was one of the first items to be considered, once the concept of a bridge across this entrance to Lake Michigan had been accepted. Professors Charles P. Berkey and Sidney Paige, of Columbia University, prepared an early report on the site geology, suggesting that "the Mackinac breccia and the elements composing it, which is the principal formation involved in this problem, has the strength required to support the proposed bridge piers with an ample margin of safety." None of the ensuing detailed subsurface investigations, starting with a comprehensive test-boring program in 1939, disproved this initial suggestion. Somewhat naturally, it took many years before designs were completed and construction began, but the hopes of the years were fulfilled in 1954.

The main suspension span is 1,140 m (3,800 ft) long, even the side spans being each 540 m (1,800 ft). To drive across this structure is a memorable experience, even for those who do not know that the main piers rest upon bedrock (the Mackinac breccia) 58.5 m (195 ft) below water level (Fig. 24.24). The piers were located after detailed test drilling at prospective sites, the occurrence of occasional small "caves" in the breccia being one determining factor. Not only did the excellent subsurface exploration give to the bridge designers all necessary information for the design of the bridge piers, but it also yielded geological information of unusual value. A preglacial valley, now filled with till, extending 105 m (350 ft) below water level, was located, as was also a secondary valley extending to

FIG. 24.24 The suspension bridge which now spans the Straits of Mackinac, Michigan.

24-36 CIVIL ENGINEERING WORKS

52 m (174 ft) below lake level, two features of the local geology which led to reconsideration of the glacial history of this part of the Great Lakes.[24.43]

Dr. Terzaghi once said that "On account of the fact that there is no glory attached to the foundations, and that the sources of success or failure are hidden deep in the ground, building foundations have always been treated as stepchildren and their acts of revenge for lack of attention can be very embarrassing." Of no group of foundations is this more true than of those for bridges. Only once in all their experiences have the writers come across a bridge foundation that was given recognition by being featured in a major advertisement. Figure 24.25 shows this, the remarkable subsurface below the Richmond-San Rafael Bridge across a northern portion of San Francisco Bay. This is another very long bridge, 6,403.5 m (21,345 ft) with its approaches. Detailed subsurface exploration, with associated geotechnical laboratory tests, indicated that bearing piles driven into the strata shown would not provide an adequate foundation. Thirty-five-cm (14-inch) steel H piles were therefore driven to bedrock, the piles ranging in length up to 52.5 m (175 ft), driving being effected in the case of 68 of the piers by means of precast-concrete, bell-shaped, template-type submerged pier units.[24.44] A total length of 175,710 m (585,700 ft) of steel piles was used, approximately half supplied by the United States Steel Corporation. Small wonder that this company featured the bridge in one of their major advertisements at the time (Fig. 24.25).

24.11 INSPECTION AND MAINTENANCE

The regular and thorough inspection of all civil engineering structures is a matter that can hardly be overemphasized. It is always important but nowhere more so than in such

FIG. 24.25 Richmond–San Rafael Bridge over San Francisco Bay. The bridge is 21,345 feet long. Stratification encountered points up the need for extra long piles. (U.S. Steel Corporation.)

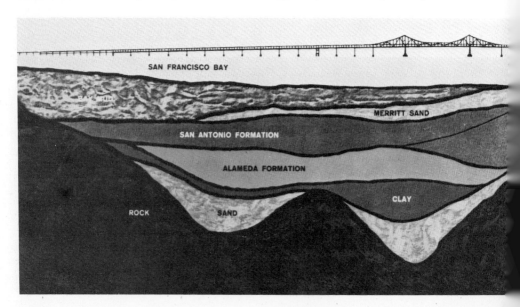

features as bridge piers, the underwater parts of which are not seen from day to day and yet are most prone to possible periodic damage. In the case of piers and abutments founded on dry land, inspection of the structures and regular checks on their positions and levels are the main requirements of inspection. Should any movement of the pier be detected, and no deterioration of the constituent material be evident, then inevitably some feature of the underlying geological strata is responsible for the trouble; and in consequence, foundation-bed conditions must be investigated anew.

An outstanding example of the benefits to be derived from inspection work is given by the famous Lethbridge Viaduct of the Canadian Pacific Railway, a steel-trestle structure 1,598.1 m (5,327 ft) long carrying track 93.6 m (312 ft) high above the bed of the Oldman River of Alberta (Fig. 24.26). The trestle bents were founded on concrete pedestals supported on concrete piles driven into the clay which overlies the whole site. Completed in 1909, the deck of the bridge soon showed signs of settlement; a line of check levels run in 1913 disclosed a subsidence of 64 mm (2½ in) at pedestal 59 south. Cracks and gopher holes in the clay below were plugged with clay, and drainage arrangements improved, but settlement continued; eventually, the pedestal had to be underpinned by caissons carried to shale rock 15 m (50 ft) below ground level, and a drainage tunnel had to be constructed. As a result of these measures, subsidence was arrested (Fig. 24.27).

Similar vigilance in track maintenance was responsible for the first detection of trouble with one of Robert Maillart's justly famous slender, reinforced-concrete bridges in Switzerland. This bridge is located at Klosters, a well-known skiing center near the eastern border (Fig. 24.28). The bridge carries the meter gauge Rhaetian Railway running from a junction with the Swiss Federal Railways at Landquart to Davos and St. Moritz. Built upon a sharp curve, the bridge is characterized by its main arch of 30-m (100-ft) span; its total length is 75 m (254 ft). The arch rests upon mass-concrete abutments; one of these

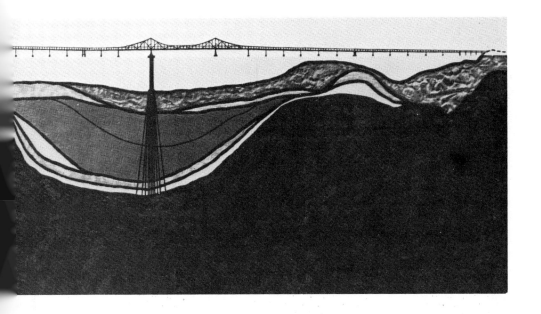

FIG. 24.26 The Lethbridge Viaduct of the Canadian Pacific Railway, Alberta, seen from the northwest.

is located on a relatively steep hillside through which the railway tunnel passes; the portal of the tunnel is close to the end of the bridge. Soon after the bridge was placed in service, one of the track maintenance men noticed a slight rise in the track at the center of the span; this was corrected in the roadbed formation, but it was soon noticed again.

This led to detailed study of the situation, and it was then found that the soffit of the arch was slowly rising. In turn, this was traced to a slow but progressive movement toward the river of the hillside abutment. Extensive borings were taken, and it was found that the movement was widespread on the sloping ground, since the underlying rock talus was in unstable equilibrium. After much study it was decided that the only possible solution was to brace one abutment against the other, even though this would interfere with the general appearance of the bridge. Therefore, a heavily reinforced concrete strut was built across the river at haunch level; the strut was of U section with a width of 2.5 m (8 ft 3 in) between the insides of the webs, or walls, as they appear to be to the uninformed visitor. (During a visit, one of the writers found that the local villagers were still enjoying to the full the access across the river thus given by what some of them took to be a footpath constructed for their special benefit.) Embedded within the strut are strain gauges, regular observations upon which have enabled the responsible engineers to check on the adequacy of their design and, incidentally, to accumulate information, which they are making publicly available, on the magnitude of the pressures developed in this unusual way.[24.46]

The inspection of piers founded below normal water levels is a more extensive operation, involving underwater work and the regular sounding of the riverbed or watercourse between piers and for some distance on either side of the bridge site. Underwater inspection in diving suits, although generally a job for the field inspecting staff, is something that should regularly be undertaken also by the bridge engineering staff. It is not necessary to stress here the value of diving experience to engineers, but the importance to an engineer of being able to see the effect of the bridge structure on riverbed stability is a special feature which may fittingly be mentioned. Examination of the pier structure is a part of underwater work, but equally so is the study of actual bed conditions, especially around the pier bases. This inspection, coupled with soundings over the whole water area adjacent to the bridge, if performed regularly, will be a constant check on the dangerous possibility of riverbed scouring. The regular nature of this inspection work must be stressed.

Naturally, inspections should be timed in accordance with periodic danger periods, e.g., those caused by regular spring-flood discharges in rivers. Inspections must also be supplemented by special surveys if any new structures are built in a waterway near enough to the existing bridge to affect it even to a slight degree. Interesting records are available of the methods and programs adopted for bridge inspection work by leading railways. It may perhaps be mentioned that in order to speed up the inspection of bridge piers in shallow water, Canadian National Railways employed on one of its inspection crews an expert high diver who carried out underwater inspection without the use of a cumbersome diving suit. Skin divers are now regularly used for this vital inspection work.

The importance of these and other features of bridge inspection work is well illustrated by two bridge failures that occurred in the United States in the late summer of 1933, both of which caused loss of life, injuries, and serious damage to stock and structures. The first occurred at the Anacostia River Bridge of the Pennsylvania Railroad near Washington,

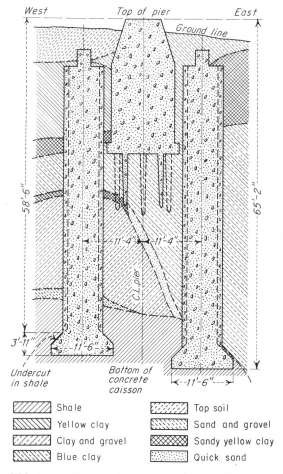

FIG. 24.27 Pier No. 59 S of the Lethbridge Viaduct, Canadian Pacific Railway, Alberta, showing the geological strata beneath it and the way in which the pier was rebuilt.

FIG. 24.28 Reinforced-concrete arch bridge at Klosters, Switzerland, carrying the narrow-gauge Rhaetian Railway to Davos and St. Mortiz, showing the special reinforced-concrete strut between abutments.

D.C., a four-span, deck plate-girder structure, which, because of the failure of the center pier, collapsed when a train was passing over it. Subsequent investigation showed that the gravel stratum on which the piers were founded had been seriously eroded; the culmination of the erosion process was probably caused by a tropical storm which was abating when the collapse occurred. The railroad company had no record of soundings or underwater inspection from the time the bridge was built in 1904 until the date of failure; the company stated that as no settlements or cracks had developed, unsafe conditions underwater had not been indicated. Some dredging had been done downstream after the completion of the bridge, but at the investigation it was stated that, even if this had not affected the riverbed at the bridge, "it was the duty of the railroad company to keep informed concerning changes that might have any effect on the safety of the bridge." [24.47]

The second accident, which also occurred in August 1933, was on the Southern Pacific line between Hargis and Tucumcari, New Mexico. It, too, was due to unusual floodwaters which undermined the east abutment of the deck plate-girder structure and thus caused the collapse of the bridge as a train was crossing. The Bureau of Safety of the Interstate Commerce Commission, in its report on the accident, attributed the failure to the fact that the embankment had not been protected against erosion and to heavy rains which had caused excessive floodwater; the position of a highway bridge (constructed 45 m or

150 ft upstream) had increased the current velocity and probably had diverted it against the railroad-approach embankments.[24.47]

Mention of such illustrative examples prompts the thought that it is always easy to be wise after the event. This cannot be disputed; and yet at the same time, it must be recalled that one of the surest ways of learning constructive methods is to study past mistakes and errors. Bridge inspection can provide information of general value only by records of the discovery of features unsuspected and of serious consequence; if some of these have been discovered too late, they may still serve as a reminder of the supreme importance of regular examination by the engineer of the foundation strata in which such confidence for the sure and certain support of bridge structures is placed.

24.12 CONCLUSION

The scouring action of rivers, when bed conditions permit this, is one of the most insidious natural phenomena with which the civil engineer has to deal. Unless the greatest care is exercised in regular inspections of bridges and other structures founded in flowing water, erosion can take place which gives no evidence at the surface and thus remains undetected. If only to demonstrate what riverbed erosion can effect, Fig. 24.29 has been included to show what test borings revealed at the site of the 17th July Bridge at North-

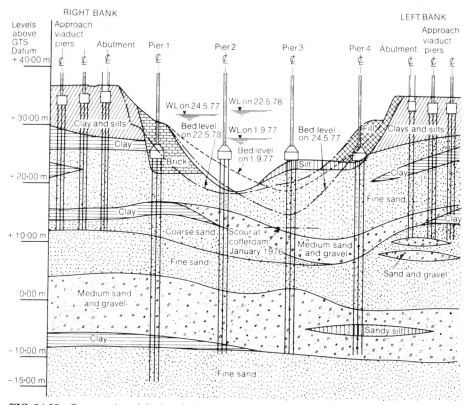

FIG. 24.29 Cross section of the foundation-bed conditions below the 17th July Bridge over the river Tigris at Baghdad, Iraq, showing the varied soil strata and the depths of scour.

FIG. 24.30 The new bridge over the river Danube at Bratislava, Czechoslovakia, as seen from the north bank, the famous castle of Bratislava being off to the right.

gate in the ancient city of Baghdad. This is a four-lane, five-span, prestressed-concrete bridge over the river Tigris, 270 m (almost 900 ft) long. Not only did the borings reveal the many changes in the riverbed, but construction operations encountered ancient remains from the period of Nebuchadnezzar.[24.48]

As a final example from the great number of bridges that could have been cited to illustrate the message of this chapter, Fig. 24.30 shows the new bridge across the river Danube at Bratislava in Czechoslovakia. Its striking design can speak for itself, but to enjoy the restaurant located at the top of the inclined tower is an experience not easily forgotten. Foundation beds under the two main supports for the bridge differ because of a fault located beneath the river. On the right bank 14 m (46 ft) of alluvial deposits overlie Neogene clay and sand. On the left bank, granitoid bedrock is overlain by 12 to 23 m (40 to 43 ft) of miscellaneous fill material. The engineering solutions to the design problems presented by these conditions show yet again the profound influence of geology on bridge design.[24.49]

24.13 REFERENCES

24.1 *British Bridges,* Public Works, Roads and Transport Congress, London, 1933, p. 173.

24.2 Ibid., pp. 175–176.

24.3 J. A. L. Waddell, *Bridge Engineering,* vol. 2, Wiley, New York, 1916, p. 1544.

24.4 P. Allen, "The Georges River Bridge, New South Wales," *Minutes of Proceedings of the Institution of Civil Engineers,* **232**:183 (1932).

24.5 D. Moran, "Sampling and Soil Tests for Bay Bridge, San Francisco," *Engineering News-Record,* **111**:404 (1933); *see also* other articles in the same journal on other aspects of this great bridge, e.g., C. H. Purcell, C. E. Andrew, and G. B. Woodruff, "Deep Caissons for Bay Bridge," **113**:227 (1934).

24.6 W. M. Scott, Winnipeg, by personal communication.

24.7 L. W. Millard, "The Mortimer E. Cooley Bridge," *Civil Engineering,* **7**:617 (1937).

24.8 "Tappan Zee Bridge A Foundation Triumph," *Engineering News-Record,* **155**:44 (14 April 1955).

24.9 E. J. Buckton and H. J. Fereday, "The Demolition of Waterloo Bridge," *Journal of the Institution of Civil Engineers,* **3**:472 (1936).

24.10 "The Vaudreuil Rigid-Frame Bridge," *The Engineer,* **150**:160 (1935).

24.11 C. J. Mackenzie, "The Broadway Bridge," *The Engineering Journal,* **17**:3 (1934).

24.12 "Bridging the St. Lawrence at the Ile d'Orléans," *Engineering News-Record,* **112**:356 (1934).

24.13 "Road Foundation Problems," *Civil Engineering and Public Works Review,* **32**:442 (1937).

24.14 H. E. Brooke-Bradley, "Bridge Foundations," *Structural Engineer* (new series), **12**:103 (1934).

24.15 P. A. Scott, and C. Roberts, "The Volta Bridge," *Proceedings of the Institution of Civil Engineers,* **9**:395 (1958).

24.16 O. H. Pilkey, "Arresting Abutment Shifting on a Bascule Bridge," *Engineering News-Record,* **108**:725 (1932).

24.17 A. G. Stermac, M. Devata, and K. C. Selby, "Unusual Movements of Abutments Supported on End Bearing Piles," *Canadian Geotechnical Journal,* **5**:69 (1968).

24.18 W. T. Everall, "Description of Several Unusual Structures Adopted in Bridge Construction," Paper no. 160, kindly supplied by the chief engineer, North Western Railway, India.

24.19 *The Strengthening of an L. M. S. Railway Bridge, British Steel Piling Co., Ltd.* Bulletin 4, 1935, geological data kindly supplied by W. H. Giles and D. C. Bean.

24.20 R. M. Finch, and A. Goldstein, "Clifton Bridge, Nottingham: Initial Design Studies and Model Test," *Proceedings of the Institution of Civil Engineers,* **12**:289 (1959).

24.21 "Elevated Roadway Adjusts to Bad Ground," *Engineering-News Record,* **178**:50 (27 April 1967).

24.22 J. L. Savage, "Earthquake Studies for Pitt River Bridge," *Civil Engineering,* **9**:470 (1939).

24.23 "Deck-Girder Railroad Bridge Has Earthquake-Resistant Features," *Engineering News-Record,* **134**:120 (1945).

24.24 L. S. Moisseiff, "Provision for Seismic Forces in Design of Golden Gate Bridge," *Civil Engineering,* **10**:33 (1940).

24.25 *British Bridges,* op. cit., pp. 186, 190.

24.26 A. W. H. Dean, "Construction of a Submergible Road-Bridge over the Nerbudda River near Jubbulpore, Central Provinces, India," *Minutes of the Proceedings of the Institution of Civil Engineers,* **239**:178 (1936).

24.27 R. Borhek, "Scouring of Foundations as a Cause of Bridge Failure," *Roads and Bridges,* **33**:50 (1943).

24.28 D. W. Smith, "Bridge Failures," *Proceedings of the Institution of Civil Engineers,* p. I, 1976, p. 367; paper reprinted without tabular material in *Civil Engineering,* **47**:59 (November 1977).

24.29 C. P. Disney, and R. F. Legget, *Modern Railroad Structures,* McGraw-Hill, New York, 1949, p. 32.

24.30 L. T. C. Rolt, "Centenary of the Royal Albert Bridge, at Saltash," *The Railway Magazine,* **103**:307 (1959).

24.31 G. B. Woodruff, "An Overturned 19,000-Ton Caisson Successfully Salvaged," *Engineering News-Record,* **106**:275 (1931).

24.32 G. O. Boulton, "Construction of the Grey Street Bridge, Brisbane, Australia," *Civil Engineering and Public Works Review,* **26**:55 (1931).

24.33 "Failure to Clean Bottom Results in Defective Pier Bases," *Engineering News-Record,* **95**:211 (1925).

24.34 Vitruvius, *The Ten Books on Architecture* (M. H. Morgan, trans.), Dover, New York, 1960, p. 162.

24-44 CIVIL ENGINEERING WORKS

24.35 "Sinking Open Cofferdams Through Glacial Drift," *Engineering News-Record,* **114**:1 (1935).

24.36 "Fighting Water in a Bridge Foundation," *Engineering News-Record,* **130**:806 (1943).

24.37 W. H. Townsend, "Underground Water Supply Complicates Bridge Pier Construction," *Engineering News-Record,* **128**:495 (1942).

24.38 "Tailor-Made Cofferdams," *Engineering News-Record,* **121**:207 (1938).

24.39 J. E. Richey, "Nature's Contribution to the Forth Bridge," *Transactions of the Society of Engineers,* **45**:275 (1954).

24.40 R. F. Legget, and F. L Peckover, "Foundation Performance of a 100-Year-Old Bridge," *Canadian Geotechnical Journal,* **10**:504 (1973); see also in same journal, F. L. Peckover, and R. F. Legget, "Canadian Penetration Tests of 1872, **10**:528 (1973).

24.41 H. Q. Golder, and A. B. Sanderson, "Bridge Foundation Preloaded to Eliminate Settlement," *Civil Engineering,* **31**:62–65 (1961).

24.42 H. Q. Golder, and G. C. Willeumier, "Design of the Main Foundations of the Port Mann Bridge," *The Engineering Journal,* **47**:22 (August 1964).

24.43 R. M. Boynton, "Mackinac Bridge," *Civil Engineering,* **26**:301 (1956); see also W. N. Melhorn, "Geology of Mackinac Straits in Relation to Mackinac Bridge," *Journal of Geology,* **67**:403 (1959).

24.44 "Problem: How to put the Richmond-San Rafael Bridge across these holes....," Advertisement in the technical press by the United States Steel Corporation, Pittsburgh, 1970.

24.45 F. W. Alexander, "Maintenance of Substructure of the Lethbridge Viaduct," *The Engineering Journal,* **17**:523 (1934).

24.46 C. Mohr, and R. Haefeli, "Umbau der Landquartbrucke der Rhatischen Bahn in Klosters," *Schweizerische Bauzeitung,* **65**:5,32 (1947).

24.47 "Two Railroad Bridge Failures Laid to Inadequate Inspection," *Engineering News-Record,* **111**:687 (1933).

24.48 M. E. Morris, and R. B. Hannant, "Some Aspects of the Design and Construction of the 17th July Bridge at Northgate, Baghdad," *Proceedings of the Institution of Civil Engineers,* vol. 66, p. I, 1979, pp. 437–456.

24.49 Milan Matula, Bratislava, by personal communication.

Suggestions for Further Reading

Geysers and Geothermal Energy by J. S. Rinehart is a 256-page volume, published in 1980 by Springer-Verlag of New York. An even more recent volume is *Geothermal Energy Development* by E. W. Butler and J. B. Pick (352 pages) published in 1982 by Plenum of New York. A very useful review of research and development in this field is *Geothermal Energy* edited by H. C. Armstrong and published by UNESCO (Paris) in 1973.

Geology in the Siting of Nuclear Power Plants, edited by Allen W. Hatheway and C. R. McLure, although referenced above, must be mentioned again, such is its value. This 256-page volume contains a set of specially prepared papers covering all major aspects of its subject. It is the fourth in the series of *Reviews in Engineering Geology* published by the Geological Society of America (3300 Penrose Place, Boulder, CO 80301). The first two volumes in the series contain general collections of papers of great value in the practice of engineering geology.

chapter 25

DAM FOUNDATIONS

25.1 Historical Note / 25-2
25.2 Failures of Dams / 25-3
25.3 Inspection and Maintenance / 25-8
25.4 Review of Dam Construction / 25-9
25.5 Anchoring Dams to Foundation Beds / 25-10
25.6 Preliminary Work / 25-13
25.7 Exploratory Work During Construction / 25-18
25.8 Soundness of Bedrock / 25-24
25.9 Possibility of Ground Movement / 25-27
25.10 Permeability of Bedrock / 25-32
25.11 Dams on Permeable Foundation Beds / 25-36
25.12 Construction Problems / 25-38
25.13 Problems with Dams in Service / 25-50
25.14 Conclusion / 25-54
25.15 References / 25-54

The construction of a dam to retain water causes more interference with natural conditions than does any other civil engineering operation. The validity of this assertion will be realized after even cursory consideration; the statement will later be amplified and explained, but of itself it constitutes a leading reason for the devotion at some length of special attention to the foundations of dams. Equally striking is the critically important function that dams perform in storing water for domestic supply, for the generation of power, for flood control, for recreation, and for irrigation. The reliance that must be placed on structures carrying out these functions, together with the fact that they exist in all parts of the world and in surprising numbers, ranging from the smallest timber-check dam to such structures as the Lloyd Barrage Dam or the Hoover Dam, provides further reason for the devotion of particular attention to the design of dam foundations. Finally, although failures of civil engineering works are always of serious consequence, failures of dams are possibly more serious than others, since they generally occur during periods of abnormal weather, often without warning, and almost always with disastrous results. Defects in foundation beds are an unfortunate factor in many dam failures, and another telling argument is thus presented to support the necessity of neglecting no single feature of foundation beds that may possibly affect the dam that is to rest upon them.

25-2   CIVIL ENGINEERING WORKS

Although the benefits to be derived from a critical study of engineering failures are duly stressed in other parts of this book, it may be noticed that constructive examples are offered, whenever possible, of works carried to successful completion with the aid of the methods and investigations described. A slight departure from this course will be made in the case of dams, and a brief review will be presented of some failures of the past. This is no descent to the almost hysterical attention devoted to dams whenever a failure does occur. At these times the pages of the popular press would almost lead one to believe that failures of dams are a regular occurrence, and on these occasions, not a word is said of the thousands of dams successfully performing their tasks, nor is the fact often mentioned that failures in many other branches of civil engineering, e.g., water purification, are equally liable to endanger the lives of people. Failures of dams are presumably "good news." It is not as news items that they will be considered here but rather as useful illustrations of some of the reasons for the failure of engineering structures and of the significance of foundation-bed conditions among these reasons.

25.1 HISTORICAL NOTE

Among dams of ancient times, that constructed by Joshua to help the Israelites cross the river Jordan is one of the earliest to be recorded, although the Marduk Dam across the river Tigris is even older, its construction having been carried out in almost prehistoric times. Built for river regulation, it lasted until the end of the thirteenth century. The first masonry dam of which there are good records was that built by Menes, first king of the first Egyptian dynasty, some time before 4000 B.C. It was constructed to divert the waters of the Nile to provide the site for the ancient city of Memphis, about 19 km (12 mi) to the north. It was 457 m (1,500 ft) long and at least 15 m (50 ft) wide. It was maintained for 4,500 years and then neglected. Another famous old dam was that near Yemen in Arabia, a dam of which little is known beyond its construction date (1700 B.C.), its great size (3.2 km or 2 mi long, 36 m or 120 ft high, and 150 m or 500 ft wide), and the fact that it failed about A.D. 300 in a flood recorded in Arabian literature.[25.1]

These few examples are cited to illustrate the longevity of dam building as a major branch of civil engineering. Further examples could be described; but as information about foundation-bed conditions is almost completely lacking in ancient records, they would not contribute to the purpose of this chapter. As an example from comparatively modern times, the Puentes Dam in Spain may usefully be mentioned. Constructed in the years between 1785 and 1791, it was 282 m (925 ft) long, curved in plan, and 45.6 m (152 ft) thick at the widest part of the base, with a maximum height of 50.1 m (167 ft), a perfectly safe design from the structural point of view. Constructed of rubble masonry and faced with cut stone, it was finished off with ornamentation befitting its standing as one of the wonders of Spain. Although the dam was apparently founded in part on bedrock, a gravel pocket about 20.1 m (67 ft) wide was encountered near the center of the site, and the founding of the dam across this gap was solved by an ingenious piled design carrying a grillage and protected by an upstream apron. This method of overcoming geological difficulties served satisfactorily just after the dam was built, but it was not reliable, and on 30 April 1802, that part of the dam above the earth pocket failed, following the washout of the underlying foundation beds. Still available is an eyewitness account of the disastrous collapse of this "plug" of masonry, which went out like a cork, leaving what appeared to be two massive bridge abutments with a gap between them 16.8 m (56 ft) broad and 32.4 m (108 ft) high.[25.2] This failure was somewhat exceptional, since other large dams constructed in the same period functioned successfully and thus inaugurated modern practice of dam design and construction.

25.2 FAILURES OF DAMS

The record of failures of dams in succeeding years provides a useful if a somewhat melancholy study. Analysis of the causes of failures indicates fairly definitely that the main reasons have been (1) failure to provide adequate spillway capacity, and (2) defective foundation-bed conditions; these two factors have accounted for the majority of all recorded failures. Spillway capacity is determined mainly from anticipated runoff from the catchment area above the dam, a subject on which geological conditions have a considerable indirect influence, as will be explained in Chap. 26. The second cause of failures noted is dependent essentially on geological features, although the specific reason for failure may vary from one case to another.

Several reviews of dam failures have been prepared, and the number of failures so listed is remarkable. Lapworth mentions that over 100 failures of dam structures due to undermining of water-bearing beds below the dam foundation occurred between 1864 and 1876.[25.3] Justin lists details of over 60 failures of earth dams alone between 1869 and 1919.[25.4] The study of failures in such numbers, although interesting and impressive, is not helpful with regard to avoidance of trouble in future work; only a few special cases will therefore be referred to in detail, since specific study of individual failures can be so helpful. In his paper, Dr. Lapworth mentions several interesting cases, including the Hauser Lake failure which bears a striking similarity to the Puentes Dam failure, since a central portion 120 m (400 ft) long, founded on water-bearing gravel, was destroyed in 1908 while the remaining sections were left intact.

An example from relatively recent practice is that of the Austin Dam, which was constructed in 1893 to provide a water and power supply for the city of Austin, Texas. Difficulties were encountered during construction. Eventually the dam was finished and placed in service; it was of masonry, 327 m (1,091 ft) long, 20.4 m (68 ft) high, and 19.8 m (66 ft) wide, resting on Cretaceous limestone, clays, and shales. The limestone strata were almost horizontal but alternated in texture, some much harder than others. It was said that the limestone was dissolved in places, giving rise to underground caverns, one of which was described by workmen but is not mentioned in engineering reports. The clays and shales were slippery and very broken, since the site was on a fault zone. Flood flows over the dam caused some erosion of weaker strata and at least one serious washout, which was repaired. This inevitably contributed to the sudden failure of the structure on 7 April 1900, when a central section collapsed completely; the dam was then overtopped by 3.3 m (11 ft) of water because of severe flood flow. The concentration of flow through the resulting gap moved two adjacent blocks, each about 75 m (250 ft) long, downstream for a distance of about 18 m (60 ft). The causes of failure are complex, but they appear to include removal by erosion of rock supporting the heel of the dam, the slippery nature of at least part of the foundation bed, and the presence of percolating water under the base of the dam.[25.5] The gap was closed and the dam placed in service again in 1915, but shortly thereafter another flood took out 20 crest gates and blocked the tailrace and turbine draft tubes with debris. Not until April 1940, were the dam and the associated powerhouse finally put into continuing service; a thorough job of reconstruction included extensive grouting and foundation rehabilitation that can really be called underpinning, even though applied to a dam.[25.6]

The failure of the St. Francis Dam will be within the memory of some readers. This great gravity dam, 61.5 m (205 ft) high and 210 m (700 ft) long, was completed in 1926 by the Bureau of Water Works and Supply of the city of Los Angeles to assist in the storage of water for that well-known California community. It was located in the San Francisquito Canyon. Curved in plan and connecting with a long, low, wing wall, it created a reservoir with a capacity of 47 million m³ (38,000 acre-feet). The dam was founded partially on schist and partially (for about one-third of its length at the southwest end) on a reddish conglomerate with sandy and shaly layers. The contact between the two rocks is a fault, generally recognized to have been long inactive. Although the schist is a relatively sound rock, the conglomerate

> ... is by no means a strong rock. A test ... gave a crushing strength of 500 pounds to the square inch.... When wet, the rock shows a considerable change, [a sample starts] to flake and crumble when placed in a beaker of water and in about 15 minutes slumps to the bottom of the vessel as a loose gritty sediment that can be stirred about with the finger.... So far as can be ascertained, no geological examination was made of the dam-site before construction began and no crushing or immersion tests were made of the conglomerate.[25.7]

Seepage was noticed through the conglomerate when the reservoir first became full (in March 1928); eventually, the dam failed on 12 March 1928, with a tragic toll of 426 lives and untold property damage (Fig. 25.1). Boards of inquiry were appointed by the state of California and by the city of Los Angeles; both agreed that the main cause of failure was the nature of the rocks under the dam. Quoting Dr. F. L. Ransome again:

> The plain lesson of the disaster is that engineers, no matter how extensive their experience in the building of dams or how skillful in the design of such structures, cannot safely dispense with the knowledge of the character and structure of the adjacent rocks, such as only an expert and thorough geological examination can provide.[25.7]

Just about 30 years later, an almost equally tragic failure occurred; 344 people were killed when the Malpasset Dam in southern France collapsed after several days of unusually severe rain (Fig. 25.2). The dam was completed in 1954; it was a thin arched dam of reinforced concrete, its greatest thickness only 6.8 m (22 ft 8 in), and its height about 60 m (200 ft). Curved to a radius of 103 m (344 ft), it was built in a narrow gorge of Le Reyran River for the department of Var to serve both irrigation and water supply. Its sudden collapse on 2 December 1959, resulted in a catastrophic flood that carried everything before it for 12 km (7 mi) downstream; most of the people who lost their lives lived in the town of Fréjus, which was in the path of the flood. The preliminary report of the

FIG. 25.1 The remains of the St. Francis Dam, California, after the structure had failed.

FIG. 25.2 The site of the Malpasset Dam on Le Reyran River in southern France just after this thin concrete-arched dam had failed.

commission of inquiry, established by the French Ministry of Agriculture, confirmed that the structural design, although daring, was sound. After reviewing all possible causes for the disaster, the commission was forced to the conclusion that the principal cause of the catastrophe was a rupture of the rock below the foundations, a rupture that induced substantial displacement, notably of the abutment, and so the destruction of the dam. The bedrock was a mica schist, reported to be sheared and jointed, with a wedge-shaped mass overlying a clay-filled seam.

In the actual wording of the conclusion of one of the official commissions, given in the original language to avoid loss of any nuances in translation:*

> La cause de la rupture doit être cherchée exclusivement dans le terrain au-dessous du niveau des fondations ... et ... la cause la plus probable de l'accident doit être attribuée à la présence d'un plan de glissement ou faille amont supérieure ... qui suivait presque parallèlement, sur une grande longueur et à faible distance, les fondations de la voûte dans la partie haute de la rive gauche. La déformabilité déjà grande du terrain de fondation a été aggravée localement par la présence de ce plan de glissement. L'ouvrage n'a pu s'adapter à cette aggravation.[25.8]

Commenting upon this tragic failure two years after the collapse, Dr. Karl Terzaghi had this to say:

> A conventional site exploration, including careful examination of the rock outcrops and the recovery of cores from two-inch boreholes by a competent driller, would show—and very likely had shown—that the rock contains numerous joints, some of which are open and filled with clay. From these data an experienced and conservative engineering geologist could have drawn the conclusion that the site is a dangerous one ... the most advanced means of rock explora-

---

*The cause of the failure is to be sought solely in the ground below the level of the foundations ... and ... the most probable cause of the accident is to be attributed to the presence of a plane of sliding, or superior upstream fault, ... which for a great length and at a slight distance followed almost parallel to the foundation of the arch in the upper section of the left bank. The susceptibility of the foundation material to deformation, already great, was increased locally by the presence of the sliding plane. The work was not capable of adapting to such aggravation.

tion at our disposal ... would have added no significant information to what was known concerning the safety of the left abutment of the dam, before the failure occurred.[25.9]

This extended reference to the failure of the Malpasset Dam has been made in view of the great concerns which the failure created in the engineering profession around the world, especially after some of the engineers responsible were brought to court to answer charges of imprudence; extended reference is made also because of the profound significance of Dr. Terzaghi's comment. No amount of the most sophisticated testing of rock (or soil) will be of effect if the basic geology of a foundation site is not first determined to be sound from the geological point of view. As further illustration of this principle of dam foundations, there may be mentioned an example of a dam failure within the boundaries of a great city. The dam was well designed and constructed but could not perform as designed when a failure of foundation beds took place. The location of this dam is of significance, since the urban development which surrounded it, particularly downstream, lent unusual importance to its safety.

Within the limits of Los Angeles, an earth dam 69.6 m (228 ft) high and 195 m (640 ft) long and containing 650,000 m³ (850,000 yd³) of soil was constructed to provide what came to be known as the Baldwin Hills Reservoir; the reservoir was intended for the storage of water coming to the city from the Owens River and Colorado River aqueducts. The site used was convenient for supplying a rapidly growing residential area in southwest Los Angeles. Construction started in 1947 and was completed in 1951. Excavation revealed a zone of Late Pleistocene or Recent normal faults, with a displacement of 18 m (59 ft), that were believed to be a part of the active Inglewood fault system. The gate tower was relocated to avoid this fault zone. The dam was naturally operated with the greatest care and

FIG. 25.3 The Baldwin Hills Reservoir, Los Angeles, California, after it had been emptied following failure of the earth retaining dam; the breach can clearly be seen.

kept under constant supervision, with special inspections made regularly. One such inspection was made on 3 April 1963, but on Saturday afternoon, 14 December of the same year, the dam failed, releasing most of the water stored in the reservoir in a disastrous flood (Fig. 25.3). Leakage that started late on that Saturday morning enabled the authorities to issue warnings. With the efficient aid of the Los Angeles Police Department, whose men on motorcycles, in cars, and in helicopters used bullhorns to alert all residents, a mass evacuation from the area below the dam was made, but despite these efforts five persons were drowned, some trapped in automobiles. Naturally great property damage was done, estimates ranging to more than $10 million.

In the ensuing legal actions, the city of Los Angeles, on its own behalf and for the Los Angeles Department of Water and Power, brought suit against a group of important oil companies, contending that the extraction of petroleum, water, and natural gas from the ground at the Inglewood oil field near the reservoir caused land subsidence that triggered the collapse of the dam, the design and construction of which were not called into question. The total claimed was $25.8 million. The matter was settled out of court for a total amount to be paid to the city of $3.875 million. Extensive studies of the dam failure were made, naturally, and these revealed—if nothing else—how complex were the geological questions that were raised by the accident.[25.10]

Finally, brief reference must be made to the failure of the Teton Dam in Idaho on 5 June 1976, not only because it is one of the most recent of major dam failures but because of the wealth of information made publicly available on all aspects of the design, construction, and failure of the dam by its builders, the U.S. Bureau of Reclamation. The frank disclosure of all necessary information by the Bureau was an example of professional engineering at its very best, but despite this, the real cause of the failure will probably never be fully known, since the collapse of the structure took with it the essential evidence. The dam, on the Teton River in eastern Idaho, was 92 m (305 ft) high above river bed and 122 m (405 ft) above the lowest point in the foundation. Its crest was about 900 m (3,000 ft) long. Built to provide flood control, power generation, recreation, and (most importantly) irrigation water, it was topped out in November 1975. Filling of the reservoir

FIG. 25.4 Remains of the Teton Dam in Idaho, after its failure.

behind the dam was almost complete when it failed. The foundation bed was volcanic rhyolite, heavily jointed and cracked, there being even some man-sized openings found beneath the dam location. Recognizing these conditions fully, the USBR carried out the most extensive grouting program that they had ever undertaken, a total of 16,000 m^3 (570,000 ft^3) being injected. Despite this, the dam failed and killed 11 people, rendered 25,000 homeless, and caused about $400 million in damages.

Two days before the failure, two small leakages downstream of the toe of the dam were noticed. Early on the day of failure, an engineer noticed a small leak about one-third of the way up the dam, close to the right abutment. Attempts were made to plug the leak but in vain, complete failure taking place later the same morning, a large part of the dam being washed out as can be seen from Fig. 25.4. Two expert commissions of inquiry were immediately established, and other studies of the failure were made. It seemed to be the consensus that the failure took place because of piping in the loesslike silt, of which impermeable material the main embankment was constructed. What caused the piping can now only be conjectured, possible causes being failure of some sort in the relatively narrow key trench across the foundation, unsealed fractures in the bedrock, or the absence of filters between the bedrock and the core material. A large area of wet soil found when the remaining part of the dam was carefully studied added still further to the complexity of determining exactly why the Teton Dam did fail. The overall message for the future, however, is clear—a reemphasis upon the necessary essential integrity of foundation bedrock and a consideration of the interaction between bedrock and core material in all earth dams.[25.11]

25.3 INSPECTION AND MAINTENANCE

It was an ironic coincidence that, in the months before the Teton Dam failure, there had been much discussion about dam safety and, correspondingly, about the regular inspection and maintenance of dams. In November 1977, the President of the United States initiated a major program of dam inspection, the U.S. Corps of Engineers being charged with the inspection of 9,000 nonfederal dams, with special reference to their safety. This program supplemented the dam inspection programs of individual states, which were expected to cover the remaining 40,000 dams (generally more than 7.5 m or 25 ft high) known to exist. Previous to this, in April 1977, President Carter had charged agencies of the United States government with ensuring the adequacy of their own dam safety programs. Most developed countries have similar ongoing programs for dam inspection, quite often inaugurated after some serious dam failure in the country concerned. In Great Britain, for example, the first relevant legislation appears to have been the Reservoirs (Safety Provisions) Act of 1930, passed after a number of dam failures. Annual inspections are now obligatory for all structures above a certain size. The restriction of such legislative requirements to dams was at first a matter of some surprise to many engineers, especially as the requirements have been usually initiated as a result of some notable failure. But the vital importance of dams, their frequently isolated locations, the fact that they are often seen regularly by very few people, and the tragic consequences of any failure of a dam have combined to counteract these early impressions.

With or without state regulations, however, regular inspection of all dam structures is an essential requirement. Naturally, examination is restricted to those parts of the dam which can be seen—together with the inspection of galleries, tunnels, and shafts that have been left open after the completion of construction. The practice in certain large dams of leaving exploratory tunnels open is eminently desirable; this means of access is invaluable. One certain guide in inspection is the leakage of water through or around the dam. Records of this will naturally have been kept regularly, and any increase, either sudden or gradual, will be a call to remedial action. Study of the leakage, the sediment it carries, and its chemical content may reveal the reason for the increase and suggest remedial measures. Such study of leakage in extreme cases when the damage is irreparable will at least permit due warning of impending trouble. Failures should become more of a rarity as the years go on, especially if, in the course of increasing the safety and reliability of dam foundations, due attention to the associated geological conditions is always regarded as a factor of the very greatest importance, not only during design and construction but also in all future maintenance and regular inspections.

The ultimate dependence of all dams, as also of all other structures, upon the geology of the foundation beds upon which they rest is so obvious that one would imagine that this is where the inspections of dams would always start. If, as has been suggested to the authors, this has not always been the case, the oversight must have been due to some administrative misunderstanding, since it is to be expected that all those engaged on this vital work will realize the vital contributions that engineering geologists can make to such work.

Reliance upon interpretation of aerial photographs as an economical way of inspecting dams carries with it a similar lack of appreciation of the importance of geology. Aerial photographs can be useful for general surveys, but they cannot take the place of detailed examination of ground conditions on foot, especially in the vicinity of dams, where occasionally quite minor details of observation may prove to be of unusual significance. Such "foot-slogging" field inspection work will be greatly assisted if complete as-constructed drawings of the dam foundations are available. In some cases such drawings will contribute directly to the understanding of later observations. Their importance, especially in regard to the future safety of the dam for which they were prepared, cannot be overemphasized.

25.4 REVIEW OF DAM CONSTRUCTION

A return may now be made to the statement with which this chapter started, and consideration may be given to the effects of dam construction. A dam is an artificial structure erected to support a waterproof membrane designed to retain water above the level that it normally occupies at the site of the dam; suitable provision is made for passing a certain calculated flow of water past the dam, through it, over it, or around it, depending on local circumstances. The membrane may be, and generally is, an integral part of the dam structure. It may be supported in several ways by such varying designs as (1) an earth-fill or rock-fill dam, in which the membrane is either on the upstream face or in the center as a core wall; (2) a gravity dam of masonry or mass concrete, in which the membrane is the upstream face of the dam itself; or (3) a reinforced-concrete dam of the arched type, multiple-arch, multiple-dome, or some other special design, in all of which an unbroken, reinforced-concrete skin serves as the waterproof membrane. It will be seen that dams can be generally grouped into two main divisions: earth- or rock-fill dams, which depend for their stability on the natural repose of unconsolidated material, and concrete or masonry dams, which depend for their stability on the structural performance of the material used for construction. The type of dam to be constructed at any location must be determined mainly from geological considerations; the actual kind of dam to be constructed, once its general type has been decided, will also be dependent to some extent on geological conditions affecting the supply of structural materials.

Since all dams retain water to a certain predetermined level, the flow of water in the watercourse being regulated is seriously affected below the dam site; the flow is generally regulated to a more uniform discharge than that given by the stream itself. In addition, the underground water conditions in the valley above the dam location are completely changed; the level of the groundwater table is raised at least to the water level of the reservoir near the water line, and changes of decreasing importance occur farther up the valley. Below the dam, the level of the water table may be lowered if normal streamflow is depleted. Between the two sides of the dam, there is thus set up a considerable difference in groundwater level. Although the waterproof membrane generally extends from side to side of the dam site, effectively isolating the two groundwater tables, this artificial condition will always exist while water is being retained.

The structure will exert unusually high unit pressures on certain parts of the underlying foundation beds. The beds will be submerged well below water level in the reservoir area and so will be subjected to hydrostatic pressure, which may be appreciable in high dams. These are the main reasons underlying the assertion that a dam causes more interference with natural conditions than does any other type of civil engineering structure. In the special case of weirs, or dams founded on pervious strata, in which the waterproof membrane merely deflects the flow of groundwater but does not stop it completely, the serious change in the underground conditions caused by the dam will be obvious.

In the case of all dams designed to seal completely the flow of water, surface and underground, down the valley which they cross, the action of the dam in resisting the pressure of the retained water gives rise to four main geological problems: (1) determination of the soundness of the underlying foundation beds and of their ability to carry the designed loading; (2) determination of the degree of watertightness of the foundation beds at the dam location and of the measures, if any, required to render the underlying geological strata quite watertight; (3) a study of the effect on the foundation bedrock of prolonged exposure to water; and (4) an investigation of the possibility of earth movements at the site of the dam and determination of the measures to be taken as a safeguard. Some of the geological problems affecting the reservoirs formed by dams may be encountered also during construction; but for convenience, they will be considered separately in the following chapter, which is devoted specifically to reservoir problems.

Dams founded on pervious foundations present their own special problems, problems mainly associated with the controlled flow of water beneath the structure. It may not be possible to make any dam foundation absolutely tight, but the slight leakage occurring with dams on so-called "impermeable" foundation beds, although associated with uplift pressure, is generally of no other consequence. For those dams on admittedly permeable foundation beds, the exact nature of the flow of water through the underlying strata is a vital part of the design; thus, accurate knowledge of the water-carrying properties of the unconsolidated materials encountered becomes a matter of importance. The accumulation of sediment above dams, deposited as the water comes practically to rest in the reservoir formed by the dam, is another associated geological problem; it will be studied generally in Chap. 41. The scouring out of exposed strata immediately below dam structures, because of the excessive velocities of flood flows discharging over or around the dam, is also a matter of importance in the design of dams; it, too, is reviewed in Chap. 41.

25.5 ANCHORING DAMS TO FOUNDATION BEDS

Many special problems naturally arise in connection with the construction of dams, so varied are the sites in location and nature and so great are the possibilities for different designs. Since special problems, interesting though they may be, do not usually have a general application, they cannot receive attention here, apart possibly from one installation which was, for a time, unique. This is the Cheurfas Dam in Algeria, constructed in 1882, with a maximum height of 32 m (107 ft) (Fig. 25.5). Soon after construction, one abutment of the dam was washed away, but it was reconstructed and served well until recent years. An examination in 1927 disclosed serious weakness; and as the most economical means of strengthening the dam, 25-cm (10-in) -diameter holes, 3.9 m (13 ft) apart, were drilled from the crest of the dam to the foundation and then to a depth of 21 to 24 m (70 to 80 ft) into the calcareous sandstone, limestone, yellowish sandstone, and argillaceous marl on which the dam is founded. Specially prepared steel cables were then anchored in these holes to the foundation strata, secured in special caps on the dam crest, placed under great tension, and then anchored; their effect was to "tie down" the dam to the foundation strata (Fig. 25.6).[25.12]

The same system of tying down a dam into its foundation beds was successfully applied to the Tansa hand-placed masonry dam after the structure was found to be in an unsafe condition.[25.13] The Tansa Dam is 52 m (173 ft) high; it serves the water supply of Bombay, India. This method was also employed to raise the spillway level of the Steenbras Dam in South Africa by 2 m (6½ ft) in order to provide a quick and economical increase in the reservoir capacity for the water supply of Cape Town. The character of the Table Mountain sandstone, upon which the Steenbras Dam is founded, is such that percussion drills could not be used successfully; holes for the post-tensioning cables had to be sunk by diamond drilling. All cables were carefully grouted into place and tested beyond their

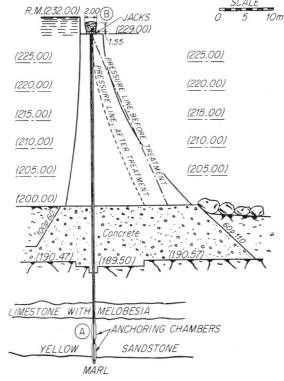

FIG. 25.5 Cross section of the Cheurfas Dam, Algeria, showing the foundation strata and the method of anchoring the dam to them.

FIG. 25.6 Close-up view of one of the anchor heads at the Cheurfas Dam, Algeria.

FIG. 25.7 Anchorages for the tying down of a lock wall forming part of the John Hollis Bankhead Dam on the Black Warrior River, Alabama.

design load; six cables only, of the total of 326, failed to meet the stringent test conditions.[25.14]

This ingenious method of strengthening old dams was pioneered at Cheurfas by the famous French engineer, Andre Coyne. With the two exceptions just noted, it took a little time for the method to be applied to other dams but the development of satisfactory strain gauges for checking on embedded anchors was one factor which resulted finally in its wider use. The first application in England appears to have been for the raising of the Argal Dam of the Falmouth water-supply system. Embedded post-tensioned anchors into the underlying bedrock permitted the addition of 3 m (10 ft) to the height of this masonry-faced, 30-year-old dam. This doubled the capacity of the reservoir which it formed, at a fraction of the cost of a new dam.[25.15]

One of the first, if not the first, applications in the United States was for the strengthening of the 50-year-old John Hollis Bankhead Dam on the Black Warrior River in Alabama, a project of the U.S. Corps of Engineers (Fig. 25.7). The 345-m (1,150-ft) -long spillway structure did not satisfy the stability requirements of the Corps when checked as part of major reconstruction work. An extensive test program was carried out, including the placing and stressing of 16 test anchors into the underlying sandstone, the record of which is a useful guide for other possible installations. Only after the satisfactory completion of tests was a contract awarded (in 1967) for the installation of the main group of 135 anchors.[25.16] Another half-century-old dam, the Ryan Dam on the Missouri River (just upstream of Great Falls), was similarly strengthened in 1970 by its owners, the Montana Power Company, by tying the old structure into the underlying Cretaceous sandstone and shale. The advantage of being able to carry out this work without interference to the associated power station was again a powerful argument in favor of using this method.[25.17] And finally, an inspection of the Conowingo Dam of the Philadelphia Electric Company, in keeping with the 1965 order of the Federal Power Commission, established the need to strengthen this major structure. A detailed geological study of the foundation beds pointed to the use of tiedown anchors as a possible solution, and this was adopted in 1977.[25.18]

These brief references to strengthening existing dams have been included since there will probably be an increasing call for such remedial works in the years ahead. Remedial work will increase if only because, as the material status of Western civilization advances, the best sites for dams in convenient and economical locations will be found to have been used up. Concurrently, advances in design concepts and construction methods are permitting the construction of dams at sites earlier found to be unsuitable. With all such developments, the need for accurate knowledge of every detail of site conditions is becoming more urgent and more important, rather than the reverse, so that the geology of dam foundations will continue to be a vital part of engineering geology.

25.6 PRELIMINARY WORK

Since the location of a proposed dam will generally be restricted by topographical, economic, and social considerations, the areas to be examined as possible sites will be fairly well defined. In general, preliminary studies will follow the lines already suggested in Chap. 10. Accurate geological sections along possible lines for the dam will be a major requirement. The nature of valleys will sometimes mean that bedrock is some distance below ground level, and therefore the determination of the rock surface across the valley will often be a first step in the preparation of the section. Geophysical methods can be of great assistance in this work, when utilized in connection with strategically placed boreholes. The existence of buried valleys must be checked carefully, not only from test-bore results but also from general studies of the local geology, for valleys 21 m (70 ft) deep have been found between two boreholes as close together as 15 m (50 ft).

The possible presence of boulders is another danger to be guarded against, especially in glacial formations; the greatest care must be taken with test drilling. The almost insuperable difficulties met during the construction of the Silent Valley Dam for the Belfast City and District Water Commission in Northern Ireland were due very probably to the assumption that granite boulders encountered by the original test boreholes were solid rock. Although a rock foundation bed had been anticipated at a depth of 15 m (50 ft) below ground level, the cutoff trench actually had to be carried to a depth of 54 m (180 ft), through running sand of such a nature that the use of the maximum working pressure of compressed air served to reduce the water level by only about a meter (Fig. 25.8). The final solution of the problems thus encountered provides a fascinating but sobering study.[25.19]

In the determination of the geological sections, local topographical detail and, more particularly, local geological features must always be given special consideration in valleys that have been subjected to glaciation. It was from surface observations at the site of the Vyrnwy Dam for the water supply of the city of Liverpool, England, that Dr. G. F. Deacon (joint engineer with Thomas Hawksley) deduced that the valley was the site of an old glacial lake, held up at one time by a rock bar. The center line of the dam, which was the first large dam to be built in Great Britain, was located for the necessary parliamentary plans along the inferred position of the rock bar (Fig. 25.9). Subsequent detailed investigation by trial holes and shafts proved the inference to be practically correct, and the dam was built as originally located. Dr. Deacon estimated that a deviation of the center line up or down the valley of only 200 m (656 ft) would have added £300,000 to £400,000 ($1,500,000 to $2,000,000) to the cost of the work.[25.20] As a contrast to this example may be mentioned a small dam known to one of the writers which was constructed without any such preliminary studies of the rock surface; as a result, several thousand dollars was wasted because the center line was only 15 m (50 ft) upstream from what would have been a better location.

FIG. 25.8 Reservoir in the Mountains of Mourne, Northern Ireland, formed by the Silent Valley Dam (a photograph taken soon after the dam, despite all difficulties, was completed.

FIG. 25.9 Excavation in progress for the Vyrnwy Dam in northern Wales for the water supply of Liverpool, England; this is the southwest end of the site, as exposed in June 1882, showing the glaciated surface of the bedrock (and, incidentally, being an early example of a "progress photo").

Whenever feasible, the rock surface should be traced right across the valley being investigated. This is not always possible, since faulting sometimes provides clefts filled to great depths with unconsolidated material. One of the authors has carried out some investigations in the northern part of Ontario, in a rock gorge barely 15 m (50 ft) wide through which passed the whole flow of a fairly large river. Solid diabase extends for a great distance on either side of the gorge, yet diamond drill holes carried 54 m (180 ft) below water level in the riverbed failed to reveal solid rock; the holes passed through boulders, gravel, and finally more than 30 m (100 ft) of compacted sand. This is typical of valley faults, quite a number of which are to be found in Canada; they always demand very careful study. Confirming what has just been said about previously unsuitable sites for dams now being utilized, 40 years after the early study of this rock gorge, further site exploration was carried out. As a result, and in view of the advances made in earth-dam design through soil mechanics research, a 51-m (170-ft) -high earth dam was designed and constructed at this remarkable site, the deep gorge being bottomed out at a depth of 69 m (230 ft) below river level, the final excavation naturally being handwork of a particularly intricate character (Fig. 25.10). The narrow gorge, in graywacke, was found to have been caused by erosion along one of the shear zones, striking northwesterly, which excavation revealed.[25.21]

One of the precepts of the engineering geologist is *never to take anything for granted* in site studies for engineering works; especially is this so for dam sites. The necessity for this was probably never better shown than in the geological study carried out on the Vltava River in Czechoslovakia for the Orlik Dam, the highest dam in the country when constructed in the mid-1960s and the highest of a series of dams built, or planned, for the development of power on this famous stream (Fig. 25.11). The dam is located 91 km (57

FIG. 25.10 The Lower Notch gorge on the Montreal River, Ontario, as revealed during excavation for the water power dam now at this location; scale can be gauged from construction plant to be seen.

FIG. 25.11 The Orlik Dam on the Vltava River, Czechoslovakia, south (upstream) of Prague.

DAM FOUNDATIONS   25-17

mi) upstream of the capital city of Prague; it is 91 m (300 ft) high and 515 m (1,700 ft) long at the crest. The geology around the proposed site for the dam was known to be generally of Proterozoic metamorphic rocks of igneous origin, forming here part of the mantle of the great Central Bohemian Pluton, which consists of a complex mixture of plutonic igneous rocks. Geological conditions at the site appeared to be favorable, the Vltava River flowing though a narrow valley in crystalline schists.

Careful geological reconnaissance showed, however, that superficial geology was misleading. The narrowing of the valley at the site selected was due, not to solid resistant rocks in situ, but to a massive rock slide of Pleistocene age. Detailed study of the site through shafts, test pits, and test boring confirmed the geological deduction. The chosen site was abandoned and, on the basis of geological advice, a new site selected 200 m (650 ft) upstream of that originally chosen, which was upstream of the slide area. The bed of the valley here consisted of schistose, altered amphibolite-rich rocks, suitable for the founding of a mass-concrete gravity dam, which may now be seen at this geologically interesting location.[25.22] (Construction of the dam raised the normal water level for a distance of 60 km (38 mi) above the dam with the result, typical for Europe, that many buildings and structures were either flooded or disturbed. Much remedial work had to be carried out, therefore, perhaps the most interesting being the underpinning of the castle of Orlik by the grouting of the supporting cliff of granodiorite and the anchoring of its exposed face with prestressed rock bolts.)

Different again were the site conditions which had to be investigated for the proposed Auburn Dam on the North Fork of the American River, planned as a part of the great Central Valley Project of California, a project of the U.S. Bureau of Reclamation. The dam was initially designed as a double-curvature, concrete arch structure, with a height of 205 m (685 ft) and a crest length of 1,245 m (4,150 ft); it would thus have been one of the largest arch-dam structures of the world. Rocks at the site consisted chiefly of amphibolite with minor metasediments, generally of a competent nature. These rocks are, however, interlayered with considerably weaker zones composed of talc schist, chlorite schist, and occasionally talcose serpentine. The importance of the proposed dam and the character of the foundation materials led the Bureau to initiate one of the most extensive site studies ever undertaken for any such structure. Designed to give complete information for the dam site and the area within 150 m (500 ft) of its axis, the investigation involved approximately 8 km (5 mi) of surface trenches, 1,065 m (3,550 ft) of exploratory tunnels (six of them), 11 shafts totaling 669 m (2,230 ft) driven from selected points within the tunnels, five raises totaling 308 m (1,028 ft), and a shaft of 45 m (150 ft) leading into one of the tunnels. Test drilling, yielding NX diameter cores, was carried out from 306 holes and totaled 26,400 m (88,000 ft).

To facilitate all this work and detailed examination of all the exposed bedrock in the foundation area, the entire site was stripped and cleaned down to bedrock. Small wonder that ten years and $280 millions were spent by the Bureau on its site investigation and design work. Occurrence of an earthquake in August 1975, 80 km (50 mi) from the dam site, raised questions about the seismicity of the proposed site and the safety of the proposed arch dam. This led to external studies and much public controversy with the result that proposed construction was postponed, the future of the project being undecided as this volume goes to press. No matter what the final decision may be, nothing can detract from the remarkable and exhaustive site investigation, an example on the largest scale of what should be the procedure for the sites of all proposed dams.[25.23]

The records made available from preliminary site studies will usually enable those in charge of the engineering work to decide on the general type of dam to be used and to begin their economic and design studies. If, for example, the rock floor cannot be reached

25-18　CIVIL ENGINEERING WORKS

by ordinary drilling methods, an earth- or rock-fill dam or a "floating" concrete structure may be necessary. If a rock bed is found to be available at reasonable depths, economic considerations will guide the designer in a choice between an earth- or rock-fill dam and one of the gravity or other structural type. The choice of the kind of dam to be used, although so closely related to geological conditions, must be left for study elsewhere, since it is essentially an engineering problem. Enough has been said to indicate the amount of geological survey work necessary before even a choice can be made. Once the choice has been made, geological work must continue, specially adapted to serve the particular requirements of the kind of dam now contemplated; and it must be carried to such a stage that, from the information so obtained, assurance can be had that construction of the dam as proposed can be completed within the estimated cost.

25.7　EXPLORATORY WORK DURING CONSTRUCTION

It has already been indicated that investigations of the geological conditions underlying the site of a dam do not cease when active construction work begins; on the contrary, the start of excavation means, in many cases, that geological study can be extended, actual rock surfaces examined, and previously formulated opinions checked. For all major dams, the services of a geological expert will be advisable, and the expert's work should most certainly be continued until all excavation has been completed and construction begun. The final inspection of the foundation bed as prepared for use will be the most critical part of the investigation work.

The start of active construction operations will generally mean that equipment and power supply will be available for the use of aids to exploration other than those possible in ordinary preliminary work. The digging of more extensive test pits is a first possibility but one that is merely an extension of work included in most preliminary investigations. A further extension is the excavation of shafts into rock and of exploratory tunnels. The purpose of such shafts and tunnels is to permit visual inspection of the actual rock structure in checking on the geological sections across the site, previously prepared, and on the soundness and strength of the rock in relation to the designed loading. In many deep gorges, which are especially suitable for dam sites but where borings cannot easily be taken in the riverbed, underwater tunnels drifted from shafts sunk in the gorge sides can prove invaluable. Exploratory excavation work of this type can often perform a double purpose by providing access to rock well below the foundation-bed surfaces; into this rock cement grout may be injected, if this is necessary to render the foundation strata watertight. In this way, grouting can be made thoroughly effective and more extensive in its influence than is possible from surface operations.

As an example of a dam at the site of which exploratory tunnels were widely used, there may be mentioned the main dam of Le Sautet water-power development in the south of France on Le Drac, one of the headwaters of the Rhône (Fig. 25.12). Located in a magnificent gorge 180 m (600 ft) deep and 0.8 km (0.5 mi) long, the dam is 124 m (414 ft) high and yet has a crest length of only 79 m (263 ft). The dam proper is an arched structure, bearing onto the walls of the gorge, but it is backed with lean concrete designed to buttress the two sides of the gorge; the resulting cross section is almost that of a gravity dam. The rock in which the gorge is located is a limestone formation, which called for most careful exploratory work. The preliminary investigations and surveys extended over a period of ten years, and 495 m (1,650 ft) of exploratory tunnels were driven. As a result of this study, it was determined that the rock was as sound as could be desired, without faults or other imperfections. Watertightness of the gorge sides was assured by an exten-

FIG. 25.12 Le Sautet Dam, powerhouse, and gorge, Drac River, France; the dam is 124 m (414 ft) high; the extensive exploration of the limestone on which the dam is founded is explained in the text.

sive grouting program; about 6,000 m (20,000 ft) of grout holes were drilled and grouted, absorbing about 2.7 million kg (6 million lb) of cement under pressures as high as 35 kg/m² (500 psi). Much of the grouting was done from the tunnels, some of which have been left open so that further grouting can be done in the future if inspection shows this to be necessary.[25.24]

Another major aid to construction is the use of special drilling machines capable of taking out drill cores up to 1.8 m (72 in) in diameter, as described in Chap. 11. The use of large-diameter drill holes at the Prettyboy Dam in Maryland was then mentioned. The highly foliated nature of the schist revealed in the foundation bed of this dam resulted in such overbreak and general rock movement after blasting operations that other methods had to be tried for the excavation of the necessary cutoff trench (Fig. 25.13). After repeated trials, a novel expedient was adopted. Wire-rope saws were rigged up between pairs of the Calyx drill holes, and with these the necessary trenching work was successfully carried out; this was probably the first occasion on which wire saws were used on a highly quartzose rock. Figure 25.14 illustrates two of the wire-saw rigs used at the site of the Prettyboy Dam.[25.25]

The continued study devoted to all the geological aspects of the construction of the Hoover Dam provides an unusually good example of the invariable requirements of dam

FIG. 25.13 General view of excavation for the Prettyboy Dam for the water supply of the city of Baltimore, Maryland.

FIG. 25.14 Wire-rope saws in use for rock excavation at the site of the Prettyboy Dam; the drill holes had been used for rock examination by geologists.

construction work—the close inspection of all rock surfaces as they are stripped and a final and thorough inspection before concreting is started. In 1928, a board of engineers and geologists recommended after much study that the dam be located at the Black Canyon site on the Colorado River. The canyon walls were formidable, defying intimate inspection; the gorge floor was buried deep below the swift-flowing water of the river, heavy boulders, and river sediment. Predictions were made, however, explorations were renewed, and active construction soon started; by 1933, the canyon floor had been cleaned up (the river flow was diverted through tunnels), and the geological formation was thus available for direct inspection (Fig. 25.15). Dr. Charles P. Berkey was a member of the original board, and he made a final inspection at this time. It revealed that every major contention and assumption made by the board five years before was confirmed by intimate examination of the gorge in the dry.

All the rock formations of the gorge were found to be thoroughly capable of bearing the load and resisting the thrusts of the dam. Stability and general watertightness of the walls and floor as well as of the four great tunnels in the canyon walls exceeded all expectations, although grouting had to be carried out to obstruct leakage and otherwise guard against the great pressure of the water to be retained in this unusually deep reservoir. The rock of the foundation beds was found not to soften under prolonged submergence. The gorge did not follow a fault zone, but a most interesting feature revealed by excavation was the existence of an "inner gorge," forming a narrow and tortuous channel roughly along the center of the main gorge at a depth of 23 to 24 m (75 to 80 ft) below the rock benches on either side (Fig. 25.16). The side rock benches were generally smooth and uniform (although showing some potholes); but the inner gorge was pitted and fluted, being generally very uneven in form and depth. This form suggested that the whole of the pre-

FIG. 25.15 General view of the site of the Hoover Dam as finally excavated, showing the canyon walls and the "inner gorge"; the photo was taken as the second bucket of concrete was being placed on 6 June 1933.

FIG. 25.16 Close-up view of the final excavation and clean-up work in the deepest section of the "inner gorge" at the site of the Hoover Dam, showing some of the river gravel deposits.

viously superincumbent mass of sediment had moved in great flood "tides," being subject to scour, whereas in the center there had been a considerable whirling action which set up "pothole erosion" at greater depth. Although this would seem hard to believe, it was proved beyond doubt by the essential continuity of river fill from top to bottom and also by the finding of a sawed plank of wood embedded in the gravel at the edge of the inner gorge, a position that it is believed could have been reached only by burial during a recent flood. On this site now stands Hoover Dam, a tribute to all involved in its planning, design, construction and maintenance.[25.26]

It will have long since been obvious to the reader that all dams, from the greatest to the smallest, have some feature of special interest in their foundations. Innumerable examples are therefore available, but one more only can be cited to illustrate how geological investigation and construction must go hand in hand. The Serre-Ponçon Dam of Electricité de France, part of the Durance development, is a notable structure located in the French Alps at what appears to be an ideal dam site on La Durance River near the towns of Embrun and Gap. The river here flowed through a narrow gorge formed in limestone formations of the Lower Lias, close to a broad valley in Embrunais black marl. So obvious was this a site for a dam that explorations started in 1913. It was soon found, however, by means of many test borings and the driving of test shafts and tunnels that, although there was no buried glacial river course, the natural valley reached a depth below present surface level of 110 m (360 ft) and was filled with alluvial material. Even these deposits were unusual, since rock scree, falling from the sides of the gorge, had become embedded with the alluvial material and in some cases was surrounded by a clay matrix. Subsurface conditions were further complicated by the presence of thermal waters at a temperature of 60°C (140°F), highly charged with calcium sulfate leached from the gypsum Lias below and seeping upward through faults and seams in the limestone. Plans for the "obvious" mass-concrete or arched dam had therefore to be abandoned, and the site was left unused until the progress of soil mechanics made an earth dam of requisite size an economic and engineering possibility. An earth dam has therefore been constructed on this interesting site; it is 600 m (1,970 ft) wide at its crest and 650 m (2,130 ft) long along the valley floor.[25.27]

This great structure, containing 13.7 million m^3 (18 million yd^3) of material, is what the visitor to this unusually beautiful site sees today. What is not seen is all the work that had to be done in sealing off the permeable alluvial material beneath the dam. Most careful exploratory work was done in advance of the final decision to proceed with construction. The preliminary work included extensive field tests and experimental grouting, the efficacy of which was observed from specially sunk exploration shafts. Construction started with the sinking of 19 parallel rows of holes, initially for exploration of the complex subsurface but primarily for use as grout holes. Only the boreholes in the four central rows were carried the full depth to bedrock, outer rows being terminated at shallower depths. Special grouts were used, the main one being a mixture of bentonitic clay and blast-furnace wet-crushed cement; other mixtures were made in some cases without cement, others without clay. The result was a grouted cutoff curtain extending up from the bedrock surface and connecting with the impervious core of the dam itself, constructed of specially selected compacted soils, its size denoted by the fact that the core alone contains 2 million m^3 (2.6 million yd^3) of soil. Thirty thousand tons of material were injected as grout. Eleven shafts, totaling 480 m (1,600 ft) in extent, were sunk for control purposes. The result is an engineering masterpiece, a massive tribute to all phases of geotechnical engineering, starting with detailed attention to the most complex geology underlying the dam (Fig. 25.17).

Small dams have their foundation problems also, problems different in degree from those herein so briefly described but just as important. Nothing can ever be taken for granted in the founding of a dam, nor in its performance, ceaseless vigilance being essential. And in the case of every dam, no matter what the problems encountered in its found-

FIG. 25.17 The Serre-Ponçon Dam on La Durance River, France, under construction; the view shows what an ideal site this location appeared to be.

25-24 CIVIL ENGINEERING WORKS

ing, *as-constructed drawings showing every detail of the rock or soil on which the dam is founded must be prepared as the final contribution of the engineering geologist to the construction program.* Record must be made in this way of all the variations from design assumptions that geological studies during construction have revealed. *Such records are absolutely essential as an aid to continued maintenance of the dam. And they can only be prepared once, i.e., before the first concrete is placed or the first load of soil deposited for compaction.*

25.8 SOUNDNESS OF BEDROCK

It is now necessary to turn from general considerations to special geological problems that may be encountered in dam construction. The essential soundness of the strata on which a dam is to be founded is a prime requirement that must be investigated. Two questions that the civil engineer desires to have answered well before construction operations are to start concern the probable nature of the rock surface as it will be exposed on excavation and the possible presence of disintegrated rock, since estimates of cost will depend on the answers. Test pits, if carried to rock surface, may give some indication of the answers to these questions; test drilling likewise may give an indication, although with certain types of disintegration the rotten rock, when still in its original position, may give surprisingly good cores, showing its true nature only when exposed to the atmosphere. Thus it is that a careful geological study is a most necessary aid in determining whether or not disintegration is to be expected. A geologist may be able to deduce this from knowledge of the essential nature of the rock type to be encountered and of possible disintegrating influences and from a detailed study of cores obtained from test drilling.

In Washington, D.C., a bed of granite was found which had decayed, to a depth of 24 m (80 ft), to such an extent that it could be removed with pick and shovel. In the state of Georgia, limestone has been found decomposed to a depth of 60 m (200 ft); in Brazil, shales have been found disintegrated to a depth of 118 m (394 ft) below surface level.[25.28] These figures indicate how serious this problem can be. Actual examples from construction practice will elucidate the problem even further. Before the start of construction of the Assouan (Aswân) Dam on the river Nile in Egypt, in 1898, it was assumed that in the deeper channels of the riverbed, where drilling could not be carried out because of the swift currents, all decomposed granite would be eroded. When the site was unwatered, it was found repeatedly that there were depths of decomposed granite at the bottom of most of the channels; the removal of this material added greatly to the cost of the work. In one season, five times the estimated quantity of rock had to be removed; for the complete structure, excavation was 100 percent in excess of that calculated. It is of some interest to note that Sir Benjamin Baker, engineer for the work, consulted Sir Archibald Geikie, then director of the Geological Survey of Great Britain, about possible fissures in the granite; consultation apparently took place in London and not at the site.[25.29] At the Sennar Dam on the Blue Nile in the Sudan, the reverse experience was encountered; the excavation necessary was less than originally calculated, and there was a consequent saving in cost. In a North American example, a cutoff wall resting on andesite had to be carried 12 m (40 ft) lower than the level originally suggested by core borings in order to reach what was finally considered to be sound rock.

An added difficulty is that of determining exactly what is meant (or desired) by "sound rock"; the term may have quite different meanings for engineer and for geologist. The engineer associates structural soundness with the strength of the rock and its impermeability, whereas the geologist may tend to consider the matter from the mineralogical standpoint. The subject is complex, therefore, and its consideration requires experience and a close cooperation between engineer and geologist. A final note of warning must be stressed with regard to rock structure near fault zones. As exemplified by the rock encountered at the Prettyboy Dam, rock adjacent to fault planes may quite probably be brecciated and possibly useless as a sound foundation rock. Normally, fractured rock can easily be detected from its appearance if not from its location, but under special circumstances this may not be easy. When work is proceeding in very cold weather, for example, the effect of frost may convert such breccia into an apparently solid mass. In the north of Ontario, one of the authors has noted this effect; diabase which had been fractured into small fragments appeared to be quite solid when exposed at a temperature of $-35°C$ ($-30°F$) but revealed its true nature (detected from the location) only when subjected for some time to the action of steam jets. This fragmentation is almost certainly the result of exposure to permafrost conditions at some time in the geological past. It is a phenomenon encountered occasionally in areas which have been glaciated, but it has probably not always been recognized. Most careful checking is essential if there is any possibility of such fragmentation in any bedrock to be used as a foundation bed.

The structural strength of rock in proposed foundation beds is a main requirement of soundness. Compressive strength is the property generally in question; dam structures (gravity, arched, or buttressed) are designed with a maximum unit pressure at the heel as one criterion for stability, and this is equated to the maximum permissible compressive stress in the rock. Although it seems probable that no dam failure has been directly due to a failure of a rock-foundation stratum in compression, this in no way minimizes the importance of this aspect of the subject. The compressive tests usually carried out deal with rock specimens in the dry, whereas part, at least, of the rock surface supporting a dam structure will be exposed continuously to water. A further consideration of rock soundness therefore must always be the effect of prolonged exposure to water. The bedrock is to be exposed not only to water, but to water under appreciable pressure. Any failure of the rock under the influence of water may therefore lead to serious structural weakness; and if the weathered rock is for any reason displaced as it weakens, as it may be by seeping water, serious damage may be done. The failure of the St. Francis Dam provides at once a notable and most tragic example.

Preliminary testing can remove almost completely any doubts entertained on this matter, as has been instanced in many investigations. For example, tests made in connection with the Madden Dam site at Alhajuela, Panama Canal Zone, may be mentioned. Among the rock formations investigated was a bluish-gray, fine-grained sandstone, probably the Gatun formation, samples of which were tested by the National Bureau of Standards and found to have a compressive strength of 245 kg/cm^2 (3,500 psi) when dry but only 60 kg/m^2 (850 psi) when wet. Similar figures for a light-gray, medium-grained sandstone also tested, probably the Caimito formation, were 160 and 38 kg/m^2 (2,300 and 550 psi), respectively. The figures quoted are extremes. Petrographic examination disclosed the presence in the sandstones of a clay mineral occurring as a film coating feldspar and other grains; abundant glauconite contributed to the weakness.[25.30] This example serves to show the importance of tests on thoroughly wet rocks; incidentally, it illustrates the troubles that may be encountered because of the unwanted presence of clay.

Another possibility of gravity dam failure—the possibility that the dam will slide—is to some extent dependent on geological conditions. Various values have been suggested for the coefficient of friction between either concrete or masonry and mortar and rock surfaces. Tests reveal that, provided a rock floor is properly cleaned and prepared, the bond obtained between concrete and rock can be so efficient that failure will take place only by a shearing fracture of solid rock mass or of the concrete. Special preparation of rock surfaces may be necessary in the case of glaciated rock that may be worn so smooth that it must be roughened artificially in order to give a good bond. Perfect bond may be obtained between concrete and the rock surface, but the rock mass itself may slide forward under the influence of the pressure transmitted from the dam foundations. Movement of this kind will take place only along planes of weakness in the rock, such as bedding planes, and only when normal resistance to movement has been in some way removed. Naturally, geological structural arrangement will be the determining factor, since movement is possible only on bedding planes that are horizontal or that dip away from the dam. This arrangement can be determined only by adequate preliminary geological study. The Austin Dam already mentioned provides an illustrative example.

There is at least one dam under which the foundation beds were "tied together" by means of grouted steel anchor rods in order to achieve increased resistance to the downstream thrust exerted on the dam foundation by the water pressure. The Harland County Dam on the Republican River in southern Nebraska rests upon Cretaceous chalk with interbedded bentonite seams up to 7.5 cm (3 in) in thickness. There is a general upstream dip of about 2 percent, and several faults cross the site diagonally. Faults were grouted along the upstream side of the spillway section of the dam, and deep drain holes were provided in the chalk under the spillway in order to reduce uplift pressures and seepage. Anchor rods were grouted inside 15-cm (6-in) -diameter drill holes, inclined upstream on a 2:1 slope, so arranged and designed as to ensure stability against sliding.[25.31]

When the foundation beds at dam sites are finally seen, after the necessary dewatering operations and subsequent riverbed excavation, it is not uncommon for unexpected geological features to be revealed. For this reason constant attention to all geological details is imperative. Few dams have probably undergone such difficulties with foundation bedrock after the site was cleared as the Bort Dam, key structure of the Dordogne power system in southwestern France. A gravity structure, curved in plan, its exact location and shape were determined by the necessity of utilizing two types of foundation rock, a sound gneiss capable of sustaining 73,000 to 97,500 kg/m² (15,000 to 20,000 psf) and a weaker mica schist with a strength of only 2,449 kg/m² (5,000 psf); a crushed fault zone between the two formations crossed the gorge immediately under the site of the dam. Great difficulties were experienced during construction, notably with incipient slips in the mica schist when exposed in the side of the gorge. A series of reinforced-concrete "rockslide protection works" had to be constructed in order to hold up great masses of the mica schist that exhibited instability before the main mass of the dam had reached them.[25.32]

Difficulties of a different kind, but equally serious, were met during the construction of the Bhakra Dam on the Sutlej River in northern India. This great Indian engineering structure is a concrete gravity dam, 510 m (1,700 ft) long, 172.5 m (575 ft) wide at its base, and 222 m (740 ft) high from the bottom of excavation. The dam is founded on sandstone interspersed with numerous seams of a claystone-siltstone formation. General practice was to excavate these seams, when encountered, to a depth of twice the width of the seam and then to backfill with concrete. One major seam was encountered, however, which was 33 m (110 ft) wide at riverbed level and inclined at 70° to the horizontal as it crossed beneath the location of the dam. The consultants recommended that the seam be excavated to a depth of about 30 m (100 ft) below the level of the riverbed. Excavation proved to be most difficult, however, and it had to stop about halfway down to the desired elevation. The open cut so formed was then backfilled with concrete; 12-ft square shafts were left at 20-ft intervals, through which the excavation was eventually resumed. Mining methods were used, even to the extent of using light blasting charges, but the excavation was taken out only by blocks; each one was backfilled with concrete before the adjoining blocks were excavated. The work was successfully extended to the desired depth and the dam was completed.[25.33]

The Warragamba Dam was the largest dam in the southern hemisphere when it was completed in 1962; it is a major element in the water supply system for Sydney, New

FIG. 25.18 The Warragamba Dam, New South Wales, Australia, for the water supply of Sydney.

South Wales, Australia (Fig. 25.18). It is located on the Warragamba River about 64 km (40 mi) from the seacoast, in the foothills of the Blue Mountains. A 15-m (50-ft) -high weir was constructed in 1940 as a temporary water storage measure, at what appeared to be the obvious site for the ultimate high dam. Detailed geological studies of the gorge in which the weir was located, and the area around, were carried out as the Second World War came to a close, geological fieldwork being supplemented by test drilling, some test holes being put down with Calyx drills 1.2 m (4 ft) in diameter, enabling the consultant geologists to examine the local sandstone rock in situ.[25.34] (The record of this investigation is an excellent example of all that was suggested in Chap. 10.)

As a result, a new site, which was in every way superior, was found 1,200 m (4,000 ft) upstream of the "obvious" site and was adopted for the mass-concrete Warragamba Dam. The dam is 120 m (394 ft) high, and yet only 350 m (1150 ft) long at its crest, dimensions which demonstrate what an ideal site was used. The gorge has been eroded in an uplifted sandstone plateau, local bedrock being Triassic sandstones and shales, generally the Hawkesbury sandstones mentioned elsewhere in this volume. A fault zone exists under the dam site; this was carefully excavated and backfilled with concrete before construction of the dam was started. Tests on samples of the bedrock were conducted at the University of Sydney and showed strengths when wet of 60 to 80 percent of corresponding dry strengths. The most unusual feature encountered was movement of the bedrock, as exhibited in Calyx holes and smaller drill holes, a lateral movement of as much as 12.5 cm (5 in) being observed. This was attributed to the unloading of the superincumbent rock. Arrangements were made for the measurement of rock stresses and strains in the bedrock adjacent to the dam, following its completion.[25.35] Publication of the results of these observations will aid all similar dam building.

25.9 POSSIBILITY OF GROUND MOVEMENT

Earth movement in its relation to dam foundations is of vital importance. The nature of dam structures is such that, unless due allowance has been made in design, any movement of the foundation beds may lead to serious structural damage. In areas subject to earthquake shocks, allowance for seismic forces must always be made, even for dams founded on the most solid of rock formations. A usual allowance is to consider an earthquake shock with a horizontal acceleration of a fraction of that of gravity; vertical acceleration is usually neglected. In connection with the application of this design requirement to the Morris Dam, built in the San Gabriel Canyon near Azusa, California, to augment the municipal water supply of Pasadena, California, it was found that the natural period of vibration of the dam as a whole (98.4 m or 328 ft high) was 0.16 seconds or less (Fig. 25.19). It was estimated that the volume of material in this dam had to be increased by 15 percent to allow for the thickening necessary to resist seismic forces.[25.36] Special dam designs have been evolved to be proof against earthquake shocks even if quite severe. There is, for example, a special type of laminated concrete paving used for facing a considerable number of rock-fill dams in Chile; the rock fills are designed so that they will not flatten under shock; the slopes adopted (1.6 and 1.8 : 1) were arrived at after a long and careful study of rockslides, many of which had been caused by earthquake shocks, in the district adjacent to the dam sites.[25.37]

Many dams, especially those of old design, have failed as the result of earthquake shock, a fact that is not surprising in view of the relatively recent progress of seismic research work. What is perhaps more surprising is that dams built at right angles across faults along which movement has taken place have not failed. Two good examples are provided by earth-fill, puddle-core dams built across the San Andreas fault in California in 1870 and 1877, respectively. The San Andreas Dam is 28.5 m (95 ft) high above stream bed and 39 m (130 ft) above the bottom of its cutoff trench; and the Upper Crystal Springs Dam is 25.5 m (85 ft) high above stream bed and 57 m (190 ft) above the bottom of its cutoff trench. Both were constructed on clay foundation beds; the fault passes almost at right angles across their crest lines. The disastrous San Francisco earthquake of 1906 reached a maximum intensity of 10 on the rossi-forel scale; movement was concentrated along the fault line. At the two dams, permanent displacement up to a maximum of 3.6 m (12 ft) took place; and although outlet tunnels around the dams were badly fractured, the dam structures themselves remained stable. Despite the appreciable movement of one-half of the dams, they remained watertight and so stand today. A large, mass-concrete gravity dam (the Crystal Springs Dam, constructed in 1877 and enlarged in 1888 and 1890) located only 0.4 km (¼ mi) from the fault was undamaged.[25.38] Earth dams in Japan

FIG. 25.19 The Morris Dam for the water supply of Pasadena, California, located over a fault in the San Gabriel Canyon.

have similarly withstood severe earthquake shocks without serious damage. The design for a dam for the water supply of Rangoon, Burma (in an area also subject to earthquakes), was based on this experience. In the absence of any clay suitable for use as a puddled core, the dam was designed with a flexible concrete core wall. The dam is 37.8 m (126 ft) high. The concrete core wall was built as a series of panels connected by means of ingenious grooved joints filled with asphalt.[25.39]

The example cited of what can happen at a fault plane is a fitting reminder of the importance to be attached to the presence at a dam site of geological faults—certain signs of past earth movement. It will be increasingly difficult to find sites for dam structures without some such defect, and engineering ingenuity has already been displayed to advantage in devising means of overcoming the difficulties in design thus introduced. At the site of the Morris Dam, California, for example, a minor fault intersects the dam foundations near the base of the right abutment in a direction almost normal to the axis of the dam. Study of the stratification of old stream-bed gravels revealed the fact that no appreciable move-

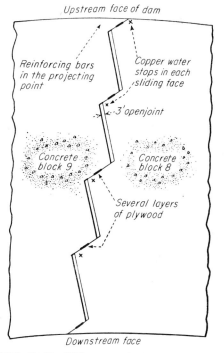

FIG. 25.20 Sliding type of joint which divides the Morris Dam (see Fig. 25.19) into two units over an old, inactive fault.

ment along this and other fault planes had taken place for a period of approximately 10,000 years. The fault received special treatment in design; an open joint with vertical sliding planes was provided in the dam structure over the trace of the fault. The joint ran from top to bottom of the dam between two of the blocks into which it was divided for construction purposes (Fig. 25.20). The four sliding planes are at an angle of 45° with the horizontal and so lie in the direction of past movement. Planes of contact are separated by a bituminous filler which can yield slightly if motion is not as anticipated. A somewhat similar type of sliding joint, but horizontal, was incorporated in the design of the 63-m (210-ft) Lake Loveland concrete arch dam near San Diego, California. The joint was installed above a shelf in the left abutment in an attempt to eliminate cracking of the concrete at this location when the dam is under load.[25.40]

Another interesting and illuminating example is provided by the foundation-bed conditions at the site of the Owyhee Dam built about 1930 by the U.S. Bureau of Reclamation to serve a large irrigation project (Fig. 25.21). The dam is 106.5 m (355 ft) high above bed level but 159 m (530 ft) high above lowest concrete level; it provides 880 million m³ (715,000 acre-feet) of storage on the Owyhee River in eastern Oregon.

> The dam site was formed by the present Owyhee River cutting a narrow gorge 400 to 600 ft deep through a flow of rhyolite which blocked the valley subsequent to the deposition of the original tuff underlying the district. The outer surface of this flow of rhyolite moving on the original tuff—probably in contact with water—chilled comparatively quickly, producing a layer of pitchstone agglomerate, roughly about 25 feet thick, extending as a contact between the two materials. Subsequent erosion has removed the upper levels of this rhyolitic flow, and the river had cut its gorge down into the remainder. This rock, which forms the abutment and

FIG. 25.21 Owyhee Dam, Oregon; a general view of the unwatered site, showing the canyon walls and the fault described in the text; (scale may be judged by the crawler crane in center foreground).

foundation material for the structure, is an extremely hard crystalline mass due to the slow interior cooling of the flow. Beneath it lies the stratum of more elastic pitchstone agglomerate as a transition material blending into the softer tuff.

A fault in the center of the streambed parallel to the canyon subsequently has occurred with horizontal displacement. Exploration of this fault zone with diamond-drill borings showed that faulted material was evident in the hard rhyolite in the form of small shattered pieces, but indicated that the disturbance lessened in the more yielding agglomerate and practically disappeared upon entering the tuff. The original plan provided for the removal of this faulted material on the line of the upstream cutoff by means of a shaft to the undisturbed tuff.... Excavation work in this cutoff shaft and general foundation excavation along the upper portion of the fault revealed fractured and loose material, which made it advisable to remove the material in the fault zone throughout the entire width of the structure and refill the crevice with concrete to insure proper foundation conditions. This possibility had been foreseen as a result of the drilling and was provided for in the specifications.... Excavation proceeded through about 83 feet of faulted and broken rhyolite, 22 feet of the agglomerate and 10 feet into the practically undisturbed tuff to El. 2,145, at a depth of about 213 feet below low water level of the river. It is of interest to note that the level of actual excavation required to reach undisturbed material below the zone of fracture was within 2 feet of the estimated depth established by the exploratory borings in the zone.[25.41]

This unusual foundation work is described in detail because of its special interest and the close relation between construction procedure and geological structure that it reveals. Practically all recorded cases of major dam construction over faults include some features of interest. Reference can be made, however, to two more examples only, which are of special interest; one is the arched foundation necessitated at the Rodriguez Dam on the Tijuana River near Tijuana, Mexico, constructed during the years from 1928 to 1930 (Fig. 25.22). The dam is of the Ambursen-buttress type, 72 m (240 ft) high above lowest foundation concrete, having a crest length of 660 m (2,200 ft) and providing a storage of 136 million m^3 (110,000 acre-feet).

Rock, which was exposed over much of the site, consists of rhyolite and granite with a fused contact, indicating watertightness on this plane. In the lower part of the gorge the rock surfaces were hard and fresh, although broken.... Cleavage planes are evident on either side,

FIG. 25.22 Rodriguez Dam, Mexico; a close-up view of the special arch incorporated into the foundation because of the presence of a fault.

dipping toward streambed. As preliminary investigation nine borings were made in streambed and about fifty test pits were excavated at higher elevations. Cores could not be obtained in the other holes but two indicated sharp clean rock chips and the remaining two showed rather definite evidence of disintegration.... Foundation excavation work uncovered bedrock at depths of about 35 feet, as had been expected, but it proved considerably broken and unsuited for buttress foundations. A more serious difficulty was the finding of a pronounced fault about 20 feet wide parallel to the stream along the sound rock of the east bank.... In addition, deeper excavation along the line of the cutoff showed that a weak, inferior bedrock could be expected to continue below the existing elevation of foundation excavation.

Further and more extensive geological inspection of the site was made in the light of foundation excavation.... Excavation in streambed over the foundation area revealed sound rock walls about 90 feet apart at the upstream toe and 50 feet apart at the downstream edge of the structure.... Excavation in the cutoff indicated that conditions could not be bettered by going deeper. The alternative was to provide a foundation structure across streambed to equalize that portion of the load which could be carried by the weaker rock and transfer the remainder to the sound rock of the walls.[25.42]

Eventually, after much study and approval by a consulting board, a concrete arch with a span across the stream bed varying from 25.8 to 22.8 m (86 to 76 ft), with the barrel 67.5 m (225 ft) long, was designed; the space between the intrados of the arch and the stream bed was built up with lean concrete to provide a support for constructing the arch and to transfer a portion of the load vertically, although no allowance for this was made in the arch design. The arch was successfully constructed, and the dam completed (Fig. 25.23).[25.42]

Finally, reference may be made to a dam incorporating in its design a considerable degree of flexibility, not in anticipation of earthquake shocks or movement over faults, but because of the possibility of differential settlement between one part of the dam and another. The Salso River Dam is located in Sicily, Italy, near Pozzillo above the Catania Plain. Part is founded on sandstone but the downstream section had to be founded on

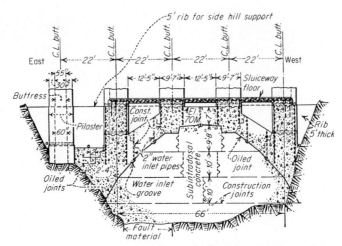

FIG. 25.23 Rodriguez Dam, Mexico; cross section of the special arch shown in Fig. 25.22.

clay. It is a gravity structure, 54.9 m (183 ft) high and 231 m (770 ft) long. The structure was built of individual blocks of concrete, suitably proportioned to give the necessary, approximately cubical shape, with sides of about 4 m (13 ft). The upstream face was formed by a continuous metal plate, keyed into the sandstone in order to provide a watertight membrane. The blocks were connected with specially designed joints, incorporating provision for some sliding in case of differential settlement when full load is against the dam.[25.43]

25.10 PERMEABILITY OF BEDROCK

The second essential requirement of a rock foundation bed for a dam is that the entire geological structure underlying the site of the dam, in addition to being strong enough to carry the designed loads, will be sound enough to provide a watertight barrier to the water impounded by the dam. Although the requirement is so obvious, dams have failed because of its neglect, and an imposing and tragic list could be produced in confirmation of this statement. It is only very rarely that major leaks through a dam foundation have been experienced. One example only need be mentioned; it has been termed "the greatest object lesson that the history of engineering foundations had to offer." The Hales Bar Dam on the Tennessee River in the United States was founded on a pure but soluble limestone, well known for its cavernous formation (Fig. 25.24). Cavities were encountered to such an extent during the excavation of the dam site that completion was delayed several years, and the cost was increased far beyond the original estimate. Several hundred thousand barrels of cement were injected into the fissures and openings in the underlying rock. The dam site is located on a syncline, the lower side of which is especially susceptible to solution water. Further trouble due to this situation was foretold, notably by L. C. Glenn, not long after the dam was put into use.

About ten years after the dam had been completed (in 1926), leakage through this limestone had become quite serious, and many unusual methods were used in attempts to provide an effective seal, including the use of various kinds of mattresses and the dumping of rocks, gravel, clay, and bales of hay. Success was achieved only by drilling a large num-

FIG. 25.24 The Hales Bar Dam while remedial work to overcome leakage beneath the dam was being undertaken by the Tennessee Valley Authority.

ber of holes to an average depth of about 27 m (90 ft) and injecting into these hot liquid asphalt, the flow of which was assisted by ingenious devices; over 11,000 bbl of asphalt were used. The process was later repeated, and leakage was eventually stopped. By 1941, however, leakage had again increased, amounting then to 48 cms (1,700 cfs), so that further remedial action had to be taken. The dam was now owned by the Tennessee Valley Authority.

The geological and engineering staffs of the TVA made careful studies, and many rehabilitation schemes were evolved. It was finally decided to form a continuous curtain of concrete along the face line of the dam by means of continuously connected Calyx-drill holes, 45 cm (18 in) in diameter. More than 200 holes were drilled to depths up to 18 m (60 ft); the total length of drilling exceeded 3,600 m (12,000 ft). The holes were lined with cement-asbestos pipe and then filled with concrete. Study of the Bangor limestone obtained from cores at this time showed a rate of solution much greater than would be expected for most limestones, a fact dramatically demonstrated by the troubles caused by leakage at the Hales Bar Dam.[25.44] Finally, TVA decided in 1963 to give up the unequal battle. The Nickajack Dam (Fig. 25.25) was designed and constructed 10 km (6.4 mi) below the Hales Bar Dam, the height of the new dam being such that it impounds water to the level intended as the impounded level behind the Hales Bar structure which is now submerged—but only after the expenditure of over $10 million in attempting to correct the imperfect geology of the original dam site.[25.45]

The first reason for having rock strata underlying dams as watertight as possible is obviously to make sure that no water escapes. Not only is this necessary from the point of view of water conservation, but it is also essential because any steady flow through a solid rock formation is certain to have some erosive action, which in all probability will gradually but steadily intensify the defective conditions causing the original leakage. The geological investigations already mentioned will indicate the underground structure to be encountered along the line of the dam, and from a study of this cross section a general idea, at least, of the effectiveness of the strata in retaining water can be obtained. Natu-

FIG. 25.25 The final solution to the Hales Bar Dam "problem" of the Tennessee Valley Authority: the Nickajack Dam, which raises the water level sufficiently to flood over the upstream Hales Bar Dam.

rally, limestone and all soft rock formations are suspect until proved to be sound; limestone formations, especially, are unusually soluble and so characterized by underground caverns or open fissures.

Although limestone causes most of the geological troubles in foundations, it was a shale and sandstone combination that presented one of the most unusual examples known to the writers of a permeable foundation bed for a dam. The Santa Felicia Dam of the United Water Conservation District of California is a 60-m (200-ft) rolled-earth dam located due north of Los Angeles on Piru Creek. It lies over sandstones and shales of Miocene age, much folded and tilted and known to be oil bearing. One oil well was actually located in the reservoir and further drilling was contemplated. The district had therefore to pay for the abandonment of this part of the oil field and for the capping of producing wells in the resevoir area; this cost $200,000. Oil at the dam site was therefore suspected, but as test drilling proceeded, oil was encountered in every one of the test borings and in the 70 grout holes drilled into the old stream bed. Gas under pressure was struck by some holes. All holes had to be sealed off, and oil seeps (of which there were many) had to be capped just before earth fill was placed over them. Anticipated water pressure would more than counteract the oil and gas pressure when the reservoir was filled, but this unusual "permeability" of the foundation strata made the construction of the Santa Felicia Dam a somewhat uncommon operation.[25.46]

In addition to the possibility of major leaks, an almost equally serious problem is presented by the danger of minor leaks, especially on the site of the dam structure itself. Springs are included in this category; if noted either before or during construction, or if preliminary investigations suggest that they may tend to form at constant planes after construction is complete, the most careful precautions must be taken to box them out when the lower courses of the dam are being built, so that they may be suitably connected with the drainage system. This last possibility is probably the most difficult to investigate before construction and the most potentially dangerous, since the slight leakage of water that may occur between rock surface and dam structure after completion will give rise to what is generally known as *uplift pressure*. Uplift pressure was not, as is sometimes suggested, neglected in all early dam designs; this is made clear by the following quotation. Dr. Deacon, describing the design of the Vyrnwy Dam to which reference has already been made, stated (in 1896) that:

> Although no visible springs of water issued from the beds of rock thus exposed, it was by no means certain that, when the reservoir was formed and the head on one side became 144 feet, springs subject to that pressure would not occur. Moreover, a mere moisture, rapidly evaporated when exposed to air might when sealed down acquire a pressure from the adjoining hills of far more than that due to the intended level of the lake . . . and the Author agreed with him [the late Mr. T. Hawksley] in thinking it desirable to provide relief drains, which so far as he is aware, had not been done in connection with any former masonry dam.[25.47]

The predictions of Messrs. Hawksley and Deacon have been generally confirmed at many dams on which tests of uplift pressure have been made. Many of these are founded on perfectly sound rock strata, so that the uplift pressures may be said to be independent, in these cases at least, of geological structure. At other dams the reverse is true, and therefore uplift pressures may rightly be classed generally as a problem associated with the geology of dam sites. Drainage has long been a matter for keen debate, but the general consensus appears to be that, provided the drains are carefully located and are of such a size and character that they cannot possibly become blocked, their installation is a wise precaution. Drainage trenches filled with crushed rock and provided with suitable outlets constitute a usual expedient. Inspection galleries inside a dam structure are often connected with the drainage system so that regular checks on the drainage can be made; and in some cases known to the authors, one of the inspection galleries is located on the rock floor itself in order that the rock surface can be periodically examined.

The geological implications of uplift pressure beneath dams, and the relevance of geological structure under dams, have been admirably reviewed by W. H. Stuart.[25.48] Starting with the conventional assumption that uplift pressure varies from full head at the toe of a masonry dam to zero at the heel, Stuart was able to present actual measured pressures beneath a number of major dams in the United States. Drainage arrangements varied but were always downstream of the grout curtain or equivalent cutoff. "Geological conditions such as the direction, dip and spacing of joints, fractures and bedding, and the location of impervious boundaries were used in designing the grout and drainage systems" of the six dams dealt with, which varied in height up to 113.7 m (379 ft). Foundation-bed conditions varied from sandstone over shale to granite, but the essentials of the drainage systems used were similar, all having 7.5-cm (3-in) -diameter drain holes spaced at from 3- to 6-m (10- to 20-ft) centers, with depths up to 18 m (60 ft), apart only from the 111-m (370-ft) Detroit Dam where they were drilled to 36 m (120 ft). Measured uplift pressures were all less than the conventional conservative assumption, but the engineering decision as to the uplift pressures to be allowed for in design must always be reached only after the most thorough study of anticipated geological conditions beneath every proposed dam. Drainage systems, as Stuart points out, are similarly dependent upon these conditions, even the smallest detail of geological structure being of possible significance. When dams have been completed and drainage systems are operating satisfactorily, regular inspection and maintenance of these facilities is essential. Plugging of drains, due to deposition from the water they convey, has happened. Inspection should detect this so that remedial measures can be taken immediately.

It will be clear that the presence of water in rock foundation beds for dams, and the possibility of flow through such bedrock, is a complex matter, just as complex as is the variation of geological conditions that may be encountered. Accordingly, in addition to study of all relevant geological information, including the detailed results of all boring, tests should be carried out whenever possible by means of the boreholes used for obtaining core samples. The use of carefully controlled and observed water-pressure tests in these holes provides a fairly reliable confirmation of underground structural soundness. If the holes hold water under the maximum test pressures, it is clear that the strata penetrated by the test holes are watertight in the neighborhood of the holes.

This qualification is a necessary reminder that water-pressure tests in drill holes are not an infallible guide; when considered in conjunction with ascertained geological structure, however, such tests can be a reasonably reliable indication of watertightness. Pressure tests of this nature were utilized in connection with the geological investigation of the Madden Dam site in the Panama Canal Zone, although in this case water was forced under pressure into flowing wells to see whether they would provide underground connection to other wells or any other possible means of escape for the test water. For tests of this kind, the use of color dyes, which will be dealt with in the following chapter in connection with leakage from reservoirs, is often of value. Tests were also carried out, during the Madden investigation, on the permeability of test specimens of the rock types encountered; the tests were made on core samples taken from drill holes. The apparatus employed was not unlike the standard type used for testing permeability of unconsolidated material; the specimens were packed into sections of steel pipe with lead wool and steam packing and subjected to a head of 5.6 kg/cm^2 (80 psi).[25.49]

This was an early example. Today, water-pressure testing in exploration drill holes may be classed as almost standard procedure in studies of dam foundations, although there is never any standardization of the conditions which such explorations will reveal. Since only rarely will bedrocks be permeable of themselves, permeability of rock foundations will usually be the result of structural features such as joints and fissures, thus pointing again to the vital necessity of geological study as a preliminary to actual exploration work.

25.11 DAMS ON PERMEABLE FOUNDATION BEDS

Discussion of dam foundations up to this point has been confined to structures constituting watertight barriers across a valley; watertightness is achieved either by founding the dam directly on impermeable rock or by having some type of cutoff wall constructed below the dam structure proper, connecting it to a solid-rock floor below. There are many sites at which dams have been required but at which solid rock exists only at very great depths below existing ground level. The unconsolidated deposits on which a dam structure must be founded at these sites present many unusual problems of design and construction. In permeable strata such as sands and gravels, the stability of "floating" dams, as they are sometimes called, depends on a predetermined safe rate of flow beneath the dam; in impermeable strata, for example clay formations, reliance must be placed on the efficiency of the clay as a seal against uncontrolled flow beneath the dam. In both cases, accurate knowledge of the nature of the unconsolidated deposits to be encountered is as essential for design as for construction.

Thus the importance of preliminary geological studies at dam sites again comes up for emphasis. Methods of investigation need not be detailed again, since general procedure and the analysis of results are similar whatever the geological conditions at the dam site may be. Novelty enters in as a result of the fact that the unconsolidated deposits now have to be regarded as constitutent parts of the structure, necessitating detailed testing prior to use in design, instead of as something to be got rid of by excavation or penetrated by cutoff trenches or sheeting. Undisturbed soil sampling must therefore be a feature of preliminary work. The soil characteristics to be investigated in connection with dams on foundation beds of unconsolidated materials are common to other branches of civil engi-

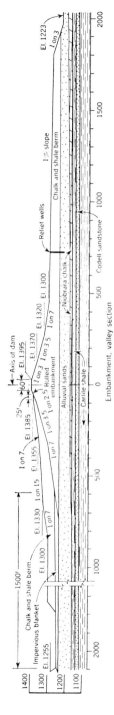

FIG. 25.26 Cross section at right angles to the axis of the Fort Randall Dam on the Missouri River, showing foundation-bed conditions.

neering work, but there is probably no other branch of civil engineering work in which close correlation of geology and soil studies is so absolutely essential.

Many dams in the Middle East, such as some of those in India, are structures built on permeable foundation beds. In earlier years, and especially in India, design of the necessary cross sections was to a large extent empirical, design rules being gradually refined in the light of accumulated experience. The Central Board of Irrigation (of India) published some notable guides. In more recent years, accurate soil testing and the development of sound theories of groundwater flow have permitted the preparation of dependable flow nets for such structures, giving more certainty to design, although design must still be based upon the results of adequate and accurate subsurface investigation for the whole site.

The Fort Randall Dam, one of the group of major river-control structures on the Missouri River, may be cited as an outstanding example (Fig. 25.26). Located in southeastern South Dakota, the dam is one unit in the Pick-Sloan Comprehensive Plan for the conservation and utilization of the water resources of the Missouri River basin. It controls the runoff from one-twelfth of the entire area of the United States. The spillway outlet works and associated powerhouse (with 40,000 kW units operating under a head of 33.6 m or 112 ft installed) are located at the northeast end of the dam, all founded on the Niobrara Chalk which underlies the entire site. The main part of the dam, however, is an embankment constructed of 20 million m^3 (26 million yd^3) of the glacial drift overburden which had to be removed elsewhere on the site. Fourteen and a half million m^3 (19 million yd^3) of chalk and shale bedrock was placed as the 450-m (1,500-ft) long downstream berm. The embankment was constructed over the natural alluvial soils in the valley floor which consist of interbedded layers of silty clays, sands, gravelly sands, and clayey gravels, naturally a pervious foundation bed.

To provide the necessary length of travel for water in this underlying formation, an impervious blanket extending 450 m (1,500 ft) upstream from the embankment was provided, varying in thickness from 3 m (10 ft) at the upstream end to 6 m (20 ft) where it joins the embankment proper. Soil in the embankment was rolled, at about optimum moisture content, in 20-cm (8-in) layers, usually with six passes of a tamper-type roller; that for the blanket was similarly placed in 30-cm (12-in) lifts. Figure 25.26 illustrates the general arrangement of this major structure, similar to that used for innumerable smaller dams on permeable foundations. The row of relief drains in the downstream berm will be noted. The use of 75- and 90-cm (30- and 36-in) -diameter Calyx drill holes for studying clay seams that were found in the bedrock was but one of the many other interesting features used during the construction of this outstanding dam.[25.50]

25.12 CONSTRUCTION PROBLEMS

It may be fitting to conclude this recital of the problems that can be encountered in the geological formations that form the foundation beds of all dams with brief notes on some geological problems that have been encountered during the construction of a few major dams. A major objective of this section is to emphasize yet again the fact that study of the geological conditions at a dam site must be continued from the start of preliminary investigations up to and beyond the final passing of the cleaned and prepared rock surface as the approved foundation bed. Observation of rock conditions must continue until all is safely covered up with concrete, masonry, or soil, as the case may be. The following notes are brief indeed, but a selection of dams from around the world has been made to illustrate the infinite variety of problems than can arise when a major dam is under construction.

Problems with Rock Foundation Beds A problem that may be regarded as essentially part of the construction operations consists in the removal of all rotten and disintegrated rock; keeping accurate records of all rock that is so removed is equally important. When finally the approved foundation bed is ready for undertaking the building of the dam, a complete and detailed survey of the bed must be made; this is one of the most critical duties of a resident engineer or geologist. Although record surveys of this kind may seem to be unimportant at the time, they can often prove to be of inestimable benefit; their value is incalculable because, if they are not accurately made when the foundation bed is ready, they can never be obtained again. The authors write on this point from personal experience of a rock-surface survey so obtained in the depth of a Canadian winter which proved to be of great and unexpected assistance when an adjoining section of a dam was constructed in the following spring.

There are some detailed matters in connection with the preparation of a rock foundation bed which call for mention. One essential of construction is to get as good a bond as possible between concrete and rock. To obtain this, the rock surface must be clean, free from all loose particles of material (even though they be rock fragments) and free from standing pools of water, however small, when concreting operations start. Opinions differ concerning the methods to be used when a glaciated rock surface is encountered, but it is generally agreed that any polished surfaces should at least be roughened by "pop shots," even if excavation to break up the surface completely is not undertaken. Failure of the Gleno Dam in Italy has been attributed to sliding of the structure on a glaciated rock surface. Preliminary geological advice will generally be able to foretell the presence of such a feature, so that an appropriate construction schedule can be prepared. The exact location and regular discharge of all springs encountered must be most carefully recorded

FIG. 25.27 Unusual pothole detected in preliminary investigations and uncovered during excavation for the foundation of the Noxon Rapids Dam, Montana; workers indicate the scale.

lest they prove to be possible sources of trouble. Any unusual springs should be immediately brought to the attention of the geological adviser so that he may examine them and advise accordingly. The detailed inspection of all unusual features, in addition to general supervision of excavation operations, is just as essential a part of the duties of the geological adviser as are the preliminary geological investigations.

Only during construction will the full extent of unusual foundation-bed features, revealed in part by preliminary investigations, be seen. During the construction of the dam for the Merwin Hydro Electric Project on the Lewis River in Washington (downstream from and associated with the Swift Dam), a buried gorge immediately below the location of the powerhouse in the 375-m (1,250-ft) dam was detected in preliminary studies, but naturally, it was not fully seen until excavation was complete. It was then found to be 21 m (70 ft) wide and 36 m (120 ft) deep, a buried river valley from preglacial times. In a bold solution, a massive concrete arch was built to span this inner gorge and on this the powerhouse of the Pacific Power and Light Company has successfully performed since the dam was completed in 1931.[25.51]

Not too far away, at the site of the Noxon Rapids Dam of the Washington Water Power Company in Montana, glacial action has had similarly significant effects upon the bed of the river, as was fully revealed only after dewatering. Figure 25.27 shows one of the potholes then discovered; it extended 24 m (80 ft) below riverbed level and required 17,000 m^3 (22,000 yd^3) of concrete before it was completely filled. The site is on indurated argillite but lies in a wide valley that was once the bottom of glacial Lake Missoula. Failure of an ice dam in glacial times, about 40 km (25 mi) above the dam site, led to strange and awesome erosional effects in the sediments already deposited, and this led to complicated soil conditions which had to be dealt with in the construction of the dam. The final solution was a mass-concrete spillway and powerhouse section, flanked by two large earth-fill wing dams.[25.52]

Geological Problems and Construction Schedules Geological features revealed during construction can sometimes affect progress schedules. During the construction of the Beechwood Dam and powerhouse for the New Brunswick Power Commission, on the St. John River 160 km (100 mi) north of Fredericton, a relatively small fault in the argillite that made up the entire riverbed was discovered during the second construction season, despite a most careful preliminary program of test drilling. Some alteration of plans was called for in order to deal adequately with the fault zone; this was readily arranged, but the change affected the schedule for the placing of earth fill in the adjacent wing dam. The delay meant that placing of soil had to proceed during Canadian winter conditions or be postponed until the following year. In an ingenious solution, an asphalt-mixing plant, hastily procured, was used to heat the soil before placement, and thus the work was finished almost as originally scheduled.[25.53]

An outstanding example of how difficult bedrock conditions can be overcome during construction is given by the experience on the Wimbleball Dam in Somerset, England. The dam is a concrete diamond-headed buttress structure (the fine appearance of which is shown in Figure 17.5) constructed to provide water storage for two of southwest England's water authorities, which are joint owners. Maximum height is 50 m (164 ft) and the crest length is 300 m (1,000 ft). Foundation beds are a combination of siltsone and sandstone. The site chosen was one of 17 possible sites, a deciding factor being environmental considerations. The site was carefully studied by expert engineering geologists; 17 test boreholes were put down. As a further precaution, 22 trial pits were later sunk and a full-scale rock loading test carried out when excavation permitted. It was "concluded that the strata . . . were cleaved, folded, and much jointed, and that the mudstones particularly weathered to clay in places." Forewarned was certainly forearmed in this case. Excavation revealed the anticipated bedrock conditions, to be seen in Fig. 25.28. In some places excavation had to be carried rather deeper than expected (to a maximum depth of 26 m or 85 ft), and a careful grouting program was carried out. Despite the extra work involved, impounding of water started at the end of 1977 so that delivery of water from the reservoir started in 1978, as originally planned. The fine record of this work, written up by those responsible, is well worthy of special study.[25.54]

Many years before, similar difficulties were encountered during the construction of one of the major rock-fill dams of the western part of the United States. The Los Angeles County Flood Control District was set up in 1915 to carry out flood-protection and water-conservation work for Los Angeles County. In 1924, $25 million of the funds raised by bond issue was allocated for the construction of a concrete gravity dam of unprecedented size in the San Gabriel Canyon. Work was begun under contract in 1929, and completion of the dam was expected within six years.

> While excavation was still under way a serious earth slip at the west abutment resulted in holding up the work for a further study of foundations. Before the end of 1929 the contract for constructing the dam was cancelled. Surveys for a new dam site developed what was thought to be a safe and satisfactory location just above high water level of the reservoir of the Morris Dam, constructed by the city of Pasadena lower down the San Gabriel Canyon. At this site it was proposed to build a rockfill dam ... to have a height of 300 feet above streambed, a crest length of 1,670 feet, and a base thickness of 900 feet at streambed level. ... A contract was awarded in the latter part of 1932. ... After that contract was made, work in the quarry progressed satisfactorily, although the unexpectedly large percentage of material that had to be rejected resulted in low yardages delivered at the dam. ... Up to October 1st, 1934, approx-

FIG. 25.28 Wimbleball Dam, Exmoor, England; a view in the excavated foundations showing something of the difficult rock conditions.

imately 3,650,000 cubic yards of material, including stripping and waste, or the equivalent of about two thirds of the total volume of the dam, had been removed from the quarry. But little more than one tenth of this material actually went into the dam.[25.55] [Quotation amended and corrected.]

After further conferences and reports, work on the new contract was temporarily held up, and the dam design was revised again to take into consideration the rock available from the quarry. On this basis, the work was satisfactorily completed in July 1937. Over 7.5 million m^3 (10 million yd^3) of material was finally placed in the dam; about 750,000 m^3 (1 million yd^3) was a sand-clay mixture used as an impervious blanket near the upstream face, and all other material was quarried rock. All loose rock fill was placed with the aid of sluicing; shrinkage approximated 6 percent. The dam is functioning satisfactorily, having already been tested by severe floods (Fig. 25.29).

A smaller rock-fill dam, known as San Gabriel Dam No. 2, was constructed about the same period. No trouble was experienced with rock supply in this case; the main difficulty involved the settlement of the rock fill when near completion, a settlement which was attributed to lack of sluicing for compaction of the fill. These brief details of a notable water-conservation project (a full bibliography for which is noted) serve to illustrate the vital importance of material supply and, in particular, of rock fill. Geological advice was repeatedly taken in connection with this work, though some of the geological difficulties were unpredictable.

Grouting as Solution to Geological Problems   Problems with karstic limestone have already called for mention. Probably the most difficult case yet encountered of karstic limestone as the foundation bed for a dam was in Turkey. The Keban Dam was, when built, the largest construction project ever carried out in Turkey. It was the first of a series of over 20 dams which constituted a major step forward in the development of the country's resources (Fig. 25.30). Located in a spectacular gorge in eastern Anatolia, just below the junction of the Firat and Murat rivers (headwaters of the great Euphrates), Keban

FIG. 25.29 San Gabriel Flood Control Dam No. 1, under construction in August 1934; concrete toe wall in the foreground and Quarry No. 10 in the background.

FIG. 25.30 The Keban Dam, when built the largest in Turkey, now in full service, following unusual difficulties during construction.

Dam is a multipurpose structure, controlling water for irrigation and flood prevention and storing water for use in the associated powerhouse equipped with eight 155-MW generators. The dam is founded on karstic limestone, the nature of which was generally known from preliminary investigations. It is a combined rock-fill and mass-concrete gravity dam, with a height of 204 m (680 ft).

Nobody would have expected, when work started, that the ultimate depth of the necessary grout curtain and concrete cutoff wall would extend to a depth of 345 m (1,150 ft), approaching twice the height of the structure. Grouting had to be carried to this depth in order to penetrate to the underlying watertight schist in which a sure cutoff was obtained. Grouting was carried out from a series of horizontal 3-m (10-ft) -square adits, spaced vertically 39 m (130 ft) apart. Five such access tunnels were excavated in the right bank and eight in the left. Vertical shafts were also used for access. Completion of the dam was delayed and further cost increases incurred when geological studies revealed the presence of a major cavity 138 m (459 ft) long, 117 m (390 ft) wide, and 30 m (100 ft) high at a depth of almost 330 m (1,100 ft) beneath the crest of the dam. Known on the job as "the Crab" because of its shape, it had to be filled with 60,000 m³ (80,000 yd³) of tremie concrete, pumped into place, and bisected by the main cutoff wall. Some idea of the scope of this problem is given by the fact that 116,000 m³ (155,000 yd³) of concrete was used in the main concrete cutoff wall, and that 69,000 m (230,000 ft) of exploratory and grout-check holes were drilled in addition to the 288,300 m (961,000 ft) of holes drilled for the grout curtain. The work was successfully completed in 1974, three years after the originally scheduled date.[25.56]

Massive grouting operations are not peculiar to the Middle East. An equally impressive job was completed in 1961 in the coastal mountains of British Columbia. The Mission Dam looks today like a normal combined earth- and rock-fill dam, but it is necessary to read the account of its construction in order to appreciate fully the continually unfolding train of problems that were encountered during its construction. A grout curtain 150 m

FIG. 25.31 The Mission Dam, British Columbia, now named the Terzaghi Dam, the commemorative plaque to be seen in the foreground; the restricted site prevents the photographing of a more general view of the dam with the plaque.

(500 ft) deep had to be installed in water-bearing alluvial materials, down to the underlying bedrock, and this for a dam which itself is only 60 m (200 ft) high. Intrepid diving work, often in mud, was necessary at an early stage. Later, two acres of polyvinyl-chloride sheeting had to be installed in order to assist with the consolidation of the clay used in construction. Dr. Karl Terzaghi was the special consultant on this dam; he regarded it as one of the most challenging projects he had ever faced. The dam was renamed the Terzaghi Dam so rightly at the Sixth International Soil Mechanics and Foundation Engineering Conference held in Montreal in 1965 (Fig. 25.31). The plaque then dedicated is now attached to a granite boulder near the crest of the dam amid superlative scenery which gives no direct indication of all the problems that construction of the dam involved.[25.57]

Preliminary Construction to Counteract Geological Problems At the other end of the spectrum, so to speak, were the almost ideal conditions available for the founding of the Karadj Dam in Iran. Located 40 km (25 mi) northwest of the capital city of Teheran and designed to be a part of the city's water supply system, this dam is located in another spectacular gorge, that of the Karadj River. The geology featured in the gorge is a chaotic mixture of igneous and sedimentary rocks, but at the most desirable spot for a dam, Nature had cooperated by providing a sill of hard and competent diorite. On this excellent rock, a double curvature, reinforced-concrete dam 177 m (590 ft) high was built in the early 1960s, a quarter of a century after the first studies of the site were made. The main problem at the site was obtaining the necessary level areas for construction plant and operations. A bench had to be excavated in the wall of the gorge, and the excavated material was deposited in a neighboring side canyon; this provided the necessary level area for the screening and concrete-batching plants, but at a cost of $250,000.[25.58]

Even more remarkable was the preliminary construction work that had to be done before the building of another of Iran's modern dams could start. This is the Dez Dam, on the Dez River in the Khuzistan area of southwestern Iran, another structure designed

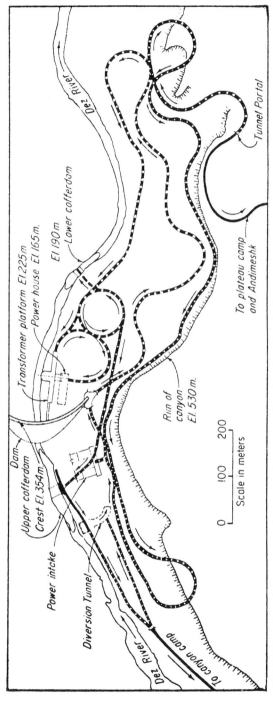

FIG. 25.32 Plan of the access road necessary for construction of the Dez Dam in Iran.

to provide irrigation water while also providing for flood control and generating power in an adjacent powerhouse. Located in a narrow gorge of the river, this dam site was not known in advance of modern surveys, being identified in the course of aerial survey work in 1956. Sides of the gorge rise about 420 m (1,400 ft) from river level to that of the surrounding plateau. A 35-km (22-mi) access road had to be constructed, in the first instance on to the plateau for transport by jeep so that site investigation could proceed. When designs were complete and the start of construction was in prospect, access to the gorge itself had to be arranged. The main trucking road was 24 km (15 mi) in length from the nearest railhead at Andīmeshk to the plateau. The only means of getting it down to the level of the gorge was by tunneling. Six km (4 mi) of road was therefore constructed in tunnel, dropping steadily towards the gorge, with a maximum grade of 8 percent. Figure 25.32 shows the quite remarkable system of spiral tunnels in the route that was so carefully planned for this unique access road. The unusual route is explained by the necessity for keeping the tunnel (which is 7.5 m or 25 ft in diameter) close to the side of the gorge so that short adits could be provided at intervals, through which excavation was disposed of economically by being dumped into the gorge below. The rock is a competent fused limestone conglomerate with no faults or cavities, in great contrast to other limestone formations which have been mentioned.[25.59]

The idea of having to drive large tunnels merely for access to a difficult construction site might at first sight seem to be so unusual as to be unique. Other examples, however, could be cited, but attention will be rather directed to another tunnel excavated merely to aid construction, but for a very different reason. The Atiamuri Dam on New Zealand's Waikato River was built as the fifth of a chain of seven hydroelectric stations on this scenic river in the country's North Island. The dam consists of a mass-concrete section 36 m (120 ft) high and 150 m (500 ft) long, founded on bedrock, and a connecting earth embankment 240 m (800 ft) long containing 350,000 m^3 (460,000 yd^3) of soil, 75,000 m^3 (100,000 yd^3) in the core alone. The earth dam is built over alluvial deposits in older beds of the river, deposits varying in depth from 15 to 36 m (50 to 120 ft) and naturally water-bearing. Groundwater level was about 24 m (80 ft) below the surface, to which depth excavation for the concrete core wall proceeded with no unusual difficulty. It proved impossible to drive steel sheet piling below this because of the presence of boulders, so the old valley had somehow to be drained so that the concrete cutoff wall could be constructed up from the level of the bedrock, a silicified breccia.

The contractor therefore sank a shaft to a depth of 37.5 m (125 ft) from a 1.8- by 2.4-m (6- by 8-ft) header at original ground level. From the bottom of the shaft a horizontal drainage gallery was then driven under the axis of the core wall for a distance of 99 m (330 ft). It was fitted with a bulkhead, and a pumping station was installed at the bottom of the shaft. Drill holes were then put down from the bottom of the excavated core-wall trench. Groundwater drained through these holes into the construction tunnel, the water table being lowered as required for excavation by controlling the amount of pumping (Fig. 25.33). A pumping rate of 950 lpm (250 gpm) held the water table steady. When the core wall had been founded on the bedrock and the necessary grouting beneath it had been done, the water table was allowed to return to its normal elevation, the tunnel and shaft were sealed off with concrete, and the dam completed. The competent character of the silicified breccia bedrock permitted this bold expedient to be successfully completed.[25.60]

Faults at Construction Sites A different kind of construction expedient was necessary during the building of the Seminoe Dam of the Kendrick Project of the U.S. Bureau of Reclamation. Two faults, normal to each other, were found to intersect beneath the center of the dam; they were uncovered only when excavation was possible after dewatering the site of the dam. The dam is a concrete structure, 78 m (260 ft) high, arched in plan

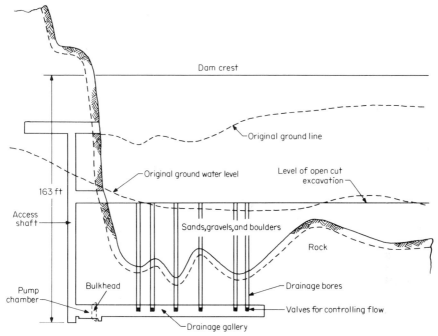

FIG. 25.33 Drainage tunnel, shown in diagrammatic form, used for the construction of the Atiamuri Dam, New Zealand.

as it is located in a narrow gorge. After the dam had been started, engineers decided to clean out the two fault seams and backfill them with concrete so that progress of the main structure would not be impeded. This involved what can only be called a mining operation; shafts were left in the concrete base of the dam through which access to the fault zones was obtained and through which all muck was removed and concrete transported for the backfilling operation. Despite the hazard of the work, 2,600 m³ (3,500 yd³) of seam material was removed and replaced with concrete without accident. Concrete placing in the main dam was completed six weeks before the last placement of concrete in the foundation.[25.61]

A similar problem had to be faced during the construction of the Folsom Dam on the American River, 32 km (20 mi) east of Sacramento, California. This is another combined mass-concrete and earth-fill dam with a total length of 7 km (4.4 mi). One of its unique features is that it was a project of the U.S. Corps of Engineers, but the associated powerhouse was built by the U.S. Bureau of Reclamation. The mass-concrete section is founded on sound granite rock, but the earth-fill part of the dam, and eight of the small saddle dams needed around the reservoir, are founded on granite, the upper part of which is decomposed, cutoff walls being carried down in each case to granite that is firm, even though weathered to a degree. During the course of construction, a fault was discovered in the sound granite on which the concrete part of the dam was being placed. The fault was traced back under some of the concrete that had already been placed (as a precaution during a high-water period). A tunnel had therefore to be excavated along the line of the fault and under the newly placed concrete (Fig. 25.34). Fortunately, the fault petered out and so backfilling it with concrete did not prove to be too difficult a job, although about 37,000 m³ (50,000 yd³) of additional concrete was required.[25.62]

FIG. 25.34 Folson Dam, California; a close-up view of the excavated fault zone, showing some of the concrete filling.

Profitable Use of Problem Materials Another unusual feature of this job was that much of the soil to be removed was old gold-dredge tailings. The general contract therefore contained a provision that 15 percent of the value of any gold found during the work had to be turned over to the U.S. government. Of more direct geological interest, it was found that the properties of the highly decomposed granite were such that it could be used as compacted fill for the impervious sections of the main earth-fill dam and the saddle dams. This use of decomposed rock as soil is a useful reminder of the origin of all soils, even though the origin is not usually so obvious as here.

This practice of using decomposed granite as earth fill is not unusual in areas of deep weathering. Newbery has conveniently recorded some interesting cases as widely separated as Australia, Malaysia, Hong Kong, and the Seychelles, in all of which decomposed granite was encountered above sound bedrock at the sites of proposed dams, with depths of weathering varying from 15 m (50 ft) to more than 30 m (100 ft). In three of the cases, the decomposed granite was successfully used as compacted soil in the construction of the earth-fill sections of combined mass-concrete and earth-fill structures.[25.63] Although, therefore, weathering of granite and other competent rocks can add to the problems of dam building, the products of weathering can sometimes be used and the problem turned to an advantage.

Problems in Construction of Cutoff Walls Cutoff walls, when necessary, as they so frequently are in the abutments to dams and under earth-fill structures, always provide the contractor with a construction challenge. Geology has contributed to this phase of dam construction also. The Mount Morris Dam, on the Genesee River in northern New York, was built on a site characterized by limestone and shale formations similar to those encountered in the area of Niagara Falls, to which reference has already been made. The dam, designed for flood control only, is a mass-concrete structure 75 m (250 ft) high, but it was necessary to carry cutoff walls well into the two abutments. The shale rock made excavation for these walls difficult, but the contractor solved his difficulty by tunneling into the walls of the gorge and then stoping (upward) into the shale, mucking out through the access tunnels. Excavation was carried vertically in this way for a distance of 10.6 m (35 ft), and the process was then repeated after the open excavation had been filled with carefully placed concrete.[25.64]

A problem of an entirely different kind had to be faced in the construction of the Merriman Dam, part of the Delaware Aqueduct project for the water supply of the city of New York. Here, a complex assembly of glacial, waterborne, and lake-deposited sands, gravels, boulders, silty clays, and glacial till had to be penetrated to depths up to 54 m (180 ft) in order to provide certain cutoff for the 60-m (200-ft) -high earth-fill dam, designed to form one of the main reservoirs of the Delaware system. It was decided that the best way to achieve this would be by the use of caissons so designed and constructed that compressed air could be applied if and when necessary. Twenty reinforced-concrete caissons were used for this purpose; they measured about 3.6 by 13.5 m (12 by 45 ft), the height varying with the level of bedrock upon which they were finally founded. Free air was used for much of the excavation, but pressures up to 2.6 kg/cm² (38 psi) did become necessary in the final stages. Figure 25.35 illustrates, in diagrammatic form only, the heterogeneous character of the glacial soils that were penetrated by this unique cutoff wall; it illustrates, as well as can be done, the variation in soil conditions that must be expected in this type of glaciated country.[25.65]

Problems with Shales Finally, it has been during construction operations that troubles due to overconsolidated shales have made themselves evident. When this type of shale is known to exist, a contractor can be forewarned and can adopt a method such as

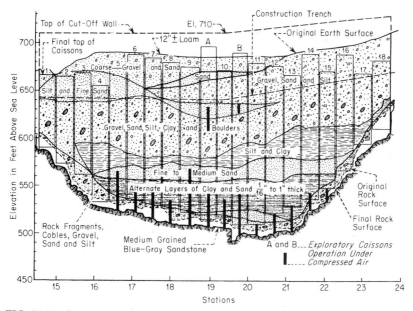

FIG. 25.35 Diagrammatic section through the soil strata encountered by the concrete caissons forming the cutoff of the Merriman Dam of the Delaware Aqueduct works for the water supply of New York.

25-50   CIVIL ENGINEERING WORKS

that followed on the Oahe Dam on the Missouri River. Here, the shale was excavated in relatively small areas; as soon as it was exposed at final grade, it was immediately covered with a bituminous coating; concrete had to be placed within 48 hours—in practice, it was usually in place within 24 hours. Only when this initial concrete slab had set were holes drilled through it and the necessary anchor bolts set in place for the purpose of holding the shale down and restricting its swelling tendency.[25.66] The same problem is often encountered in areas featured by shale as the local bedrock, another example being the Green Mountain Dam, a unit of the Colorado-Big Thompson Project on the Blue River, a tributary of the Colorado. Here the local Morrison shale caused the same difficulties; as a result, the bedrock was protected with an asphalt-bitumen emulsion covering applied within half an hour of the shale's exposure to the air.

25.13 PROBLEMS WITH DAMS IN SERVICE

Occasionally, despite the greatest care taken in design and construction, trouble—usually in the form of excessive leakage—may develop after a dam has been placed in service and is retaining water. There have been cases where trouble has been experienced because of concrete deterioration, but the usual cause is some inadequacy of the underlying geology, especially so in the case of excessive leakage. Stress has already been placed upon the vital necessity of regularly inspecting all dams and of executing all necessary work for their maintenance in first-class condition. It may be helpful to indicate the steps taken with one or two dams when trouble was encountered or, in the first place, suspected.

The Mulholland Dam is located in Weid Canyon, above and in plain view of Hollywood, California. It is a mass gravity structure, even though curved in plan (due to site requirements), containing 130,000 m³ (175,000 yd³) of concrete. It was built in 1924–1925 as a part of the water-supply system of Los Angeles. After the failure of the St. Francis Dam, suit was taken in August 1928 by local residents against the owner, the City of Los

FIG. 25.36 The Mulholland Dam, Los Angeles, seen just after earth fill had been placed against the downstream face, as described in the text.

Angeles, to force the abandoment of the reservoir formed by the dam, on the ground that it constituted a menace to life and property. Several inspections of the dam were made by groups of recognized experts, including geologists, and their reports were almost unanimously favorable, one emphatic endorsement of the geology of the site being that of Dr. Charles P. Berkey, one of the great pioneers of engineering geology. Despite these opinions, and even though the lawsuit was dismissed, the City of Los Angeles yielded to public opinion, to a degree, and in 1934 placed 220,000 m³ (300,000 yd³) of earth fill against the downstream face of this fine concrete structure, landscaping the sloped surface of the fill with trees and shrubs so that it would be practically concealed from Hollywood (Fig. 25.36). The cost of this "remedial" work was $250,000. The dam stands today, still effectively hidden, its stability justifying the studies of the underlying geology made half a century ago.[25.67]

Quite different was the experience met with at Ewden Valley works for the water supply of the city of Sheffield, Yorkshire, England. Construction of the two impounding earth dams in the Ewden Valley began in the year 1913; the respective reservoirs are known as the Broomhead and the More Hall; the former is for water supply and the latter for compensation water purposes only. The Broomhead Dam is about 297 m (990 ft) long at the top of the embankment, and overflow level is 27 m (91 ft) above lowest drawoff level (Fig. 25.37). The configuration of the valley disclosed the presence of ancient slips near the dam site, and studies and test borings indicated a very disturbed condition of the underground strata, which consists of shale, sandstones, and grits with overlying yellow clay. Excavation for the cutoff trench confirmed this deduction; the trench had to be carried to a depth of 36 m (120 ft) below stream level before the strata were found in undisturbed horizontal positions. Underground water flow in a band of grit was encountered and was tapped and drained as the trench was concreted up.

Geological and specialist engineering advice was constantly taken as the work proceeded; but after the start of impounding water in May 1928, increased flow below the dam was observed, which indicated leakage from the reservoir. Further geological investigations were made (also in connection with a serious landslip), including many more borings; as a result of these investigations, a program of cementation was initiated. In about six months (that is, up to July 1930), 6 million kg (14 million lb) of cement was injected, which reduced the leakage from 11.1 to 7.2 mld (3 to 1.9 mgd). This latter figure has been still further reduced by additional grouting work carried out when the reservoir

FIG. 25.37 Spillway of the Broomhead Dam, Yorkshire, England.

was drained and also by sluicing silt behind sandbag dams formed where rock outcrops were found in the bed of the reservoir. Leakage still persists through unknown underground passages but has steadily decreased through the years; the works are a most striking example of the possibility of leakage occurring through rocks normally sound but locally distorted and fissured, despite most careful engineering work and geological investigation.[25.68]

Very different were the conditions at the site of the Wolf Creek Dam on the Cumberland River in south-central Kentucky, where a major dam was built between 1941 and 1951. The dam was a project of the U.S. Corps of Engineers; somewhat naturally, construction had to be discontinued between August 1943 and September 1946. The structure consists of a mass-concrete section 539 m (1796 ft) long, carrying ten tainter spillway gates, the dam tying in to the left wall of the valley and occupying the area of the old stream valley. A rolled-earth embankment, 1,182 m (3,940 ft) long, adjoins the mass-concrete structure, extending to the right abutment in the valley wall. The dam has a maximum height of 77 m (258 ft) and can impound 7.5 billion m^3 (6.1 million acre-feet) of water. Geological formations at the site range in age from Mississippian to Ordovician, and include limestone and shale of different formations. The shale is at a shallow depth beneath the spillway section of the dam and passes beneath the Leipers limestone which forms the main foundation bed. Groundwater has dissolved the rock along the joints and bedding planes of the Leipers formation, resulting in a karstic structure. This was well recognized when the dam was built, the cross section of the dam including a deep cutoff trench into the bedrock; from the bottom of the trench grouting was carried out.

After satisfactory performance for 16 years, the dam exhibited the first sign of possible trouble in October 1967 when a muddy flow was noticed in the clear waters of the tailrace

FIG. 25.38 Wolf Creek Dam; a view showing construction operations in progress for the construction of the cutoff wall.

from the powerhouse associated with the dam. Within a few months a sinkhole was observed in the embankment near the downstream toe. An extensive exploration program was therefore undertaken, and high piezometric pressures were found under the dam, indicating leakage through the limestone. Water level in the reservoir was lowered 6 m (20 ft) for a period of eight months, and an extensive program of grouting with cement was undertaken, being completed in June 1970.

Extensive studies were then undertaken to monitor the performance of the embankment to see if further steps for ensuring its safety were necessary. A board of consultants concluded that such steps were necessary, grouting alone being regarded as insufficient for a permanent remedial measure. Accordingly, it was decided that a positive cutoff wall extending well into the bedrock below was necessary in the embankment. Two contracts, totaling almost $100 million, were awarded in 1975 and 1977, respectively, for the construction of a concrete cutoff wall, starting at the mass-concrete section of the dam and extending along the axis of the embankment at about its center point, with an additional wall adjacent to the switchyard close to the dam (Fig. 25.38). The slurry-trench method was chosen, using first circular holes varying from 1.3 to 1.2 m (52 to 47 in); these were sunk with special clamshell buckets, bentonite slurry filling the holes until the concrete was placed. These primary holes were 1.3 m (4 ft 6 in) apart, the space between them being filled with secondary concrete elements once the primary elements were set. In this

FIG. 25.39 A beautiful dam in an ideal setting, typifying the significance of geology in the founding of all dams: the Escales Dam on the Ribagorzano River in northern Spain.

25-54 CIVIL ENGINEERING WORKS

way, a permanent concrete wall was provided through 60 m (200 ft) of compacted soil and 24 m (80 ft) into the underlying limestone. The work was completed in 1979.[25.69]

25.14 CONCLUSION

There are few civil engineering works which so attract public attention as do dams, large or small. There are few other works that so often seem to be a cooperative effort between man and Nature. To see a thin-arched concrete dam in a narrow river gorge or a massive concrete dam in a great river valley spanning between exposed rock walls is to experience a sense of aesthetic satisfaction that is not always aroused by the works of the engineer. Figure 25.39 illustrates but one striking example of the "aesthetics of dams," if the term may be allowed. It shows the impounding dam for the Escales hydroelectric project in northeastern Spain. It is a gravity structure, 125 m (410 ft) high and 200 m (650 ft) long on its crest. Water impounded above the dam is led through tunnels to a large underground power station in which three 12.500-kW units operate under a head of 117.5 m (385 ft). The dam is located on the Ribagorzano River which has its headwaters in the Pyrenees. Flanked by supporting alternating strata of Cretaceous marls and limestone, which dip almost vertically, this dam—as do so many others in similar locations—shows a sense of fitness with its surroundings that makes it a fine symbol of the interrelation of geology with the design and construction of major civil engineering works.

25.15 REFERENCES

25.1 C. Prelini, "Some Dams of the Ancients," *Engineering News-Record*, **87**:556 (1921).

25.2 L. J. Mensch, in a letter, *Engineering News-Record*, **100**:674 (1928).

25.3 H. Lapworth, "The Geology of Dam Trenches," *Transactions of the Institution of Water Engineers*, **16**:25 (1911).

25.4 J. D. Justin, "The Design of Earth Dams," *Transactions of the American Society of Civil Engineers*, **87**:1, 133–134 (1924).

25.5 Lapworth, op. cit., p. 23; see also T. U. Taylor, *The Austin Dam*, U.S. Geological Survey Water Supply Paper no. 40, 1900.

25.6 C. McDonough, "Historic Austin Dam Rebuilt," *Engineering News-Record*, **124**:845 (1940); see also in same journal, G. L. Freeman and R. B. Alsop, "Underpinning Austin Dam," **126**:180 (1941).

25.7 F. L. Ransome, "Geology of the St. Francis Dam Site," *Economic Geology*, **23**:553 (1928).

25.8 "Malpasset Dam on French Riviera Fails," *Civil Engineering*, **30**:59 (1960); see also "Preliminary Report on the Malpasset Dam," *The Engineer*, **209**:812 (1960); see also *Rapport Définitif de la Commission d'Enquête du Barrage de Malpasset*, Paris, August 1960.

25.9 K. Terzaghi, "Failure of the Malpasset Dam," *Engineering News-Record*, **168**:58 (15 February 1962).

25.10 W. E. Jessup, "Baldwin Hills Dam Failure," *Civil Engineering*, **34**:62 (February 1964); see also press reports of legal decision; see also D. H. Hamilton and R. L. Meehan, "Ground Rupture in the Baldwin Hills," *Science*, **172**:333 (1971).

25.11 "Teton Dam Failure," *Civil Engineering*, **47**:56 (August 1977); see also same journal, **49**:78 (January 1979).

25.12 A. Coyne, "The Construction of Large Modern Dams," *Structural Engineer* (new series), **15**:70 (1937).

25.13 "Tensioned Cables Will Strengthen Threatened Bombay Water Supply Dam," *Engineering News-Record*, **147**:54 (5 July 1951); see also same journal **153**:39 (30 September 1954).

25.14 S. S. Morris, "Steenbras Dam Strengthened by Post-Tensioning Cables," *Civil Engineering,* **26**:75 (1936).

25.15 "Coyne Prestressing Gives New Strength to an Old Dam," *Engineering News-Record,* **168**:42 (19 April 1962).

25.16 F. G. Thompson, "Strengthening of John Hollis Bankhead Dam," *Civil Engineering,* **39**:75 (December 1969).

25.17 "Record Tendons Anchor Old Dam to Foundation," *Engineering News-Record,* **186**:20 (11 February 1971).

25.18 E. A. Marcinkevich and A. Z. Ephros, "Dam Anchored to Foundation to Withstand Floods," *Civil Engineering,* **48**:55 (April 1978).

25.19 G. McIldowie, "The Construction of the Silent Valley Reservoir, Belfast Water Supply," *Minutes of Proceedings of the Institution of Civil Engineers,* **239**:465 (1936).

25.20 G. F. Deacon, "The Vyrnwy Works for the Water-Supply of Liverpool," *Minutes of Proceedings of the Institution of Civil Engineers,* **126**:24 (1896).

25.21 "Rock Bottom is Way Down at Lower Notch Dam," *Engineering News-Record,* **184**:24 (4 June 1970); see also *Lower Notch Generating Station,* Hydro Electric Power Commission of Ontario, Toronto, 1971.

25.22 Q. Zaruba, "Geology of the Orlik Dam Site," *Water Power,* **17**:273 (1965).

25.23 L. R. Frei, "Auburn Dam: Foundation Investigation, Design and Construction," *Field Trip Guide Book,* Association of Engineering Geologists, 1976 (rev. April 1976 as separate paper, U.S. Bureau of Reclamation); see also "Professional Roles Debated in Auburn Arch Death," *Engineering News-Record,* **202**:18 (1 February 1979).

25.24 R. A. Sutherland, "French Build High Dam in Narrow Limestone Canyon," *Engineering News-Record,* **115**:706 (1935).

25.25 H. A. Johnston, "Underground Exploration with Calyx Drills," *Engineering News-Record,* **120**:436 (1938).

25.26 C. P. Berkey, "Geology of Boulder and Norris Dam Sites," *Civil Engineering,* **5**:24 (1935); see also "Gorge Excavation Confirms Geological Assumptions," *Engineering News-Record,* **111**:761 (1933).

25.27 J. Cabanius and R. Maigre, "The Serre-Ponçon Dam: The Durance Development," *Travaux,* no. 286, August 1960, p. 43.

25.28 W. H. Twenhofel, *A Treatise on Sedimentation,* 2d. ed., Williams & Wilkins, Baltimore, 1932, p. 17.

25.29 M. Fitzmaurice, "The Nile Reservoir Assuan," *Minutes of Proceedings of the Institution of Civil Engineers,* **152**:71 (1903).

25.30 F. Reeves and C. P. Ross, *A Geologic Study of the Madden Dam Project, Alhajuela, Canal Zone,* U.S. Geological Survey Bulletin 821, 1931; see also F. H. Kellog, "Clay Grouting at Madden Reservoir," *Engineering News-Record,* **109**:395 (1932).

25.31 S. C. Happ, "Treatment of Chalk Foundation with Bentonite Seams, Harlan County Dam, Nebraska," *Bulletin of the Geological Society of America,* **61**:1568 (1950).

25.32 W. G. Bowman, "The French Touch in Dams," *Engineering News-Record,* **146**:33 (10 May 1951).

25.33 K. L. Rao, "Engineering Problems in Recent River Valley Projects in India," *Proceedings of the Institution of Civil Engineers,* **11**:1 (1958).

25.34 L. L. Waterhouse, W. R. Browne, and D. G. Moye, "Preliminary Geological Investigations in Connection with the Proposed Warragamba Dam, N.S.W.," *Journal of the Institution of Engineers of Australia,* **23**:74 (April-May 1951).

25.35 T. B. Nicol, "Warragamba Dam," *Proceedings of the Institution of Civil Engineers,* **27**:491 (1964).

25.36 S. B. Morris and C. E. Pearce, "Concrete Gravity Dam for Faulted Mountain Area," *Engineering News-Record,* **113**:823 (1934).

25.37 "Earthquake-Proof Dam in Chile," *Engineering News-Record*, **107**:725 (1931).

25.38 "Three Dams on San Andreas Fault Have Resisted Earthquakes," *Engineering News-Record*, **109**:218 (1932); see also N. A. Eckart, "Development of San Francisco's Water Supply to Care for Emergencies," *Bulletin of the Seismological Society of America*, **27**:3 (1937).

25.39 "Flexible Core for Dam in Earthquake Country," *Engineering News-Record*, **125**:279 (1940).

25.40 G. E. Goodall, "Horizontal Joint Put in Arch Dam to Prevent Cracking," *Engineering News-Record*, **137**:884 (1946).

25.41 "Foundation Procedure at Owyhee Dam," *Engineering News-Record*, **106**:178 (1931).

25.42 "Unique Cutoff Construction and Arched Foundation Features of Rodriguez Dam," *Engineering News-Record*, **105**:600 (1930); see also C. P. Williams, "Foundation Treatment at Rodriguez Dam," *Transactions of the American Society of Civil Engineers*, **99**:295 (1935).

25.43 "Joints Let Dam Flex with Soil Movements," *Engineering News-Record*, **161**:32 (6 November 1958).

25.44 G. W. Christians, "Asphalt Grouting under Hales Bar Dam," *Engineering News-Record*, **96**:798 (1926); see also in same journal, "Stopping a River under a Dam," **127**:654 (1941); see also J. W. Frink, "Solution of Limestone beneath Hales Bar Dam," *Journal of Geology*, **53**:137 (1945).

25.45 F. P. Lacy and W. W. Engle, "Nickajack Dam—Replacement for Hales Bar," *Civil Engineering*, **35**:50 (July 1965); see also "TVA Gives Up on Hales Bar Dam," *Engineering News-Record*, **170**:26 (18 April 1963), includes bibliography of ENR articles on the Hales Bar Dam.

25.46 J. Hinds and N. S. Long, "Oil No Boom in Foundations at Santa Felicia Dam," *Civil Engineering*, **25**:220 (1955).

25.47 Deacon, op. cit.

25.48 W. H. Stuart, "Influence of Geological Conditions on Uplift," *Proceedings of the American Society of Civil Engineers*, vol. 87, SM6, Paper 3008, 1961, p. 1.

25.49 Reeves and Ross, op. cit.

25.50 H. J. Hoeffer, "Fort Randall Dam to Provide More Storage on Missouri River," *Civil Engineering*, **22**:474 (1952).

25.51 "Merwin Hydro Project," *Guide for Western (U.S.) Tour*, International Congress on Large Dams, New York, 1958, p. 156.

25.52 K. O. Strenge, "Noxon Rapids Dam Meets Extraordinary Geologic Conditions," *Civil Engineering*, **29**:474 (1959).

25.53 I. D. MacKenzie and E. L. Brown, "Geological Features and Foundation Treatment at the Beechwood Development," *The Engineering Journal*, **42**:54 (1959); see also H. McFarlane, *The Beechwood Earth-Fill Dam*, tech. memo. no. 63, National Research Council Associate Committee on Soil and Snow Mechanics, Ottawa, 1960.

25.54 K. T. Bass and C. W. Isherwood, "The Foundations of the Wimbleball Dam," *Journal of the Institution of Water Engineers and Scientists*, **32**:187 (May 1978); see also, same publication, D. Battersby, K. T. Bass, R. A. Reader, and K. W. Evans, "The Promotion, Design and Construction of Wimbleball," **33**:399 (September 1979); and, as a more "popular" account, see "Crushed Rock Wreaks Wimbleball Havoc," *New Civil Engineer*, no. 223, 16 (16 December 1976).

25.55 "San Gabriel River Flood Control," *Engineering News-Record*, **114**:113 (1935); see bibliography p. 116.

25.56 W. G. Bowman, "Turkey's Keban Dam is Deeper Than Tall," *Engineering News-Record*, **180**:78 (11 April 1968); see also in same journal, "First Power at Keban—Three Years Later," **193**:21 (10 October 1974).

25.57 K. Terzaghi and Y. Lacroix, "The Mission Dam," *Geotechnique*, **14**:13 (1964); see also "Three Phases of Mission Dam," *Engineering News-Record*, **166**:33 (29 June 1961).

25.58 W. G. Bowman, "Iran's Two Big Dams Promise a Better Life; (1) Karadj Dam: A Difficult Job But No Hitches," *Engineering News-Record*, **166**:38 (16 March 1961).

25.59 W. L. Voorduin, "Iran's Dez Dam," *Civil Engineering,* **30**:50 (August 1960).

25.60 "Dam Trench Drained from Below," *Engineering News-Record,* **161**:47 (23 October 1958).

25.61 "Mining under Seminoe Dam," *Engineering News-Record,* **122**:490 (1939).

25.62 F. Kochis and H. A. Johnson, "Faulted Foundation Complicates Construction of Folsom Dam," *Civil Engineering,* **24**:659 (1954); see also "Dam Building in the Gold Country," *Engineering News-Record,* **151**:39 (23 July 1953).

25.63 J. Newbery, "Dam Foundations on Decomposed Granite," *Proceedings of the Third International Congress, International Association of Engineering Geology,* vol. 1, sec. III, 1978, p. 169.

25.64 "Upward Mining of Cut-off Trenches," *Engineering News-Record,* **145**:44 (20 July 1950).

25.65 "Caissons for a Cut-off Wall," *Engineering News-Record,* **125**:761 (1940); see also in same journal, **127**:426 (1941).

25.66 R. Smith, "At Oahe: Contractor Licks Shale by Building Backwards," *Engineering News-Record,* **163**:42 (23 July 1959).

25.67 "Mulholland Dam Backed by Earthfill Against Downstream Face," *Engineering News-Record,* **112**:558 (1934).

25.68 J. K. Swales, "The Broomhead Reservoir," *Water and Water Engineering,* **36**:565 (1934); and an in-person communication from G. T. Calder, Yorkshire Water Authority, Southern Division.

25.69 J. Kellberg and M. Simmons, "Geology of the Cumberland River Basin and the Wolf Creek Damsite, Kentucky," *Bulletin of the Association of Engineering Geologists,* **14**:245 (1977); see also F. B. Couch and A. L. R. diCervia, "Seepage Cut-Off Wall Installed Through Dam Is Construction First," *Civil Engineering,* **49**:62 (January 1979).

Suggestions for Further Reading

Large dams create such international interest that it is not too surprising to find that the International Congress on Large Dams (generally known by its acronym ICOLD) is in some ways unique among international engineering organizations. Having its origin in the World Power Conference, it is now an active and esteemed organization in its own right, with headquarters located at 151 Boulevard Haussmann, 75008, Paris. There are active local committees in participating countries; the one for the United States is located in the Engineering Societies Center, 345 East 47th Street, New York, NY 10017.

The regularly held congresses of ICOLD are arranged to address themselves usually to three leading questions selected in ample time for the preparation of papers by those with special interest in them for presentation at the next congress. Now numbering well over fifty, these questions range over the whole subject of dam design and construction. Of special interest to users of this volume is Question No. 53 covered at the 14th congress: *Influence of Geology and Geotechnics on the Design of Dams.*

The volumes of ICOLD proceedings contain a wealth of information on all aspects of dams and geology. Reference to ICOLD publications can always be a useful starting point when the first preliminary steps have to be taken for a new dam-building project. The national registers of major dams, the preparation of which ICOLD encourages, provide an invaluable comparative guide, both nationally and internationally.

As with other branches of civil engineering, the major civil engineering societies give attention to the design and construction of dams. Indicative of this active interest is the publication by the American Society of Civil Engineers of the proceedings of two conferences sponsored by the Engineering Foundation. The first, published in 1973, is a 948-page volume called *Inspection, Maintenance and Rehabilitation of Old Dams;* the second, published in 1974, is a comprehensive 472-page volume on *Foundations of Dams.* In a similar way, the Institution of Civil Engineers (U.K.) published in 1981 the record of a conference on *Dams and Earthquakes* as a 304-page book.

Major dam-building agencies have enriched the literature of civil engineering by preparing some outstanding volumes in their work. In 1949, for example, the U.S. Government Printing Office published TVA Technical Report 22 on *Geology and Foundation Treatment: Tennessee Valley Authority Projects,* following the most active period of TVA dam building. The U.S. Bureau of Reclamation has similarly shared its experiences in dam building, notably in a magnificent multivolume report on the building of the Hoover (Boulder Canyon) Dam; Bulletin 1 of Part III of this deals with *Geological Investigations.* At the other end of the scale, so to speak, is the Bureau's well-known book *Design of Small Dams* which has been so helpful to many since first it was published. While its name was the Water and Power Resources Service, the Bureau published in 1980 a notable report by R. B. Jansen called *Dams and Public Safety.*

To indicate the widespread availability of information on dam foundations, mention may be made of a paper to be found on pp. 197–216 of part (volume) 13 of *Proceedings of the 24th International Geological Congress* (Montreal) by D. Stefanofic and S. Swmak. It is a most useful review called "The Role of Geophysics in Dam Construction," naturally with special reference to practice in eastern European countries.

chapter 26

RESERVOIRS AND CATCHMENT AREAS

26.1 Leakage from Reservoirs / 26-2
26.2 Detection of Leakage / 26-8
26.3 Prevention and Elimination of Leakage / 26-9
26.4 Secondary Effects of Reservoir Flooding / 26-11
26.5 Landslides / 26-12
26.6 Silting Up of Reservoirs / 26-13
26.7 Methane in Reservoirs / 26-18
26.8 A Perched Reservoir / 26-19
26.9 River Diversions / 26-21
26.10 Conclusion / 26-22
26.11 References / 26-22

The function of dams, in general, is to retain water above the elevation at which it normally stands. This impounding may vary from the control of water levels within a few feet of normal elevation by a regulating weir to the creation of the immense lake now behind Gouin Dam in northern Quebec, on the St. Maurice River. Apart from specially restricted cases such as low-head regulating weirs, a reservoir will be formed by a dam to serve as an artificial storage basin for the water that is retained. Usually the dam structure forms but a small part of the periphery of the reservoir; its bottom and the greater part of its sides (if this distinction may be used) will consist of the natural crust of the earth. A superficial assumption is that, provided the dam site is given due attention with regard to watertightness, all will be well. If this were always the case, this chapter would not be necessary.

The construction of a dam and the subsequent impounding of water behind it cause more interference with natural conditions than do almost all other works of the civil engineer. The problems created at the dam site are common to the whole reservoir above this site, although to varying degrees. In brief, a difference will be set up in the level of the water table corresponding to the height of the dam between the two sides of the dam; all material in the bed of the reservoir and especially close to the dam will be subjected to considerable hydraulic pressure; all flooded areas will in the future be subject to the action of water upon them; and finally the groundwater level up the valley in which the reservoir is located will be directly affected by the rise in water level, generally for a considerable distance away from the actual shore line of impounded water. In consequence, there will be a tendency for the impounded water to find some means of escape through any weaknesses that may exist in the structure of the ground. Materials in the sphere of influence of the impounded water and liable to be affected by exposure to water may fail to retain their former stability; landslides are one possible direct result of failure. Features such as wells that come within the area of influence of the reservoir and are dependent on previous groundwater levels will be correspondingly affected. Cursory consideration will show that all these matters depend fundamentally on the geological structure of the reservoir basin. A study of this geology is therefore an essential complement of geological investigation at a dam site, since only by this means can possible dangers be foreseen and suitable precautionary measures be taken in advance.

Another important reason for the careful study of the geology of reservoir sites is that an accurate estimate of the capacity of a reservoir can thus be made; this requirement often has associated with it the determination of the necessary height of dam. The figure must often be determined purely from the point of view of elevation when maximum working head is the criterion, as in the case of water power plants. In other cases, especially with water-supply reservoirs, effective storage capacity is the determining factor. For a reservoir formed by a dam and a basin of impervious strata, the calculation of this capacity is a simple matter, and reservoir capacities are often calculated on this assumption. If attention is given to the geological formations underlying the basin, it may be found that large masses of pervious rocks are in contact with the impounded water and are so arranged that they can drain out (at least to some extent) as the reservoir water level is lowered. With the aid of reasonably careful preliminary geological investigations, an estimate of the additional storage thus available can be obtained, with consequent benefit to the economic aspect of the scheme.

26.1 LEAKAGE FROM RESERVOIRS

Leakage from reservoirs is obviously a potential source of trouble. Surprising though it may seem, attention has not always been given to the possibility of leakage developing. Of the many examples that can be cited to support this suggestion, there need be mentioned only the abandonment of the Jerome Reservoir in Idaho and of the Hondo Reservoir in New Mexico as a result of excessive and uncontrollable leakage which occurred after the impounding dams had been built. That structures of magnitude should have to be abandoned, not because of defects of design or construction but because of the undetermined nature of local geological structure, is surely most telling evidence of the need of adequate study of every feature connected with the geology of a reservoir site.

It is, perhaps, not without significance that the two most remarkable examples of the interrelation of geology and engineering encountered by the writers have involved reservoirs. One of these cases is that dealt with in the penultimate section of this chapter, in which it will be seen that geological study was applied with constructive effect. The other example is, on the contrary, one of the most singular cases imaginable of what can result from the neglect of geology. Unfortunately, it must be described anonymously, but the lesson it tells does not suffer from this lack of identification. There is standing today in the northeastern part of North America a fine, buttress-type, reinforced-concrete dam about 12 m (40 ft) high and 540 m (⅓ mi) long, constructed in 1910 and still in good condition, even though it has never retained water. It was built by an owner who spurned professional advice. A cursory reading of a geological report then extant would have shown that one side of the valley crossed by the dam, one buttress therefore of the structure and one complete side of the intended reservoir, consisted of a glacial moraine made up of small boulders. As could have been foretold, the reservoir leaked like a sieve as soon as impounding of the water commenced.

Not willing to be beaten by geology, the owner had a vast area paved with an asphalt-coated, reinforced-concrete slab; later he had a cutoff wall, in places taken to a depth of 24 m (80 ft) below ground level, carried up the valley from the dam. All to no avail; the reservoir still leaked to such an extent that the intended generation of the power from the water that was to have been impounded had to be given up. Penstock and powerhouse were dismantled. The dam stands today in mute testimony to what the neglect of geology can do. But as has happened before, the beavers who inhabited the stream that was to be dammed by the concrete structure did what the owner could not achieve. They built their own series of dams and did retain water; one of the beaver dams was carefully constructed around a sinkhole, the leakage through which was thus stopped by Nature's own engineers.

Failure by excessive leakage may take place because of defects in the structural arrangement of the underlying strata, e.g., a fault or excessive fissuring; the failure of one or more of the rock types to stand up to exposure to water; the dip of pervious rock strata which drains water away from a reservoir site; or the absence of adequate impervious barriers at critical points in the topographic perimeter of the reservoir area. As the dam site in a reservoir will usually be at the deepest part of the valley utilized to form the basin, the impounded head will be greatest at this location. Therefore, structural geological defects and the existence of rocks that may be affected by prolonged exposure to water are of special consequence and of greatest potential danger in the vicinity of the dam. Especially is this true of doubtful rocks such as those which are to any degree soluble in water. Rock salt is naturally the most dangerous of all; but as it is not of frequent occurrence, it is not of great significance. Gypsum, although less soluble than rock salt, is seriously affected by exposure to water and so is potentially dangerous. Of unusual significance in all reservoir investigations is the presence of limestone. Although it is the least permeable of the rocks mentioned, its relatively wide distribution renders it comparatively well known. This general familiarity, although not perhaps breeding proverbial contempt, does lead to the tendency to form incorrect judgments regarding its soundness, especially when considered in conjunction with the relatively solid appearance of many outcrops.

All limestones are at least partially soluble in ordinary water. If through the movement of water the resulting solutions can be removed from contact with the rock and replaced by fresh water, solution will continue. In all limestone formations subject to movement of underground water, therefore, special precautions must be taken to make sure that no potentially serious zones of weakness exist. The well-known caverns found in many limestone districts are the result of water action. Sometimes these are the result of the action of flowing streams, which for a part of their courses flow underground; parts of Derbyshire, England, are typical of similar well-known scenic areas to be found in many parts of the world. Some limestone caverns are the result of periodic movements of underground water tables. In consequence, a close study of existing underground water levels, although always of value in connection with the investigation of reservoir sites, is of special value in the case of those formed by limestone strata. Many examples from practice testify to the importance of close study of any limestone formation encountered in reservoir work.

The troubles at the Hales Bar Dam were due to cavernous limestone. The pioneer work on grouting with clay at the site of the Madden Reservoir in the Panama Canal Zone was occasioned by the careful studies of the underlying limestone strata which had been made in the preliminary geological investigations; all points of weakness were effectively sealed up before the reservoir was filled. The Malad Reservoir, near Malad City, Idaho, represents another example. A dam 132 m (440 ft) long and 21 m (70 ft) high was projected to impound water, and construction was started in 1917. Work had to be stopped when the crest had reached a height of only 15 m (50 ft); although the level of the water being impounded had risen only 3 m (10 ft) above normal stream level, leakage was occurring at a relatively high rate. Investigation suggested that the leakage was occurring through a bed of soluble limestone 150 m (500 ft) long and 7.5 m (25 ft) wide in outcrop on the side of the valley near the dam but effectively screened from view by a talus slope. These examples include cases previously noted in order to illustrate the close relationship between the geological conditions affecting dam sites and those germane to reservoirs and catchment areas.[26.1]

Leakage from reservoirs, as distinct from that under dams, will be so remote from the normal practice of civil engineers that it may possibly be regarded as something of a "freak" occurrence. Unfortunately, the accumulated records of civil engineering practice include all-too-many cases of such leakage. As recently as the late 1960s, for example, another case developed in Greece. The Greek government in 1960 built an earth-fill dam, 27 m (90 ft) high and 350 m (1,148 ft) long, on the Perdika River in the Ptolemais region in the north of the country. It was for the water supply of adjacent industries. The dam was satisfactorily completed and the necessary 8-km (5-mi)-long pipeline was installed. But the reservoir remained dry, despite heavy spring rains. Studies revealed a previously undetected limestone outcrop in the bed of the reservoir.[26.2] Intensive studies of the situation were made by German and Greek experts, but no immediate solution was found. It is understood that the problem is now receiving further attention.

There are repeated references in this Handbook to karstic conditions encountered in limestone, the name being derived from the unusual topography—due to solution of limestone—in the high land that separates the Dalmatian coast of Yugoslavia from the interior of the country. The natural beauty of the area is in striking contrast to the innumerable problems that karstic conditions have caused in civil engineering works. If only because of this link with Yugoslavia, brief mention must be made of two dams successfully built in the heart of the Karst country, on the Trebisnjica River, well known to geomorphologists as the "disappearing river." For much of its length, it flows underground in solution cavities in the local limestone. Many studies have been made of its underground course but this is still not completely known. Following many years of study of the river after it appears in a normal watercourse near the lovely town of Dubrovnik, Yugoslavian engineers considered that a dam could safely be built near Bileća, about 17.5 km (11 mi) downstream of the river's first appearance above ground, with another diversion dam 13 km (8 mi) farther downstream at Gorica. Water is diverted by the latter structure into a 16-km (10-mi)-long tunnel whence it flows into the Dubrovnik powerhouse to generate 108,000 kW. (Fig. 26.1)

The main Grancarevo Dam is a concrete arch structure, 121 m (403 ft) high and 337 m (1,115 ft) long at its crest. It impounds more than one billion m³ (one million acre-feet), or 43 percent of the annual runoff from the watershed above. Bedrock is Lias limestone, heavily seamed on the left bank, surface cracks around the dam site being clay-filled. A total of 3,600 m (12,000 ft) of test borings preceded the design and construction of the dam. Construction methods were strictly controlled under the general contract, charges of explosives for excavation in the foundation rock being limited, as but one example, to 100 kg (220 lb), being often much less than this. The final clean-up excavation was carried out by hand tools in order to eliminate vibrations. Much jetting of rock slopes in the river banks was carried out. From the foundation trench, which went 19.5 m (65 ft) below the riverbed, a 165-m (550-ft)-deep grout curtain was injected, using a clay-cement-bentonite mixture. Despite all precautions, two small slides occurred during construction, but these were readily corrected; 350 observation points were established around the dam, with electrical connections to a central control station for maintaining a continuous watch on the performance of the dam and reservoir.[26.3]

FIG. 26.1 Another reservoir in karst country, far from Jugoslavia; the upper view shows the Liulang Cave in Yunnan Province, China, and in the lower view is seen the small water power plant operated by the flow through this unusual reservoir.

26-6 CIVIL ENGINEERING WORKS

There may be finally mentioned the Lone Pine Reservoir in Arizona which has never yet been used, even though the 30-m (101-ft) -high earth-fill dam intended to form it was completed in 1934–1936. Built as a WPA project, the dam was the highest earth-fill dam in Arizona when dedicated. Unfortunately, no geological advice was taken prior to the construction of the dam at an "obvious" site noted by early settlers. Geology beneath the reservoir is complex, with subsurface salt deposits possibly responsible for sinkholes in an adjacent area. The main rock types are basalts, limestones, and sandstones, even the structure of which tends to increase the probability of leakage. Water level reached a high mark of 22 m (74 ft) on 6 April 1936 but receded to the 13 m (43 ft) mark by 8 May; further filling proved impossible. A full account of the geological difficulties has been published.[26.4]

The possibility of leakage through a pervious stratum dipping away from the reservoir valley is best explained by means of a simple diagram (Fig. 26.2). This is not an unusual geological section; and although the figure is diagrammatic only, many actual sections of a similar type can be found. The inclined structure of pervious and impervious strata will naturally have one result—any water in contact with the pervious beds in valley A (as, for example, water impounded in the valley by a dam with crest level about X) will flow underground until it is discharged into valley B. Discharge might take the form of springs. The structural arrangement shown is as simple as possible; it is capable of much variation without the fundamental source of leakage being changed. For example, the lower pervious stratum might not outcrop in valley B but remain underground with no natural outlet for the entrapped water if the surface stratum is impervious. Artesian pressure would thus be set up, and a potential supply of underground water would be provided.

Finally, there always exists the possiblity that leakage may take place through some permeable stratum at a low point in the ground encircling the reservoir. Rarely will the topography be such that there is no indication of any low spot in the perimeter of the reservoir site. Any depressions will inevitably call for special geological study, since the fact that they are a departure from regular topographical features is an indication that some unusual process has been at work at some time in the past. Glacial action is frequently responsible for depressions—"saddles," they may be called, if they take that particular form between two valleys. A section through a simple saddle in which the rock level has been worn down well below surface level, pervious deposits making up the observed level of the ridge at this location, is shown in Fig. 26.3. If water is impounded above "critical WL" and the broken line extending through "pervious material" is the necessary underground hydraulic gradient for flow through that material, it is clear that leakage will occur from valley A into valley B. In addition to causing loss of water from

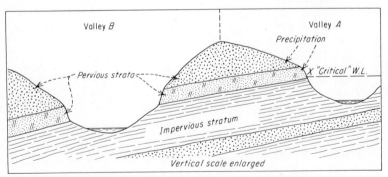

FIG. 26.2 Simplified geological section illustrating possible leakage from a reservoir.

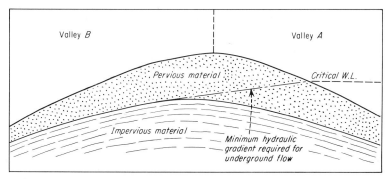

FIG. 26.3 Simplified geological section illustrating another possible means of leakage from a reservoir.

the reservoir, this underground flow of water may cause trouble in valley B if there is any instability there of unconsolidated deposits which might be intensified by the presence of excessive underground water.

There are available all too many descriptions of losses from reservoirs due to this cause. One of the most remarkable examples is that given by the Cedar Reservoir in the state of Washington, briefly noted in a number of references which prompted J. Hoover Mackin to undertake a careful study of the case many years after it had first given trouble. In 1914 the city of Seattle constructed a concrete dam in a gorge of the Cedar River, a tributary of the Snoqualmie River, with a crest length of 239 m (795 ft) and a total height above bedrock of 65 m (217 ft). Before the dam was built, the city was advised by its consultants of the danger of severe leakage around the northeast abutment, but this advice was neglected. Filling of the reservoir started in 1915, but before it had proceeded far, leakage was reaching a volume of 110 mld (30 mgd). Despite this, filling was slowly continued; leakage increased and finally washed away the small town of Moncton, whose 200 inhabitants had to be relocated. A new board of consultants was appointed, and test holes were put down. The findings of this board were just the same as those of the earlier consultants, namely, that the northeast abutment of the dam was part of a large glacial moraial deposit of open-textured gravel, occupying a deep and capacious preglacial valley of the Cedar River.

Much effort was exerted and large amounts of money were spent in attempts to seal the reservoir. Presumably to test the efficiency of the sealing operations, water was allowed to rise again in the reservoir toward the end of 1918. Leakage again appeared and increased rapidly. Around midnight on 23 December, a great earth movement occurred on the eastern part of the northeast face of the moraial embankment (i.e., the part of the face of the moraine removed from the dam), approximately 1,800 m (6,000 ft) from the dam. Well over 764 thousand m³ (1 million yd³) of material was washed out by the rush of water; the initial discharge when the failure occurred was estimated at between 85 and 570 cms (3,000 and 20,000 cfs). The resulting flood rushed down a small creek valley to the Snoqualmie River, taking out the small town of Edgewick, sawmills, and other property and destroying some track of the Milwaukee Railroad; good fortune alone obviated a major loss of life, for no train was passing at the time. Subsequent legal actions dragged on for the next decade, but eventually the city of Seattle had judgment rendered against it of over $300,000. The dam still stands, but somewhat naturally, no attempt has again been made to use it for its intended purpose of impounding water. Mackin's paper on the failure gives a detailed treatment of the geology of the moraial ridge and provides a rational explanation of the exact happenings at the time of failure.[26.5]

26-8 CIVIL ENGINEERING WORKS

26.2 DETECTION OF LEAKAGE

Leakage from a reservoir will generally be so obvious as to call for no special aids in detection. Sudden increases in streamflow below a dam site without a corresponding increase of rainfall will be a direct indication of some escape of water; and the appearance of springs, boils, or other surface indications of underground water in locations previously dry and stable will also indicate unsatisfactory conditions. Minor leakage may be more difficult to trace, especially in connection with the investigation of solution channels in limestone. The use of dyes in cases of this kind can be invaluable; small quantities of dye can be added to the moving water at a convenient station and samples taken from the suspected source of leakage. Various chemicals have been used for this purpose; permanganate of potash is perhaps the most common; in some cases, ordinary bluing can be used. The use of bacteria has been suggested; but for practically all special work, a chemical indicator is used. Many chemicals are thus employed, among them fluorescein and its sodium salt, uranin. These give water a characteristic color which can be detected even when diluted to the extent of 1 part in more than 10 million parts of water.

Even with the use of such sensitive dyes, however, leakage sometimes cannot be traced. An interesting, although possibly unique, example which may be mentioned is a small water-power installation in Cephalonia, one of the islands of the Ionian archipelago (Greece). The undershot wheel of the installation is operated by the flow of seawater from the Mediterranean *into* the island (Figure 26.4). The strange stream so formed eventually disappears from view in the caverns and fissures of the limestone of which the island is there constituted. During British occupation of the island, quantities of petroleum were poured into the seawater as it passed the mill, and the island was surrounded by boats whose crews were on the lookout for any signs of emerging petroleum. None appeared.[26.6] If any of the investigators had known their Greek history, they might have found the answer to the problem, since the ancient Greeks knew of a large underground cavern accessible from the other side of the island in which there is a brackish lakelet, rediscov-

FIG. 26.4 Undershot waterwheel in Cephalonia, Greece, operated by the flow of the sea *into* the island.

ered in the 1960s. A 13-km (8-mi) underground channel connects the inlet at the mill with the lakelet, as related by Lawrence Durrell in his interesting book, *The Greek Islands*.[26.7]

Another of the peacetime applications of atomic energy is placing into the hands of the geologist and engineer a new tool of great potential for the tracing of subsurface water flow. The use of radioactive isotopes for such a purpose is an obvious use of these newly developed chemicals. There are many difficulties in the way of their widespread adoption: the possibility of interaction with the aquifers, their relatively high cost, biological and radiation dangers, and the varying half-lives of different isotopes. The converse of these difficulties provides, of course, the desirable features of the use of isotopes for tracing groundwater movement.

A number of isotopes have been successfully used in this way. One of the early applications was in an isolated location in the Libyan desert of Egypt where Sir C. S. Fox checked on the watertightness of a proposed reservoir site by using isotopes to show that groundwater was not moving, rather than the reverse.[26.8] Rubidium chloride was specially prepared for this application at Harwell, the British Atomic Energy Research Establishment, flown out from London to Cairo, driven immediately to the site, and put to use within little more than a day after its preparation. Similar applications have been made in several countries; interesting applications using a cobalt isotope are reported from Israel, for example, where a radioactive hexacyanocobaltate was found to be the most suitable isotope for use with Yarkon River water in the limestone aquifer in the Lydia well field. Naturally, background for freshly pumped water has to be determined before the radioactive material is introduced.[26.9] The single-well pulse technique has been found to be reasonably effective and economical, even though results so obtained have to be interpreted with care if the aquifer being studied is extensive. Expert assistance must still be utilized for all such uses of radioactive materials, but they do hold promise of extending very greatly subsurface water investigations.

26.3 PREVENTION AND ELIMINATION OF LEAKAGE

If the preliminary geological investigations already described are applied to the study of reservoir and catchment areas when a dam construction project is planned, most of the possible sources of leakage from the reservoir should normally be discovered. The various possibilities of leakage have been noted, and the nature of the means of leakage will be some indication of whether or not it can be prevented by anticipatory remedial works. In the case of strata dipping across a valley into adjacent areas, little, if anything, can be done to interfere with the natural direction of underground water flow. When saddles occur with low spots in the bedrock elevation, as indicated in Fig. 26.3, cutoff dams can be constructed to seal effectively the open waterways thus available for underground flow. The simplest type of structure will suffice for this purpose, provided a cutoff wall of some type is carried to the rock floor between the points at which it ceases to be above reservoir water level.

In other cases of possible leakage, especially where localized fissuring of rock is encountered, grouting may effect a satisfactory seal if carefully carried out in close association with complementary geological study. Grouting operations may similarly be employed for remedial work if leakage is discovered after a dam has been completed and the storage of water within a reservoir has begun. As it is seldom that a reservoir can be emptied completely, or even partially, for more than a short time after it has been placed into active service, remedial work to eliminate leakage from a full or partially filled reservoir is often an extremely difficult matter. Another method of reducing and possibly eliminating leak-

26-10 CIVIL ENGINEERING WORKS

age is to employ natural sealing material, such as clay or the silt content of the watercourse feeding the reservoir, to block up the openings in the reservoir through which leakage is occurring, when the leaks are definite and not too widespread.

If the exact location of the area of weakness is known, sluicing or dumping suitable material into the reservoir immediately above this area may often be effective. Even mattresses or fascines, similar to those used in river-training work, may be used if the area of weakness can be accurately located. The process may be a natural one. W. L. Strange has described how the Much Kundi Tank in Bijapur, India, formed by a masonry dam on schistose rocks, was sealed and made tight after 12 years of natural silting.[26.10] In the discussion of Strange's paper, mention was made of a Yorkshire (England) reservoir, formed by a 24-m (80-ft) dam, which leaked as much as 2 mld (528,000 gpd) but which silted up and sealed itself tight in two years. The permanence of any self-sealing operation will obviously depend on the nature of the leakage. If the sealing is completely in basalt and similar rocks, then it may be permanently effective; but if it is only partially successful in a limestone formation, then the constant flow of the small amount of leakage still persisting may be sufficient to enlarge existing leakage channels by further solution and cause a recurrence of serious trouble.

A variety of other methods has been used in achieving economical sealing of reservoirs that leak. There is a record of a reservoir in the Italian Apennines on the river Ripa, formed by a gravity-arched concrete dam 47 m (154 ft) high, leaking through a crevasse in the rock floor and through pervious rock; the leak was eventually sealed by successive applications of a cement-sand mixture applied with cement guns. The extent of the operation is indicated by the fact that the sealing operations cost over six times as much as the original dam.[26.11] The use of asphalt as a sealing medium by the TVA has already been noted in connection with the Hales Bar Dam. The TVA successfully used the same system in sealing the reservoir formed by the Great Falls Dam on the Caney Fork River in central

FIG. 26.5 The Great Falls project, now of the Tennessee Valley Authority, on the Collins River, Tennessee, just below its confluence with the Caney Fork River.

Tennessee. The dam was built in 1915–1916 and redesigned and raised to a height of 28 m (92 ft) in 1925. Due to the local topography, the reservoir formed by the dam was curved in plan; as a result, it comes within 180 m (600 ft) of the Caney Fork River when the latter is some 1,500 m (5,000 ft) downstream of the dam (Fig. 26.5). It was possible, therefore, to drive a tunnel across this narrow neck and thus achieve an operating head of 44 m (147 ft) for the small powerhouse that was built on the river.

Built originally by a private power company, the project was taken over by the TVA in 1939. Leakage had been observed since 1925, all along the bank of the Caney Fork River nearest to the reservoir, between the dam and the powerhouse. When studied by the TVA, this was found to amount to 10 percent of the water that could have been used for power generation; the leakage therefore represented a considerable economic loss, and the leakage was increasing in volume each year. It showed itself at the junction of the Upper Fort Payne formation, a seamy limestone, and a lower cherty massive limestone and was evident at 14 locations along the river. The success of grouting with asphalt at the Hales Bar Dam encouraged the TVA to experiment with the same process at one of the well-defined leaks on this project. The result was successful, no recurrence of the flow which was stopped taking place within six months.[26.12]

26.4 SECONDARY EFFECTS OF RESERVOIR FLOODING

The function of an impounding dam is to store a predetermined quantity of water and, in the case of a water-power scheme, to raise the top level of the impounded water to a certain specified maximum. Areas of dry land must be flooded as the water rises, and the flooding will often entail extensive civil engineering construction of what may be termed a secondary nature. Buildings may have to be relocated, and sometimes whole settlements may have to be moved. Transportation routes, which often follow valleys, may traverse land that will be flooded when the reservoir fills. Relocation work can sometimes attain major proportions. In the planning of all such work, the effect of the ultimate maximum water level must be considered. This type of work will often be located on hillsides, and the possibility of landslides is one that must always be given special attention.

Close to the home of one of the authors runs the Ottawa River. Early in the 1960s this major river was damned by the mass-concrete Carillon Dam of Hydro Quebec, at a point 96 m (60 mi) below the city of Ottawa. Although no major route relocations had to be undertaken except in the immediate vicinity of the dam, a major bridge had to be reconstructed and much rehabilitation work carried out along both banks for many kilometers above the dam. Small landslides occurred, attributed to the slight rise in the water level, 80 m (50 mi) above the new barrier. This is typical when any dam is built in developed areas. The example of the Kinzua Dam (Chap. 2) showed how extensive route relocation can be.

Underground water levels will be materially affected in reservoirs, the top levels of which are appreciably above normal river elevation, and this change in a natural ground condition may have widespread effects. Building foundations, especially those on clay, must be carefully studied. Underground water-supply systems will clearly be affected. If there is any possibility that minor seepage, apart from appreciable leakage, will reach catchment areas other than that in which the reservoir is situated, attention must be given to this contingency lest it ever develop serious proportions. Peculiar problems may often arise due to special local circumstances. In a paper on dam construction in New South Wales, Australia, mention was made of the selection of a dam site for impounding water for the supply of the city of Sydney. Six sites were chosen. The first selected for construc-

26-12 CIVIL ENGINEERING WORKS

tion was on the Cataract River and had a catchment area above it of 13,700 ha (53 mi²). The geological formation of the Cataract catchment consists of Hawkesbury sandstone overlying coal measures, and the area is bounded on its upper end by a range of high hills rising abruptly a short distance back from the ocean. Coal seams outcrop on the side of the hills facing the ocean and dip with a fairly rapid inclination underneath the Cataract catchment. The seams have been worked from the face of the hills for some time, and coal mining is carried out in the area.[26.13]

It will be seen that there arises the interesting double problem that coal mining might interfere with dams and reservoirs and that seepage from reservoirs might interfere with mining operations. The second appears to be the more probable, although no trouble of either kind has yet been reported throughout this whole area. The New South Wales Mines Department has laid down marginal zones around areas of stored water in which board-and-pillar extraction of coal is permitted. The Mines Department permits board-and-pillar extraction under stored water, but total extraction of coal beneath stored water and the marginal zones is forbidden, except with the permission of the secretary for mines. The widths of the zones and the sizes of boards and pillars are fixed by the Mines Department. The situation is complicated by the unusually complex local geology, which includes two series of freshwater sedimentary rocks of the Triassic system.

26.5 LANDSLIDES

The possibility of landslides developing around reservoirs, when the water level is raised, has been mentioned. Changes in groundwater level and consequent changes in pore pressure in soils are secondary effects of dam construction which must receive careful attention as a basic part of design. The growing number of pumped storage projects, not only for peak power generation but also for water storage, adds special urgency to the investigation of any possibility of major landslides into the impounded water of a reservoir. The very nature of such projects necessarily involves relatively rapid changes in water level, both up and down, and this requires certainty, in advance, as to the stability of all slopes that are going to be affected.

The vital necessity for this study of potential slides around reservoirs is, unfortunately, confirmed by one of the most tragic of all failures involving a civil engineering structure in modern times, the overtopping of the Vaiont Dam in Italy as a result of a massive rock slide in the reservoir on 9 October 1967. The slide precipitated several hundred million cubic meters of soil and rock into the reservoir, creating a wave estimated at 90 m (300 ft) high. This overtopped the 257-m (858-ft) -high arched dam, the resulting flood below the dam destroying several villages and a town of 4,600 people, almost 3,000 losing their lives (Fig. 26.6). Engineers responsible were arraigned in the Italian courts, adding still further to the tragic aspects of this dreadful disaster.

The dam still stands, since it was only slightly damaged as a result of the slide and ensuing wave; it remains a monument to the skill of arched dam designers, but its distinction as the world's highest arched dam, when it was built, has long since been forgotten in the light of the failure in the reservoir. Located in a very narrow gorge of the Vaiont River, a tributary of the Piave River in northeastern Italy, the dam was completed in 1960. It is really a domed structure, keyed into the competent rock of the sides of the gorge. It is only 3.4 m (11 ft) thick at its crest and 22.7 m (75 ft) thick at its base. It was carefully instrumented so that its performance could be checked. It stands today retaining mainly rock and soil. The area in which the dam was built is characterized by a thick section of sedimentary rocks, predominantly limestone, with interbedded marl and clay seams. The entire area had been subjected to (geologically) recent movements. Slow creep

FIG. 26.6 The Vaiont Dam in northern Italy, after the tragic overtopping and flood consequent upon the great slide, the scar of which is plainly visible.

movement of part of the reservoir wall had been noted but the final movement was sudden. A helpful review of the disaster shows how vital was the local geology.[26.14]

As this Handbook goes to press, investigations are continuing in the mountains of British Columbia into the stability of the Downie Slide, since its toe will be flooded when the Revelstoke hydroelectric project becomes operational (Fig. 26.7). This ancient slide contains about 1.5 billion m^3 (1.9 billion yd^3) of rock; it is thought to have taken place between 9,000 and 10,000 years ago but with no high velocity sliding involved. It is on the west bank of the Columbia River about 65 km (40 mi) north of the town of Revelstoke. Disturbed material is evident for about 2.5 km (1.5 mi) along the river. The slope of the slide surface is about 18°, except near river level. The license for the construction of the project has been issued with the understanding that water level in the reservoir will be restricted to 9 m (30 ft) below full design pool level until stabilization measures have been completed and are operating satisfactorily.

26.6 SILTING UP OF RESERVOIRS

In all but exceptional cases, the construction of a dam or weir across a stream bed will raise the water level upstream of the axis of the dam and thus increase the cross-sectional

FIG. 26.7 The ancient Downie Slide, British Columbia, detailed study of which preceded completion of the Revelstoke water power project on the Columbia River.

area of the stream and decrease the average velocity. This will automatically eliminate a part of the suspended load of the stream, the load falling to the bottom of the stream or reservoir. Although all the bed load may move down toward the dam, it will eventually be caught in the basin formed by the dam structure. There exist more than a few records of large dams rendered either ineffective or useless by the unscheduled and rapid filling of the reservoir behind them with transported material. To mention but three of these, reference may be made to the so-called New Lake Austin on the Colorado River in Texas, which lost 95.6 percent of its capacity in 13 years; La Grange Reservoir on the Tuolumne River in California, which lost 83 percent in 36 years; and the Habra Reservoir on the Habra River in Algeria, which lost 58 percent in 22 years. These figures are quoted from an extensive list of silted reservoirs compiled many years ago by J. C. Stevens.[26.15] Reference is also made in this paper to the unusual siltation record for the Lake McMillan Reservoir of the Pecos Irrigation District in the United States, which lost 55.5 percent of its capacity between 1894 and 1920. Since the latter year, no further accumulation of silt has taken place because an extensive growth of tamarisk (salt cedar) in the flatlands immediately above the reservoir has so reduced the stream velocity that the main part of the silt load is deposited before the reservoir is reached.

An essential part of preliminary work in connection with the construction of all dam structures is consequently an investigation of the amount of sediment normally transported by the stream to be dammed. It may often be found that the quantity of material carried is negligible; but in other cases, further and more detailed studies will be necessary. These will inevitably include a careful correlation of information on the quantity of debris transported and its geological origin, so that the nature of the resulting deposit may be anticipated. This information is necessary in view of the engineering problems involved. These may be classified into two groups: (1) possible methods of removal of the transported material either before it is deposited in the reservoir or after it has come to rest behind the dam; and (2) the effect that deposits of material, if they cannot be avoided, will have on the stability and on the utility of the structure being planned.

Detailed discussion of these problems is mainly an engineering matter, but their essential dependence upon geological conditions will be apparent. The second problem relates mainly to the effect of silt and other retained debris upon mechanical and other moving parts integral with a dam structure, since under usual circumstances the deposition of silt against the face of a dam does not increase the normal pressure upon it. The possibility of removing sediment that is trapped in a reservoir is, however, a matter of much wider significance. For very large, mass-concrete structures, it may be practically impossible to include in the design any means for passing silt through the dam after retention; the problem then becomes one of economics with relation to the anticipated useful life of the reservoir. For smaller structures that are known in advance of construction to be so located that they will retain silt, bypass arrangements for occasional sluicing of the retained material can sometimes be incorporated into the design of the dam. Alternatively, arrangements can be contemplated, or even made after construction when necessary, for the removal of retained silt by sluicing or dredging. And in extreme cases, when sediment is unexpectedly retained by a dam, the structure may be breached in order to make the reservoir again serviceable.

The high silt content of the Rio Grande and the Colorado rivers is well known; examples of major dams upon these two rivers may therefore appropriately be mentioned. Elephant Butte Dam is on the Rio Grande River a short distance north of the junction of Mexico and the states of Texas and New Mexico. It was completed in 1916; its drainage basin has an area of 67,000 km^2 (25,923 mi^2). Probably because of the international character of the river, it has been continuously under measured observation for many years, and so there are available records of the silt load coming into Elephant Butte Reservoir that can be correlated with the measured deposition of silt on a scale probably unequaled for any other major reservoir. Estimates based upon these observations suggest that the useful life of the reservoir after its first filling will be about 150 years, unless some unanticipated change in river regime or other natural conditions should take place.[26.16]

On the Colorado River, Hoover Dam has created one of the largest of man-made lakes, Lake Mead, already mentioned several times in this book because of its outstanding geological significance. All the silt load of the Colorado River is naturally held up as the water of the river passes the great dam, so that studies of sedimentation have been actively pursued ever since the dam first began to retain water. Estimates of the useful life of the reservoir have naturally varied, but one of the most detailed suggests that the reservoir will not be filled with sediment to its crest level until about 445 years after the first closing of the gates controlling the bypass tunnels on 1 February 1935.[26.17] When the magnitude of Hoover Dam is recalled, this estimate of 445 years is still a sobering figure, even though human thinking is not normally accustomed to look ahead beyond the current century. The estimate is in keeping with the suggestion that the useful life of the reservoirs constructed in North America in the second quarter of this century, which probably exceed in number all those built previously, averages about 200 years.

The cleaning out of filled reservoirs is a normal part of the operation of flood-control dams, such as those of the Los Angeles County Flood Control District (Fig. 26.8). Reservoirs for water supply, however, experience the same sort of problem, although usually at a slower rate. The San Fernando Reservoir of the Los Angeles Bureau of Waterworks and Supply, for example, lost 24 percent of its capacity in about 20 years. Arrangements were made for dewatering the reservoir for a period of 38 days, during which a major sluicing operation was undertaken; about 15,000 m^3 (20,000 yd^3) of sediment were sluiced out of the reservoir into a specially constructed debris basin; it was found that about 69 m^3 (90 yd^3) could be moved in this way per man-day.[26.18] Philadelphia faced a similar problem with its Queen Lane sedimentation basin in which 450,000 m^3 (500,000 yd^3) of silt had accumulated during its first 50 years of operation. The basin had to be kept in operation, and so on the basis of satisfactory experience with the same problem at the city's Torres-

26-16 **CIVIL ENGINEERING WORKS**

FIG. 26.8 Sluicing operations in progress upstream of the Santa Anita Dam, removing flood debris retained by this arched concrete dam of the Los Angeles County Flood Control District, California.

dale plant, it was decided to remove the silt by an electrically operated floating dredge. The dredge had to be assembled at the site and launched into the basin (that used at Torresdale was floated in from the Delaware River). The large accumulation of silt is accounted for by the fact that the Schuylkill River water, with its high silt content that has already been mentioned, is treated in the Queen Lane plant by chemical coagulation so that, to some extent, the silt deposition here is premeditated, although still derived from riverflow.[26.19]

More unusual was the experience with a 36-m (120-ft) arched concrete dam spanning the narrow Cat Creek Canyon in an isolated part of Nevada, constructed in 1931 by the U.S. Bureau of Reclamation to serve the water supply needs of the naval ammunition depot at Hawthorne, with an impounded volume of 190 million L (50 million gal). On 21 July 1955, a half-hour thunderstorm caused a flash flood in Cat Creek so intense that it cleaned off all erodible material from the catchment area above the dam, depositing it behind the dam and almost completely filling the reservoir. After 27,000 m^3 (35,000 yd^3) had been removed by excavation with some difficulty, the reservoir was again filled with debris and silt after heavy rains. The decision was made to breach it at its lowest point, since no sluiceway had been considered necessary in the original design. This difficult operation was successfully completed early in 1956, and a 1.2- by 1.2-m (4- by 4-ft) hole was blasted through the concrete, close to the bed of the canyon. For half an hour after the breach was made, little came through the hole, but then the mud plug shot out with a roar and in the course of a few hours, 38,000 m^3 (50,000 yd^3) of silt and water rushed through; the water effectively washed the silt downstream. Sluicing then completed the job. A sluice gate was attached to the face of the dam, utilizing the hole as the sluiceway, so that future deposits of silt (unless of a phenomenal character) can readily be removed.[26.20]

Sometimes economics will dictate a temporary solution rather than such a drastic, even though permanent, solution as that just described. The United Water Conservation District of Santa Paula, California, built the Santa Felicia Dam on Piru Creek in 1955, forming a pleasant body of water soon called Piru Lake. The dam is an earth-fill structure 83.8 m (275 ft) high; the reservoir has a capacity of 123.3 million m^3 (100,000 acre-feet). Water is stored in the lake during the rainy season and released during the summer months for irrigation and groundwater replenishment. The lake is much used for recreational purposes. Its outlet was through a vertical outlet pipe, embedded in concrete on the floor of the reservoir and extending (with its trashrack intake) about 11 m (37 ft) above the valley floor (Fig. 26.9).

During 21 years of operation, silt and debris were carried into the reservoir by floodwaters to such an extent that the side walls of the intake were blocked. Remedial measures were essential. Lowering the water level of the lake was undesirable for a number of reasons, including damage to the fish in the reservoir and loss of recreational facilities during the period of drawdown. As an alternative, an ingenious and economical solution was developed by the staff of the district. A 13-m (42-ft) reinforced-concrete pipe extension to the outlet was constructed; its ends were sealed and it was floated into place. Based on

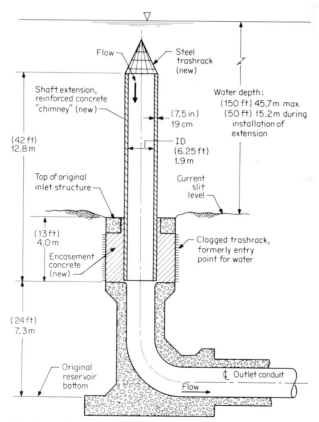

FIG. 26.9 Santa Felicia Reservoir, California; cross section through the original intake structure showing also the extension fitted to this, as described in the text.

experiments with models, it was successfully tipped up so that it floated in a vertical position, being then lowered into place, by varying the floatation, on top of the existing intake into which it was sealed. When secured in place, it gave an 8.7-m (29-ft) extension to the intake which, it is estimated, will take care of sediment accumulation for the next 30 years. Then another problem will have to be faced, unless it has proved possible in the interim to control the inflow of silt.[26.21]

The deposition of sediment in reservoirs is so obvious a result of dam construction that there remains little more to be said about it. Silt contents of streams and rivers can be determined at the time of design; predictions can be made of the probable continuation of this amount of silt reaching the dam after its construction; the effect of the dam upon the riverbed downstream can be estimated and the possible effects of the degradation of the stream bed investigated; and finally the economics of the entire project can be evaluated in making the final decision regarding building. A key factor in such a study would be the probable finite useful length of life for the reservoir. Behind all such mundane proceedings, however, is the tragic significance of the origin of most of the sediment that is to be trapped in the reservoir. To some extent, the sediment may be derived by normal riverflow from the beds of the streams and rivers that bring it to the dam site. In almost all cases, however, most of the silt load will represent topsoil that has been washed off the surface of the land in the watershed above the dam. No civil engineer should be able to look upon a silt-laden stream without being disturbed by the critical problem of soil erosion.

Accordingly, reference should be made to Chap. 49 in which the whole question of the interrelation of the works of the civil engineer and the protection of the environment is reviewed. Here it is sufficient to note that the caustic comments which one sometimes reads about the way in which "dams built by engineers get silted up very quickly" should, in almost all cases, be directed not at dam designers and builders, but at those responsible for poor land-management practices. Much progress has been made in the control of soil erosion but much still remains to be done. All such work demands the attentive interest of civil engineers and engineering geologists since upon its success depends the long-term usefulness of many dams and other structures.

26.7 METHANE IN RESERVOIRS

Although it is, fortunately, a very rare occurrence, the accumulation of methane gas in the bottoms of reservoirs must be briefly noted since, when it does occur, the consequences can be, and have been, serious indeed. Accidents due to this unusual cause have probably happened in years past, but the first record of the presence of methane due to the storage of water in reservoirs appears to have been in Soviet literature, notably in connection with hydroelectric stations on the Volga River. Other gases accompanied the methane, the origins of which are discussed in a general paper by Kuznetsov.[26.22] A very serious accident did occur in the construction of the Furnas hydroelectric project in Brazil during the closure of the diversion tunnels. An explosion caused unusually complex difficulties with the closure of the tunnels. It was attributed to the presence of methane, which had accumulated in the organic matter in the bed of the reservoir and which was released with flow into the tunnels, becoming explosive when mixed with air. The accident and notable remedial work were described to the Institution of Civil Engineers. In the discussion of the paper, an account was given of another similar accident, involving serious loss of life, at the Akosombo Dam on the Volta River in Ghana. Other cases were cited, even of methane in excavations for bridge piers, while a Canadian example was mentioned in which a pontoon used for the removal of closure gates on a hydroelectric scheme was swamped by a large bubble when the first lift occurred.[26.23] The possibility of methane being generated in submerged organic matter, as in the bottoms of reservoirs, should therefore be kept in mind in all relevant cases.

26.8 A PERCHED RESERVOIR

Between 1946 and 1948, the Hydro Electric Power Commission of Ontario completed its Aguasabon power plant on the north shore of Lake Superior, about 210 m (130 mi) east of Thunder Bay. It has an installed capacity of 40,000 kW generated under a head of 87 m (290 ft). Water is brought to the small powerhouse, located on Terrace Bay, through a 4.5-m (15-ft) -diameter tunnel, which leads to twin steel penstocks. The Aguasabon River enters Lake Superior about 3 km (2 mi) to the east, its water having been impounded by a concrete dam about 2.5 km (1.5 mi) from the mouth. (See Fig. 26.10) Because of unusual local topography, the reservoir so formed not only extends up the river in the usual way but floods through a narrow gorge just above the dam into a large, basin-shaped area to the west of the river, extending to within 800 m (0.5 mi) of the shore of the lake and yet

FIG. 26.10 Powerhouse and surge tank of the Aguasabon water power project, northern Ontario; Lake Superior is in the foreground and Blue Jay Reservoir is to be seen in the background, its water level 87 m (290 ft) above the level of the lake.

87 m (290 ft) above it (at reservoir water line). Superficial examination suggested that a rock ridge might continue under the overburden between the reservoir site and the lake, even though the ridge was covered with a mantle of soil. The usual careful preliminary test drilling of the Ontario Hydro, however, penetrated to depths well below lake level along this ridge and found nothing but sand and gravel; the presence of many boulders eliminated any possibility of constructing a cutoff wall (even to such a depth) between the adjacent rock outcrops.

Study of the natural basin directed attention to a small pond of water (Blue Jay Lake) in its center, a pool which could be retained only by underlying impervious material. The glacial history of this part of northern Ontario suggested that this material might be a layer of glacial silt similar to other glacial lake deposits in this part of the Canadian Shield. Trenching around the edge of the reservoir area and very careful test drilling (with a minimum number of holes actually penetrating the blanket) revealed a continuous bed of compact and almost impervious glacial silt over the entire bed of the basin, extending—most fortunately—to just about the intended top water level for the reservoir. Somewhat naturally, tests were made to determine by penetration the depth of the silt layer, but where this was studied it was found that, although in the middle of the reservoir area a thickness of some 18 m (60 ft) existed, the silt layer thinned out to a relatively few meters toward the upper edge. With the existence of this natural reservoir lining known, planning of the project could be completed. The plant was built, and the reservoir was filled. There was a slight increase in the level of the groundwater in the sand and gravel almost 90 m (300 ft) *beneath* the reservoir, but this gradually leveled off, and the plant has been in continuous operation ever since its opening, dependent for its water supply upon this natural "perched" reservoir.[26.24]

FIG. 26.11 The Kenney sloping-core rock-fill dam on the headwaters of the Fraser River, British Columbia, just as the impounding of water was starting. The final water level is such that water is diverted through a tunnel to the Kitimat power development of the Aluminum Company of Canada, Ltd., on the Pacific coast.

26.9 RIVER DIVERSIONS

Another special feature of the Aguasabon project is that its construction was made economically possible only when the normal flow of the Aguasabon River was increased by diverting into it part of the flow of the Kenogami River, which flows into Hudson Bay, by what is known as the Long Lac Cutoff. Diversions of water from one watershed into another are infrequent, but those that do exist are all of unusual interest. Sometimes, such as the diversion of water from one side of the continental divide to the other to augment the water supply of Denver, they are major engineering projects carried out without any special assistance from geology. They always involve a variety of works, dams, new canals, tunnels (usually), the diverted water flowing into existing reservoirs. They fit conveniently, therefore, into no one chapter of this Handbook but are mentioned here because a number of the interbasin river diversions carried out in Canada (at least 19 to date) have been possible because of geology. (See Fig. 26.11.)

In the recession to the north of the last ice sheet from Canada, a number of major glacial outwash channels were created. Now covered with vegetation, these can be detected by careful geomorphological study. Some have been used as early transportation

FIG. 26.12 Long Lac diversion dam of Ontario Hydro which raises water level sufficiently to divert some flow that would normally reach Arctic waters into Lake Superior and the St. Lawrence system.

routes (see Chap. 31). It was by using old glacial outwash channels north of Lake Superior, between the great river basins of northern Canada and the relatively narrow area of land draining into the north shore of Lake Superior, that it was possible to divert Arctic waters into Lake Superior. Additional power was generated as the diverted water flowed into the lake and thereafter as it augmented the flow of the St. Lawrence on its way to the sea over Niagara and down the international section (Fig. 26.12). There are few locations where geology can provide such a convenience, but these few are so important that this brief reference seems warranted.[26.25]

26.10 CONCLUSION

It can be seen that the formation of reservoirs by the construction of dams raises many questions that are not at first apparent. Let is be said again that few, if any, works of the civil engineer have such a profound effect upon the environment as does the formation of a new reservoir, especially if it is one of great extent. Not only are there all the possible effects already briefly noted but, in the case of deep reservoirs, minor earthquake shocks may result if the surrounding geology is susceptible to such changes in load. This is a matter which will be discussed in Chap. 45. Already the subject of reservoir-induced seismicity (RIS) has become a recognized field of study among seismic experts.

The hydrological cycle in the vicinity of all new reservoirs will be affected by the change in the water level of what had been the river; the cycle is affected not only because of the disturbance of the previously stable groundwater situation but also because of the increase in local evaporation from the surface of the reservoir. In temperate climates, these changes may be slight, but in hot climates, the change in evaporation may prove to be serious. Many rivers receive part of their flow from underground sources that often seep into the normal riverflow below water level. With a change in water level, such as will result from the building of a dam, the groundwater situation will be so altered that the previous underground flow into the river may even be reversed. Doubts were expressed about similar possibilities prior to the construction of the High Dam at Assouan, but it is yet too early to say whether its beneficial effects have been counterbalanced by less desirable results.[26.26] This is an extreme case, but the formation of every new reservoir requires the most wide-ranging study of all possible indirect effects, almost all of which will be dependent either directly or indirectly upon the underlying and surrounding geology.

26.10 REFERENCES

26.1 "Porosity of Reservoir Prevents Water Storage," *Engineering News-Record,* **96**:561 (1926).

26.2 "Leak in Greece; Reservoir Site Surveyed by Germans Won't Hold Water," *Engineering News-Record,* **178**:37 (4 May 1967).

26.3 "Disappearing Yugoslavian River Is Harnessed for Hydro Projects," *Engineering News-Record,* **176**:26 (24 March 1966).

26.4 G. A. Kiersch, "Geologic Causes for Failure of Lone Pine Reservoir, East Central Arizona," *Economic Geology,* **55**:854 (1958).

26.5 J. Hoover Mackin, *A Geologic Interpretation of the Failure of the Cedar Reservoir, Washington,* Engineering Experiment Station Bulletin no. 107, University of Washington, Seattle, 1941.

26.6 "A Curious Water Wheel," *The Engineer,* **142**:590 (1926).

26.7 L. Durrell, *The Greek Islands,* Faber, London, 1978, p. 49.

26.8 C. S. Fox, "Using Radioactive Isotopes to Trace Movement of Underground Waters," *Municipal Utilities*, **90**(4):30 (April 1952).

26.9 E. Halevy, A. Nir, Y. Harpaz, and S. Mandzel, *Use of Radioisotopes in Studies of Groundwater Flow*, Second United Nations Conference on the Peaceful Uses of Atomic Energy, Paper 15/P/1613, 1958.

26.10 W. L. Strange, "Reservoirs with High Earthern Dams in Western India," *Minutes of Proceedings of the Institution of Civil Engineers*, **132**:137 (1898).

26.11 B. C. Collier, "Sealing a Leaking Reservoir in the Italian Appenines," *Engineering News-Record*, **108**:293 (1932).

26.12 L. A. Schmidt, "Reservoir Leakage Stopped at Outlet," *Engineering News-Record*, **134**:65 (1945).

26.13 L. L. Waterhouse, W. R. Browne, and D. G. Moye, "Preliminary Geological Investigations in Connection with the Proposed Warragamba Dam, N.S.W.," *Journal of the Institution of Engineers of Australia*, **32**:74 (1951); and information from the Metropolitan Water, Sewage and Drainage Board, Sydney, N.S.W..

26.14 G. A. Kiersch, "Vaiont Reservoir Disaster," *Civil Engineering*, **34**:32 (March 1964); see also in same journal, "Vaiont Slide—Seven Years Later," **40**:86 (June 1970).

26.15 J. C. Stevens, "The Silt Problem," *Transactions of the American Society of Civil Engineers*, **101**:207 (1936).

26.16 J. C. Stevens, "The Future of Lake Mead and Elephant Butte Reservoirs," *Transactions of the American Society of Civil Engineers*, **111**:1231 (1946).

26.17 "Filling of Lake Mead with Silt Estimated to Take 445 Years," *Engineering News-Record*, **145**:34 (6 July 1950).

26.18 Removing Reservoir Silt by Sluicing Operations," *Engineering News-Record*, **127**:20 (1941).

26.19 "Dredge Cleans Waterworks Basin Without Interrupting Services," *Engineering News-Record*, **137**:314 (5 September 1946).

26.20 C. F. Mobley, "Muck-Filled Reservoir Behind 120 Foot Dam is Cleared," *Engineering News-Record*, **160**:40 (15 May 1958).

26.21 E. O. Bengry and W. T. Caltrider, "Reservoir Outlet Extended Above Silt to Prevent Clogging," *Civil Engineering*, **48**:81 (September 1978).

26.22 A. M. Kuznetsov, "The Phenomenon of Gas Formation in the Foundations of Concrete Dams," *Gidrotekh Stroit*, **36**(10):33 (1965); issued as National Research Council of Canada Technical Translation no. 1310, S. H. Bayley (trans.), Ottawa, 1967.

26.23 W. M. MacGregor and F. H. Lyra, "Furnas Hydro-Electric Scheme, Brazil: Closure of Diversion Tunnels," *Proceedings of the Institution of Civil Engineers*, **36**:21 (1967). With valuable discussion.

26.24 R. F. Legget, "A Perched Reservoir in Northern Ontario, Canada," *Geotechnique*, **3**:259 (1953).

26.25 R. Kellerhals, M. Church, and L. B. Davies, "Morphological Effects of Inter-Basin River Diversions," *Canadian Journal of Civil Engineering*, **6**:19 (1979).

26.26 A. A. Ahmed, "Recent Developments in Nile Control *and* An Analytical Study of the Storage Losses in the Nile Basin, with Special Reference to Aswan Dam Reservoir and the High Dam Reservoir (Sadd-el-Aali)," *Proceedings of the Institution of Civil Engineers*, **17**:137, 181 (1960).

Suggestions for Further Reading

Since reservoirs are created by dams, most publications dealing with dams contain some information relative to the reservoirs they form. In particular, the publications of the International Congress on Large Dams (ICOLD) give due attention to problems with reservoirs. One of the active working commissions of the Congress has written a paper called *Sedimentation in Reservoirs.*

In view of the very great care that is being taken with the investigation of the ancient Downie slide in British Columbia (before flooding of the proposed reservoir into which it will impinge is permitted), another account of it may be a useful reference. "Structure and Stratigraphic Setting of the Downie Slide, Columbia River Valley, B.C." by R. C. Brown and J. E. Psutica is to be found in vol. 17 (p. 698) of the *Canadian Journal of Earth Sciences* (1980), published by the National Research Council of Canada, Ottawa.

Little has been published on watersheds, significant though they are in relation to river flow and floods. One of the few recent publications is *Watershed Management,* a 789-page volume published in 1975 by the American Society of Civil Engineers.

chapter 27

GROUTING

27.1 Historical Note / 27-2
27.2 Grouting with Clay / 27-2
27.3 Grouting with Asphalt / 27-4
27.4 Grouting with Cement / 27-5
27.5 Special Applications / 27-13
27.6 Grouting with Chemicals / 27-16
27.7 The Grand Rapids Project, Manitoba / 27-18
27.8 Conclusion / 27-20
27.9 References / 27-20

Grouting has become one of the most widely used specialist techniques in the practice of civil engineering. It is so essentially an engineering operation that the devotion of a full chapter to the subject in this Handbook might at first appear to be unwarranted. It is, however, almost always carried out to overcome some defect in foundation beds, to improve upon natural geological conditions; as a result, it is completely dependent upon geology for its success. Without, therefore, going into any detail about equipment or methods, the uses of grouting in civil engineering work will be reviewed, with special emphasis upon its interrelation with geology.

Grouting may be generally described as the injection of suitable material into certain sections of the earth's crust, under pressure and through the medium of specially drilled holes, in order to seal any open fissures, cracks, cavities, or other openings in the strata encountered. Cementation is a word applied to grouting in some cases when a slurry of portland cement (either alone or mixed with a proportion of sand) is used as the sealing material; it is also used in a proprietary connection. In addition to cement, clay, asphalt, and various combinations of chemical solutions may be used to form the necessary seal. It will have been noticed that grouting called for frequent mention in the treatment of dam foundations. Necessary watertightness of such foundations makes grouting an invaluable tool in the great majority of dam foundations, but today it is utilized also in many other branches of civil engineering work. Chemical grouting, in particular, is frequently called into use for difficult foundation conditions in confined spaces, such as in urban tunnel work. Grouting is a subject, therefore, of the greatest importance in relation to the uses of geology on civil engineering projects.

27.1 HISTORICAL NOTE

Few, if any, of the specialist processes in civil engineering have had their history so well documented as has grouting. In two masterly papers published in *Geotechnique* in 1960 and 1961, Rudolph Glossop, a British foundation engineer of wide experience, presented a fascinating review of the slow development of the "injection process," as grouting is, perhaps, more correctly called.[27.1] All interested in the historical background of grouting should consult these classic papers. Glossop's researches make clear that the process was invented, in the first instance for the repair of masonry, in France in 1802 by Charles Berigny, although he did not publish any written description until 1826. A notable paper published in 1837, also in France, by Raynal dealt with the use of injections for the repair of damaged masonry. This report was noticed by an officer of the (British) Royal Engineers, Lieut. William Denison, who published an abstract of the paper in the *Professional Papers of the Corps of Royal Engineers* in 1840. This seems to have been the first reference to the injection process in English. Denison had seen a simple gravity system of injection used about 1830 on the masonry of the locks of the Rideau Canal in Canada, almost within sight of where these words are being written.[27.2]

First mention of grouting in American engineering literature appears to have been by William E. Worthen, who related that he had used it in 1854 to strengthen a masonry pier on a bridge of the New Haven Railway line, having read some years before of "the French way of injecting under foundations," quite possibly in Denison's paper since the volumes of the Royal Engineer papers became well known. One of the great pioneers of modern grouting was W. R. Kinipple, a British engineer, who extended its use considerably. He was consultant on the use of grouting beneath some of the early dams on the river Nile in Egypt. The first use of grouting with cement to strengthen bedrock seems to have been by Thomas Hawksley, a notable British engineer, who in 1876 grouted the foundation rock beneath an earth dam at the Tunstall Reservoir. In 1877 and 1878 he followed this up by using grouting on the Cowm Reservoir of the Rochdale Corporation. The corresponding pioneer use of grouting into rock in the United States appears to have been in 1893 for the New Croton Dam for the water supply of New York.

Civil engineers were slow in adopting this new aid, but gradually its potential was realized, with the result that today it is part of standard practice. There have been many useful reports upon grouting, notably by committees of the American Society of Civil Engineers. Some references to the most useful of these are given at the end of this chapter. Cement is almost the universal material used in grouting, but clay and asphalt have also been used to a limited extent. Some applications of these unusual materials will, therefore, be briefly noted first, and then some geologically significant examples of grouting with portland cement will be presented.

27.2 GROUTING WITH CLAY

The use of puddled clay to fill openings in rock is such an obvious expedient that it was probably used from the earliest days of modern civil engineering, without any record of this being made. One of the first recorded uses of clay was on the New Croton Dam, already mentioned, where the contractors suggested it. Clay was forced into fissures in the rock down a 7.5-cm (3-in) -diameter pipe, using a small pile drive, as much as 1,460 kg (3,520 lb) of clay being forced into one hole.

Clay was apparently first used on a major scale in the sealing of the limestone foundation beds underlying the Madden Reservoir in the Panama Canal Zone. The limestone formations encountered are badly weathered in central localities and contain under-

FIG. 27.1 Typical view in limestone cavern at the site of the Madden Dam, Panama Canal, prior to sealing with clay grouting; reprecipitated calcite lining on walls may be seen.

ground caverns and fissures, which render them far from the watertight barrier required to form the bed of this reservoir. Cement grouting was considered impracticable because a loamy sand with which the cement would not bond lined the caverns and crevices. Considerations of expense ruled out the possibility of using asphalt, and so the board of consultants for the project suggested the use of clay. A valuable series of experiments was carried out, the results of which demonstrated that the best effects were obtained with mixtures of clay and water containing about 55 percent water. Grout of this consistency traveled up to 15 m (50 ft) underground and penetrated seams as narrow as 13 mm (0.5 in). It was found that the mixture had to be prepared with the aid of blungers, such as those that are used in the ceramic industry; screens were also necessary. On this job, 7.5- and 17.5-cm (3- and 7-in) -diameter drill holes were used, and the pressure varied with the consistency of the grout (Fig. 27.1). Excavation of small pits at surface blowouts showed up several well-grouted seams, and other specific test results demonstrated the special utility of this method of grouting for the cavernous limestone encountered.[27.3] Although clay does not give such a perfect seal as cement, for example, it can be employed where cement cannot be used and therefore is a useful tool of the foundation engineer.

Another example of a somewhat similar kind was the use of a fine, red sandy loam, having a high clay content, mixed with a small proportion of cement (varying up to a mixture of 1 part cement to 6 parts soil) as grouting material for sealing an old earth-fill dam near Huguenot, New York, which supplies the Port Jervis community with water. No information on subsurface strata was available when the dam was raised, and serious leakage subsequently developed. To stop this, an extensive test-boring and grouting program had to be carried out to seal boulder and gravel strata previously unsuspected. The case is a good example of the expense that can be caused by lack of adequate construction records.[27.4]

French engineers have further developed the use of clay for grouting in recent years by adopting a finely powdered clay in suspension, the clay so selected that it will have some thixotropic action. Such clays must have high liquid limits; sodium bentonitic clays have been found to be most satisfactory.

27-4   CIVIL ENGINEERING WORKS

FIG. 27.2 Limestone abutment of the Great Falls Dam on the Collins River, Tennessee, just below its confluence with the Caney Fork River (*top*) before and (*bottom*) after the extensive grouting program carried out with asphalt and cement grout in 1946.

27.3 GROUTING WITH ASPHALT

Aqueous emulsions of bitumen have been used for grouting, especially when the main function of the grout was to reduce permeability rather than to fill open voids. Asphalt was used for the latter purpose in connection with the Hales Bar Dam, as has been noted. It was also used for forming the necessary cutoff wall at the Claytor Dam on the New

River near Radford, Virginia. Founded on stratified gray dolomite with shale and chert present in seams, the dam was located over some large solution channels that naturally had to be filled. After a study of various possibilities was made, the asphalt-sand mixture was decided upon. Mixed in a standard construction premix plant, the asphalt was, in general, dumped into place, but its consistency was such that it penetrated into fine seams and acted as a satisfactory grout. It is believed that asphalt has also been used in grouting operations carried out in the U.S.S.R.

In sealing rock at the Great Falls development, as described in the last chapter, the Tennessee Valley Authority utilized asphalt grouting in association with a backing of a cement-grouted curtain quite successfully (Fig. 27.2). The Authority has used similar grouting at the Timms Ford Dam. It was found desirable not to use asphalt grouting in the vicinity of actual structures. Quite naturally, its use makes any additional grouting difficult to execute, and it does not enhance the bearing capacity of rock so grouted, as cement grouting will do. For special locations, however, and under emergency conditions, it is one type of grouting that can be used effectively.[27.5]

27.4 GROUTING WITH CEMENT

Grouting with cement is the most usual method adopted, and civil engineering literature contains references to its successful application in many different types of work. The method depends essentially on the fact that if portland cement is mixed in a slurry with water, it will, in course of time "set up" hard and bond satisfactorily with other materials, such as rock, with which it may be in contact. Originally, the grout was simply poured into open cavities and allowed to set freely. Eventually, the idea of forcing the grout into normally inaccessible cavities by the use of pressure was developed. The origin of cement grouting is a matter of conjecture; it seems probable that the Romans were at least familiar with the idea, even though they had no means of using pressure for injection. Pressure grouting was certainly used to some extent in the nineteenth century as already noted, but not until James Greathead invented his grouting machine at the end of the century for assisting in the construction of tube railway tunnels in London did it become a regular construction operation.

One of the first dam foundations to be effectively sealed by grouting was that of the Kinder Embankment of the Stockport Corporation, England. During the construction of this work (from 1903 to 1905), the bedrock was found to be so fissured and faulted that it could not support the originally intended masonry dam; a satisfactory shale bed was eventually found at a depth of 54 m (180 ft) below ground level. In the construction of the trench to this depth, much trouble was encountered with water. The drainpipes installed to tap the water, as concreting advanced, were carried up until water ceased to flow from them and were then grouted up under pressure, a practice now frequently followed.[27.6] One of the first dam projects on which pressure grouting was used in the United States was the Estacada Dam constructed in Oregon in 1912. Although so early an application, no less than 10,200 m (34,000 ft) of drill holes were used for the grouting operation, using diamond and Calyx holes and a Canniff air-stirring grouting machine, then in common use in the United States. The dam was built on volcanic breccia which was very permeable. As showing how well the basic principles of grouting were already appreciated, this extract from the description of the work is significant: "The general idea provided for drilling a double row of holes of an average depth of 50 ft under the heel of the dam across the entire valley to and under the shore abutments and the subsequent forcing into each of these holes of grout of such consistency as to percolate through the entire substructure."[27.7]

British practice contains many interesting examples of cementation as applied to dam construction, including several cases in which its use has materially reduced the necessary depth of cutoff walls. Precementation has been another feature of British practice; the term is used to indicate the cementation of underlying strata before dam excavation is begun. The object of precementation is twofold: (1) to consolidate effectively (by means of the cement grout) the material that has to be excavated and (2) to obtain the benefit of the superincumbent weight of undisturbed strata over the fissures being sealed. This weight is necessary to prevent vertical lifting of rock strata above fissures which the pressure on the grout tends to cause, movement that has been readily observed in a number of cases. Precementation can be carried out, even through surface strata of unconsolidated material, if the drill holes used are suitably cased, but this introduces the risk of leakage occurring at the junction with solid-rock strata; it is always advisable, therefore, to strip unconsolidated materials before cementation is undertaken.

Considerable savings can be effected by the careful application of this method, as was clearly brought out in its use at the Scout Dike Dam site in England. This is an interesting example, especially since vertical shafts were used from which horizontal drill holes, in addition to the usual vertical drill holes, were made. Work on the dam, which is an earthfill structure forming part of the Barnsley Corporation Waterworks, Yorkshire, England, was started in 1924. Underlying the site are alternating beds of sandstone and shale; a bed of permeable sandstone outcrops in the stream throughout the length of the reservoir, and below this is a tight bed of shale. The sandstone bed was found, by means of trial borings, to be almost horizontal and uniform in thickness along the line of the dam. In order to avoid taking the trench down to the depths that the position of this bed necessitated (over 24 m or 80 ft at the ends of the dam), cementation was resorted to, and afterward the trench was taken down only to the first sound shale or sandstone encountered.

A series of vertical boreholes 7.5 m (25 ft) apart was put down into the impervious shale underlying the sandstone encountered at stream level, and cement was injected into them. Later, when the trench was excavated, shafts were sunk 60 m (200 ft) apart to the same level as the bottom of the boreholes, horizontal holes were drilled between them, and cement was again injected. In this way, any vertical joints possibly missed by the boreholes were sealed (Fig. 27.3). All holes were drilled with a view to obtaining cores, and the evidence thus obtained was a valuable addition to preliminary information. It was found during the progress of the work that grouting pressures of not more than 28 kg/cm^2 (400 psi) had undoubtedly lifted strata, so that the limiting pressure was reduced to 7 kg/cm^2 (100 psi). Careful estimates of cost were kept which proved beyond doubt the great monetary saving achieved by this precementation program.[27.8]

Cementation under high pressure was carried out for a very different purpose in the construction of the Abitibi Canyon water-power dam in northern Ontario. This large structure, located on a river flowing into Hudson Bay, is founded on gabbro which preliminary test borings showed to be free from seams and fissures. When the site of the dam in the canyon was unwatered, it was found that along the stream bed there was a zone of rock, which, for a width of about 12 m (40 ft), had been altered in composition and texture by possible faulting and by the passage of intrusive gases through the rock when molten. Since it was not known whether or not this zone would be impervious to water, a shaft was sunk to a depth of over 30 m (100 ft) below river level at the upstream face of the dam. No water was encountered, and the rock appeared to be sound; but it was decided to carry on exploratory work by drilling from the bottom of the shaft and, when porosity or seams were found, to use cement grout at pressures not less than the full static head from the reservoir with water at top level. This course was followed, and the excavation was concreted up so that construction of the main body of the dam could continue; a smaller shaft was boxed out, measuring 3 by 3.6 m (10 by 12 ft), with an enlargement at the bottom (Fig. 27.4). From this shaft, holes were drilled, and grouting was carried out

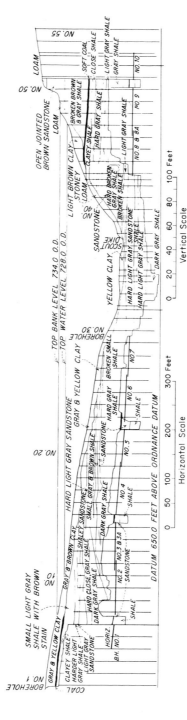

FIG. 27.3 Geological section along the trench of the Scout Dike Dam, Barnsley, England, showing boreholes used for precementation.

FIG. 27.4 Abitibi Canyon power project, northern Ontario (now a part of the Ontario Hydro system); a general view of excavation for the dam, looking upstream and showing the main cofferdam, diversion tunnel intakes, and deep shaft for excavation of unsound rock.

at 3-m (10-ft) stages in each hole, using pressures up to 17.5 kg/cm^2 (250 psi). In addition, grouting was carried out in old exploratory drill holes in the canyon walls and from inspection galleries left in the dam; a total of 1,527 bags of cement were used with satisfactory results.[27.9]

The foregoing examples have been cited because of their major geological interest. It has been, however, in connection with the construction of large dams in the United States during the second quarter of this century that perhaps the greatest progress has taken place in developing the techniques of cement grouting. Notable have been the practices developed by the Tennessee Valley Authority, to mention but one of the several large American engineering organizations that have contributed to this advance. The TVA's experience in cement grouting was born of necessity, developed because of the poor foundation conditions, usually in limestone, available for several of their large dams.

Early experience with the foundations for the Norris Dam may be cited in view of its extent and pioneer interest. The dam is a gravity structure located on the Clinch River about 130 km (80 mi) above its confluence with the Tennesse River; it is 85 m (285 ft) high and floods an area of 14,000 ha (34,200 acres) at normal pool level. The entire valley above and below the dam site is underlain by Knox dolomite; in many locations there are underground caves and porous beds. The dolomite was described in official reports as "a comparatively hard and substantial rock ... quite massive, compact and sound. Some jointing has occured ... [with] occasional seams, of importance in regard to possible leakage." Complete exploration of the dam site confirmed previously expressed opinions of consultative boards to the effect that the foundation rock itself is characterized by massive, thick strata of excellent rock interrupted at definite intervals by horizontal seams which are partly open, partly clay filled, and for a portion of the area in close contact. A comprehensive grouting program was laid out covering the entire base area of the dam, spillway apron, and powerhouse. The grouting treatment was divided into two parts: shallow, low-pressure grouting covering the entire area of the foundation and deep, high-pressure grouting to form an impermeable curtain under the heel of the structure. On the reservoir rim, the work involved the determination of the location of those portions in need of treatment to prevent excessive leakage and the grouting of the parts found to be faulty.

Spacing of the drill holes through which grouting was done is shown in Fig. 27.5. A system of interlocking patterns was chosen for the shallow grouting. All seams were washed out by reversing air and water flow between adjacent holes prior to grouting. One complete pattern was first drilled and grouted to a depth of 6 m (20 ft). Following this primary grouting, a system of 12-m (40-ft) holes, evenly spaced between those first drilled, was superimposed upon the original pattern, and the same sequence of operations was repeated. After the dam had been built to a height of 30 m (100 ft), deep foundation grouting holes were drilled from the lower gallery to establish a deep grouting curtain. This curtain is located approximately 1.8 m (6 ft) downstream from the axis of the dam. The work of drilling and grouting the holes 3 m (10 ft) apart in the galleries was divided into three parts. First, groups of three holes on 30-m (100-ft) centers between groups were drilled, washed, and grouted, after which groups of three holes halfway between the first and second groups were similarly treated. This left space for two holes between the first and second groups and these were drilled last. In general, a refusal pressure of 10.5 kg/cm^2 (150 psi) was used in the curtain grouting.[27.10]

At almost every one of the major TVA structures, new experience was gained in the art of grouting in limestone. At the Chickamauga Dam, for example, founded on a fissured and cavernous limestone overlain by 12 to 15 m (40 to 50 ft) of overburden, ranging from good, sandy clay loam to mixed gravel and finally to limestone slabs and clay, subsurface conditions were found to be so variable that most of the original grout holes on 30-m (100-ft) centers missed the worst areas. One large cavern was found as much as 27 m (90 ft)

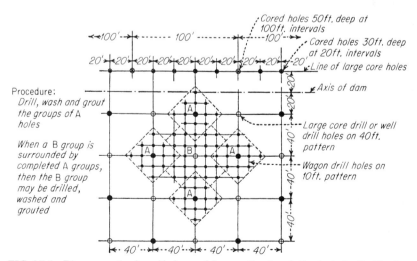

FIG. 27.5 Diagrammatic plan of holes used for grouting foundation beds for the Norris Dam, Tennessee Valley Authority.

below normal river level. Because of the size of caverns to be filled, grouting was developed in which a mixture of sand, bentonite, and cement was used.[27.11] Published records make the invaluable experience thus gained available to the public.

Records exist not only for dams of the TVA but also for structures built by the U.S. Bureau of Reclamation. Brief reference to one of these, the Hoover Dam, must be made, since the work then done was one of the great pioneer grouting projects. All grouting work was specified in detail with regard to materials, pressures, and procedure. The contractors, Six Companies, Inc., developed special equipment for this as well as for other parts of the work. Wherever concrete was to be placed against rock, advance provision was made for forcing grout into rock fissures and seams, if any existed. While concreting was in progress, special pipe connections were installed through which grout could later be forced, when the dam was carrying full water load, in order to seal effectively the contact between concrete and rock. By means of carefully planned drill holes, the attempt was made to provide a grout cutoff curtain in the bedrock across the dam site, and to assist in attaining this objective, grout holes were drilled from the outlet tunnels up to 45 m (150 ft) deep. After the construction of the dam had advanced appreciably, grouting was also carried out from the inspection galleries left in the body of the dam. Checks on the efficacy of the grouting were made by taking out cores of grouted rock; these showed satisfactory results. Pressures used varied up to 70 kg/cm^2 (1,000 psi); the average quantity of cement used (neat cement was used throughout) was 0.8 sack per ft of hole for all the longer holes.[27.12]

To describe further examples of the grouting of foundation beds below dams would involve somewhat tiresome repetition since, although every case has its special features, the purposes in every case are identical—to reduce permeability of the bedrock or to give it increased strength. A few more cases will, therefore, merely be summarized. The application of grouting to the foundation beds under the Beechwood Dam of the St. John River in New Brunswick is of special importance because of the care with which it was described

FIG. 27.6 The completed Beechwood water power project on the St. John River, New Brunswick, the intensive grouting beneath which is mentioned in the text.

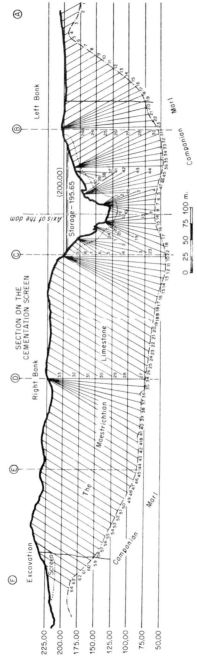

FIG. 27.7 Cross section showing the extensive grout curtain formed at the site of the Foum el Gherza Dam in the Sahara zone of Algeria to seal the marl and limestone strata shown; sulfate-resisting cement was used for the structure.

27-11

in a published paper. The dam rests on Silurian shales, limestones, and sandstones, intensively folded and so with many open joints. A program of grouting, involving the injection of 10,665 bags of cement through 2,883 holes over a total length of 32,388 m (107,691 ft), was carried out successfully (Fig. 27.6). The account of this work demonstrates the necessary close interrelation of grouting procedures and bedrock geology.[27.13]

In other lands, too, grouting has become an almost essential part of the practice of dam construction. Many modern European dams, especially the daring arched dams that now distinguish many of the spectacular gorges in high mountains, are assisted in their functioning by hidden grout curtains carefully placed in the surrounding rock. The St. Pierrett-Cognet Dam, an important unit in the Drac development in France, may be considered typical. Although the dam is 75 m (246 ft) high, the gorge in which it is located is so narrow that the entire structure contains only 38,000 m³ (29,000 yd³) of concrete. The gorge is in a highly fissured Bajocian Lias, reasonably watertight but subject to surface weathering. The rock surrounding the dam absorbed 300 tonnes of cement at the rate of 30 kg/m³ (55 kg/yd³) of finished curtain.[27.14] Figure 27.7 illustrates clearly the extent of the grouting necessary to seal marl and limestone strata at the site of the Foum el Gherza Dam in Algeria.

In northeastern Iraq, the great Dokan project is a combined irrigation, water-supply, flood-control, and water power development; the main retaining structure is an arched concrete dam, 116 m (380 ft) high, across the Lesser Zab River. The rock succession at the dam site starts at the surface with a marl, under which is thinly bedded limestone, and then dolomite; an important "contact zone" occurs between the weathered surface of the dolomite and the limestone. The dam is located almost at the top of an anticlinal fold, and the contact zone is exposed in both abutments. Careful preliminary work showed the necessity for extensive grouting; this had to be carried out, eventually, over a total length (including the dam) of 2,451 m (8,000 ft). Special tunnels were excavated in both abutments to facilitate the grouting work, which was carried out to depths up to 60 m (197 ft), where a pressure of 40 kg/cm² (5,700 psi) was used. Table 27.1 summarizes the main statistics of this grouting operation and shows comparable figures for four other dams, two of which have already been mentioned. It is not, therefore, surprising to find that the grouting work for the Dokan Dam, including the two tunnels, cost £2.85 million ($15 million).[27.15]

From relatively insignificant beginnings, pressure grouting, or cementation, has developed into an important feature of all major construction operations. The actual performance of grouting operations is essentially an engineering matter, but assessment of the need and successful application depend on an accurate knowledge of underlying geological structure. Grouting must therefore always be a cooperative endeavor of engineer and geologist, and the success of the operations described in the foregoing typical examples has been dependent to a large extent on such joint action.

TABLE 27.1 Some Comparative Statistics of Grouting in Dam Foundations.[27.15]

Project	Area of curtain, ft²	Total drlling, ft	Total injection, tonnes
Dokan	4,528,000	601,000	77,766
Hoover Dam	1,116,000*	190,981	13,250
Camarasa (Spain)	3,102,000*	169,402	186,000
Chickamauga (U.S.)	228,000*	477,427	59,500
Bin-el-Ouidane (Morocco)	1,453,000	121,000	27,122

*Figures are approximate.

27.5 SPECIAL APPLICATIONS

Grouting has proved a most useful aid in other types of engineering work, notably in mining and particularly in the sinking of shafts where water-bearing strata have to be penetrated. A pioneer in this work was Albert François. In the early years of the century, François worked out and developed the use of a special direct-acting pump capable of developing 210 kg/cm^2 (3,000 psi) pressure; the pump had flexible and reliable mechanical controls and permitted the use of cementation pipes sealed into the ground and the employment of special chemical solutions, lubricating in their action, to ensure the cementation of all fine cracks encountered. Development work has naturally taken place along general lines other than those associated with this particular process, so that it can now be said that cementation is an accepted part of mining engineering practice; cementation programing can be accurately planned and controlled if considered with proper relation to the geological structure concerned. Not only has grouting been applied to mining in a positive way but also, if it may be so expressed, in a negative way. Trouble was experienced with an underground fire, believed to be of coal covered up when the track was laid, along a 3.2-km (2-mi) section of the (old) Midland Railway main line near Hasland in Derbyshire, England. The fire was extinguished by a special application of cement grouting. Admittedly this was an unusual case, but it does serve to illustrate the flexibility of the technique.[27.16]

Building foundations have occasionally required the use of grouting in the supporting beds, one example being the Havana Hilton Hotel, which is founded on coral bedrock. This relatively porous rock was strengthened by the injection of over 1 million gal of Intrusion-Prepakt grout.[27.17] This is one of the several respected specialist names in the grouting field, the Prepakt process consisting of injecting grout, of a propriety composition, into crushed rock aggregate already in place—in effect, making concrete in situ. This scheme has long been used (as for example in a bridge over the New Croton Reservoir before the end of the last century), but the composition of the specialist grout and the accumulated experience in its use combine to make the system one of great value today for some especially difficult foundation jobs.

Grouting may even have a direct effect upon design. In West Germany a new highway bridge, 1,110 m (3,700 ft) long, was planned across the valley of the Kocher River, south of Stuttgart, to carry the Geislingen Autobahn (Fig. 27.8). Geological strata in the valley sides and bottom consist of shell-limestone, which early geological reports suggested was not strong enough to carry the loads that piers for a new bridge would impose. Accordingly, a cable-stayed bridge with a main span of 650 m (2,165 ft) was proposed as an alternative design. Further examination of the local geology, and a reevaluation of the bedrock, suggested that if the rock was strengthened by cement grouting, a more economical box-girder design, supported on eight piers, could be used. The revised design, made possible by grouting, required free-standing piers 176 m (588 ft) high. The limestone in the valley sides and floor was in three formations, the lowest one the strongest of the three but still in need of strengthening. Shafts were therefore sunk for three of the main piers so that they could be founded, with all the other piers, on the lowest limestone formation. All the foundation beds were then grouted with cement (Fig. 27.9). By this change in design a saving of over $20 million was effected.[27.18]

The temporary structures (cofferdams) that have to be used in many civil engineering projects to retain water temporarily during the construction period, notably for dam foundations and the founding of bridges, must have watertight strata beneath them if they are to be effective. Grouting has therefore been of assistance in some very large cofferdams. One of the most extensive cofferdam operations utilizing steel sheet-pile cell units was that carried out during the construction of the Grand Coulee Dam in Washington. The

FIG. 27.8 Bridge over the Kocher River, south of Stuttgart, West Germany, on the Geislingen Autobahn, the design of which was changed to that here seen because of the use of grouting to improve foundation-bed conditions.

Columbia River was diverted first into one half of its original bed and then into the other by cofferdams which were about 900 m (3,000 ft) long. The piles were driven generally into a consolidated glacial deposit which was difficult to penetrate, resulting in splitting and deformation of the steel piling.

When excavation was carried out within the cofferdams, a sand seam 10 to 25 cm (4 to 10 in) thick was encountered; it was exposed across the excavation at an elevation just below the general elevation of the bottoms of the steel piles. In the west cofferdam, no trouble was caused by this seam; but in the east cofferdam, owing to some complicating factors, including the deformation of the bottoms of steel piles, a leak developed through the seam and reached a flow of 132,500 lpm (35,000 gpm). Much trouble resulted, including the wrecking of a small part of the cellular dam; some of the steel piles were split from top to bottom. The emergency was met by a combination of engineering skill and construction ingenuity, the record of which is of great interest to all engineers; the leak was finally stopped without serious interruption of the main dam construction operations. The only feature of this emergency work that need here be mentioned is that 10-cm (4-in) well drills were used to locate the main passage through the sand seam that the leak was taking, and cavities were found from 30 to 60 cm (1 to 2 ft) deep. This passage was successfully plugged, despite the great flow through it, by the use of a special grout; ordinary

FIG. 27.9 One of the pier foundation pits for the Kocher River bridge, West Germany, showing the results of grouting.

FIG. 27.10 The Jim Woodruff Dam as it is today, the remarkable use of grouting during its construction being described in the text.

cement grout washed out before it could set. Four hundred batches of this grout were used to plug the leak described; the grout was pumped through an 18-cm (7-in) concrete pipeline and forced into the passage under a pressure of 2.5 kg/cm^2 (35 psi).[27.19]

A problem of quite a different sort had to be faced with the cofferdam for the navigation lock at the Jim Woodruff Dam, Chattahoochee, Florida (Fig. 27.10). Porous Tampa limestone underlay the area that had to be dewatered for construction of the lock. Earlier experience at a site 300 m (1,000 ft) downstream had shown the problems that could result from leakage through this rock. Detailed site studies, including the use of Calyx drill holes for actual inspection of the rock in place, showed the necessity for a grout curtain in the rock before the necessary cofferdam was constructed. After much experimentation, a sanded-cement grout with two special admixtures was found suitable, and a grout curtain 847 m (2,824 ft) around the perimeter of the area to be excavated was successfully injected. A small quantity of lampblack was added to the mixture so that the efficacy of the grouting could readily be seen when excavation was carried into the limestone near the sides of the excavation. Some minor leakage occurred but no more than could easily be controlled by pumping. Grouting, therefore, is not a process for application only in permanent installations. It is a specialist process of wide application when the local geological conditions call for its use in the first place and facilitate its successful application in the second.[27.20]

27.6 GROUTING WITH CHEMICALS

Useful as grouting with cement has proved to be in civil engineering practice, it has some limitations. Because cement has to be forced into the interstices that have to be sealed, there is a limiting size of joint that can be so filled, cement being ineffective for finer openings. Cement grouting has rarely been too successful in soils, and yet water-bearing alluvial materials are often encountered, especially in shallow tunnel work. The idea of using chemicals for grouting attracted inventive minds for many years, but the first successful system was that patented in 1925 by a Dutch engineer, H. J. Joosten. The basis of his system was to inject successively a solution of sodium silicate (waterglass) and then a strong electrolytic saline solution. When the two solutions meet, they react and produce a soft gel which, with the absorbed film, results in a hard, dehydrated gel that binds together the particles with which it is in contact. There have been many variations since this first patent was granted, and many other systems have been patented, so that the application of chemical grouting is today almost always a proprietary operation. There have been a few failures, naturally, but the system has proved so valuable in so many difficult cases that some of these successes must be briefly recounted.

One of the first applications of the Joosten process in North America was for sealing off leaks that had developed in 1941 in the Chicago Freight Tunnel, but prior to this, a number of major installations in Europe had been aided by the system, notably works carried out by the British contracting firm of John Mowlem and Co. In 1949 an inclined tunnel was being sunk for access to a new Alpha mine of the Shuler Coal Company in Illinois. A 1.2-m (4-ft) -thick stratum of water-bearing sand, the water under a 15-m (50-ft) hydraulic head, had to be penetrated. Once the sand was encountered and pumping had proved useless in controlling the water, the Joosten process was applied. Eight grouting operations were carried out successively, each about 2.5 m (8 ft) long, forming the necessary impermeable seal in the sand; tunneling followed and was successfully completed into the competent underlying coal bed.[27.21]

Another system of chemical grouting was successfully used in preparing the old tunnels which were incorporated into the Pennsylvania Turnpike. One end of the excavation for the Chase Manhattan Bank Building in New York City, where singularly difficult soil conditions were encountered (the New York "bull's liver," underlain by boulders and sand), was successfully completed after the application of chemical grouting. In driving a new major tunnel under the river Thames at Dartford, England, portland-cement grouting was used from an earlier pilot tunnel to solidify the fractured zone found at the top of the well-known Chalk, which is here the bedrock. A stratum of Thames gravel, with an open structure, overlies the Chalk. This was successfully solidified in advance of excavation by the use of one of the more recent combined-chemical grout treatments, the grout used containing portland cement, bentonite, silicates, and other chemicals.[27.22] A similar process was used to form one of the largest grout curtains ever injected, that under the Assouan High Dam on the river Nile, Egypt. The curtain extends to a depth of 240 m (835 ft) below river level, the V-shaped gorge of the Nile being 492 m (1,640 ft) wide.[27.23]

Most applications of chemical grouting, however, are on smaller jobs such as the construction of a 42-m (139-ft)-long tunnel for electrical cables on the south bank of the Monongahela River at Pittsburgh. The tunnel had to be driven on an incline, finishing in a cofferdam 6.9 m (23 ft) below the normal river level and passing under busy railway tracks. A chemical-cement grout with special bulking properties was used in advance of tunneling through the porous mixture of slag and cinders. Despite some difficulty in the lower part of the tunnel, the job was successfully completed, thanks to this chemical grouting.[27.24] Finally, there may be mentioned the use of grouting for a small building foundation, an application of unusual significance.

All schools in California built before 1933 must by law be investigated to see if they meet seismic safety requirements; if they do not, the potential hazards must be eliminated. Roosevelt Junior High School, built about 1930, was one of many schools of the San Francisco Unified School District so affected. It is a three-storey building of mixed construction type, with spread footings beneath columns. Test borings revealed 1 to 2 m (3 to 6 ft) of loose fine sand and silty sand, underlain by loose-to-medium-dense windblown sand, commonly called dune sand. Strengthening of the

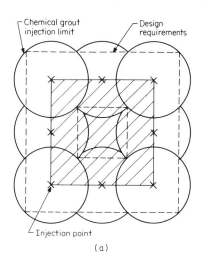

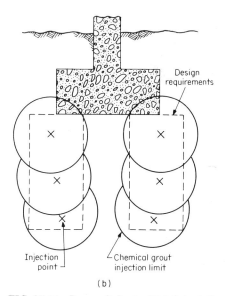

FIG. 27.11 Roosevelt Junior High School, San Francisco, California; the use of chemical grouting beneath its foundations, as here shown, gave its structure the requisite safety against seismic shock.

27-18 CIVIL ENGINEERING WORKS

foundation was imperative. Alternative designs were considered. Grouting beneath each foundation with chemical grout was the solution adopted. A proprietary organic resin grout was employed by the successful contractor, working to a performance specification prepared by the engineer. As with all performance specifications, testing had to be carried out to prove the results achieved. This gave samples which met the strength requirements; thanks to the care with which the preliminary work had been carried out, grouting following a precise program under each footing (Fig. 27.11). It was estimated that the use of chemical grout, costing less than $250,000, saved as much as $1 million when compared with standard underpinning methods.[27.25] The widespread potential of thus strengthening existing foundations will at once be obvious, as will also the necessity for having complete subsurface information if the method is to be applied successfully.

27.7 THE GRAND RAPIDS PROJECT, MANITOBA

Any use of the term "largest in the world" invites immediate rebuttal, so let it merely be said that one of the most extensive grouting projects ever carried out was for the foundations of the dams for the Grand Rapids hydroelectric project in Manitoba. The project is of importance if only because it could not have been carried out had it not been for the advances in recent years in geotechnique, alike in grouting methods as in the use of soil mechanics, which facilitated the selection, design, and placing of soils in no less than 26 km (16 mi) of dikes. (This is not an invention of the authors, but a statement to one of them by the president of Manitoba Hydro, the owner, while the plant was being constructed.)

Grand Rapids was the name given to the rapid stretch of water down which the Saskatchewan River, draining the Canadian prairies, used to flow into Lake Winnipeg, about 400 km (250 mi) north-northwest of Winnipeg. The natural fall is about 22.5 m (75 ft), which could be increased to 36 m (120 ft), in view of the local topography, for power generation, provided the local bedrock could be made watertight and the many km of low dikes necessary could be safely constructed of soil available in the locality. Studies were first made in 1911 but discontinued in view of the limitations already indicated. They were resumed in 1953 when it was known that the necessary techniques were available. Con-

FIG. 27.12 The Grand Rapids water power project of Manitoba Hydro, on the western shore of Lake Winnipeg, showing how the Saskatchewan River is controlled by dam, powerhouse, and dikes.

struction started in 1960, the plant being commissioned in November of 1963 (Fig. 27.12). It generates 340,000 kW in three units tied in with the Manitoba Hydro transmission system.

The entire area of the project and its reservoir is in glaciated terrain. Precambrian bedrock underlies the site at a depth of almost 150 m (500 ft) below ground level. It is overlain by Ordovician and Silurian rocks, consisting generally of fragmented dolomitic limestone and massively bedded dolomites. Minor formations of calcareous shales and sandstones were encountered in the extensive preliminary subsurface investigations. The most significant finding, however, was a thin layer of argillite, containing sandstone lenses and clay seams and underlying the entire site, clearly of assistance in providing a useful barrier against any vertical movement of groundwater. Its location determined much of the major grouting program which had to be undertaken. Over 4.35 million m^3 (5.7 million yd^3) of soil and 1.37 million m^3 (1.8 million yd^3) of rock were used to construct the 26 km (16 mi) of dikes required, to an average height of 7.5 m (25 ft), maximum height being 30 m (100 ft), top level being 3 m (10 ft) above normal forebay level.

The fractured nature of the bedrock was reflected in many sinkholes and solution channels found near the necessary line of the dikes. Extensive grouting was, therefore, the key to the success of the project. A preliminary contract for 120 experimental grouting holes was awarded in 1959. Good results were obtained, so that it was possible to proceed with the major program which eventually involved 600,000 m (2 million ft) of drill holes, extending over the 25 km (16 mi) of dikes and the mass-concrete headworks, structures, etc., the total being equivalent to a curtain wall of grout 29 km (18 mi) long. Holes were usually on 5-ft centers, primary holes being drilled at 20-ft centers, generally to a depth of 61 m (200 ft), secondary holes to 36 m (120 ft), and tertiary holes to 18 m (60 ft). Twelve different grout mixes were used, carefully selected on the basis of detailed preliminary examination of the rock to be penetrated. NX diamond-drill holes were regularly used to check on the efficacy of the grouting. Sinkholes, when accessible, were cleaned out and backfilled, the largest taking 4,200 m^3 (5,500 yd^3) of low-strength concrete for this purpose. Special consolidation grouting was undertaken under all mass-concrete structures (Fig. 27.13).[27.26]

To go further would involve too much detail. The job was successfully completed and the project has operated successfully from the day of its commissioning, thanks to the

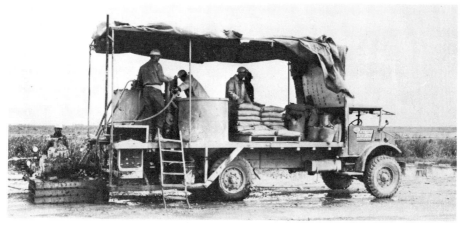

FIG. 27.13   Portable grouting plant as used on the Grand Rapids project, Manitoba.

27-20 CIVIL ENGINEERING WORKS

attention given so meticulously to the geology underlying all the works and the expert use of the technique of grouting with cement, carried out in complete liaison with detailed geological studies.

27.8 CONCLUSION

Many of the projects described in this volume were founded on geological formations so sound that grouting was not necessary to ensure safety and stability. This is a reflection of general civil engineering practice. Grouting is only a means of improving upon imperfections of Nature when rocks or soils do not have the sound, strong properties required for the purpose to which they are to be adapted by the civil engineer, generally as foundation beds carrying substantial loads. When geological conditions are known with certainty and when these show that grouting is necessary, then it can be a most useful technique for achieving satisfactory foundation conditions. Its application is always a job for the experienced expert. Poor grouting may be worse than none at all. But the expert needs to know exactly what subsurface conditions are before he can apply his skills; geological investigations must always precede and accompany all grouting operations, if those operations are to be successful.

28.9 REFERENCES

27.1 R. Glossop, "The Invention and Development of Injection Processes," *Geotechnique*, **10**:91 (1960) and **11**:255 (1961).

27.2 W. Denison, "Notes on Injecting Cement or Hydraulic Lime into Leaky Joints of Masonry," *Papers on Subjects Connected with the Duties of the Corps of Royal Engineers*, **4**:208, 209 (1840).

27.3 F. H. Kellog, "Clay Grouting at Madden Reservoir," *Engineering News-Record*, **109**:395 (1932).

27.4 E. A. Willbur, "Grouting Checks Leakage in Earthfill Dam," *Engineering News-Record*, **115**:499 (1935).

27.5 H. S. Slocum, "Asphalt Cutoff at Claytor Dam," *Engineering News-Record*, **128**:490 (1942); see also in same journal, L. A. Schmidt, "Reservoir Leakage Stopped at Outlets," **134**:65 (1945); with information kindly provided by TVA.

27.6 A. A. Barnes, "Cementation of Strata below Reservoir Embankments," *Transactions of the Institution of Water Engineers*, **32**:42 (1927).

27.7 H. S. Rands, "Grouting Cutoff for the Estacada Dam," *Transactions of the American Society of Civil Engineers*, **78**:447 (1915).

27.8 J. R. Fox, "Precementation of Reservoir Trench," *Transactions of the Institution of Water Engineers*, **32**:69 (1927).

27.9 M. Taylor, "Cementation of Abitibi Dam Foundations," *Canadian Engineer*, **66**:5 (1934); and information received from R. L. Hearn, Queenston.

27.10 "Thousands of Holes Grouted under Noris Dam," *Engineering News-Record*, **115**:699 (1935).

27.11 "Corewall Grouting at Chickamauga Dam," *Engineering News-Record*, **122**:551 (1939).

27.12 V. L. Minear, "The Art of Pressure Grouting," *Reclamation Era*, **27**:56 (1927); see also "Extensive Rock Grouting at Boulder Dam," *Engineering News-Record*, **114**:795 (1935).

27.13 I. D. MacKenzie and E. L. Brown, "Geological Features and Foundation Treatment at the Beechwood Development," *The Engineering Journal*, **42**:54–62 (1959).

27.14 P. Volumard and R. Dubost, "The Saint-Pierett Dam (The Drac Development)," *Travaux*, **286**:5 (August 1958).

27.15 G. M. Binnie, J. G. Campbell, N. H. Gimson, P. F. F. Lancaster-Jones, and G. A. Gillott, "The Dokan Project: The Flood Disposal Works and the Grouted Cut-off Curtain," *Proceedings of the Institution of Civil Engineers,* **14**:157 (1959).

27.16 "Fire Under Track near Chesterfield," *The Railway Magazine,* **108**:807 (1961).

27.17 H. C. Boschen, "Havana Traffic Tunnel Built in the Dry under Almendares River," *Civil Engineering,* **23**:447 (1953).

27.18 "Site Geological Review Slashes Bridge's Cost," *Engineering News-Record,* **200**:24 (27 April 1978).

27.19 "Contractors Win River Battle," *Engineering News-Record,* **119**:13 (1937).

27.20 F. A. Robeson and W. E. Webb, "Cofferdam Grouting at Jim Woodruff Dam," *Engineering News-Record,* **145**:35 (6 July 1950).

27.21 "Chemical Treatment Cures Sick Tunnel," *Construction Methods and Equipment,* **31**:66 (September 1949).

27.22 J. Kell, "Pre-treatment of Gravel for Compressed Air Tunnelling under the River Thames at Dartford," *Chartered Civil Engineer,* Institution of Civil Engineers, London, March 1957, p. 3; see also "River Gravel Solidified for British Tunnel," *Engineering News-Record,* **159**:58 (14 November 1957).

27.23 W. G. Bowman, "Record Grout Curtain Seals Nile's Leaky Bed," *Engineering News-Record,* **180**:22 (29 February 1968).

27.24 M. C. Behre, "Chemical Grout Stops Water," *Civil Engineering,* **32**:44 (September 1962).

27.25 E. D. Graff and E. G. Zacher, "Sand to Sandstone: Foundation Strengthening with Chemical Grout," *Civil Engineering,* **49**:67 (January 1979).

27.26 J. R. Rettie and F. W. Patterson, "Some Foundation Considerations at the Grand Rapids Hydro-Electric Project," *The Engineering Journal,* **46**:32 (December 1963).

Suggestions for Further Reading

In the decade preceding the publication of this Handbook, interest in grouting among engineers in North America increased at a remarkable rate. This is indicated by the publication in 1982 by the American Society of Civil Engineers of a volume of no less than 1018 pages called *Grouting in Geotechnical Engineering,* which reports the proceedings of a specialized conference held in New Orleans.

Some complete volumes have been published on the subject, one of the first being *Grouting in Engineering Practice* by R. Bowen; its second edition of 285 pages was published in 1981 by the Halstead Press (John Wiley & Sons) of New York. There has also been much activity in professional committee work, where experience is shared so helpfully, notably within the Geotechnical Engineering Division of ASCE; some of the reports have appeared in the Society's GT Proceedings.

chapter 28

WATER SUPPLY

28.1 Historical Note / 28-2
28.2 Sources of Water Supply / 28-3
28.3 Relation of Geology to Runoff / 28-3
28.4 Water Quality / 28-5
28.5 Water Supply from Riverflow / 28-8
28.6 Water Supply from Impounding Reservoirs / 28-10
28.7 Water Supply from Groundwater / 28-15
28.8 Replenishment of Groundwater / 28-21
28.9 Some Unusual Sources of Supply / 28-25
28.10 Some Geologically Significant Systems / 28-27
28.11 References / 28-35

The appearance of a relatively short chapter dealing with water supply may seem strange at first sight, but the subject is presented in this way in order that its importance may be duly emphasized and that the relation of geological considerations to water supply may be clearly defined. All too often the "geology of water supply" is featured in a manner not in keeping with the importance of the engineering side of water-supply work, to which geology can be no more than a serviceable aid. That geological conditions have a profound effect on sources of water and on the means by which water is supplied for public use cannot be denied. Geological conditions are, however, a part of the natural order and cannot well be altered, whereas the task of the civil engineer in designing a water-supply system is the utilization for the public good of the particular natural conditions with which he has to deal, often a task of great difficulty.

Much can be written, even in summary form, about the importance of water supply; it can hardly be overemphasized. It is a prime necessity in public health, an essential to cleanliness, and a fundamental requirement of almost all modern manufacturing processes. Indeed, a satisfactory supply of pure water is one of the engineer's greatest gifts to the public and at the same time one of his greatest responsibilities. So vital is the maintenance of a public water supply that no possible risks can be taken in connection with the necessary engineering design. Few cities are in the fortunate position of having a duplicate water supply system to safeguard them against serious trouble if anything should interfere accidentally with the functioning of the main supply. Unusual precautions have to be taken, therefore, in providing temporary storage facilities; for this reason, conservatism in water engineering design and practice is often justified. Correspondingly,

28-2 CIVIL ENGINEERING WORKS

in no branch of civil engineering are geological considerations more important than in connection with water supply: the most careful preliminary geological investigations are always absolutely essential. The civil engineering works required for water-supply projects include tunnels, dams, canals, reservoirs, grouting, and buildings, all subjects considered in other chapters of this Handbook. In this chapter, general reviews only will be presented of complete water-supply systems, with special emphasis on the problems associated with groundwater supplies.

28.1 HISTORICAL NOTE

The importance of water supply is reflected in the attention that was paid to water-supply systems by some of the earliest engineers. The inscription of the Moabite stone, dating from the tenth century B.C., contains references to water conduits and cisterns. There is evidence to suggest that reservoirs for water supply existed in Babylon as early as 4000 B.C. The Bible contains not a few references to water-supply systems; some of these systems were of considerable magnitude, and parts of them have remained in use up to the present day. Notable among these were the waterworks for the city of Jerusalem, which were installed in Solomon's newly conquered city about 900 B.C. and showed highly developed technical skill. They were extended by King Hiskia in 700 B.C.; an inscription records that his engineers successfully bored a tunnel almost a mile long by working from the two ends. The Romans also augmented the supply, and parts of the original installation are still used today to supply the Mosque of Omar. Works in Asia Minor at Priene (dating about 350 B.C.), a town of only 5,000 inhabitants, are hardly excelled by those of any modern town, since every house was connected by earthenware pipes to mains supplied from a large and steady spring on the hillside above the town.

The waterworks of ancient Rome are widely known and appreciated, but the fact that the supply to the city is estimated to have exceeded 360 mld (80 mgd) is not perhaps so generally realized. The principal sources of the Roman water supply were springs in the great beds of limestone in the valley of the river Amio. Many of the springs fed into one of the 11 aqueducts that were constructed to carry the supply to the Imperial City, those monumental undertakings even the ruins of which testify to the engineering skill of their Roman builders. Fortunately, a contemporary description of these works exists, written by Sextus Julius Frontinus, water commissioner of Rome in A.D. 97. This earliest of engineering authors was an enthusiast, commenting in one place thus: "Will anybody compare the idle Pyramids, or those other useless though much renowned works of the Greeks, with these aqueducts, with these many indispensable structures?" This has a modern touch, as indeed has much else of what Frontinus wrote; his descriptions are worthy of study by all engineers interested in the history of their art. It will be found that he does not neglect geological features.[28.1]

The Middle Ages were not devoid of their water-supply engineers. The modern emancipation of women is possibly reflected in the fact that the first post-Roman artificial water supply system in England appears to have been that installed by St. Eanswide, of whom it is reported that "she haled and drew water over the hills and rocks against nature from Swecton a mile off to her Oratoria at the sea-side." The good lady in question was the daughter of Eadbald, king of Kent (A.D. 616–640), and was the first prioress of St. Peter's Priory near Folkestone.[28.2] So the fascinating story can be unfolded; it will be found related in detail in more appropriate places, but enough has been presented to show that the water engineer of today is following in a great tradition and that in some respects at least, modern problems were reflected in these ancient works.

28.2 SOURCES OF WATER SUPPLY

This plural subheading is almost a misnomer, since the ultimate source of practically all water supply is the fall of rain; the use of the plural implies that the natural cycle may be broken into in this way for classification purposes. The estimation of the quantity of rain that will fall upon an area is always of importance in water-supply studies, whether the supplies are to be drawn from catchment areas, lakes, rivers, wells, or springs, the features usually referred to as sources of water supply. The quantity depends on many local factors, including the physical configuration of the district, its elevation above sea level, its relation to mountain masses and prevailing winds, and the proximity of the sea. There appears to be some reason for believing that the presence of forests also has an influence on rainfall.

In the study of what happens to rainfall when it reaches the ground, geological conditions assume importance. Part of the rain is evaporated; part will run off the surface; part will be absorbed by the vegetation on which it falls; and part will eventually find its way into the subsoil. Cursory consideration will show that all four processes depend to some extent on the geological nature of the ground on which the rain falls and that only a portion of the rainfall on any area is available for use. That which runs off the surface to form the flow of streams and rivers is clearly of special interest to water engineers. The available runoff naturally includes also that part of the water which eventually finds its way into watercourses by seepage through the local ground formation. Geology has therefore an appreciable effect on the relation of runoff to rainfall.

The most satisfactory method of evaluating probable runoffs is by means of stream gauging in conjunction with the study of rainfall records. If a measure of actual streamflow is thus obtained, the various reductions in quantity of water from the amount falling as rain are automatically taken into account. This does not mean that study of the geology of the catchment area can be neglected; it cannot, unless long-term streamflow records are available, since after varying intensities and durations of rainfall, geological peculiarities may affect the relation of runoff to rainfall to a varying degree and consequently may interfere with the regularity of stream discharge. The loss of water through evaporation into the atmosphere and absorption by plant life is of interest to the water engineer only insofar as it represents a quantity of rainfall only indirectly for human use. That which percolates into the ground, though it represents an amount that has to be determined as a loss in the rainfall-runoff relationship, goes to replenish the great volumes of underground water which are of such vital importance to life on the earth's surface. Clearly, the amount that thus enters the ground is directly dependent on the geological nature of the exposed surface and on the type of vegetation growing thereon.

28.3 RELATION OF GEOLOGY TO RUNOFF

The calculation of rainfall losses is a vitally important part of hydrological study in connection with all engineering works dependent on assured water supply. Studies of the matter are to be found in many standard works, and many empirical formulas have been evolved to suggest a general relation between rainfall and runoff for catchment areas of different sizes and different climatic conditions. Of the factors affecting rainfall losses, two are fundamentally dependent on geological structure. It will therefore be seen that, although general formulas are undoubtedly useful in limited application, they must be interpreted in the light of information on the local geology if they are to be effective and of use.

Loss of rainfall by percolation to underground pervious strata is the first of these two factors. Consideration of this possibility leads inevitably to a general study of groundwater. In all but the most exceptional watersheds (such as those used at Gibraltar), some portion of the rain will fall on pervious material and so sink below ground surface. There it will eventually join that large body of water which is permanently held in the interstices existing in the underground strata, taking its part in the slow but steady movements of the contents of this underground reservoir. It is water derived in one way or another from this source that constitutes the flow in the corresponding watercourse during periods of dry weather, when the flow obviously cannot be the direct consequence of rainfall. This "dry-weather flow," as it is termed in engineering practice, is the criterion on which fundamental calculations regarding water supply must be based. The determination of the minimum possible dry-weather flow at any gauging station is a critical hydrological investigation, since it will often be impossible to measure this by actual gauging before a water-impounding project is started. Records at the sites of works that have been constructed can usually be obtained, and these constitute a valuable guide to similar cases. It is in determining the similarity of catchment areas in this connection that geological structure must be considered.

D. Halton Thomson once compiled a study of more than 30 records of various dry-weather flows for British streams, and these showed discharges varying from 0.004 cms per 1,000 ha (0.055 cfs per 1,000 acres) on the river Alwen in North Wales (drainage area 2,520 ha or 6,313 acres) to 0.14 cms (2.00 cfs) gauged on the river Avon, Scotland (drainage area 2,020 ha or 5,000 acres). The ratio of the two records from these catchment areas, almost identical topographically and in respect to climate, is 1 : 36. This is a surprising variation, and the explanation is wholly geological. Superficially, both areas might be regarded as impervious; the geology of this part of the Alwen Valley consists of Silurian shales and grits covered with boulder clay; that of the Avon is wholly granite. Based on this general comparison, it would seem that the dry-weather flows from the two areas should be almost equal. Further study discloses the fact that although the Alwen catchment area is watertight, as might be expected, and thus provides little or no underground storage, the granite over which the Avon flows is so fissured and decomposed that vast quantities of water are held in its interstices. The ample underground storage thus provided is drawn upon by the stream during persistent dry periods.[28.3]

Another interesting example is afforded by flow records of two different points on the river Exe in Devonshire, the lower one just above the range of tidal influence and the other near the headwaters of the river above the junction of two important tributaries with the main stream. Table 28.1 shows that the unit dry-weather flow for the whole area is almost four times that of the upper part of the area alone; the reason for the variation is, again, solely geological. About one-eighth only of the upper part of the area is formed of pervious rocks, the New Red sandstone, and the remainder is Devonian measures and igneous rocks; thus, practically no underground storage exists. The basins of the two tributaries that join the main river below the upper gauging station are largely formed of the pervious New Red sandstone, and in consequence, they are both well served with underground stor-

TABLE 28.1 Flow Records of Two Different Points on the Exe River [28.3]

	Upper Exe Basin	Whole of Exe Basin
Drainage area	620 km² (240 mi²)	1,200 km² (461 mi²)
Average annual rainfall	1.2 m (47 in)	1.1 m (42 in)
Dry-weather flow (September 1911 average)	154 l per 100 ha (22 cfs per 1,000 acres)	610 l per 100 ha (87 cfs per 1,000 acres)

age to the extent that their unit dry-weather flows are so high that they raise the figure for the whole catchment area to that indicated in the table.[28.4]

The examples quoted are by no means unusual, and they therefore illustrate vividly the effect that geological conditions have on dry-weather flows. This effect is felt to varying degrees for all stages of streamflow. Even during torrential rainfall, a pervious surface stratum will rarely be so completely saturated with water that rain falling on it will all run off, giving a flow equivalent to 100 percent of the rainfall. This is confirmed by the unique conditions that combined to cause the catastrophic floods early in 1936 in the eastern part of North America. Unexpectedly early rains fell on catchment areas that were still frozen hard, and thus constituted catchment areas almost "ideal," if considered from the standpoint of impermeability. The resulting runoffs caused floods that broke all known records and in some cases exceeded the calculated "1,000-year maximum." That such floods do not normally occur is due to the regulating effect of geological formations, which determine to some degreee the course of water from the time it falls as rain until the time it reaches the watercourse to join with the streamflow. The importance of this effect can hardly be overemphasized.

The figures quoted illustrate how geological conditions alone can explain some variation in runoff records. By means of similar studies, not only can records be compared, but the accuracy of recorded flows can be checked as possible limiting cases. Unfortunately, there appears to have been all too little published on this important aspect of applied geology. Whenever runoff records have to be correlated with rainfall, the importance of the geology underlying the watershed must be kept in mind if meaningful and useful correlations are to be made.

28.4 WATER QUALITY

Rainfall absorbs gases and floating solid particles from the air before it reaches the ground. Rainwater is therefore by no means pure; it is often definitely acidic, especially during thunderstorms, when nitric acid may be formed. There is currently much public concern about acidity in rainfall ("acid rain") caused by industrial pollution. Small portions of carbon dioxide are always absorbed. Consequently, in flowing over rock formations of various types, rainwater has a slight chemical effect on them, and the characteristics of lake and river water can therefore vary considerably. Although the detailed examination of water intended for use as domestic supplies is now a matter for specialist study, more particularly from the bacteriological aspect, it is one with which the civil engineer has to be generally familiar. Knowledge of the geological conditions causing the various impurities in water is a prerequisite for such familiarity.

Many upland gathering grounds, by reason of their topographical arrangement and the underlying geological structure, are covered with a layer of peat for much of their total area. Rain falling on peat deposits and running off to become streamflow inevitably takes with it traces of the peat and of the organic acids found in peat deposits. It thus becomes brackish, often slightly discolored, and usually slightly acidic. If the acidity is pronounced, the water may be slightly plumbosolvent, a grave matter in England and other European countries where lead piping is still in use for water-supply purposes. Similarly, such water may have a serious effect on concrete surfaces exposed to it. Investigations in Scotland have shown that the only normal types of open conduit lining that successfully resist the action of moorland water are Staffordshire blue brick (an engineering brick of high quality) and aluminous-cement concrete. Ordinary portland-cement concrete of varying consistencies is seriously affected.[28.5]

The brackish taste and discoloration are objectionable from the point of view of the user of the water. Both can be removed by modern filtration and treatment methods. In many cases, recourse has been had to stripping off all the peat from the area to be submerged by impounded water, often at great cost. Thus, a total volume of 3 million m^3 (4 million yd^3) of peat was removed from the site of No. 5 Reservoir of the Boston, Massachusetts, waterworks. The bottom of the reservoir formed by the Silent Valley Dam in Northern Ireland was found to be covered with peat, in some places 9 m (30 ft) deep. The peat was not removed, but the valley floor was covered with a layer of clean sand 60 to 90 cm (2 to 3 ft) in thickness. An area of 286,000 m^2 (342,000 yd^2) was covered in this way with sand obtained from local glacial deposits.[28.6] It will be seen, therefore, that thorough examination of the surface condition of any proposed reservoir site is a necessary part of preliminary investigation work.

Hardness, temporary and permanent, is perhaps the most widely recognized of possible impurities. It has been said that one can gain a general appreciation of the geology from which a water supply comes, when paying a first visit to a locality, by simply washing one's hands. Hardness is entirely geological in origin, since, in surface waters, it is the result of contact with rocks containing calcium or magnesium carbonate or calcium or magnesium sulfate. If a river is known to flow over limestone or dolomitic deposits, the source of the contamination can easily be found. Sometimes the strata in the bed of the river do not suggest the cause of the salt content. The river Derwent in Derbyshire, England, for example, flows over grit beds, and yet its water is hard. The hardness is due to the fact that the streamflow is fed by tributary streams which come down from mountain limestone deposits. Study of geological features both in riverbed and over catchment area will suggest the type of hardness and the necessary treatment.

Although generally troublesome, hardness of water is not always a disadvantage, as some beer drinkers may know. It is usually the mineral content of water that gives special local flavor to pale and bitter ales; hard water is desirable for their manufacture. For example, ales brewed at Burton-on-Trent, England, are made from hard water obtained from valley gravels and Keuper sandstones. It was once carefully calculated that in one year drinkers of Burton ales consumed 160,000 kg (350,000 lb) of solid gypsum in the process of assuaging their thirst. Soft water is required, on the contrary, for stout. Waters used in the north of Scotland and elsewhere for a similar, although stronger, purpose are also dependent on their minute mineral content for their peculiar efficiency (Fig. 28.1). Just in case one or two readers may have some interest in this rather specialized aspect of water quality, it may be recorded that one famous company which operates several distilleries once arranged for each of the distilleries to use water from one of the others, an erudite research project of some appeal. It was found that, although the resulting Scotch whisky produced was, in every case, excellent, it "was entirely different from that produced before [they] temporarily exchanged the waters."[28.7] It would take even more than geology to explain this result. One erudite writer on this vital topic even goes so far as to say that there is a difference between Scotch made from water that "comes off granite through peat" and that made from water "off peat into granite."[28.8] Further research is indicated.

More prosaic is the record of a study from South Africa of five pairs of streams in Natal. Care was taken in selecting the streams to ensure that they were unpolluted and geologically homogeneous and that the two streams in each pair were geographically far apart. Samples were taken for analysis from all ten streams in both the dry and rainy season. The report on this investigation presents a table which demonstrates clearly the influence of the varying geological formations over which the streams ran. Total dissolved solids varied from 314 ppm (for the Dwyka formation) to 38 ppm (for the Beaufort formation). Breakdown of the chemical content into percentages of Ca, Mg, Na, K, CO_3, SO_4, Cl, and SiO_2 reflect the profound influence of the geology of the catchment areas.[28.9]

FIG. 28.1 The distillery at Bunnahabhain, Isle of Islay, Scotland, illustrating the typical "granite and peat" topography that is conducive to the distilling of Scotch whisky.

The subject is a vast one, with a correspondingly extensive literature. It is only in the case of what may be called "natural waters" that geology will so often be found to exercise some influence over quality. In these modern days, unfortunately, impurities which are not geological in origin may be found in water intended for water-supply purposes. It was found some years ago, for example, that the Los Angeles area had to face serious problems because of the contamination of its groundwater resources by the wastes from oil wells. Trouble started when sumps were excavated along the east bank of the Los Angeles Flood Control Channel for the purpose of reclaiming the oil which remained in the wastewaters from oil-well drilling. These sumps handled more than 9.5 mld (2.5 mgd) of wastewater, containing 9,000 ppm of chlorine or 13,000 ppm of sodium chloride. Adjacent wells penetrating the local sands and gravels soon showed contamination; some had to be abandoned and others deepened and cemented in to seal off the source of contamination. Tests showed conclusively that the bottom and sides of unlined ditches and sumps were not sealed off by the drilling mud and oil emulsions carried by the wastewaters.[28.9]

This example is typical of the kind of problem that must be faced whenever groundwater is being used that may have contact with possible sources of industrial waste contamination. In the past, a major source of such pollution has been the wastewater from coal mines, especially that from disused mines which are not being actively maintained. Some states have had to take very strong legal action in order to eradicate the grave danger which arises from the interaction of air and water upon the sulfur-bearing compounds exposed in coal-mining operations. The problem has been aggressively tackled, and the results have been highly satisfactory.[28.10]

The other side of this interdependence is well illustrated by the widely quoted "Montebello incident" in California. The wastes from a plant making a weed killer commonly called 2-4-D were discharged into the sewers of the city of Alhambra. After passing through the city's sewage-treatment plant, the wastes were discharged as a small part of the effluent into the Rio Hondo River about a mile upstream from an interconnection with a subterranean groundwater basin. Within a short time, 11 wells serving 25,000 people were so seriously affected that even after cutting off the source of contamination, large expenditures were necessary to treat the water in order to make it safe again for public use.[28.11]

28.5 WATER SUPPLY FROM RIVERFLOW

Probably the simplest and most straightforward method of obtaining a supply of potable water is to abstract it directly from the flow of a river or from a freshwater lake; it is also probably the first method ever used, if imagination may be allowed to range back to prehistoric human activities. Today, some of the greatest cities in the world secure their water supplies in this way—London, from the river Thames; Montreal, from the St. Lawrence; Chicago and other North American lake cities, from the Great Lakes. What relation does geology have to this branch of waterworks engineering? There is, first, the indirect influence that geological conditions have on the rate of riverflow. The local geological structure at the site of the water intake will affect to a considerable degree the type of structure adopted to provide the necessary intake. At Toronto, for example, one of the intakes for the city's duplicate water-supply system was designed as a tunnel in the local shale for a distance of 990 m (3,300 ft) out from the lakeshore, finishing in a vertical shaft leading up to a series of precast concrete and steel pipes laid on the lake bed in a trench, at the ends of which are located the actual intakes, suitably screened. The location of the necessary control buildings for intake conduits from rivers or lakes adjacent to a watercourse or shoreline, will often introduce unusual geological conditions in the foundation beds that have to be used.

The abstraction of water from a river may be undertaken in a manner similar to that followed in Toronto; in general, Montreal and London obtain their supplies in this way. In some instances, however, when a river is bordered by porous beds of sand or gravel, advantage is taken of this geological feature, and the water is abstracted through the porous stratum and not directly. This method has many advantages, the most important of which probably is that the water is naturally filtered by its passage through the porous stratum; the use of artificial, slow sand-filter beds in other water-supply schemes is a special adaptation of this process. Many such installations are to be found on the continent of Europe on some of the highly industrialized rivers. The cities of Berlin and Frankfurt in Germany and Göteborg in Sweden obtain at least a part of their water supplies in this unusual way, as do some cities in the great Ruhr industrial area of Germany. Hamburg, Germany, in addition to obtaining water that has filtered through from a riverbed, uses two irrigating channels as additional "distributors"; river water is pumped into the channels which are located to the north and south of the boreholes from which the city supply is finally obtained.[28.12] The city of Breslau, Germany, has used a similar practice. The devices used to abstract the water from the porous strata vary, but in some cases they are similar to the well points used on construction projects to dry up water-bearing ground. Thus, Düsseldorf, Germany, uses borings 21 m (70 ft) apart with gravel-packed suction pipes 30 cm (12 in) in diameter, the yield of each being about 110,000 lph (28,800 gph).[28.13]

This interesting method is not confined to Europe. The city of Des Moines, Iowa, has long obtained its water supply in this way from the Raccoon River which flows through a valley about a mile wide consisting of impervious material with a superficial deposit of water-bearing sand. Collecting galleries consisting of reinforced-concrete pipes 120 cm (4 ft) in diameter are located parallel to the river bank and about 60 m (200 ft) away from it. They are located in this way in order to obtain fairly uniform percolation; a minimum good-quality supply of over 1 mld (300,000 gpd) is obtained (Fig. 28.2).[28.14]

A somewhat more unusual installation of this kind is that which supplies the city of Kano in northern Nigeria, Africa, an important center with a population of over 100,000.

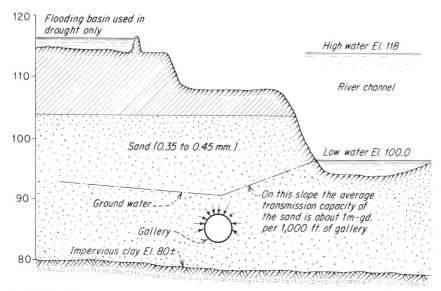

FIG. 28.2 Diagrammatic cross section showing intake arrangements for the water supply of Des Moines, Iowa.

Prior to the construction of new waterworks, the necessary supply had been drawn from wells dug in the local laterite. As this was unsatisfactory, a new installation was completed. Five large intake "wells" in the form of reinforced-concrete cylinders, three with internal diameters of 2.7 m (9 ft) and two with internal diameters of 4.5 m (15 ft), respectively, were sunk by grabbing from the interior to depths of between 10.5 and 15.6 m (35 and 52 ft). The wells rest on rock and penetrate the sand which constitutes the local riverbed; they are located in the riverbed over a stretch of about 3 km (2 mi) (Fig. 28.3). Intake pipes of 10-cm (4-in) diameter are installed in two of the wells and porous concrete blocks in three; these are located near the rock surface through which water is drawn from the surrounding sand; the sand acts as a filter and also as a storage reservoir during dry seasons. A daily flow of over 2 million L (600,000 gal) was obtained.[28.15]

To include here reference to a water supply from groundwater flowing beneath a dried-out riverbed may appear to be stretching the title of this section rather far, but the inclusion is literally correct. A somewhat similar example from North American practice is to be found in Virginia. The city of Harrisonburg was led to the consideration of augmenting its water-supply system on account of a water shortage during the great drought of 1930. Surveys and investigations finally led to the conclusion that the supply could be obtained from groundwater flow in the valley of the Dry River, from the normal streamflow of which the city already drew its main supply. It was found possible to extend an existing concrete diversion dam across the riverbed for a distance of about 270 m (900 ft) through a relatively level part of the bed of the valley and as far as a steep rock cliff. This necessitated a thin, reinforced-concrete wall, varying from 30 to 60 cm (12 to 24 in) in thickness and keyed into the underlying fine-grained, closely cemented Pocono sandstone rock for a distance of 40 cm (16 in). The entire wall was backfilled on the upstream side with selected stone, obtained to some extent from excavation. At the location of the lowest level of the rock, a suitable concrete collecting gallery was constructed upstream of the dam, from which a 35-cm (14-in) supply main leads to the city. The valley bed consists of products of disintegration and erosion varying in size from fine sand to large boulders; and in this

28-10 CIVIL ENGINEERING WORKS

FIG. 28.3 Sinking intake well No. 3 for the water supply of Kano, Nigeria, in dried-out riverbed.

deposit, 12 underground streams in definite courses were encountered; these streams carried such an amount of water as to cause difficulties during excavation. This flow is now permanently collected in a perforated concrete pipe laid under the stone backfilling and connecting with the collecting gallery (Fig. 28.4). An average flow of about 3,600 lpm (700 gpm) is obtained from this unique installation, which has performed satisfactorily since its completion.[28.16] A similar idea has been used in southern France.[28.17]

28.6 WATER SUPPLY FROM IMPOUNDING RESERVOIRS

When water for public use cannot be obtained directly by abstraction from river or lake, an obvious alternative is to utilize a distant source of supply such as the streams found in mountain areas, impounding them artificially and conveying the water thus obtained by means of aqueducts to the cities for which it is required. Although these schemes are always bold engineering conceptions, they are not a development of modern engineering, as a glance at the records of the Roman aqueducts will make clear. These ancient water-supply systems have now been eclipsed in size and conception by those of modern times, but they will always retain their interest. Water is conveyed today for city use in aqueducts hundreds of kilometers long; that for the supply of Los Angeles, California, brings

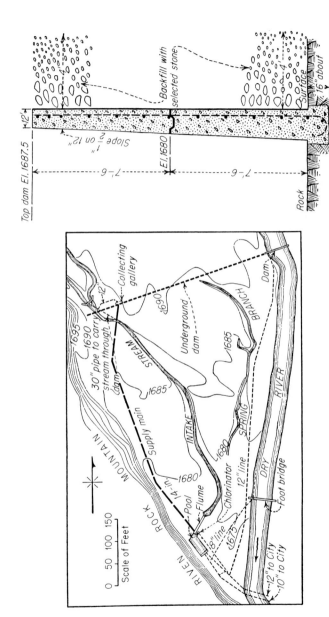

FIG. 28.4 Underground dam for the water supply of Harrisonburg, Virginia.

28-11

water 400 km (250 mi) to the city from the Owens Valley, 70 km (43 mi) of this being made up of 142 tunnels of varying length. This installation is but one of a distinguished group of major engineering enterprises of this type, which in the United States include the Hetch Hetchy scheme supplying San Francisco and the Catskill and Delaware systems for New York; in Great Britain they include the Vyrnwy scheme supplying Liverpool, the Rhyader supplying Birmingham, the Haweswater for Manchester, and those that supply other inland cities of England.

A typical water-supply scheme of this type consists of a catchment area, part of the rainfall on which is impounded behind a storage dam. From the reservoir formed by the dam, water is led away through an intake structure into the main aqueduct in which it is conveyed to a point on the city distributing system, generally a storage reservoir. It will be seen that the engineering problems introduced by the works mentioned are not peculiar to water-supply engineering; all of them are important parts of many water-power schemes, for example. Insofar as the relation of geology to the works indicated is concerned, no unusual features are introduced by their use for water supply purposes. The problems that they introduce can therefore be treated generally.

Although the engineering works for water supply derived from distant sources introduce no unique problems in relation to geology, individual examples of this type of project have provided the engineer with most valuable information on the general geological questions then introduced. Especially notable was the work of this nature during the construction of the Vyrnwy Dam for the water supply of Liverpool, England. The Ewden Valley works, also in England, have similarly been mentioned as providing most valuable information to engineers because of the geological conditions encountered. These and other English and European water-supply systems have involved works of great magnitude, but some of the North American water-supply schemes are far larger and may well be briefly mentioned.

The Catskill system for the city of New York, to which extended reference has already been made, is one such example. It includes two great tunnels, one going under the Hudson River 366 m (1,200 ft) below water level, and two great dams; the close study of geology made in connection with this and other associated works was to a large extent pioneer work in this field. The Board of Water Supply for the City of New York continued to correlate geological study with its contract work and applied this particularly to the new Delaware River system, to supplement still further the water supply of the great metropolitan area of New York. The Delaware River aqueduct consists of 137 km (85 mi) of tunnel, in six sections, the longest of which is 72 km (45 mi) long; the predominating rock types are shales, slates, sandstones, gneisses, and schists.

San Francisco receives its water supply from an impounding reservoir system of great magnitude; the Hetch Hetchy water supply and power scheme was put into service in 1934. Water is obtained in a catchment area in the Yosemite National Park and conveyed to the city through an aqueduct 250 km (155 mi) long, 130 km (82 mi) of which consists of tunnels, the remainder in pipelines. Several dams are also included; the largest is the O'Shaughnessy (impounding) Dam, which is 129 m (430 ft) high. It is founded on granite which was seen to contain many large and deep potholes when uncovered by excavation. The tunnels were driven through varying rock strata, including granite, slate, and recent sedimentary formations. In these last, much trouble was encountered because of "quicksand" and excessive pressure from chlorites, methane gas, and sulfureted hydrogen; an explosion of methane was the cause of a fatal incident. Despite all difficulties, the tunnels were completed; in some the granite rock was so satisfactory that the tunnels are in use without concrete lining.[28.18]

The Colorado River Aqueduct of the Metropolitan District of Southern California, which conveys water a distance of 385 km (240.5 mi) from the Colorado River to the Los

Angeles area for the use of the 13 cities which constitute the metropolitan district, is an even greater project. The aqueduct was designed for a peak flow of 45.5 cms (1,605 cfs), or approximately 40 million mld (1,000 mgd). There are five pumping stations having a total lift of 485.4 m (1,618 ft). The aqueduct consists of 147 km (91.9 mi) of tunnels, 100 km (63.4 mi) of open canal, 89 km (54.7 mi) of conduit, and 47 km (28.9 mi) of siphons. Excavation in the aqueduct totaled about 19 million m^3 (25 million yd^3) and 2.7 million m^3 (3.5 million yd^3) of concrete were used. Of the total tunnel footage, 58 percent was supported by structural lining; 14 percent was coated with gunite; and the remainder was unlined.

These figures will give some idea of the magnitude of the project, but its unique nature will fully be realized only when it is considered that the eastern part of the aqueduct is located in an arid desert region, construction operations in which could begin only after the building of 237 km (148 mi) of surfaced highway, 750 km (471 mi) of power lines, and 290 km (180 mi) of water-supply lines. The aqueduct passes through the desert as a concrete-lined open canal, and the lining is watertight, as the surface of the ground consists generally of alluvial deposits. It crosses three earthquake faults which are classed as active; flexible pressure pipelines across these locations have been considered desirable. Excavation for the Parker Dam, the impounding reservoir from which the supply of water is obtained, and the construction of the tunnels were affected by unusual geological conditions. (See Fig. 28.5.) It is not possible to describe these works in detail, but, as is the case so fortunately with many leading civil engineering projects, full descriptions are available in technical journals to which reference can be made by those interested. In these descriptions repeated references are made to geological features that affected the work.[28.19]

FIG. 28.5 Eagle Mountain pumping plant on the Colorado Aqueduct, California, which lifts water 131 m (438 ft) out of the total lift on the aqueduct of 485 m (1,617 ft), showing typical terrain traversed by this water-supply system.

Even in such a relatively small country as Great Britain, water-supply systems interconnecting different river basins have now been necessary. The outstanding example is the Kielder project in northeastern England, an undertaking of the Northumbrian Water Authority. It was conceived as an answer to steadily growing water demands in the area served by the Authority, a rise from 635 mld (140 mgd) in 1961 to 910 mld (200 mgd) in 1971 suggesting a possible demand of 1,440 mld (317 mgd) by 1981 and 1,970 mld (433 mgd) by the turn of the century. Careful studies suggested that one large storage facility would be more economical than five or six smaller reservoirs, even though the one reservoir necessitated a means of transferring water between the three river basins involved. The Kielder Dam, which will be western Europe's largest earth-fill dam, was therefore designed. Its start and the progress of construction were interfered with far more by environmental concerns, voiced at two bitter public hearings, than by any geological factors, although the dam site and the source of fill material naturally involved the usual close attention to geology. The transfer conduit is a 38-km (24-mi) -long tunnel, 2.8 m (9.5 ft) in internal diameter. It starts at the river Tyne 24 km (15 mi) upstream of the city of Newcastle upon Tyne and runs generally south, linking the Tyne watershed with those of the rivers Wear and Tees and passing under the river Derwent, a tributary of the Tyne. Supplies will therefore be adjusted to demand over this large, generally industrial area. Even though the project is not complete as this Handbook goes to press, enough has already been published to show that the final published records will add a notable chapter to the record of water-supply engineering and to the influence of geology on water-supply undertakings.[28.20]

Unavoidably, these references to water-supply schemes involving the use of reservoirs and interconnecting conduits have had to be brief in the extreme, so varied are the works involved and so many the associations of such works with geology. It is, therefore, almost invidious to have to mention in the same staccato manner the greatest of all such projects, the California Water Plan. Despite the consistent operation of the major California water-supply projects already mentioned, further supplies were necessary for the great urban populations in the southern part of this large state. Over 70 percent of the natural streamflow occurs in the northern third of the state, and yet the great demand for water is in the south. Accordingly, in 1951 the state legislature authorized the construction of the California Water Project, by the Department of Natural Resources, to bring northern water in excess of local needs to the south. Later a special Department of Water Resources was established to be responsible for this and other water concerns of the state. In 1960 a $1.75-billion bond issue was approved by the voters. The total project cost amounted to $2.3 billion. Construction occupied more than the whole decade of the sixties (Fig. 28.6).

Most parts of the vast undertaking were operational by 1973. These included 1,075 km (684.5 mi) of aqueducts, mostly in canal but with 33 km (20.5 mi) in tunnel; power plants, operated by falling water in the aqueducts, capable of generating over 5 billion kWh; pumping plants requiring for their operation almost 13 billion kWh; and no less than 21 major dams with a total volume of over 204 million m^3 (267 million yd^3) of material, the Oroville Dam alone containing 61 million m^3 (80 million yd^3) of soil. Since the area in which these many works had to be constructed includes some notable faults and some active seismic regions, design studies have added much to knowledge in these two fields. Construction practices also presented many notable advances, one of the most unusual being the flooding of ponds in the San Joaquin Valley in order to induce consolidation of the unusual soils there in advance of canal construction. The literature of civil engineering and of engineering geology has been enriched by the records already published of this outstanding water supply system, named as the "Outstanding U.S. Civil Engineering Achievement of 1972" by the American Society of Civil Engineers. Merely an introduction is given in the references noted.[28.21]

FIG. 28.6 Outline plan of the main features of the California Water Plan.

28.7 WATER SUPPLY FROM GROUNDWATER

Wells and boreholes constitute the principal means of obtaining a supply of water from underground sources. They may be considered together, since they are similar in principle and action. As a matter of convenience, wells are generally used for small depths and boreholes for greater depths. Unless they pass through very stable strata, wells (using this term to denote both types of hole) will be cased in some way throughout the depth of all strata, except that from which the supply is to be obtained. Occasionally wire screens are used to protect that part of the well into which the water is being drawn. Another common precaution is to fill the excavated hole with specially selected coarse gravel through which the discharge pipe is led. This may decrease the yield of the well by about 5 percent, but it will prevent plugging.

In wells that penetrate strata having a low specific yield, it may be necessary to

28-16 CIVIL ENGINEERING WORKS

attempt to secure a greater exposed surface in the pervious strata than is given around the perimeter of the well shaft. This is easily effected by driving adits into the strata from suitable locations in the well. It will generally be a wet construction operation; but it has the advantage that as soon as water problems become too troublesome, the object of the adit is achieved. The Romans used adits in wells that they excavated in Great Britain; today, adits are quite common in wells in the Chalk of southern England and in the Bunter sandstones. In the Canary Islands adits alone are sometimes used as water-collecting galleries; they are driven into steeply sloping lava beds in order to obtain the water that seeps through this porous rock. The *karez* of northern India and the *keghriz* of Iran are primitive types of a special kind of adit of considerable interest but restricted application (Fig. 28.7). The planning and building of these unusual collecting galleries appear to be a matter of intuition rather than calculation, and the work is generally hereditary. Soviet engineers working in the Caucasus have tried to construct *keghriz* scientifically but without success.[28.22]

Fortunately, the usual type of well installation does not call for anything beyond ordinary scientific skill. When the excavation has tapped the water-yielding stratum, a body of relatively free water will be available which can be drawn upon as required. The level of this water may vary from time to time quite apart from the variation due to pumping. When the water table is near the surface, barometric pressure may affect the level, and even temperature has been found to have some effect because the capillary properties of the strata above are affected by a change in temperature. Water levels in adjacent wells may not coincide if there is a marked change in geological conditions between them. A striking example occurred at St. Andrews, Scotland. A farm drew its supply from a borehole 8.4 m (28 ft) deep in sands and gravel. An architect sank a well 18 m (60 ft) from the borehole in order to supply some new cottages; only a trace of water was found, although

FIG. 28.7 *Karez* near Quetta, Pakistan.

the well was sunk to a depth of 9 m (30 ft) in the gravel. This was in all probability due to variation in level of the underlying impervious boulder-clay floor.[28.23]

When pumping is started in a well that has been satisfactorily completed to a free-water surface, the water level will naturally fall and continue to do so until the hydraulic gradient in the surrounding material is such that flow will take place to the well to equalize the amount withdrawn. It can be shown theoretically that for uniform material, the curve of the gradient is a parabola. As this will be the same all around the well, the phenomenon known as the *cone of depression* is obtained. The intersection of a cone of depression with the water table will trace out a circle which will mark the range of influence of the pumping operation. It will be clear that the effective yield of any other well located within this range will be interfered with to some extent. Such interference is a critical matter in built-up city areas where groundwater resources are being taxed to the limit. It is surprising to note how wide these ranges of influence may be in pervious strata. In Liverpool, England, the effect of pumping has been noticed as much as 3 km (2 mi) away from the well in which pumping was taking place; the intervening ground was largely pervious red sandstone.

The wells so far considered have been those which are ordinarily encountered, having no unusual geological features. In addition, not only artesian wells but also a few special types of wells may be noted briefly; although they are not of common occurrence, they are often of great importance in some localities. There are, in the first place, shallow wells which depend for their supply on a *perched water table,* i.e., a body of groundwater perched high above the zone of saturation by means of an impermeable stratum overlying the main body of porous material. Some striking examples are found on Long Island, New York, where a thin layer of almost impermeable till separates a pervious surface deposit from the main body of pervious strata, and the main water table may be as much as 75 m (250 ft) below this perched water.[28.24]

Australia provides numerous examples of two other types of unusual wells. *Soak wells* is the name that has been given to the shallow wells used for obtaining groundwater that collects, after running off exposed sloping surfaces of impervious weathered granite, in the accumulation of decomposed or disintegrated granite at the foot of these slopes. The products of decomposition vary and often yield quartz grains in a matrix of kaolin; from such a deposit no water supply can be obtained. But if the borehole is continued, a water supply will often be obtained from the body of water trapped between the bottom of this deposit and the top of the completely fresh granite; the transitional space is often open enough to yield water itself and to serve as a drain for the overlying kaolin.

The second type of unusual well encountered in parts of Australia (and elsewhere) is known as a *tray well* (Fig. 28.8). This is really a case of tapping a localized body of perched groundwater, a body of water perched not on a continuous impermeable stratum but on a lenticular bed of impervious material often of limited extent. A good example is furnished by a well at Challner, where a saucer-shaped bed of impervious clay with some ironstone nodules about it, located halfway down through a deep stratum of sand which overlies solid granite, collects a small quantity of freshwater as it percolates through from the surface. This supply is tapped by the Challner Well; endeavors were made to increase the supply by boring further, but the bed of clay was pierced and the freshwater started to empty into the sand below. Fortunately, the well was saved by plugging the hole. Three hundred and five meters (10 chains) away from this well is another, known as Nugent's Well, the bore of which encountered no water above the thin clay layer, so that it had to be extended to the main body of groundwater overlying the granite directly. This second well is also unusual: it cannot be extended right down to rock surface because the water immediately above the rock has a strong saline content, which makes it quite unsuitable for domestic use. Only the thin layer of freshwater that tops this saltwater layer can be

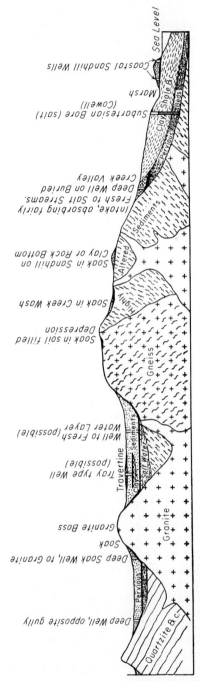

FIG. 28.8 Idealized diagrammatic geological section for the counties of Jervois, Buxton, and York, South Australia, showing the formation of tray wells, and the dependence of these and other water supply features upon geological structure.

tapped for use.[28.25] Salt water may be present for two general reasons: (1) encroachment of seawater into wells located near the seacoast and (2) contamination of inland well waters with excessive mineral content. In both cases, as soon as the salt content passes a certain limit, the water ceases to be fit for human use.

The contamination of inland groundwater supplies with excessive salt content, such as is found in parts of Australia and in the Sudan, has in the past been attributed to the leaching out of decomposition products of overlying or adjacent rocks. Consideration of the chlorine content of samples investigated in South Australia disproved this supposition and pointed to the fact that saline content is due to insufficient drainage for carrying off the salts brought down by rain. Continued evaporation of water gradually concentrates these dilute salt solutions, with the results already indicated. (Decomposition products probably contribute to the actual salt content found.) As evidence of what rainwater may contain, it may be mentioned that in England the average chlorine content of rainwater is about 2.2 ppm, equivalent to 3.62 parts of salt. Rain falling at Land's End during strong southwest winds blowing in from the sea has been found to contain one hundred times this amount.[28.26] As can be imagined, the problem is a complex one and is further complicated by varying local conditions. The greatest care must always be taken to avoid driving a well into the saltwater layer; and in the case of correctly driven wells, care must be taken to prevent the possibility of overpumping. This, when it occurs, introduces what can only be called "man-made geological problems," which are considered in Chap. 48.

There is little need to describe any of the usual borehole-pumping installations which supply water so economically to so many municipalities. Two unusual developments in this field, however, will serve to illustrate the complete dependence of all underground water supplies upon the local geology, as is also shown in Chap. 8. The Province of Saskatchewan embarked in 1963, jointly with the government of Canada, on a study of the province's groundwater resources. An area of 259,000 km^2 (100,000 mi^2) in the southern part of Saskatchewan had been studied in detail by 1970. By deep drilling combined with geological study, the team of investigators, led by E. A. Christiansen, discovered a number of buried valleys previously unsuspected. These are the same preglacial valleys which have been frequently mentioned in other chapters, but in the prairies they are located nowhere near the surface, being found at depths such as 150 m (500 ft) below the surface. As is usual, the valleys were found to be filled with glacial deposits, some till, but much sand and gravel, and this was water-bearing.[28.27] The discovery, therefore, was not just of buried valleys but of buried rivers which can eventually be tapped as sources of water supply. It has been estimated, for example, that the buried watercourse (sometimes known as the Ancestral Missouri) found in the Estevan Valley could supply an estimated 200 to 400 mld (50 to 100 mgd) from several wells; this would be worth at least $10 million. Water quality is such that treatment would be needed before use, but when water is scarce, this is a small penalty to have to pay. Observation wells have now been installed at almost 50 locations, and Saskatchewan will have eventually a very clear picture of its deep underground water resources.

The second special case is the latest development in water supply of London, England, one of the best-known and oldest-established systems for any major city of today. The city obtained its water, at first, from the great aquifer provided beneath it by the Chalk of the London Basin. When additional supplies were needed, water was abstracted from the river Lea and the river Thames, the latter now providing the major part of the metropolitan supply. All who fly in to London's Heathrow Airport fly right over some of the impounding reservoirs necessary for balancing demand with supply. Still more water is necessary, but the Metropolitan Water Board is not allowed to take any further water from the Thames, above Teddington Weir, when the riverflow is less than 643.5 mld (163 mgd), and this means usually for much of the summer. Instead of building still more reservoirs to hold water throughout the months of low flow, the Thames Conservancy (responsible for water above the weir) embarked on an experimental program of utilizing groundwater to augment low water flows.

The Conservancy selected an area of 1,000 km^2 (386 mi^2) in Berkshire in which flow two small streams, the Lambourne and the Winterbourne. After most intensive preliminary geological study, and with arrangements made for a continuing ecological study of all effects at the surface, a three-year program was started; the program involved pumping groundwater into the streams at times of low flow in the river, and so of low rainfall (Fig. 28.9). Preliminary calculations had suggested that the water table, thus lowered, would probably be restored to its normal elevation with the advent of winter rains, in view of the permeability of the Chalk underlying the streams. Because of the low water table when the rain came, some of the water from heavy rains would percolate into the ground to replenish the groundwater supply, instead of all the excess flowing down the streams. It will be seen, therefore, that the hydrological cycle is not interfered with, except in the timing of input to and discharge from the groundwater reservoir beneath the streams. It was estimated that over a ten-year period a net gain in the flow of the river Thames of up to 378.5 mld (100 mgd) at the end of a severe dry period could be obtained in this way, at an estimated cost of from one-third to one-quarter that of the conventional balancing reservoirs.[28.28]

Even from this capsule account of this inspired concept, it can be seen that success of the experiment was entirely dependent upon the most detailed investigation of the underlying geology, even to extensive sampling of the Chalk in order to check its permeability so that it could be determined whether the water it held was stagnant or moving. The experiment was quite successful; the scheme is being implemented; and a similar scheme is already in operation in the catchment area of the Great Ouse River in eastern

FIG. 28.9 Thames River Valley, England; pumping station for experimental recharging of tributary streams to the river Thames.

England.[28.29] When the local geology is suitable, this concept of adjusting the timing of the flow into and out of the ground to increase low flows in rivers used as a source of water supply is one that could be widely adopted with consequent great economy and no deleterious effects on the environment.

28.8 REPLENISHMENT OF GROUNDWATER

All over the world, water supplies for human use are being obtained from groundwater, from springs when they exist, from wells when the water table is close to the surface, and by pumping when the water table is at such a depth that water cannot otherwise be obtained. And all over the world, there are today all-too-many cases where there has been excessive pumping, generally described as "over-pumping," with a variety of serious consequences, such as seawater intrusion at locations on seacoasts and subsidence of overlying ground, problems which are dealt with elsewhere in this volume. Most of the water pumped out of the ground for human use is disposed of as "waste." It is a natural concept that if some of this waste could be reclaimed and returned to the ground, it could be reused and a better balance thus be obtained in the use of the great natural resource which groundwater represents. Excess surface waters might be used in the same way. This has been done, now to a steadily increasing degree, in many parts of the world.[28.30]

It would appear that the first attempts to replenish groundwater were made in 1881 at Northampton in England; water was fed back in this case into the Lias limestone. Experiments were conducted in the London Basin by the East London Water Company as early as 1890. In North America, the Denver Union Water Company may have spread water over the gravel cone of the South Platte River near the mouth of its canyon in 1889.[28.30] It was not until well into the present century, however, that the need for replenishment of groundwater became so urgent that major schemes were investigated and ultimately put into operation. Los Angeles, with its great concentration of population, has been the center of much activity in this field. Experiments were first started in the San Fernando Valley as early as 1931, and spreading grounds were gradually developed that could affect one-third of the groundwater supply for the city of Los Angeles. After five years' successful operation, the system took on a definite character. At the end of this period, of a possible 73 ha (180 acres), 19 ha (42 acres) were in use for this purpose, through which 3 cms (106 cfs) of water were fed to the valley bed. Basins 30 by 120 m (100 by 400 ft) were separated by dikes 1.2 m (4 ft) high and 4.5 m (15 ft) wide. To each basin a 15-cm (6-in) pipe supplied water to a depth of 15 cm (6 in) during filling operations. The water was then left to stand and seep into the ground; this took between 8 and 24 hours, depending on the state of the beds, which had to be harrowed and cleaned of silt at intervals of ten days. Percolation went on at the rate of between 0.9 and 3.0 m per day (3 and 10 fpd), depending on the state of the beds. The water reached the North Hollywood pumping station, some distance down the valley, one year later and was there pumped out for use through eighteen 50-cm (20-in) wells 120 m (400 ft) deep.[28.31]

Since this early work, great progress has been made; today the Los Angeles area probably has more examples of groundwater recharge than any other area in the world (Fig. 28.10). Because of the continuing serious depletion of groundwater supplies, estimated to amount to 865 million m³ (700,000 acre-feet), a special water-replenishment district has now been created to coordinate many local efforts. The area includes 29 cities with a combined population of 2.5 million people. Costs will be met by levying pumping assessments on water companies and large water users. Between 50 and 75 new wells have been drilled for recharging purposes along 17.5 km (11 mi) of coastline as part of the "Barrier Project" designed to limit further encroachment of seawater into this valuable area.[28.32] Some prog-

FIG. 28.10 Restoring water to the land in California; the Rio Hondo spreading grounds for groundwater recharge of the Los Angeles Flood Control District, near Pico Rivera.

ress has been made also in the Los Angeles area in the use of reclaimed sewage-plant effluent for recharging groundwater supplies; naturally this operation is under strict control.[28.33]

The city of Amarillo, Texas, has similarly improved its groundwater situation by recharging from the surface. The practice is, however, not restricted to the United States. In Europe, too, falling groundwater levels have created similar needs. In Sweden there are 20 municipalities that obtain their water supply from underground sources now supplemented by surface recharging. And in London, England, the Metropolitan Water Board has used old shafts and adits to put water back into the Chalk that serves as such a useful underground reservoir for the great metropolitan area. This program was operational long before the groundwater pumping project described in the last section. In this particular case, it was possible to measure the amount of the recharged water that was obtained by later pumping. Only about 40 percent was so reclaimed, although in other recorded cases full return of all recharged water has been obtained. Interesting administrative questions are raised by the missing 60 percent of the London water, but they may be left for disentanglement by legal authorities[28.34] The important thing is that some of the debt incurred by a too-profigate use of one of Nature's great bounties is at last beginning to be repaid.

It is believed that the recharging of groundwater reservoirs is today being practiced to some degree in every state of the United States. In some cases recharge is achieved by flooding areas that are underlain by permeable soils so that percolation down to the water table takes place directly. In others, the same result is achieved by discharging water into wells, whence it makes its way through permeable soil or rock to the main body of groundwater. In some cases "raw water," as from riverflow, is used for recharge; in some, storm water which is collected separately from sewage is used; and in an increasing number of cases, treated sewage is approved for use by the appropriate public health authorities, after all requisite testing has been done, with assurance that it is suitable for the purpose. With the increasing attention today being given to the necessity for all sewage to be treated adequately (even though this may be done in sewage-treatment plants disguised by such modern names as "environmental protection installations"), there will be correspondingly an increasing quantity of safe wastewater available for groundwater recharge.

Indicative of the even broader attention now being given to this matter is the fact that the Denver Board of Water Commissioners, after citizens had rejected a $200-million bond issue intended for major extensions to their transmountain supply system, turned to the idea of extending the work they had already done on reclaiming water. They had previously experimented with direct use of purified sewage effluents after additional treatment involving three further processes—dual filtration using activated carbon, chemical treatment by adding lime or alum, and biological treatment by nitrification. The possibility of using such water for industrial purposes, and diverting potable water, previously used by industry, for human use, was seen as a real possibility. Accordingly, if to these processes one adds the natural filtration that percolation into the ground will give to water thus recharged, the idea of using purified sewage effluent, an idea which to some seems revolutionary at first, takes on a quite practical aspect.[28.35]

The city of Chanute, Kansas, under strict control of the state's health department, was one of the pioneers in this conservation of water after the drought of 1952–1957, at the end of which their water-supply river, the Neosho, ceased to flow.[28.36] The town of Minot in north-central North Dakota, has an interesting recharging scheme in operation, a small dam having been constructed across the Souris River in order to divert some riverwater into a special basin and canal system, in the bed of which gravel-packed, 76.2-cm (30-in) -diameter wells were installed, through which the water percolates into the underlying Minot aquifer. Total cost of the entire installation was less than $200,000, as compared with an estimated $12 million for one of the other alternatives, a supply pipeline from the Garrison Reservoir.[28.37]

On a vastly greater scale, the country of Israel has in operation a major project for the use of treated sewage to supplement their quite critical national water supply. The initial wastewater reclamation scheme was expected to add 12 percent to the total supply. Special lagoons were constructed, the treated water being allowed to percolate into the ground, being reclaimed (after an estimated stay of one year in the ground) by pumping from wells (Fig. 28.11). Naturally, the engineering details of this installation are complex, with the most stringent controls in the interest of public health, but it is mentioned thus briefly to demonstrate how extensive the practice of recharging groundwater, and the associated use of "wastewater," has become.[28.38]

Finally, mention must be made of one of the most enlightened multipurpose developments involving groundwater recharge of which the writers have knowledge. It relates to Coors beer, well known in the western part of the United States. The Coors brewery is located to the west of Denver, on Clear Creek in the outskirts of the municipality of Golden. Springs on the large property of the Coors Company supply the water for making their well-known beverage. Much more water is required in their plant for industrial purposes, however, and this is obtained from Clear Creek. The bed of this mountain-fed stream consists in this area of deep deposits of gravel, 18 m (60 ft) deep in places, which were found to be gold-bearing. Most of the gold was garnered from it many years ago, the

FIG. 28.11 Spreading ground for the infiltration of treated effluent for recharging groundwater, Tel Aviv, Israel.

gravel now proving its further worth as concrete aggregate. Large gravel-mining operations are to be found adjacent to the creek, for which a master plan of reclamation and development has been prepared. Immediately below the brewery, gravel is being mined under strict control so as to leave carefully shaped shallow pits. When available gravel has been removed, these pits are allowed to fill with water from the creek and so form attractive looking small lakes. The water percolates through the gravel whence it is reclaimed for plant use after this period of natural filtration (Fig. 28.12). The combination of gravel

FIG. 28.12 General view of the way the Adolph Coors Company is using worked-out gravel pits as spreading grounds for water recharge near Golden, Colorado.

mining, water conservation, and development of public recreational facilities in this area is a vivid example of what can be done when cooperation between different responsible agencies can be ensured, with farsighted policy making on the part of management, all based on a thorough knowledge of the quite unusual local geology.[28.39]

28.9 SOME UNUSUAL SOURCES OF SUPPLY

Brief reference may be made to one or two unusual sources of water supply which have been adopted in some parts of the world. The supply to the community living at the foot of the Rock of Gibraltar is one such example. The great rock consists of a hard and compact Jurassic limestone which provides no natural supply of springwater. For strategic reasons, the potable water supply has had to be obtained within the fortress area, and so the supply was limited for many years to the rainfall on the surface of the rock itself. This is steeply sloped on the east side, and the town is located on the west; a catchment area of 10 ha (24 acres) was constructed on the east side, using corrugated iron sheets laid on 40° sand slopes. The sheets were secured by means of timber purlins and piles creosoted under pressure. An additional catchment area of 6 ha (15 acres) was secured by utilizing steep rock slopes on the west side (Fig. 28.13). Rockfalls frequently damage the east catchment area, but because of the careful watch that is maintained, trouble is quickly located; washouts of considerable areas have nevertheless occurred. An additional supply of potable water was found by means of test borings sunk in the sand of the flat tableland to the north of the rock. Two wells were later dug and these supply about half of the total consumption. The brackish water supply to the town for all purposes other than drinking is obtained from shallow wells in the same area; it is kept separate from the potable supply. All drinking water is stored in reservoirs tunneled out of the rock; these have a total capacity of 63.5 million L (16.8 million gal); the reservoirs are generally in the form of galleries, leading off from two main construction adits, driven from west to east at a level of roughly 112 m (375 ft) above sea level. An additional corrugated-iron catchment area of 4 ha (10 acres), together with two new reservoirs inside the rock, has been added to supplement the existing water supply for shipping.

Fig. 28.13 The "ironclad" catchment area on the Rock of Gibraltar.

This interesting example shows how geological conditions distinctly unfavorable to the type of water-supply system that might have been expected (springs, wells, or other methods of underground extraction) have been adapted to give an almost perfect, although miniature, example of a surface water supply. It is of interest to note that the water thus obtained at Gibralter is used without treatment, and although it is rather "flat," it is quite palatable and consistently safe.

In parts of Australia, "ironclad" catchment areas are sometimes used. These are extensions of the "100 percent" catchment areas provided by such impermeable surfaces as the roofs of farm buildings. Corrugated iron sheets laid on ground specially graded to give a convenient fall to special storage tanks have provided a large number of ironclad catchments at relatively low cost. All the scanty rainfall in these locations is thus obtained for use; an area of 2,400 m² (26,000 ft²) has been found sufficient for the average farm.[28.40]

The use of corrugated iron can hardly be classed as a geological feature; but in other parts of Australia (notably Western Australia), a similar result is achieved by the use of rock catchment areas. At some places outcrops of granite are remarkably solid and so shaped that small concrete walls can be constructed around them in such a way that all rain falling upon them can be led to suitable storage tanks. Sometimes, holes excavated in the granite are used as the storage tanks.[28.41] In areas where no solid rock is available for use in this rather unusual way, Australian engineers have ingeniously waterproofed catchment areas whose surface consists of sandy soils. Attempts to render the ground surface impermeable and to promote full runoff of the sparse rainfall were first made in 1935 in Western Australia; emulsified bitumen was used on carefully graded natural soil. The experiments proved successful, and a small catchment area located at Narrogin in Western Australia was then treated in this way. Two areas of 30 and 18 acres, respectively, were treated by carefully rolling the cleared soil and then sealing the soil's surface with 2.2 L/m² (0.4 gal/yd²) of emulsified bitumen. Local rainfall averages 45 cm (18 in) per year of which only about 5 percent used to run off, to be retained in the previously constructed reservoir. Suitable drains and surface channels now ensure complete runoff from the "waterproofed" catchment to the reservoir where it is retained for public use.[28.42]

Another unusual type of surface water supply is provided by the dew ponds that are still found in Sussex and other parts of southern England and probably also in other parts of the world. Some of these dew ponds seem to be natural formations, possibly finished off to some extent by the farmers who use them, but they are occasionally made artificially. The water deposited in them as dew is the result of condensation of the atmospheric moisture on the cold surface provided in the pond. The construction of a pond has been described as follows:

A basin-shaped hollow is excavated in an open space, well exposed to damp sea winds. The hollow is covered by a layer of straw and twigs, or other non-conducting material, about 18 inches thick. On this is laid a continuous bed of puddled clay, about 2 feet thick, which in its turn is covered by a layer of broken stone. The object is to provide a surface of stone and clay which rapidly grows cold at night, and the dew thus collected is caught by the layer of puddle clay, and conducted to the central pond. Such a prepared area, about 200 feet in diameter, under favourable conditions, will keep the pond in the centre about 20 feet in diameter, and say 3 feet maximum depth. So far as can be gathered, in default of systematic observations, the yield is about 0.01 inch per night during the summer, over the prepared area.[28.43]

The geological aspect of this source of supply is concerned with the provision of an impermeable basin, and the surface deposits of clay found in the south of England are well suited to this. The ponds illustrate vividly the direct relation of water supply to atmospheric moisture; the water thus obtained so directly is of admirable quality.

The city of Juneau in Alaska has been built on a small coastal plain on Douglas Sound, a section of the 1,600-km (1,000 mi) -long inland passage that distinguishes the northwest coast of North America. The restriction of the site by mountains is well shown by the fact that, even in this isolated location, a large steel bridge has had to be constructed across the channel to give access to building sites on Douglas Island. Water supply for the city was obtained from Gold Creek well above the town and from wells in Last Chance Basin through which the creek flows. Cold-weather problems caused many difficulties, and so a new supply system had to be developed. A tunnel had been driven in 1898 by a mining company that intended in this way to drain the gravels of the Last Chance Basin in the expectation of discovering gold. Their efforts were defeated when the heading broke through into the gravels, the tunnel being flooded. Another company tried in 1902 to wash the gravels into the tunnel. Some success was achieved, but a flash flood in 1905 plugged the tunnel and brought operations to a halt. The tunnel was then abandoned and became clogged with debris, but it was still connected with the water-bearing gravels. Having been driven through sound basalt-gabbro rock, the tunnel itself was found to be quite sound when it was investigated as a possible conduit for water. The tunnel was cleaned out, a concrete connection with the water-bearing gravels was constructed with proper controls, and connection to the city reservoir was made at the lower end of the tunnel that now serves as the major aqueduct. Gravity flow from wells in the gravels is normally sufficient, but water can also be pumped when necessary. The large diameter of the tunnel results in low velocities, and so the rock around the tunnel acts as a heat exchanger, warming the water slightly from its initial temperature of about 5°C (41°F) throughout the year, thus eliminating the previous problems with winter freezing.[28.44]

28.10 SOME GEOLOGICALLY SIGNIFICANT SYSTEMS

So varied are the conditions encountered in the development of public water-supply systems that the foregoing general review of some of the influences of geology on all such systems may usefully conclude with brief descriptions of some projects which have featured unusual geological conditions or have involved unusual geological investigations for their success. To select such a short list is to invite criticism for omissions, but if the following examples will be taken merely as typical, they will at least illustrate how vital are geological studies to the prosecution of water-supply engineering.

London, England One has only to glance at a geological section beneath London (Fig. 8.4) to realize that few cities of the world have been so fortunate in their geological setting. Although in the past, wells into the Chalk have supplied an appreciable percentage of the city's water needs, today only about 14 percent comes from this source, about the same amount from the flow of the river Lea, and almost three-quarters of the total now used from the flow of the river Thames. The London Basin remains, however, as a truly remarkable urban geological phenomenon; its use is strictly controlled. For London's utilization of the surface waters, many outstanding works have been necessary—tunnels, embankments for reservoirs, replenishment facilities in the Lea Valley, and the most recent development just described.

Long Island, New York Long Island constitutes the coastal plain of the states of New York and Connecticut, separated from the mainland by Long Island Sound and isolated geologically from it with respect to its water-bearing beds. The island supports a population of several million people, and in recent years a large part of its water supply has been obtained from groundwater. Crystalline bedrock outcrops in some of the western

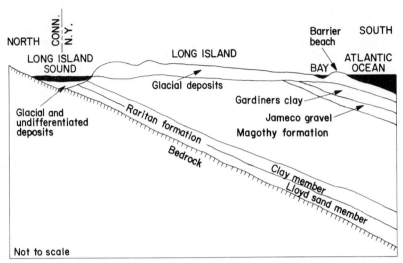

FIG. 28.14 Simplified cross section through Long Island, New York, indicating water-bearing horizons.

parts of the island, and the rock surface dips to the east so that in places it is at least 600 m (2,000 ft) below the surface. Overlying the rock is a great deposit of stratified Cretaceous sands, clays, and gravels; the dip of the strata decreases toward the surface (Fig. 28.14). At the surface, stratified sand and gravel drift deposits and two moraines give rise to a fairly regular surface topography; the highest ground elevation is about 126 m (420 ft) above the sea. These unconsolidated deposits constitute the great underground reservoir from which has been drawn probably the greatest concentrated groundwater supply utilized in North America. As the unconsolidated strata do not cross the sound to the mainland, the groundwater underlying the island is obtained only from the rain falling on it, and, consequently, the relation of this body of freshwater with the surrounding seawater is a delicate one. The rainfall averages about 105 cm (42 in) per year, of which possibly one-half reaches the zone of saturation. Opinions differ as to how much of this water can be obtained for use; but from the records available, it is known that over 750 mld (200 mgd) have been pumped for domestic and industrial use.

Unfortunately, this pumping has been concentrated in developed areas, and a serious situation with regard to saline contamination has been created. Under the older part of Brooklyn, the water table dropped to such a depth that water under Brooklyn became unfit for drinking purposes. Before this, however, the situation had come under strict control, with the result that by 1947 virtually all pumping of groundwater had stopped, Brooklyn and the area around obtaining all their water supply from the New York Board of Water Supply. Even though so much of the Brooklyn area was paved, natural intake of rain in the areas around led to a slow but steady restoration of original water levels. Even this most satisfactory consequence of controls raised some problems. This rising water table came back to an elevation above that which existed when some public works in Brooklyn had been built, most notably the Newkirk Avenue subway station, so that remedial work had to be undertaken. This experience illustrates vividly the interdependence of groundwater, local geology, paving of surfaces, and overpumping.[28.45]

The example is so important that reference may here be made to a notable publication of the U.S. Geological Survey and the New York State Water Resources Commission, an *Atlas of Long Island's Water Resources*.[28.46] This shows the interesting situation now existing on Long Island, from almost natural conditions at the north end of the island, through partial paving of surfaces and some pumping, to the recent restoration of natural subsurface conditions beneath Brooklyn. All who face major groundwater problems can gain much by a study of the Long Island situation, now so well controlled and recorded.

South Coastal Basin of California Southern California is well known in many parts of the world for reasons other than engineering. What is not so well known is that part of the water supply to this area is obtained in a singularly interesting manner from underground sources. The South Coastal Basin of California comprises the Los Angeles metropolitan area and territory to the east, mainly agricultural but including 26 incorporated cities. The area consists essentially of the coastal plain, separated from the desert hinterland by three ranges of mountains generally parallel with the coast. In these mountains rise three major streams: the Los Angeles, the San Gabriel, and the Santa Ana, the valleys of each of which are well defined. Geologically, the area is most complex, with the result that there are 37 fairly distinct basins, although the boundaries of these are not always evident at the surface; many are fault planes. The slope of the valleys is quite steep, but that of the coastal plain is relatively flat; the junctions of valleys and plain provide almost perfect examples of discharge cones, as is well illustrated in Fig. 28.15.

Rainfall in the mountains is high but irregular; thus, the discharge of the streams is "flashy." Their courses through the coastal plain are often dry, the dry-weather flow percolating into their beds to replenish the groundwater stored in the unconsolidated deposits forming the plain. These deposits are varied in the extreme, although generally coarse. Faulting has complicated their distribution, but the area as a whole constitutes a great underground reservoir. It has been calculated that its capacity, in the 30-m (100-ft) layer immediately below outflow level, is no less than 85 billion m³ (7 million acre-feet), of which

FIG. 28.15 Part of the area providing Southern California's great underground storage basin; the apex of the detrital cone of the San Gabriel River is shown as the river emerges on to the south coastal plain.

only about 60 percent has so far been used. The total surface area of the basins is 3,400 km² (1,305 mi²), and the average specific yield of the unconsolidated strata is 8.4 percent by volume. Despite the high cost of pumping, the groundwater is used not only for domestic purposes but also for agricultural irrigation. So great has been the demand that the water level has dropped very seriously. It is to augment this supply that some of the great aqueducts already mentioned have been constructed.[28.47]

Vancouver, British Columbia In Vancouver, western metropolis of Canada, serving a population of almost a million, is the Greater Vancouver Water District, established in 1926. It has been given complete control over catchment areas of three small rivers with an area of 58,500 ha (145,000 acres). They are located immediately to the north of Burrard Inlet on which the city is located, in the Pacific Range of the Coast Mountains that includes peaks of almost 1,500 m (5,000 ft) within sight of the city (Fig. 28.16). The high rainfall ensures good runoff, and this is stored for city use in reservoirs created by dams on the small rivers. Large pipes convey the water by gravity from the reservoirs to the water district, the crossing of Burrard Inlet being effected by a notable pressure tunnel, the construction of which was a difficult operation due to the geological conditions encountered. Since the water comes from catchment areas underlain by igneous rocks, its quality is remarkable—hardness rarely exceeding 10 ppm, total dissolved solids only 23 ppm—so that chlorination is all the treatment that is required. This wonderful water supply is really fortuitous, since the city was founded not as a result of any careful city planning but in connection with the terminus of the Canadian Pacific Railway on Canada's west coast. There are few major urban areas throughout the world that are so fortunate as to have a gravity supply of pure water available within 16 km (10 mi) of the civic center. The growth of Vancouver has certainly been aided by this good fortune.[28.48]

Sydney, Australia Water supply to the great metropolitan area of Sydney, New South Wales, now with a population of well over 2 million, has an unbroken history since the

FIG. 28.16 Vancouver, British Columbia; a general view of the city, looking northeast, showing the adjacent Coast Range mountains in which are located reservoirs for the city's water supply.

days of the first settlement. It is recorded that on 26 January 1788 "sufficient ground was cleared for the encamping of the officer's guard and the convicts who had been landed ... at the head of the cove, near the run of fresh water, which stole silently along through the very thick wood...." The Tank Stream, thus so pleasantly described, supplied water to the little settlement from 1788 to 1826. "It still flows, a dark and hidden stream, beneath the shops and offices of modern Sydney," but it had become polluted before it was abandoned as the main water supply in 1826. Groundwater was next used, a supply being obtained by tunneling for "Busby's Bore" in a corner of Hyde Park, but this supply also became insufficient. In 1886 the first water was brought to the city from the Nepean watershed to the south, collected by means of a weir across the Nepean River at Pheasant's Nest. This was the beginning of a series of notable collecting projects, using larger catchments, all further removed from the city, and involving the construction of many notable civil engineering works. Geology added its complications since valuable coal seams underlay some of the dam sites needed for the water supply projects. The N.S.W. Mines Department laid down stringent limitations upon coal mining both in the vicinity of the dams that were built on such unusual sites and beneath the reservoirs formed by the dams; no troubles, either of leakage or settlement, have been reported. The last of these major projects was the Warragamba scheme that included the building of the largest dam in the southern hemisphere, the dam across the Warragamba River. This great structure, 120 m (394 ft) high but only 330 m (1080 ft) long at its crest because of the precipitous gorge in which it is built, has already called for mention, not only because of its size but also because the sandstone rock on which it is founded caused complications requiring most expert geological advice during construction. Observations on this dam since its construction are adding an important chapter to the records of engineering geology.[28.49]

Hong Kong The small crown colony of Hong Kong, with a total area of less than 103,000 ha (255,000 acres), has a rapidly increasing population that is already over 3 million. It has good rainfall, but this naturally varies year by year. Its water problem is therefore primarily one of storage, since the supply obtained from runoff from the rocky hills that make up so much of the colony has now been supplemented by water from China. The Chinese have reversed the flow of the Shih Ma River and pump its water through eight pumping stations, aided by six large reservoirs, to a point just north of the Hong Kong border. The colony is taking 67,500 million L (18,000 million gal) of this water per year, giving the Chinese almost £1,000,000 ($3,000,000) in foreign exchange. Storage, however, is difficult since there are no natural large reservoir sites amid the Hong Kong hills. Every advantage has been taken for the building of small storage reservoirs, but for major storage it was finally decided to convert an inlet from the sea to hold freshwater instead of seawater. The Plover Cove project has converted an area of 1,110 ha (2,740 acres) into a main freshwater reservoir. Intensive study of the geology of the seabed revealed about 10.5 m (34.4 ft) of soft clay overlying 1.5 m (4.9 ft) of stronger clays, beneath which is bedrock of granite in various stages of weathering. Dredging removed all the soft clay, the stiffer clay being left as the bottom of the reservoir to act as a seal against penetration of seawater once the reservoir was filled. The main dam is 2,070 m (6,800 ft) long and 45 m (148 ft) high, but with only 16.7 m (55 ft) above water level (Fig. 28.17). Geologically it is interesting because of the extensive use made of decomposed granite, and granodiorite, from the hillsides overlooking the famous harbor. So rapidly has the demand for water increased that an even larger project at High Island, involving still further use of the seabed as a reservoir site, is now in use.[28.50]

Ogden Valley, Utah Possibly the most unusual water-supply system of its kind is that supplying the city of Ogden. In 1914 the city started drilling wells in the artesian system

28-32 CIVIL ENGINEERING WORKS

FIG. 28.17 The upper part of the Plover Cove reservoir dam, Hong Kong, the larger part of which is below water level.

of the lower Ogden Valley, created by the existence of silts and gravels beneath an impervious bed of 21 m (70 ft) of varved clay. By 1933 a total of 51 wells had been drilled in what was called "Artesian Park"; the flowing wells are illustrated in Fig. 28.18. In 1935 and 1936 the U.S. Bureau of Reclamation constructed the Pine View Dam at the lower end of Ogden Valley; the reservoir so formed floods Artesian Park, which was the source of the city's water supply. Before the dam was built, however, a contract was negotiated between the city and the United States government providing for the capping of the artesian wells and the construction of the necessary collecting system, which would naturally be flooded by the water in the reservoir. The city therefore reserved the right to have the reservoir drained if ever there appeared to be trouble with its water system. The reservoir

FIG. 28.18 The flowing artesian wells of the Ogden city water supply before they were submerged beneath the water of the Pine View Reservoir; a photograph taken probably about 1925.

has been drained for this purpose, and repairs have been executed. Some of the damage that has been detected appears to have been due to the consolidation of the clay stratum because of the extra load of the water impounded by the dam. In 1955, therefore, special precautions were taken before the dam was raised by 8.7 m (29 ft) to give additional storage. Flexible connections were used to minimize the effects of future ground movement. The system has been described as a "three-layer" water system, an unusual but not inaccurate title for an unusual but successful engineering project.[28.51]

Sheikhdom of Kuwait Kuwait, although only 1.6 million ha (4 million acres) in extent, has become well known because of the great quantity of oil obtained from beneath its surface, now over 2 million bbl of petroleum per day. Located at the head of the Arabian Gulf, it has an average annual rainfall of only 8 cm (3 in). This all comes during the winter months of November to May, sometimes as intense local storms but more generally as light winter showers. Temperatures range from 51 to $-7°C$ (123 to 19.4°F). Post-Miocene rocks underlie the generally inhospitable surface of the country, the northern part being generally a gravelly plain. The Dibdibba formation, thought to be of Pleistocene age, consists predominantly of sand and gravel and serves as an aquifer, as do also underlying formations of limestone. The oil reserves are in more deeply underlying Cretaceous sandstones and limestones. Regional dip of the strata is to the east and northeast. Settlers in the area, prior to the recent oil developments, were able to get water from shallow wells; dug wells were also used, but all the water so obtained was brackish due to contact with some of the local rock formations.

Drilling for oil started in 1936, and further brackish water supplies were discovered in this way. Two major sources were thus developed, the water being distilled for potable use but used directly for other purposes. In 1960, fresh groundwater was surprisingly discovered by "a rig drilling brackish-water wells for the Basra Road construction [being] accidentally sited on the wrong location when moved during a sand storm." Careful studies revealed that this freshwater is surrounded and underlain by brackish water. The details of the local geological structure are such that, from inflow at the surface following winter rains, some water is retained in the Dibdibba cemented sandstone before being rendered brackish. Now, 32 mld (8.5 mgd) are obtained from this source, which is here described to emphasize the existence of groundwater beneath even the most unexpected areas.[28.52]

Honolulu, Hawaii The eight inhabited islands that comprise the main part of the state of Hawaii are world-famous for their beauty and their equitable climate. In a more restricted sense, they are known throughout the world by geologists as unusually interesting examples of volcanic-island formation. Since they have been developed for the extensive production of sugar cane and pineapples, for defense, and in more recent years as almost ideal tourist resorts, an increasing amount of engineering work has been necessary. This has included the development of what are some of the most unusual water-supply systems of the world. The islands are formed of volcanic lava, a rock that is generally highly permeable. Since the islands are blessed with good rainfall and as the generally rugged terrain exposes the rocks, water seeps into the ground continually and so gives excellent local supplies. But because the rocks are so permeable, their open pore structure is also filled with seawater beneath the surface of the sea, and so deep below the islands. The conjunction of the fresh and saltwater held in the rocky foundations to the Hawaiian Islands is one of the most remarkable examples of a hydrogeological phenomenon that is becoming of increasing importance.

The phenomenon is generally known as the *Ghyben-Herzberg effect*, since its nature was first recognized by W. Badon Ghyben of the Netherlands in 1889 and apparently quite independently by B. Herzberg in Germany in 1901. As with many such discoveries in science, other scientists had noticed the phenomenon but had not described it so clearly as the two men mentioned, whose names are given here since readers may come across them in reading about groundwater. The situation with which their names are now associated is, in principle, very simple; it is readily visualized. Since seawater, because of its salt content, is about one-fortieth heavier than freshwater, it follows that the freshwater will, if undisturbed, "float" (so to speak) on the top of the seawater. The depth to which the fresh water will extend below sea level will be 40 times the height of the lens of freshwater. There will be no dividing line between saltwater and fresh, but experience has shown that the actual occurrence of both types of water in a permeable formation, such as the lava rocks of Hawaii, does approximate to this theoretical pattern when undisturbed.

Inhabitants of Oahu, relatively few in number until recent times, relied upon the flow of surface streams for their water supply until 1879, when the great reservoir of groundwater underlying the island was discovered. Because of impermeable surface deposits, the groundwater was then under artesian pressure in many parts of the island. This was tapped by the first well drilled (in 1879), a well that was the forerunner of the more than 400 wells that had been drilled by 1910. By 1923, the groundwater level in the vicinity of Honolulu had dropped 6.9 m (22 ft 6 in) below its original level, which was only 12.6 m (41 ft 4 in) above sea level. A few wells had been "salted up" by intrusion of seawater; the delicate balance of sea and freshwater beneath the island was in danger of being seriously disturbed. In 1929 the Honolulu Board of Water Supply was established to control and manage the water supply to the city. In 1959, through a merger with the former Suburban Water System, the board was given control over all water in the island of Oahu.

The key to successful conservation of this almost unique water-supply system was the adoption, starting in 1936, of a type of well first used about the turn of the century on the island of Maui for the supply of irrigation water for sugar-cane plantations. *Maui wells,* as they are now widely known, consist essentially of horizontal tunnels excavated in the local volcanic rock at carefully selected locations, generally just below the known freshwater levels (Fig. 28.19). Access to the tunnels is usually obtained by means of inclined shafts, sometimes by vertical shafts, in which the necessary pumps and discharge mains

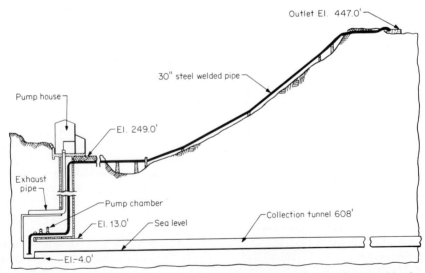

FIG. 28.19 Simplified section of a typical Maui-type well as used in the Hawaiian Islands.

FIG. 28.20 View inside one of the Maui wells serving Honolulu, Hawaii, giving an impression of the flow of water through the porous rock. The photograph was taken before the lower part of the tunnel was allowed to fill up.

can be located. By pumping at controlled rates from the supply which accumulates in the tunnels, a water supply is, so to speak, "skimmed off" the reservoir of freshwater so delicately balanced upon the underlying saltwater. The surface level of the sea is depressed beneath the island by the overlying freshwater to the extent that for every foot of freshwater in the rocks above sea level, almost 40 ft is found below sea level. The rapid rise of saltwater consequent upon depletion of the freshwater reservoir will be obvious from these figures. By the use of the horizontal skimming tunnels of the Maui wells, an assured supply is now being obtained for the city of Honolulu and the other rapidly developing communities on Oahu (Fig. 28.20).[28.53]

It would be difficult to imagine a more vivid example than the Maui wells of Hawaii of the constructive application of the fundamentals of an unusual local geological situation for "the use and convenience of man" (to use the famous phrase from the 1818 charter of the Institution of Civil Engineers). To stand in the complete silence at the end of one of the collecting tunnels in Oahu; to watch the crystal-clear water in the lower part of the tunnel; to know that it is slowly seeping from the surrounding basalt, a volcanic rock the character of which is itself geologically interesting; and to think of the delicate balance existing between the clear freshwater thus being collected for use and the great expanse of seawater deep in the porous rock below one's feet is to have one of the most moving experiences possible for one interested in the proper appreciation of geology in all the works of the civil engineer.

28.11 REFERENCES

28.1 C. Herschel, *The Two Books on the Water Supply of the City of Rome of Sextus Julius Frontinus,* Dana Estes, Boston, 1899, p. 10.

28-36 CIVIL ENGINEERING WORKS

28.2 Lambarde, *Perambulation of Kent*, London, 1826; quoted in *Meteorological Magazine*, **63**:63 (February 1928).

28.3 D. H. Thomson, "Water and Water Power," *Transactions of the Liverpool Engineering Society*, **44**:105 (1923).

28.4 Ibid.

28.5 W. T. Halcrow, G. B. Brook, and R. Preston, "The Corrosive Attack of Moorland Water on Concrete," *Transactions of the Institution of Water Engineers*, **33**:187 (1929).

28.6 G. McIldowie, "The Construction of the Silent Valley Reservoir, Belfast Water Supply," *Minutes of Proceedings of the Institution of Civil Engineers*, **239**:465 (1936).

28.7 *Annual Report of Distillers-Corporation, 1969*, Seagrams, Montreal, 1970, p. 27.

28.8 D. Daiches, *Scotch Whisky*, Deutsch, London, 1969.

28.9 B. Harmon, "Contamination of Groundwater Resources," *Civil Engineering*, **11**:345 (1941).

28.10 "New Look at Acid Mine-Drainage," *Engineering News-Record*, **164**:61 (19 May 1960); see also reports of the Ohio River Valley Water Sanitation Committee, Cincinnati.

28.11 "Groundwater Pollution in California Points to Industrial Waste Discharge," *Engineering News-Record*, **137**:785 (1946).

28.12 J. Bowman, "German Waterworks Practice," *Water and Water Engineering*, **38**:665 (1936).

28.13 Discussion of Gourlay paper, *Minutes of Proceedings of the Institution of Civil Engineers*, **237**:530 (1935); see ref. 28.15.

28.14 "The New Filter Gallery at Des Moines," *Engineering News-Record*, **65**:468 (1912); and information from the general manager, Des Moines Water Works.

28.15 H. J. F. Gourlay, "The Water Supply of Kano, Northern Nigeria," *Minutes of Proceedings of the Institution of Civil Engineers*, **237**:454 (1935).

28.16 A. B. McDaniels, "Groundwater Cut-off Wall Provides New Water Supply," *Engineering News-Record*, **113**:757 (1934).

28.17 "Underwater Dam Brings Fresh Hopes to Arid Areas," *New Civil Engineer*, **119**:53 (21 November 1974).

28.18 "Hetch Hetchy Water Supply," *Engineering News-Record*, **113**:129 (1934), and publications of the City of San Francisco.

28.19 J. Hinds, "Colorado River Water for California," *Civil Engineering*, **7**:573 (1937); see also in same journal, **5**:2 (1935).

28.20 "Dam and Three-River Hookup Give Britain Flexible Water Tap," *Engineering News-Record*, **198**:22 (16 June 1977).

28.21 H. O. Banks, "California's Water Plan," *Civil Engineering*, **30**:52 (December 1960); see also in same journal, H. G. Dewey, "California's State Water Project—on Schedule," **36**:48 (January 1966), and W. R. Gianelli and R. B. Jansen, "California Water Project," **42**:78 (June 1972).

28.22 R. D. S. Thompson, "Capturing Water in the Desert," *Engineering News-Record*, **120**:327 (1938) *(karez)*; "L'Alimentation en eau des régions désertiques de la Perse et du Caucase," *L'Eau*, Paris, 1933, p. 86 *(keghriz)*.

28.23 G. A. Cummins, "Underground Water Circulation," *Water and Water Engineering*, **38**:319 (1936).

28.24 O. E. Meinzer, *Outlines of Methods for Estimating Groundwater Supplies*, U.S. Geological Survey Water Supply Paper no. 638(c), Washington, 1932, pp. 99–144.

28.25 R. Lockhart Jack, *Geological Structure and Other Factors in Relation to Underground Water Supply in Portions of South Australia*, Geological Survey of South Australia Bulletin 14, Adelaide, 1930.

28.26 B. Smith, "Geological Aspects of Underground Water Supplies" (Cantor Lecture of the Royal Society of Arts), *Water and Water Engineering*, **38**:223 (1936).

28.27 L. Quigley, "Saskatchewan Finds Water for Tomorrow," *Water and Pollution Control,* **104**:50 (November 1966).

28.28 *Thames Conservancy Water Plan: Pilot Scheme Successfully Completed,* Thames Conservancy, Reading, May 1970; see also "Borrowing Water for Thirsty London," *New Civil Engineer,* **67**:32 (22 November 1973).

28.29 H. van Oosterom, R. A. Downing, and F. M. Law, "Development of the Chalk Aquifer in the Great Ouse Basin" (informal discussion report), *Proceedings of the Institution of Civil Engineers,* **57**, Part 2, pp. 187–190 (March 1974).

28.30 A. T. Mitchelson, "Conservation of Water through Recharge of the Underground Supply," *Civil Engineering,* **9**:163–165 (1939).

28.31 D. A. Lane, "Artificial Storing of Groundwater by Spreading," *Journal of the American Waterworks Association,* **28**:1240–1251 (1936).

28.32 "L.A. to Expand Groundwater Recharge," *Engineering News-Record,* **163**:28 (3 December 1959).

28.33 F. B. Laverty, "Recharging Groundwater with Reclaimed Sewage Effluent," *Civil Engineering,* **28**:585 (1958).

28.34 E. S. Boniface, "Some Experiments in Artificial Recharge in the Lower Lea Valley," *Proceedings of the Institution of Civil Engineers,* **16**:325 (1959).

28.35 "Denver Moves to Fill Future Water Supply Gap," *Engineering News-Record,* **190**:68 (22 March 1973).

28.36 Paschal Grimes, city engineer of Chanute, by personal communication; see also *Journal of the American Waterworks Association,* **50**:1021 (August 1958).

28.37 W. A. Pettyjohn, "Artificial Recharge: A Potential Solution for Many Water Problems," *Water Well Journal,* **28**:48 (September 1974); see also by same author, "Design and Construction of a Dual Recharge System at Minot, North Dakota," *Groundwater,* **6**:4 (July–August 1968).

28.38 "Israel Turns to Sewage for Water," *Engineering News-Record,* **183**:42 (6 November 1969).

28.39 W. R. Hansen, "Geomorphic Constraints on Land Development in the Front Range Urban Corridor, Colorado," *Urban Geomorphology,* Special Paper no. 174, Geological Society of America, Boulder, 1976, p. 85 (with useful references).

28.40 L. R. East, *Water Supply Problems in Australia,* Selected Paper no. 141, Institution of Civil Engineers, London, 1933.

28.41 R. Lockhart Jack, *Some Developments in Shallow Water Areas in the North East of South Australia,* Geological Survey of South Australia Bulletin no. 11, Adelaide, 1925; see also other bulletins issued by the Survey.

28.42 J. W. Young, "Bituminous Surfacing Treatment of Portion of the Water Supply Catchment at Narrogin, Western Australia," *Civil Engineering and Public Works Review,* **36**:548 (August 1941).

28.43 P. A. M. Parker, *The Control of Water,* 2d ed., Routledge, London, 1925, p. 207.

28.44 "Miner's Mistake is Modern Blessing," *Engineering News-Record,* **168**:37 (8 March 1962).

28.45 Chief Engineer J. T. O'Neill, by personal communication, 1977.

28.46 P. Cohen, O. L. Franke, and B. L. Foxworthy, *An Atlas of Long Island's Water Resources,* New York Water Resources Commission Bulletin no. 62, 1968.

28.47 "Underground Water Storage in California's South Coastal Basin," *Engineering News-Record,* **115**:733 (1935).

28.48 Greater Vancouver Water District folder, 1958.

28.49 *Water for Sydney,* Metropolitan Water, Sewerage and Drainage Board, Sydney, N.S.W., Australia, 1959.

28.50 "Hong Kong Stores Water in the Sea," *Engineering News-Record,* **172**:26 (9 January 1964).

28.51 "Artesian Well Water Supply for the City of Ogden from the Bed of Pine View Reservoir," *Reclamation Era,* **24**:208 (August 1938); see also R. R. Woolley, "Artesian versus Surface Supply—Ogden River Project," *Civil Engineering,* **11**:536 (1941).

28.52 R. E. Aten and R. E. Bergstrom, *Groundwater Hydrology of Kuwait,* American Society of Civil Engineers Reprint 152, from Water Resources Conference, March 1965.

28.53 See reports of the Honolulu Board of Water Supply; see also "Skimming Fresh Water off Salt," *Engineering News-Record,* **156**:47 (15 March 1956); and in same journal, N. D. Stearns "Wells for the Water of Hawaii," **118**:450 (1937).

Suggestions for Further Reading

Despite the importance of geology in all aspects of water supply, there appears to have been only one book in English dealing specifically with the subject. This is *The Geology of Water Supply* by H. B. Woodward, published by Edward Arnold & Co. of London in 1910. Despite the years that have elapsed since its publication, it is still worth reading if a copy can be located.

There are available today a number of fine books that include in their contents useful comments on the geological aspects of water supply. *Water,* the 1955 Year Book of the U.S. Department of Agriculture, must again be mentioned as a most useful general reference. The *California Water Atlas,* edited by W. L. Kahrl and published by the state of California in 1979, is a masterpiece of book production, packed with information about water in California including a most useful bibliography. It is well worth studying by all users of this Handbook, alike for the beauty of its presentation as for the brilliant picture it gives of the water situation in a state where water is so critical. It is to be hoped that it will act as an incentive for the production of comparable volumes for other areas.

So important has water become around the world that there is now a monthly periodical devoted to *World Water.* This is published by Thomas Telford, Ltd., the publishing arm of the Institution of Civil Engineers (U.K.); its editorial office is at 201 Cotton Exchange, Old Hall Street, Liverpool 3, England.

In this Handbook, the authors make clear their high regard for the imaginative pioneer groundwater recharging project of the Thames Conservancy, one that presents much potential for water conservation where geological conditions are favourable for the "balancing" of groundwater supplies. The Institution of Civil Engineers published in 1978 a 242-page volume of proceedings of a conference on the *Thames Groundwater Scheme* from which further information can be obtained. A most useful 50-item bibliography on this concept and its applications will be found on p. 171 of volume 33 of the *Journal of the Institution of Water Engineers and Scientists* (U.K.); this was kindly drawn to the attention of the authors by A. Hunter Blair, principal engineer of the Anglian Water Authority.

chapter 29

CANALS

29.1 Historical Note / 29-2
29.2 The Canal Age / 29-2
29.3 The Panama Canal / 29-5
29.4 The St. Lawrence Seaway / 29-9
29.5 Canal Locks / 29-11
29.6 Some European Canals / 29-14
29.7 References / 29-15

The history of transportation routes presents a vivid picture of the development of civil engineering—from the first efforts of primitive peoples to bridge a stream with a fallen tree trunk to the conception and construction of modern express highways with all associated works. Beaten tracks to drinking places were probably the first artificially constructed pathways. Communication between settlements was a natural development and thus gradually arose the idea of main transportation routes. In early historical records, many references to practical road building are to be found. The road work of the Roman Empire was the culmination of such development, and some Roman roads are still in use today. Many centuries had to pass before road construction regained the eminence that it held in Roman times. This latest advance has not yet been halted and continues as the virile construction practice of today. The nineteenth century saw the road challenged by the railroad as the main artery of transport. In this present era, the road has more than answered the challenge, and both now play their part in the general transportation scene. Many problems, both economic and social, face the railroads as they adjust themselves to their competitive position. The advantages the railroads present, however, are still so marked that they will long continue to play a vital part in general transportation.

Road, rail, and now air transport have become so familiar that it is easy to forget that before the beginning of the industrial revolution, and indeed for several decades after its start, transport by water was by far the most important means of movement. Beginning with the primitive local use of rivers and slowly extending into the use of lakes and inland seas, the record of water transport is continuous throughout the whole range of human history. The use of the ocean itself for travel presents one of the most fascinating chapters in the story of civilization. Even so limited a part of this story as the provision of the necessary marine works is of such importance in the history of engineering that a separate chapter must be devoted to it. It will be convenient to consider the use of rivers as trans-

29-2   CIVIL ENGINEERING WORKS

portation routes in connection with other aspects of river engineering; this branch of water transport will therefore be considered in Chap. 33. There remain for consideration canals, those artificial waterways necessary to join natural waterways or to give access from the sea to inland points. Canals played an important part in the growth of modern transportation, and although the building of new canals is now a rare occurrence, the canals in use today are so vital to the human economy and present such interest from so many points of view, including the geological, that brief reference to them will be the subject of this chapter.

29.1   HISTORICAL NOTE

The extension of inland waterway systems and the penetration of narrow land barriers between seas by means of canals have attracted the attention of master builders throughout the ages. A canal connecting the Nile and the Red Sea is said to have been begun in the fourteenth century B.C.; work on the project was abandoned by Necho about 610 B.C. after 120,000 men had lost their lives in the excavation work. According to Strabo, the canal was finished by Ptolemy II, who is said to have constructed in it locks with movable gates. The many canals constructed in what is now Iraq are familiar to students of early history. Xerxes, in his war against the Greeks, constructed a canal across the isthmus of Mount Athos; trouble was encountered because the sides (up to 15 m or 50 ft deep) were excavated at too steep a slope.[29.1] The Isthmus of Corinth was a continuing challenge to the early Greeks, even though the Corinth Canal was not finally completed until as recently as 1893. A crossing of the isthmus is said to have been first conceived by Periander of Corinth, one of the Seven Sages. With the facilities available to him, he could not possibly have constructed a canal, but in order to give the merchants of Corinth an eastward outlet, he built the best alternative to a canal, what might be termed the first marine railway. This was the *dioclos,* a paved roadway upon which vessels could be portaged from the Gulf of Corinth to the Saronic Gulf, 6.4 km (4 mi) away.

A legend to this effect has long persisted; in 1957 Greek archaeologists found undoubted remains of the *dioclos,* still in good condition where it had not been cut away for the modern canal or crossed by railway tracks. It was 3.9 m (13 ft) wide and built of heavy rectangular blocks of the local limestone. Evidence that it was built in the time of Periander, probably about 606 B.C., is found in the letters carved into some of the limestone blocks; the letters are from the Corinthian alphabet which was in use in the seventh century B.C. Periander's idea of constructing a canal was taken up by later rulers. One of them, however, Demetrius Poliorcetes, was dissuaded from starting the work by his priests, who claimed that the sea level in the Gulf of Corinth was higher than that in the Saronic. Nero came nearest to completing the project in these early years. He himself inaugurated the work in A.D. 67, cutting the first sod with a golden spade. With the aid of 6,000 slaves (sent by Vespasian from Judea) he did complete 2,000 m (2,200 yd) on one side of the isthmus and 1,450 m (1,600 yd) on the other. Four months later the work was stopped, probably because of the revolt of Vindex, but when modern engineers resumed the work in the nineteenth century, they used exactly the same cuts. The canal is 6.4 km (4 mi) long, 21 m (70 ft) wide, and 8 m (26 ft) deep (Fig. 29.1). It was seriously damaged during the Second World War but was soon back in service.[29.2]

29.2   THE CANAL AGE

It is often forgotten that modern civil engineering really got its start in the building of canals. As a prelude to the building of artificial waterways, much good work was done in

FIG. 29.1 The Corinth Canal, Greece.

improving navigation on rivers in Europe, and especially in France and Great Britain, starting as early as the seventeenth century. When the need for interconnections between rivers was seen, the first canals were conceived, one of the first being the Midi Canal in France, still in use. In England, the first canal was the St. Helen's Canal, but it was the Duke of Bridgewater's Canal, opened in 1757, that marked the start of what was to be the canal era.

It was in the building of canals that the first notable British and French engineers gained their experience, one of the most notable, if also one of the least known, being William Jessop, "from his experience and abilities looked upon as the first engineer of the kingdom" when he was appointed in 1792 as principal engineer for the famous Grand Junction Canal of Great Britain. Canal fever was soon passed to the United States and to Canada, and many truly remarkable works were carried out in the early decades of the nineteenth century. But the coming of the railway dashed the hopes of the canal enthusiasts, the canal age coming to an end about 1840. Many of the canals built in those exciting days are still in use; some are even being restored to use. Canal building, however, will only rarely come into the province of the civil engineer of today, despite its fascinating history, but this reminder of the century during which so much canal building was done is a necessary prelude to brief reference to some important modern canals.

Some indication of the enthusiasm that canal building generated in the early nineteenth century is given by this quotation from Goethe, the appearance of whose name in this volume is entirely appropriate since, as the "universal scholar," he had a very lively interest in geology. Eckermann reports him to have said in 1827:

29-4 CIVIL ENGINEERING WORKS

So much, however, is certain that, if they succeed in cutting a canal that ships of any burden and size can be navigated through it from the Mexican Gulf to the Pacific Ocean, innumerable benefits would result to the whole human race. But I should wonder if the United States were to let an opportunity of getting such a work into their own hands escape. . . . it is absolutely indispensable for the United States to effect a passage from the Mexican Gulf to the Pacific Ocean and I am certain they will do it. Would that I might live to see it!—but I shall not. I should like to see another thing—a junction of the Danube and the Rhine. But this undertaking is so gigantic that I have doubts of its completion, particularly when I consider our German resources. And thirdly, and lastly, I should wish to see England in possession of a canal through the Isthmus of Suez. Would I could live to see these three great works! It would be well worth the trouble to last some fifty more years for that purpose.[29.3]

In this, as in so many other things, Goethe was a man with vision. Others shared his enthusiasm for canals, and many canal works were started in the nineteenth century, although some were abandoned long before actual completion. In most of these cases, it seems to have been a neglect of geology that led to the abandonment of projects that appeared superficially to be practicable. The "pathology of canals" would form an interesting, if sobering, study in engineering geology. Let brief reference to but one example suffice, with corresponding apologies to the Emerald Isle in which so much fine engineering work has been carried out. T. Mellard Reade in a paper presented in 1888 records the following case.

Another well-known piece of stupidity was the attempt to make a canal between Loughs Mask and Corrib in the West of Ireland. This was one of the relief works planned during the great famine, and it is not known who was responsible for it. A canal between the two loughs would have been of some utility, but the strata in which they lie is carboniferous limestone much jointed, and the communication or drainage between the two waters is by underground channels. The engineers, Royal or otherwise, went bravely to work, the canal was cut in the limestone rock, the water turned in, when lo! it all quickly disappeared. The abandoned cutting may now be seen sparsely overgrown with grass.[29.4]

FIG. 29.2 Entrance to the Rideau Canal at Ottawa, built in 1826–1832 as a military waterway joining the Ottawa and St. Lawrence rivers.

This appears to be an unfortunate reference to the work of the Royal Engineers (of Great Britain), since some of their early canal building was notable pioneer civil engineering. Outstanding is the Rideau Canal, built between 1826 and 1832 as a military canal from Ottawa to Kingston in eastern Canada. It is 197 km (123 mi) long, and its construction involved the canalization of two rivers and several lakes by means of 47 locks and over 50 dams; it was a masterly demonstration of what the Royal Engineers could do in virgin forest (Fig. 29.2). Lt. Col. John By, the engineer in charge, must have had an innate sense of sound geological understanding, since both the location of the canal and the selection of masonry for locks and dams could not have been better done today.[29.5]

29.3 THE PANAMA CANAL

Superficially, the excavation of a regular water channel along a selected route appears to be a simple operation. It is only when the details of design and construction are considered that the true complexity of the work becomes evident. Canal bed and canal banks should be impermeable; if permeable, the leakage must be predictable with certainty. Canal banks, if necessary, must be stable. Bridge foundations adjacent to the canal must be sound and secure; similarly, the foundation of all lock and control structures must be susceptible to rational design and expeditious construction. Geology is of fundamental importance in all these problems, as can be seen at a glance. Adequate knowledge of all geological features is therefore an essential prerequisite to canal construction.

The problems and complexities that geology can introduce to canal construction have nowhere been better illustrated than in the construction of the canal which Goethe so brilliantly foresaw—that across the Isthmus of Panama. It was, however, the French who started to build the Panama Canal at the instigation of Ferdinand de Lesseps, following his success in constructing the Suez Canal. That no serious problems had arisen in this earlier work, opened for use in 1869, was due to the satisfactory local geology, but it was, in a way, geology (combined with dread tropical diseases) that finally defeated the French at Panama. The story of the Panama Canal has been brilliantly told by David McCulloch in his book *The Path between the Seas*.[29.6] The only criticism that can be made of this fascinating record of the long and tangled story is that, although geology and especially that of the Culebra Cut is mentioned, its significance is not made clear. With the suggestion that interested readers should consult McCulloch's fine book for the overall story, the following summary of the geological aspects of this major construction battle will speak for itself.

The Panama Canal, now one of the most important waterways of the world, is located just north of the equator. It crosses the Isthmus of Panama where the great continental divide dips to its lowest elevation, about 94 m (312 ft) above sea level. The route of the canal follows existing river valleys and crosses the divide in the great Gaillard Cut; it is 64.5 km (40.27 mi) long. Construction was started in 1882 by the French company which operated until 1889; a reorganized company started work in 1894 but was eventually bought by the United States of America in 1902, under whose auspices the canal was completed in 1914. The canal was officially opened on 12 July 1920.

Despite the magnitude of its engineering features, attention must here be restricted to the canal's geological interest. This is typified, perhaps, by the following figures of excavation carried out up to 30 June 1959:[29.7]

	M³	Yd³
Excavated by French	22,800,000	29,908,000
Excavated by Americans	449,000,000	588,181,200
Excavated in Gaillard Cut	144,500,000	189,832,700
Excavated in Gaillard Cut (attributed to slides)	58,300,000	76,228,000

29-6 CIVIL ENGINEERING WORKS

This last surprising figure is at least a clue to the difficulties encountered in the excavation of the great cut penetrating the continental divide. The Gaillard Cut is 14 km (8.75 mi) long. The geological formations encountered during the construction of the canal are mainly sedimentary beds of shale and sandstone, irregular masses of intruded basaltic and agglomeritic dikes and plugs, and volcanic tuff. Their distribution is somewhat complicated and is seriously affected by frequent faults. The greater part of the Gaillard Cut is through rock strata; although these strata are variable, slides in them caused little trouble. In the Culebra section of the cut, however, a synclinal trough about 1.6 km (1 mi) wide at canal level crosses the cut directly at its deepest portion. The trough is filled with a "fine-grained sandy clayey formation" called cucaracha; it is structurally weak, and its repeated failure led to most of the slides that interfered so seriously with the construction of the canal.

These slides began at least as early as 1884, just after the start of excavation by the French, and they continued until long after the completion of the canal. One slide of over 380,000 m^3 (500,000 yd^3) happened overnight; others were of considerable duration. They were of varied types and included upheavals of the floor of the cut, described thus by Colonel Gaillard:

> Most of the slides of the past year [1911–1912] were breaks resulting from the failure of an underlying layer of rock of poor quality due to the pressure of the enormous weight which crushes the underlying layer, forces it laterally and causes it to rise up and heave in the bottom of the cut. The heaving at times is 30 feet.

An accompanying photograph (Fig. 29.3) shows such an upheaval in the form of an island which appeared overnight out of a depth of 10 m (30 ft) of water.

Figure 29.4 represents a cross section through the Gaillard Cut as it exists today and as it was originally projected. From a study of this, coupled with recognition of the structural instability of the cucaracha formation, one can appreciate that the slides were due to the inherent weakness of this material in standing up to the loading to which it was subjected by progress of the excavation, intensified in some places by the extra loading of excavated material dumped too near the canal. The unsatisfactory nature of the cucaracha formation was reported on in 1898 by the French geologists Bertrand and Zurcher, and it has been suggested that their opinions were not utilized. It is therefore of significance to note that the final report of the committee of the National Academy of Sciences

FIG. 29.3 A landslide on the Panama Canal, 18 and 19 September 1915, showing an island formed overnight in 10 m (30 ft) of water.

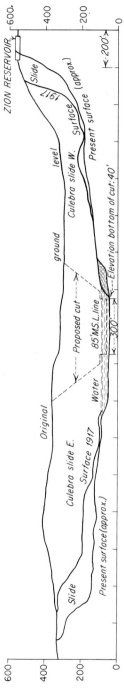

FIG. 29.4 Cross section through the Gaillard Cut, Panama Canal, showing excavation as originally proposed and that actually carried out.

29-8 CIVIL ENGINEERING WORKS

appointed by the President of the United States to study the Panama Canal slides (following the closing of the canal in September 1915 because of blockage by slides) included this statement:

> The Committee regrets that the United States engineers in charge of the canal excavation have not had the benefit, from the outset, of the best available technical advice in regard to the proper slopes for the canal banks, based on a thorough study of the rocks in the banks, and in regard to the character of the slides. Dr. Howe, who was attached to the canal force under Engineer John F. Stevens, 1906–7, before much progress had been made in excavation, was occupied mainly in the preliminary geological study of the canal route and in special problems which were pressing at that time. Dr. Hayes, the first geologist called upon to examine the Culebra slides, was sent to the canal in 1910, on the request of President Taft, but remained there for a short time only. On his recommendation a geologist [Mr. MacDonald] was attached to the canal corps and for three years rendered valuable assistance, although the great slides were already past control. He was the first to recognize the general character of the deep-seated deformation which characterizes the Culebra slides and explains the upheaval of the bottom of the cut, which was a feature of some movements.

The report of this committee and its valuable appendixes constitute a document of great engineering and geological interest and illustrate vividly the necessity of the application of geological study to all problems such as those which the construction of the Gaillard Cut involved.[29.8]

The Mr. MacDonald mentioned so appreciatively in the report was a Canadian geologist on the staff of the U.S. Geological Survey when selected for this challenging post. In later years he returned to his native Nova Scotia and became professor of geology at St. Francis Xavier University, but he returned to the Canal Zone in 1939 to help with wartime investigations and died there in 1942.

Careful maintenance has naturally been a continuing feature of the operation of the canal. In the course of this work, a surveyor noticed in 1938 a small crack at the top of Contractors' Hill, a dominating feature of this critical part of the canal. The crack was regularly observed; in 1954, it had increased to such an extent that remedial action was clearly necessary. On the recommendation of consultants, a drainage tunnel, 1.5 by 2.1 m (5 by 7 ft), was quickly excavated in order to drain groundwater from the vicinity of the

FIG. 29.5 Excavation on Contractors' Hill, Panama Canal, required to stabilize this part of the Culebra Cut, as described in the text.

main crack, since this water was one of the contributory causes. A contract was awarded for the removal of about 1.9 million m³ (2.5 million yd³) of the Pedro Miguel agglomerate forming the hill; the excavation was intended to relieve the pressure on the underlying cucaracha shale. Most unusual expedients were necessary in carrying out controlled blasting in order to cause as little disturbance as possible; the work was successfully completed before the end of 1955; the finished appearance of the cut is shown in Fig. 29.5.[29.9] The Second World War taxed the capacity of the great canal to the limit, and much study has been given to possible means of enlarging the facility it provides. Eventually, there will probably be more construction at Panama; canal construction here and elsewhere will continue its challenge to the cooperative efforts of engineers and geologists.

29.4 THE ST. LAWRENCE SEAWAY

The other major ship canal of North America, built jointly by Canada and the United States of America half a century after the Panama Canal, is that known as the St. Lawrence Seaway. The full Seaway includes improved river channels, canalized sections of the St. Lawrence, crossings of the Great Lakes with much local dredging, and two canals proper. After half a century of political delays, the ceremonial first sod was turned on 10 August 1954. Less than five years later, the Seaway was officially opened by the Queen of Canada and the President of the United States. The international power station had been delivering power since 1958, the first vessel having sailed through the completed Iroquois Lock on 22 November 1957, little more than three years after the start of work. Total cost of the Seaway and the associated power development was over $1 billion, shared internationally. At one time more than 22,000 men were at work between Montreal and Lake Ontario. Over 500 professional engineers alone had worked in the offices of the four principal design agencies, and there were many smaller offices also involved. The St. Lawrence Seaway must surely rank as one of the very greatest of all civil engineering projects. (See Fig. 29.6.)

The overall concept of the project will be familiar—the provision of an 8.1-m (27-ft) waterway, capable of handling oceangoing vessels, from the head of the navigable St. Lawrence River at Montreal to Lake Ontario, and thence into the Upper Lakes. This involved new locks and a canal from Montreal to Lake St. Louis; locks at Beauharnois; two lift locks and a guard lock in the international section of the St. Lawrence, designed integrally with the international powerhouse which utilizes almost the full fall from Lake Ontario to Lake St. Francis to generate over 1.5 million kW; the Welland Ship Canal (opened in 1932) providing, with some deepening, the link between lakes Ontario and Erie; all combined with much dredging of channels and the construction of a large number of ancillary works, some of which (such as the Long Sault Dam) are major structures in their own right by any normal standard.[29.10]

By international agreement, the two lift locks in the international section of the river were located on the United States side. All the earlier, smaller locks had been built by Canada and were located on the Canadian side, where geological conditions are favorable. The Eisenhower and Snell locks are notable structures, with the same lock dimensions as for other Seaway locks—24 m (80 ft) wide, 258 m (860 ft) from pintle to pintle, and a minimum of 10 m (30 ft) of water over sills. They are located near the eastern end of the Wiley-Dondero Canal, a straight channel almost 16 km (10 mi) long, all but the western end being a major new cutting excavated through dry land, with a bottom width of 133 m (442 ft). This involved almost 15 million m³ (20,000,000 yd³) of excavation, in advance of which extensive test boring, soil sampling and soil testing, and a demonstration test pit were carried out. Serious difficulties were encountered with this excavation work because

FIG. 29.6 The St. Lawrence Seaway and Power Project; crowds on top of the dike of glacial till adjacent to the Long Sault of the St. Lawrence which had been dried up, on 1 July 1958, just prior to the admission of water by blasting the main cofferdam, marking the end of the 4-m (14-ft) canals and the start of the Seaway.

of the properties of the till and marine clay involved; these difficulties are described in Chap. 19.

The Welland Canal is the fourth canal to bypass Niagara Falls, the first having been opened in 1829. Major improvements to the relatively simple original canal resulted in the second (opened in 1845) and then the third canal (opened in 1887). The fourth Welland Canal was started in 1912 but, because of the First World War, it was not finished until 1931, being officially opened in 1932. It followed a new route for part of its length; its planning involved engineering quite inspired for the early years of this century (Fig. 29.7). It is about 42 km (26 mi) long, its total drop of 99.5 m (326.5 ft) being achieved in only seven locks (compared with 40 locks in the first canal). Each of these was built 258 m (860 ft) long by 24 m (80 ft) wide, this great increase over any dimensions for all other Canadian canals having been possibly influenced by the final designs for the locks on the Panama Canal.

Throughout its course, it crosses ten distinct geological formations of the Ordovician and Silurian measures, including dolomites, limestones, sandstones, and shales. Of these, the Queenston shale gave unusual trouble during excavation, disintegrating as it dried out in the atmosphere. The practice developed, therefore, of so excavating it that a thin cover was left in place until just before concreting began; the cover was then removed after having thus protected the final excavated surface. Trouble was experienced here also, a quarter of a century before the remainder of the Seaway was built, with excavation in glacial till, but this was eventually overcome and the canal finished as planned. The Welland Canal of today is just as it was finished in 1931, good maintenance having ensured its unimpeded service. The dimensions of its locks fixed the dimensions of all other locks on the St. Lawrence Seaway. Construction of a 13-km (8-mi) bypass during the 1970s eliminated some awkward impediments to navigation, reducing normal passage through the canal by about one-half hour. This work involved some interesting problems with groundwater control.[29.10]

FIG. 29.7 The twin locks on the Welland Canal which take this major ship canal up the Niagara escarpment.

The Welland Canal was, however, only one part of the St. Lawrence Seaway. It will already be clear that this billion-dollar project was in many ways a unique construction undertaking. Excluding what had already been done on the Welland Canal, dry excavation amounted to 126 million m^3 (166 million yd^3) with 27 million m^3 (35 million yd^3) of dredging in addition. Over 19 million m^3 (25 million yd^3) of material were used in the construction of the necessary dikes to form the navigation canal between Montreal and Beauharnois. Corresponding figures for the international power project built integrally with the Seaway are 62 million m^3 (82 million yd^3) of dry excavation, 9 million m^3 (12 million yd^3) of dredging, and 13.5 million m^3 (18 million yd^3) of material in dikes. In preparation for this work, 1,700 test borings were put down in the Canadian section alone, totaling 9,600 m (32,000 ft) in soil and 12,000 m (40,000 ft) in bedrock. These were supplemented by many test pits, auger holes, and special investigations. The result of this most careful subsurface study was reflected in the equally impressive progress with construction, all major works being completed within less than five years. Fortunately, much of the experience gained involving geological factors has been recorded and published. Especially notable was the use of glacial till as the impermeable central parts of the miles of dikes required; no serious problems were experienced with this most economical use of excavated material.[29.11] Problems were encountered with groundwater, the most troublesome being always in bedrock excavation. Sandstone, especially that underlying the upper lock at Beauharnois, produced the greatest inflow into excavated areas. All problems were overcome, however, with no serious delays in construction, testimony again to the value of careful advance geological investigation and thorough subsurface investigation.

29.5 CANAL LOCKS

The civil engineering works necessary for the completion of major canals include those common to other types of projects and so call for no special mention. But the foundation of locks is of particular interest because of the varying hydrostatic conditions to which the completed structures will be subjected. When locks are founded upon solid rock, few problems usually arise. Some locks, however, must be founded on soil, and this necessitates special care in design and construction. As an example of somewhat unusual conditions ingeniously utilized, there may be mentioned the main sea lock of the Ijmuiden Canal, Holland, a diagrammatic cross section through the foundation of which is shown in Fig. 29.8. The local freshwater supply is obtained from groundwater trapped beneath the top clay stratum shown in the diagram. There was grave danger that if construction operations interfered with this underground reservoir, the supply might be affected in quantity and possibly in saline content by ingress of seawater. As a result of thorough subsurface exploration, the conditions shown in Fig. 29.8 were determined, and this permitted the complete enclosure of the space occupied by the lock—394 m (1,315 ft) long by 48 m (161 ft) wide—by a wall of steel sheet piling. This course was followed as indicated in the diagram, and the work was successfully completed. It constitutes an outstanding example of the adaptation of local geological features as a definite part of a scheme of construction.[29.12]

Entirely different were the geological conditions beneath one of the very few lock walls to fail in North America. When the Wheeler Dam and lock were built on the Tennessee River, near Athens, Alabama, in 1936 by the U.S. Corps of Engineers, the usual, most careful subsurface investigations were made all over the lock site. The installation was taken over by the Tennessee Valley Authority when that agency was established. The lock performed well and with no evident trouble until June 1961. Excavation was then in progress for a new and larger lock when 131 m (436 ft) of the landward wall of the original lock suddenly moved outwards. Prior to this, staff of the TVA had also conducted most detailed subsurface studies for the foundations of the new lock and had found nothing amiss. The lock failure was naturally subjected to most critical study, and it was found to have been due to sliding on a thin (1- to 6-mm or $\frac{1}{16}$- to ⅜-in-thick) seam of clay. This seam occurred in the 15-cm (6-in) stratum of shale which is a part of the Fort Payne formation underlying the site, most of which is a limestone, argillaceous to crystalline, fine- to somewhat coarse-grained, and all practically horizontal. Careful test drilling had

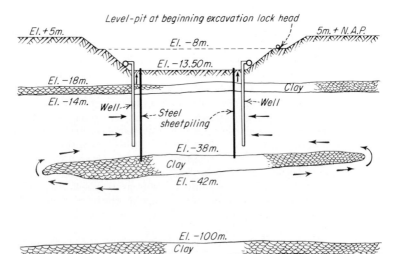

FIG. 29.8 Subsurface conditions at the site of the Ijmuiden Lock of the North Sea Canal, The Netherlands.

failed to detect the thin clay seam during both the subsurface investigations by the two prestigious engineering organizations noted.[29.13] Subsequent public discussion of the failure, sensibly constructive in nature, included a number of suggestions for improving observations made with test holes, reflected today in modern practice as outlined in Chap. 11.[29.14]

Why the Wheeler Lock should have performed so satisfactorily for 25 years and then suddenly failed remains something of a mystery even as it points, yet once again, to the vital importance of sound and regular maintenance based on periodic rigid inspections of all civil engineering structures. This necessity was well shown by some of the maintenance work carried out on the Panama Canal in 1958. Inspection had shown some trouble with underlock culverts in the Pedro Miguel locks. The lock floors were found to be perforated with many (4-in) weep holes, but no record could be found from the days of the Panama Canal Company to indicate what they were for. It was said that they had been installed in order to relieve hydrostatic pressure in the bedrock because notable uplift of the rock had been experienced during the long construction period. The uplift was almost certainly due to rebound in the bedrock following the deep excavation required for the lock chamber, but this was a phenomenon unrecognized at the time of construction. The bedrock was the Culebra formation which is a weak rock; shear failures were found when the rock

FIG. 29.9 Major maintenance work in progress on the west side of the Pedro Miguel Lock in the Panama Canal, February 1965.

29-14   CIVIL ENGINEERING WORKS

was uncovered during the lock rehabilitation work. The openings in the rock thus caused had permitted water to circulate, and this had washed out some weathered material, resulting in voids which had, in turn, affected the culverts. All was corrected and a complete restoration job done with a minimum of interference with the canal operation.[29.15] (Fig. 29.9)

29.6 SOME EUROPEAN CANALS

Because of the relatively extensive use of inland water navigation in Europe, starting with the river Rhine and other major rivers but now extended to inter-river waterways, major canal construction has been a feature of European civil engineering for many years. Major canals are still under construction in Europe, and so this brief reference to two of these waterways is warranted.

The Albert Canal is a masterpiece of Belgian engineering, completed just before the Second World War; because of its critical location, it was a vital factor in some of the war's most crucial fighting. The purpose of the canal was, indeed, twofold: it served as a defense barrier and as a means of conveying barges with a capacity of 2,600 tons from the sea to the important industrial area around Liège. It is 130 km (80 mi) long and has only six locks (as compared with 23 in the smaller canal which it replaced); these give a drop of 55 m (184 ft) from the river Meuse to the Antwerp docks. Excavation amounted to 68 million m^3 (89 million yd^3). Geology was a determining factor throughout its route; two features only can be mentioned here. Near Lanaye, at the Liège end of the canal, an open cut ranging from 19.5 to 63.6 m (65 to 212 ft) in depth proved to be necessary. This had to be carried out through a formation of tufa and chalk, a rock combination which hardens on exposure to the atmosphere. Steep slopes, as shown in Fig. 29.10, were therefore possible because of this rock characteristic. Steam shovels were used to remove clay and gravel overburden and also to remove the rock, which was so soft when first exposed that it could be removed with liquid-air cartridges. Pneumatic tools were used for dressing the finished slopes; the good effect can be seen from the illustration in Fig. 29.11. Berms, 0.9 to 1.8 m (3 to 6 ft) wide, were broken out at intervals of 10 m (33 ft). Another of the major cuts on the canal is near Briegden, some kilometers "downstream" from Lanaye. Here careful preliminary exploration had revealed difficult conditions at a depth of 39 m (130 ft), with waterlogged sand and clay; to have avoided these strata would have necessitated two extra locks in order to raise the canal level above them. The bold decision was made to stay with the original concept of a deep cut, using special drainage installations to take

FIG. 29.10   The cutting at Lanaye on the Albert Canal, Belgium, during construction.

FIG. 29.11   The Lanaye cutting on the Albert Canal, Belgium, as completed and in use.

care of groundwater, special construction methods throughout, and most carefully designed side slopes. The great cut, over 45 m (150 ft) deep, was successfully completed, and it has continued to perform well.[29.16]

Scheduled for completion in 1985 is the Main-Danube Canal being constructed in West Germany, another of the inspired dreams of Goethe. Known also as the Europa Canal, this 170-km (106-mi) inland waterway will complete a great network of waterways which will permit inland water navigation from the North Sea to the Black Sea. The section from the river Main, an important tributary of the Rhine, to the famous city of Nuremberg was completed in 1963 at a cost of $400 million. In the remaining section between Nuremberg and the Danube River, the canal will climb to its peak elevation of 325 m (1,065 ft) above the Main and then drop 285 m (951 ft) down to the Danube. Locks have a standard length of 186.9 m (623 ft) and a width of 11.7 m (39 ft) to accommodate the typical European-sized barge. Some of these are self-propelled; others travel in pairs pushed by a powerful tug. Water conservation is essential in view of the traffic which the canal is expected to carry. Water from locks being emptied is therefore pumped into storage ponds adjacent to the canal whence it can be returned to the locks when they are again filled. This sudden loading and unloading of the foundations of the storage ponds has caused some problems with settlement of soil, but these are design problems which will be cleared up long before the great waterway is complete.[29.17]

Some of the earliest accurate observations of landslides in clay were made on early French canals by Alexandre Collin in the middle of the nineteenth century. Slides impeded the construction of the Kiel Canal, such a vital link between the Baltic and the North Seas. Construction of European canals has, therefore, enriched the literature of civil engineering and assisted in its advance with special reference to waterway construction. Some canals will still be built, although they will inevitably be infrequent enterprises and so a small minority of the major civil engineering works of the future. When new canals are built, however, they will be perhaps even more dependent upon geology than any other work in the practice of civil engineering, apart only from that involved in water supply. Fortunately there is now available a wealth of experience upon which canal designers of the future will be able to draw.

29.7   REFERENCES

29.1   F. J. Bennett, "The Influence of Geology on Early Settlement and Roads," *Proceedings of the Geologists Association,* **10**:372 (1888).

29.2   *The Times,* London, 20 August 1937.

29-16　CIVIL ENGINEERING WORKS

29.3　J. P. Eckermann, *Conversations of Goethe with Eckermann,* Everyman ed., Dent, London, 1930, pp. 173–174.

29.4　T. M. Reade, "The Advantage to the Engineer of a Study of Geology," *Transactions of the Liverpool Engineering Society,* **10**:36 (1888).

29.5　R. F. Legget, *Rideau Waterway,* rev. ed., University of Toronto Press, Toronto, 1972.

29.6　D. McCulloch, *The Path between the Seas,* Simon and Schuster, New York, 1977.

29.7　Statistics from the engineering and construction director, Panama Canal Company, Canal Zone.

29.8　Report of the Committee of the National Academy of Sciences on Panama Canal Slides, *Memoirs of the National Academy of Sciences,* vol. 18, Washington, 1924.

29.9　H. M. Arnold, "Taking the Menace out of Contractors' Hill," *Engineering News-Record,* **154**:34 (24 February 1955).

29.10　W. Grothaus and D. M. Ripley, "The St. Lawrence Seaway, 27-ft Canals and Channels," *Proceedings of the American Society of Civil Engineers,* vol. 84, WW2, Paper no. 1518, New York, 1958; and in the same series, L. H. Burpee, "Canadian Section of the St. Lawrence Seaway," vol. 86, WW1, Paper no. 2420, 1960; see also R. F. Legget, *The Seaway,* Clarke Irwin, Toronto, 1979.

29.11　F. L. Peckover and T. G. Dustin, "The St. Lawrence Seaway: Soil and Foundation Problems," *The Engineering Journal,* **41**:69 (September 1958).

29.12　J. A. Ringers, "Construction of the New Ijmuiden Lock," *Engineering News-Record,* **104**:769 (1930).

29.13　"Clay Seam Wrecked Wheeler Lock," *Engineering News-Record,* **168**:19 (4 January 1962); see also "Horizontal Clay Seam Triggered Wheeler Lock Failure," *Civil Engineering,* **32**:89 (March 1962).

29.14　Useful discussion in correspondence columns of *Engineering News-Record,* **168**:6 (1 March 1962) and **169**:6 (9 August 1962).

29.15　F. H. Lerchen, "Explosive Buckling of Floor Corrected in Panama Lock Overhaul," *Civil Engineering,* **29**:158 (1959).

29.16　"War Christens Belgium's Albert Canal," *Engineering News-Record,* **124**:729 (1940).

29.17　"German Canal Climbs Toward European Link," *Engineering News-Record,* **200**:61, 62 (22 January 1978).

Suggestions for Further Reading

Somewhat naturally, modern reading with regard to canals consists almost wholly of descriptions of the few new canals that have been built and of historical accounts of early canals. The American Canal Society, with its useful *Bulletin,* serves the interests of those so historically minded.

Two classic works on canals are still well worth reading today: the great report on the landslides on the Panama Canal by the special committee of the National Academy of Sciences, referenced above, and the record prepared well over a century ago of observations on landslides at French canals by Alexandre Collin. This notable work has been translated by W. R. Schriever and published in English by the University of Toronto Press in 1956.

chapter 30

ROADS

30.1 Route Location / 30-4
30.2 Climate / 30-6
30.3 Drainage / 30-8
30.4 Materials / 30-12
30.5 Construction / 30-15
30.6 Roads as Dams / 30-20
30.7 Some Geologically Significant Roads / 30-22
30.8 What Road Construction May Reveal / 30-27
30.9 References / 30-28

The Story of the Road is the title of a book (by J. W. Gregory) which traces the broad outlines of highway development through the ages, a book that is mentioned at the outset of this chapter since so little can here be said about the background of this particular application of geology.[30.1] Roman road building was always carried out with intuitive geological appreciation. At the height of the Roman Empire, over 80,000 km (50,000 mi) of first-class highways had been built and were in use by Roman legions. Julius Caesar was at one time curator of the Via Appia and is said to have expended vast sums of his own money upon its improvement. On the South American continent, however, was to be found the greatest of all early roads. This road, built by the Incas, was 6,400 km (4,000 mi) long and stretched from Quito in Ecuador to Tucuman; it was 7.5 m (25 ft) wide, and much of its surface was paved with bitumen. A second road paralleled the main road for almost 3,200 km (2,000 mi) along the coast. Much of this great road can be seen today; some of it is still in use. All who are interested in the historical background of modern engineering have a treat in store for them if they have not yet read Prescott's *The Conquest of Peru*, a volume that can well be termed supplementary reading to any study of the relation between geology and road building.

It was not until the dawn of the industrial era that road building again assumed importance; Britain and France led the way in this early branch of modern civil engineering. John Metcalfe (Blind Jock of Knaresborough) was probably the first of the pioneers. Sir Walter Scott, through the character of Wandering Willie in *Redgauntlet*, has shown how this gifted man overcame his blindness even to the extent of appreciating geology in his road-building work. It was not until 60 years later that Thomas Telford and James Loudon McAdam started their notable road work, the results of which are still to be seen in

30-2   CIVIL ENGINEERING WORKS

Great Britain today. Although neither of these great men wrote very much, the writings they did leave show clearly that geology (even though not so called) was a vital part of their thinking about roads.[30.2]

North America had its road pioneers also, quite naturally long after those of Europe in view of the differing pace of development between the Old World and the New. It was as early as 1906, for example, that Dr. C. M. Strahan, then county engineer of Clark County, Georgia, started his experiments on the desirable properties of soils for road-building purposes, although his first published statement on this work did not appear until 1914. This pioneer research work of Dr. Strahan paved the way in North America for the more general scientific study of soils for engineering purposes, now so well known as soil mechanics, as was noted in Chap. 6. Just a year after the start of this work in Georgia, the Wisconsin Geological Survey (in 1907) was granted $10,000 by the state legislature for investigations and experimental work in road building.[30.3] Such progress was made in comparable studies in Illinois that in 1927 the Illinois Geological Survey organized its own areal and engineering geology division, one of the first such units to be established.[30.4] In other states, similar progress was made. In Kansas, for example, the state highway commission appointed a chief geologist as early as 1937, a position later occupied by S. E. Horner, who made notable contributions to the applications of geology to highway engineering.[30.5]

Today, there is probably no branch of civil engineering in which the potential contributions of geology are more generally appreciated than in the field of road design and construction. There are even long-established annual meetings devoted to this one subject; for instance, annual symposia on geology as applied to highway engineering have been held in the eastern states since 1950. At such meetings, those engaged in this kind of professional work meet and share their experiences. Special note must be made of the annual meetings of the Transportation Research Board (for many years known as the Highway Research Board) of the U.S. National Academy of Sciences; the records of the board, together with other T.R.B. publications, contain a wealth of information on the uses of geology in highway engineering. Possibly the ultimate recognition is given by the fact that the American Society for Testing and Materials (ASTM) has published a standard test method for surveying and sampling highway subgrade soils.[30.6] There are some who dislike seeing any geological procedure reduced (if that be the word) to the status of a standard procedure, but if the ASTM document is taken as a guide, for it is so intended, then it can be of real assistance to those who have to direct the work of others in this field.

With all this valuable information widely available for use, there might seem to be little need for a chapter such as this in this Handbook. But the information has to be used; interesting as it is, it is of no value if left between the covers of books. Accordingly, cases have been selected for the sections which follow to illustrate both the sound application of geology in highway engineering and what can happen if it is neglected. As a cautionary tale, so to speak, Fig. 30.1 has been included to show a very simple example, one from many years ago but illustrating the type of problem that can be met with if preliminary subsurface investigation is not properly carried out for highway construction. The section shows the subsurface conditions, determined after the trouble had happened, between two lakes now connected by a small stream. It is clear that the highway could not have been located more unfortunately, since it is directly above the ancient glacial valley that at one time existed between the two lakes. The existence of this valley could have been determined by a combination of test borings and geological investigation, and about 7.5 m (25 ft) of fill over this section of road could have been saved.[30.7]

Fortunately, such cases are now the exception rather than the rule. It will be appreciated that in the actual design of a new highway many of the principal branches of civil

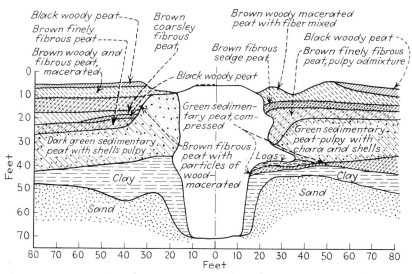

FIG. 30.1 Cross section through an unusually deep fill; preliminary borings would have indicated local subsurface conditions and so suggested an alternative location.

engineering work may be involved, such as the construction of bridges, the driving of tunnels, and the stabilizing of slopes, all of which are common to other fields of work and which are herein considered in separate chapters. The design of the roadbed and the procurement of construction materials are, however, two special features of highway engineering. It might be thought that sound and solid rock would provide an admirable roadbed; however, this is not always so, since solid rock is of varying strength and durability, and the existence of joints often adds a further complication. The softer rocks such as chalk constitute fair roadbeds when dry, but they can be troublesome when wet. Clay is similar; it is extremely unreliable when wet because many clays expand as they increase their moisture content. Probably the best of all roadbeds is provided by well-packed coarse sand and gravel resting on some type of solid substratum. This ideal case is not often met, and the engineer has to contend with actual ground conditions varying from fissured igneous rock to peat bogs.

In this branch of civil engineering only, the wheel may be said to have turned full circle since there are now available highway maps showing the geology of the areas covered by the maps. Ontario has issued, for example, through its departments of Natural Resources and of Transportation and Communications (formerly, Highways), a fine highway geological map of Ontario. There is reluctance to go further and erect signs on the sides of highways drawing attention to significant geological features, since there would be a danger of such interest being shown in these features that attention might be diverted from safe driving, possibly with disastrous results. There are a few such signs, but usually so arranged that they can be viewed only by travelers on foot or in parked cars. A better approach has been the publication of handbooks in which geologists (and engineers) explain and interpret the geology of the land adjacent to main highways for the information and enjoyment of those who travel these routes by bus or automobile. One of the earliest handbooks known to the writers is that issued in 1958 by the West Texas Geological Society. This is a well-produced, quarto-sized guidebook to the geology (and also the history) of the country traversed by U.S. highways 90 and 80 from Del Rio to El Paso. A concise road log is well supplemented by a brief description of the areal geology and notes on the wildlife and plants to be seen along the highway.[30.8] The guide may well serve as a model to other groups who contemplate the same sort of public service.

30.1 ROUTE LOCATION

In land transportation, there are two main problems to be faced in the design and construction of both roads and railways: (1) the selection of the route to be followed and (2) the provision of a roadbed and a bearing surface of reasonably permanent stability and capable of carrying intended traffic. The final choice in both will today depend largely on economic aspects, but geology may often have a considerable bearing on the final solutions. In very early days, road location was a matter of adapting for vehicular use older pathways that had been formed by primitive peoples as their tracks through the forests and along river banks. Roman road construction marked a significant break from this primitive practice; Roman engineers usually laid out their roads, upon which the greatness of their empire depended, as straight as possible, a practice that continued in France until the eighteenth century. When changes in direction had to be made, this was generally done on high ground or at station houses; curves were used only infrequently. The Fosse Way in England, to cite one example, is about 320 km (200 mi) long (joining Lincoln and Axminster), but the greatest departure from a straight line between the two end points is only 10 km (6 mi). There has been much discussion of the reason for this almost invariable practice of route selection; one suggestion (not very well-founded) is that the Romans could not construct vehicles that could easily negotiate curves. The Romans only rarely used anything but embankments for easing the grades on their roads; there does seem to be some foundation for believing that this was done deliberately to promote snow clearing by the wind, a practice regarded by many as a feature of modern road design.[30.9]

Roman road building, however, for centuries stood alone. It took the coming of the railway to direct attention to the basic problems of economical route selection, of severely limiting grades, and of keeping curvature to a minimum. In flat country, selection was largely a matter of convenience and of land availability; the route chosen was selected to give the shortest possible convenient link between the centers to be joined, although deviations were made if necessary to obtain improved river crossings. In hilly country, however, many alternatives could be considered. The ridge route up a valley, for example, might be chosen in preference to the valley route; the climb to the ridge would more than compensate in convenience for the numerous river crossings usually necessary if a valley route was used. Such selections were inevitably influenced by the underlying geology, since it determines the local topography.

In areas which have been glaciated, it will often be found that glacial landforms have influenced the location of early roads. It is surprising how frequently the older towns in Ontario, and in similar glaciated areas, will be found to have a "Ridge Road" as a well-established older highway, the name clearly indicating the original use of a prominent geomorphological feature, always a glacial landform. Ridge Road, between Stirling and West Huntingdon, north of Trenton at the east end of Lake Ontario, is one excellent example, since it so clearly follows a well-defined esker. Special glacial landforms should, therefore, always be kept in mind in the location of new roads in glaciated areas, since such landforms may provide useful locations not immediately obvious. One relatively recent example was the location of Ontario's major divided highway (Route 401) leading from Ontario cities to the province of Quebec and Montreal. Between Port Hope and Brighton, the highway was located on an old beach of the former Lake Iroquois, instead of the more obvious location along the shore of Lake Ontario. Better grades were obtained as were excellent supplies of sand and gravel needed for construction, even though numerous bridges were necessary for crossing the streams and gullies that had eroded courses in the old beach deposits. And as recently as 1973, one of the new roads into northern mining areas of Ontario had its location changed in the planning stage back to an adjacent esker which it followed as far as practicable; it then became possible to utilize a morainal surface for the continuation to the mining area.[30.10]

The building of superhighways has resulted in many similar vivid examples of the effect of local geology upon road location and upon the costs of construction within the limits thus determined. The similarity of modern road location with the long-established practice of railroad location is nowhere better illustrated than by the construction of the Pennsylvania Turnpike, much of the route of which utilized an abandoned railway right-of-way. Even though this pioneer, modern toll highway was opened to traffic on 1 October 1940, after a remarkably short construction period of only two years, the Pennsylvania Geological Survey published a notable guidebook to the geology of the entire highway as soon as possible after the end of the Second World War. This 54-page description gives a good general account of the geology of the region through which the road is located and a succinct description of the special engineering features affected by the geology along the highway, especially in the old railroad tunnels; subsurface investigations carried out for the new road are also described. The guidebook concludes with a "geologic itinerary" from one end of the turnpike to the other; the whole book is well illustrated by almost 40 photographs.[30.11]

New roads will still be built, not only in the less-developed parts of the world but also in North America; new railways will also still be built. What techniques are now available for final route selection after a general course has been determined? Here is found change indeed, and current techniques contrast greatly with the laborious work of field location practiced until the 1930s. Today, office study of topographical and geological maps is a first requirement, and almost always a possibility, even though the scales of maps available for more isolated areas are still small. Concurrently, the detailed study of aerial photographs of the area through which the proposed route must run is now a well established and invariable aid that gives a different pattern to route location.

Preliminary work for the Ohio Turnpike Project No. 1 may be cited as an early example of the application to a modern highway undertaking of the procedure summarized above. This first-class, modern dual highway is 390 km (241 mi) long and cost about $1 million per mile; almost 30 million m^3 (40 million yd^3) of excavation was involved, with a corresponding volume of fill, and over 5.8 million m^2 (7 million yd^2) of concrete paving. A complete set of consecutive aerial photographs to a scale of 1 : 10,000, covering a strip of territory 2.4 km (1.5 mi) wide within which it was known the road would run, was first studied. A band 600 m (2,000 ft) wide was then selected for more detailed study. All available pedological and geological maps and reports relating in any way to the selected route were studied concurrently. Thus all main soil types and soil boundaries, drainage patterns, etc., could be marked upon the photographs; the photographs were viewed stereoscopically and terrain-interpretation techniques were used. The Ohio Geological Survey provided some unpublished material on abandoned glacial shorelines which, when applied to the photographs, proved to be unusually accurate. This is a good example of the use of unpublished geological information that is sometimes to be found if approach is made to the appropriate authorities. Preliminary correlation of pedological soil characteristics with engineering soil properties, assisted by study of information provided by the Ohio State Highway Testing and Research Laboratory, meant that a good approximation to soil conditions along the route was obtained merely by office work.

Careful checklists were prepared for field study. These greatly facilitated the final surveys in the field, which did not have to be so extensive because of the preliminary correlation of all existing information through the vivid aid of aerial photographs. Before field studies commenced, 34 stream crossings and 114 road and railway crossings were similarly studied in detail in the office; the study was again aided by state highway department information obtained from adjacent locations. Twenty-two contracts were then awarded to 11 contractors for the requisite test-boring program, the main lines of which had been carefully and accurately determined by means of the office studies. Eventually 18,013 m (60,405 ft) of earth borings and 2,644 m (8,812 ft) of rock borings were put down at a total cost of $441,785, a figure that may usefully be compared again with the total cost for the entire project of over $200 million. All rock cores and soil samples were passed to the state geologist for safekeeping and public use, a good example of the reciprocal service that now so happily characterizes cooperative geological-engineering investigations.

Some special problems were encountered with glacial tills and with road construction over swamplands, but the only really unusual feature of the project was that part of the route passed through an area well known for coal strip-mining operations. Land of potential value for strip-mining operations had to be carefully surveyed before being preempted for road construction. Forty special prospect borings were put down to supplement surface observations (which included study of adjacent strip-mining operations) in order to provide the necessary information upon which land valuations could be based. It was found that, for the Mahoning County area, the average quantity of coal to be obtained by strip mining could be estimated as 125 tons per acre per inch of seam. It was also found that strip mining of coal could be regarded as economical if a maximum of 60 cm (2 ft) of overburden was removed for every inch thickness of merchantable coal. Although these figures are directly applicable only to this area of Ohio, they may be useful as a general guide. Cash payments for land taken over for the turnpike project were made on the basis of the factual evidence thus assembled, despite the claims made by landowners in the manner usual on such occasions. The full record of these preliminary geological and soil studies for the Ohio Turnpike is a valuable guide to the best of modern practice in route selection.[30.12]

With railways and roads now so commonplace in the more developed countries of the world, it should not be forgotten that surveys for many of the pioneer routes of the New World in particular (especially for railways) provided the first information about the local geology of the areas traversed. There are fortunately on record many papers giving invaluable geological information as by-products of such preliminary engineering work. A singularly useful paper of this sort was published as early as 1888 [30.13]; one of the most recent examples is the record prepared of the geology along the Alaska Highway, a wartime emergency route that penetrated the relatively unknown Canadian Northwest.[30.14] Its construction was a remarkable wartime achievement; many geological problems were encountered then and have had to be dealt with since in its maintenance and improvement as a fine civilian highway.

30.2 CLIMATE

Because of the widespread nature of road construction operations and the fact that they always involve disturbance of natural ground surfaces, they are unusually susceptible to the vagaries of the weather. Even in North America, there are areas in which a working season of only a few months is available each year because of high precipitation in the form of snow and rain. Slow progress on the Trans Canada Highway through the Rocky Mountains, for example, was attributable to this cause. Even in an area such as Michigan, the weather has affected the progress of road construction. During the building of the Detroit Industrial Expressway, large, wide cuts in brown and blue glacial clays were necessary. When the stratum of weathered brown clay was exposed to freezing weather, its laminated character led to its rapid surface disintegration and subsequent sloughing when wet by spring rains; this interfered appreciably with construction progress.[30.15]

More unusual, perhaps, was the effect of heavy rain upon the rock used as the subbase for Florida's Sunshine State Parkway, completed in 1957. Soon after its opening, trouble was experienced between Stuart and Fort Pierce at the northern end of the road because of "alligatoring" of the pavement. This was traced to the fact that, during paving operations, heavy and unseasonable rainfall (up to 52.5 cm or 21 in) was experienced. This had the effect of seriously wetting the Belle Glade, or Okeechobee, rock, a somewhat unusual type of lime rock which here formed the base of the road; it became "greasy" when wet and under load, and this led to the pavement failures. Tests upon this rock in its dry state had been quite satisfactory; possibly the proverbial sunshine of Florida contributed to the apparent neglect of the possibility of change in properties when the rock was wet.[30.16] This can be taken as another salutary reminder of the importance of climate in relation to geology.

In no other branch of engineering is the importance of climate more to be remembered than in highway engineering. As a first step even in preliminary designs, the *hythergraphs* for the different areas through which a new highway is to pass should be critically exam-

FIG. 30.2 All that is left of the Parkgate Station (Alberta) of the old Grand Trunk Pacific Railway built in 1915. It is now engulfed in moving sand dunes, whose progression down the valley of the Athabasca River is due to the prevailing wind: a vivid illustration of the possible effect on transportation routes of one climatic factor, wind.

ined. This is but a first step. Studies should be made of the groundwater depletion situation along the selected route. The influence of climate upon materials exposed in the vicinity of the highway must be investigated, especially if any are to be used for construction. Particularly is this true in nonglaciated areas. Significant work in this field has been done in South Africa, notably by H. H. Weinert. For the Ventersdorp supergroup of andesitic and basaltic lavas, with which are associated some pyroclastic rocks, three quite different soil profiles are found; these profiles correspond with the three well-recognized climatic zones of semiarid, subhumid dry, and subhumid moist. Following a wide-ranging study of the interrelation of climate and the weathering of exposed rocks, Weinert has demonstrated "that mechanical disintegration is the predominant mode of rock weathering in areas where his climatic N-value is greater than 5, whereas chemical decomposition predominates where the N-value is less than 5." This N-value is determined by dividing 12 times the evaporation during the month of January (summer in South Africa) by the annual precipitation.[30.17] The guide noted is for South Africa only, but the concept is one worthy of detailed study in all areas featured by similar climatic conditions. (Fig. 30.2)

30.3 DRAINAGE

Drainage is of supreme importance in all highway work in order to prevent waterlogging of the subgrade and consequent trouble. Drainage methods vary; they must inevitably be suitably related to the nature of the materials to be drained, the physical properties of which—in the case of unconsolidated materials—can be tested in the same way as can those of other soils. The underlying geological principle of keeping the water table sufficiently low so that gravity flow through the material can take place is applied in various ways: with "self-draining" porous roadbed materials by means of suitable cross-sectional road design and with impermeable materials by the provision of suitable artificial drainage channels or by the use of a layer of sand or gravel.

It is not often that the reverse process has to be undertaken, but this has been necessary in at least one instance. A concrete roadway in Ellis County, Kansas, was found to be settling unevenly and causing high joints between adjacent slabs. Investigations showed that the differential movement was due to changes in the moisture content of the subgrade soil occurring at joints and cracks. This was successfully corrected, and the concrete slabs were releveled by introducing water into the subgrade material through an installation of well points jacked into place about 25 cm (10 in) below the slab from the roadway shoulder.[30.18]

Drainage can only be successful if the design of drainage facilities is carried out when full and detailed knowledge is available of the geological strata beneath the roadbed and in the area surrounding the road. If carried out when a road is initially constructed, adequate drainage facilities represent only a very small part of total road cost. If, however, initial drainage design is inadequate, and drainage facilities have to be added when a road is in service, before their inadequacy has had the far more serious effect of damaging the road itself, then costs can be astronomical. These truisms might seem to be out of place in a modern handbook since the need for good drainage has been recognized, by some, from the very earliest days of road construction. But as recently as 1979, a leading North American engineer had this to say, inter alia:

> Pavements designed without good internal drainage are doomed to early failure the day they are completed, because failure mechanisms are built right in. . . . On a national basis it would be impossible to place an exact monetary value on the total loss caused by the undrained pavement practice, but it is clearly rising to near catastrophic levels . . . [and using the writer's estimates] the potential loss [in the United States] that can be blamed on poor drainage is about $200 billion for those 15 years [1976 to 1990] or a little over $15 billion annually.[30.19]

Harry Cedergren, in this arresting statement, reminds his readers of the statement of John McAdam, made as long ago as 1820: "The erroneous opinion ... that a road may be sufficiently strong, artificially, to carry heavy carriages though the subsoil be in a wet state ... has produced most of the defects of roads in Great Britain." Those words may not have traveled over to North America at the time they were voiced, but the reader may wish to try to date the following statement published in North America: "Drainage is an affair of primary importance in roadmaking, and requires much skill to execute in a proper manner ... it may even be necessary to carry the process of draining far beyond the area of the road.... In this country, the roads sustain much injury from heaving up by frost. This would be, in great measure, prevented by adopting a better system of drainage...." These words come from a small 42-page pamphlet published in Toronto, Upper Canada (now Ontario) in 1841. The author was a remarkable man, both a civil engineer and geologist (just as was William Smith), named Thomas Roy. Little is known of his private life although he served as the equivalent of the city engineer of the fledgling city of Toronto during his final years (he died in 1842), but the whole of his pamphlet on road building can be read with profit today, a century and a half later.[30.20]

The warnings were there for all to read, but throughout the century and a half since there has been probably more trouble on highways due to inadequate drainage facilities than from any other cause. Even on so fine a modern highway as the New York Thruway, serious trouble was experienced, apparently because of inadequate drainage, on the Berkshire Extension very soon after it was opened. The area in which the trouble evidenced itself is underlain by glacial till overlying horizontal thin strata of clayey silt and fine sand. Because of heavy rainfall during working seasons, trouble was initially experienced with earth moving in the necessary cuts. The roadway consists of a 23-cm (9-in) -thick reinforced-concrete pavement on a base course of select granular material. Cracking and "pumping" did serious damage to the pavement, parts of which had to be removed. Material below the base course was then removed, some to a depth of 3 m (10 ft), and then replaced before the roadway was rebuilt.[30.21]

This case was one of the few reported, but it is regrettably typical of widespread trouble due to inadequate drainage, even on modern highways. Design of the drainage system is an engineering matter. All that can be said here is that unless drainage systems are designed with full appreciation of the geological conditions over which a road is located, they may not perform as intended. Drainage facilities are essentially simple, a variety of types of pipe being now available to meet almost any combination of ground conditions. Exceptional drainage measures are discussed in Chap. 18. The use of sand drains alone may be again mentioned, since they will sometimes provide a solution to an otherwise serious drainage problem. Used in a simple form by early French engineers, sand drains were developed in their modern application in the 1930s, in the first instance for road work on the Pacific Coast.

Water, as is so often the case, is the prime cause of trouble in another important phase of road building, i.e., when the route selected lies over terrain that is so unsatisfactory that a road cannot be built directly upon it. Peat bogs, swamps, marshes, muskeg, and tidal flats are examples of such a condition; all are relatively new geological deposits which have not consolidated to any extent and which consequently contain a high percentage of water. Study of the local geology and an adequate program of test borings will usually reveal the extent of these deposits, since they are not the "limitless sinkholes" sometimes popularly imagined. In early civil engineering work, construction across such materials was generally tackled by dumping solid fill material along the route until it finally stopped settling or, alternatively, by laying down mattresses of brushwood or similar material to serve as artificial foundation beds for solid fill which was then deposited on them. The crossing of Chat Moss in 1827 by the original Liverpool and Manchester Railway in England, with George Stephenson as the engineer, was one of the earliest notable examples of the use of this device. These ideas are sometimes used today; but with accurate information on the extent of the poor material, other methods can often be applied to better advantage. Replacing the soft material with dry fill by means of jetting is a new method; jetting of peat with pressures up to 2,000 g/cm^2 (100 psi) is carried out underneath the dry fill, which is allowed to settle as the disturbed peat is displaced laterally.

Displacement of unstable deposits by blasting is an even more widely used method; blasting with high explosives is carried out either beneath dry fill which has been dumped in place or well down in the soft material in front of the fill deposit. Figure 30.3 is a photograph of an early example of this work; its dependence on accurate data regarding subsurface conditions will be clear. The work illustrated, on the notable dual highway constructed from Baltimore to Havre de Grace, Maryland, was necessitated where the route crossed swamp areas (Fig. 30.4). The surface vegetable mat was first removed by blasting with 50 percent nitroglycerine, and the fill was then advanced and blasted into position; a 40 percent gelatin explosive which has a slow heaving effect was used.[30.22] This type of construction operation has been considerably developed in North America in more recent years.

At the other extreme from the problem posed by swamp conditions is the problem of building roads through loess, from which water must be kept away if its stability is to be maintained. Travelers record that in the interior of China, ancient roads are to be found in the loess country which forms so large a part of that great land. The roads have been worn deep into the dry, friable loessal soil by the passage of wagons through the centuries; the wind has blown away the dust formed by the steady disintegration of the loess by wagon wheels. Pumpelly, for example, records having seen such roads worn 15 m (50 ft) below the general level of the country traversed.[30.23] Widely distributed throughout the world, from New Zealand in the south to the northern parts of the United States and the U.S.S.R., loess has now been carefully studied in relation to its occurrence on highway routes. Splendid exposures of this interesting soil are to be found on roads in Iowa. During the relocation of U.S. Highway 127 near Magnolia, the Iowa State Highway Commission experimented with a new type of terraced design for deep cuts in the loess, cuts which involved the movement of 610,000 m^3 (800,000 yd^3); some cuts were 24 m (80 ft) deep. Terraces 4.6 m (15 ft) wide were cut between vertical faces each 4.6 m (15 ft) high; the terraces were graded slightly toward their inner edges so that precipitation could be taken

FIG. 30.3 Blasting operations in progress for fill consolidation over swampland near Ha-Ha Branch, U.S. Route 40, Maryland.

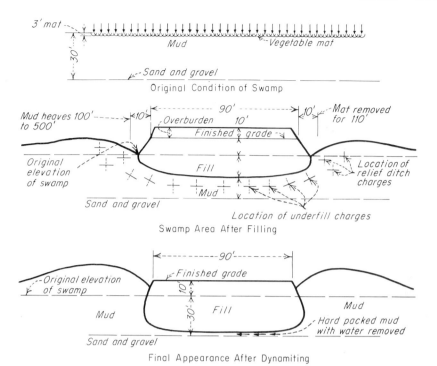

FIG. 30.4 Diagrams illustrating the method followed in consolidating fill material over swampland during the construction of U.S. Route 40, Maryland.

FIG. 30.5 Vertical cuts in loess just east of Vicksburg, Mississippi, showing how well it will stand if protected from water.

30-12 CIVIL ENGINEERING WORKS

care of by a properly installed drainage system (Fig. 30.5). The loess here is typical in texture; it is composed of angular to subangular quartzose and feldspathic silt and fine sand with montmorillonite as the binder. Calcite is also present but not as a binder. Examination of specially prepared thin sections under a microscope shows an open structure with little interlocking of the grains.[30.24]

30.4 MATERIALS

The materials used in road making are almost all used in their natural state (cement being the main exception), and so the geology of road materials is important. Nobody recognized this more than the Romans. Typical of their construction methods was the use of a base course of sand upon which four separate layers of masonry were laid with the addition of a simple kind of concrete. In Italy, the top layer (the *summa crusta*) was usually constructed of lava cut into hexagonal shapes so carefully that the joints could scarcely be seen. The Romans always selected the hardest stone they could readily procure in each district served by their roads, just as did the Incas in their almost incredible road construction down the Andes in the northwestern part of South America.[30.25]

Among the road-making materials used regularly by the Romans was the waste from iron forges, especially the cinders from the forge fires. Roads so constructed were known as *viae ferriae*, a name that is said to be the origin of the French term *chemins ferrés* and almost certainly the origin of the English expression *road metal*. So recent have been the basic changes in road construction that there will be users of this Handbook who can recall hearing this old term actually in use. It derives from the use of crushed stone that dates from the start of modern road development in France and Great Britain in the early years of the nineteenth century. Thomas Telford and James Loudon McAdam (whose name is memorialized in "macadamized" roadways) were the two great British pioneers. Although differing in their ideas on roadway design, each in his own way had a thorough appreciation of the importance of geology in road planning and in the selection of natural materials for road making.

If only for its historical interest, the importance that McAdam himself gave to the geology of the rock he used is well shown in these words from his own notes (1823):

> Flint makes an excellent road, if due attention be paid to the size, but from want of attention, many of the flint roads are rough, loose and expensive. Limestone when properly prepared and applied makes a smooth solid road and becomes consolidated sooner than any other material; but from its nature it is not the most lasting. Whinstone is the most durable of all materials; and wherever it is well and judiciously applied, the roads are comparatively good and cheap.[30.26]

Among the principles that McAdam laid down were the following: (1) no stone should exceed 2.5 cm (1 in) in any major dimension; (2) all stone used should be clean; (3) the foundation should not be solid rock or hard pavement (a matter on which Telford had other opinions); (4) the camber used should not be too great; and (5) the rock when laid should possess a natural water bond. McAdam's ideas were based on his practical experience; his selection of rocks suitable for road use was similarly empirical. Subsequent detailed study and microscopic examination of rocks available for road work generally confirmed his original ideas.

The necessity for good wearing surfaces for city streets to resist the abrasion of steel-rimmed wheels and to give adhesion for horses, the universal source of earlier motive power, led to the development of special pavements using both natural and manufactured materials in the form of granite setts or paving brick. Granite setts, usually 10-cm (4-in) cubes, used to be one of the main products of all granite quarries. Although generally superseded, they may still be seen in use in the older parts of industrial cities, and quite frequently even on country roads, in the older parts of Europe, such as in central Bohemia. In Prague may also still be seen slices cut from large trees and trimmed to hexagonal shape forming the roadways of ancient courtyards—an early example of acoustical control. When naturally shaped hexagonal blocks were readily available, as from quarries, in columnar-jointed basalt, they were sometimes employed directly, a practice that may still be seen in use today for the revetments along the banks of the Rhine, some of the hexagonal blocks so used coming from Czechoslovakia.

With the turn into the twentieth century came also the beginnings of great changes in traffic upon roads and streets, especially in cities, and a corresponding change in the previous, somewhat empirical, approach to the design and construction of roadways. The proper use of soil is a development of the last century; it appears that investigations initiated in 1906 by Dr. C. M. Strahan, then county engineer of Clark County, Georgia, were among the earliest attempts to apply to road work some of the facts concerning soil behavior that were then just being observed. Dr. Strahan's first public statement was made in 1914. At the first conference of state highway-testing engineers held in Washington in February 1917, the matter received considered attention, and from that time forward advance leading to today's well-established discipline of geotechnique can be noted.

Soil, concrete, and asphalt are the three major materials used in modern road work. Asphalt will not generally be as familiar as the others to engineers; it is a truly remarkable material. The word is properly used to describe solid or semisolid native bitumens and bitumens that are obtained by refining petroleum, which melt upon the application of heat and which consist of a mixture of hydrocarbons and their derivatives. A vast technology has been developed to serve all users of bitumens and bituminous mixtures, but it will have been noted that, whether obtained directly in Nature or extracted from petroleum, asphalts are ultimately derived from geological formations. The naturally occurring asphalts are of unusual geological interest. Important deposits of pure bitumen are found near the Dead Sea and in the world-famous Asphalt Lake in Trinidad. Refined bitumens are obtained from a wide variety of crude petroleums. Of special geological interest also are the asphalt-bearing rocks that are found in many parts of the world and that may be used, after crushing, directly in highway work. Many of the streets of Paris have been paved with such material. From the numerous examples of its use in North America, the practice of the Texas Highway Department may be cited.

About 160 km (100 mi) west of the city of San Antonio is found a porous and highly fossiliferous limestone known as the Anacacho formation, a member of the Gulf series of Cretaceous rocks. The formation covers a wide area, but that part with the highest concentration of natural asphaltic content is found in southern Uvalde and Kinney counties. In the 1890s asphalt was actually extracted from this rock for use in the paint and rubber industries. It was used, after crushing and screening, in the construction of streets and sidewalks in San Antonio as early as the turn of the century. It was put to wider use in highway construction for paving in 1912. This was the start of a gradual increase in its use in road work to such an extent that the Texas Highway Department has been using more than half a million tons of it every year since the early 1950s (Fig. 30.6).

According to Dr. T. S. Patty, "the asphalt content of the presently mined limestone ranges from less than 1% to about 14% ... [but] quarry operations coupled with plant practices are used to blend proper proportions of the 'lean' rock with the 'rich' to insure an asphalt content between 5 and 9% (by weight) for the commonly used asphaltic paving materials shipped for highway use from the Uvalde sources."[30.27] In order to activate the native asphalt in the limestone so as to render the mixture suitable for direct use, a flux oil, consisting of a nonvolatile oil containing about 26 percent of oil-produced asphalt, is

FIG. 30.6 Rock-asphalt quarry and processing plant, Uvalde County, Texas.

used. So important has this source of material become that the highway department operates its own field-testing laboratory at the Uvalde rock-asphalt mines. Known reserves of this valuable "industrial mineral" are sufficient to last into the twenty-first century; it is anticipated that the material may well come to be substituted for other kinds of aggregate, the supplies of which are rapidly becoming exhausted in parts of Texas. Similar records are available for other areas that have the good fortune to have natural rock-asphalt deposits. In regional geological studies related to the urban planning of new towns, therefore, this is one mineral that should always be kept in mind by those responsible for the studies, since it can be of such convenience as well as of great economic benefit.

Economics will usually dictate the use of material as close as possible to the road location; study of the local geology is an essential prerequisite to deciding what local materials can be thus used. In some cases, unusual materials have been put to good use after necessary laboratory studies of their properties. In Liberty County, Texas, for example, seashells obtained from reefs in Galveston Bay have been successfully used, when mixed with sand, to form a most satisfactory subbase for main road construction.[30.28] More unusual, but of potentially wider application (with supplies of good road-building material steadily decreasing), is the practice of calcining in place the clay which forms the natural ground at or adjacent to a road location. Apparently this idea was first tried out in Australia (an original patent having been granted to L. H. R. Irvine of Sydney, New South Wales), where local adobe clay, as found in parts of New South Wales and Queensland, was successfully treated in this manner. This clay is one that can be satisfactorily calcined by means of wood-fired down-draft furnaces of the air-gas-producer type; the resulting product is bricklike in texture. The untreated clay (which softens quickly when moist) is used as a binder in places where termite mounds (providing antbed material) are not available; the resulting road is stable in wet weather, and the stability is achieved at such relatively low cost.[30.29] In more recent years, the same technique has been used with the black cotton soils of the tropics. In both the Sudan and Nigeria, experiments have been conducted on the effect of calcining this difficult soil, but this has so far been carried out in specially built kilns adjacent to airfield-construction jobs.[30.30]

With the rapid increase in road construction in developing parts of the world, so many of these being in tropical regions, there is much activity in the investigation of local residual soils in relation to road construction. Further work on the possibilities presented by calcining soils in place for secondary road construction would seem to be warranted. Merely as an indication of the activity in this general field, there may be mentioned work done at the Road Research Laboratory in the United Kingdom, work on which this progress report was made:

> The volcanic soils have a varied clay mineralogy reflecting environmental factors and derivation from a wide range of volcanic rocks. Common clay minerals which may be present are halloysite, disordered kaolinite, goethite, montmorillonite, hematite and vermiculite. These greatly influence the properties of soils such as the East African red clays in which aggregation of clay results from flocculation by ionized iron and cementation by iron oxides with resulting decrease in plasticity. The results of plasticity tests on volcanic soils investigated tend to plot in three groups: (i) basaltic soils which are inorganic with high plasticity and which are often red, well-drained and commonly contain kaolinite or halloysite; (ii) doleritic soils with low-intermediate plasticities; (iii) ash and tuff soils with high plasticity which are often black, poorly-drained and commonly contain montmorillonite. Maximum dry densities of these soils are low, usually in the range 0.96–1.44 Mg/m^3 (60–90 lb/ft^3), but they may be used for fill and sometimes for the sub-base. The soils can usually be stabilized satisfactorily with cement or lime and used for the base. Weathered volcanic rocks might be expected to provide an economic source of road-making material. Their use is limited, however, since they are generally mechanically weak and/or have been affected by chemical weathering which may not be readily apparent in hand specimen. Decomposing diorites have been used in the Virgin Islands for sub-bases and cement-stabilized bases while, in Grenada, well-graded unweathered agglomerates provide material for bases and any fines are non-plastic rock powder. Screened agglomerate has been used as surface-dressing aggregate. Volcanic soils, locally known as "tiffs," from St. Lucia [have been] studied by the Road Research Laboratory. They have low clay contents, are non-plastic and are sensitive to moisture changes; when wet they become spongy and unstable.[30.31]

These uses of unusual materials involve the association of design with construction methods, a feature commonly called for in the practice of civil engineering, particularly in highway work. Accurate foreknowledge of the materials to be met in excavation for road cuts will not only assist in accurate cost estimating by engineers, but also in the preparation of bids by contractors based on carefully planned construction methods; expenditure of money on careful preliminary work pays good dividends here, as always, because of the layout of highway-construction jobs, stretched along a reasonably straight alignment. The moving of material from cuts into fills provides opportunity for especially economical earth moving if the nature of the soil is accurately known. The possibilities of sluicing as a means of combined excavation and transportation should not be overlooked when the soil to be moved is sand; gravity is still an efficient servant of the builder.

30.5 CONSTRUCTION

Construction methods are naturally the domain of the civil engineer, but even here geological information will always be helpful, will sometimes be imperative, and may occasionally be critical. In the placement of fill, for example, modern methods require rigorous control over the compaction of the layers in which soil must now be placed at or close to optimum moisture content. Preliminary geological study can here assist in suggesting the uniformity that is to be expected in the natural deposits of soil to be used as fill. In the excavation of cuttings, accurate geological knowledge is clearly essential before any contract can be awarded, and continued observation of the strata revealed as excavation pro-

30-16   CIVIL ENGINEERING WORKS

gresses is also essential to ensure that further excavation will be in materials as expected—or to forewarn, should any previously unsuspected change in the geology be found. And when troubles develop with a finished road, or as a highway approaches completion, geology may be able to provide the explanation and suggest the solution.

Placing fill by pumping has been mentioned. This is an old-fashioned method that is still applicable when local conditions are suitable and the geology of road location and fill availability are favorable. This was well shown during the construction of Interstate Route 10 along the Gulf Coast at a location about 50 km (30 mi) southeast of Baton Rouge. The preferred location crossed many low-lying areas that contain unstable surface soils, including swamp deposits. The Louisiana Department of Highways made careful subsurface exploration of these conditions, despite the difficulties of the site where tidal waters often covered the ground surface. It was determined in this way that down to about 6.6 m (22 ft) unsuitable material could be removed and replaced with a suitable load-carrying soil at reasonable cost. In some cases, this "exchange of materials" can be effected by surcharging the undesirable material with the sound fill, transformation of the material in the roadbed being sometimes assisted by blasting. In this case, however, after it was found that a stratum of dense clay having a good load-bearing capacity existed throughout the site at a reasonable depth for a distance of 32 km (20 mi), it was concluded that direct replacement of the poor soils would be an economical solution.

Contract documents including all necessary controls over excavation methods, fill selection, and placement were therefore prepared, and a contract was awarded. An ideal granular material was found in the bed of the Mississippi River, which almost parallels this part of the Interstate but is about 16 to 24 km (10 to 15 mi) away. Clearing was an extensive operation in itself, but when this was finished a dragline dipper dredge was floated in to the site. It prepared an initial access channel along the 32 km (20 mi) of road and was followed by a large hydraulic dredge which did the actual mucking. The muck-water slurry which this large machine pumped out was deposited on the right-of-way perimeter and in other disposal areas. Excavation was carried out to depths of about 3.3 m (11 ft) below the surface. Fill placement then started, the granular fill being pumped from the bed of the Mississippi by a 75-cm (30-in) hydraulic dredge. With the aid of booster pumping stations, the fill was pumped to the road site, even though this meant pumping the sand-water mixture as much as 24 km (15 mi). An energy dissipater was used at the discharge point, deflecting the discharge upward. The coarser particles fell into place in the embankment; the finer silt-size particles floated out into the open excavated area whence they were removed by a smaller 25-cm (10-in) hydraulic dredge. For the main pumping operation a solid-to-waste ratio of between 14 and 18 percent was obtained. Careful studies showed that the method could be economical for distances up to 40 km (25 mi). The cost worked out to be about $2 million per mile as compared with $4.5 million a mile if a structure-supported design had been used for the road.[30.32] (Fig. 30.7)

Almost at the other end of the spectrum, so to speak, was the experience on a section of Britain's M4 motorway in Glamorgan, South Wales. For the 14.3-km (8.9-mi) Bridgend northern bypass section, from Pencoed to Stormy Down, a careful subsurface study was made as usual. Preliminary study of geological maps indicated that cavities in the underlying rock would be encountered. Local swallow holes confirmed this. Bedrock was known to be a limestone conglomerate overlying shales, sandstones, and thin coal seams. Air-percussion test drilling was therefore carried out to depths of 20 m on a 20-m (65- by 65-ft) grid pattern. In the areas suspected of being the worst, featured by surface depressions up to 6 m (20 ft) deep, a 5-m (16-ft) grid was used. Results were inconsistent, a model of the borehole results displaying no pattern to the voids. It required actual observations within the test holes, using borehole TV cameras, to show that the voids between the limestone blocks were generally clay filled, being open only when percolating water had

FIG. 30.7 Moving glacial till by sluicing. Requiring only simple equipment, it is an economical method of excavation when grades permit.

eroded the clay. Special flexible provisions were therefore incorporated into the contract documents so that the contractor could be suitably paid for all the special work required in the filling of individual cavities with aggregate, once they were exposed as excavation proceeded. Drainage provisions on this job were even assisted, to a degree, by use of old swallow holes (Fig. 30.8).[30.33]

There are other problems that may not be readily identified even if preliminary investigations are properly carried out with full appreciation of the local geology. A somewhat extreme example will be described in more detail since it illustrates so clearly what can happen even with excavation in "solid rock." The California Division of Highways decided to route the Southern Freeway, as a part of the network of modern highways serving the central area of San Francisco, through the Potrero Hill area of the city above an existing tunnel of the Southern Pacific Railroad, one-half a kilometer (1,600 ft) long, built in 1906. The tunnel had been lined with unreinforced-concrete walls and invert, with a roof arch of six courses of brick over the two tracks. Late in 1966 the general contractor for the road started his excavation and had completed it by the following September. In order to protect property adjacent to the big cut, an arched and rock-bolted retaining wall up to 16.5 m (54 ft) high was constructed along the east side of the excavation. All this was in the local Franciscan formation (through which the tunnel had been excavated) that consists of folded strata of shales with sandstone boulders and some serpentine present, the rock being badly fractured although quite solid. The maximum depth of the excavation for the road was about 18 m (59 ft), and the minimum clearance over the tunnel was 8 m (26 ft).

One might imagine that with this clearance, and in solid rock, all would have been well, but even before excavation was complete, owners of buildings adjacent to the cut had noticed cracks in walls and roofs; cracks appeared next in sidewalks; and then water and gas mains began to fracture. By the time the excavation was finished, the tracks in the railroad tunnel were found to be "rough," or slightly out of alignment, and movement was detected in the concrete walls and invert of the tunnel and in the brick roof arch. Inclinometers were installed to measure the movement that was obviously taking place in the

30-18 CIVIL ENGINEERING WORKS

FIG. 30.8 Typical rock as exposed during careful preliminary work on the Bridgend bypass section of Motorway M4 in Glamorgan, South Wales.

tunnel, the serious nature of which can be appreciated from the fact that the tunnel is the only access to the San Francisco terminus of the Southern Pacific system. A maximum rate of movement of 0.6 mm (0.02 in) per day was observed, the entire western side of the tunnel moving inward. Studies were being made concurrently of possible solutions to the problem; the method adopted is shown in Fig. 30.9. Repairs had to be carried out so that they would not interfere with rail traffic and would cause minimum interference with the highway.

Using three large, truck-mounted drill rigs, a total of sixty 1.07-m (42-in) -diameter holes—half being inclined holes drilled to depths of 27.4 m (90 ft)—were drilled to a total length of 1,450 m (4,770 ft). Into the holes were inserted 0.9-m (30-in) -wide flange steel beams weighing 342 kg/m (230 lb/ft). These were immediately concreted in place with pumped concrete made with high-early-strength cement. The steel struts between the pairs of piles were then inserted in specially excavated trenches and tightly wedged in place before being embedded in weak concrete, which then constituted a good foundation for the road surfacing. A decrease in the movement was first observed after only nine frames had been installed. One month after the completion of this unusual rehabilitation job, movement had virtually stopped. No serious maintenance problems with the tunnel have since been experienced. The concept of the strut design proposed by engineers of the railroad company was the transfer of the load that was clearly actuating the movement of

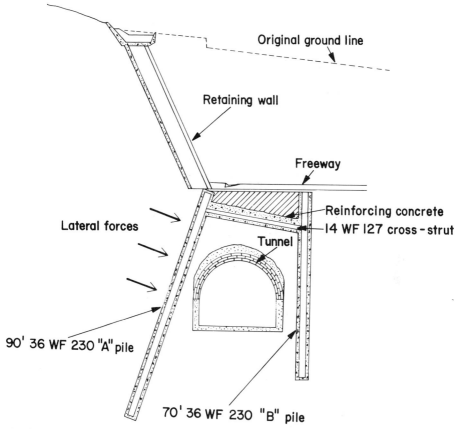

FIG. 30.9 Simplified cross section through new freeway cutting and the Southern Pacific Railroad tunnel, San Francisco, California, showing the remedial measures described in the text.

the west wall of the tunnel by means of a strong enough framework to the continuous body of rock on the east side of the tunnel. The movement of the rock appeared to have been the result of the release of internal stresses in the rock by the deep excavation for the road, a salutary reminder of the state of stress in which "solid rock" deep below ground surface always exists.[30.34]

Finally, brief mention must be made of one aspect of highway construction that has become increasingly recognized in recent years—the erosion of soil into streams and other bodies of clear water from areas disturbed by highway excavation. Previously regarded as "something that could not be avoided," this spoilation of clear water is now a problem that must receive attention. Rainfall running off newly placed fill or discharging from drainage channels in cuttings in process of excavation will naturally carry with it some of the soil over which it flows. Knowledge of local rainfall patterns is really essential prior to the start of highway construction for a number of reasons, not the least of which is to gain some indication of what arrangements should be made for controlling runoff from the work in progress before it causes trouble. The matter has been receiving close attention from all major road-building agencies.

Typical were experiments conducted by the Pennsylvania Department of Transportation, jointly with the U.S. Geological Survey, during the construction of Interstate High-

FIG. 30.10 Scammonden Dam near Huddersfield, Yorkshire, England, with motorway M62 crossing the crest; the dam is of rockfill with a clay core and, when built, was the highest in England.

way 81. Four adjoining small watersheds were monitored. Results showed clearly that clay-sized particles were those most responsible for sediment carried off new works, even though the clay content of the soils in the construction area was only 20 percent.[30.35] There is now available useful and helpful information as to the best and most economical methods for controlling runoff from construction jobs. This note is, therefore, one of warning about a feature of road-building operations which must never be forgotten, in the control of which climatic information and knowledge of the local geology are essential.

30.6 ROADS AS DAMS

It has been found in a few locations that a new road that must be located on an embankment can, with a slight change in route, serve a double purpose. One of the most notable examples of this dual use was some new road construction under the supervision of the engineer and surveyor of the county council of West Riding in Yorkshire, near the city of Huddersfield, England.

The new Lancashire-Yorkshire Motorway, for the construction of which the West Riding County Council was the agent for the British Ministry of Transport, connects the

Great North Road up the east coast of England with the Stretford-Eccles Motorway (Road M62) in the west. Passing through the Pennines involved some very heavy engineering work, including rock cuts as much as 46 m (150 ft) deep and an embankment as high as 65 m (213 ft). This very high embankment was adjacent to the city of Huddersfield, and so advantage of its construction was taken to use the embankment as a rock-fill dam to impound water for the Huddersfield Water Undertaking (Fig. 30.10). This most economical dual project was therefore under two engineering jurisdictions, that of the county for the excavation and all road work and that of consulting engineers for the impounding dam. For the latter use, broken or crushed rock had to have certain very specific properties in the way of density, durability, strength, and grading. On the other hand, excavation of the great quantity of rock from the roadway cutting would be desirably carried out using the heaviest equipment possible, if it could be done without blasting, and with strict limitation if blasting was necessary. Accordingly, very extensive field trials of both excavation (using different methods) and fill placement were carried out at the site of the works in order to ensure that the rock could be successfully used and to obtain information that would permit the preparation of an adequate specification for the main contract.

The trials were carried out at the western end of one of the deepest of the proposed cuttings, a cutting that would have a maximum depth of 46 m (150 ft). The trial embankment was constructed 1.2 km (4,000 ft) farther to the west. The trial drilling and excavation was confined to the top 18 m (59 ft) of the cutting. The strata to be traversed by the new road form part of the Millstone Grit series of rocks, one of the great divisions of the carboniferous rocks of Great Britain. The trials were carried out in a sandstone known as the Midgley Grit, which is generally massive in character but with many open joints. The work was carried out on a day-labor basis by an experienced general contractor under unusually close supervision. Most careful records were taken of all aspects of the trials, even including cinephotography, from the results of which a sound film was made for displaying to contractors when they came to tender on the main contract. The jointing in the rock permitted the use of massive rippers hauled by 385-hp tractors, but the resulting grading of excavated rock was not satisfactory (Fig. 30.11). Multiple-row and single-row drilling and blasting were tried using a variety of explosives. Corresponding variations were used in the deposition of the broken rock in the test embankment and for its compaction in place.

Laboratory analyses using standard and special tests were run on samples of the excavated rock. These showed that the rock was not suitable for coating with tar or bitumen, that it could be used as aggregate for concrete of specified low-compressive strength only, and that the rock, although not acceptable for use as a subbase in road construction, could be used to within 15.2 cm (6 in) of the surface of the subbase. The results of the major field trials cannot usefully be summarized, but they are clearly set out in the excellent paper that was published soon after these trials were carried out. This reference can be recommended for careful study to all who have to face similar problems in the use of large quantities of excavated rock for embankment construction or, indeed, for any purpose other than dumping to waste.[30.36]

It should be added, to avoid any misunderstanding, that it is only very rarely in the conduct of civil engineering work that full-scale trials of this kind have to be carried out. The process of design based upon the accumulated experience of the profession and the steady development of sound theoretical methods is so far advanced that, when materials are under the control of the design engineer, a project can be designed completely in the design office with certainty that when constructed it will perform as intended. When dealing with the ground, however, the materials are not under the control of the designer, who has to accept what is found and "make the best of it." This is why foundation engineering

30-22 CIVIL ENGINEERING WORKS

FIG. 30.11 Rock-ripping test in progress in preliminary full-scale tests prior to the construction of Motorway M62 near Huddersfield, Yorkshire, England.

will always remain an *art*, depending on experience and the ability to assess, right on the job, the best way of dealing with soil and rock conditions as revealed by preliminary investigation. Correspondingly, when large quantities of material have to be excavated, most careful assessment of the results of preliminary investigations is essential, and, when necessary, as in this case from Huddersfield, actual full-scale trials must be carried out to remove all uncertainty.

If only to show that the Huddersfield road-dam is not unique, there may be noted a slight change that was made in the planning of the I-80 Clearfield bypass road in Pennsylvania. It was to be built to the south of the borough, adjacent to its South Park area; this area had experienced a serious drainage problem which was caused by the runoff from a steeply sloping area of over 120 ha (300 acres) south of the site of the new highway. Accordingly, the highway embankment was redesigned so that it could act safely as a dam which would hold up stormwater after heavy rainfall in a basin that could cover as much as 2 ha (5 acres), discharge being controlled by the limited capacity of the 60-cm (24-in) culvert beneath the road. In this way an area of 40 ha (100 acres) was made available for development without the danger of further flooding.[30.37]

30.7 SOME GEOLOGICALLY SIGNIFICANT ROADS

All round the world there are today remarkable examples of the influence of geology upon road design and construction. Since it has been possible to mention only so few in this short chapter, it seems desirable to include summary reference to a few more examples, if only as a tribute to all that has been done in the frequent acceptance of geology as the major determinant in highway location. The state of California contains many notable examples, such as a rather large cut carried out by the Division of Highways on U.S. Highway 101 near Dyerville Bridge over the south fork of the Eel River in Humboldt County.

Careful preliminary investigation showed that the cut would be in the Yager formation of Upper Cretaceous age; the material to be encountered would be interbedded sandstone and shale with numerous small zones of conglomerate. The upper 12 to 15 m (40 to 50 ft) were weathered, badly jointed, and fractured. On the basis of the preliminary information, the cut was designed to have a 1 : 1 slope with 10-m (30-ft) -wide horizontal berms at 18-m (60-ft) intervals (Fig. 30.12).[30.38]

Major cuttings are often geological display cases even though, in the course of time, the original geological exposure may become concealed by the products of weathering or vegetation. Those who use the Carquinez crossing at San Francisco today probably never realize that the great cut through which they drive, 105 m (350 ft) deep, involved the excavation of 6.5 million m³ (8.5 million yd³) of soil and rock, about half the excavation for the Panama Canal. Figure 30.13 is a view of what may literally be termed a typical cutting on the interstate highways that now serve the United States so well, this being an exposure near McAdoo, Pennsylvania, of the Pottsville formation with gray conglomerate overlain by red sandstone and shales with a syncline clearly in view. In contrast, Interstate Highway 70, as it goes through the beautiful Vail Pass in Colorado, was specifically designed to blend in with the landscape, with the result that its remarkable engineering and geological features are not at all obvious as one drives over it today.

FIG. 30.12 Major cut on U.S. Highway 101 in Humboldt County, California, adjacent to the south fork of the Eel River; top of cut 144 m (480 ft) above road; scale can be judged by size of automobile on road.

FIG. 30.13 Typical large rock cut on Interstate Highway 81, south of McAdoo, Pennsylvania, showing a syncline in the Pottsville formation with gray conglomeratic sandstone overlain by red sandstones and shales.

In many cases, the geological interest of a highway is beneath the surface. There are few better examples of this aspect of geology in highway design than on Ontario Highway No. 11 between Fort Frances and Atikokan. It crosses Rainy Lake on what is known as the Rainy Lake Causeway, 4.5 km (2.8 mi) long, costing $4.5 million. The road "leapfrogs" from island to island at this quite beautiful location, with water depths between islands up to 15 m (50 ft) deep. Local geology is typical of Precambrian Shield country so that there was sound bedrock available all along the causeway, even at considerable depths, but the rock was overlain in the channels by varved clays which created their own problems. Ice thrust had to be allowed for in the design of the bridge piers. Compressed air and a bubbling system had to be used during construction.[30.39] (Fig. 30.14)

South America includes some outstanding highways such as the Cochabamba-to-Santa Cruz road in Bolivia. Five hundred km (312 mi) long, much of it through rugged terrain, the road is located in the rain forest area, so fog is a local climatic feature. But trucks can now make the journey along this Andean highway between the two cities in 15 hours, whereas in earlier days the trip could take up to a month.[30.40] Perhaps the most remarkable road in the Andes, excepting only that of the Incas, is the famous spiral highway at El Inferno.

Spain has built some notable roads, one of the most recent being that between Bilbao and San Sebastian on its northern rocky coast. Although only 117 km (73 mi) long, it cost $270 million. In advance of construction an engineering-geological map, to a scale of 1:1,000, was prepared showing the full route. Bedrock consists of sedimentary rocks including thinly bedded limestone, marl, sandstones, and shales, rocks that can often cause problems. By skillful design, the length of the road carried in tunnels and over bridges was reduced to about 12 percent of the total, but this involved cuts up to 69 m (230 ft) deep and fills up to 79 m (262 ft) high. The four-lane toll highway is now a geological exhibit in its own right.[30.41]

Swiss highways are generally well known for their striking features, but it is not always

ROADS 30-25

FIG. 30.14 The Alaska Highway crossing a large alluvial fan, the result of erosion; British Columbia, near Mucho Lake, mileage 467.

realized that northern Italy has roads of comparable spectacular interest. The "Highway of the Sun" is one, especially the section from Florence to Bologna running through the Apennines with their most complex geology, much of the exposed bedrock being susceptible to weathering (Fig. 30.15). In view of this and the extremes of local climate, the design was the reverse of that adopted for the Spanish road, tunnels and bridges being

FIG. 30.15 Typical view on Italy's Highway of the Sun between Florence and Bologna.

used in preference to big cuts and fills. Seventeen km (11 mi) of the total length of 82 km (51 mi) consists of structures. These include 45 bridges and viaducts, totaling 11 km (7 mi) in length, and 25 tunnels, 6 km (4 mi) in length. Four of the tunnels are artificial structures, erected to protect the road from possible landslides, a real hazard because of the character of the sedimentary rocks. For the same reason, such elaborate protective devices as massive retaining walls are to be seen all along the highway.[30.42]

Landslides are one of the most widespread of all hazards to be encountered in road design. The new highway in the crown colony of Hong Kong along the seacoast of the New Territories had to be designed and constructed in the face of unusually complex geology, bedrock being generally distorted granite, with much disintegration of the solid rock in places. Seven bad slides, usually on old and weathered joint planes, occurred even during construction. The rugged terrain and complex geology necessitated cuts up to 59 m (198 ft) deep and viaducts up to 39 m (130 ft) above ground surface.[30.43] It is not often that one encounters the expression "miserable geology." It is the only appropriate term to describe the local geology encountered by this highway!

All major highways are not new, as the Roman roads and the road of the Incas show well. Some famous routes were pathways rather than roads, none more famous than the old "Silk Route" from the Far East, in use for at least 2,000 years and described by Marco Polo in the thirteenth century. Pakistan and China have now replaced with a modern road that section of the old trail through the great mountain ranges which separate their two

FIG. 30.16 Typical view on the Karakoran Highway, part of an eventual International Asian Highway from Turkey to China.

countries, the Himalayas, Karakorum, and Hindukush. Under construction for 20 years, this highest road in the world was completed in 1978. With a ruling gradient of 5 percent, the Karakorum Highway (as it is known) is 800 km (500 mi) long, its highest point 4,500 m (15,000 ft) above the sea. Its varied geology must be imagined from Fig. 30.16. It will eventually form a vital part of the International Asian Highway crossing Asia from Turkey to China and Singapore, its total length 22,400 km (14,000 mi).[30.44]

30.8 WHAT ROAD CONSTRUCTION MAY REVEAL

The ways in which the works of the civil engineer can benefit the advance of geological knowledge will be reviewed in Chap. 50, but this chapter would not be complete if it did not invite attention to the fact that the building of roads, since this always involves work over a large area of ground, probably presents more opportunities for unusual subsurface discoveries than any other construction operation. Two examples must here suffice.

Dover, England, has been an important center for over 2,000 years. It has long been a center of interest for British archaeologists, one of the most eminent of whom, Sir Mortimer Wheeler, predicted in 1929 that remains of the ancient Roman port and fortress would be found beneath the buildings of the present-day town to the west of one of the hills which feature the townscape. In 1970, the British Ministry of Transport began construction of the York Street bypass road designed to give better access to the present port facilities. The route selected crossed the site, as suggested by Sir Mortimer Wheeler, of the old Roman port. With the cooperation of the Ministry, the town council, the Ministry of the Environment, the Pilgrim Trust, and others, all under the "umbrella" of the Council for Kentish Archaeology, a great company of volunteers was assembled and allowed to excavate, by hand, over the critical area after existing buildings had been demolished. During 90 days of nonstop work, 5,000 tons of soil and concrete were removed by hand. Not only was the finest and most complete Roman fort ever found in Great Britain uncovered but other remains going back to neolithic times 4,000 years ago were also revealed. Design of the road was changed so as not to interfere with this fascinating discovery, part of the road being raised 1.8 m (6 ft) above the intended grade, so that there is now a magnificent archaeological display in Dover, thanks to road construction and inspired local archaeological interest.[30.45]

Roman ruins will not be found in North America, but unusual or special geological features may be revealed by excavations for roads. This was recognized in the Federal Highway Act of 1957 with its far-sighted provision, which makes it possible for federal funds to be applied toward the salvage of historical, archaeological, and paleontological specimens when they occur within the limits of federal highway projects. It was under this provision that the extra excavation was made adjacent to Interstate Highway 71 as it enters the city of Cleveland, Ohio, near Brookside Park (Fig. 30.17). Many must have blamed the obvious extra excavation at this location on "bad engineering," whereas it was a cooperative effort between the highway authorities and the Cleveland Natural Science Museum; as a result of this effort, the museum obtained what must be the world's most extensive collection of Devonian fossil fishes.[30.46] A civil engineer, the late George B. Sowers, then commissioner of engineering, City of Cleveland, had brought these unusual fossils to the attention of the museum nearly 40 years previously when he found them in an excavation. The construction project enabled the museum to obtain its first specimens.

All such cooperative efforts in the interest of science stem from an appreciation of the value of historical or geological exposures arising from engineering work. It is surely incumbent upon all geotechnical engineers to nurture such appreciation among their

FIG. 30.17 Extra excavation adjacent to Interstate Highway 71 at Brookside Park, Cleveland, from which fossil Devonian fishes were obtained in quantity.

staffs so that unique opportunities for adding to the store of human knowledge, without interfering with necessary engineering work, shall not be lost.

30.9 REFERENCES

30.1 J. W. Gregory, *The Story of the Road*, MacLehose, London, 1931.

30.2 J. L. McAdam, *Remarks on the Present System of Road Making*, 7th ed., Longman, Hurst, Rees, Orme and Brown, London, 1823.

30.3 E. E. Bean, "Economic Geology and Highway Construction," *Economic Geology*, **16**:215 (1921).

30.4 G. E. Ekblaw, "Twenty-five Years of Engineering Geology in Illinois," *Transactions of the Illinois Academy of Sciences*, **46**:7 (1953).

30.5 Memoir of S. E. Horner, *Bulletin of the Geological Society of America*, **65**:119 (1954).

30.6 "Standard Recommended Practice for Investigating and Sampling Soil and Rock for Engineering Purposes," *1980 Annual Book of Standards*, pt. 19, American Society for Testing and Materials, Philadelphia, 1980, p. 104 (and other standards in the same volume).

30.7 V. R. Burton, "Fill Settlement in Peat Marshes," *Public Roads*, **7**:233 (1927).

30.8 *West Texas Geological Society Road Log, Del Rio-El Paso*, West Texas Geological Society, Midland, 1958.

30.9 A. P. Gest, *Engineering (Our Debt to Greece and Rome)*, Longmans, New York, 1930.

30.10 R. F. Legget, "Glacial Landforms and Civil Engineering," in D. R. Coates (ed.), *Glacial Geomorphology*, State University of New York, Binghamton, 1974, p. 351.

30.11 A. B. Cleaves and R. C. Stephenson, *Guidebook to the Geology of the Pennsylvania Turnpike: Carlisle to Irwin*, Pennsylvania Topographic and Geologic Survey Bulletin G 24, 1949.

30.12 C. W. A. Supp, "Geological and Soils Engineering on the Ohio Turnpike Project," *Proceedings of the Fifth Conference on Geology as Applied to Highway Engineering*, Columbus, 1954, p. 61.

30.13 F. J. Bennett, "The Influence of Geology on Early Settlements and Roads," *Proceedings of the Geologists Association*, **10**:372 (1888).

30.14 C. S. Denny, "Late Quaternary Geology and Frost Phenomena along Alaska Highway, Northern British Columbia and South-eastern Yukon," *Bulletin of the Geological Society of America*, **63**:883 (1952).

30.15 "Depressing a Highway in Unstable Soil," *Engineering News-Record*, **133**:409 (1944).

30.16 "Repair of Florida Pike Forced by Rain Effect," *Engineering News-Record*, **158**:27 (21 March 1957).

30.17 C. W. Thornthwaite, "An Approach toward a Rational Classification of Climate," *Geographical Review*, **38**:55 (1948); see also as an excellent application, B. R. Schulze, "The Climate of South Africa according to Thornthwaite's Rational Classification," *South African Geographical Journal*, **40**:31 (1958).

30.18 H. Allen and A. W. Johnson, "Adding Water to Subgrade Levels Up Pavement," *Engineering News-Record*, **113**:464 (1934).

30.19 H. R. Cedergren, "Poor Pavement Drainage Could Cost $15 Billion Yearly," *Engineering News-Record*, **200**:21 (8 June 1976).

30.20 T. Roy, *Remarks on the Principles and Practice of Road Making As Applicable to Canada*, Bowell, Toronto, 1841.

30.21 "Poor Drainage Blamed for Failure of Two-Year-Old Road Pavement," *Engineering News-Record*, **165**:36 (28 July 1960).

30.22 N. L. Smith, "Notable Dual Road Completed," *Engineering News-Record*, **121**:45 (1938).

30.23 H. B. Woodward, *The Geology of Soils and Subsoils*, E. Arnold, London, 1912.

30.24 "Town Road Carved through Loess Bluffs," *Engineering News-Record*, **125**:277 (1940); see also C. S. Gwynne, "Terraced Highway Side Slopes in Loess, Southwestern Iowa," *Bulletin of the Geological Society of America*, **61**:1347 (1950).

30.25 W. H. Prescott, *The Conquest of Peru*, Dent, London, 1942, p. 39.

30.26 McAdam, op. cit.

30.27 T. S. Patty, "Industrial Mineral Utilization by the Texas Highway Department," unpubl. paper, 1970.

30.28 E. R. Young and R. T. Pinchback, "Sand-Shell Admixture as Flexible Road Base," *Civil Engineering*, **11**:286 (1941).

30.29 "Heat Treatment of Soils as Base for Road Construction," *Commonwealth Engineer*, **21**:399 (1934).

30.30 "Airfield Construction on Overseas Soils, a Symposium," *Proceedings of the Institution of Civil Engineers*, **8**:211 (1957).

30.31 D. C. Cawsey, "Volcanic Road-Making Materials in the Tropics," *Proceedings of the Geological Society of London*, **1662**:13 (1970).

30.32 J. W. Starring, "Sand Fill Pumped 15 Miles for Interstate Construction," *Civil Engineering*, **41**:44 (February 1971).

30.33 "Cavity Filling Precedes M4 Earthworks," *New Civil Engineer*, **305**:16 (17 August 1978).

30.34 J. P. Nicoletti and J. M. Keith, "External 'Shell' Stops Soil Movement and Saves Tunnel," *Civil Engineering*, **39**:72 (April 1969).

30.35 W. G. Weber and L. A. Reed, "Sediment Run-off During Highway Construction," *Civil Engineering*, **46**:76 (March 1976).

30.36 H. M. Williams and J. N. Stothard, "Rock Excavation and Specification Trial for the Lancashire-Yorkshire Motorway, Yorkshire (West Riding) Section," *Proceedings of the Institution of Civil Engineers*, **36**:607 (1967).

30.37 S. P. Singh, "Highway Embankment Doubles as Dam," *Civil Engineering*, **49**:80 (April 1979).

30.38 "Big Freeway Will Link L.A., San Diego, Mexico," *Engineering News-Record*, **161**:30–39 (27 November 1958).

30.39 M. A. J. Matich, A. Rutka, and P. F. Andersen, "Foundation Aspects of the Rainy Lake Causeway," *The Engineering Journal,* **46**:23 (November 1963).

30.40 "Spectacular New Highway Opened to Traffic in Bolivia," *Civil Engineering,* **25**:47 (January 1955).

30.41 "Spanish Cut-and-Fill Operation Reduces Cost of Mountain Road," *Engineering News-Record,* **192**:37 (21 February 1974).

30.42 "Structures Dominate Highway of the Sun," *Engineering News-Record,* **166**:40 (1 June 1961).

30.43 "Hong Kong Roadbuilders Battle Rain, Geology," *Engineering News-Record,* **197**:21 (5 August 1976).

30.44 "World's Highest Highway Through Himalayas Links China and Pakistan," *Civil Engineering,* **49**:58 (April 1979).

30.45 *Roman Dover—Britain's Buried Pompeii,* CIB (Kent) Archaeological Rescue Corps, Dover, 1973.

30.46 W. E. Scheele, "Fossil Dig," *The Explorer,* **7**:5, (1965).

Suggestions for Further Reading

Highway Engineering is the one branch of civil engineering which, in North America, has long recognized the place which geology must occupy in design and construction. This is clearly shown by the many excellent relevant papers to be found in the proceedings and other publications of the (old) Highway Research Board, now the Transportation Research Board, originally established by the National Academy of Sciences.

There have been held annually since 1950 special Symposia on Geology in Highway Engineering, originated largely by W. T. Parrott of Virginia, stimulating smaller gatherings also with a fine record of proceedings. These have usually been published in relatively simple form by the highway departments and/or universities of the state in which the meetings were held. For ease in referring to these most valuable records, the following list of the locations of the Symposia has been compiled with valued assistance from J. W. Guinee of the staff of the Transportation Research Board:

Year	Location	Sponsoring Highway Dept.	Year	Location	Sponsoring Highway Dept.
1950	Richmond	Virginia	1966	Ames	Iowa
1951	Richmond	Virginia	1967	Lafayette	Indiana
1952	Lexington	Virginia	1968	Morgantown	West Virginia
1953	Charleston	West Virginia	1969	Urbana	Illinois
1954	Columbus	Ohio	1970	Lawrence	Kansas
1955	Baltimore	Maryland	1971	Norman	Oklahoma
1966	Raleigh	North Carolina	1972	Hampton	Virginia
1957	State College	Pennsylvania	1973	Sheridan	Wyoming
1958	Charlottesville	Virginia	1974	Raleigh	North Carolina
1959	Atlanta	Georgia	1975	Coeur D'Alene	Idaho
1960	Tallahassee	Florida	1976	Orlando	Florida
1961	Knoxville	Tennessee	1977	Rapid City	South Dakota
1962	Phoenix	Arizona	1978	Annapolis	Maryland
1963	College Station	Texas	1797	Portland	Oregon
1964	Rolla	Missouri	1980	Austin	Texas
1965	Lexington	Kentucky	1981	Gatlinburg	Tennessee

chapter 31

RAILWAYS

31.1 Railway Location / 31-1
31.2 Railway Construction / 31-3
31.3 "Maintenance of Way" / 31-10
31.4 Rockfalls / 31-14
31.5 Special Hazards / 31-16
31.6 Relocation of Railways / 31-24
31.7 Subways / 31-26
31.8 Conclusion / 31-32
31.9 References / 31-32

Railways still play a vital part in land transportation, even though the average citizen may have lost sight of this because of the all-pervading attachment to the automobile for personal travel. The volume of freight handled by railways is at an all-time high. Passenger traffic has always been subsidiary to the carriage of freight, but even passenger travel is making a slow comeback in many countries of the Western World. In developing countries, new railways are being built, and this aspect of "development" is liable to continue for some time. The critical energy situation in which the whole world now finds itself lends added importance to travel and transport by rail, in view of the economy in fuel which it represents. Especially is this true of urban travel where the automobile must soon take second place to public transportation systems, subways being preeminent among these in all large urban areas. Accordingly, it is not the misplaced enthusiasm of "railway buffs" (and both the writers admit to belonging to this unusual group) that leads to the inclusion of this chapter in this Handbook but rather a realistic appraisal of the position of railways today. New main lines may be constructed only in some more isolated parts of the world, but existing lines all over the world have to be maintained, and in some cases, improved, to carry a mounting volume of traffic. And in the major cities of the world, subway construction may be expected to provide some of the greatest of construction challenges for many years yet to come.

31.1 RAILWAY LOCATION

New main railway lines are still being built in Canada, notably by the British Columbia Railway, but most new railway construction will be in less well-known parts of the world.

31-2 CIVIL ENGINEERING WORKS

The methods adopted for the location of new lines are almost identical with those adopted for road location, recognition of significant geomorphological features, and especially of glacial features in glaciated regions, being a first and important step. The records of railway locations are replete with graphic examples of such geological determinants; to select even one example is not easy. Figure 31.1, however, shows what may perhaps be called a classical case. The view shows how the main Gotthard line of the Swiss Federal Railways makes its way across Lake Lugano. This lake very effectively bars the direct route that could otherwise have been followed in the direction of Como. The lake, opposite the town of Lugano, is over 270 m (900 ft) deep. Swiss engineers found the solution to their location problem by noticing that Melide, a village at the base of Monte San Salvadore, the conical peak that is visible from Lugano, was located on a rather odd-looking promontory in the lake. This suggested that it might be a part of a glacial moraine. Soundings in the lake were therefore taken opposite the village, and a ridge was found, running across the lake at quite a shallow depth, clearly the remainder of the suspected moraine. An embankment was, therefore, easy to construct; it is this that is seen in the photograph; the embankment carries not only the railway but also a main highway. The railway line curves sharply at the end of the embankment to follow a winding route along the foot of the mountains until it reaches the border station of Chiasso.[31.1]

Glacial landforms will probably be only rarely encountered in railway construction of the future even though they have influenced recent railway construction in the northern part of North America, as Fig. 31.2 shows quite clearly. Attention should, therefore, be rather directed to the use of aerial photography as a location tool, briefly touched upon in Chap. 10. As there noted, the art of aerial-photo interpretation has now developed to such an extent that it has its own voluminous literature to which recourse may be had for detailed accounts of techniques. There are few applications in which the value of this procedure is more in evidence than in the selection of routes for highways and railways. Field surveys are still necessary, but these can now be restricted to the route selected by means of aerial photographs, with the certain knowledge that the route is the best of all available. Survey work can thus be speeded up and carried out with more thoroughness than would be desirable if final choice of route depended upon preliminary surveys them-

FIG. 31.1 The Melide Causeway across Lake Lugano, constructed on top of a submerged glacial moraine, carrying the Gotthard line of Swiss Federal Railways on its way to Italy.

FIG. 31.2 Low-altitude aerial view of old glacial lake beaches south of the present-day Great Slave Lake, Northwest Territories; one of the beaches was used for the route of the Great Slave Lake Railway, seen in the right background.

selves. Soil and rock surveys along the route, and in adjacent areas for the provision of necessary fill or roadbed material, must similarly be carried out, but again, the conduct of such surveys will be greatly aided by the preliminary information gained from a study of aerial photographs, especially if this can be done in association with available geological and pedological maps.

Illustrative of the saving in time now possible is the record of the preliminary surveys conducted for the construction of an extension of the Nigerian Railway into Bornu Province in central Africa, a project which was assisted by the World Bank. Here aerial photographs were not available and so had to be specially taken. A total length of 980 km (611 mi) was covered by aerial-strip photography (to include possible alternative routes); the work involved just over 50 hours of flying time and was done in 11 working days. Mapping of the route selected by means of the aerial photographs, a distance of 710 km (443 mi), was completed in 15 months; the only unusual geological feature involved the unavoidable use of a section of route over black cotton soil, one of the most difficult of tropical soils with which to deal in engineering work. Despite the isolated tropical location of the work, the total cost of aerial photography and ground mapping was approximately £70,000 ($350,000) or £173 ($865) per mile of final alignment.[31.2]

31.2 RAILWAY CONSTRUCTION

The building of railways throughout the world has probably provided more vivid examples of the interrelation of geology and civil engineering than have most other types of construction. All the determinants of highway construction are present in railway work, and there are the added imperatives of achieving strictly limited grades and curvature. Only with the most modern type of superhighway is road building coming to approach the same criteria as has railway construction for more than a century. It is interesting to reflect that even the practice in modern highway work of utilizing as much as possible of excavated material was a common feature of early railway construction work, at least once with quite surprising results. George Stephenson, that great figure in early British engineering history, was, among his many assignments, engineer for the North Midland line. He therefore designed and supervised the construction of the Clay Cross Tunnel (north of Nottingham). During the excavation of the tunnel, first-class coal seams were encountered, the potential of which was quickly realized by Stephenson, with his usual acumen. He interested the "Liverpool Party" in his find, and with their help he was able to form the Clay Cross Colliery Company, one of the most successful of British colliery enterprises, which was in continuous operation for over a century. Stephenson erected coke ovens for the processing of some of the coal and became so interested in the local industry that developed around his colliery that he bought Tapton House, a beautiful country residence near Chesterfield, where he died in 1848.[31.3]

All interested in British Railways know that Swindon was for almost a century the headquarters of the justly famous Great Western Railway; what is not so well known is that the stone houses and railway buildings that distinguish the town were all constructed of limestone excavated from the famous Box Tunnel. The Lockwood Viaduct on the (old) Lancashire and Yorkshire Railway's Huddersfield-to-Penistone line was built with 23,000 m³ (30,000 yd³) of stone excavated from the cutting at Berry Brow Station and then quarried by masons at the adjacent site of the viaduct (Fig. 31.3).[31.4] Possibly even more remarkable is the fact that the clay excavated from the long Penge Tunnel of the Southern Railway in the southern outskirts of London was formed into bricks, fired in a brickmaking plant specially constructed just outside the tunnel, and then used for local houses and railway buildings. These early British railway engineers knew their local natural building materials and also their engineering economics. Their example may be useful today in developing countries.

There are few major railway lines that, somewhere along their routes, do not display striking exposures of local bedrock and that do not exhibit the ingenuity of location engineers in circumventing natural geological obstacles. A complete volume could readily be produced containing nothing but summary accounts of such examples. Each reader of these words can probably picture some example, personally known, that could be here

FIG. 31.3 The Lockwood Viaduct of British Railways, near Huddersfield, England, showing the cutting, rock from which was used for the masonry in the viaduct.

recorded. But brief mention of just a few examples must suffice; they are to be considered against the background of railroad economics and the fact that it is still possible to construct and maintain a main-line railway more economically than a first-class highway. It must also be remembered that there are many railroad lines that penetrate country where there are still no roads. Although this part of the picture is changing, it adds special interest to the geological significance of pioneer railroad construction.

Possibly the most remarkable of all examples is the stretch of the Denver and Rio Grande Western Railway through the Royal Gorge of the Arkansas River in south-central Colorado. The normal limitations of railroad location dictated that the line had to come up this gorge, starting just above river level. The gorge is over 300 m (1,000 ft) deep, its sides precipitous in the extreme, cutting through upturned Precambrian strata. The character of the granite put tunneling out of the question. The narrow confines of the gorge, coupled with the character of the rock, made bench excavation also impracticable. The responsible engineers therefore took the unusual (and it is believed unique) course of actually suspending a part of the line over the river by means of special steel suspender structures. Figures 31.4 and 31.5 illustrate the result of this bold piece of engineering, long one of the scenic attractions of travel through Colorado.[31.5]

At the other extreme is the use of a natural tunnel by the Bristol-Appalachia section

FIG. 31.4 Royal Gorge on the Arkansas River, Colarado, through which runs the single-track line of the Denver and Rio Grande Western Railroad.

FIG. 31.5 Unusual steel suspension structure used to carry the Denver and Rio Grande Western Railroad line through the narrowest part of the Royal Gorge, Colorado.

of the Southern Railway System through Powell Mountain in Virginia. The tunnel is 236 m (788 ft) long and is a natural erosion feature in the local limestone; it was virtually unknown until 1880 when J. H. McCue found it during railway survey work. It was then filled with driftwood, stones, and dirt; when cleaned out, it gave a perfect solution to this particular piece of railroad location; the walls were already smoothed by past water action, and the cross section, 36.5 m (120 ft) wide and 27.4 m (90 ft) high at the east portal, was more than enough to provide easy alignment for the projected rail line, merely by the use of an easy reverse curve within the tunnel (Fig. 31.6). This unique tunnel is still in use, but no passengers now have the opportunity of passing through it, as the line is now used for freight traffic only.[31.6]

Even in so small a country as England, but undoubtedly because of its unique and varied geology, there are many railway lines of geological significance. The Dore and Chinley line of the old Midland Railway (now British Railways) is but one example; it was an important rail link in the Peak district of north-central England. It is unusual in that one-quarter of its entire length is underground, the Totley and Cowburn tunnels being in part responsible for this unusual feature. These two long tunnels, both of them through limestone, provided great contrast during construction; the Cowburn Tunnel was practically dry, but the Totley Tunnel workings were so wet that there developed a local saying that

FIG. 31.6 A train of the Southern Railway emerging from the natural tunnel through Powell Mountain, Virginia.

all the workmen on the job (back in 1892) possessed the miraculous power of Moses. Every time a workman struck a rock, water seemed to spring out of it; a flow of over 22,600 lpm (5,000 gpm) occurred at one heading.[31.7]

It would be very difficult to select any one of the railways of Switzerland for mention here; there are so many that provide classical examples of how geology determines railway location. Attention may rather be directed to one of the several railways of South America that provide some of the most exciting routes now in public use. In Peru, for example, the Peruvian Central line from Callao and Lima to Oroya via the Rímac Valley, was an epic piece of construction; its operation is a continuing masterpiece of railway work. It has a summit of 3,862 m (12,873 ft) above sea level, but most of the rise to this elevation takes place in 117 km (73 mi). It is not surprising, therefore, that it has grades as steep as 1 in 22, but despite this the line is worked by adhesion. It required 67 tunnels, 62 bridges, and 11 zigzags in order to penetrate to the rich Montana mineral district with its great copper mines at Cerro de Pasco.[31.8]

So the list could continue with reference to such well-known features as the spiral tunnels on the main line of the Canadian Pacific Railway through the Rocky Mountains near Field, British Columbia; the cable-worked inclines up the Serra do Mar of the (old) São Paulo Railway of Brazil, with their elaborate protection works; the famous Rimutaka

31-8 **CIVIL ENGINEERING WORKS**

incline of New Zealand and the great tunnel that now replaces it. No such record as this, however, brief as it must be, would be complete without reference to two railways of North America, the White Pass and Yukon route joining White Horse, Yukon Territory, to the sea at Skagway, Alaska, and the Alaska Railroad connecting Seward and Fairbanks. Because of the route that it follows, the White Pass line is located generally on solid rock. The same cannot be said of the Alaska Railroad, however. Since much of it is over relatively level terrain, it runs over permafrost and presents the inevitable maintenance problems. On each of these lines are to be found sections that can best be described as illustrations of "geology in the raw" (Fig. 31.7).

Modern excavations, tunnels, bridges large and small, all are regular features of new railway construction and call for no special comment here. All the remarkable resources and equipment now available for civil engineering operations have been applied in railway work. One notable example alone can be noted. A new main line, eventually to link Amsterdam with Leiden, is being constructed by Netherlands Railways, so planned that it will also provide a rail link with the well-known Schiphol Airport from Amsterdam while augmenting commuter services. This remarkable main railway line already runs beneath the airport. The tunnel which carries it is 6 km (3.7 mi) long; its construction has been spread over a number of years, but for its first section the slurry-trench method was used to form the sides of the necessary cut, "cut-and-cover" being the overall procedure. Soil conditions include the usual high water table, surface soil being clayey sand overlying sandy clay, with a 45 cm (18 in) layer of peat about 6 m (20 ft) below the surface, below which is hard sand. Chemical consolidation of soil was used for a short section, but the bulk of the work has been carried out by driving steel sheet-pile walls to form the necessary trench, concrete for the invert slab being placed under water (Fig. 31.8). Tension piles were also necessary, to resist the hydrostatic uplift. Prestressed piles of special design were used for this purpose. The entire project, based on sound knowledge of the local geology, is a masterpiece of modern civil engineering.[31.9]

FIG. 31.7 Train on the Alaska Railroad, near mileage 350, showing subsidence of track due to thawing of frozen soil.

FIG. 31.8 Schiphol Airport, near Amsterdam, the Netherlands, showing new rail line under construction, here in a steel sheet-piled trench.

Even so specialized a technique as soil freezing has been used in modern railway construction. When the new Central Station of Canadian National Railways in Montreal was formed by a major enlargement of the old Tunnel Station, in association with the great Place Ville Marie building complex, it was necessary to rearrange the southern end of the Mount Royal Tunnel which terminates at the north end of the new station. A central concrete dividing wall between two tracks had to be removed and the two concrete arches over the tracks replaced with a single concrete arch over both. The tunnel is in limestone bedrock, here 9 m (29.5 ft) below ground surface, the overlying soil being compact glacial till overlain by a stratum of silt and then by stiff brown clay. Immediately above the 52.5 m (175 ft) of tunnel that had thus to be reconstructed was an exceptionally busy street with the usual underground maze of sewers, conduits, and pipes and multistorey buildings on either side. The silt was known to be a troublesome material; difficulties had already been encountered with it when it was found beneath the water table, as was the case here.

After every possible construction expedient had been looked into carefully, it was decided to freeze the silt and clay, remove this material when frozen, and place the new concrete arch while the ground was still frozen. The city of Montreal authorized occupation of the street for the construction period. Figure 31.9 shows the general layout, the freezing pipes being driven vertically downward from the surface well into the till; the pipes were connected at the surface with the necessary pipe network that was fed from a central refrigeration plant specially installed on the street for this purpose. The work was done in the winter (as is now regular practice in Canada), and the cold weather naturally aided the freezing operation; 1,528 m^3 (2,000 yd^3) of clay and silt were frozen satisfactorily in 30 days. Excavation was carried out above the existing arches at an average rate of a 45-cm (18-in) advance per day, but all soil had to be drilled and blasted (since it was frozen) under controlled conditions, as rail traffic continued to use the tracks below. When

FIG. 31.9 Freezing shallow ground above the south portal of the Mount Royal Tunnel of Canadian National Railways, Montreal, showing proximity of buildings.

excavation was complete, the old arches were used as a working platform for the erection of the new arch and were removed only in the final stage of this unique operation.[31.10]

31.3 "MAINTENANCE OF WAY"

This old term, still used as a railway title on many lines in the form "engineer of maintenance of way," is one full of meaning. The apparently simple inspection of railway lines in regular service by one or two men riding on a trolley looks so simple and insignificant that its vital importance has sometimes been lost sight of, especially when budgets are tight. The long experience of a senior track foreman, however, is essential to the maintenance of a stable track, his eye able to discern the first sign of trouble, long before an accident results. Accordingly, "maintaining lines of communication" is a phrase as potent with meaning for railway engineers as for highway engineers. Local failure or obstruction of older transportation routes did not usually involve the safety of travelers, since the trouble could generally be seen in advance of slow traffic and so avoided. In the case of railways and modern highways, however, human lives are involved if the routes cannot be kept clear and intact. Thus it is that track maintenance and inspection are such important parts of railway work. Bridge foundations constitute a special hazard; in Chap. 24 will be found some comments on this matter and a note on two accidents due to bridge failures. Tunnels always require special care in inspection, with due regard to the stability of rock exposed in unlined sections. Embankments always require careful attention, especially in the period just after their completion and in the vicinity of bridge abutments. Cuttings are similarly important, as was shown in Chap. 19. The trouble most readily imagined in cuttings is that due to landslides of adjacent unconsolidated material; all-too-many occurrences of this kind have taken a toll of human lives.

As an extreme case, there may be mentioned a statement made in 1894 by the chief engineer of the (old) London and North Western Railway of England that "there was neither a bank nor cutting between Euston (London) and Rugby that had not slipped at one time or other"; this statement deals with the first 160 km (100 mi) of one of the main railway lines of England.[31.11] Railways have been very general sufferers from landslides, both in natural undisturbed material near to which railway lines have been located (as on hillsides, for example) and in cuttings and fills constructed to carry the roadbed. It is perhaps safe to say that no railway line has suffered more than (although some of the South American lines may have suffered as much as) the Hill Section of the (former) Assam-Bengal Railway in India. The Hill Section is 183 km (114¼ mi) long and was opened to traffic in February 1904; its construction was undertaken largely for military reasons, and it has been described by a viceroy of India as a "millstone round the neck of the Indian Finance Department." From the very beginning of operation, slips and washouts gave trouble; in ten years, about £100,000 ($500,000) had been expended in necessary maintenance, and the line had been closed on several occasions. In 1915, a further disaster overtook the line—excessive rainfall, culminating in a fall of 66 cm (26 in) in 48 hours. The section was subsequently closed for two years while remedial work was carried out at a still further cost of over £225,000.

Disregarding the financial aspects of the work, all engineers can appreciate the heroic efforts of the engineers concerned with the maintenance of this line in the face of the difficulties that the quoted figures suggest. The troubles were due in large part to the local geology; the rocks are of the Tertiary measures and consist mainly of alternating beds of carbonaceous shales and sandstones so affected by earth movements of the past that they possess little durability and break up into small fragments on exposure to the atmosphere. The shales vary from rock as hard as slate to material with the consistency of clay, and the sandstones vary from rock sand to first-class building stone. A detailed record of some of the major difficulties created by this material is fortunately available; the following account illustrates a notable case of slope readjustment—first by way of a slide and subsequently by means of remedial measures coupled with other features of interest.[31.12]

Figure 31.10a represents a cross section at a point on the line near the south portal of the Chamartalla Tunnel; Fig. 31.10b shows how this was after the slip in 1913. The cutting in the rock was at first taken out to a steeper slope, but after many falls of rock and earth had occurred, the slope was cut back to that shown. Revetment was also built to protect the newly exposed shale from exposure to the atmosphere. The retaining wall shown at river level was the cause of much trouble. When rebuilt in 1899, its foundation was on hard shale at riverbed level, and no erosion was anticipated, but in 1902, and again in 1908, the river in flood flow undercut this wall to a depth of 3.6 m (12 ft) and eroded the hard shale completely; the wall had to be rebuilt again. In 1913, a great mass of material slipped from the hillside above onto the tracks; the load was so great that the retaining wall was forced into the river, and the whole railway formation was carried away for a length of 45 m (150 ft). Communication was restored temporarily within 15 days. Permanent reconstruction introduced many problems, but the final solution is shown in Fig. 31.10b; the design of the covered way was so prepared that it would offer a minimum

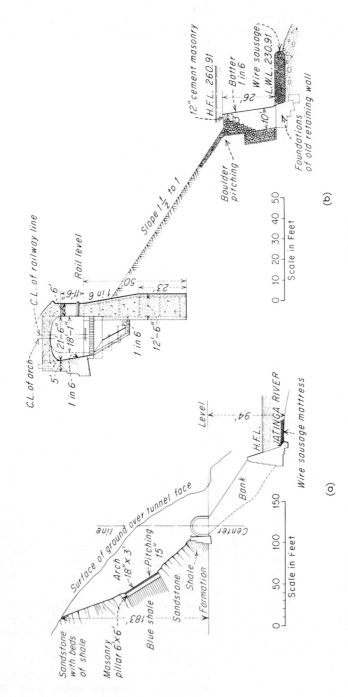

FIG. 31.10 Landslide remedial work on the Hill section of the Assam-Bengal Railway, India; (a) cross section near Chamartalla Tunnel in 1913, before landslide; and (b) covered way structure, erected following the 1913 slip and undamaged in 1915.

obstruction to further slides, which were intended to pass over it. A sloped cushion of earth fill and hand-packed stone was to be permanently maintained on top at the slope indicated; the slope below the track was trimmed as shown. The foot of the retaining wall was protected with wire-sausage mattresses to secure it against future scouring. The piers for the covered way, as will be seen, had to be carried to a depth of 15 m (50 ft) below track level in order to reach a solid foundation bed, since rock above this level was badly fissured. In 1915, when the major disaster to the line took place, a slip passed over the covered way without causing any damage. Fortunately, engineers are seldom faced with such difficulties as these, but the case is of great interest, because it shows how an exceedingly serious situation can be remedied by judicious work on the affected slope.

In great contrast, but illustrating well the engineering problems that railway maintenance involves, there may be noted some of the difficulties met with in maintaining the Alaska Railroad at the other side of the world from India. Probably the most troublesome part of this truly pioneer line is where it crosses the Alaska Range (miles 322 to 385). The central Alaskan earthquake of October 1947 triggered a most serious landslide in this area. This disrupted traffic for several days and led to a request to the U.S. Geological Survey for assistance with the track troubles in the area. (The report summarizing the results of this study is a memorable document, and the record it presents makes other railroad engineering problems pale into relative insignificance.[31.13]) Not only does the line have inevitable troubles with thawed ground because of the permafrost under the foundation for the roadbed, but there are corresponding problems with seasonal frost heaving in the frost-susceptible soils beneath the track that have thawed out from their original condition. Slumps and earth flows along the Nenana Gorge had to be regularly dealt with; these are caused by slow movement of rock debris on the underlying steep slopes as much as a few hundred meters deep. They move the entire railroad down toward the river at rates varying from a few meters per year to a few meters per hour. Near mile 351, an ancient landslide, long ago stabilized by being perennially frozen, is slowly becoming active again as the frozen ground thaws out because of the inevitable disturbance of local surface conditions.

At the north end of a tunnel near Garner, the railroad crosses the lower part of a large talus cone which is slowly creeping toward the adjacent river under the increasing load of debris falling from the cliffs above. Large blocks of schist constitute much of this debris; some are at the heads of steep chutes that lead down through the cliffs to the track, and these blocks occasionally break loose. The largest, at the time the USGS study was made, was 30 m (100 ft) long on its side and was creeping at the rate of 60 cm (0.2 ft) per year toward the edge of a precipice over which it will eventually fall onto the track below. It is not, therefore, altogether surprising that many geologists have recommended that the only thing to do about this part of the Alaska Railroad is to realign the track permanently. Meantime, the engineers responsible for this vitally important rail link continue their really epic work of keeping the traffic moving in the face of geological problems probably without parallel anywhere else in the world.

Most railway maintenance work has no such excitement as these special cases involved. In comparison, the work may appear to be pedestrian—constant and regular inspection of track, especially after heavy rains, with keen observation to detect the slightest sign of possible trouble. Incipient landslides can usually be anticipated by such observations. Signs of "pumping" under track will indicate a failure of drainage facilities. These and many other details will be in the experience of the maintenance-of-way men upon whose lonely work so much depends. They will know the geology along their sections of track instinctively, even though they may not realize this. Even the provision of new ballast, when necessary, has a very definite geological input, since it is not just a case of using the most economical crushed rock that is available, but crushed rock that is geologically suit-

31-14   CIVIL ENGINEERING WORKS

able for the special use that service as ballast involves. Even on this detail of railway maintenance there is now a useful and extensive literature available for consultation.[31.14]

31.4 ROCKFALLS

Rockfalls constitute a special hazard in the maintenance of railway track wherever the track runs in rock cuts or adjacent to steep rock faces. Although such faces, and the sides of cuts, may be perfectly safe and stable when completed, exposure to weathering influences may eventually lead to some loosening of rock on joint planes, or on other surfaces of weakness, with ultimate dislodgement of rock fragments, some of which may be large enough to cause real trouble. Peckover, in an admirable review of this problem, has shown that for western Canada there is a definite correlation between precipitation (rainfall in general) and the incidence of rockfalls. Again, regular and careful inspections, coupled with a detailed and continuing study of weather records, are a first requirement.[31.15]

Many railways have stretches of line that require special maintenance procedures, such as multiple tracks of the Pennsylvania Railroad along the Ohio and Monongahela rivers in the vicinity of Pittsburgh. The alternating strata of sandstones and shales exposed above these tracks constitute a continuing hazard; as a result the railroad has a special team of trackmen who patrol the most critical 9.6 km (6 mi) of track day and night. This is naturally an expensive procedure. In many locations, therefore, and where rail traffic is not heavy, mechanical or electrical guard devices have been installed. One of the first of these was on the Callander and Oban line of the (old) Caledonian Railway in the Highlands of Scotland; where the line passes along the north side of the Pass of Brander, it is subject to frequent rockfalls. As long ago as 1881, a special 2.7-m (9-ft) wire fence was erected alongside the line, connected to 14 signals located throughout the critical length of 5 km (3¼ mi); the fence was so arranged that if a falling rock cleared the fence, it should not land on the track, whereas if the rock touched the wires one or more signals would be put in the danger position. The severance of a wire would also cause an electric bell to ring in the nearest signal box and in the houses of the local track-maintenance men. The speed of the trains is severely restricted, especially at night; it was this fact that minimized the damage done in 1947 when a train was derailed at this location by a large boulder that did land on the track, despite all the precautions described.[31.16]

A more recent example of the same general type of protection is found on the other side of the world, in Australia. On the main line of the Western Australian Government Railways between Perth and Northam, the only remaining section of single line was duplicated in 1944 in order to double-track the entire route. The new line became the down main line. So that its grade might be eased somewhat, it was constructed on a course which deviated a little from the location of the existing line. Construction involved a deep cut in granite, some 450 m (1,500 ft) long, in which protection from rockfalls was essential in view of the character of the granite exposed after the cut had been excavated. Steel fences, 2.5 m (8.5 ft) high with supports 9.9 m (33 ft) apart, were therefore erected on both sides of the cut. The wires forming the fences are in continuous circuits that are energized; since the line is controlled by automatic track-circuit signaling, a simple but effective system of electrical interlocking was installed which makes the protection system completely automatic (Fig. 31.11). If a wire is broken by falling rock or if a battery or circuit fails, two relays become de-energized, and the signal controlling the line in the cut is placed at danger. An indicator in the Swan Valley signal box shows the operator there that a rockfall (or other failure) has occurred; the control signal can only be reset with a special key at the signal itself.[31.17]

The former Great Northern Railroad system (of the United States) has had to main-

FIG. 31.11 Electrical slide-detection fence on the Swan View deviation cutting of Western Australian Government Railways, on the down eastern main line in the Darling Range.

tain 36,000 m (120,000 ft) of protective fencing of this general kind. Peculiar to this system, however, is the fact that this total includes about 12,000 m (40,000 ft) of "mud-slide fence," which consists of two wires stretched near the ground so that they will be interfered with if a mud slide, as distinct from a rockfall, should occur. Nearly all this fence is installed between Seattle and Vancouver, British Columbia, on the G.N.R. coast line. The system also uses protective fences for detecting snow slides through the Rockies and the Cascade Mountains. The railway also experiences unusual difficulties in maintaining its warning fences in the Glacier Park area, not from any of the causes that might normally be expected, but because elk break the wires. "When a big bull comes to the fence on his way up or down the mountain side, he will wind the wires in his horns and then charge,

FIG. 31.12 Special protection works to minimize rockfalls, on Austrian Federal Railways.

31-16 CIVIL ENGINEERING WORKS

taking the wires with him. It is a #10 copperweld insulated line wire, and he usually has to drag it along with him until the horns fall off in the early spring."[31.18]

Equally as important as these warning devices are the measures that can be taken to prevent rockfalls from occurring. This requires careful (and sometimes dangerous) examination of the rock faces from which rock may fall. Rock bolting of dangerous sections is a first possibility as a preventive measure. Equally important is the scaling of rock slopes and faces in such a way that no water can accumulate anywhere on them, thus reducing the possibility of serious weathering. In some cases, rock faces that might cause trouble can be underpinned by suitably designed and located rock buttresses. An extreme case of concrete protection work is shown in Fig. 31.12. This illustration comes from the author of a wide-ranging review of rockfalls and necessary protection work in Europe and North America; Peckover's report will prove of much assistance to those who face similar problems.[31.19]

31.5 SPECIAL HAZARDS

There are other hazards to the safety of railways that must be reviewed, even though they are not so general in nature as are rockfalls and landslides. One very serious problem in some mountainous districts is the possibility of mud runs—streams of mud and rock debris washed down from the lower slopes of bare mountainsides by the runoff that follows torrential rains. In general, these are experienced widely in tropical countries and are especially severe in sections of South America and in India. The extreme climatic conditions promote rapid disintegration of exposures of relatively weak rocks such as shales, and the concentrated rainfall acts as an efficient transporting agent. If mud runs reach bridge openings under a railroad designed to take normal streamflow, the consequent restriction of movement will block the openings, and thereafter control of the runs will be difficult. It is reported that several sections of the Bolivia Railway have had to be relocated because of this; in one place a town adjacent to the railway had to be completely relocated because the original site was buried deep under the unusual detrital cone formed by blockage of a large mud run at a bridge opening. This bridge, originally 3 m (10 ft) above stream-bed level, is now located in a cutting (Fig. 31.13).[31.20]

FIG. 31.13 Arque Bridge on the Antofagasta and Bolivia Railway, showing two of the original 40 spans (which have to be kept open) and the accumulation of rock debris above the original valley floor.

Even more serious is the danger of damage to transportation routes, railways in particular, by avalanches. Snow may not usually be thought of as a geological agent, but all who have seen an avalanche, or examined the results of one, will know that moving snow has quite remarkable power for terrain modification. Here Switzerland must be mentioned, for the Swiss have led the world not only in the development of avalanche-protection works but, in more recent years, in research at the Swiss Institute for Snow and Avalanche Research at the head of the Pasern funicular railway at Davos. From the start of railway construction in Switzerland, avalanches have been of constant concern to Swiss engineers. The building of the great Alpine tunnels, with their portals necessarily located at the heads of steep valleys, merely served to intensify the need of avalanche-protection works.

The case of the Lötschberg Railway may be cited as an example. This line, which includes the Lötschberg Tunnel, gives a convenient short route between Berne and the Simplon Tunnel to the south. For almost the entire stretch of line from Goppenstein (where the line emerges from the tunnel) down to Brigue, the line is carried high up on the rocky sides of the Lonza and Rhône valleys, where avalanches are frequent. Fortunately, the paths of regular avalanches can readily be detected on wooded slopes, for they will strip off all major vegetation in their paths. Engineers responsible for the Lötschberg line were therefore able to design protection works, mainly deflecting walls but with some notable snowsheds also, in conjunction with the construction of the railway line itself. In

FIG. 31.14 Avalanche-deflection structure of the Schintigraben in the Lonza ravine on the Lötschberg Railway, Switzerland.

31-18   CIVIL ENGINEERING WORKS

the Lonza Valley, in particular, there must be a record number of avalanche works in relation to the length of line protected. More than 1,000 protection walls have been constructed along this route; some are 12 m (40 ft) high and located as much as 2,550 m (8,500 ft) above sea level. More than 10 million trees have also been planted as natural protection up mountain slopes as high as 1,950 m (6,500 ft) (Fig. 31.14). The value of all this work is shown by the fact that the Lötschberg Railway has never been closed for any appreciable length of time since it was first opened in 1913, despite the fact that its route lies across so large a number of well-recognized avalanche paths.[31.21]

A little to the east of the line just described runs the Furka-Oberalp Railway. Since it runs through a relatively uninhabited part of the country, its owners do not attempt to keep open that part of the line from Oberwald up to the Furka Tunnel and then downhill to Realp near Andermatt during the winter season. On its way down from the Furka Tunnel, this line passes through a very deep valley much subject to avalanches. During the first winter after completion of the line, a bridge that had been constructed here across the Steffenbach Gorge was swept away by an avalanche. Study showed that the bottom of the gorge was a regular path for avalanches of unusual intensity. With typical Swiss ingenuity, therefore, a new bridge was designed that would not be so destroyed when not in use in the winter. Its three deck spans are designed in such a way that the central span can be hinged on the end of one of the side spans, and the three spans, with their steel supporting trestles, can then be swung into positions of rest against the two abutments of the bridge, to stay there out of danger throughout the winter (Fig. 31.15). When the line has to be opened in the spring, the sections of the bridge are unfolded, swung back, and secured together in place. This entire operation (as well as the corresponding dismantling in the fall) takes no more than one working day for a trained group of workers if the weather is favorable.[31.22]

These are but two examples of Swiss railway engineering; they are typical of the boldness in location and design that has produced in this little mountainous country (only 362

FIG. 31.15 The Steffenbach Bridge on the Furke-Oberalp Railway, Switzerland: *(a)* as erected for use in the summer; *(b)*, *(c)*, and *(d)* being dismantled in stages at the end of the summer operating season; *(e)* folded up during winter closure of the line to permit the passage of avalances.

(b)

(c)

(d)

(e)

by 220 km or 226 by 137 mi in size) over 5,600 km (3,500 route mi) of railway. Snowsheds for avalanche protection are almost a commonplace on the lines at higher altitudes; one of the longest serves both a highway and the Furka-Oberalp Railway at the head of the Oberalp Pass, 2,001 m (6,670 ft) above the sea. Even more impressive, however, are the long series of snowsheds, in places almost continuous, that protect the railway line that now joins Oslo and central Norway with Bergen on the western coast, probably the most remarkable of all European railway lines outside Switzerland.

Snowsheds are called for also in North American railway practice. One of the most extensive installations was on the old route of the Canadian Pacific Railway where it penetrated the Selkirk Mountains through Rogers Pass. So troublesome were the avalanches in this area that, when the time came to renew the original snowsheds, it was decided to relocate the line even though this meant building the 8-km (5-mi) -long Connaught Tunnel. It now seems probable that much of the early trouble with avalanches in the pass was due to the indiscriminate burning off of almost all vegetation by the early railway builders. Avalanches still occur in the pass; they have been intensively studied in connection with the building of the Trans Canada Highway, which uses much of the old railroad right-of-way through the pass and for which a special series of avalanche-protection works has been designed (Fig. 31.16).

Finally, mention must be made of the danger due to floods. Protection against floods is largely a civil engineering matter, depending on the accuracy with which preliminary flood predictions can be made for the design of necessary waterways beneath roads or railways. In Chap. 24 mention is made of the scouring of bridge foundations and the precautions to be taken against this hazard. Reference will here be made to a case which is of special geological significance and which involved one of the most tragic of all accidents in the history of railroading (Fig. 31.17). Late on Christmas Eve, 1953, the main part of the Wellington-to-Auckland Express in New Zealand was swept away when it plunged into the swirling waters of Whangaehu River at the site of what had been the Tangiwai Bridge. In this accident, 151 persons were killed; 20 bodies were never found. The main finding of the official board of inquiry was that:

> The accident was caused by the sudden release from the Crater Lake on Mount Ruapehu through an outlet cave beneath the Whangaehu Glacier of a huge mass of water which was channeled down the Whangaehu River carrying with it a high content of ash from the 1945 eruption and blocks of ice due to the collapse of large volumes of the glacier. This flood, which can properly be called a "lahar," proceeded down the mountain as a wave, uplifting huge quantities of sand, silt and boulders. It was most violent and turbulent and of great destructive effect. It destroyed portions of the railway bridge at Tangiwai before the arrival of train No. 626, which was engulfed when proceeding across the bridge.

No blame was attached to the engineers or to the train personnel involved; the tragedy was one of those that can only be described, with reverence, as an "act of God."[31.23]

It is impossible to summarize the technical aspects of the accident; those interested are referred to the unusually complete report of the board of inquiry and to an associated publication. It must be admitted that the combination of an active volcano and a nearby glacier, the discharge from which led directly down a relatively short river crossed by the railway, is unusual, but it would appear that similar damage to communication routes has

FIG. 31.16 The result of an avalanche at Cougar Creek, near Glacier, British Columbia, on the main line of Canadian Pacific Railway, despite the existence of protection works above the line.

FIG. 31.17 Tangiwai Bridge over the Whangaehu River, North Island, New Zealand, (a) carrying a fast freight train on the day before the disaster; (b) the scene of the disaster on Christmas Eve, when the bridge was engulfed while carrying a crowded passenger train of New Zealand Government Railways.

been done in a few other locations from similar discharges, although, fortunately, without such dreadful loss of life. The board suggested that the geological term *lahar* should be better known by civil engineers. It is defined thus in their report:

    A lahar is a type of mudflow that occurs in volcanic areas. Lahars may be formed by the waters of crater lakes being released by the collapse of the crater wall or by volcanic eruption, by the melting of snow and ice by volcanic heat, or by the action of rain on volcanic ash deposited on the steep slopes of volcanoes. They usually pick up large quantities of volcanic ash and other debris, and form a thick slurry that on account of its high density may carry even enormous boulders for many miles across fairly flat country once the initial momentum has been gained.

The potential danger of lahars in all volcanic areas, therefore, is an added hazard to be carefully considered by those responsible for the design of transportation routes.

The original Tangiwai Bridge consisted of three deck plate girder steel spans supported on concrete piers. Figure 31.17a (which, by chance, was taken on the day before the accident) shows its general appearance. It has since been completely replaced by a new bridge consisting of two 36-m (120-ft) steel through-truss spans, with one mid-river concrete pier. New Zealand engineers have designed and installed a warning system against any future floods that, in some respects, must be unique (Fig. 31.18). It consists of a 6-m (20-ft) -high concrete pier located in the river 12.8 km (8 mi) above the bridge. A series of lead electrodes is exposed, at progressively increasing heights, spirally around the pier; the electrodes are suitably connected to five separate circuits with a somewhat elaborate system of warnings that includes even a device that displays a warning signal actuated by the *speed* of a sudden rise in the water level in the river, since it is the danger of sudden flash floods that is regarded as unusually serious. The lead electrodes were selected for use after it was found that the water of the Whangaehu River was highly

FIG. 31.18 Special river-level warning tower on the Whangaehu River, 13 km (8 mi) above the Tangiwai Bridge, erected following the disaster of 1953.

31-24 CIVIL ENGINEERING WORKS

FIG. 31.19 Blockage of the main line of the Canadian Pacific Railway due to the descent of the jökulhlaup described in the text.

acidic; its low pH value permits the design of a warning system that is as close to perfection as engineering design can achieve. If, for example, power is cut off for any reason, the system can still operate for more than 24 hours from its own battery system. The warning panel is located at Waiouru station, 10 km (6½ mi) away from the river pier and 13 km (8 mi) from the bridge site.[31.24]

Lest it be thought that lahars are so exceptional that they are merely a matter of special interest, it must be recorded that a somewhat similar occurrence, although without volcanic action, took place on the main line of the Canadian Pacific Railway. On 5 September 1978, just as a freight train was leaving the first of the famous spiral tunnels, a mass of soil and water swept down from the mountains above. The locomotive was derailed and damage was done to both train and track, but fortunately no lives were lost. It was estimated that possibly 175,000 m^3 (230,000 yd^3) of debris were washed onto the track by the outflow of possibly 1 million L (250,000 gal) of water from the base of the Cathedral Glacier (Fig. 31.19).[31.25] Wherever glaciers come close to railway tracks or are so located that flows from them can affect rail tracks, this is another hazard that must be kept in mind.

31.6 RELOCATION OF RAILWAYS

Although few new main railway lines are now being constructed in North America, there are many necessary railway relocations, either to assist maintenance or as a part of the major replanning that is now a feature of many of the larger urban communities throughout the continent. An example of the former was the relocation of more than a mile of the (former) Great Northern double-track main line north of Seattle, in view of the continuous trouble with rockfalls along the old location. After the necessary subsurface investigation, it was found that a new location seaward of the old one could be used, despite the exposure to the sea of the new embankment that relocation would involve. The new route is, on the average, 30 m (100 ft) away from the old, the maximum separation being 50 m (165 ft). Specially selected rock was brought from a quarry in the Cascade Mountains, 100 km (65 mi) away, and dumped from a temporary timber and steel trestle. Seven 1.8-m (72-in) concrete culverts were installed in the fill to allow for tidal flow in and out from the lagoon formed between the old and the new tracks. The fact that the contract was for a sum of $1.25 million (in 1958) shows the monetary cost of the trouble caused by such simple hazards as rockfalls.[31.26]

Many of the urban relocations of railways consist in the adaptation of outlying railway yards and freight lines for the passenger service that was once so conveniently provided to a downtown terminal or through station. A few are more extensive operations than this. The most extensive in recent years of which the writers have knowledge is the complete rearrangement of railway lines in the vicinity of the new bypass on the Welland Canal at the eastern end of the Niagara peninsula in Ontario. This 13.6-km (8.5-mi) -long new ship canal was constructed in order to speed up operations on the Welland Canal, one means of achieving this result being the elimination of the road and rail crossings of the original canal, all but one on bridges. Six bridge crossings were therefore eliminated when the new bypass was opened for use, but this necessitated the relocation of almost 160 km (100 mi) of railway.

All roads and all rail lines were combined in a master plan, the highway crossings of the bypass being made in two new tunnels. All rail lines were grouped in such a way that they could be combined with the highways in the design of one of the two new tunnels, both of which were built in the dry before the newly excavated canal prism had been completed and flooded (Fig. 31.20). The combined railway-highway tunnel consists of a reinforced-concrete box section 32 m (106.5 ft) wide and 9 m (30 ft) deep. With 9 m (30 ft) of water in the new canal, this meant that the invert of the new tunnel was 22.5 m (75 ft) below ground level. Because of the limiting grade for the three railway lines approaching and leaving the tunnel, the necessary approaches had to be each 4 km (2.5 mi) long, so that the total excavation amounted to 12.5 million m^3 (16.4 million yd^3). Fortunately, local geological conditions were favorable. Bedrock is generally dolomitic limestone overlain by up to 24 m (80 ft) of soil, usually a thin stratum of till followed by glacial lake clay, the upper portions of which are generally weathered, with the usual 60 cm (2 ft) of topsoil and organic matter. Construction procedures, although on a massive scale, were not unusual apart only from the extensive system that had to be installed for control of

FIG. 31.20 The Townline Road-Rail Tunnel beneath the Welland Canal Bypass, Niagara, Canada, opened for use in 1972–1973.

31-26 CIVIL ENGINEERING WORKS

groundwater, since the bedrock was charged with water under subartesian pressure, the free water table being often close to the surface. The main general contract for this work amounted to over $31 million, and this was for only one unit in a railway (and highway) relocation project.[31.27]

31.7 SUBWAYS

Construction of rapid-transit subways is now in progress or under consideration in many of the major cities of the world, clear indication of the planning attention that is being given to the essential movement of large numbers of people efficiently and with a minimum of pollution. It is axiomatic that the closer such railways can be to ground surface, the more convenient will they be and probably the more economical. This means, however, that the subsurface conditions to be encountered must be accepted rather than chosen, as would be the case with deep tunnels. In almost all cases, excavation will be in surficial deposits, possibly also in rock, this combined operation always presenting its own peculiar difficulties on construction. A further problem is that, with location close to ground surface, the cut-and-cover method of construction will sometimes prove to be the most desirable and economical, if permission to disrupt traffic temporarily can be obtained from civic authorities. Despite all problems, and the high capital costs involved, subway construction may be expected to continue, probably at an accelerated rate, as the traffic problems of major cities become more acute, and the energy situation more critical.

The Geology of London and Subway Construction The city of London, England, is preeminent in its use of subways, not only for passenger traffic but also (unknown to the general public) for such special purposes as the conveyance of mail from the central main post office to railway terminals, the "post office tube" being a subway in its own right. For those interested, a retired chief engineer of London Transport has compiled a succinct but fascinating book on the first half century of the London "tubes," from which one can see clearly how fortunate London has been in its underlying geology, so suitable for tube railway construction.[31.28] The geological section in Fig. 8.4 illustrates this.

The London clay is a stiff, fissured, and overconsolidated Eocene clay. This is the correct geological-engineering description of the material that is of such importance to all subsurface work in London. The fact that it is of Eocene (geological) age shows that it is much older than the clays normally encountered in all the glaciated parts of the world. During the excavation of tunnels in the clay, when handwork is necessary, the clay is cut off in large slices that have the consistency of a medium cheese. The slabs can be easily handled. The sight of these large pieces of this material being thrown so easily into mucking cars must always impress interested observers as about the easiest form of excavation possible. Its very advantages from the point of view of excavation, however, pose other problems when structures within the clay are disturbed. During the construction of the great Shell Building on the south shore of the river Thames in the center of London, for example, the presence of tube railways under the site added to normal construction difficulties.

The complex of buildings in this center is one of the largest of such aggregations of buildings in London, its site having an area of 3 ha (7.4 acres). Construction of a basement under the complete site is some indication of the extent of the project. Beneath the site of one of the main buildings, and lying very close to the location of the 20-storey tower were the twin tubes of the Bakerloo Tube (as this part of the Underground is known), with the two tubes one above the other, the top one having only 1.5 m (5 ft) of clearance below the basement slab for the large building (Fig. 31.21). Although the building load

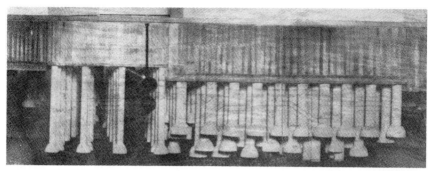

FIG. 31.21 Photograph of a model of the foundations under the Shell Center, on the south bank of the river Thames, London, showing measures adopted to minimize interference with existing tube railway tunnels.

could have been carried on the clay, that part overlying the two Bakerloo tunnels was founded on cylinders with belled-out bottoms, excavated in the clay to a depth well below the bottom tube and then filled with concrete suitably reinforced, to avoid overstressing the tubes (and for other structural reasons).

There was the further danger that the tubes might be forced somewhat upward by the rebound of the clay when the deep excavation for the building was complete, again in the light of the characteristics of the clay. To guard against this, the excavation over the tubes was carried out in strips each 9 m (30 ft) long. The concrete foundation slab was placed in each strip as soon as it had been excavated and before the adjacent strips were excavated. As can be imagined, the tube tunnels were most carefully instrumented throughout the lengths that ran through the excavation. Through this instrumentation, in itself an unusual and most expert foundation study, the tubes were kept under constant surveillance. The job was completed without any disturbance to traffic in the tubes and with eventual movements so small as to be negligible.[31.29]

Although much of the excavation for London's subways has been in "free air," compressed air has sometimes to be used, with all the special precautions (and extra cost) that this involves. It is a measure to be avoided if at all possible. This possibility will be evident only from a most detailed geological study of the ground around, above, and below the line of the proposed tunnel. An almost unsuspected danger with compressed air was experienced during the construction of the new Victoria line in London, the first tube railway to be constructed there for 30 years. At one location where compressed air was being used in the Woolwich and Reading water-bearing sands that lie beneath the London clay, it had the effect of forcing into a neighboring tunnel, also under construction, the air that had been trapped throughout geological time in the voids of the sand. This long-stagnant air was deficient in oxygen and could have led to fatalities, although fortunately it did not in this case.[31.30] The same phenomenon has been noted in other tunnels in Melbourne and Seattle, for example, where two excavations were being worked in close proximity, one under air and in porous material.

Freezing of soil is always an expensive construction technique, but when geological conditions are so bad that any other solution is questionable, its use may prove to be truly economical, especially when time is considered. It has even been used on construction work for the London Underground, where construction is usually thought of as always being in the easily handled London clay. During the construction of the modern Victoria line, subsurface exploration showed that the upper half of the inclined tunnel for the access escalator at Tottenham Hale Station would have to be excavated through dense, water-bearing sandy gravel that contained some cobbles. Few worse combinations can be imagined for excavation in confined quarters such as this station involved. The use of a chemical means of consolidating the gravel (another well-tried method) was first considered, but detailed test boring revealed the existence of lenses of silt within the gravel, material that would be questionable for chemical grouting. Freezing was therefore adopted as the method for consolidating the ground so that excavation could safely be carried out. A relatively new system using liquid nitrogen, instead of the more usual cold brine solution, was tried—with some initial misgivings but with complete success. The rate of freezing, so critical on this job, was retarded by the flow of the groundwater through the gravel, but eventually excavation could proceed, working a full face in the inclined tunnel. The time saved by the use of liquid nitrogen more than compensated for its increased cost.[31.31]

Consideration of Preexisting Foundations in Subway Construction Difficulties encountered in subway construction due to the foundations of existing buildings are not peculiar to London. In almost every case of the building of a modern urban subway railway system, great problems have had to be faced either with the underpinning of existing structures or in constructing the new tunnel so as not to interfere with existing foundations close to the tunnel route. Brussels, capital of Belgium, is well launched into the building of a 70-km (44-mi) double-track subway system. Early problems on this system included tunneling under a famous 68-year-old ceremonial masonry arch (weighing 26,000 tons), resting on poor foundations, and tunneling under two existing large brick conduits used for the flow of the river Senne and for overflow storm water from the city's sewers.[31.32] The reverse of this happened in Paris, France, when the foundation for what was then (1972) Europe's tallest office building, rising 197 m (656 ft) above ground level, had to be constructed around and under an operating subway of the Paris Metro.[31.33] In São Paulo, Brazil, the presence of many multistorey buildings, flush with the street lines adjacent to the city's subway, made mandatory the use of a shield for excavation.[31.34]

Even more remarkable was the construction of the tunnels for the Stephansplatz Station on the Vienna subway, Austria. As usual, these tunnels were located beneath the center of a main road on which St. Stephen's Cathedral is located (Fig. 31.22). The cathedral is regarded as Vienna's greatest cultural monument. The Virgin Chapel, dating back to 1367, is located almost flush with the street line. Foundations of the cathedral vary, but some rest on loessal soil. The most intensive study of the geology of the cathedral surrounds was made. The revealed soil section included (from ground level down): loess-loam, broken slate, sandy loam and gravel, and broken slate and pebbles, with the water table about 15 m (50 ft) below the surface. Intensive studies were made of construction methods that could be safely used without damage to the cathedral. Suffice it to say here that the methods adopted included special grouting, the use of compressed air at appropriate pressures for the shield excavation, and the use of low-vibration bored piles to support station structures. As can be well imagined, the most careful measurements were taken throughout the entire construction period, which was successfully completed with no damage to the historic structure, such settlements as did occur being measured as a few millimeters.[31.35]

The Hydrogeology of Rotterdam and Subway Construction Few of these lines encountered geological conditions that had a determining effect upon design or construction, the geology of their routes having to be accepted and dealt with in what may be called standard ways. There is one subway, however, the design and construction of which were entirely controlled by the local hydrogeology. All who know the modern city of Rotterdam in the Netherlands know what a delightful blending of the old and the new it now

FIG. 31.22 The west front of St. Stephen's Cathedral, Vienna, Austria, showing the proximity of the exit escalator from the new subway railway line.

presents. The new city center is one of the finest of all examples of enlightened modern city planning. Visitors with any sense of history know that this entire area has been completely rebuilt since the end of the Second World War, following its devastation by low-flying bombers at the start of the war. There is now in use the first section of a rapid-transit subway system, including a second notable tunnel under the Nieuwe Maas, that gives ready access to the business area, while not adding any vehicular traffic to the already crowded central section of the city.

The subsurface of Rotterdam must be outlined in the first instance so that this remarkable construction operation can be the better appreciated. The most important feature of the geology of Rotterdam is the existence at a depth of from 16 to 18 m (52.5 to 59.25 ft) below the local datum level, equal to average sea level, of a stratum of medium-to-coarse sand containing some gravel, a Pleistocene deposit brought down by the rivers Rhine and Meuse. It is naturally water-bearing; pumping from it in recent years for air-conditioning and cooling installations in city buildings has been extensive. It extends to a depth of more than 40 m (130 ft) in the eastern part of the city, diminishing to about 30 m (100 ft) in the western area. Upon this excellent foundation bed, all major buildings and structures in Rotterdam are founded with every confidence, through the medium of piles. Above the sand is a stratum of alluvial clay about 1 m (3.25 ft) thick, followed by a layer of organic peat, and a final stratum of marine clay with a total thickness of about 2 m (6.5 ft). Thence to the surface is a variety of shallow clay and peat deposits, with man-made fill naturally present to varying degrees. It is on these varied deposits that roadways and light buildings have to be founded.

Since much of the outlying part of Rotterdam is of very recent development, the observant visitor can usually see some roadways being reconstructed as the slow settlement of the organic peat beneath them has to be offset by filling and rebuilding. The devices of sand drains and preloading are used to accelerate the process of consolidation. Because the general level of the city is below sea level, as with so much of the Netherlands, the

31-30 CIVIL ENGINEERING WORKS

Dutch dikes against the sea being justly world-famous, the groundwater throughout the entire area of the city is close to ground level, never more than 2 m (6.5 ft) below the surface. Fortunately, it is reasonably constant in elevation, and so no difficulties are experienced with deterioration of the timber piles so widely used all over the city. But any deep excavations necessarily interfere very seriously with the water table. Herein lay the great problem to be faced in the building of the subway system.

All the rapid-transit lines on the north side of the Nieuwe Maas, i.e., in the central part of the city, had to be constructed in tunnel. Once the line emerges to the south from the tunnel under the Maas, it becomes an elevated line and will so continue, with surface portions, as it is extended into the suburbs. Of the initial 7.6 km (4.7 mi), 3.1 km (1.9 mi) had to be in tunnel, with the remaining 4.5 km (2.8 mi) on a prestressed reinforced-concrete superstructure. Bottom level of the tunnel section had to be at about 11 m (36 ft) below ground level, the combined twin tunnels requiring a structural width of about 10 m (33 ft). A tunnel of these general dimensions had to be constructed right through the central city area, close to many large buildings founded on piles, with its terminal adjacent to the new Central Railroad Station. Tunneling through the varied soil strata to be encountered would have been difficult and hazardous, especially with the high water table to be handled. Excavation of open trenches for construction by conventional cut-and-cover methods would have involved a massive pumping operation for each section of trench for a five-year period and great disturbance of groundwater conditions around each trench, even with the most expert use of well points and controlled pumping. The decision was therefore made, in order to avoid pumping, to build even these "inland" sections of the subway tunnel as precast concrete tubes and to float them into place right in the center of the city. This bold operation was successfully completed and the first section of the subway was opened in 1968.

FIG. 31.23 Subway under construction in the heart of Rotterdam, the Netherlands, almost all the buildings seen having been built after the Second World War; the excavated trench for the subway structure between rows of steel sheet piling can be clearly seen, and to the right of center the yard for the fabrication of tunnel sections.

The large tunnel sections were constructed in three "dry docks," two in the central city area and one on an island in the Maas about 5.6 km (3.5 mi) away. Two complete sections could be built in each dock at once. The inland sections measured 45 to 60 m (147 to 197 ft) long, 10 to 25 m (33 to 79 ft) wide, and 6 to 10 m (20 to 33 ft) deep. Sections for the river part of the tunnel, which was also built by sinking tubes, were 90 m (295 ft) long, 10 m (33 ft) wide, and 6 m (20 ft) high. Some of the inland sections had to be curved. They weighed up to 5,090 tons, but most fortunately this vast load could easily be carried by water. While the tubes were being built, trenches of the appropriate size were excavated within single rows of steel sheet piling. Pumping was carried out only for the initial excavation. Installation of the necessary steel bracing was easily completed. The trenches were then allowed to fill up with water, excavation was completed with clamshell buckets, and specially designed concrete piles were then driven from a floating pile driver all over the bottom of each trench to provide the bearing medium for transmitting the load of the tunnel section down to the underlying sand. At intervals, the great tubes were ballasted with water, their ends having been carefully sealed, and they were then floated along the excavated trench in sequence. When in final position, they were sunk into place and connected with the adjoining tubes by most ingenious sealing mechanisms. Fill was placed around them and on top, the steel piling was removed, and the street surfaces were restored so well that, even shortly after the work had been finished, it was difficult to imagine that sections of railway tunnel had indeed been floated into place along these selfsame roadways (Figs. 31.23 and 31.24).[31.36]

Casual visitors to Rotterdam during the five-year construction period may have thought to themselves that the Dutch people were enlivening the great city by a modern type of canal with somewhat unattractive steel sides. All visitors who knew anything about building, however, knew that they were looking at one of the boldest urban construction

FIG. 31.24 Another view of the Rotterdam subway under construction, looking towards the main railway station and subway terminal; this view gives an idea of the scale of the workings by comparison with automobiles on adjacent streets.

operations ever carried out. They must have shared the admiration felt by one of the authors as he was privileged to follow the necessarily slow progress of this remarkable work during annual visits to Rotterdam throughout the period of construction. The project will long remain a classic example of fine engineering and excellent construction, taking advantage of unusual subsurface conditions that appear at first sight to be so unfavorable for tunnel construction. Details of the construction techniques employed, especially for the critical river crossing, have been made available in the literature of engineering by those responsible. All that need be said here, in addition, is that construction throughout Holland is complicated by subsurface and groundwater conditions not dissimilar to those in Rotterdam. For the construction of the similarly interesting Amsterdam subway, with similar geological conditions, an alternative method was used. Sections of the subway tunnel were built as concrete caissons on the surface and then sunk into place, with the necessary use of compressed air.[31.37]

31.8 CONCLUSION

The record of modern subway building is in the great tradition of railway construction. Another distinguished chapter is being added to the record as this Handbook is prepared for the press in the steady progress being made on the Washington subway, already in partial service in the nation's capital. The complex geology in which it is being constructed would require a full volume for adequate description; it is to be hoped that such a record will be prepared.

Great efforts are being made, especially in the United States but in other countries also, to reduce the cost of tunneling, one reason for this concern being the knowledge that more subways are going to be needed if the traffic problems of major cities are to be solved. There is much written and said today about the innovations needed in tunneling, one complaint being that the tunneling practice in New York (where 50 percent of all tunnels in the United States are said to be located) is conservative, and yet is slavishly followed elsewhere in the United States. Innovations are needed, as always; tunnel costs should be minimized to the maximum extent possible. But the fact that the local geology always determines the final result is a fact of life that is mentioned all too rarely in these current discussions. One of the authors has heard an address by one of those leading the search for reduced tunneling costs in which geology was not mentioned once.

In all railway work, and today in all considerations of subway railway construction within the borders of great cities, adequate knowledge of the geology along all routes being studied is the first requirement. Unless the geology along the route selected is known with reasonable certainty, then all other considerations may be "of nothing worth." As the subways of the world proliferate—as they must do—urban geology will present challenges such as have not yet been seen. Upon the answer to those challenges will depend the success of the railways yet to be built.

31.9 REFERENCES

31.1 *The Times*, London, 20 August 1937.

31.2 Rolf Emerson, "A Project for Extending the Nigerian Railway into Bornu Province," *Proceedings of the Institution of Civil Engineers*, **12**:153 (1959).

31.3 C. H. Ellis, *British Railway History 1830–1876*, G. Allen, London, 1954, p. 93.

31.4 A. C. O'Dell, *Railways and Geography*, Hutchison, London, 1956.

31.5 Ibid., p. 55.

31.6 "A Natural Tunnel in Virginia," *The Railway Magazine,* **78**:319 (1946); see also H. P. Woodward, "Natural Bridge and Natural Tunnel, Virginia," *Journal of Geology,* **44**:604 (1936).

31.7 C. H. Ellis, *The Midland Railway,* Ian Allan, London, 1953, p. 98.

31.8 O'Dell, op. cit., p. 130.

31.9 "Rail Tunnel Underpasses a Runway in Holland," *Engineering News-Record,* **178**:42 (20 April 1967); and information from Ir W. de Steur, chief engineer (Concrete Building), Netherlands Railways.

31.10 G. J. Low, "Soil Freezing to Reconstruct a Railway Tunnel," *Proceedings of the American Society of Civil Engineers,* vol. 86, CO3, Paper no. 2639, 1960.

31.11 Francis Fox, *Sixty-Three Years of Engineering,* J. Murray, London, 1924, p. 47.

31.12 T. R. Nolan, "Slips and Washouts on the Hill Section of the Assam-Bengal Railway," *Minutes of Proceedings of the Institution of Civil Engineers,* **218**:2 (1924).

31.13 C. Wahrhaftig and R. F. Black, *Engineering Geology along Part of the Alaska Railway,* U.S. Geological Survey Professional Paper no. 293-8, 1958, p. 69.

31.14 G. P. Raymond, "Railroad Ballast Prescription: State of the Art," *Proceedings of the American Society of Civil Engineers,* vol. 105, GT 2, Paper no. 14397, 1979, p. 305.

31.15 F. L. Peckover, *Treatment of Rock Falls on Railway Lines,* American Railway Engineering Association Bulletin no. 653, Chicago, 1976, p. 471.

31.16 *The Railway Magazine,* **104**:735 (1958).

31.17 The secretary for railways, Perth, Western Australia, by personal communication.

31.18 J. R. Thomas, Great Northern Railway, by personal communication.

31.19 See ref. 31.15.

31.20 S. W. F. Morum, "The Treatment of Mud-Runs in Bolivia," *Journal of the Institution of Civil Engineers,* **1**:426 (1935).

31.21 C. J. Allen, *Switzerland's Amazing Railways,* Nelson, New York, 1953, p. 24.

31.22 Loc. cit., pp. 25-26.

31.23 *Tangiwai Railway Disaster: Report of the Board of Inquiry,* Wellington, New Zealand, 1954; see also N. E. Odell, "Mount Ruapehu, New Zealand: Observations on its Crater Lake and Glaciers," *Journal of Glaciology,* **2**:601 (1955).

31.24 "Rising Water Completes Circuit, Sounds Flood Alarm," *Engineering News-Record,* **161**:58 (16 October 1958).

31.25 L. E. Jackson, "A Catastrophic Glacial Outburst Flood *(jökulhlaup)* Mechanism for Debris Flow Generation at Spiral Tunnels, Kicking Horse River Basin, British Columbia," *Canadian Geotechnical Journal,* **16**:806 (1979).

31.26 "Railroad Crosses Bay to By-Pass Rock-Slide Danger," *Engineering News-Record,* **160**:52 (9 January 1958).

31.27 "Wet Site Turned into Dry Home for Massive Tunnel," *Engineering News-Record,* **187**:22 (4 November 1971).

31.28 H. G. Follenfant, *Reconstructing London's Underground,* London Transport, London, 1974.

31.29 "Digging Foundation around Bridge Approach and Subway Tunnels," *Engineering News-Record,* **162**:60 (18 June 1959).

31.30 H. D. Morgan and J. V. Bartlett, "The Victoria Line: Part Three: Tunnel Design," *Proceedings of the Institution of Civil Engineers,* suppl. vol., Paper no. 7270 S, 1969, p. 388.

31.31 J. A. M. Clark, G. S. Hook, J. J. Lee, P. L. Mason, and D. G. Thomas, "The Victoria Line: Part Four: Some Modern Developments in Tunnelling Construction," *Proceedings of the Institution of Civil Engineers,* suppl. vol., Paper no. 7270 S, 1969, p. 397.

31.32 "Subway Builders Snake Tunnels Under Old Arch and Conduits," *Engineering News-Record,* **190**:26 (14 June 1973).

31-34 CIVIL ENGINEERING WORKS

31.33 "Europe's Tallest Building Encases an Operating Subway," *Engineering News-Record,* **189**:22 (13 July 1972).
31.34 "Tough Conditions Challenge São Paulo Subway Builders," *Engineering News-Record,* **201**:23 (26 October 1978).
31.35 A. Dollerl, A. Hondl, and E. Proksch, *The Construction of the Vienna Underground Railway and Measures Taken to Protect St. Steven's Cathedral,* D. A. Sinclair (trans.), Technical Translation no. 1900, National Research Council of Canada, Ottawa, 1977.
31.36 G. Plantema, "Rotterdam's Rapid Transit Tunnel Built by Sunken-Tube Method," *Civil Engineering,* **35**:34 (August 1965); and by personal communication from Dr. Plantema.
31.37 D. Halperin, "The Sinking of the Amsterdam Metro," *Civil Engineering,* **45**:92 (September 1975); see also "Amsterdam Subway Built on Surface Then Sunk into Water-Laden Soil," *Engineering News-Record,* **187**:31 (19 August 1971).

Suggestions for Further Reading

Railway engineering is a branch of civil engineering that is fortunate in having had for a long time its own specialized engineering society. This is the American Railroad Engineering Association with its headquarters at 2000 L Street, Washington, D.C. 20036. Reference to a paper in its publications will be found above, indicating clearly the Association's appreciation of geology in railroad engineering work.

Construction of subways will necessarily be a feature of the development of larger cities of the world. It may be helpful, therefore, to add to the references above an account of some of the most difficult urban railway tunneling ever carried out, this being in Melbourne, Australia. A concise account of this remarkable work was presented to the American Society of Civil Engineers at their San Francisco meeting in October 1977; the 30-page paper by A. M. Petrofsky and J. A. Ramsay, *Underground Construction Down Under,* was published as ASCE Preprint No. 2971.

Many problems of railway design, construction, and maintenance are similar to those in highway engineering work. This is perhaps reflected in the change of name of the Highway Research Board (of Washington, D.C.) to the Transportation Research Board. Reference to some of the publications listed at the end of Chapter 30, especially those of the TRB in recent years, will be found of use in railway work.

chapter 32

AIRFIELDS

32.1 Site Study for Mirabel Airport / 32-3
32.2 Drainage / 32-4
32.3 Cut and Fill / 32-6
32.4 Some Special Cases / 32-7
32.5 Airfield Cover / 32-9
32.6 References / 32-9

So phenomenal has been the growth in civil aviation all over the world in the last half century, especially since the end of the Second World War, that there is probably no branch of civil engineering work about which the public is so impatient and responsible officials so frustrated as the construction of airfields—the building of new fields where necessary and the enlargement and improvement of existing facilities. Singapore had to build a completely new airport within 20 years of the opening of the former one; Washington, D.C., was in almost the same position. Cities such as Chicago have had to construct duplicate fields. And all-too-many fine landing facilities, built apparently for the foreseeable future but a few years ago, are proving to be inadequate and incapable of expansion to meet the needs of larger planes in regular commercial use. In all this rather urgent design and construction work, involving extraordinary sums of money judged by any standards, geology is not the determinant that it usually is in other kinds of construction.

Modern airfields have to provide very large, reasonably level areas without serious impediments to flying in the vicinity. Desirable flat areas, when available, owe their flatness to their geology. Typical are the floors of old lakes or seas, usually clays, and alluvial plains often underlain by coarse materials. The areas must be capable of being thoroughly drained at all times; and above all, they must be as conveniently situated, with respect to the cities they serve, as is possible. In view of these requirements, the geology of potential sites almost always has to be accepted; convenience of access and economic availability of reasonably level ground are the basic determinants. In only one case that has come to the authors' attention has a location, otherwise suitable, had to be turned down because of unfavorable geological conditions; in this location, the presence of an extremely sensitive marine clay gave promise of unavoidable settlements of long duration that were quite unacceptable. Some airfields have had to use sites involving construction over filled ground; examples include the La Guardia Airport at New York and the airport at Baltimore. In such cases, serious design problems have to be met and solved, but they are more in the nature of difficulties in soil mechanics rather than in the field of geology.[32.1]

Despite this overall situation, the geology of airports is still of great importance. It is imperative that preliminary geological and subsurface investigational work be well done and that it take into account adequately all the area upon which construction is to take place, together with any adjacent areas which may influence drainage of the field. In the first place, determination of the amounts of cut and fill material and its exact character will be necessary for the preparation of estimated costs, of tentative time schedules, and ultimately, of contract documents for actual construction. Correspondingly, accurate subsurface information is essential for the proper design of pavements (especially because of the large wheel loads now commonplace in design standards) and of foundations for the structures that will serve the port. Finally, and perhaps of greatest importance, adequate drainage facilities can only be properly designed and installed if the soils to be encountered, their drainage characteristics, and the local groundwater conditions are known with accuracy. These are not "exciting" applications of geology; they are, nonetheless, vital and call for the same degree of careful attention as does the most spectacular service provided by geology for tunnels, dam foundations, or landslide correction.

Landing-field areas have to be graded to a given level, drained, and provided with suitable runways. Design of the latter is comparable to highway design; similar materials are used with similar design requirements. When turf is used to surface a field, special attention must be given to the type of grass and to the spreading of a good top layer of soil if the local material is unsuitable for good growth. Drainage is the counterpart of all such work; it will often prove to be the most difficult part of airport work. The fact that the landing field has to be practically level constitutes a leading problem and necessitates the closest attention to the gradients adopted for drains, in order to keep trench excavation to a minimum. A soil survey is therefore essential. This can often be carried out with the aid of hand-boring outfits, provided the overall geology warrants their use. Once this information is available, the type of drainage system necessary can be decided upon. If porous materials underlie the site, the system can be simple; if clay or similar material is found, an elaborate system of field drains and main drains may be necessary. Installation of drains must be undertaken with great care; backfilling (of selected porous material) up to within 15 cm (6 in) of the surface is essential. Surface material should be as uniform as possible over the field and therefore should cover all drains and refilled soft spots, the discovery of which in preliminary test borings can be one of the most important contributions of the civil engineer to this phase of transportation engineering. United States figures suggest that one accident out of every eight in aviation is due to landing-field defects; even in this branch of work, therefore, adequate preliminary investigation of ground conditions can be of real avail.

All the items of work so far mentioned are common to other branches of civil engineering. There is, therefore, nothing of special note in the design and construction of airfields beyond the vast area always involved, the great costs inevitably associated with installations of such size, and the necessity for the closest possible attention to all details of design, such as drainage facilities and pavement cross section, not only because of the extensive use that has to be made of the basic designs once they have been established, but also because of the imperative need of absolute safety in the performance of airport runways in particular. Accordingly, this brief chapter can be regarded, in a way, as a reminder of the importance of accurate subsurface investigation and of the significant correlation of drainage facilities with the underlying geology. As an illustration of the extent of areal studies for major airports, a summary account of the site studies for the Mirabel Airport, near Montreal, will be presented (Fig. 32.1). This is one of the most recent entirely new major airports to be constructed in North America. Although its location (especially in relation to the existing Dorval Airport of Montreal) and the necessity for such a large facility have been matters of discussion in Canada, there has been no

FIG. 32.1 General view of Mirabel Airport, a new international facility 40 km (25 mi) from Montreal, Quebec.

question about its engineering, the excellence of which may be deduced from the following summary.

32.1 SITE STUDY FOR MIRABEL AIRPORT

Selection of the site near Montreal, from a number of widely spaced alternatives, involved preliminary terrain studies. Once the site had been selected, more detailed study commenced with the aid of all available aerial photographs, all available information on the geology of the region centering around Ste. Scholastique, a small town northwest of Montreal, and a general appreciation of the local hydrogeology. An area of 35,500 ha (88,000 acres) was expropriated so that control could be exercised on all environmental questions relating to the operation of the new facility. Of this great area, 1,600 ha (3,900 acres) was required for the first phase of the operational area, which will eventually utilize 6,700 ha (18,500 acres).

The site consists generally of rolling country, flattening towards the east. Most of the area is a clay plain at about 75 m (250 ft) above sea level, underlain by the Leda (Champlain) marine clay so characteristic of the St. Lawrence Valley. The clay has a desiccated crust of from 1.5 to 2.4 m (5 to 8 ft) and is, in turn, underlain by glacial till resting on bedrock. The hills which produce the broken terrain are generally composed of till. Local depressions in the till were found to be filled with uniform fine sand and organic material. The water table is generally between 1.5 and 3.0 m (5 and 10 ft) below the surface (Fig. 32.2).

The larger area was the subject of a wide-ranging interdisciplinary ecological study, but full consideration was given from the start to agricultural features, forestry, ecology, and the recreational aspects of the land being preempted for the actual construction of the airport. With the surface geology of the area known in general terms, the first step in the detailed study was an intensive program of auger borings, all over the site, to a depth of from 50 to 75 cm (20 to 30 in), disturbed soil samples being secured for initial identification tests. Concurrently, seismic surveying was carried out over the site in order to determine the depth to bedrock.

32-4   CIVIL ENGINEERING WORKS

FIG. 32.2 Soil conditions as exposed during construction of services for Mirabel Airport, Montreal; the straw is for protection during cold-weather construction.

All field-test results were fed into a computer (located in Montreal) through which a contour map of the bedrock was prepared. Based on this and the results of the shallow borings, a major program of deep soil borings was then developed. Two hundred holes were put down with a large machine-mounted drill rig, on the average one to each 2.5 km (1 mi^2), but irregularly spaced in accordance with preliminary information given by the shallow borings. Soil samples were taken in every hole with a side sampling device and the resulting undisturbed samples subjected to a full suite of soil tests. Only when all the resulting information was ready in usable form did it provide the basis for the engineering design of runways, taxiways, and other features of the airport and for the layout of drainage facilities. Construction started in 1973 and Mirabel Airport became operational in 1975.[32.2]

32.2   DRAINAGE

Drainage of airports is one feature of their design the importance of which cannot be overstressed. The large area of natural ground that has to be covered up with pavement for runways and taxiways, quite apart from that area utilized for airport buildings, makes such a considerable change to the local hydrogeology that the drainage system (as it must be called) for an airport is a major part of the overall design, mundane though it may seem to be. It is not within the province of this Handbook to deal with design details of such systems, vital as they are. Reference can be made to standard works on drainage facilities for this information. But to illustrate how a sound knowledge of the geology underlying the site of a proposed airfield can affect the design, there may be mentioned a case dating back to the years of the Second World War.

During the war years, a small airport was constructed at Bowling Green, Kentucky. With an area of only 107 ha (265 acres), providing four 1,170-m (3,900-ft) runways, it was not a large project, but its design was carried out with the usual care exercised by the U.S. Corps of Engineers, the constructing agency. It was found that the site was located in what was known locally as "sinkhole country"; the bedrock is a cavernous limestone (the upper stratum of which is known locally as "Cathedral rock"), overlain at the airport site by 4.5 to 6.0 m (15 to 20 ft) of dense red clay. Existing sinkholes were located and other potential downward drainage channels were surveyed; it was thus possible to grade the field so that 12 manholes, leading directly into the underlying limestone, would take care of all surface drainage. Manholes were connected to precast concrete pipe lengths which led down to the Cathedral rock, through which 30-cm (12-in) -diameter holes were drilled to the deeper and more cavernous limestone. Surface grading toward the manholes sufficed for most of the surface drainage, but some 600 m (2,000 ft) of French drains were used in flatter areas in order to conduct drainage to the nearest manholes. The airport has now been in use for three decades, and the unusual drainage system has worked quite satisfactorily. It illustrates what can be done by utilizing constructively a local geological condition that is ordinarily regarded as undesirable.[32.3] (Fig. 32.3.)

Since airports serving coastal cities are frequently located on low-lying land not suitable for other purposes, subsoils are sometimes of poor quality, with high water contents. In such cases, not only is surface drainage important but also subsurface drainage, in order to induce and control settlement of such soils, especially under runways and taxiways. Sand drains have proved their great utility for this purpose in a number of cases. There need be only mentioned the immediate postwar rehabilitation of both the Newark and La Guardia airports serving New York, taken over by the Port of New York Authority

FIG. 32.3 Bowling Green Airport, Kentucky, showing sinkholes *(right front)* some of which were used for drainage of the runways etc.

32-6   CIVIL ENGINEERING WORKS

in 1947. Three-quarters of the original La Guardia Airport was located over a deep deposit of marine organic silt. Six meters (20 ft) of ashes and debris had been placed over this in 1938, but differential settlement of the early runways had taken place. As part of the major rehabilitation program carried out by the Authority, sand drains were installed (at 4.5-m or 15-ft centers). Conditions at the Newark Airport were similar; up to 7.5 m (25 ft) of soft and highly compressible peat and organic silt (an old marsh deposit) overlies sand and varved clay, bedrock being at about 20 m (65 ft) below the surface. Sand drains were again used to consolidate the organic materials, spaced at 3.0- to 4.2-m (10- to 14-ft) centers, depending upon the thickness of the compressible soil. It must be stressed that the use of sand drains was only one feature of the extensive rehabilitation programs carried out at both these ports, work which has gone on, to a degree, up to the present and which has included some notable civil engineering features. Sand drains showed their usefulness, however, as they have in many similar locations.[32.4]

32.3 CUT AND FILL

After drainage, the estimating of quantities for cut and fill and the planning and execution of such work probably reveal the necessity of adequate geological study more than anything else in airfield work. Much of what was said about open excavation in Chap. 19 applies directly to airport construction; since excavation does not usually extend to great depths, even though large quantities are often involved, few additional comments are called for. One example may usefully be cited, however, to illustrate the extent to which airfield excavation may have to proceed. Admittedly the case is an extreme one, but there are not a few airfields in North America now in regular use that called for construction operations on an almost comparable scale.

One of the many smaller airfields constructed in the United States as the Second World War came to a close, with civil aviation looming large as an important postwar development, was that built to serve the metropolitan district around Charleston, the county seat of Kanawha County and the state capital of West Virginia. Kanawha Airport was constructed for the Kanawha County Court; it is located in the rugged, hilly country northeast of Charleston. The severity of the local terrain necessitated moving slightly more than 7.4 million m^3 (9.7 million yd^3) of material in order to give the level area necessary to accommodate the main runway, which is 1,800 m (6,000 ft) long, and subsidiary runways of 1,740 and 1,560 m (5,800 and 5,200 ft), respectively. The maximum difference in elevation between the highest point in a cut section and the toe of the deepest fill was 135 m (450 ft); one fill alone extended 69 m (230 ft) from its toe to runway level. Alternating layers of shale and sandstone were encountered; the absence of groundwater eliminated any real problems with excavation. Benches filled with rock from 0.8 to 6.1 m^3 (1 to 8 yd^3) in size were used to support all slopes steeper than 1 in 3. About one-half of all excavation had to be drilled and blasted, but good fragmentation was obtained; drop weights were used for reducing large rock fragments to manageable size for handling. The Harrison Construction Company of Pittsburgh used a grand total of 226 pieces of earthmoving equipment in carrying out this mammoth piece of "terrain reconstruction" (Fig. 32.4).[32.5]

Since flying provides lines of communication all over the globe, it calls for the construction of airports in many of the far places of the world. Frequently, airport construction projects have given rise to unusual experiences with geological conditions not normally encountered in the general run of civil engineering work. Some of those who served during the Second World War with the Seabees will long remember the unusual airfield construction on Iwo Jima. This small Pacific island, measuring only 9 by 5 km (5½ by 3 mi), is of

FIG. 32.4 Grading work in progress at Kanawha Airport, Charleston, West Virginia.

recent volcanic origin. An active volcano provides the main outcrop of solid rock on the island (Mount Suribachi); the general plateau constituting the main part of the island, at an elevation of 102 m (340 ft) above the sea, consists of volcanic ash in two main forms— a loose black cinder commonly called "black sand" (which was found to be magnetic) and a consolidated buff-colored ash known as "sandrock." The black sand was easy to handle and formed inexpensive and stable fill material. The buff sandrock was similarly well suited for use as a "stabilized" surfacing for airstrips and roadways for initial construction requirements. This latter material was found to have an in situ moisture content greater than that required for optimum compaction, compaction that gave field densities of only 1,150 kg/m^3 (72 lb/ft^3). Modern earth-moving equipment proved quite satisfactory for military airfield construction. Possibly the most unusual features of this isolated wartime job were that the temperature of the volcanic soils exposed in borrow pits ranged from 43 to 98°C (110 to 208°F) and that the temperature of groundwater was about 57°C (135°F); these were vivid reminders of the proximity of recent volcanic action.[32.6]

32.4 SOME SPECIAL CASES

In the steady development of her remaining colonial territories, Great Britain has pioneered in the building of airfields in a variety of terrains, notably in tropical regions. An unusually comprehensive group of papers summarizing some of this experience with the construction of airfields in the tropics was presented in London in 1957 to the Institution of Civil Engineers.[32.7] Since the six papers are in themselves condensed in form, they cannot readily be summarized. Suffice it to say that they provide a most useful introduction to some of the construction problems encountered with the tropical black and red soils, lateritic soils, and alluvial sands and silts found in tropical regions, with some reference to construction methods that have been found satisfactory with these most difficult of materials. The references presented with the papers are in themselves an excellent

32-8 CIVIL ENGINEERING WORKS

guide to some of the literature then available on residual soil formation, treatment, and use. The entire presentation is a fine example of the sharing of experience and information that is at once so striking and so valuable a part of professional civil engineering practice. Naturally there have been excellent conferences since this one, references to which will be found at the end of this chapter, but attention is directed to this relatively early meeting since it related to the special problems of developing countries.

A typical example of the necessary extension of a smaller airport, work on which is still in progress as this volume goes to press, is that which serves the salubrious island of St. Thomas, one of the Virgin Islands. The existing Harry S. Truman Airport was built in 1950. Its single runway is only 1,380 m (4,600 ft) long, and this naturally limits the size of plane that can use it. The increase in tourist traffic has meant the provision of a "shuttle service" of small planes from one of the neighboring islands having airports with longer runways. As is so often the case, the St. Thomas Airport is adjacent to the sea so that it has been possible to plan the extension of the existing runway by the construction of an additional length of 720 m (2,400 ft) on an embankment in the sea. Local bedrock is quartz keratophyre, a medium-to-hard rock, with some diabase dikes. A further improvement is the removal of Sara Hill, a hill of the same rock type that is at present partially in line with the runway, in order to give a safer approach. Main excavation is of Cabritaberg Hill, adjacent to the existing runway (Fig. 32.5). Drilling is being done with the aid of modern

FIG. 32.5 Harry S. Truman Airport, St. Thomas, Virgin Islands; Atlas Copco drills at work on rock excavation, with runway in background.

all-hydraulic crawler drills. Rock is being moved partially by truck but also by means of a specially constructed belt conveyor. Some rock has also to be placed from barges in locations in the sea inaccessible in other ways. When the job is completed, 5.8 million m^3 (7.4 million yd^3) will have been moved in the ways stated, the volume being a good indication of the large quantities of material involved in the contruction of even relatively small airport facilities. In this case, both the adjacent seabed and the rock available for use as fill were geologically satisfactory, but this can only be assured if preliminary site studies are adequately carried out.[32.8]

32.5 AIRFIELD COVER

Finally, note must be made of a major problem when, as is usually the case, all or parts of an airport surface have to be formed by either the deposition of new fill or the removal of existing ground, in both cases resulting in an exposure of bare soil to the elements. If sand is the fill, again as is often the case, it can cause trouble when dry if blown about by wind. Accordingly, stabilization of bare surfaces must be carried out as soon as possible after construction and thereafter maintained by suitable procedures. Nothing has been found better than natural vegetation, generally grasses, for this purpose, although other measures, such as the use of bituminized fabric or of light metal grid, have been used in time of war or emergency. Grasses, however, are now the almost universal means of surface stabilization. Selection of the grasses to be used will be dependent on the nature of the soil to be covered and on the local climatic conditions. When these are known, useful advice can be obtained from the U.S. Department of Agriculture, as is indicated in Chap. 49. Merely to indicate the extent of the problem, when New York's John F. Kennedy Airport was constructed (originally as Idlewild), 50 million m^3 (65 million yd^3) of sand were deposited over the selected site, resulting in an area of 2,000 ha (4,900 acres) that had to be stabilized. It was found that a mixture of 10 percent poverty grass and 90 percent beach grass was the most desirable combination for the condition there. Each case must be considered on its own, naturally, but beach grass (*Ammophila arenaria* and *A. breviligulata*) will be found to be useful under a variety of local conditions.

32.6 REFERENCES

32.1 N. W. McLeod, "Influence of Geology on the Design and Construction of Airports," in R. F. Legget (ed.), *Soils in Canada,* rev. ed., University of Toronto Press, Toronto, 1965, p. 195.

32.2 G. Y. Sebastyan, "Design and Construction of the New Montreal International Airport," paper presented to the American Society of Civil Engineers, Montreal, July 1974.

32.3 "Underground Channels Utilized for Airport Drainage," *Engineering News-Record,* **130**:498 (1943).

32.4 J. M. Kyle and M. S. Kapp, "Sand Drain Applications by the Port of New York Authority," *Proceedings of the American Society of Civil Engineers,* vol. 80, Separate 456, 1954.

32.5 "10,000,000 yd. Earthmoving Job Levels Hills for West Virginia Airport," *Construction Methods,* **186**:86 (January 1946).

32.6 F. B. Campbell and W. K. Chase, "Seabees Encounter Unusual Soils on Iwo Jima," *Civil Engineering,* **15**:505 (1945).

32.7 "Airfield Construction on Overseas Soils, a Symposium," *Proceedings of the Institution of Civil Engineers,* **8**:211 (1957).

32.8 "V.I. Runway Project Gets Under Way," *Civil Engineering,* **49**:30 (May 1979).

Suggestions for Further Reading

Literature on airports is relatively recent but it has increased rapidly since the end of the second world war. There are now regularly held national and international conferences on airport design and construction, the proceedings of which contain much useful information, but almost wholly with regard to aboveground features. Typical was the sixth World Airport Conference, the proceedings of which, under the title *Airports for the Community*, were published by the Institution of Civil Engineers (U.K.) in 1979.

Deep foundations are only very rarely encountered in airport design, and so foundation bed problems almost always relate to surface soils. Accordingly, the most useful sources of information for airport subsurface design are likely to be the relevant geological survey reports and those of agricultural soil surveys. Search should always be made for information linking the locally recognized and named agricultural soils with engineering soil properties, such as has been done in some states in connection with highway engineering, either by the state geological survey or the state highway department.

chapter 33

RIVER-TRAINING WORKS

33.1 Historical Note / 33-2
33.2 The Mississippi River / 33-3
33.3 Deltas and Estuaries / 33-10
33.4 River Dynamics / 33-12
33.5 Small-Scale Works / 33-13
33.6 Bank Protection / 33-17
33.7 Canalization of Rivers / 33-18
33.8 Some Other Works / 33-23
33.9 References / 33-24

For more than 3,000 years men have been trying to "tame" some of the great rivers of the world. The first civil engineering works of any magnitude were in all probability river-training works. Today, the training of rivers is still an important branch of civil engineering, so diverse in its character, so widespread in its practice, and so specialized for each individual river that brief reference only can be made to it here, despite the fact that the meandering of rivers, which river-training works are designed to control, is a part of the natural order, one phenomenon in the broad field of physical geology. Decisions regarding design must necessarily be empirical; and as no two rivers are alike in all particulars, the local application of the results of experience elsewhere is often fraught with uncertainty and is liable to misinterpretation.

River-control work is essentially the control of a natural riverflow within a certain well-defined course, usually that which the stream normally occupies. Since deviations from this course will most probably occur during periods of flood flow, control work is often a matter of flood regulation. In addition to this, the prevention of erosion of river banks during periods of normal flow constitutes another important branch of control work. Finally, the improvement of rivers in connection with civic or industrial development, such as the straightening of a riverbed in connection with town planning and the canalization of a stretch of river to facilitate navigation, is another type of river regulation that is often of importance. In undertaking work of this nature, an engineer is dealing with a fundamental geological process, and all designs must be based on an acceptance of the natural characteristics of riverflow. In the past, a rigid restriction of this work to assumed natural conditions has led to major controversy, especially with regard to the control of the Mississippi River, one of the outstanding river-control projects of the world; special attention is given to it later in this chapter.

In the ordinary practice of civil engineering, river-training work will be usually on a relatively small scale. Much can be learned, however, from experience with the larger rivers of the world, for which reason a brief historical note follows. Special attention may be invited to the work of British engineers in India. This work was essentially empirical in nature, based on the accumulation of experience throughout the century of modern India's development. Carefully recorded in many papers to the Institution of Civil Engineers (London), this experience was synthesized in some notable books of an earlier day. *The Control of Water* by P. A. Morley Parker is an excellent example. It is still of value today, but copies have well-nigh disappeared. The Central Board of Irrigation (of India) in its first annual report, issued in 1935, provided a fine summary of Indian river-training work and also a summary of world views, at that time, of desirable procedures in this important branch of civil engineering.

33.1 HISTORICAL NOTE

Four of the major river systems of the world have well-developed floodplains long occupied by human settlements; it is natural that these four should have been prominent in the long history of river-training work. They are the Mississippi, the Nile, and two Chinese Rivers, the Yangtze and the Yellow (or Hwang Ho). Other rivers of comparable size are the Amazon, the Congo, the Lena, and the Brahmaputra, but only the last of these has seen development comparable to that which has been carried out on the two great rivers of China for 30 centuries or more. The Yangtze River is 5,100 km (3,200 mi) in length; it is navigable for a longer distance than any other river in the world. Its flow on 19 July 1915, was recorded as 71,600 cms (2,531,692 cfs), the greatest flow ever recorded for one river channel, much in excess of the maximum flood flow on the Mississippi. In 1871, the level of the Yangtze rose to the phenomenal height of 82.5 m (275 ft) above normal in the Wind Box Gorge, 1,750 km (1,100 mi) from its mouth. About 180 million people live in its watershed. Its floods have caused untold tragedies in loss of life down through the ages. Dikes have been manfully built in attempts to control its channel; the word "manfully" is used in its literal sense, since 50,000 people were at work on one dike alone in repairing flood damage after an inundation in April 1927. Today, the Yangtze is one of the most completely diked rivers of the world.[33.1]

Even more remarkable is the Yellow River, "China's Sorrow," 4,000 km (2,500 mi) long, with at least 100 million people living in its basin; its name is derived from the high content of loess that it carries to the sea and out into the sea, sediment being visible far from its mouth. At one time the river ran north by Tientsin and at another it emptied in central China into the Yangtze River, near Chekiang, several hundred km to the south. For 700 years it discharged eastward into the Yellow Sea, but in 1852 it broke its banks at a point more than 480 km (300 mi) from its mouth, formed a new channel across the province of Shantung, and finally a new mouth in the Gulf of Chihli, almost 480 km (300 mi) from its old mouth; it occupies this position today. The sediment that it carries is unusually fine, so that if the velocity, which averages 2.4 mps (8 fps), drops below 0.9 mps (3 fps), silt is deposited rapidly, the waters overtop the existing dikes, and a new page of sorrow starts. It was an old Chinese tradition that restless dragons stirred the mud in the rivers, causing great calamities. Until early in this century, foreign engineers working in China often had their work delayed while the river dragons were propitiated.[33.2]

Chinese manpower, coupled with expert engineering advice, controlled the Yellow River when it broke its banks again in 1935 near the towns of Tung Chuang and Lin Pao

Chi. Four million people were affected by the resulting flood through a breach in the dike system 2.4 km (1.5 mi) long. The ancient Chinese contraction method was used to close the gap; the ends of the breach were first built up strongly, new dike structures were then built out gradually from both ends, and the final closure was made with the aid of immense, stone-filled wire sausages and willow-fascine mattresses. As many as 25,000 men were employed at the peak of the work, but almost every operation was by hand, even the tamping of the soil placed in the dike; ingenious stone "flappers" lifted by eight men with attached ropes suggested that the Chinese had a good appreciation of the necessity of soil compaction long before it became a well-established technique in modern soil mechanics.[33.3] This case illustrates a development of the Chinese river-training work that has an unbroken record through the centuries of controlling two of the most unruly rivers of the world.

India has also had its share of sorrows caused by rivers on the rampage, particularly the Brahmaputra. This great river left its normal course spectacularly toward the end of the eighteenth century, seriously affecting the ancient city of Mymensingh. With the development of irrigation works of magnitude by British engineers in the nineteenth century, river-training works became an established part of Indian engineering. It was in 1887 that J. H. Bell advanced new proposals for training the Chenab River at Sher Shah where a large new bridge was proposed. He suggested concentrating the low riverflow in a relatively deep channel by means of low embankments, soon known far beyond India as "Bell's bunds." The same idea was further developed by Sir Francis Spring and has since been amended and improved.

33.2 THE MISSISSIPPI RIVER

Nowhere, perhaps, has this precept been better illustrated than in the history of the control of the "Father of Waters," America's greatest river, and its lusty tributary, the Missouri, so commonly known as "Big Muddy." The Mississippi is fed by a drainage area of 3,220,000 km^2 (1,243,000 mi^2) and has a total length of about 3,840 km (2,400 mi). Its average discharge at Vicksburg is about 14,000 cms (500,000 cfs); flood discharge is as high as 56,000 cms (2 million cfs). In general profile, the river is flat and slow-moving for the greater part of its length. The valley of the Mississippi follows the general form for such river valleys. Flood waters, before confinement by levees was effected, covered a much greater area than the normal streambed, at times as much as 77,500 km^2 (30,000 mi^2). The first attempts were made in the early nineteenth century to confine the river during flood periods by raising the natural levees with articifial embankments, thus raising the next flood water level and promoting deepening of the channel through this restriction. In consequence, no local addition to a levee could really be made without all levees being raised similarly. The first levee, built at New Orleans in 1717, was 1.2 m (4 ft) high. In 1932, the average height was more than 3.9 m (13 ft), and 4,000 km (2,500 mi) of artificial levees were being maintained on the river. All this work, up to 1932, was based on the ideas of maintaining the existing channel of the river.

In June 1932, after years of discussions, a new policy was inaugurated under the direction of Gen. H. B. Ferguson, newly appointed president of the Mississippi River Commission. Another characteristic of the river was adopted for use; the natural process whereby the river makes cutoffs at narrow necks between bends, at long intervals of time, was expedited by making artificial cutoffs with the aid of powerful dredges. These cutoffs have materially reduced the length of the river and have lowered its flood stages. Supplementary dredging is done in the reaches between cutoffs to assist the river in restoring the steepened gradients to normal. A famous example of a natural cutoff is at Vicksburg, Mississippi, where in 1863 during the Civil War, General Grant tried to isolate the city from the river by cutting a canal through a narrow neck of land. This attempt was not successful, but in 1876, the river made the change for itself. Fortunately, the town of Vicksburg was able to maintain contact with the main river through the diversion of another stream; it is now well known to engineers as the location of the large-scale model of the river on which most valuable experiments have been made at the Waterways Experiment Station (of the U.S. Corps of Engineers). This shortening of the river was started, initially, at 12 sharp bends (Figs. 33.1 and 33.2). When all these cutoffs were fully developed, the total shortening they effected amounted to nearly 185 km (116 mi) in a stretch of 530 km (330 mi). Associated with this work, relief floodways were provided, designed to allow floodwaters to spill over the natural banks but only under control and in specially selected locations. Reservoirs have been created on tributary streams by the construction of some important dams; although this work is of relatively minor importance on a river as large as the Mississippi, regulation by this means on smaller streams is often extremely valuable. Finally, continued bank-protection work, involving mattress designs of much engineering ingenuity, has been carried out.[33.4]

This great program of river training, to a new plan, had just got well under way when there occurred the remarkable flood of 1936 and 1937, the greatest ever to flow down the Mississippi to the sea without any break in the levees. With the engineering organization for the river-training work already in the field and at strategic offices in the great valley, it was possible to make a detailed study of the entire flood to a degree that had not previously been attempted. This study gave much invaluable new information about the Mississippi, all recorded conveniently for future use. The flood directed attention to gaps in available knowledge, and a major program of investigation was started shortly afterward;

FIG. 33.1 The site of the Sarak Island cutoff on the Mississippi River, when right-of-way had been cleared and the first dredge cut started. The main river flows from right to left at top and from left to right at bottom.

FIG. 33.2 The Sarak Island cutoff on the Mississippi River in initial operation; some of the riverflow still uses the old riverbed, although this had started to silt up when the photograph was taken.

the investigation was directed particularly into the geological features of the valley dependent on the great river. The work resulted in one of the most comprehensive engineering-geological studies ever made, taking in the entire valley not only by means of surface observations but with the aid of the most complete collection of boring records that could be assembled. At least 16,000 boring records were made available by engineering organizations operating in the valley; the results of the extensive test-boring programs carried out in the search for oil near the Mississippi delta have added more valuable information to this steadily growing fund of knowledge (Fig. 33.3). The results have been summarized in a number of papers and in a comprehensive volume by H. N. Fisk, who was directly responsible for much of the geological part of this notable undertaking.[33.5]

As with all such studies, the results have proved of mutual benefit—to engineering, in providing the essential background for the continuing improvement of the great system of training works for the Mississippi, and to geology, in revealing hitherto unsuspected aspects of the Pleistocene history of the valley sediments. It has been shown, for example, that there has been considerable subsidence of the earth's surface in the delta area under the increasing weight of the deposited sediment; this fact accounts for the limited extension of the delta deposits within recent decades. Jetties constructed at the mouths of the Mississippi in 1875, for example, have not had to be extended since then. Corresponding elevation of adjacent areas on dry land appears to have occurred. The investigation continues, but it already constitutes one of the most fascinating of all interrelations of geology and engineering.

Of the many aspects of the geology of this interesting area, one more only can be mentioned; this is the most dangerous, potentially, of all. For some time the future of the lower 480 km (300 mi) of the Mississippi River has been threatened by the possibility of the "capture" of its flow by the Atchafalaya River through what has long been known as the "Old River diversion." The Atchafalaya River, although only a minor waterway in comparison with the Mississippi, is still the third largest American river that flows into

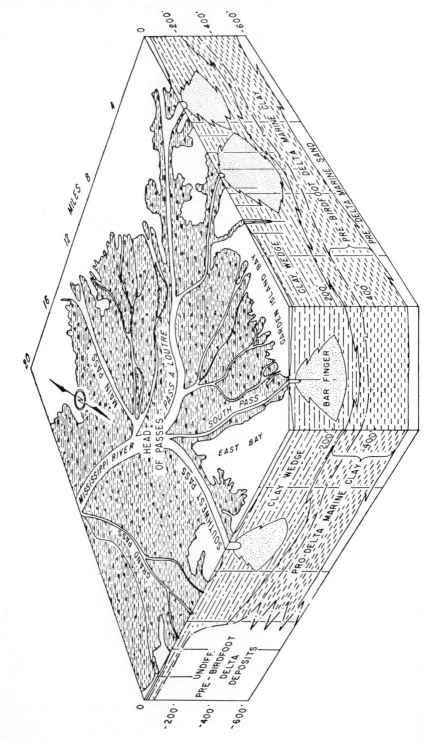

FIG. 33.3 Diagram of the Mississippi River delta (a good example of the use of a "block diagram" to illustrate geological features).

the sea. The Old River connection is 9.6 km (6 mi) long; its location is shown in Fig. 33.4. In 1956, the Old River took 23.5 percent of the flow of the main river, diverting it into the Atchafalaya; it was estimated that in the absence of any control measures, this diversion could reach 40 percent by 1975. At that stage it would constitute a real threat to the lower section of the main river, since the lower flow would promote the silting up of the main channel, with probably disastrous effects upon the major port of New Orleans. A master control plan was prepared by the U.S. Corps of Engineers, and the first stages of it are already complete. The total plan will take from eight to ten years to complete and will cost about $47 million. The essential parts of the plan are shown in Fig. 33.4. Two control structures were built on the west bank of the Mississippi, upstream from Old River, a navigation lock connecting the two rivers, with the corresponding connecting channels. Additional levee work was involved, and the Old River itself is now sealed off, with all the new works functioning as planned (Fig. 33.5). This vast project, so critical in its conception, planning, design, and construction, is necessitated merely to control the further development of a relatively simple, if unusually large, case of natural "river capture."[33.6]

Although improvements were still being made to the Mississippi River and Tributaries Flood Control Project (the official name of MR&TP), it showed its worth in the terrible flood of early 1973. Starting in October 1972, a phenomenal period of eight months of heavy precipitation throughout the valley of the great river foretold an unusual flood. Flood stage on the river was reached in the St. Louis area on 11 March. On 15 March the lower part of the valley was hit by one of the most severe storms on record, up to 27.5 cm (11 in) of rain falling in 30 hours. Inevitably, a record flood developed, the worst since 1937 south of Cairo, and between Cairo and St. Louis a flood of record. For 90 days the river was out of its banks at Vicksburg. All along 1,900 km (1,200 mi) of river, emergency works were taken in hand, direction centralized in the office of the president of the Mis-

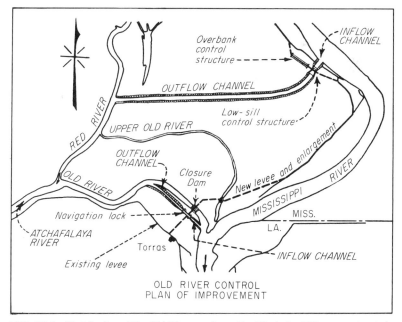

FIG. 33.4 Sketch plan of the Old River works on the Mississippi River, constructed for control purposes.

FIG. 33.5 General view of the Old River works on the Mississippi River in 1974, after repair of damage caused by the 1973 flood.

sissippi River Commission at Vicksburg, direction being from the offices of the district engineers of the U.S. Corps of Engineers at St. Louis, Memphis, Vicksburg, and New Orleans. Two of the emergency major spillways had to be used, successfully, but some damage was done to the low sill structure at Old River. About 6.7 million ha (16.5 million acres) of land were flooded and 69,000 people were made temporarily homeless. Economic losses were estimated at about $1 billion. It was estimated, however, that without the works of the MR&T Project, damages would have been more than $15 billion, and 6 million ha (14.5 million acres) of additional land would have been flooded, with what loss of life could not be envisaged, whereas very few lives were lost as a result of the 1973 flood (Fig. 33.6). It gave invaluable experience to those responsible for the great river, showed that the works so carefully planned were sound, but led to a program of raising the height of about 1,080 km (800 mi) of main-line dikes.[33.7]

Brief reference has been made to the main tributary of the Mississippi, the Missouri River. For more than 100 years, the Missouri has been a troublesome river to those who live near its banks; its floods sometimes reach catastrophic proportions. It is the chief source of the load of sediment in the Mississippi, whose load, although often talked about, is minor in degree compared to that of its tributary. In flood flow, the silt content of the Mississippi rises to 5,000 ppm. That of the Missouri, by comparison, may rise as high as 20,000 ppm, a figure exceeded only by such rivers as the Colorado and the Rio Grande, which may reach as high as 40,000 ppm at times of high flood. Two plans for the amelioration of Missouri floods were prepared, one by the U.S. Corps of Engineers, submitted in 1943, and one by the U.S. Bureau of Reclamation, released in 1944. Although having the same objective, the approaches were somewhat different; reservoirs on the main stream were a prime factor in the first scheme and "upriver" works a leading feature of the second. The two schemes were combined; the Missouri Basin Inter-Agency Committee was formed in 1945 to implement the policies of the Federal Inter-Agency River Basin Committee (Fig. 33.7). The former committee consists of four representatives of the United States government and four from the states in the Missouri Basin. Steady progress has

FIG. 33.6 A typical view showing the extent of the 1973 flood on the Mississippi River.

been made with many parts of the overall valley plan, and reference to some of the major engineering works appears elsewhere in this book.[33.8]

As typical of the troubles that the Missouri can cause, there may be mentioned just one example, its change to a new course in the vicinity of Decatur, Nebraska, in 1946. A new $2 million bridge had been located at a site on the deserted bed; it was decided to take advantage of the absence of the river to build the new bridge in the dry, with considerable economy (Fig. 33.8). It was estimated that $400,000 would be saved in this way, with the river to be returned to its old course when the bridge was built. But the Missouri, as so often in the past, did not cooperate, and major construction work had to be undertaken in order to get the river back into its old channel and to keep it there; much of the

FIG. 33.7 View of the Missouri River, upstream from mile 440 in Buchanan County, Missouri, showing channel-stabilization works along both banks.

FIG. 33.8 New bed for the Missouri River under a highway at Decatur, Nebraska; the river was shortly after diverted under the bridge and the highway continued as an embankment across the old riverbed.

anticipated saving was thus lost. But the task was completed in 1956, and the bridge that was once on dry land is now serving as was intended, with water flowing between its piers.[33.9]

33.3 DELTAS AND ESTUARIES

The valley of the Nile provided annually a vivid example of the silt-bearing function of rivers; as a result, the fertility of that cradle of civilization has been long renowned. Records taken at the famous Roda gauge near Cairo (which has been maintained for many hundreds of years) show that the average sediment deposit over the valley is about 10 cm (4 in) per century. The corresponding formation of deltas at the mouths of several of the major rivers of the world, for example the Ganges and the Mississippi in addition to the Nile, is one of the most obvious and generally appreciated demonstrations of the transporting power of rivers. The somewhat similar phenomenon of the existence of bars at the entrances to tidal estuaries into which quite small rivers may discharge is a telling reminder of the great transporting power of the sea. Both natural features introduce engineering problems of the greatest magnitude, problems which, although simple in essence, are extremely complicated in detail, having provided a fertile field for discussion for at least 3,000 years.

The building up of large areas of new land in deltas provides land for development work of various kinds, and this involves structural foundation problems, often of unusual complexity. It is estimated that the delta of the Rhône in France (under observation since 400 B.C.) has been growing at the rate of 11 m (36 ft) yearly; the corresponding rate for both the Danube and the Nile is 3 m (10 ft) per year.[33.10] It may be noted that the physiography of the mouth of a river is a factor of great importance in determining the existence of deltaic conditions, since rivers such as the Severn in England and the Amazon in South America, although they carry large quantities of silt, exhibit no special deltaic features; the funnel-shaped estuaries promote self-maintenance of the main channels. One of the largest deltas in the world is also one of the most unusual; this is the delta at the mouth of the Mackenzie River in Arctic Canada (Fig. 33.9). It is 240 km (150 mi) long and up to 80 km (50 mi) wide, is underlain by permafrost, and is enclosed by high land to east and west; the main load of sediment in the river is deposited in a great underwater delta in the Beaufort Sea.

The geological conditions affecting the formation of deltas are well recognized; the load of the stream is deposited on the riverbed because the velocity, and therefore the carrying capacity, is reduced when the streamflow merges with the water of the sea or lake. Part of the load which might still be held in suspension may be deposited by flocculation of the colloidal material by the salt solution provided by the seawater. As a result, deposits are formed in regular beds, and they constitute some of the newest of sedimentary deposits. These may reach surprising depths; for example, a boring put down near Venice on the delta of the river Po in Italy reached a depth of 150 m (500 ft) without penetrating the bottom of the deltaic beds. It will be appreciated that, although a delta is popularly imagined to be above water level, a deltaic formation is essentially an underwater geological structure, a fact influencing its engineering significance. It may also be emphasized that, although a delta may be the formation of a single stream, breakages of the natural levees above the river mouth may possibly cause the main stream to branch out into the sea. The Nile delta and that of the Ganges are two excellent examples of this special feature.

The magnitude of the engineering problem created by the deposition of the sediment load of rivers in estuarial water where a delta has not formed is well shown by the experience of the port of Baltimore, Maryland, the only one of the original group of towns on Chesapeake Bay that has developed into an ocean port. Designated a port by the General Assembly of Maryland as early as 1706, it has developed into one of the greatest of American ports; the harbor proper is the parent estuary of the Petapsco River and the arms formed by its minor tributaries. In the course of 200 years, the head of navigation has been pushed steadily seaward for a distance of 11 km (7 mi); depth at low water at one well-defined location decreased from 5.1 m (17 ft) in 1845 to about 15 cm (6 in) in 1924. It is not surprising to find that dredging was first started (by the federal government) as early as 1836. It is said that Baltimore in 1783 saw the development of the first "mud

FIG. 33.9 Typical view in the delta of the Mackenzie River on the Arctic coast of the Northwest Territories of Canada; all the ground to be seen is perennially frozen.

machine," a simple type of drag dredge operated by horsepower. In the course of a century, the United States government spent about $17 million in dredging 85 million m³ (111 million yd³) of sediment from Baltimore harbor, and the work of maintenance dredging still continues. There is no doubt about the origin of the material deposited in the harbor and then removed at such expense by dredging. Soil erosion in the watershed of the Petapsco River is well recognized; measurements taken of the corresponding silt load carried by this small river agree reasonably well with the amount of material known to be deposited in the harbor. [33.11] If only for strictly economic reasons, therefore, civil engineers must take cognizance of the menace of soil erosion. More will be found on this subject in Chap. 49.

33.4 RIVER DYNAMICS

It must be repeated that most civil engineering work involving rivers is on a scale considerably less than that of the work on the major rivers so far considered. Despite the difference in scale, however, the way in which great rivers—and especially the Mississippi—have been controlled is not only of interest, and even of inspiration, but also of value in showing how Nature can be assisted by well-planned engineering works. This is a salutary reminder that *any* engineering work on *any* river, large or small, must be carried out against consideration of the river as a whole and as a vital element in the hydrological cycle. Rivers are part of the dynamic natural system, each one with some features that are unique. Engineering work at a river bend, for example, can only be carried out properly after possible effects downstream have been evaluated and with full consideration of all upstream features, including careful study of all flow records available so that the annual variation of flow can be estimated. Without consideration of the river system as a whole and a review of its performance throughout the twelve-month cycle, the design and construction of training or diversion works at any one location may prove to be inadequate at least, and possibly disastrous.

River dynamics is, therefore, a term of real meaning. When looking at a section of smooth-flowing water, one may all too easily forget that this is but a part of an intricate system, fed by rain either directly by runoff or indirectly through seepage of groundwater. In both cases, geology is the basic determinant. Once the initial stream has formed, its progress to its mouth will again depend on the geology along its course, rock exposures creating rapids, erodible soils adding to the load of sediment it carries, flow over limestones even leading sometimes to the temporary disappearance of the stream into underground channels. And the action of the moving stream, throughout its course, is one of the most potent of the natural forces now at work on the surface of the earth as an integral part of the geological cycle: denudation—deposition—earth movement—denudation. . . . It is always salutary in river work to keep firmly in mind that the carrying power of moving water varies as the sixth power of its velocity. The process of erosion then becomes more meaningful.

The study of river systems is now well established. Some of the basic concepts recognized today were first advanced by a notable hydraulic engineer, Robert E. Horton; in more recent years Luna Leopold, for many years chief hydrologist of the U.S. Geological Survey, has been a leader in this field. References to some works of these men will be found at the end of this chapter. Typical of modern approaches to river hydraulics is the concept of the four orders of streams. A stream that has no tributaries, such as one in the headwaters of a river system, is considered to be of the first order. When two first-order streams unite, then a stream of the second order is formed. Third-order streams are formed by the union of two second-order streams, and the fourth order, the final classification, is the designation given to the watercourse after two third-order streams have joined (Fig. 33.10). This may seem somewhat artificial, but it is found that, for typical river systems, both the number of tributaries and their total length bear a linear relationship with their orders of magnitude.[33.12]

When river systems are considered in a continental, or even worldwide, context, then different types of rivers are readily recognized. Major variations in type depend on the geological age of the streams. Relatively young rivers begin, ideally, as streams flowing over level plateaus and eroding their courses into the material of which the plateau is composed. This will lead to very rapid development and spectacular scenery, the Grand Canyon being a superb example. When the landscape formed in this way reaches its ultimate state of development, the rivers enter a mature stage, irregularities in their beds being smoothed out. Correspondingly, old age in river development sets in when the coalescence of flat and gently sloping surfaces of adjacent systems begins to take effect, working up from coastal regions at the expense of outstanding bluffs and divides.

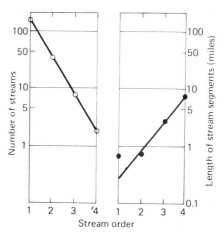

FIG. 33.10 Diagram of average stream length and number of streams, as a function of stream order, for Seneca Creek above Dawsonville, Maryland (drainage area 260 km² or 100 mi²).

These are very broad concepts. Of more direct moment as a necessary prelude to any consideration of river-training work at bends is a lively appreciation of the somewhat complex motion of water flowing around a bend. As a stream starts flowing round a bend, the surface water, since it is faster than that at the bottom of the stream, will move towards the concave bank. It will tend to erode the material of which the bank is formed and as it does so will be deflected downward. Water at the bottom must move upward in a compensating movement. The combination induces in the flowing water a sort of helix movement of individual water particles. Viewed in cross section, water will tend to flow across the stream bed away from the concave and eroding bank towards the other side of the stream, depositing some of its load from the erosion as it does so. Just as concave banks tend to be eroded, so convex banks tend to accumulate sediment and grow outwards, in keeping with the erosion of the facing bank. On a large scale, and when considered over a number of years, it is this water action that explains the meandering of rivers when flowing in flat plains.

The civil engineer, in the ordinary course of everyday practice, will probably encounter this natural phenomenon when unexpected flood flow down a stream or river has caused serious erosion at a critical bend. Here, an appreciation of the basic hydraulics is essential before remedial works are carried out. Erosive forces will have increased because of the enlargement of the bend. Protective works must, therefore, take this into account. Once the immediate danger has been overcome, the overall situation must be carefully reviewed before any permanent works are decided upon, all against a background of the fundamentals of river hydraulics, guides to which will be found listed at the end of this chapter.

33.5 SMALL-SCALE WORKS

The simplest form of river-training work is the restoration of a stream channel after havoc has been wrought to it by sudden flooding. Such experiences will always reveal the weak

33-14 **CIVIL ENGINEERING WORKS**

spots in stream routing, such as at bends where significant erosion may take place. There is an immediate temptation to eliminate such trouble spots in rehabilitation work, but if further trouble is to be avoided, the best possible course to follow is to try to reproduce earlier natural conditions as closely as possible. Stream beds must be cleaned out and all debris moved by the flood removed; eroded ground must be replaced, if possible, and bank-protection work installed to prevent, as far as possible, any further erosion at the same bends. This will naturally be on concave banks, although all too often one can see bank-protection works carried out, at needless expense, on convex banks where, in all normal cases, things can take care of themselves.

All this appears very simple, and hardly engineering. It is simple, as so many natural processes are, but unless every operation is carried out with full appreciation of the natural processes involved, then mistakes can be made even with such apparently minor works. The U.S. Soil Conservation Service has, through its notable work down the years, accumulated much experience in this kind of training work and has usefully recorded much of it in valuable publications, a guide to which is given at the end of the chapter. Keller has published a most worthwhile general review in small compass, a useful initial guide. He draws attention to a 394-page report on the subject of channel modification prepared by consultants for the Council on Environmental Quality, any work on streams now being subject to review not only from the engineering point of view but also in relation to the environment and to the ecology of the stream.[33.13]

Keller gives, as a useful example of emergency stream-relocation work, the case of Hat Creek in Nelson County, Virginia. Hurricane Camille (1969) changed this relatively small stream into a raging torrent, which departed from its normal course and destroyed buildings and parts of a highway. The Soil Conservation Service worked with the Virginia Highway Department in urgently needed rehabilitation work, moving the stream back to its prestorm position and repairing and restoring its banks and adjacent slopes. The damage and the success of the rehabilitation work were well shown in photographs, one of which is reproduced as Fig. 33.11. It must be stressed that this emergency operation was carefully planned and designed to cooperate with Nature in restoring the stream to its original state, a very different matter from the unplanned work all too often carried out in the panic conditions following a flood, heavy construction equipment being used for massive earth moving without regard to previously existing natural conditions.

Floods and flood-protection works are considered in Chap. 46, but they must also be briefly mentioned here. Phenomenal rainfalls cannot be prevented, so floods are a part of

FIG. 33.11 Severe damage to the banks of Hat Creek, Nelson County, Virginia, following Hurricane Camille in 1969; the original channel was obliterated.

FIG. 33.12 Natural stream control, a typical small beaver dam.

the natural order. But with increasing deforestation and urban development, the incidence and magnitude of floods have been increasing as a direct reflection of interference with the flood protection—river training—that Nature provides if man does not interfere. Vegetation is one of the best of all runoff controls; if it is removed, the rate, and so the volume, of runoff will increase, enhancing the possibility of floods. Animal life along smaller streams makes its own contribution to river control. Beavers (so rightly called "Nature's own engineers") build dams which act as perfect barriers, almost always found to be located at points on a stream where there is a reasonable area for ponding immediately above the dam (Fig. 33.12). If beavers are displaced, either naturally or by advancing development, their dams will deteriorate and eventually disappear, leaving the stream without its earlier controls.

Following the example of beavers was the admitted basis of a major stream-control project carried out near Salt Lake City just before the Second World War; its message is worth recording briefly even today. L. M. Winsor, then district engineer at Salt Lake City for the Bureau of Agricultural Engineering of the U.S. Department of Agriculture, had noticed in 1922 that removal of beaver dams in a stream above the town of Nephi had resulted in much trouble at a control dam in the town due to filling of the dam's reservoir with gravel. Townspeople confirmed that beaver dams had been removed from the stream. Stream velocity had therefore increased, and with increased velocity came increased carrying power, resulting in the movement and subsequent deposition of gravel previously in the stream bed. A simple "beaver dam" was built at a cost of about $3,000, and this was so effective that it eliminated the need for the $5,000 previously spent each year in removing the unwanted gravel. Others had, naturally, used this same idea for river training previously. The origin of the idea will never be known, but this was at least an early application of a barrier dam as a natural river-training work.

The Garfield plant of the Kennecott Minerals Company (Utah Copper) is located 35 km (22 mi) west of Salt Lake City; in 1939 it represented an investment of $12 million, having been extended since it was built about 1900. It was located close to the mouth of a small canyon. Floods coming down the steep stream bed in the canyon caused trouble at the plant, and a number of protective works were installed. A flood of unprecedented magnitude swept down the canyon in July 1927 and caused serious damage not only to existing protective works but to the plant itself. Mr. Winsor was consulted about the problem. On the basis of his earlier experience, he recommended that a barrier dam be constructed right across the mouth of the canyon. This was constructed as an earth dam, almost 1.6 km (1 mi) in length and founded on alluvial soil. The dam varies in height to about 30 m (100 ft) at the crossing of the stream bed, after rebuilding from an original height of 17 m (57 ft). A carefully designed spillway was constructed over the stream bed, curved in plan, rubble masonry (using boulders from the canyon debris) being used as the building material. The pool this created behind the dam reduced the water gradient from the 12 percent slope in the natural canyon to the small slope necessary for normal streamflow. Sand and gravel, brought down by small floods, naturally accumulated behind this simple barrier dam to such an extent that by 1937, none having been removed, the fill had risen in places to the crest of the dam. Despite this, a flood estimated at 425 cms (15,000 cfs) passed over the spillway safely, even though it carried much debris and boulders which, had the reservoir not been filled, would have been retained there (Fig. 33.13). As a flash flood, this peak flow lasted only 20 minutes, but no damage was done to the plant or its surrounds. These small control works are still in effective operation.[33.14]

Small works such as those described would normally be carried out today only in relation to the control of the stream or river as a whole. In some parts of North America, the need for dealing with river-control problems on a watershed basis, rather than in accordance with purely artificial political boundaries (important as these are for other purposes) is now recognized. One of the most notable examples is that followed by the government of the province of Ontario, which has established conservation authorities for almost all rivers, large and small, in the southern part of the province. The work of these bodies is financed jointly by the province and by all the municipal units (on a proportional basis) in the watershed of the respective rivers, each municipality having representation on the authority. Members-at-large are usually also appointed to each authority, the more

FIG. 33.13 Debris control at the Garfield smelter plant of Kennecott Minerals Company on north slope of the Oquirrh Mountains at the mouth of Kessler Canyon, initiated after a serious flood in 1927; work follows natural lines by using small check dams; this view looks towards one spillway.

detailed work of which is carried out through committees, one of which is always concerned with flood control and river training. Technical advice is available through the provincial Department of Natural Resources; consultants are used for study and design of all major works, the execution of which is carried out by contract. Engineers, and geologists, are among the hundreds of citizens serving as members of these local authorities, giving their time voluntarily in the interest of conserving the river in the watershed of which they reside. The conservation authorities are mentioned as a good example of what should be a widespread practice and as representing an ideal way of studying and carrying out river-training works on a modest scale. It has been well said that the work of these authorities "represents the best value for the dollar" of any public expenditure in Ontario.[33.15]

33.6 BANK PROTECTION

The importance of bank-protection work must be emphasized, since it enters into so many phases of river-control work. Methods of protection range from the construction of permanent retaining walls in front of natural soil slopes for preventing the initial formation of soil gullies to the use of expedients such as mattresses of brushwood, dumped rock, and even sandbags in cases of emergency. In the evolution of the many successful types of protection, due consideration has inevitably had to be given to the nature of the river action at the location. For example, on many sharp bends, the use of a wall of deep-driven, interlocking steel sheet piling has often proved successful where other expedients have failed, because it enabled additional scour of the riverbed to take place in front of the wall until a condition of equilibrium was attained. In the case of levee protection, many ingenious designs of flexible mattresses built of different materials, including reinforced concrete, take up initial variations of the riverbed and also any alteration caused by further minor erosion at the edge of the mattress. Flexibility is also the keynote of the "wire-sausage" type of protection adopted so successfully in India and China; cages of wire are filled with loose stones and so placed that in case of the failure of one sausage the superimposed cages will automatically move down to take its place. Another idea has been the planting of low-growing willow trees at or about normal water line, so that the foliage may impede the flow at flood periods and therefore reduce its eroding power. Throughout all this work, however, as in all satisfactory river-training work, the basic concept has been to cooperate as closely as possible with the natural process of riverflow, to control it within necessary limits without attempting to usurp the place of Nature.

The utility for small-scale bank protection of the old "wire-sausage" idea may well be stressed, especially as there are now commercially available ready-made wire-netting baskets to which the name *gabion* is sometimes applied. The idea is a very old one. There are records of its use by the (British) Corps of Royal Engineers well over a century ago. Like many simple ideas, however, it has not been as widely applied as one might expect. The ease of installation and general utility of such rock-filled wire boxes almost invariably leads, after an initial trial, to extended use. Figure 33.14 shows an imaginative use of gabions for drainage control adjacent to an interstate highway project. Examples are slowly becoming more widespread.

For large-scale bank protection, interlocking steel sheet piling is one desirable solution, as has been noted, because of the resistance it provides to further scouring. It involves, however, the use of construction equipment by experienced operators, since the driving of this type of piling is a specialized task which, if not done well, may lead to trouble. Alternatively, large blocks of rock (of a suitably resistant kind) can be used, if carefully placed and after consideration of the possible consequences of setting such impediments

33-18 CIVIL ENGINEERING WORKS

FIG. 33.14 An imaginative use of gabions (rock-filled wire-net boxes) for control of drainage flows along Interstate 88, adjacent to the Schenevus exit in Otsego County, New York.

in a stream bed. Where no rock is economically available, artificial "rocks" can be fabricated of concrete, plain or reinforced.

The use of large rocks, natural or artificial, is a feature of many branches of civil engineering work, such as the facing of the upstream slopes of earth dams, the protection of cofferdams against scour, shore protection, and the difficult task of closing off a narrowing gap in rock-fill cofferdam construction. There are available, therefore, scattered references to the precautions to be taken and guides to the selection of rock of suitable nature and of suitable size. The latter is a matter of the greatest importance since if a mistake is made, rock may disappear downstream when the stream velocity exceeds a critical value, with all the economic loss that this represents. Figure 33.15 is a useful guide, prepared by the U.S. Corps of Engineers, North Pacific Division, and based, as can be seen, on widely distributed experience.[33.16]

33.7 CANALIZATION OF RIVERS

So wide-ranging has been the use of the world's rivers that it is easy to forget that the canalization of rivers (already referred to in Chap. 29) is a form of river training. For the efficient operation of the necessary impounding dams, spillways, and associated navigation locks and powerhouses, the rivers involved must naturally be under complete control. Many storage reservoirs on subsidiary streams have been developed primarily to control the flow down rivers that are used commercially following canalization. Figure 33.16 is a useful reminder of the way in which some major rivers have been canalized throughout much of their length. The advance in civil engineering practice is well shown by the reduction in the number of dams across the Ohio River from the first 46-lock system, the reduction representing a saving of at least 50 hours in the time necessary for sailing from Pittsburgh to Cairo (Illinois).[33.17]

There are examples of such river-training works all over the world, ranging from one of the earliest in North America—the complete canalization of the small Rideau and

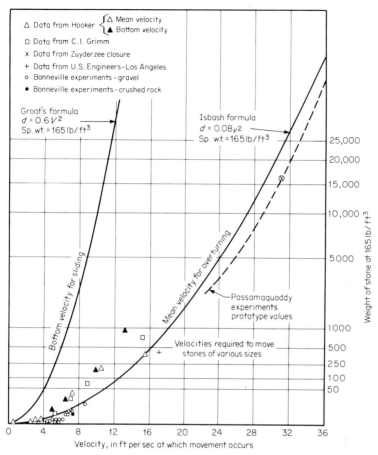

FIG. 33.15 Graph illustrating the stability of rocks in flowing water, based on the actual examples indicated.

Cataraqui rivers in Ontario to form the Rideau Canal, constructed in 1826–1831—to the mighty works required for the canalization of the upper reaches of the Mississippi River, now extending to the center of Minneapolis. The upper reaches of the Rhine, where it runs between France and West Germany, provide another quite spectacular example; and even some of the tributaries of the Rhine have now been similarly controlled. One of these is the lovely Moselle River, known to all lovers of good wine.

Every structure thus erected across a river, or as part of the navigation or power facilities associated with canalization impounding dams, involves geological problems, since all such structures are founded on geological strata. This one branch of civil engineering therefore provides almost innumerable examples which could be cited in support of the thesis of this Handbook. Some have already been mentioned. Two more must suffice for brief mention here, the first from the Moselle River of West Germany, used for navigation certainly since Roman times 2,000 years ago. On 26 May 1964, a new chapter in its long history was inaugurated; on that day it was opened throughout as a fully navigable waterway. This followed the signing of the Moselle Treaty in 1956 by France, Luxembourg, and West Germany, the three nations agreeing to harness and regulate the river, sharing the

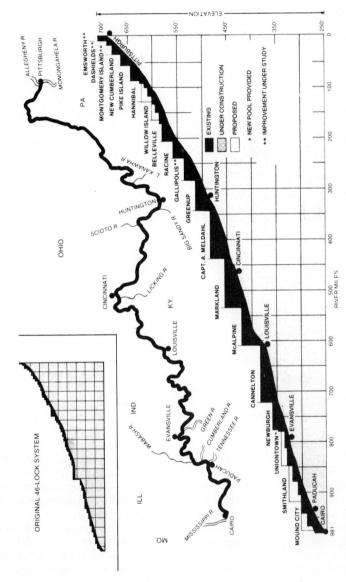

FIG. 33.16 "Engineers' staircase up the Ohio River"; diagram illustrating the control of the Ohio River by locks and dams.

FIG. 33.17 Construction of one of the control drains parallel to the river bank on the Moselle River (a tributary of the river Rhine).

200 MW of power that would thus be generated. Locks permit the movement of 3,500-ton-capacity tugs for towing loaded barges. Eleven dams and associated works were necessary, nine located in Germany and two on the international section of the river between Luxembourg and West Germany.

In the relatively short length of river (only 271 km or 170 mi) involved in the canalization project, the water level drops 90 m (296 ft). Conditions, therefore, were much different from those on major North American projects such as the Ohio. The construction of most of the 11 dams raised the water level of the river close to or even over the top levels of the previous river banks. Many ancillary works were therefore necessary for protecting riverside towns and villages from flooding, for adjusting water supply and sewerage systems, and for providing necessary clearances under bridges. As illustrating the sort of problem that this type of work can create, Fig. 33.17 shows diagrammatically one of the groundwater problems. The water table in the ground adjacent to the river, before canalization, sloped (as always) down to the normal water level, the slope dependent upon flow and the permeability of the ground. This condition had existed for centuries and so river-bank buildings, such as the warehouse indicated, had foundations designed to suit. The basement of this particular warehouse was used for the storage of Moselle wines. With the rise in water level consequent upon the construction of the adjacent downstream dam, the water table would have risen to such an extent that storage would have been impossible. All along the Moselle, therefore, special provisions had to be made in such locations to maintain groundwater conditions in their previous state. One solution is illustrated, the construction of an impermeable barrier (by the slurry-trench and allied methods) and then the provision of a drainage facility behind this, so located that it would maintain groundwater at or very close to its original level. In more serious cases and where no well-established wine storage was involved, fill had to be placed to raise ground levels; in a few cases villages and buildings had to be moved. And all this work was due merely to changes in the level of groundwater.[33.18] (Fig. 33.18.)

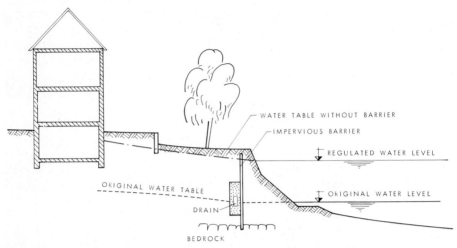

FIG. 33.18 Diagrammatic representation of the effect of raising water levels in the Moselle River and the effect on river-bank buildings without the remedial drainage works shown.

One of the most complex construction operations in the canalization of the upper Mississippi was the completion of the control of the river into the harbor of Minneapolis by the building of two locks adjacent to St. Anthony Falls (Fig. 33.19). The upper lock had to be fitted into the famous Stone Bridge of the (then) Great Northern Railroad. Developments along both banks of the river made design unusually difficult, but design work was aided by the use of a large model of this stretch of river in the adjacent St. Anthony Falls Hydraulic Laboratory of the University of Minnesota. Further complications were provided by the local geology. Existence of the famous falls is due to this geology, a stratum of resistant Platteville limestone overlying the deep stratum of St. Peter sandstone, which so distinguishes the subsurface of the Twin Cities, the limestone bed thinning out to zero just above the falls. A thin stratum of impervious siltstone within the St. Peter formation further complicated the foundation designs. Upon excavation, the St. Peter formation was found to be featured by horizontal seams which, when dewatering commenced, permitted the flow of water through the stratum. Eventually a well-point system had to be installed because of the inflow. Installation of the well points was itself a critical operation, since they had not to penetrate the impervious silt stratum, but this, and other problems, was overcome successfully, and the two locks are now an additional interesting feature of the St. Anthony Falls area.[33.19]

Works such as these are a far cry from the "dredging over sandbars and removal of snags" with which the U.S. Corps of Engineers commenced their work of controlling the Ohio River a century and a half ago. Today river-control works for the canalization of major streams involve some of the most expensive and complex of construction operations. Much of the experience gained in carrying out these works in North America has fortunately been recorded in the literature so that there are many useful guides available. It cannot be too strongly stressed, however, that every such project is unique in itself and may present previously unforeseen problems or hazards, as the experience at the Wheeler Lock so clearly demonstrated. The most accurate possible knowledge of the subsurface throughout the entire site is the first and essential requirement, knowledge of the hydrology of the river being equally important preliminary information, especially for the planning of cofferdam design and construction.

FIG. 33.19 St. Anthony Falls on the upper reaches of the Mississippi River at Minneapolis, Minnesota, with the lock and control works mentioned in the text.

33.8 SOME OTHER WORKS

Brief as it must be, this chapter would not be complete without at least a passing reference to Niagara Falls. This majestic natural feature is itself the result of the local geology. The steady erosion of the falls upriver, well recorded in written historical works, is vivid evidence of the inexorable cycle of Nature. The United States and Canada have been consulting regularly, ever since 1878, about measures to be taken to preserve the unique beauty of the falls. Water power development at the falls started in 1877 and added to problems of controlling the river. Major remedial works were contemplated by the 1941 exchange of notes between the two neighboring countries, and in 1943 a causeway leading to an artificial island, from which a rock-fill submerged weir extended almost to the Ontario shore, was constructed jointly by the two countries in an unusually interesting operation. The weir helped to preserve the beauty of the falls while permitting the abstraction of extra water for wartime power development.[33.20] The falls are constantly under international review and further training works are carried out when seen to be necessary. So important is this work that the American Falls were "dried up" in 1969 so that a detailed inspection of them could be made in the dry (Fig. 33.20).[33.21]

Nor should the special problems of the river Rhine, the greatest of European waterways, be forgotten. The effects of dredging upon the great port of Duisburg were described in Chap. 2. Construction of the delta works by the Netherlands is touched upon in Chap. 35 in relation to Dutch land reclamation. Closely associated with this project is the construction of three river-control dams on the Lower Rhine, which parallels the Waal River

FIG. 33.20 Niagara Falls; the American Falls dewatered for inspection and testing in 1969.

and is one of the waterways through which Rhine water reaches the sea. The dams will influence flow to the Zuider Zee through the Yssel River, the whole network of these rivers, all now trained by man, being one of the world's outstanding examples of applied hydrogeology. Pollution of the Rhine is now coming under control, assistance in this vital work coming also from river-training works on tributaries of the main river, especially in the coal-mining districts of the Ruhr. Here the Rheinische Braunkohlwerke, A.G., popularly known as the Rheinbraun, has carried out river-training works, even river relocations and the moving of complete communities, all in relation to the winning of coal.[33.22] The works of the Rheinbraun are so extensive, so well planned and integrated that this chapter may well conclude by this brief tribute to an organization which should be consulted by all facing similar problems.

33.9 REFERENCES

33.1 H. Chatley, "The Hydraulics of Large Rivers," *Civil Engineering and Public Works Review*, **33**:59 (February 1938).

33.2 "Flood Prevention and Other Hydraulic Problems in China," *The Engineer*, **166**:278 (9 September 1938).

33.3 O. J. Todd, "A Runaway River Controlled," *Engineering News-Record*, **116**:735 (1936).

33.4 D. O. Elliot, *The Improvement of the Lower Mississippi River for Flood Control and Navigation*, 3 vols., U.S. Waterways Experiment Station, Vicksburg, Miss., 1932.

33.5 H. N. Fisk, *Geological Investigation of the Alluvial Valley of the Lower Mississippi River*, Mississippi River Commission, Vicksburg, Miss., 1944; see also G. H. Matthes, "Paradoxes of the Mississippi," *Scientific American*, **184**:19 (April 1951).

33.6 "New Control Dams Remove Threat to Ol' Mississippi," *Engineering News-Record,* **164**:32 (7 January 1960).

33.7 C. C. Noble, "The Mississippi River Flood of 1973," in D. R. Coates (ed.), *Geomorphology and Engineering,* Dowden, Hutchinson & Ross, Stroudsburg, Pa., 1976, p. 79.

33.8 R. C. Crawford, "Flood Control Plan for the Missouri River," *Civil Engineering,* **14**:413 (1944); see also same journal, **16**:64 (1946).

33.9 "Bringing the River to the Dry-Land Bridge," *Engineering News-Record,* **154**:25 (12 May 1955).

33.10 C. R. Longwell, A. Knopf, and R. F. Flint, *A Textbook of Geology: Part I, Physical Geology,* Wiley, New York, 1932, p. 65.

33.11 L. C. Gottschalk, "Effects of Soil Erosion on Navigation in Upper Chesapeake Bay," *Geographical Review,* **35**:221 (1945); see also "Sedimentation in a Great Harbor," *Soil Conservation,* **10**:3 (1944).

33.12 L. B. Leopold, *Water: a Primer,* Freeman, San Francisco, 1974, p. 63.

33.13 E. A. Keller, "Channelization: Environmental, Geomorphic and Engineering Aspects," in D. R. Coates (ed.), *Geomorphology and Engineering,* Dowden, Hutchinson and Ross, Stroudsburg, Pa., 1976, p. 115.

33.14 W. S. Adamson, "Engineers Copy Animal's 'Plan' to Curb Elements," *Salt Lake City Tribune,* 8 January 1939; and I. M. Winsor, "The Barrier System of Flood Control," *Civil Engineering,* **8**:675–678 (1938).

33.15 A. H. Richardson, *Conservation by the People: The History of the Conservation Movement in Ontario to 1970,* University of Toronto Press, Toronto, 1974.

33.16 B. E. Torpen, "Large Rocks in River Control Works," *Civil Engineering,* **26**:587 (1956).

33.17 "New Stairway up the Ohio Has Fewer Steps," *Engineering News-Record,* **194**:32 (8 May 1975).

33.18 K. J. Melzer, *Harnessing the Moselle River—An Example of Interaction with the Environment,* Preprint no. 1858, American Society of Civil Engineers, New York, 1972.

33.19 D. R. Cady, "Foundation Difficulties and High Water Overcome" *Civil Engineering,* **24**:505 (1954).

33.20 N. Marr, "Flow Diversion at Niagara Falls," *Civil Engineering,* **13**:321 (1943); see also same journal, **13**:359, 403 (1943).

33.21 *Preservation and Enhancement of Niagara Falls,* International Joint Commission, 1953.

33.22 "Open Pit Miners Move River, Restore Land," *Engineering News-Record,* **195**:16 (6 November 1975); "Can Political Tide Clean Rhine-Sewer," *New Civil Engineer,* **212**:16 (30 September 1976).

Suggestions for Further Reading

Reflection suggests that, rather than giving a long list of references, as indicated in the text, the authors should direct the attention of users of this Handbook to *Water: A Primer* by Luna B. Leopold. This excellent brief book, published by W. H. Freeman of San Francisco in 1974, is a lucid introduction to the general principles of hydrology and will serve as a guide to further study.

chapter 34

MARINE WORKS

34.1 The Tide, Waves, and Currents / 34-2
34.2 The Earth Beneath the Sea / 34-5
34.3 Typical Coastal Problems / 34-7
34.4 Docks and Harbors / 34-10
34.5 Breakwaters / 34-15
34.6 Coastal Erosion / 34-17
34.7 Littoral Drift / 34-22
34.8 Maintenance of Tidal Estuaries / 34-28
34.9 Dredging / 34-32
34.10 Some Submarine Problems / 34-35
34.11 Offshore Structures / 34-37
34.12 Conclusion / 34-42
34.13 References / 34-43

The Sea Around Us is not only the title of a beautifully written and most interesting book by Rachel Carson but it is also a literal description of the place that the ocean occupies in the life of man.[34.1] When it is remembered that almost three-quarters of the surface of the globe is covered by the sea, the fact that it does surround all the inhabited lands will be the better appreciated, even though it may be difficult to realize this when one lives in midcontinent, almost 3,200 km (2,000 mi) from the nearest coastline. On the other hand, no resident of the British Isles lives more than 112 km (70 mi) from the sea; residents of smaller islands frequently see the sea every day of their lives. It was along seacoasts that some of the first human settlements developed. The first long-distance travel was by sea. For many centuries, ocean and coastal travel was the only means of mass transport. The Mediterranean Sea, in particular, may be regarded in many ways as one of the cradles of human history. Along its shores some of the earliest of man-made harbors were constructed. Some still remain. Others have long since disappeared, vivid reminders that the construction of marine works is always a battle between man and Nature, requiring for human success a singularly careful appreciation of the natural forces that have to be contended with and knowledge as accurate as is possible of the site conditions and of the local geology.

Seawater has a high salt content, but it is "salt" in a peculiar way; the composition of the solid materials that the seawater contains differs appreciably from the average salt content of the freshwater entering the sea from rivers. The difference is accounted for

34-2 CIVIL ENGINEERING WORKS

biologically—the animal life of the sea uses some dissolved salts but not others. The Red Sea is the saltiest part of the ocean; its mineral content reaches 40 parts per thousand. Along the Atlantic coast of North America the salt content of seawater varies from 33 to 36 parts per thousand; these are typical limits for most of the oceans. The temperature of seawater varies from a low of $-2°C$ (28°F) in polar seas to a high of 37°C (98°F) in the Persian Gulf. Only when the temperature of the sea is over 21°C (70°F) can coral reefs, one commonly recognized type of marine growth, develop. These remarkable formations, which are the result of the gradual accumulation of the skeletal remains of organisms with a high content of calcium carbonate, are thus generally confined between latitudes 30°S and 30°N.

34.1 THE TIDE, WAVES, AND CURRENTS

The movement of the sea is more important to the civil engineer than the specific properties of seawater. Never does one see the ocean at rest. In motion, as at the height of a great storm, it presents one of the most majestic of all natural phenomena. The wind is the chief factor in the formation of waves and currents, but the regular movement of the tides is perhaps the greater determinant in the design and construction of the marine

FIG. 34.1 Slipway in the harbor of Saint John, New Brunswick, (on the Bay of Fundy) at low tide.

works of the civil engineer. These three types of motion constitute some of the "great forces of Nature" that it is the task of the civil engineer to tame, even as down through the ages they have been amongst the more important agencies of physical geology, molding the shape and form of the earth along all its borders on the sea.

Local currents, near harbors and in estuaries, will affect the layout and siting of marine works, but when once determined by hydrographic surveying, the currents must then usually be accepted as a part of the natural order and integrated into design. Wave action, however, is of far more serious consequence. In general, waves are of two types: swell waves and storm waves. Both are generated by the action of the wind on the surface of the sea, storm waves more directly than swell waves. Much study has been given in recent years to the mechanics of wave motion. There are many convenient references, but again the civil engineer will usually be more concerned with the effects of wave action than with its detailed character. Of major importance is the maximum force that is to be expected from wave action at the location of any proposed works. This problem appears to have been first studied by Thomas Stevenson (the father, incidentally, of Robert Louis Stevenson) in connection with the design of harbor works in Scotland. He propounded the first empirical rules for determining the probable forces of waves, based upon the length of the *fetch* at a particular location—the distance of uninterrupted open sea in the direction of prevailing wind. Thomas Stevenson was the third member of five generations of this

FIG. 34.2 The same slipway as in Fig. 34.1, at Saint John, New Brunswick, at high tide.

famous family, all of whom were in charge of the building of the lighthouses of Scotland, a record unique in world engineering and one which has been well told in a book entitled *A Star for Seamen*.[34.2]

Those who have never seen the sea, particularly at the height of a great storm, find it difficult to imagine the force that the sea can exert. Even well-authenticated figures tend to seem unrealistic. As an indication of what waves can be and do, it may be noted that, although ocean waves do not normally exceed 7.5 m (25 ft) in height in midocean, there is a well-established record of a storm wave in the Pacific with a height of 33.6 m (112 ft).[34.3] One of the writers has witnessed storm waves in the Atlantic over 15 m (50 ft) high. Approaching a coastline, waves will naturally be reduced in size, but the potential force of breaking waves is remarkable. One of the stormiest parts of all the oceans is the Pentland Firth, a stretch of water separating the northern coast of Scotland from the Orkney Islands. Close to the eastern end lies the town of Wick with its small harbor, famous in the annals of civil engineering. When its breakwater was destroyed in 1872, there was clear proof that blocks of masonry weighing up to 1,350 (long) tons had been moved intact by the sea. A new pier was constructed, but in 1877 it was also destroyed by a great storm; accurate observations showed that a block of masonry weighing 2,600 (long) tons had been carried away from its original location.[34.4] One of the lighthouses guarding this stormy Scottish coast (at the western end of the firth) is at Dunnet Head; the windows around the light of this high structure, which stands atop a 90-m (300-ft) cliff, are frequently broken by stones tossed aloft by storm waves breaking on the coast below. A door, 58.5 m (195 ft) above the sea, near the top of the lighthouse at Unst, the most northerly of the Shetland Islands, has been broken open at the height of a storm. So the tall but true tales of the force of the sea can be continued. It is small wonder that the design and construction of marine works provide some of the supreme examples of the art of the civil engineer.

The movement of the tide is a natural phenomenon of an entirely different but equally majestic character, involving the simultaneous movement of all the water in the oceans. Under the combined attraction of the moon and the sun, but chiefly the moon, the waters of the oceans tend to move away from the earth; this action, when combined with the rotational motion of the three celestial bodies, results in the rhythmical movements known as the tide. Normal tides have a dual cycle of about 24 hours and 50 minutes (the extra 50 minutes are most significant for the scheduling of marine-construction operations). As the moon waxes and wanes each lunar month so do the tides vary from neap (or low) tides, twice a month when the pull of the sun and the moon are opposed, to spring (or high) tides, also twice a month when the two bodies are in line and so pulling together on the waters of the sea. In almost all countries of the world which have tidal coastlines, there are now regularly published tide tables that give particulars of tidal ranges and times for many months ahead; such publications are naturally of vital importance to marine constructors.

With so simple a basic explanation for the movement of the tides, it is surprising to find so great a variation in tidal ranges along the coasts of the world. Possibly the most remarkable variation is that along the coastline of the northeastern part of North America. At Nantucket, near Boston, the range is less than 30 cm (1 ft). As one goes north into Canada, the range steadily increases until, in the Bay of Fundy, between New Brunswick and Nova Scotia, some of the highest tidal ranges of the world are experienced; spring tides reach a height of 16.2 m (54 ft) in Minas Basin (see figures on previous two pages). This lovely spot is only about 640 km (400 mi) from Nantucket. Tides even higher than the famous Fundy tides have been observed in the Canadian Arctic. Such strange variations as the fact that many coastlines have almost negligible tides, that some locations have only one tide each day instead of two, and that other locations have a tidal period of an even 12 hours can be explained only after meticulous oceanographic study. This is a branch of learning that attracts the interest of many civil engineers, but in the practice of civil engineering what is essential for all marine work is accurate knowledge of local tidal conditions. This is to be obtained directly from published tide tables for locations listed in such tables and by interpolation from them for other points, always with a check (by means of a correlation of readings from a local tide gauge with those for a listed port) for the location of any new marine works of any magnitude.

Civil engineering works upon the large lakes of the world may properly be called "marine works," even though they are in freshwater. There are no tides in the ordinary sense on such lakes, but on certain very large, shallow lakes, mass movements of water occur that may be more troublesome than the tide. These movements are known as *seiches*. Such movements of water, resulting in rapid changes in water level, may be caused by changes in barometric pressure or, more commonly, by changes in the wind. The free water surface is subject to a shearing force as the wind moves over it; the transfer of energy results in water displacement, the magnitude of which depends on the depth of the lake, the wind speed, and the fetch over which it blows. Lake Erie, the shallowest of the Great Lakes, is renowned for the seiches that distinguish it. Differences of 4 m (13.5 ft) have been observed between the levels of the two ends of the lake, at Buffalo and Toledo, with wind "setups" (as they have come to be known) of as much as 2.5 m (8.4 ft) in the harbor of Buffalo. The effects upon harbor structures can well be imagined. One indirect result is the development of currents at intermediate points along the lake; the harbor of Conneaut experiences currents as high as 30 cm (1 ft) per second at its entrance in response to a major seiche.[34.5]

34.2 THE EARTH BENEATH THE SEA

It is difficult to refrain from using the title of yet another book to describe this short treatment of some features of the geology of the ocean bed which are of importance to civil engineers in the prosecution of their work. *The Earth Beneath the Sea* is the title of a pioneer volume by F. P. Shepard, one of the leaders in the relatively new study of submarine geology.[34.6] Not only does this title describe a graphically written account of current knowledge of the seabed and of methods of underwater investigation, but it is also a telling reminder that there does exist "solid ground" (for want of a better term) beneath the surface of the sea. To see a stretch of wide beach at low tide, as in the Bay of Fundy, helps one to appreciate the fact that the solid crust of the earth is continuous over the entire globe, even though large parts of it are shielded from view by the waters of the sea. Since there was little tide around the shores of those countries in which civilization took root, it is not surprising that strange and misleading ideas about the seabed persisted for so long. Only in the last few decades has even a general impression of the character of the ocean bed been obtained, a statement well supported by the fact that only about 15,000 deep soundings had been taken over the entire globe when echo sounding was introduced on a general scale immediately following the First World War.

The first deep-sea sounding, to a depth of 4,430 m (2,425 fathoms), was obtained by Sir James Clark Ross in 1840 during one of the early voyages of exploration to the Antarctic. He used a weight on the end of a manila line, and this crude method had to suffice until 1870 when Lord Kelvin used piano wire for the same purpose and so extended appreciably the depths that could be plumbed. Prior to this development, fewer than 200 soundings had been made throughout the entire Atlantic area. The 15,000 soundings, already noted as having been taken by mechanical means, represented about one sounding for every 15,500 km² (6,000 mi²) of sea; thus it has been only through the medium of indirect depth determinations, notably the use of echo-sounding devices, that a real beginning has been made in determining the overall configuration of the bottom of the sea. Immense mountain ranges, correspondingly great "deeps," and gigantic submarine canyons, the dimensions of which surpass any to be seen on dry land, have now been located. The irregularity of the seabed is well recognized, but perhaps the most unusual underwater physiographic feature of all is the wide extent of the *continental shelf.*

This remarkable area has been defined officially as the "zone around the continents, extending from low-water line to the depth at which there is a marked increase of slope to greater depth"; this outer slope is called the *continental slope.* The average depth to the shelf is 130 m (72 fathoms), the greatest depth about 365 m (200 fathoms). The widest parts of the shelf are those bordering glaciated lands; the Arctic coasts have the broadest shelf of all, the average width being about 67 km (42 mi), the average slope about 1.9 m to the km (10 ft to the mi). Vividly demonstrated on all small-scale ocean charts, the continental shelf is clearly of great practical significance to fishermen, to mariners, to those concerned with cable laying, and in relatively recent years, to those in search of petroleum. Civil engineering structures for both oil drilling and defense purposes now stand upon the shelf at selected locations off the American coast. Some North American lightships have been replaced by tower-type structures founded on the shelf.

It requires but the most general appreciation of geology to realize that this submarine shelf must bear some relation to geological phenomena of the recent period. Changes in the level of the sea have already been mentioned. There is an almost certain connection between such changes and the continental shelf as it exists today. The latest determinations suggest that the level of the sea rose from about 78 m (260 ft) below present level to possibly 6 m (20 ft) below this level between 17,000 and 6,000 years ago. Movements of the sea level in the last 6,000 years are not so well defined, but it is almost certain that the level is rising today very slowly; there is good evidence to suggest that the level rose about 10 cm (4 in) along the Atlantic coast of the United States between 1930 and 1948. Those who find it difficult to visualize the concept of the changing level of the sea might refer to the many examples of "raised beaches" to be found in all parts of the world; these are vivid evidence of a much higher sea level in the past.[34.7] The reverse process appears to be closely related to the existence of the continental shelf that is today such an important feature of submarine geology. Correspondingly, fragments of peat and artifacts of the Stone Age which have been brought up in nets from the Dogger Bank in the North Sea by fishermen show clearly that this famous fishing ground was once dry land. The shallow nature of this famous sea has proved to be providential for the erection of the offshore drilling platforms required for exploration for, and extraction of, the oil found beneath its bed, as will be seen in a later section of this chapter.

Interesting and scientifically significant as such general geological considerations are, they are of only indirect importance to the civil engineer, who works to a different time scale. Explanation of the formation of the shelf, however, is a prerequisite for study of its composition; its surface characteristics, particularly, constitute a matter of real importance for the engineer. Following in the trail of underwater depth determinations has been the start of a study of the sediments that constitute the seabed. Already it is known, in a general way, that the sediments on the shelf consist predominantly of sand, some mud, and a small percentage of gravel, whereas the percentage of mud is much increased in the deposits on the continental slope. The different types of mud that constitute so much of the deep ocean bottom have already been delineated in a general way; more information about them is steadily being accumulated. They raise some interesting questions, such as why sediments in the Atlantic Ocean reach a depth of 3,600 m (12,000 ft), whereas those in the Pacific and Indian oceans are never more than 300 m (1,000 ft) deep. The existence of manganese nodules on the sea floor has been known since the time of the pioneer expedition of *H.M.S. Challenger,* but their presence is still unexplained.

It was during the *Challenger* expedition that the first advance was made in the study of the nature of the seabed. It had been traditional for mariners to have a lump of tallow on the underside of their sounding leads, to which samples of bottom sediments would adhere and thus be brought to the surface for study. The scientists of the *Challenger,* however, were able to obtain short cores of the material forming the seabed. The practice of undersea core boring, although fraught with many mechanical difficulties, has steadily progressed since then, and continuous cores can now be obtained from great depths. Study of the sediments so retrieved from the bottom of the ocean is opening up new phases of geological study. Concurrently, there has been steady advance in personal inspections of the sea bottom—first by means of diving equipment, with the most complicated of which depths of 150 m (500 ft) can be reached by a diver; then with special units, such as William Beebe's bathysphere, with which a depth of 908.4 m (3,028 ft) beneath the surface was reached in 1934; and finally by the use of the U.S. Navy's bathyscape, the *Trieste,* in which Jacques Picard and Lt. Don Walsh reached a depth of 10,734 m (35,780 ft) at the bottom of the "Challenger Deep" on 23 January 1960.[34.8]

Investigations at such great depths are of only academic interest to the civil engineer, but methods of submarine investigation at shallower depths are of great practical importance. The standard diving suit will long remain the chief aid to such personal inspection; depths of about 15 m (50 ft) can be readily penetrated with only a minimum of training when expert supervision is available, as it must be for all such underwater inspection. The value to civil engineers of personal inspection of the geology of the seabed at the site of the works for which they are responsible is inestimable. On one job, one of the writers was told that the dredging contractor was being hampered by the presence of "boulders as big as cottages." It required only a short survey in a diving suit (once the contractor had recovered from the shock of being asked for the use of his diving outfit) to see that the expression "cottages" was a poetic exaggeration; the efficacy of the dredging that had been done was, however, similarly demonstrated by this personal inspection. The modern practice of skin diving naturally presents great possibilities for use in connection with the marine work of the civil engineer. The extension of its use can be a significant factor in giving all engineers who are involved in marine work a better appreciation of the physiography and the geology of the earth beneath the sea.

34.3 TYPICAL COASTAL PROBLEMS

So wide is the range of civil engineering problems that come within the general subject of marine works that those who have had no opportunity for engaging in this branch of professional practice may be puzzled even by the variety of the major sections into which this chapter is divided. It may be helpful, therefore, if a brief summary is given of some of the problems that have been experienced on one short length of seacoast. The fact that the shoreline of the British Isles is never more than 112 km (70 mi) from any inland point has already been mentioned. When it is recalled that there are few, if any, parts of the world where such varied geological conditions are found in so small an area as that of Great Britain, a glance at some happenings along the coastline of England and Wales may be useful (Fig. 34.3). The coast of Scotland, especially in the far north, has had its own fill of problems, but, in view of limitations of space, they must be left for readers to study on their own.

The White Cliffs of Dover will be generally familiar as one of the striking features of this coastline, which presents so many truly beautiful natural scenes, but it is only one. The whole coastline has been studied for centuries and so has been often described; two books of special relevance are *The Coastline of England and Wales* and *The Sea Coast,* both by J. A. Steers.[34.9] The best starting point, however, is the third (and final) report of

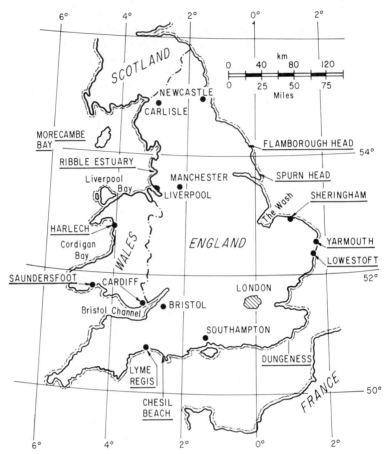

FIG. 34.3 Outline map of England and Wales showing some locations mentioned in the text.

the Royal Commission on Coast Erosion.[34.10] Although published in 1911, this masterly document is still of great value and utility. One of the many historical lessons underlined in this notable volume is that, without the most careful planning and design, engineering works on one stretch of coast may cause serious trouble on adjacent sections.

Morecambe Bay is a leading feature of the coastline of northwestern England; at low tide much of the area of the bay is dry sand, although "quick sands" are found in some parts. In earlier days, stagecoaches drove across the bay when the tides were suitable; a troop of cavalry similarly used it as a shortcut during the First World War. George Stephenson proposed in 1837 the construction of a great wall across the bay that would have enclosed 16,000 ha (40,000 acres). It was not built, but the project is again under renewed study. About 405 ha (1,000 acres) were reclaimed from the bay when the Furness Railway was built along its northern shore, embankments across the many little estuaries serving a double purpose. The sand of the bay is part of the coastal deposits in this part of England. Sand dunes around the estuary of the river Ribble have had to be stabilized by vegetation. Some of the dune grasses found as natural protection of these dunes were not plants native to Great Britain. It was found that they had grown from seeds swept out of ships in the neighboring port of Liverpool after cargoes of grain feed for poultry had been brought in from the United States during the First World War.

The centuries-old battle to keep open and to improve the entrance to the port of Liverpool will be mentioned later. If one here turns westward and travels along the coast of north Wales, one sees the dubious results of running (necessarily) a main road and a main railway line in close juxtaposition to an exposed rock shoreline. The striking coastal scenery of the island of Anglesea leaves no doubt that this engineering location was dictated by geology, even though it has led to ceaseless maintenance work along most of the coast. On the west coast of Wales, Harlech Castle is a notable feature, built in 1286 by King Edward I. It is now more than 800 m (0.5 mi) from the sea but must have been built originally right on the coast, silting and the change in course of a small river having altered completely this stretch of shore. *Sarns* are an unusual feature to the south of Harlech, causeways linked to the legend of the lost land of Cantref-y-Gwaelod, a large settled area said to have been reclaimed by the sea through neglect of dike maintenance by a renowned drunkard. Even the best of engineering works cannot withstand such mistreatment.

The Bristol Channel is featured by some fine man-made ports. One small port, not now used, is that of Saundersfoot, the remains of a fine shingle beach here not being a natural feature but the result of dumping ballast from across the sea when the port was in use. During excavation in 1928 for one of the more recent harbors, that at Port Talbot, a stone was found, in an upright position but 6 m (20 ft) below the surface, bearing the date 1626, clear evidence of sand movement along this coast. Even more remarkable is the history of the church at Penard; built about 1270 the church and neighboring castle were inundated by advancing sand dunes in 1528 and remained covered until rediscovered in 1861.

Along the south coast of England are signs of many major landslides such as that which took place near Lyme Regis on Christmas night, 1839, precipitating 8 million tons of rock into the sea, forming what is known today as Dowland's Chasm. Chesil Beach, one of the most remarkable beaches of the world, leads to the Isle of Portland, for centuries source of building stone that is world-famous. The beach is 29 km (18 mi) long, its most remarkable feature being that the shingle of which it is composed is graded consistently from pea size at its northwestern end to coarse gravel (up to 7.5 cm or 3 in in diameter) at the Portland end. Dungeness, farther to the east, is the largest shingle foreland in Great Britain. Figure 34.4 shows clearly the stages in its growth, probably since Neolithic times. Ninety-eight percent of the shingle here is flint (from nodules in the neighboring Chalk). Its seaward growth has averaged about 4.8 m (16 ft) a year throughout the last four centuries at least.

Shoreline erosion has been the great problem along the east coast of England. Records show that erosion was a recognized problem as early as 1391, when a jetty had to be built at Cromer to protect fishing boats; the work was not successful. Protection works have been necessary ever since, William Smith (amongst many other early engineers) being consulted about the problem. A commission of inquiry reported as early as 1846. The twin ports of Yarmouth and Lowestoft give good evidence of the effects of engineering works on neighboring coasts. The great expanse of The Wash is a "museum" of efforts to control coastal erosion, remains of Roman works having been found, a salutary reminder to civil engineers engaged on coastal works today that they are following in a long tradition.

Immediately north of the Humber, between Spurn Head and Flamborough Head, there is what is perhaps the most famous example in the world of an eroding coastline. The coast consists mainly of boulder clay in cliffs that reach 30 m (100 ft) in height. Despite this, the entire stretch of coast south of the resort town of Bridlington has been eroding steadily since Roman times, with a general average loss of almost 1.8 m (6 ft) per year. It has been estimated that the equivalent of a strip of land 4 km (2½ mi) wide along this

FIG. 34.4 Aerial view of Dungeness on the south coast of England, showing the trend of the shingle ridges.

entire coast has been washed into the sea since the Romans left England. *The Lost Towns of the Yorkshire Coast* is the melancholy title of a volume that presents a detailed review of this extreme case of coastal erosion.[34.11] The material so eroded accumulates at Spurn Head, where it provides the inverse problem of steady extension of this headland at a rate of about 12 m (40 ft) per year; this land growth has already necessitated many moves of the lighthouse that stands on the point guarding the entrance to the Humber.

The instability of this section of coast was well shown by the attempt in the late nineteenth century to construct a new dock at Sutton Bridge, a venture in which the Great Northern Railway had a substantial financial interest, since it would have provided a most convenient port connection for that important rail line. The dock was completed and opened with due ceremony on 14 May 1881, but it was discovered on the very next day that leakage was taking place from the entrance lock. A thorough inspection of the site was made, and expert advice taken, but the consensus was that the site was so unsuitable that reconstruction was impracticable. It is said that the only ship which entered the dock on its one day of operation brought a load of timber from Norway and sailed the next day with a cargo of coal. The entire project was abandoned. The Sutton Bridge golf course now occupies the site; a side of the abandoned dock forms one of the bunkers for the course. It is not unreasonable to imagine that consultation with a Roman engineer who worked at this location many centuries before might have obviated one of the few really complete failures in the history of modern civil engineering.[34.12]

34.4 DOCKS AND HARBORS

Dock and harbor engineering has always been a leading branch of civil engineering; its history goes back to the very earliest of all civil engineering works. Because of the inevitable interference with natural processes caused by dock construction and the understandable lack of knowledge of such processes on the part of early constructors, many ancient harbors, although initially successful, eventually proved unsuccessful for reasons that can only be classed as geological. One of the most famous of all ancient ports was that at Tyre, a city often mentioned in the Bible, second port of the Phoenician empire; it was founded on an island and included two harbors protected by rock-fill breakwaters. In 332 B.C., Alexander the Great destroyed the city after building a solid causeway to the island. The causeway interfered with the local movement of coastal sand, and the channel soon filled up; the site of Tyre is now on a peninsula. The great Roman harbor of Ostia, having marine structures which even today command respect, finally was silted up by sediment brought down by the Tiber, despite the ingenuity of the Roman engineers in designing structures to offset this sedimentation; Ostia's site is now 2.4 km (1.5 mi) from the sea.[34.13] Throughout the intervening 2,000 years, engineers have had to give ceaseless effort to overcoming similar difficulties. If today the abandonment of a harbor is almost unheard of, this indicates no modification of the action of natural processes along seacoasts but only the availability of dredging equipment and other devices capable of dealing with the great volume of sediment carried to sea by rivers or moved along the coast by the sea.

Modern dock and harbor practice has included some of the most notable civil engineering works of recent years. The extensive dock system of Liverpool, England; the steady development of the port of New York; and the many and varied engineering works in the gulf and tidal portion of the River St. Lawrence—these and many similar developments testify to the magnitude of dock and harbor works. Geological features affect all construction of this kind to some degree, although usually in one or more of the ways described in the other chapters of this book. Thus a notable underwater geophysical survey was carried out in the harbor of Algiers, using an electrical method, in order to obtain information on the position of a rock surface covered by superficial deposits. For the construction of the rock-fill breakwater at the new harbor of Haifa, Israel, a large quarry had to be opened up; it is a striking reminder of the similar construction methods adopted at the adjacent harbors, Tyre and Alexandria, over 2,000 years before.[34.14]

The ports on both sides of the English Channel form a particularly interesting group of harbors, catering to one of the busiest international exchanges of traffic to be found anywhere. The long-discussed channel tunnel will have, as its main objective, the improvement of this cross-channel traffic. One of the major reports upon this tunnel project was completed in 1930, but the British government decided, shortly thereafter, not to proceed with the project, not for technical reasons but because of national policy. The Southern Railway Company of England thereupon decided to proceed with a train-ferry scheme connecting Dover with the other side of the channel. The ferry terminal at Dover was designed as an enclosed and watertight dock in which the train-ferry vessels could berth at all stages of the 7.5-m (25-ft) tide, the water level being raised or lowered by pumps to the berthing level (Fig. 34.5). The geological formation of the seabed at the site consists of the Lower Chalk Measures. Work and borings previously carried out in the vicinity and the existence of the underwater headings for the proposed tunnel, which were within a distance of a few hundred meters and which had been practically dry for 50 years, suggested that the chalk would be of a solid and homogeneous nature. It was therefore proposed that the work should be carried out in the dry within a cofferdam constructed of steel piling driven into the chalk.

This did not prove possible, as the hardness of the chalk limited the penetration of the piles, but a slight modification of the design permitted the work to go ahead. When pumping of the enclosed area was started, it was "found that, although a head of from 10 to 20 feet could be sustained, the difference in pressure . . . caused an inflow of water through the sea-bed in the immediate vicinity of the works greater than the pumps could discharge." Usual methods of sealing were tried, but all proved unsuccessful. Small bags filled with permanganate of potash were placed by divers in fissures in the seabed outside the cofferdam, and the color showed up all over the enclosed area. Further consideration

34-12 CIVIL ENGINEERING WORKS

FIG. 34.5 Dover Harbor on the southern coast of England, the train-ferry dock mentioned in the text to the left of center.

showed that increased pumping might enlarge the fissures and make matters worse. Special methods, such as the use of freezing, were contemplated, but eventually the entire project had to be carried out in the wet instead of in the dry; the placing of large quantities of underwater concrete proved to be a notable construction operation. Various explanations of the unusual state of the chalk were advanced, the most interesting of which was that the chalk encountered consisted not of solid chalk but of ancient rockfalls from adjacent chalk cliffs, consolidated by the passage above it, over a long period of time, of the littoral drift of gravel for which the adjacent coast is noted. A drill hole put down some kilometers away for the construction of the Beachy Head lighthouse showed solid chalk, although similar difficulties were encountered during construction.[34.15] Another explanation, based on wider experience with the surface of the Chalk of England when exposed in excavation, is that the shattering of the rock near the surface was the result of fossil permafrost conditions which probably existed when much of Great Britain was under ice cover.

The construction of the Dover ferry terminal was one of the most difficult marine undertakings of modern times. Almost at the other extreme, from the geological point of view, was the building of a port on the Great Lakes to serve a new taconite (low-grade iron ore) mining development. (Civil engineering works on major lakes of the world, particularly on the Great Lakes, are marine works in every sense except that the water to be built in is fresh and not salt.) Taconite Harbor is located on the northern shore of Lake Superior, 128 km (80 mi) northeast of Duluth and 120 km (75 mi) due east of the taconite plant at Hoyt Lake. Two islands were conveniently located in the lake at this point, about 450 m (1,500 ft) offshore. When the islands were connected and extended, it was possible

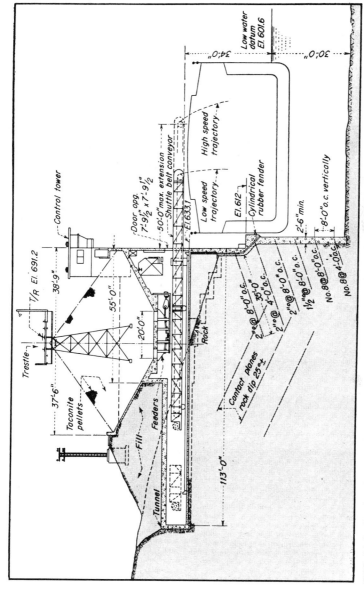

FIG. 34.6 Section through the main wharf of Taconite Harbor, Lake Superior, showing the use of rock anchors.

34-14   CIVIL ENGINEERING WORKS

to construct a large enough breakwater to give complete protection to the area planned for the loading facilities. The local bedrock is a fairly sound amygdaloidal basalt, laid down in successive surface flows, with a strike parallel to the shoreline and a dip of about 25° toward the lake. After careful study it was decided to take advantage of this favorable geological formation and excavate the rock to the shape required for the finished wharf, covering the excavated rock face with a relatively thin wall of reinforced concrete after anchoring the rock with long steel dowels against possible movement on the dip toward the lake. Figure 34.6 shows the finished cross section, and Fig. 34.7 is an aerial view of the work under construction; the rock excavation was carried out within the area enclosed by a cellular cofferdam. Almost 750,000 m^3 (1 million yd^3) of rock was removed, excavation starting with line drilling at 30-cm (1-ft) centers. Some evidence of contact planes was noticed as rock was removed from the vertical face; some of the planes were water-bearing, and some contained fragmented material. Extensive grouting was carried out both before and after excavation, and all anchor rods (the arrangement of which was designed after detailed geological study) were carefully grouted into place.[34.16]

It is only very infrequently that local rock conditions can be adapted for harbor design so effectively as at Taconite Harbor, but it is similarly only seldom that a harbor has to be developed from scratch, so to speak, without taking advantage of some natural feature that requires only modification to form the harbor desired. Most of the major harbors of the world have developed from the initial use of a natural harbor. Some of the most famous—the harbors at Sydney, Australia, and St. John's, Newfoundland, the shelter provided by San Francisco Bay, and the dock facilities of London and New York—are supreme examples. These and many other harbors can quite aptly be described as "shelters provided by Nature." This old expression is really another way of saying "shelter provided by local geological conditions." A study of the geological factors responsible for major American harbors was published (by the U.S. Geological Survey) as early as 1893.[34.17] In most cases, it has been the flooding of land by the rise of the sea that has led to the natural sea-filled depressions that today form such splendid harbors as those mentioned, and many other worldwide ports.

There is probably only one harbor in the world that engineers formed by imitating Nature in this way, i.e., by flooding a natural depression that could thus be connected to

FIG. 34.7 Aerial view of Taconite Harbor, Minnesota, showing the breakwater formed by two offshore islands in Lake Superior.

FIG. 34.8 The beautiful harbor of Ischia in the Gulf of Naples, Italy, showing the artificially formed entrance.

the sea. This unique example is in the Mediterranean, on the north coast of the little island of Ischia (described by Ludwig Bemelmans as his "Cinderella Island"), lying immediately to the west of Naples.[34.18] Porto d'Ishia is today one of the safest harbors on the Gulf of Naples, but until the 1850s it was a landlocked crater lake, separated from the sea by a narrow neck of land. Italian engineers spent two years excavating an opening through this neck. The resulting harbor is seen in all its beauty in Fig. 34.8, filled as usual with the colored sails of local boats carrying cargoes of wine, fruit, and vegetables. In this case, it required relatively little work to convert a natural feature into a perfect harbor, but usually major civil engineering work is necessary to form a new harbor. Possibly the greatest challenge to the civil engineer engaged in dock and harbor engineering comes when natural conditions do not provide all the shelter needed and protective works, such as breakwaters, have to be constructed, almost in defiance of the elements.

34.5 BREAKWATERS

From the very earliest times attempts have been made to develop harbors of refuge by the construction of protective breakwaters in locations that provide no natural shelter for shipping. One of the very earliest examples of the potential dangers in this branch of civil engineering is provided by the history of port works in the Gulf of Ephesus. Strabo has recorded the history of this early harbor construction, going back for many centuries B.C. Under King Attalus Philadelphius of Pergamum, "engineers" constructed a breakwater across the end of the gulf in an attempt to provide a protected harbor, but the breakwater interfered with the natural channel and accelerated silting so greatly that the main objective of the work was not achieved. Many years later, Roman engineers attempted to rectify the mistake, but it was too late and the natural harbor was ruined. Today, it is difficult to imagine that the sea once extended 24 km (15 mi) into the Gulf of Ephesus.[34.19]

Similar examples, extending in time throughout the intervening 3,000 years, could be quoted, but all would demonstrate that in the construction of breakwaters the greatest possible care must be taken to investigate fully all possible interference with natural geological conditions *before* construction is started. As recently as 1939, a small breakwater was constructed at Redondo Beach, California, that soon revealed itself to be a serious impediment to the natural littoral drift along the adjacent stretch of coast, and expensive remedial measures have been necessary since that time. This is but one example of many that could be cited, all of which would give solid support to the suggestion of J. B. Schijf, a leading Netherlands coastal engineer, that in all coastal engineering "[you should] be sure to put off to tomorrow what you do not absolutely have to do today."[34.20]

This is a cryptic way of pointing out that in the design and construction of breakwaters, unusually careful preliminary study *must* be made of all phases of the local geology, static and dynamic, before design is finalized and construction begins. Not only must the seabed be investigated fully with respect to its stability, bearing capacity, and ease of removal if dredging is involved, but the adjacent stretches of coast must be studied in order to determine local currents, littoral drift, and any features that may in any way be affected by the proposed structure. This is a procedure that can take years, even for a small harbor. One of the writers has published a summary account of one such investigation for which he was responsible at Forestville in the Gulf of St. Lawrence. The investigation was for an initial harbor development on a small scale (although since greatly enlarged); despite this, the preliminary field studies extended over a period of more than two years (Fig. 34.9). Methods of investigation are but little different from those treated generally in Chap. 10; the added difficulty of working in tidal waters is often the main variation.[34.21]

The associated problems of coastal erosion and littoral drift are so important that they call for special mention in the sections that follow. The character of the seabed at the site of a proposed breakwater does not usually present unusual problems; the action of the sea itself generally results in satisfactory bed conditions, at least in relatively shallow water, although these conditions must always be carefully investigated. The main problem of design is to ensure stability against the anticipated action of the waves that will strike the proposed structure. With modern knowledge of wave mechanics, the availability of model studies as a guide to design, and the abundance of meteorological records now on hand for most coastal regions where construction may be anticipated, design is today rather more determinate than it was in the early years of modern civil engineering practice. Often, however, the old empirical design rules still have their use.

The provision of large blocks of suitable rock is usually a major construction problem and frequently leads to quarrying operations on a large scale; this work also calls for full and careful preliminary geological investigation. Where no suitable rock is economically available, structures of mass concrete may have to be used, leading to critical structural design requirements. Alternatively, artificial "rocks" may be used, such as the patented

FIG. 34.9 Start of the construction of a breakwater in the Gulf of St. Lawrence at Forestville, Quebec.

"tetrapods," successfully developed in the first instance by French engineers for use at harbors in North Africa, but now used in such widely spaced locations as Kahului Harbor on the Hawaiian island of Maui, at Crescent City, California, and at another Pacific coast location shortly to be mentioned. Still another alternative for locations where rocks of only limited size can be obtained is to construct the breakwater to the usual mound form with grout pipes embedded in it. Through these, special cement grout can be forced as soon as the rock is stabilized, thus reversing the usual procedure of fragmenting large masses of rock into smaller pieces for handling by recementing small pieces into a solid mass of artificial conglomerate large enough to withstand all anticipated seas. This procedure was successfully followed at the small Quebec harbor of Forestville already mentioned, and at the adjacent port of Baie Comeau.[34.22]

Geological processes have to be remembered and kept ceaselessly under observation throughout construction and long afterward in all breakwater work, as the records of civil engineering make clear. One of the largest and most important "artificial" harbors formed by large breakwaters is that at Valparaíso in Chile, South America. The base for one of the main breakwater extensions here was formed by depositing sand from dredgers in water up to 48 m (160 ft) deep onto a seabed consisting of black clayey silt. Deformation of the entire seabed in the vicinity of the work caused slides of the deposited material, and over 50 percent of the material deposited, as compared to theoretical quantities, was "lost." It was found, during deposition of the sand, that everything of a light character—"shells, mud and things of that sort"—was washed out of it; the sand that reached the bottom was so dense that an anchor fluke would not enter it; this provides an interesting example of the principles of sedimentation in operation on a major scale.[34.23]

34.6 COASTAL EROSION

There can be few engineers who are not familiar with a stretch of coast on ocean, lake, or inland sea that is being eroded at a rate noticeable in the course of a few years or possibly months. Although sandspits and bars give some evidence of corresponding accretion, the balance always appears to be against the land in this constant battle with the sea. Rockbound coasts and cliffs are similarly affected, although their increased resistance as compared with that of unconsolidated beach deposits usually renders erosion on rocky shorelines of small immediate consequence. This relative unimportance is emphasized by the fact that coastlines distinguished by continuous rock outcrops are not so favorably placed for development, either for pleasure or for commerce, as are coastlines formed of unconsolidated material. Although the process of erosion is similar for all types of shore and although the erosion of rock cliffs often presents geological features of unusual interest, attention will be devoted more particularly to the erosion of low-lying lands.

Despite the fact that the erosion of coastlines is a part of the natural geological cycle, the protection of coastlines against erosion has become a worldwide problem. Steady increase in land values, development of large industrial estates on coastal lands, the natural desire of home owners to have building sites near the water—these and other factors make the problem one of steadily increasing economic importance every year. All around the coasts of Great Britain are to be seen substantial coastal-protection works. The same picture is developing on many North American coasts; the concern of the United States is indicated by the formation of the Beach Erosion Board as a special agency of the federal government. In Japan, with its long coastline of 27,000 km (16,214 mi) for a land area of only 370,000 km² (142,338 mi²), the problem is equally serious. From Australia, South America, India, and many other lands come reports of engineering studies of how to protect valuable coastal lands from the erosive force of the waters of the sea or of great lakes.

Even the crudest estimate of the cost of coastal-protection works would be so astronomical as to defy belief. The following are typical figures for individual projects: $26 million for works on the coast of Long Island, $1.5 million to protect merely about 600 m (2,000 ft) of the tracks of the Pere Marquette Railroad along the shore of Lake Michigan, $250,000 for studies alone to investigate what Rio de Janiero should do to preserve its beaches. Such figures could be duplicated hundreds of times over and would still fail to indicate the full magnitude of this branch of civil engineering work which depends so completely upon a full appreciation of the geological conditions of the coast that has to be protected.

To those unfamiliar with the shorelines of inland lakes, the fact that coastal erosion is also a serious problem on lakeshores may be surprising. The natural forces that have to be contended with, however, are just the same as those along the seacoast. In the case of the Great Lakes, storm damage can equal that on many of the seacoasts of the world. Because of the intensive industrial and residential development around the shores of lakes Ontario, Erie, Huron, and Michigan, coastal erosion on these lakes is probably as serious today as anywhere in the world. Intensive studies are in progress by U.S. and Canadian agencies, designed to limit the damage to the (literally) billions of dollars' worth of property fronting on these lakes. It has been estimated that the loss of land and of property on the Lake Erie shore in the state of Ohio alone, during a period of merely 20 years, amounted to almost $9 million. Study of long-term records shows a clear correlation between rapid shore erosion and years of high lake-water levels. These, in turn, are related to cyclical variations in precipitation. All that is involved in this major branch of remedial civil engineering work is one of the simplest of geological processes, but a process exercised on a grand scale.[34.24]

The erosive action of pounding waves and swirling water is always obvious, so that the cause of coastal erosion need not be discussed in detail. Generally, it may be said that the essentials of the action are now understood, since the direct hydrodynamic forces exerted by waves have been successfully measured and the erosive effect of eddies is at least generally appreciated. When the nature of a beach is such that waves carry with them sand or gravel in suspension, the resulting intensified action in eroding the standing beach will readily be appreciated. Wave action is only one factor in erosion; the determining factor is the geological formation of the coastline that is exposed to wave action. On rock-bound coasts characterized by regular strata of appreciable thickness, erosion will generally be uniform. When strata of varying types having differing resistance to abrasion and solution are exposed, differential erosion will result, often with somewhat fantastic effects, such as the formation of natural arches. Only when erosion threatens a stretch of rocky coast that has been developed right up to the edge of cliffs is the civil engineer's skill and ingenuity called upon. The cause of erosion will usually be obvious after even a preliminary survey, provided this has been made with due regard to the geological formation of the coastline.

A section similar to that shown in Fig. 34.10 might be found in many places. The section is through a small cliff close to the famous old church of Lyme Regis in Dorset, England. The sketch is based on one included in *Sixty-Three Years of Engineering,* an entertaining volume of reminiscences by Sir Francis Fox, who carried out the necessary remedial works.[34.25] The alternating layers of shale, being much softer than the limestone beds, were easily eroded by the sea waves, and the alternate limestone beds were left overhanging. When the overhang became too great, a fracture of the limestone bed resulted, and renewed erosion of the shale took place. Other forces at work in completing the disintegration were the action of heavy seas, which lifted the projecting layers of limestone and so loosened them from their beds, and slips in the upper layers of unconsolidated materials due to seepage of groundwater along the face of the receding cliff. The remedy successfully adopted was the construction of a continuous wall protecting the underlying layers of shale from further erosion and thus stabilizing the entire cliff face.

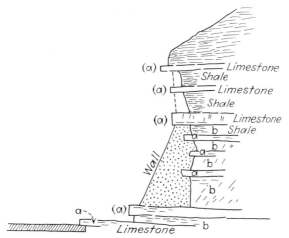

FIG. 34.10 Simplified geological section of sea cliff at Lyme Regis, England, showing protection wall constructed to halt erosion.

Shores consisting of unconsolidated material can rarely be regarded as stable; even prevailing winds sometimes affect their stability. If any development work is to be carried out adjacent to the high-water line on such beaches, consideration of general beach conditions is essential. Protection walls will be a frequent recourse, especially if the works are to project below the high-water mark. Foundation-bed conditions for these retaining structures must be given unusual attention. Scour at the foot of protection walls will always be a possibility to be seriously considered; allowance must be made for every change in beach regime that can be visualized. Cutoff walls of steel sheet piling below reinforced-concrete superstructures provide a convenient and effective method of construction, ensuring protection against scouring. Provided the wall structure does not encroach far below high-water mark, the existence of this wall should not affect the stability of the beach. If a bulkhead wall has to be constructed either within the tidal range on the beach or below the low-water line, then study must be made of the possible consequences of construction.

This leads to the study of beach formation—an extensive branch of physical geology and a perennial subject for debate among engineers responsible for coastal works. Of necessity, study must be largely empirical, since local conditions are of unusual significance (Fig. 34.11). In general, two leading types of coastlines may be recognized: the shoreline of emergence and the shoreline of submergence, the names being descriptive. The first is the result of a gradual uplift of the seabed; the floor of the sea emerges to form a flat coastal plain consisting of unconsolidated deposits. Wave action will then erode the edge of the emergent landmass, carrying this away from the water line to form an offshore barrier beach which will gradually work nearer to the original shoreline until ultimately the two will merge.

Shorelines of submergence are the result of a gradual depression of a coastline; the original coastal area is submerged beneath the sea, which will thus stretch far inland and be in contact with an eroded land surface. The initial characteristic is extreme irregularity of outline; features such as fjords and deep bays are indicative of submerged valleys. Erosion by the sea will continue, and its first effect will be a tendency to straighten the shoreline, a process in which strata that are harder than the surrounding beds will offer

34-20   CIVIL ENGINEERING WORKS

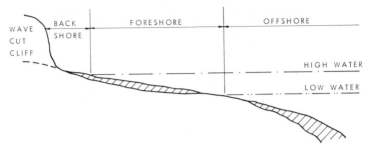

FIG. 34.11 Diagram of essential elements in ocean beach formations.

greater resistance and consequently give rise to special features such as stacks, arches, and small offshore islands. The products of erosion will be rapidly washed into deep water in initial stages of development, but eventually a small beach will form at the foot of the cliffs. If littoral currents exist, they will tend to sweep these beach deposits along the coast, past headlands, to be deposited in the deeper water of bays, forming what are, in effect, submarine embankments. With continued action, these formations will reach water level; they become spits or hooks and in some cases are of such great extent that they almost completely block a whole bay. Eventually, a submergent shoreline will lose its irregularities and acquire a beach formation. After this formation has reached equilibrium its development will be identical with that of a shoreline of emergence. When a shoreline has reached this stage of stability, erosion of the coast or cliff will tend to proceed regularly.

Movement of the eroded materials will be caused by currents and waves; the general tendency will be for the finer material to be carried into deeper water and for the coarser material to form the main beach deposit. The nature of beaches may vary considerably from place to place; but as a general rule, gravels will be found within 1.6 km (1 mi) from shore in depths of not more than 9 m (30 ft) of water; sand rarely extends beyond a depth of 90 m (300 ft). Figure 34.11 illustrates diagrammatically an ideal beach formation which has become reasonably stable. Studies have shown that the size of sand grains or gravel is related to the slope of a beach—the larger the sand grain the steeper the slope.

The shores of the Great Lakes provide striking examples of coastal features, some classical in form. Two of the most notable are the *compound recurved spits* that provide the harbors of Toronto, Ontario, and of Erie, Pennsylvania. They are shown in outline form in Figs. 34.12 and 34.13. It requires but little imagination to visualize that these two major projections from the normal lakeshore have been caused by the interaction of movement of material along the shore (but in opposite directions in the two cases) and the flow from a river, deflecting this material away from its mouth. Nature has thus provided two of the finest harbors on the Great Lakes, although much work has had to be done to maintain and develop them, especially at Presque Isle (a most descriptive name). At Toronto, the eastern entrance was created by a storm early in this century. Once formed, it was kept open and is now maintained by dredging. Knowledge of the drift coming from the Scarborough Cliffs to the east was put to good effect by a contractor for one of the great coal docks in the harbor. Plans for the dock required a large quantity of sand fill. The low tenderer for the work reckoned on dredging the sand from a point just outside the east entrance with the anticipation that he would have a continually replenished supply, necessitating little if any moving of his dredge. His geological knowledge paid off; once his dredge had made its initial cut, it did not have to be moved again, but was provided with all the fill required for the job by the orderly process of "geology in action."

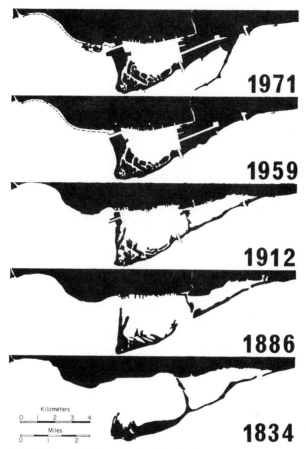

FIG. 34.12 Sketches showing the development of the harbor of Toronto in Lake Ontario, protected by Toronto Island formed by littoral drift from the east (i.e., from the right).

Erie Harbor is notable in that it was the first harbor project ever undertaken by the U.S. Corps of Engineers, the original act authorizing its improvement having been passed as early as 3 March 1823. The shape of the peninsula has changed appreciably through the intervening years; there is now a continuous record of the civil engineering works that have been necessary to keep the narrow connecting neck of land unbroken as a convenient access to the great park that has been developed on the spit proper. Many surveys have been made and many measurements of erosion taken. Eventually, it was decided that natural replenishment of the beaches that were being so regularly eroded was insufficient, despite all that had been done in the way of constructing groins and other protective works (at an expenditure of several million dollars). In 1955 tenders were received for the dredging of over 3 million m³ (4 million yd³) of sand to form an artificial beach; this work was carried out in 1955 and 1956 at a total cost of about $2.5 million.[34.26]

The maintenance of the Presque Isle shoreline is a fine example of cooperative effort in the careful study of the natural forces involved; even with this sound approach, some idea of the costs entailed is to be gained from the two figures quoted. As an example of

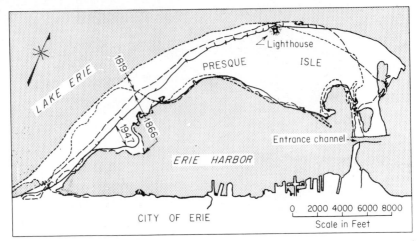

FIG. 34.13 The Harbor of Erie, Pennsylvania, in Lake Erie, another compound recurved spit formed by littoral drift.

what costs may be involved without due initial consideration of the geological implications of engineering work, there may be mentioned the destruction by a hurricane of more than one-half of U.S. Highway 90 between Pass Christian and Biloxi, Mississippi, in 1915; damage was estimated at $13 million. After the highway had been replaced, the local county authority constructed a seawall 38 km (24 mi) long, adjacent to the road, at a cost of $3.4 million. The beach disappeared completely in a very short time; fill behind the wall escaped, and the highway was again threatened. The U.S. Corps of Engineers was asked to help in the restoration work, but before it could provide its expert assistance, another hurricane occurred and did more damage. The estimated cost of the repair work eventually carried out was about one-third of the original total cost of the wall. The beach was replaced by pumping. It is estimated that 25,000 m³ (32,500 yd³) is lost from the beach every year, but in view of the large amount of pumped fill placed in 1951, routine maintenance has been sufficient to keep it in good condition.[34.27]

34.7 LITTORAL DRIFT

It is not necessary to see the works of the civil engineer to appreciate the existence of the extensive movement of sand in regular and steady manner along many stretches of coast, a movement generally known as *littoral drift*. Its existence is obvious, owing to the regular accretion of products of erosion in not a few coastal locations. At river mouths, conditions are especially favorable for its observation; one peculiar result of the interaction of drift and estuarine conditions is the *barachois* (a French word meaning sand and gravel bars), which are a prominent feature of certain parts of the Atlantic coast of Canada. The records of physical geology contain many examples of great interest based on observations of coastal conditions and development. Valuable information is also presented in studies undertaken in conjunction with engineering works, such as the extension of Rockaway Point, Long Island, a sandspit near New York (Fig. 34.14). The point of this spit has advanced at an average rate (at least between 1835 and 1927) of more than 60 m (200 ft) per year. When the progression of the point was halted by a rock-mound jetty, 230,000 m³ (300,000 yd³) of sand was found to have accumulated behind it in the course of little more than a year.[34.28]

FIG. 34.14 Accumulation of sand at the east jetty, Rockaway Point, Long Island, New York.

With the advent of man-made structures projecting out into the sea, littoral drift has come to be of great practical importance, usually as a difficult problem to be faced but occasionally as a factor beneficial to coastal development. On any section of coast where there is appreciable drift (and this includes most sand beaches), the projection of any solid structure below mean tide level will result in the building up of sand or other drift material on one side of the structure and a corresponding, although not necessarily equal, erosion of material from the other side. If the drift is of any magnitude, the serious effects of the action will be readily apparent, especially in the case of works protecting harbor entrances. The possibility of trapping drifting sand by deliberately stopping littoral drift in order to build up a good beach section will be equally clear.

The coasts and major harbors of India have provided some of the most valuable information yet available on this important matter, much having been obtained as the result of bitter experience which has been faithfully recorded for the benefit of the profession generally. Thus it has been found that at Madras on the east coast of India, where the sand travels from north to south, approximately 1 million tons of sand pass any given spot on the coast each year. Similar conditions exist on the east coast of South Africa; at Durban, in particular, great difficulties have been experienced in maintaining the harbor entrance.

Much of this accumulated experience was applied to the design of the harbor of Vizagapatam between Madras and Calcutta on the east coast of India in the late 1920s. Initial

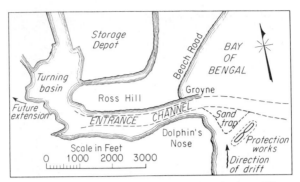

FIG. 34.15 Sketch map of the entrance to the harbor of Vizagapatam, India, showing arrangement used to deal with littoral drift.

studies suggested that no trouble would be encountered with drift, but the start of dredging operations at the harbor entrance soon demonstrated that conditions similar to those at Madras existed. Eventually, it was found that (1) the volume of sand traveling along the coast amounted to 1 million tons per year; (2) the travel took place largely in shallow water and was generally confined to a zone 180 m (600 ft) wide; and (3) the travel was primarily the result of wave action, although sometimes it was accentuated by shore currents. At Madras and Durban, erosion corresponding to the drift deflected by harbor works had occurred, and hundreds of acres of agricultural land had been lost in this way. At Durban, 535,000 m³ (700,000 yd³) of sand was required to restore an eroded foreshore to its original state. This aspect of littoral drift also had to be considered in connection with the Vizagapatam Harbor entrance. Eventually, a scale model was constructed by means of which the action of waves of varying degree on the coastal section was noted. This gave valuable indications of what protective works could be effectively carried out; the final solution is shown in Fig. 34.15.

The breakwater (consisting, incidentally, of two scuttled ships loaded with rubble) ensures deflection of the littoral drift between it and the coastline; therefore, the drift has to fall into the dredged sand trap on the inner side of the entrance channel; from here the sand is excavated by a dredge working under the lee of the protective breakwater (Fig.

FIG. 34.16 Protective breakwater, formed by two scuttled steamboats, at Vizagapatam Harbor, India, the purpose of which is described in the text.

34.16). This discharges the sand onto the other side of the entrance channel, whence it is gradually moved away by the erosive drift action.[34.29]

This example from engineering experience in India has been cited not only because of its intrinsic interest but because it illustrates a rather perplexing but certainly unfortunate break in communication between civil engineers of different countries. British engineers working in India for more than a century after the start of the development of that great subcontinent recorded in a quite remarkable way their varied experience, generally in Indian and British publications. Their papers on coastal engineering were notable, but only rarely does one see reference to them in North American publications. As recently as 1959, a paper to the American Society of Civil Engineers (ASCE) contained the statement that "by-passing . . . was suggested for the first time at the Navigation Congress at Brussels in 1935." There cannot have been anyone at that conference who had heard of Vizagapatam.

In another recent ASCE paper, however, the following quotation from a paper to the (British) Institution of Civil Engineers given by Sir Francis Spring in 1912 is cited with appreciation. It is worthy of repeated study by all concerned with coastal engineering work.

> The chief lesson to be learned from a study of sand-travel at Madras appears to be that it is absolutely necessary, when an engineer is called on to advise about a work situation, on a coast where there is any suspicion of travelling sand, mud, or shingle, that he should be allowed adequate time to make observations, conducted with due precautions, of the directions and causes of travel; and that if he should arrive at the conclusion that such travel is likely to affect the proposed works in the near or distant future, he ought at least to acquaint his employers with what he conceives to be the broad general facts of the case and their probable financial effects, whether on the proposed works or on other works or interests.[34.30]

This sound advice of many decades ago could have saved untold trouble and expense if it had been more widely known and heeded.

The hydrodynamic aspects of littoral drift have now been carefully studied and the general character of the movement is well understood. The influence of the wind has been shown to be dominant. This is well illustrated on stretches of coast where the direction of drift is found to change, generally at a well-defined location. In the case of the north shore of Lake Ontario (in the vicinity of Whitby), this location can be linked directly with the direction of the prevailing wind on the lake and the corresponding fetches on the Ontario shore. In the case of the west shore of Lake Michigan, the change appears to occur in the vicinity of 55th Street, Chicago. The source of the material constituting the drift always provides a geological problem of interest and significance. If there are cliffs that are obviously being eroded in the vicinity of the area of accretion and in the direction from which the drift is coming, they will clearly be a first source. Not so obvious, but probably more important, is the sediment brought down by rivers and streams discharging near the area under study. It has been estimated, for example, that the discharge of streams and rivers into the Pacific Ocean between Point Conception and the Ventura River amounts to almost 1.5 million m^3 (2 million yd^3) of sediment annually, and much of this moves steadily along the coast. If the supply from either of these sources—cliffs or streams—is interfered with, as can so easily be done by a misplaced engineering structure built on the beach, trouble may soon develop.

There are many stretches of sand beach, the progressive erosion of which will interfere with either recreational facilities or developed property; these beaches must therefore be conserved as much as possible. When erosion is the result of continued littoral drift, the use of groins is often the most satisfactory solution. *Groins* are wall structures generally built at right angles to the shoreline and extending below low-water line; their action

34-26 CIVIL ENGINEERING WORKS

interferes to such an extent with the travel of drift that the sand is trapped against the walls and then gradually accumulates until a stable beach is formed. The shores of many parts of the British Isles have been protected in this way, as has much of the low-lying eastern seaboard of the United States. Erosion of the coasts of Florida and New Jersey in particular has been severe; records of the New Jersey coastline show an average recession of about 60 cm (2 ft) per year (45.9 m or 153 ft between 1840 and 1920), and figures for Florida are similar. Problems arising on these coasts were among those considered and investigated by the U.S. Beach Erosion Board; as a result, specific recommendations regarding groin location were drawn up. It was found, for example, that groins need not extend higher than the level of uprush at high water; construction at right angles to the coastline was found to be beneficial. A spacing of 1½ to 4 times the length from the bulkhead or shoreline was tentatively suggested, the longer spacing being successful when the volume of drift was large, and vice versa.

Almost all older installations of groins used walls that were solid in construction, providing a complete barrier to material coming into contact with them. In more recent years, the practice has developed in some locations of constructing groins with "windows," short sections of wall constructed to a lesser height than the main section of the groin, with the intention of allowing some of the drift to pass the groin while retaining sufficient to maintain the beach. Another method of achieving the same objective is to use permeable groins, of which there are a number of types now available, some patented. A notable installation of this type of groin is at Sheridan Park, Wisconsin, on the west coast of Lake Michigan, where erosion was taking place at the rate of 60 cm (2 ft) per year until corrective measures were undertaken in 1933. The groin installation then constructed has served well in the intervening years.[34.31] It is illustrated in Fig. 34.17.

It must be emphasized, however, that the solution to each individual coastal problem is to be found only by making the most exhaustive field studies at the site and by comparing the situation thus determined with the experience of others who have had to deal with similar problems. No two coastal-protection problems will be exactly similar; possible local variations must continually be kept in view. Careful study of available records is, however, as useful in this branch of applied geology as anywhere. The Beach Erosion Board of the United States built up an impressive record of publications now available for consultation. The experience of other nations should not be forgotten, particularly the records of British engineers, since they have now documented experience extending back well over a century for protection works along a coastline presenting most varied geolog-

FIG. 34.17 Permeable groins in front of Sheridan Park, Cudahy, Wisconsin, showing their effect after 40 years' service.

ical conditions. Figure 34.18 is included to show a typical example of British coastal-protection work; the paper in which it was first reproduced can be recommended as a guide to good British practice.[34.32]

The ingenious solution adopted at Vizagapatam for bypassing the sand brought along the coast by littoral drift is but one example of the application of this idea whenever a harbor entrance has to be kept open on a coast featured by littoral drift. As a further example, the bypassing of the entrance to Lake Worth may be briefly described. The inlet gives access to the harbor of Palm Beach, some 113 km (70 mi) north of Miami; it was constructed by private interests in 1945; its 90-m (300-ft) navigable channel is protected by rock-mound jetties. Sand started to accumulate against the north jetty soon after completion, and beaches between the Lake Worth inlet and that at the South Lake Worth inlet (25 km or 15½ mi to the south) were correspondingly depleted; naturally, the effect was serious in view of the popularity of this area.

Various studies were made, and large quantities of sand were pumped into stockpiles on the beach on several occasions. Observations showed that the beaches could be replenished, at least in part, by periodic nourishment at the rate of about 150,000 m³ (200,000 yd³) a year. Successful operation of a small bypassing sand-pumping plant at the south inlet, with a capacity of 60,000 m³ (80,000 yd³) per year, installed in 1937, focused attention upon the possibility of using a larger pumping installation. Studies proved the suitability of this solution, but local difficulties (such as not drawing any sand from that

FIG. 34.18 Coastal-protection works at Sheringham, Norfolk, England, showing clearly the action of the groins.

34-28 CIVIL ENGINEERING WORKS

FIG. 34.19 Sand-bypassing installation at the entrance to Palm Beach Harbor, Florida, the pumping plant transferring sand brought by littoral drift from the right, across the harbor entrance, to the beach off left.

already built up around the north jetty and making sure that local residents would not hear any noise from the sand movement in the suction or discharge pipes) made the task of the consulting engineers somewhat difficult. A monolithic concrete pumphouse was constructed, equipped with a 400-hp electric motor-pump installation with a 30-cm (12-in) suction inlet and 25-cm (10-in) -diameter discharge, all protected by an L-shaped groin (Fig. 34.19). The plant was completed in 1958; its total cost of about $500,000 is an indication of the intrinsic value of the beaches that the plant is designed to maintain by merely pumping sand from one side of a 270-m (900-ft) wide inlet to the other—vivid evidence of the dynamics of coastal protection and of the ceaseless action of the sea on the coast.[34.33]

34.8 MAINTENANCE OF TIDAL ESTUARIES

The formation and persistence of bars at the entrance to many tidal river estuaries has long been a matter of keen debate and investigation. Bars may be said to have been the object of some of the earliest hydraulic research of which record exists; the Abbot Castelli included them in the subjects that he investigated in the years around 1660. Other early Italian and French engineers devoted considerable attention to them, and many explanations of their existence have been advanced in the intervening years. Bars in tidal estuaries and river mouths may be divided into two general classes: (1) those consisting of hard material not affected by the scouring action of normal currents passing over them, and (2) those of sand, shingle, or other unconsolidated material, which, while retaining their general features, are constantly subject to alteration from effects caused by winds, waves, and varying currents. In addition, temporary bars of sand or shingle, such as the famous Chesil Bank, are occasionally heaped up by the action of waves in heavy gales and afterward displaced by currents.

An example of the first type of bar is that which existed at the entrance to Aberdeen Harbor, Scotland. It consisted of boulder clay and when removed by dredging was formed again by the piling up of sea sand from the neighboring shore. Continued dredging finally gave a permanent channel by the removal of all the sand exposed in the vicinity as it filled into the bar channel. This left exposed only boulder clay, which was not eroded to any appreciable degree by tidal currents.[34.34] Many examples of the second type of bar could be given, since details of such bars have been presented in connection with the many discussions of the general subject before technical societies. These discussions would appear to have reduced to a certainty the cause of the formation of such bars from out of the wealth of controversy that has raged about the question. It seems almost certain that sand and shingle bars across the mouths of tidal estuaries are formed by the interaction of tidal and other shore currents, although local topographical detail has considerable indirect effect. The influence of the discharge of river water into these estuaries appears to be small. An interesting fact is that, although the exact location of a river discharge channel in the sandbanks of its estuary may change appreciably, a sandbar will always exist at the entrance to any new channel, if such was the case in the original channel.

It may be fitting to describe what is at once one of the most remarkable and one of the best known of these natural impediments to navigation, that at the entrance to the river Mersey in England, on which stands Liverpool, one of the country's great ports. The river is relatively small—only 90 km (56 mi) long and draining only 4,500 km^2 (1,724 mi^2). It has a high tidal range (up to 9.6 m or 32 ft), and the lower part of the river is therefore swept by large volumes of water on every tide. The actual mouth is located in what may be called the "corner" of Liverpool Bay, in which there are certain well-defined sandbanks, some of which are high and dry at low water. Through these, the river maintains a regular channel with a depth of between 9 and 15 m (30 and 50 ft) below low-water level. The lower section of the channel has undergone considerable change of position in the last 100 years, limiting positions being 4.8 km (3 mi) apart. In every case, the channel finished in a large sandbar; that which existed before dredging started rose directly out of 15.9 and 13.8 m (53 and 46 ft) of water at low tide in the landward and seaward directions, respectively, giving a depth at the bar itself between 2.1 and 5.1 m (7 and 17 ft). Flood conditions in the upper reaches of the river appear to have the effect of increasing the available depth of water.

The interference which this obstruction meant to navigation can well be appreciated; but it was only in 1890, after exhaustive studies, that dredging across the bar started experimentally. The success obtained in deepening the bar brought into prominence other shoals in the main approach channel, and it was foreseen that if the depth of this channel throughout its length was to be commensurate with the deepening that had been effected on the bar, considerable further dredging would be necessary. For some years, therefore, extensive and systematic dredging was practiced; and since the desired depth in the main channel was achieved, the natural tendency to deposit on the former site of the bar has been easily countered by occasional dredging (Fig. 34.20). Inside the bar, the channel has been stabilized by the construction of large rock revetments or training walls.[34.35]

Although dredging is often an essential part of improvement work, it deals only with an effect, whereas the training walls involve an attempt to modify a cause of silting or shoal formation. The general function of training walls is to train a river in a permanent course in which its own velocity will have the necessary scouring action. The use of training works is a more hazardous operation than the initiation of dredging, since it involves large expenditures and is liable to make matters worse instead of better if it is not successful. As an illustration, the development of the entrance to New York Harbor may be mentioned. Four long shoals obstructed the main ship channel and Gedney Channel in providing access into the lower bay from the ocean. One proposed form of improvement was by dredging, maintaining this channel, should it be necessary, either by periodic dredging or by contracting the entrance by the construction of a dike running across the shoals from the Coney Island side. The estimated cost for obtaining the dredged channel

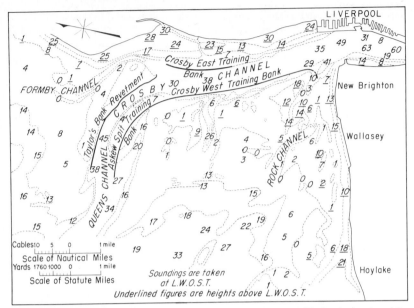

FIG. 34.20 Liverpool Bay and the estuary of the river Mersey, England, showing ship channel and training walls.

was almost $1.5 million, with a total cost of improvement, should the contraction works be necessary, estimated at between $5 million and $6 million. Between 1884 and 1890, after much discussion and some trials of dredging, the proposal for the dike was discarded and the channel dredged to 9 m (30 ft), saving approximately $4 million. Since that time a new channel 600 m (2,000 ft) wide and 13.5 m (45 ft) deep has been dredged from the main entrance to New York Harbor. Practically no maintenance is required on this channel owing to the normal scour action of the lower section of the Hudson River.[34.36]

A sidelight on the second point mentioned, and an incident of general interest in connection with tidal rivers, is provided by some observations made of the river Garonne, France, at the end of the last century. A steamer was sunk by collision at the mouth of the river opposite Le Verdon and rested with her keel at the bottom of the channel; masts and funnel alone showed at low water. On examination, it was found that the vessel was completely buried in sand and that the sandbank extended 91 m (300 ft) fore and aft and 45.7 m (150 ft) on either side. Subsequently, a second examination showed that the sandbank had completely disappeared; the hull was unsupported at the ends and rested on sand only in the central section. Further study showed that the sandbank formed and disappeared with every tide; the first examination had been made at the end of the ebb tide and the second at the end of the flood; the ebb deposited the sand, and the flood removed it. No more striking evidence of sand travel with tidal currents can be imagined; it serves also to illustrate what troubles can be invoked by a training wall not correctly located.[34.37]

Many rivers and river entrance channels have been successfully developed and deepened by the judicious use of training works. Liverpool is a case in point, but of more importance (since regular dredging is not involved) is the mouth of the Mississippi River, improved during the period from 1875 to 1879 by Gen. James B. Eads (at his own expense until the improvement had definitely satisfied governmental requirements). The method

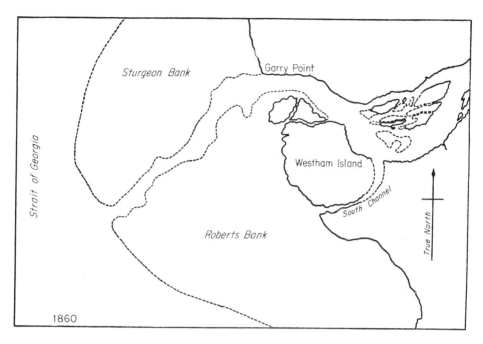
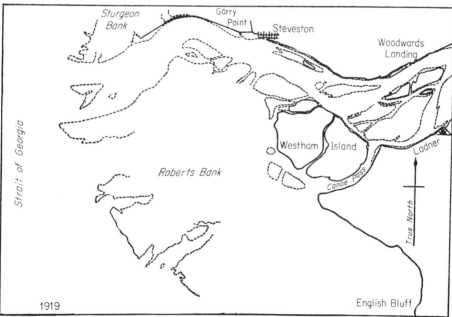

FIG. 34.21 Sketch maps showing the development of the Fraser River delta, British Columbia, and the changes in river channels between 1860 and 1919.

adopted by this distinguished American engineer was to concentrate the scour of the current in one of the "passes" of the river mouth by two parallel jetties of fascine work and thus to make the water the agent for effecting the improvement of its own course. A full account of this important example of river-training work is available in a volume that has long been of value and interest to engineers.[34.38]

There may also be mentioned the case of the Fraser River in British Columbia, which, although not perhaps so important commercially as some rivers mentioned, is yet one of the great rivers of the Pacific coast of North America. It has a length of 1250 km (790 mi) and drains an area of 238,000 km^2 (91,700 mi^2); the average yearly flood discharge is 10,700 cms (380,000 cfs). The river passes through drift deposits and is therefore fairly heavily charged with sediment. This has resulted in the formation of an unusual delta, the outbuilding of which occurs because the river currents dominate the tidal currents and because the wave action is not great (Fig. 34.21). The tides and tidal currents have had the effect of giving an unusual form to the seaward part of the delta and of introducing unusual and complex problems into the successful training of the estuary of the river. Borings have disclosed the depth of the deltaic formation as at least 240 m (800 ft), and evidence has also been found of the formation of calcareous sandstone in relatively recent years in the delta. The main navigation channel is maintained by extensive training walls and by dredging; in dredging a bar at the mouth of one of the arms of the delta, a deposit of this sandstone had to be cut through by the dredge. The training works are in the charge of engineers of the Department of Public Works, Canada. In connection with their work, the engineers have cooperated with the Geological Survey of Canada; a most useful report was one early result of this cooperation.[34.39]

Local conditions affecting tidal estuaries are so variable that detailed study of each case is usually required. In recent years, specially constructed scale models have been used in many such studies; this is one of the most interesting and important branches of civil engineering in which models have proved of assistance. By the use of the principles of hydraulic similarity, it is possible to construct a model of an estuary to such a distorted linear scale that observation in a vertical direction can accurately be made without the use of a model too large in plan. By extension of the similarity to the scale of time, tidal effects occurring over long periods of years can be reproduced in a matter of days. This branch of model work appears to have been first used in France in 1875, when a model was constructed to study the action of certain proposals for improving the river Garonne between Bordeaux and the sea. The first real attempt to apply scientific principles in the design of models was made in 1885 by Prof. Osborne Reynolds at the University of Manchester, and his work is being continued today in leading hydraulic-research laboratories throughout the world. It must be emphasized, however, that even the best of model studies are only a supplement to the most meticulous field investigations; Nature always has to be observed with the greatest care before any change of her natural order is contemplated.

34.9 DREDGING

Dredging is the term usually applied to any process of open excavation under water. Inevitably it involves the use of floating equipment, many of the larger dredgers now in use being full-sized oceangoing vessels. The simplest form of equipment consists of barge-mounted mobile cranes, followed by a great variety of craft ranging up to the largest type of suction dredge. These usually discharge into carrying vessels alongside, but some dredges are self-contained, equipped with their own holds for carrying dredged material to disposal areas. Whatever the equipment may be, it is clear that a thorough knowledge of the material to be dredged is a prerequisite for any successful dredging operation.

Records of dredging practice contain descriptions of many projects with unusual geological interest. One of these, a combined flood-protection and navigation project for waterways in Florida, was initiated in 1930 to prevent a repetition of the disastrous floods of 1926 and 1928. Included in the project was the construction of 106 km (66 mi) of levees around Lake Okeechobee. The material for the levees was to be obtained from the bed of the lake. Much of Florida is geologically new country; the bed of the lake was still in process of formation. Bed materials vary widely, and uniform deposits are exceptional; but under the recent deposits are strata of marl and limestone of varying thickness. The recent deposits often contain considerable quantities of seashells, sand, and other material so finely ground as to be almost colloidal. From this brief description, it will be seen that the material was most unfavorable to dredging operations. Almost all the contractors who started on the work elected to use draglines for excavation, but later experiments with a powerful hydraulic dredge proved so successful that other dredgers were brought in. Special cutters had to be used at the end of the suction pipes; however, good progress was made at low cost, and the resulting levees were more satisfactory than those made with the draglines. The marl and limestone layers had to be broken initially by drilling and blasting, but the dredging equipment handled all material thereafter. The relatively new deposits proved their geological youth by the way in which they could be disintegrated.[34.40]

This example emphasizes the fact that dredging is really only open excavation carried out under water; the submarine character of the work permits the use of floating equipment with its attendant economies. The basic principles upon which the work must be carried out remain unchanged. Adequate preliminary investigation is even more important for this kind of work than for excavation work in the dry, since the material to be removed cannot be seen in advance. Removal of unconsolidated material usually revolves around the selection and the availability of suitable floating equipment for the dredging work. Suction dredgers fitted with cutter heads for disturbing the material to be excavated are perhaps the most common type in wide use; ladder dredgers are excellent for special purposes, and large dipper dredgers (in effect, excavating shovels mounted on barges) are suitable for shallow and heavy work. Of special importance is the possible presence of boulders. Suction dredgers should always be fitted with a rock catcher in the suction line; this is a simple trap device for catching all boulders and rock fragments, which may be sucked up the discharge line, before they get into the pumping unit and possibly cause damage to the impeller. One of the writers has had experience on two large dredging jobs, one in an inland lake and the other in the Gulf of St. Lawrence, where the only real problems encountered arose from the presence of boulders; in each case boulders were anticipated, but they still caused trouble with the dredging machinery.[34.41]

For underwater rock excavation two methods are in general use; one involves underwater drilling and subsequent blasting with explosives, and the other makes use of a floating rock breaker. In both methods, but especially when a rock breaker is to be used, accurate knowledge of the bedding and the nature of the rock will prove to be of great value, not only in planning construction methods but also in estimating construction schedules and in specifying the work suitably in relation to possible overbreak. Many remarkable examples of underwater rock excavation testify to the difficulty of the work and its uncertainty.

One of the earliest as well as one of the most remarkable major examples of underwater rock work was the removal of Flood Rock, a rocky ledge in Long Island Sound, New York; 61,000 m^3 (80,000 yd^3) was removed in this operation in 1885. The work had to be continued in later years, and a modern channel was completed only in 1920. Another interesting piece of work was carried out in New York in 1922 when a subaqueous reef in the East River was removed to improve navigation. The work had to be carried out under the water of a tidal river with a swift current and to a final depth which was only 4.5 m (15 ft) above the roof of a busy subway tunnel. So accurately was the drilling and blasting of the crystalline Manhattan schist carried out, however, that the project was successfully completed without damage to the tunnel or interference with subway traffic. A feature of special interest was the extensive use of field seismographs to check on the extent of the vibrations caused by the blasting.[34.42]

The importance of the nature of the rock to be excavated in underwater work is well illustrated by experience obtained during rock-dredging operations in the harbor of St. Helier, Jersey, one of the Channel islands. A rock breaker equipped with a 10,000-kg (22,000-lb) ram was used for breaking the rock of the local seabed which varied from a syenitic granite to a true diorite. In the harbor area, the rock was badly fissured; but outside the harbor entrance channel, it was a purer and sounder diorite. The contractors estimated that the latter material cost five times as much to break as the former, even though the two were found in such close proximity.[34.43] The type of rock may also dictate the excavation method to be adopted, as is shown by a dredging contract carried out at Sunderland, England. Over 76,400 m³ (100,000 yd³) of rock was removed; work was carried to a maximum depth of 13.2 m (44 ft) without the use of blasting or a rock breaker. This was possible because all the rock was a loosely bedded limestone, varying from a few centimeters to 60 cm (2 ft) in thickness, and it was broken up by the dipper of a large bucket dredger digging underneath the upper projecting stratum on the working face.[34.43] Some of the most extensive work in removing underwater rock was necessitated by the increase in the depth of channels in the St. Lawrence River associated with the international Power and Seaway Project of the 1950s. Much of the dredging fleet of the entire Great Lakes area was assembled for this operation; some new equipment was developed. Figure 34.22 shows the rather spectacular 20-frame drill rig used by Marine Operators for rock drilling in the Amherstburg Channel of the Detroit River in 1957.[34.44]

Under the exigencies of war, the same kind of equipment, but on a vastly different scale, was frequently developed from existing land equipment. The Seabees of the U.S. Navy, for example, were faced with the job of removing 17,000 m³ (22,000 yd³) of coral, ranging from soft to hard, in depths of water up to 5.4 m (18 ft), at a harbor in the Solomon Islands. Standard Navy pontoons were lashed together to form a barge large enough to mount three standard 10-ton cranes with 10.5-m (35-ft) swinging leads hung from 12-m (40-ft) booms. Pile-driving equipment was used with the leads to drive steel pipes into the coral, in which dynamite charges were then set; the barge was moved 150 m (500 ft) away for the firing and then returned so that the blasted rock could be removed by a 1½-

FIG 34.22 Multiple-drill barge for rock excavation at work in the Detroit River near Amherstburg, Ontario, in connection with the St. Lawrence Seaway and Power Project.

cubic-yard clamshell bucket, operated by a 25-ton crawler crane sitting on the bow of the work barge. Radio equipment, today not infrequently used on large construction jobs, was an innovation on this job; installed by the Signal Corps, it enabled the men on the barge to maintain contact with the shore and also with the crew at work surveying the channel for which the blasting was carried out. In blasting through coral reefs in Samar, Philippine Islands, the Seabees adopted an alternative, and even simpler, technique—that of placing dynamite charges right on the surface of the coral, which was to be removed with the help of divers, and firing the charges without the aid of any equipment. Using 60 percent dynamite when available, they developed a pattern of blasting that could be varied according to the type of reef and the character of the coral. Although uneconomical in the use of explosive (requiring just over 2.4 kg/m^3 or 4 lb/yd^3 of dynamite as compared with an average working figure of 0.6 kg/m^3 or 1 lb/yd^3), the method was quick and was used when excavation had to be carried out without any suitable equipment near at hand.[34.45]

34.10 SOME SUBMARINE PROBLEMS

Marine works designed and constructed by civil engineers are usually close to or on the shore; even those few isolated structures built offshore are necessarily in relatively shallow water. Just as engineers and geologists now study the details of the land beneath the sea, so also are engineers beginning to consider what may properly be called "submarine problems." Mining operations, such as those in the Cape Breton coalfield of Nova Scotia, have been conducted beneath the seabed for many years. Tunnels beneath the sea or under tidal rivers, although still not numerous, are not now the novelty they were when the Brunels, father and son, succeeded in completing the first underwater tunnel under the river Thames at Wapping in 1843. A more recent underwater tunneling operation was carried out for an entirely different purpose; this was for placing the explosive charge necessary for the removal of Ripple Rock in the coastal waters of British Columbia.

Ripple Rock was the name given to the dual-hump-shaped summit of a well-rounded "hogsback" located near the south end of Seymour Narrows in the beautiful inland passage between Vancouver Island and the mainland of Canada. Long an impediment to navigation, the rock became a hazard of increasing danger; its peaks reached to within 2.7 and 6.0 m (9 and 20 ft) of low water in a navigable channel marked by strong tidal currents and difficult to navigate even without the added problem of avoiding Ripple Rock. Elaborate test drilling showed that the rock underlying the adjoining channels, and forming Ripple Rock itself, is generally fresh basalt, with some andesite and a few bands of breccia. After much study, the engineers in charge decided to remove the rock by tunneling out to it from an adjacent island and then placing a suitable charge of explosive, large enough to blast off the entire twin peaks. A shaft, 2.1 by 5.4 m (7 by 18 ft) and 171 m (570 ft) deep, was first sunk on Maud Island. From this, a 2.1- by 2.4-m (7- by 8-ft) tunnel was driven under the channel to a point under the center of the rock. Two 2.1- by 4.5-m (7- by 15-ft) raises, each about 90 m (300 ft) long, were then driven up to within 12 m (40 ft) of each summit. Finally, about 966 m (3,220 ft) of coyote workings were driven under the rock, 1.5 by 1.2 m (5 by 4 ft) in section. A total charge of 1,240,000 kg (2,736,324 lb) of "Nitramex 2H" was then placed throughout the workings and detonated on the morning of 5 April 1958; the charge represented 4.1 lb of explosive per lb of rock and water to be removed (Fig. 34.23). The operation was entirely successful, and the hazard to navigation was completely removed at a cost somewhat in excess of $2.5 million. Scientists from a number of Canadian research and scientific organizations took many geophysical observations over a wide area at the time of the blast, so that this largest of civilian man-made explosions was very fully documented.[34.46]

34-36   CIVIL ENGINEERING WORKS

FIG. 34.23   The destruction of Ripple Rock, Inner Channel, British Columbia, on 5 April 1958.

This was a carefully planned and premeditated earth movement beneath the sea. Other submarine earth movements completely unplanned and quite unexpected have made their occurrence felt by their effects upon engineering structures. Many underwater slides have been noted by Dutch engineers in connection with their extensive coastal engineering work. Koopejan reported 224 major slides in a 65-year period involving the mass movement of about 2.3 million m³ (3 million yd³). Dr. Terzaghi has reported upon a slide in clean sand and gravel forming an underwater delta in Howe Sound, British Columbia, which took out part of a wharf and warehouse. More serious, however, have been major submarine slides in some of the fjords of western Norway, even though the physical damage they have done is relatively small. Damage of a very serious nature occurred on 18 November 1929, when more than 20 breaks developed in trans-Atlantic cables south of Newfoundland on the edge of and down the continental slope from the Grand Banks. The breaks occurred very shortly after a serious earthquake. The times and locations of the breaks followed an orderly sequence down the continental slope, suggesting a speed of about 90 km/h (60 mph), gradually decreasing along the 560 km (350 mi) over which breaks occurred, but giving an overall average velocity of about 37.5 km/h (25 mph). Kuenan has suggested that the breaks were caused by turbidity currents, but Terzaghi suggests rather that they were the result of temporary liquefaction of sediments forming the seabed.[34.47] Figure 34.24 is a simplified section along the route of one of the trans-Atlantic cables. Even allowing for the distorted scale, it typifies the remarkable configuration of this particular part of the ocean floor so that the hazards of cable laying and maintenance can well be imagined.

Cable laying may not strictly be classified as civil engineering, but the importance of the geology of the ocean floor in relation to cable laying will be obvious. A new application of oceanographic studies has therefore developed; underwater photography is a useful tool.[34.48] The significance of the movements of the seabed, already mentioned, in all such work is clearly shown by the experience on record of undersea slides at the mouth of the Magdalena River, Columbia, South America, in August 1935. The slide carried away 480 m (1,600 ft) of the western end of a breakwater and most of the long-term river bar; that same night, several hours later, tension breaks occurred in the submarine cable located 24 km (15 mi) offshore in 1,280 m (700 fathoms) of water. When the cable was brought up

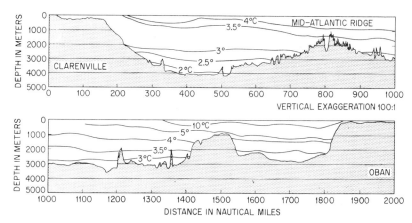

FIG. 34.24 Profile across the North Atlantic Ocean from Clarenville, Newfoundland, to Oban, Scotland, showing isotherms of mean sea temperature.

for repair, grass of the type that grew near the jetties was brought up with it.[34.48] Clearly, the geology of the seabed is of vital concern to civil engineers engaged upon marine works. This concern has taken on entirely new proportions in the 1960s and 1970s as the worldwide search for oil has extended beneath the surface of the sea, now often far from land. Offshore structures, always of interest, have therefore taken on a new importance.

34.11 OFFSHORE STRUCTURES

Lighthouses come immediately to mind as the oldest type of offshore structure. All who have sailed the seas will be familiar with some well-known "light" in isolation on a hidden shoal or rocky islet, a mere glance at which will show how dependent is the stability of the structure upon the geology of the location it has been built to indicate. The Pharos of Alexandria, built by Sostratus of Cnidus in the reign of Ptolemy II between 283 and 247 B.C., was probably the most famous of all early lighthouses. Built on a small island near the mouth of the Nile, it served until demolished by an earthquake in the thirteenth century. Of lighthouses in use today, that at Cordonan, near the mouth of the Gironde River in the Bay of Biscay, is the oldest. The first light at this location was constructed as early as 805 A.D., the second by the Black Prince; the present structure was built between 1584 and 1611. With more recent additions it is now 62 m (207 ft) high. (It is of some interest to note that the light it showed was produced until the eighteenth century by a fire of oak logs; a coal fire was used for many years after that; this is a reminder of the relative youth of modern technology such as is now applied in illumination.)

Possibly the most famous of all lighthouses is that guarding the Eddystone rocks, 22.5 km (14 mi) off the coast near Plymouth, England, a landmark (seamark?) in the English Channel familiar to all who have sailed up this historic seaway. The first light on the rocks was a timber structure built between 1695 and 1698 by Henry Winstanley but tragically washed away by a great storm in 1705. It was replaced by "Rudyard's Tower," built of oak but destroyed by fire in 1755. Then John Smeaton undertook one of his outstanding works, the construction of the first stone Eddystone Lighthouse, one of the truly pioneer structures of civil engineering. Not only did Smeaton study the geology of the rock upon which he had to build, and of the quarry from which he obtained his finished masonry, but he also studied carefully the records of Roman civil engineering. He followed Roman practice in the use of wooden wedges to fasten together the dovetails by which he interconnected each course of masonry and, possibly of more significance, in the use of pozzolan cement. Smeaton actually sent to Italy for the ingredients of the cement so that he could be sure he was using the material so well proved by the Romans. Started in 1756, the tower was first lit in 1759; it served continually until 1882 when it, in turn, was replaced by the structure to be seen today (Fig. 34.25). This was built by J. N. Douglas and is located 36 m (120 ft) south-southeast of Smeaton's tower; one reason for constructing the new tower was that Smeaton's had been undermined somewhat by the sea.[34.49]

The story of almost every lighthouse, certainly those located offshore, has its geological interest. Even those founded on sand shoals instead of on bedrock are of interest. Perhaps the most famous of these guards the entrance to the river Weser in Germany; it was built between 1881 and 1885 and exemplifies one of the first uses of iron caissons. Studies of the seabed suggested the necessary use of a mattress covered by loose rock rubble in order to eliminate dangers of possible erosion; the sand within the caisson was removed by pumping and replaced with concrete. Fortunately, there are available interesting records of lighthouse building, study of which is a very sobering experience for all who are to be concerned with marine construction. Supplementing the service of lighthouses has been, and still is, the even more isolated service of lightships to mark shoals and other shallow water features where a lighthouse of normal design could not be constructed. The pioneering efforts of the oil industry in constructing platforms (usually supported on steel tower structures) have already enabled a number of lightships to be replaced by permanent structures, firmly secured into the seabed beneath.

Small, offshore trestle structures for drilling close to shore had been used for many years; these naturally led to the use of more permanent steel structures, and these were steadily increased in size, in the depth of water in which they were built, and in the distance from shore at which they were located. The modern era may be said to have started in 1947 when the first steel oil-drilling structure out of sight of land was constructed off

FIG. 34.25 The Eddystone lighthouse in the English Channel, off Plymouth, England; the "stump" of John Smeaton's pioneer structure of 1759 may be seen on the right.

the shore of Louisiana. This was constructed in a depth of water of 6 m (20 ft); another soon followed in 15 m (50 ft) of water. Both were supported and secured by piles driven as deeply as possible into the seabed with the largest pile hammers available. Geotechnical information, in addition to a general knowledge of the geology of the seabed at all selected locations, was clearly necessary for installations of this sort. *Submarine geotechnical engineering,* now an important discipline, may be said to have started at this time.[34.50]

In view of the success of these early major offshore oil-drilling platforms, the U.S. Air Force planned to build five somewhat similar structures in shallow water along the Atlantic coast; only four were built. Somewhat naturally they were dubbed "Texas towers," their function being to support radar installations and the staffs necessary to operate them. The towers were designed only after a most thorough investigation of the rather scanty information then available concerning maximum wind and wave conditions in the open ocean. They were designed for the force of 18-m (60-ft) waves, but the platforms were set at a height to clear the crests of 27-m (90-ft) waves. They were designed for depths of water from 16.5 to 54 m (55 to 180 ft), a fact which is a reminder of the existence of the continental shelf, even as far offshore as the George's Bank Tower (about 160 km or 100 mi from the nearest land). To investigate the seabed at the selected locations, borings were put down from a patented barge arrangement, spudded to the seabed, an arrangement which was workable for water depths up to 24 m (80 ft). Soil conditions were determined to depths of 45 m (150 ft) beneath the sea floor. Generally, about 3 m (10 ft) of loose sand was found underlain by compact sand as far as the borings went, but borings at two locations encountered strata of medium-to-stiff clay interbedded with the sand at depths of 18 to 24 m (60 to 80 ft) below ocean bottom. Towers were not placed at locations where clay was encountered. At a possible location on Brown's Bank, off the southern tip of Nova Scotia, the water depth of 42 m (140 ft) made boring impracticable. The sea floor was investigated, therefore, by means of underwater photography, undertaken by the Woods Hole Oceanographic Institute; this technique showed that the surface of the ocean bottom consisted of sand and gravel with boulders up to 1.2 m (4 ft) in diameter.[34.51]

Valuable oceanographic information was obtained as an incidental service from the Texas towers, but their initial good service was marred by damage to Tower 4 by both Hurricane Daisy in 1958 and again by Hurricane Donna in 1960, the tower collapsing in January 1961 with serious loss of life. There have been a few corresponding failures with oil-drilling rigs, especially in bad storms, but the careful study made of all such happenings has aided the steady progress of safe design. Accordingly, the U.S. Coast Guard commenced in 1964 to replace some of its lightships along the Atlantic coast with offshore steel towers. Eight such towers have now been in service since the late 1960s, giving most satisfactory service. When it was decided to replace a lightship with a tower, a detailed survey of the surrounding seabed was first carried out, to a radius of 1.6 km (1 mi). Any sharp changes in the surface of the bottom were thus detected. Borings were then taken at selected locations, to a depth of at least 48 m (160 ft) unless bedrock was encountered above this. Towers are used only when the materials in the seabed are sand and/or gravel, but not when it is all clay. Typical is the Chesapeake light station, now standing in 19.2 m (64 ft) of water, 21 km (14 mi) east of Cape Henry, Virginia (Fig. 34.26). Soils in which the 97.5-cm (33-in) -diameter pipe piles were driven consist mainly of sand, with some sand-clay and clay with shell fragments, one layer of cobbles being encountered.[34.52]

The most remarkable developments, however, have taken place in the design and construction of offshore platforms for oil drilling, initially in the Gulf of Mexico but today in locations all round the world. Installations in water as deep as 61 m (200 ft) were made in the 1960s. This was remarkable at the time, but in 1976 the Hondo platform was installed in Santa Barbara Channel off the coast of California in 259 m (850 ft) of water. Twelve structures, designed to withstand massive ice pressures, were erected as early as 1964 in

34-40 CIVIL ENGINEERING WORKS

FIG. 34.26 Chesapeake Bay lightship and the permanent structure that has now replaced it, off the coast of Maryland and Virginia.

Cook Strait off the coast of Alaska. It has been in the North Sea, however, that some of the most remarkable developments have taken place, pride of place probably going to the Ekofisk complex, a 93-m (304-ft) -diameter, reinforced-concrete structure weighing 444 million kg (489,000 tons) and floated into place after being built in Norway (Fig. 34.27). It rests safely in 70 m (230 ft) of water.[34.53]

These are engineering achievements, but all depend upon knowledge of the seabed for their safe performance. The geology of any one part of the bed of the ocean gradually becomes known as the proprietary information gathered by individual operators is released for public use, as has been done most generously. Most of the geological formations encountered are of Pleistocene age. Unusual features that have been encountered have included gas-charged sediments beneath the slope of the shelf and soft-grained calcareous sand in Bass Strait between Victoria and Tasmania, Australia.

Even when records of the local geology are available, geotechnical investigations are necessary in great detail for every installation. A fine review of the development of this special branch of geotechnique has been published by Focht and Kraft.[34.50] There will also be found useful records in the volumes of proceedings of the many conferences on offshore engineering, now being held in most of the countries involved in offshore work; a guide to these will be found at the end of the chapter.

Indicative of the importance attached to geology in all the preliminary investigations for offshore structures is the fact that in 1975 the (British) Institute of Geological Sciences, the name by which the Geological Survey of Great Britain is now known, embarked on a detailed study of a selected area in the North Sea, due east of the city of Aberdeen, Scotland. The study served as a supplement to their previous mapping (to a scale of 1:250,000) of the United Kingdom's continental shelf. Seabed samples, obtained by a variety of methods, were tested at sea and in the laboratory for basic geotechnical parameters. An engineering assessment of the Quaternary sequence on the shelf was made possible through these studies, geotechnical results being shown on small-scale maps for

FIG. 34.27 The Ekofisk oil-drilling structure being fitted out before being towed to its location in the North Sea.

depths up to 5 m (16 ft) and as tables of values for depths up to 250 m (820 ft). The detailed work east of Aberdeen has been completed over an area of 400 km^2 (155 mi^2), and a map of all relevant geotechnical information has been produced by computer analysis. The great value of this undersea geological mapping can well be imagined.[34.54]

If only to show that there are other ways of carrying out offshore drilling, there may be mentioned that on the other side of North America, off the Pacific coast of the United States between Santa Barbara and Ventura, there now stands one of the few man-made islands of any magnitude that have been constructed in the open ocean. Rincon Island was constructed for the Richfield Oil Corporation as an alternative to an open tower of the type previously described, other examples of which have been built in the Gulf of Mexico for supporting oil-drilling rigs (Fig. 34.28). The island will permit the owner company to explore, with modern "slant" drilling techniques, a large proportion of an offshore oil lease granted to it by the state of California. At the island site, water varies from 12.3 to 14.4 m (41 to 48 ft) in depth. Extensive site investigation, carried out by a variety of methods, revealed that the bottom consists of a silty sand grading into sandy silt, increasing in thickness with increasing depth of water, which varies at the island site from 4.2 to 7.5 m (14 to 25 ft). Slope of the ocean floor was found to be 3 percent. Recent shale or siltstone underlies the sediment. Sand movement on the sea floor at the site was found to be negligible; thus, the construction of a solid island structure was a feasible proposition. The completed island has an area of 2.5 ha (6.3 acres) on the ocean bed, with a working area of 0.45 ha (1.1 acres) out of a total area at elevation 16 of 0.9 ha (2.2 acres). The island was formed by means of rock embankments enclosing sand fill; the rock faces are protected from the action of the sea by 1,130 tetrapods having a total weight of 35,000 tons. Rock was obtained from a quarry on Santa Cruz Island, previously used for the construction of the Santa Barbara breakwater; it is Eocene sandstone of the Cold Water formation. It proved to be variable as quarried, but a satisfactory supply, which passed all requisite tests, was obtained. The island was completed in less than two years in 1958.[34.55]

34-42 CIVIL ENGINEERING WORKS

FIG. 34.28 Rincon Island, man-made for the support of oil-exploration installations off the coast of California.

34.12 CONCLUSION

This chapter must end as it began—with a reminder of the sea's awesome power, which dominates all marine design. There are few better ways of doing this than by recounting the tragedy of Hallsands, a name that civil engineers concerned with marine works should always remember. Hallsands was the name of a small fishing village on the south coast of Devonshire in England. It was located above a fine shingle beach and backed by a steep rock cliff. Records showed that the shingle had been stable for at least a century, when, in 1897, the great dockyard at nearby Devonport had to be enlarged. Large quantities of material were required for the construction work. Permission was given to the responsible contractor to dredge shingle from below the beach at Hallsands. The villagers, 126 people living in Hallsands' 37 houses, protested, but it was not until 1902 and after the removal of 660,000 tons that dredging was stopped. But the damage had been done. Despite construction of a protection wall, the beach level had dropped by 5.7 m (18 ft); without natural protection, heavy waves washed right over the wall.

Some residents of the village moved away, but most remained in their homes until, between 25 and 27 January 1917, there occurred a combination of strong northeast winds and high spring tides which led to such violent storm conditions that 24 of the remaining 26 houses disappeared into the sea.[34.56] The village of Hallsands has disappeared, not from natural causes but because of the work of man carried out without full appreciation of the power of the sea.

34.13 REFERENCES

34.1 Rachel L. Carson, *The Sea Around Us,* Oxford University Press, New York, 1951.

34.2 C. Mair, *A Star for Seamen,* J. Murray, London, 1978.

34.3 Carson, op. cit., p. 122.

34.4 T. Stevenson, *The Design and Construction of Harbours,* 3rd ed., A. & C. Black, London, 1886, p. 49.

34.5 I. A. Hunt and L. Bajorunas, "The Effect of Seiches at Conneaut Harbor," *Proceedings of the American Society of Civil Engineers,* vol. 85, WW2, Paper no. 2067, New York, 1959.

34.6 F. P. Shepard, *The Earth beneath the Sea,* Johns Hopkins, Baltimore, 1959.

34.7 R. Fairbridge, "The Changing Level of the Sea," *Scientific American,* **202**:70 (1960).

34.8 "How Far to the Bottom," *The New Scientist,* **7**:515 (3 March 1960).

34.9 J. A. Steers, *The Coastline of England and Wales,* Cambridge University Press, New York, 1946; see also by same author, *The Sea Coast,* Collins, London, 1953.

34.10 *Third Report of the Royal Commission on Coast Erosion,* H. M. Stationery Office, London, 1911.

34.11 T. Sheppard, *The Lost Towns of the Yorkshire Coast . . . ,* Brown, London, 1912.

34.12 "Sutton Bridge Dock," *The Railway Magazine,* **96**:422 (1955).

34.13 C. R. S. Kirkpatrick, "The Development of Dock and Harbour Engineering," Vernon Harcourt Lecture before the Institution of Civil Engineers, London, 1926.

34.14 E. J. Buckton, "The Construction of Haifa Harbour," *Minutes of Proceedings of the Institution of Civil Engineers,* **239**:544 (1936).

34.15 G. Ellson, "Dover Train-Ferry Dock," *Journal of the Institution of Civil Engineers,* **7**:223 (1937).

34.16 A. de F. Quinn, "Great Lakes Port for Shipping Taconite Built by Ore Industry," *Engineering News-Record,* **156**:38 (18 October 1956).

34.17 N. S. Shaler, *The Geologic History of Harbors,* pt. II, U.S. Geological Survey 13th annual report, 1893, p. 99.

34.18 L. Bemelmans, *Father, Dear Father,* Viking, New York, 1953, p. 143.

34.19 W. H. Hunter, *Rivers and Estuaries,* Longmans, London, 1913.

34.20 J. B. Schijf, "Generalities on Coastal Processes and Protection," *Proceedings of the American Society of Civil Engineers,* vol. 85, WW1, Paper no. 1976, New York, 1959.

34.21 R. F. Legget, "Development of a Pulpwood Shipping Harbour, Forestville, Quebec," *The Engineering Journal,* **36**:1287–1294 (October 1953).

34.22 W. Johnston, "Consolidation of Rock Embankment to Prevent Wave Erosion," *Roads and Bridges,* **83**:67 (September 1945).

34.23 W. F. Stanton and A. G. Le Clercq, "The Improvement of the Port of Valparaiso and Extension of the Breakwater," *Minutes of Proceedings of the Institution of Civil Engineers,* **233**:109 (1931).

34.24 G. W. White and W. H. Gould, "Erosion of Lake Erie Shore," *Engineering Experiment News,* **17**:3 (1945).

34.25 Francis Fox, *Sixty-Three Years of Engineering,* J. Murray, London, 1924, p. 167.

34.26 F. H. Forney and G. A. Lynde, "Beach Protection Engineers Attempt to Outwit Nature at

Presque Isle Peninsula," *Civil Engineering,* **21**:508 (1951); see also in same journal **28**:172 (1958).

34.27 F. F. Escoffier, "Harrison County (Mississippi) Artificial Beach," *Transactions of the American Society of Civil Engineers,* **123**:817 (1958).

34.28 Beach Erosion Studies by Federal Board, *Engineering News-Record,* **111**:281 (1933).

34.29 W. C. Ash and O. B. Rattenbury, "Vizagapatam Harbour," *Journal of the Institution of Civil Engineers,* **1**:235 (1935).

34.30 Francis Spring, "Coastal Sand Travel near Madras Harbour," *Minutes of Proceedings of the Institution of Civil Engineers,* **194**:153 (1913).

34.31 E. A. Howard, "Permeable Groins of Concrete Check Beach Erosion," *Engineering News-Record,* **114**:594 (1935).

34.32 J. Duvivier, "Coast Protection: Some Recent Work on the East Coast 1942–52," *Proceedings of the Institution of Civil Engineers,* vol. 2, pt. II, London, 1953, p. 510.

34.33 C. Senour and J. E. Bardes, "Sand By-passing Plant at Lake Worth Inlet, Florida," *Proceedings of the American Society of Civil Engineers,* vol. 85, WW2, Paper no. 1980, New York, 1959.

34.34 W. H. Wheeler, *Tidal Rivers,* Longmans, London, 1893, p. 145.

34.35 Ibid., p. 365; see also C. Peel, *The Mersey Estuary,* Manchester and District Association of the Institution of Civil Engineers, 1926.

34.36 Wheeler, op. cit., p. 191; also information from Mr. Billings Wilson, assistant general manager, The Port of New York Authority, by personal communication.

34.37 Wheeler, op. cit., p. 142.

34.38 E. McHenry (ed.), *The Addresses and Papers of James B. Eads,* Slawson, St. Louis, Mo., 1884.

34.39 W. A. Johnson, *Sedimentation of the Fraser River Delta,* Memoir no. 125, Geological Survey of Canada, Ottawa, 1912.

34.40 "Cheap Levees by Hydraulic Fill at Lake Okeechobee," *Engineering News-Record,* **115**:81 (1935).

34.41 Legget, op. cit.

34.42 E. Burr, "Remove Subaqueous Ledge above Rapid Transit Tunnel," *Engineering News-Record,* **89**:1021 (1922).

34.43 R. D. Gwyther, "Improvements to St. Helier's Harbour, Jersey," *Minutes of Proceedings of the Institution of Civil Engineers,* **238**:100 (1934).

34.44 P. C. Nyzer and H. E. Hill, "Scheduling Equipment for Great Lakes Channel Dredging," *Civil Engineering,* **27**:470 (1957).

34.45 C. Phillips, "Blasting Submarine Coral," *Engineering News-Record,* **136**:779 (1946).

34.46 E. Dolmage, E. E. Mason, and J. W. Stewart, "Demolition of Ripple Rock," *Transactions of the Canadian Institute of Mining and Metallurgy,* **61**:382 (1958).

34.47 K. Terzaghi, "Varieties of Submarine Slope Failures," *Proceedings of the 8th Texas Conference on Soil Mechanics and Foundation Engineering,* Austin, 1956.

34.48 C. H. Elmendorf and B. C. Heezen, "Oceanographic Information for Engineering Submarine Cable Systems," *Bell System Technical Journal,* **36**:1047 (1957).

34.49 John Smeaton, *A Narrative of the Building and a Description of the Construction of the Edystone Lighthouse with Stone . . . ,* Hughs, London, 1791 (an engineering classic).

34.50 J. A. Focht and L. M. Kraft, "Progress in Marine Geotechnical Engineering," *Proceedings of the American Society of Civil Engineers,* **103**(GT10):1097 (1977).

34.51 P. C. Rutledge, "Design of Texas Towers Offshore Radar Stations," *Proceedings of the 8th Texas Conference on Soil Mechanics and Foundation Engineering,* Austin, 1956.

34.52 J. V. Ruffin, "Steel Offshore Towers Replace Lightships," *Civil Engineering*, **35**:72 (November 1965).

34.53 B. C. Gerwick and E. Hognstad, "Concrete Oil Storage Tank Placed on North Sea Floor," *Civil Engineering*, **43**:81–85 (August 1973).

34.54 Information in a personal communication from the Institute of Geological Sciences.

34.55 J. A. Blume and J. M. Keith, "Rincon Offshore Island and Open Causeway," *Proceedings of the American Society of Civil Engineers*, vol. 85, WW3, Paper no. 2170, New York, 1959.

34.56 Allison Wilson, "The Lesson of Hallsands," *New Scientist*, **45**:311 (12 February 1970).

Suggestions for Further Reading

During the decade prior to the publication of this Handbook there has been what can only be described as an explosion of information about marine works. Much of this relates to what is now called "ocean engineering," a field greatly stimulated by the rapid extension of offshore drilling. Some of the new publications now available recording the proceedings of the numerous conferences on the subject now held in several countries will be found to contain valuable material about the geology of the seabed and its influence on the design and construction of the new types of marine works called for by this new activity. The last reference cited above is a good illustration. The paper by Focht and Kraft is an admirable review in relatively short compass of the geotechnical aspects of offshore structures.

One of the largest of all technical and scientific meetings now regularly held is the Offshore Technology Conference, the 14th of which was held in Houston in May 1982. It is organized by the Society of Petroleum Engineers (6200 North Central Expressway, Drawer 64705, Dallas, TX 75206) with many cosponsors. Oceans '82 was the name given to the 1982 conference sponsored by the Marine Technical Society and the Council on Ocean Engineering of the Institute of Electrical and Electronic Engineers, again with many cosponsors. Over 300 papers were scheduled for presentation.

The American Society of Civil Engineers has held regular conferences on the subject of ports. It has published proceedings of the regular international conferences on *Coastal Engineering*, typical being the three volumes (of 3090 pages) for the 16th conference in 1979. It has now started regular conferences on *Coastal and Ocean Management*. The Institution of Civil Engineers (U.K.) has also published a number of valuable records in this field: *Dredging* (124 pages, 1975), *Design and Construction of Offshore Structures* (184 pages, 1977), *Sea Defence and Coast Protection Works* (240 pages, 1978), *Maintenance of Maritime Structures* (252 pages, 1978). It should not be forgotten that the earlier proceedings of these two leading civil engineering societies contain many valuable papers on marine works which are still well worthy of study.

Indicative of international interest was a conference organized as a workshop by NATO. Its proceedings, *Marine Slides and other Mass Movements* edited by S. Saxov and J. K. Nieuwenhuis, were published in 1982 as a volume of 188 pages by Plenum of New York.

Although for many years *Dock and Harbour Engineering*, published in London, was virtually alone in the periodical field, it has now been joined by a number of new journals, all of value in their several ways. *Coastal Engineering*, a quarterly published by Elsevier Scientific Publishing Co., Amsterdam, with E. W. Bijku as editor-in-chief, is of special relevance to matters discussed in this chapter.

Volume 3 of a comprehensive work called *The Sea* deals with *The Earth beneath the Sea*, the editor being M. N. Hill; this 963-page volume was published in 1963 by Wiley Interscience of New York. *Coastal Engineering* is a two-volume work by R. Silvester, published in 1974 by Elsevier of New York. *Beaches and Coasts* is one of the valuable Benchmark Paper series published by Dowden, Hutchinson and Ross of Stroudsburg, Pennsylvania, edited by C. A. M. King. *Our Changing Coastline* is the title of a 579-page volume by F. P. Shepherd and H. R. Wanless, published by McGraw-Hill in 1979, remarkable for its use of aerial photographs to illustrate the inexorable work of

Nature. *Oceanography: a View of the Earth* by M. G. Gross appeared in a third edition of 498 pages from Prentice-Hall in 1982.

Finally, if only as a reminder that the Great Lakes present their own marine problems, there may be noted two reports, copies of which are available from the government of Ontario (880 Bay Street, Toronto M7S 1NB) entitled *Great Lakes Shore Management Guide* and *Great Lakes Shore Processes and Shore Protection*.

chapter 35

LAND RECLAMATION

35.1 Zuider Zee Reclamation / 35-1
35.2 Some Smaller Examples / 35-6
35.3 Disposal of Excavated Material / 35-9
35.4 Landfill with Waste Materials / 35-12
35.5 Conclusion / 35-14
35.6 References / 35-15

The man-made offshore island just mentioned in Chap. 34 and those described in the chapter which follows are but two small examples of another part of civil engineering as practiced on or near coastlines—the reclamation of areas previously under water for use as developed land. In the constant battle against erosion on some coasts, the nature of some protective works permits their installation actually in the sea (or lake), which then provides a safe area in which fill can be placed and new land thus formed. The concept of "winning land from the sea" has been extended far beyond this simple basic idea, but in every case it is essential that full information on the subsurface conditions be accurately known in advance of any planning. It is also essential to assess the natural processes, such as littoral drift and local currents, which may affect the completed works. Examples of land reclamation from the sea are to be found in many countries. The Back Bay reclamation scheme in Bombay, India, is a notable case; the airport runway at Hong Kong provides a dramatic example. Airports in the United States and Canada have been similarly extended over land that was recently under water. Many dock and harbor projects have included the reclamation of land for port service areas, often utilizing material obtained from dredging approach channels to the dock facilities. Interesting and significant as are all such works, they pale into insignificance when compared with the work of Dutch engineers in the protection and extension of the land area of the Netherlands.

35.1 ZUIDER ZEE RECLAMATION

Forty percent of the Netherlands, known throughout history as part of the Low Countries, lies below mean sea level. Some 1,920 km (1,200 mi) of dikes protect this lovely and gracious land from the inroads of the sea, dikes built up over the centuries and the subject of many stories. Steady lowering of the level of the land, combined with a corresponding

35-2 CIVIL ENGINEERING WORKS

rise in the level of the North Sea, has given rise to a continuing battle against the encroachment of the sea into the fertile low-lying plains of Holland. Roman records mention a small freshwater lake in the area known through recent history as the Zuider Zee. Historical records of the struggle go back for at least 1,000 years, when the first dikes were built between high mounds that had been constructed as safety measures for use in times of high water. *Poldering,* or the reclaiming of land from the sea, started as early as A.D. 1200. Windmills, for pumping out areas so reclaimed, came into use about 1600. But the battle had been a losing one until very recent years. It is estimated that, from 1200 until quite recently, the sea claimed 567,000 ha (1.4 million acres), whereas reclamation along the coast and by pumping won back less than 526,000 ha (1.3 million acres). In the year

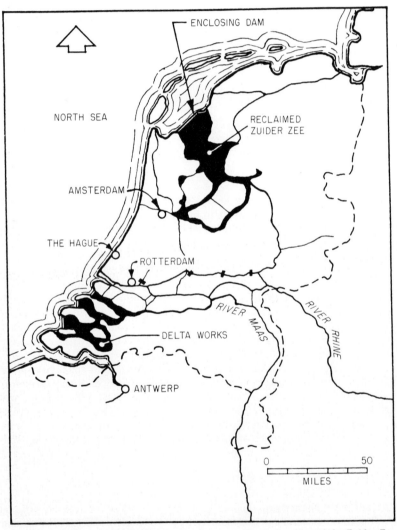

FIG. 35.1 Map of the Netherlands showing generally the location of the Zuider Zee project and the area to be enclosed by the delta works.

1421 a dreadful storm destroyed 65 villages through inundation and took the lives of 10,000 people.

In 1918 the Dutch government passed an act setting in motion the greatest land-reclamation project that the world has yet seen—the reclamation of 222,000 ha (550,000 acres) in the Zuider Zee and the transformation of this great bay into a freshwater lake with an area of almost 400,000 ha (1 million acres) (Fig. 35.1). The incidence of nine major floods in the course of the previous 40 years and the loss of over 243,000 ha (600,000 acres) spurred the Dutch nation to this great effort. When the project is complete, as it will be well before the end of the present century, the area of arable land in Holland will have been increased by 10 percent, the total area of the country by 7 percent. Of the land thus far reclaimed, 85 percent has proven to be high-quality agricultural soil. The task is being carried out in stages. The Wieringermeer Polder, the first to be drained, was put into use in 1930; it has an area of 20,000 ha (49,000 acres). The North Eastern Polder, 48,200 ha (119,000 acres), was dried out in 1942 and Eastern Flevoland, 54,000 ha (133,000 acres), in 1957 (Fig. 35.2). The Markerwaard and Southern Flevoland, having 61,000 and 40,500 ha (150,000 and 100,000 acres), respectively, are scheduled for future draining.

The entire project is being carried out along lines first proposed by a great civil engineer, Dr. Ing. C. Lely, a fine monument to whom is seen by all who visit the Zuider Zee works. Key to the entire scheme was the construction of the great enclosing dam across the northern end of the Zuider Zee. About 32 km (20 mi) long and 90 m (300 ft) wide at its crest, this earth embankment, built with the aid of fascine mattresses for protection of the sea floor, was constructed between 1927 and 1932 in the open sea. Pumping started immediately upon its completion, and the enclosed water was soon converted into the IJsselmeer, a vast, freshwater lake. Within this, further dikes were then built so that the areas they enclosed could be pumped out and the land now known as the *polders* could be reclaimed for cultivation (Fig. 35.3). The enclosing dam had the effect of pushing the

FIG. 35.2 Reclamation of the Zuider Zee, the Netherlands; the circular dike of Eastern Flevoland, before the polder was dried out, with temporary Lelystad on right.

35-4　CIVIL ENGINEERING WORKS

FIG. 35.3 Pumping the sand core of a major dike by dredgers for a main dike of the Zuider Zee works; deposition is between two impervious dikes of boulder clay.

sea back 85 km (53 mi) from the southern end of the old Zuider Zee; the shoreline of the country was thus reduced by 300 km (186 mi). The freshening of the water enclosed soon had the effect of leaching out much of the salt content of the soils to be drained. Samples of water showed salt contents of 1.39 g/L in 1943 and 0.82 g/L in 1955. Draining, cultivation, and settlement of the polders followed as quickly as possible after the drying out was complete. Today there are thriving communities, lovely buildings, tree-lined roads, and many comfortable and happy homes where but a few years ago there was the sea (Fig. 35.4). Excavation carried out in connection with the work of settlement has yielded many archaeological treasures dating back many centuries and fossil relics dating from the last glacial period; these are indirect geological results of a great geological-engineering undertaking.[35.1]

The first day of February 1953 is a date that will long be remembered by the Dutch people and, indeed, by many in other countries bordering on the North Sea, for on that day occurred a storm coupled with high-water levels that rivaled the great storm of 1421 in the havoc it wrought. In the Netherlands, 133 villages were seriously damaged by inundation and 1,800 people died. The cost of physical damage amounted to at least $400 million. Many dikes were breached. Reacting characteristically, the Dutch people put into effect, through legislation passed just 20 days after the flood, another great reclamation scheme known as the Delta Plan. The aim of this is to block off by permanent barrages, with all necessary ancillary structures, sluice gates, and pumping stations, four of the main channels of the Schelde estuary. Two main channels will be left open to serve the ports of Rotterdam and Antwerp, but these will be isolated by suitable structures from the impounded area, the water in which will gradually become fresh as the existing saltwater is pumped out and replaced from the rivers that feed into the estuary. No new land will be won, but existing fertile land will be protected and future danger from flooding averted.

Strengthening and raising existing dikes is naturally a major part of the plan; one of the islands upon which work is already being carried out is Walcheren, which achieved such ill fame as a battleground in the Second World War. In repairing the Walcheren dikes blasted by Allied bombing, Dutch engineers tried a new method, making use of some floating concrete caissons, left over (so to speak) from the Normandy landing operations of the war. Put to this constructive use, these wartime units proved serviceable (as they have frequently proved to be in North American harbor construction). The use of similar units is being followed in the implementation of the Delta Plan but on a much larger scale. Work on actual closure of some of the channels started in 1967; completion is expected in 1985, some time having been lost while the method for closing off the Ooster Schelde was debated nationally. The final decision was to leave it open to tidal action, a reversal of earlier plans made in response to fears for the ecology of the area to be enclosed. Estimates of cost for this great project have naturally varied greatly as planning has progressed, the latest figure available being a total of probably $3 billion. It is mainly a masterpiece of engineering, on a grand scale, geological conditions being generally straightforward, the seabed being sand, with more than enough sand available for pumping for construction of the artificial islands, dikes, and major dams required. The seabed has to be protected and a variety of mattresses have been tried, all based on some of the most extensive and elaborate model testing ever carried out in advance of a major civil engineering project.[35.2] And Dutch and German engineers already have their eyes upon an

FIG. 35.4 The new town of Emmeloord, the Netherlands, entirely built on reclaimed land.

even greater reclamation scheme, this time for the winning of new land, impounding the northern part of the Zuider Zee by using the Frisian Islands as the basis for an immense enclosing dam that will far exceed the existing dam in the Zuider Zee in size and daring of conception.[35.3]

One of the most remarkable features of both the Zuider Zee reclamation work and the Delta Plan has yet to be mentioned. Although land reclamation is an important element in the Zuider Zee work and shore protection a vital part of both schemes, another common feature of the greatest importance is the conversion of great areas of saltwater into corresponding areas of freshwater—the IJsselmeer and the new lake that will be formed in the Schelde estuary. The necessity for this arises from the difficulties already being experienced in Holland, despite its annual rainfall of 67.5 cm (27 in), in meeting the demands for water of the growing number of inhabitants of this heavily populated little country. To think that there is a water shortage in a land in which it is almost impossible to walk more than a few kilometers without being in sight of some body of water is, perhaps, as dramatic a piece of evidence as can be found to emphasize the critical place that water has come to occupy in the world.

The Zuider Zee works and the Delta Plan are both so vast that they put into the shade other Dutch land reclamation works, which elsewhere would be major enterprises. A new steam power plant for meeting demand anticipated from Eastern Flevoland, for example, was built on an artificial island with an area of 11 ha (28 acres) constructed in the waters of the IJssel Lake. The lakebed consists of mud overlying compact sand. Dredges first cleared a depth of 6 m (20 ft) of mud from the site and then moved into the lake to dredge sand. This was brought to the site and deposited directly from barges until the original depth of water of 9.9 m (33 ft) was reduced to about 1.5 m (5 ft). A stone retaining wall was then constructed around the perimeter of the planned island, sand being pumped into the area thus enclosed. About 1.5 million m^3 (2 million yd^3) of sand fill was thus used to form the island, suitable protection being provided to prevent erosion of its banks. This was a "small" Dutch job.[35.4]

With similar geological conditions to assist economical construction, the port of Rotterdam increased its already extensive facilities in the late 1960s by creating a new port area built out into the North Sea opposite the Hook of Holland. The new port was developed entirely on new land, about 2,200 ha (5,500 acres) being formed on the west side of the mouth of the Maas River by pumping, at a total cost of about \$330 million. The ultimate cost of about \$35,000 for each new acre was regarded as reasonably economical.[35.5] Looking at this busy area today, it is difficult to appreciate that just a few years ago it was open sea. With such examples in use, it is perhaps not too surprising to find that plans have been developed for a 10-km^2 (4-mi^2) area in the North Sea near The Hague to be constructed along similar lines for providing homes for 40,000 people, such is the pressure on land use in the Netherlands.[35.6]

35.2 SOME SMALLER EXAMPLES

Just as with other branches of civil engineering, the great size of the land reclamation projects of the Netherlands permitted the use of large and sophisticated equipment for construction operations which are essentially quite simple, capable of being carried out on a small scale with the simplest of equipment. Brief accounts of major projects, however, not only show what can be done when geological conditions are favorable (as they are in Holland) but can serve as a reminder of the possibilities which quite small sites, and demands, may present. Although the projects so far described have all been carried out in water, there are also extensive land reclamation schemes which have made usable land of what was otherwise unusable, such as swampland. Interference with wetlands can only be done after proper ecological studies, but when approved and carried out with care, such interference can be socially valuable.

As an illustration of a combined dredging and filling operation which illustrates this well, there may be mentioned the preparation of an area for a new central market in San Juan Bay, Puerto Rico. The 40-ha (100-acre) site adjoins the mouth of the Puerto Nuevo River. When the river was dredged and rerouted in order to obviate troubles with flooding, some of the area later developed for the market was used as a spoil area for dredged material. The site was an old mangrove swamp; soft organic silt extended from the surface to firm residual clay 7.5 to 12.0 m (25 to 40 ft) below. The area was therefore first completely cleared, and a 1.2-m (4-ft) sand blanket was pumped over it; this was used as a working area, and 33,500 sand drains, 45 cm (18 in) in diameter, were installed to predrain the area. In order to speed up the required consolidation, the areas required for specific and immediate building projects were overloaded (i.e., covered with an excess amount of sand fill, which was later removed when the additional weight had caused the necessary consolidation of the silt). The amount of overloading was predetermined on the basis of the time of dredging, the duration of the sand-drain operation at each site, and the date at which it was required for building purposes. The extent of this site operation is indicated by the cost of almost $6 million for the first two phases, $912,000 for the river-diversion work, and $4.9 million for the fill and stabilization work.[35.7]

Quite different, but still land reclamation, was the construction of the Century III shopping mall at West Miffling, Pennsylvania, 19 km (12 mi) from Pittsburgh. Swampland was again used, but this time the fill for the necessary stable site was obtained from a slag dump (of the U.S. Steel Corporation, which paid most of the costs). The site overlies the famous Pittsburgh vein, honeycombed with tunnels of old coal workings. Excavation was carried down to the bottom of the seam; remaining tunnels were filled with a mixture of fly ash and cement; over this initial fill a pad of slag was placed and compacted with vibrating rollers to give adequate bearing for the 190 shops and 6,000 parking spaces, all at a cost of $100 million when the mall was complete.[35.8]

One of the best-known areas of marginal land in North America, not yet extensively used, is that known as Hackensack Meadows, a large area of low-lying land between the cities of Newark and New York, bounded on the east by the high land of the New Jersey Palisades. All travelers going south from New York by road, rail, or air (if visibility permits) see this vast area of swampland that has been thus described:

> ... once a glacial lake, the Meadows is an inhospitable mix of sand, silt and clay topped with a layer of black organic mud, green reeds and swamp grass. For generations the area has been used as a refuse dump for the New York-New Jersey metropolitan area, so a thick layer of garbage blankets much of the marsh.[35.9]

Much test boring has been done over the swamp. Borings have revealed the existence of two pre-Pleistocene buried valleys cut more than 30 m (100 ft) below present sea level into the underlying Triassic sandstone, one valley extending under the city of Newark, the other underlying the eastern margin of the Meadows. Light buildings have been erected on the Meadows in its northern section. Test borings for these revealed between 1.8 and 2.1 m (6 to 7 ft) of highly compressible surface tidal deposit, "meadowmat," or fibrous peat, overlying a thin stratum of stiff varved clay below which is a deep (7.5- to 22.5-m or 25- to 75-ft) deposit of soft-to-firm varved clay. Compact glacial till then overlies the sandstone bedrock. On these foundation beds one-storey buildings have been founded, and have performed well, by being placed on specially selected and carefully placed fill resting directly on the stiff varved clay, after removal of the fibrous material (Fig. 35.5). This shows what can be done, based on careful study of the properties of the

FIG. 35.5 Building on the Hackensack Meadows, New Jersey; the Environmental Center at an early stage of construction.

materials of the foundation beds, with a minimum disturbance of natural conditions, a tribute to the work of the Hackensack Meadowlands Development Commission.[35.9]

Already the Meadowlands sports complex has been completed; it utilizes 200 ha (488 acres) of which 54 ha (130 acres) is a landscaped area for the parking of 20,000 automobiles and 400 buses. Reclamation was achieved by cofferdamming the race track and stadium areas and placing sand fill. This was initially dredged in the open sea from the Ambrose Channel, barged up the Hudson River, and then pumped to the site. Future plans cover the orderly development of 8,100 ha (20,000 acres).[35.10]

Reverting briefly to new land formed by reclaiming areas initially under water, this type of reclamation is not confined to seacoasts. A singularly interesting example is to be

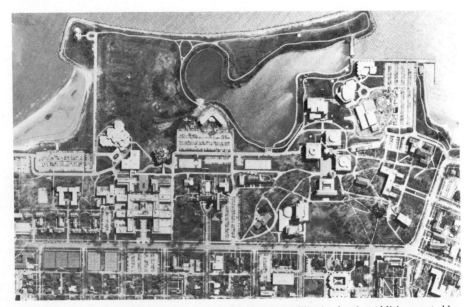

FIG. 35.6 The campus of Northwestern University, Evanston, Illinois, showing addition created by land reclamation from Lake Michigan.

found at Evanston, Illinois, where, in 1963, Northwestern University constructed a major landfill, extending the campus property into the waters of Lake Michigan (Fig. 35.6). Unfortunately, this appears to have deflected littoral drift along the lakeshore, with the result that within a few months of the completion of the landfill, the shortest of the city of Evanston's water intakes (on the bed of the lake) began clogging with sand. Dredging gave temporary relief, but eventually a new intake had to be constructed, and one of the existing intakes extended into the lake, both now extending 1,750 m (5,350 ft) offshore. This experience is a salutary reminder of the interrelation of natural processes, especially in the open water of sea or lakes.[35.11]

Finally, there may be mentioned a quite straightforward example of land reclamation, this one in the harbor of Toronto. With the anticipated growth of traffic in this important inland port of Lake Ontario (anticipation more than realized once the St. Lawrence Seaway was completed), it was decided to increase available storage area by reclaiming 9 ha (22 acres) of open water. A wall of interlocking steel sheet piling was driven all around the area selected at the eastern end of the harbor. When this was complete, the area so enclosed was filled with sand pumped from the open water of the lake through a floating pipeline passing through the eastern entrance, the sand so pumped being soon replaced by the littoral drift from the Scarborough Bluffs, which here distinguishes coastal conditions.[35.12]

35.3 DISPOSAL OF EXCAVATED MATERIAL

The problem of disposing of material from excavations is one that is steadily increasing in complexity as cities grow larger and available sites for the easy dumping of material diminish in number and size. It is a problem governed to a large extent by economics, due especially to the high cost of trucking. In earlier days, a simple solution was to find an unused area of land, an old quarry or even an unused ravine, and merely end dump the rock or soil into it from trucks. Uncontrolled dumping of this kind led to much trouble. In one city that has the good fortune to have many lovely ravines around it, some of these natural beauty spots were thus lost to public use. What was a real problem for a time was the occurrence of slides of this dumped material because of instability of the steep slopes. But today with public control over so many dumping operations, this is a situation that would not normally arise. End dumping on private ground is still widely practiced, however; owners see an easy and cheap method of obtaining what they consider to be good building lots as depressions are filled in and level areas developed. Unless fill is properly placed, in layers and compacted, there may be continued settlement of land so formed due to the slow, natural compaction of the fill, quite apart from any settlement that the load it induces may cause on underlying weak soils. A first requirement for the disposal of fill is, therefore and as usual, accurate knowledge of the subsurface conditions.

A combination of factors has led many cities to consider accepting the responsibility for disposal of excavated material, since it can often be desirable fill, conveniently available to downtown areas with their extremely high land costs. The procurement of 9.5 ha (23.4 acres) of new land in downtown Manhattan, the value of which must be left to the imagination, was effected by the controlled dumping of excavation from the World Trade Center in the manner shown in Fig. 35.7.[35.13] Lest it be thought that this is a new and revolutionary idea, it must be recorded that when the first railway into Canada's capital city of Ottawa was built from the St. Lawrence River, at Prescott, one of the great early civil engineers of Canada, Walter Shanly, used all the fill from a big cut at the Prescott end to obtain 3.2 ha (8 acres) of new land in the St. Lawrence River; the fill was dumped into an area enclosed by about 300 m (1,000 ft) of wooden piling.[35.14] This was in 1852, but the idea has been used on a large scale only in recent years.

FIG. 35.7 One tower of the World Trade Center, New York, approaching completion with filled ground in Hudson River in background.

The city of Milwaukee has reclaimed 15 ha (37 acres), most conveniently located close to its downtown area, from Lake Michigan, and this acreage will be used for harbor and marina developments. It is not surprising to find that, always excepting the special case of the Netherlands, Japan probably leads the world in the creation in this way of new land for cities. A national program of reclaiming no less than 47,000 ha (183 m²) around its major ports, at an estimated cost of $7 billion, was announced in 1970. Much of this will probably be material pumped from the adjacent seabed, but for many years Japan, with its desperate shortage of land, has been using also its urban waste materials for filling. As but one example, two areas in Tokyo harbor at Kawasaki were filled between 1957 and 1963 with 24 million m³ (31.5 million yd³) of soil, most of it hydraulically placed by pumping but with the upper 1-m (3-ft) thickness provided from excavated material, ash, and slag, giving a total area of 3.8 million m² (4.5 million yd²). And this was but the start of a major filling program at this location (Fig. 35.8).[35.15]

An alternative use of excavated material is being followed by the city of Toronto. Its existing harbor (shown in Fig. 34.12) is incapable of any further extension within the wonderfully sheltered Toronto Bay, and so plans were developed for a new section of the harbor to the east of the present area. It had to be out in the open water of Lake Ontario, and so protection was essential. A new breakwater was planned that is 4,920 m (16,000 ft) long, giving an eventual two-thirds increase in harbor capacity. As a result of long-term advance

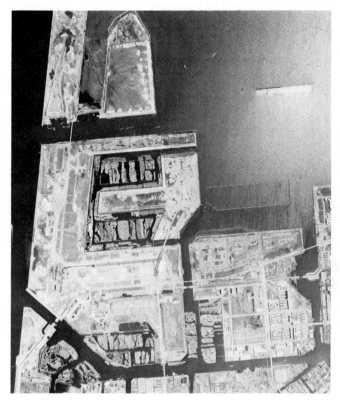

FIG. 35.8 Reclamation of land in the harbor of Tokyo, Japan, using fill material from the city; shown are some completed areas and two under construction.

planning, the new breakwater is being constructed rather more slowly than is usually the case but is costing the public practically nothing (Fig. 35.9). It is being constructed entirely of fill from excavations in the central part of this rapidly growing city. Contractors are not charged for the privilege of dumping, and so they save in their operations by the relatively short haul out to the breakwater. The Harbor Commission has to provide a bulldozing operation for control of the fill, but this is estimated to cost not more than $50,000 for the entire job, which contains about 13 million m³ (17 million yd³) of material—all obtained conveniently and for nothing.

A large consignment of French wines was even contributed as fill to the breakwater by the Liquor Control Board of Ontario, but only because the wine had been accidentally frozen and so was unfit for the palates of their customers. Blocks of concrete and rock are not used for the breakwater but are broken up and crushed in a special installation from which a steady supply of crushed rock for surfacing harbor roads is obtained. Geology helps, for the locally excavated glacial soils and shale bedrock are suitable materials for this unusual use, and the bottom of Lake Ontario, formed of the same materials, provides an excellent foundation bed. There must be many cities that could use their valuable materials from excavations in similar constructive ways, with no disfigurement of the landscape and to public economic benefit, but only as a result of careful advance planning and provided the local geology ensures material that is suitable and locations for its dis-

35-12 CIVIL ENGINEERING WORKS

FIG. 35.9 Toronto's new breakwater in Lake Ontario, entirely constructed from material excavated in the central city area; similar projects are being constructed to east and west of the main harbor area.

posal that can be used safely. The relatively small expenditure needed for the necessary subsurface investigation can pay dividends many times over, even for this unusual phase of modern urban development.[35.16]

35.4 LANDFILL WITH WASTE MATERIALS

Throughout the developed world, the disposal of domestic and industrial waste is steadily becoming a major municipal problem. It is a far cry from the earlier, simpler use of a hidden garbage dump, since solutions now involve major engineering works. The extent of the problem is but rarely realized. Estimates by the U.S. Public Health Service suggest that the total amount of household, commercial, and municipal wastes in the United States in one year as the sixties came to a close was more than 250 million tonnes, to which must be added more than 550 million tonnes of agricultural wastes and more than 1.1 billion tonnes of wastes from the mineral industry. The cost of removing such waste material in the United States has been estimated to be $4.5 billion annually (80 percent of the total being trucking costs), an amount that increases every year.[35.17] Here geology has been able to assist in some locations. New ecologically and environmentally acceptable sites have been selected, always after detailed geological investigation. Plans have even been made for transporting waste by rail to disposal sites in the desert.

In Great Britain, municipal officials have no opportunity of sending their garbage off into the desert in the hope that it can be forgotten. They have to adopt other methods. On the south coast stands the city of Portsmouth. An area of 164 ha (405 acres) of Portsmouth Harbor, long known as Paulsgrove Lake, situated between the mainland and Hornsea Island, has now been enclosed from the sea by the construction of relatively simple embankments made of Chalk excavated from a nearby hill. Refuse is now being dumped under controlled conditions within this large area. It is estimated that it will hold more than 2.25 million tonnes, but plans are already being developed for an incinerator plant that will be in operation when the area is about half full (Fig. 35.10). The residue from the incineration process will be so much less bulky that the life of the area for disposal

FIG. 35.10 New disposal area for garbage and waste, in Portsmouth Harbor, England; embankments enclose areas to be reclaimed; material (Chalk) from new road constructed in the background is being transported by belt conveyor to form foundation material at the new building site.

will be increased from the original estimates of 20 years to perhaps 50 years. In addition to providing 161 ha (400 acres) of usable new land, this disposal site will also provide new road access to Portsmouth. This was a part of the initial planning that naturally included a careful program of subsurface exploration to ensure that the site was quite suitable. The results of advance predictions based on the geological survey that was first made were confirmed by 28 test borings and 4 probes. Very recent marine deposits were found to form the surface, overlying earlier alluvium and then the Upper Cretaceous Chalk.[35.18]

The same practice, on a more limited scale, was started in 1932 by the city of Liverpool, one of England's major ports, on the river Mersey. At a location with the delightful name of Otterspool, to the south of the city, 17.3 ha (42.5 acres) were enclosed by a seawall, and embankments were made from the sandstone rock excavated from the first Mersey Tunnel. Some years after tipping was finished, the area had been converted into a waterside park with good landscaping, including gardens and a waterfront promenade (Fig. 35.11). In 1955 a further 28.2 ha (70 acres) were reclaimed in a similar way, and here it was estimated that 6.1 million m^3 (8 million yd^3) could be deposited.[35.19] Across the Irish Sea, the port of Dublin in Eire (Ireland) has been regularly reclaiming 6 to 8 ha (15 to 20 acres) a year using domestic refuse and excavated material under controlled tipping. Suitable embankments are first built. Dumped material is compacted by the simple means of dropping a 3-tonne weight from heights up to 6.1 m (20 ft). After compaction, a gravel top is added, and this new land is used for any loading up to 5.47 kg/m^2 (130 lb/ft^2).[35.20]

There are, therefore, solutions to this most pressing problem, but they involve sound, long-term planning and an acceptance before any plans are finalized. The crowded nature of Japan's cities suggests that here, too, the problem must be pressing. It is, and there are some lessons to be learned from the Japanese attack on the problem. Eighty percent of all the solid waste of Tokyo is burned in incinerators. Even so, this generates large quantities of burned waste that, together with the 6,800 tonnes daily that are not burned, is used for fill in areas controlled just as have been described. Power is even generated from

FIG. 35.11 Otterspool esplanade on the river Mersey, Liverpool, England, a new seawall having created an enclosed area used for disposal of domestic wastes, now finished off as this pleasant park.

the incinerators, the 2,400 kW of by-product power being used for refuse plant operation. In the extensive reclamation program already mentioned, new land is in part being filled with refuse and incinerator waste. Dream Island in Tokyo Bay, an enclosed area of 45.3 ha (112 acres), has been formed entirely in this way, solid waste, incinerator residue, and clean fill being placed in alternate layers. It will be used as a public park and for the site of another incinerator. One commercial development in Japan has attracted worldwide attention and possibly points the way to future attacks on the garbage problem. The Tezuka Kosan Company has developed a satisfactory process for compressing garbage under unusually high pressures, forming rock-hard blocks. Wire mesh is used to contain the rubbish at the start of the process. After being compressed under a force of more than 2,400 tonnes, the finished blocks are dipped in hot asphalt and are then used for a variety of purposes, such as forming simple embankment walls. The high pressures used are said to prevent rotting and the formation of methane. Perhaps most importantly, a "waste" material is being turned to a constructive purpose, as it is, of course, when properly used for desirable filling. This economy of material is something that will become increasingly vital.[35.21]

35.5 CONCLUSION

This brief review of such an apparently simple matter as winning land from the sea has led to consideration of a variety of topics, some with social overtones. Accordingly, it seems desirable to bring the review to a close by emphasizing, yet again, the fact that, although the operations described are essentially simple in character, they will be successful only if, before any work is done, accurate knowledge of the subsurface of the area to be covered and of the characteristics of the material to be used for fill has been obtained, studied, and applied in design. The Bombay Back Bay land-reclamation scheme in India was mentioned briefly at the outset of this chapter. It was initiated in the mid-1920s, an imaginative and simple project for giving the great city of Bombay some much-needed new land. The area to be reclaimed was surrounded by a rock-fill dike, and a large dredger was commissioned to fill this area with material pumped from the adjacent seabed. The great project was completed; few residents of the area today probably realize that they are living on reclaimed land. Before completion, however, a great deal of trouble was experienced because of the neglect of one apparently minor factor, the settling characteristics of the material obtained from the seabed. This was unusually fine grained and so did not settle quickly, with the result that much material was lost by seepage through the open structure of the surrounding rock-fill dike. This happened to be one of the first major engineering works with which one of the authors had any direct contact. The lesson it taught has never been forgotten.

35.6 REFERENCES

35.1 *From Fisherman's Paradise to Farmer's Pride,* Netherlands Government Information Service, 1959 (a good outline of this entire project).

35.2 J. B. Schijf, J. J. Dronkers, and H. A. Ferguson, "The Delta Project—a Symposium," *Transactions of the American Society of Civil Engineers,* **125**:1290 (1960); see also E. O. Hauser, "Holland's New Weapon against the Sea," *Saturday Evening Post,* 17 January 1959; see also *The Delta Plan,* Information Department, Ministry of Transport and Waterstaat, The Hague, Netherlands.

35.3 K. P. Blumenthal, "Some aspects of Land Reclamation in the Netherlands," Proceedings of 10th Conference on Coastal Erosion, Tokyo, 1966, pp. 1331–1359.

35.4 "Dutch Build a Power Plant on a 28-Acre Island That Wasn't There," *Engineering News-Record,* **175**:104 (18 November 1965).

35.5 "Rotterdam Expands Riverfront with Stone and North Sea Sand," *Engineering News-Record,* **179**:26 (2 November 1967).

35.6 "Dutch Landfill for 40,000 Planned," *Engineering News-Record,* **200**:18 (6 April 1978).

35.7 S. S. Cooke-Yarborough, "Making a Market out of a Swamp," *Engineering News-Record,* **163**:48 (13 August 1959).

35.8 "Slurry, Compaction Firm up Wasteland for Building," *Engineering News-Record,* **201**:15 (26 October 1978).

35.9 H. L. Lobdell, "Settlement of Buildings Constructed in Hackensack Meadows," *Proceedings of the American Society of Civil Engineers* (New York), vol. 96; SM4, Paper no. 7398, 1970, p. 1235; see also K. Widmer and D. G. Parillo, "Pre-Pleistocene Topography of the Hackensack Meadows, New Jersey," in abstract vol. for annual meeting at Pittsburgh, 1959, Geological Society of America, Boulder, Colo., p. 140A.

35.10 "The Hackensack Meadowlands Project," *Civil Engineering,* **47**:52 (May 1977).

35.11 "Landfill Forces Intake to Extend Its Reach," *Engineering News-Record,* **194**:24 (6 February 1975).

35.12 *Marginal Way Coal Dock, Toronto,* Bulletin of British Steel Piling Company, London, 1937.

35.13 M. S. Kapp, "Slurry-Trench Construction for Basement Wall of World Trade Center," *Civil Engineering,* **39**:36 (April 1969).

35.14 S. R. Elliot, *The Bytown and Prescott 1854–1979,* Bytown Railway Society, Ottawa, 1979, p. 6.

35.15 "Japan's Pollution Grows with Population," *Engineering News-Record,* **185**:26 (17 December 1970).

35.16 J. H. Jones, *A Bold Concept for the Development of the Toronto Waterfront,* Toronto Harbor Commission, January 1968; see also *Report on Coastal Engineering Studies and Design in the Toronto Area,* Engineering Department, Toronto Harbor Commission, 1968.

35.17 "Solid Wastes Pile Up While Laws Crack Down and Engineers Gear Up," *Engineering News-Record,* **182**:28 (12 June 1969).

38.18 M. E. Pitt, "Portsmouth Harbour Reclamation Scheme," *Proceedings of the Institution of Civil Engineers,* **47**:157 (1970); see also for discussion, **49**:539 (1971).

35.19 *Otterspool Riverside Promenade,* City of Liverpool, May 1970.

35.20 P. M. O'Sullivan, "Recent Development Works in Dublin Port," *Proceedings of the Institution of Civil Engineers,* suppl. vol. London, 1970, p. 153.

35.21 "Japan's Pollution Grows . . . ," op. cit.

chapter 36

PROBLEMS OF COLD REGIONS

36.1 Geology in Cold Regions / 36-2
36.2 Permafrost / 36-3
36.3 Logistics / 36-8
36.4 Aerial Photography / 36-10
36.5 Test Drilling / 36-10
36.6 Problems with Rock / 36-12
36.7 Problems with Soil / 36-13
36.8 Problems with Ice / 36-18
36.9 Materials / 36-20
36.10 Maintenance / 36-20
36.11 Conclusion / 36-21
36.12 References / 36-22

The cold regions of the world are popularly thought of as "the Arctic," but today the Antarctic must also be included in any treatment of areas of abnormal cold. The southern tip of South America, and the many islands to be found in that climatically inhospitable part of the world, although well to the north of the Antarctic circle, also share the problems of cold-weather building. Mountain tops in more temperate realms can experience what is normally thought of as "arctic weather," there being permafrost, for example, at the top of Mount Washington in New England and in the higher parts of the western mountains. In all these areas, developments are taking place involving some civil engineering operations. Weather stations "on the roof of the world," services and facilities for mining operations in northern parts of North America and the Soviet Union, support services for scientific stations in Antarctica—all these, and more, suggest that some attention must be given in this Handbook to the special problems that do arise when geological conditions in cold regions have to be considered as preliminary to the execution of civil engineering works.

Starting with the wartime establishment of Point Barrow on the northern coast of Alaska, activities in this largest state of the union now include the operation of a major oil pipeline, the construction of which marked a real advance in cold-weather construction. Construction of joint weather stations at five locations in Canada's Queen Elizabeth Islands, in the years immediately following the end of the Second World War, was but the

36-2   CIVIL ENGINEERING WORKS

start of engineering activities in the Canadian north which have included the construction of an entirely new town (Inuvik) in the Mackenzie delta and active drilling for gas and oil on land in the Arctic islands and in the waters of the Beaufort Sea. Norway and the U.S.S.R. share the northern coast of Europe, and the long coastline of Siberia is the northern rim of the continent of Asia. Here port works, vast mining operations, and explorations for petroleum products are carried out. Coal mining on Spitzbergen and volcanic problems in Iceland are further special features of the cold regions of the northern hemisphere, while explorations in Antarctica and scientific studies in Tierra del Fuego have become well known. These are all exciting developments which have attracted worldwide attention.

It is, therefore, difficult to deal with cold weather problems in small compass, but this must be done. Readers, therefore, will please note that this chapter is in no way even a capsule review of cold-weather construction. It is, on the other hand, the briefest possible survey of the special problems relating to the study of geology as a vital part of the site investigations that are the essential preliminary to all civil engineering design and construction for such areas. This will inevitably give a distorted picture of cold-weather developments in general. Guides to useful reviews of all aspects of such developments are given in some of the volumes cited at the end of this chapter, to which reference should be made for this more complete coverage.

36.1 GEOLOGY IN COLD REGIONS

The basic character of geology does not change merely as one moves north or south into the cold regions of the world. The overall structure of the earth's crust is continuous all over the surface of the globe. There is, however, one major and vital factor that distinguishes the cold regions—wherever there is water present, it is in the form of ice. Ice, therefore, has to be regarded as a geological material, just as is water in all temperate climates. And the physical facts related to the cooling of water and its transformation into solid ice, with an increase in volume, must ever be kept in mind in all geological work carried out in cold regions. This is at once so simple and so obvious that the emphasis here given to the importance of ice might seem to be misplaced. Any study of developments in cold regions, however, will very quickly reveal records of troubles that are directly attributable to neglect of the presence of ice; so it is that the presence of ice cannot be overemphasized. Some of the more important consequences of the solid nature of water in cold regions will be dealt with in later sections of this chapter.

One of the surprises of the overall geological picture of northern regions (and even, to a minute degree, of Antarctica also) is the fact that a small part of northern Canada, for example, was not subject to recent glaciation. Soils encountered in these northern regions, therefore, may be residual (Fig. 36.1). Since this nonglaciated area extends almost as far south as Whitehorse, the capital city of the Yukon Territory, it is not just a matter of academic interest but one of practical significance. Weathering processes are similar to what is experienced in more temperate climates, but the unusual extremes of climate may sometimes have unexpected effects. For the larger part of the cold regions under review, winter conditions (with the ground surface frozen) will persist for at least eight months of the year, if not longer. The other four months are characterized by relatively rapid rises in mean daily temperatures (daily temperatures sometimes going as high as 33°C (91°F) on the Arctic coast) with increasing length of daylight until, north of the Arctic circle or south of the Antarctic circle, at least one period of 24 hours will be all daylight.

Such vegetation as there is will, therefore, grow very rapidly during these short periods of "summer," sometimes with surprising results. In the search for a suitable site for the new northern Canadian town of Inuvik, on the eastern edge of the delta of the Mackenzie

FIG. 36.1 Residual soil in northern Canada! Weathered granite, an exposure near Whitehorse, Yukon Territory, in a small unglaciated area.

River, the most obvious of all the sites studied was on the western edge of the delta, on alluvial fans fed by small streams running off (in summer) the lower slopes of the Richardson Mountains, which here form the rim of the delta. Within a distance of 2.4 km (1.5 mi) from where bedrock could be seen to be disintegrating into fragments to be washed away and ground up into gravel-and-sand-sized fragments, there was an area admirably suited for the town site (Fig. 36.2). All surface indications pointed to a fine subsurface of such granular material. As a check, a test hole was drilled in the center of this area. This penetrated no granular material at all below the surface layer but rather a continuous stratum of ice-rich organic silt, rendering the site quite useless for the proposed purpose. This unusual weathering is thought to be due to the rapid growth of grasses in the long summer daylight and then to equally rapid decay at the end of summer, providing an additional agency of weathering.[36.1] The lesson is clear. In cold regions especially, one can take nothing for granted in subsurface geology.

36.2 PERMAFROST

Air temperatures in the cold regions of the world are not appreciably lower than those in central continental regions such as the Midwest of the United States and the Canadian prairies. The prevalence of high winds, however, makes far northern and southern cli-

36-4 CIVIL ENGINEERING WORKS

FIG. 36.2 Alluvial fans in the delta of the Mackenzie River, Northwest Territories, formed of disintegrated rock from the mountains in the immediate background, and yet containing no sand or gravel.

mates inhospitable (to put it mildly). More significant, however, is the duration of winter in extreme latitudes. This is well shown by any study of degree-days such as that illustrated in Fig. 36.3, degree-days being determined for heating purposes by the difference between the average air temperature for any location and 18°C. The summation of these values for a complete winter gives a good indication of the relative heating load (for example) to be considered for buildings. A corresponding figure (below 0°C) is used as an index of freezing.

It follows that, with an increasing total of degree-days for the winter, the corresponding average annual air temperature will gradually fall. As was explained in Chap. 9, this average air temperature will be close to, but not exactly equal to, the constant temperature of the ground for the same location, as measured at a depth of about 15 m (50 ft). When, therefore, the average annual air temperature falls below 0°C, the ground at the location will also be very close to freezing temperature. As one moves north, average air temperature will steadily decrease; so also will the constant ground temperature. When the steady temperature of the ground is below freezing temperature, the condition of permafrost exists. In the Far North this condition can extend to depths of over 500 m (1640 ft).

The word *permafrost* is one of those terminological inexactitudes which quickly become accepted in normal language, even though not literally correct. It was coined during the Second World War by Dr. Simeon Muller and was very quickly adopted in general usage to describe what should more accurately be called perennially frozen ground.[36.2] Any suggestion of absolute permanence is impossible, and improbable in view of the slowly

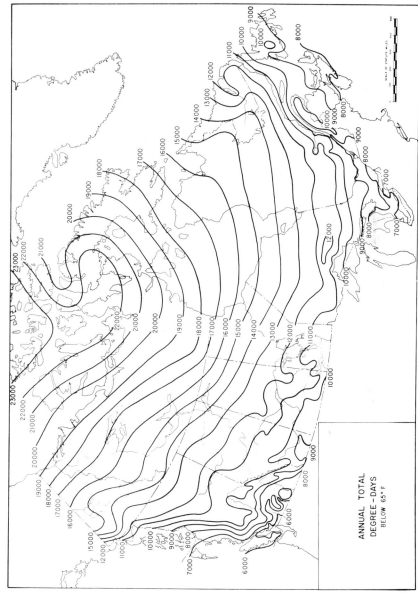

FIG. 36.3 Degree-days below 18°C (65°F), showing increase in the duration of cold toward the north.

changing world climate. But the name *permafrost* has stuck, despite valiant efforts to use more correct terms, and the word will be used in this chapter in the now widely accepted sense.[36.3] It must be stressed that, stricly speaking, the word applies only to a condition of the ground and not to any material. In popular talk and writing about cold regions, it will often be found that the word is misused to describe the ground material that causes most trouble when permafrost is encountered (see the following paragraph), but this wrong usage will not be followed here.

If the ground consists of solid bedrock, or dry sand and gravel, the condition of permafrost will cause no unusual problems in engineering work, although the material will be cold and special precautions will have to be taken in placing concrete, for example, in contact with it. The possible presence of ice must always be kept in mind as will be explained later. When, however, ground consists of any water-bearing soil—such as fine sand, silt, or clay—then the permafrost condition will result in the contained water being ice. When undisturbed, such frozen soil will be quite stable ("as hard as rock" when first encountered), as the well-established terrains in the Arctic and Antarctic testify. If, however, the condition of the ground is disturbed and its temperature regime changed, as by the removal of the usual organic cover, then the ice may melt and transform the soil from a stable to an unstable material in relatively short time. This possibility is the main problem to be faced in regions underlain by permafrost when any engineering works have there to be undertaken.

This is no small local feature. More than one-half of the land area of Canada and one-half of the U.S.S.R. are underlain by perennially frozen ground, as is more than three-quarters of Alaska. The entire Antarctic continent is similarly underlain by perennially frozen ground. Figure 36.4 shows in broad outline the limits of permafrost in the northern hemisphere. It will be clear that in the Far North (or the "Far South"), permafrost is to be expected as the normal condition of the ground. Correspondingly, in the developed part of Canada, as also along the southern coast of Alaska, permafrost will not be present. In between, however, there is a great area which is, in effect, a transitional zone, much of it characterized by discontinuous permafrost. The presence of the condition of permafrost in this zone will depend on the local geology, local drainage, the character of the ground cover, the nature of the vegetation, and the orientation of the ground surface, i.e., southern-facing slopes may be free in summer, while northern-facing slopes may exhibit the frozen condition. In all such borderline cases, the temperature of the ground will be at or close to freezing point, the factors indicated being sufficient to elevate ground temperature near the surface slightly, perhaps enough to place it above freezing point. If undetected, the presence of ice in such areas can cause much trouble. Identification of permafrost is, therefore, an essential part of site investigation in cold regions.

Fortunately, there are now available helpful guides. The *Permafrost Map of Canada* was a pioneer effort in this direction, first issued in 1967 jointly by the Division of Building Research of the National Research Council of Canada and the Geological Survey of Canada.[36.4] There are now similar guides for Alaska, China, and the U.S.S.R., China having a permafrost map showing that condition to exist in the northwest of the country and in Tibet. Such maps are general guides only. Examination of the ground, in the first place from the air, is essential for a true assessment. Fortunately, aerial photographs taken when the ground is clear of snow are now available for almost all relevant northern areas. By careful photo interpretation a fairly good idea can be obtained as to whether ground is underlain by permafrost or not. Actual conditions must be checked on the ground, on foot.

As indicating the utility of aerial photographs the case of the selection of the site for Inuvik, the new governmental center and town in northwestern Canada, may be cited. The government of Canada decided, in 1953, that a replacement of the existing settlement of Aklavik (located in the Mackenzie River delta) was essential; aerial photographs of the

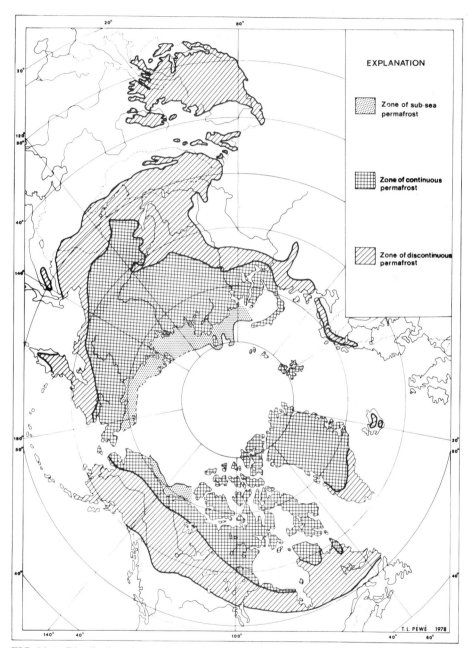

FIG. 36.4 Distribution of permafrost in the northern hemisphere.

entire delta area, 240 km long by 80 km wide (150 mi by 50 mi) were examined in Ottawa during the following winter by members of the site-survey team, and six possibly suitable sites were selected. In the field, during the ensuing short open summer season, these were quickly reduced to four. Test drilling was carried out at these four sites (one of which was mentioned in the previous section), and the site finally approved for the new town of Inu-

36-8   CIVIL ENGINEERING WORKS

vik was in this way readily selected. There was permafrost at all sites, but the aerial photographs permitted an accurate assessment of surface geology to be made in an office 4,800 km (3,000 mi) away, before any fieldwork was undertaken (Fig. 36.5).[36.5]

Patterned ground, as illustrated in Fig. 36.6, is one characteristic of perennially frozen, ice-bearing ground, especially in low-lying areas. Study of vegetation (if any) will give a good indication of the type of soil at and near the ground surface. The presence of white or silver birch, for example, is an almost certain indication that the underlying ground consists of sand and gravel, whereas the presence of black spruce in northern areas will suggest that the ground is ice-bearing silt. These are merely indications of the guides that are available to the detection of the permafrost condition and the usual soils associated with it. In all cold regions, the presence of permafrost must be assumed until positive proof is available that the ground is not so perennially frozen. Unfortunately, all too often the approach to northern work has been the reverse of this—permafrost was not anticipated until it made itself inescapably obvious.

36.3   LOGISTICS

This subject heading will strike an odd note, but logistics is an essential part of all work in far northern or southern regions. In the first place, the distances involved are very

FIG. 36.5   Inuvik, the new town in the delta of the Mackenzie River, Northwest Territories, its location selected after study of aerial photographs of the entire delta.

FIG. 36.6 Patterned ground, such as is widely found in permafrost areas, this view being on Banks Island, in the Queen Elizabeth archipelago, Canada.

great. Access to Antarctica is clearly a matter of intricate logistics, the distance from New Zealand (one of the usual jumping-off points) being about 4,200 km (2,600 mi). Journeys into northern lands are similarly formidable long-distance journeys, from Edmonton to the Arctic coast of Canada being about 2,080 km (1,300 mi), for example. Air transport makes the time element very different today from what it was before the advent of flying, but the distances remain the same, and the costs involved are great.

Site studies in the north can only properly be carried out when the ground surface itself can be seen, and this means work in the field in the very short open season of summer. In the Far North this may be a matter of a few weeks only; throughout the north "four months summer, eight months winter" is a common saying, and it typifies the limited duration during which normal fieldwork (such as described in Chaps. 10 and 11) can

FIG. 36.7 Transportation on the Mackenzie River, Northwest Territories, a string of loaded barges being pushed downstream to the Arctic coast in the main channel of the river delta.

36-10   CIVIL ENGINEERING WORKS

be carried out. If distances, cost, and short summer seasons are thought of together, then the term *logistics* becomes one of real meaning, even in relation to the subject matter of this Handbook. All planning must be done (except for emergency or wartime operations) at least one year ahead. It may require two years' advance planning for arranging the transport of any goods that cannot be taken into the north by air, such are the limitations of normal freighting into the north of Canada. Alaska and the Yukon do have year-round access from the sea, but the western and eastern Arctic of Canada do not. The shipping service on the Mackenzie River serves the western Arctic, with a three-to-four-month operating season (Fig. 36.7). The annual flotilla of cargo vessels which leaves Montreal in July provides the only access other than by air to the eastern Arctic. These are factors, in addition to geological matters, that must be considered when planning site investigations for all cold-region developments.[36.6]

36.4   AERIAL PHOTOGRAPHY

It follows that speed in carrying out fieldwork, when the ground is clear of snow, is another logistical factor of cold-region work. Accordingly, aerial survey work in cold-region investigations assumes an importance even greater than it has for normal site investigations. There are now available aerial photographs of all the northern part of the North American continent (and doubtless also for the U.S.S.R.), so that a good start at site studies can be made without going into the north, a convenience well illustrated by the initial selection of possible sites for the new town of Inuvik in the Northwest Territories. The utility of aerial-photo interpretation and guides to its practice were given in Chap. 10, but for convenience the most useful of the guides are also listed at the end of this chapter.

Illustrations were provided in this earlier chapter of detailed aerial studies of pipeline routes in Alaska and Canada. This technique was used to survey the route of the Alaska pipeline, 950 km (590 mi) of which are underlain by permafrost. A photo-mosaic of the proposed 1,260-km (790-mi) route, a strip 3.2 km (2 mi) wide, was prepared, and the location of all borings marked on it. In this way, valuable time was saved, as the coordinated work was carried on during the short summer seasons.[36.7] For the proposed pipeline down the Mackenzie Valley in Canada (not yet built), an airphoto terrain study was carried out over an area 3,200 km (2,000 mi) in length and from 4.8 to 48 km (3 to 30 mi) wide. This overall survey enabled those responsible to narrow down the choices of possible routes, final checking of "ground truth" being done by simple inspection and some test boring on the ground.[36.8] These were two major jobs, but the techniques used are equally applicable to much smaller projects in cold regions, "time being always of the essence" (to use the old phrase), so that sites may be properly investigated while the ground is clear of snow.

36.5   TEST DRILLING

In the quest for gas and oil in cold regions, quite remarkable feats have been performed in deep drilling on land and in Arctic waters. Drilling techniques have been developed and advanced, permitting exploration to great depths below the surface (up to 4,000 m or 13,000 ft in the Canadian north). If soil is encountered in such holes, near ground surface, it is merely a nuisance to be penetrated as quickly as possible so that "real" drilling in bedrock may proceed. It has been possible, with the cooperation of the owners, to utilize some deep dry holes, filled with diesel oil, for the insertion and withdrawal of probes to measure deep ground temperatures and so to further permafrost studies.

Test drilling for site studies, however, remains on a very different scale, small, easily transported drill rigs being used for all this work, methods followed being similar to those used in more temperate regions, with one vital difference. If test drilling is in frozen, water-bearing soil and undisturbed samples are required, special precautions must be used to preserve the frozen condition of the soil, the use of warm wash water, for example, being clearly undesirable. Core drilling, as in rock, is the normal procedure. Suitable refrigerated containers must be available at the drill site, and soil samples must be placed in the containers as soon as the samples are removed from the ground, so that they may be tested later in their natural condition in a laboratory. Ice content and ice fabric are two essential characteristics to be determined, in addition to the regular tests. Those with experience have recorded much useful information of this aspect of cold-weather site study (Fig. 36.8).

It will be appreciated that the hand methods used in ordinary site studies—such as the digging of test pits and the hand drilling of shallow holes—will not be possible in the north, except to a very limited degree. At the end of summer, shallow thawing will have taken place in frozen soils, even beneath an organic cover, so that hand excavation of a pit can be carried out to the bottom of the *active layer* (as the zone of summer thawing is called). Below this, excavation will be in the equivalent of rock, the combination of ice and soil being almost always rocklike. Air hammers or similar mechanical equipment will be necessary for drilling extensive pits, which will sometimes be necessary for special installations, even though, as will shortly be explained, one of the basic precepts of engineering in cold regions is to "preserve the permafrost" to the maximum extent possible.

FIG. 36.8 Test drilling under Arctic conditions, near the delta of the Mackenzie River, Northwest Territories.

36-12 CIVIL ENGINEERING WORKS

Advance planning will again be seen to be essential, and on a time scale quite different from that usually followed, since, if jackhammers will be necessary, they must be arranged for well in advance for reasons already delineated. Test drilling for site studies, therefore, becomes a part of the overall project, and its planning must be coordinated with that for the actual construction work that is to follow.

36.6 PROBLEMS WITH ROCK

The ordinary characteristics of rock in cold regions will be similar to what they are in more temperate areas. Only when concrete has to be placed, or pipelines installed, in contact with "frozen rock" will the prevalent low temperature be of importance. There is, however, one special feature that must always be anticipated, the possible presence of water in the form of ice. If bedrock is competent and sound, with no joints, seams, or fissures whatever, then there will be no possibility of visible ice being present. If, however, and as is often the case, the bedrock is featured by jointing, in the case of igneous rocks, or has well-defined bedding planes and possibly seams, in the case of sedimentary rocks, then ice may be present. In the normal process of ice formation in confined spaces, once some ice has formed it will tend to grow, if there is a supply of water available for this. Accordingly, although most occurrences of ice within bedrock will be limited in thickness, it is possible to encounter massive ground ice that has formed in open fissures well below the rock surface. Once again, when bedrock is encountered at northern or other cold-region sites, nothing can be taken for granted. Every possible means must be taken to ensure that it is indeed "solid rock," with no ice present, if it is to be incorporated in any way in an engineering project.

There is a further problem that is occasionally met. If the joint system in a bedrock exposure tends in a vertical or near-vertical direction, or if the bedding planes of sedimentary rock have been uptilted from their original horizontal position, then there is the possibility that water may penetrate along such joint planes in summer periods and, in due course, freeze. The mechanics of this action may be complex, but the phenomenon of

FIG. 36.9 "Growing rocks" at Churchill, Manitoba.

FIG. 36.10 Frost effects on jointed phyllite near the east end of Spider Lake, Northwest Territories, the blocks having been forced upwards along joint planes.

"growing rocks" is clearly due to the collection and freezing of water in joints. "Growing rocks" is the name applied to blocks of bedrock, seemingly in place but clearly separated from adjacent blocks, which have a vertical or near-vertical orientation.[36.9] Figure 36.9 shows one of the better-known examples, at Churchill, Manitoba. Although south of the 60th parallel, and so not in the Northwest Territories of Canada, Churchill is indeed a cold region, as all who have visited it during winter months can testify. (It has been the cold-weather testing area for the Canadian armed forces). It will be seen that the "growing rocks" appear to be quite solid; they can readily be mistaken for solid, immovable bedrock (Fig. 36.10). But they are not. Individual blocks of this bedrock mass are moving upward, very slowly but quite definitely. Accordingly, they cannot be used, even if they were in a suitable location, as a foundation bed for any structure, nor can they be used as benchmarks!

36.7 PROBLEMS WITH SOIL

Without going into any explanation of the existence of permafrost in cold regions, or of the climatic and environmental factors that have led to the formation of the widespread peat deposits of northern regions (called in Canada, *muskeg*, an old Indian term), it must be observed that in cold regions such mineral soil deposits as there are will usually be hidden from normal view by the organic cover. This mat of closely interwoven organic fibers is a remarkable material. Its fabric is such that it may contain as much as 1500

FIG. 36.11 The result of disturbing the muskeg cover over perennially frozen coarse silt ("bull's liver" when thawed) near Uranium City, Saskatchewan.

percent of its own solid weight as water in summer, as ice in winter. The thermodynamics of its protective function in preserving frozen soil forms a most interesting study. Here it must suffice to record that, if this organic protective cover is once removed, or even damaged, then under the influence of summer sunshine the frozen condition of the underlying soil will be changed, usually with lamentable—and irreversible—results such as are shown in Fig. 36.11.[36.10]

Water-bearing soils as found in cold regions (sands, silts, and clays) have generally a very recent geological history. They are almost invariably unconsolidated and usually have high moisture contents. They must have been frozen within a short time (in geological terms) after deposition, which was from lakes of glacial outflows. Varved soils may sometimes be seen in cold regions, their water content concentrated, in the form of ice lenses, at the bottom of each varve. The exact mode of deposition and freezing of such "soils in the raw" is a relatively unexplored part of Pleistocene geology, but interest has been heightened by the discovery, in such frozen soils in the U.S.S.R., of complete mammoths, their flesh intact, some with vegetation still in their mouths.[36.11] Massive ice is also found in deposits of sand and gravel (Fig. 36.12).

Cold-region engineering has to accept the reality of such soils, whatever their exact history—and with all the unfortunate consequences if their presence and character are not determined in preliminary site study. As can be seen from Fig. 36.11, once such soils are exposed their excess water content quickly turns them into "soup." In the lower right-hand corner of the view the soil can be seen flowing towards the adjacent drainage ditch. Exposure of the soil in this case was due to the excess of zeal of a good foreman, but one inexperienced in northern road construction. He thought that the careful covering with brush of the muskeg along the line of the road under construction and the careful deposition of fill over the brush, all as protection of the permafrost condition, was just a waste of time. He had a good bulldozer and wanted to show what it could do. He soon found out. It took a full year before the soil shown had stabilized sufficiently to be reworked.

Figure 36.13 shows another common effect of thawing soil in cold regions, "drunken buildings" in northern Canada and Alaska being a sure sign that they were erected before proper cold-region building methods were fully understood (after the Second World War). The case shown is typical of early buildings, the heat from which in winter had penetrated the ground beneath with inevitable consolidation of the underlying soil. The most famous

FIG. 36.12 Cavity in sand and gravel caused by melting of ground ice after removal of surface cover near Inuvik, Northwest Territories.

FIG. 36.13 Old store building in Dawson City, Yukon Territory, differential settlement being due to thawing of the underlying frozen soil.

case was that of the first building in the Canadian north to be constructed with a full concrete basement, just prior to 1939. This was a fine attempt to provide comfortable living quarters; unfortunately, building on perennially frozen ground was then something not yet fully appreciated. The house was built on unconsolidated silt, the frozen nature of which had not been realized when the site was cleared for building, so solid is such soil in its frozen state. The furnace was turned on with the advent of winter. By Christmas serious settlement of the building could be observed by eye. By the end of the heating season, the whole house had settled about 45 cm (18 in). Had heating continued, the house would have eventually disappeared below ground. The furnace was removed and the basement filled in; the house may still be seen.[36.12]

Today, there is naturally no excuse for any such misadventures. For operations such as road building, once the location has been selected, the greatest care must be taken to ensure that no construction equipment moves on to the area to be covered by the road until the organic cover has been fully protected with interlocking layers of brush, laid in place by hand. Crushed-rock, or sand-and-gravel, fill may then be placed, by end dumping from equipment working over the brush, to the necessary depth that has been calculated in advance. In some northern roads, sheets of synthetic insulating material are now being used, placed horizontally below the road's wearing surface; experimentation is active. The success of these basic procedures is well shown by an early example, the runway of the airport at Inuvik in the Mackenzie delta. The best possible site was chosen, but this was over the worst kind of perennially frozen, ice-rich soil conditions. The runway was constructed as just described in 1956–1958, necessary minimum depth of crushed-rock fill being calculated in advance as 2.4 m (8 ft) (Fig. 36.14). Temperature recording devices were installed in ground and fill. Twenty years' experience has shown that, if anything, the design was somewhat conservative, the 0°C isotherm having moved upwards into the fill. The runway, having been paved in 1969, continues to handle heavy traffic admirably.[36.13]

When buildings have to be erected over frozen soils, they can be built in the same way,

FIG. 36.14 Major airstrip at Inuvik, Northwest Territories, under construction; the muskeg cover is being carefully preserved; initial rock fill was placed during severe winter conditions, remaining fill being placed on top of it during the following short summer season.

i.e., on gravel or crushed-rock pads placed on brush-protected muskeg cover. This type of construction is naturally limited to small buildings. In one extreme case, a special refrigeration plant was built as part of a building complex in order artificially to keep the underlying soil frozen, despite the heat loss from the building.[36.14] In the case of the vast majority of buildings that have to be put up in cold regions, when the only foundation material available is perennially frozen, water-bearing soil, the invariable method is to build them on piles. These piles, timber or steel but reinforced-concrete if necessary (as always in the U.S.S.R.), are not driven into the ground but placed there in holes which have either been drilled into the frozen material or (in earlier days) steamed out with steam jets, the annular space around the piles being filled with "mud" and allowed to freeze. Figure 36.15 shows a typical installation in the Canadian north, the space left between the floor of the building and the ground surface being clearly visible. This is to permit winter-cold access to the ground surface beneath the building, further to ensure that the frozen nature of the foundation beds is maintained despite the heat losses from the building. All building services (potable water supply, power cables, disposal of wastewater, etc.) must naturally be brought to and from buildings above ground. The "utilidors" seen in Fig. 36.15 are the structures specially built for this purpose.[36.15]

To go further would involve detailed discussion of engineering design methods. What has been said is to illustrate the unusual and extreme methods that have to be taken in cold regions, north or south, in Alaska, Canada, and the U.S.S.R., to provide modern facilities for living in these inhospitable parts of the world, when preliminary site studies have shown that the local geological condition is frozen, water-bearing soil. Only on the basis of the most meticulous site studies, carried out despite all the difficulties of such work in cold regions, can any northern designs be prepared, and only then against a background of solid experience with cold-region problems. In only one way can this frozen condition of soil be turned to some advantage. Occasionally, some excavation in such soils

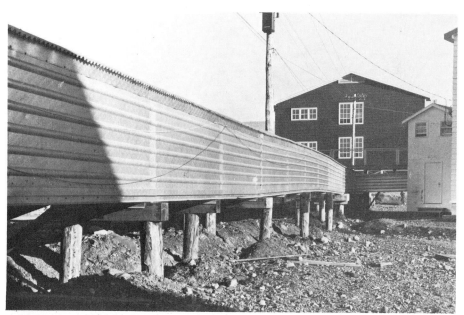

FIG. 36.15 Typical view of northern buildings on piled foundations, frozen in place, and insulated "utilidors" for conveying services to individual buildings.

36-18 CIVIL ENGINEERING WORKS

must be carried out. One would imagine that drilling and blasting would be essential, but this is not always the case. If the exact nature of the soil is known and there is time to let Nature assist, the area to be excavated can be deliberately exposed to summer sunlight, and thawing will take place. The thawed soil can then be easily scraped off and removed, the thawing continuing until enough layers have been removed in this simple way to reach the required depth. But this is small compensation indeed for the mass of problems that must be faced in all cold-region civil engineering work, problems that can only be solved on the basis of accurate knowledge of every facet of the local geology.

36.8 PROBLEMS WITH ICE

The idea of considering ice as a material may at first seem strange, but reflection will show that it is a solid material, differing from those in common use in that its properties are so clearly dependent upon temperature. The properties of ice in various forms and from differing origins are now well known. It is, for example, a plastic material susceptible to deformation under long-term loads of appropriate magnitude, as demonstrated so well by the flow of glaciers. At low temperatures ice has appreciable structural strength.

FIG. 36.16 An ice wedge, a frequent form of ground ice in northern regions; this wedge, 4 m (13 ft) wide at its top, is in 40,000-year-old Pleistocene sediments on Garry Island, Northwest Territories.

The possibility of its presence in cold regions in the form of ground ice, in both soils and rock, has already been noted. This possibility must always be kept in mind. If the permafrost condition is maintained beneath all structures, just as has been described, then any ground ice that is present beneath the site will remain as ice and perform as expected. It will be impossible to determine with any certainty that ground ice is not present, since this would require test borings at impossibly close spacing. Figure 36.12 shows a typical exposure of ground ice, movement of adjacent soil exposing it to view. Figure 36.16 shows another common manifestation of ice in cold regions, an ice wedge that has formed as a patterned surface developed on the ground. One has only to glance at photographs such as these or, better still, to see actual exposures of ice in the field to develop a healthy respect for the fact that the possibility of ice in the ground is something that can never be forgotten in cold regions.

Ice can be put to good service in all such areas. Winter snow roads can be improved by judicious icing. Ice bridges can be formed across rivers, for use during the hard months of winter. Some outstanding examples were used on the initial access road for the James Bay hydroelectric development in northern Quebec (Fig. 36.17). In the winter of 1971-1972, eight major ice bridges were constructed on the temporary 720-km (450-mi) winter access road that was necessary to facilitate the building of advance facilities and of the permanent access road with its fixed bridges. The ice crossings of rivers ranged in length up to 630 m (2,100 ft), this being the length of the bridge over the Waswanipi River. All crossings were carefully designed on the basis of laboratory strength tests of ice. Once river ice had formed, the bridge crossing areas were flooded to form layers of ice above that formed in the river. Final ice thicknesses ranged from 125 to 175 cm (50 to 70 in). The bridges performed admirably.[36.16]

Correspondingly, sea ice, once it had been tested and in view of the knowledge now available about the properties of ice and its performance under load, has been used in the

FIG. 36.17 An ice bridge, providing a winter crossing of this river on the access road to the James Bay water power development in northern Quebec, in advance of the construction of a permanent bridge crossing.

form of artificial islands for the mounting of large drill rigs for oil and gas exploration beneath the seabed. This use of ice circumvents the necessity for constructing actual offshore islands. Typifying these advances, which lie on the borders of geology, is the existence of the International Glaciological Society, a learned society which brings together the growing number of scientists and engineers who are concerned with ice and its uses.

36.9 MATERIALS

Brief note must be made of the use of natural materials for engineering works in cold regions. Local materials occupy a far more important place in construction in such areas than is usually the case, because of the logistics of transporting heavy construction material the great distances necessary from normal places of supply. There is, therefore, always a temptation to use local materials without further thought, whenever they are available. This is a temptation that must be firmly resisted, and even more meticulous geological study and laboratory investigation than usual must be carried out before any local materials are adopted for construction. If anything does go wrong, as it can if unsuitable materials are used, the consequences can be serious indeed. Sand and gravel, for example, are often available in northern areas that have been glaciated. Because of their mode of deposition from glaciers, these granular soils may not have anything like the grading required for use as concrete aggregate, despite their superficial appearance. Testing is essential; addition of extra-fine material is sometimes found to be necessary. Even bedrock must be initially suspect if it has to be excavated for use as concrete aggregate. One of the most unusual cases on record of alkali-aggregate reaction was encountered in just about the last place in the world where it might have been expected—at Alert, the small Canadian establishment at the northern tip of Ellesmere Island, the most northern permanent community in the world. Here it was found that subgraywackes reacted with portland cement in some concrete structures. Studies showed that materials other than the layered-structured silicates were also affected by the alkali in the cement.[36.17] The warning must again be repeated—in cold-region engineering, nothing can be taken for granted when local geology or local natural materials are involved.

36.10 MAINTENANCE

Regular inspection and maintenance of all civil engineering structures is imperative. Fortunately, this is now well recognized as a vital part of professional practice. Nowhere is regular inspection more necessary than with works in cold regions, particularly for structures that are founded (in ways that have been described) on perennially frozen, water-bearing soils. The permafrost condition under such structures *must* be maintained intact. For important structures, temperature measuring and recording devices may be installed at suitable points in the foundations. Once installed, such instruments must be carefully maintained and serviced. Despite the fact that, if the foundations have been well designed, there will be no appreciable changes in ground temperature after the initial disturbance, readings must be carried on with diligence throughout the life of the structure, just in case something untoward should occur, undetected, which might endanger the stability of the ground. This sort of routine work is tedious; every effort must be made to get those responsible for it to realize its vital importance.

For smaller structures, the use of instrumentation may not be warranted. Visual inspections, necessary in all cases even if supplementing temperature measurements, must then be relied upon. These too may become "routine drudgery" to the unthinking;

their importance must be kept in mind. After the snow has gone, special inspection must be made to see if any standing water remains. If so, drainage facilities must be corrected to ensure that this water is removed and does not return. Not only must all structures be inspected but so too must ground conditions, as evidenced on the surface, all around developed areas. If any significant changes are observed, then remedial maintenance work must be put in hand immediately. The ground environment in all cold regions is sensitive, especially so when the ground consists of soil. That environment must be conserved with all due care and diligence.

36.11 CONCLUSION

The warning at the outset of this short chapter must be repeated; this chapter is not a guide to northern building. The writers have attempted to show how, in the cold regions of the world, all the precautions relative to the geology underlying any site to be used for civil engineering projects described in earlier chapters must still be observed, with unusual attention given to the special features that distinguish the earth's surface in these areas. Such are the developments now current in cold regions that the approaches outlined in this chapter have long ceased to be merely of academic interest. They are in use by all experienced in northern (or southern) design and construction. Islands have been built off the northern Canadian coast, in the shallow waters of the Beaufort Sea, upon which great drilling installations have been erected in the search for gas and oil (Fig. 36.18).[36.18] More pipelines for gas and oil will be built, alike in the U.S.S.R. as in Canada and Alaska. An experimental pipeline has already been installed in the seabed beneath the Arctic ice, linking two of the Queen Elizabeth Islands of Canada.

It is greatly to be hoped that there will be a moratorium on any further road building in the northern parts of North America, in the interests of protecting the fragile natural environment. As new mines are developed, new railways may be necessary. The Soviet Union is showing the world what can be done in this direction by the construction of the

FIG. 36.18 Issungnak Island, constructed by Imperial Oil, in the shallow waters of the Beaufort Sea, off the Arctic coast of Canada, to provide a base for oil-drilling operations.

duplicate railway line across southern Siberia, the 2,240-km (1,453-mi) Baikal-Amur Railway being well advanced as this volume goes to press. It is estimated that it may cost $15 billion, with up to 100,000 workers engaged upon its construction, much of which is over ground underlain by permafrost.[36.19] For all these and similar works, knowledge of the geology underlying the sites to be used and the routes to be followed is the prime essential.

36.12 REFERENCES

36.1 R. F. Legget, R. J. E. Brown, and G. H. Johnston, "Alluvial Fan Formation near Aklavik, N.W.T., Canada," *Bulletin of the Geological Society of America,* **77**:15 (1966).

36.2 S. W. Muller, *Permafrost or Permanently Frozen Ground and Related Engineering Problems,* 2nd ed., U.S. Geological Survey Special Report, Strategic Engineering Study No. 62, Washington, 1945.

36.3 K. Bryan, "Cryopedology—The Study of Frozen Ground and Intensive Frost-Action with Suggestions on Nomenclature," *American Journal of Science,* **244**:622 (1946).

36.4 R. J. E. Brown, *Permafrost in Canada,* Map 1246A, NRC no. 9769, Division of Building Research, National Research Council, and the Geological Survey of Canada, Ottawa, 1967.

36.5 C. L. Merrill, J. A. Pihlainen, and R. F. Legget, "The New Aklavik—Search for the Site," *The Engineering Journal,* **43**:52 (January 1960).

36.6 K. B. Woods and R. F. Legget, "Transportation and Economic Potential in the Arctic," *Traffic Quarterly,* October 1960, p. 435.

36.7 R. A. Kreig and R. D. Reger, "Preconstruction Terrain Evaluation for the Trans-Alaska Pipeline Project," in D. R. Coates (ed.), *Geomorphology and Engineering,* Dowden, Hutchinson and Ross, Stroudsburg, Pa., 1976, p. 55.

36.8 J. D. Mollard, "Airphoto Terrain Classification and Mapping for Northern Feasibility Studies," in R. F. Legget and I. C. MacFarlane (eds.), *Proceedings of the Northern Pipeline Research Conference,* NRC no. 12498, National Research Council of Canada, 1972, p. 105.

36.9 D. H. Yardley, "Frost-Thrusting in the Northwest Territories," *Journal of Geology,* **59**:65–69 (1951).

36.10 I. C. MacFarlane (ed.), *Muskeg Engineering Handbook,* University of Toronto Press, Toronto, 1969.

36.11 "Study Begins of Baby Mammoth," *Nature,* **273**:485 (1978); see also I. T. Sanderson, "Riddle of the Frozen Giants," *The Saturday Evening Post,* 16 January 1960, p. 39.

36.12 R. F. Legget, "Permafrost in North America," *Proceedings of the 1st International Permafrost Conference,* Purdue University, 1963, p. 4.

36.13 C. L. Merrill, J. A. Pihlainen, and R. F. Legget, op. cit.

36.14 K. Terzaghi, "Permafrost," *Journal of the Boston Society of Civil Engineers,* **39**:1–50 (1952).

36.15 G. H. Johnston, *Permafrost Engineering—Design and Construction,* J. Wiley, Toronto, 1981.

36.16 L. M. Lefebvre, "Winter Roads and Ice Bridges," *Civil Engineering,* **49**:54 (December 1979).

36.17 J. E. Gillott and E. G. Swenson, "Some Unusual Alkali-Expansive Aggregates," *Engineering Geology,* **7**:181 (1973).

36.18 "Arctic Drilling Island Built During the Winter," *Engineering News-Record,* **192**:16 (13 June 1974).

36.19 "Soviet Railroad Conquers Terrain and Cold," *Engineering News-Record,* **200**:28 (6 April 1978).

part 4

SPECIAL PROBLEMS

chapter 37

LAND SUBSIDENCE

37.1 Major Natural Movements / 37-4
37.2 Subsidence due to Pumping / 37-6
37.3 Subsidence due to Mining / 37-14
37.4 Conclusion / 37-22
37.5 References / 37-23

Many of the examples from civil engineering practice so far described in preceding chapters have shown clearly that the crust of the earth is not the rigid, unyielding stratum that is often popularly imagined. The crust consists of materials which have properties exactly similar to those of all other solid (and liquid) materials. They react to stress and exhibit strain. They are not immovable, as many natural phenomena clearly show. In this and the chapters which follow, therefore, some of the special features of the earth's crust and its constituent materials will be reviewed in relation to the works of the civil engineer, again with examples from design and construction practice to the extent that is possible.

Vertical displacements of the ground may first be considered. It is common knowledge that such movements sometimes follow the occurrence of earthquakes. After one earthquake in New Zealand, for example, the town of Karamea, located on a deltaic formation, sank 60 cm (2 ft). Some ground subsidence, however, is not sudden, as is the case after such catastrophic natural events, but quite gradual, so gradual that it has only been with the advent of precise leveling that sinking has been observed and recorded. A subsidence of 37.5 cm (15 in) in the vicinity of Kosmo, Utah, was determined in this way; it followed an earthquake which occurred on 12 March 1934, but the sinking was not evident to the eye.[37.1]

Some examples of ground subsidence are quite local; others involve vast areas of land and sea. An unusual case of a local distortion is what has been called the "Palmdale Bulge." This upward movement of about 84 km² (32.4 mi²) of ground in southern California was discovered by staff of the U.S. Geological Survey in 1976. It extended from near Point Arguello eastward to the Arizona border, with its center in the Mojave Desert about 64 km (40 mi) north of Los Angeles. It is, therefore, located in the vicinity of the San Andreas fault; scientists at first thought that there was some relation between the two. Once observed, it was kept under close observation. Check levels showed that parts of the bulge had risen by more than 20 cm (8 in) by 1974, but then the rising stopped and the land began to subside, much of the bulge being back to its "normal" elevation, if not slightly below, by 1979. An increase in the emission of radon from some wells in the area added to the questions raised by the bulge. As this volume goes to press, it remains "an unexplained and perplexing phenomenon."[37.2]

More significant to the engineer, although from a long-term point of view, are the crustal movements affecting large areas of land in Scandinavia, northern Canada, and elsewhere, and the corresponding changes in the respective levels of sea and land. These movements might be regarded as of academic interest only, but when it is realized that one can see evidence of changes in sea level in works built only a few hundred years ago, then it will be appreciated that this is one natural phenomenon that must receive at least brief notice in this volume.

37.1 MAJOR NATURAL MOVEMENTS

The coasts of Italy provide evidence of unusual interest. To the west of Naples at the ancient city of Pozzuoli there stand today the ruins of the Temple of Serapis. Surprisingly, the floor of the temple is now beneath the waters of the sea, and so the few remaining columns stand on a watery base. Close examination of the columns will show that they are marked up to a height of 6 m (20 ft) above present sea level with small holes. These will be found to be the holes bored by a type of clam that can penetrate marble. Similar clams are found in adjacent bluffs to a height of 7 m (23 ft) above present sea level. They are a marine organism and so must have been alive in the sea when they did their boring into the columns (Fig. 37.1). The only explanation that fits these facts is that the ground on which the temple was built must have subsided after construction to such a depth as would have permitted the clams to do their boring at the height on the columns that can be seen today. Thereafter the level of the land, relative to the sea, must have risen again but not to its original elevation. There is some evidence to show that part of this uplift took place about the year 1500. The fact that similar evidence of changing relative levels is not found on all other coasts shows that this must have been a local phenomenon, but it was not an isolated occurrence, since in the Mediterranean there have been discoveries of ancient cities now hidden beneath the sea, providing a fertile exploration area for scuba divers. The Temple of Serapis is of special interest since it was first described by Sir Charles Lyell in his famous *Principles of Geology,* published in 1828, the first textbook on geology as it is known today and a precursor of the work of Charles Darwin, who always paid tribute to the contributions made to his thinking by Lyell. *Principles of Geology* is a book well worth reading today.

Evidence of a more general character is to be found in the New World. Diving has revealed the existence of a forest of at least 1,000 trees on the seabed off Panama City, Florida, at a depth of 18 m (52 ft). The trees stand in a sandy bottom deposit; their trunks have been worn away, but as measured at the level of the sand a typical trunk is 20 cm (8 in) in diameter. Radiocarbon dating has indicated an age of about 36,500 years before the present time; samples of a peat deposit associated with the trees gives a corresponding age of 40,000 years.[37.3] This example, which has been studied in some detail (by diving), leaves no doubt that the sea level today must be a good deal higher than it was tens of thousands of years ago, i.e., still in the recent glacial period. Evidence of a different sort, but equally convincing, is to be found in many raised beaches. In the Canadian Arctic, for example, there are to be seen on the east side of Pelly Bay three raised beaches, all of which have yielded samples of organic material that have been accurately dated. This study shows that the beaches have the following ages in years before the present (BP):

First beach: 52.5 m (172 ft) above present sea level formed at 7,160 BP
Second beach: 87 m (286 ft) above present sea level formed at 7,880 BP
Third beach: 162 m (530 ft) above present sea level formed at 8,370 BP

A little simple arithmetic will show that, to begin with, the sea was changing relative to the land at a rate of 17.5 cm (7 in) per year, the change later slowing to about 5 cm (2 in) per year. [37.4]

These examples are given first, since the bald statement that the level of the sea has risen about 100 m (330 ft) since the end of the last glacial period might appear to some readers to be a piece of science fiction. When it is remembered, however, that the ice cover in North America was perhaps 3,000 m (10,000 ft) thick at its maximum, extending from the Far North to as far south as St. Louis, the water released as it melted must have had a profound effect upon the sea. Evidence from around the world, similar to the last two examples noted, shows that it certainly did. Not only did the volume of water in the sea increase from the melting of the ice mass, but the land itself rose as the tremendous load of ice was removed from it, the rise naturally varying with the previous superimposed weight. And this process is still going on. This is the significant aspect in the present context of this excursion into geological theory.

It has been estimated, for example, that if all the ice in Antarctica and in Greenland that can be seen today were to melt because of amelioration of world climate, the level of

FIG. 37.1 Ruins of the Temple of Serapis at Pozzuoli, Italy, showing marks of marine-clam borings on the columns, as noted in text.

the sea would rise at least an additional 70 m (230 ft). Fortunately, the time scale of geological change is such that this eventuality can be regarded today with scientific interest but with no immediate concern. On the other hand, there is now available evidence obtained from careful study of the readings of accurate tidal-level gauges throughout the world that shows that the crust of the earth is still rising very slowly in those areas that carried the greatest loads of ice something like 11,000 years ago. There is in Amsterdam in the Netherlands a famous tidal gauge for which records are available from the year 1682. The records from this and other long-established gauges leave no doubt about the rise, especially around the Baltic. At Scandinavian ports, for example, rise of the land relative to the sea within recent time can be measured in centimeters. The most remarkable rises, however, are found in the north of Canada, along the Arctic coast and more particularly in Hudson Bay. Tidal records there show that at the northern port of Churchill, Manitoba, the land is still rising at a rate of almost 25 mm (1 in) per year. Because Hudson Bay is a relatively shallow body of water, it is possible to imagine the bed of the bay becoming dry land, since extrapolation of the records already obtained of the rise of ground level in this vast region indicates a probable ultimate total rise of more than the depth of the bay.[37.5]

These changes in the level of the land, even on Hudson Bay, are still very small, even though within the next century the status of Churchill as a port is probably going to be affected, just as have been some of the smaller fishing harbors in Scandinavia already. It must also be remembered, however, that just as the regions once loaded with ice are still rising, so other parts of the crust are correspondingly sinking by small amounts. The eastern coast of North America, for example, starting in Nova Scotia and extending into the Gulf of Mexico, as well as many parts of the shores of the Mediterranean, shows an annual subsidence of up to 2.5 mm (0.1 in) per year. Again, the changes are so small as to be insignificant for all practical purposes, but they present one further prospect for the future. If the Great Lakes are considered, it will be realized that they lie between the region of maximum rise in ground level and that of small subsidence of ground levels along the coast. There is, therefore, in progress a slow warping of the levels of the land around the lakes. This was first noticed as early as 1853 and has been studied carefully since 1894, records of water levels in the lakes being available from 1860. It is possible, therefore, to extrapolate from the records already obtained to show the changes that are actually occurring. Maximum calculated movement is on the north shore of Lake Superior, the amount being 0.53 m (21 in) per century, decreasing to 30 cm (12 in) at Sault Ste. Marie, 15 cm (6 in) at the north end of Green Bay, and zero at Lake Erie. Within the century, therefore, harbors on Lake Superior will almost certainly be affected. Taking a longer range view, it has been calculated that within the next 3,000 years, this warping of the continent will make Lake Michigan drain into the Mississippi River instead of into the St. Lawrence Valley, just as it did in glacial times. This will create problems for the International Joint Commission of that time.[37.6]

37.2 SUBSIDENCE DUE TO PUMPING

Ground movements so far discussed can be termed "natural" in character, but there are types of subsidence for which people are responsible. Settlement of ground due to subsurface mining operations has long been recognized; it is briefly discussed in the next section. In more recent years, however, potentially more serious subsidences have been created by the removal of water and oil from the ground. In certain areas, subsidence from these causes has become so serious that vast sums of money have already been expended for remedial engineering works. Ground movement of this kind can be likened to a large-scale demonstration of the process of consolidation, now so well recognized in soil mechanics.

The winning of water, oil, and gas from beneath the crust of the earth in some areas carries with it the penalty of land subsidence. Unlike subsidence caused by the mining of solid materials, subsidence caused by the withdrawal of fluids will rarely be sudden but will usually be so slow as to be imperceptible under ordinary observation. It will usually cover so wide an area that even when extensive subsidence has been caused it will grade so gradually toward the undisturbed ground as to be difficult to appreciate unless there is some man-made structure to act as a telltale. When it is realized that in one well-known case subsidence from this cause has reached a total of 7.8 m (25.5 ft) at the center of an area of 5,200 ha (12,300 acres), all of which has been affected to varying degree, then it will be realized that this is probably a more serious problem than the better-known subsidences due to the mining of solid materials. It is a problem of current importance in many parts of the world. Inevitably it will continue to be so, probably becoming even more serious than it is already, in view of the increasing demand for water and the never-ceasing hunt for new sources of gas and oil.

Once the matter is mentioned, the basic mechanism of subsidence can be imagined in general terms, even though the detailed analysis of some examples is complex indeed. Fluids are to be abstracted from beneath the crust. If the fluids merely filled the open pores of porous rocks or soils at ordinary pressures, then abstraction of the fluids should cause no essential change in volume. If, however, there is any free fluid held in the ground under pressure until released by the drill hole or by pumping, then the possibility of underground movements can be readily visualized. The ground will tend to adjust itself to such changes of pressure just as it will because of any other change in applied stress. If, in addition, this underground pressure has been supporting claylike soils, its release may permit such a redistribution of underground stresses that the clay will begin to consolidate, with consequent decrease in volume. In these, and other ways, very slow movements of soil and rock can take place after considerable changes in the volume and pressure of entrapped fluids. If the local geological conditions are appropriate, these changes can show at the surface by a general but very slow subsidence.

The pumping of groundwater to supply a part of the demands of the population of London was mentioned in Chap. 28, as was the existence of the great stratum of London clay beneath the great city. The result has been exactly such a general ground subsidence as is here outlined, even though very few Londoners would probably believe that this has taken place. Precise leveling in London has permitted accurate assessments to be made of the relatively small settlements that have taken place. The latest estimate of maximum settlement since 1865 is 2.14 cm (0.85 in), occurring near Hyde Park, decreasing to almost negligible amounts at the Crystal Palace in the south and the Brent Reservoir in the north. There has been, inevitably, a corresponding drop in the level of the water table beneath the city, this being an indicator of the cause of the subsidence. The drop started about 1820, when the head of artesian water was as much as 9 m (30 ft) above sea level; today the level has dropped in places to as low as 90 m (300 ft) below the same datum.

This pattern is to be found repeated all around the world: a natural supply of groundwater beneath the site of a city, a start at pumping to use this natural source of essential water, overpumping (and there is no other term that can be used) leading to serious lowering of the original water table and destruction of artesian conditions (if they existed to begin with), with increased pumping costs, possibly contamination from seawater if in coastal locations, and—of concern here—subsidence of the overlying land surface if the underlying geological conditions are conducive to this. Fortunately, there are still some locations where the water table has remained, or been kept, sensibly constant, but these cases are the exception rather than the rule.

37-8 **SPECIAL PROBLEMS**

In North America, a pioneer hydrologist, Dr. O. E. Meinzer, noted an early case of this type of ground subsidence in connection with his groundwater investigations in the Dakotas and in the Goose Creek oil field in the United States. He was apparently the first to treat this matter in a scientific manner, showing how the subsidence could be accounted for by the abstraction of water from underlying beds.[37.7] The abstraction of groundwater is responsible for one of the best-known cases of subsidence, that in the Santa Clara Valley, about 32 km (20 mi) south of San Francisco, California. The area affected is 520 km^2 (200 mi^2), and the maximum settlement has amounted to slightly more than 1.5 m (5 ft); this occurred in the business district of the city of San Jose, which has a population of 80,000. The subsidence was first noted during the progress of a survey in 1920. The results quickly became serious, as the area involved includes a part of the shoreline of San Francisco Bay, which had to be protected by dikes. Settlement had virtually ceased by the end of 1937, when restrictions on well-pumping operations raised the groundwater level by 15 m (50 ft). The permanent subsidence, however, remains.[37.8]

More serious has been subsidence in the city of Long Beach, adjacent to Los Angeles in southern California. The largest oil field in California, and the second richest in the United States, was found under the coastal part of Long Beach—unfortunately, it might almost be added, in view of the damage it has caused. The first wells were drilled in December 1936, and the first settlement was noted shortly thereafter. Since the area in which settlement was first detected was located between two well-known faults, it was at first thought that the settlement might be related to the aftereffects of the 1933 earthquake in this area. As settlement continued to increase, however, it was soon found to be directly related to the withdrawal of oil, gas, and groundwater from the underlying strata. By 1950, an elliptical area, including much of the harbor and measuring 8 by 5 km (5 by 3 mi), was seriously affected; the maximum settlement was 3.3 m (11 ft), and this was

FIG. 37.2 View inside the U.S. naval dockyard, Long Beach, California, showing the retaining wall needed, as a result of land subsidence, to keep out the sea.

increasing at the rate of about 0.5 m (1.5 ft) per year. Horizontal displacements were also recorded; a 1,800-m (6,000-ft) base line had been shortened by 1.5 m (5 ft). First remedial work to engineering structures had to be undertaken at about this time; a major bascule bridge was jacked up 22.5 m (7.5 ft), and protection walls and other works were constructed at the naval dockyard that was close to the area of maximum settlement (Fig. 37.2). Maximum settlement had increased to 5.1 m (17 ft) by 1952, with an anticipated ultimate maximum of 7.5 m (25 ft). Repressurizing of the strata from which the oil and gas were being taken was considered at that time to be impracticable for technical as well as for legal reasons, since so many owners were involved in the oil operation (including the city of Long Beach itself).[37.9]

By 1959, an area of 52 km² (20 mi²) was affected, and maximum settlement had reached 7.8 m (26 ft); settlements of as much as 0.6 m (2 ft) had occurred 14.4 km (9 mi) from the center of the disturbed area. Since the U.S. Navy's $175 million dockyard was at the center of the area, it is not surprising that the federal government had to take steps to remedy the situation; thus, an injunction was sought for the closure of the field. Over 800 million bbl of oil had been obtained from the field, and known reserves amounted to about half this amount. An emergency crash program was therefore initiated under the direction of a Subsidence Control and Repressurization Administration for the Long Beach Harbor Commission. Eight special pumping plants were built, with an installed capacity of 142 mld (37.5 mgd) (capable of increase to 167 mld, or 44 mgd), which pump saltwater out of special source wells; after treatment, the saltwater is forced back into the oil-bearing strata, under pressures ranging from 60 to 87 kg/cm² (850 to 1,250 psi). Cost of the plants, financed by the city, have been recovered through a surcharge paid by oil operators on each barrel of water pumped underground. By the end of 1959, almost 75 mld (20 mgd) were being pumped underground; it was then estimated that the subsidence under the naval installations would be halted in about six months.[37.10]

It is naturally in connection with coastal developments that land subsidence can be most serious. One interesting solution to the problem thus created was adopted in connection with the protection of a stretch of shoreline in Venezuela. The area around Lake Maracaibo near Lagunillas in this South American state is extensively developed as an oil field. The ground is swampy, and so a drainage program was initiated in 1923, following which it was noticed that the ground was slowly subsiding. The rate of subsidence was about 30 cm (12 in) annually, so that before long, properties of the operating oil companies were in danger of flooding. Various methods of protecting the lakefront were tried, but finally a flexible type of reinforced-concrete retaining wall, with a sloping face secured by raking piles at the back, was evolved and constructed in 1932. The wall was subjected to a slight earthquake in 1933 but showed no signs of damage.[37.11]

Land subsidence due to pumping is by no means confined to the Americas. Europe presents one of the most tragic cases of all, the settlement of Venice. This world-famous city, located at the head of the Adriatic Sea, dates back to the seventh century (Fig. 37.3). Some of the population of the adjacent mainland then moved out as a safety measure onto a cluster of small islands, known locally as *barene,* in the center of a large lagoon. Extending along the shore for 56 km (33 mi), 10 km (6 mi) in width, the lagoon has an area of 545,000 ha (1.35 million acres). Three local rivers were diverted, a sea-protection wall was built with three inlets for the tide, and the city gradually developed by building upon piles driven into the "mud" of the lagoon, the accumulation of river silt that creates the *barene.* Only within the past century has Venice been connected with the adjacent mainland, first by a railway bridge and in 1935 by a highway bridge. Its isolated position was well used in establishing its preeminence as a medieval mercantile power; its position as a renowned center of art is known to all. Problems of high water, however, have plagued the city for centuries, although only in recent decades to an alarming degree. Combina-

37-10 SPECIAL PROBLEMS

FIG. 37.3 The beauty of Venice, showing the surrounding lagoon.

tions of changes in atmospheric pressure, heavy rain, high tides, winds, and mass movements of the water of the Adriatic Sea (known as seiches) have caused periods of unusually high water *(acque alte)* from time to time. One of the most serious of these periods was in 1966. It is from this year that most serious study has been given to the future of this noble city, study that has been well publicized around the world.

Records of 58 *acque alte* during the 100 years prior to 1966 showed that 48 had occurred in the last 35 years, but 30 in the last 10 years. Two additional factors were recognized as being of increasing importance—the gradual rise in the level of the sea and the progressive subsidence of the bed of the lagoon. These two more permanent features, when considered in combination with the natural forces just listed, help to explain the growing frequency of flooding at Venice. Nothing can be done, naturally, about the general rise in sea level at Venice (an amount of 15 cm, or 6 in, in a century having been observed), but the subsidence of the ground beneath the lagoon appears to be due to human activity, which it should be possible to control.

Subsidence has been increasing, from 1 mm (0.004 in) per year in the period 1909–1925 to 5 mm (0.02 in) per year in the period 1953–1961. A number of reasons have been discussed. Some think that movements of the underlying bedrock are responsible; others, that the extraction of natural gas from under the delta of the Po River may have had some effect. A more probable cause, however, since it is a phenomenon found elsewhere in the world, is the consolidation of the soil upon which the city rests as a result of excessive pumping of groundwater from aquifers beneath. Yet a further factor is the recent construction of a large industrial development about 3.2 km (2 mi) from the city on the shore of the adjacent mainland, imposing new loads on the sediments of the lagoon. There appears to be some correlation between the observed subsidence and the pumping from the 7,000 wells that have been in use. Drilling for gas and oil in the vicinity of Venice has now been prohibited. Efforts are being made to reduce the pumping of groundwater.[37.12] These are but the start of the major rehabilitation project embarked upon by the government of Italy in October of 1971, with an initial planned expenditure of $400 million.[37.13] Some imaginative engineering suggestions for a permanent solution to the settlement of the great city have been advanced. All involve record-breaking operations. But the treasures which the old city contains may ultimately require that permanent rehabilitation be undertaken.

Mexico City calls for extended reference in even so brief a recital as this. Mexico City lies in the west-central part of the Valley of Mexico, a closed basin at a general elevation of about 2,500 m (8,200 ft) above the sea, surrounded by high mountains. The relatively flat part of the valley, including the site of the city, is underlain by alluvial deposits, but beneath these is a most complex geological structure dating from the end of the Cretaceous period, when deformation of the local limestone was accompanied by great volcanic action, action that has continued at a decreasing rate almost to the present. Even as recently as 2,000 years ago, a basaltic lava flow drove the inhabitants from the valley. Subsoil therefore includes soft, fine-grained deposits of volcanic ash and water-transported sediments, as well as sand and gravel with interbedded clayey silt. The latter constitute a fine aquifer beneath the city; naturally as the city grew so did pumping of water from this convenient source. The overlying beds of silts and clays are, however, highly compressible and so are susceptible to any change in moisture content. The inevitable result was that with the pumping from the deep aquifer came also serious subsidence at the ground surface. Demands from an increase in population from half a million in 1895 to 1 million in 1922, and to 5 million in 1960, led to a corresponding decrease in the pressure in the aquifer, originally artesian. The heavy stone buildings of the original Aztecs and later of the Spaniards naturally caused some early settlement because of the weak character of some of the underlying soils (despite the use by the Aztecs of deep pillars as foundations), but general ground subsidence seems to have started at the end of the last century.

Records taken on four benchmarks in the center of the city show a general subsidence of about 1.5 m (5 ft) between 1900 and 1940, but then a rapid increase, with total settlement of the general ground level of as much as 7 m (23 ft) by 1960. All observant visitors to this interesting city can see evidence of the subsidence in older buildings and, more dramatically, in the protrusion of old well casings originally flush with the surface but now sticking up by as much as 6 m (20 ft) into the air (Fig. 37.4). The skill with which Mexican engineers can deal with unusual local geological conditions in their foundation designs was well shown in Chap. 22. It remains to be noted that water is now brought into the city in order to reduce the pumping of groundwater, now most strictly controlled. Recharge wells are being installed and other conservation measures taken. The ground subsidence must remain, but the further development of the city will continue without the same problems that this caused in the past, now that the subsurface conditions are so fully appreciated and understood.[37.14]

In Japan, on the island of Honshu, serious ground subsidence has taken place as a result of pumping of groundwater at Tokyo, Nagoya, and Osaka, to mention three important cities. In Osaka the subsidence is attributed to the consolidation of an alluvial clay stratum 30 m (100 ft) thick. Between 1885 and 1928, when pumping was much less than in more recent years subsidence was from 6 to 13 mm (0.23 to 0.51 in) per year. Since then, however, pumping has increased concurrently with population, with the result that as much as about 3 m (10 ft) of subsidence has taken place in some areas. If there are any skeptical students of this general phenomenon, they may be convinced by the fact that ground subsidence at Osaka practically stopped in 1944–1945 when the city was being subjected to very heavy bombing in the closing stages of the Second World War and pumping correspondingly stopped (Fig. 37.5). Tokyo experienced the same temporary lull in the record of its ground subsidence for the same reason during its wartime ordeal of bombing. Even more serious is the situation in the city of Niigata, Japan, where serious ground subsidence has taken place due to the withdrawal of methane gas and groundwater. The city is located on the coast, so that protection from flooding by the sea presents special problems. A large chemical industry has grown up around the availability of the methane, and so intensive study of the situation has naturally been carried out. Repressuring of the strata affected is being carried out through wells.[37.15]

FIG. 37.4 Mexico City, Mexico; a well casing, the top of which was once at ground level, illustrating dramatically the 6 m (20 ft) of ground movement that has taken place; the group are members of ASTM Committee D18 on Soils for Engineering Purposes.

Even the great state of Texas must be mentioned in this summary account. It was the withdrawal of oil, gas (at an initial pressure of 70.3 kg/cm² or 5 psi), and water from the Goose Creek oil field in Texas that led to the first case of ground subsidence from this cause that appears to have been described in the literature. The area affected was about 4 km (2.5 mi) long by 3 km (1.9 mi) wide, and subsidence amounted in 1925 to more than 0.9 m (3 ft). The area affected conformed with the area being tapped by the producing wells. Because of the subsidence, some land was flooded permanently. The state of Texas claimed title to this submerged land, but the courts did not sustain the claim on the ground that the subsidence was caused by an act of man.[37.16]

One further example must be described briefly, this being the area around the modern city of Houston, Texas. The pumping of oil, gas, and water from beneath this rapidly developing area had affected (even by 1962) an area of almost 180,000 ha (445,000 acres), with subsidence of at least 30 cm (12 in), maximum settlements exceeding 0.9 m (3 ft), the area including not only the city of Houston but also the communities of Pasadena, Baytown, and Bellaire. The area is underlain by the Beaumont clay; the San Jacinto Monument is within the area of general subsidence. An unusual feature is the evidence of faults at the surface, faults in the underlying oil-field strata that show themselves by differential subsidence at the surface. Sudden changes in elevation of as much as 60 cm (24

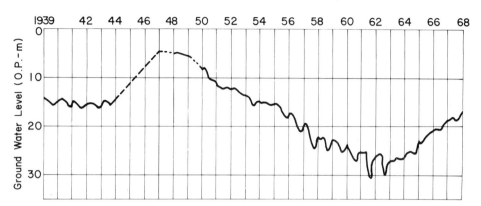

FIG. 37.5 Record of groundwater level beneath Kujoh Park, Osaka, Japan, showing the effect of wartime interference with pumping during 1943–1948.

in) are evident along some roads. Asphaltic roadway surfaces have performed well under these unusual distortions, but road maintenance is naturally a problem. Among the many local problems reportedly caused by the subsidence was the fracture of a natural gas supply line to an industrial plant. This incident, which occurred fortunately on a Sunday afternoon so that no lives were lost, must not be magnified, but it does illustrate the urban problems that follow subsidences such as have been herein so briefly summarized. The faults in the Houston area are particularly troublesome; local practice in laying sewers across known fault lines has been to use twin pipes, the active sewer pipe being placed within a larger pipe for a distance of 150 m (500 ft) across the line of the fault as a precautionary measure.[37.17]

It must be emphasized that overpumping of groundwater does not necessarily cause ground subsidence in the area above. This will happen only if the local geological conditions are such that strata exist that will be affected by the withdrawal of fluids from the ground in their vicinity. Although normally these strata will be soils, and especially clays, withdrawal of fluids from rock strata that act as aquifers can have exactly the same effect, since the groundwater conditions in all overlying strata will be so seriously affected. Precise leveling (by which most of these subsidences have been detected in the first instance) within the city of Savannah, Georgia, has revealed subsidence of as much as 10 cm (4 in) that can be directly attributed to a great increase in pumping groundwater from the Ocala limestone and associated limestones of Eocene age that underlie the city.[37.18]

In Memphis, Tennessee, on the other hand, a decline in the artesian head of more than 30 m (99 ft) within confined aquifers, also of Eocene age, does not appear to have resulted in any appreciable subsidence of the overlying ground. This is not to defend overpumping, since groundwater resources should always be conserved to the extent that is possible, through controlled pumping and recharging when necessary, but rather is meant to emphasize that the use of groundwater, even with excessive drawdowns, does not necessarily mean trouble with ground subsidence. The possibility always exists, however, and can be determined in its probability only by detailed study of the relevant geology and hydrogeology. If pumping of groundwater is used as a water-supply system, drawing from beneath an urban area, every effort must be made to check regularly on local ground levels to ensure that preliminary geological predictions are correct. Limitations upon such pumping, although superficially annoying, can have a multiple beneficial result, not only in limiting the danger of subsidence but in conserving groundwater, eliminating danger of contamination of the supply in coastal locations, and even (in some cases) maintaining satisfactory foundation-bed conditions.

37.3 SUBSIDENCE DUE TO MINING

Land subsidence due to pumping is familiar to those resident in, or near to, areas so affected, but it is a geological problem little appreciated elsewhere. Subsidence due to mining operations, on the contrary, is well known. Some statements that one reads in the popular press almost suggest that settlement is always to be expected in the wake of mining activity. The results of earlier mining operations provide many examples to confirm this suggestion. A great deal of mining has always taken place in igneous rocks, commonly called "hard rock." If mining excavations are not too large, such rocks are competent enough to support themselves without trouble. Correspondingly, a good deal of mining today is carried out deep beneath the surface of the ground, so deep that excavations have no influence at all on the ground surface. For more shallow mining operations, especially in the less competent rocks, there are today in most of the developed countries of the world official regulations as to how such work may be carried out, the regulations serving as social controls designed to minimize damage to surface property in all normal cases. Typical is the requirement in some areas that all excavations must be backfilled with inert material after ore has been removed, this being one of the ways in which large quantities of sand and gravel are now being used up. In some mines it is possible to use the mine waste material for this purpose, thus solving two problems simultaneously.

Examples from the Old World Mining was not always so carefully conducted as it is, in general, today. Europe presents many classical cases, one only of which can be described. It shows that trouble with mining subsidence is no new thing and illustrates well the interrelation of groundwater and underground excavations. The ancient town of Kutná Hora is located in eastern Bohemia, some 65 km (40 mi) to the east of Prague, Czechoslovakia. Its development defies all the rules that usually governed the growth of medieval towns. The discovery of silver-rich veins in the local rocks toward the end of the thirteenth century led to the start of settlement. For several decades people were attracted to the locality, living in small villages close to the deposits, villages that were later amalgamated into one town. At the start of the fourteenth century, silver mining was well established. In 1528 coins known as "Prague groschen" were minted in Kutná Hora, by hand, and were in widespread use as currency. Kutná Hora had then a population of 60,000 and had become the second largest town of Bohemia. At the turn into the fifteenth century the King of Bohemia promulgated the first European codification of mining laws *(Jus regale montanorum)* for use at Kutná Hora; it was soon in use in many other countries. Silver mining started to decline at the end of the sixteenth century and today is nonexistent. But the town remains, with its problems, and one can still visit one of the old mine workings and even see the ancient coins minted by hand just as they were made over four centuries ago.

The natural conditions of the area around Kutná Hora were not at all favorable for the founding of a large town. There is no local waterway of importance, no convenient transportation routes come near, and the physiography of the terrain around is not suited to easy defense in time of war. The town is located in a broad depression, the topography of which is naturally controlled by the geological structure of the area. To the north the depression is closed by three hills made up of crystalline schists, chiefly gneiss. On the hill slopes and over the bottom of the depression are relics of Cretaceous marine sediments, mainly sandy limestones and marlstones. A small stream (the Vrchlice) runs through the area, its bed cut down into a canyon-shaped valley in the local crystalline rocks. Mining activity in these same rocks greatly disturbed the continuity of the bedrock underlying the old town. The presence of old waste dumps of excavated gneiss and of heaps of slag from the old foundries complicates many surface features. The old mine workings, the full extent of which, as also their exact location, cannot now be determined with any accuracy, have led to surface subsidence in many parts of the town. The subsidences are quite irregular, in location and in timing, but many beautiful old buildings and many of great historic interest have been damaged. Temporary supports are used to the extent that is possible, and no new building is now permitted above the mined-out area (Fig. 37.6).

The hydrogeological conditions beneath the town and its surroundings were also strongly affected by the old mines. Originally, the local bedrock was generally impermeable, but mining seriously disturbed this condition. Water in the mines was drained into numerous sumps and galleries and along open joints in the rock, then pumped to disposal at the surface. At the present time, the mines are flooded, and so the level of the groundwater is the same as the level in the Vrchlice stream. Surface water infiltrates into the old mine workings. As the groundwater flows through the fractured bedrock and over old spoil heaps, it is enriched with soluble sulfates, and this makes it quite unsuitable for use, for example, in making concrete. Many of the wells originally used lost their water as the mines were developed and began to interfere seriously with groundwater conditions. Eventually, the town lost its established supply of water from wells, and, as early as the fifteenth century, a town water supply had to be installed. Water was brought to the town in wooden pipes from nearby supplies in the Cenomanian sandstones and led into several fountains, one beautiful example of which, illustrated in Fig. 37.7, still stands in one of the old city's squares. It must be almost unique as an architectural gem that is the result of untoward mining experience.[37.19]

FIG. 37.6 Kutná Hora, Czechoslovakia; supporting ancient buildings damaged by ground settlements due to mining.

37-16 SPECIAL PROBLEMS

FIG. 37.7 Kutná Hora, Czechoslovakia; the ancient water fountain for water distribution; note the granite setts in roadway; the figure on the left is Dr. Quido Zaruba.

Somewhat naturally, it is in the Old World that mining subsidences most frequently occur; there are on record many more almost classical cases of unexpected subsidence when old mine workings were encountered. Richey records a case where old mine workings were found only when an excavator at work on a building site disappeared into shallow workings that had been penetrated; the workings were covered only by a thin roof of solid rock, boulder clay, and other deposits. The site had to be opened up and the old workings filled in before construction could proceed. Old coal workings, previously unsuspected, were even encountered beneath the site for a new science building at the University of Glasgow, Scotland, and had to be fully investigated; necessary additional roof support had to be provided before the new building could be commenced.[37.20]

A somewhat extreme, but nonetheless interesting, case may be briefly noted to show what mining subsidence can really do. The region around Barrow-in-Furness, immediately to the southwest of the lovely English Lake District, has been long famous for its iron mines. The local "ironstone" has been mined from underground workings for many decades, and earlier workings have never been fully surveyed. On 22 September 1892, locomotive No. 115 (a 0-6-0 tender engine) of the Furness Railway was shunting some iron-ore wagons into a siding at the railway yard at Lindale, near Barrow, when the driver felt the ground beneath his engine start to fail. He tried to reverse the engine, but the ground actually opened up before he could do this. He was able to jump off but was badly shaken; his fireman jumped also and was unhurt. As the cavity increased, first rails disappeared into it and then the locomotive and its tender (Fig. 37.8). By the time the breakdown gang with their large crane reached the site, they were able to rescue only the tender. The locomotive was never seen again.[37.21]

Examples from the New World   It must not be thought that land subsidence due to mining operations is something that is confined to the older countries of Europe. North

America has its full share of examples, even though its mines are, by comparison, relatively so new. As recently as 1968, for example, serious property damage took place in the borough of Ashley, 6.4 km (4 mi) south of Wilkes-Barre, Pennsylvania, streets, bridges, water and gas lines, homes, and churches being affected in various ways. Legal action came before the courts in 1970.[37.22] Wherever there exist old mine workings at relatively shallow depths beneath the ground, the possibility of danger exists. Subsidence may take place suddenly as a result, for example, of slow but steady erosion of material in the workings by groundwater. Operations on the ground surface can sometimes affect the stability of these underground openings. The construction of foundations for new structures over mined areas is a matter that always requires unusual care in site investigation. The greatest caution must be observed, therefore, in urban development in any area below which mining is known to have taken place.

A score of urban areas could be used to illustrate some of the problems. The district around the city of Pittsburgh, Pennsylvania, may be cited as an example, typical but not at all unusual. Coal was first mined in Pittsburgh in 1759 by British soldiers on Mount Washington overlooking Fort Pitt, site of the present city. Those who know this busy center will know also that the area around is hilly, the floodplain elevation being about 200 m (660 ft) above sea level, the tops of the surrounding hills about 360 m (1,180 ft) above the same datum. The main coal seam, named the Pittsburgh, is at the bottom of the local Monongahela formation, but it has been eroded, in the past, over one-third of the area. Other seams are also mined locally from as much as 180 m (590 ft) below the surface. The Pittsburgh seam outcrops on hilltops in the city area but then, because of the dip of the beds, goes to depths of over 190 m (620 ft) to the south of the city. It has been extensively mined, rights to mine it being originally purchased with waivers against all surface damage. Starting in 1957, one company, recognizing the social implications of their operations, guaranteed surface safety from subsidence if approximately 50 percent of the coal under a private property was purchased by the home owner and left in place. This practice has been extended. In one 10-year period, one company that had made such a guarantee provided support for 635 homes and had to repair only 2 percent of the total. In 1966 there was enacted the Bituminous Mine Subsidence and Land Conservation Act of the Commonwealth of Pennsylvania, which is designed to prevent undermining that would damage public buildings or noncommercial property. There has been accumulated in the Pittsburgh area, therefore, a large volume of experience with a wide variety of subsidence problems. Much of this has been placed on the record and is now available for the guidance of others facing similar problems.[37.23]

FIG. 37.8 The cavity formed on the (old) Furness Railway at Lindale, near Barrow, Lancashire, England, into which a locomotive sank and disappeared in 1892.

FIG. 37.9 Pittsburgh, Pennsylvania; photograph taken with a borehole camera showing aggregate filling of an old coal working, in contact with the roof of the old excavation.

Accurate knowledge of all subsurface conditions and especially of the extent of old mine workings is a starting point for all new use of land in this area. Plans of mine workings are helpful, if available, but this is not always the case with the older mines. Ingenious methods have been developed for examining old workings, if they cannot actually be visited and surveyed directly. One method has been to drill 15-cm (6-in) -diameter holes from the surface into the mine workings, then to lower a special rotating borehole camera down the hole; with appropriate fittings, the camera can be operated from the surface to photograph the area around the bottom of the hole (Fig. 37.9). In this, and in other ways, it is possible even today to prepare good maps of the outline of old workings and then to relate these to the locations proposed for new structures. When this was done for a proposed new school, for example, it was found that safer foundation conditions would be obtained if the building was moved about 240 m (800 ft) from the site originally proposed. Arrangements were made with the local authorities to exchange the original site for a new one at the correct location, park land being involved, so that the school was finally built with 19.8 m (65 ft) of rock between the foundation and the mined-out Pittsburgh coal seam, 70 percent of which still remains in place and will not be touched.[37.24]

Methods of Avoiding Subsidence When sites cannot be changed, then other methods to avoid subsidence must be used. They are necessarily expensive, since new supports may have to be provided deep below the surface to take the place of the support originally given by the coal when in place. Various engineering methods can be used to achieve this result, starting with the construction of solid concrete piers or even walls directly beneath the foundation locations for the structure to be erected above. Alternatives include varieties of grouting the ground, in some cases sand and gravel being introduced into the old workings and then grouted up into a solid and load-bearing mass. Precautions such as these require engineering judgment for their selection and use, but this can be exercised only on the basis of complete geological information about the subsurface conditions. In some areas it is necessary to consider such provisions for worked-out coal seams as deep as 45 m (148 ft), but the necessity for subsurface rehabilitation work will vary considerably, depending on local geological formations. An alternative procedure, if the danger of subsidence is remote, is to include in the building design provisions for jacking it back into place if settlement does take place after construction.

This has been done in some cases in Pennsylvania, but it has become a common practice in the Midlands of England, where large areas are underlain by old coal workings. The very simple device of founding large tanks on three supports instead of the usual four has been used in some places. The Nottinghamshire County Council pioneered with a simple design for light school buildings that included in the structural-steel frameworks adjustable joints that could be used after the buildings were completed to make quite appreciable changes in line and elevation, should subsidence take place beneath the building. A three-storey school building constructed as an articulated steel frame using this "clasp" system was almost complete when coal mining beneath it caused a differential settlement, from one end of its 51-m (168-ft) -long concrete basement slab to the other, of over 30 cm (12 in). Located at Heanor in Derbyshire, England, the school was built over the valuable Piper coal seam that was worked out in the summer of 1963. Since there had been a maximum settlement of 0.93 m (3 ft) in an existing school building, the new building was designed with settlement in view. The concrete base slab was articulated in two sections, 15 m and 36 m (50 and 118 ft) long, respectively, and 12 m (40 ft) wide. One end settled 52 cm (21.4 in) and the other 16.5 cm (6.5 in). All joints, including that in the roof, worked as intended, and "no damage which would have rendered the building unfit for use had it been completed" resulted (Fig. 37.10).[37.25] Such provisions suffice, of course, for small subsidences only, but local experience and knowledge of all underground conditions—including the geology—will serve as a sound guide to the adoption of such design features.

What has given new urgency to this problem in coal-mining areas is the trend toward the construction of high, multistorey buildings with their increased foundation loadings. Again, this is a worldwide movement, and so the special problem of founding such tall buildings over old coal workings has received much attention. An unusually comprehensive review of the matter has been published in Great Britain.[37.26] This excellent paper points out that there were (when it was written) probably 400 to 600 multistorey buildings being constructed in the United Kingdom each year, of which possibly one-third were being built in areas overlying the famous Coal Measures that provided one of the foundations for Britain's industrial eminence in the earlier part of the Industrial Age. It is further observed that the loads imposed on the underlying foundation beds by these large buildings are similar to those imposed by many high dams. The paper, to which reference should be made by all who have to face similar problems, confirms what has already here been said about the vital necessity for accurate subsurface information, the detailed investigations of the old mine workings beneath any building site, and the careful selection of the exact location of buildings based on the study of this information. Many cases are described. A group of three multistorey blocks in Leeds, for example, was relocated and reoriented by varying amounts to avoid two old mine shafts and a third possible shaft, as well as to fit better to the two underlying coal seams, one of which had been worked out and one of which was still untouched.

A particularly difficult problem had to be faced in connection with the erection of a 16-storey block of 126 flats for the county borough of Gateshead in northeast England. It was known that the gently sloping site is underlain by glacial till on top of nearly horizontal strata of productive Coal Measures. Mine records were available only for coal workings below a depth of 90 m (300 ft). It was not known if the upper coal seams had been worked in much earlier days. A test borehole revealed six coal seams in the upper 90 m (300 ft), three of which appeared to have been worked from the cavities encountered in drilling; shallow workings were also found under part of the site. Cavities 90 to 120 cm

37-20 SPECIAL PROBLEMS

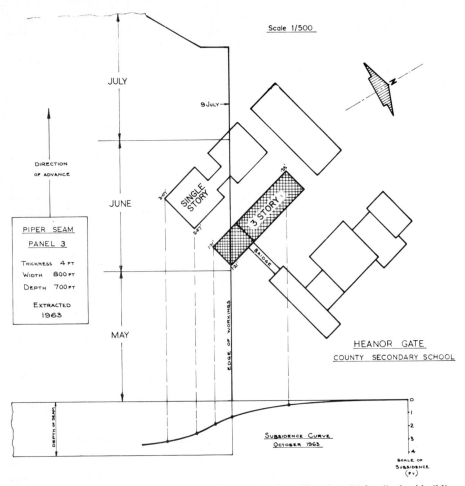

FIG. 37.10 Heanor, Derbyshire, England; diagrammatic plan of location of "clasp" school building with articulated foundation slab, in relation to coal workings.

(35 to 47 in) high were found by further drilling, but some had collapsed and were now filled with loosely packed debris fallen from the roof.

It was decided after most careful study to consolidate the shallow workings and the topmost of these old seams, the other two being at depths of 45 and 72 m (147 and 236 ft), respectively. To do this, holes were drilled around the perimeter of the site and grouted up to provide an enclosing dam within which the remainder of the grouting could then proceed. Pressures up to 15.5 kg/cm^2 (220 psi) had to be used, but assurance was gained that the upper 18 m (59 ft) of rock under the building foundation had been well consolidated. A rigid concrete-raft foundation under the complete building block was decided upon. Arrangements were included in the structural design for some jacking in the future if, despite all the precautions, subsidence of the supporting ground still took place. These details are thus briefly recorded to indicate that, even on a most difficult site, with normally inaccessible, worked-out coal seams now rubble filled, it is possible with modern engineering and construction techniques to achieve a sound design for a 16-storey building, provided, as always, due attention is given to obtaining, in the first place, an accurate picture of the exact geological conditions beneath the site. (Figure 37.11 illustrates a cross section of a site similar to that described here.)

Even gold mining has to be considered in relation to ground subsidence in some areas. Johannesburg is world-famous for its gold ore. In earlier days, this was found almost at the surface and within the limits of what is now the great modern city. The old workings have long since been unused, but they are still there. They necessitate severe restriction of modern building in a great wide strip of land (the old reef) running through the city from east to west immediately to the south of the main city center. Ground movements still take place, and so this area has limited capability of development. It was, however, ideally located for road construction when a great system of freeways was planned in the early sixties. Land for the east-west freeway, most conveniently located (as can be imagined) to serve central Johannesburg, was acquired at relatively low cost. From the planning point of view this was all to the good, but the design and construction of this modern highway presented most unusual problems. Ground movements over the old workings had to be included as definite possibilities, and yet a wide overhead modern highway had to be constructed with certainty of performance. Adding to the complications was the fact that some gold still remains in some of the old shallow workings. Although then uneconomic to remove, it had to be left available for removal in case the price of gold should go up. The engineers responsible were able to lay down limits of movements to be allowed for in design, given strict control over any future mining. But these restrictions are considerable: a 5-cm (2-in) change in grade in 30 m (100 ft), longitudinal movements of up to 95 mm (2.5 in) in 30 m (100 ft) in any individual span, and a maximum cumulative movement of 25 cm (10 in) in a group of spans. The engineering solution is one of the greatest

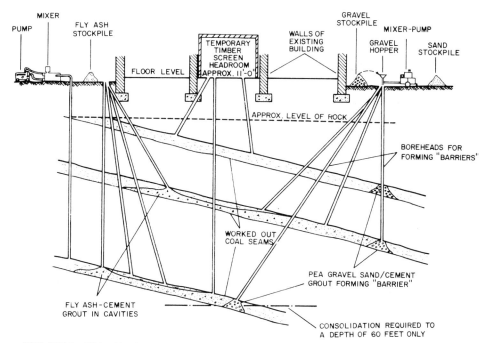

FIG. 37.11 Kirkcaldy, Scotland; diagrammatic cross section showing coal seams and injection boreholes used for filling old mine workings in order to stabilize the building site above.

interest—the supporting piers, which weigh as much as 600 tonnes, are so mounted on bearings that, if and when necessary, they can be jacked either vertically or horizontally to compensate for any ground movement that may displace them. Few of the motorists who now use this fine highway can ever realize upon what an unusual structure they are riding.[37.27]

In summary, if subsidence due to mining operations of the past is a possibility in any area to be developed for urban purposes, study must be made of the probable extent of any subsidence that may occur. Special requirements for the design of all structures can then be prepared and designs completed with every certainty of satisfactory performance of the structures when built, provided adequate maintenance and inspection is afforded. If there is any doubt as to the possible existence of old mine workings under any area to be developed, every possible source of information must be followed up to determine whether old workings do exist or not. Preliminary site investigations must be planned and executed with this possibility as one underground feature to be watched for. In new areas, and for new cities, the possible existence of valuable ores, coal, or sand and gravel must be determined in advance of any detailed planning. Once the local geology is known and the exact location of any mineral deposits has been determined, then study can be made about their extraction. If the deposits are close to the surface it may be possible to extract them before the area is used for building. If they are at shallow depths but not available by surface working, then regulations can be drawn up in advance governing their future extraction, so that surface subsidence will be under strict control at all times. There are many fine examples of such regulations that can be drawn upon for guidance, those of the Dortmund Board of Mines (West Germany) being an excellent example. If mining is to be deep below the ground, it will be possible to determine in advance with reasonable certainty what chance there is of any future subsidence and provision made accordingly in the urban planning process. Correspondingly, the shafts necessary for deep workings can be located in advance so as to fit with the overall plan.

37.4 CONCLUSION

The possibility of land subsidence due to either the pumping of oil and/or water or to mining operations is now so well recognized that most developed countries have well-drafted regulations governing the continuation of such operations already in progress, and the commencement of any new projects in areas where this hazard may be encountered. This is part of the general attention now being given to town planning, which must, naturally, be conditioned by any possibility of land subsidence. In areas far removed from urban development, however, the same attention is being given to the possibility of land settlement which may affect civil engineering works. Especially notable have been the investigations carried out by and for the Metropolitan Water, Sewage and Drainage Board of Sydney, Australia. Coal beds underlie some of the catchment areas utilized by this notable agency and so under some of its dams. In cooperation with the government of New South Wales, rigid controls are exercised over mining in all areas in which the board has interest. Although coal measures exist under the Warragamba Dam, for example, no mining has been, or will be, carried out under the dam or under its reservoir area.

All such regulations are naturally based upon a full appreciation of all aspects of the local geology, this being but one example of the absolute necessity for geological study as a preliminary to any and all land planning. There are all-too-many examples of trouble due to land subsidence that was not anticipated when urban development was initiated, the case of Niigata in Japan being one of the most remarkable. In his notable first volume, *Engineering Geology of Southern Africa,* Brink gives some illuminating examples of land subsidence due to both deep and shallow mining operations and usefully stresses that, despite all advances in theoretical approaches, subsidence is still unpredictable to a degree, dependent as it is upon geology, with all the variety in underground conditions that geological features may introduce.

The attention now being given to this subject is well illustrated by the number of meetings, local and international, now being held about land subsidence. A guide to some of the more useful proceedings of such meetings is given at the end of this chapter. Special note may be made of the conference sponsored by UNESCO, held in Japan in 1969, dealing only with subsidence due to pumping; to the report published in 1959 by the Institution of Civil Engineers in London, England, on problems of mining subsidence; and to the *Subsidence Engineer's Handbook,* published in 1975 by the National Coal Board of Great Britain. By way of contrast, and as a fitting conclusion to this chapter with its unavoidable recital of unfortunate examples of land subsidence, the reader is reminded that knowledge of subsidence mechanisms in some parts of the world is now so far advanced that it was possible to lower the harbor of Duisburg, in West Germany, by predetermined amounts, in order to restore requisite water depths at its wharves. This was accomplished by mining coal under strict control from beneath the harbor area (see Chap. 2). Land subsidence is not all bad, therefore. It may be expected that as unwanted land subsidence is reduced by the exercise of geologically determined controls, so also may desirable subsidence be achieved, in the few places where it is needed, by equally constructive applications of geological knowledge.

37.5 REFERENCES

37.1 "Land Subsidence of 15 in Follows Earthquake in Utah," *Engineering News-Record,* **114**:322 (1935).

37.2 "Rising Land May Presage Earthquake," *Engineering News-Record,* **184**:11 (26 February 1976); and subsequent news reports.

37.3 G. Shumway, G. B. Dowling, G. Salsman, and R. H. Payne, "Submerged Forest of Mid-Wisconsin Age on the Continental Shelf off Panama City, Florida," in abstract vol. for annual meeting at Cincinnati, Ohio, 1961, Geological Society of America, Boulder, Colo., p. 147A.

37.4 D. Wilkinson, *The Arctic Coast,* Illustrated Natural History of Canada Series, NSL Natural Science of Canada, Toronto, 1970, p. 50.

37.5 R. W. Fairbridge, "The Changing Level of the Sea," *Scientific American,* **202**:70 (May 1960).

37.6 F. W. Clarke, "Some Implications of Crustal Movement in Engineering Planning," *Canadian Journal of Earth Sciences,* **7**:628–633 (1970).

37.7 O. E. Meinzer, "Compressibility and Elasticity of Artesian Aquifers," *Economic Geology,* **23**:263 (1928).

37.8 E. E. Stohsnet, "Santa Clara Valley Subsidence Has Now Reached 5 Feet," *Engineering News-Record,* **118**:479 (1937).

37.9 C. H. Neel, "Surface Subsidence of a Naval Shipyard," *The Military Engineer,* **49**:432 (1957).

37.10 "Water Buoys Land That Sank as Oil Was Removed," *Engineering News-Record,* **163**:26 (12 November 1959).

37.11 J. J. Collins, "New Type Seawall Built for Subsiding Lake Shore in Venezuela," *Engineering News-Record,* **114**:405 (1935).

37.12 C. Berghinz, "Venice Is Sinking into the Sea," *Civil Engineering,* **41**:67 (March 1971).

37.13 Reuter's press reports in North American newspapers, 22 and 23 October 1971; see also, S. Fay and P. Knightley, *The Death of Venice,* Deutsch, London, 1976.

37-24 SPECIAL PROBLEMS

37.14 F. Mooser, "Informe sobre la geologia de la cuenca del Valle de México y zonas colindantes" (Report on the geology of the Valley of Mexico and adjacent area), Comision Hidrologia de la Cuenca del Valle de México, Oficina de Estudios Especiales, 38 pp., 1961.

37.15 J. F. Poland and G. H. Davis, "Land Subsidence due to Withdrawal of Fluids," in D. J. Varnes and G. A. Kiersch (eds.), *Reviews in Engineering Geology,* vol. II, Geological Society of America, Boulder, Colo., 1969, p. 216.

37.16 A. S. Allen, "Geologic Settings of Subsidence," in D. J. Varnes and G. A. Kiersch (eds.), *Reviews in Engineering Geology,* vol. II, Geological Society of America, Boulder, Colo., 1969, p. 305.

37.17 P. Weaver and M. M. Sheets, *Guide Book for Excursion No. 5,* for annual meeting at Houston, November 1962, Geological Society of America, Boulder, Colo.

37.18 G. H. Davis, J. B. Small, and H. B. Counts, "Land Subsidence Related to Decline of Artesian Head in the Ocala Limestone at Savannah, Georgia," in abstract vol. for annual meeting at Pittsburgh, 1958, Geological Society of America, Boulder, Colo., p. 27A.

37.19 Q. Zaruba and K. Hromada, "Techniko-Geologicky Rozbor Území Města Kutné Hory" ("Geologico-Technical Surveying of the Town of Kutná Hora"), *Svasek 9,* Geotechnica, Prague, 1950; and information in private communications from Dr. Zaruba.

37.20 J. E. Richey, "Surface Effects of Mining Subsidence," *Transactions of the Society of Engineers,* **55**:95 (1952).

37.21 News report in the *Lancaster Guardian* for 24 September 1892, noted in *The Railway Magazine,* April 1958; and personal communication from E. H. Scholes, Lancaster.

37.22 *The New York Times,* 12 February 1970, p. 49.

37.23 C. I. Mansur and M. C. Skouby, "Mine Grouting to Control Building Settlement," *Proceedings of the American Society of Civil Engineers,* vol. 96; SM2, Paper 7166, New York, 1970, p. 511.

37.24 R. E. Gray and J. F. Meyers, "Mine Subsidence and Support Methods in the Pittsburgh Area," *Proceedings of the American Society of Civil Engineers,* vol. 96; SM4, Paper 7407, New York, 1970, p. 1267.

37.25 F. W. L. Heathcote, "Movement of Articulated Buildings on Subsidence Sites," *Proceedings of the Institution of Civil Engineers,* **30**:347 (1965); see also **33**:492, 1966.

37.26 D. G. Price, A. B. Malkin, and J. L. Knill, "Foundations of Multi-storey Blocks on the Coal Measures with Special Reference to Old Mine Workings," *Quarterly Journal of Engineering Geology,* **1**:271 (1969).

37.27 "Elevated Roadway Adjusts to Bad Ground," *Engineering News-Record,* **178**:50 (27 April 1967).

Suggestions for Further Reading

Mining authorities throughout the world have published useful guides to the prevention and control of surface settlements due to mining operations. One of the most helpful of these in English is *Subsidence Engineer's Handbook,* a 1975 publication of 111 pages published by the Mining Department of the National Coal Board of England; it is exceptionally well illustrated. Many conferences on the subject have been held in recent years, in view of public interest in the problem. Typical of the more local meetings was the second Conference on Ground Movement and Structures, under the auspices of the University of Wales and the Institution of Structural Engineers held in Cardiff in 1980; one of the papers in the proceedings is "Subsidence over Abandoned Mines in the Pittsburgh Coalbed" by R. W. Bruhn, M. G. Magnuson, and R. E. Gray. UNESCO organized as one of the several recent international meetings a Workshop of Ground Subsidence held in Mexico in September 1979. The proceedings are available from UNESCO, 7 Place de Fontenoy, 75700 Paris. In North America, the American Society of Civil Engineers published *Evaluation and Prediction of Subsidence* in 1979, a 594-page report on one of their conferences. It must be stressed that these are but a few of the many valuable recent meetings and publications on mining subsidence, but reference to them will be at least a start at tapping the extensive and valuable information on this critical matter that is now available.

chapter 38

SINKHOLES

38.1 Sinkholes in Limestone / 38-3
38.2 Sinkholes due to Salt / 38-5
38.3 Sinkholes due to Pumping / 38-8
38.4 Karst Country / 38-9
38.5 Detection and Remedial Works / 38-12
38.6 References / 38-13

Sinkhole is a word that is self-explanatory. In geology the term is applied to any marked, localized depression in the surface of the ground that has obviously been caused by removal of underlying material. Sinkholes are usually roughly circular in plan but not necessarily so. They may range from small holes 0.3 to 0.6 m (12 to 24 in) in diameter to large depressions 100 m (330 ft) or more in diameter. They are usually the result of underlying limestone but again not necessarily so. The large depression now occupied by Crater Lake in Saskatchewan, about 32 km (20 mi) south of Yorkton, is thought to be due to solution of underlying salt beds; other sinkholes due to this cause are a feature of the Canadian west.[38.1] Culpepper's Dish, a well-known feature in England, has a circumference of almost 180 m (600 ft); its sides slope to a depth of about 45 m (150 ft). A district in Jamaica is known as the "Cockpit Country" because of the sinkholes found there in the white limestone; they vary from shallow basins to pits 150 m (500 ft) deep. An interesting case occurred in connection with the disposal of sewage from the city of Norwich, England; sewage was conveyed to a farm at Whitlingham, and the land on which it was discharged was soon covered with holes from 0.9 to 1.5 m (3 to 5 ft) in diameter and of varying depth; the washing out of sand galls in the local chalk formation had resulted in these sinkholes in the surface material.[38.2]

Usually, however, it is the slow solution of limestone by acidic water that eventually results in holes that suddenly evidence themselves at the surface, either directly or by the collapse of the soil or other material above the cavity. The slowness of this action must be emphasized, especially since the dissolving water is mainly rainfall, with its minute content of carbon dioxide turning it into a very weak acid. It has been estimated, for example, that the rainfall in Kentucky (where many sinkholes are found) will dissolve a layer of limestone 1-cm thick in a period of 66 years.[38.3] Humic acids in soil also assist with the process of weathering. Sinkholes are a common feature of the landscape in many regions throughout the word in which limestone or dolomite is the local bedrock. Parts of

Spain are characterized by sinkholes, some due to the solution of gypsum and the consequent collapse of overlying soil. Readers who saw either of the 1970 films *Patton* or *Cromwell* might find it hard to realize that they were both filmed (almost simultaneously) on the Urabasa Plain in Spain. One of the operational difficulties of the respective directors was the occasional disappearance of tanks *(Patton)* or horses *(Cromwell)* into the "underground caves," as they were described, new sinkholes in this context.[38.4]

In England, the county of Cheshire is distinguished in its landscape by *meres*, small shallow lakes, yet another evidence of sinkholes, in this case due to the removal of the underlying salt about which more will be said later. And in many parts of the United States, more particularly toward the south, there are great areas in which sinkholes are either evident or are a potential hazard to development. Many project sites located on carbonate rocks in the area served by the Tennessee Valley Authority, for example, have solution cavities, the first phase of sinkhole formation, as one of the characteristic features of the subsurface. In all limestone and dolomitic areas, therefore, site exploration must be conducted on the assumption that sinkholes may be present, every effort being made to determine the soundness and continuity of bedrock near the surface.

Because of their mode of formation, natural sinkholes (as distinct from those caused by human activity below ground) will be irregular in occurrence. The solution channels in carbonate rocks may start out along joint planes but thereafter develop in quite heterogeneous fashion. There can be no certainty, therefore, about the absence of sinkholes on building sites in "sinkhole country." The only practicable solution, after preliminary site studies, is to sink a test hole at the site of every important column or load-bearing member, a costly procedure at first sight but a most desirable "insurance procedure." One has only to think of such an accident as occurred in Akron, Ohio, in 1969 when the collapse of part of the roof of a six-year-old department store caused one death and injuries to ten people, the collapse being attributed to settlement of one column that had been founded over an undetected sinkhole, to realize how essential is such detailed subsurface investigation if there is any possibility at all of sinkholes beneath a building site.[38.5]

The extent of underground solution channels in limestone was shown by the careful study made of subsurface conditions at the site of a new paper mill for the Bowaters Company at Calhoun, Tennessee. Knox dolomite is the underlying bedrock. In some places 10 percent of the rock by volume had been dissolved away by percolating waters. Cavities were found to extend to depths of more than 60 m (200 ft) below the level of the adjacent Hiwassee River.[38.6] Many more examples could be cited to show that buildings and other structures can be safely and economically founded on carbonate rocks, despite solution channels and the sinkholes that evidence these at the surface, if adequate subsurface investigations are first carried out, planned in detail in cooperation with careful study of the local geology.

Although they can cause difficulties in some aspects of urban development, and especially in foundation design and construction, sinkholes can sometimes be turned to good effect. The fact that the solution channels in carbonate rocks that lead to the formation of surface sinkholes are caused by the flow of water indicates that there must be some underground outlet for the water that moves through the rock. If the local geology is studied carefully and the underground course of percolating water determined with accuracy, sinkholes may provide a convenient means of disposing of surface waters. This can be done with safety and assurance only with the information that geological study can give. When underground conditions are known with certainty, however, excellent drainage systems can be developed by leading surface drainage through appropriately designed channels to one or more sinkholes, the sides of which must naturally be finished off properly to ensure their stability and successful performance when drainage water comes pouring into them.

The city of Springfield, Missouri, has used this system in a number of locations, turn-

FIG. 38.1 Natural sinkhole converted for use as a drain for surface water by the City of Springfield, Missouri.

ing this hazard into a civic convenience (Fig. 38.1).[38.7] The airport of Bowling Green, Kentucky, is completely drained in this way, all drainage ditches leading to selected sinkholes that were incorporated into the drainage system by appropriate design.[38.8] In road construction in southern England, there have been a number of cases when surface drainage, both permanent and temporary, has made use of the "swallow holes" or "soak holes" that exist in the Chalk bedrock that has called for such frequent prior reference. Being also a carbonate rock, the Chalk is sometimes featured by solution cavities and channels, disappearing streams being one of the more interesting evidences of this.[38.9] The regular use of sinkholes for drainage purposes is naturally for the conduct of surface waters only, but even they may contaminate either the groundwater with which they may come into contact or the streamflow into which they will eventually discharge after their underground journey. It is not surprising to find that, in the past and before modern sanitary controls, the existence of convenient subterranean cavities provided an open invitation to economically minded persons and organizations for disposal of their sewage and other wastewaters.[38.10]

38.1 SINKHOLES IN LIMESTONE

As an increasing number of records of sinkholes have been published, it has become abundantly clear that they occur most generally in country underlain by limestone, with some

38-4 SPECIAL PROBLEMS

due to underground solution of dolomites and gypsum. Examples are known with other bedrock types, but (with the exception of rock salt) they are unusual. Most sinkholes appear to form in residual soil ranging in thickness from 12 to 30 m (40 to 100 ft), according to a valuable study by Williams and Vineyard.[38.11] These authors report that 46 out of a total of 97 catastrophic surface failures in the state of Missouri, recorded since the 1930s, were caused in some way by human activity. These actions were, naturally, not deliberate, but they had the effect of triggering the formation of a sinkhole which might eventually have formed from natural causes. Williams and Vineyard record, as a notable case from their own state, a highway failure in Pulaski County in February 1966. It was initially triggered by water discharging from the downstream end of a box culvert under the road, which concentrated surface drainage at the one spot, but traffic vibrations appeared to have been a contributory factor to the collapse of the road, a collapse occurring over the roof of a cave that was slowly enlarging upward in residual soil. They also record that highway construction in Hannibal caused small surface depressions in a limestone formation through which the road was being built; in this way the formation was found to have a maze of caves. Three boys entered one of these openings and were never seen again.

Several states have experienced troubles with sinkholes either during highway construction or, more seriously, after a road has been completed with unsuspected cavities beneath it. As but one example, a bad sinkhole developed on Bullfrog Valley Road near Hershey, Pennsylvania, between 24 February and 2 March 1979, completely wrecking the road structure.[38.12] It started as a small depression on one side of the road but gradually spread, eventually leading to the complete collapse of the road, as one motorist discovered to his chagrin when he went around a safety barrier (Fig. 38.2). Near Tarpon Springs, Florida, a highway bridge on U.S. Route 19 collapsed suddenly in December 1968, unfortunately killing one person and leading to injuries to five others. Sinkhole formation in the Anclote River, which was crossed by the bridge, was blamed for the accident.[38.13]

Even more remarkable was the collapse of a timber bridge on a highway over Piscola Creek in the state of Georgia in 1976. The bridge was to be replaced and test drilling for

FIG. 38.2 An effective barrier to smooth driving; road destruction caused by a sudden sinkhole development.

FIG. 38.3 Highway bridge over Piscola Creek, Georgia, after part of it had disappeared into a sinkhole that developed unexpectedly; the sinkhole was satisfactorily sealed soon after this photo was taken.

a new bridge was in progress. As part of the site investigation a 15-cm (6-in) hole had been put down to a depth of about 27 m (90 ft), apparently without encountering any underground cavity, although the site is underlain by dolomitic limestone. The drilling crew left the site in perfect shape on a Friday afternoon, but by the next day a local farmer noticed that a hole almost a meter in diameter had opened up, and by Sunday the damage was so serious that the bridge had to be closed. Figure 38.3 shows the final appearance of the bridge, part of which actually disappeared into the sinkhole which had developed to a considerable size as a result of water flowing from the creek down the exploratory drill hole into a deep underground cavity.[38.14]

This sad recital could continue at great length but all to the same effect—a record of sinkholes developing in limestone country with little if no warning, causing serious damage to engineering structures. There are even records of residential housing developers who were warned of the possibility of sinkholes on land which they planned to subdivide, but who, because there was no initial evidence of this at the surface, neglected the warnings and then were faced with serious damage to their property after their changes to the natural drainage of the area had aggravated sinkhole formation. (Fig. 38.4.) In all work in limestone country, therefore, a special feature of site investigation must be the most thorough search for any evidence of sinkhole formation anywhere near the site. Methods to be followed will be described at the end of this brief chapter.

38.2 SINKHOLES DUE TO SALT

Attention has so far been concentrated on sinkholes in limestone country, but the same phenomena are also experienced in areas from under which salt is being extracted, either by mining or pumping. Although not nearly so widespread as limestone bedrock, massive

38-6 SPECIAL PROBLEMS

FIG. 38.4 View from the air of the large sinkhole which suddenly developed at Winter Park, Florida, on 8 May 1981, after a prolonged dry spell, destroying one home and adjacent property.

deposits of common salt are not uncommon. Since salt is soluble in water, the potential danger can readily be seen. Salt is not so strong a material as limestone, and so subsidence can take place at the surface, even though pillars of salt have been left as supports in mined-out areas. The plastic flow of such pillars of salt has been reported in salt mines in Kansas. Removal of salt by solution is widely practiced and is clearly a basically simple procedure and therefore undoubtedly economical. It cannot, however, be controlled in extent to the same degree as solid mining, with consequent possible uncontrolled subsidence. Some years ago serious subsidence developed very suddenly in Windsor, Ontario, where salt has long been obtained from the extensive beds underlying this part of southern Ontario and adjacent Michigan. Salt is extracted by pumping, the brine going directly to an adjacent chemical works. It was thought that the subsidence was due to collapse of old, unused salt workings at the 270-m (890-ft) level, but whatever the cause, a hole measuring about 150 by 110 m (500 by 360 ft) and about 8.1 m (27 ft) deep suddenly opened up and filled with water from the workings below. Two buildings were lost and much material damage was done (Fig. 38.5).[38.15]

One of the most famous of all salt deposits is that in the county of Cheshire, England, since it was in use even in pre-Roman times when the local springs were a source of salt for human use. The Romans introduced the more efficient process of evaporation from lead pans. There follows an unbroken period of almost 2,000 years of use of this valuable natural resource. The actual mining of the salt followed the discovery of beds of salt during test drilling for coal carried out as early as 1676 "in the liberties of Wm. Mayberry."

FIG. 38.5 Damage to property of the Canadian Salt Company, Windsor, Ontario, after the development of the sinkhole described in the text.

The extent of its use is shown by records of shipments of "white salt" rising from 17,900 tonnes in 1759 to 41,000 tonnes by 1782. By the year 1820, as much as 190,000 tonnes were being delivered by barge to the nearby port of Liverpool for further shipment. It is not, therefore, surprising to find that a volume of no less than 1,206 pages, *Salt in Cheshire*, was published in 1915. Nor is it surprising to find that throughout the area of Cheshire overlying the salt deposits there is a wide variety of evidence of land subsidence, of which the Cheshire meres are but one example (Fig. 38.6). There are two main beds, the upper one generally mined through shafts, the lower by the pumping process. When old mines collapse, funnel-shaped holes normally form, but the pumping gives generally no such localized subsidence. In more recent years, quite naturally, scientific methods have been applied to the control of all mining and especially of pumping. By maintaining accurate records and leaving concentrated brine in position in the salt beds when necessary, subsidence can be controlled within limits.[38.16]

FIG. 38.6 Tatton Mere, Cheshire, England, typical of Cheshire meres, water-filled depressions caused by solution of underlying salt beds.

38.3 SINKHOLES DUE TO PUMPING

Subsidence at ground surface due to pumping brine from the underground is understandable. It is rather more difficult to appreciate that the pumping of groundwater can cause not only gradual subsidence of large areas (as described in Chap. 37) but may even cause sinkholes to appear at the surface with dramatic, and possibly catastrophic, suddenness. Naturally, this will be only if local geological conditions include hidden cavities. A well-recorded case happened at Hershey, Pennsylvania, 14.4 km (9 mi) east of Harrisburg, near the plant of the Hershey Chocolate Company. A neighboring quarry from which high-calcium Annville limestone was obtained was kept dry by a continuous pumping program. The rate of pumping was suddenly doubled in May 1949. Groundwater levels were immediately affected in an area of about 2,600 ha (6,400 acres), including the site of the chocolate company's plant. Over 100 sinkholes appeared, some of them causing serious trouble. Groundwater-recharging operations were then started. Although these restored groundwater levels to some extent, they also increased the pumping load of the quarry company. Legal action followed, but the recharging was not stopped. The quarry company therefore started a program of grouting around their operating site, and something like the original conditions were restored.[38.17] The case well illustrates the complete interdependence of groundwater conditions over a considerable area when ground conditions permit free underground flow.

Far more serious, since human lives were involved, is an example from South Africa. Johannesburg, founded and developed in connection with the world-famous local gold mining, has seen the location of active mining change from within city limits to new mines to the west of the city. As mining has followed the gold-bearing reefs, operations have had to go deeper, and since the ground is water-bearing, pumping has had to be extended. In 1960 a major dewatering program was initiated in the Far West Rand mining district some 65 km (40 mi) to the west of the city. A thick bed of Transvaal dolomite and dolomitic limestone (up to 1,000 m or 3,300 ft thick) overlies the Witwatersrand gold-bearing beds. Thick, vertical syenite dikes cut across the dolomite and divide the great bed into compartments, one of which, the Oberholzer, is at the center of the new mining area. Large springs flowed in the vicinity of these dikes prior to 1960, but the big pumping program soon dried these up and then started to lower the entire water table in the compartment. Between 1962 and 1966 eight sinkholes larger than 50 m (165 ft) in diameter and deeper than 30 m (100 ft) had appeared, together with many smaller ones. In December 1962 a large sinkhole suddenly developed under the crushing plant adjacent to one of the mining shafts; the whole plant disappeared and 29 lives were lost (Fig. 38.7). In August 1964 a similar occurrence took the lives of five people as their home dropped 30 m (100 ft) into another suddenly developed sinkhole.[38.18]

This is the briefest of summaries of a most complex situation, the circumstances of which are such that it is improbable that any similar occurrence will develop. For those who wish to know more about the case, reference may be made to Dr. A. B. A. Brink's outstanding book, *Engineering Geology of Southern Africa*.[38.19] An excellent summary of the sinkholes which developed between 1962 and 1975 is there given, the last entry being of a sinkhole, 20 m (66 ft) wide and 7 m (23 ft) deep, that opened up under a railway track just as a train was approaching. Although coaches were derailed, fortunately no lives were lost. Dr. Brink summarizes the studies that have been made of these occurrences and outlines the probable explanations, the basic cause being the existence of a cavity in the residual soil above the dolomite, the roof of which collapses because of some change in underground conditions, thus creating the sinkhole. In cases where the same sort of action occurs underground, with no rupture at the surface and with a rather slow subsidence over a small area, the periphery characterized by tension cracks in the ground, the name *doline*

FIG. 38.7 Sinkhole which engulfed the crusher plant at the West Driefontein mine, South Africa, in December 1962.

is used in South Africa; the term indicates what is a very limited example of compaction subsidence.

Naturally, South African authorities now exercise the strictest control over all building proposals in areas possibly subject to the development of sinkholes. A 30-m (100-ft) grid of geophysical stations must be investigated using the gravimetric method, and appropriate drilling of exploratory boreholes is mandatory before a township may be proclaimed; and these are minimum requirements. Protective methods are in force to guard the stability of existing structures in the sinkhole region, these including the absolute control of all water at the surface, exploratory drilling, and the installation of ingenious telescopic benchmarks which will indicate directly any differential ground movement at the various depths at which units of the benchmark are installed. Reference to Fig. 38.8 will, perhaps, lend credence to the suggestion that, based upon personal inspection of two of these sinkholes by one of the writers, there are few more vivid manifestations of the possibly critical interplay of unusual geological conditions with human activities.

38.4 KARST COUNTRY

Sinkholes have been known for a long time, naturally occurring examples from time immemorial and those due to man's underground operations of some sort for many decades. It has been, however, only in the years since the end of the Second World War that widespread attention has had to be given to them. This has followed upon the quite remarkable developments in major road building, in the construction of dams, and in the

FIG. 38.8 The edge of a sinkhole that developed rapidly in the area west of Johannesburg, South Africa, after deep pumping for mine drainage had started.

rapid expansion of urban areas. Sinkholes have caused serious troubles in a number of areas in all these branches of engineering work, as also in many others. Concurrent with these developments there has been a corresponding increase in professional circles in the exchange of information about such troubles. There has gradually grown up an appreciation of the fact that sinkholes do not occur haphazardly but are confined, in general, to certain well-delineated regions, characterized by the presence of limestone or dolomite as the bedrock at or near ground surface. To such areas the name *karst* has now been widely applied. (Fig. 38.9.)

The name derives from a beautiful part of Yugoslavia in the Dinaric Alps adjacent to the Adriatic coast which has long been distinguished for its multitudinous natural sinkholes. Both the name *karst* and the derivative *karstic regions* are now in common usage in describing areas in which sinkholes do occur naturally or may be induced by the works of man. Several states, including Alabama, Georgia, Missouri, and Pennsylvania to mention just a few, have such areas within their boundaries. As indicating the widespread interest in, and concern about, such areas, the International Association of Engineering Geologists arranged a special "symposium-by-correspondence," which was published in its Bulletin no. 12 in 1975.[38.20] Organized by Professor Calembert of Belgium, the symposium includes contributions from the following countries: South Africa, West Germany, Belgium, Brazil, Canada, Spain, France, Great Britain, India, Ireland (Eire), Italy, Switzerland, Czechoslovakia, the United States, and Yugoslavia (the order being alphabetical in French).

Further evidence of current interest, and of the availability of useful information on karst, is given by the symposium held by the Association of Engineering Geologists (U.S.) as a part of its annual meeting of 1978, appropriately enough in Hershey, Pennsylvania,

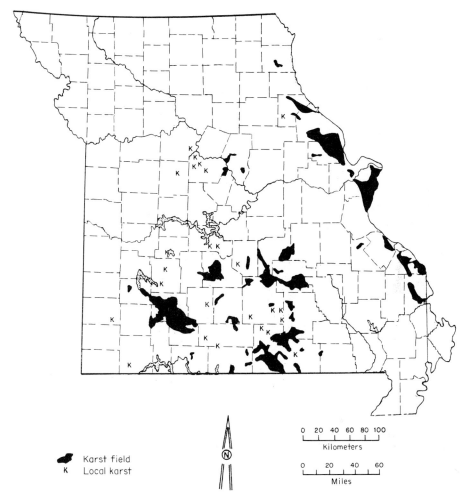

FIG. 38.9 Map of the main karst fields of Missouri, showing the wide extent of this geological phenomenon even in this one state.

which has already been mentioned in this chapter.[38.21] Useful information was then presented about the karst terrain of the Hershey Valley, about problems encountered in mining in such areas, and about the use of remote-sensing methods and geophysical methods for detecting hidden cavities in karstic areas. In a review of the experience of TVA with karstic conditions in dam foundations, A. D. Soderberg stated that of a total of 27 dams which the Authority has constructed, 15 were built on limestone foundations. Seven of their eleven fossil-fuel power plants and four of the six nuclear power plants are similarly founded on limestone bedrock.[38.22] Some of the experiences of the TVA with cavities found in some of these foundations have been related elsewhere in this book.

With Yugoslavia naturally taking pride of place, it is worth recording that Czechoslovakia has an appreciable area of karstic terrain, which has been under investigation for many years by both official geological agencies and by amateur (speleological) groups. All such areas are now specially protected by the state for scientific purposes, having been declared "natural or geological reservations," supervised by the State Institutes of Natural Reservation in Prague and in Bratislava.[38.23] Some of the caves in the karstic areas have yielded valuable archaeological remains. Engineering works in these areas have been characterized by imaginative designs.

The most spectacular karstic areas of which the writers have knowledge occur in China. Figure 38.10 gives merely a glimpse of the remarkable karstic scenery to be seen in that vast land. A volume entitled *Karst in China* (in English), if it can be consulted, will be found to contain a fine selection of photographs illustrating not only some of the resultant scenery but also some of the engineering works carried out by Chinese engineers in these difficult regions.[38.24]

38.5 DETECTION AND REMEDIAL WORKS

It will now be clear that there must be an additional factor in site investigation when it is known that the underlying bedrock is limestone or dolomite. Every precaution must be taken to ensure that sinkholes will not prove to be a problem. Study of and inquiry about the regional geology will usually reveal some indication of any natural sinkholes that may be in the area. Study of aerial photographs of the site and its vicinity will take on a new dimension, since existing natural sinkholes will usually be distinguishable. Expert geological advice should naturally be taken, especially in nonglaciated areas where residual soils may overlie the bedrock. It is in these soils that the most serious cases of sinkholes have occurred. Exploratory test drilling for detailed site investigation must be supervised with unusual care so as to detect the slightest indication of underground cavities or even channels, which, with any change in groundwater conditions, might possibly cause trouble at the surface.

Beyond this, indirect methods must be relied upon, geophysical methods as a start. All four major types of geophysical exploration have been tried, with varying degrees of success. South African experience, as recorded by Brink, has suggested that the gravimetric method has the best chance of success, electrical resistivity and seismic methods having proved of limited use for South African conditions, with electromagnetic methods giving some useful results in the detection of joints. Experimentation is proceeding in many places with other, even more indirect, methods.

FIG. 38.10 Karst topography in China; the "Stone Forest" at Lunan in Yunnan Province; scale may be judged by the human figures on the pathway left of center.

The use of microwave radiometers for determining the thickness of dust on the surface of the moon was investigated in 1965, and this led to the idea that this technique might be used for subsurface investigation on a wider scale. An experimental survey was therefore conducted in an area near Cool, California, to see if the existence of hidden caverns in the local Calaveras limestone could be detected in this way. The necessary equipment was housed in a mobile laboratory and on a 1½-ton flatbed truck, all the work being carried out adjacent to an existing highway. Soil cover varied from 3.6 to 18 m (12 to 59 ft). No locations of any cavities were known when the survey started, but the microwave radiometer results were checked later, and in almost every case they had correctly detected caverns. The significance of this work is considerable, for it gives promise of detection of a serious natural hazard with certainty and at low cost.[38.25] There are few areas of research work in the earth sciences that present more challenging opportunities than the search for more positive means of detecting subsurface cavities without the need for closely spaced test drilling.

Sinkholes will still occur. As merely one recent example, the "December Giant," a sinkhole 97 m (325 ft) long, 90 m (300 ft) wide and 36 m (120 ft) deep, developed without warning in Shelby County, Alabama, on 2 December 1972, fortunately in an isolated location in thick woods, although a resident of the nearest house was startled by a house-shaking rumble and the sound of trees breaking.[38.26] In many cases, the site must be rehabilitated. A first requirement is careful exploration of the hole to determine its depth, the cause if possible, and what connection there is between the sinkhole and underlying strata. Only when this has been done, and the possibility of further movement eliminated with certainty, can any attempt be made to rehabilitate the area that has been devastated. Remedial methods are an engineering matter, once all geological details are known with certainty, but it may be noted that practice in Missouri includes the sealing with concrete of any open joints in exposed rock surfaces and the use of rock filling, based on the concept of a graded filter. The thrust of this approach is to block off the main opening in the bedrock that has permitted water to drain downward and thus caused a sinkhole to form. Once the joints have been sealed, the void can be filled by normal methods, following standard practice for the compaction of all soil so used, with special attention given to the control of all surface drainage when natural ground level is again restored.

38.6 REFERENCES

38.1 E. A. Christiansen, "Geology of the Crater Lake Collapse Structure in Southeastern Saskatchewan," *Canadian Journal of Earth Sciences,* **8**:1505 (1971).

38.2 *The Engineer,* **51**:123 (1881).

38.3 R. F. Flint, C. R. Longwell, and J. E. Sanders, *Physical Geology,* Wiley, New York, 1969, p. 250.

38.4 K. Dzeguze, press report in *The Toronto Globe and Mail,* 12 December 1970.

38.5 "Sinkhole Causes Roof Failure," *Engineering News-Record,* **183**:23 (11 December 1969).

38.6 L. F. Grant, "Solution in Bedrock at the Calhoun, Tenn., Plant of Bowaters' Southern Paper Corp.," *Bulletin of the Geological Society of America,* **67**:1751 (1956).

38.7 E. E. Brucker, "Geology and Treatment of Sinkholes in Land Development," *Missouri Mineral Industry News,* **10**:125 (July 1975).

38.8 "Underground Channels Utilized for Airport Drainage," *Engineering News-Record,* **130**:498 (1943).

38.9 F. G. H. Blyth, *A Geology for Engineers,* Edward Arnold, London, 1945, p. 211.

38-14   SPECIAL PROBLEMS

38.10 "Unusual Flood in Ohio Town," *Engineering News-Record,* **119**:40 (1937).

38.11 J. H. Williams and J. D. Vineyard, "Geologic Indicators of Catastrophic Collapse in Karst Terrain in Missouri," *Transportation Research Record,* no. 612, 1976, p. 31.

38.12 A. R. Geyer and A. A. Socolow, "A Sinkhole Swallows a Road Before Our Very Eyes," *Pennsylvania Geology,* **10**:2 (1979).

38.13 "Sinkhole Drops Bridge and Piers," *Engineering News-Record,* **182**:15 (2 January 1969).

38.14 Commissioner Moreland, Georgia Transportation Department, by personal communication; see also "Sinkhole, Enlarged by Drillholes, Swallows Old Bridge," *Engineering News-Record,* **194**:50 (20 March 1975).

38.15 *The Toronto Globe and Mail,* and other daily newspapers for 21 February 1954; see also R. Terzaghi, "Brinefield Subsidence at Windsor," *3rd Symposium on Salt,* N. Ohio Geological Society **2**:298-307 (1969).

38.16 A. J. Calvert, *Salt in Cheshire,* Spon, London, 1915.

38.17 R. M. Foose, "Groundwater Behavior in the Hershey Valley," *Bulletin of the Geological Society of America,* **64**:623 (1953).

38.18 J. E. Jennings, "Building on Dolomite in the Transvaal," *The Civil Engineer in South Africa,* **8**:41 (January 1966); and information from Professor Jennings.

38.19 A. B. A. Brink, *Engineering Geology of Southern Africa,* vol. 1, Building Publications, Pretoria, 1979; see Chapter VII.

38.20 L. Calembert (ed.), *Engineering Problems in Karstic Regions,* Bulletin No. 12, International Association of Engineering Geology, Krefeld, W. Germany, 1975, p. 93.

38.21 R. M. Foose, "Engineering Geology in Karst Terrain, a Symposium," *Bulletin of the Association of Engineering Geologists* (a special issue), vol. 16, no. 2, Summer 1979, p. 353.

38.22 A. D. Soderberg, "Expect the Unexpected; Foundations for Dams in Karst," *Bulletin of the Association of Engineering Geologists* (a special issue), vol. 16, no. 2, Summer 1979, p. 409.

38.23 A. Droppa, P. Rysavy, and F. Skrivanek, "Organisation of Karst Investigations in Czechoslovakia," *Ceskoslovensky Kras,* no. 17, 1966, p. 112 (with English summary).

38.24 J. Silar, "Development of Tower Karst of China and North Vietnam," *Bulletin of the National Speleological Society,* **27**:35 (1965); see also *Karst in China,* Shanghai People's Publishing House, 1976.

38.25 J. M. Kennedy, "A Microwave Radiometric Study of Buried Karst Topography," *Bulletin of the Geological Society of America,* **79**:735 (1968).

38.26 "The Hole Truth," *Missouri Mineral News,* **13**:81 (May 1973).

Suggestions for Further Reading

Sinkholes are such isolated and local phenomena that they have not yet attracted the degree of attention which would have resulted in extensive coverage in publications. Since many of them occur in karstic country, useful information is to be obtained from geological publications dealing with such terrains, now being produced in several countries. International interest is indicated by the fact that one of the working commissions of the International Association of Engineering Geology has as its subject Engineering Geology in Karstic Areas. The bulletin of this Association, published in Krefeld, West Germany, is a valuable source of information on many of the subjects dealt with in this Handbook. A local publication which has wide significance is a report prepared for the City of Springfield, Missouri, in 1977 by W. C. Hayes and W. G. Russell, entitled *Urban Development in a Karst Terrain.* It could be helpful to other cities that face the problems that Springfield has dealt with so successfully.

chapter 39

LANDSLIDES

39.1 Slope Stability / 39-3
39.2 Natural Landslides / 39-5
39.3 Landslides in Sensitive Clays / 39-8
39.4 Mud Runs / 39-10
39.5 Causes of Landslides / 39-11
39.6 Preventive and Remedial Works / 39-16
39.7 References / 39-24

The term *mass wasting* is sometimes used in academic circles to denote in general one of the most widespread and effective of natural actions in the geological cycle—the movement of rock and soil under the action of gravity once they have been displaced from their normal positions. In more popular language the term *landslide* is used for the same purpose, and will be so used here, even though strictly speaking it should be restricted to movements of soil when the structural stability of a slope is disturbed. It is so used in the various classifications of earth and rock movements (mass wasting) which have been developed down the years. One pioneer effort was that of C. F. Stewart Sharpe in his book, published in 1938, entitled *Landslides and Related Phenomena*. In more recent years David Varnes has gone further in his subdivision of these evidences of ground instability. His latest system of classification was published in one of the notable volumes on landslides now available, but the authors prefer an earlier version of Varnes's table, which is reproduced, by permission, as Table 39.1.[39.1] This chapter deals with true landslides and related movements; that which follows deals with rockfalls, which are the second most troublesome type of ground movement affecting engineering works.

There has been a wealth of information published about landslides. It is the one subject touching both geology and engineering which has attracted attention ever since the start of modern civil engineering. As early as 1846, a volume dealing only with landslides on French canals was published. It will be noted later, but it was only the start of a steady stream of publications on landslides which still continue to appear. A notable volume was one published just before the close of the nineteenth century called *Slopes and Subsidences on Public Works*. In more recent years there have been a number of splendid and helpful volumes. Accordingly, all that can be done within the confines of this short chapter is to present some of the principles involved, with just a few examples. The chapter, therefore, represents nothing like a complete overview of landslides and their prevention, but

TABLE 39.1 Classification of Landslides: Types of Movement Experienced by Various Types of Material

Consolidated material	Falls	Slides	
		Few units	Many units
Bedrock	Rockfall	Rotational slump Planar block glide	Rockslide
Soil	Soilfall	Rotational slump Planar block glide	Debris slide Failure by lateral spreading

Unconsolidated material	Flows	
	Dry	Wet
Rock fragments	Rock fragment flow	
Sand and silt	Sand run Loess flow	Rapid earthflow Sand or silt flow
Mixed	Debris avalanche	Debris flow
Mostly plastic		Slow earthflow Mudflow
Complex material	Combined forms of movement	

Source: Adapted from reference 39.1.

since every landslide is in some respect unique, possibly this general treatment will be useful as an introduction to more detailed studies to which the list at the end of this chapter will be a guide.

All types of landslides are associated with movement of material constituting a part of the earth's crust, movement caused fundamentally by gravity and taking place because of some inherent instability in the arrangement of the materials concerned. All types are dependent completely on the nature of the materials involved and on their relative arrangement, in other words, on the local geology at the site of the slide. All types may occur naturally; all may develop during the course of civil engineering work. If they occur naturally, they must be regarded as an inevitable part of the general geological cycle, since they contribute to the erosion of parts of the earth's surface and thus are important factors in the processes that are developing topographical features of the world today. If they occur during civil engineering construction, they betray some interference with the natural stability of the part of the earth's crust in which work is being carried out. In all cases, although the exact cause of the slide may be difficult to determine, they will be due to either one or a combination of several natural causes, which can be determined if investigations are pursued in the proper directions.

It follows that although naturally occurring slides may be classed as "acts of God," if by that term is implied a part of the natural order of the universe, they cannot be so classified if the term is meant to suggest that the cause of a slide is a mystery. Slides do not develop their essential instability suddenly (if those few due to earthquake shocks are excepted); the movement that develops is merely the indication that a critical point has been passed. It should be possible therefore to anticipate many landslides, if the necessary preliminary investigations are made, that is to say, if the local geology is studied with unusual care.

It is to be noted that, as a part of the geological cycle, landslides are a regular and normal feature of landscape development. A few notable natural landslides will therefore be briefly described, but since these obey the basic laws of slope stability, an introduction to this theoretical background will first be presented.

39.1 SLOPE STABILITY

The main types of earth slopes which present problems of control are the sides of cuttings, the sides of deposits of fill material, and natural hillsides that have to be used in connection with civil engineering construction. Whenever such slopes are composed of material that does not occur in bedded strata, the determination of stability becomes a matter of investigating the mechanical stability of a mass of unconsolidated material conforming to the outline of the slope under consideration. Once the physical properties of the material are known, it should be possible, by application of the laws of mechanics, to determine whether or not a specified slope will be stable. Today, this is a well-accepted part of studies in soil mechanics. If an earth slope consists of coarse-grained soils—sand or gravel—then the maximum stable slope will correspond with the angle of internal friction of the material, readily determined by simple tests. There are necessary precautions to be taken to protect such slopes, but since they will be self-draining, there is rarely any question about their stability. Slopes of fine-grained soils, and especially of clay, are entirely different.

As early as 1846, the fundamentals of the stability of clay slopes were recognized by a gifted French engineer, Alexandre Collin.[39.2] He made field studies of extensive clay slips at 15 locations, primarily on the canals then being constructed with such vigor in France. He saw that they were all deep, rotational slips, generally on a cycloidal surface, starting with a tension crack at the top of the slope (Fig. 39.1). He carried out the first shear tests

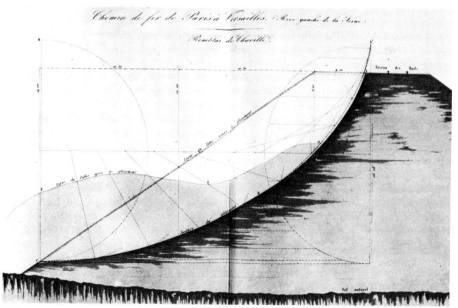

FIG. 39.1 A typical plate from Alexandre Collin's book, *Landslides in Clay* (published in Paris, France, in 1846), showing typical cycloidal form of failure surface.

39-4 SPECIAL PROBLEMS

on clay, in order to determine the properties of the soils which he might then apply in the mathematical analyses that he also carried out. In every sense, he was one of the great pioneers of soil mechanics, but his work was, for the most part, forgotten for almost a century, though it was mentioned in connection with the studies made of the Panama Canal slides.[39.3] These studies were notable, as were those carried out in connection with slides on the Kiel Canal, on the Swedish State Railways and in the harbor of Göteborg, Sweden. From these beginnings came the modern theoretical approaches to slope stability which are outlined in texts on soil mechanics.

Not until 1956, however, did a translation of Collin's book of 1846 become available in English. Reading the book is a sobering experience, for it is clear that Collin recognized the importance of quick shear tests in comparison with slow shear tests, the importance of what is now called "undisturbed sampling," and above all, the dependence of all theoretical studies of slope stability upon the local geological structure. This dependence is an essential counterpart of the most meticulous theoretical stability calculations. To say this is not to belittle the value of the theoretical approach; it is essential. The assumptions made in calculations, however, must always be checked in the field, and any geological peculiarities of the site must be carefully investigated so that the relevance of mathematical computations may be realistically assessed (Fig. 39.2). The presence of and the possible future changes in groundwater must, in particular, be carefully considered. Geology, therefore, must play a part in all investigations of slope stability.

Such studies can often be aided by observing natural slopes that have been formed, and have proved to be stable, under the same general conditions as those being considered in connection with engineering work. The correlation of natural and artificial slopes has been outlined by Ward.[39.4] Skempton and DeLory have published an intriguing account of the natural slopes found with the London clay.[39.5] Other regional studies are available, and they confirm what, well over a century ago, Collin had noted:

> Must we not, in going from the observation of fills and of cuts made by man to the observation of mountains whose sides are or seem to be immense slopes of natural fills or cuts created by upheaval or erosion, find in one as in the other material traces of the permanent force which so acts on all bodies of our planet as to submit them to its immutable laws.

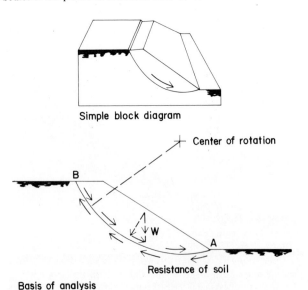

FIG. 39.2 Elements of the stability of clay slopes.

It has taken a long time for Alexandre Collin's clear guides to the stability of clay slopes to be appreciated. Despite all recent advances in geotechnique, his words are still worth reading today.

39.2 NATURAL LANDSLIDES

Natural landslides may affect the work of the civil engineer in many ways. Those which have taken place in the past may be responsible for topographical features with which the engineer now has to deal. Many major slides have blocked up river valleys, and the resulting constrictions give dam sites which at first appear to be admirable. In view of the nature of the material in slides, and its unconsolidated and often disturbed condition, sites of this kind may be very far from ideal. Dams have, however, been successfully founded at such sites. Some interesting examples have been given in a paper by W. G. Atwood.[38.6] One of the examples Atwood cites–the Farmers' Union Reservoir on the Rio Grande River in Colorado—blocks off only one-quarter of the width of the true valley in the local lava flows; the remaining three-quarters is dammed effectively, although with small leakage, by a great mass of landslide material (Fig. 39.3). The incidence of another great landslide mass upon a major dam location is mentioned in Chap. 25; the Bonneville Dam site was

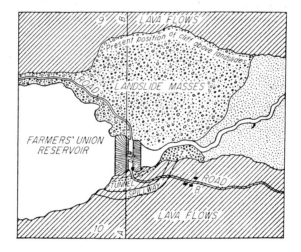

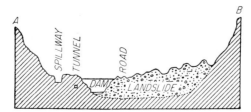

FIG. 39.3 General plan and section of the site of the Farmers' Union Dam (earth and rock fill, with concrete core wall) on the Rio Grande River, Colorado, showing relation of dam to landslide area.

created by the deflection of the Columbia River by an ancient landslide. Even more remarkable is the Cheakamus Dam in British Columbia, not only founded on the debris from an old rock slide but constructed of selected debris material.

It will be clear, therefore, that recognition of ancient landslides is an important part of geological survey work. Local topographical detail will often be a helpful guide, but the composition of the unconsolidated mass will be a more certain one. Several distinguishing features may be mentioned: the rock fragments will be angular and neither polished (like fragments in glacial deposits) nor rounded (like stream gravel); all material will obviously be the same as that found in adjacent strata (unlike the mixed materials in glacial deposits); and the disposition of the different materials in the mass will be irregular. The motion of landslides, both natural slides and those caused by engineering work, may vary from an almost imperceptible rate to what may be appalling suddenness. The great Gros Ventre slide in Wyoming filled a river valley for 2.4 km (1.5 mi) of its course with a natural dam 300 m (1,000 ft) wide at the top with an average depth of 54 m (180 ft). Eyewitnesses stated that the slide, the total volume of which was estimated at 38 million m³ (50 million yd³), occurred in less than five minutes. It was later washed out by the natural flow of the river.[39.7] Figure 39.4 shows a typical section through the main slide from which it can be seen that the slide was essentially of the structural type; slipping occurred on a bedding plane. The flow of the Thompson River in British Columbia was similarly completely blocked in October 1881 by a slide estimated to contain 60 million m³ (79 million yd³) and caused by collapse of a small, amateur-built irrigation dam on a terrace above the incised river valley; the resulting dam of slide debris was 48 m (160 ft) high until overtopped by the impounded flow of the river.[39.8]

One of the greatest landslips of which there is any record took place in Pauri Garhwal, India, toward the end of the last century. Strictly speaking, it was a rockfall, but it is always referred to as the Gohna landslide. Rising in the high Himalayas, the Birehiganga River is normally a small stream but is subject to high flood flows. It is an important tributary of the Ganges. Reports coming out of the valley of this small river in the late fall of 1893 told of "a mountain having fallen." When the district surveyor visited the village of Gohna, he found that on 22 September a vast mass of rock had slipped down a distance of almost 1,500 m (5,000 ft), creating such a dust pall that villages around could not tell for some time what had happened; the dust settled to a depth of several inches for miles around. Another slip occurred on 18 October 1893; the final natural dam thus created across the river valley was 294 m (980 ft) high above the bottom of the stream bed, 3.2 km

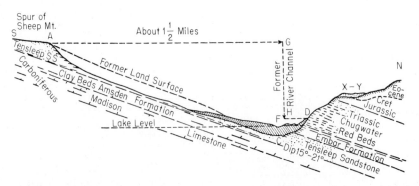

FIG. 39.4 Geological section across Gros Ventre Valley, Wyoming, showing the landslide of June 1925.

FIG. 39.5 Stereoscopic vertical aerial view of an old flow slide in the Leda clay of the Ottawa River valley, a flow slide not discernible at ground level.

(2 mi) long and 1.6 km (1 mi) wide at its base. The slide consisted of pulverized and broken dolomite, varying from hand-size boulders to blocks of 57 m^3 (2,000 ft^3) or more.

The British superintending engineer for the district decided that nothing could be done until the accumulating riverflow overtopped the great dam. The timing of this was carefully estimated, and all possible precautions were taken to avoid loss of life when the dam was overtopped. A telegraph line was run down the valley from the observation station at the dam so that ample notice of the flood could be given. In every town and village below the dam, stone pillars were erected to indicate the point below which it would be unsafe for people to remain when the flood occurred. Early on the morning of 25 August 1894, almost one year after the slide, the water began to trickle over the top of the dam. Within 24 hours, the level of the artificial lake had dropped 117 m (390 ft), and it was estimated that 28 million m^3 (10 billion ft^3) of water had rushed down in a gigantic flood. Great material damage was done; towns and villages (including the whole of Srinagar) were swept away. Because of the precautions taken, however, no lives were lost except for a fakir and his family who were destroyed not by the flood but by a slip in the face of the remaining barrier when they returned to their home after they had been unwillingly and forcibly removed.[39.9]

Fortunately, the techniques now available through aerial photography have provided a quick and reasonably certain method of seeing where old landslides have taken place. By examining aerial photographs, preferably stereoscopically, a very large number of old, previously unsuspected landslides have been found in the St. Lawrence Valley region of Canada and in Norway and Sweden. Figure 39.5 shows one example that is located close to the boundaries of the city of Ottawa. Although not more than 1.6 km (1 mi) from the office used by one of the authors for many years, neither he nor any of his colleagues had

39-8 SPECIAL PROBLEMS

ever suspected its existence until they first saw this aerial photograph, so completely had nature shielded with vegetation this disruption of several centuries ago. Subsequent studies of the slope in which the slide occurred revealed the old land surface, so that it was possible to measure and analyze the slope stability. Rigorous examination of the best aerial photographs available in a search for any evidence of old landslides must be, therefore, a vital part of the study of regional geology in almost all site investigations.

39.3 LANDSLIDES IN SENSITIVE CLAYS

Some of the largest natural landslides for which there are records are those in regions characterized by sensitive clays. Clays as encountered in Nature inherently vary a great deal. They will range all the way from stiff clay that has almost the character of shale to clay so soft that it can well be described as mud. Of particular concern in connection with landslides are those clays that appear to be stiff but are unusually sensitive in character. When clays of this type that are exposed with a sloping face are disturbed at the toes of the slopes, slides may be started. Unlike normal landslides, slides in sensitive clays can develop from small beginnings to massive earth movements, the movement taking the form of an actual flow as the excess water inside the clay is released and the initially solid material is converted into a liquid mud to an extent sufficient to carry along large blocks of the material still in its solid state. These flow slides are a familiar feature in Scandinavia and in the region of the St. Lawrence and Ottawa valleys in eastern Canada. Many examples in recent years have been studied intensively and recorded in geotechnical literature.

Of special interest in this context is the existence of a small village municipality in the province of Quebec with the significant name of Les Éboulements (one of the French words for landslides). Located 104 km (65 mi) downstream from Quebec City in the broadening section of the River St. Lawrence, it is situated on the consolidated fan caused by one of these flow slides of long ago (Fig. 39.6). Some terrible examples were recorded in

FIG. 39.6 Les Éboulements, Quebec, a village called by one of the two French words for "landslides," situated on the stable fan created long ago by a major flow slide.

FIG. 39.7 View from a helicopter shortly after the disastrous landslide on 4 May 1971 at St. Jean-Vianney in the Lake St. John district of Quebec, in which 31 lives were lost; the "flow" nature of the slide is clear.

early Scandinavian and Canadian history, slides such as that at Notre Dame de la Salette (on the Lièvre River near Ottawa) that took 33 lives and removed 2.4 ha (6 acres) of land. In more recent years, and affecting a prosperous small city, the slide at Nicolet, Quebec, on the Nicolet River, was serious and could have been much more so had it not stopped providentially just before undermining the local cathedral. It took place on 12 November 1955, just before noon, and in a very few minutes a volume of about 165,000 m³ (216,000 yd³) had flowed out from what had been a solid bank into the Nicolet River. Three lives were unfortunately lost; damage was estimated at several million dollars.[39.10]

One of the most serious of all slides in the Leda clay took place at St. Jean-Vianney, in the Lake St. John district of Quebec, shortly after 10 P.M. on 4 May 1971. The slide took place on the bank of the Rivière aux Vases, involving some houses in a recent residential development. It is estimated that about 6.9 million m³ (9 million yd³) moved during the slide from an area of 268,000 m² (320,000 yd²). Unfortunately, the slide carried 40 homes to destruction and 31 persons to their deaths (Fig. 39.7). The slide appears to have taken place within the crater of a much larger slide of the same type that took place about 300 years ago. Reports of the extensive investigation of this slide are most valuable geotechnical documents.[39.11]

Although fortunately localized within those areas underlain by sensitive clays, these

39-10 SPECIAL PROBLEMS

flow slides are mentioned not only because of their special character but because they illustrate well an important aspect of slide detection. Of the slides mentioned, two started from purely natural causes with no human intervention. The exact cause of the Nicolet slide has never been accurately determined, although natural erosion of the river bank must have been involved; some local construction activity may have triggered it. The slide at St. Jean-Vianney appears to have started because of erosion of the toe of a bank by spring flow of the Rivière aux Vases. Instability of the local soil had previously been recorded.

39.4 MUD RUNS

In the classification of landslides, a large grouping is designated *flows*. Although much less frequent than the types of landslides so far described, earth movements consisting of soil or rock in the form of streams that either are flowing or have obviously flowed are encountered in many localities; they can cause serious problems for the engineer. Most interesting geologically are the flows associated with rock-talus slopes that sometimes take on the character of *rock glaciers,* a descriptive name applied to such flows in geological literature. Only rarely do engineering works encounter this type of flow, but a section

FIG. 39.8 Mud run of newly excavated glacial clay, which was dumped from a dragline bucket; the material is flowing into the partially drained basin of Steep Rock Lake, Ontario.

of the main line of the Canadian Pacific Railway was located across the debris fan at the foot of such a flow near Mount Stephen. Mud slides cover the track on the average of once every year and so constitute a continuing maintenance problem.

Mixtures of rock and mud are more serious because of the havoc they can create, even though they are normally restricted to barren areas that may be subjected to sudden high rainfalls. *Mud runs,* as they have come to be called, are generally confined to tropical countries; they are especially severe in India and in parts of South America, where they are locally known as *huaicos.* Extreme climatic conditions will promote rapid disintegration of relatively weak rocks such as shales; the corresponding concentrated rainfall acts as an efficient transporting agent. If mud runs, for example, reach bridge openings that have been designed to take normal streamflow (of water), the consequent restriction of movement may block the opening and the results may be serious. Several sections of the Bolivia Railway had to be relocated for this reason. Careful study of the regional geology will usually reveal evidence of previous mud runs in areas where they are prevalent, almost always in mountainous terrain.

The slides in sensitive clays may, on occasion, result in what appear to be mud runs, when the excess moisture content in the clays is sufficient to turn the solid clay temporarily into a fluid condition. Figure 39.8 is an example of one such flow, so fluid that, when an extra load of excavation was dumped at its upper end, the toe of the flow would respond to this extra load within a few minutes. These special clays derive their sensitivity from their geological origin, the marine clays through deposition in saltwater. This is evidenced by the fact that their natural moisture content in situ will be greater than their liquid limits. Whenever this condition is found in site investigation, the greatest care must be taken in avoiding any disturbance of the soils beyond what is essential, lest such a result as that shown in Fig. 39.8 result.

39.5 CAUSES OF LANDSLIDES

From what has already been said, it can be seen that there are many ways in which a landslide may be started. This makes it almost impossible to classify the causes of landslides. An almost invariable feature is water, or the lack of it, so that once again the vital importance of climate to civil engineering operations becomes evident. Many natural slopes are in a potentially unstable condition so that their failure can readily be triggered. Earthquake shocks are always a potent cause of movement. The process of erosion is another common cause; it may act in one of several ways. Differential erosion of strata of varying stability may leave overhanging material of a harder stratum, which will eventually break away and cause a slide. Erosion of the toe of a slope of unconsolidated material may remove the essential support from the material above, which will start to move downward until stability is restored again. This will happen more easily on sloping bedding planes, which are always a source of possible weakness when bordering on any natural or excavated sloping face.

Fault planes constitute another frequent cause of slides, if the planes are so arranged that they isolate blocks of material which are thus left free to move; the work done in opening out the Cofton Tunnel is an example of slides due to this cause. The equilibrium of material in its natural position may sometimes be affected if unduly heavy loads are placed upon it. This is often the case with excavation work; the dumping of excavated material too near to the cut being made is responsible for many slides. The slides on the Panama Canal may be included among these, although the slides there were due to a combination of causes, of which this was only one. Erection of buildings on unconsolidated material will sometimes have the same effect. The growing of carnations has been blamed for serious landslides near Menton, France, that took 11 lives. Local farmers uprooted so many olive trees in order to get ground on which to plant carnations (a more valuable crop) that the stabilizing effect of the tree roots was destroyed and the results were disastrous.

Probably the most important factor of all is a change in groundwater conditions. This may be caused by interference with natural drainage conditions, excessive evaporation from normally damp ground, or an increase in groundwater due to excessive rainfall. This last, probably the most common way in which groundwater conditions are affected, is especially serious because excessive rainfall will also increase surface runoff, which may result in erosion of material at the toe of a slope and so intensify sliding tendencies. Rarely will surface erosion of itself lead to trouble; the landslides that are occasioned by intense rainfall are generally due to a corresponding change in groundwater conditions. If, however, surface erosion strips off vegetation and leaves bare ground exposed to future rainfall, this may lead very quickly to an increase in groundwater and so to slides. The presence of water underground has three main effects: the water will increase the effective weight of the material that it saturates; it will create appreciable pore pressure; and it will tend to weaken many materials, including the weaker kinds of rock and unconsolidated materials with any clay content. The combination of one or more of these effects with other consequences of heavy rainfall will show why many slides occur in wet weather and why drainage is so often an effective remedy for sliding.

The presence of excessive groundwater is so generally realized as a potent cause of landslides that it may be difficult to imagine a slide caused by drought conditions. One such case, however, led to the wrecking of two trains, although the volume of material in the slide was only about 300 m^3 (400 yd^3). The slide occurred on the Richmond, Fredericksburg, and Potomac Railway in Virginia in 1933. It took place in a cut 300 m (1,000 ft) long at a point on the west side where the cut was about 24 m (80 ft) deep. Excavation was through sedimentary deposits of fine material derived from local metamorphic beds but containing no clay; deposition had been irregular, and induration had taken place to such an extent that the railway engineers classified the material as rock. They had found it difficult to excavate with a steam shovel and able to stand at a slope of about 60° in the lower part of the cut, flattening off to about 45° in the upper part. Three distinct strata were evident in the cut; the slide occurred from the third one and impinged on the tracks with the result noted. The strata were considerably jointed; expert opinion was that the slide was due to "the actual or incipient joint planes that parallel the west face of the cut (which) are the result of general shrinkage of the exposed beds because of excessive loss of moisture during a long dry period."[39.12]

All the causes so far listed could be illustrated from North American experience. Figures 39.5 and 39.7 illustrate just two examples. Since many readers will be familiar at least with reports of such slides, some further examples may well be given from two countries that may not be as familiar to most readers, each of which has its own special problem areas for landslides. New Zealand is one of the most interesting countries from the geological point of view, and a most pleasant place in other ways. Local names, derived from the original inhabitants, have a music all their own. One of the country's most lovely lakes, with the musical name of Waikaremoana, was created by a majestic landslide that probably happened in historic time, for this is suggested by a fanciful but poetic explanation of the lake's formation, traditional with the Maoris.[39.13]

Within a distance of less than 32 km (20 mi), in the vicinity of the city of Dunedin on the southeast shore of the South Island, a remarkable series of engineering troubles occurred, all caused by ground movement. Dr. W. N. Benson has described them in two notable papers.[39.14] Cornish Head is a good starting point: it is a notable headland that is itself scarred by major slides of solid rock, caused by the yielding of the local Burnside mudstone over which occur beds of basalt and conglomerate. Not far away is the now-disused Puketeraki Tunnel of the New Zealand Railways. Constructed in 1878, the tunnel is 155 m (516 ft) long on a gradient of 1 in 66. It was excavated through the Caversham sandstone that is here located just over the Burnside mudstone. The tunnel, although this was not realized at the time, was actually driven through a slump scarp, movement of which has been aggravated in recent years by surface streams working through the overlying strata into the mudstone below the tunnel. Trouble with movements was experienced soon after the tunnel was opened. Starting in 1932 measurements of movements were regularly taken; it was soon found that the tunnel was actually being twisted by a movement of the entire rock mass that it penetrated. By 1934, a resultant total movement of up to 45 cm (17.9 in) had been measured. With the geological cause of the trouble then realized, it was decided that the only course to pursue was to abandon the tunnel; the line was relocated in a cutting excavated between the tunnel line and the edge of a sea cliff nearby (Fig. 39.9).

A few miles down the coast is the site of a large mental hospital, 240 ha (600 acres) of beautifully situated rolling land which appear to be an almost ideal location for this therapeutic purpose. When the area was first proposed for hospital use, Sir James Hector (then director of the New Zealand Geological Survey) warned that "the clay would move—like any other plastic substance—with an almost molecular motion." His warning was disregarded, and it is easy to realize why as one stands on the upper part of the site, looking out over the sea, with good "firm" clay in evidence all around as a foundation material. Unfortunately, as later geological studies have shown, the entire site is underlain not only by the Burnside mudstone but, beneath it, by the Abbotsford mudstone, an equally unreliable rock, with a major fault zone running across the building area. The weak character of the mudstones had led to severe slumping, so that much of what appears to be solid ground is merely surface evidence of successive slump scarps. Serious movements have therefore affected the hospital from its earliest days; the first main building had to be completely demolished a few years after its erection because of serious and irremedial cracking. Downhill movements of entire buildings have been measured in feet; differential settlements of many inches have accompanied such mass sliding. So serious has the instability of the entire site proved to be that it is to be abandoned as soon as

FIG. 39.9 Abandoned Puketeraki Tunnel near Palmerston, on the main South Island line of New Zealand Government Railways, showing the relocated line in open cut on the left.

39-14 SPECIAL PROBLEMS

other buildings can be provided. Unfortunately, further serious trouble with slips was encountered in this area in the summer of 1979, a new housing development being adversely affected.

It is therefore not surprising to find many and quite extensive landslides and mudflows (from the disintegrating mudstones) in the Dunedin area, although it should be added that these are evident only to the keen observer; the natural beauty of the area distracts normal attention from all such geological peculiarities. Even in the areas of bedrock much more competent than the mudstones, troubles have been experienced. About 40 km (25 mi) west-southwest of Dunedin is a lovely little valley, the Waipori Gorge, in which is located a small municipal water power plant, fed by water from an impounding reservoir at the head of the gorge. The water is led to the plant through a tunnel that originally terminated at a point 204 m (680 ft) above the riverbed level at the powerhouse site. Three 1.1-m (42-in) and one 1.5-m (60-in) -diameter, steel pressure pipes were installed on anchor blocks; the concrete was keyed into blocks of the local schist protruding from the steep surface of the side slope to the valley, which was known to be in solid schist. So solid did the rock foundation to the anchor blocks appear to be that no doubts at all were entertained when the plant was put into service. In 1929, however, following the thawing of an unusually heavy snowstorm, engineers in charge of the plant noticed what appeared to be slight movements of some of the anchor blocks, following some slumping of the slope surface. Careful measurements were instituted, and the movement was confirmed; the whole slope proved to be unstable; a maximum downward movement of 12.3 cm (4.85 in) was detected. The movements of the different blocks were irregular, and this increased the complexity of the problem. It was then found that the anchor blocks were founded not on solid rock but on a mixture of rock debris, clay, and very large fault blocks of schist that had been mistaken for solid rock. The penstocks were therefore replaced by a continuation of the pressure tunnel, but even in this construction, difficulty was experienced when weathered schist was found to extend far below the surface of apparently solid rock.

Czechoslovakia is a lovely land with more than its fair share of landslides, especially in the Carpathians of Slovakia. So numerous are these land movements that some years

FIG. 39.10 The 1960 slide at Handlová, Czechoslovakia; the first indication of the ground displacement at the head of the slide.

ago a survey resulted in no less than 9,100 being registered. Results of these extensive studies have been published in English, one of the most notable books on landslides being by Drs. Zaruba and Mencl of Czechoslovakia.[39.15] Handlová is an important coal-mining town in central Slovakia. It is bordered on the south by the small Handlová River, beyond which there is most pleasant, well-treed pastureland sloping upward from the river bank. In the town is a large power station, fired with brown coal and therefore emitting much fly ash. Prevailing winds carried the ash to the south, and so the character of the grazing land was gradually changed by the surface layer of ash that slowly accumulated. In some parts, grazing had to be given up and the land was plowed. This immediately allowed rain to percolate into the surface. This percolation disturbed the very delicate groundwater condition that existed beneath the sloping ground. Handlová was subject to unusually high rainfall in 1960 (1,045 mm or 41 in as compared with the long-term average of 689 mm or 27 in). The combination of all these circumstances brought a rise in the water table beneath the pastureland, which, coupled with the geological conditions, led to a start of earth movement early in December 1960 (Figs. 39.10 and 39.11). Accurate measurements were initiated by 22 December; at one point a movement of almost 150 m (460 ft) had taken place within a month. And a volume of about 20 million m^3 (26 million yd^3) of clay, clayey silts, and rock debris was moving downhill, threatening to engulf the town.

The slide was very quickly brought under control by a spendid example of teamwork, with the result that by the end of January 1961 most of the movement had been greatly reduced, but not before 150 houses had been destroyed by the inexorable movement of this huge mass of soil (Fig. 39.12). Drainage was the key to the control measures, as it is so commonly in landslide remedial works. Today the great hill to the south of Handlová is again stable, the soil resting, but not now moving, on the underlying argillaceous shale rock that was so easily affected by the excess groundwater. To sit on its grassy slopes and to look at the town below, which could so easily have been so seriously damaged, and all because of fly ash interfering with sheep grazing, is to be very forcibly reminded of the dual importance of clean air and slope drainage. Flow from the drainage tunnels and pipes that were so hastily installed is now quite small, so limited that one has to have a very

FIG. 39.11 The major landslide at Handlová, Czechoslovakia; progressive slump of the scarp of the slide due to movement at the toe.

FIG. 39.12 Damage to buildings at the toe of the Handlová landslide, Czechoslovakia.

real appreciation of the mechanics of soil movement to appreciate what that small quantity of water meant before it was thus extracted from the soil through which it flows.[39.16]

39.6 PREVENTIVE AND REMEDIAL WORKS

The Handlová slide emphasizes the importance of drainage as a means of slide control, a subject that must be briefly noted since slides do still occur, despite all the knowledge that is now available (but not always used) that can lead to their avoidance. Landslide control is naturally work that can be undertaken only on the basis of most expert advice and after full study and determination of the real cause of the earth movement. Sometimes "unloading" the slope in movement will be called for. In other cases of small slides, special retaining structures may be necessary to hold the slopes so that the area below it can safely be used. In extreme cases of very valuable land, attempts may be made to stabilize the soil by chemical means, if local conditions are suitable. On construction works, freezing of a large mass of moving soil has been successfully accomplished, the freezing plant being left in operation while construction continued. The solution must be chosen by the engineers responsible to fit the particular case, after all the local factors are fully known. Landslide investigation is a complex and difficult task. In the majority of cases, water will be found to be the culprit, and drainage works, to get the water out from the critical locations, will often be the only work necessary.

This is unspectacular work, and it can therefore be misunderstood unless the conditions under the ground being drained are appreciated. In some cases, full-sized tunnels have had to be built; tunneling within the slide material can often be a critical operation. Large drainage pipes are another means of achieving the desired ends, if the pipes can be safely and efficiently installed. A more recent development has been the use of quite small pipes, at frequent intervals, perforated so that they will collect water easily and either forced or jetted into place from the face of the slope (Fig. 39.13). This method eliminates the danger of work in the slide material and can be remarkably effective. It is, however,

FIG. 39.13 Horizontal perforated 5-cm (2-in) steel drain pipes as installed on the Trans Canada Highway west of Banff, Alberta, for slope drainage and stabilization.

mundane and undramatic in appearance; it depends for its success upon the accurate location of the drainage pipes in relation to the local geology and especially to the surface of sliding. Coupled with it, and with all other remedial works, is the necessity of controlling, if possible, the ingress of water into the slide area. Surface drainage is here an essential. This is work that can be seen, but it is ameliorative rather than remedial. Both types of work are important and can well be regarded with respect by those who benefit from the control of unstable soil that they provide.

Consideration of the overall stability of the entire mass of material involved must always be of cardinal importance. If due regard is paid to stability, such remedial measures as removal of material at the toe of a moving slope and the construction of either pile or crib retaining walls at the toe of a slide will not be adopted, except possibly as temporary emergency measures. In a few isolated slides of very small magnitude, these measures may be effective, but the general result of toe removal will be to lessen the resistance to movement and so to increase the extent of the slide instead of stopping it. The use of piles, in particular, is usually clear evidence that those responsible for the remedial works do not appreciate the basic character of earth movement in landslips; the idea of retaining a deep-seated movement of a large mass of soil by means of slender piles is on a par with King Canute's efforts to stem the tide. The vibration caused by the operation of pile driving may even make matters worse if the soil in question is unduly sensitive.

39-18   SPECIAL PROBLEMS

Piles are useful elements of civil engineering design when local soil conditions demand their use; remedial work on landslips is not such an occasion.

The use of retaining structures may also betray a neglect of the fundamentals of engineering design. Retaining structures should be designed for a predetermined load, which they are to transmit to a foundation bed of known capacity. If constructed to stop a slide movement, unless the slide is of very small extent, they will have to withstand the load induced by movement of the whole slide—generally unpredictable with any accuracy—and will transmit the load, in many cases, to a lower part of the material affected by the slide. To illustrate the frequent failure of such measures to control slides would be to labor the point too strongly; examples will be found in several of the references given. There are two possible exceptions to this general unsuitability of retaining walls. First, they may be used in a cutting, both banks of which are unstable and both banks of which can be held by retaining structures connected by solid struts across the bottom of the cutting. This measure sometimes proves economical; but usually it is more satisfactory to achieve stability of the slopes of the cutting and thus to dispense with the need of such walls. Second, small walls may be satisfactorily used not so much as retaining structures but as protection for the lower parts of slopes and as provision for drainage.

It seems to be possible to provide an exception to almost every general comment about landslides, their prevention and restoration. Figure 39.14 provides just such an exception to what has been said, in general, about the use of retaining structures. It shows a new main building in the extensive Coors brewery at Golden, Colorado, which was specifically designed to act as an abutment for the pressure exerted by a potentially unstable hillside, which here abuts onto a main road. The combination of building, road tunnel, and retaining structure is a pleasing example of geological-engineering design.

Slope readjustment is another important remedial measure, which really amounts to doing under control and in a limited way what a slide will do automatically if it is allowed to take place. Now that the stability of slopes can be theoretically determined with the aid of the techniques of soil mechanics, landslides caused by excessive slopes should steadily decrease in number. In earlier years, however, when civil enigineering did not

FIG. 39.14 A major building of the extensive plant of the Adolph Coors Company at Golden, Colorado, linked to a retaining wall to give stability to the sloping ground above.

have the benefit of modern soil studies, the excavation of cuts with excessive slopes was a potent cause of earth movement; correspondingly, slope adjustment was frequently the only solution.

Drainage is one of the main remedial measures. Surface drainage is generally envisaged as necessary in most engineering work, including that dealing with slides. In a few cases, surface drainage may be effective; generally, however, it will be useless because the critical location in a slide is the plane on which sliding is taking place—usually well below the surface except at the extremities of the slide. At times, surface drainage may even be harmful. To be effective in preventing or remedying slides, drainage must intercept the groundwater which is tending to promote the instability. Sometimes a very small amount of water can cause extensive slides. Examples from actual practice have shown that features such as thin beds of sand, only 2.5 to 5.0 cm (1 to 2 in) thick and only slightly water-bearing, and even a 2.5-cm (1-in) seam of decomposed coal, can lead to extensive slides of the superincumbent material. Drainage can be applied effectively, therefore, only when those in charge fully know and understand the cause of the slides and the source of the groundwater that is contributing to movement. This knowledge can be obtained by a judicious combination of geological survey work and test boring; needless to say, such work can be carried out far more easily and to much better effect before, rather than after, any movement has taken place.

In recent years, drainage has been widely recognized as essential to the maintenance of stable slopes and the avoidance of much trouble with earth movements. Many ingenious variations of standard drainage methods have been developed for use with unstable slopes. Highway departments, in particular, have special equipment and specially trained staff for such drainage work. In this branch of work, highway engineers have been able to profit from the experiences of railway engineers and to take advantage of all that modern soil studies can contribute. Highway construction in California now provides many notable examples of excellent drainage installations; the unusual troubles encountered with earth movements in that part of the United States are demonstrated by the fact that the cost of landslide-stabilizing works have, on occasion, amounted to more than one-third of the total annual state expenditure on highway maintenance.[39.17] Horizontal drains of conventional design can often be used; horizontal drains jacked into place, in order to avoid further disturbance of slopes, are a variant. The use of perforated pipes for this purpose is widespread, but almost every type of pipe and culvert has found application in some example of slope stabilization. There are now, fortunately, many hundreds of examples to show the efficacy of this preventive measure. To illustrate almost all phases of such work, an example will be cited that constitutes probably one of the worst individual landslide situations ever to be faced and solved by civil engineers.

The main line of British Railways connecting the ports of Folkestone and Dover (previously a part of the Southern Railway) runs along the south coast of England, which is characterized by the widely known White Cliffs of Dover. Between the Martello Tunnel near Folkestone and the Abbotsford Tunnel nearer Dover is a 3.2-km (2-mi) stretch of coast known as the Folkestone Warren. It has suffered extensive landslips from the earliest historical times, possibly since the breaching of the Straits of Dover in Neolithic times. The double-track railway line runs in a shallow cutting in the undercliff; the main cliffs rise about 150 m (500 ft) above sea level. General conditions at the Warren in 1951 are shown in Fig. 39.15. Figure 39.16 shows the appearance of part of it in 1915. In this year one of the most serious landslides of recent times occurred; another had taken place in 1877. Ever since the line was opened in 1844, however, those responsible for its maintenance have been plagued by earth movements; major remedial work was put in hand in 1948 following one of the most intensive programs of subsurface exploration of which the writers have knowledge.

This work was started in 1939 but had to be stopped because of the war. Between 1940

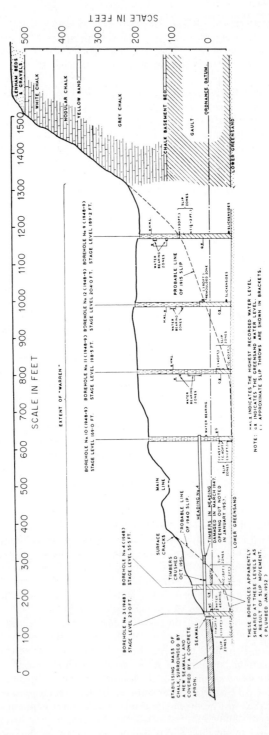

FIG. 39.15 General cross section of the ground conditions at the Folkestone landslide area, Kent, England, showing something of the test borings and remedial works carried out.

FIG. 39.16 Folkestone Warren landslide, Kent, England, of 1915, showing the train caught in the west end of the slip.

and 1948, the position of the shear surface responsible for the major movement was clearly demonstrated in two of the timbered drainage headings that had been constructed before the war. The movement continued slowly throughout the war years, but fortunately, no major slide occurred. An extensive program of boreholes provided the basic information upon which an explanation of the movements and the main plan for the necessary remedial work could be based. This case, however, was one of the many in which engineering studies alone were not enough; the engineers in charge worked in the closest cooperation with officers of the Geological Survey of Great Britain, and their joint efforts eventually provided a reasonably complete solution to the century-old problem. In view of the discontinuities created by earlier slips, which had caused gaps in the geological sequence, it was necessary to make a detailed study of the fossils found in the samples of the Gault formation obtained in boreholes in order to establish with certainty the stratigraphical succession in each particular boring. It would be hard to imagine a better example of the complete interdependence of such engineering and geological studies.[39.18]

The Warren owes its existence to the facility with which northeasterly dipping Gault clay and chalk marl, forming the lower 30 m (100 ft) of the Lower Chalk, are eroded by the sea. External factors affecting the earth movements have therefore been the effects of heavy rainfall and the erosive action of the sea. The Gault has marked swelling characteristics, so that any disturbance creates increasing trouble as the newly exposed material comes into contact with water. Figure 39.15 is a typical cross section resulting from the investigations carried out between 1948 and 1950; the figure shows one of the drainage headings and the foreshore loading. The investigations showed clearly that the landslips penetrate to the base of the Gault and that failure had been generally confined to a plastic sheet of the Gault immediately overlying a thin layer of phosphatic nodules embedded in a concretionary grit, popularly known as the "sulfur band." When the cause was known, remedial works of a major nature were put in hand. These included a new drainage heading more than 210 m (700 ft) long, constructed as a 2-m (6.5-ft) shield-driven tunnel, lined with precast concrete segments, whose purpose was to lower permanently the groundwater level in the disturbed area. In order to limit further erosion by the sea, about a kilometer (0.5 mi) of the critical part of the coast was stabilized by weighting with chalk, held in place by concrete walls and slabs, and by carefully located groins, constructed at right angles into the sea for added protection (Fig. 39.17). It has recently been shown that the great slide of 1915 was almost certainly caused by interference with littoral drift by the construction of new works at Folkestone Harbor, to the west. This intensified erosion at the site of the slip.[39.19]

The Folkestone Warren landslips are unique, but the records that the engineers asso-

FIG. 39.17 The weighted foreshore at Folkestone Warren, Kent, England, this being one of the remedial works; the visitors are from the 1957 International Conference on Soil Mechanics and Foundation Engineering then being held in London.

ciated with the recent investigations have published are invaluable guides even for remedial work on a much smaller scale. The case epitomizes successful drainage work in connection with landslides, except that it was carried out after a slide had occurred instead of before, as can often be done. The subsurface was adequately explored; the cause of the slide was found; and the main contributing cause was dealt with by a carefully planned drainage system. These steps characterize most preventive and remedial work in connection with landslides, although the exact method of dealing with the main cause of sliding may vary. When the slide is caused by normal groundwater, special methods may sometimes be adopted in addition to regular drainage systems. Well points may be used to advantage in some places.

One large hillside in California is being maintained in its present state only because of the successful operation some years ago of an extensive hot-air pumping system which circulated warm air through tunnels excavated in the clay bank in order to dry out groundwater that was constantly seeping toward the hill face. This unusual scheme has worked so well that it has been possible to fill all the tunnels with gravel and discontinue drying, while leaving the general arrangement so fixed that drying can be resumed in the future if this is ever necessary (Fig. 39.18). The drying operation was carried on from 1933 to 1939, and it has not yet proved necessary to repeat the process.[39.20] Admittedly, this is an extreme example but it is included in this brief review to show what can be done, and how economically serious may be even relatively small slides.

Landslides are so numerous, so varied in type and size, and always so dependent upon special local circumstances that the foregoing review has necessarily had to be rather general. It is hoped that it will at least provide useful background for more detailed consideration of the "landslide situation" in specific localities (Fig. 39.19). Many urban areas are spared from this particular geological difficulty. Others have more than their fair share, as was indicated in the case of many municipalities in Czechoslovakia. Southern California has achieved an unenviable distinction in this connection, since around the Los Angeles area in particular small landslides associated with hillside development became such a serious local problem that a special "grading ordinance" was passed in 1952. It has since been amended and strengthened. Geologists in this region have shared their hard-won experience in a number of most useful publications; as a result, public convenience has been enhanced and recognition of the importance of engineering geology ensured.[39.21]

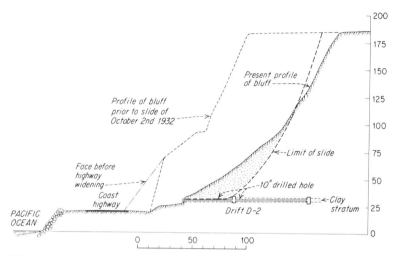

FIG. 39.18 General plan and typical cross section of the landslide area at the Palisades, west of Los Angeles, California, showing the arrangements used for air drying the unstable clay.

In South America, there are areas where landslides have been really troublesome, one being in the city of Santos, major seaport of Brazil. Heavy rains (so common a cause) led to a series of most serious soil and rock slides in March 1956 that resulted in more than 100 deaths and many millions of dollars worth of damage. Around London, England, there have been a few most troublesome slides occurring long after the slopes that failed had been formed. On the old main road between Manchester and Sheffield in central England,

FIG. 39.19 Mud run covering the main line of the Canadian Pacific Railway, near Field, British Columbia, caused by water escaping from a glacier in the mountains above the line.

there is one section that has provided almost continuous maintenance problems where it passes over a geologically old slip scarp on Mam Tor, known locally by the telling name "Shivering Mountain"; the scarp has been the cause of a number of recent landslides. So the record could continue—all around the world. There is probably no part of civil engineering that calls more clearly for the joint efforts of engineering geologists and engineers trained in geotechnical work than does work in relation to landslides—but always, to the maximum extent possible, *before* they occur.

If a landslide *does* occur the following course of action might well be prescribed:

1. Don't panic.
2. Don't rush out and "drive piles."
3. Study all water exhibited at site and remove free water.
4. Determine, or review, local geology.
5. If possible, view site from air and study aerial photos.
6. Study former landslides that can be located, if any.
7. If a slip-slide, try to locate evidence of failure surface.
8. Ask "Why did the slide happen?" before seeking solution.
9. Study local climate records, long-term and recent.

And only then:

10. Put down any necessary exploratory borings.
11. Design solution, using slope stability calculations.
12. Install necessary drainage facilities and/or stabilizers.
13. Trim and finish slopes, planting suitable vegetal cover.
14. Prepare best possible record and ensure its deposition where it will be regularly used in inspection and maintenance.

39.7 REFERENCES

39.1 D. J. Varnes, "Landslide Types and Processes," chap. 3, in E. B. Eckel (ed.), *Landslides and Engineering Practice*, Highway Research Board Special Report no. 29, National Academy of Sciences, Washington, D. C., 1958, p. 20.

39.2 A. Collin, *Landslides in Clays* (1846), W. R. Schriever (trans.), University of Toronto Press, Toronto, 1956.

39.3 *Report of the Committee of the National Academy of Sciences on Panama Canal Slides*, Memoir XVIII, National Academy of Sciences, Washington, D.C., 1924.

39.4 W. H. Ward, "The Stability of Natural Slopes," *Geographical Journal*, **105**:170 (1945).

39.5 A. W. Skempton and F. A. DeLory, "Stability of Natural Slopes in London Clay," *Proceedings of the Fourth International Conference on Soil Mechanics and Foundation Engineering*, **2**:378 (1957).

39.6 W. G. Atwood, *Relation of Landslides and Glacial Deposits to Reservoir Sites*, U.S. Geological Survey Bulletin no. 685, Washington, D.C., 1918.

39.7 F. E. Emerson, "180-ft. Dam Formed by Landslide in Gros Ventre Canyon," *Engineering News-Record*, **95**:467 (1925); see also W. C. Alden, "Landslide and Flood at Gros Ventre, Wyoming," *Transactions of the American Institute of Mining and Metallurgy*, **76**:347 (1927).

39.8 R. E. Stanton, "The Great Landslides on the Canadian Pacific Railway," *Minutes of Proceedings of the Institution of Civil Engineers*, **132**:1 (1897).

39.9 Information received from various Indian sources through the late Lt. Gen. Sir Harold Williams, Roorkee, India.

39.10 C. B. Crawford and W. J. Eden, "Nicolet Landslide of November 1955, Quebec, Canada," *Engineering Case Histories No. 4,* Geological Society of America, Boulder, Colo., 1963, p. 45.

39.11 F. Tavenas, J.-Y. Chagnon, and P. La Rochelle, "The Saint Jean-Vianney Landslide: Observations and Eyewitness Accounts," *Canadian Geotechnical Journal,* **8**:463 (1971); see also R. F. Legget and P. LaSalle, "Soil Studies at Shipshaw, Quebec: 1941 and 1969." *Canadian Geotechnical Journal,* **15**:556–564 (1978).

39.12 G. E. Ladd, "Bank Slide in Deep Cut Caused by Drought," *Engineering News-Record,* **112**:324 (1934).

39.13 M. Ongley, "Waikaremoana," *New Zealand Journal of Science and Technology,* **14**:173 (1932).

39.14 W. N. Benson, "Landslides and Their Relation to Engineering in the Dunedin District, New Zealand," *Economic Geology,* **41**:328 (1946); see also by same author, "Landslides and Allied Features in the Dunedin District in Relation to Geological Structure, Topography and Engineering," *Transactions of the Royal Society of New Zealand,* vol. 70, pt. III, Wellington, 1940, p. 249.

39.15 Q. Zaruba and V. Mencl, *Landslides and Their Control,* Elsevier, Prague, 1969.

39.16 Ibid., p. 42.

39.17 T. H. Dennis and R. J. Allan, "Slide Problem: Storms Do Costly Damage to State Highways," *California Highways and Public Works,* **10**:1 (23 July 1941).

39.18 A. M. Muir Wood, "Folkestone Warren Landslips: Investigations, 1948–1950," and N. E. V. Viner-Brady, "Folkestone Warren Landslips: Remedial Measures, 1948–1954," *Proceedings of the Institution of Civil Engineers,* pt. II, **4**:410 (1955); see also J. N. Hutchinson, E. N. Bromhead, and J. F. Lupini, "Additional Observations on the Folkestone Warren Landslides," *Quarterly Journal of Engineering Geology,* **13**:1 (1980).

39.19 J. N. Hutchinson, "Assessment of the Effectiveness of Corrective Measures in Relation to Geological Conditions and Types of Slope Movement," in *Symposium on Landslides and Other Mass Movements,* International Association of Engineering Geology Bulletin no. 16, 1977, p. 131.

39.20 R. A. Hill, "Clay Stratum Dried out to Prevent Landslips," *Civil Engineering,* **4**:403 (1934); and information in a personal communication from Mr. Hill.

39.21 R. Lung and R. Proctor (eds.), *Engineering Geology in Southern California,* Association of Engineering Geologists, Glendale, Calif., 1966.

Suggestions for Further Reading

After this chapter had been completed, there appeared a fine account of one of the most remarkable landslides of North America, one that has been moving for hundreds of years and which has caused maintenance problems on a mainline railway for a century and similar problems more recently on a main highway. This is the Drynoch Slide of British Columbia, which is 5.3 km long. It is described in *Engineering Geology and Geotechnical Studies of the Drynoch Landslide, B.C.* by D. K. Van Dine, Report No. 79-31 of the Geological Survey of Canada (34 pages) published in 1980.

Selection from the many useful publications now available on landslides is difficult, but the following titles include the best treatments known to the authors. Superseding the excellent volume published by the Highway Research Board, and referenced as 39.1 above, is *Landslides: Analysis and Control,* edited by R. L. Schuster and R. J. Krisek, Special Report No. 176 of the Transportation Research Board (National Academy of Sciences), a fine volume of 234 pages published in Washington, D.C., in 1978. Almost a companion volume to it is *Landslides,* the third of the series of Engineering Geology Review volumes of the Geological Society of America published in 1977. The editor of this excellent collection of papers, which is unusually well illustrated, with 278 pages, is D. R. Coates.

SPECIAL PROBLEMS

The International Association of Engineering Geology held a special conference on landslides in Prague, Czechoslovakia, in 1977, a truly international gathering including delegates from China. The papers then given were published in no. 16 of the Association's *Bulletin* (from Krefeld, West Germany) and they constitute a volume comparable with the two just mentioned, some of the papers being among the most stimulating discussions of landslides known to the authors.

It is the view of the authors that quite the outstanding recent book on this subject is *Landslides and Their Control* by Q. Zaruba and V. Mencl, both of Czechoslovakia. Written in English and published by Elsevier, it has been out of print but a second edition has been prepared and is due to appear at about the same time as this Handbook. It is unique in its approach, because it combines a full geological treatment of landslides with detailed geotechnical studies, thus epitomizing the necessary linking of geology with geotechnique as urged in Chapter 6.

There are naturally a very large number of individual papers on landslides in a wide variety of books and journals. An unusually good example of a general review paper is one by Dr. Karl Terzaghi which appears on page 83 of the volume published by the Geological Society of America in honor of Dr. Charles P. Berkey and so known as the *Berkey Volume* (1950); it was edited by Sidney Paige.

International interest in landslides is reflected in the existence of yet another Working Commission of the International Association of Engineering Geology. No landslide is yet known to have crossed an international border, but there must be few (if any) countries of the world in some part of which landslides are not a regular occurrence. International studies such as this Working Commission is carrying out can prove to be of widespread interest and benefit.

Since landslides constitute one of the most widespread and serious of all natural hazards, it may be appropriate to mention here the "Procedures" to be followed by the U.S. Geological Survey in implementing, for its part, the requirements of Section 202 of the Disaster Relief Act, passed by the Congress of the United States in 1974. Proposed "Procedures" were published in the Federal Register, vol. 42, no. 70, for Tuesday, 12 April 1977, on pages 19292 to 19296. These proposals outline clearly the manner in which Notices of Potential Hazards shall be prepared and used. Quite apart from its specific purpose in guiding the U.S. Geological Survey, this document is of great interest for all concerned with natural hazards; it includes a useful bibliography.

chapter 40

ROCKFALLS

40.1 Some Major Rockfalls / 40-1
40.2 Some Smaller Rockfalls / 40-5
40.3 Minor Rockfalls / 40-11
40.4 Niagara Falls / 40-13
40.5 Rockslides / 40-14
40.6 Avalanches / 40-16
40.7 Conclusion / 40-19
40.8 References / 40-23

Rockfalls constitute one of the major subgroups into which *mass wasting* is usually divided. They can have profound influence upon the works of the civil engineer and so warrant this special attention. Reference to Table 39.1 will show the relation of rockfalls to other types of ground movement, the relative frequency of which varies widely, some types being quite rare. It is always difficult, in such general considerations, to draw a hard-and-fast dividing line between soil and rock, and so the suggested categories of ground movement to some extent overlap. The term *rockfall* is, however, capable of little misunderstanding; precipitate collapses of solid rock form, therefore, the main subject matter of this chapter. Rockslides will also be considered, and a short final section is devoted to avalanches (an "orphan subject"), disastrous failures of another natural material, ice, especially after heavy snowfall in temperate mountain regions. Civil engineers should be at least familiar with avalanches, and with the information now available about them, since ice is a vital part of the natural order with which the broad scope of this Handbook is concerned.

40.1 SOME MAJOR ROCKFALLS

Rockfalls, in general, are more determinate and understandable than other types of earth movement. They occur when a mass of rock becomes detached from surrounding bedrock in some way and thus becomes free to move downward, if its position will permit. The loosening of blocks of rock will almost always be associated with such features as bedding planes, joints, cleavage, or a local fault zone or plane. The immediate cause of loosening

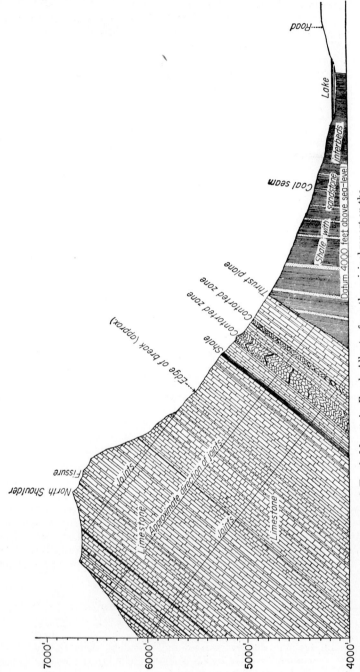

FIG. 40.1 Geological section through Turtle Mountain, Frank, Alberta, from the original report on the great rockfall.

will be a change, which will usually be the result of weathering, in the material adjacent to the planes of possible movement. Natural rockfalls are evident on any steep mountain face; talus slopes are a direct result. Many are of great extent, but large rockfalls will generally take place in uninhabited mountain regions and so be of interest mainly to the geologist. One great rockfall was not so located. Figure 40.1 shows Turtle Mountain in Alberta, which is composed of Paleozoic limestones overthrust upon a syncline of Mesozoic shales, sandstone, and coal beds. In the early morning of 20 April 1903, the crest of this great mountain broke off and slid to the valley below, taking at least 70 lives, damaging much property, and burying nearly 2,200 m (7,000 ft) of the Crow's Nest Railway. The catastrophe was closely investigated by Canadian geologists and engineers; the general conclusion was that the major cause was failure of the rock along the joint planes which are a prominent feature of the limestone formation, although operations in a coal mine located at the foot of the mountain may have contributed to the rockfall. Sixteen miners at work in the mine when the disaster took place were rescued, almost miraculously, after digging themselves an escape way through the soft coal seam to where it outcropped on the mountain side.[40.1]

The interest in this tragic happening, and the relative availability of the site, has led to a number of further studies of the disaster, one of the most recent being that by Cruden and Krahn in 1973.[40.2] The advances made in studying the local geology are well shown by comparing Fig. 40.1 with Fig. 40.2, the latter being one of the illustrations supporting Cruden and Krahn's explanation. Principles of rock mechanics have also been applied in another study of the "Frank slide" (as it is still known) by Krahn and Morgenstern, this analysis indicating factors of safety close to unity.[40.2] Although the authors caution against

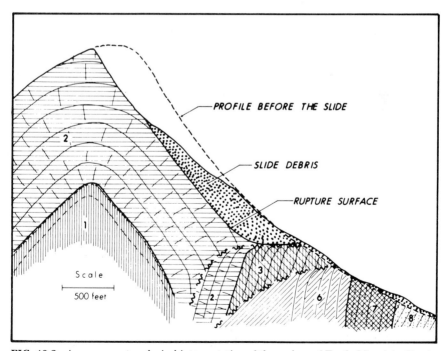

FIG. 40.2 A more recent geological interpretation of the geology of Turtle Mountain, Frank, Alberta, at the site of the great rockfall.

extrapolating their results even to small rock slopes, it is significant that what was formerly regarded as an "act of God" can now be subjected to rigorous analysis based on geological study, sampling, and sound theory.

Western Canada was the scene of another of the very great rockfalls of recent times, this being now known as the "Hope landslide," even though it was, in fact, a rockfall combined with a rockslide. It took place just before dawn on 9 January 1965 in a narrow valley some 16 km (10 mi) to the east of the town of Hope, British Columbia, and 160 km (100 mi) to the east of Vancouver, through which runs a relatively new transmountain highway connecting Hope with Princeton. The original road is now buried to a depth of up to 78 m (240 ft) under the pile of 130 million tonnes of rock that came down from the steep northern side of the valley. Four lives were lost; almost miraculously, passengers in a bus and other automobiles escaped. The rock involved is massive-to-slightly-schistose metavolcanics with intrusive sheets of felsite that dipped nearly parallel to the 30° slope of the mountainside. Rainfall did not appear to have been a contributing factor, but two small earthquake shocks that were associated with heavy snow avalanches and some landslides appear to have been the triggering mechanism. Today, a new road has been constructed over the top of the great pile of fallen rock. A parking lot and observation area have been constructed (Fig. 40.3). To stand here and see the bare face of the rock from which this

FIG. 40.3 Sign at the lookout point adjacent to the new road that has been constructed on top of a great depth of broken rock (60 m or 200 ft), at the Hope rockfall, British Columbia, the original highway being beneath the rock pile.

vast quantity fell so quickly is a most remarkable experience for any traveler who has any appreciation of geology at all.[40.3]

There have been other major rockfalls, one of the most tragic of which took 115 lives at Elm, Switzerland. It started as two small rockslides on each side of a quarry excavated into a mountainside. Almost immediately, a vast mass of rock detached itself, crashed down, and flowed across the valley, traveling over 1.6 km (1 mi) before it stopped. About 10 million m³ (13 million yd³) of rock descended an average of 435 m (1,450 ft) vertically in an elapsed time of 55 seconds.[40.4] Another major rockfall (the Blackhawk landslide) has been found between the San Bernardino Mountains and the Mojave Desert, covering an area of 3.2 by 8.0 km (2 by 5 mi), with a thickness up to 30 m (100 ft). It is suggested that the flow mechanism by which this vast amount of breccia traveled so far was a layer of compressed air, trapped by a sudden and precipitous fall of rock from Blackhawk Mountain. The same explanation would rationalize the phenomenal movement of solid rock in the disaster at Elm.[40.5] For those interested in these majestic natural catastrophes, a comprehensive review of the great rockfalls of modern times which have generated fast-moving streams of rock debris (called *sturzstroms* in Europe) was published by Hsu in 1975.[40.6] Eisbacher has published an interesting review of his studies of major natural rockfalls in the mountains of western Canada.[40.7]

40.2 SOME SMALLER ROCKFALLS

The rockfalls encountered in the ordinary practice of civil engineering are, of course, on a very different scale from that of these great natural phenomena. Only very rarely will vast failures be encountered, even in the construction of transportation routes in mountainous areas. Rockfalls of smaller extent are, however, not uncommon along transportation routes that are adjacent to natural rock cliffs and in cuttings excavated through rock. An example is illustrated in Fig. 40.4, which shows the reinforced-concrete protective

FIG. 40.4 Reinforced-concrete protective works at Vriog Cliff, near Barmouth, Wales, on the western region of British Railways.

"tunnel" constructed by the former Great Western Railway Company near Barmouth on the west coast of Wales. Vriog Cliff rises to a height of 450 m (1,500 ft); road and railway are benched into it as they are carried along this stretch of coast. On 4 March 1933, a train was swept off the tracks at this location by a slide that happened after a severe blizzard was followed by a period of thaw. The slip took place at a reversed fault in the local Ordovician strata, consisting here of dark flags and slates, one or two beds of limestone, and abundant volcanic ash. The protective works were constructed, after a complete survey of the adjacent cliff, in order to obviate any possibility of trouble from slides that might develop in the future because of continued weathering of the exposed rocks.[40.8]

Difficulties in maintaining the British Railways line between Dover and Folkestone have become well known. As if the great landslides at Folkestone Warren were not enough, this same line has been the scene of major rockfalls at the Dover end, when the chalk through which it passes has become weakened, particularly after heavy rains. Between the Abbotscliff and Shakespeare tunnels, the line is protected by a seawall 3.2 km (2 mi) long. Chalk cliffs abut onto the line. They are so steep that they cannot be climbed by human beings, but some years ago wild goats, belonging to a butcher, disported themselves on the rocky faces and troubled the railway engineers by frequently displacing loose pieces of chalk which then fell on the track. Eventually the goats were all shot by a local marksman, under the auspices of an inspector of the Royal Society for the Prevention of Cruelty to Animals. It was at this point that a rockfall of 19,000 m^3 (25,000 yd^3) took place in November 1939; watchmen had been posted, however, since trouble was anticipated. Trains were stopped just before the fall occurred. The tunnels restricted the size of earth-moving equipment that could be brought to the site to clear the tracks, so it was not until 7 January that the line was again opened for traffic.[40.9]

Further along this scenic coast to the east is the ancient port and city of Dover. The White Cliffs of Dover are world-renowned and are almost symbolic of stability. Unfortunately, their very whiteness is an indication of their geological character—they are composed of the familiar Chalk of southern England, a soft rock containing layers of flints. This leads to differential weathering; occasionally the flints fall away to the ground. In the mid-sixties this hazard became very serious for the houses that, at one location, are located close to the foot of the cliff. An inspection of the cliffs above the city of Dover was made with the idea of removing all loose flints. Inspection revealed a rather more serious situation, the Chalk overhanging at some places; in others, lumps of fissured Chalk were found to be badly weathered. A thorough rehabilitation job was therefore ordered by the British Ministry of Works. The appointed contractor installed a 10-ton derrick on an old wartime concrete structure at the top of the cliff, and a thorough scaling of the cliff was carried out, with all due safeguards provided for the houses below (Fig. 40.5). Portable radios greatly assisted in coordinating work on the cliff, the watchmen being below and the crane operators and other personnel at the top of the cliff.[40.10]

Hydroelectric powerhouses are often located in rocky gorges and so are sometimes liable to damage by rockfalls; this is always an important feature to be watched for in design. It is also one of the many factors that have led to the widespread adoption of underground power stations, free from all such danger. Usually, if probable rockfalls can be detected, it will prove economical to remove the loose rock under control, with the aid of skilled "rockmen." If this cannot be done, then some form of protective works may be constructed so that rockfalls will not cause serious damage. This type of protection work is most common on transportation routes, but the work illustrated in Fig. 40.6 is a good example of what can be done, and of what had to be done, in this direction to protect a powerhouse.

The hazard of falling boulders was dramatically shown by the damage done to the Southern California Edison Company's Big Creek Plant No. 1 on 6 April 1946; a 5-ton boulder, never recognized as a potential menace, became lubricated by melting snow and rain and slid down the slope on which it had been stationary. It bounced against the steel penstock lines, denting one so seriously that, under the water pressure of 14 Kg/cm^2 (200 psi), the pipe ripped open and released a flood of water that soon picked up a full load of

FIG. 40.5 Work in progress scaling the White Cliffs of Dover, England, showing protective wire mesh, with contractor's equipment at the top of the cliff.

soil and boulders from the fill around the penstocks. The resulting mud and boulder stream poured down the hillside and entered the powerhouse below. Damage was especially severe because the operators, although able to shut down three units, were caught by the flood before they could shut down the fourth.[40.11]

One of the most catastrophic of all rockfalls, although fortunately one that caused no great loss of life, was that in the Niagara Gorge on 7 June 1956. The sequence of layered limestones and shales making up the Niagara formation calls for several references in this volume. The rocks are well exposed in the upper part of the gorge, and the background they provided to the old Schoellkopf hydroelectric station of the Niagara Mohawk Power Company, clearly seen from the Rainbow Bridge, was familiar to many. On the date mentioned, serious leakage developed early in the day, and water seeped into the powerhouse from the rock wall above. Pumping was in progress when, at 5:10 P.M., 38,000 m³ (50,000 yd³) of rock fell on top of the powerhouse, demolishing almost two-thirds of it and unfortunately killing one man who was unable to escape (Fig. 40.7). The station developed 300,000 kW and this was automatically cut off. Estimates of the damage ranged from $100 million down to $20 million.[40.12] One of the main results of the disaster, no official report upon which appears to have been published, was the decision to proceed with the much larger power development of the New York State Power Authority, which has its station in the lower gorge.

40-8 SPECIAL PROBLEMS

FIG. 40.6 Concrete guard walls protect the Calderwood powerhouse of the Aluminum Company of America on the Little Tennessee River, Tennessee.

Exposed rock faces of doubtful stability can often be made safe by the modern practice of rock bolting. The basic idea is simple—that of drilling a hole into rock suspected of instability and inserting into this a long steel bolt, suitably wedged at its end so that, through driving, it can be securely fastened in the hole, thus anchoring the outer layer of rock to that in which the bolt is keyed. Frequently practiced in the past on a small scale,

FIG. 40.7 The Schoellkopf power station at Niagara Falls, New York, after its destruction by a rockfall from the cliffs above.

rock bolting has in relatively recent years assumed major importance. Today it is a widespread practice not only for anchoring exposed rock faces, but for supporting the soffits of unlined rock tunnels and underground powerhouses. Notable advance in the technique of rock bolting has been made in Australia on the Snowy Mountains project and in France where a well-illustrated volume has been published on this single topic.[40.13]

Illustrative of an extensive application is the use of rock bolts as a precautionary measure to anchor the exposed faces of the local pink sandstone at the Glen Canyon Dam on the Colorado River in Arizona. Thoroughly explored by the U.S. Bureau of Reclamation, which was responsible for the entire project, the local geological conditions are almost ideal; the sandstone was the only rock encountered, and it was always in sound condition. Some 7,600 m³ (10,000 yd³), however, had to be removed from the canyon walls, and this led to some spalling, with potential danger from more (Fig. 40.8). On the west side, where

FIG. 40.8 Installing rock bolts in the sandstone cliffs at the site of the Glen Canyon Dam on the Colorado River, Arizona.

spalling was worst, 1,000 mine-roof bolts, 1.8 and 2.4 m (6 and 8 ft) long and 1.9 cm (¾ in) in diameter, have been installed to "pin" the rock back to itself.[40.14]

A similar precautionary measure was taken in 1953 at the Nevada valvehouse of the Hoover Dam, also on the Colorado. Although no movement of the rock had been noticed, it was decided to "pin back" a slab 54 by 36 m (180 by 120 ft) by means of 349 five-cm (two-in) anchor bolts grouted into place in 7.5-cm (3-in) holes. The work was carried out

40-10 SPECIAL PROBLEMS

from a spectacular piece of steel scaffolding; its appearance was a vivid reminder of the scale of this great engineering project.[40.15] A similar operation was carried out at the Howard A. Hanson Dam on the Green River in Washington, but because of the clay-filled seams encountered during drilling for the placing of rock bolts, the necessary grip could not be obtained with just the usual wedges. Accordingly, larger holes were drilled (6.2-cm or 2½-in holes for 2.5-cm or 1-in bolts) so that they could be cleaned out by blowing. All bolts were grouted into place and then tightened up against large steel washers with torque wrenches. In all, 1,056 bolts were installed; their average length was 6.2 m (20.8 ft), the longest being 12 m (40 ft).[40.16]

The use of rock bolts of various types is now so well established and such common practice that it is probably not necessary to give further examples. Figure 40.9 illustrates,

FIG. 40.9 Scaffolding on the right abutment of the Kukuan Dam in Taiwan for giving access to tunnels, with rail reinforcing, instead of rock bolts, used to stabilize the alternating sandstone and quartzite.

however, a somewhat unusual case when time did not permit the extensive use of rock bolts to stabilize a steep rock face, as had been intended, the local geology assisting in the development of an alternative solution. The scene is on the right abutment of the concrete arch dam of the Kukuan water power development of the Taiwan Power Company. The V-shaped gorge in which the dam is located has alternating layers of sandstone and quartzite, with some thin strata of slate, forming the right abutment but dipping at 60°.

Naturally there was talus at the bottom of the gorge, and sound rock had to be obtained for founding the dam. It was necessary to support rock strata higher up the side of the gorge before further excavation; rock bolts were planned for this purpose but time was pressing. As an alternative, it was decided to drive short tunnels into the steeply sloping rock face; the tunnels were roughly 1.5-m (5-ft) square in cross section, ranging from 6 to 15 m (20 to 50 ft) deep. They were heavily reinforced with old railroad rails and had grout pipes embedded in the concrete with which they were filled so that precautionary grouting could later be carried out. In this way, and since the tunnels were driven at right angles to the bedding planes, the outer strata were securely "tied together," and the face was stabilized. The accompanying view shows scaffolding which might not, perhaps, secure approval from some North American safety inspectors, but it is commonplace in the East. Its use here is a tribute to the agility and courage of the local labor in Taiwan.[40.17]

Reference must finally be made to the rockfalls that have characterized the steep and rocky coasts of Norway. The topography of the western part of this country is widely known from the many photographs of its most beautiful scenery. Much of the fjord country consists of metamorphic rocks of Cambrian and Silurian age; most rockfalls occur in the older metamorphic rocks of Caledonian origin. Records show that about 200 major rockfalls took place between 1640 and 1900 and that 50 have taken place since the start of this century. Most of them occur in the spring or the fall of the year, being clearly related either to freezing conditions or to excessive runoff. Over 350 people are known to have been killed in rockfalls; many of this number perished as a result of the six catastrophic rockfalls on record. These occurred generally in narrow fjords, where the fall of rock causes great waves in the narrow waterways and the consequent tragic flooding of little villages that are at once so picturesque and so vital a part of the Norwegian economy. Descriptions of these tragic happenings are available.

Study of the records of such rockfalls and careful study of rock conditions in the vicinity of known falls have elucidated the causes of such accidents, so that preventive measures are now possible. The Norwegian Geotechnical Institute has been engaged in this important work and has designed and installed special extensometers for measuring slight differential rock movements (such as across an open seam) that may be anticipated prior to a major rockfall. Since these must be installed in isolated locations such as on the tops of steep cliffs above deep fjords, the Institute has coupled its extensometers to automatic radio transmitters, from which messages which will indicate any change in rock conditions are sent at regular intervals to observer stations at the bottom of the valleys. Figure 40.10 shows such an installation on the mountains Flofjell, some 20 to 30 km (12 to 18 mi) north of Loen, approximately 1,000 m (3,300 ft) above sea level. This is on the Nordfjord about 180 km (110 mi) north of Bergen.[40.18]

40.3 MINOR ROCKFALLS

One hesitates to describe any rockfall as a "minor" incident, since even small stones falling from a great height can wreak much damage. The term is used, however, to indicate the fall of individual pieces of rock, harmless if they do not hit anything, but dangerous if they do. It is in the regular inspection and maintenance of civil engineering works, and especially of transportation routes, that they are most commonly encountered. All motorists will be familiar with the "falling rocks" signs that are to be seen on main highways where this danger exists. Trimming of rock cuts sometimes becomes necessary if falling rocks become too troublesome. Following the example of railways, some road authorities have had to install safety fences to protect road users from this hazard. On the Nuuanu Pali Highway in Hawaii, for example, a road running between Honolulu and Windward

FIG. 40.10 Automatic rock-movement gauge installed by the Norwegian Geotechnical Institute on the Flofjell, north of Loen, Norway, at the head of the Nordfjord (Lake Strynsvannet is to be seen in the background).

on the island of Oahu, a double row of strong steel fencing has had to be installed at one critical location, with good effect. In Quebec City, even the historically famous rock known as Cape Diamond, so familiar a sight to all who have sailed up the River St. Lawrence, occasionally drops rocks on the narrow street that winds along the shore below it. City authorities have installed rock baskets to trap any such falling boulders.

These are but two of the precautionary measures that may be taken in areas where rockfalls are to be expected. As with other such design features, they are in themselves relatively so simple that they might seem to have no place in such a Handbook as this. Anyone who has seen, however, the damage that can be done by a falling boulder will need no reminder that such simple features are in no way to be spurned. Even simpler is the idea of excavating an extra width of road or railway right-of-way in order to leave an area at the foot of the rock face, but separated from the road or railway, sometimes by a low concrete wall, onto which base rock may fall and be retained in safety. The effectiveness of this practice is often evident to travelers through the accumulation of fallen rocks that gradually develops in such areas.

At the other extreme of protective measures are the electrically operated protection fences that are used for particularly dangerous stretches of road or rail. These are illustrated in Chap. 31. When the area from which rocks may fall is localized, another alternative is to cover the site with strong steel-wire mesh, adequately secured. An outstanding example of this practice was carried out when Interstate 70 (a fine highway mentioned elsewhere in this book) was opened through Clear Creek Canyon, west of Denver. In the Floyd Hill-Idaho Springs section of the road, twin one-way, two-lane tunnels were constructed through a jutting promontory dipping at about 45° across the line of the road. With weathering on exposed bedding planes, some rock loosening was to be expected. As a part of the tunnel contract, therefore, the Colorado Department of Highways included the installation of a protective cover of chain-link fence consisting of No. 9 galvanized wire and secured into place by 252 rock bolts, 1.8- to 2.4-m (6- to 8-ft)-long, on 2.7-m (9-ft) centers. These rock bindings were high above the road. Their installation was carried out ingeniously by workmen in specially constructed working platforms suspended from the end of a 52-m (172-ft) boom of a mobile crawler crane.[40.19] After well over a decade of service, this "rockfall guard" has proved its usefulness (Fig. 40.11).

FIG. 40.11 Chain-link drape attached to rock above portals of tunnels on Interstate 70 near Denver, Colorado, as a safety measure to control rockfalls.

40.4 NIAGARA FALLS

This chapter would not be complete without at least passing reference to Niagara Falls, at once one of the best-known examples in the world of rockfalls and a vivid example of the natural process of erosion to be dealt with in Chap. 41. This majestic natural phenomenon has been under regular observation since the falls were first seen in 1678 by an intrepid pioneer missionary priest, Father Hennepin. It is known, therefore, that since that time the gorge has receded up the Niagara River by about 300 m (1,000 ft). Modern studies suggest that the full length of the gorge (from its start near the natural escarpment at Queenston) has eroded in about 12,000 years, a figure which brings home the relative speed (geologically) of this particular example of erosion by running water. The erosion is actually caused by the relative weakness of the shale strata which underlie the surface Lockport limestone, the shale weakening under the action of water and periodically caus-

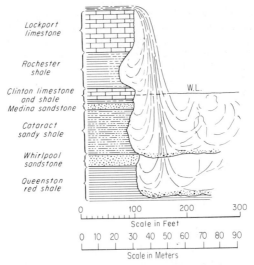

FIG. 40.12 Geological section through Niagara Falls, illustrating the alternation of strata which assist erosion.

ing the superincumbent limestone to break off in huge slabs (Fig. 40.12). It is these that are such a prominent part of the spectacle of the falls.

The gorge has been formed by the retrogression of the Horseshoe, or Canadian Falls. It was only when this erosion passed what is today called Prospect Point that the American Falls began to form, with the corresponding formation of Goat Island as it is today. Since then, erosion has taken place at both falls, as the great piles of rock slabs at the feet of the two cataracts testify so vividly. The International Joint Commission, under whose jurisdiction the falls come, became so concerned about the future of the American Falls that a special study was carried out for them in the early 1970s by the American Falls International Board. Appendix C to the fine report that emanated from this inquiry, the appendix dealing with geology and rock mechanics, is a singularly valuable document in relation to a major example of rockfalls. The American Falls were dewatered in 1969 so that rock conditions at the lip could be examined in situ by the U.S. Corps of Engineers and their consultants, an extra tourist attraction being thus provided for a short period. Figure 33.20 shows the American Falls in this unaccustomed condition; it serves as a vivid reminder of the mechanics of rockfalls.[40.20] (Fig. 40.13.)

40.5 ROCKSLIDES

Rockfalls are fairly obvious hazards to civil engineering works. Rockslides are not, but may be of equally serious importance. The term is applied to masses of loose rock which, either alone or in association with soil, are slowly moving under the force of gravity along some buried inclined surface of competent bedrock. There are cases (as in the Carpathian Mountains) where immense blocks of rock, separated by jointing but having all the appearance of bedrock, are actually slowly moving along an underlying stratum of clay. More common, however, are masses of fragmented rock on sloping hillsides which, in the course of geological time, have become so consolidated that they appear to be solid rock but which are nevertheless moving slowly.

FIG. 40.13 Examining exposed rock formations at Niagara Falls, Ontario, in connection with design of remedial works, now complete.

Travelers in Switzerland who use the attractive, narrow-gauge Rhaetian Railway to go up into the mountains from the junction at Chur, either for skiing or to visit the famous Institute for Snow and Avalanche Research at Davos, pass through the pleasant town of Klosters. Here the railway crosses a small river on a curved bridge, entering a short tunnel almost immediately. If one can stop at Klosters and detrain, then one can see this splendid example of the slender, reinforced-concrete bridges designed by the justly renowned Robert Maillart, fitting perfectly into its lovely surroundings. Closer inspection of the bridge will show that, although it appears to be an arched bridge with one clear span, there is a reinforced-concrete beam between the two abutments, so designed that it serves as a most convenient footbridge between the two parts of the village. So it was thought to be when first erected; it may still be so regarded by residents who have not inquired as to its real purpose. After the bridge was placed in service, the track upon it was naturally inspected regularly. An alert permanent-way employee noticed one day that the track in the center of the bridge appeared to have risen slightly. Further inspections showed that the center of the bridge was indeed rising very slowly. When a full investigation was made, it was found that the southern abutment was founded on rock talus that, although apparently a stable foundation bed, was actually moving very slowly downhill. The only solution was to brace one abutment against the other, and this was done by means of the "footbridge." Embedded in it were the necessary instruments for measuring the strain induced by this unusual rockslide, and these confirmed the basis used for the design of this unusual remedial measure (see Sec. 24.11).[40.21]

One of the most interesting cities in England from the geological point of view (and others) is the historic seaport of Bristol. Even though Avonmouth is today the major seaport, some vessels still come up the scenic gorge of river Avon to the still-active Bristol docks. Just after passing under the famous Clifton Suspension Bridge, ships pass through a part of the gorge where the north bank is quite precipitous, leaving room for only a narrow roadway between the river and the foot of the cliff. On the top of the 60-m (200-ft)-high cliff is a well-known hotel and valuable residential property. So steep is the cliff that a funicular railway was installed in the late nineteenth century in a specially built tunnel running obliquely from near the hotel to the road below. With developments in transportation, the Clifton Rocks Railway, as it was known, fell into disuse but the tunnel remained. In 1956 it was noticed that cracks had appeared between the lower portal and the Carboniferous limestone of the cliff. This limestone consists of thick beds of heavily jointed rock separated generally by thin layers of red clay, probably derived from rock weathering. The beds dip roughly parallel to the face of the gorge at an angle of 33°. A full-scale investigation was launched.

Test drilling confirmed the existence of a bed of clay about 1.5 m (5 ft) thick, first seen when a retaining wall near the portal was removed at the start of the rehabilitation work. It seemed clear that very large blocks of the limestone were tending to slide downhill on this clay surface. Test boreholes revealed that the groundwater was standing 1.2 m (4 ft) above the top of the clay layer, whereas the underlying limestone was dry, water draining into it even down the small test holes. After careful analysis had been made, it was decided to put down six gravel-filled boreholes as far as the underlying limestone; the boreholes were to act as permanent drains, thus keeping the top of the clay stratum reasonably dry. In addition, strengthening of the tunnel portal and an associated retaining wall and tying them by steel rods to the limestone bedrock was also carried out, in case any further movement did tend to develop. It seemed clear that the disused tunnel had been acting as a drain for water moving through the upper limestone along the upper surface of the clay stratum, since borings showed no water above the clay some distance from the tunnel. That the tunnel was disused may have concealed for some time any change in drainage conditions within the tunnel. The remedial measures were effective, and no further work has had to be done to maintain the stability of this particularly important section of the wall of the Avon Gorge.[40.22] (Fig. 40.14).

The lessons from these two cases, so different and yet so similar in basic cause, are so obvious that they need no elaboration except, perhaps, for the observation that both illustrate, yet again, the vital importance of regular inspection in the course of what is unhappily called "routine maintenance" of engineering structures. The existence of the disused tunnel at Bristol shows how quickly such features can be forgotten, or at least neglected. Once again it can be seen that inspection must comprehend not only the examination of structures as such but the geological conditions upon which their stability ultimately depends.

40.6 AVALANCHES

Another natural hazard occurs when yet another material, snow, slides down inclined surfaces. The word *avalanche* has already been used in its broadened sense to describe the descent of groups of rocks, but the name was originally restricted to the description of sudden descents of masses of snow down steep natural slopes. Avalanches can happen only in mountainous country and might, therefore, appear to have little to do with engineering

FIG. 40.14 Rock-protection tunnel approaching completion beneath the Clifton Suspension Bridge, Bristol, Avon, England.

development. But with the steady encroachment of tourist resorts in mountain fastnesses and the continual search for new mineral deposits in mountainous country, planning for such developments cannot neglect the danger of avalanches (Fig. 40.15). The destruction of a new mining camp in the mountains of British Columbia in 1965, with the loss of 26 lives, and the tragic accidents at ski resorts in France and Switzerland during the winter of 1969–1970, despite their advanced protective services, were grim reminders of the toll to be paid if avalanches cannot be controlled. Fortunately, their occurrence is restricted to areas with very steep slopes that are subjected to heavy snowfall, but they certainly warrant a brief reference if this volume is to serve adequately to show how many and varied are the aspects of geology that must be considered in engineering developments around the world.

History has its own grim reminders of the havoc that avalanches can cause to human settlements, such as the virtual destruction of the village of Leukerbad in Switzerland in 1718, with the loss of 52 lives, and the destruction of many buildings in the well-known and beautifully situated town of Davos on 23 December 1919, with the loss of several lives. One is forced to conclude that avalanches were then regarded as "acts of God" that just had to be accepted, as they were, for example, in the maintenance of the Canadian Pacific Railway through Rogers Pass in the Selkirk Mountains of British Columbia. Here they became so serious a problem that the 8-km (5-mi) Connaught Tunnel was completed in

FIG. 40.15 Destruction of the Hotel Drei Könige at Andermatt, Switzerland, by an avalanche in January 1951 (Goethe stayed in this hotel when on his way to Italy).

1916 to provide a new route for the railway. The pass was left to revert to its original wild state until it came to be used, in the late fifties, for the Trans Canada Highway. It was by then possible to plan avalanche control well ahead of the building of the highway; a research officer lived in the pass through three preceding winters and measured every avalanche as it occurred, providing consequent on-the-spot defense-work planning (Fig. 40.16).[40.23] This followed from contacts with the Swiss Snow and Avalanche Research Institute high up on the Pasern above Davos in eastern Switzerland. This famous station was a pioneer in the scientific approach to avalanches, having been established in 1936. The director and his staff have shared fully their experience with others interested around the world, and especially in North America. Interestingly, the station was established by and still comes under the direction of the Swiss Forestry Department, since avalanches have done so much damage to the reforestation program of which Switzerland is so justly proud, even though the station's activities cover protection against avalanches for all types of property, and, above all, for people, especially skiers.

Although the first study of works for protection against avalanches appears to have been carried out as early as 1880, such efforts were personal and spasmodic until the years after the First World War. The first classification of avalanches—and there are many varieties—appears to have been that of E. C. Richardson in 1909 in a book on skiing. Thereafter it was the skiers who led the way toward the truly scientific study of these winter hazards, in Germany, Austria, Switzerland, and England. Publication in 1936 of *Snow Structure and Ski Fields* by Gerald Seligman, founder and honorary president of the International Glaciological Society and editor of its excellent journal until his death, marked a real turning point. Today, the mechanisms of avalanche flow are well understood; the climatic and terrain factors that can combine to create incipient avalanches are

FIG. 40.16 Tupper No. 1 snowshed on the Trans Canada Highway in Rogers Pass, British Columbia, its location and design based on a three-year study of avalanches in the pass.

known; field observations at any prospective building site, carried out through at least one or two winters (but preferably more) will indicate locations in which defense works are desirable.

Defenses against avalanches include the construction of snowsheds (effective but expensive), the use of mounds of soil and rock to deflect the flow of avalanches from critical places, and the deliberate initiation of avalanches by mortar gunfire from carefully selected gun emplacements (Fig. 40.17). All three methods are used for the Rogers Pass section of the Trans Canada Highway. This major road has not been closed for more than an hour or two (usually after controlled gunfiring) since it was opened in 1962, despite annual snowfalls as high as 15.5 m (50 ft).[40.24] It is mentioned only to show that avalanches can now be controlled, even under extreme conditions, although there can never be any absolute certainty about the incidence of avalanches, so many are the factors that can very quickly combine to create a hazardous situation. Constant vigilance is again necessary, with regular patrols of expert observers. Today, this practice is becoming more widespread because of the popularity of skiing, but the significance of avalanches in relation to any engineering works close to steep mountain slopes must never be forgotten.

40.7 CONCLUSION

Until relatively recent years, rockfalls have received little attention in the normal practice of civil engineering. There have been, naturally, exceptional individual cases in which the possibility of a rockfall was fully integrated into design, such as some of the cases cited in this chapter. In general, however, rockfalls were usually regarded as "acts of God" (to use that lamentable phrase again), and, if on a small scale, their rectification and the carrying out of remedial works was left to maintenance personnel. This neglect is well illustrated by the lack of any useful written records in English until recent years. There are now, fortunately, some useful guides, all of which inevitably make clear the absolute dependence upon the local geology of all aspects of rockfalls, rockslides, and other rock movements. Knowledge of geology cannot ensure prediction of rockfalls with certainty, but a

FIG. 40.17 Avalanche starting its descent (after initiation by gunfire) on the Laurie avalanche path between Revelstoke and Glacier, British Columbia, Canada.

study of local geological conditions can make prediction the more certain and provide knowledge of the causes of falls, knowledge without which remedial methods may be ineffective (Fig. 40.18).

Studies that have been made have already revealed some significant and useful information. Figure 40.19 from a notable paper by Peckover and Kerr, for example, shows the close correlation of rockfall incidence with climate.[40.25] The graph summarizes experience over a 37-year period on the Yale subdivision of Canadian National Railways through the Rockies (and incidentally is a tribute to the accurate record-keeping on CNR maintenance work). It will be seen that the maximum frequency of rockfalls occurs between November and March, when the mean temperature is at or slightly below freezing point, and is almost certainly linked with the freeze-thaw cycles that occur at such times.

Since rockfalls can have such serious consequences in urban areas, it would be useful if the engineering-geology maps now being developed for some urban areas could indicate the probability of rockfalls. There is already a good example of what can be done in this direction. The remarkable work in correlating local geology with urban development in the Front Range area of Colorado, in and around Denver, has already been mentioned appreciatively. Some of this work finds expression in the series of maps that have been prepared for the Golden Quadrangle. The third map in this set, by Howard E. Simpson,

FIG. 40.18 Protection work on Swiss Federal Railways against rock and snow slides.

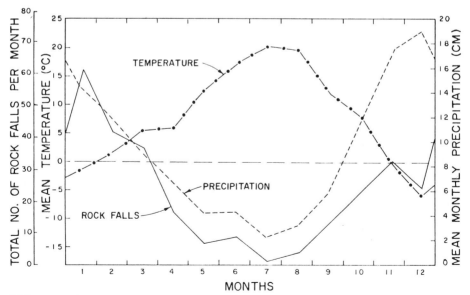

FIG. 40.19 Chart illustrating the interrelation of climate and rockfall occurrences on the Yale subdivision of Canadian National Railways, British Columbia.

40-22 SPECIAL PROBLEMS

shows areas of potential rockfalls in the Golden Quadrangle, Jefferson County, Colorado.[40.26] The approach road to the city of Golden from the east runs between two remarkable mesas, North and South Table Mountains. One has only to glance at these to see evidence of persistant rockfalls. Simpson has mapped these, and other falls along the mountain front to the west, and studied all available local records to get an historical perspective of this hazard. The results are clearly shown, in color, on his map, which is accompanied (on the same sheet) with a helpful explanation. Any development within this quadrangle should, therefore, never proceed without reference being first made to this useful map. It will be a long time before the existence of such maps becomes commonplace, but this desirable advance in making geological information on specific problems conveniently available for public use has at least started. This map can serve as a useful guide and helpful reminder to all concerned with engineering works in any area where rockfalls may occur, a matter necessitating care in design and *regular and expert inspection* once a project is complete.

Finally, there is one Rock which is so well known around the world that this chapter, short as it is, would not be complete without some reference to it, especially as some rock strengthening is in progress there as this volume goes to press. This is Castle Rock, on which Edinburgh Castle is built, providing all visitors to the Scottish capital city a stirring sight from Princes Street (Fig. 40.20). The rock consists (geologically) of a plug of fresh, hard basalt emplaced in Carboniferous sediments. These consist of alternating sandstones, shales, and marls. They may be seen to the east of the rock, underlying the famous esplanade. Ice movement has been from the west and this has exposed the basalt on the northern, western, and southern flanks of the plug, leaving grooves in the exposed bedrock.

FIG. 40.20 Edinburgh Castle, Scotland, and its famous rock, some reinforcement of which has been necessary.

When a plan of Castle Rock is examined, it can be seen that it constitutes a classical example of "crag-and-tail" structure, a feature commonly encountered in areas over which ice has moved. It is regrettable to have to add that the great rock is not quite so firm and immovable as it appears to be at first sight. There have been some landslides in the till on the northern face of the tail, and occasional falls of blocks of the basalt have taken place in recent years. To assist with the necessary corrective measures, a detailed engineering-geological study of the rock was carried out in 1966-1967. A fine record of this study has been published and can be commended to all who face problems with the stability of exposed rock faces.[40.27]

40.8 REFERENCES

40.1 *Report of the Commission Appointed to Investigate Turtle Mountain, Frank, Alberta*, Memoir no. 27, Geological Survey of Canada, Ottawa, 1912.

40.2 D. M. Cruden and J. Krahn, "A Re-Examination of the Frank Slide," *Canadian Geotechnical Journal*, **10**:581 (1973); see also J. Krahn and N. R. Morgenstern, "Mechanics of the Frank Slide," *Rock Engineering for Foundations and Slopes*, vol. 1, American Society of Civil Engineers, New York, 1976, p. 309.

40.3 W. H. Mathews and K. C. McTaggart, "The Hope Slide, British Columbia," *Proceedings of the Geological Association of Canada*, **20**:65 (1969).

40.4 E. B. Eckel (ed.), *Landslides and Engineering Practice*, Highway Research Board Special Report no. 29, National Academy of Sciences, Washington, D.C., 1958.

40.5 R. L. Shreve, "Geology of the Blackhawk Landslide, Lucerne Valley, California," in abstract vol. for Vancouver meeting of the Cordilleran Section of the Geological Society of America 1960, p. 41.

40.6 K. J. Hsu, "Catastrophic Debris Streams *(Sturzstroms)* Generated by Rockfalls," *Bulletin of the Geological Society of America*, **86**:129 (1975).

40.7 G. H. Eisbacher, "Cliff Collapse and Rock Avalanches *(Sturzstroms)* in the Mackenzie Mountains, North-Western Canada," *Canadian Geotechnical Journal*, **16**:309 (1979).

40.8 R. Carpmael, "Noteworthy Engineering Work on Vriog Cliffs, Near Barmouth," *Great Western Railway Magazine*, March 1938, p. 118; and H. H. Reynolds, unpubl. paper.

40.9 C. S. Lee, "Chalk Falls between Folkestone and Dover," *The Railway Magazine*, **66**:531 (1940).

40.10 "Chalk Cliffs Are Being Stabilized at Dover," *The Engineer*, **225**:627 (1968).

40.11 "Penstocks and Power House Damaged as Result of Falling Boulder," *Engineering News-Record*, **136**:934 (1946).

40.12 "Tremor Triggers Niagara Rock Slide," *Engineering News-Record*, **156**:27 (24 June 1956).

40.13 A. Huson and A. Costes, *Le Boulonnage des Roches en Souterrain*, Editions Eyrolles, Paris, 1959.

40.14 "Preventing Spalling at Glen Canyon Dam," *Civil Engineering*, **28**:958 (1958).

40.15 "Slab is Anchored to Wall of Canyon," *Engineering News-Record*, **151**:25 (17 December 1953).

40.16 H. E. Christman, "Bolts Stabilize High Rock Slopes," *Civil Engineering*, **30**:98 (1960).

40.17 A. E. Niederhoff, "Concrete 'Dowels' Anchor Cliff," *Engineering News-Record*, **165**:45 (20 October 1960).

40.18 L. Bjerrum and F. Jorstad, *Stability of Rock Slopes in Norway*, Norwegian Geotechnical Institute Publication No. 79, pp. 1-11, 1968.

40.19 "Steel Drapes Guard Twin Tunnels," *Engineering News-Record*, **167**:30 (14 September 1961).

40.20 S. S. Philbrick, "Horizontal Configuration and the Rate of Erosion of Niagara Falls," *Bulletin of the Geological Society of America*, **81**:3723 (1970).

40.21 C. Mohr and R. Haefeli, «Umbau der Landquartbrucke der Rhatischen Bahn in Klosters,» *Schweitzerische Bauzeitung*, **65**:5, 32 (1947); see also R. Haefeli, C. Schaerer, and G. Amberg, "The Behaviour under the Influence of Soil Creep Pressure of the Concrete Bridge Built at Klosters by the Rhaetian Railway Company, Switzerland," *Proceedings of the Third International Conference on Soil Mechanics and Foundation Engineering*, **2**:175 (1953).

40.22 D. J. Henkel, "Slide Movements on an Inclined Clay Layer in the Avon Gorge in Bristol," *Proceedings of the Fifth International Conference on Soil Mechanics and Foundation Engineering*, **2**:619 (1961).

40.23 P. A. Schaerer, "Planning Defences against Avalanches," *Canadian Geotechnical Journal*, **7**:397 (1970).

40.24 P. A. Schaerer, "Relation Between the Mass of Avalanches and Characteristics of Terrain at Rogers Pass, B.C., Canada," *Publication No. 104*, International Association of Scientific Hydrology, Paris, 1971, pp. 378–380.

40.25 F. L. Peckover and J. W. G. Kerr, "Treatment and Maintenance of Rock Slopes on Transportation Routes," *Canadian Geotechnical Journal*, **14**:487 (1977).

40.26 H. E. Simpson, *Area of Potential Rockfalls in the Golden Quadrangle, Jefferson County, Colorado,* Map no. 1761C, U.S. Geological Survey, Washington, D.C., 1971.

40.27 D. C. Price and J. L. Knill, "Engineering Geology of Edinburgh Castle Rock," *Geotechnique*, **17**:411 (1967).

Suggestions for Further Reading

Barry Voight has placed the profession in his debt by assembling in two large volumes (851 pages in 1978 and 868 pages in 1980) a fine collection of information called *Rockslides and Avalanches,* published by Elsevier of New York. Many of the papers thus assembled have relevance to the problems with which this chapter deals.

A summary of the information obtained by the dewatering of the American Falls at Niagara will be found in Appendix C to the Final Report of the American Falls International Board of the International Joint Commission. (This most important but little-known body, with equal United States and Canadian membership, is responsible to the respective governments for the solution of all problems related to boundary waters between the two countries.)

Three Norwegian authors, R. Schach, K. Carshol, and A. M. Heltzen, have prepared a useful small book of 96 pages, published in 1979 by Pergamon Press, called *Rock Bolts,* the devices that can be so helpful in securing loose rock before a fall occurs.

chapter 41

EROSION AND SEDIMENTATION

41.1 Erosion by Streamflow / 41-2
41.2 Erosion Below Dams / 41-5
41.3 Transportation by Streamflow / 41-8
41.4 Silting of Reservoirs / 41-12
41.5 Erosion on Construction Projects / 41-13
41.6 References / 41-14

The phenomenon of erosion and the complementary action of sedimentation can probably be regarded as the geological processes of most importance in civil engineering work, even when considered only insofar as they are operating at present. If the past is considered and even cursory thought is given to the difficulties introduced into construction work by the existence of buried valleys and by the variation of unconsolidated sediments and glacial deposits, the significance of these processes will be considerably increased. They constitute a vital part of the geological cycle. By the glaciation now taking place in arctic regions, just as by the cutting of solid rock by the wind-driven sand of tropical deserts, parts of the earth's crust are steadily being eroded. Other parts are similarly being built up. In combination, these processes disturb the static equilibrium of the earth's crust, with the result that the erosion and sedimentation of today are contributing to earth movements of the future.

The civil engineer is repeatedly faced with problems caused by erosion: on the seacoast, the action of the sea may be destroying valuable developed property; at bends in rivers, the current steadily encroaches upon bankside property or adjacent transportation routes; and in other areas, the combined action of wind and rain erodes fertile topsoil. Similarly, the deposition of silt behind many artificially constructed barriers to riverflow, the silting up of slow-flowing rivers, and the formation of bars in tidal estuaries present special problems in the solution of which the engineer is called upon to play a leading part.

Even in this brief introductory note civil engineering works of such important types as river-training work, coastal protection, and soil conservation have already been suggested. All have a common feature, since in them the work of the engineer is concerned with an attempt to conserve or develop physical features that are constantly exposed to natural

41-2   SPECIAL PROBLEMS

forces tending to alter them in some way. These forces will persist despite all the efforts of the engineer. Hence, in civil engineering work connected with erosion or silting, the engineer not only must face structural problems but must also consider the effect that the resulting structures will have on the future action of these natural forces. In all these considerations, a knowledge of the fundamental geological processes at work is essential; their neglect may result in the failure of works structurally quite sound.

41.1   EROSION BY STREAMFLOW

The flow of water in rivers and streams is perhaps the best recognized of the major causes of erosion. As the material eroded must be carried away by streamflow, complementary problems of silt deposition arise whenever the flow of a silt-laden river is interfered with in any way. Many and diverse engineering problems are thus created; the correct solution of these problems requires an appreciation of the general geological characteristics of riverflow as a necessary background. The life history of rivers, although essentially a study in physical geology, is therefore a matter that should be familiar, at least in general outline, to all engineers engaged on river works. Visualizing a complete profile of a river from source to mouth, one can imagine the river gradually decreasing in slope as it nears its mouth, increasing in volume and cross section on its way. This theoretical profile is not universal, although many of the large rivers of the world, such as the Amazon and the Ganges, conform to it. Rivers rising in mountainous country near the sea may be steeply graded from source to mouth, whereas other rivers, which have their sources in relatively low-lying ground, will have a flat profile throughout and corresponding characteristics.

The curved profile is one that a river will constantly tend to produce, and it is therefore a fundamental concept underlying all river studies; changes in gradient are achieved through the erosive power of running water and of the debris that it transports. Thus the steepest part of the theoretical profile may be regarded as the youthful stage of a river, the central section its middle-age period, and the flat gradient at the lower end its old age; the meandering of slow-flowing streams across surrounding plains is another characteristic of this last stage. All references to time apply to the geological time scale; but despite this, rivers in all stages of development may be seen today, and many show clearly their immediate past history. The canyons of the Colorado River in the United States, the Cheddar Gorge through which the river Avon flows in the south of England, and similar scenic river features present vivid evidence of progressive deepening due to erosion.

The main interest of the civil engineer in river history is in connection with the erosive processes that have been, and still are, at work. Weathering of the exposed rock surfaces can be regarded as the starting point in the train of erosion that eventually finishes in the depths of the sea or of an inland lake. Gravity and the wind will have the effect of bringing the products of weathering down from the hillsides. After rainstorms, the flows in even the smallest gullies will start the work of transporting this debris downward with the stream. The erosive action of streams, although starting in this way, can be generally classified as three separate processes: (1) transportation, (2) corrasion, and (3) solution. The start of the transportation of material by a stream has just been described. This is a process easy to visualize and in some cases to watch, and it is one the general nature of which can be readily appreciated. Solution is also a type of action easy to understand.

The term *corrasion* has been used to denote the combined erosive action, mechanical in nature, of the flow of water and of the material that it is transporting along the bed of a stream. The action has been likened to the passing of a continually moving file over the stream bottom. A direct analogy can be seen in the practice of cutting solid rock with wire saws, quartz grains being usually introduced beneath the moving wire rope, to be carried along by it as it passes over the rock. The wire rope is similar in action to the riverflow, and the quartz grains to the material being transported by the stream. Corrasion will clearly depend on the velocity of the streamflow, on the amount and nature of the load that the stream is transporting, on the plan and profile of the stream course, and on the geological nature of the stream bed.

The direct influence of stream velocity on erosive action is a feature that needs no elaboration. The load of a stream is also of great importance, since the erosive action of water itself flowing in a straight and regular course over solid rock is small indeed. When it is carrying along sand and gravel, that portion of the load which travels on the bed of the stream will be constantly in irregular movement, bumping against the bottom and intensifying eddies. In this way, the resistance of the bed material to movement will be broken down, and erosion will gradually take place. A striking although rather localized example of this action is provided by the potholes which are often found in the beds of swift-flowing streams; the presence of pebbles or even boulders at the bottom of such holes is a telling reminder of their mode of formation (Fig. 41.1). During the construction of the Assouan Dam in Egypt, potholes were found in the unwatered riverbed area, 1.2 to 1.5 m (4 to 5 ft) in diameter, up to 6 m (20 ft) deep, and eroded out of the hardest granite.[41.1]

The plan and profile of a stream naturally affect its flow and so inevitably are related to its corrasive action. Bed profiles have been discussed in a general way; the increased

FIG. 41.1 An unusual example of a deep pothole exposed by rock excavation for railway construction in northern Quebec.

41-4 SPECIAL PROBLEMS

erosion liable to take place in the more steeply graded section of a stream bed will be easily appreciated, especially since the erosive power of a stream depends not directly on the velocity but on some power of the velocity higher than the square. Another feature of a stream profile may be the existence of waterfalls. These will often be due to geological conditions influencing bed erosion.

No better example can be quoted than that provided by Niagara Falls, where the Niagara River drops over the Niagara escarpment in two waterfalls, 50 m (167 ft) high, world-famous for their majestic beauty and of peculiar interest to engineers in view of the water-power developments associated with them. It will be found that this world-famous waterfall was briefly described in the preceding chapter in relation to rockfalls which characterize both parts of the great cataract. Niagara Falls is mentioned again, however, to emphasize the complete interrelation of natural processes such as riverflow, erosion, and rockfalls (in this example). Devotion of this separate chapter to erosion and sedimentation is, therefore, in one way, an artificial subdivision. These associated processes have been highlighted in this way because their very familiarity can lead to their neglect, even though they affect so many types of civil engineering work. All projects involving rivers or streams are liable to be affected by erosion or silting, or both, and this applies not only to the great rivers already mentioned but to every stream, large or small.

If only by way of contrast to the rivers already mentioned, it may be observed that the relatively short rivers of the South Island of New Zealand carry some of the highest loads of sediment of any rivers, as can so clearly be seen by the observant traveler who journeys by road or railway down the east coast of this lovely island. The Waimakariri River, which flows into the sea near Christchurch, is less than 160 km (100 mi) long with a catchment area of only 2,460 km² (950 mi²), and yet, as Fig. 41.2 shows, it can transport over a quarter of a million tons of sediment in a day at times of high flood. Much of this load comes from its small tributary streams flowing swiftly out of the valleys of the Southern Alps.[41.2]

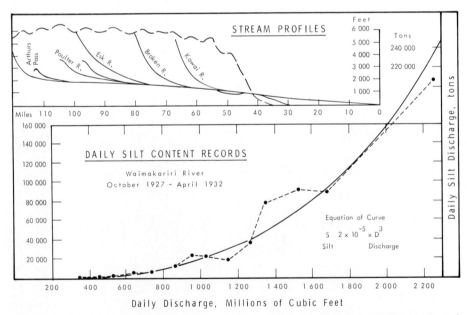

FIG. 41.2 Relation between the average daily water discharge and amount of sediment transported by the Waimakariri River, New Zealand.

41.2 EROSION BELOW DAMS

Erosion that is induced by man's interference with natural riverflow through the construction of dams can have consequences that go far beyond recognition in terms of money. The concentration of water flow and the high head under which it may take place render the hydraulic design of spillway channels always a delicate matter, and the proportioning of the artificial structure must be most carefully done. Unfortunate experiences have led engineers to undertake comprehensive research work, and there now exists in engineering literature a series of most valuable papers dealing with the hydraulic solution of the problems thus raised; many of these are based on the use of model structures.

An early example is a paper by Butcher and Atkinson which describes extensive research work on models undertaken as a result of certain problems of erosion that had arisen in Sudan at the Sennar Dam controlling the Blue Nile.[41.3] The dam, located about 320 km (200 mi) south of Khartoum, is founded on a ridge of solid rock, but the riverbed immediately below the dam consists of boulders and rock of doubtful quality. No masonry aprons were originally provided; and after the first year's operation, it was found that considerable erosion had taken place in this part of the riverbed. A glass-sided observation tank with models of dam cross sections was employed in a way similar to that used by Dr. Rehbock in his original experiments on this problem carried out in 1905 and 1909 at the Technical University of Karlsruhe. The necessity of using some type of sill at the foot of the outfall apron of a dam was demonstrated in Butcher and Atkinson's experiments, as it had been previously during Dr. Rehbock's experiments, when the patented dentated sill that now bears his name was evolved. This detail of spillway design is mentioned because of its importance when geological studies show that the rock immediately below the overflow apron is liable to be affected by the intensity of direct flow; the use of some type of sill often reduces the eroding effect of the discharge flow.

In the discussion of this paper, some interesting examples were cited of erosion in New Red sandstone and in clay and shale beds adjacent to sluices at some of the locks of the Manchester Ship Canal. Since these openings did not have to pass large volumes of water, they illustrate how this problem of erosion may be of extreme importance even in relatively small structures. At least one major structure, the Islam Weir of the Sutlej Valley project, has failed from this cause (Fig. 41.3). The weir is one of four similar structures on

FIG. 41.3 The Islam Weir, Sutlej Valley irrigation project, Punjab, as seen from upstream following the failure of six bays on 19 September 1929.

the Sutlej River in the Punjab which constitute diversion dams for a large irrigation project. Consisting of 24 openings, each of 18-m (60-ft) span, it is a floating type of dam which rests on a sand foundation and is protected by upstream and downstream aprons. Although the head on the dam was only 5.4 m (18 ft), downstream scour in two years had amounted to 2 m (6½ ft); and in 1929, six bays of the structure collapsed. The official report of this accident is a most valuable publication.[41.4]

Numerous examples of the erosion of rock below dam spillways could be cited; in most instances the erosion developed only after the dam structure had been in use for some time. In some cases, although admittedly not in all, trouble of this kind can be foreseen if a detailed geological survey of the area immediately downstream from a spillway is made before construction begins. Careful study must be made not only of the effect of water and of concentrated flow on the rock but also of the structure of the rock. Besides the more usual erosion processes which will operate with increased intensity below a spillway discharge, there always exists the possibility that concentrated flow will loosen blocks of the rock at its more definite joints and bedding planes, especially if these contain any disintegrated material or product such as clay. This loosening would not normally be serious, but the velocities that occur in a spillway discharge may be so high that blocks of rock of great size will thus be transported downstream bodily. They provide a dangerous "tool" for the streamflow to use in further erosion and leave exposed a further area of bedrock on which the concentrated flow will impinge. Such a condition existed on the Susquehanna River in Maryland, in the case of the fissured granite below the Conowingo Dam, the joints of which were filled with clay. Remedial measures include aprons and protective works and also the provision of deep stilling pools, as in the case of the Calderwood Dam (Fig. 41.4).

FIG. 41.4 Calderwood Dam on the Little Tennessee River, Tennessee, showing the special overflow pool which "cushions" the fall of spillway discharge, thus controlling erosion.

Erosion below spillways is no "respecter of persons" or even of the importance of dams; remedial works have been carried out on many of the best-known dam structures in North America and elsewhere after erosion was detected below their spillways. The spillway at Grand Coulee Dam had a carefully designed spillway bucket, formed of concrete and designed to provide a cushion of water at the foot of the spillway proper. First flooded in

1937, the bucket was examined by divers in 1943. Variable erosion, even in the surface of the first-class concrete used for construction, averaged about 5 cm (2 in) throughout the length of the spillway and included potholes (in the concrete) formed by the action of boulders that had gotten into the bucket; one of granite had a diameter of 60 cm (24 in). Remedial action served to correct the erosion thus discovered which, although surprising, did not affect in any way the stability of the great structure.[41.5]

The erosive power of swiftly moving water was well illustrated by the damage done to the spillway of the Bonneville Dam on the Columbia River, the repair of which cost $750,000. The spillway has a dentated sill which was designed after most careful study; the concrete baffles are protected by steel armor plates. Cavitation around the baffles had completely removed some of the steel plates and eroded the concrete to depths up to 55 cm (22 in).[41.6] Erosion of shale bedrock to a depth of 6.6 m (22 ft) and the necessary placement of 38,000 m^3 (50,000 yd^3) of concrete to replace eroded concrete and shale provide further evidence of this aspect of erosion from experience with the Lake Waco Dam in Texas. More than $1 million had to be expended in finally replacing the old timber apron and rock-filled cribs used for many years to protect the historic St. Anthony Falls on the Mississippi River at Minneapolis after serious erosion had been caused by an unprecedented flood in 1952. A new reinforced-concrete apron was designed following hydraulic model studies and installed before erosion had proceeded too far into the soft St. Peter sandstone which constitutes the bedrock at the famous falls and which is mentioned elsewhere in this book.[41.7]

As a typical example of another interrelation of dam construction and erosional processes, there may be mentioned the Wilder Dam on the Connecticut River, completed in November 1950 by the New England Power Company. The dam was built at a famous site known in earlier years as "White River Falls," where Rogers' Rangers saw their raft broken by the turbulent rapids. Integral with the dam is a powerhouse equipped with two 16,500-kW generators operating under a head of 15 m (51 ft). The dam is a mass-concrete structure 630 m (2,100 ft) long. The necessary excavation revealed a deep pothole 4.5 m (15 ft) below the bottom of the river. Amid the debris filling this pothole an old piece of iron rail was found. Those responsible for the work had the rail analyzed by the Carnegie-Illinois Steel Corporation, and the analysis suggested that the rail had been made between the years 1845 and 1850. At that time the first railway was under construction up the west side of the river; this suggests the origin of the rail, but its location so far below the riverbed only a century later provides vivid evidence of the erosional power of swift water.[41.8]

Brief reference must be made to another type of erosion below dams, the retrogression of riverbeds for some distance downstream. This follows from the retention in the reservoir formed by the dam of the normal sediment load carried by the river. Water discharged over or through a retaining dam will therefore be relatively clear water. If the nature of the riverbed over which it then flows is such that it is easily eroded, the flowing water will gradually pick up a load corresponding to that which it had above the reservoir, and degradation of the riverbed will follow. In the case of the riverbed below the Hoover Dam, this action was fully anticipated. Most useful studies were made, and reported, in advance of the construction of the dam, but this has not always been done, certainly not before the Hoover Dam studies.[41.9] These revealed actual failures of structures, and near failures, because there was no advance recognition of this perfectly natural process.

One of the first cases to be observed and to be fully described in the literature was the failure of a debris barrier on the Yuba River about 22 km (14 mi) above Marysville, California. Constructed in two "lifts," this simple barrier trapped debris coming down the river, but it failed in March 1907, possibly because of retrogression of the riverbed below. This bed erosion did take place, and was recorded, although there was some doubt as to whether it was actually the direct cause of the failure. More modern cases were studied, not only in North America but also in Europe, all to the same effect, the inexorable action of natural bed erosion reasserting itself when flowing water had been artificially relieved of its normal load of sediment. Today, it is possible to estimate this erosive action and so to take all necessary remedial measures, once the problem has been appreciated.

41.3 TRANSPORTATION BY STREAMFLOW

When rivers reach the state described as that of old age, they will have a bed profile that approaches the minimum possible if the discharge of the river is to reach the sea through the cross section of the natural waterway. This hydraulic criterion is related to the critical velocity for silt and scour, since there is one particular condition at which a regulated stream will neither erode the exposed material in its bed nor deposit the silt that it carries in suspension. This silt-and-scour criterion is a condition of delicate hydraulic balance, and many problems of river engineering, especially those connected with river-development work involving some alteration of natural features, may be seriously complicated by the continued presence in a riverbed of solid material being transported down the river as a product of erosion. The matter has been the object of much engineering research work; records of detailed investigations by engineers go back at least to 1851 and appear in increasing numbers in engineering publications since that time. Concurrently, the approach to what was originally a relatively simple geological process has changed from that of generalized observation to a detailed and sometimes complicated analytical mathematical investigation.

The quantitative study of transportation by running water is now rightly regarded as an important branch of the science of hydraulics, with which this book is not directly concerned, but the basis of all such analytical study is the fundamental geological process of erosion. As an extreme example of how serious the matter may become, an observation of the San Juan River in Colorado may be mentioned; in flood the river appears to be a moving stream of red mud and proves on test to be 75 percent by volume composed of silt and red sand. Admittedly, this is an extreme case, but the fact that careful estimates suggest that as much as 250 million tons of silt annually pass the lower end of the canyon section of the Colorado River, and that the Mississippi River carries almost twice this amount of silt into its delta each year, will show that silt transportation is a process of magnitude.

In many parts of the world, the regular deposition of silt so transported is beneficial to fertile areas; the Nile Valley of Egypt is perhaps the best-known example. In other parts of the world, deposition of silt transported by riverflow during flood periods can be injurious; New Zealand, South Africa, and the Malay peninsula are seriously affected in this way. As an example of what may result from this cause, the town of Kuala Kubu in Malaya may be mentioned. The town had to be completely abandoned in 1929 and reestablished on a new site some kilometers away, because the original site was buried several meters deep in river-deposited silt.[41.10]

The source of the material transported by riverflow is usually confined to the catchment area of the particular river. The quantity of material transported and the regularity of its movement are therefore dependent on the geological characteristics of the catchment basin. In preliminary studies of a river prior to the prosecution of development or remedial work, a rough survey of the geology of the drainage area is therefore an essential requirement. Coupled with the complementary topographical survey, this will indicate the stage of development that the river has reached above the point considered, the natural or initial causes of erosion (such as disintegration of steep rock slopes, wind action, surface soil erosion), and probably the main source of material transported by the river. This information is useful mainly as a guide to further studies, since the effect of transportation by water is to reduce practically all material, whatever its initial state, to particles fine enough to be classified as either fine sand or silt. For example, the vast quantity of sediment transported by the Colorado River is actually rock flour mixed with some organic matter and ground so fine by water action that in grain size it is truly described as silt.

A more detailed study and one of immediate value to the engineer is the securing of samples of river water together with the silt they carry and the quantitative analysis of these in order to obtain an estimate of the average quantity of material being transported. For this purpose, one of the several types of sediment sampler now available may be used. Most such samplers consist essentially of a closed container to which river water can be admitted at some determinate point in the cross section of the river. Results given by sampling present only a partial picture, since transportation is not a simple matter of suspension but a threefold process. Described in the simplest way, the means of transportation may be said to be a rolling along the riverbed of the heaviest particles, an intermittent bumping up and rolling of the next grade of particles, and finally the transportation in suspension of the finest particles of the river's load. The total load is therefore divisible into two parts: the bed load is that transported by the first two types of movement, and the suspended load is that moved by the third. The suspended load is capable of accurate measurement by means of samples, but no really satisfactory method of measuring bed load has yet been devised for general application.

The study of the several types of movement is what has naturally attracted the attention of investigators, and the correlation of the results of mathematical analysis with experimental results has provided engineering literature with many notable papers. The engineer is interested primarily in the results of theoretical investigations and is especially interested in the way in which these results may be applied to specific problems. The transport of material in normal river section is not usually a concern, but whenever a river section has to be artificially enlarged or reduced, with consequent changes in velocity, the engineer should be able to calculate in advance how these changes will affect the transportation of material, especially in relation to the scouring out of bed material if this is unconsolidated. An associated problem is that faced by engineers engaged on irrigation work who have to design their canal systems so that the bulk of the transported load in the irrigation water will be deposited before the water reaches the final feeder systems. The deposition of river silt on irrigated farmland has in some instances been found to have injurious effects on fertility. Elaborate precautions have therefore to be taken in order to remove silt before river water is admitted to a canal system.

One installation of this type is at the entrance to the All-American Canal which takes off from the Colorado River about 480 km (300 mi) below Hoover Dam and serves the Imperial Irrigation District (Fig. 41.5). Although Hoover Dam naturally holds up practically all the material transported by the Colorado River as far down as the dam site, the scouring action of the river is such that in the 19 km (12 mi) immediately downstream from the dam, the riverbed scoured out to an average depth of about 90 cm (3 ft) in the first five months following the closure of the gates at the head of the diversion tunnels at the Hoover Dam site. This was not an unexpected result, since preliminary design work for the canal intake works had been based on an assumed total load of 54.5 million kg (120 million lb) of sediment per day in the maximum anticipated diversion of 425 cms (15,000 cfs) of water. Estimates suggested that the cost of removing this amount of sediment from the irrigation system, if the load were deposited in the canals, would amount to $1 million per year; a desilting installation at the canal intake proved therefore to be an economic possibility. The installation eventually provided consists essentially of six settling basins, 81 by 231 m (269 by 769 ft), with a rated capacity of 57 cms (2,000 cfs); the maximum velocity in the basins is 0.07 m per second (0.24 fps); and the deposited silt is removed by

FIG. 41.5 Imperial Dam on the Colorado River, and the desilting works at the entrance to the All-American Canal, prior to admission of river water.

power-operated rotary scrapers. The efficiency of desilting at normal flow has been found to be 80 percent by the engineering staff of the U.S. Bureau of Reclamation, which is responsible for the installation.[41.11]

After almost a quarter of a century's operation, however, changes were found to be necessary. Silt which accumulated in the settling basins had been dumped into the Colorado River, below the canal intake, and transported by normal river action downstream, some of it traveling as far as the delta of the river. Increased demand for Colorado River water after 1957, coupled with low flows, led to some deterioration of the riverbed downstream of the desilting works and so the problem of disposing of the unwanted silt became acute. It is expected that continuing river improvements above the Imperial Dam will eventually lead to a decrease in the sediment load at the dam, but, in the meantime, a solution had to be found for disposing of the silt collected. It was finally decided to install a suction dredge in a swampy area just below the desilting works, but some distance from the river, and to sluice silt down a new sediment-disposal channel leading to the dredge location (Fig. 41.6). Once the dredge had dug a collecting basin, silt would accumulate in this and then be dredged out to be discharged on to the surface of the swamp, which would thus be converted into dry land.[41.12]

In India, the main problem faced by irrigation engineers is sometimes the reverse of that indicated by the example just cited, since the silt transported by Indian rivers is not always injurious to agriculture. It is present in such quantities that the general practice has developed of designing Indian irrigation works so that the silt and sand are deposited in the channels on individual farms where they can easily be cleared out by the peasant farmers. Thus a main canal will be so "designed as to its section and slope that it will be able to carry forward its full quota of sand silt without either deposition or erosion."[41.13] These words are quoted from a publication of the well-known irrigation engineer, R. G. Kennedy; as a result of prolonged investigation, Kennedy developed a formula for the

FIG. 41.6 Dredging work of the U.S. Bureau of Reclamation, similar to that described in the text, this operation being at Toprock Marsh in association with the U.S. Fish and Wildlife Service.

conditions of flow described, a formula that now bears his name and is widely used. Kennedy's work was but one step in a large series of studies carried out in many parts of India devoted to the problem of so-called "stable flow" and closely related to river-training work. In this work, hydraulic studies come into close contact with the geological background; the grain size of the sediment is an all-important factor. An appreciation of the geological origin of the silt will go far toward assisting in the correct application of the results of such hydraulic research.

The more usual problem faced by civil engineers in connection with the material transported by running water, however, is the removal of the solid material. Not only must sediment be removed in some cases, as at the Imperial Dam desilting works, but coarser debris must sometimes be removed for the protection of machinery through which the debris-laden water is to pass, such as the turbines in water power plants. It is only in mountainous areas with streams of occasional high velocity that this becomes a serious problem. A notable desanding installation is a part of the Marlengo hydroelectric station on the river Adige in the southern Alps where the mountainous terrain necessitated building the installation underground; at the adjacent Tel station, there is a corresponding installation. The building of these two special plants led to an immediate decrease in the cost of maintenance and a corresponding increase in the amount of power generated.[41.14]

The debris carried by streams and rivers in flood is a familiar sight in almost all parts of the world. Excessive floods sometimes necessitate debris-removal operations in their train, but only rarely do engineering works have to be specially constructed to take care of flood debris. One important area where such works are necessary is that around the city of Los Angeles. The Los Angeles County Flood Control District has developed an extensive series of debris-collecting basins which serve to protect the lives of residents in the great metropolitan district of Los Angeles and the valuable property that now lies in this

41-12 SPECIAL PROBLEMS

vast developed area so close to the foothills of the San Gabriel Mountains. Some of the basins are relatively shallow, formed by low-lying dikes; others are located in narrow valleys, formed by dams, some of which are notable in size and design.

Flood debris has to be removed regularly from the collecting basins; this is usually done by sluicing into the normal streamflow, but in some cases it is done by excavation with earth-moving equipment. Because of the character of the watershed and the heavy flood flows that occur in this area, an average amount of debris (soil, gravel, and boulders) of 18,000 m^3/km^2 (60,000 yd^3/mi^2) for one flood has been found to collect; occasional measurements register twice this amount. Actual observations of flood flows have disclosed flows of 85 percent solid material, the consistency of sloppy concrete. These figures for material eroded from steeply sloped mountain watersheds may be contrasted with the silt content of the Mississippi River that rarely exceeds 0.5 percent (5,000 ppm). Even that of the Missouri River, "Big Muddy," rarely exceeds 20,000 ppm or about 2 percent.[41.15]

41.4 SILTING OF RESERVOIRS

Implicit in much of what has already been said is that, whenever a dam is constructed across a flowing stream, large or small, the decrease in velocity of the stream will result in the deposition of most, if not all, of its load of sediment in the upper part of the reservoir. So important is this matter that, although it was dealt with in Chap. 26, it must be at least mentioned in this chapter also. Figure 41.7 is a reminder of the magnitude of the problem

FIG. 41.7 Reservoir of the old Furnish Dam, Umatilla River, Oregon, silted up after only 16 years of service; the dam has now been destroyed and the surface of the silt beds used for farming.

and one way of dealing with it, the provision in a dam structure of outlet works through which silt can be sluiced whenever the water level in the reservoir can be lowered for this purpose. The best solution of all is to control the initial deposition of silt. Some success has been achieved in some reservoirs by the development of suitable vegetation along the banks of the stream before it enters the reservoir, and of the reservoir banks also, in an effort to reduce velocities in such a way that silt will be deposited in the delta formed by the stream as it enters the still water of the reservoir, thus preventing the sediment from working its way down into the effectively useful part of the reservoir. The problem is a serious one; it will come up for yet further mention in Chap. 46.

41.5 EROSION ON CONSTRUCTION PROJECTS

It might be thought that the problem of dealing with erosion and silting is something encountered only in connection with large projects such as dam building. On the contrary, it is a problem that may be encountered, if it is recognized, by almost every civil engineer engaged on construction work. The operations of construction almost always involve disturbance of the natural ground surface at the construction site. Once natural surface cover is removed with the associated topsoil, soil will be exposed. If on a sloping surface, the first rainstorm will start the work of erosion and sediment-laden water will run off to the nearest drainage ditch and thence into whatever watercourse the ditch leads to.

In times past, such spoliation of natural streams was perhaps regarded as inevitable and of little consequence, especially since the action would stop when construction was complete. Such an attitude is not acceptable today. Every effort must be made to prevent any such sediment-laden water from reaching natural streams, even to the construction of special sedimentation ponds. In a masterly review of this "mundane" subject in urban and suburban areas of Maryland, Wolman and Schick record sediment loads from construction sites varying from a few thousand to a maximum of 55,000 tons/km^2 (140,000 tons/mi^2) per year, whereas the highest comparative concentration from agricultural lands was 390 tons/km^2 (1,000 tons/mi^2) per year.[41.16] Analysis of building sites in the same region in Maryland showed that 50 percent were open for eight months, 60 percent for nine, and 25 percent for more than one year. The problem is therefore a serious one, calling for attention on every building site other than those enclosed by other urban constructions. Many states and other major agencies have now enacted strict regulations governing runoff from all soil exposed during construction. Civil engineers concerned for the environment will not need such imperatives for attention to an aspect of erosion and silting all too often neglected in the past but now a normal precautionary measure in the layout of construction projects.

The problem is especially serious during railway and highway construction during which, inevitably, large areas of land are exposed without their natural cover. State highway departments have therefore been in the lead in this widespread modern approach to the problem of controlling erosion and silting on construction sites. The U.S. Soil Conservation Service can provide valuable advice through its field officers and by means of its valuable publications, a guide to which will be found following this chapter. Exact methods will depend on the detailed layout and soil conditions at each site, so generalizations are not too helpful. In addition to sedimentation ponds, there may be mentioned as useful aids the construction of diversion ditches along steep slopes to get runoff under control before it reaches any possible outlet to streams; the building of low fences which will trap silt being carried off slopes; "tracking" exposed slopes by running tracked vehicles along them on contour so as to form roughly horizontal grooves which will hold water; the placement of sandbags as temporary dams across smaller drainage ditches to be in use for a short time only; and the covering of large exposed slopes with special fabrics to protect them until vegetation gets well established as permanent protection (Fig. 41.8).[41.17]

There is, perhaps, no better way of drawing this short chapter to a close than by reminding readers that almost every construction job may be the scene of erosion and

41-14 SPECIAL PROBLEMS

FIG. 41.8 Settling basins on the Annapolis Tidal Power project at Annapolis Basin, Nova Scotia; through these all pump discharges must pass before entering the waters of the Basin in order to eliminate suspended matter.

silting. However small the occurrence may seem to be, it should be controlled. And as one watches soil being eroded, perhaps during heavy rain, and then being carried away, one is witnessing on a small scale one of the great natural processes, a vital part of the geological cycle.

41.6 REFERENCES

41.1 M. Fitzmaurice, "The Nile Reservoir, Assuan," *Minutes of Proceedings of the Institution of Civil Engineers,* **152**:71 (1903).

41.2 W. N. Benson, "Notes on the Suspended Load of the Waimakariri River," *New Zealand Journal of Science and Technology,* **27**:421 (1946).

41.3 A. D. D. Butcher and J. D. Atkinson, "The Causes and Prevention of Bed Erosion, with Special Reference to the Protection of Structures Controlling Rivers and Canals," *Minutes of Proceedings of the Institution of Civil Engineers,* **235**:175 (1934).

41.4 *Report of the Islam Enquiry Committee into the Failure of the Islam Weir, S.V.P., on September 19, 1929,* Lahore, India, December 1939.

41.5 K. B. Keener, "Spillway Erosion at Grand Coulee Dam," *Engineering News-Record,* **133**:41 (13 July 1944).

41.6 J. C. Stevens, "Scour Prevention Below Bonneville Dam," *Engineering News-Record,* **118**:61 (1937); see also in same journal, **154**:36 (21 April 1955).

41.7 A. V. Deinhart, "Chute Spillway Preserves St. Anthony Falls," *Civil Engineering,* **26**:12 (1956).

41.8 Display at Wilder Dam, Hanover, N.H.

41.9 E. W. Lane, "Retrogression of Levels in River Beds Below Dams," *Engineering News-Record,* **112**:836 (1934).

41.10 S. Fortier and H. F. Blaney, *Silt in the Colorado River and Its Relation to Irrigation,* U.S. Department of Agriculture Bulletin no. 67, 1928.

41.11 C. P. Vetter, "Why Desilting Works for the All-American Canal?" *Engineering News-Record,* **118**:321 (1937).

41.12 "Dredge Rids River of Silt from Irrigation Water," *Engineering News-Record,* **172**:38 (14 May 1964).

41.13 R. G. Kennedy, *Hydraulic Diagrams for Channels in Earth,* 2 ed., Edinburgh, 1907.

41.14 H. Dufour, ≪Le Dessableur, Les Turbines et La Production D'énergie de L'usine de Marlengo,≫ *La Houille Blanche,* Paris, January-February 1936, p. 1.

41.15 P. Baumann, "Control of Flood Debris in San Gabriel Area," *Civil Engineering,* **14**:143 (1944).

41.16 M. G. Wolman and A. P. Schick, "Effects of Construction on Fluvial Sediment, Urban and Suburban Areas of Maryland," *Water Resources Research,* **3**:451 (1967).

41.17 G. Dallaire, "Controlling Erosion and Sedimentation at Construction Sites," *Civil Engineering,* **46**:73 (October 1976).

Suggestions for Further Reading

Since erosion and sedimentation are parts of the geological cycle, it is natural that useful references to them can be found in all books dealing with physical geology and geomorphology. There is readily available, therefore, a great amount of helpful information to all who have access to a geological library. One example may be mentioned: *Fluvial Processes in Geomorphology* by L. B. Leopold, M. C. Wolman, and J. P. Miller, a useful volume published by Freeman of San Francisco in 1964.

The application of the principles governing erosion and sedimentation has been implicit in much hydraulic engineering. It was the recognition of soil erosion as a national problem and of the necessity for soil conservation, however, that brought increased attention to this matter. The establishment of the U.S. Soil Conservation Service was a major step forward. This remarkable Service has now built up a wealth of experience, much of it relevant to engineering control works. Its list of publications contains many useful references to the problems discussed in this chapter. It should be consulted by all who face decision making in stream-control work.

As with other major organizations, the Soil Conservation Service has its own *Technical Engineering Manual* prepared for internal use by its staff. Although this is not publicly available, it may be mentioned that Section 3 (for example) deals with sedimentation. Issued in 1968 and reprinted in 1971, this is a most useful document, as will be found if the privilege of looking at a copy can be obtained.

The work of the U.S. Geological Survey necessarily includes study of erosion and sedimentation, and its list of publications is well worthy of consultation when problems in this field are faced. One example from the list is Circular 601 E of 1970, a study entitled *Sediment Problems in Urban Areas* by H. P. Guy.

The reference to sand dunes may surprise some users of this Handbook who had previously regarded them as exclusively a coastal feature. One of the most extensive inland areas of sand dunes in North America is described in No. 12 of *Publications in Botany* of the National Museum of Canada, published in 1982. This is the first part (on land and vegetation) of *The Lake Athabasca Sand Dunes of Northern Saskatchewan and Alberta, Canada,* by H. M. Raup and G. W. Argue.

chapter 42

PROBLEM SOILS

42.1 Origin of Soils / 42-2
42.2 Paleosols / 42-2
42.3 Weathering and Some Unusual Soils / 42-3
42.4 Lateritic Soils / 42-5
42.5 Collapsible Soils / 42-6
42.6 Black Cotton Soils / 42-6
42.7 The Pebble Marker / 42-7
42.8 Boulders / 42-9
42.9 Loess / 42-10
42.10 Loose Sands / 42-11
42.11 "Bull's Liver" / 42-12
42.12 Multiple Tills / 42-12
42.13 Loose Tills / 42-13
42.14 Sensitive Soils / 42-13
42.15 Swelling Clays / 42-15
42.16 Conclusion / 42-17
42.17 References / 42-18

Throughout this volume there have been repeated references to rocks and soils, as their uses in civil engineering or performance as foundation beds have been discussed. Geological origin has always been mentioned when appropriate and at least general classifications given for the materials of the earth so mentioned. In only a few cases have any unusual characteristics of soil or rock had to be mentioned and then, usually, only incidentally. There are, however, a number of types of soil and classes of rock which have some properties out of the ordinary. In every case, the properties that make the materials unusual are due to geological causes.

It may, therefore, be helpful to bring together, in this chapter and that which follows, brief accounts of soils and rocks which depart from the normal pattern of behavior on occasion and which, therefore, may prove troublesome if not recognized in advance of carrying out construction. It must be emphasized that the notes which follow deal only with the significant geological aspects of these unusual materials, with only incidental reference to their performance. Each type mentioned has naturally been the subject of intense investigation; references to useful relevant publications are given. In one or two cases, an explanation of the geological peculiarity of soils or rocks has already been men-

42-2   SPECIAL PROBLEMS

tioned briefly in preceding chapters. It seems desirable, even at the risk of minor repetition, to bring all together in these two chapters.

42.1 ORIGIN OF SOILS

The geological origin of soils was described generally in Chap. 5. Here the various processes of weathering were outlined, whereby solid rock is usually disintegrated first into rock fragments and eventually into soils. As one general means of subdividing soil types, the size of the constituent particles is a useful yardstick, this leading to the concept of sand-, silt- and clay-sized particles, the generic names—*sand, silt,* and *clay*—being used to describe soils which consist of particles predominantly of the one range of size. If the products of rock weathering are not moved from their original position, the soil is designated as a *residual soil*. If natural causes, such as running water, the movement of glaciers, or even the wind, move the products of weathering, the result is a *transported soil*. Study of the local geology will almost always indicate whether a deposit is that of a residual or transported soil. In warmer climates, the majority of soils are residual, transported soils predominating in more temperate regions. In areas which have been glaciated, most (but not necessarily all) soils encountered will have been transported from the location of the bedrock from which they have been derived. These processes of rock disintegration and then the transport of the products of weathering are normally thought of as geologically recent events. So it is for almost all soils normally encountered, but it is a process that is as old as geological time itself. Accordingly, in some parts of the world there may be encountered soils which, although they look just the same as recent soils, are actually old (in geological terms) and may therefore have unusual properties.

42.2 PALEOSOLS

Most soils of the far geological past have been transformed through lithification processes into rocks such as conglomerates, sandstones, siltstones, claystones, shale, or slate. The origin of these rocks is easy to recognize and to appreciate. Some clays, however, because of geological events not now to be determined, have remained as clays throughout the millenia of geological time since their deposition. In many parts of the world they may be found over considerable areas. Great Britain and Czechoslovakia are two areas where such ancient soils are found; in each country, engineering problems due to the unusual character of these soils have been encountered.

The Upper Lias clay of England may be cited as an example. It is of Lower Jurassic age and thus was formed about 150 million years ago. It has most of the usual characteristics of normal clays but has a stiff fissured structure because of its geological history. If water can enter the fine fissures in such clays, after they have been exposed by engineering excavation, for example, slow swelling will result, and this may exhibit itself in unusual ways. Chandler has reported on a study of a recent landslide, on a slope of only 9°, in Upper Lias clay in what used to be the county of Rutlandshire.[42.1] The slope on which this failure took place was similar to that found in the eastern Midlands of England and typical of natural slopes in this material, many of which must therefore be only just stable. (Fig. 42.1) Skempton has given some information about a bad slide that took place within recent years in a railway cutting that had been stable for 80 years.[42.2] This was in the Weald clay of England, which is of Lower Cretaceous age, and so possibly only about 100 million years old. It also is fissured. It may be difficult to distinguish this ancient fissuring in small samples examined in the laboratory. Geological identification of such clays in the

FIG. 42.1 Exposure of the Middle Lias clay near Grantham, England, similar to that described in reference 42.1.

field will aid in providing advance warning of the unusual properties they may have. The clays mentioned are marine in origin and may contain well-preserved fossils.

In some of the Cretaceous and Tertiary rocks of Czechoslovakia, claystones and clayey shales that contain montmorillonite are encountered. They are mentioned since they occur in so many parts of the world. Swelling of adjacent strata may lead to the formation of tensile cracks in the shale strata, even though these are quite stiff. Weakening of the shale will follow because of the access thus given to percolating water, and the shales may change into clays with corresponding loss of strength. Mencl has reported a slide caused in this way that did not take place until ten months after the excavation, into which the slide moved, had been completed.[42.3]

42.3 WEATHERING AND SOME UNUSUAL SOILS

Long-term exposure of rock surfaces to the action of the elements will eventually lead to disintegration of at least some of the constituent minerals of which the rocks are formed, the overall process being known as weathering. Some details of the changes that can thus take place in minerals were given in Chap. 4. If the rock disintegration products remain in their natural position, they come to be residual soils (Fig. 42.2). These can be subjected to all the tests now usual in geotechnical work and so can be grouped in the same way as other soils. Their geological origin and their position in relation to their parent material may well result in unusual characteristics. Thus, for example, when impure limestones weather, if the impurities are quartz and clay minerals, these will not normally be affected by the action of rain-water and so will remain undissolved. In this way, they will form the basis of the soil mantle which gradually develops over an exposed limestone surface. Red earths of this kind, known widely as *terra rossa*, are found to cover much of the cavernous limestone of the Karst area in Yugoslavia, near the Adriatic coast, the red color derived

FIG. 42.2 Typical exposure of residual soil, grading from the solid in situ granite up to surface soils; a view in Hong Kong.

from the insoluble iron hydroxides formed from the original iron compounds in the limestone.[42.4]

The two major determinants of the characteristics of residual soils can readily be seen to be the climate to which bedrock has been subjected over a long period of time and the geological nature of the bedrock. These factors will be found to be duly emphasized in the volumes now available dealing only with *weathering*; one especially useful book on this topic is by Ollier.[42.5] General correlation with overall climatic patterns is usually suggested in such detailed studies of this important natural process. Workers in South Africa have gone further and worked out statistical correlations between the products of weathering and the climatic data now available. Weinert's work is notable.[42.6] He has shown that mechanical disintegration of rock is the predominant mode of weathering in areas where the N-value is greater than 5, while chemical decomposition predominates where this value is less than 5. Weinert's *N-value* is the ratio of evaporation during January and annual precipitation, multiplied by 12. Figure 42.3 illustrates how Weinert's studies have permitted regional generalizations about the nature of residual soils in southern Africa.

The geology of southern Africa is so complex and yet has been studied in such detail in relation to civil engineering works that attention may be directed (again) to a notable volume by Brink, *Engineering Geology of Southern Africa*.[42.7] The first volume of this twin-volume work deals with the first half of the geological time scale; this is but little

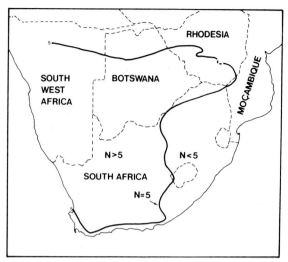

FIG. 42.3 Weinert's N-line as applied to southern Africa.

more than half of the Precambrian era in North American geological terminology. In stressing the vital importance of the geological origin of residual soils, Brink says that

> there is a world of difference between the behaviour of sandstones (and other lithological types) of different age and different mineralogical compositions, by virtue of their different modes of origin. The residual soils to which they give rise on decomposition will likewise possess distinctive engineering properties which are predictable, even if only at a broad level of generalisation, if the stratigraphic origin of the parent material has been correctly identified and recorded.[42.8]

This introductory statement is well illustrated by the fact that in the course of his first volume, Brink describes no fewer than 12 distinct variations of residual soils, all from different types of bedrock—from coarse-grained soils formed by the mechanical disintegration of granite to clays with high rates of consolidation derived from the chemical weathering of mica schists, to mention but two examples. *Wad* is the name given to a clay soil left as one of the residues of the weathering of dolomite; it is the most highly compressible soil known to occur in the Highveld of South Africa and is related to some of the catastrophic sinkhole failures noted in Chap. 38.

42.4 LATERITIC SOILS

Although further generalizations about residual soils in the brief compass of this chapter will not be very helpful, not even with regard to the depths of weathering to be expected, which may range up to 30 m (100 ft) or even more, some reference to lateritic soils (or laterite) is warranted. In tropical regions where there is heavy rainfall during the wet seasons, followed by hot, dry periods, one well-recognized process of weathering is the formation of weak acidic solutions during wet seasons, from rain and groundwater, resulting in the leaching of rocks. During dry seasons, these solutions will be concentrated, leading to the deposition of the least soluble solids. Products so produced will include hydroxides of iron and aluminum, silica, and various carbonates and sulfates. Some of these will be redissolved in the next wet season, but the hydroxides of iron and aluminum will usually remain at or near the surface. The result is a reddish-brown deposit, commonly known as *laterite*.

Bauxite is one such lateritic soil; its predominant content is alumina, the main ore from which aluminum is obtained. There are large deposits of bauxite in Africa and Jamaica. Strata of laterite may extend to depths of a few meters, the surface layer often being a hard crust. In some parts of India, the laterite will be quite soft below this crust. Particle size may vary through the clay-silt-sand range. Some laterites can be cut into bricklike shapes and dried in the sun to give reasonably satisfactory building bricks, this use being the source of the name laterite. Lateritic soils are to be found in the extreme southeastern United States, central America, much of Brazil, central Africa, India and countries to the east of it, and in parts of northern and western Australia.[42.9]

42.5 COLLAPSIBLE SOILS

One somewhat limited type of residual soil can cause such serious problems in engineering if not detected during the course of site investigation that it, too, must be mentioned. In some warm, humid regions, the weathering of granite may proceed to considerable depths. The quartz content will remain unaltered, yielding sand grains; the mica will normally remain intact; but the felspars may become kaolinized. The kaolin will be colloidal, with particles so small that, in regions of high rainfall, they may be leached out, leaving a spongy, sandy mass of micaceous silty sand. When dry, such a soil will appear to be quite stable, but, if saturated, the limited amount of colloidal kaolin remaining will lose its ability to link the solid particles. The result may be that the structure of the "open-work" sand is destroyed when the sand is compressed, the sand grains collapsing, resulting in a denser soil. If this happens under a loaded foundation, the result may be serious structural failure. This collapsible grain structure can be recognized in laboratory studies of thin sections, if suspected in the first place after study of the local geology. Differential settlement of a large, reinforced-concrete water tower at White River in the eastern Transvaal was found to be due to this cause. Rectification of the tilt in the tower was achieved by inducing settlement beneath the two unaffected legs of the tower, following a detailed geotechnical investigation.[42.10]

42.6 BLACK COTTON SOILS

Another widely distributed group of residual soils is that known as *black cotton soil, black turf* being another popular name; other terms used (in India) are *grumosols* and *regur soils*. These are black or dark-gray subtropical clays. They include some of the most expansive clays encountered, for example, in southern Africa. Expansive clays are a worldwide problem, in both residual and transported soils, with the result that much study has been, and is being, given to them. The clay mineral montmorillonite is a major component of black cotton soils. Their black color is thought to be due to organic matter, but there are divided opinions on this matter. So serious have been the problems arising from their expansive properties in South Africa that a field research site for full-scale experimentation has been developed by the Council for Scientific and Industrial Research (of South Africa) at Onderstepoort, publications from which are of much value.[42.11] It has been found that the parent material for these soils is usually mafic igneous rock, especially basalt, and that pH values and liquid-limit values are almost always high. Table 42.1 is a reproduction

TABLE 42.1 Some Soil Properties of Black Expansive Clays from Various African Countries[42.12]

Country	Parent material	Clay %	Silt %	Sand %	Liquid limit	Plasticity index
South Africa	Bushveld gabbro	45	21	34	84	51
Ethiopia	Basalt	56	38	4	109	81
Ghana	Hornblende garnet-gneiss	50	40	10	99	70
Kenya	Basalt	62	8	22	104	70
Kenya	Basalt	67	28	4	103	60
Morocco	Alluvium from basalt	54	32	14	56	32
Nigeria	Calcareous rocks	68	26	6	66	41
Tanzania	Alluvium	40	50	6	58	42
Zambia	Basalt	50	20	30	50	33

of the essential parts of a more extensive table in Brink's volume, to which reference should be made for further information and for references for each of the cases listed.[42.12] The wide geographical coverage in this table, which relates only to Africa, is remarkable. Black soils of the same general type are found in most subtropical areas, notably in India and Indonesia, as well as in parts of the U.S.S.R. (Fig. 42.4).

FIG. 42.4 Typical exposure of dried black cotton soil, from India.

42.7 THE PEBBLE MARKER

In areas featured by residual soil cover and subject to heavy rains, some surface soils will inevitably be washed away from their natural location, forming alluvial deposits after settling from the water which has moved them. Accordingly, there are many areas of the

42-8   SPECIAL PROBLEMS

tropical and subtropical world where a thin layer of transported soil overlies residual soil. In some locations, notably in southern Africa, Australia, and parts of South America, the separation of the two is very clearly marked by a thin layer of pebbles. The pebbles themselves may be alluvial gravel, possibly colluvial gravel (angular) moved by sheet erosion, or even of organic origin. The age of the gravel will vary greatly. In some locations, archaeological remains may be associated with it, thus aiding age determination.

Jennings has thus described its importance:

> First, it represents a stratum of free drainage which must be sealed off in certain forms of construction such as dams or, if drainage is required, it may be retained and be usefully employed for providing a free flow of water. Secondly, it indicates the level below which soil behaviour may be approximately predicted from other experience with similar decomposed rock types. . . . if the country rocks are of types which allow one to accept the principle that the degree of weathering will decrease with depth, then considerable subsoil information is provided to a great depth without the need for deep and expensive boreholes.[42.13]

It will be clear that this detail of soil profiles in regions featured by residual soils can be of great significance. The pebble layer is usually quite distinct, as may be seen from Fig. 42.5.

FIG. 42.5   A pebble marker of quartz gravel separates a transported soil of sandy colluvium from underlying compressible residual soil formed by decomposition of Precambrian andesite in situ; Johannesburg, South Africa.

42.8 BOULDERS

Boulders are a source of potential trouble in civil engineering construction, especially when they are encountered buried in soil. Accordingly, this brief note has been "buried" in this chapter, between notes on residual and transported soils, as a reminder of the great trouble that boulders can cause if not anticipated. In residual soils, they will be large blocks of country rock, which, for one reason or another, has not decomposed in the same way or at the same rate as the surrounding rock. Boulders are more common in decomposed granite than in other parent rocks of residual soils. They have caused much difficulty in excavation work in Hong Kong, as but one example, and as can be readily imagined from Fig. 42.6.

FIG. 42.6 A deep cutting in Hong Kong showing residual soil derived from local bedrock with boulders of unchanged rock.

In transported soils they will be merely very large fragments of bedrock in the process of gradual mechanical disintegration, leading to gravel, then sand, and finally silt and clay. They are a common feature of glacial till (Fig. 42.7). They are sometimes found embedded in glacial-lake and marine clays, coming to these unusual positions probably by being dropped from ice "rafts" floating in the glacial lakes in which the clay has been deposited. This is well shown by the occasional boulder that is found embedded even in varved clays.

Boulders are usually too small to be detected by geophysical means. Test borings, penetrating so small a fraction of the soil being investigated, cannot be expected to encounter boulders except by what must be described as almost pure chance. Accordingly, the geological aspects of site investigation assume special significance. A thorough study of the geology of the area which includes the building site will indicate with reasonable certainty how the soils encountered have come to occupy their present position, in the case of transported soils, and in the case of residual soils how the soils have been formed and from what bedrock they have been derived. In all such cases, aided always by a search of the

42-10 SPECIAL PROBLEMS

FIG. 42.7 Typical boulders—erratic blocks—as found in many glacial tills, this being on a construction site in Ontario, Canada.

area for any actual examples of boulders, it will be possible to indicate whether boulders are a probability in the subsurface. If they are, the civil engineer can include in the contract documents a suitable clause indicating that boulders *may be* encountered when excavation (or other subsurface operation) proceeds; contractors may then indicate a provisional amount which may be used for payment if and when boulders are encountered. If they are not, little is lost; if they are, trouble and costly argument will be avoided.

42.9 LOESS

Loess is the most widespread of aeolian soil types. It is often found exposed with almost vertical faces, usually in areas of low rainfall. Microscopic examination shows the soil particles to consist generally of fresh minerals such as quartz, feldspar, calcite, or mica, but with other material acting as a binder, so that the soil has a relatively hard texture when dry. The clay mineral montmorillonite is one such binding material; calcium carbonate is also found in some types of loess. This binder can give loessal soils considerable dry strength, sufficient to sustain the overburden of 100 m (330 ft) or more of dry soil. The addition of water generally destroys the binding action of whatever has been holding the individual grains together, and the internal soil structure will literally collapse. The previous open texture of the soil mass will be destroyed as the water breaks the bond between adjacent grains and then facilitates their moving together. Even with loess in place, the addition of water to the surface will so destroy the strength of the soil that progressive collapse of its structure will result, possibly with serious settlement at the surface (Fig. 42.8). One well-recorded case showed a settlement equal to 10 percent of the original thickness of loess. Lest this be thought to be exceptional, there is on record an example from Kamloops, British Columbia, where 15 buildings all showed the results of ground settlement beneath them, settlement due to the collapse of loessal soil.[42.14] One is shown in Fig. 42.9. Even the watering of gardens can have serious and unexpected results if the underlying soil, unsuspected, is loess.

Distribution of this unusual soil is quite widespread, not only in the United States but also in the U.S.S.R., New Zealand, and especially in China, where its erosion and transport

FIG. 42.8 Test plot in the San Joaquin Valley, California, for determining subsidence features following wetting of loess; photo taken after 14 months of operation.

FIG. 42.9 Damage to a building foundation caused by settlement resulting from collapse of loessal soil because of uncontrolled drainage, in Kamloops, British Columbia.

by water has given rise to the well-known names of the Yellow River and the Yellow Sea. Roads across loessal country have demonstrated unusual results of its exposure through centuries, Pumpelly having seen roads that were worn 15 m (50 ft) below the surrounding ground surface because of erosion by the wind of the fine loessal particles after disturbance by traffic on the road.[42.15] Once loess is recognized, there are now well-tried ways of dealing with it. These include, naturally, "capturing" all surface water liable to come into contact with loess, trimming excavations in loess (such as road cuts) with faces as close to vertical as possible, and using benches when necessary.

42.10 LOOSE SANDS

One would imagine that sand is one soil type of which there could not be any unusual variety. This is almost true but not quite. Almost all sands encountered in nature are well compacted, sand grains being packed together at something approaching the theoretically closest arrangement, giving bulk densities in the range of 2,000 kg/m^3 (125 lb/ft^3). Occasionally, however, sand will be encountered which, because of some peculiarity of its deposition, has a density much less than this. Such loose sands will look just the same as medium or dense sands, but their low relative density can be detected by a simple field test. This is a test which should always be carried out if there is any possible question about the nature of a sand stratum, since the low density indicates a loose packing of sand grains, a particle structure that can be changed either by vibration (as from a machinery foundation) or by movement of water (as from drainage).

Once a low-density sand has been detected, there are various methods available for compacting it to bring its density up to a more normal value, thus automatically increasing its shear strength, which will be low because of the low density. Compaction with heavy weights is the simplest method but one with severe limitations. The driving of piles will set up vibrations in the sand, with consequent compaction. Blasting with small shots of explosive is yet another way of inducing vibrations in the soil, and there is a patented method involving a special vibrating tool which is used in association with forcing water into the soil. By one or other of these construction techniques it is possible to decrease the porosity of such loose sands from as much as 45 to 35 percent. There are a variety of geological causes that may be responsible for the loose packing of sand, causes that can now only be assumed. The engineering geologist should always keep in mind the possibility of even sand having this unusual property, recommending field testing wherever it appears desirable.

42.11 "BULL'S LIVER"

This rather vulgar term is used on construction to indicate an unusual form that silts may assume. Fine-grained silt is occasionally found in a saturated condition and with its particle size and distribution such that, on exposure, it has a glossy appearance. It will also be frequently brown in color, and in this way the term *bull's liver* has developed quite naturally as a vivid description. Occasionally, the wet silt will have such a consistency that a piece of it, similar in size to the calves' livers to be seen in butchers' shops, can be held up between one's fingers, apparent cohesion giving it the necessary strength. This is dangerous soil. If disturbed or vibrated, it can readily assume a "quick" condition (or *liquefy*, to use a common but rather inaccurate expression), with consequent loss of all strength and the usual problems of dealing with a flowing soil. On the other hand, if it is confined in its natural position, especially if in association with sand through which slow drainage can take place, it can prove to be an acceptable bearing stratum for reasonable foundation loads. A typical example of this soil is often encountered during excavations, or test borings, in the vicinity of New York City, where the deposits in glacial Lake Flushing include varved deposits of fine sand, silt, and clay, the silt frequently being in the characteristic state of "bull's liver." Piles used to be employed for building foundations over such soils, but extensive geotechnical studies have led to the use of raft foundations, with complete success and great economy, the varved soils not being disturbed.[42.16]

42.12 MULTIPLE TILLS

One would not expect to find unusual soils among the various soil types that may be the result of glacial transport. Even here, however, geology can influence the character of glacial tills. In many glaciated areas, the bedrock has been subjected to several traverses of ice sheets, advancing and retreating. Each of these sheets may be expected to deposit till on the rock surface. Generally, such deposits will be nicely cleaned off by the advance of the next glaciation, so that the till exposed today is that of the most recent glaciation. Sometimes, however, some till from an earlier glaciation may remain in place. A common reason is the packing of till against a prominent rock exposure over which the ice sheet must move.

A typical example is the escarpment of the Niagara Peninsula, over which Niagara Falls tumbles. This is located almost at right angles to the southward movement of the great ice sheets of Pleistocene glaciation. Accordingly, along the base of the escarpment, two tills may be encountered, the older buried beneath the later (Wisconsin) till.[42.17] The significance of this unusual geological feature is that, primarily because of the ice movement, the lower till may be very tough, sometimes having a rocklike consistency when dry, although it will disintegrate in the presence of water. It is in its dry state that it has to be excavated, however, and this has caused much trouble on construction. Use of the penetration test in connection with trial test boring holes, when carefully done, can now detect any buried till which has properties significantly different from that at the surface. And study of the local glacial geology will indicate, usually with reasonable certainty, the probability of there being two tills present (Fig. 42.10).[42.18]

42.13 LOOSE TILLS

At the other extreme are glacial tills occasionally encountered close to the ice centers from which the great ice sheets of Pleistocene times emerged; these tills have an unusually low density and are therefore porous. If disturbed in the presence of water (as after heavy rains), they can lose all their strength and very quickly take on the form of sloppy mud. Densities as low as 2,000 kg/m^3 (125 lb/ft^3) have been encountered in Labrador for till that was clearly deposited as ground moraine, the Labrador ice center being not far removed from the location in question.[42.19] Although well graded in so far as constituent soil particles are concerned, some of these tills are found with natural moisture contents considerably higher than the optimum moisture contents determined by the standard ASTM test. Any attempt to "work" such tills, as for compaction in a subgrade, will almost automatically lead to trouble.

The good grading of these loose tills renders them fairly impermeable, so that any mechanical operations, such as compaction, will increase internal pore pressures, thus subverting the compacting effort and developing a "spongy" character. Till with naturally low moisture content must be added to achieve the desired results. The operation of heavy construction equipment over such material, when the natural water content is excessive, can lead to the condition shown in Fig. 42.11. The fact that this scene is in glaciated terrain, all the soil in view being glacial till, will be surprising to all who have encountered only "normal" tills. Admittedly, the occurrence of such loose tills is unusual, confined (it is believed) to areas close to the origins of ice movement, but that it *can* happen serves to emphasize the almost infinite variety of geological conditions affecting soils.

42.14 SENSITIVE SOILS

In geotechnical work, the term *sensitivity* is applied to soils when the ratio of their strength in an undisturbed state to their strength after they have been disturbed and "remolded" at a constant moisture content is found to be more than 64. There have been frequent references to such soils in this volume, but they are so important and can cause

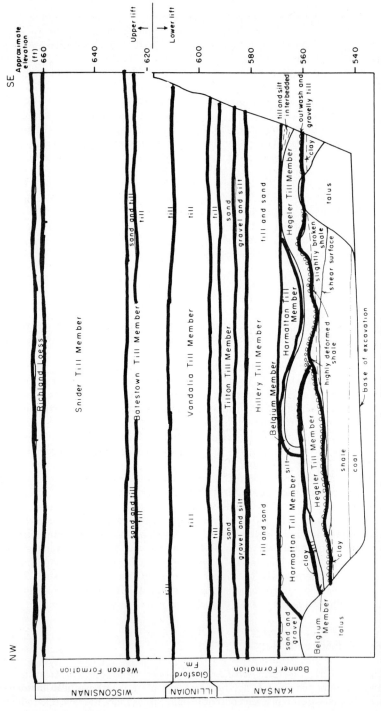

FIG. 42.10 Diagrammatic representation of an exposure of multiple tills in Ohio.

FIG. 42.11 Difficulties with construction equipment in loose till at mileage 201 on the Quebec, North Shore and Labrador Railway.

such trouble if not recognized that this brief further note is warranted. In general, sensitive soils are found to have their origin in deposition in a marine environment during the retreat of the ice sheets. Particle sizes are in the silt-clay range. They are found in the northern part of the U.S.S.R. and the southern part of Scandinavia, in New Zealand, and in the St. Lawrence and Richelieu valleys of North America.[42.20] The glacial history of such clays is a sure guide to their sensitive and unusual property.

They are distinguished, when examined with an electron microscope, by an open mineral fabric, a characteristic that was deduced long before the advent of such modern instruments by study of their properties as soils. Their natural moisture content will usually be found to be higher than their liquid limit, a most surprising feature until their open fabric is recalled (Fig. 42.12). Once disturbed, this open fabric is destroyed; free water is released, the process being irreversible; this may lead to the previously solid soil being converted into a flowing stream of "mud" which, when it dries, will have quite normal properties. The danger, therefore, of disturbing such soils during construction operations will be obvious, as will be the necessity, so often neglected, of ensuring that permanent works are so located that nothing can disturb the sensitive soils on which they are founded. These soils can be a perfectly satisfactory foundation bed if, but only if, undisturbed.

42.15 SWELLING CLAYS

It has been estimated that damage to physical property in the United States of America, due to swelling soils, now amounts to $2.25 billion every year.[42.21] Another estimate puts the figure at well over $1 billion. Damage from this cause is known to be experienced in every state of the Union and probably in every province of Canada. It is commonplace in many other countries, troubles in South Africa (for example) having already been mentioned. If search were made, it would probably be found that there are few, if indeed any, parts of the world where this unusual propensity of some clays does not evidence itself

FIG. 42.12 The sensitivity of Leda clay, both specimens being originally identical, that on the right having been remolded (but with no addition of water).

when modern structures such as paved roads or reinforced-concrete foundations are erected. It is, therefore, a most serious problem.

The fact that swelling clays are encountered in all parts of the United States shows that they can occur either as residual or as transported soils. South Africa's swelling soils are all residual; Canada's are all transported. The common factor is the mineralogical content of the clays. If one of the products of rock decomposition is one of the smectite minerals, either montmorillonite or saponite, then there is a possibility of trouble. These minerals have a layered structure; if water enters into the individual minerals, it can vary the basal spacing and swelling will result. The most extreme case is that of bentonite, a material which occurs in nature. Its swelling property has been recognized, and it has been utilized to form drilling muds and other heavy liquids employed in construction and mining operations. The main constituent of all true bentonites is the mineral montmorillonite.

There are other types of swelling soil, such as those containing pyrite and some alkalis, but it is the montmorillonite clays that are responsible for most of the troubles in civil engineering work. The swelling characteristic of the soil can be detected by laboratory tests. It will be found in such tests that swelling clays will also tend to have high plasticity and low permeability, being strong when dry but having low strengths when wet. Such clays may understandably be still in their natural locations or may have been formed by the weathering of rock fragments during or after transportation. They must be anticipated, therefore, in all regions.

The damage they are able to cause stems from the great pressures which they can exert as they swell when water has gained access to them, pressures up to several tons per square foot having been reported. If swelling is anticipated, structures can sometimes be designed to resist such pressures, but this is costly and uncertain, while many structures, such as road pavements, cannot be so designed. The access of water is often due to the changes in the hydrogeological balance at ground surface, which will be disturbed when, for example, a concrete slab is placed on the ground thus sealing the surface against normal air-water interaction. Much damage, therefore, has been reported from housing projects in which houses have been constructed founded on slabs-on-ground, an economical method of construction if site conditions are appropriate, appropriate conditions include, naturally, the absence of swelling clays. These are, however, engineering matters. The geologist must be aware of the potential of trouble if swelling clays are encountered and include appropriate warnings, when needed, in all site assessments.[42.22]

42.16 CONCLUSION

The foregoing staccato-like notes on some unusual soils merely touch the highlights of the more commonly encountered unusual soils, with associated warnings for both residual and transported soil terrains. The wide extent of the "swelling soil problem" is indicated by the number of conferences that have been held on this one subject alone. There has even been published a volume of over 200 pages dealing merely with pavements on expansive clays.[42.23] A thorough understanding of the formation of soils, the mineralogy of soils, and the fundamentals of soil physics is really essential for any detailed study of unusual soil types. These basic topics have also been the sole subject of an even larger recent volume, one of which can be warmly recommended to all who wish to pursue this subject further.[42.24]

The mechanical properties of soils are now well understood, the discipline of soil mechanics having made great progress since its formal recognition in 1936. The geology of soils has not yet advanced quite as far, and yet these two aspects of soil study are quite interdependent. Detailed studies of soils have an interest all their own, as is reflected by the part which soil studies have played in what has now been properly called *forensic geology*. A volume on this erudite aspect of geology has now been published,[42.25] and it is not surprising to find that it contains much about soils, including (naturally) reference to Dr. Watson's analysis of Sherlock Holmes's knowledge of geology—"practical but limited; tells at a glance different soils from each other"—knowledge Holmes put to use in a number of cases.[42.26] Long before this, however, as early as 1873, George P. Marsh reported the solution of a criminal case through use of the unusual character of some of the soil underlying the city of Berlin.[42.27]

At the other extreme from such intriguing uses of small soil specimens is the knowledge now being acquired of the vast deposits of soil beneath the oceans. Deep borings in the oceans beds are revealing information about the nature of what has so commonly been thought of as "ooze," some findings having already contributed to the broadening of Pleistocene geology. These investigations have also disclosed that erosional features such as the buried valleys on land (so often mentioned in this volume) are found also in seabeds. In the North Sea, geotechnical investigations have revealed erosional valleys up to 3,000 m (10,000 ft) wide and more than 75 m (250 ft) deep, infilled with sediments much closer to those normally consolidated than seabed deposits elsewhere. Some glacial tills have been encountered deep in the sea that are very heavily overconsolidated.[42.28] The problems that such materials pose for offshore engineering are paralleled by the geological questions they raise, confirming the combined approach to the study of soils that is always so essential and to which this short chapter has been a mere introduction.

42.17 REFERENCES

42.1 R. J. Chandler, "Lias Weathering Processes and Their Effect on Shear Strength," *Geotechnique*, **22**: 403 (1972).

42.2 A. W. Skempton, "Long Term Stability of Clay Slopes," *Geotechnique*, **14**:77 (1964).

42.3 V. Mencl, in a personal communication.

42.4 A. Holmes, *Physical Geology*, 2d ed. Nelson, London, 1965, p. 394.

42.5 C. Ollier, *Weathering*, Longmans, London, 1975.

42.6 H. H. Weinert, "Climate and the Potential Performance of Weathered Dolerites in Road Foundations," *Proceedings of the Second Southern African Regional Conference on Soil Mechanics and Foundation Engineering*, 1959.

42.7 A. B. A. Brink, *Engineering Geology of Southern Africa, vol 1*, Building Publications, Pretoria, 1979.

42.8 Ibid., p. 33.

42.9 Mary McNeill, "Lateritic Soils," *Scientific American*, **211**:96 (1964).

42.10 Brink, op. cit., p. 72.

42.11 C. M. A. DeBruijn, "Moisture Redistribution in Southern African Soils," *Proceedings of the Eighth International Conference on Soil Mechanics and Foundation Engineering*, **2.2**:37–44 (1973).

42.12 Brink, op. cit., p. 286.

42.13 J. E. Jennings, A. B. A. Brink, and A. A. B. Williams, "Revised Guide to Soil Profiling for Civil Engineering Purposes in Southern Africa," *Transactions of the South African Institution of Civil Engineers*, **15**:3–12 (1973).

42.14 E. L. Krinitsky, and W. J. Turnbull, *Loess Deposits of Mississippi*, Geological Society of America Special Paper no. 94, 1967, typical of many excellent regional accounts; see also R. M. Hardy, "Construction Problems in Silty Soils," *Engineering Journal*, **33**:775 (1950).

42.15 H. B. Woodward, *The Geology of Soils and Subsoils*, E. Arnold, London, 1912.

42.16 J. D. Parsons, "New York's Glacial Lake Formation of Varved Clay and Silt," *Proceedings of the American Society of Civil Engineers*, GT6, Paper no. 12218, 1976, p. 605.

42.17 R. F. Legget, "Till in Engineering," unpublished paper, May 1980.

42.18 G. W. White, "Engineering Implications of Stratigraphy of Glacial Deposits," *Proceedings of the Twnety-Fourth International Geological Congress*, sec. 13, 1972, p. 76.

42.19 W. J. Eden, "Construction Difficulties with Loose Glacial Tills on Labrador Plateau," in R. F. Legget (ed.), *Glacial Till*, Royal Society of Canada, Ottawa, 1976, p. 391.

42.20 C. B. Crawford, "Engineering Studies of Leda Clay," in R. F. Legget (ed.), *Soils in Canada*, 2d ed. Royal Society of Canada, Ottawa, 1965, p. 200.

42.21 K. A. Godfrey, "Expansive and Shrinking Soils—Building Design Problems Being Attacked," *Civil Engineering*, **48**:87 (October 1978).

42.22 D. E. Jones, and W. G. Holtz, "Expansive Soils—The Hidden Disaster, *Civil Engineering*, **43**:49 (August 1973); see also W. G. Holtz and S. S. Hart, *Home Construction on Shrinking and Swelling Soils*, National Science Foundation, Washington, D.C., 1978.

42.23 G. Kassiff, M. Livneh, and G. Wiseman, *Pavements on Expansive Clays*, Jerusalem Academic Press, Jerusalem, 1970.

42.24 *Expansive Soils* (Proceedings of a conference held in Denver, June 1980), American Society of Civil Engineers, New York, 1981.

42.25 R. C. Murray and J. C. F. Tedrow, *Forensic Geology*, Rutgers University Press, New Brunswick, N.J., 1975.

42.26 A. Conan Doyle, *A Study in Scarlet*, in *The Complete Sherlock Homes*, vol. 1, Doubleday, New York, 1953, p. 12.

42.27 G. P. Marsh, *The Earth as Modified by Human Action*, Samson Low, Marston, Low, and Searle, London, 1874, p. 141.

42.28 M. E. Milling, "Geological Appraisal of Foundation Conditions, Northern North Sea," *Proceedings of Conference, Ocean International* (Brighton, England), Society for Underwater Technology, London, 1975, p. 310.

Suggestions for Further Reading

That the emphasis in this Chapter placed upon the problems created by expansive soils is not incorrect is borne out by the fact that the American Society of Civil Engineers has an Expansive Soils Research Council. In 1981, the Society published the proceedings of an international conference on the subject held in Denver. The proceedings, entitled *Expansive Soils,* is a two-volume work with a total of 900 pages.

The international nature of these problems is well illustrated by the fact that one of the first major meetings held to discuss the topic was in South Africa. Papers from this Symposium on Expansive Clays were published in *Transactions of the South African Institution of Civil Engineers* in issues at the end of 1957 and in early 1958. The publication of *Pavements on Expansive Clays* by G. Kassiff, M. Livneh, and G. Wiseman of the Faculty of Civil Engineering of the Technion in Haifa, Israel, is yet further evidence of the widespread nature of the problems.

A book on another aspect of the same basic problem is *Foundations on Expansive Soils* by F. H. Chen, published by Elsevier of New York as No. 12 of their Geotechnical Engineering series. It is a volume of 280 pages published in 1979.

Lateritic soils have been receiving concerted attention, notably at a special Seminar on the Engineering Properties of Lateritic Soils held in August 1969, prior to the 7th Congress of the International Society of Soil Mechanics and Foundation Engineering, at the Asian Institute of Technology in Bangkok, Thailand. Also, Elsevier published (in 1979) a book called *Lateritic Soil Engineering* by M. D. Gidigasu in the same series as noted above, a volume of 540 pages.

The geology, mineralogy, properties, and uses of bentonites are well covered in a book of 256 pages called *Bentonites,* published by Elsevier in 1978, the authors being R. E. Grim and N. Guven.

chapter 43

PROBLEM ROCKS

43.1 Swelling Shales / 43-2
43.2 Rocks Under Stress / 43-3
43.3 Cavernous Limestone / 43-5
43.4 Anhydrite / 43-5
43.5 Gypsum / 43-7
43.6 Rocks Reactive with Cement / 43-7
43.7 Shales and Bacteria / 43-8
43.8 Kaolinization of Bedrock / 43-8
43.9 Shattered Bedrock / 43-9
43.10 References / 43-12

In site investigation for civil engineering works, it is always most satisfying to find that a clear exposure of solid, competent-looking bedrock is available as a foundation bed or as the material to be excavated. When good sound rock is excavated, it is a perfectly natural assumption that this fine-looking material should not be wasted but used as required on the works, often as aggregate for concrete. But, just as with soils, it must be remembered that "rock" is not just one simple material but a great variety of different materials, all composed of a combination of minerals. Often visible to the eye in coarse-grained rock, the minerals are still there in the finest-grained rocks, even though they cannot be distinguished. And the history of the geological conditions to which the rock has been subjected in the long course of its existence may have affected its properties; even this change may not be discernible in macroexamination.

Rocks cannot just be accepted without question, therefore, when they are encountered on sites for civil engineering works. Just as with soils, most rocks will be found to be perfectly satisfactory materials for the purposes in view, but this cannot be assumed. They must be studied with great care, and their geological and mineralogical character, as well as their history, must be determined before any assumptions as to their use can be made. And again, just as with soils, rock may be occasionally met with that is not all that it appears to be. The following notes provide an outline guide to some unusual rocks that may be encountered.

43.1 SWELLING SHALES

Shales are laminated sedimentary rocks formed by the consolidation of clay; they may therefore be expected to contain the usual suite of clay minerals which so distinguish clay soils. The fact that some shales have the property of swelling when exposed is not, therefore, too surprising, but it may be most disturbing for those responsible for structures affected by such rock movements. One large area featured by such swelling clay-shales is the Midwest of North America, on both sides of the international border, centered in Montana and Saskatchewan. The chief "culprit," now well recognized in civil engineering and engineering-geological circles, is known as the Bearpaw shale. It was responsible for extensive damage to structures before its properties were fully recognized. It has also affected the geomorphology of river banks where it is exposed to view, notably those of the South Saskatchewan River along which are to be seen successions of minor (natural) landslides, all caused by progressive failure of this material, failure associated with its swelling propensity (Fig. 43.1).

FIG. 43.1 Aerial view of slumping due to swelling of the Bearpaw shale on the bank of the South Saskatchewan River, Saskatchewan.

The shale contains the mineral montmorillonite, which is at least partially responsible for the swelling that distinguishes the Bearpaw formation when, for example, it is exposed to the air in excavations. Detailed studies, however, have shown that another reason for the swelling, almost certainly, is that this Late Cretaceous marine-deposited clay-shale exhibits rebound when overburden is removed from it, rebound from the compression that it underwent when under the intense loading of superincumbent ice sheets. It has been estimated that loadings of the order of 1.0 to 1.5 million kg/m² (100 to 150 tons/ft²) may have been caused by the ice at its maximum thickness.[43.1] When this property is recognized in advance (as is now usually the case), appropriate measures can be taken in design. Such measures include provision for holding down anchors embedded in concrete slabs that must be placed on the exposed shale, as in the spillway for the Oahe Dam on the Missouri River.[43.2]

Another area that is plagued with problems due to expansive shales is that running east from the city of Cleveland along the shores of Lake Erie. Here the local shale contains, unfortunately, the mineral pyrite, in some cases as much as 4.5 percent by weight. The pyrite occurs in different forms, from microscopic-sized particles to crystals of up to 6 mm (¼ in) in diameter. When exposed to air, as by excavation of a foundation area, the pyrite oxidizes and absorbs water, forming iron sulfate and sulfuric acid. In so doing, it expands appreciably and therefore will exert pressure on any structure that has been placed upon it. Troubles with heaving floors because of this have been experienced at a number of buildings in the Cleveland area, including some at Western Reserve University, the Case Institute of Technology, and the Nela Park plant of the General Electric Company. One solution to the problem has been to arrange for all excavation in such shale to be stopped a few inches above final elevation. All exposed shale is then sprayed with a bituminous coating. The final few inches of excavation are taken out, in small sections, immediately before concrete is placed. Oxidation is thus prevented and heaving obviated.[43.3] (Fig.43.2)

FIG. 43.2 Heave of concrete floor due to underlying expanding shale.

Shales should, therefore, be suspect in engineering work. There are some that behave quite normally and give rise to no trouble. Watch must be kept, however, for shales with montmorillonite as a major constituent and preliminary tests made if there is any possibility of swelling. This is a further illustration of the importance of clay mineralogy to civil engineering work, a subject with which engineers should be at least generally familiar.

43.2 ROCKS UNDER STRESS

Long-term, very slow expansion of bedrock has been experienced on some civil engineering works, especially in the northeast of the North American continent. Other examples are slowly coming to light, so that the problem, although often unrecognized, is in all probability widespread, even worldwide. The expansion is a result of the gradual release of locked-in stresses in the rock. The stress in the Bearpaw shale, just mentioned, can readily be explained by reason of the load of ice that this rock is known to have supported. The stresses causing slow horizontal expansion are such that the greatest principal stress is in a horizontal direction. When excavation is carried out in such rocks, the restraint provided by the rock which is removed is eliminated, and either heaving or horizontal expansion of the rock may occur. This is not unknown in quarrying operations (Fig. 43.3). For example, one night in 1969, in a limestone quarry in Missouri, a 1.2-m (4-ft)-thick bed of competent Salem limestone fractured as it buckled, rising almost 60 cm (24 in) and cracking for a length of 90 m (295 ft). The limestone rock at the bottom of the deep canal for the new power station of Niagara, New York, gave evidence of similar rock heaving. Such expansion is, therefore, a perfectly natural result of releasing internal stresses in rock and so can usually be anticipated.[43.4]

FIG. 43.3 Rupture in Devonian Unadilla shale in a quarry in East Chenango County, New York, probably due to release of internal stress.

Stresses of this type are now being recognized in a number of locations in southwestern Ontario and northern New York, in shales and limestone. The principal stress is always in, or close to, a horizontal plane. Its magnitude is indicated by the rupture of a 1.8-m (6-ft)-thick, reinforced-concrete box tunnel wall at Welland shortly after construction (Fig. 43.4).[43.5] There is available a record of 70 years of rock movement in the wheel pit of the Canadian Niagara Power Company's plant at Niagara Falls.[43.6] Similar long-term rock movements have been reported from the Lockport, New York, area. Records are available also of similar movements in Australian sandstone at the Warragamba Dam,[43.7] and elsewhere in Australia.[43.8] Similar experience in Malaysia has also been reported.[43.9]

There seems to be a possibility that these generally unrecognized stresses, now evidenced by resulting rock movements which can be observed and measured, may be in some way connected with plate tectonics, the concept of which has revolutionized the science of geology in the last two decades.[43.10]

The study of in situ rock stresses is a slowly unfolding new field of exploration in geotechnique. Fortunately, good instrumentation has been installed in some crucial locations. It must be recognized that, in some areas, this may be another unusual feature of bedrock to be watched for. No channel of inquiry regarding local experiences should be left unexplored. The experience in one of the Niagara Falls power plants, starting at the turn of the century was, for example, known to a few but not generally recognized as of wide significance until comparatively recently. (Fig. 43.5)

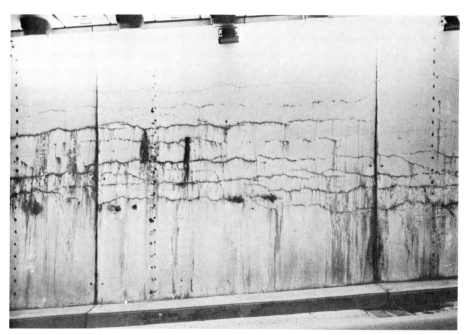

FIG. 43.4 Fracturing of a (new) reinforced-concrete tunnel wall due to rock pressure, Thorold, Ontario.

43.3 CAVERNOUS LIMESTONE

Short though it must be, this listing of unusual features that may be encountered in rocks would not be complete without brief reference to cavernous limestone. There have been many references in this volume to problems occasioned in civil engineering work because caverns in limestone were not recognized in advance. The fact that limestone is a rock soluble in water that is even slightly acidic in nature must, therefore, be emphasized yet again. This solubility is a property of limestone that must never be forgotten in site investigations. When this very slow action is set in the framework of geological time, then the existence of caverns in some limestones can be the better appreciated. Limestones that have been exposed to atmospheric conditions for a relatively short period, in geological terms, may not be affected by solution cavities. Limestones in recently glaciated areas are in this category. But the possibility of cavities in limestones must never be out of mind in site investigations. In karstic areas, search for cavities will be a normal part of site studies, but even where karstic conditions have not been recognized, the possibility of solution channels must never be neglected. If only as a reminder, the sometimes tragic sinkholes that have developed in the West Rand in South Africa, can be cited again. Problems with foundations of some of the outstanding dams of the TVA were the result of solution cavities in the local limestones. And many smaller jobs have had their cost greatly increased because unexpected openings were found in what had been assumed to be "solid bedrock."

43.4 ANHYDRITE

When sediments are deposited in a marine environment, considerable quantities of calcium sulfate combined with water may be precipitated. It is in this way that the extensive

43-6   SPECIAL PROBLEMS

FIG. 43.5 Offset drill holes along preexisting fault planes in the Hebron formation of schists and gneisses, as exposed during the construction of Connecticut State Route 11, near Colchester.

deposits of gypsum that are found today have been formed. Under the action of great pressures, as can now be imagined to have existed in earlier geological times, and correspondingly high temperatures, the combined water in gypsum may be driven off, leaving pure calcium sulfate, or *anhydrite*. This mineral, usually in the form of a compact, whitish rock, is dangerous. It will readily combine again with water to form gypsum and in so doing will expand rapidly, with an increase in volume in some cases of over 50 percent. Fortunately, anhydrite does not occur in nature very widely, but it does occur and so must be watched for. Its occurrence near the surface will naturally be in dry terrain; but it can also be encountered in wet terrain, deep below the surface, protected from any contact with water until disturbed by human activity.

There is on record the occurrence of a loud boom heard late one night in the town of Paint Rock, Texas. It was found the next day to have been due to a massive rock uplift on a nearby ranch. The uplift, amounting in places to 3 m (10 ft) had taken place along a stream bed in which water had been standing; the stream bed was underlain by anhydrite.[43.11] The driving of some Alpine tunnels was seriously interfered with when Triassic rocks which contain anhydrite were encountered; there have been similar experiences on tunnels in France.[43.12] Indicative of the concern over such problems was a 1975 symposium devoted solely to engineering-geological problems of tunneling in sulfate-bearing rock, held under the auspices of the Swiss and German national groups of the International Association of Engineering Geology; the six main papers from this meeting have been published and form a useful guide to problems with this unusual rock.[43.13]

43.5 GYPSUM

Deposits of gypsum are much more frequent than those of anhydrite, fortunately, since the problems with gypsum are not as dangerous as those anhydrite can create. There are problems with gypsum, however, notably because of its solubility in water. It has been found that, in running water, up to 2,400 ppm of gypsum may be dissolved, compared with a maximum figure of 400 ppm of calcium carbonate. Solution channels are liable to be encountered, therefore, in rocks containing any gypsum. Leakage may result if structures impounding water are built on foundation beds which contain any gypsum. A recent case was in the subsurface at the Red Bluff Dam in Texas, where an extensive grouting program had to be undertaken following solution of gypsum beneath the toe of the dam and associated riprap.[43.14]

Problems with gypsum are probably more widespread than is generally realized, as was well brought out by James and Lupton in a review of the solution processes for both anhydrite and gypsum.[43.15] These authors have guided the writers to a somewhat remote paper of great value, published in Venezuela. Calcano and Alzura summarize a wide-ranging study of gypsum in dam foundations which they carried out in connection with the foundations of the El Iriso Dam in their own country. They give details of 12 cases, mainly dams, in the foundations of which gypsum was encountered; they describe the associated difficulties and the remedial measures adopted. They conclude with some helpful recommendations, which center around the need for the most detailed subsurface investigation of any site at which gypsum is encountered or is even suspected.[43.16] Remedial measures stem from the necessity of keeping water away from the gypsum when a site involving gypsum cannot be avoided, the use of carefully designed and installed cement grout curtains being one of the most effective means. It follows that periodic inspection of all such structures, at more frequent intervals than is usual, is imperative in order to ensure that the gypsum remains intact, unaffected by water.

43.6 ROCKS REACTIVE WITH CEMENT

When materials for use in the making of concrete were considered in Chap. 16, attention was directed to the fact that there are some rock types which may react with the alkali content of portland cements, with ultimate deterioration of the resulting concrete. This note is a reminder of this serious possibility and of the widespread troubles that have arisen from such reactions in the past. In areas where concrete is already in use, careful study should be made of any evidence of "concrete cracking" which might indicate that alkali-aggregate reaction is taking place, even though generally unsuspected. When rock has to be used for concrete aggregate in an area where concrete has not previously been used, then every type of rock proposed as aggregate should be subjected to rigorous testing before use. There are useful ASTM standard tests for detecting unsound aggregate. Beyond these, however, and in all cases of doubt, petrological examination of the proposed aggregate should be carried out by experts in this field, since in recent years there have been encountered cases of aggregate reaction which were not shown up by ASTM tests. Some argillaceous dolomitic limestones, for example, have been detected as potential troublemakers. There are now available useful reviews of this important subject to serve as guides to those who have cause to investigate it (see end of Chap. 16).

43.7 SHALES AND BACTERIA

The idea of bacteria having anything to do with solid rock will appear to most readers to be something that might have strayed into this volume from *Alice in Wonderland*, unless they have encountered the very successful use of autotrophic bacteria in the refining of uranium ore. It was passing acquaintance with this unusual ore-refining process that led to the solution of a problem that was found to be due to the action of such bacteria in black shale, quite the most unusual "rock problem" with which the writers are acquainted. A brief description of the problem and its solution is, perhaps, the best way of presenting this aspect of unusual rocks that must be watched for in civil engineering works involving shale.

A multistorey commercial building in Ottawa experienced heaving of a well-built concrete basement floor some years after the building had been completed and placed in service. The foundation was well-designed, a crushed-rock layer underlying the reinforced-concrete floor slab. Tile drainpipes were embedded in the crushed rock to ensure that if any water did accumulate, it could drain safely away. Despite this, parts of the floor did rise and continued to rise after being placed under observation. Every known solution was studied and discarded. Study of records of the foundation bedrock, the local black Billings shale, a well-known and competent rock, revealed that it might be traversed by a small fault under the building. Excavation at an adjacent building site confirmed this.

It was then that the idea of possible bacterial action occurred to the chief investigator. Bacterial experts were consulted; they confirmed that such bacteria could oxidize pyrite if there was any pyrite in the shale, provided environmental conditions were appropriate—a moist atmosphere and a temperature of between 30 and 35°C (86 and 95°F). The floor was studied again; the necessary temperature was there, since emergency electrical generating sets were located on the basement floor, giving rise to high air temperatures beneath when operating. The moisture could have come *in* through the drainage pipes due to the unusually high atmospheric relative humidities that feature Ottawa climate in high summer. And the Billings shale was known to have some pyrite in it.

With all the conditions necessary for bacterial action present, the owners were persuaded to excavate a hole in the beautifully finished floor slab. At a depth in the shale of about 50 cm (20 in), the characteristic yellow color of the mineral jarosite was uncovered, and oxidation of the pyritic content of the black shale was confirmed (Fig. 43.6). The shale was also found to be disintegrated in the vicinity of the fault. The solution was easy, once the cause of the trouble had been located. A concentrated solution of potassium hydroxide was poured into the excavation, thus neutralizing the acidic condition and removing the environmental conditions favorable to the bacterial action.[43.17] The only unsolved problem was how the bacteria managed to find this ideal combination of conditions for their activity. It must readily be admitted that the combination of a fault in pyritic shale under a sound concrete slab in the presence of moisture and a high temperature will only very rarely be encountered. But even without this ideal combination of conditions, other cases of oxidation of pyrite in shale have been encountered, at least one other building in Ottawa on the Billings shale having been affected. The possibility is, therefore, yet another example of unusual rock conditions that must be kept in mind.

43.8 KAOLINIZATION OF BEDROCK

Weathering of bedrock, as has been repeatedly noted, is to be expected in unglaciated areas and especially in warmer environments. One would not expect to find it in northern regions, except where glaciation has not taken place, as is the case in the western part of the Yukon Territory. Kaye has discovered, however, almost certain evidence of the weath-

FIG. 43.6 Looking down into test pit excavated in the floor of a modern building in Ottawa, founded on black shale in which significant mineralogical changes took place along an old fault.

ering of bedrock in the Boston area of Massachussetts.[43.18] Test boreholes in the Boston urban region had revealed puzzling whitish claylike material near the bedrock surface. As the area covered by such subsurface explorations gradually extended, the same thing was encountered over a steadily increasing area. "Soft rock" had also been encountered in some of the major tunnels which now underlie the city. Detailed laboratory studies of the "soft" material showed that it was kaolinite, a basic product of rock weathering. The cause of the weathering is uncertain, and yet determination of the cause is important from the engineering point of view. If due to surface weathering processes, then it would be restricted to shallow depths, but, as noted, it is found (in tunnels) at depths up to almost 100 m (330 ft). This suggests that the weathering is due either to hydrothermal processes acting upon the fine-grained diabase in which it is found, as well as in fine-grained volcanic rocks, or to deep penetration of groundwater, with consequent weathering. This unusual feature is not peculiar to the Boston area. Deep weathering has been encountered in Idaho and the piedmont area of the southeastern United States. So limited is the information yet available that the matter is one still requiring much investigation, but it is another unusual feature of bedrock that may sometimes be encountered.

43.9 SHATTERED BEDROCK

In central Europe the effects of deep freezing during periglacial (permafrost) conditions have long been recognized. During the last glacial period, glaciation of the Alps was more extensive than it is today, but between two of the glaciated areas there was a belt roughly

43-10 SPECIAL PROBLEMS

500 km (310 mi) wide which was free of ice, had a dry climate, and was subject to very cold temperature. Under these conditions, soils and bedrock were affected in glacial periods by strong physical weathering and deep perennial freezing. Slope detritus of today testifies to the intensive weathering. Permafrost is indicated by relic ice wedges, now soil-filled, and by the loosening and contortion of shallow rock, especially in layered and schistose rocks, a condition well described as *shattered*. Figure 43.7 shows a typical exposure of this shattered rock in Czechoslovakia.[43.19]

FIG. 43.7 Weathered sandy marl, the rock fragmentation being the result of periglacial action, the area having been at one time frozen; a view in Czechoslovakia.

It seems probable that this same condition is more widespread than has been realized. Patterned ground, a sure indication of the permafrost condition, has been detected from the air not only in England but also in southern Ontario (Fig. 43.8). There is now good evidence that at the time of the last glacial period, permafrost was extensive in England. It must also have been widespread in North America. One of the most telling experiences suggesting that shattered rock due to permafrost is not confined to central Europe was the experience gained in sinking two shafts for a new cable tunnel across the river Thames in England, between Tilbury and Gravesend. Excellent subsurface investigations were carried out, permitting the award of a contract for the work, tunneling to be at an elevation below the bed of the river so that work could be carried out in "free air."[43.20]

When, however, the sinking of the necessary working shafts reached the surface of the Chalk bedrock, at the expected elevation, the rock was found to be badly disintegrated to a depth of up to 6 m (20 ft). Its shattered condition made excavation difficult and held up the progress of the work. The tunnel is in that part of England thought to have been subjected to permafrost conditions. It would appear, therefore, that in all probability the shattered condition of the Chalk was due to its exposure to perennial freezing.[43.21] In Canada, exposed bedrock has been found which, although at first sight appears to be solid, is actually shattered into angular fragments still so tightly held in their natural position that little or no surface weathering has taken place. This shattering is in argillite, a layered rock (just as in the European experience).[43.22] Figure 43.9 shows a typical view of this shattered rock after it has been disturbed, actually by excavation for use as "gravel." Accordingly, the possibility that apparently solid bedrock in northern areas has actually been

FIG. 43.8 Large-scale patterned ground near Woodstock, Ontario, as seen from a low-flying aircraft, believed to have been formed in a permafrost environment in this region prior to the retreat of late Wisconsin ice from this area, about 13,000 years ago.

FIG. 43.9 Fractured argillite, unweathered on fractures except close to the surface, believed to have been caused by a permafrost environment, Grand Manan, New Brunswick.

43.10 REFERENCES

43.1 R. Peterson, "Rebound in the Bearpaw Shale, Western Canada," *Bulletin of the Geological Society of America*, **69**:1113 (1958).

43.2 "At Oahe: Contractor Licks Shale by Building Backwards," *Engineering News-Record*, **163**:42 (23 July 1959).

43.3 "Structures Don't Settle in This Shale; But Watch Out for Heave," *Engineering News-Record*, **164**:46 (4 February 1960).

43.4 H. P. Cerutti, "Twin Conduits for Niagara," *Civil Engineering*, **30**:50–53 (July 1960).

43.5 C. F. P. Bowen, F. I. Hewson, D. H. Macdonald, and R. G. Tanner, "Rock Squeeze at Thorold Tunnel," *Canadian Geotechnical Journal*, **13**:111 (1976).

43.6 A. D. Hogg, "Some Engineering Studies of Rock Movement in the Niagara Area," *Engineering Case Histories No. 3*, Geological Society of America, Boulder, Colo. 1959, p. 1.

43.7 T. B. Nicol, "Warragamba Dam," *Proceedings of the Institution of Civil Engineers*, **27**:491 (1964).

43.8 D. G. Moye, "Rock Mechanics in the Investigation and Construction of T1 Underground Power Station, Snowy Mountains, Australia," *Engineering Case Histories No. 3*, Geological Society of America, Boulder, Colo., 1959, p. 13.

43.9 J. Newbery, "Engineering Geology in the Investigation and Construction of the Batang Padang Hydroelectric Scheme, Malaysia," *Quarterly Journal of Engineering Geology*, **3**:151 (1970).

43.10 M. L. Sbar, and L. R. Sykes, "Contemporary Compressive Stress and Seismicity in Eastern North America: An Example of Intra-Plate Tectonics," *Bulletin of the Geological Society of America*, **84**:1861 (1973).

43.11 G. Brune, "Anhydrite and Gypsum Problems in Engineering Geology," *Engineering Geology* (now the *Bulletin of the Association of Engineering Geologists*), **2**:26 (1965).

43.12 K. F. Henke, "Engineering Geological Problems of Tunnelling in Sulphate Bearing Rock, Summary of a Symposium," *Bulletin of the International Association of Engineering Geology*, **13**:37–70 (June 1976).

43.13 Ibid.

43.14 V. S. Reed, "Grouting of the Red Bluff Dam, Reeves and Loving Counties, Texas," in abstract vol. for annual meeting of Association of Engineering Geologists, Hershey, Pennsylvania, 1978, p. 15.

43.15 A. N. James and A. R. R. Lupton, "Gypsum and Anhydrite in Foundations of Hydraulic Structures," *Geotechnique*, **28**:249 (1978).

43.16 C. E. F. Calcano and P. R. Alzura, "Problems of Dissolution of Gypsum in Some Dam Sites," *Bulletin of the Venezuelan Society of Soil Mechanics and Foundation Engineering*, Caracas, July-September 1967.

43.17 E. Penner, J. E. Gillott, and W. J. Eden, "Investigation of Heave in Billings Shale by Mineralogical and Biogeochemical Methods," *Canadian Geotechnical Journal*, **7**:333 (1970).

43.18 C. A. Kaye, *Kaolinization of Bedrock of the Boston, Massachusetts Area*, Professional Paper no. 575-C, U.S. Geological Survey, Washington, 1967, p. C 165.

43.19 Q. Zaruba, "Frozen Ground Phenomena of Pleistocene Age and Their Significance in Engineering Problems," *Proceedings of the Eighteenth International Geological Congress*, pt.13, London, 1952, p. 264.

43.20 C. K. Haswell, "Thames Cable Tunnel," *Proceedings of the Institution of Civil Engineers*, **44**:323 (1969).

43.21 R. B. G. Williams, "Permafrost in England During the Last Glacial Period," *Nature*, **205**:1304 (1965).

43.22 I. E. Higginbottom and P. G. Fookes, "Engineering Aspects of Periglacial Features in Britain," *Quarterly Journal of Engineering Geology*, **3**:85 (1970); see also R. F. Legget, "Glacial Geology of Grand Manan Island, New Brunswick," *Canadian Journal of Earth Sciences*, **17**:440–452 (1980).

chapter 44

FAULTS AND JOINTS

44.1 The Nature of Faults / 44-2
44.2 The Effects of Faults / 44-2
44.3 Some Major Faults / 44-3
44.4 Detection of Faults / 44-6
44.5 Faults and Tunnels / 44-8
44.6 Faults and Dam Foundations / 44-10
44.7 Faults and Nuclear Plants / 44-11
44.8 Evidence of Fault Movement / 44-13
44.9 Joints / 44-14
44.10 References / 44-15

Faults are fracture surfaces along which bedrock has been displaced, involving relative movement of the two sections of the rock that has been fractured from a few centimeters to many kilometers. They were mentioned briefly in the summary outline of geology presented in Chap. 4 and there illustrated. There have been many references in succeeding chapters to faults affecting some of the engineering works described. Faults have attracted an increasing amount of attention during the 1970s because of the strict control of sites intended for the construction of nuclear power plants, the occurrence of faults beneath such sites being clearly undesirable if the faults are, or are liable to become, active. This interest appears to have suggested, in some circles, that faults had not previously been given due attention on civil engineering works, but this is clearly a serious misapprehension.

Faults have been recognized as potential sources of danger on engineering works ever since the geology of construction sites was first given adequate attention. Controls over faults in rocks underlying nuclear power sites are more stringent than have been necessary for other works, although faults beneath dam foundations have always been regarded most seriously and designed for accordingly. In all probability it has been the publicity surrounding the opposition to the building of nuclear power plants that has focussed undue attention upon faults in the relevant geological formations. In itself this is probably not a bad thing, but it is regrettable that this attention should have had the side effect of creating wrong ideas about the continuing and detailed attention to faults in all foundation beds utilized in modern civil engineering. It seems desirable to bring together in this

44-2   SPECIAL PROBLEMS

chapter a summary of the main features of faults and their possible effects on civil engineering works and to present guides to recognizing them in site investigations.

44.1 THE NATURE OF FAULTS

Faults are fractures that result from failure of the rocks involved to withstand the great pressures exerted upon them. Earthquakes are the expression of release of such pressures. Earthquakes may also cause movement on existing faults in the vicinity of the seismic action, but there is today no invariable link between such earth movements and the existence of faults. Indeed, it can be said that earthquakes sometimes are the result of sudden movements along faults, the two certainly being closely associated and evidencing (again) that the materials of the earth's crust are susceptible to stress, strain, and rupture in exactly the same way as are all other solid materials.

Faults assume different geometrical forms. Movement on the fault plane is called the *slip*. If this is up or down the dip of the rock, the result will be a *dip slip;* if at right angles, then a *strike slip* will result. With dip slips, further subdivision relates to *normal* faults, in which movement is down the dip, and *reverse,* or *thrust,* faults in which movement is up the dip. With strike slips, relative displacement as viewed from the ground surface will dictate whether they are *right-lateral (dextral)* faults, or *left-lateral (sinistral)* faults. It is not surprising that strike slips are sometimes called *wrench*, or *tear,* faults.

These are the main terms necessary for describing faults in such a way as to indicate clearly how the main movement on the fault has taken place. The only other terms to be noted (from the many now current in detailed geological discussions of faults) are the names used to describe the two faces of a fault—the *hanging wall* is that which overlies the other or *foot wall,* faults very rarely being vertical, so that inclination of the fault plane to the vertical is almost universal. Surfaces of the fault planes will often be found to be highly polished as a result of the movement, in which case they are referred to as *slickensides*. It is finally to be noted that, although a fault plane has so far been referred to, in some cases, and especially so in larger faults, the rupture has taken place not on a single surface but in a zone of material, all of which has been seriously distorted; the term *fault zone* will, therefore, sometimes be encountered.

44.2 THE EFFECTS OF FAULTS

The main effect of a fault is to displace either one, or both, of the rock masses lying on either side of it. In most of the smaller faults such as are encountered on engineering works, the displacement will not be large, in some cases only a few centimeters. But in some of the great and ancient fault systems around the world, displacements of up to 550 km (350 mi) have been determined (this for the San Andreas fault, shortly to be mentioned). Great movements along faults have naturally had profound effects on local geological structures and so on surface features, which, correspondingly, will affect engineering works built upon them. Fault zones will be filled with ground-up material, usually called *gouge,* which may prove troublesome on construction. More serious, however, is the frequent effect of fault planes in acting as channels along which groundwater can penetrate. It is for this reason that the occurrence of faults in rock that has to be tunneled can be so very serious, especially if undetected in advance of excavation. Detection of faults along the proposed lines for tunnel routes is therefore an unusually important part of site investigation. If faults which might prove to be troublesome are located in advance, it may be desirable to relocate the tunnel from the line first proposed. This was done, for example, on tunnels forming part of the great Catskill Aqueduct for the water supply of New York, with most fortunate results. If site investigation suggests that a fault does underlie a site, the extra expenditure of putting down drill holes inclined so as to penetrate the supposed fault is well warranted.

The fracturing which faults represent is a normal part of the dynamic process of geological development. Faults have occurred throughout the geological time scale; some, therefore, are of great age, while others may be seen in Pleistocene deposits, being, correspondingly, very young indeed in geological terms. New faults are usually to be seen following earthquakes, as was well evidenced after the Hebgen earthquake of 1959. The age of a fault is related, in some ways, to its present character, i.e., whether it is an *active* fault or a *dead* fault, *competent* and *passive* faults being alternative names, clearly indicating whether or not more movement on the fault plane is to be expected in the foreseeable future. This is naturally a most important characteristic of faults that may impinge on engineering works; it has become of extreme importance in connection with the siting of nuclear power plants. It is, therefore, a matter which will call for further attention in Sec. 44.7 of this chapter. But the faintest possibility of movement taking place upon a fault plane under any engineering structure is one that has to be avoided at all costs, even to the extent of abandoning a proposed building site.

44.3 SOME MAJOR FAULTS

Throughout the world there are a number of major fault systems that have had such a marked effect upon local topography that they are known and generally recognized far beyond geological circles. In the Highlands of Scotland, for example, the Great Glen, perhaps the most prominent feature of this lovely area, is the scene of many romantic tales of the past. It is also the route followed by the Caledonian Canal, one of the great pioneer canals of Great Britain, linking the North Sea with the Atlantic (Fig. 44.1). The glen is the expression of a broad belt of shattered rock forming the fault zone of earth movement that took place in late Devonian and early Carboniferous times, with a lateral displacement of about 104 km (65 mi). This is part of the geological complexity of this part of

FIG. 44.1 A view in the Great Glen of Scotland, showing the Caledonian Canal which traverses the glen from Fort William to Inverness and exemplifies use of a fault valley for transportation.

44-4 SPECIAL PROBLEMS

Scotland. During the last two centuries, about 60 minor earthquakes or earth tremors have taken place along the Great Glen, vivid reminder that it is a great fault and of the stresses in the adjacent rocks.

Pride of place, however, must be given to a fault that, on the one hand, has been studied in more detail than any other major fault and, on the other, has had more effect on engineering works than any other, the San Andreas fault of California. It is the master fault of a great network of faults in California, indicative of extensive earth movements here many eons ago which still continue. The San Andreas fault extends on land for a total distance of about 1,000 km (620 mi) until it runs into the sea off the northern coast of California. It extends deep into the earth a distance of at least 32 km (20 mi). The total accumulated displacement along the fault may be as much as 550 km (340 mi), and this movement continues at a rate of about 5 cm (2 in) per year, as disclosed by precise surveying techniques. If the fault has been similarly active during 100 million years of existence, a displacement of over 500 km (300 mi) is quite understandable. The movement is so slow that many residents of California who live close to the fault are not even aware of its existence. During the 1906 San Francisco earthquake, however, a movement of more than 1 m (3.3 ft) (6 m or 20 ft being the maximum) occurred, as was shown by the distortion of roads and structures that crossed the fault; in each case, the ground to the west of the fault moved relatively northward.[44.1]

Figure 44.2 shows generally the location of the great fault, and it will be seen that it

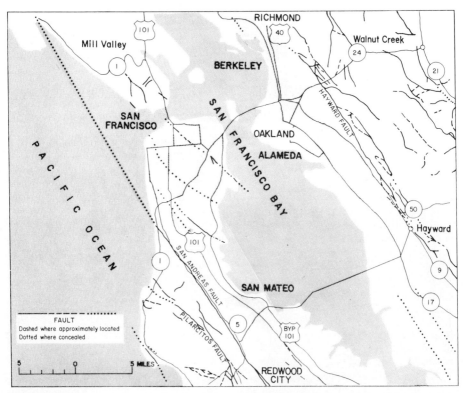

FIG. 44.2 Outline map showing the San Andreas and other faults in the vicinity of San Francisco, California.

passes immediately to the south of the great concentration of population around the city of San Francisco and within a few kilometers of the San Francisco Airport. Its location is known with accuracy; it has been photographed from a high-flying jet aircraft using side-scanning radar in a manner that gives promise for the detection of other faults in other areas using this modern technique (Fig. 44.3). There is today, therefore, no possible excuse

FIG. 44.3 The San Andreas fault in California, as viewed from a height of 20 km (12.4 mi) through the medium of side-scanning radar.

for any structure that would be damaged by excessive movement to be built across the fault, as has happened in the past. Roads must naturally cross the fault, but if movement does occur here it can readily be corrected. Difficult though its location is from the point of view of urban development, the San Andreas fault is most conveniently located for geological study. It is, therefore, being intensively studied, just as it has been under careful observation for many years; it still presents many unresolved scientific questions, the solutions to which will certainly aid in determining the best way to deal with such geological features when they are encountered in other areas.[44.2]

One of the subsidiary faults in California, running roughly parallel with the San Andreas fault, is that which passes through Oakland and Berkeley on the eastern side of San Francisco Bay. It is related to the earthquake of 1836 and to the well-known Hayward earthquake of 1868. Some structures and railways had been built across the fault as early as 1866, and so it was possible to determine the fault slippage that then took place. Detailed studies of recent movements have been carried out in the Niles district of Fremont, located almost at the south end of San Francisco Bay. The evidence suggests that most of the recent movement took place between 1949 and 1957 and that no appreciable movements have taken place since then. Six sets of railway tracks cross the fault near Fremont almost at right angles and so serve, although not by design, as excellent indicators of movement. They are built over the low-gradient alluvial fan of Alameda Creek, a surface underlain by over 90 m (300 ft) of sand, gravel, and clay. Movements are, therefore, modified by this soil cover, and good maintenance of the tracks naturally conceals the long-term movements. Surveys suggest, however, that the oldest sets of tracks, built about 1910, have suffered a total displacement of 21 cm (8.3 in). An average movement of 4 mm (0.2 in) per year has been recently reported, but movement at a much greater rate seems to have taken place for relatively short periods. Railway personnel have reported as much as 4 cm (1.6 in) in a period of two or three days. The zone of deformation on the railway tracks is from 6 to 24 m (20 to 80 ft) wide, another result of the shielding given by the soil overburden.[44.3]

New Zealand calls for some mention in even such a brief review of major earth movements as this. It has two major fault systems, one in the North Island and one in the South. The so-called "Alpine fault" of the South Island runs down the center of the scenic Southern Alps and is known to form some 500 km (310 mi) of the boundary of the Indian-Pacific plate, now well recognized in plate tectonics. Despite the prevalence of earthquake shocks in New Zealand, no major earthquake has taken place on or close to the central part of this major fault for at least the last 150 years. Detailed geological study has traced the occurrence of earlier large earthquakes in its vicinity at intervals of about 500 years, as well as many smaller earthquakes during historic times.[44.4] The Wellington fault in the North Island is closely linked with earthquakes and will call for further mention in the next chapter.

In the subcontinent of India, J. B. Auden has identified three major areas of faulting, with associated dike formation, by a detailed study of drainage patterns which were found to be determined to a large extent by the earth movements.[44.5] Faults are widespread in the 259,000 km^2 (100,000 mi^2) of the Deccan plateau (Hyderabad). Similar conditions are found in the Bombay Konkan and in the Gujarat along the Tapti River east of Mandvi. Dr. Auden's attention was directed to these features in part through geological studies he was called upon to make, while with the Geological Survey of India, of tunnel and dam sites. Finally, in view of the remarkable earthquake records of China, and of the great earthquakes that have taken place and have been recorded in that vast country, it is almost certain that major fault systems exist there. As Chinese geological and engineering experience is shared with Western lands, information about Chinese fault systems may be anticipated.

44.4 DETECTION OF FAULTS

These major fault systems are easily recognized and are already well known in general. The faults with which the civil engineer is usually concerned will be, by comparison, very small indeed. Small though they be, they can dictate major changes in design and cause untold trouble on construction, if not detected during site investigation. It is in the course of the geological mapping of a proposed building site that the existence of fault(s) will usually be revealed. As the results of study of the local geology are correlated with the records of test borings and test drilling, plotted and, when necessary, demonstrated on a model of the subsurface, any major fault will almost certainly be shown up by a break in the continuity of strata which alone can explain the results obtained. Determining the actual location of the fault may not be easy, especially if there is substantial soil cover over bedrock. Here is where more detailed examination of surface features becomes necessary.

It is only rarely that a fault will be well exposed at the surface. Figure 44.4 is an example known to the writers, one that has achieved minor fame as a local scenic spot. Search for any such direct evidence is, obviously, a first step in the determination of faults under a site being studied. It will be indirect evidence, however, upon which reliance will usually have to be placed. Examination of aerial photographs of a considerable area embracing the site under detailed study is naturally a logical step. This may not yield any useful results, but it is always essential as an initial study. In some cases *lineaments* may be observed, clearly defined lines of varying length which cannot always be distinguished on the ground. They may be due to a number of causes, one of which is a slight change in vegetation along the line of a fault due to a change in the soil (and possibly in groundwater conditions) resulting from existence of the fault plane, or zone, beneath. Naturally, watch should be kept for any such lineaments during study on the ground. Aerial photographs are now such a useful tool that it is worthwhile spending some time in their examination

FIG. 44.4 A well-exposed fault that has become a scenic attraction; Red Point fault on Grand Manan, New Brunswick, in which older Green Head strata are brought in contact with younger Triassic rocks on the left.

before detailed fieldwork is undertaken. The necessity of studying an area extending far beyond the limits of the proposed building site cannot be overstressed, since if any faults are present they will almost certainly involve an appreciable area.

On the ground, the first special feature to be looked for is the existence of any fault scarps, near-vertical exposures that may have been caused by movement on a fault plane. These are regularly seen after earthquakes, Fig. 44.5 being a typical example. In the course of time, weathering and other geological processes will gradually erode this distinctive type of feature but it is so sure a guide that every unusual difference in ground elevation should be investigated. The earth tremors which must accompany fault formation will usually trigger landslides in any incipient unstable slopes. All evidences of landslides should therefore be looked at carefully, and, if any alignment of slides becomes evident, the possibility of an associated fault with the same orientation must be entertained. Correspondingly, any unusual alignment of other geomorphic features should be suspect until it can be ascertained definitely that it is not an expression of an underlying fault.

Interference with groundwater conditions has already been mentioned as one of the important effects of faults. It is not an invariable result, but in all areas where the water table is reasonably close to the surface, it is a possibility worthy of close study. If, for example, a sudden change in the elevation of the water table is detected, in wells or cased

44-8 SPECIAL PROBLEMS

FIG. 44.5 Trace of the Red Canyon fault, as exposed after the Hebgen earthquake, looking west towards Culligan Ranch, showing typical damage to buildings.

boreholes, this may very well be because a fault plane is acting as a groundwater barrier. Surface expressions of groundwater conditions, such as the existence of springs, are always worthy of special attention. If, for example, springs are found in a regular alignment, the possibility that this may reveal the alignment of an underlying fault is worth looking into. And if a stream bed is observed with sudden offsets, clearly not due to any change in the geology of the stream bed, this may well prove to be a good indicator of ground movement due to an unseen fault.

This is all tedious work, but so is any investigation of ground conditions if it is to be well and properly completed. It is, therefore, always well to remember that the job may have been done before, if the area in question has been mapped in the ordinary course of geological survey work, without thought (at the time) for the possible engineering implications of the results. Earlier in this volume two examples were cited of trouble that occurred exactly where a geological report of 1946 had clearly shown faults to exist within the limits of the city of Ottawa.[44.6] In one of these cases, the old report was not consulted, and an extra cost of something like $50,000 was the result. In the other case, the report was consulted and gave invaluable information, which was confirmed in the field and which helped to solve an unusual and difficult problem.

44.5 FAULTS AND TUNNELS

There are few major tunnels in the world, the construction of which was not affected by the problems created by the faults encountered. Not only can faults effect a sudden change of rock type, often necessitating changes in tunneling methods, but rock in the vicinity of the fault is liable to be fragmented and so troublesome, often for some distance on either side of the fault itself. If a fault is approached by driving into the hanging wall, rock failures in the tunnel roof as the fault is approached may be serious. Probably the most serious of all potential troubles is the possibility that a fault will be water-bearing and so liable to flood the tunnel workings when it is penetrated. Even when the existence of a fault is known in advance (as should always be the case), the exact character of the water troubles it may create can never be predicted with certainty. And despite the most careful and detailed study of the subsurfaces to be penetrated, in advance of the start of tunneling, sometimes these known problems cannot be avoided by rerouting the tunnel, many tunnel routes being fixed by considerations of factors other than the geology to be encountered.

One example only need be cited to illustrate the serious nature of troubles with faults in tunneling. The San Jacinto Tunnel was probably the most difficult part of the great 630-km (392-mi) Colorado River Aqueduct constructed by the Metropolitan Water District of Southern California. One of 38 tunnels in this major water-supply project, it is 21 km (13 mi) long and penetrates the San Jacinto Mountains. Rock penetrated is mainly granite with subordinate masses of metamorphic rocks, including schists and quartzite. Shafts were used for access to working headings. Soon after excavation had started, working out from the Potrero shaft, a fault was encountered which allowed a flow of water, estimated to be 28,400 lpm (7,500 gpm), as well as over 760 m^3 (1,000 yd^3) of rock debris, to enter the heading. The flow could not be stopped; eventually the 245-m (815-ft) shaft was filled with water to within 48 m (160 ft) of the surface, a vivid demonstration of the interconnection of the fault zone with a large volume of groundwater, either held under subartesian pressure or filling voids in the bedrock for more than 180 m (600 ft) above the tunnel line. The headings were ultimately dewatered; a detailed geological survey along the line of the tunnel was conducted. Through this study 24 additional faults were located as being probably on the tunnel line.[44.7] Even with this foreknowledge, troubles continued, mainly with inflows of water, a maximum pumping rate of 150,000 lpm (40,000 gpm) having to be provided, with some groundwater encountered under a pressure of 42 kg/m^2 (600 psi). The tunnel was eventually completed, but only after some relocations at the sections giving the worst trouble; it has been in continuous service since 1939 (Fig. 44.6). It provides today a vivid reminder of the vital importance of determining as accurately as possible in advance of tunneling the location and nature of all faults that may be encountered on every tunnel line.[44.8]

FIG. 44.6 Leakage into the San Jacinto Tunnel as a result of the penetration of water-bearing faults.

44-10   SPECIAL PROBLEMS

44.6 FAULTS AND DAM FOUNDATIONS

The necessity for sound and solid foundation beds upon which to found a dam is so obvious as to require no elaboration. Fortunately, rock conditions can be inspected before use, once all overburden has been removed and the foundation area has been cleaned off. Faults thus exposed can be subjected to close scrutiny and instrumented, if necessary, to confirm that they are indeed inactive or passive (Fig. 44.7). Design requirements will dic-

FIG. 44.7 Cleaning out a fault beneath the Beechwood Dam of the New Brunswick Power Commission, a good example of the care and attention given to such geological features.

tate whether the site can be used as intended or not. It is, accordingly, most desirable that the presence of any faults along the line of the dam be determined before any excavation is undertaken.

There is yet a further danger, one that has occasionally caused trouble after a dam has been completed and put into service. Extension of the fault upstream of the dam may possibly include a section that will admit water, especially under the increased head that the formation of the reservoir will create, and, in extreme cases, this can result in leakage beneath the dam. Some significant examples of this unusual condition, located generally in the northwest of the United States, are to be found in a useful paper by Esmiol.[44.9]

As but one example of the trouble that faults can cause in dam construction, the Mattupatti Dam in Travancore-Cochin, south India, may be cited. The dam was planned to be only 37.5 m (125 ft) above its foundation bed. Excavation revealed what were thought to be gravels in the foundation, but this material was actually weathered pegmatite. Further excavation showed a width of 21 m (70 ft) in which there were three faults, filled with clay gouge and chlorite schist. Before excavation was completed down to competent rock, it had reached 39 m (130 ft) below stream level, a distance greater than the originally intended total height of the dam.[44.10]

44.7 FAULTS AND NUCLEAR PLANTS

So crucial is the necessity for absolute safety at all nuclear plants, notably nuclear power plants, that consideration of faults at sites proposed for such projects since about 1960 has already probably involved more study, discussion, and debate than had been given to the subject of faults in all previous time. Typical of recent activity was the full-day, well-attended joint meeting of the Geological Society of America's Engineering Geology Division and the Association of Engineering Geologists in Seattle, Washington, in November 1977; the meeting was devoted entirely to a consideration of capable faults (the term now widely used to describe active faults). There are more than 50 references to faults in the volume titled *Geology in the Siting of Nuclear Power Stations,* published in 1979 by the Geological Society of America.[44.11] Beyond directing attention to the vital importance of one detail of structural geology in this special branch of civil engineering, therefore, this section can merely serve as a guide to the latest useful publications in this specialist field.

One of the desirable by-products of this attention to faults at possible sites for nuclear plants has been the "spillover" into other branches of civil engineering. Table 44.1 is but an indication of the regulatory requirements now current in the United States with regard to faults which may exist at the sites of dams, hospitals, liquid natural gas facilities, and

TABLE 44.1 Comparison of U.S. Fault Criteria for Critical Structures

Agency*	Facility	Terminology	Activity criteria
NRC	Nuclear power plants	Capable fault	• 35,000-yr singular movement • 500,000-yr multiple movement • Macroseismicity • Structural relationship of faults
COE	Dams	Capable fault	• 35,000-yr movement • Macroseismicity (magnitude 3.5) • Structural arrangement of faults
DOT	Liquid natural gas facilities	Surface faulting (active fault)	• 35,000-yr singular movement • 500,000-yr multiple movement • Structural arrangement of faults
VA	Hospitals	Active fault	• 10,000-yr movement
BR	Dams	Active fault Inactive fault Indeterminate fault	• 100,000-yr movement • Distribution of seismicity
EPA	Hazardous waste facilities	Active fault	• Movement in lifetime of waste toxicity

*Agencies:
NRC   Nuclear Regulatory Commission
COE   Corps of Engineers
DOT   Department of Transportation
VA   Veterans Administration
BR   Bureau of Reclamation
EPA   Environmental Protection Agency

Source G. A. Robbins, L. A. White, and T. J. Bennett, "Seismic and Geologic Siting Regulations and Guidelines Formulated for Critical Structures by U.S. Federal Agencies," *Report of the GSA Committee on Geology and Public Policy,* Geological Society of America, Boulder, Colo., August 1979, pp. 8–11, by permission.

44-12 SPECIAL PROBLEMS

hazardous waste facilities. It will be noted that the terms *capable* and *active* fault are used interchangeably, another of the terminological complications that seem to plague so much of applied science; there is virtue in both terms, few faults being literally active at the present time, many faults being capable of activity in the foreseeable future.

Determination of the potential activity of a fault starts with a determination of its geological age. A variety of sensitive methods have been developed for this purpose, generally related to the laboratory dating methods that have been developed so remarkably in recent years. Merely to indicate the extent of this part of the current study of faults, there is in the GSA volume just referred to a paper dealing with dating techniques in fault investigations; the paper is itself 12,000 words long and is accompanied by a list of over 40 references.[44.12] So has the subject developed in just a few years.

It is, therefore, impossible to go into any detail here, but the continuing importance of field investigations must be stressed. All the methods normally used in site investigations must be utilized, with study of aerial photographs for an extensive area around the proposed site but a first and essential step. Going beyond what is usually possible at sites proposed for "normal" civil engineering projects, the necessity for actually seeing what lies beneath the surface has led to the use of trenching at proposed nuclear sites (Fig. 44.8). Carried if possible down to the rock surface, trenches, even if they penetrate only

FIG. 44.8 Trenching at a test reactor site near Pleasanton, California, to determine the hazard posed by the Verona fault, part of a program which involved the opening up of about 4,000 m (13,000 ft) of trenches.

soil, can still yield information of value and give some indication of the possible presence of active faults, since recent activity would be displayed in Pleistocene deposits. Trenches must be excavated and well and safely braced by experienced personnel, after careful location by the engineering geologist. As excavation proceeds, logging must be accurately done as soon as access to the trench can be gained. The information so obtained can be an important complement to that gained through more usual methods of subsurface exploration, often providing useful confirmation of features that would otherwise be only indirectly obtained.[44.13]

44.8 EVIDENCE OF FAULT MOVEMENT

Ground movement along and across the San Andreas fault is not too difficult to visualize, so well recognized is the fault itself. It will be more difficult for many to appreciate that this is not something peculiar to the San Andreas fault and that evidence of activity of quite small faults, far removed from any major features, may be encountered in engineering work. Studies for a new pipeline in the Mojave Desert revealed repeated cracking of a nearby highway. It was thought that this might have been due to movement on the Rosamond fault. Pipeline design was modified to allow for flexible-joint pipes across the zone where further cracking might occur.[44.14] A number of faults have been identified in and around Houston, Texas, some 50 having been now recorded. Rates of displacement have varied up to 34 mm (1.3 in) per year. In many locations water mains and other critical pipelines are laid within a larger enclosing pipe to allow safely for any future movements.[44.15]

It has been suggested that the slow earth movement (on a fault) that has been observed between the cities of Tucson and Phoenix, Arizona, which has already produced a 16-km (10-mi)-long scarp, 30 cm (12 in) high, may be due to local pumping of groundwater.[44.16] It is thought that it is not the result of seismic stresses, and it has not itself triggered any microseisms. It has, however, caused problems in the maintenance of highways and railroads. Active faults, therefore, are not an academic invention but a very real feature of the earth's crust, fortunately not encountered very often on civil engineering works, but definite enough as a possibility that they are one further feature of the crust that must be kept clearly in mind in all site studies for engineering works (Fig. 44.9).

FIG. 44.9 The Wasatch fault of Utah, as it crosses the mouth of Bell Canyon on the Wasatch Front.

44.9 JOINTS

Joints are fractures in rock along which no displacement has occurred. They will be a familiar feature to all who have examined even a few rock outcrops. Associated with joints are the bedding planes of sedimentary rocks, surfaces indicating some interruption with the original process of deposition and now planes of weakness (usually) along which movement may take place if, for instance, beds originally horizontal, but tilted by subsequent earth movements, are then interfered with by human activity. In New York City, for example, serious rockslides have occurred into new excavations when the Manhattan schist, widespread as bedrock and steeply inclined to the vertical, has had the capacity of its bedding planes weakened in some way. In one notable case, a block with a volume of 460 m^3 (600 yd^3) slid after a bedding plane with a dip of 65° had been weakened by tree roots, possibly aided by frost action (Fig. 44.10).[44.17] Bedding planes must, therefore, be

FIG. 44.10 Rockslip on joint planes in broken Manhattan schist, New York, the dip being 65°, movement precipitated by the action of tree roots.

watched carefully in preliminary studies, especially if now steeply inclined. Joints, however, will usually be vertical in sedimentary rocks which still have their bedding planes horizontal, and so at right angles to the stratification. They occur in almost all types of rock, however, and if undetected can sometimes cause trouble.

Amongst the more interesting types of jointing is the polygonal pattern (often hexagonal) observed in basalts and other fine-grained igneous rocks, as well as in mud when it dries quickly under the influence of solar heat. Shrinkage is clearly here the cause. In granites, however, there are often three sets of joints, the main set being called the *rift* and the secondary set the *grain*, with a *sheet* structure as possibly a third. One common cause of jointing is the folding which distinguishes many rocks. The convex upper surface of even a gentle fold creates stresses in the rock which evidence themselves by the fracture of the rock after severe strain. Study of local structural geology will be a normal part of site investigation on geological work; evidence thus gained of any folds will suggest the possibility of jointing.

Some joints appear to be due to tension or shear stresses. Some have been filled with molten magma and so now present themselves as dikes. It is, however, the open or unfilled joint that can be most troublesome, and some formations are notorious for their open-jointed structure. Dealing with them is one of the most widespread reasons for grouting programs, carefully designed and injected portland-cement grout usually sealing up joints that are open enough to take any grout, thus converting the jointed mass into solid rock.

Joints in rock are a benefit to all quarrymen; they may assist correspondingly with excavation of rock. But this is small compensation for the minor problems which joints can create, especially if they are open.

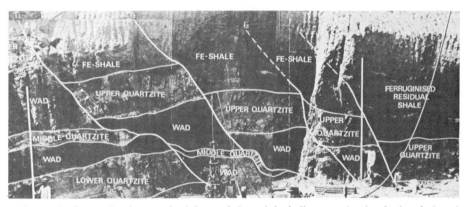

FIG. 44.11 Composite photograph of the north face of the bulk excavation for the foundation of the Randfontein mill building, South Africa, showing the displacement of the Black Reef strata resulting from four step faults.[42.7]

44.10 REFERENCES

44.1 *San Andreas Fault* (one of a popular leaflet series), U.S. Geological Survey, Washington, 1965.

44.2 W. R. Dickinson and A. Grantz (eds.), *Proceedings of Conference on Geologic Problems of the San Andreas Fault System,* Geological Sciences vol. 11, Stanford University Publications, Stanford, 1967.

44.3 L. S. Cluff and K. V. Steinbrugge, "Hayward Fault Slippage in the Irvington-Niles Districts of Fremont, California," in abstract vol. for annual meeting of Cordilleran Section of the Geological Society of America, Reno, Nev., 1966, p. 27; see also in same volume, M. G. Bonilla, "Slippage on Hayward Fault Indicated by Deformation of Railroad Tracks in the Niles District of Fremont, California," p. 22.

44.4 J. A. Langbein, "The Manapouri Power Project, New Zealand," *Proceedings of the Institution of Civil Engineers,* **50**:311 (1971).

44.5 J. B. Auden, "Erosional Patterns and Fracture Zones in Peninsular India," *The Geological Magazine,* **91**:89 (1954).

44.6 A. E. Wilson, *Geology of the Ottawa-St. Lawrence Lowland, Ontario and Quebec,* Memoir no. 241, Geological Survey of Canada, Ottawa, 1946.

44.7 L. H. Henderson, "Detailed Geological Mapping and Fault Studies of the San Jacinto Tunnel Line and Vicinity," *Journal of Geology,* **47**:314–324 (1939).

44.8 B. C. Leadbetter, "Driving An Extremely Difficult Tunnel," *Engineering News-Record,* **121**:669–673 (1938).

44.9 E. E. Esmiol, "Seepage Through Foundations Containing Discontinuities," *Proceedings of the American Society of Civil Engineers* vol. 83, SM1, Paper 1143, New York, 1957.

44.10 Auden, op. cit.

44.11 A. W. Hatheway and C. R. McClure (eds.), *Geology in the Siting of Nuclear Power Stations* (reviews in *Engineering Geology,* No. 4), Geological Society of America, Boulder, Colo., 1979.

44.12 P. J. Murphy, J. Briedis, and J. H. Peck, "Dating Techniques in Fault Investigations," in A. W. Hatheway and C. R. McClure (eds.), *Geology in the Siting of Nuclear Power Stations,* Geological Society of America, Boulder, Colo., 1979, pp. 153–168.

44.13 A. W. Hatheway and F. B. Leighton, "Trenching as an Exploratory Method," in A. W. Hatheway and C. R. McClure (eds.), *Geology in the Siting of Nuclear Power Stations,* Geological Society of America, Boulder, Colo., 1979, pp. 169–195.

44.14 J. R. Keaton, "Evidence for Possible Fault Creep Near Rosamond, Kern County," in abstract vol. for annual meeting of Association of Engineering Geologists in Seattle, 1977, pp. 28, 29.

44.15 W. M. Reid, "Active Faults in Houston, Texas," in abstract vol. for annual meeting of Geological Society of America in Dallas, 1973, pp. 777, 778.

44.16 News note in *Civil Engineering,* **49**:93 (June 1979).

44.17 J. Feld, Report on International Conference on Rock Mechanics, *Civil Engineering,* **36**:59 (December 1966).

Suggestions for Further Reading

Faults are such a vital part of the structure of the earth's crust that their geological aspects can be studied in any comprehensive book on geology. Fortunately there are some excellent guides to the treatment of faults encountered on engineering works. *Geology in the Siting of Nuclear Power Plants,* edited by A. W. Hatheway and C. R. McLure (the fourth of the Geological Society of America's Reviews in Engineering Geology series), published in Boulder, Colorado, in 1980, has an excellent treatment of faults which, although written with respect to nuclear power plants, has general application. In 1976 a comprehensive paper on faults in dam foundations, based on North American experience, was published but in the British journal *Geotechnique* (**24**:367); it is "Potentially Active Faults in Dam Foundations" by J. L. Sherard, L. S. Cluff, and C. R. Allen. Those who know it regard it as a classic paper.

A record of valuable experience is *Passing Faults on the Delaware Aqueduct* which, although published in 1941, is still of value; it is to be found in *Engineering News-Record* **50**:40 (31 July 1941).

chapter 45

VOLCANOES AND EARTHQUAKES

45.1 Volcanoes / 45-3
45.2 Earthquakes / 45-6
45.3 Seismic Regulations / 45-16
45.4 Earthquake Prediction / 45-18
45.5 Man-made Earthquakes / 45-19
45.6 Tsunami / 45-20
45.7 Conclusion / 45-21
45.8 References / 45-22

The relatively minor types of earth movement, or mass wasting, that have so far been discussed will have shown that the crust of the earth is composed of ordinary solid materials which react to the stresses induced in them, generally in a manner similar to structural materials which may be tested in a laboratory. This has been said in other chapters; the statement must be repeated here as a prelude to considering briefly such major earth movements as earthquakes, and the associated phenomenon of volcanic action. The latter is clearly related to the existence of hot springs and other releases at high temperatures from the crust, evidencing the great heat that exists well below the earth's surface. Earthquakes, on the other hand, are now widely understood to be related to the geologically slow movement of the massive plates of which the solid crust of the earth is made up. The whole subject of plate tectonics has been one of the most exciting advances in the history of geology. Continental drift was first suggested over a century ago, and seriously promoted at the start of the present century by Taylor and Wegener. The concept which holds that the continents in the southern hemisphere can be fitted together to give what was called Gondwanaland, although ridiculed for many years, is now so well regarded that it has even called for mention in this volume dealing only with applied geology. There is today little doubt in geological circles that "the continents are drifting," that the great plates are moving, and this explains many geological features previously puzzling—including earthquakes.

Earthquakes and volcanoes are a part of the natural dynamics of the world. They are usually catastrophes when they occur, all too often causing loss of life and limb. Engineering works are inevitably affected, and the activities of the engineer are frequently a

45-2 SPECIAL PROBLEMS

vital part of necessary rescue and ameliorative operations. The catastrophes have to be accepted as a part of the natural order of things, but, as will shortly be seen, measures to prevent harm to people, by prediction and forewarnings, are now being actively pursued. Planning, for example, must always be carried out so as to avoid any structures over faults. Active volcanoes are naturally well known, and under constant observation, but even within recent years, two new volcanoes have appeared, one in Mexico and one in Iceland. As this Handbook was in preparation, Mount St. Helens erupted in Washington (Fig. 45.1). Correspondingly, the most active seismic regions of the world are now well recog-

FIG. 45.1 It can happen here. Eruption of Mount St. Helens, as seen on 18 May 1980.

nized, and great efforts are being made in searching for methods of earthquake prediction. The Chinese have special interest in this subject and have had some initial success.

North Americans naturally think first of the west coast of the continent as an area of known seismic activity, but there have been serious earthquake shocks also in and near the east coast. Both Iceland and New Zealand are distinguished by hot springs, faulting, volcanoes, and earthquake shocks, the former land associated in some way with the mid-Atlantic ridge and the latter with similar evidences of crustal movement beneath southern oceans. New Zealand, a particularly lovely place with most hospitable people, is a fascinating area for the geologist since it is geologically so "new." Even the capital city of Wellington is located on the site of severe earthquake movement that was observed in 1855 by early settlers. A paper by an eminent local geologist bears the title, "For How Long Will Wellington Escape Destruction by Earthquakes?"[45.1] On the other hand, one of the country's most lovely lakes, with the musical name of Waikaremoana, was created by a majestic landslide that probably happened in historic times, as is suggested by a fanciful, poetical explanation of the lake's formation traditional with the Maori.

Indicative of the "world-watch" on these catastrophic happenings was the establishment of the Center for Short-Lived Phenomena in January 1968 by the Smithsonian Institution. Located in Cambridge, Massachusetts, it is served by about 3,000 registered correspondents in more than 150 countries who report to the center immediately upon short-lived natural events and who then follow up such occurrences for as long as necessary. With the aid of modern communications, the center is able to issue warnings, when this is necessary. It issues summary reports annually, a typical issue being a 300-page book listing 113 events in 51 countries during the one year. Geological events numbered just under 50, including 19 major earthquakes and 22 volcanic eruptions, tsunami, and associated events. To the extent that is now possible, the center is an early-warning station for natural catastrophes. And its work has fostered a new appreciation of the relative frequency of volcanic and earthquake action.

45.1 VOLCANOES

There is still, in some quarters, a little scepticism about the theories of geologists and geophysicists regarding the composition of the central part of the globe. Their determinations suggest that the central core has a density of 10.72 g/cm^3 and is surrounded by the mantle with a density of only 4.53 g/cm^3, although this is higher than the density of the thin outer crust, with its average density of just less than 3 g/cm^3. Some who have never seen a volcano in eruption find it hard to imagine the molten magma which emerges from its cone, with the concomitant evidence of release of high pressure and great heat. Users of this volume will have no such uncertainty, but even they may be surprised by looking at Fig. 45.2, which shows molten magma being ejected from an ordinary borehole. This was a hole sunk for extraction of thermal heat at Namafjall in Iceland. The hole was only 1,138 m (3,730 ft) deep and was 16 cm (6.25 in) in diameter. Ten people, fortunately, witnessed the event (on 8 September 1977), which lasted from 15 to 25 minutes, the incandescent column of magma shooting 15 to 25 m (50 to 80 ft) high. It was accompanied by other evidences of volcanic action but is illustrated here to emphasize yet again the intricate mechanism that controls our earth.[45.2]

History contains many records of volcanic eruptions, some with devastating results. Destruction of the ancient cities of Pompeii and Herculaneum by the eruption of Mount Vesuvius in 79 A.D. is now well known, following the discovery of the ruins of Herculaneum in 1748 and the wide publicity given in words and pictures to the remarkable excavations that have revealed so much of the buildings and streets of the old cities. Vesuvius and Stromboli in Italy, Hekla in Iceland, and Mont Pelée in Martinique are the most notable of the volcanoes that have built up cones around the vents in the earth's surface through which molten lava has been emitted from the molten magma beneath the crust. Mont Pelée was in violent eruption in 1902 when the town of St. Pierre was completely destroyed. Another tremendous natural explosion of recent times was the eruption of Krakatoa in the Strait of Sunda, Indonesia, in 1883. The volcano of Paricutín in Mexico, which started in level farmland in 1943, is of the same conical type (Fig. 45.3), whereas the new Icelandic volcano, Surtsey, started by emission of lava beneath the sea from a long fissure. The fact that the Hawaiian Islands are the eroded tops of volcanoes rising

45-4 SPECIAL PROBLEMS

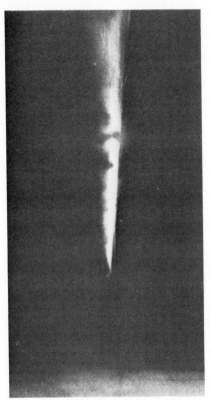

FIG. 45.2 (a) Spurt of magma, estimated to be between 15 and 25 m (50 and 80 ft) high, ejected from borehole at Namafjall, Iceland, about midnight on 8 September 1977. Photo on right (b) taken from same spot on the following morning.

FIG. 45.3 Paricutín, Mexico; the new volcano appeared first as a small eruption in 1943.

from the depths of the Pacific Ocean is well known and is evidenced by the generally mild activity of the still-active peaks.

This short list includes the best known of the relatively few volcanoes that have been active in recent times. Their existence is so well known that their influence on any urban development in their vicinity can be taken for granted. Equally limited in distribution today are those areas in which hot water and associated gases, including steam, escape from the crust of the earth through cracks or hot springs. The ways in which this supply of subterranean heat can be harnessed for public use were touched upon in Chap. 23, the city of Reykjavík in Iceland being preeminent in this connection. Here a minor form of volcanism is being turned to good effect.

Engineers in Italy can never forget the influence of volcanic action in their country, as was demonstrated in the early 1970s by a sudden rise in the ground level in the city of Pozzuoli, well known at least by name as the location from which pozzolan was first obtained in quantity by Roman builders. The movement started in 1969, and within six months the ground had risen in places as much as 0.7 m (2.25 ft), continuing to rise at the rate of about 10 mm (0.4 in) per day. It is believed that underground movement of molten lava, the area being very close to Mount Vesuvius, is the cause of this reversal of the usual form of ground movement, settlements being almost the only kind of movement normally encountered. By the spring of 1970, the Italian Ministry of Public Works had ordered 9,000 of the city's 65,000 residents to evacuate their homes. A new city is being built at a cost in excess of $12 million to house the displaced townsfolk. This most unusual movement may also be responsible in part for serious ground movements that have affected an area close to the adjacent city of Naples. Serious subsidences have taken place, with accompanying slides and even opening up of cracks in the earth's surface, with consequent damage to property. Heavy rains in the fall of 1969 are believed to have been a contributing factor, causing the infiltration of groundwater into a geologically unstable area, but the subterranean volcanic action is thought to be at least partially responsible.[45.3] The world as a whole is fortunate in that such evidence of the instability of the thin crust on which the welfare of the entire earth depends is so very limited. To see molten lava pouring forth from an active volcano, and—in the case of some of the Hawaiian volcanoes, for example—pouring back again into the earth through crevasses, is a sobering experience. To find complete mountains, such as Paricutín in Mexico, formed within a few years on what was previously level farmland, is a reminder that the earth is not the fixed and permanent sphere of the poets.

Observatories now exist on, or near, a number of the more active volcanoes, where these are close to developed areas; from these observatories warnings are issued when any renewed activity can be predicted. Emergency measures exist and can be put into operation by responsible local authorities for minimizing damage and eliminating risk to lives. Attention has also been given to volcanoes, in well developed areas, that have not been active recently but which might well resume activity. Mount Ranier in the state of Washington may be cited as an example. Its last significant eruption was between 120 and 150 years ago, with some minor eruptions reported at the end of the last century. One of the remarkable "miscellaneous" geological maps published by the U.S. Geological Survey shows vividly the potential hazards that would be created by future eruptions of Mount Ranier.[45.4] It is accompanied by useful explanatory notes by its author, D. R. Crandell, and by a small map showing the extent of tephra deposits more than 2.5 cm (1 in) thick from eruptions of the last 7,000 years, and the probable extent of mudflows and lava flows for this same period. The main map shows high-, moderate- and low-risk areas for mudflows and lava flows and for tephra (air-borne deposits). A detailed study has been made by Janet Cullen of what impact a major eruption of Mount Ranier would have on public service delivery systems in the Puyallup Valley, one of those shown as a hazard area on the Crandell map.[45.5] Preparation and publication of such vitally important information is as far as the geologist can go; further action is in the hands of public authorities.

45.2 EARTHQUAKES

Earthquakes can be among the most terrible and devastating of all natural phenomena affecting the surface of the earth, and so the lives of many people. They have real relevance to engineering. Although fortunately not frequent in occurrence, earthquakes can wreak such havoc when they do occur that it is surprising to find that their scientific investigation in the Western World is of relative recent date. In China, on the other hand, records of earthquakes exist which cover a period of 3,000 years down to the present, earthquake prediction being today one of the highest priorities in Chinese scientific endeavor. The Lisbon earthquake of 1775 is really the first of which any scientific description exists in the West. The first seismograph was not developed until the 1880s, following the pioneer seismic studies of Robert Mallet, who published his first paper in 1862. His description of the great Neapolitan earthquake of 1857 (published in 1862) provided for the first time a sound basis upon which the science of *seismology* could be founded.[45.6] Mallet's papers include many excellent accounts of damage to buildings; the engineering effects of earthquakes have been a stimulus to the more fundamental study of seismic forces throughout the relatively short history of this branch of science.

Some Examples of Major Earthquakes From the Chinese records that have recently become available, it is known that something like 10,000 earthquakes have occurred in that large country in the last 3,000 years, 530 of them causing major disasters, the worst one in Kansu Province in 1556, when more than 820,000 people were killed, all recorded by name.[45.7] The earthquake that devastated Tokyo, Japan, on 1 September 1923, will be within the memory of some readers; it destroyed 140,000 lives and caused damage estimated at $3 billion. The earthquake that devastated Quetta, India, in 1935, took a toll of at least 35,000 lives in a few seconds. The earth movement that causes the quaking may be either volcanic or tectonic, the latter type being more common. In many instances, the permanent deformation of the earth's crust can be observed. Thus the San Francisco earthquake of 18 April 1906 caused movement along 435 km (270 mi) of the San Andreas rift, the great fracture zone which runs almost parallel to the California coastline of western America. A maximum horizontal movement of 6.3 m (21 ft) was afterward found. No vertical movement occurred; but in the Yakutat Bay (Alaska) earthquake of 1899, a vertical movement of 15 m (50 ft) took place over a large area, completely altering the local topography and creating new waterfalls on local watercourses.[45.8]

Some readers may recall the earthquake in Chile in 1960. The next major disaster took place in Skopje, Yugoslavia, on 26 July 1963, when an earthquake of great intensity demolished most of the city. The center of the earthquake action, the *epicenter* in scientific terms, was between 6 and 10 km (3.8 and 6.2 mi) northwest of the city, which therefore felt almost the full force of the vibrations set up by this sudden geological disturbance. About 2,000 people lost their lives; about 85 percent of the city's population were left homeless. The city had grown very rapidly between 1949 and the time that disaster struck, with the result that the multistorey, unreinforced masonry buildings that had been used to provide much-needed accommodation suffered severe damage.[45.9] In 1967, an earthquake of only moderate intensity occurred about 48 km (30 mi) east of Caracas, Venezuela, but close enough to cause considerable damage in this rapidly growing city. The loss of life was limited to an estimated total of 277, about 200 of whom were killed by the complete collapse of five buildings (Fig. 45.4). Of the thousand multistorey buildings,

VOLCANOES AND EARTHQUAKES   45-7

FIG. 45.4 Caracas, Venezuela; wreckage of the San José building by the earthquake of 1967, typical of the widespread damage throughout the city.

ranging in height from 10 to 30 storeys, then in use in Caracas, 180 suffered serious damage. The structural results of this earthquake were therefore studied in detail, as is now done for all major earthquakes, and the results have been published in a series of notable and useful reports.[45.10]

North America is not immune from these natural hazards. The San Francisco earthquake of 1906 has already been mentioned. In more recent years, there were two quakes of moderate intensity in Churchill County, Nevada, on 16 December 1954, fortunately not close to any large city but serious enough to be felt in six states. Surface fault displacements occurred throughout a zone 32 by 96 km (20 and 60 mi), as much as 6 m (20 ft) vertical displacement occurring, with lateral movement along faults of as much as 3.6 m (12 ft).[45.11]

Lest it be thought that all North American earthquakes take place in the western part of the continent, it may be pointed out that one of the most potentially active seismic areas of the continent is the St. Lawrence Valley and adjoining parts of New England, Ontario, and Quebec. There is a long record of earthquake activity in this region, fortunately with no serious quakes having taken place in very recent times. On 28 February 1925, however, an earthquake of relatively low intensity took place and affected a surprisingly wide area, centered in the region of Quebec City, where serious damage was done to heavy, reinforced-concrete columns supporting a large grain elevator. Most fortunately, no lives were lost directly, but there were several deaths due to associated shock. On 2 May 1944 a small earthquake occurred west of Montreal, its epicenter not far from the seaway city of Cornwall, Ontario. One of the authors was in Cornwall on the day following the earthquake and examined the damage done by this minor earth shaking; it amounted to about $1 million. No lives were lost, again fortunately and almost miraculously, because of the late hour at which the disaster took place. Had it been earlier in the day, the col-

45-8 SPECIAL PROBLEMS

lapse of many chimneys and parapets would almost certainly have caused a number of deaths, many of them among children, since school buildings were affected.[45.12]

The same fortuitous timing led to the very small loss of life in the most recent of major earthquakes in North America, that which took place at 5:36 p.m. on Good Friday, 27 March 1964, at Anchorage, Alaska. It released twice as much energy as the 1906 San Francisco earthquake, and it was felt over an area of 130 million ha (320 million acres). It left 114 people dead or missing, but had it taken place on an ordinary working day, or even earlier in the day on that Good Friday, the possible loss of life is disturbing even to consider. It was estimated that it would take the better part of a billion dollars to rehabilitate the damaged public and private property. Uplift and subsidence affected an area of at least 9 million ha (22 million acres), these vast earth movements taking place about a well-defined zone that can be regarded as a "hinge" (Fig. 45.5). To the east of the hinge zone,

FIG. 45.5 Anchorage, Alaska; building subsidence due to the Fourth Avenue slide caused by the Good Friday, 1964, earthquake.

land levels went up as much as 2.25 m (7.33 ft), while to the west, subsidence of as much as 1.62 m (5.33 ft) took place. In the country around, thousands of avalanches were triggered, as were massive rockslides. One rockslide occurred 217 km (134 mi) away from the epicenter, at the southern tip of Kayak Island, and unfortunately killed a Coast Guardsman stationed at the Cape St. Elias lighthouse.[45.13]

The Alaska Railroad was violently disrupted, most of the bridges on the affected 300-km (185-mi) stretch being seriously damaged. Of unusual importance were the deep-seated landslides that took place along the adjacent coasts, taking out much of the waterfront property at the cities of Valdez and Seward. At Valdez a 10,815-ton vessel was being unloaded at the docks, which were accordingly crowded with people. As soon as the tremors started, the entire dock area began to sink into the sea; 30 lives were lost. At the town of Homer, 256 km (160 mi) southeast of the epicenter, the small harbor disappeared into a "funnel-shaped pool," as it was described by a witness. An area on a sandbar extending out from the town subsided from 1.2 to 1.8 m (4 to 6 ft), most of this being due to the consolidation of a 140-m (460-ft)-thick stratum of alluvium by as much as 75 cm (30 in), and the further settlement of the underlying bedrock by about 60 cm (24 in). But the effects of the earthquake were felt far beyond Alaska. Incredible though it may seem, the water level in wells to which automatic level recorders had been fitted in both Winnipeg and Ottawa, 3,000 and 5,000 km (1,850 and 3,100 mi) away, respectively, from the epicenter, showed violent oscillations six and eight minutes after the first shock took place at Anchorage.[45.14]

Many pages could be filled with similar summary notes on just the main features of this great earthquake, but enough has been said to show that it was truly a remarkable natural phenomenon. It has been studied in great detail, probably in far more detail than any previous occurrence of this kind. The results have been published in many notable papers and reports, to which readers who wish more information may be directed. A few words more must be said, however, about the effect of the earthquake on the city of Anchorage. Damage estimated at $200 million was caused, including the destruction of 215 residences and severe damage to 157 commercial properties in this leading Alaskan city. Nine people lost their lives. Most of the casualties and the worst damage resulted from some serious landslides triggered by the earthquake, at Turnagain Heights, L Street, and 14th Avenue. Each of these areas is underlain by a gray silty clay, with some lenses of sand, well known locally as the "Bootlegger Cove clay" (Fig. 45.6).

FIG. 45.6 Anchorage, Alaska; damage in the Turnagain Heights area due to major sliding caused by the Good Friday, 1964, earthquake.

Geologists on the staff of the U.S. Geological Survey had studied the geology of the Anchorage area in 1949. A preliminary report on this work and a map were made available in 1950. The final printed report was published in 1959 as Geological Survey Bulletin No. 1093.[45.15] This valuable report was therefore publicly available for use four years before the earthquake took place. The Bootlegger Cove clay is described in the report and delineated in its extent. Its physical properties were described, and conditions under which

45-10   SPECIAL PROBLEMS

landslides could be activated in the clay are described under a section dealing with slumps and flows. Local engineers used this report in their soil and site investigations in the Anchorage area. The Alaska Road Commission and the Alaska Department of Highways both used the report, but for a variety of reasons it was not used as a background for planning in Anchorage.

> The planning department was relatively new and its early problems concerned more pressing matters. The report is a general treatment and did not zone or classify the ground except by geologic map units. The map is not a document the planners can use directly without interpretation by a geologist. There were no geologists on the planning staff.[45.16]

The fact remains that the report was publicly available but apparently was not used in the planning of this rapidly developing city. It is idle to speculate what its use in planning might have done, since the damage by the earthquake has taken place and cannot be undone. USGS Bulletin no. 1093 has already taken its special place in the literature of engineering geology, not only as a record of fine work on the geology of an important urban area but also as a report that should have been better known beyond purely geological circles.

Cause, Measurement, and Effect of Earthquakes   It must be emphasized that the movements along faults are not the result of earthquakes; on the contrary, they are usually the cause of the quakes that follow them. (Figure 45.7 maps occurrences of major earthquakes in 1970.) The great masses of material involved in the movements naturally set up dynamic effects of magnitude. In the New Zealand earthquake of 1929, several observers witnessed the earth's surface undulating to such an extent that a ripple seemed to pass along the ground under observation. The main result, however, is the initiation of earthquake waves (or, more accurately, vibrations) within the crust of the earth. These travel great distances, and it is because of them that a localized movement of the earth can have disastrous effects over a wide area. Earthquake waves are recorded by means of seismographs, delicate instruments similar to those used in seismic prospecting but housed in special buildings and far more sensitive. Modern seismographs are, indeed, so sensitive that the instrument in the Bidston Hill Observatory near Liverpool, England, regularly records the rise and fall of the tide in the adjacent Liverpool bay and estuary, which cause slight movement of the rock on which the observatory is founded.

Seismologists have distinguished three types of waves set up by earth movements; two of these travel through the earth, and the third at the surface. All three can be distinguished on the seismograms obtained from seismographs, and by means of suitable calculations, the origin of the waves, or epicenter, can be determined. The area affected by earthquake waves is often extensive; in the case of the Assam, India, earthquake of 1897, it is estimated that an area of 4.55 million km^2 (1.75 million mi^2) was thus influenced. Amplitudes and periods of vibration vary; the factor of most significance in engineering is the horizontal acceleration that the vibrations tend to produce. Various scales have been suggested and used to denote the differing intensities of earthquake shocks. One of the most recent is the modification of the Mercalli scale suggested by C. F. Richter.[45.17] This provides such a clear guide that it is reproduced in full in Fig. 45.8.

Not only is geological action the cause of earthquakes, but local geological features can have a marked effect on the local results of earthquake shock. This is evidenced by the fact that earthquake shocks are not felt in deep mines located within seismic areas. Since the radiating vibrations travel at different speeds in different materials, it is to be expected that effects in rock and unconsolidated material will be different. This is found in practice; earthquakes cause much more trouble in areas of unconsolidated materials than in those with solid rock exposed at the surface. Greater amplitudes of vibrations are possible in the former than in the latter, and consequently, greater accelerations are to be expected in the softer materials. In the San Francisco earthquake, the maximum acceleration recorded in marshy ground was about 3 m (10 ft) per second per second; in corresponding rock outcrops, a figure as low as 27 cm (0.89 ft) per second per second was observed. Similar records have been obtained in other earthquakes.

It follows that in civil engineering work in seismic areas, structures should be founded wherever possible on solid bedrock. Another reason for this is that under the action of the transmitted vibrations, unconsolidated material will fracture and be displaced much more easily than will solid rock. Actual experience with buildings and engineering structures in severe earthquakes is in line with this suggestion. When the local geology is such that a rock foundation bed cannot possibly be used, foundation design must take into consideration the unusual forces that may be exerted upon the structure during an earthquake; a raft type of foundation, so designed that it will withstand upward forces in addition to the usual downward loads, is one type that has proved satisfactory, as have also continuous foundations carried on piles driven so that they rest on an underlying rock stratum.

In built-up city areas, the shock from an earthquake is liable to fracture water mains; and as fire is always a consequent hazard, water-distribution systems have to be designed and constructed with special care in seismic areas. Underground conditions may be so affected that groundwater distribution will be changed; old springs may stop flowing and new ones may be formed. Unconsolidated deposits that are below the surface may react in strange ways. Fill material will very often be affected; the intense vibration may compact it to an extent not possible by any normal means. In the New Zealand earthquake of 1929, all made ground sank to some extent. One fill which had been in place for six years sank almost 1 m (3 ft), although it was only 6.6 m (22 ft) high; the grass on the surface was not displaced.[45.18]

A feature which is sometimes of serious consequence is the effect that an earthquake shock can have at exposed vertical or sloping faces, such as the steep banks of rivers or those of artificial cuts. The action of the vibrations on reaching a sloping face has been likened to the movement of the last billiard ball of a row that is struck sharply at the other end. Although hardly an accurate picture, this may give the reader an appreciation of the severe landslides and rockfalls that so often occur on such faces after passage of an earthquake shock. The damage caused by earthquakes to major engineering excavations, especially tunnels, that are adjacent to lines of major earth movement can be disastrous. The Arvin-Tehachapi earthquake in California on 21 July 1952 affected the area crossed by a steeply graded main line of the Southern Pacific Railroad on which there are 15 tunnels. Four of these were on a loop south of Caliente, adjacent to the White Wolff fault, upon which serious movement took place. Damage to the tunnels and the track within them was so serious that three of them had to be transformed into open cuts. The damage was economically irreparable; the total cost of this work was about $1 million.[45.19]

Much work has been done in correlating damage to buildings, the stiffness of building frames, and ground conditions. In a valuable review by Seed and Idriss, for example, they conclude that "short period structures develop their highest damage potentials when they are located on relatively shallow or stiff soil deposits and long period structures when they are constructed on deep or soft soil deposits."[45.20] During the 1970s much attention was given in geotechnical circles to the problems caused by the *liquefaction* of fine-grained soils as a result of earthquake shocks. The name has been questioned as terminologically incorrect, but it has come to be accepted and is now in common usage. It is best defined as "the sudden large decrease of shearing resistance of a cohesionless soil, caused by a collapse of the structure by shock or strain, and associated with a sudden but temporary increase of the pore fluid pressure. It involves a temporary transformation of the material into a fluid mass." The damage that could be caused to structures founded on such soils,

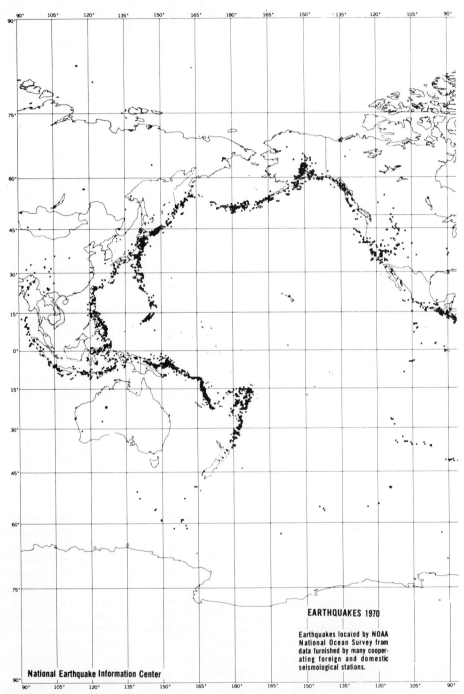

FIG. 45.7 Major earthquakes throughout the world during 1970 (the last such compilation available in 1981).

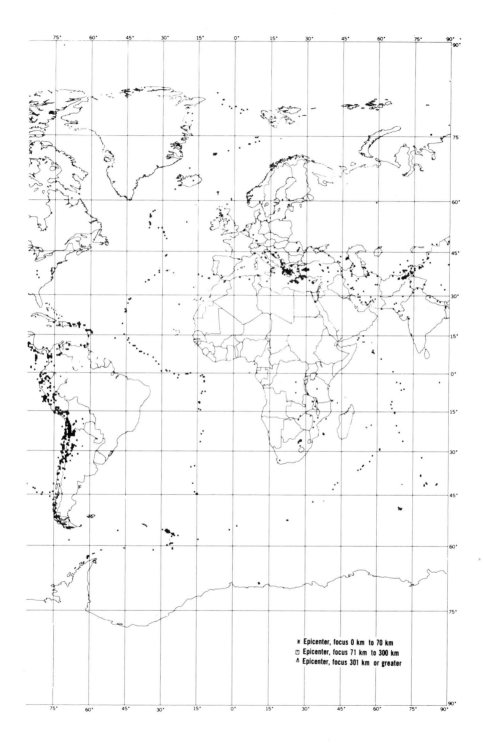

I	Not felt; Marginal and long-period effects of large earthquakes.
II	Felt by persons at rest, on upper floors, or favorably placed.
III	Felt indoors; Hanging objects swing; Vibration like passing of light trucks. Duration estimated. May not be recognized as an earthquake.
IV	Hanging objects swing. Vibration like passing of heavy trucks; or sensation of a jolt like a heavy ball striking the walls. Standing motor cars rock. Windows, dishes, doors rattle. Glasses clink. Crockery clashes. In the upper range of IV wooden walls and frame creak.
V	Felt outdoors; direction estimated. Sleepers wakened. Liquids disturbed, some spilled. Small unstable objects displaced or upset. Doors swing, close, open. Shutters, pictures move. Pendulum clocks stop, start, change rate.
VI	Felt by all. Many frightened and run outdoors. Persons walk unsteadily. Windows, dishes, glassware broken. Knicknacks, books etc., off shelves. Pictures off walls. Furniture moved or overturned. Weak plaster and masonry D cracked. Small bells ring (church, school). Trees, bushes shaken (visibly, or heard to rustle CFR).
VII	Difficult to stand. Noticed by drivers of motor cars. Hanging objects quiver. Furniture broken. Damage to masonry D, including cracks. Weak chimneys broken at roof line. Fall of plaster, loose bricks, stones, tiles, cornices (also unbraced parapets and architectural ornaments—CFR). Some cracks in masonry C. Waves on ponds, water turbid with mud. Small slides and caving in along sand or gravel banks. Large bells ring. Concrete irrigation ditches damaged.
VIII	Steering of motor cars affected. Damage to masonry C, partial collapse. Some damage to masonry B; none to masonry A. Fall of stucco and some masonry walls. Twisting, fall of chimneys, factory stacks, monuments, towers, elevated tanks. Frame houses moved on foundations if not bolted down; loose panel walls thrown out. Decayed piling broken off. Branches broken from trees. Changes in flow or temperature of springs and wells. Cracks in wet ground and on steep slopes.
IX	General panic. Masonry D destroyed; masonry C heavily damaged, sometimes with complete collapse; masonry B seriously damaged. (General damage to foundations—CFR.) Frame structures, if not bolted, shifted off foundations. Frames racked. Serious damage to reservoirs. Underground pipes broken. Conspicuous cracks in ground. In alluviated areas sand and mud ejected, earthquake fountains, sand craters.
X	Most masonry and frame structures destroyed with their foundations. Some well-built wooden structures and bridges destroyed. Serious damage to dams, dikes, embankments. Large landslides. Water thrown on banks of canals, rivers, lakes, etc. Sand and mud shifted horizontally on beaches and flat land. Rails bent slightly.
XI	Rails bent greatly. Underground pipelines completely out of service.
XII	Damage nearly total. Large rock masses displaced. Lines of sight and level distorted. Objects thrown into the air.

FIG. 45.8 Mercalli Scale of Earthquake Intensities[45.17]

during an earthquake, can well be imagined; records of earthquake damage contain all-too-many examples (Fig. 45.9). During the 1906 San Francisco earthquake an "earth-flow in sandy soil [was observed] at the base of the San Bruno Mountains.... The sand and water forming this slide came out of a hole several hundred feet long and 46 m wide, flowed down the hill ... carried away a pile of lumber and knocked the powerhouse from its foundations."[45.21]

The importance of all these features to civil engineering design and construction will be clear. In seismic urban areas, local regulations usually govern the allowance to be made in structural design for the possibility of earthquake shock; a common rule is that a structure must be able to withstand a horizontal force caused by a horizontal acceleration of one-tenth that of gravity. In other areas, however, regulations will give no guide to the engineer, who will then have to study the past seismic history of the locality to see if any special features have been revealed. Local geology will be a necessary complementary study. The advisability of getting foundation beds of rock has been noted. Similarly, level ground should be selected in preference to undulating or hilly ground. Marshy ground should be particularly avoided, and the use of fill material should be kept to a minimum. Close vicinity of civil engineering works to natural cuttings such as riverbeds, canals, and coasts should be avoided so far as this is possible.

Many of the features that have been mentioned to show the interrelation of earthquakes and engineering works were dramatically demonstrated by the effects of the Hebgen Lake (Montana) earthquake of 7 August 1959. By a fortunate chance, two mem-

FIG. 45.9 Effect of soil liquefaction on a department store building in Niigata, Japan, following an earthquake in 1964.

bers of the engineering geology branch of the U.S. Geological Survey were at work in the vicinity, one above and one downstream from the Hebgen Dam adjacent to which the major damage was done. There are available, therefore, detailed accounts of the geological aspects of the earthquake to associate with the seismological studies made possible through correlative seismogram analyses. Hebgen Dam is an earth-fill structure on the Madison River in southwestern Montana; the structure has a concrete core wall, having been built in 1915 by a power company. It retains 370 million m³ (300,000 acre-feet) of water. The epicenter of the quake appeared to be close to the dam; major movement occurred along a well-recognized fault paralleling the lake formed by the dam and passing within 210 m (700 ft) of the right abutment. A vertical displacement of about 1.5 m (15 ft) took place on the fault. The dam was badly fractured and settled as much as 1.2 m (4 ft), but it did not fail, even though overtopped at least four times as a result of the major waves set up in the reservoir by the ground movement. Far more serious was the massive slide which took place 11 km (7 mi) downstream from the dam and which buried an unknown number of vacationers. At least 33 million m³ (43 million yd³) of rock fell into the river gorge, completely blocking the riverflow; the slide area was 1.6 km (1 mi) long, and the debris reached 120 m (400 ft) above the level of the stream bed. The U.S. Corps of Engineers quickly initiated emergency action, and a relief spillway was soon excavated;

45-16 SPECIAL PROBLEMS

FIG. 45.10 Hebgen Lake, Montana; visitors viewing the commemorative plaque at Hebgen Lake, formed by the earthquake of 1959, following dedication ceremony on 17 August 1970.

this permitted the river to resume its flow, but the slide and the lake it created remain as grim reminders of the damage that an earthquake can cause (Fig. 45.10).[45.22]

45.3 SEISMIC REGULATIONS

Earthquakes will continue to take place throughout the world; they cannot be stopped. Engineering and other structures will continue to be needed, even in areas where the possibility of earthquakes happening is known to exist. What can be done? Fortunately, today, a great deal, in view of the mass of information collected about earthquakes throughout this century and the many instrumental records obtained by seismic observatories around the world. Figure 45.7 shows the occurrence of major earthquakes around the world in 1970. It will be seen at a glance that, although earthquakes take place all over the world, they are to a large extent concentrated in certain well-defined areas such as the west coast of North and South America (the meeting places of continental plates). This pattern is repeated every year. It suggests that it should be possible to define, in a general way, areas in which earthquakes are more liable to occur than others.

It will have already been observed that it is almost impossible to describe major earthquakes even briefly without mentioning some of the engineering consequences as well as their geological significance. There is, indeed, a large overlap between the interests of both disciplines, but, in general, it is the task of the geologist-geophysicist to delineate the areas in which there is risk of seismic action, and of the engineer to develop design and construction methods which will ensure the maximum degree of safety for structures in those areas. The meeting place of the two disciplines is the joint effort required in delineating and classifying areas of seismic risk and then in developing the factors to be used in structural design to ensure maximum safety. These specific requirements for structural design and construction are today prominent features of the building regulations now in use in all developed parts of the world, administered by appropriate regulatory bodies such as municipal governments or state or provincial agencies. There are few such well-recognized regulations that do not now contain specific requirements with regard to seismic design.

As an example, the National Building Code of Canada may be cited. It is a nationally recognized set of building regulations, prepared by a representative national committee and in use, in a variety of ways, throughout the Dominion. Accompanying each revised edition of this code is a seismic map of Canada, the latest available being featured as Fig. 45.11. It will be seen that the whole country is divided into four numbered zones. Zone 0

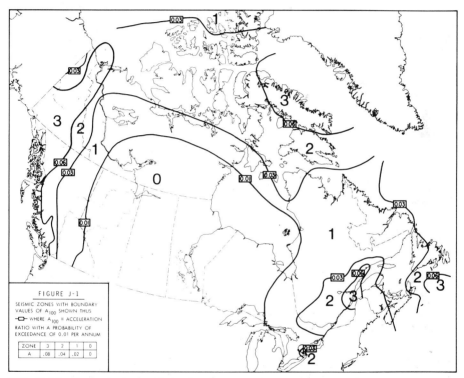

FIG. 45.11 Zones of seismic risk for Canada, featured in the *National Building Code of Canada,* 1980 Supplement (typical of guides that can be prepared from records of earthquakes).

indicates those parts of the country in which no earthquakes are to be expected. Zones 1 to 3 are areas with increasing probability of earthquakes occurring, with corresponding increasing probability of damage being experienced if an earthquake should occur. The map was prepared by scientists of the Canadian Department of Energy, Mines and Resources, with advice from a national committee of experts and in association with officers of the National Research Council of Canada.

The record of every earthquake that has ever happened in Canada and of which there is any written record was carefully examined; the first one dates from 1545. When all the records had been assembled, a computer study of a suitable sample was made to determine the probability of earthquakes occurring at selected locations once in 100 years.

45-18 SPECIAL PROBLEMS

From the results of this extensive analysis, the seismic map was prepared. A *seismic regionalization factor* has been developed for each of the three zones of possible activity. The values of this factor for all the main municipalities of Canada are given in one of the technical supplements to the Canadian *Code*. They are thus available for use by all responsible for the design of buildings and other structures throughout the Dominion. The values are inserted in the formulas that must be used to allow for earthquake loads in structural design, and they can be relied upon as being as accurate as is possible, based on the records of actual earthquakes of the last 300 years.[45.23]

There is another factor that must be used in building design, this being what is called the *foundation factor*. Experience throughout the world has shown that much more damage is caused to structures by earthquake shocks when they are founded on soil than when they are directly in contact with firm bedrock. An expert Japanese committee has suggested, based on the extensive experience that the Japanese have unfortunately accumulated in view of the long history of earthquake shocks in their country, that the effect of differing ground conditions can be expressed by the following factors:[45.24]

Marshy land	1.5
Alluvial ground	1.0
Diluvial (older and more consolidated) ground	0.7
Tertiary rock	0.4

There is still much uncertainty about the actual values that can be attached to different sorts of ground, even though the major difference between soil and rock is well attested. This is one of the many matters under investigation by seismologists and allied workers in all those countries that suffer regularly from earthquake shocks. The Canadian Code, noted above, uses the factor of 1.5 for "soils [of] low dynamic shear modulus such as highly compressible soils" and a value of 1.0 for all other soils.

The particular example of the Canadian *Code* was chosen since it features a seismic zone map in wide use throughout a large country; the principles which it embodies are international. There is, naturally, worldwide sharing about earthquake occurrences and earthquake engineering. Every major earthquake, no matter where it may occur, is now studied by teams of experts and the results are made generally available. Particularly valuable studies were made in California after the San Fernando earthquake of 9 February 1971. One of these showed that most of the surface rupturing occurred along well-recognized faults of the Santa Susana-Sierra Madre fault zone.[45.25] Another was concerned with damage to highways and bridges, the serious nature of liquefaction failures being notable.[45.26] Serious though the effects of earthquakes always are, each one now contributes to the store of world knowledge, and so to public benefit.

45.4 EARTHQUAKE PREDICTION

On 4 February 1975 an earthquake took place in Liaoning Province, China. Five hours *before* the quake took place, a general warning had been issued; shops had been shut in the town of Yingkow, and there had been a general evacuation in the counties of Yingkow and Haich'eng.[45.27] This dramatic event showed how far Chinese scientists have progressed in their major study of earthquakes. It reinforced the attention already being given to the possibilities of earthquake prediction in the United States, the U.S.S.R., and some other countries. The problem is most complex, but some progress appears to have been made.

VOLCANOES AND EARTHQUAKES   45-19

Attention is even being given to the "old wives' tale" that some animals will, by their behavior, give clear indication of a coming earthquake. Studies are being aided by the development of small seismic shocks by human activity, as can now be done.

It seems probable that, eventually, it should be possible to predict with some degree of accuracy the pending occurrence of a major earthquake. But when this problem is solved, it will bring others in its train. Who is to be responsible for the mass evacuation that should, presumably, be made? If this is not done under strict control, traffic problems may prove to be far more serious than any problems the earthquake might create. And since earthquakes are no respecters of locale, who will be able to say that it will be safer to be on a highway or in open country (if you can get there) than in a well-designed and constructed seismic-proof building? And what if the predicted earthquake does not take place? Will not this have most serious consequences for any further attempt to get people to move when it is again thought necessary? The problems are endless! They all lie outside the scope of this volume, but they require mention in the hope that efforts of geologists and engineers in the direction of even safer seismic-proof buildings will not be relaxed.

45.5. MAN-MADE EARTHQUAKES

Mention of man-made earthquakes in the preceding section may be thought by some readers to be an aberration, but it is not. The Hoover Dam on the Columbia River was completed in 1935. Precise leveling has shown that the land around the lake formed by the dam subsided in places as much as 17.5 cm (7 in) in the first 20 years of operation. Not surprisingly, the change in the state of stress in the underlying rocks caused by the huge new load of water (estimated to be 35 thousand million tons) and evidenced by the subsidence, created minor seismic shocks, the first being felt in 1936, and the first being recorded in 1938 after seismographs had been installed around the reservoir. Since then, a very large number of minor shocks (the largest of magnitude 4 on the Richter scale) have been recorded.[45.28] The same experience has been found around some other major reservoirs, such as that formed by the Kariba Dam in southern Africa and that by the Kremastá Dam in Greece. These events serve to emphasize the absolute necessity for detailed geological study of the area to be covered by the waters of all major reservoirs to ensure that, if minor seismic shocks are thus induced, no serious damage will result. It is now common practice to install a small network of seismological stations around such major reservoirs as a part of the permanent installation.

Rather more remarkable was the experience at the Rocky Mountain Arsenal of the U.S. Army in 1962 and the years immediately following. Liquid wastes from the manufacturing processes in this plant seeped downward from receiving ponds into a shallow aquifer. This seepage appeared to have contaminated groundwater to such a degree that apparently damage was done to some crops irrigated by the water. As an alternative solution, a lined reservoir 43.3 ha (107 acres) in extent was constructed, together with necessary observation wells all around it to ensure its safe operation. Its capacity was limited, however, and so to accommodate the extra liquid wastes, a very deep wastewater well was sunk. Extending to a depth of 3,420 m (11,200 ft) below the surface, the well is believed to be the deepest ever sunk for such a purpose. It was designed to dispose of the waste liquid by seepage into the 300-m (1,000-ft)-thick Fountain formation of porous sandstone that is overlain by an impermeable stratum of shale, 900 m (3,000 ft) thick, which would act as a safe seal. The well extends to the underlying Precambrian granite and starts with a diameter of 60 cm (24 in) at the surface. Together with a treatment plant to remove colloidal solids from the liquid before it was put down the well, the facility was estimated to have cost over $1 million. It went into operation in March 1962. In April of the same year, observations disclosed the start of a series of minor earthquake shocks in the Denver area, over 700 being registered by November 1965. These were attributed to the effect which the vast quantities of liquid being pumped into the well under pressure had on the complex rock structure into which the well had been drilled. At times the quantity was almost 11.4 million L (3 million gal) per month. Many difficult questions arose when this unusual situation was studied in depth, but there was no doubt about the correlation of the pumping and the incidence of the earthquake shocks. Pumping into the well was stopped in February 1966.[45.29] As is often the case, this rather unfortunate experience had beneficial results since the same idea, under strict control, has been used as part of the experimental program for earthquake prediction.

Not too far removed from such man-made seismic shocks is the possibility of detecting incipient failures in excavations in rock by the use of very sensitive instruments, able to pick up the microseisms that are then often created. The Seismitron is one excellent proprietary instrument specially developed for this purpose.[45.30] It can be seen that all the topics discussed in this chapter are closely interrelated, from the largest of earthquakes to the smallest microseism, all being demonstrations of the reaction of solid rock to stresses and consequent strains. The "solid earth" is solid indeed, although it is not composed of any mystical materials but of quite ordinary matter, the limited properties of which can never be forgotten in any civil engineering work.

45.6 TSUNAMI

Figure 45.7 shows that earthquakes, quite naturally, occur beneath the sea as well as on dry land. Earthquakes on land and near seacoasts will inevitably affect the solid ground beneath the sea within the radius of influence from their epicenters. Underwater disturbances may create severe upheavals of the seawater in the vicinity, and these will form great surface waves. These are known by the Japanese name *tsunami*, a much more accurate title than the name *tidal wave* that one still sometimes hears. Quite unlike the ordinary waves on the ocean, tsunami may extend for 160 km (100 mi) in width and be as much as 1,000 km (600 mi) from crest to crest. In the deep waters of the Pacific Ocean they can travel at the amazing speed of 800 kmh (500 mph). Since they may be from 6 to 18 m (20 to 60 ft) in height, their devastating effect when they hit a shoreline is not difficult to imagine. Because their effect has been observed by skilled scientists, a good deal is known about their action.

The *tsunami* caused by the Good Friday Alaskan earthquake struck all the adjoining coastline of Alaska with destructive force. They were recorded on tide gauges as far away as Japan, Hawaii, and California, where damage was caused. The *tsunami* created by the eruption of Krakatoa in Indonesia in 1883 raced across the Pacific and created heavy seas even as far away as San Francisco. They are much more common in the Pacific Ocean than elsewhere, seldom occurring in the Atlantic Ocean. Japan has been subjected to 15 destructive *tsunami* since 1596, eight of them being disastrous. The Hawaiian Islands are subjected to a serious *tsunami* about once every 25 years. A particularly bad one struck the Islands in 1946 (Fig. 45.12); it was fortunately observed by a number of scientists and led directly to a study of the possibility of detecting *tsunami* in advance of their reaching a coastline. An ingenious instrument was developed by an officer of the U.S. Coast and Geodetic Survey, and a warning system centered in Honolulu was established to correlate the results obtained from the instruments that have now been installed at critical Pacific locations. There have been 15 *tsunami* in the Pacific since 1946, but only one (in 1952) of any serious magnitude. The warning system worked well. Not a life was lost, and damage in the Hawaiian Islands amounted to only about $800,000. It has been stated that "there is little doubt that the warning system saved lives and reduced the damage."[45.31]

FIG. 45.12 Hilo, Hawaii, damage done to residential buildings by the *tsunami* of 1 April 1946.

As with earthquakes, however, *tsunami* will continue to occur; they cannot be stopped, even though they may be detected in advance. As was shown in 1952, this can lead to the saving of lives, but nothing can save fixed property from the inexorable force of the sea. Urban development close to shorelines on any ocean coast must, therefore, be considered in relation to the possibility of damage by *tsunami*. If this is a probability, buildings must be firmly anchored to the ground or raised above it, if indeed, they have to be placed close to the shoreline and cannot be relocated on higher ground. One of the most percipient questions that one of the authors ever received in the course of his work in directing building research in Canada was about the possibility of *tsunami* in the Atlantic Ocean; the question came from a planner in Newfoundland, charged with responsibility for developing the new fishing villages that have been created in "the oldest Dominion" by the amalgamation of isolated small groups of fishermen's homes. This intriguing social operation often meant floating complete houses across some of the many bays that feature the rockbound coast of the island to a carefully selected site for a new village.

The few records of *tsunami* in the Atlantic were then traced; the most serious *tsunami* was apparently that caused by the Lisbon earthquake of 1 November 1755, the waves following which persisted for about a week and reached as far as England. Despite the low risk, all buildings in the new villages of Newfoundland are located so that they are well above high-tide level. This is fortunate since, in studies of the history of seismology in Canada, Dr. Anne Stevens found that there had been a serious *tsunami* in recent years on the Burin Peninsula on the south shore of Newfoundland. On 18 November 1929, a major earthquake with a magnitude of 7.2 took place in the Grand Banks of this part of the Atlantic, about 270 km (170 mi) south of the peninsula and 400 km (250 mi) south of the capital city of St. John's. It was felt throughout eastern Canada and in the New England states.[45.32] The resulting *tsunami* reached the coastline almost exactly at high tide, striking with great force the southern end of the Burin Peninsula and the west side of Placentia Bay. With no warning at all, 27 lives were lost. Actions of great heroism were common. Almost 100 buildings were destroyed and untold damage was done to fishing craft and equipment. So "it can happen here" is an old tag not to be disregarded by any residents living close by the sea.

45.7 CONCLUSION

Those words had just been written when Mount St. Helens erupted, attracting the attention of the world and bringing home to all in North America that active volcanoes are not found only in countries far away. Preliminary warnings enabled preparations to be made for necessary evacuations, but lives were still lost, unfortunately, and physical damage was widespread and of great extent. Figure 45.1 is a vivid picture of the eruption as seen from the air. Figure 46.2 is less dramatic, but it shows the aftermath of the main eruption, the devastation of a beautiful valley. Scenes of similar havoc were to be seen around much of the mountain.

When one sees such photographs of an occurrence of our own time, it becomes a little easier to appreciate what some of the great eruptions of the past must have been like. Human imagination, however, is still insufficient to enable one to grasp what the prehistoric eruption of Santorin must have been like. The remains of the small island, lying due north of Crete, are still beautiful, but the wide bay around which one may sail today (29 km or 18 mi around the inner rim) is probably the crater of the eruption, now believed to have taken place in the Bronze Age, which "blew the head off" the island. The *tsunami* which resulted must have caused devastation all round the Greek islands. It is now thought that quite possibly the rain of volcanic ash settling on Crete—100 km (60 mi) away—may be the explanation of the disappearance of the Minoan civilization. This is a sobering possibility. It serves well to bring this brief review to a close.[45.33]

45.8 REFERENCES

45.1 C. A. Cotton, "For How Long Will Wellington Escape Destruction By Earthquakes?" *New Zealand Journal of Science and Technology*, **3**:229 (1921).

45.2 G. Larsen, K. Gronvold, and S. Thorarinsson, "Volcanic Eruption Through a Geothermal Borehole at Namafjall, Iceland," *Nature*, **278**:707 (1979).

45.3 L. Yokoyama, "Pozzuoli Event in 1970," *Nature*, **229**:531 (1971).

45.4 D. R. Crandell, *Map Showing Potential Hazards from Future Eruptions of Mount Ranier, Washington*, Map L-836, U.S. Geological Survey, Washington, 1973.

45.5 J. M. Cullen, "Impact of a Major Eruption of Mount Ranier on Public Service Delivery Systems in the Puyallup Valley, Washington," Departments of Geography and Urban Planning, University of Washington, Seattle, 1978.

45.6 R. Mallet, *The Great Neapolitan Earthquake of 1857*, 2 vols. 1862.

45.7 J. T. Wilson, "Mao's Almanac: 3,000 Years of Killer Earthquakes," *Saturday Review*, **54**:60 (19 February 1972).

45.8 C. R. Longwell, A. Knopf, and R. F. Flint, *A Textbook of Geology: Part I, Physical Geology*, Wiley, New York, 1932, p. 296.

45.9 W. E. Kunze, M. Fintel, and J. E. Amrheim, "Skopje Earthquake Damage," *Civil Engineering*, **33**:56 (December 1963).

45.10 H. J. Degenkolb, "Seismic Tremor [Shook] Caracas Severely," *Engineering News-Record*, **179**:16 (24 August 1967); the American Iron and Steel Institute has published a more detailed account of this earthquake in volume form.

45.11 D. B. Slemmons, "Structural and Geomorphic Effects of the Dixie Valley-Fairview Earthquakes of December 16, 1954, Churchill County, Nevada," in abstract vol. for annual meeting in New Orleans, Geological Society of America, Boulder, Colo., 1955, p. 94A.

45.12 R. F. Legget, "Earthquake Damage at Cornwall," *The Engineering Journal*, **27**:572 (1944).

45.13 E. B. Eckel, *The Alaska Earthquake, March 27, 1964: Lessons and Conclusions,* Professional Paper no. 546, U.S. Geological Survey, Washington, 1970.

45.14 J. S. Scott and F. W. Render, "Effect of an Alaskan Earthquake on Water Levels in Wells at Winnipeg and Ottawa, Canada," *Journal of Hydrology*, **2**:262 (1964).

45.15 R. D. Miller and E. Dobrovolny, *Surficial Geology of Anchorage and Vicinity, Alaska*, Bulletin no. 1093, U.S. Geological Survey, Washington, 1959.

45.16 E. Dobrovolny and H. R. Schmoll, "Geology as Applied to Urban Planning: An Example from the Greater Anchorage Area Borough, Alaska," *Proceedings of the Twenty-Third International Geological Congress*, sec. 12, 1968, p. 39.

45.17 C. F. Richter, *Elementary Seismology*, Freeman, San Francisco, 1958.

45.18 F. W. Furkert, "The Effect of Earthquakes on Engineering Structures," *Minutes of Proceedings of the Institution of Civil Engineers*, **235**:344 (1933).

45.19 "Arvin-Tehachapi Earthquake," *Mineral Information Service*, vol. 5, no. 9, San Francisco, September 1952.

45.20 H. B. Seed and I. M. Idriss, "Influence of Soil Conditions on Building Damage Potential During Earthquakes," *Proceedings of the American Society of Civil Engineers*, Paper no. 7909, **97**:(ST2):639 (1971).

45.21 H. B. Seed, "Soil Liquefaction: Cyclic Mobility and Evaluation for Level Ground," *Proceedings of the American Society of Civil Engineers*, Paper no. 14,380, **105**:(GT2):201 (1979).

45.22 J. L. Sherard, "What the Earthquake Did to Hebgen Dam," *Engineering News-Record*, **163**:26 (10 September 1959); see also K. R. Barney, "Madison Canyon Slide," *Civil Engineering*, **30**:72 (August 1960).

45.23 *National Building Code of Canada 1980*, 8th ed., National Research Council, Ottawa, 1980; see suppl. P.

45.24 *Report of the Special Committee for the Study of the Fukui Earthquake, 1948*, National Research Council of Japan, 1950.

45.25 R. J. Proctor, R. Crook, M. H. McKeown, and R. L. Moresco, "Relation of Known Faults to Surface Ruptures, 1971 San Fernando Earthquake, Southern California," *Bulletin of the Geological Society of America*, **83**:1601 (1972).

45.26 A. L. Elliot, "The San Fernando Earthquake: A Lesson in Highway and Bridge Design," *Civil Engineering*, **42**:95 (September 1972).

45.27 R. M. Hamilton, "The Status of Earthquake Prediction," *Earthquake Prediction—Opportunity to Avert Disaster*, Circular no. 729, U.S. Geological Survey, Washington, 1976.

45.28 J. P. Rothe, "Fill a Lake, Start an Earthquake," *The New Scientist*, **39**:75 (11 July 1968).

45.29 D. M. Evans, "Man-Made Earthquakes in Denver," *Geotimes*, **11**:11 (May–June 1966); see also, in same journal, **13**:19 (July–August 1967).

45.30 F. D. Beard, "Microseismic Forecasting of Excavation Failures," *Civil Engineering*, **32**:50 (May 1962).

45.31 J. Bernstein, "Tsunamis," *Scientific American*, **191**:60 (August 1954), a good general review.

45.32 "Earthquake and Tidal Wave Disaster," *The Newfoundland Quarterly*, **29**:20 (December 1929); see also, in same journal, G. Jones, "The South Coast Disaster of 1929," **71**:35 (January 1975).

45.33 L. Durrell, *The Greek Islands*, Faber, London, 1978.

Suggestions for Further Reading

Despite their relative infrequency, volcanic eruptions received considerable attention in the seventies and eighties, even before the eruption of Mount St. Helens. This is indicated by the number of recent books on the subject. One of the first of these was *Volcanoes and Earthquakes: Geological Violence* by G. B. Oakeshott, published in 1979 by McGraw-Hill. 1981 saw two more books on the subject published, one being *Volcanoes* by R. and B. Decker (244 pages) published by Freeman of San Francisco; the second, *Volcanoes of the World* by T. Simicin, L. Siebert, L. M. McClelland, D. Bridge, A. N. Ewhall, and J. H. Latte, was published by Academic Press of New York. A comprehensive (844-page) report on *The 1980 Eruption of Mount St. Helens, Washington*, by P. W. Lipman and D. R. Mullineaux, has been published by the U.S. Geological Survey (USGS), who had previously published (as Bulletin 1383 C) *Potential Hazards from Future Eruptions of Mount St. Helens Volcano* by D. R. Crandall and D. R. Mullineaux, a report which was issued in 1978!

45-24 SPECIAL PROBLEMS

Alaska's Good Friday Earthquake of March 27, 1964: A Preliminary Geologic Evaluation by D. A. Grantz, G. Plafker, and R. Kachadoorim, Circular C491 of 1964, is another USGS publication released prior to the major report referred to above. The Geological Survey is naturally very active in the field of earthquake studies, maintaining a special Office of Earthquake Studies and issuing a regular *Earthquake Information Bulletin*.

"Earthquake Prediction" was treated by Frank Press in a percipient paper with this title in *Scientific American* for May 1975 (pages 14–23). An abstract of an early review of "man-made earthquakes" was published in *Bulletin of the Geological Society of America* **67**:1763 under the title "Reservoir Loading and Local Earthquakes in the Lake Mead Area," the authors being D. S. Carter and F. H. Werner (of USGS). Circular 688 (1973) of the Survey, by T. C. Youd, deals with *Liquefaction, Flow and Associated Ground Failure*.

As this Handbook goes to press, there have already been seven World Conferences on Earthquake Engineering, the proceedings of which are of great value. The eighth conference is to be held in San Francisco in July 1984, the host organization being the Earthquake Engineering Research Institute of Berkeley, California. The active work of this institute is yet another indication of the attention now being given to earthquakes.

Probabilistic and statistical studies are part of this widespread activity, as are the public risks involved. Indicative of these more general studies are two publications: an Open File Report (No. 76-416) of the USGS in 1976 by S. T. Algermissen and D. M. Perkins called *A Probabilistic Estimate of Maximum Accelerations in Rock in the Contiguous United States*, with an accompanying map, and *Natural Hazard Risk Assessment and Public Policy: Anticipating the Unexpected* by W. J. Patak and A. A. Atkisson, a 300-page book published in 1982 by Springer-Verlag of New York.

chapter 46

FLOODS

46.1 *Jökulhlaup* / 46-2
46.2 Mud Runs / 46-3
46.3 Catchment Areas / 46-4
46.4 Causes of Floods / 46-6
46.5 Flood Protection / 46-8
46.6 Special Flood-Control Measures / 46-10
46.7 Floodplains / 46-14
46.8 Conclusion / 46-16
46.9 References / 46-17

Of all the natural catastrophes that can beset mankind, major floods are in some ways the most distressing. One can help to fight a fire; damage is done by an earthquake, but reconstruction work can start quickly; a tornado passes and causes great damage, but quiet is quickly restored. But to see a great river burst its banks, floodwaters creeping ever higher and higher with nothing to be done until the peak is passed, this tries the endurance of even the most stoical among us.

Storms along seacoasts are a part of living by the sea. Tsunami can do great damage, as was seen at the end of the last chapter. But the sea can cause trouble to urban communities located on seacoasts quite apart from the rare occurrence of tsunami. When high spring tides coincide with violent winds, the combination can cause phenomenal highwater levels that, coupled with the waves created by the high winds, can lead to disaster conditions. The eastern side of the great delta of the Ganges, then in eastern Pakistan, was devastated on 12 November 1970 by just such a combination of typhoon winds and high tides. The catastrophe may well have been the most serious disaster of its kind in history. Early estimates placed the number of dead at over a million, but the full extent of the tragedy will never be known. All who have flown over this heavily populated, flat-lying area, intersected by the seemingly innumerable channels into which the great river divides, will long carry in their minds the thought of what the disaster must have meant to those living on their little farms on the many islands, with no possibility of escape in the face of such mighty natural forces. And the use of such areas for settlement can really only be understood by those who have seen what the concentration of population in the Indian subcontinent means in human terms.

46-2 SPECIAL PROBLEMS

Similar combinations of extremes of natural occurrences in other parts of the world can usually be predicted and necessary precautionary measures taken. Throughout North America, citizens are now used to reading reports of hurricanes forming over the Caribbean, with disaster warnings issued well in advance by the responsible authorities to all who live in the coastal and inland areas that may be affected. Evacuations are now accepted as essential under extreme conditions. One hundred fifty thousand people were evacuated from the delta area in Louisiana in 1965, for example, because of the onslaught of Hurricane Betsy. The evacuation is said to have probably saved 50,000 lives, since the whole area was completely inundated. These unusual combinations are not limited to tropical areas, even though it is in such areas that hurricanes develop. The North Sea has an unenviable reputation for bad storm conditions, and these have combined with high spring tides to wreak disaster on European coasts. History records one such tragedy in the year 1421, but the much worse inundation that occurred on 1 February 1953 will never be forgotten by those on the southeast coast of England and, more particularly, those in the coastal area of the Netherlands who lived through that critical period; 133 Dutch villages were seriously damaged and 1,800 people died. Damage was estimated to be at least $400 million. Dikes were breached and seawalls on the coasts of England badly damaged (Fig. 46.1). Reacting with characteristic courage, the Dutch people passed legislation within 20 days of the flood, initiating the great Delta project that has been under construction since that time and that will ultimately benefit the whole of this most interesting country.[46.1]

FIG. 46.1 Harwich, England; an indication of the damage wrought by the North Sea flood of 1 February 1953; a Royal Navy diver is about to descend in search of a manhole cover through which drainage might be assisted.

46.1 JÖKULHLAUP

The Delta Plan has been mentioned earlier in this volume; so also have the catastrophic floods that can result from sudden failure of ice that has retained water on the surface of a glacier. Such repetition is unavoidable and even desirable, if only to lend support to the *geological* significance of floods and the responsibilities that this involves. The failure of ice that retains water in lakes on glaciers can take place in a number of ways, but melting of critically located ice is usually the root cause. In Scandinavia this is termed *jökulhlaup* (an Icelandic word). Whatever it is called, the phenomenon should be known and understood by all concerned with any development in the foothills below glacier-covered mountains. A tragic example of what can happen is the mud-water avalanche that took place on 10 January 1962, due to failure of a glacier on Peru's highest mountain, devastating the high Huaylas Valley in the Andes, destroying several villages, and taking over 4,000 lives.[46.2] An even worse disaster took place in 1970 in the valley of the Río Paria, killing probably 50,000 people and destroying much of the old Spanish colonial city of Huaráz.

Peru has now established the Corporación Peruana del Santa, with its headquarters at Carhuaz. Annual surveys of glacial lakes are undertaken, and any known hazardous conditions are carefully investigated and, if possible, ameliorated before any real danger develops. Dynamiting the exits from glacial lakes and even tunneling to lower the level of critical lakes are two measures already used. It can be seen, therefore, that when natural hazards even as serious and unpredictable as the *jökulhlaup* are recognized, measures can be taken to provide warning to all those living below the danger zones so that the necessary precautionary measures can be taken. Fortunately, such areas are rare throughout the world, but they warrant this attention here since they demonstrate so vividly the fact that geological conditions are not permanent and static, but dynamic.

46.2 MUD RUNS

It is not only these violent discharges from glacial lakes that can precipitate debris, as well as water, on to the flat areas on to which they eventually discharge. Rivers and streams in more normal terrain can be equally destructive if their watersheds permit them to pick up large loads of debris, usually denoted as *mud,* since heavier particles will quickly be dropped as stream velocity decreases, whereas finer soil particles can be carried great distances.

Pictures of the devastation of the beautiful city of Florence and many of its art treasures by the disastrous flood on the river Arno in November 1966 will be vivid recollections for all art lovers. Many readers may have had personal experiences of the mess that can so quickly be caused by mud deposited by floodwaters where it is not wanted. All these occurrences are reminders of the transporting power of running water and of the fact that this movement by water of soil and rock is an important part of the geological cycle.

There are some parts of the world where such mass movements of solid material occur regularly, and not in the orderly and controlled way featured for so many centuries in the Nile Valley. In India and in South America *mud runs,* as these mass movements are usually called, are of common occurrence. So common are they in certain parts of South America that they have been given the local name of *huaicos.* If they take place in open country they may do little harm, but if they encounter in their path any obstruction such as bridge piers or other man-made works, real trouble may result. If they flow as far as developed urban areas, the damage they cause may be considerable. Since there must naturally be a convenient source of disintegrated rock or soil to be moved, mud runs are usually encountered in hilly or mountainous country. Landslides, large and small, are a potent source of "mud," especially if they block the normal flow down watercourses.

In July 1953 a remarkable mud run took place 64 km (40 mi) to the northwest of Needles, California, following a series of small landslides near the Von Trigger Spring.

FIG. 46.2 Ruin of a lovely valley following the eruption of Mount St. Helens, a view taken on 4 June 1980.

Streams fed by the heavy rainfall carried material from the toes of the slides out of the Lanfair Valley and downstream beyond the Von Trigger Spring for a total distance of at least 72 km (45 mi). Although not observed, it is believed that the mud run extended as far as Cadiz Dry Lake, which would have meant a total travel of almost 100 km (62 mi).[46.3] This case is mentioned, even though no serious damage was involved, to indicate how far mud runs extend at times of high flood. In Japan, the local volcanic soils provide in many places ideal material for mud runs. The Nirasaki mudflow is a well-known feature extending from the slopes of the volcano Yatsuga-take to the town of Nirasaki, a distance of 20 km (12 mi), and it is regularly replenished in times of high rainfall.[46.4]

46.3 CATCHMENT AREAS

Although the victims of flooding, and those responsible for associated emergency relief and rehabilitation measures, naturally think only of the floodwaters they can see, all overall considerations of flooding must consider the whole watershed of stream or river and in a framework of time. It is in order to emphasize this obvious—but so frequently neglected—factor in flooding that the foregoing unusual cases have been presented first. In these cases, just as in what may be called "normal floods," the catchment area must be considered in association with rainfall for a full assessment of flood problems.

A catchment area that was formed of a continuous exposure of impermeable rock would yield at any gauging station in the watercourse a runoff equivalent to the rainfall on the total area above the station less the amount lost by evaporation. In the usual catchment area, however, other losses must also be considered; the most important of these are the water that is absorbed by vegetation, the water that is temporarily absorbed into the underground water reservoir formed by any pervious rocks underlying the area, and the water that may be deflected into adjacent catchment areas because of peculiarities of geological structure. The calculation of rainfall losses is a vitally important part of the hydrological study done in connection with all engineering works dependent on water. Studies of the matter are to be found in many standard works, and many empirical formulas have been evolved to suggest a general relation between rainfall and runoff for catchment areas of different sizes and different climatic conditions. Of the factors affecting rainfall losses, two of those mentioned are fundamentally dependent on geological structure. It will therefore be seen that although general formulas are undoubtedly useful in limited application, they must be interpreted in the light of information on the local geology if they are to be effective and of use.

Loss of rainfall by percolation to underground pervious strata is the first of these two factors. Consideration of this possibility leads inevitably to a general study of groundwater. In all but the most exceptional watersheds (such as those used at Gibraltar), some portion of the rain will fall on pervious material and so sink below ground surface. There it will eventually join that large body of water which is permanently held in the interstices existing in the underground strata, taking its part in the slow but steady movements of the contents of this underground reservoir. The determination of the minimum possible dry-weather flow and the probable maximum flood at any gauging station is a critical hydrological investigation, since it will often be impossible to measure them by actual gauging. Records at the sites of works that have been constructed can usually be obtained, and these constitute a valuable guide to similar cases. It is in determining the similarity of catchment areas in this connection that geological structure must be considered.

D. Halton Thomson many years ago compiled a study of more than 30 records of various dry-weather flows for British streams, and these show discharges varying from 0.004 cms per 1,000 ha (0.055 cfs per 1,000 acres) on the river Alwen in north Wales (drainage area 2,520 ha or 6,313 acres) to 0.14 cms (2.00 cfs) gauged on the river Avon, Scotland (drainage area 2,020 ha or 5,000 acres). The ratio of the two records from these catchment areas, almost identical in respect to topography and climate, is 1:36. This is a surprising variation, and the explanation is wholly geological. Superficially, both areas might be regarded as impervious; the geology of this part of the Alwen Valley consists of Silurian shales and grits covered with boulder clay; that of the Avon is wholly granite. Based on this general comparison, it would seem that the dry-weather flows from the two areas should be almost equal. Further study discloses the fact that although the Alwen catchment area is watertight, as might be expected, and thus provides little or no underground storage, the granite over which the Avon flows is so fissured and decomposed that vast quantities of water are held in its interstices. The ample underground storage thus provided is drawn upon by the stream during persistent dry periods.[46.5]

Another interesting example is afforded by flow records at two different points on the river Exe in Devonshire, the lower point just above the range of tidal influence and the other near the headwaters of the river above the junction of two important tributaries with the main stream. The unit dry-weather flow for the whole area is almost four times that of the upper part of the area alone; the reason for the variation is, again, solely geological. About one-eighth only of the upper part of the area is formed of pervious rocks, the New Red sandstone, and the remainder is Devonian measures and igneous rocks; thus practically no underground storage exists. The basins of the two tributaries that join the main river below the upper gauging station are largely formed of the pervious New Red sandstone, and in consequence, they are both well served with underground storage to the extent that their unit dry-weather flows are so high that they raise the total flow figure for the whole catchment area. (Comparative figures are given in Table 28.1.)[46.6]

The examples quoted are by no means unusual, and they therefore illustrate vividly the effect that geological conditions have on riverflow. This effect is felt to varying degrees for all stages of streamflow. Even during torrential rainfall, a pervious surface stratum will rarely be so completely saturated with water that rain falling on it will all run off, giving a flow equivalent to 100 percent of the rainfall, but it is possible.

This is confirmed by the unique conditions that combined to cause the catastrophic floods early in 1936 in the eastern part of North America. Unexpectedly early rains fell on catchment areas that were still frozen hard and that thus constituted catchment areas almost "ideal" if considered from the standpoint of impermeability. The resulting runoffs caused floods that broke all known records and in some cases exceeded the calculated "1,000-year maximum." That such floods do not normally occur is due to the regulating effect of geological formations, which determine to some degree the course of water from the time it falls as rain until the time it reaches the watercourse to join with the streamflow. The importance of this effect can hardly be overemphasized. The figures quoted illustrate how geological conditions alone can explain some variation in runoff records. By means of similar studies, not only can records be compared, but the accuracy of recorded flows can be checked as possible limiting cases.

46.4 CAUSES OF FLOODS

Every flood has some unique aspects, necessarily so since Nature is involved. It is possible, however, to delineate the main causes of floods, which always result from a combination of several elements. The role of geology of the catchment area involved should now be clear. But the surface area of the catchment can be affected by climatic factors, frost in the ground, snow on the surface, and the relative position of the water table. Precipitation, in the form of both snow and rainfall, is the vital factor, not only its amount but also the duration of its fall (in the case of rain) and the possible combination of rainfall and snowmelt in spring periods. These are all natural factors, operative whether humans are resident in the catchment area or not. Man, has, however, introduced another disturbing element by cutting down vegetation, notably trees, and so facilitating runoff from the ground surface. There are some who, with an excess of zeal for conservation, suggest that human activity is entirely responsible for modern floods. Historical records show clearly, however, that this is not the case, although there can be no doubt but that human activities have intensified the flood problem. In all basic studies of floods, it is the overview that is so necessary, all factors being considered and weighted in the search for the best solution to the "flood problem" for any particular watershed (Fig. 46.3). The cost and extent of so-called "structural flood-protection measures" are alone warrant for the detailed study that flood problems are now receiving, quite apart from the possible danger to human lives. The following outline is suggestive only, a guide to the major factors to be considered in all local flood investigations.

The fact that climate has called for consideration yet again in this Handbook may first be noted, in view of the general disregard of this important natural feature. Serious springtime floods in the northeastern part of North America will often be found to be related to the timing of the first frost in the ground. If the first frost comes, and persists, before the first snowfall, the ground surface may remain frozen throughout the winter, leading to rapid runoff in the spring when the snow melts. Such a condition is especially serious for any watercourse of the northern hemisphere that runs northward, the Red River (of the North) being an almost classical example. Rising in the Dakotas, the river runs almost due north through Manitoba, discharging into Lake Winnipeg. It has been the scene of most serious floods, intensified by the fact that warmer temperatures may lead to runoff in its upper sections while its lower reaches are still frozen (Fig. 46.4).

FIG. 46.3 Aerial view of some of the damage at Putnam, Connecticut, caused by the Quinebaug River in August 1955, following unusual heavy rainfall; the normal river channel is to be seen in the lower left corner.

FIG. 46.4 Typical street scene in suburban Winnipeg, Manitoba, during the 1950 flood of the Red River.

Far greater are the corresponding problems on the Mackenzie River of northwest Canada, seventh largest river of the world. It rises in the Rocky Mountains to the west of Edmonton but discharges into the Arctic Ocean in the far northwest corner of Canada. Its headwaters are, therefore, "running full" in the spring while the lower stretch of the river is still hard-frozen. Thus, in some of its lower reaches, the normal high-water level in the spring is always 18 m (60 ft) above normal river level, remaining at this level until "the ice goes out," as it does quite suddenly. This condition is well recognized by the relatively few residents of that part of the Canadian north. Preparations are made accordingly, all buildings in the affected stretches of the great river being mounted on skids and moved up, well above the high-water level, when the open season of navigation closes in the preceding fall.

The depth of snow lying on the ground of the northern parts of North America as the spring approaches will determine, when combined with springtime air temperatures, the rate and amount of spring runoff. Regular snow courses are now surveyed, at intervals each winter in northern watersheds liable to serious flooding so that estimates of spring runoff can be made. The same precaution cannot be taken with rainfall, which has to be accepted when it comes. But the state of the ground on to which rain falls can be, and must be, studied in critical watersheds as spring approaches, to see what the groundwater conditions are. Here the valuable work of Thornthwaite again calls for appreciative mention, the groundwater-depletion concept being vital in flood prediction. Longtime rainfall records permit preparation of statistical estimates to show the probable "once-in-100-year" rainfall (or a corresponding figure) for use in all engineering designs which impinge on waterways. It is impossible, however, to estimate in advance the rate at which rain will fall. That this can be a serious factor was shown by the floods in early May 1973 in the Greater Denver area, a valuable account of which has been published.[46.7] Here up to 11.13 cm (4.38 in) of rain (a new 100-year record) fell over the Denver area, an extreme amount in itself, but it fell gradually, starting as a light drizzle on a Saturday evening, the rain falling steadily all the next day and so saturating the ground. Extensive damage was done to property by the resulting flood.

The Denver flood was due to ordinary rainfall. Far worse floods have been due to the catastrophic rainfall that often accompanies the hurricanes that are experienced almost every year on the mainland of North America. Their courses are now traced and so warnings are regularly issued, with consequent less risk of loss of life, although some deaths do still unfortunately occur. The magnitude of the flood flows cannot, however, be estimated. Damage can be very serious indeed, despite the precautionary measures that the early-warning system makes possible. Hurricane Agnes in 1972 produced what was probably the greatest disaster of its kind ever to be inflicted upon the United States, with over 100 lives lost, 387,000 people evacuated from their homes, and an estimated cost of over $3 billion in physical damage.[46.8]

46.5 FLOOD PROTECTION

Discussion reverts here to "normal" floods, while accepting (reluctantly) the near-impossibility of providing protection against the ravages of hurricanes. It is understandable that from the earliest times of settlement, efforts have been made to minimize damage from floodwaters in urban areas adjacent to watercourses. Straightening stream beds liable to flooding may give some relief to adjacent areas, even though it may increase storm damage downstream. Erection of protective dikes, to confine streams within limits, is a comparable development, effective if well and carefully designed but liable to cause upstream damage if not. Building of small dams to retain water in flood time above developed areas was a logical development, effective in part if such dams are well built and efficiently operated, but possibly disastrous if not. These simple, "amateur" ideas have been developed down the years into major structural flood-protection measures, such as the great dike systems along the banks of the Mississippi, the concrete floodwalls in many important city areas, and the large retaining dams in the headwaters of rivers subject to regular flood flows. Some of the structures described earlier in this Handbook were built for flood-control purposes, or as multipurpose dams, one function of which is provision for storage of floodwaters to reduce downstream flows.

Although today it is generally recognized that such structural measures can be only a part of the full answer to flood protection, there has been much argument down the years between those who favored these measures and those who insisted that "little dams" and soil water conservation measures in the watershed generally were the real answer. Both sides were right, to varying degrees in different locations, the known occurrence of major floods before any land development showing that the one without the other may not be enough. The importance of "watershed engineering," however, cannot now be questioned. One of the major pioneer efforts in this direction was the establishment of the Muskingum Watershed Conservancy District in Ohio in 1933 and the associated leadership provided by A. E. Morgan, the District's chief engineer. It comprises 18 counties, amounting to about one-fifth of the land area of the state. Started with major federal grants, it has been self-supporting for many years.[46.9] It is still visited by people from all over the world who come to see how watershed conservation measures can have multiple benefits, of which flood amelioration is one of the most important. Similar work has been done by the Tennessee Valley Authority. On a much smaller scale are the conservation authorities established by the government of Ontario for almost all the small rivers of southern Ontario. Serious flooding on the Ganaraska River, entering Lake Ontario at Port Hope, was one of the sparks which generated this remarkable development, the Ganaraska Authority being established in 1947. Forty such authorities now control conservation measures over a total area of less than 260,000 km^2 (100,000 mi^2), so these are indeed small-scale operations, but worthy of close attention by all who face similar flooding problems on smaller rivers.[46.10]

Of great significance was the use of both methods of flood control in the reconstruction in 1978 of the flood-control dam which formed the Charles River basin between Boston and Cambridge. The U.S. Corps of Engineers was responsible for the replacement of the old earth-fill dam, originally constructed in 1910, with a new concrete structure, incorporating a new ship lock and pumping station (Fig. 46.5). This followed the $5 million flood loss in 1955 after Hurricane Diane. Of unusual importance, however, was the asso-

FIG. 46.5 New flood-control works and lock at the mouth of the Charles River, Boston-Cambridge, Massachusetts.

46-10   SPECIAL PROBLEMS

ciated controlled use of 8,000 ha (20,000 acres) of marsh and swamp area ("wetlands") in the upper part of the watershed, the area serving as a natural sponge to slow up runoff.[46.11]

Geological details have not been included in the foregoing capsule comments on flood-prevention measures, although they are implicit in all. The geology of watersheds has already been touched upon. That geology must be known with exactitude for the proper planning of all upstream engineering, as well as for the assessment of the foundation conditions beneath all structures erected as flood-control works. Despite all such measures, floods will still occur when the natural phenomena leading to flood conditions combine to the appropriate degree. In extreme cases, it may prove necessary to move completely settlements that cannot possibly be fully protected. This was done in 1979–1980 in the case of the small town of Field in northern Ontario, which was devastated by flooding of the Sturgeon River in 1979. In June of that year, the government of Ontario announced that they would assist the residents of the small community to move all their buildings, not destroyed by the flood, to a new site on higher ground.[46.12] This is an extreme measure, rarely capable of execution in view of cost. Some alternative measures will be briefly described.

46.6 SPECIAL FLOOD-CONTROL MEASURES

The city of Pullman, Washington, with a population of 22,000, has suffered three serious floods in the course of this century, the last in 1948. A proposed flood-protection program was prepared along conventional lines, but its estimated cost ($2.5 million) was more than the city was willing to pay. A citizens committee was appointed and, with the aid of expert advice, developed an alternative plan. The committee recommended the formation of a county flood control district; the implementation of an early-warning system for floods, of controls over the use of the floodplain of the Palouse River, and of special watch, prior to floods, over all watercourses, culverts and bridge openings within the city limits; the formulation of emergency flood plans; and the use of flood insurance. A flood in 1972 showed the efficacy of the new plan.[46.13]

An urban drainage and flood control district has been established in the Denver region of Colorado and is engaged in preparing master plans for about 20 drainage districts. An associated agency is the Platte River Development Committee, one of the first projects of which has been the development of Confluence Park at the junction of the South Platte River and Cherry Creek, converting what was a real eyesore into a pleasant recreational area and providing for flood-flow control (Fig. 46.6).[46.14] Quite different are the problems faced by the Albuquerque Metropolitan Flood Control Authority, since that agency has to deal with arroyos coming out of the Sandia Mountains, dry for 99 percent of the year but when in flood capable of doing great damage. After much study, a master drainage-management plan was prepared, calling for the use of natural channels, improved and lined channels in residential areas, detention ponding, and floodplain control.[46.15]

These examples are symptomatic of what smaller cities are doing, against the geological necessity of considering floods as watershed problems. Some very large cities have had to make more drastic steps. One of the first of these was the construction of the Red River Floodway between 1961 and 1965 to protect the major Canadian city of Winnipeg, following its devastation by the 1950 flood on the Red River. Studies suggested that the only solution was to provide an alternative flood route around the city. This was excavated on the east side of the river, being 48 km (30 mi) long from its intake south of the city to its junction with the river well downstream. It has a peak capacity of 1,700 cms (60,000 cfs); several uses, during bad floods on the river, have confirmed this design feature. Excavation in the local gray clay amounted to over 76 million m^3 (100 million yd^3), or about 40

FIG. 46.6 River improvements at the confluence of the South Platte River with Cherry Creek, this pleasing area now known as Confluence Park, in Denver, Colorado.

percent of all the excavation on the Panama Canal. Bridges crossing the floodway numbered ten; their foundations required special attention. The local groundwater condition was naturally disturbed but came to new equilibrium within a year of the completion of construction (Fig. 46.7).[46.16]

The city of Chicago has long had a serious flood problem, in part because of the paving over of so much of the 930 km² (360 mi²) of the Greater Chicago area. Average annual flood losses have amounted to $31 million, with damage every year to an average of 50,000 homes.[46.17] Long and intense study has naturally been given to this great problem, and many solutions have been suggested. That finally adopted, and under active construction as this volume goes to press, involves one of the greatest of all deep-tunneling projects. The basic concept is to provide large, deep, underground caverns for the temporary stor-

FIG. 46.7 The Winnipeg Floodway, Manitoba, in use, looking south (toward the city of Winnipeg), the Red River being on the right, flowing towards Lake Winnipeg.

age of polluted storm waters, with associated collection systems and pumping stations. The Metropolitan Sanitary District of Greater Chicago will evenutally spend $5.6 billion on the complete system, which will include no less than 200 km (125 mi) of major tunnels (Fig. 46.8).[46.18]

FIG. 46.8 One of the completed Chicago main drainage tunnels; the scale can be gauged by the standard rail tracks.

It will be appreciated that it is quite impossible even to summarize this vast, imaginative project, so complex are the engineering problems involved. The geology, however, is relatively straightforward. It was naturally explored in great detail before tunneling started. The city is underlain by up to 60 m (200 ft) of Pleistocene deposits overlying Silurian, Ordovician, and Cambrian strata, fortunately almost horizontal. Most of the major underground excavation will be in the Platteville dolomite, an Ordovican rock with good tunneling characteristics, except where it includes small sections of shale.[46.19] The tunnels are up to 90 m (300 ft) below ground surface and up to 10.5 m (35 ft) in diameter. Mechanical moles are being put to excellent use for much of the tunneling; some trouble with clay seams has been encountered. There are few better examples anywhere in the world of how the geology underlying a great city can be put to such constructive use.

Finally, the protection in England of Greater London, including the ancient city, from the danger of flooding from the sea must be mentioned. The Thames Barrage is yet another massive, imaginative concept—a dam built across the Thames River well below the main part of London, equipped with large, movable gates which can be raised into position to provide protection from any unusually high tidal surges if and when these occur. The "flood risk area" covers 116 km^2 (45 mi^2) and includes about a quarter of a million dwellings, 50 subway stations, and 35 hospitals. The estimated value of assets that might be flooded is $8 billion. The famous Thames embankments are an expression of early flood-protection measures, but the barrage was finally decided to be necessary. It is located in a part of the river known as Woolwich Reach, 10 km (6 mi) below the well-known Tower Bridge. Four main channels, for normal flow and all shipping, each 60 m (200 ft) wide, are being provided between the main piers. Construction of the barrage is in full swing as this volume goes to press (Fig. 46.9).

FIG. 46.9 Aerial view of the Thames Barrage, England, under construction in December 1980; the historic river flows from left to right, Greater London being to the left.

So important is the project, and so complex the engineering which it involves, that a full-day conference was held at the Institution of Civil Engineers in London on 5 October 1977 at which all major aspects of the design were reviewed. The proceedings of this meeting form a volume of unusual interest. The long history of flooding in London, starting with a flood in the year 1099, is first reviewed. A singularly informative paper by Fookes and Martin describes the geology of the Woolwich Reach site, which had been selected after a long search process. The paper outlines the very detailed studies made of the site geology and of the geotechnical properties of all the materials that would be encountered, studies that cost £108,634, (less than $300,000) and revealed significant information. The northern end of the barrier will be founded on Thanet sand and the southern end on the well-known Chalk of southern England, and site studies showed that near its upper surface the Thanet sand had been reworked. Some frost shattering of the upper part of the Chalk was found at the southern end of the site, but not where the Thanet sand was in

46-14 SPECIAL PROBLEMS

place over the Chalk (Fig. 46.9). The effect of the barrage on the existing regime of the Thames Estuary was also reviewed before a series of papers describing in detail the main features of the design of the structure. Attention is directed especially to the paper by Fookes and Martin, since it is a masterly account of a most thorough site investigation.[46.20]

46.7 FLOODPLAINS

Frequent mention of floodplains will have already indicated that their use in urban development usually leads to major flood problems. Floodplains are a perfectly normal, natural geological feature of most river valleys. They are always flat, usually with good topsoil (from river deposits); their location close to open water makes them most desirable building sites, especially for residential construction. They are, indeed, excellent building areas apart only from the overriding danger of flood at times of high riverflow. Despite this clear and obvious danger, almost every city which has a river within its boundaries will be found to have flood-prone buildings on the floodplain.

The experience of Rapid City, a well-settled community of about 50,000 in western South Dakota, should be known to every engineering geologist and civil engineer (Fig. 46.10). On the night of 9 June 1972 a phenomenal meeting of a cold front and warm moist

FIG. 46.10 Rapid City, South Dakota; the memorial to the victims of the 1972 flood now standing in the park that has been developed on the floodplain.

air over the Black Hills resulted in an extreme rainfall, as much as 37.5 cm (15 in) falling in some places in six hours. Rapid Creek, from which the city takes its name, reached a flood flow of 1,410 cms (50,000 cfs). Building had been permitted on the Creek's lovely floodplain, for the Pactola Reservoir, 22 km (14 mi) upstream from the city, had given a false sense of security. But the worst of the rain fell in the area between the reservoir and the city. Two hundred and thirty-eight people lost their lives during that dreadful night. Property damage was estimated to be $128 million. The area flooded was almost identical with that covered by a flood in 1907. The U.S. Geological Survey had published 1:24,000 maps of the Rapid City area *before* the flood; the coincidence of the flooded area with the alluvium plotted on the maps is significant. It is shown in Fig. 46.11, reproduced from a paper describing the experience in some detail; it shows the location of bodies recovered after the flood; five were never found.[46.21]

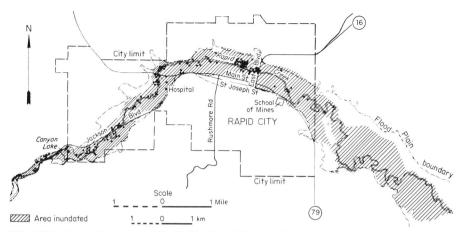

FIG. 46.11 Generalized map of Rapid City, South Dakota, showing the area inundated in the 1972 flood, and the area mapped as alluvium by the U.S. Geological Survey; black dots show where bodies were recovered after the flood.

This was one of the very worst such experiences, but it is far from unique. On 15 October 1954 Hurricane Hazel penetrated as far as the Toronto area in southern Ontario. Again through an unusual combination of meteorological circumstances, extreme rainfall took place over a small area, as much as 27 mm (1.06 in) in one hour at one gauge, a total of 177 mm (7 in) in 24 hours at Brampton. This was, however, only slightly more than a previous maximum of 173 mm (6.90 in) in 1915. The small Humber River, with adjacent streams such as Etobicoke Creek, rose to catastrophic heights. Residential building had been permitted in the borough of Etobicoke on the pleasing floodplain. Eighty people lost their lives, again in the night; damage was estimated at $25 million. The local municipality soon passed the necessary ordinances prohibiting all building on this floodplain.[46.22] There has since been continuing pressure to relax this provision, despite the awful warning of October 1954.

Herein lies the problem. Floodplains provide some of the superficially most desirable areas for residential development in all municipalities blessed with a river within their boundaries. And yet they are the most dangerous areas. It is difficult to determine the magnitude of the problem, but a survey of 1971 showed that there were then 5,200 cities and towns of significant size in the United States, part, or even all, of which were located in floodplains.[46.23] Fifteen states had then adopted regulations regarding the use of floodplains, and 360 municipalities had already adopted floodplain regulations. Progress has been made since then, but even allowing for maximum progress, the problem remains a serious one; pressure to use floodplains, despite all that can be explained about their danger, will long persist. Prairie du Chien, a pleasant town of 6,000 on the Mississippi River in Wisconsin, when surveyed some years ago had 48 buildings in the lower part of the town that needed removal, 157 that should have been relocated, and 33 that should have been raised, if they were to be safe from the 100-year flood for that part of the great river. Its valley, 2,400 m (1.5 mi) wide, is here confined between high cliffs, with the river normally in two channels, each 300 m (1,000 ft) wide.

The problem is at last receiving attention in larger centers, if only because of the emphasis now being given to town planning. The problem is just as serious, in character if not in degree, along smaller streams and in undeveloped areas. Action is being taken here also. Typical are the powers that have been given by the government of Ontario to the small river conservation authorities of the southern part of the province. The author-

46-16 SPECIAL PROBLEMS

ities have been encouraged and financially assisted to have the floodplain limits of their respective rivers accurately surveyed and the floodplains plotted on maps that are made publicly available. Once their individual regulations are approved, the authorities have the power to restrict or prohibit the placing of any new fill in the floodway, to prohibit construction that impinges on the floodway, and to control through necessary permits any proposed alteration to the natural course of their river or its tributary streams.[46.24] Many other authorities now have similar powers to deal with this development problem.

With every such advance in the practical appreciation of this vital geomorphological feature, the dangers of loss of life and of damage to property due to floods will be reduced (Fig. 46.12). Much, however, remains to be done. Guard will always have to be kept, even

FIG. 46.12 Looking northwest over the floodplain of the Columbia River at Vanport, Oregon, during the flood of May–June 1948; the railway embankment to left of center had been acting as a levee until it was breached, affecting 20,000 people and causing 13 deaths.

over floodplains that have been cleared of buildings and are under strict control. They provide admirable parkland, flooding of which can be permitted. Other uses, not endangered by flooding, have even included small golf courses. But residential construction should most desirably be completely eliminated from all floodplains in North America. As a recent mayor of Rapid City, South Dakota, has well said: "It is stupid to sleep in the floodplain."[46.25]

46.8 CONCLUSION

This chapter, dealing as it does with so vast a subject, has been necessarily general in nature. Brevity inevitably renders discussion of the flood problem somewhat simplistic at first glance. Some might think that the matters discussed have little to do either with civil engineering or geology—but not those readers who have seen a great river in flood or who have been called upon to assist with flood-rehabilitation work (Fig. 46.13). The flow of

FIG. 46.13 Typical view of damage following the 1977 southwestern Pennsylvania flood (a 500-year record), a reminder of what floods can do.

water down streams and rivers is a vital part of the dynamics of geology and an important aspect of that area of the science often called geomorphology. The relevance of geology to runoff has been explained. The geological implications of floodplains, so easily recognized by all familiar with physical geology, are clear-cut. The significance of flood flows to all civil engineering works either on or near rivers is so obvious, once thought is given to it, that it requires no elaboration.

Beyond all such direct interconnections, however, is the fact that civil engineers and geologists are informed citizens, singularly well-informed citizens, about all the physical aspects of town and country planning. Users of this Handbook will, therefore, have a dual interest in the impact of geology on all aspects of planning and on the environment. Accordingly, the four following short chapters provide at least an introduction to aspects of the relation of geology to civil engineering that are rapidly assuming increasing importance but which go some distance beyond the strictly professional subjects with which all preceding chapters have been concerned. To some the chapters will convey nothing new, but it is hoped that they will prove of interest even so, and of real use to those who have not yet come to appreciate that geologists and civil engineers do have concerns about the environment, and have already done a good deal about it.

46.9 REFERENCES

46.1 This summary prepared from a number of sources: e.g., *The Netherlands and the Water*, Netherlands Ministry of Transport and Waterstaat, The Hague, 1967.

46.2 L. Stowe, "The Ordeal of Huaráz," *Reader's Digest*, **94**:63. (June 1969).

46.3 J. F. Mann and R. O. Stone, "Von Trigger-Cadiz Mudflow of July 1953," in abstract vol. for annual meeting at Los Angeles, Geological Society of America, Boulder, Colo. 1954, p. 77.

46.4 A. C. Mason and H. L. Foster, "Extruded Mudflow Hills of Nirasaki, Japan," in abstract vol. for the annual meeting at Los Angeles, Geological Society of America, Boulder, Colo., 1954, p. 78.

46.5 D. H. Thomson, "Water and Water Power," *Transactions of the Liverpool Engineering Society,* **44**:105 (1923).

46.6 *Ibid.*

46.7 W. R. Hansen, "Effects of the May 5–6, 1973, Storm in the Greater Denver Area, Colorado," Circular no. 689, U.S. Geological Survey, Washington, 1973.

46.8 R. R. Parizek, "Agnes, Aftermath and the Future," in abstract vol. for the annual meeting at Minneapolis, Geological Society of America, Boulder Colo., 1972, p. 621.

46.9 B. C. Browning, "The Muskingum Watershed Conservancy District," in J. Forman and O. E. Fink (eds.), *Water and Man,* Friends of the Land, Columbus, Ohio, 1950, p. 213.

46.10 A. H. Richardson, *Conservation by the People; the History of the Conservation Movement in Ontario to 1970,* University of Toronto Press, Toronto, 1974; floodplain regulations published by individual authorities and publicized in the respective local newspapers.

46.11 F. Notardonato and A. F. Doyle, "Corps Takes New Approach to Flood Control," *Civil Engineering,* **49**:65 (June 1979).

46.12 "Town to Move to High Ground," *Ottawa Journal,* 26 May 1979.

46.13 H. D. Copp, "Pullman Tackles Its Flooding Problem," *Civil Engineering,* **42**:42 (August 1972).

46.14 K. R. Wright and W. C. Taggart, "The Recycling of a River," *Civil Engineering,* **46**:42 (November 1976).

46.15 H. F. Bishop, "Flood Control Planning in Albuquerque," *Civil Engineering,* **48**:74 (April 1978).

46.16 F. W. Render, "Geohydrology of the Metropolitan Winnipeg Area as Related to Groundwater Supply and Construction," *Canadian Geotechnical Journal,* **7**:243 (1970).

46.17 R. Lanyon, "Flood Plain Management in Metropolitan Chicago," *Civil Engineering,* **44**:79 (May 1974).

46.18 *The Chicago Underflow Plan,* Metropolitan Sanitary District of Greater Chicago, Chicago, December 1972; see also "Chicago Pioneers in Correcting Water Pollution, Controlling Floods," *Civil Engineering,* **48**:58 (July 1978); see also F. C. Neil and F. E. Dalton, "Water Resources for Chicago—History and Future," *Proceedings of the American Society of Civil Engineers,* Paper no. 15269, **106**(WR 1):173 (1980).

46.19 T. C. Buschbach and G. E. Heim, "Primary Geologic Investigations of Rock Tunnel Sites for Flood and Pollution Control in the Greater Chicago Area," *Environmental Geology Notes No. 32,* Illinois Geological Survey, Urbana, May 1972.

46.20 *Thames Barrage Design* (record of a conference in London, 5 October 1977), The Institution of Civil Engineers, London, 1977.

46.21 P. H. Rahn, "Lessons Learned from the June 9, 1972 Flood in Rapid City, South Dakota," *Bulletin of the Association of Engineering Geologists,* **12**:83 (1975).

46.22 C. C. Boughner, "Hurricane Hazel," *Weather,* **10**:200 (1955).

46.23 J. E. Goddard, "Flood-Plain Management Must Be Ecologically and Economically Sound," *Civil Engineering,* **41**:81 (September 1971).

46.24 Richardson, op. cit.

46.25 C. R. Wall, "Conference Focusses on Flood Plain Management," *Civil Engineering,* **45**:20–21 (October 1975).

Suggestions for Further Reading

In its water resources work, the U.S. Geological Survey (USGS) has naturally been active in studies of floods. Typical of its publications in the field are two Circulars of 1970: No. 601 D by H. E. Thomas and W. J. Schneider with the intriguing title *Water as an Urban Resource and Nuisance,* and No. 601 C, *Flood Hazard Mapping in Metropolitan Chicago,* by J. R. Sheaffer, D. W. Ellis, and A. M. Spieker.

part 5

GEOLOGY AND THE ENVIRONMENT

chapter 47

GEOLOGY AND PLANNING

47.1 Elements of Planning / 47-3
47.2 "Design with Nature" / 47-5
47.3 Planning and Geology / 47-6
47.4 Natural Hazards / 47-10
47.5 Some Examples / 47-11
47.6 Terrain Analysis / 47-18
47.7 Geological Information / 47-19
47.8 The Future of Planning / 47-21
47.9 References / 47-22

The concept of physical planning for the development of all areas put to human use, both urban and rural, is now happily accepted throughout the world. Indiscriminate development is coming to be a thing of the past. There is a growing realization that planning all land development, with full consideration of all social implications, is quite essential if land is to be well used and the towns and cities of the future are to be pleasant places in which to live. The old tag that "God is not making any more land" has finally entered into public consciousness, so that one does not have to argue, as was the case in former years, that all land must be used to best social advantage, development being subject to public scrutiny and control.

The happy results of urban planning are to be seen in many a "New Town," of which there are now a growing number, and in the redeveloped older parts of existing cities. In many of the latter cases the juxtaposition of old, unplanned parts of cities with redeveloped areas, with judicious use of open spaces and design with human needs always in mind, is the best of all demonstrations of the desirability of sound planning and land control. And all such planning is concerned, ultimately, with the use of the ground, both at the surface and below it to any depth that may in any way affect the execution of plans—in other words, with geology. Geology and planning are, therefore, inseparably associated, a fact that unfortunately has not always been fully appreciated.

47.1 ELEMENTS OF PLANNING

There have been countless definitions of planning, but in essence it is the analysis of human needs in any urban or rural area and the orderly synthesis of physical features which will help to achieve the goals that have been developed. Many disciplines are therefore involved in all comprehensive planning exercises. This chapter is concerned with one major aspect only of the physical aspects of planning. In order to set what follows this section in its proper context, Table 47.1 has been prepared (based on much more extensive matrices, such as those of Pollard and Moore) to show the many professional interrelations that must inevitably enter into even the smallest exercise in planning.

It is to be noted that the diagram emphasizes the physical parts of planning. The first three items, and the last two, would naturally call for considerable extension in any complete picture of the planning process. It is, however, with the planning, design, and implementation of physical facilities and the use of land that this chapter is concerned. The chart should be examined with this clearly in mind. Correspondingly, the disciplines involved with the physical expressions of planning are indicated, and not practitioners. Those who bear the honored title of *"Planner"* always have special training in one or other of the related disciplines, but, whatever their background training, it is clear that they

TABLE 47.1 Summary of Professional Disciplines in Urban and Regional Planning

Studies	Disciplines				
	Civil engineering	Architecture	Landscape architecture	Geology	Other
Economic base					x
Population					x
Social factors					x
Geology: soils	x	•	x	x	•
Climate	x	x	x	x	•
Flood problems	x	•	x	x	•
General land use	•	•	•	x	x
Housing	•	x	•	x	x
Schools	x	x	•	x	x
Hospitals	x	x	•	x	x
Public buildings	x	x	•	x	x
Libraries	x	x	•	x	x
Fire and police	x	x	•	x	x
Parks and recreation	•	•	x	x	x
Commercial development	x	x	•	x	x
Industrial development	x	x	•	x	x
Roads and traffic	x	•	•	x	•
Public transportation	x	•	•	x	x
Parking	•	x	•	x	x
Power supply	x	•	•	x	•
Water supply	x	•	•	x	•
Sewerage	x	•	•	x	•
Central heating	x	x	•	x	x
Waste disposal	x	•	•	x	x
Urban renewal	•	x	•	•	x
Historical aspects	x	x	•	•	x

x Primary interest
• Possibly active interest

Source Adapted from a much larger tabular statement in W. S. Pollard and D. W. Moore, "The State of the Art of Planning," *Proceedings of the American Society of Civil Engineers*, vol. 95, UP1, April 1969.

must have a lively appreciation of the essential place of civil engineering, architecture, landscape architecture, and geology in all planning work. That geology has a bearing on every aspect of physical planning should be no surprise to those who use this Handbook; planning has been implicit in a great part of the contents of this Handbook. This should not be surprising, since every civil engineering work is planned in itself and in its relation to the physical environment and any associated works. This basic relationship between civil engineering and planning, especially urban planning, must be stressed. When a town or region formally embraces planning as one of its essential services, it is sometimes possible to hear, and to read, comments which suggest, even if they do not actually state, that all previous public works in the planning district have been built to no plan at all. Some mistakes may have been made in the location of "preplanning" works, but they are exceptions to the generally ordered and planned developments of which public works are a part.

Today all cities, most towns, and many smaller communities have official plans into which all new developments must fit. Such plans are not static, "one-shot" efforts but are constantly under review and revision. It is therefore incumbent for civil engineers to have a clear overall picture of the planning process, if only to assist them in the preliminary planning of the works for which they are responsible. Rather than being puzzled by some of the more abstruse modern definitions of planning, they can take inspiration from these words of Lewis Mumford in one of his early notable books:

> The task of regional planning ... is to make the region ready to sustain the richest types of human culture and the fullest span of human life, offering a home to every type of character and disposition and human mood; creating and preserving objective fields of response to man's deeper subjective needs. [47.1]

Planning involves at its outset the wise use of land. Land is not just white space on the pieces of paper on which the planner first sketches initial ideas and, later, more mature ones. Land is the surface expression of geological processes working through the ages. Today's environment is the result of many natural modifying influences over a very long period. Geological, hydrogeological, climatic, and other natural processes have combined to yield the land surface of today. Development of this land must, therefore, be planned with a full realization of the natural forces which have brought it to its present state, especially the geomorphological processes. Planning must take into account the essentially dynamic character of Nature, so that the works of man will fit into the natural pattern and not disturb the delicate inherent balance any more than is essential. To conserve the soundness, the productive power, the beauty, and the natural equilibrium of any area being developed should be the overriding objective of all planners, so that future generations will not be deprived of the privileges that Nature has given to those of today.

Geology, clearly, must therefore be the starting point of all planning. This will be so obvious to users of this Handbook that they may question the necessity of even saying so. But it is still possible to complete degree courses in urban and regional planning without having taken any courses in geology at all! Even civil engineering handbooks on planning have appeared without any specific reference to geology in the planning process. The situation is changing, but all who have any association with official planning work should ensure that geology does occupy its proper place, so that planning may get started in the right way and mistakes may be avoided.

47.2 "DESIGN WITH NATURE"

Lest it be thought that the foregoing views are those only of civil engineers and geologists, it must be recorded that, although admittedly many publications on planning do not recognize the place of geology in the planning process, some important publications do. Their number is increasing, but mention may be made here of just one, a notable volume published first in 1969. It is graced with the title *Design with Nature*. Its author, Professor Ian McHarg, is, significantly, a landscape architect and planner, head of the department of landscape architecture and regional planning at the University of Pennsylvania. His most attractive volume (which has a foreword by Lewis Mumford) shows, by means of a number of vivid examples from the author's professional practice, how geology can and must be used in the planning process. Professor McHarg has developed the idea of using a series of base maps, each of which shows one of the major significant factors that must be considered together in planning for any region. One of these is geology. The maps, printed on transparent material, are then superimposed the one on the other to see where the conflicts arise. The positive guidance thus given so vividly (color being an essential element in the plotting) enables proper allocation of space, logically arranged for public benefit and without contravention of any of the natural barriers geology can sometimes create. [47.2]

Professor McHarg calls his main subdivisions "physiographic obstructions" and "social-cost values." The former include slope, drainage, bedrock, geology, and soils. In commenting on a study of Staten Island, New York, the author says:

> It is a special place—its geological history made it so. Silurian schists form the spine of the island, but the great Wisconsin glacier of Pleistocene time left its mark, for there lies the evidence of the terminal moraine. There are glacial lakes, ocean beaches, rivers, marshes, forests, old sand dunes, and even satellite islands.... The serpentine ridge and the diabase dyke of Staten Island can only be comprehended in terms of historical geology. The superficial expression of the island is a consequence of Pleistocene glaciation. The climatic processes over time have modified the geological formations, which account for the current physiography, drainage, and distribution of soils.

These are the words of a planner who looks upon geology in planning in exactly the way that the authors have suggested. In an earlier day, a great thinker said the same thing with economy of words. Over every planner's desk and drawing board could well be mounted words of Francis Bacon (1561–1626): "NATURE TO BE COMMANDED MUST BE OBEYED."

47.3 PLANNING AND GEOLOGY

Geology has so far been used as a general term only. It is now necessary to illustrate how geological conditions influence the planning of all land use, from the few acres acquired for a new industrial plant to a large region such as a county, from a new village to the extension of a major urban community. The principles to be applied do not vary, although the particular features of different sites will. Accordingly, all that can here be done is to provide a listing of the main geological features that must be considered at the start of all planning work. Some features will apply to all sites; others may be encountered only occasionally.

For all sites a consideration of the *general picture of the geology underlying the site* is the first essential; the general picture would comprise the depth to bedrock, the type and structure of bedrock, and the nature of the overlying soils. For all parts of North America there are now available geological maps. For many regions they may be small scale only, but efforts must be made to obtain the most detailed geological maps available. Study of the geological sections which accompany almost all such maps will be the most satisfactory way of introducing those engaged in the planning work, who are not familiar with geology, to the subsurface conditions which geological maps reveal, conditions that affect almost every aspect of planning.

A consideration of *climate* is the second basic requirement. There should be access to the longest available weather records for the nearest recording station to the area being planned or if there is no nearby station, to the best set of records from the station most suitable for comparative purposes, as recommended by meteorological experts. Temperature records will be important, especially for winter months in relation to heating loads, but precipitation records are also of vital importance.

Rainfall records will be intimately related to *groundwater conditions.* Groundwater could well be called the geological Cinderella of planning. It is not normally seen and so is not as obvious a factor as soil and bedrock (if it outcrops), but the effect of groundwater conditions on land development can sometimes be disastrous if the conditions are not recognized in advance and if the variation in water level throughout at least one annual period is not checked. Annual records alone will show if there are any "wet areas" or any intermittent springs in the land being planned: records will also show the varying depth at which the water table may be encountered when construction starts in implementation of the plan.

These conditions are but a part of a site's *hydrogeology,* which must be carefully reviewed—i.e., the interrelation of the soil and rock under the site with water, in all its manifestations. Natural drainage must be studied to see how it will be interfered with. Any running streams must be investigated, especially with reference to extreme high and low flows, for which determination long-term runoff records are desirable. If none are available, early efforts must be made to initiate flow recording for assessment in relation to the best long-term precipitation records that can be obtained.

Flood conditions call for special considerations, even on small streams, but especially so on any larger watercourses that are included in the site. Floodplains, in particular, must be delineated and their reservation for non-residential purposes ensured from the very outset of planning.

Natural drainage must be determined for initial planning purposes, and every consideration must be given to the major changes that the paving of roads and streets and the covering of ground with buildings will cause. It is so easy to overlook the fact that all such areas will then have 100 percent runoff, giving a drainage volume far in excess of any natural drainage; carefully designed precautions are therefore essential. There are useful guides to the volumes of water to be handled at various stages of urbanization; most notable is a good general guide by Leopold. [47.3]

Not only must water be disposed of, but *water supply* must be considered at a very early stage in planning. If water can be obtained from river or lake, the location of the necessary intake and treatment plant will be a key element in planning. If groundwater conditions are such that a supply can safely be obtained by pumping, this will require a much more detailed study of groundwater conditions than might otherwise be necessary. Here the source of supply will not be just "pumping out of a well"; the entire groundwater regime will be affected; experts must be consulted about any possibility of land subsidence resulting from even moderate pumping.

The *soils* over the site must be accurately surveyed, sampled, and tested, so that their overall geotechnical character can be known; more detailed studies will be necessary later at each individual building site. The overall survey will show if there are any unusual soils present, such as expansive clays which can cause problems with ground heaving, especially when covered up with impermeable material such as concrete, the ground-air-moisture interface being thus disrupted. Regional geology must be studied to determine the possibility of the presence of any sensitive clays, such as the Leda clay of the St. Lawrence, Ottawa, and Richelieu valleys. The landslide at St. Jean-Vianney, Quebec, is a telling reminder of the tragedy that can result from slides in these soils.

Evidence of *landslides* that have taken place naturally on the site must be diligently searched for; they can be useful indicators of unstable soil conditions which may be present elsewhere without having revealed themselves through actual slides. In the same way, exposed rock faces must be examined for any signs of rock instability or of previous *rockfalls*.

In the survey of soils over a site, special attention must be given to any deposits of sand and gravel. If testing of this material indicates that it is suitable for use as concrete aggregate, then it becomes a resource of value, the controlled use of which should be considered before any detailed planning is done. There are all-too-many deposits of valuable sand and gravel, now covered up by buildings within the limits of North American cities, that should have been exploited before the buildings were erected, but which are now lost for beneficial use. When the extent of such deposits is known, planning can take into account the areas that will be left after excavation of the deposit has been completed. *Sequential land use* should become a term of real meaning.

If there is no sand and gravel on the site, then another part of geological reconnaissance must be a search for sand and gravel adjacent to the site or, if there is none to be found, the location of convenient areas where bedrock of suitable quality may be quarried. The cost of transporting heavy and bulky concrete aggregate is now so high that every effort must be made at an early stage of planning to locate suitable supplies of natural material for use in the concrete that today is almost always such an important all-purpose building material.

The *depth to bedrock* over the site is another important geological feature that must be determined. Preliminary test boring will give some idea of this, but for a complete picture, the use of geophysical surveying methods, in association with test boring, must be considered. This requirement applies, naturally, only if bedrock is relatively close to the surface. For the Midwest, where the bedrock may be hundreds of meters below the surface, soil studies will usually suffice. If bedrock is close to the surface, its presence can influence appreciably the type of structure that can be used on the site. It may also affect the necessary installation of underground services. (If this point seems too elementary, let it be recorded that a housing development in a city that must be nameless was planned without this factor being considered; all the water mains and sewers had to be subsequently installed in carefully excavated rock trenches, drilled and blasted at a cost that can be imagined, one that was not included in the preliminary estimates.)

Some services may have to be located so deep underground as to require *tunneling*. Location of the bedrock surface is essential before any tunnel work can be considered, as is also a thorough knowledge of the character of the rock(s) that will be encountered. Tunnels will probably come up for review when consideration is given to the provision of sewers for collection of liquid wastes, and companion sewers for collection of storm water from drainage facilities. In all modern communities these liquids must be properly treated before disposal; this involves major collector sewers, frequently of such size that they must be in tunnel. The difficulties resulting from the installation today of such sewers, neglected at an earlier time in older towns, will be all too familiar to those who have had to live in, or to drive through, such towns while main streets are disrupted by the necessary trench excavation or tunnel head shafts.

In even general studies of the bedrock underlying a site to be planned, it is essential to find out (if possible) if there are any *faults* underlying the site. Their presence will give warning of the necessity for specially detailed subsurface investigations of all installations adjacent to the fault line. Faults may affect tunneling and the use of any *underground space* that may be included in the site plan. If there are any *natural caverns* under the site, they will have to be considered in planning, and put to use if appropriate. Only in exceptional cases will it be possible to mine the underground space beneath an urban development for limestone to be used for building purposes, as at Kansas City, Missouri. This possibility should be kept in mind, however, in geological studies for planning purposes. The necessity for energy conservation will almost certainly make the use of underground space, for appropriate purposes and when local geology is suitable, a steadily increasing feature of urban development.

Finally, the presence of faults will be a reminder of the necessity for studying carefully the *seismic risks* that the site carries with it. Fortunately, there is now available in many countries information, on a national scale, as to the varying seismic risks throughout the country in question, based on detailed study of records of earthquakes that have taken place. Zones of different seismic-risk intensity are now delineated by responsible authorities. Such initial guides are not, of themselves, enough. Study must be made of any local records of earthquakes or earth tremors throughout the period of recorded local history. Nothing can be left to chance in ensuring that the necessary information is available for inclusion in building design of the requisite factors for earthquake resistance.

It is to be noted that all the geological features listed relate to the ground that is to be developed, and they must be considered before any physical planning can be started. A review of these features will show the actual characteristics of the area to be planned; the land cannot then be regarded as a great "blank space" on which the planner can exercise creative imagination with no regard for any physical factors. Even from this mere listing, it can be seen that consideration of the geology of an area to be planned is an essential preliminary to all physical planning. This is the first phase of geological involvement in land development. When specific structures are being considered after adoption of a plan, detailed investigations must be made of every site; here geology is again an integral part of each study. And finally, when construction starts, as has been seen so frequently in this Handbook, the geology of the site must be studied in detail as it is revealed by excavation, in order to ensure that design assumptions are correct and that no unusual and unsuspected features are revealed.

It can also be seen that in initial study of the geology of an area to be planned, the function of geological investigations is wholly constructive. When development starts, sound geological assessments can assist in minimizing flood damage, in preventing trouble from landslides or rockfalls, in ensuring that valuable sand and gravel deposits are not covered up and so lost for all time, and in assisting in the determination of seismic risks, so that adequate precautions can be taken in the siting and structural design of buildings. In other words, geology provides as much insurance as possible against the damage that can result from *natural hazards,* some of which, in the inexorable order of Nature, will happen sooner or later at almost all sites.

Finally, if the assistance which geology can so clearly give to physical planners is to be fully effective, those responsible for site geological studies must be able to present their information clearly and simply to co-workers, who may be, initially, ignorant of geology and geological terms. Language used, therefore, must be as simple and straightforward as possible. Simple terms (such as *soil* and *rock*) should be used whenever appropriate, geological terminology being avoided as much as possible. Maps should similarly be prepared with planning in view, and not as examples of the amount of geology that can be shown on one sheet. The great utility of geological sections should always be kept in mind. Above all, it must be remembered that geology is here a means to an end and not the end in itself. This vital matter of the *transfer of information* is at last being recognized in geological circles, as evidenced by the 1979 publication of a useful brochure on this subject by the Geological Society of America.[47.4] And the Committee on Geological Sciences (of the National Academy of Sciences), in its excellent book, *The Earth and Human Affairs,* has said:

> Knowledge of the earth ought to be available at all governmental levels where environmental concerns are likely to come up for consideration. We must find ways for reliable, complete,

and understandable information to be efficiently and promptly provided for persons in government and the public at large. [47.5]

47.4 NATURAL HAZARDS

Natural hazards have just been mentioned. They are part of the order of Nature and cannot be avoided. If they are recognized in advance of planning, however, much can be done, with the special aid of geology, to mitigate their effects, not always with regard to property but assuredly so with regard to human lives. The extent of this damage is seldom fully realized. It was estimated in 1974 that the annual average value of property damage in the United States due to natural hazards—of which earthquakes and floods are the most serious—was on the order of $10 billion.[47.6] The estimate was prepared in connection with a special conference held in 1973 for discussions of natural hazards, convened jointly by the American Society of Civil Engineers and the Engineering Foundation.

Widespread efforts are being made to gain public recognition of the serious nature of these potentially great losses while they can be considered logically and dispassionately, rather than in the panic atmosphere which, quite naturally, persists after a serious natural disaster has taken place. Typical was a special issue of the house organ of the professional engineers of British Columbia in which a strong appeal was made for the institution of a provincial natural hazards policy.[47.7]

If it is thought that this matter is being overemphasized, another official figure can be cited. In 1973, an important and comprehensive report was published in California showing that a careful estimate would put the value of damage from natural hazards in the state of California between the years 1970 and 2000 at $55 billion. This figure was contrasted with a corresponding estimate of $50 billion for urban and wildland fire losses in the state during the same period. It was further estimated that of the $55 billion loss, 98 percent would be due to earthquake shocks, loss of mineral resources, landslides and floods. Of special interest is the estimate that $38 billion of the total loss could be avoided by the application of current loss-prevention methods, in which geology naturally looms large. And the estimated cost of these measures was $6 billion, giving a benefit:cost ratio of 6.2:1. The report, part of a study of a master plan of urban geology for California, was prepared under the direction of the state geologist.[47.8] It is a landmark document, but it has one regrettable feature: the hazards are not called "natural hazards," but "geological hazards."

This latter term has come into use in relatively recent years. There is even a book (a good one) which includes the term in its title. [47.9] The writers have unthinkingly used it themselves, one of them in print, but they have now come to see what an unfortunate term it is, tending to destroy public appreciation of the fact that it is the constructive use of geology that can minimize damage due to natural hazards. It is easy to see how this lamentable term came into use, since the great natural hazards do have geological origins. This is especially true in California, so it is not surprising to find that a paper titled "The Influence of Geologic Hazards on Legislation in California" was presented to an international audience by a notable geologist of that state.[47.10] The term is even used in some otherwise notable and constructive legislation in Colorado. In 1967, there was held in Utah the Governor's Conference on Geologic Hazards, an admirable and helpful meeting apart from its name.[47.11] At this meeting, hazards discussed included construction hazards such as dam and reservoir failures and trench cave-ins, failures that are strictly engineering in nature, although, again, geological features may contribute to them.

Having come, even if rather late, to realize the harm that the use of the term "geological hazards" is doing, especially to appreciation of the place that geology must occupy in all sound planning, the writers urge all readers to do what they can to correct its use when they encounter it. The term is not used elsewhere in this Handbook, nor will it appear in anything further which the authors may write. It is greatly to be hoped that it will disappear as quickly as it has come into use, being replaced by the correct term, *natural hazards*—to the alleviation of which geology has so much to contribute.

47.5 SOME EXAMPLES

Some examples of the uses of geology in planning must be given, if only to illustrate what has already been suggested as the place that geology must have in planning and to indicate some of the different approaches taken to its use. In so short a chapter as this, a mere sampling must suffice. From the many cases now fortunately available, five will be mentioned—one major city, three larger territories, and one small, essentially rural area. To select one city out of the many in North America that do now recognize geology in all their planning work invites invidious comparisons. Fortunately, the choice is simplified, since the capital city of the United States offers an example, reference to which involves no territorial preferences.

Washington, D.C. In January 1968, the Metropolitan Washington Council of Governments (COG) published a well-printed brochure entitled *Natural Features of the Washington Metropolitan Area,* subtitled appropriately *An Ecological Reconnaissance.*[47.12] This beautifully produced 50-page booklet was financed, in part, through an urban planning grant from the Department of Housing and Urban Development under the provisions of the (U.S.) Housing Act of 1954, a welcome indication of the interest of federal government in this type of approach to planning. The purpose of the report, as stated is

> to present and analyze, under one cover, certain basic data essential to an understanding of the natural environment of the Washington Metropolitan Area. Within this urban region, approximately 25 square miles per year—or about one per cent of the metropolitan area annually—will be converted from farm, swamp, and hillside to subdivision, street and park between now and the year 2000.

The metropolitan region that includes Washington is 610,000 ha (1.5 million acres) in area, measuring about 112 km (70 miles) wide and 89 km (55 miles) long. It is situated between the Atlantic Ocean and the Appalachian Mountains. Its geology is therefore varied, including several geological provinces, covering a period of geological time of as much as 600 million years. The first map in this impressive publication shows the generalized geology of the area (Fig. 47.1).

A concise text accompanies this first map and explains in outline the four main provinces—the coastal plain, the Piedmont province, the Triassic lowland, and the Blue Ridge complex—each of which is clearly indicated on the accompanying map. It is further explained how geology influences some of the key factors in urban development, 11 factors being listed, with their significance. Typical are slope stability, foundation conditions, excavation characteristics, groundwater features, and suitability for septic tanks. Maps that follow show generalized industrial mineral distribution (this being a $25 million-a-year industry in the area), ground elevations, the slope of the ground, soil distribution, streams and drainage basins, the presence of floodplains, groundwater distribution, woodlands—all with supporting statistics and explanatory notes—and, finally, a general map that is a composite of all the natural features previously delineated. This is obtained by superposition (not unlike the procedure employed by Professor McHarg) and is accompanied by notes on the policies that must be considered in the further development of the

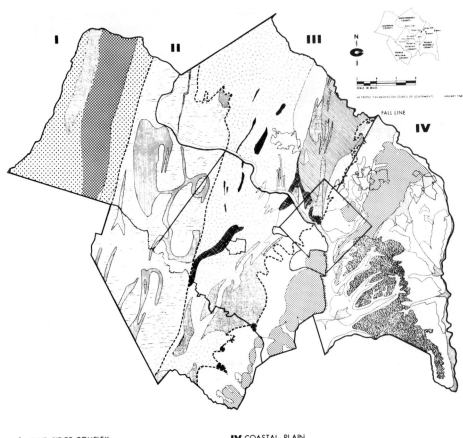

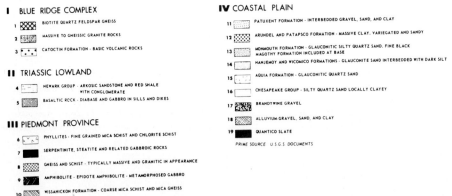

FIG. 47.1 Generalized geology of the Washington metropolitan area, based on U.S. Geological Survey publications and prepared for planning use.

area, and their relation to the natural environment so excellently illustrated in the report. It is impossible even to glance through these pages without being made aware of the profound influence of geology upon the natural features of the area.

The information thus assembled was drawn from many sources. The geology of the Washington area has been studied by the U.S. Geological Survey (the national headquarters of which has always been in the area) since 1882. Dr. N. H. Darton, of the Survey, started his observations in 1891 and was able to publish a masterly report in 1950, summarizing his studies over a period of 50 years.[47.13] There must be few, if any, cities that have had the benefit of such devoted attention to their geology from one individual for half a century. His work has been carried on by other members of the Survey staff. In 1961 a detailed engineering-geologic study was started of the 112,500 ha (278,000 acres) of the inner urban area; it is yielding a series of modern geological maps, on a scale of 1:24,000, that depict not only the major lithologic units but also the general engineering properties of these units and the mineral resources of the area.

The San Francisco Bay Area The San Francisco Bay Area must be given pride of place in considering the planning of regions in relation to geology, since here were prepared two of the pioneer engineering-geology maps of the U.S. Geological Survey. In 1957 there was published a map showing the "areal and engineering geology of the Oakland West Quadrangle, California," by Dorothy H. Radbruch, and in the following year a corresponding map of the "geology of the San Francisco North Quadrangle, California," by Schlocker, Bonilla, and Radbruch.[47.14] Both sheets present not only excellently reproduced geological maps of the land area shown but long descriptions of the geology and tables giving generalized descriptions of the engineering properties of the earth materials encountered. These early tables gave such information as permeability, slope stability, and foundation characteristics. Geological sections complete the quite remarkable coverage that is presented on these single sheets (Fig. 47.2).

The origin of the project may well be quoted from the legend on the San Francisco sheet

> [The San Francisco Bay] area was selected because its strategic location, both as the focal point of development for a large part of the west coast and as the gateway to trans-Pacific commerce, has made it one of the outstanding centers of continuous expansion and construction in the nation. The study in no way pretends to supplant detailed site studies. Rather it tries to supply an accurate background picture of the lithology and of the geologic processes that change or modify the earth materials involved. As much of the quadrangle that does not lie beneath the bay is hidden by streets, buildings and other man-made structures, geologic observations were confined largely to undeveloped lots and to current excavations for utility lines and building foundations and had to be supplemented by data obtained from boreholes and earlier foundation construction. Several thousand logs of boreholes drilled by private firms as well as federal, state, county and municipal agencies provided data that were invaluable in filling out the geologic story of the area.

This remains today an excellent summary of what maps for engineering-geological purposes can and should indicate. This early work was supplemented on a much larger scale by a major joint project of the Geological Survey and the U.S. Department of Housing and Urban Development. Although all the initial objectives of this master study may not have been met because of a change in priorities for funding, *Program Design,* published in 1971 and describing the operation, can still be rightly described as a classic document in its field.[47.15] This 120-page document is a complete outline of the many factors involved in studying the geology of this great area from the point of view of assisting urban planning.

FIG. 47.2 San Francisco, California; aerial view of the city showing generally the area covered by the engineering-geological map mentioned in the text.

The area of the Bay region is 19,920 km² (7,416 mi²), almost the same as the total area of Israel. Population of the nine counties covered by the project is over 5 million. Indicative of what has been done in the development of the region is the reduction in the water area of San Francisco Bay from 1,750 to 1,030 km² (680 to 400 mi²), 725 km² (280 mi²) having been filled or diked. There is today, therefore, a wealth of information in the more than 70 maps and reports already published on the geology of this important region, information available to assist all local planning work. The 1971 volume will long serve as a guide to others faced with planning similar large-scale areal geological studies.

Denver, Colorado Denver is the center of a rapidly growing conurbation located in a magnificent natural setting close to the foothills of the Rocky Mountains; the significant geology of the region is obvious to even the most casual observer. With the city of Boulder about 32 km (20 miles) to the northwest, and with many satellite communities around Denver itself, there are many common problems to be faced in the steady development of this important area (Fig. 47.3). Early in 1967, a controversy developed over the proposed subdivision of a small hillside tract near Boulder. City officials, in the face of strong objections on the grounds of the hazards attached to soil and geological conditions along this eastern margin of the adjacent Front Range, sought for specific terrain information. They turned to the U.S. Geological Survey, as a result of which a manuscript copy of a geological map of the Boulder quadrangle was promptly placed in open, or public, files, so that it could be officially consulted. This was a geological map, however, and so questions inevi-

FIG. 47.3 Denver, Colorado; general view of the city with its imposing backdrop of the Rocky Mountains; the development of the area below the Front Range is mentioned in the text as illustrating the vital uses of geology in planning.

tably arose about the interpretation of the geology that it showed in relation to problems of local land development.

Almost coincidentally with this specific request, the Denver Regional Council of Governments (DRCOG)—formerly the Inter-County Regional Planning Commission—requested the U.S. Geological Survey to conduct a regional planning study of geological conditions as they affect the use of land along the eastern margin of the mountains near Denver. A cooperative engineering-geological research project was therefore started in August 1967, this being the first such project shared by the USGS and a regional planning organization. Initially, the project was financed cooperatively among the regional planning commission, several city and county governments in the region, and USGS. After July 1968 it was financed by the Denver Regional Council of Governments and the USGS.

This cooperative study covered an area of 116,000 ha (287,000 acres). The resulting engineering-geological maps, together with accompanying reports, were first released in preliminary form by 1971.[47.16] Each map covers an area of 14,200 ha (35,000 acres). The two objectives of the program are

> to supply DRCOG with engineering geology information for planning use in a part of the Denver metropolitan area that is rapidly becoming urbanized, mainly by augmenting existing geological information with engineering geology information ... [and] to explore how geologic information can be made more understandable and useful to more non-geologists, in part by developing methods of representation of map information, text arrangements, and verbal expression.

The second objective can be seen to be a research project in the preparation of engineering-geological maps, so that the whole program has a significance that extends far beyond the immediate needs of the Denver area, important as they are.

A helpful summary of the Front Range Urban Corridor Project has been published by Hansen in a GSA symposium volume on urban geomorphology.[47.17] As with almost every such regional project, although the results obtained are of prime and immediate benefit to those responsible for the planning of the region in question, the published results are of far wider significance, while what may be called the "by-products" are sometimes unexpected. Possibly associated with this project was the inclusion in the state (of Colorado) land subdivision regulations of authority to establish land-subdivision criteria dealing with "earth testing" and other geotechnical matters. And the city of Lakewood, one of Denver's suburban communities, has now included in its planning ordinance a requirement for a site report by an engineering geologist. The City of Boulder has a full time geologist on its staff.

Allegheny County, Pennsylvania Different again were the problems faced in relating geology with planning for Allegheny County, Pennsylvania. With an area of 1,886 km^2 (727 mi^2), the county is centered around the city of Pittsburgh and so has no less than 60 percent of its area urbanized. The summary report of the extensive geological investigations carried out by the U.S. Geological Survey in the 1970s for the Appalachian Regional Commission (with other agencies also involved), contains this eloquent summary of how the county in the space of 200 years became:

> one of the most important industrial centers in the United States, largely because of an original abundance of environmental resources: rivers and streams for transportation and water supply, timber for construction material, fuel and charcoal for the early iron furnaces, iron ore and limestone for the iron and steel industry, coal for fuel and coking, oil and gas, and others. These resources strongly counterbalanced the unfavourable configuration of the land surface which is characterized by streams flowing in narrow steepsided valleys at levels commonly 300 feet and locally more than 600 feet below intervening ridges. . . .

And the legacies of two centuries of development are similarly well summarized:

> On the positive side is a varied, vital society, relatively stable in population, having established institutions and a broad economic base. Negative aspects include environmental problems. . . . [the principal, in addition to air and water pollution, being] flooding, landsliding, and mine subsidence.[47.18]

Landsliding is so serious a local problem that special attention was given to landslides in the geological studies. More than 2,000 recent and prehistoric landslide areas were mapped in this one county. Slopes that may be susceptible to significant landsliding underlie about 15 percent of the county. The final results of the geological planning studies were published in the form of a series of specially prepared maps, of which no less than three relate to landsliding.[47.19] These have been supplemented since the original publications by further maps, of which the latest, by Pomeroy, is one of adjoining Beaver County and shows more than 3,000 landslide areas, most of them small but still affecting all local planning (Fig. 47.4).[47.20] The summary report of the Allegheny County studies is a succinct document, of much general interest; the foregoing quotations are taken from it.

St. Lawrence County, New York Finally, and on an entirely different scale in terms of the urban development concerned, the remarkable work done in St. Lawrence County, New York, must be mentioned. As an example, it is far from being unique, but it serves to show what can be done in an essentially rural area. The county occupies the extreme northwestern corner of the central part of New York State, and is thus bordered by the international section of the St. Lawrence River. It is a large county, (7,280 km^2 or 2,800 mi^2); it has no cities within its boundaries but does have four towns, such as Potsdam and Canton, with populations of more than 10,000. Most of the settlements are small towns

FIG. 47.4 Head scarp of a landslide at Washington, Pennsylvania, one of those shown on the landslide map mentioned in the text.

with populations numbered in hundreds. The county does have a planning board, with its office in Potsdam. With the aid of a relatively small financial grant from the federal government (through the Department of Housing and Urban Development under the Comprehensive Planning Assistance Program authorized under Section 701 of the Federal Housing Act of 1954 as amended) and under the Comprehensive Planning Assistance Program of the New York Department of State, it was possible for the planning board to engage a small staff.

Geology was studied first, possibly because the county then had the good fortune to have a geologist, in the capacity of an interested and concerned citizen, as a member of the planning board. This permitted the preparation of a general land-use map for the county as a basis for all future planning. Then similar land-use plans, with appropriate explanations, were prepared for a number of the towns and hamlets in the county, each printed in bright colors, sometimes even with a small generalized map of land-use areas, delineating in broad terms the major land uses recommended. That for the town of Fowler is printed on one piece of paper (28 by 41.5 cm or 11 by 16.5 in), with concise and simple explanantions for the development of the hamlet printed on the back. That for the town of Macomb (population about 1,000) is in the form of a 48-page, pocket-sized, typewritten pamphlet, well reproduced, with four pages of good photographs, the land-use map, again in vivid colors and easily read, fixed to the inner back cover (Fig. 47.5). The title of this little 1976 publication is well chosen—*Macomb; Plan for the Future.*[47.21]

For large cities, for intensely developed regions, for small cities and towns, and for quiet rural areas, the needs are the same if future development is to follow sound lines and build communities good to live in and convenient to work in. A basic appreciation of geology is the starting point for the land-use map which can provide the framework for all detailed development, natural hazards being anticipated from geological studies, precautionary measures ensured through good regulations and sound engineering. Every effort must be made to have such information conveniently available for all human settlements.

FIG. 47.5 Cover of one of the series of simple but effective town plans prepared for St. Lawrence County, New York.

47.6 TERRAIN ANALYSIS

There are many parts of the world, including some parts of North America, where the large areas involved and the lack of more than a mere outline of the regional geology prohibit detailed geological planning such as has just been described. Here the judicious application of aerial-photo interpretation has been providing answers in an increasingly significant manner. Such skills have been developed in interpreting terrain features from aerial photographs that it has proved possible to prepare maps with all important geological features indicated, upon the basis of which the usual requirements for good land planning can be developed. The most recent example and one of the most extensive known to the authors will be briefly outlined.

In the development of its municipal regulations, the province of Ontario has stressed the need for every municipality in the province to have a land-use plan. The province has a total area of 950,000 km² (365,880 mi²), about two-thirds of which can be called "northern Ontario," with some major communities on and close to the Great Lakes but generally sparsely settled, small pioneer communities being widely separated, often by somewhat inhospitable country. The provincial Department of Northern Affairs accordingly arranged with the Department of Natural Resources for the Ontario Geological Survey to conduct an engineering-geology terrain study of 370,000 km² (143,000 mi²), covering the area generally between 46° and 51°N. Time was pressing. Arrangements were therefore made to have the work carried out by contract with five professional firms, all working in cooperation, coordinated by one of the firms and under the general direction of the chief of the Engineering and Terrain Geology Section of the Survey. Approximately 12,000 aer-

GEOLOGY AND PLANNING   47-19

ial photographs of the area were available (the entire land area of Canada having been surveyed from the air a number of years ago, to scales of between 1:30,000 and 1:70,000). An agreed-upon system of symbols was developed, the combination of color (to show landforms), letters (to indicate topographic features), and symbols (to indicate special terrain features) giving a clear and readily understood guide to landforms and terrain character (Fig. 47.6).

Once the base maps were completed, it was then possible to prepare from about a quarter of them a derivative map showing such features as sand and gravel deposits, groundwater supplies, and excavation suitability. In less than three years the job was done, at a total cost of about $1.5 million, the result being a set of 102 maps, each accompanied by a simple explanatory booklet, conveniently packaged, the whole presentation being designed so as to be convenient and understandable for the informed layman. Derived maps will be prepared as required, many being already available. The remarkable project well illustrates how, with the aid of modern technology (here in the form of good aerial photographs) and expert professional skills, the geological factors necessary for the planning of even "wild and inhospitable places" (although much of the Precambrian country of northern Ontario is far from that, being attractive to tourists) can be obtained economically and efficiently.[47.22]

47.7 GEOLOGICAL INFORMATION

The few geological reports that were available for northern Ontario were naturally studied prior to the aerial-photo interpretation exercise just described. Published geological reports will always provide the starting point for all geological contributions to planning. Rarely will there be found geological maps to a scale large enough to allow the maps to be used directly for planning purposes. But the geological maps that are available will give the general geological picture of the region in which the planning area is located. More detailed geological surveys of this area will thus be materially assisted. Appendix B to this Handbook is a guide to geological surveys throughout North America and, in a more general way, in the rest of the world. Officers of such surveys will be willing to assist with such applications of their work and to suggest sources, in addition to their own publications, which may be of service.

Such additional publications will be varied indeed. There must be few major urban areas in North America which do not have some local guides to their geology, at least in general terms. Some typical examples were given in Chap. 13, but these were a small sampling only of the remarkable local geological records which will often be found if search is made. One such record only need be noted here (again) as typifying the sort of information that may be found. The *Saskatoon Folio*, as it is known throughout Canada, and indeed beyond, is a fine compilation prepared by a group of volunteer workers, public financing being provided only for the printing job in preparing the work for publication.[47.23] An area of 324,000 ha (800,000 acres), centered on the city of Saskatoon, is described with the aid of 54 maps and diagrams, the maps showing features such as bedrock, surficial geology, pedology, climate, and geotechnology very clearly, with useful descriptive accompanying notes. If every city had a corresponding compilation, the job of urban planners would be immeasurably assisted.

A general guide—*Earth-Science Information in Land-Use Planning* is its title—is one of the most useful by-products of the San Francisco Bay Region Environment and Resources Planning Study (the full title), to which tribute has already been paid with an all-too-brief statement of its scope and achievements. Published by the U.S. Geological Survey but written by three consultants, this 28-page brochure is even more useful than

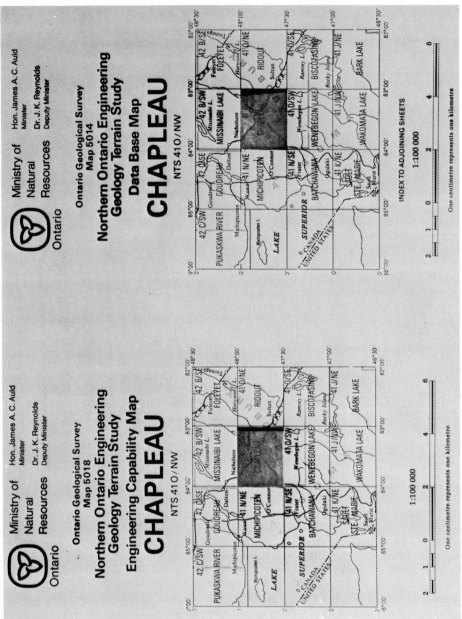

FIG. 47.6 Sample titles of the series on engineering geology, and first derivative maps prepared for northern Ontario, as described in the text.

its main title would suggest.[47.24] It is subtitled *Guidelines for Earth Scientists and Planners,* and helpful suggestions for the best use of geological information in the planning process occupy the latter part of the text. Suggestions as to sources of earth science information come first however, even the inevitable acronym EST appearing on the first page. As another introduction to the uses of geology in planning, this small but valuable publication is warmly commended.

Finally, for all urban planning it is natural to think that the local civic authorities will have an accumulation of geological information, most probably in the city engineer's office, from the records of the works, both public and private, that have been carried out within their city limits and which have revealed through test borings and/or excavation the geology of their sites. Surprisingly, and most unfortunately, few cities do have such banks of earth science information. This strange void exists despite the fact that plans have to be registered with all civic authorities before building permits can be granted, while engineering records of the city's own public works are part of the civic archival material, or should be. This is the general situation, however, one that stands in need of urgent correction. The preparation of the first engineering-geological map of San Francisco was greatly aided by the assembly of just this sort of information. The Twin Cities, Minneapolis and St. Paul, now have an invaluable guide to their subsurface through maps prepared with the aid of records of about 4,000 test bores and 900 water wells.[47.25] Their example should be widely followed.

47.8 THE FUTURE OF PLANNING

The future of urban and regional planning is assured as a vital, and indeed essential, prerequisite to the best use of land for all human purposes. Planning can only be effectively carried out against the background provided by thorough knowledge of the subsurface of the area under consideration, i.e., of its geology. Implementation of plans requires more detailed subsurface investigations at the sites of all structures and for underground works such as tunnels. When construction is carried out, constant watch is necessary so that the actual details of geological conditions, previously deduced from surveys and test programs, can be determined and recorded for future benefit. When this positive information about the subsurface is fed back into the planning process, and made publicly available for all to use, the public will benefit, planning will be improved, and economy will be served.

It is, therefore, incumbent upon all geologists to be appreciative of the uses to which their science can be put in the public service. Those charged with geological studies for planning purposes must be prepared to present their information in the simplest possible way, conscious that geology is here a means to an end but appreciative also of the benefits that may come as their deductions are tested when construction starts. A mere glance at Table 47.1 will be a reminder of the intimate relation between civil engineering and planning, this areal exercise being in one way an extension of the planning that is a part of every civil engineering design. Special responsibility rests upon the civil engineer to ensure that all the information about the subsurface that his works provide is not "buried" in filing cabinets but is placed in responsible hands for use in future planning work. It can thus be made available publicly for consultation by those interested in local geology and by professionals whose work may be speeded and assisted by such information. And if a city does not yet have a proper system for collecting, assembling and utilizing information on its own subsurface, civil engineers should be in the lead in making sure that such a system is developed and supported—before it is too late.

47.9 REFERENCES

47.1 Lewis Mumford, *The Culture of Cities,* Harcourt, Brace, (New York), 1938, p. 336.

47.2 I. McHarg, *Design with Nature,* Natural History Press (American Museum of Natural History), New York, 1969; see also by same author, "Ecological Values and Regional Planning," *Civil Engineering,* **40**:40. (August 1970).

47.3 L. Leopold, *Hydrology for Urban Land Planning—A Guidebook on the Hydrologic Effects of Urban Land Use,* Circular no. 554, U.S. Geological Survey, Washington, 1968.

47.4 N. W. Rutter (ed.), *Geological Information—Problems in Transfer from Scientist to Policy Maker,* Special Report, Geological Society of America, Boulder, Colo., 1979.

47.5 *The Earth and Human Affairs* (report of Committee on Geological Sciences of the National Academy of Sciences, Ian Campbell, chairman), Canfield, San Francisco, 1972.

47.6 J. H. Wiggins "Towards a Coherent Natural Hazards Policy," *Civil Engineering,* **44**:74. (April 1974).

47.7 K. G. Farquharson, S. O. Russell, and N. A. Skermer (eds.), *Risks and Hazards.* (special issue of *The B.C. Professional Engineer*), vol 27, no. 1, Vancouver, January 1976.

47.8 J. T. Alfors, J. L. Burnett, and T. W. Gay, *The Nature, Magnitude, and Cost of Geologic Hazards in California and Recommendations for Their Mitigation,* Bulletin no. 198, California Division of Mines and Geology, Sacramento, 1973.

47.9 B. A. Bolt W. L. Horn, G. A. Macdonald, and R. F. Scott, *Geological Hazards,* Springer-Verlag, New York, 1975.

47.10 I. Campbell, "The Influence of Geologic Hazards on Legislation in California," *Bulletin of the International Association of Engineering Geology,* **14**:201 (1976).

47.11 *Governor's Conference on Geological Hazards in Utah, 14 December 1967,* Special Studies no. 32, Utah Geological and Mineral Survey, Salt Lake City, 1970.

47.12 *Natural Features of the Washington Metropolitan Area,* Metropolitan Washington Council of Governments, Washington, D.C., 1968.

47.13 N. H. Darton, *Configuration of the Bedrock Surface of the District of Columbia and Vicinity,* Professional Paper no. 217, U.S. Geological Survey, Washington, 1950.

47.14 D. H. Radbruch, *Areal and Engineering Geology of the Oakland West Quadrangle, California,* Map no. 1-239, U.S. Geological Survey, Washington, 1957; see also J. Schlocker, M. G. Bonilla, and D. H. Radbruch, *Geology of the San Francisco North Quadrangle, California,* Map no. I-272, U.S. Geological Survey, Washington, 1958.

47.15 *Program Design, 1971, San Francisco Bay Region Environment and Resources Planning Study,* U.S. Geological Survey and U.S. Department of Housing and Urban Development, Menlo Park, Calif., 1971.

47.16 M. E. Gardner, *Preliminary Engineering Geologic Map of the Boulder Quadrangle, Boulder County, Colorado,* U.S. Geological Survey, (open file), Washington, 1968.

47.17 W. R. Hansen, "Geomorphic Restraints on Land Development in the Front Range Corridor, Colorado," in D. R. Coates (ed.), *Urban Geomorphology,* Special Paper no. 174, Geological Society of America, Boulder, Colo., 1976.

47.18 W. R. Wagner, et al., *Geology of the Pittsburgh Area,* General Geology Report no. G 59, Pennsylvania Geological Survey, Harrisburg, 1970.

47.19 R. P. Briggs, *Environmental Geology, Allegheny County and Vicinity, Pennsylvania—Description of a Program and Its Results,* Circular no. 747, U.S. Geological Survey, Washington, 1977.

47.20 J. S. Pomeroy, *Map Showing Landslides and Areas Most Susceptible to Sliding in Beaver County, Pennsylvania,* Map no. I-1160, U.S. Geological Survey, Washington, 1980.

47.21 *Macomb—Plan for the Future,* St. Lawrence County Planning Board, Potsdam, N.Y., 1976 (a typical example).

47.22 N. A. Roed and D. R. Halett, *Chapleau Area, Districts of Algoma and Sudbury,* Northern Ontario Engineering Geology Terrain Study 80, accompanied by Map no. 5014 (Data Base Map) and Map no. 5018 (Engineering Capability Map), Ontario Ministry of Natural Resources, Toronto, 1979, are typical examples.

47.23 E. A. Christiansen (ed.), *Physical Environment of Saskatoon, Canada,* Saskatchewan Research Council, Saskatoon, and National Research Council, Ottawa, 1970.

47.24 W. Spangle et al, *Earth-Science Information in Land-Use Planning: Guidelines for Earth Scientists and Planners,* Circular no. 721, U.S. Geological Survey, Washington, 1976.

47.25 R. F. Norvitch and M. S. Walton (eds.), *Geologic and Hydrologic Aspects of Tunneling in the Twin Cities Area, Minnesota,* Miscellaneous Investigations Series, Map I-1157, U.S. Geological Survey, Washington, 1979.

Suggestions for Further Reading

Unfortunately, literature on the subject of planning is still relatively sparse. Planners in general, with a few notable exceptions such as Professor McHarg (see above), have yet to accord recognition to the key place that geology must occupy in all aspects of urban and regional planning. Engineering geologists, on the other hand, are well aware of the relevance of their work to the basic requirements of planning. This was shown, for example, by the holding of a conference in April 1970 at the University of Southern Illinois on Engineering Geology in Relation to Urban Land Use and Planning. Typical of the papers then given was one by H. C. Schlicker called "Engineering Geology in the Northwest."

The Geological Society of America, through its Committee on the Environment and Public Policy, has recognized the importance of geology in the planning field. For example, the Committee has published a separate report called *Geological Constraints in the Urban Environment,* the result of a panel discussion held in 1975.

The United States Geological Survey has long recognized the place geology has in planning. Three of its publications illustrating this concern are: *Seismic Safety and Land Use Planning,* Professional Paper 941 B by M. L. Blair and W. E. Spangle (1978); *Flood Prone Areas in Land Use Planning,* Professional Paper 942 by A. G. Waananen et al. (1977); and *Hydrogeology for Urban Planning,* Circular 554 by L. B. Leopold (1968).

Engineering interest is indicated by the fact that the biennial meeting of civil engineering societies of the United States, Great Britain, and Canada (ASCE, Inst. C.E., and CSCE) held in Torquay, England, was devoted to consideration of Planning and the Civil Engineer. The *Proceedings* were published by the Institution of Civil Engineers in 1982 as an 176-page volume.

Although appreciative reference has been made to the book *Geological Hazards,* it may usefully be referenced again as a reminder of some of the hazards, dependent upon geology, that should always be considered in general approaches to planning. Authors of this notable volume (published by Springer-Verlag of New York in 1975) were B. A. Bolt, W. L. Horn, C. A. Macdonald, and R. F. Scott.

chapter 48

MAN-MADE GEOLOGICAL PROBLEMS

48.1 Large Dams / 48-2
48.2 Diversion of Water / 48-4
48.3 More Problems with Water / 48-6
48.4 Pollution / 48-7
48.5 Waste Disposal / 48-9
48.6 Problems with Methane / 48-13
48.7 Waste from Mines / 48-14
48.8 Conclusion / 48-16
48.9 References / 48-17

"One-Tenth of the land surface of England and Wales is covered with buildings, railways, roads and the like. In the past twenty years about 650,000 acres have been lost to farming."[48.1] Similar statements, but with amended figures, could be made for most of the countries of the Western World. Even in so large a land as the United States, more than one percent of the total land area has been utilized for streets and roads alone, and that figure makes no allowance for developed urban areas apart from the space used for streets.[48.2] Only when viewed in this broad perspective does the significant change that man has caused in the character of the surface of the earth become clear. Much of human activity in changing the face of the earth has been beneficial, "the crooked [have been] made straight; the rough places plain." There have naturally been associated problems, perhaps the most obvious being the change in the disposition of rainfall when it all runs off a paved area without reaching the ground as it normally would do, a problem which Benjamin Franklin had noted and so commented upon in his will (1790).[48.3]

In preceding chapters reference has been made to the large open excavations made necessary by some engineering projects, to tunneling through the depths of mountains, to execution of major works on seacoasts that interfere with the action of the sea, to the draining of lakes, to the construction of water-retaining dams and so the formation of artificial reservoirs. All these are activities that affect the surface of the earth, and so they are, in effect, geological operations. In their local setting, the projects may appear immense in scope and significant in their effect upon natural conditions. Even the greatest of civil engineering projects has only to be viewed from the air, however, to be put into perspective; it will then appear to be a relatively small thing when compared with the unaided operations of Nature. Viewed in such perspective, civil engineering projects can be seen merely to "scratch the surface."

The remarkable achievements of civil engineering are not to be belittled; all that is suggested is that the scale of comparison being used must always be kept in mind. The catacombs of Rome, for example, originally developed as quarries, have a length of 885 km (550 mi), yet a visitor to this ancient city may never realize that such excavations are so close at hand. The catacombs of Paris, not so well known, required the excavation of 10 million m³ (13 million yd³), about four times the volume of the Great Pyramid. Excavation of the "wet docks" of the port of Liverpool, England, still the greatest continuous area of enclosed dock space in the world, required the excavation of 4.6 million m³ (6 million yd³) of soil, from which, incidentally, 30 million bricks were manufactured for use in local buildings. Typical of more general figures that can be cited in this connection is the volume of about 30 billion m³ (40 billion yd³) of soil and rock that is estimated to have been excavated in Great Britain for all purposes prior to the start of the First World War, a figure that has today been greatly exceeded.[48.4]

These are impressive figures when considered in the light of the quantities normally dealt with in engineering works. When such excavations are seen in broader context, however, then their effect, in a geological sense, can be seen to be almost negligible. The materials removed are inert; their removal seldom has any indirect effects. With mines, however, the situation is often different; and the disposal of waste material from mines has, in the past, created serious problems. Much attention is now being given to the disposal of all waste materials, but problems remain, essentially geological in character even though caused by human activity. This Handbook would not be complete without some recognition of the worst of these problems, even though this recognition must be given in little more than a listing. There are lessons still to be learned, however, and so even a listing may be of some value.

48.1 LARGE DAMS

The construction of large dams does interfere with previously existing natural conditions, having indirect as well as direct and clearly visible results. Groundwater conditions will be immediately changed upstream, and possibly downstream, from the dam. The reservoir area lying beneath the elevation of the crest level of the dam will be flooded and removed, more or less permanently, from human sight and use. All vegetation in the flooded area will be changed; trees are now almost always removed before flooding. Animal and insect life will be displaced from the reservoir area. And a new load, the weight of the retained water, will be placed upon the rocks that form the bed of the reservoir.

These factors might appear to be of slight importance until they are considered in relation to very large structures. Hoover Dam, for example, is 218 m (726.4 ft) high; the new lake which it formed, Lake Mead, is 184 km (115 mi) long, with an area of 66,000 ha (163,000 acres) and an estimated volume of 38 billion m³ (31 million acre-feet) (Fig. 48.1). Interference with biological conditions was not here an important matter because of the character of the country that was flooded. But some surprising effects have been observed in the operation of the reservoir, such as the discharge of muddy water through the outlet gates within eight months of closing the gates to start the filling. This demonstrated clearly the existence of density currents in the waters of the new lake and pointed the way to a new appreciation of sedimentation processes. Silting of the reservoir was naturally anticipated and had been allowed for in the design of the dam. Somewhat unexpected, however, was the change in the chemical balance of the water within the reservoir. It has been found that large quantities of salts, particularly gypsum, are going into solution at

FIG. 48.1 Lake Mead, created by the Hoover Dam on the Colorado River.

a rate of more than 1.5 million tons per year. On the other hand, it is estimated that as much as 800,000 tons of salts, mainly calcium carbonate, are being deposited on the floor of the reservoir every year.[48.5]

Drastic interference with biological conditions was the first major effect of the closing in 1959 of the discharge gates of the Kariba Dam on the Zambesi River in Zimbabwe (Rhodesia): this is a 126-m (420-ft) -high, arched concrete dam across the Kariba Gorge (Fig. 48.2). In an underground power station, fifteen 100,000-kW generators, of which six were initially installed, are developing power for use throughout Zimbabwe. This outstanding civil engineering project is further distinguished in that the dam has created the largest of all man-made lakes, Lake Kariba, 280 km (175 mi) long, impounding 160 billion m³ (130 million acre-feet). Much of the flooded area was covered with lush tropical vegetation and inhabited by a singularly varied wildlife. Operations to rescue wild animals trapped on artificially formed islands as the water gradually rose constituted an epic zoological adventure. More permanent effects upon local fauna and flora were anticipated, as well as some effect upon the equilibrium of the adjacent country. Unusually complete arrangements were made for careful scientific observation of the continuing effects of this great dam upon the country around it.[48.6]

Even before construction of the High Dam at Assouan started, there had been published warnings of what some of the serious effects might be from building this great bar-

FIG. 48.2 Kariba Dam on the Zambesi River, Africa, with its vast reservoir filling for the first time.

rier across the river Nile.[48.7] But construction went ahead and the dam was completed. It is yet too early to see what long-term effects it will have, but already brickmaking and agriculture in the vast valley downstream have been interfered with, since the annual replenishment of silt at the time of the spring flood has been stopped. The salinity of the eastern part of the Mediterranean Sea has been affected; so also has the local fishing, again because of the absence of the usual spring flood from the Nile. And flow patterns in the Suez Canal have been altered.[48.8]

48.2 DIVERSION OF WATER

It has been in association with the building of dams for the generation of water power and for water supply that some of the most dramatic changes to natural patterns have been made; these patterns are disturbed when water is diverted from one watershed to another, usually by means of tunnels. The Big Thompson scheme of the U.S. Bureau of Reclamation diverts water from the Pacific to the Atlantic watersheds in the vicinity of Denver, Colorado. In northern Ontario, the Long Lac and Ogoki projects divert water from the Arctic watershed of the Nelson River into Lake Superior and the St. Lawrence watershed; the water thus gained from Hudson Bay flows through powerhouses at Sault Ste. Marie, Niagara Falls, and Cornwall-Massena before reaching sea level below Montreal. In northern British Columbia, another large Canadian power project takes water from the upper reaches of the Fraser River and, by means of an impounding dam and long tunnel, brings it quickly to the sea through the Coast Range by way of the great Kemano powerhouse. And in Australia, as explained in Chap. 2, the Snowy Mountains project diverts water through the Great Dividing Range so that it may usefully generate power before being made available for irrigation purposes in one of the great fruit-growing districts of the world.

These are water-diversion schemes of vision and great extent, the design and construction of which called for civil engineering skills of high order. Although they affect the flows in the rivers from which water is abstracted, they have all been so designed that there will be no permanent injury to the depleted watersheds and but little interference with local natural conditions. Far different are some civil engineering projects that have been long under discussion but have as yet failed to come to realization. Prominent among such schemes is the plan to divert water from the Mediterranean Sea into the Qattara depression of the Western Desert of Egypt. This 144- by 300-km (90- by 190-mi) depressed section of the earth's crust lies between the Egyptian-Libyan border and the Nile Valley, less than 320 km (200 mi) from Cairo. Its lowest point is 131 m (436 ft) below sea level. The project, under discussion for many years, depends essentially upon excavating a 64-km (40-mi) -long channel from the coast near El Alamein in order to admit seawater to the depression. A power installation of 225,000 kW is anticipated, but it is also hoped that seawater in the depression will change freshwater conditions and so permit farming in the area. Key to the ultimate economic feasibility of the scheme is the excavation of 290 million m^3 (380 million yd^3) of soil and rock. If and when the project is undertaken, it will indeed be a civil engineering job with major geological overtones, changing the entire character of a singularly interesting, if unusual, part of this historical section of northern Africa.[48.9] A new estimate of 1974 placed the probable cost at over $1 billion, and the power potential as high as 10,000 MW, if pumped storage is used.[48.10] A more recent proposal contemplates a similar vast canal between the Mediterranean and the Dead Sea, the possible consequences of which stagger the imagination.[48.11]

Civil engineering projects carried out for quite different purposes have had similar diversionary effect. One example is found where embankments have been constructed in the sea to connect mainland to islands or islands to one another, thereby interfering with the normal pattern of tidal currents to some extent. Most of the projects of this kind are small and so perhaps may not be significant in a geological sense, although some have had serious effects in causing accumulations of moving sand. One of the largest of recent projects was the construction of a rock-fill embankment to form a causeway between Cape Breton Island and the mainland of Nova Scotia in northeastern Canada (Fig. 48.3). The chief purpose of this work was to give better rail and road access to Cape Breton than was possible with the previously existing ferry service. A bridge could have achieved the same purpose, even though its piers would have had to be unusually deep, but a $23 million embankment was built by direct tipping of rock working from the shores, with a navigation lock at one end to take care of the shipping that had regularly used the Strait of Canso. Currents through this strait were quite strong. These have now been stopped, since

FIG. 48.3 The Canso Causeway across the Strait of Canso, Nova Scotia, linking Cape Breton with the mainland.

the embankment provides a complete barrier to flow. It will be many years before the possible effects of the causeway upon marine conditions and upon the climate can be evaluated.[48.12]

The causeway and the adjacent parts of the Strait of Canso are already under observation so that these long-term effects of the embankment upon the regional ecology will be recorded. One totally unexpected result has already been beneficial. Construction of the embankment naturally stopped the tidal movements through the strait and therefore the wintertime movements of broken sea ice. This ice now accumulates at the western side of the causeway (the Gulf of St. Lawrence side), but the eastern (Atlantic) side remains clear of ice all winter. There are depths of 30 m (100 ft) of water within 220 m (750 ft) of the shore in the long fjordlike strait between the open Atlantic and the causeway, and so the beginnings of a major year-round, very deepwater port have already been developed. The first such wharf serves a major oil refinery; it has already had the largest oil tanker in the world berthed alongside.

Other diversions have also had side effects. As a floodway around the city of Winnipeg, Manitoba, designed to carry the peak of bad floods on the Red River (of the North), a large canal has been constructed. Normally it is almost dry and grass-covered, the word "almost" having to be used since parts of the canal invert are below the local water table. There is, therefore, usually a steady seepage of groundwater into the canal, normal flow being about 9,100 lpm (2,000 gpm). When a major spring flood has to be diverted in part into the floodway, the water level in it quickly rises to an elevation higher than the local water table and so flow is reversed, water seeping *from* the canal into the surrounding ground. This was anticipated, and the possiblity of contamination of the groundwater was considered. A careful groundwater survey was made, with continuing observations at, and after, flooding; some contamination has been detected near the canal intake, but this was quite restricted and did not prove serious.[48.13]

An increasing volume of water, in total, is being diverted from rivers and lakes for use in cooling for power stations. In earlier days, the environmental effects of this were probably not realized, but it is now recognized that *thermal pollution* (as it has rightly been called) is already a serious problem. If it is above a critically low limit, the increased temperature of water thus diverted for cooling purposes can have serious ecological effects after its return to lake or river. The increasing number of nuclear power plants has compounded the problem, since these plants reject 40 percent more waste heat than fossil-fuel stations of the same capacity. Many regulatory authorities now restrict any such increases in temperature to limits such as 2.8°C (5°F) for rivers and streams, 1.7°C (3°F) for lakes and reservoirs, and 2.2°C (4°F) for estuarine areas, with no rise in temperature permitted for specially sensitive streams (such as those with good trout and salmon fishing).[48.14] A first result has been that preliminary studies for major power stations now must include an investigation of the probable effect of the use of adjacent water for cooling purposes; some notable surveys have been described. A second result is that in a steadily increasing number of cases, in both new plants and existing plants, large cooling towers are being constructed on land in order to limit thermal pollution. Acceptance of this inevitable addition to standard power-station design in North America can be dated from early 1970.

48.3 MORE PROBLEMS WITH WATER

"Water," as Izaak Walton wrote so many years ago, "is the eldest daughter of creation, the element upon which the Spirit of God did first move. The water is more productive than the earth. Nay, the earth hath no fruitfulness without showers or dews, for all the herbs and flowers and fruits are produced and thrive by water." The civil engineer, who knows well the place of water in the natural order, may not always appreciate to the full that the uses to which water has been put have seriously depleted many of the most extensive supplies of groundwater in all parts of the world. The engineer cannot be blamed for the overpumping that has unfortunately characterized so many water-supply systems; but he does inherit the problems that overpumping creates and has to face the challenge of designing and carrying out necessary remedial works. All-too-many examples could be cited of the sometimes disastrous results of uncontrolled pumping from natural groundwater reservoirs. To illustrate what has taken place in the past in many countries, some older figures may be quoted as typical:

1. Records obtained from the area of the Great Australian Artesian Basin disclosed that the discharge of 233 bores in New South Wales (taken as examples) decreased by 39.92 percent between 1914 and 1928, i.e., from about 380 to 230 mld (100 to 60 mgd).[48.15]
2. In the case of the Great Dakota Artesian Basin in the United States, the artesian pressure at Woonsocket, South Dakota, dropped from 17.5 kg/cm^2(250 psi) in 1892 to 2.5 kg/cm^2(35 psi) in 1923.[48.16]
3. In the Paris syncline, France, the level of the water table dropped about 75m (250 ft) in 93 years.[48.17]
4. In the Chicago district in the United States, the drop in the level of the groundwater surface has averaged about 1.2 m (4 ft) per year.[48.18]
5. In the London Basin, England, the underground water level is known to have dropped since about 1878 at the rate of about 45 cm (18 in) per year.[48.19]

Depletion of groundwater is, therefore, no new thing. One might think that, with these (and many more) examples from earlier years, there would have been some amelioration of this violation of one of the greatest of natural resources. But, as in so many human affairs, things had to get worse—as shown by examples given elsewhere in this Handbook—before they got better. Today there is, generally, an appreciation of the vital need to protect all groundwater reserves. Regulations limiting withdrawals are in widespread use. And in a growing number of locations, water is now being "recharged" into the subsurface, workers of today making valiant efforts to undo the harm done by the workers of yesterday. It is not surprising to find that even this procedure is no new thing. It would appear that the first attempts to replenish groundwater were made in 1881 at Northampton in England; water was fed back in this case into the Lias limestone. Experiments were conducted in the London Basin by the East London Water Company as early as 1890. Today the practice is widespread, an interesting development being the use of treated sewage effluents for recharging, a combination of solutions to two man-made geological problems.[48.20](Fig. 48.4)

48.4 POLLUTION

The fact that treated sewage can safely be used in this way is, perhaps, the best of all indications of the progress made in dealing with the pollution of water. Testing and controls are now so well developed that pollution of water to be used for any human purpose can quickly be detected, and limitations upon use put into effect. The necessity for exercising such controls, as at all-too-many urban swimming beaches, shows that much yet remains to be done in the battle against water pollution, but, together with air pollution, it is now so well recognized as a social blight that this brief mention must suffice.

FIG. 48.4 Water discharging from the Glendora Tunnel of the Metropolitan Water District of Southern California into the normally dry San Gabriel River to recharge groundwater supplies of the Upper San Grabriel Valley Municipal Water District.

Just as public control over *water pollution* appears to be making substantial progress, a new source of pollution appears. The demand of motorists for clear streets in winter periods, in all northern parts of the continent, has led to a phenomenal increase in the use of salts on highways for snow and ice melting. When snow and ice along highways melt, and run off to drains and so to adjacent waterways, they carry much of this salt with them and this then serves as a new pollutant. Much of the salt that does not stay on the highway, as all northern users of this Handbook will know, accumulates on automobiles from which it is regularly washed off in car-wash establishments. Drainage from these modern service stations is therefore a concentrated solution of the deicing salts used, especially if the carwash recirculates its wash water, as some apparently do. If this drainage discharges into a small watercourse, the resulting pollution can be serious indeed.[48.21] With every advance in modern convenience, therefore, there seems to be at least one new problem, and all too many of these are in the nature of man-made geological problems.

This is even true in connection with *air pollution*. By providing the surfaces upon which the products of mechanical engineering operate, the civil engineer is very indirectly responsible for one of the most significant, until quite recent years, of all changes in the physical state of the Western World—the steady increase in the amount of carbon dioxide in the atmosphere. It has long been known that this contamination of the atmosphere affects the amount of solar radiation reaching the surface of the earth.[48.22] Recent climatic changes are held by some to be directly related to the products of combustion, the effect of which upon the atmospheric conditions in the many city areas is, unfortunately, all too well known. The incidence of radioactive material in the atmosphere from atomic developments has now created greater and more immediate problems, the full significance of which is still uncertain. Admittedly, these are long-term problems, yet they are now recognized and, by some, regarded as already serious.

48.5 WASTE DISPOSAL

Another modern problem of increasing seriousness is the disposal of waste materials. It was estimated in 1977 that the United States will soon be handling no less than 3 billion tonnes of solid wastes in one year. Sixty-three percent of this will be from mines (much of it from coal mines), 22 percent from agriculture, 9 percent from industrial sources, 5 percent from municipal sources, and about 1 percent from sewage plants.[43.23] That 5 percent from municipal sources does not appear to be much; yet it represents at least 150 million tonnes a year, other estimates putting the quantity as high as 250 million tonnes. Few North American cities have yet been able to take the logical, but politically unattractive, step of building plants for the conversion of domestic wastes into heat and power. The example of Montreal, with a most efficient plant of this kind, and other cities that have taken a lead in this development, must be followed by an increasing number of cities if energy conservation is ever to have real meaning. Even then, there will still be the continuing problem of dealing with material that will not readily disintegrate or burn—in the United States something like 7 million old automobiles, 50 billion cans, 30 billion bottles each year. But in the meantime, disposal of waste to landfills will have to continue. And here is where geology comes in.

The old-fashioned "garbage dump" has been the butt of many a joke, but it was at least a start at communal disposal of domestic wastes. With the rapid increase in the volume of waste to be disposed of, from the "affluent society" of North America, the old garbage dump has become the local "land disposal site." Initial problems of finding areas suitable for this essential public service were followed by efforts to ensure that disposal sites were sanitary—thus "sanitary landfills"—tidy, and so operated that new garbage was quickly covered up with soil, giving new open land to the municipality (Fig. 48.5).

FIG. 48.5 Modern sanitary landfill of a major Canadian municipality, showing heavy-tracked "packer" crushing newly deposited waste material, which is covered with stockpiled soil at the end of each day. Geology under the site had been studied, and groundwater conditions beneath it are monitored regularly through permanent observation wells around the site.

48-10 GEOLOGY AND THE ENVIRONMENT

This was all to the good, but only rarely was it realized that the rain falling on garbage would seep through it, become contaminated, and, unless geological conditions were very favorable, seep eventually into the groundwater beneath the site and be carried away by the steady but slow movement of most bodies of water beneath the ground.

Here was a geological problem of complexity, and one not readily appreciated except by those who fully understood and appreciated groundwater hydrology. Fortunately, it came generally to be recognized, although not before serious trouble had resulted in some localities because the source of pollution was not detected. Detailed studies of groundwater movement in the vicinity of garbage disposal areas have led to a useful body of knowledge which can now be drawn upon before any new site is selected. A new word has even been coined—*malenclave,* the zone of contaminated groundwater below a disposal site (Fig. 48.6). Studies of sites in Iowa suggest that steady-state conditions may be

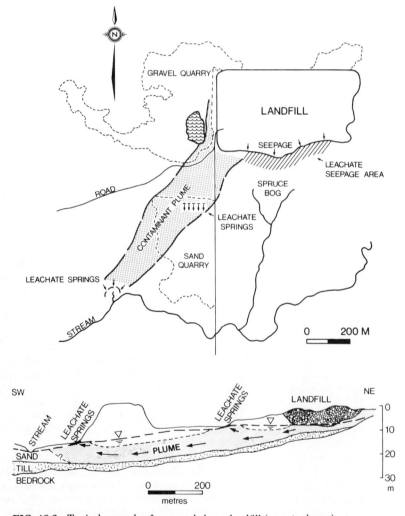

FIG. 48.6 Typical example of seepage below a landfill (an actual case).

reached in about three years, one of the sites investigated having a malenclave 2,100 m (7,000 ft) long, 1,350 m (4,500 ft) wide, and in places of over 18 m (60 ft) deep.[48.24]

Correspondingly, regulations are now in operation in most states and provinces strictly limiting the use of land for disposal sites until the necessary *geological* surveys have been made in addition to the usual topographic surveys and studies of the social effects of site location. Much useful work has been done in educating the public about the geological implications of disposing of garbage, so that the overall situation can be said to be under control, even though much work has still to be done, especially at older sites. Typical of the new attitude toward this problem is the title of a useful paper by Grover Emrich, "What Is a Geologist Doing in a Health Department?"[48.25] It will be appreciated that that geologist has a real job to do.

There are associated problems. The first is the use to which completely filled and soil-covered disposal sites shall be put. They are often conveniently located, and attractive for development, but if possible they should be reserved for use as parkland or for other recreational purposes so that two further problems do not arise—the founding of structures upon such material and the danger of methane gas. These problems have to be faced, however, when it is found that old disposal sites underlie areas that have been purchased for active development. If preliminary subsurface investigations are properly conducted, and the history of the sites investigated, previous use for garbage disposal can be determined, original ground level can be located, and then foundation design studies carried out in the usual way. Methane problems are so serious that they will be considered separately later.

Disposal of the semisolid residue from the treatment of sewage in municipal plants is a civic problem even less well realized than the disposal of garbage. Sewage treatment plants are now being constructed in ever-increasing numbers in delayed response to the demand for a reduction in the pollution of waterways. These plants produce an effluent that is of such quality that it can be discharged safely into running water or lakes, but there remains in the plant, for disposal in some other way, the sludge that collects in the treatment tanks. Burning is one way of disposing of the sludge, since it does contain combustible material. In seaboard cities, dumping in the sea is an alternative way of getting rid of the sludge, but even this easy means of disposal is now being questioned, since pollution of the sea is under increasingly critical review. In relatively recent years as much as 150,000 m^3 (174,000 yd^3) of sludge from New York's sewage plants were deposited annually at a specially assigned location at the head of the (underwater) Hudson Canyon in the continental shelf.[48.26] Deep-sea deposition is being increasingly required. Possibly the best of all means of disposing of sewage sludge is the method practiced now for many years by the City of Milwaukee in the preparation of their well-known fertilizer, Milorganite.

Disposal of solid wastes from industry is often a difficult and costly matter. It is something that industries have to include as a part of their overall operations. Much research has been done in finding better ways of using wastes; today this search for economic solutions to the waste problem is being intensified. In some cases, such as the disposal of the sludge obtained from the refining of bauxite for the manufacture of aluminium, there is no solution at present other than to provide storage ponds in which the sludge can settle out from the water with which it is processed. Satisfactory methods of providing this storage have been developed, costly though they can be.

Geology is implicit in the provision of all sites for such disposal purposes, as is always the case, but the somewhat unusual geological implications of this subject relate to the materials that can be made from some industrial waste products. In the manufacture of steel, for example, there are large quantities of slag to be disposed of. The use of treated slag as a concrete aggregate has been mentioned, but it is also widely used as fill, when its high specific gravity is no disadvantage. As with natural materials used directly for fill, knowledge of the properties of the slag proposed for fill is essential before its approval for use. It has been found, for example, that certain types of slag from the open-hearth system of steelmaking have the undesirable property of expanding in the course of time. If placed under a building, such material can cause very serious trouble because of this expansion, as some cases on record have shown.[48.27]

Correspondingly, the large quantities of fly ash that derive from the operation of large, fuel-fired steam power stations present a massive problem of disposal; extensive research programs are under way in many countries in seeking uses for this interesting material, another man-made geological product. It is being used as fill in old mine workings, as a filler with bituminous mixtures, as a constituent of stabilized soil mixtures, as a medium for water treatment, especially in neutralizing acidic soils, and for the manufacture of bricks and concrete. The quantities of fly ash that are today being produced are immense. British Railways has built four complete special trains merely for transporting fly ash from just three steam power stations in the English Midlands to the Fletton area. Here the ash is being used to fill up worked-out clay pits and an old reservoir in a major land-rehabilitation project that is estimated to last for 35 years.(Fig. 48.7).[48.28]

Similar remedial works are badly needed in many parts of the world where the landscape has been disfigured by massive excavation (as for coal or iron ore obtained from open pits) or in the mining of other types of ore that have been handled hydraulically. California faces unusual problems from this cause, stemming from the hydraulic placer mining for gold carried out in years before 1884 when the U.S. Circuit Court of Appeals prohibited uncontrolled deposition of hydraulic-mining debris. Despite the years between, such debris has been responsible for reducing the capacity of reservoirs, such as that behind the Combie Dam, by as much as 125 hectare-meters (1000 acre-feet). Dredging gold-bearing gravels has been carried out more recently on almost every gold-bearing river in northern California. Some of the dredged tailings were used in the construction of the great Oroville earth dam. An extensive research program has been initiated seeking

FIG. 48.7 Fly ash being used as fill on the Baldock Bypass of the Great North Road (Highway A1) of England. The fill was placed against bridge abutments. A relative compaction of 90% was achieved.

ways of manufacturing building materials from this old waste material, and some success has been achieved.[48.29]

Geology is involved also in the disposal of liquid industrial wastes, since one solution to this problem has been to inject them under pressure through wells into deep-lying strata. There are today possibly as many as 300 deep injection wells in use in North America, some for "untreatable" industrial wastes, although the majority are for the more acceptable disposal of brine from oil fields. Typical injection wells are at least 1,000 m (3,300 ft) deep. The geological implications will at once be obvious. Only after the most searching geological investigation can this method of disposal be followed, the possibility of contamination being one that must be eliminated beyond all doubt.[48.30]

48.6 PROBLEMS WITH METHANE

Problems with methane gas, as encountered in fills composed of refuse, have been mentioned, as has the infrequent possibility of encountering this gas in groundwater because of decomposition of buried natural organic material. It is a particularly insidious gas since it has no smell; it can form a highly explosive mixture with varying percentages of air (oxygen), while being lethal to human beings at much lower concentrations than would be explosive. Fortunately, there are instrumental means of detecting the gas, and these should always be employed if there is any possibility of methane being present in the subsurface of any site upon which building is proposed. If there is any possibility of methane being generated by the disintegration of domestic refuse beneath an area now covered, that site should be rejected as unsuitable for any structures which would enclose space.[48.31]

The most difficult problems, and occasional tragedies, arise when buildings have been erected over old disposal sites without the possible danger of methane having been recognized. Planners of residential developments must be especially wary, since there is no simple way of getting rid of the methane other than waiting for some years. There are all-too-many records of explosions in buildings, usually in basements, because of accumulation of unsuspected methane. Once recognized, the danger must be tackled without delay. There are means of providing vents from the source of the methane, and some details of design can minimize the danger of explosion, but these can only be applied after a most thorough exploration of the subsurface, the job that should have been done before building was carried out. Once safety precautions are instituted, they must be vigilantly maintained. Close to the home of one of the writers is a school, built over an old disposal area, which had to be vacated in the depth of one winter of heavy snowfall, since drifting snow had plugged up steel culverts being used as vents from the fill beneath. There is no general solution to this difficult problem; each case will be unique, dependent upon local subsurface conditions.

Even more remarkable was a case in Florida in late 1969, believed to be the first of its kind but presenting strange implications; 32 explosions took place in some precast, prestressed, reinforced hollow concrete piles that had been driven into place for the new Buckman Bridge near Jacksonville. The contractor carried out the work exactly as required but, as was normal custom up to that time, did not remove the cardboard forms from inside the hollow piles. Water got into the piles by wave action before the piles were capped. The court ruled that the plans for the bridge had failed to call for venting of the hollow piles, bacteria in the water had reacted with the cardboard of the inner forms, and the explosions had resulted. And the bacteria? They were found to be due to pollution of the river water. It may well be said that this is not a geological problem, but if water, probably the main geological material of the earth's crust, can be so polluted as to cause an accident like this, geologists and civil engineers can feel just as responsible as all other citizens, if not more so.[48.32]

48.7 WASTE FROM MINES

Waste from mines has long been deposited in large piles adjacent to mines, often to the disfigurement of the local landscape—a matter at last receiving attention from local authorities and mining companies. In some cases, slips have occurred in the sloping sides of these small, man-made mountains, sometimes with disastrous results. The Aberfan disaster in south Wales will still be hauntingly familiar to most readers; it was essentially a geological failure. Slips in refuse banks often are, and so warrant inclusion in this review of the problem that geology can create with human help.

It was at 9 A.M. on the foggy morning of 21 October 1966 that the No. 7 tip above the village of Aberfan failed and created a flow slide that rushed down the valleyside at a speed of nearly 32 km/h (20 mph), engulfing a farmhouse, destroying the village school, and finally wrecking some homes in the village. In all, 144 people died, of whom 116 were children. As is always the case in such circumstances, a most complete public inquiry was held by the British government, the inquiry tribunal consisting of a learned judge supported by a prominent civil engineer, an expert in geotechnical work. Their published report is a masterly document. It is accompanied by two companion volumes in which are printed eight detailed technical reports, covering every aspect of the disaster, naturally including the geology of the tip site, which was of critical importance.[48.33] Tragic though the disaster was, it has led to this constructive result, as major accidents properly investigated and reported upon always should. In the past it has been all too easy to neglect the hazards that spoil dumps can present, despite a long history of serious accidents that they have caused.

This neglect was brought out at the Aberfan inquiry when attention was directed to a paper entitled "The Stability of Colliery Spoilbanks," written by a mining engineer and geologist of wide experience, G. L. Watkins, and published in the (British) journal *Colliery Engineering* in November 1959, seven years before the Aberfan disaster, but apparently not taken to heart. In the third and fourth lines of the paper, Watkins sets forth his purpose: ". . . to discuss the stability of colliery spoilbanks, especially since in wet weather many slips have occurred."[48.34] It is, of course, always easy to be "wise after the event." It will never be known whether a reading of this paper by anyone connected with the Aberfan tip might have averted this disaster, but the existence of the paper is a telling reminder of the availability of almost limitless information, to be found and used by those who look for it. This is as true of geological matters as of any other field of science or engineering. Watkins' paper was reprinted in November 1966, exactly seven years later, in full, in the same journal in which it had originally appeared. The fact that this is the only case that the authors have ever seen of a paper's being completely reprinted well illustrates the importance that was attached to it at the time of the Aberfan inquiry and supports the intent of the foregoing comments.

Since the tragedy of Aberfan will never happen again, it is to be hoped, it would not be profitable to repeat all the details, not even of the local geology, which was somewhat complex. It may be useful, however, merely to record that the Merthyr Vale Colliery, from which the material in the tip came, first produced coal in 1875. A number of tips were created for disposal of the colliery waste, all close to Aberfan. In 1944, a considerable slip occured in No. 4 tip, and a large part of the tip traveled 480 m (1,600 ft) down the mountainside toward the village. The No. 7 tip, the one that failed, was started in 1958. Its height in 1966 was estimated to be 60 m (200 ft) from tip to toe. The tipping point was

156 m (510 ft) above the village and about 600 m (2,000 ft) from the nearest house. Once the slide started it developed into a flow slide, not unlike those in sensitive clays. One of the contributing factors appeared almost certainly to have been artesian water from the underlying jointed Brithdir sandstone. The importance of geology in this investigation is shown by the fact that two of the eight official technical reports, now published, are by geologists; the companion geotechnical reports naturally touched upon the local geology.

Geologists, therefore, have naturally been involved with the inquiries into the stability of mine tips, especially at coal mines, that have been carried out in other countries since the Aberfan disaster. In the United States, the U.S. Geological Survey undertook such an inquiry. The results were reported by William E. Davies in 1967 in a paper that is so important that nothing better could be done here than to quote in full from the author's own summary:

> Following the disaster at Aberfan, Wales, 60 waste banks were studied in the bituminous region of Virginia, West Virginia, and Kentucky. The banks occur as long mounds and ridges both across valleys and on their flanks, as isolated cones and ridges on flat ground, and as flat valley fills. The banks are as much as 800 ft. high and 1 mile long. Some are used as dams to contain settling ponds for wash water from coal processing plants. The banks consist primarily of coal, shale, and sandstone initially of cobble and boulder size, but the material is broken in dumping and quickly slakes to plates and chips as much as ½ inch across. Of an estimated 1500 banks in the 3 states, 600 are on fire or have burned, converting the waste to red dog, a weakly fused mass of angular blocks and coarse platy ash. Many of the banks are subject to failure from a variety of causes: slippage along shear planes within the bank, saturation of materials by heavy rains, explosion of burning banks, overloading of foundations beneath banks, overtopping and washout of banks that dam valleys, rock and soil slides on valley walls and hillsides breaking through banks, deep gullying, and excavation of toes of banks. In the last 40 years in the 3 states there have been 9 refuse bank failures claiming at least 25 lives; of the 60 banks examined, 38 now show signs of instability.[48.35]

It is regrettable to have to record that a waste embankment built after this report was published did fail, with the consequent loss of 125 lives, this being the Buffalo Creek disaster in West Virginia in February 1972 (Fig. 48.8). Perhaps the best comment to be made is still that by one of those connected with the official inquiry: "We feel that the tragedy was doubly sad because the basic cause was a misdirected and ill-fated attempt to improve the environment by clearing the stream. The dam's failure is an example of the results of

FIG. 48.8 Remains of the Buffalo Creek No. 3 waste dam, as seen from the right abutment, near Saunders West Virginia.

hastily conceived legislation—legislation requiring instant conformance without technological assurance that its regulations can be safely met.[48.36]

48.9 CONCLUSION

Human activities have indeed created many geological problems, and the end is not yet. Once recognized, as has been seen, the problems have been tackled and usually solved, regulations often being invoked to prevent any recurrence. But new developments bring new problems in their train. High-tension, direct-current transmission of power, for example, is now in use, at least experimentally. Instead of using a multiple-wire system, long-distance DC transmission involves single-cable (monopolar) transmission, with continuous return using the ground or the sea. A 500-km (310-mi) -long line in New Zealand uses the sea for return. But the molten core of the earth is a good conductor of electricity, and so it is thought that this may make the system practicable also on land. Investigations are now current into the many associated geological problems.[48.37]

It is, however, the use of radioactive material that is already providing perhaps the greatest challenges to geology that the world has yet seen. The associated problems have been so well publicized, and are the subject of so much public discussion, that mere mention of them here must suffice. Again, the central geological problem is the disposal of waste material. The extent of the problem was indicated by the fact that the total volume of all effluents from the various chemical processing plants at the Hanford Works, Washington, from January 1944 to June 1957, amounted to 29.6 billion gal, of which 26.7 billion gal were process cooling water and 2.9 billion gal were low-level radioactive waters.[48.38] These were discharged into the underlying groundwater basin, the water table of which is as much as 105 m (350 ft) below ground level. The considerable ion-exchange capacity of the formation through which the wastes percolate results in the removal of measurable radioactivity before the wastes reach the groundwater. One of the most extensive groundwater surveys yet undertaken constantly checks the effects of the Hanford waste-disposal procedure on groundwater conditions throughout the surrounding area of over 1,560 km^2 (600 mi^2). At Oak Ridge, Tennessee, pits have been constructed in a shale formation to serve as storage basins for the wastes, with some concentration of the waste by evaporation. This is but the beginning.

The use, as fill, of waste materials from uranium-processing plants, with a slight radioactive content, has now been recognized as the source of serious problems. About 500 homes and schools in Grand Junction, Colorado, were found to be built over such material.[48.39] Pockets of similarly radioactive material, apparently derived from the local processing plant, were found throughout the town of Port Hope, Ontario.[48.40] Unacceptable concentrations of the gas radon (produced by the decay of radium, itself the result of the decay of uranium) were found in a number of homes at Elliot Lake, Ontario, location of important uranium ore mines. Later studies showed that the major source of this radon was the glacial sand on which the town was built. Comparison with sand from other locations showed a ratio of 5:1 in the content of emanating radon.[48.41] Now that this problem has been recognized, the necessary remedial measures have been taken, but it is a matter that will require constant vigilance to avoid recurrences.

The dismantling of outmoded or unneeded nuclear facilities highlights the problems that the use of radioactive material creates. A small reactor at Elk River, Minnesota, one-tenth the size of more recent comparable plants, became surplus and had to be demolished, and the radioactive materials resulting from the operation of the facility had to be safely buried. The job was done, under strict control, but this apparently simple task cost $6.2 million, approximately one-half of what it had cost to build the plant originally.[48.42]

It is, however, the disposal of spent radioactive fuel from the steadily increasing number of nuclear power plants, and of similar waste material, that is posing the most urgent problems. Geology is supposed to provide the answer. Spent fuel in almost all existing stations is being safely stored in temporary facilities. As these are slowly used up, the pressure to find long-term solutions is mounting. In every country using nuclear power, this vital problem has received intense study. Many commissions of inquiry have looked into it; many excellent reports have been produced. In almost every case, geologically suitable burying places are among the solutions discussed, but these are still under discussion, with numerous field studies in progress when local inhabitants do not object (as many have).

The problem is so complex, technically and socially, and is under such intense public consideration and discussion as this book goes to press, that the writers can do no better than conclude with a brief extract from the report which they have found to be the most helpful of those known to them, the sixth report of the British Royal Commission on Environmental Pollution (Chairman Sir Brian Flowers), which deals with nuclear power and the environment.[48.43] In discussing the disposal of "the highly active waste which arises from fuel reprocessing," the Royal Commission says:

> We are confident that an acceptable solution will be found and we attach great importance to the search; for we are agreed that it would be irresponsible and morally wrong to commit future generations to the consequences of fission power on a massive scale unless it has been demonstrated beyond reasonable doubt that at least one method exists for the safe isolation of these wastes for the indefinite future.[48.44]

48.9 REFERENCES

48.1 *The Countryman* **56**:540 (1959).

48.2 E. C. Young, "The Interaction between Technical Change on the Farm and Technical Change in Marketing and Distribution," in R. Dixey (ed.), *Proceedings of the Ninth International Conference of Agricultural Economists,* Oxford University Press, London, 1956, p. 166.

48.3 N. M. Blake, *Water for the Cities,* Syracuse University Press, Syracuse, 1956.

48.4 R. L. Sherlock, *Man's Influence on the Earth,* Butterworth, London, 1931.

48.5 C. P. Vetter, "Silt Problems in Lake Mead and Downstream on the Colorado River," paper read at American Geophysical Union meeting, 8 February 1952.

48.6 D. L. Gough, "Some Scientific Opportunities at Kariba Lake," *The New Scientist,* **7**:1278 (19 May 1960).

48.7 A. A. Ahmed, "Recent Developments in Nile Control" and "An Analytical Study of the Storage Losses in the Nile Basin, with Special Reference to Aswan Dam Reservoir and the High Dam Reservoir (Sadd-el-Aali)," *Proceedings of the Institution of Civil Engineers,* **17**:137; 181 (1960).

48.8 S. A. Morcos, "Effect of the Aswan High Dam on the Current Regime in the Suez Canal," *Nature,* **214**:901 (1967); see also in same journal, **218**:759 (1968).

48.9 "Cairo to study Qattara Hydro," *Engineering News-Record,* **164**:53 (10 March 1960).

48.10 "$1-billion Qattara Lake Plan Revived," *Engineering News-Record,* **192**:10, (28 March 1974).

48.11 V. Rich "Better Med than Red," *Nature,* **276**:123 (1979).

48.12 G. Vilks, C. T. Schafer, and D. A. Walker, "The Influence of a Causeway on Oceanography and Foraminifera in the Strait of Canso, Nova Scotia," *Canadian Journal of Earth Sciences* **12**:2086–2102 (1975).

48.13 F. W. Render and P. Fritz, "Groundwater Contamination Aspects of Diverting River Water

Through the Red River Floodway," in abstract vol. for annual meeting of the Geological Association of Canada at Waterloo, 1975, p. 843.

48.14 G. E. Dallaire, "Thermal Pollution Threat Draws Near," *Civil Engineering,* **40**:67–71 (October 1970).

48.15 *Report of the Fifth Interstate Conference on Artesian Water,* Government Printer, Sydney, N.S.W., 1929, p. 9.

48.16 O. E. Meinzer, "Progress in the Control of Artesian Water Supplies," *Engineering News-Record,* **113**:167 (1934).

48.17 P. Lemoine, R. Humery, and R. Soyer,"L'Appauvrissement de la Nappe des Sables Verts de la Région Parisienne," *Compte-Rendus,* Paris, 23 May 1934, p. 1870.

48.18 C. B. Burdick, "Ground Water as a Source of Supply," *Engineering News-Record,* **105**:398 (1930).

48.19 "The Falling Water Level of the Chalk under London," *Water and Water Engineering,* **35**:440 (1933).

48.20 A. T. Mitchelson, "Conservation of Water through Recharge of the Underground Supply," *Civil Engineering,* **9**:163 (1939).

48.21 G. M. Wulkowicz, and Z. A. Saleem, "Chloride Balance of an Urban Basin in the Chicago Area," in abstract vol. for the annual meeting in Dallas, Geological Society of America, Boulder, Colo. 1973, p. 871.

48.22 H. Shapley (ed.), *Climatic Change,* Harvard University Press, Cambridge, Mass., 1957, p. 166.

48.23 "Solid Wastes Pile up While Laws Crack Down and Engineers Gear Up," *Engineering News-Record,* **183**:28 (12 June 1969).

48.24 "Sanitary Land Fills, The Latest Thinking," *Civil Engineering,* **43**:69 (March 1973), a useful review.

48.25 G. H. Emrich, "What Is a Geologist Doing in a Health Department?" unpubl. paper given at annual meeting in Milwaukee, Geological Society of America, 1970 (from Bureau of Sanitary Engineering, Pennsylvania Department of Health).

48.26 M. G. Gross, "New York Metropolitan Region—A Major Sediment Source," *Water Resources Research,* Elmsfold, N.Y., 1970.

48.27 C. B. Crawford, and K. N. Burn, "Building Damage from Expansive Steel Slag Backfill," *Proceedings of the American Society of Civil Engineers,* **95**(SM6), 1325 (1969); for discussion, see **96**: (SM4), 1470 (1970).

48.28 H. B. Sutherland and P. N. Gaskin, "Factors Affecting the Frost Susceptibility Characteristics of Pulverized Fly Ash," *Canadian Geotechnical Journal,* **7**:69 (1970).

48.29 T. C. Hansen, C. W. Richards, and S. Mindess, "Sand-Lime Bricks and Aerated Lightweight Concrete from Gold Mine Waste," *Material, Research and Standards,* **10**:21 (August 1970).

48.30 H. Bouwer, "What's New in Deep-Well Injection," *Civil Engineering,* **44**:58 (January 1974).

48.31 I. C. MacFarlane, "Gas Explosion Hazards in Sanitary Landfills," *Public Works,* **22**:76–78 (May 1970).

48.32 "Florida Bridge Pile Explosions Traced to River Water Bacteria," *Engineering News-Record,* **186**:14 (4 February 1971).

48.33 *Report of the Tribunal Appointed to Inquire into the Disaster at Aberfan on October 21st, 1966,* H. M. Stationery Office, London, 1967; and *A Selection of Technical Reports Submitted to the Aberfan Tribunal,* 2 vol., H. M. Stationery Office, London, 1969; see also a masterly review by R. Glossop in *Geotechnique,* **20**:218 (1970).

48.34 G. L. Watkins, "The Stability of Colliery Spoilbanks," *Colliery Engineering,* **36**:493 (1959); reprinted in same journal, **43**:459 (1966).

48.35 W. E. Davies, "Geologic Hazards of Coal Refuse Banks," in abstract vol. for annual meeting at New Orleans, Geological Society of America, Boulder, Colo. 1967, p. 43.

48.36 "Buffalo Creek Dam Disaster, Why It Happened," *Civil Engineering,* **43**:69 (July 1973).

48.37 L. C. Bechtold and R. H. Jahns, "Geologic Environment of HVDC Ground Currents as Related to Buried Pipelines," Paper no. 42 for conference at Cleveland, National Association of Corrosion Engineers, Houston, 1968.

48.38 W. H. Bierschenk, "Hydrological Aspects of Radioactive Waste Disposal," *Proceedings of the American Society of Civil Engineers,* Paper no. 1835, vol. 84, SA6, 1958.

48.39 H. P. Metzger, "Dear Sir, Your House is Built on Radioactive Uranium Waste," *New York Times Magazine,* 31 October 1971, p. 14; see also *Proceedings of the American Society of Civil Engineers,* **98** (SM4); 430 (1972), for useful discussion of Paper no. 8382 by M. W. Jackson.

48.40 *The Globe and Mail* (Toronto), news reports in issues for 15, 16, and 17 January 1976.

48.41 B. L. Mooney and A. G. Scott, "Radium Content of Elliot Lake Sand Deposits Compared with Other Ontario Sands," in abstract vol. for annual meeting in Toronto, Geological Association of Canada, Toronto, 1978, p. 459.

48.42 J. Fialka, "Plutonium Spill Detected from Radioactive Wastes Buried in Kentucky," *The Globe and Mail* (Toronto), 8 March 1976 (a copyrighted report from *The Washington Star*).

48.43 Royal Commission on Environmental Pollution (Chairman, Sir Brian Flowers), *Sixth Report; Nuclear Power and the Environment,* H. M. Stationery Office, Cmnd 6618, 1976.

48.44 Ibid., p 81.

Environmental geology is a term which the authors do not use, since they regard it as a tautology, geology *being* the study of the environment. A number of useful books, however, have appeared with this term in their title. One of the most recent is *Environmental Geology,* a 731-page volume by D. R. Coates published by John Wiley & Sons in 1980. It contains some valuable information on man-made geological problems as does *Industrial geology,* another new term, now used as the title of a 344-page book by J. L. Knill, published in 1978 by the Clarendon Press of Oxford University.

The American Society for Testing and Materials has indicated its interest by publishing *Environmental Standards Updated* in 1982, containing standards prepared by eleven of its active committees.

Civil Engineering Problems of the South Wales Valleys was the title of a conference held in Cardiff, South Wales, in April 1969, the *Proceedings* of which form a notable volume of 147 pages related to man-made problems, published by the Institution of Civil Engineers in 1970. A paper called *Colliery Spoil Tips after Aberfan* was presented to this Institution in 1972 and proved to be of such significance that it was published as a "separate" of 98 pages in 1973.

Proposed Guidelines for Landfill Disposal of Solid Wastes, issued by the U.S. Environmental Protection Agency, are to be found in the *Federal Register* for 26 May 1979. *Summary of Findings on Solid Waste Disposal Sites in Northern Illinois* by G. M. Hughes, R. G. Lomax, and R. N. Farvolden, is No. 45 of the Environmental Geology Notes series of the Illinois Geological Survey; a number of the Notes deal with solid-waste disposal and other man-made problems. The underground disposal of waste is the concern of another of the Working Commissions of the International Association of Engineering Geology; publications on this subject may be expected in the Association's *Bulletin* in the future.

chapter 49

CONSERVING THE ENVIRONMENT

49.1 Restoration of Land / 49-2
49.2 Slope Protection / 49-4
49.3 Stabilization of Sand Dunes / 49-6
49.4 Sand and Gravel Pits / 49-8
49.5 Mining and the Environment / 49-10
49.6 Environmental Impact Statements / 49-12
49.7 Problems with Fauna / 49-13
49.8 Flood Prevention / 49-16
49.9 Soil Erosion / 49-18
49.10 Soil Conservation / 49-19
49.11 Conclusion / 49-21
49.12 References / 49-22

In September 1948 there was held at the Institution of Civil Engineers in Westminster, London, a conference entitled *Biology and Civil Engineering*. The Institution is the acknowledged senior of all civil engineering societies of the world. It might be expected to be somewhat conservative in nature, and yet it was the host to this first conference ever held on the interrelation of civil engineering and biology. It was an international gathering with speakers from New Zealand, South Africa, and the Netherlands, as well as from Great Britain. Papers were delivered on such subjects as soil erosion and conservation, the influence of vegetation on floods, and the use of vegetation in stabilizing sand dunes and artificial slopes, all illustrated by what civil engineers had done around the world in promoting the growth of vegetation, from grasses to trees, to enhance the environment of their works. The date of the meeting is to be noted; it took place more that 30 years ago, long before the voice of the so-called "environmentalist" was heard in the land, long before environmental impact statements were a legal requirement. And the works described at the meeting were in no way unique, being rather typical of sound civil engineering practice.[49.1]

One of the speakers had this to say:

> If a man cuts his hand, Nature immediately begins to restore the protective layer. Experienced medical aid can help to speed up the process by inducing conditions favourable to Nature. When man damages the surface of the earth in a comparatively small way, Nature

steps in and creates a protective layer, first of grasses, then shrubs, and finally trees. The trouble is that man, in his haste to utilize his handiworks, is unwilling to wait the relatively long time that these unaided natural processes take. It is here that the biologist can be of great assistance to the engineer by creating conditions which remove the element of chance and enable Nature to proceed rapidly from one stage to the next.[49.2]

This statement can serve well as an introduction to the brief review which this chapter presents of the ways in which civil engineers, often aided by biologists, cooperate with Nature in utilizing vegetation for the protection of the land on which they work. Engineers have not been the despoilers of the land, as so many shrill voices of today maintain. Sometimes they are not able to do all they wish in the way of land restoration when works are complete, but this is almost always due to shortage of funds for implementing the complete plan, owners and financiers, rather than designers, often being the culprits. And geologists? Geology *is* the study of the environment, and geomorphology the more detailed study of the processes that have given us the landforms of today. In times past, geologists may not always have been as vocal about the spoiling of the land as they might have been, but this, too, is changing. Together, geologists and civil engineers, combining their experience and their knowledge of the land, can be in the forefront of those who may rightly be called "stewards of the land," stewards whose vigilance was never more needed than it is today.

49.1 RESTORATION OF LAND

Many major civil engineering operations necessitate the disturbance of the natural surface of the ground, frequently the scale being such that the topsoil cannot be scraped off and stored to one side for replacement when the job is done (standard practice on small jobs). Airport construction is a typical case. New York's John F. Kennedy Airport (built as the Idlewild Airport) was constructed by pumping 50 million m^3 (65 million yd^3) of sand from Jamaica Bay. The area of 2,000 ha (4,900 acres) thus formed was vulnerable to the winds that almost continually sweep this coastal location. Fortunately, because of the experience gained in controlling neighboring beach areas, especially Jones Beach, engineers were able to meet the problem at the airport. They sowed a mixture of 10 percent poverty grass and 90 percent beach grass all over the airport. By means of a machine planter 20,000 grass roots could be planted on 1 acre in two hours; if the job had been done by hand, this would have been a full day's work for 20 laborers.[49.3] It is of interest to note that an attempt was made at Idlewild to control the sand by spraying with oil. As soon as the volatile content of the oil evaporated, however, the residual bitumen dried out, cracked, flaked, and blew away. Repeated treatments of even the most satisfactory oil are always necessary if the more certain control given when the aid of Nature is recruited cannot be achieved.

Poverty grass and common beach grass have just been mentioned; they should more properly have been referred to as *Ammophila arenaria* with *Ammophila brevilgulata* and *Andropogon virginicus* var. *littoralis,* since here the civil engineer is using the results of the work of the botanist, to whom such semantic exactitude is as common as engineering terminology is to the engineer. There may be some engineers to whom reference to botanical terms in a book such as this may appear out of place, but not to those who have seen for themselves the immense value of the use of living agencies in engineering work. To speak of biology and civil engineering is not, therefore, to link the names of mutually exclusive disciplines but to suggest one of the most potent of all scientific combinations in the practice of civil engineering. In fact, it is not going too far to say that if Nature can be so utilized as to restore the conditions originally disturbed by engineering works, the best possible remedy will have been achieved and, in some cases, the only practicable solution.

This was well illustrated by the example of the Steep Rock Iron Mine at Steep Rock Lake in western Ontario. To gain access to the valuable iron ore beneath the lake bottom, the 26-km^2 (10-mi^2) body of water had to be drained. About 640 ha (1,600 acres) of old lake bottom were thus exposed to the air. As the black ooze of lake-bottom deposits dried out, lighter organic matter was blown away by the wind, leaving exposed the compact glacial-lake clay which has such interesting properties. This material had to be protected, especially where it had to be trimmed into slopes and benches for access to the ore body. Biologists were consulted and considerable quantities of recommended grass seed were sown. Probably because of the completely inert nature of the soil so long covered up by lake waters, the seeding failed to be very effective. But within two years Nature had taken a hand and natural seeding took place (Figure 49.1). Within three years there was a good vegetal cover over almost the entire area of the old lake bed; within five years small bushes were growing.

FIG. 49.1 Bed of Steep Rock Lake, Ontario, being reclaimed after the lake was drained; seeded vegetation in foreground, naturally seeded vegetation beyond this, and bare soil in rear.

Here time was available for allowing Nature to take its course. For smaller areas, speed of rehabilitation is usually important and topsoil can often be replaced; here the best biological advice is desirable, with the possible necessity of fertilizing the ground to be stabilized. Bodies such as the Soil Conservation Service will willingly give advice, not only about desirable grasses and shrubs but also about trees, when these can be incorporated into rehabilitation work. Two years should see tree seedlings well established; within 20 years they will present sturdy windbreaks.

The problem has become increasingly serious as more dredged material has had to be deposited on land. Much study is being given to this disposal problem. If dredged material is carefully placed, and tended to after placement, much benefit can result. The U.S. Corps of Engineers took an enlightened step when, in connection with dredging in the James River, Virginia, an artificial island was formed near Windmill Point, 80 km (50 mi) upstream from Norfolk. An area of 6.5 ha (16.25 acres) was enclosed and filled by dredg-

FIG. 49.2 Windmill Point artificial island in the James River, Virginia, during construction; it is now a wildlife refuge.

ing. After the soil had settled, surplus water was returned to the river (Fig. 49.2). Within a matter of months, a thick cover of wetland vegetation had developed; within two years a healthy wildlife was established on the newly formed island.[49.4]

The U.S. Corps of Engineers dredges 249 million m^3 (326 million yd^3) every year, so its problem is no small one. Accordingly, a number of other demonstration projects have been successfully developed. Upland habitat has been established on a disposal site at Nott Island on the Connecticut River. A 9.3-ha (18.25-acre) site has been developed on the waterfront of the harbor of Buffalo, New York, and has already attracted birds, ranging from waterfowl to songbirds, and small mammals, in addition to thick vegetation.[49.5] Areas such as these can compensate, to some degree, for some of the wetlands that have been lost because of uncontrolled development in earlier years.

49.2 SLOPE PROTECTION

When major excavations are carried out or embankments built, sloping sides are frequently an essential part of the finished outline. When well designed, on the basis of relevant geotechnical information, the slopes will be stable, but they need protection, especially if the exposed soil is easily erodible and local rainfall is appreciable. Accordingly, civil engineers have long used vegetation as an immediate supplement to the excavation process. An early method (used with success in California) was the laying of bundles of brush cuttings (for which the old English name of *wattles* is used) at the outer edge of each successive layer of rolled fill, with special compactive effort applied near the brush. The wattles served to give mechanical stability to the slopes until such time as the plantings of western rye grass had time to become well established.[49.6]

Another means of ensuring initial stability in the sloping sides of embankments or fills is to construct simple plank-and-post barriers. Properly located, these can serve as drainage ditches at intervals down a slope, thus eliminating or minimizing the raveling that would otherwise take place during and after heavy rain. Another approach has been to lay out lines on the contours of the slope, driving short timber posts along these lines and then installing wattles in a horizontal position. This work can be carried out while the cut, or fill, is being finished, so as to give stability to the slope just before selected grass seed is planted upon it. Once grasses have established themselves, then sturdier plants and even small trees can be planted to give permanent cover. Still another means of promoting vegetal growth is to cover the slope with one of the recently available proprietary types of open-work matting, well secured in place. These "geotechnical fabrics" can be remarkably successful in holding soil in place until such time as grass seeds can germinate and begin to establish their small root systems, which eventually, when interlocked into sod, give the best of all permanent surfaces to even the largest slopes (Fig. 49.3).

FIG. 49.3 Sloping face of reconstructed cutting reinforced with fabric on Motorway M4, 20 miles west of Reading, England.

It must be emphasized that the use of vegetation on soil slopes is no substitute for proper geotechnical design, upon which the basic stability of all slopes will depend. Vegetation is essential, however, for protecting the delicate surface of such stable slopes. Once in place, it must be properly maintained. If only to show what lack of maintenance can do, and as an indication of the sort of situation which engineers are sometimes called upon to clear up, the escarpment at Whitehorse, capital of the Yukon Territory, may be mentioned. Its desecration is one of the saddest examples of land spoliation known to the writers. The small city is unique in that it has a first-class airport (one of the wartime Northwest Staging Route stops) within 0.8 km (0.5 mi) of the civic center. The airport is located on a fine plateau-like terrace, 60 m (200 ft) above the level of the town, which is located on the bank of the famed Yukon River. Between the town and the terrace is a steep escarpment, winding down the valley, originally covered with a stand of magnificent pine, as shown by early photographs. At the time of the first airport construction, someone, now unknown, cut an access trail up to the plateau through the trees. When this was enlarged into an access road, the trees gradually disappeared with the inevitable sloughing that occurred after occasional heavy rainstorms. Even drainage from the airfield, and some of its buildings, was discharged over the brow of the escarpment, so that when, in 1953, a very heavy spring rainstorm occurred, a series of small mud runs took place right into the center of the town and even into its cemetery (Fig. 49.4). Measures were then taken to stop the misuse of the escarpment, the access road having been long since abandoned because of the slipping; finally remedial measures were put in hand. It will be many

49-6 GEOLOGY AND THE ENVIRONMENT

FIG. 49.4 Mud run due to erosion of escarpment in background, after removal of natural vegetation (of which the trees still to be seen were typical); rehabilitation work is now in progress; at Whitehorse, Yukon Territory.

years before these efforts can even begin to replace the original vegetal cover, the removal of which was tragic indeed.

49.3 STABILIZATION OF SAND DUNES

The importance of the wind in structural engineering is self-evident; wind loads on structures are sometimes the maximum loads to be considered in design. Wind is also important to the civil engineer, however, as an agent of erosion; its natural role in the work of erosion is speedily evident on any unprotected land. Some of the main evidences of its work are drifting sand dunes. Although these are prominent in desert regions, they often present troublesome local problems on seacoasts with sandy beaches. The travel of coastal sand has wrought great damage in Europe, especially on the shores of the Bay of Biscay. At Liège in France, for example, sand dunes have been moving for two centuries at the rate of 24 m (81 ft) per year, necessitating two successive reconstructions of the local church. The geologist can but observe this phenomenon. The botanist, however, is here of assistance, and much has been done in the development of plants, grasses, and shrubs that will take root in sand and provide a matting of root material which will effectively bind the surface together.

In a few places, where they serve as a storage ground for rainwater, wind-formed dunes have an economic value; in western Texas, for example, rangers get their water supplies from wells put down at the summits of large dunes. Reference to sand dunes in the west of Texas will be puzzling to those who associate dunes only with seacoasts. It is on coastlines that they are most in evidence, but large areas of dunes—generally of recent origin, sometimes the result of improper land practices—are now to be found in the Great Plains area and especially in New Mexico, Colorado, Texas, Kansas, Wyoming, and the Dakotas. Some of the largest are on the south shore of Lake Athabasca in northern Saskatchewan. On the Pacific coast dunes are found all the way from Baja California to southern Alaska. The ravages of forest fire, which removes all humus from the ground surface and leaves it open to the action of the wind, and the uncontrolled grazing of cattle are two prime causes of many modern dune areas. In open country, the presence of dunes may not be a menace, but in all-too-many areas they have encroached upon valuable land or tend to do so. Thus, the need of expert control measures can be seen. Although "mechanical" means of control, such as fences and brush mats, have their uses, their effect is temporary only. The use of vegetation is essential for the complete control of drifting sands; when properly used, vegetation has succeeded in solving some of the most serious problems of wind erosion.

An inland example of such control was provided when the U.S. Corps of Engineers constructed the John Martin Dam across the Arkansas River in southeastern Colorado. Flooding of the reservoir so formed was going to submerge many miles of the main line of the Santa Fe Railroad. On the south side of the river was an area of 500 ha (1,200 acres) of sand dunes which everyone wished to avoid if possible. On the other side of the river, however, purchase of 27 km (17 mi) of right-of-way through valuable irrigated farmland, as well as the construction of two large bridges, would have been necessary. The Soil Conservation Service was asked to assist; it did, and to such good effect that the dune area was stabilized so well that the crack trains of the Santa Fe have been safely using the relocated track, running right through the dunes, without any trouble at all; a substantial saving of money from the public purse was achieved in the process. The principal grasses used were little and big bluestem, sand reed grass, blue grama, switch grass, sand dropseed, and sand lovegrass. Some trees were also planted after the grass had taken root (Fig. 49.5).[49.7]

The most extensive sand dunes are those to be found along the low shorelines of the world, bordering the oceans and major lakes. Work of engineers of the Netherlands in preserving their fragile coastline down through the centuries is well known; some account of it was given at the meeting mentioned at the start of this chapter. Some grasses, usually sedges, can be remarkably tolerant of high salinity, open exposure, and lack of humus and groundwater and thus will take root on sand dunes, giving the surface ultimate stability and paving the way for more varied vegetation. Notable work of this kind has been done along the northwestern coast of the United States, large areas having been reclaimed so successfully that they are now well-developed recreational areas.

There are, however, dangers here too, since dunes along the coast are natural phenom-

FIG. 49.5 Rehabilitated sand dunes on the south side of the Arkansas River, near the reservoir formed by the John Martin Dam, Colorado, showing the effective use of selected grasses.

49-8 GEOLOGY AND THE ENVIRONMENT

ena, their formation and constant movement the result of the interaction of complex natural forces. If they are stabilized, this may have results quite unexpected and certainly never intended. Stabilization of dunes along the coast of North Carolina, for example, has been called in question because of the changes it appears to have caused.[49.8] The desirability of dune stabilization is not questioned here, but rather the way in which it has been done, in relation to the action of natural forces along the coast. This is a further reminder of the great sensitivity of coastal processes, one of the most interesting of all branches of geomorphology and, at the same time, for the civil engineer engaged in coastal control projects, one of the most challenging. If Nature can be recruited as the main stabilizing factor, economy will usually be served and more permanence assured.

49.4 SAND AND GRAVEL PITS

The sand-and-gravel business in North America is now so large and so widespread that there are few municipalities that have not felt its impact. In earlier days, deposits of sand and gravel were worked out without thought being given to the rehabilitation of the pits that were left. Happily those days are gone. Most regulating agencies now have requirements for the complete restoration of worked-out pits, if the areas in question are not to be otherwise used for some suitable purpose in accordance with a predetermined plan arising from what is now known as *sequential land use*. Some provinces of Canada require the deposit of specific sums per acre used, in one case as much as $500/acre, which is set aside for payment for later rehabilitation work. The industry itself has accepted its responsibilities for conservation of the environment. Pits are being operated in an orderly fashion, and careful planning of restoration work is a regular part of development (Fig. 49.6). Useful guides to the rehabilitation of pits have been published by governmental bodies and industry alike.[49.9]

FIG. 49.6 Golf course with trout ponds in what used to be a sand and gravel pit; a development of Steed and Evans, Ltd., of Fonthill, Ontario.

Typical of the modern approach to this environmental problem is the 10-year plan developed for the mining of 2.5 million tons of high-quality gravel located in the floodplain of Boulder Creek, Colorado, close to the city of Boulder and adjacent to the White Rocks area, famous locally as a sensitive nature reserve. A significant sidelight on the need to develop this deposit is the fact that over half the sand and gravel deposits in the vicinity of Boulder have been lost, because of unplanned development. The Flatiron Companies worked with a number of local groups, including environmentalists, in drawing up the 10-year plan, with the result that, despite some lingering opposition, such general agreement was obtained that the local board of county commissionners unanimously granted the necessary use permit (Fig. 49.7). Seven stages for development are envisaged; rehabilitation work includes the formation of lakes, wet meadows, forest, and woodlots.[49.10]

FIG. 49.7 Reclaimed sand and gravel pits at Boulder, Colorado, part of the development described in the text.

The cooperation of landscape architects in such environmental planning can be helpful, as can be consulation with horticulturists who are familiar with local growing patterns. The results of this kind of work, when well planned, can be spectacular, as Fig. 49.6 well shows. The cost is a small price for the producer to pay, in view of the resulting public benefits. Inquiry will often reveal research work at local agricultural stations that can be applied directly to pit rehabilitation. The University of Guelph, Ontario, has been a leader in such work, one of their experiments being conducted in the floor of a large industrial gravel pit.

A further danger to the environment arises from unrestricted excavation of sand and gravel from stream beds. Material from active stream beds has the advantage of being washed clean and is slightly rounded; it is, therefore, a tempting resource. But its unrestricted removal may have serious effects on the regime of the stream, especially downstream of any gravel-mining operation. Scour may be increased, with possible danger to bridge structures. Even stream channels normally inactive (such as occur in areas of low rainfall), from which excessive amounts of gravel are mined, can be the scene of trouble when unusually heavy rains result in flows that utilize channels normally dry. An experience in 1969 in Tujunga Wash, California, gave vivid evidence of this unusual danger.[49.11]

49.5 MINING AND THE ENVIRONMENT

Desecration of the landscape by opencast mining (as for coal), by quarries, and by other strip-mining activities, has long been the object of protests from those who are concerned about the environment, rightly so since many worked-out areas, not rehabilitated, were a major eyesore. Some still are. A report by the U.S. Geological Survey of 1978 is a useful reminder that subsurface mining may have more serious effects on surface environments than strip mining, if land restoration is well done in the latter case. Figure 49.8 is an illustration of the subsidence pits due to underground workings at the Monarch mine near Sheridan, Wyoming, which was abandoned in 1914.[49.12] In other parts of this Handbook, examples have been given of serious major subsidences at the surface due to underground mining activities. Attempts are naturally now being made to limit such harmful results.

The problems of surface mining remain, however, although today almost every state in the Union has regulations requiring restoration work to some degree. The first attempts to remedy the blight of surface spoil heaps were made voluntarily by industry in the Midwest in the 1920s. The first state regulations were promulgated by West Virginia in 1939, by Illinois in 1943, by Indiana in 1941, and by Pennsylvania in 1945. Thirty-eight states had necessary regulations by 1975.[49.13] Regulations generally relate to new mining ventures and now form a background to the desirable preplanning of such projects, in which all the requisite disciplines usefully participate. Local governmental agencies are, naturally, becoming increasingly involved. There still remain areas, long since mined out, which are in need of rehabilitation.

The problems are many, ranging from control of erosion off spoil banks to the restoration of farmlands uprooted by mining operations (Fig. 49.9). Geology is implicit, in the fact that local geological structure has resulted in the presence, close to the surface, of mineable ore or coal. Geological considerations can greatly assist in the planning of ameliorative measures. Another publication of the U.S. Geological Survey (of 1976) presents

FIG. 49.8 Subsidence pits resulting from the collapse of underground workings at the Monarch mine near Sheridan, Wyoming; the mine was abandoned in 1914, the photograph taken in 1975.

CONSERVING THE ENVIRONMENT   49-11

FIG. 49.9 Rehabilitation work in progress at the Bel Aire coal mine at Gillette, Wyoming, by the Amax Coal Company.

a most helpful matrix of information on the reclamation requirements of all states. The fact that there are columns for 24 different requirements and provisions is an indication of the complexity of the actions now possible in ensuring that mined-out land will be restored to the state of a desirable environment.[49.14]

In Great Britain, opencast coal mining is now widespread. All of this is under the National Coal Board, the concern of which for rehabilitation is indicated by the fact that the Board has its own forestry adviser. Opencast mining started only in 1942 but rapidly increased until 1958, when 14 million tons of coal were thus mined. Since then, opencast mining has been controlled and reduced, a recent annual production being only 7 million tons. Restoration of land so used has been an integral part of the board's program from the start (Fig. 49.10). In a useful review of their policies, it is stated that average costs, apart from earth moving, which is paid for as part of the mining operation, came to about

FIG. 49.10 Farming operations on rehabilitated land, restored after open-pit coal mining by the National Coal Board of England.

49-12 GEOLOGY AND THE ENVIRONMENT

FIG. 49.11 Recreational area in Dayton, Ohio, created out of rehabilitated sand and gravel pits.

$600 (£ 230) per acre.[49.15] This included such items as cultivation, manuring, and cropping; installation of permanent drainage and of water supply to fields; and restoration of ditches, fences, roads, etc. The list is a good indicator of the varied operations necessary in this particular part of the work of ensuring that the disfigurement of the environment through surface mining will not, in the future, be more than a temporary inconvenience.

The certainty that there will be an increase in coal-mining activity in the near future, as one response to the crucial energy situation, lends special significance to a three-year, $3 million research project funded by the state legislature of Iowa in 1976. Primary goals were the development of economical and satisfactory strip-mining methods and a study of the economics of cleaning Iowa coal. For the former purpose, a 16-ha (40-acre) site with two coal seams beneath it was used, coal reserves amounting to 135,000 tons. Overburden averaged 3 m (10 ft) of loess and glacial till, 6 m (20 ft) of oxidized shale, 0.9 to 1.5 m (3 to 5 ft) of siltstone, and 9 m (30 ft) of black pyritic shale. Rippers and scrapers were used to remove these materials, after careful husbanding of topsoil.[49.16] Similar work has been done elsewhere, so there is now available a useful body of knowledge to assist all who face similar problems. (Fig. 49.11.)

49.6 ENVIRONMENTAL IMPACT STATEMENTS

In a laudable move to assist efforts being made to conserve the environment, the Congress of the United States of America in 1970 enacted Public Law 91-190, the National Environmental Policy Act. Section 102 (2) (C) of this act requires that a statement of the probable environmental impact of a project, and alternatives to it, must be prepared in advance of project approval for all federal agencies and actions, and for all local and state projects which use federal funds. The Council on Environmental Quality has issued guidelines for the preparation of the quickly named *environmental impact statements,* many thousands of which have now been prepared. Their overall effect has been beneficial and helpful, even though the legal mandate requiring them created, in some quarters, the idea that this preliminary consideration of the environment was something new. As has already been seen, this is certainly not the case, but the legal requirement has made certain that the environment can never again be neglected in the prosecution of any public works for which federal money is used. The indirect effects of the Statements have also been beneficial.

Supplementing the guidelines just mentioned is another useful publication of the U.S. Geological Survey (the interest of which in all environmental matters will be evident). *A Procedure for Evaluating Environmental Impact* presents a matrix as a sort of "super checklist" for all who have to prepare Statements. Table 49.1 is a mere outline of this large chart. It can be seen that the complete chart contains columns for no less than 98 actions, or proposed works, which may affect the environment, and almost the same number of items characterizing the environment as it is. Naturally, not all boxes on the matrix will call for completion for any one project, but the idea that almost 8,500 combinations of factors are even involved in these Statements about the environment and public works is clear indication of the complexity of preparing concise and useful documents.[49.17]

If the Statements are prepared by expert professionals, they can be excellent safeguards; tribute has already been paid to their value. It must be added, however, that there have been criticisms expressed, and with reason, about some Statements (possibly a small minority) that have not been prepared quite as well as they should have been. Consideration of such inadequate Statements adds still further to the considerable expense which these legal requirements have added to the total cost of the public works involved. Some have been described as "[in parts] erroneous, irrelevant, misleading, and grammatically poor."[49.18] This is a serious indictment, but when one comes across in a Statement such a geological observation as "The total volume of rock in Michigan Basin is approximately 108,000 cubic miles," then one has cause to stop and realize that the mere legal requirement of a Statement is of little worth unless the Statement is thoroughly well prepared.

The critics just quoted advance sound and helpful guides as to what Statements should contain; they stress the need for general information about geology, expressed in understandable language, and explanations of geology, again expressed in understandable language, of mineral and water resources involved, and of "geological," i.e. natural, hazards, with a list of measures for providing protection against all hazards and undesirable problems. In this, they support the contents of the USGS guide already mentioned,[49.14] and the Council's guidelines. All civil engineers and engineering geologists who have occasion to be involved in any way with the preparation of environmental impact statements should ensure that these guides are followed, so that the resulting documents may be concise, sensible, and helpful, rather than impediments to the proper appreciation of the importance of protecting the environment when land is interfered with for public benefit.

49.7 PROBLEMS WITH FAUNA

It has been the attention drawn to some endangered species of wildlife, in connection with environmental impact assessment, that has awakened public awareness of the fact that the environment includes not only the land, vegetation, and forests, but also the living insects, mammals, and fishes that share with humans the delights of living in the earth's environment. Philosophical questions of real difficulty are raised when endangered species

TABLE 49.1 An Outline of an Information Matrix for Environmental Impact Statements

	Proposed Actions Which May Cause Environmental Impact									
Existing Characteristics of the Environment		Modification of regime	Land change & construction	Resource extraction	Processing	Land alteration	Resource renewal	Changes in traffic	Waste treatment	Chemical treatment	Accidents
	Earth	13 / 6	19	7	15	6	5	11	14	5	3
	Water	7									
	Atmosphere	3									
	Processes	9									
	Flora	9									
	Fauna	9									
	Land use	9									
	Recreation	7									
	Aesthetics & human interest	10									
	Cultural status	4									
	Man-made facilities										
	Ecological relationships	7									

Number of individual items in each group to be found in the original Table.

Source Authors' summary of a very complete table given in L. B. Leopold, F. E. Clarke, B. B. Hanshaw and J. R. Balsley, *A Procedure for Evaluating Environmental Impact* (Reference 49.14).

have to be considered. The civil engineer, however, has to deal occasionally with fauna in ways that do not give rise to any such basic problems.

In Australia great difficulty was experienced by the owner of a private landing airstrip on Trefoil Island in Bass Strait because of the burrowing habits of a flock of muttonbirds (shearwaters) who objected to interference with their normal habitat. A maintenance worker had to go over the strip every day to ensure that it could be used for landing. The same problem was encountered on another offshore island in Australia on which an attractive 9-hole golf course had been laid out. The club decided some years after the course was opened to extend it to 18 holes by using adjacent dunes and sandy land between the island and the coast. Grading started in the winter with the filling in of a network of burrows excavated by a colony of muttonbirds which had just left on their annual migration. The new fairway was almost ready for official opening in the late spring when the muttonbirds returned, to be most annoyed by the disturbance of their traditional burrows. Being creatures of habit, they proceeded to excavate them again in exactly the same locations, despite the change in appearance of their habitat. This played havoc with the new course, but with speedy work the mess created by the birds was rectified the next day after they had left for their daily fishing flights. At dusk, the birds returned, and methodically redug their burrows. The battle proceeded for some days, but the birds finally won. The course still has only nine holes.[49.19]

It will be said, rightly, that golf-course construction scarcely ranks as civil engineering construction, but similar problems have been encountered during the construction of embankments for highways; burrowing animals such as muskrats and woodchucks, displaced from their normal habitat by the placement of fill, can cause problems. This does not happen frequently, but cases are on record in which woodchuck burrows have been found in new fills, and measures had to be taken to ensure that they did not endanger the embankment.[49.20] Bears have well-established areas over which they range; they too object to man's intrusions. In experimental field research work in the mountains of western Canada, carefully installed instruments have been ripped up and displaced with consequent destruction of the records being taken—by bears on the prowl. Local fauna should not be forgotten, therefore, when work has to be done in areas still in a wild state. Even in settled areas, the unexpected may happen. Brink records a case in South Africa where the entire floor of a small reservoir formed by an earth dam was carefully lined with well-sealed plastic sheets to solve a serious leakage problem. This lining proved ineffective when local herons started perforating the plastic sheets with their beaks in their search for frogs which had been their accustomed diet in this location.[49.21]

There is one animal that is a conservationist in its own right, even though its activities sometimes adversely affect engineering works. The beaver has sometimes been called "Nature's own engineer." To all who have studied the construction of a large beaver dam, this suggestion is no mere poetic fancy. The way in which the beaver regularly dams up swift-flowing streams has been almost exactly followed in many check-dam projects for the control of stream erosion. Notable examples of this type of work were used in soil-conservation work in Utah by L. M. Winsor, district engineer of the U.S. Department of Agriculture at Salt Lake City, who freely acknowledged his debt to the beavers he had seen at work in this area.[49.22] In some districts, beavers have been too efficient in their work. In northern Ontario, for example, engineers of the Canadian National Railways regularly have to blast away beaver dams on streams adjacent to their main line across northern Ontario and Quebec, dams constructed with no regard, unfortunately for the C.N.R., to the relation of the crest level of the dams and the elevation of the rail right-of-way. Flooding is a regular occurrence; since beavers are persistent workers, with unlimited patience, the struggle has been a long one.

49.8 FLOOD PREVENTION

The work of beavers in controlling the flow of the streams on which they build their dams naturally called for mention earlier in this Handbook, when river control was considered. Floods are now so vital a problem that another chapter was given over to a review of the geological aspects of floods and flood control. The problem is a serious one; even though the same amount of water flows down rivers, it does not flow at the same rate now as in times past; there are increasingly great variations between flood flow and low flow. Reference to Roman writings shows clearly that, for certain European rivers, such as the Durance and the Seine in France, it can be stated with certainty that their flow was much more uniform 2,000 years ago than it is today. And there can be little double of the reason for the change: the cutting down of forests, with the consequent removal of the forest litter that acts as such a regulator of runoff, followed by the erosion of topsoil—all constituting interference with Nature's own regulation of runoff.[49.23] This is no new idea; many writers, with vision ahead of their times, have warned of the serious effects that would follow the indiscriminate removal of forest cover; Bernard Palissy, among others, wrote on this subject in the sixteenth century.

Lest it be thought that this is a situation peculiar to Europe, consideration may well be given to this quotation:

> If the summer floods in the United States are attended by less pecuniary damage than those of the Loire and other rivers of France, the Po and its tributaries in Italy . . . it is partly because the banks of American rivers are not yet lined with towns, their shores and bottoms which skirt them not yet covered with improvements whose cost is counted by millions and, consequently, a smaller amount of property is exposed to injury by inundation. But the comparative exemption of the American people from the terrible calamities which the overflow of rivers has brought in some of the fairest portions of the Old World, is, in a still greater degree, to be ascribed to the fact that, with all our thoughtless improvidence, we have not yet bared all the sources of our streams, not yet overthrown all the barriers which nature has created to restrain her own destructive energies. Let us be wise in time, and profit by the errors of our elder brethren![49.24]

These words were written in 1863 by George P. Marsh in a notable book first called *Man and Nature*, later reissued as *The Earth as Modified by Human Action*, long since forgotten, unfortunately, but containing much information and sage advice of great value still today (even though couched in typically involved Victorian prose). This great man of Vermont was a century before his time. His fine book was reprinted to mark its centenary; it should be known to all civil engineers and geologists, since it contains equally percipient comments on all aspects of conservation.

Of the floods on American rivers in recent decades and of the immense and varied engineering works undertaken to combat them, there is no need to write here. Nor need words be wasted in ascribing blame for what has happened to the land—of the United States and of Canada, in particular, although many of the countries of the Old World have their problems also with floods and the companion scourge, soil erosion. Countries like South Africa, New Zealand, and to a limited extent Australia, know well how too-speedy development of a new land can bring its train of troubles in uncontrolled water and loss of soil.[49.25] The engineer, who has not been any more responsible for this rape of the land than any other citizen, should yet be more appreciative of what it has meant and of what must be done to remedy its dire results. Any civil engineer who has been engaged upon construction works, the purpose of which was to put water back into its original surroundings because earlier man-made works had disturbed the natural balance, is forced to wonder why such things can happen, why the serious effects of the original work—such as indiscriminate land clearing—were not realized at the time. It is obviously a universal question, one which finds an answer in these words from Shakespeare's *Much Ado about Nothing:*

> For it so falls out
> That what we have, we prize not to the worth
> Whiles we enjoy it; but, being lack'd and lost,
> Why, then we rack the value, then we find
> The virtue that possession would not show us.

Some floods have always happened; historical records make clear that, although ancient floods were perhaps not as severe as modern floods, major rivers on which settlements developed did rise to flood heights from time to time. Equally clear, however, is the increased frequency of floods today, and their greater intensity, due inevitably to human interference with the surface cover of the watersheds. When flooding is considered as a watershed problem, with all the associated geological considerations, then the construction of small dams across gullies in order to check flood flows and reduce flood velocities on tributary streams can be seen to be a leading type of work, one demanding cooperative endeavor (Fig. 49.12). This sort of small-scale engineering is often called "upstream engineering"; the term has sometimes been used in a somewhat derogatory manner. If it is only remembered that all such conservation work has as a major objective the conservation of invaluable topsoil, the reducing of floods being of unusual significance and value, much of the fruitless discussion between the proponents of "big dams" and "little dams" will be avoided. Both measures are needed; applied intelligently and cooperatively, they can do much to conserve soil and reduce floods. The U.S. Forest Service has shown clearly what a significant contribution to flood prevention can be made by the ground cover of watersheds, as summarized in its record of 40 years of experimentation, *Mountain Water.*[49.26]

FIG. 49.12 Small masonry dam for stream and erosion control in Contra Costa County, California.

49.9 SOIL EROSION

The sediment in streams and rivers, that retained behind dams, that which builds the great deltas of the world, and that which slowly fills up river estuaries until removed by costly dredging—all this sediment is not just "silt." Almost all of it has once been topsoil, gracing the land served by the river or stream, the soil in which vegetation large and small has had its roots. The possible erosive action of surface flow over rock is negligible during a period of relatively few years; but erosive action on unconsolidated materials may be far from negligible. When vegetation covers the ground on which rain falls, the binding effect of plant root systems, in addition to the protection afforded by foliage, will generally result in the rain reaching the actual ground strata gradually. Alternatively, after heavy rains, water may flow over the surface vegetation at such a speed and in such a manner that little erosion of the ground will take place. It is only when vegetation is lacking that erosion can become serious. And vegetation can so easily be disturbed by the hand of man through uncontrolled tree felling, improper cultivation practices, and abuse of the land instead of its nurture (Fig. 49.13).

Although the widespread effects of soil erosion have attracted public attention in North America (as in other Western countries) in relatively recent years only, soil erosion is no new thing. The deserts of north China, of Iran, of Iraq, and of North Africa were not

FIG. 49.13 The beginnings of soil erosion, a view of idle bean land in Ventura County, California, before control measures were initiated.

always deserts; this is clear from the records of history. They have become deserts within the span of recorded history because of the gradual but steady depletion of the topsoil which once covered them, as the demands made upon the soil by expanding civilizations increased. The glory that was Carthage is now a waste of sand, a waste caused by soil exhaustion, crop failures, land abandonment, and the encroachment of the desert, and not merely as a result of warfare. The Adriatic took its name from the ancient port of Adria, located between the Po and the Adige, accessible to large vessels in the time of Ceasar Augustus, but now 22 km (14 mi) inland; the coast has been silted up by soil washed down the two rivers. No such disaster should overtake any modern port, if only because man is now well aware of the forces of Nature, their power, their interrelation, and the necessity of their proper control in the interests of human development. Vast areas of China resemble gigantic battlefields, scarred deeply in man's battle with Nature in the great loessal region; Nature makes short work of carving deep channels into any land abandoned in the unequal fight.

In the world today, there are few "developed" countries that have not felt the effects of soil erosion to some degree, although the effects have been less profound in older countries with well-established agricultural practices and equitable temperate climates, such as France and Great Britain. In South Africa, soil erosion was described by Field Marshal Smuts as a scourge second only to war. Even in so pleasant a land as New Zealand, large areas have been eroded. In South Australia, to take just one of the Australian states, it has been said that over 2,600 km^2 (1,000 mi^2) that were once pastoral country are now desert. In Canada, the first soil erosion was noticed at Indian Head, Saskatchewan, as early as 1875; since then all-too-much agricultural land has been lost. The Magdalen Islands, once a fertile group in the center of the Gulf of St. Lawrence, have had all their topsoil removed; only sand, not a stick of timber, now remains on the islands. And in the United States, it has been estimated that 50 million acres of good cropland have been ruined almost permanently, with another 50 million acres almost as seriously depleted, and an equivalent total area affected to some degree. Overall figures are so astronomical as to be almost unbelievable; two, however, may be noted. It has been estimated that as much as 5.4 billion tons of topsoil have been lost in one year in the United States; the total cost of the soil erosion itself and associated damage to engineering structures was said to be well over $3.8 billion some years ago.[49.27]

49.10 SOIL CONSERVATION

It is the damage done to engineering structures by soil erosion that usually first attracts the attention of the civil engineer to this widespread public problem. There may be some who, like one of the writers, have been forced to think of the matter because of participation in the design and construction of engineering works necessitated by the effects of soil erosion. He was privileged to be associated with the design of the Shand Dam on the Grand River in central Ontario (an earth dam formed of compacted glacial till). Its purpose was to regulate the flow of the Grand River in order to reduce flooding, but more particularly it was to give a minimum low-water flow in order to remove unsanitary conditions resulting from an almost negligible summer flow. When one has seen more than $1 million spent to counteract the misuse of land, "soil erosion" becomes very much more than a mere phrase, and the engineering significance of soil erosion is at once apparent.[49.28]

Full discussion of the cause of soil erosion and of the best methods to be followed in soil conservation is complex, since so many factors are involved (Figs. 49.14 and 49.15). Fortunately, there are now in most countries active agencies whose responsibility it is to

FIG. 49.14 The rape of the land; soil erosion by wind and water on morainal land, once well forested, in the valley of the Ganaraska River, Ontario.

FIG. 49.15 Restoring the land; the beginning of reforestation on morainal land in the valley of the Ganaraska River, Ontario.

deal with the problem. The U.S. Soil Conservation Service is an acknowledged world leader in this vital field. Conservation measures necessarily vary widely, but a fundamental feature is that all designs shall aid natural processes. Preventing runoff over bare soil is a basic requirement, contour plowing being a first step always recommended on farmlands. But exactly the same principle can be applied on engineering works, when, before the end of construction, bare soil has to be exposed to rainfall, especially if finished surfaces are sloped. Promotion of vegetal growth by selection of proper seeds, careful planting, and mulch protection are methods similarly applicable to land restoration and protection of land prior to the finishing of engineering projects. And so in the hands of civil engineers rests responsibility for doing all that can be done through the medium of their undertakings to conserve all renewable natural resources. While accepting, without recrimination against the past, the state of the land in which they work, engineers must, so far as possible, act in conformity with the operation of natural forces, and must always act in accord with all that the processes of geology can show them in their work.[49.29] (Fig. 49.16.)

FIG. 49.16 The land preserved: rice paddies in Kanagawa Prefecture, Japan, showing soil conservation practices long followed by the Japanese.

49.11 CONCLUSION

Conserving the environment is one of the great challenges now facing the human race. Much harm has been done in the past, but the problems have been recognized in time. Civil engineers have not been blameless, but they certainly have not been the spoilers of the land so often popularly imagined. Today, they can lay claim, with many others, to being "stewards of the land." They should be in the forefront as supporters of soil-conservation measures, knowing that every bit of soil retained on the ground is soil kept out of reservoirs, rivers, or harbors; they should be the first to encourage sound reforestation projects, knowing that these are one further means of controlling runoff and so of assisting in river and flood-control work; in excavation work, they should see that bare soil is never left to the mercies of the elements but is protected as quickly as possible by the natural cover vegetation so fully provides; in studies of groundwater, they should always remember the natural balance and see to it that their works do not permanently impair this condition; in the construction of dams, they should appreciate that these structures are only a partial answer to river control, an essential part in many cases, but fully effective only in association with what has been called "upstream engineering"; in river-training schemes, they must let Nature "have her head" under control, remembering again that there is a natural balance, even of stream regimes, that must not be too seriously disturbed; and in bank- and shore-protection work, rather than attempting the impossible, they must fit protection works, insofar as is possible, into the natural order. In all operations, they should be working with Nature and not against her, remembering that "Nature never deceives you, if you watch her long enough." And so, never hesitating to be an imitator of natural processes as they carry out works for the use and convenience of their fellow beings, they are, in fact, geological agents in the very best sense. They may well recall that they have the support of great figures of all ages, from many of whom appropriate words could be quoted to bring this chapter to a close. Let Thomas Carlyle, in his own inimitable, crusty style, have the last word: "Nature keeps silently a most exact Savings Bank and official register correct to the most evanescent item ... and at the end of the account, you will have it all to pay, my friend, there's the rub."

49.12 REFERENCES

49.1 *Biology and Civil Engineering* (Proceedings of the Conference held at the Institution of Civil Engineers, September 1948), Institution of Civil Engineers, London, 1949.

49.2 B. J. J. Moran, "The Use of Vegetation in Stabilizing Artificial Slopes," *Biology and Civil Engineering,* Institution of Civil Engineers, London, 1949, p. 113.

49.3 "Grass Stabilizes Sand on 4,900 Acres of New York's Idlewild Airport," *Civil Engineering,* **18**:22 (1948).

49.4 L. H. Hunt, "Corps Developing Wildlife Habitat to Solve Disposal Problems," *Civil Engineering,* **48**:101 (September 1978).

49.5 Ibid.

49.6 "Erosion Control on a Road Fill," *Engineering News-Record,* **85**:148 (1939); see also, in same issue, p. 53.

49.7 H. H. Bennett "Stilling the Dunes," *Soil Conservation,* **16**:106 (1950).

49.8 R. Dolan, "Barrier Dune System Along Outer Banks of North Carolina: A Reappraisal," *Science,* **176**:286 (1972).

49.9 Typical useful guides are V. P. Ahearn, *Land Use Planning and the Sand and Gravel Producer,* National Sand and Gravel Association, Silver Spring, Md., 1964; and A. M. Bauer, *A Guide to Site Development and Rehabilitation of Pits and Quarries,* Industrial Mineral Report no. 33, Ontario Department of Mines, Toronto, 1970.

49.10 H. F. Bishop, and B. E. Hanna, "Gravel Mining and Land Development Can Go Hand in Hand," *Civil Engineering,* **49**:65 (February 1979).

49.11 W. B. Bull, and K. M. Scott, "Impact of Mining Gravel on Urban Streambeds in the Southwest United States," *Geology,* **2**:171–174 (1974).

49.12 E. A. Imhoff, T. O. Friz, and J. L. LaFevers, *A Guide to State Programs for the Reclamation of Surface Mined Areas,* Circular no. 731, U.S. Geological Survey, Washington, 1976.

49.13 *Evaluation, Review and Coordination of Federal and Federally Assisted Programs and Projects,* Circular no. A-95, Office of Management and Budget, Washington, 13 January 1976.

49.14 L. B. Leopold, F. E. Clarke, B. B. Hanshaw, and J. B. Balsley, *A Procedure for Evaluating Environmental Impact,* Circular no. 645, U.S. Geological Survey, Washington, 1971; see also E. E. Whitehurst, *James River Habitat Site,* U.S. Corps of Engineers, Norfolk, 1977.

49.15 E. Brent-Jones "Methods and Cost of Land Restoration" from *Transactions of the Institute of Quarrying, Quarry Managers' Journal,* October 1971, p. 1.

49.16 L. V. A. Sendlein, "Land Restoration: The Iowa Experiment," in abstract vol. for annual meeting at Denver, Geological Society of America, Boulder, Colo., 1976, p. 1097.

49.17 Leopold, Clarke, Hanshaw, and Balsley, op. cit.

49.18 P. Lessing and R. A. Smosna, "Environmental Impact Statements—Worthwhile or Worthless?" *Geology,* **3**:241–242 (1975).

49.19 "Beaten by Birdies," *The Countryman,* **57**:56 (1960).

49.20 G. J. Doucet, J.-P. Sarrazin, and J. R. Bider, "Use of Highway Overpass Embankments by the Woodchuck *(Marmota monax),*" *Canadian Field Naturalist,* **88**:187 (1974).

49.21 A. B. A. Brink, *Engineering Geology of Southern Africa,* vol. 1, Building Publications, Pretoria, 1979, p. 79.

49.22 L. M. Winsor, "The Barrier System of Flood Control," *Civil Engineering,* **8**:675 (1938).

49.23 G. P. Marsh, *The Earth as Modified by Human Action,* 2d ed. of *Man and Nature,* Scribner, New York, 1874; facsimile ed., Harvard University Press, Cambridge, Mass., 1965.

49.24 Ibid., p. 241.

49.25 G. V. Jacks, and R. O. Whyte, *Vanishing Lands,* Doubleday, New York, 1939 (still a good general review).

49.26 A. R. Croft, and R. W. Bailey, *Mountain Water,* U.S. Department of Agriculture, Forest Service Intermountain Region, Ogden, Utah, 1964.

49.27 C. B. Brown, "Sediment Complicates Flood Control," *Civil Engineering,* **15**:83 (1945) (gives some useful detailed figures).

49.28 R. F. Legget, "An Engineering Study of Glacial Drift for an Earth Dam near Fergus, Ontario," *Economic Geology,* **37**:531–556 (1942).

49.29 H. H. Bennett, *Soil Conservation,* McGraw-Hill, New York, 1939 (a classic in its field).

Suggestions for Further Reading

George Perkin Marsh's great book of a century ago, referenced above, can well be the starting point for any further reading about conservation. Reading this book is a sobering experience, since admiration for what Marsh saw so clearly over one hundred years ago is mixed with regret that his warnings were not heeded. *Soil Conservation* by H. H. Bennett (also referenced above) is the modern equivalent; it is still a most useful reference volume for all civil engineers. Another landmark volume in this field was *Conservation of Ground Water* by H. E. Thomas, published in 1951 by McGraw-Hill; it was a pioneer contribution to current public awareness of the necessity for conservation of renewable natural resources.

For specific information on conservation measures in relation to civil engineering works, the publications of the U.S. Soil Conservation Service of Washington, D.C., are unrivalled; a list of these should be a ready reference in all civil engineering offices.

Descriptions of successful conservation projects are now to be found in a wide variety of publications. If only to illustrate this point (and so introduce the name of another fine journal into this Handbook), attention is directed to an article called "Cotswold Lakeland" published in a British quarterly journal well described by its name, *The Countryman.* In the spring issue for 1980 (**85**:103), this article describes how a lovely area of 14,000 acres in southern England has been transformed into the Cotswold Water Park, 3,500 acres being covered by water, with 500 acres a national nature preserve. Working of gravel started in the area over fifty years ago. This park provides another fine example of what can be done when the will is there.

chapter 50

GEOLOGY AND THE CIVIL ENGINEER

50.1 Test Borings / 50-2
50.2 Geotechnical Studies / 50-4
50.3 Open Excavation / 50-5
50.4 Bulldozers / 50-7
50.5 Tunnel Shafts / 50-8
50.6 Tunnels / 50-9
50.7 Highways / 50-11
50.8 Archaeology / 50-13
50.9 Conclusion / 50-14
50.10 References / 50-17

The purpose of this Handbook has been to demonstrate the vital importance of geology in all civil engineering design and construction, and to guide those who use it to sources of more detailed useful information. The volume would not be complete without at least brief acknowledgment of the reciprocal contributions to the science of geology which civil engineering has been able to make. This has been implicit in almost all preceding chapters. Every civil engineering construction project being unique, each one will reveal some previously unknown information about the earth, just as will every subsurface investigation by means of bore and drill holes, or through geophysical surveying. Much of this information will be of no unusual significance, but this will never be known in advance. It therefore behooves all engineering geologists and civil engineers in any way connected with construction to be constantly on the alert in case their work reveals information that will contribute usefully to geological knowledge or to archaeological findings. Some examples have already been noted, but only incidentally. This brief concluding chapter, therefore, focusses on this little-appreciated aspect of construction site observations in the hope that the few examples that are given will encourage all who read these words to take a new interest in what they see when excavation work starts on a new project, and this despite the fact that their major interest will lie in what is to be built in the excavation.

This reciprocal link between civil engineering and geology is not something that has been recognized only recently. William Smith (1769–1839), long since recognized as the "Father of British Geology," practiced and usually signed himself as a civil engineer. He had an extensive practice (for that time) on canal construction and drainage works. As he traveled the roads of England, on horseback, inspecting canal works and other early construction work, he observed the geology so exposed, and he was thus able to prepare the famous "Map of the Strata." His first map, published in 1815, was awarded a special prize and can be regarded in a way as the first real geological map; so the link with civil engineering was there at the start.[50.1]

One of the great pioneers of geology, as it is known today, was Charles Lyell (later Sir Charles) of Scotland. He visited North America for the first time in 1841. He landed in Boston, off one of the very early steamboats, and within a few hours had observed:

> Several excavations made for railways, ... through mounds of stratified and unstratified gravel and sand, and also through rock, [which] enabled me to recognize the exact resemblance of this part of New England to the less elevated regions of Norway and Sweden....[50.2]

Throughout his travels on the North American continent, travels which extended as far as Pittsburgh, he was on the lookout for excavations. He was guided around Niagara Falls and in the Toronto area by "Thomas Roy, the Civil Engineer," who had presented a paper to the Geological Society of London in 1837 on the various levels of the Great Lakes; Roy's paper was based on information he had assembled while carrying out railway surveys.

Canal construction dominated construction at that time. Professor Eaton, of Rensselaer Polytechnic, had observed geology along the Erie Canal and had published a report on this in 1824.[50.3] Lyell met him and discussed his work. During Lyell's second visit (1845–1846), he saw the excavation for the ill-fated Brunswick Canal, between the Altahama and Turtle rivers. This was 14 km (9 mi) long and was even then regarded as a rash undertaking, the only beneficial results being the finding of fossil remains in the excavation by Hamilton Coupar, who collected no less than 45 different species of echinoderms.[50.4] All this was well over a century ago. The record continues throughout all the intervening years. Workers of today, therefore, are following a good tradition when they remember that excavations may yield geological information or specimens of value.

50.1 TEST BORINGS

Every time a test boring or drill hole is put down for site investigation, the civil engineer is doing what many geologists wish they could do, if only funds permitted, in order to check on the accuracy of their deductions from observations on the surface. It is small wonder, then, that records of test borings taken strictly with an engineering purpose in view often yield information of value to geological studies. Typical was the use made by J. T. Hack of records of test borings put down at 14 locations in Chesapeake Bay in connection with projected engineering works.[50.5] Hack was able not only to use the descriptions of the materials encountered but also to correlate the recorded penetration resistances with the main types of soil found in the bed of the bay, varying from gravel and gravelly sand to river mud. The bay is 280 km (175 mi) long and 40 km (25 mi) wide, with only shallow water, usually less than 18 m (60 ft) deep. When penetrated by borings, Recent and Pleistocene deposits revealed depths, in places, up to 60 m (200 ft). From his knowledge of the bay and study of all the boring records assembled, Hack was able to prepare a profile of the earlier river system, now submerged beneath the bay, thus adding a useful chapter to local geology (Fig. 50.1). In a somewhat similar way, study of the test borings put down across Mackinac Straits (between lakes Huron and Michigan) in a study for bridge foundations revealed the existence of a deep, preglacial river valley which added significantly to knowledge of the geological development of the Great Lakes (Fig. 50.2).[50.6]

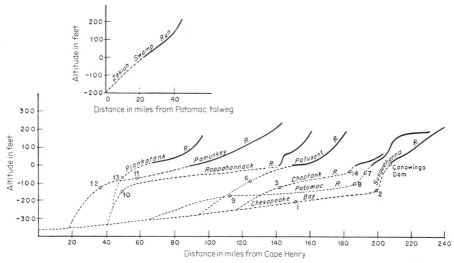

FIG. 50.1 Old stream beds in Chesapeake Bay, as determined by J. T. Hack.

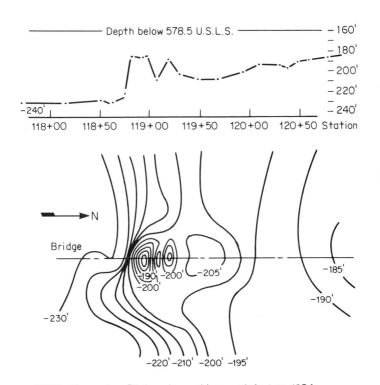

NOTE: Plan and profile based on probings made in June, 1954.

FIG. 50.2 Submerged "sea mount" revealed by study of boring results along the line of the Mackinac Straits Bridge, Michigan.

50-4 GEOLOGY AND THE ENVIRONMENT

FIG. 50.3 Quinnipiac River Bridge, New Haven, Connecticut, the test borings for which produced geologically useful samples.

Many test borings have been put down in recent years in connection with undersea oil fields and the erection of offshore structures. Fisk and McClelland collected many of the records from such test drilling in the Gulf of Mexico off the mouth of the Mississippi and carefully analyzed the correlation they made. This added a new chapter to knowledge of the geology of the continental shelf in that location. At a depth of 49 m (164 ft), for example, a sharp increase in soil shear strength was noted in one location, clearly marking the top of the weathered surface of the late Pleistocene formation known to underlie the more recent deltaic deposits brought down by the great river.[50.7]

Use made of samples from test borings in the bed of the Quinnipiac River, Connecticut, by three scientists of the U.S. Geological Survey, demonstrated another sort of by-product from preliminary work for a bridge. The samples were made available by the state highway department through an interested soils engineer (Fig. 50.3). It proved possible to carry out carbon-14 dating of some organic matter that was found in samples from depths between 9 and 12 m (30 and 40 ft) below sea level; pollen analyses were also made. Study of the results obtained indicated a steady rise of sea level with respect to the level of the land at New Haven at an average rate of slightly more than 1.8 mm (0.07 in) per year, a figure that agreed with some similar estimates.[50.8] Even micropaleontology has been studied with the aid of soil samples taken for engineering works, Lankford and Shepard studying foraminifera in 131 samples from six test borings taken across the Mississippi River, with significant findings in regard to facies interpretation.[50.9]

50.2 GEOTECHNICAL STUDIES

Modern geotechnical studies now have a good background of theory and a well-developed suite of laboratory testing procedures which greatly facilitate engineering foundation design. The full potential of these studies for geology does not appear yet to have been fully realized, but some useful applications have been made. Especially notable was the use made by Hubbert and Rubey of the concept of pore pressure and of the theories of Karl Terzaghi in their investigations of overthrust faulting.[50.10] Using the results of simple "soil indicator" tests of glacial soils from the bed of glacial Lake Agassiz, Rominger and Rutledge established five stratigraphic units in an early investigation which still does not appear to be as well known to geologists as it might be.[50.11] Harrison, in a similar way, used the results of consolidation tests to study probable ice pressures exerted upon glacial sediments near the edge of the ice mass.[50.12] Kenney developed a significant correlation between geological evidence of eustatic sea-level movements and the results of consolidation tests on soils from Boston, Nicolet, Ottawa, and Oslo.[50.13]

Misiaszek, in work the record of which is still not published, studied the records of about 200 test borings taken on the campus of the University of Illinois. By study of the natural moisture contents of undisturbed soil samples, and the records of the penetration resistance obtained during the test boring and sampling, he was able to determine the interfaces between separate till deposits.[50.14] The troubles that have been experienced with buried tills, related elsewhere in this volume, make this a significant link between geotechnical work and geological interpretations, one well worthy of much further development.

50.3 OPEN EXCAVATION

If only to show the widespread importance of examining open excavations, the writers may perhaps be forgiven for noting first that one of them was able to examine a relatively small excavation for a railway underpass structure in the Ontario city of Guelph.[50.15] It was, from the engineering point of view, a very ordinary type of grade-crossing separation project (Fig. 50.4). But beneath strata of two tills, stratified sand and silt and a weathered diamicton were found. These well exposed an underlying lens of organic material which was

FIG. 50.4 The Guelph paleosol site, Victoria Road, Guelph, Ontario; the paleosol is approximately at the level of the wet band across the middle of the cut.

50-6 GEOLOGY AND THE ENVIRONMENT

found to be resting on a paleosol developed on sand and gravel. Pollen content, fossils found in the organic material, and carbon-14 dating showed a date of over 45,000 years, corresponding to the local Port Talbot period. The find was so significant geologically that three continuously cored test holes were drilled to supplement the information that the excavation had revealed, temporarily—for the exposure was soon covered up by the permanent concrete structure. This is usually the case with engineering works, so geologists

FIG. 50.5 Aerial view of Boston, Massachussets, looking toward Logan International Airport, showing the buildings whose foundation excavations were studied by Kaye, as related in the text.

must be ready to make their examinations of excavations at short notice and with no undue delay.

Quite different were exposures during the excavation for Units 2 and 3 containment structures at the San Onofre nuclear generating station in San Diego County, California, features not shown by the careful preliminary test boring. Features observed when excavation was proceeding included several vertical and near-horizontal conjugate shear surfaces in three directions.[50.16] These were most carefully examined, excavation making this readily possible, with the result that the shear surfaces were found to be older than 70,000–130,000 years, the features revealed being of no consequence to the safety of the power plant.

An accumulation of observations of a number of excavations for building foundations on Beacon Hill, Boston, enabled Kaye to develop a new explanation for the existence of this notable feature of the city, long described as a drumlin. Kaye suggests that it is really an end moraine formed by a glacial readvance. Excavations revealed a complex of sand, gravel, and clay, instead of the anticipated till, with thrusting and other deformations clearly formed within the ice in the terminal zone. Slumping had occurred on deglaciation, forming complex secondary structures on the lower flanks of the ridge. This is a particularly revealing case, since it illustrates how patient and meticulous observations of relatively small excavations in the center of a great city can lead to a complete revision of a geological explanation that has long been accepted without question.[50.17] (Fig. 50.5.)

Finally, and in complete contrast to this case from Boston, the excavation of the Elbe-Seiten Canal in West Germany yielded information about glacial till of unusual value. The canal links the river Elbe with the Mittelland Canal. Its excavation permitted Ehlers and Stephan to examine an almost complete profile through the Pleistocene deposits of lower Saxony. Typical was a profile at km 60.6 (between Esterholz and Eelzen) which showed up to 8 m (26 ft) of grayish-brown sandy basal till, 4 m (13 ft) of fine sand, and then blackish-gray basal till. Time permitted study of the till fabric in both till deposits, and this facilitated correlation with other observations. The lower surface of the upper till showed elongated "ribs," a feature of interest in connection with current studies of ice movements in relation to till.[50.18]

50.4 BULLDOZERS

Special note must be made of two discoveries due to the alert observations of bulldozer operators, as a tribute to the men involved and in the hope that their example may be brought to the attention of other operators of these versatile automotive units, now so frequently used for small or shallow excavation operations. In the Soviet Union, many interesting finds have been made in frozen silt in regions of permafrost which has been uncovered by human activity. Exposure to solar radiation gradually thaws this material from its long-term frozen condition. Some perfectly preserved mammoths have thus been revealed to human sight, so well preserved that, in some cases, undigested vegetation has been found in their bodies. In the Kolyma-Berezovka permafrost region of Siberia, in 1978, an alert bulldozer operator noticed something which aroused his curiosity; careful operation of the machine gradually revealed a baby mammoth only seven to eight months old, perfectly preserved, naturally a find of the greatest scientific interest.[50.19]

The state highway department of Connecticut has already been mentioned, appreciatively. A new testing laboratory for this active organization was to be erected in Rocky Hill, a small community to the south of Hartford. A contract for the building was awarded; construction started with clearing of the soil on the site overlying bedrock. On 24 August 1966, Edward McCarthy was the operator of the bulldozer engaged on this operation. He

50-8 GEOLOGY AND THE ENVIRONMENT

FIG. 50.6 Dinosaur tracks first noticed by a bulldozer operator and saved for public benefit, now preserved in Dinosaur Park, Hartford, Connecticut.

noticed some strange marks when he got down to bedrock and called them to the attention of the engineer and architect on the site. (Fig. 50.6). The possible significance of the marks was realized. The director of the state geological survey and other scientists were notified. A 24-hr police guard was put in effect after one visitor had started up the contractor's bulldozer in his enthusiasm to procure a specimen. By 13 September the governor of the state announced that the 3.1-ha (7.5-acre) site would be set aside as a state park, and this was only three weeks after the initial find. The park is now popularly known as Dinosaur Park, since the strange marks that the bulldozer operator had so fortunately noticed were perfectly preserved tracks of dinosaurs (Fig. 50.7). The highway department had to find an alternative site for its laboratory.[50.20]

50.5 TUNNEL SHAFTS

It is often necessary to sink large-diameter shafts to give access to tunnel workings, many such shafts being used as working shafts, housing hoisting equipment and other services needed for tunnel construction. As they are sunk, they naturally reveal the subsurface conditions previously made known through test borings in a way that permits close examination of the strata exposed in an even better way than can be done in Calyx-drill holes. Shaft sinking, therefore, should always engage the attention of geologists, even as it is always very closely watched by those responsible for the works, since it is always a critical operation. Blank has usefully summarized a number of observations made on the surface of the gneiss which underlies Long Island, New York, some of them made during the sinking of shafts for the Board of Water Supply of New York. One of these showed a thick zone of decayed rock above its surface where this was buried under glacial deposits. In contrast, relatively fresh surfaces were found when bedrock was exposed at the surface by glacial action. In addition to the decayed rock, or saprolite, a concretionary zone was

FIG. 50.7 Plaque erected by the National Park Service at Dinosaur Park, Hartford, Connecticut.

found, especially in Shaft 14A for City Tunnel No. 2, and this has been recognized as an ancient laterite.[50.21]

In the sinking of 6.1-meter (20-ft) diameter shafts for a cable tunnel under the Thames River, to the east of the city of London, between Tilbury and Gravesend, the strata encountered were just as preliminary test boring had suggested, except that when the upper surface of the well-known Chalk (the bedrock under London) was reached, it was found to be badly disintegrated to a depth of about 6 m (20 ft). This made shaft sinking difficult, but the task was successfully completed, although with some delay. Geologically, however, the engineering difficulty was a matter of much interest, since it is entirely possible that the disintegration was the result of a permafrost condition during the last glacial period in the British Isles.[50.22] Such disintegration can be seen exposed in Czechoslovakia and probably in Canada. It is an unusual geological feature that should never be forgotten as a possibility in glaciated areas.

50.6 TUNNELS

The possibilities that tunnels present for study of geology throughout their complete lengths are so obvious that one would think there would be no need to do no more than mention "tunnels" in this chapter, leaving it at that. Unfortunately, all-too-many tunnels have been excavated, and then lined, without any record having been made of the revealed geology. Engineering records will naturally be made for all tunnels, but unless there is some geological input to the preparation of these official records, they may not prove adequate if ever the geology encountered by the tunnel driving has to be studied in detail. It was but natural that in the chapter on tunnels in this Handbook, the importance of geo-

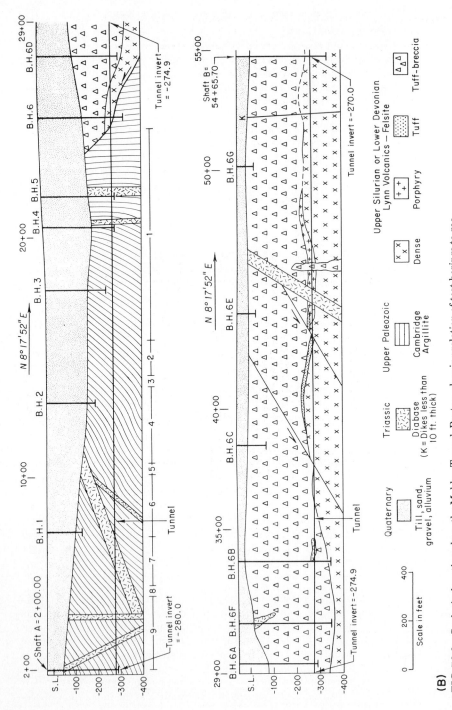

(B) **FIG. 50.8** Geological section along the Malden Tunnel, Boston, showing relation of test borings to geological structure.

logical records of all tunneling operations should be emphasized. Reference may be made to Chapter 20 for some examples.

As but a reminder, and as a further tribute to what has been done in Boston, brief mention may again be made of the detailed examination of the major tunnels now beneath this seaboard city of Marland Billings, his students and associates, and recorded in a series of valuable papers presented to the Boston Society of Civil Engineers (Fig. 50.8). It was a happy thing that Professor Billings was able to synthesize his findings in a broadly based paper, "The Geology of the Boston Basin," in which the contribution to the overall picture of the geology of the Boston Basin made by observations in the tunnels is clearly delineated.[50.23] There are few cities in which tunnel geology has been so well recorded, but special note must be made of the outstanding records prepared of the geology penetrated by tunnels for the water supply of New York City, initially by Professor Berkey of Columbia University, one of the great pioneers of engineering geology in North America.

50.7 HIGHWAYS

There are few civil engineering works which have so close a relation with geology as do highways. Since they extend over great distances, every major highway project may be expected to encounter a variety of geological conditions. Even though highways rarely involve any construction activity deep below the surface of the ground, the cuttings necessary to give the uniform grades demanded in modern superhighways necessarily provide a multitude of geological exposures. Not only are these of interest to passing motorists, but they frequently reveal details of local geology that would otherwise be unknown.

Useful guides, and even highway maps, showing the geology along major highways are now available; some of these indicate unique exposures that the highway construction has given to geology. Construction of the great system of interstate highways all over the United States has added greatly to the geological interest of U.S. roads. One exposure of significant geological interest is on Interstate 70 some kilometers west of the city of Denver. When the state highway department announced that their route for this fine highway involved making a deep cut through a local feature of some note, known as the "Hogsback," there was an immediate outcry from the active local environmentalists. Progress was almost stalled until, by mere chance, a senior member of the staff of the U.S. Geological Survey happened to sit at dinner next to one of the leaders of the opposition to the Hogsback Cut. He was able to explain what value the cut would have for geology and how it could be developed as an educational feature of great public interest and value. Opposition was thus turned into support. The cut was completed and the highway department finished it off in convenient benches, with good parking facilities and clear, useful signs (Fig. 50.9). The result is that the Hogsback Cut is now one of the most interesting of scenic viewpoints around Denver—and that is high praise indeed, in view of what geologists of the Denver area have been able to do in helping the public to appreciate the interesting geology of their area. The special map showing the big cut, prepared and published by the Colorado School of Mines, is an indication of this.[50.24]

The Hogsback Cut at Denver is but one of several such special geological exposures on major U.S. highways, of value geologically, of interest to the public. Another excellent example is to be found on the Red Mountain Expressway at the entrance to Birmingham, Alabama (Fig. 50.10).[50.25] Yet another is on Interstate 71 at the entrance to the city of Cleveland, Ohio, just as the freeway passes Brookside Park. Here a large open excavation is to be seen at the side of the road, clearly made when the new highway was built but not for any very obvious purpose. Indeed, many who see the big hole may blame it on "bad engineering." It is naturally no such thing but one of the best of all examples of the use

FIG. 50.9 The Hogsback Cut on Interstate Highway 70 outside Denver, Colorado, shown without the varied colors that add so much to this geological exhibit.

made of a far-sighted provision in the U.S. Federal Highway Act of 1957. This act makes it possible for federal funds to be applied to the salvage of historical, archeological, and paleontological specimens when they occur within the limits of federal highway projects. At Brookside Park, Interstate 71 goes through one of the most prolific deposits of Devonian fossil fishes known in North America, the existence of which was first noticed by the then city engineer of Cleveland, G. Sowers. Accordingly, the extra excavation was made in the process of procuring for the Cleveland Museum of Natural Science one of the largest collections of such fossils to be found anywhere today.[50.26]

FIG. 50.10 Rock cut on the Red Mountain Expressway at Birmingham, Alabama, with the city in the background, a fine geological exhibit now preserved for public benefit.

50.8 ARCHAEOLOGY

It will have been noticed that the 1957 Federal Highway Act mentions archaeological specimens. Although these are not, strictly speaking, geological in nature, their discovery is always associated with excavation in appropriate geological deposits, and they are so important that brief note must be made of their significance in connection with the works of the civil engineer. Again, highways come first to mind since they penetrate so much territory and, in many cases, follow old trails along which relics of earlier human occupation may be found. Some significant finds have, indeed, been made in the United States, such as the discovery in 1916, during construction of the Seventh Avenue Subway in New York City under what is now Cortlandt Station, of the remains of the bow and keel of a sailing ship, believed to be the *Tigjer,* a Dutch ship burned in 1614.[50.27] Unfortunately, it was possible to save only sections of the bow and keel at that time. Today, with the legal powers now available, better recovery would have been possible.

Somewhat naturally, it is in Europe that archaeological finds in connection with highway construction are of prime importance. In 1970 in England, for example, a team of local archaeologists discovered the remains of a complete Roman bathhouse near Welwyn, one that could be dated with some certainty about 300 A.D. This remarkable discovery was, unfortunately, right on the preferred line of a major new road from London to the north. After protracted negotiations between local archaeologists and the respective departments of the British government, it was agreed that a permanent protective cover, carrying the road over the Roman ruins, would be officially provided, if archaeologists would share the cost of exhibiting the bathhouse to the public.[50.28] Travelers from London to the north who choose to go by road, therefore, pass over the unique exhibit, in most cases probably with no realization that it is there. But it is now well preserved (Fig. 50.11).

FIG. 50.11 Metallic "tunnel" to protect the remains of the Roman bath house discovered on the route of the Welyn bypass road, England, the subgrade of which can be seen in the background.

50-14 GEOLOGY AND THE ENVIRONMENT

Even more remarkable is the experience of driving out of Dover, on the south coast of England, after crossing the English Channel on one of the justly famous ferry services. After leaving the fine new dock provided for the ferries, the motorist notes that the new highway through a crowded part of the ancient city has a surprisingly steep grade over what appears to be a "hump" in the road, just after the coast is left (Fig. 50.12). If one stops, it will be found that this hump is there to protect even more remarkable Roman ruins some of which were right on the line of this new road. The ruins include part of the original Roman harbor (with a lighthouse) and a Roman villa with painted decoration still intact. The probable location of these remarkable remains of a civilization of almost 2,000 years ago was suggested many years ago, through a process of gifted deduction, by a famous archaeologist, Sir Mortimer Wheeler. When the new road was planned, this prediction was remembered. Consultations were held with the appropriate authorities by interested archaeologists, and the result was this unusual road construction right over one part of one of the most remarkable of all Roman remains anywhere to be found.[50.29]

FIG. 50.12 New road through Dover, England, leading from the docks, the strange hump being over the Roman ruins, discovered as recounted in the text; the painted Roman villa, now on public view, is in the building on the far left.

50.9 CONCLUSION

A painted Roman villa might seem to be a long way from the general subject of this Handbook. Far from it! It is really symptomatic of all that is herein advocated. Its existence in the subsurface was deduced from sound knowledge of the region around; its actual presence was detected by carefully prosecuted subsurface exploration; its historic value was immediately realized; civil engineering design and construction were changed to protect it in its environment; and an accurate record of every detail of the villa and the surrounding geology was made before excavation was completed. That procedure is what this book is all about, although without the somewhat romantic addition of a Roman ruin to enliven the invariable interest of the subsurface.

People were involved, as they have been in every example that has been given—some of them well versed in geology, others coming to an appreciation of its importance through what they have seen of its contribution to sound civil engineering construction, and a growing number (it is hoped) discovering the pleasures which some knowledge of geology can give to the joy of living, in that it makes the beauty of scenery more meaningful, adds interest to so many of the materials of everyday, and opens the mind to great questions of the ages, as Teilhard de Chardin, and others, have shown so well.[50.30] It was with sound reason that Charles Kingsley a century ago called geology "the people's science."[50.31] It could so well be every child's introduction to the wonders of science. It should be part of

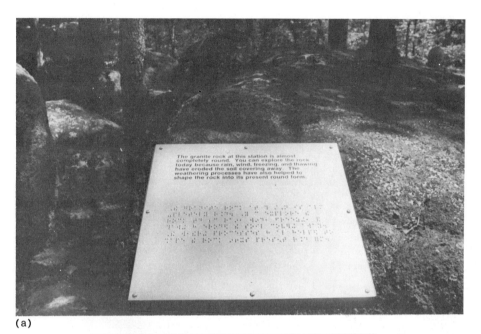

FIG. 50.13 (a) Entrance to the Braille Trail in Elephant Rocks State Park, near Graniteville, Missouri, arranged so that those who have lost their sight can "see" with their hands; (b) a close-up of the plaque at the entrance to the Trail, with its message in English and in Braille.

the scientific background of every civil engineer and an instinctive guide to all that is seen when sites are opened up.

One of the most moving geological exhibits known to the writers is to be found near Graniteville, Missouri, not too far from the geographical center of the United States. Here is located Elephant Rocks State Park, now a pleasant area of 50 ha (124 acres) from which granite was once quarried for a variety of early uses, including the piers of the famous Eads Bridge at St. Louis. The abandoned quarry was given to the State by a geologist, John S. Brown, the State assuming responsibility for it in March 1970. And in this park is a "Braille Trail," specially arranged so that those who have lost the gift of sight may still enjoy the delights of geology (Fig. 50.13). A series of knotted ropes guides blind visitors along a mile-long trail through an oak and hickory forest, circling the famous Elephant Rocks. Twenty-two special signs explain, in braille and in English, the significance of the billion-year-old igneous rocks which handicapped visitors "see" with their hands (Fig. 50.14).[50.32]

FIG. 50.14 Some of the boulders which give Elephant Rocks State Park its name, so easily "seen" with the hands.

"Sight is a faculty; seeing is an art." This is one of the many percipient sayings to be found in that splended, century-old book by George Perkins Marsh, *The Earth as Modified by Human Action*.[50.33] It is the art of seeing that it is hoped the use of this Handbook will, in a small way, develop still further in those who use it: the art of seeing the significance of every detail of the surface geology of each site they are called upon to investigate and of the subsurface geology, as their test borings supplement geological deductions and gradually reveal a three-dimensional picture of rock, soil, and water; the art of seeing the exact interrelationships of the geology of the site and the requirements of the structure under design and of seeing the correlation of what excavation actually reveals with what was expected; the art of seeing the geological significance of all that is displayed by excavation, never forgetting the possible scientific importance of what their work has disclosed and always, without question, accepting as a challenge to their capabilities as civil engineers all the complexities presented by the geology of the site. Then they will be able to accept gladly, and adopt as their own, those words which John Milton, in *Paradise Lost*, put into the mouth of the Angel (with the "contracted brow"):

> *Accuse not nature, she hath done her part;*
> *Do thou but thine. . . .*

50.10 REFERENCES

50.1 J. Phillips, *Memoirs of William Smith LLD,* J. Murray, London, 1844.

50.2 C. Lyell, *Travels in North America in the Years 1841–2, with Geological Observations in the United States, Canada, and Nova Scotia,* 2 vols., Wiley and Putnam, New York, 1845.

50.3 C. Eaton, in letter from James Hall to A. Barrett (9 June 1839), from Archives of New York Geological Survey.

50.4 Lyell, op. cit., p. 6.

50.5 J. T. Hack, "Submerged River System of Chesapeake Bay," *Bulletin of the Geological Society of America,* **69**:817 (1957).

50.6 W. N. Melhorn, "Geology of Mackinac Straits in Relation to Mackinac Bridge," *Journal of Geology,* **67**:403 (1959).

50.7 H. N. Fisk and B. McClelland, "Geology of the Continental Shelf off Louisiana: Its Influence on Offshore Foundation Design," *Bulletin of the Geological Society of America,* **70**:1369 (1955).

50.8 J. E. Upson, E. B. Leopold, and M. Rubin, "Postglacial Change of Sea Level in New Haven Harbor, Connecticut," *American Journal of Science,* **262**:121 (1964).

50.9 R. R. Lankford and F. P. Shepard, "Facies Interpretations in Mississippi Delta Borings," *Journal of Geology,* **68**:408–426 (1960).

50.10 M. K. Hubbert and W. W. Rubey, "Mechanics of Fluid Filled Porous Solids and Its Application to Overthrust Faulting"; and W. W. Rubey and M. K. Hubbert, "Overthrust Belt in Geosynclinical Area of Western Wyoming in Light of Fluid Pressure Hypothesis," *Bulletin* of *the Geological Society of America,* **70**:115; 166 (1959).

50.11 J. F. Rominger and P. C. Rutledge, "Use of Soil Mechanics Data in Correlation of Interpretation of Lake Agassiz Sediments," *Journal of Geology,* **60**:160 (1952).

50.12 W. Harrison, "Marginal Zones of Vanished Glaciers Reconstructed from the Preconsolidation-Pressure Values of Overridden Silts," *Journal of Geology,* **66**:72 (1958).

50.13 T. C. Kenney, "Sea-Level Movements and the Geologic Histories of the Post-Glacial Marine Soils at Boston, Nicolet, Ottawa, and Oslo," *Geotechnique,* **14**:203 (1964).

50.14 E. T. Misiaszek, "Engineering Properties of the Champaign-Urbana Subsoils," unpublished PhD. thesis, University of Illinois, 1960.

50.15 P. F. Karrow, R. J. Hebda, and E. W. Presant, "Interstadial (Middle Wisconsinan?) Paleosol and Fossils from Guelph, Ontario," in abstract vol. for the annual meeting at Toronto, Geological Association of Canada, Toronto, 1978, p. 432.

50.16 G. S. Hunt, H. G. Hawkins, and J. D. Scott, "Significance of Geologic Features Exposed During Excavation of Units 2 and 3, San Onofre Nuclear Generating Station, San Diego County, California," in abstract vol. for annual meeting at Syracuse, Northeastern Section, Geological Society of America, Boulder, Colo., 1975, p. 77.

50.17 C. A. Kaye, "Beacon Hill and Moraine, Boston: New Explanation of an Important Urban Feature," in D. R. Coates (ed.), *Urban Geomorphology,* Special Paper no. 174, Geological Society of America, Boulder, Colo., 1976.

50.18 J. Ehlers and H. J. Stephan, "Forms at the Base of Till as an Indication of Ice Movement," *Journal of Glaciology,* **22**:347 (1979).

50.19 "Study Begins of Baby Mammoth," *Nature,* **273**:485 (June 1978).

50.20 *Thirty-Fourth Biennial Report of the Commissioners of the State Geological and Natural History Survey,* Bulletin no. 104, Connecticut State Geological and Natural History Survey, Hartford, 1971, p. 3; see also *Dinosaur State Park,* leaflet of the Connecticut State Park and Forest Commission (no date).

50.21 H. R. Blank, "Fossil Laterite on Bedrock in Brooklyn, New York," *Geology,* **6**:21 (January 1978).

50.22 C. K. Haswell, "Thames Cable Tunnel," *Proceedings of the Institution of Civil Engineers,* **44**:323 (1969); see also R. B. G. Williams, "Fossil Patterned Ground in Eastern England," *Biuletyn Peryglacjalny,* **14**:337 (1964).

50.23 M. P. Billings, "Geology of the Boston Basin," in *Studies in New England Geology,* Memoir no. 146, Geological Society of America, Boulder, Colo., 1976.

50.24 L. W. LeRoy and R. J. Weimer, *Geology of the Interstate 70 Road Cut, Jefferson County, Colorado,* Publication no. 7, Department of Geology, Colorado School of Mines, Golden; and E. McKee, Denver, by personal communication.

50.25 P. E. LaMoreaux and T. A. Simpson, "Birmingham's Red Mountain Cut," *Geotimes,* **15**:10 (October 1970).

50.26 W. E. Scheele, "Fossil Dig," *The Explorer,* **7**(3): (1965); and articles in succeeding issues.

50.27 F. Salvadori and N. Cortes-Comerer, "Ecology of the Past—A New Responsibility," *Civil Engineering,* **47**:61–65 (June 1977).

50.28. "British Save Roman Bathhouse," *Civil Engineering,* **47**:64 (June 1977).

50.29 B. Philip, *The Roman Painted House at Dover* (1977) and *Roman Dover, Britain's Buried Pompeii* (1973), Kent Archaeological Rescue Unit, Dover Castle, Kent, Eng.

50.30. P. Teilhard de Chardin, *The Phenomenon of Man,* Harper and Row, New York, 1961, p. 142.

50.31 C. Kingsley, *Town Geology,* Daldy, Isbister, London, 1877.

50.32 "Missouri's First Braille Trail," *Missouri Mineral News,* **13**:190 (1973).

50.33 G. P. Marsh, *The Earth as Modified by Human Action,* 2d ed., Scribner, New York, 1874; facsimile ed. Harvard University Press, Cambridge, Mass., 1965.

appendix A

GLOSSARY OF GEOLOGICAL TERMS

This glossary is merely an introduction to geological terminology. Compound names such as *gabbro-syenite* are not generally included, and no mention is made of the names of minerals—whether used to describe minerals only or, as in such cases as *serpentine,* rocks as well. Reference may be made to Table 3.1 for particulars of the more common minerals. Names in the geological "timetable" are not included. Many words in normal use have been adopted literally for geological use, such as *analogy,* and there are naturally many commonly used words, such as *avalanche,* that would appear in a complete glossary of geology but which are not included here. The glossary is intended to provide the engineer with a guide to the unfamiliar terms that may be encountered in geological reading.

Aa A Hawaiian term for basaltic lava flows typified by a rough, jagged, spinose, clinkery surface.

Accessory A term applied to minerals occurring in small quantities in a rock, and whose presence or absence does not affect its diagnosis.

Acidic A descriptive term applied to those igneous rocks that contain more than 66 percent SiO_2; a term contrasted with intermediate and basic.

Adobe A name applied to clayey and silty deposits found in the desert basins of southwestern North America and in Mexico, where the material is extensively used for making sun-dried brick.

Aeolian A term applied to deposits whose constituents have been carried by, and laid down from, the wind.

Agglomerate Contemporaneous pyroclastic rocks containing a predominance of rounded or subangular fragments greater than 32 mm in diameter, lying in an ash or tuff matrix and usually localized in volcanic necks. The form of the fragments is in no way determined by the action of running water.

Alluvium The deposits made by streams in their channels and over their floodplains and deltas. The materials are usually uncemented and are of many kinds and dimensions.

Andesite A volcanic rock composed essentially of plagioclase, together with one or more of the mafic minerals, biotite, hornblende, and pyroxene.

Anorthosite A plutonic rock composed almost wholly of plagioclase.

Anticline A fold, generally convex upward, the core of which contains the stratigraphically older rocks.

Aphanitic A term referring to rocks that are so fine-grained that the individual minerals cannot be distinguished by the naked eye.

Arenaceous Like or pertaining to sand. An example is arenaceous limestone or a sandy limestone.

Argillaceous An adjective meaning clayey and applied to rocks containing clay.

Ash (volcanic) Uncemented pyroclastic debris consisting of fragments mostly under 4 mm in diameter.

Authigenic Formed where found (used of mineral particles of rocks formed by crystallization in the place they occupy).

Basalt A microlithic or porphyritic igneous rock of a lava flow or minor intrusion, having an aphanitic texture as a whole or in the groundmass and composed essentially of plagioclase and pyroxene, with or without interstitial glass.

Basic A term applied to igneous rocks having a relatively low percentage of silica; the limit below which they are regarded as basic is about 52 percent.

Batholith A huge intrusive body of igneous rock, without a known floor, in contrast to a laccolith, which rests on a floor. Like other intrusive bodies, batholiths become accessible to human observation only as a result of their exposure by erosion.

Bedding plane (of rock) Refers to the plane of junction between different beds or layers of rock.

Bentonite The plastic residue from the devitrification and attendant chemical alteration of glassy igneous material, usually volcanic tuff or ash. It swells greatly with the addition of water and forms a milky suspension.

Boulder A detached rock mass, somewhat rounded or otherwise modified by abrasion in transport; 256 mm has been suggested as a convenient lower limit for diameter.

Boulder clay A tenacious unstratified deposit of glacial origin, consisting of a hard matrix (usually rock flour) packed with subangular stones of varied sizes; *till* is the preferred name.

Breccia A clastic rock made of coarse angular or subangular fragments of varied or uniform composition (and of either exogenetic or endogenetic origin).

Calcareous An adjective applied to rocks containing calcium carbonate.

Chalk A fine-grained, somewhat friable foraminiferal limestone of Cretaceous age occurring widely in Britain and in northwestern Europe.

Chert A more or less pure siliceous rock, occurring as independent formations and also as nodules and irregular concretions in formations (generally calcareous) other than the chalk. The fracture is generally conchoidal. When worked, it is called *flint*.

Clastic A term applied to rocks composed of fragmental material derived from preexisting rocks or from the dispersed consolidation products of magmas.

Clay A fine-grained, unconsolidated material which has the characteristic property of being plastic when wet and which loses its plasticity and retains its shape upon drying or when heated.

Clay mineral One of a group of layered hydrous aluminosilicates commonly formed by weathering. Characterized by small particle size and the ability to absorb large quantities of water and of exchangeable cations.

Cleavage (1) The property of minerals whereby they can be readily separated along planes parallel to certain possible crystal faces; (2) the property of rocks, such as slates, that have been subjected to orogenic pressure, whereby they can be split into thin sheets.

Cobble A rock fragment between 64 and 256 mm in diameter, rounded or otherwise abraded in the course of aqueous or glacial transport.

Conchoidal A term used in descriptive mineralogy to describe the shell-like surface produced by the fracture of a brittle substance.

Concretions Nodular or irregular concentrations of certain authigenic constituents of sedimentary rocks and tuffs.

Conglomerate A cemented clastic rock containing rounded fragments corresponding in their grade sizes to gravel.

Crystal A body, generally solid but not necessarily so, whose component atoms are arranged in definite space lattices; crystal faces are a commonly developed outward expression of the periodic arrangement of atoms.

Current- or Cross-bedding A structure of sedimentary rocks, generally arenaceous, in which the planes of deposition lie obliquely to the planes separating the larger units of stratification.

Detritus Fragmental material, such as sand and mud, derived from older rocks by disintegration. The deposits produced by the accumulation of detritus constitute the detrital sediments.

Diabase A rock of basaltic composition, consisting essentially of labradorite and pyroxene, and characterized by ophitic texture. Rocks containing significant amounts of olivine are olivine diabases. In Great Britain basaltic rocks with ophitic texture are called *dolerites,* and the term *diabase* is restricted to altered dolerites.

Diastrophism The process or processes by which the crust of the earth is deformed, and by which continents and ocean basins, plateaus and mountains, flexures and folds of strata, and faults are produced.

Diorite A phanerocrystalline igneous rock composed of plagioclase and mafic minerals such as hornblende, biotite, and augite; hornblende is especially characteristic.

Dip The angle of inclination of the plane of stratification with the horizontal plane.

Dolerite American authors use the term *diabase* in a sense synonymous with dolerite. In Great Britain, dolerite is the name used for an igneous rock occurring as minor intrusions, consisting essentially of plagioclase and pyroxene, and distinguished from basalt by its coarser grain and the absence of glass.

Dolomite A carbonate rock, consisting predominantly of the mineral dolomite, such as dolomitic limestone and magnesian limestone, in which dolomite and calcite are present, the latter predominating; it is also called *dolostone.*

Dike An injected tabular intrusion, cutting across the bedding or other parallel structures of the invaded formation.

Endogenetic A term applied to geological processes originating within the earth and to rocks that owe their origin to such processes.

Epigene A general term for geological processes originating at or near the surface of the earth.

Epigenetic A term now generally applied to ore deposits of later origin than the rocks among which they occur.

Essential minerals The essential constituents of a rock are those minerals whose presence is, by definition, necessary.

Exfoliation The "peeling" of a rock surface in sheets due to changes of temperature or to other causes.

Exogenetic A term applied to geological processes originating at or near the surface of the earth and to rocks that owe their origin to such processes.

Fault A fracture of the earth's crust accompanied by displacement of one side of the fracture (in a direction parallel to the fracture), resulting in an abrupt break in the continuity of beds or strata.

Felsite An igneous rock, with or without phenocrysts, in which either the whole or the groundmass consists of cryptocrystalline aggregates of felsic minerals; quartz and orthoclase are those characteristically developed.

Felsitic A term applied to the cryptocrystalline texture seen in the groundmass of quartz felsites and similar rocks.

Ferruginous An adjective applied to rocks with a prominent iron content.

Flint A more or less pure siliceous rock composed mainly of granular chalcedony together with a small proportion of opaline silica and occurring in nodules, irregular concretions, layers, and veinlike masses. The fracture is conchoidal. Flint is *chert* worked by man.

Foliation   A structure represented most characteristically in schists and due to the parallel disposition of layers or lines of one or more of the conspicuous minerals of the rock; the parallelism is not a direct consequence of stratification.

Foot wall   The wall rock beneath an inclined vein or fault.

Formation   A term applied stratigraphically to a set of strata prossessing a common suite of lithological characteristics.

Gabbro   A phanerocrystalline rock consisting of labradorite, or bytownite, and augite (generally diallage).

Gneiss   A foliated or banded phanerocrystalline rock (generally, but not necessarily, felspathic and of granitic or dioritic composition) in which granular minerals, or lenticles and bands in which they predominate, alternate with schistose minerals, or lenticles or bands in which they predominate. The foliation of gneiss is more "open," irregular, or discontinuous than that of schist.

Granite   A phanerocrystalline rock, consisting essentially of quartz and alkali feldspars with any of the following: biotite, muscovite, and amphiboles and pyroxenes.

Granitic   A term applied to irregularly granular textures such as that of a nonporphyritic granite.

Gravel   An unconsolidated accumulation of rounded rock fragments predominantly larger than sand grains. Depending on the coarseness of the fragments, the material may be called *pebble gravel, cobble gravel,* or *boulder gravel.*

Grit   A course sandstone of angular grain. Variation from the grain-size range of from ½ to 1 mm in diameter should be indicated as *coarse grit, fine grit,* etc.

Groundwater   A term that should be applied only to the water in the zone of rock fracture below the earth's surface, in which interstitial water occurs, i.e., in the zone of saturation.

Gumbo   A provincial term used to describe local varieties of sticky, highly plastic clay soil.

Hanging wall   The wall rock above an inclined vein or fault.

Hardpan   A term to be avoided, in view of its wide and essentially popular local use for a wide range of materials.

Hypabyssal   A general term applied to minor intrusions, such as sills and dikes, and to the rocks of which they are made, to distinguish them from volcanic rocks and formations, on the one hand, and plutonic rocks and major intrusions such as batholiths, on the other.

Hypogene   A general term intended to include both plutonic and metamorphic classes of rocks, i.e., a term used for rocks formed within the earth.

Igneous (Rocks)   A general term for those rocks which have been formed or crystallized from molten magmas, the source of which has been the earth's interior.

Inclusion   A general term for foreign bodies (gas, liquid, glass, or mineral) enclosed by minerals; also extended in its English usage to connote *enclosures* of rocks and minerals within igneous rocks.

Indurated   A term technically restricted to compact rocks that have been hardened by the action of heat. Occasionally the term is applied also to sediments hardened by pressure and cementation.

Inlier   An area or group of rocks surrounded by rocks of younger age.

Intermediate   A term applied to rocks intermediate in silica content between the basic and the acidic groups.

Intrusive   A term applied to igneous rocks that have cooled and solidified from magma under the cover of the rock masses of the outer shell of the earth; they therefore become accessible to view only after they have been exposed by erosion of the overlying rock.

Isostasy   The principle that above some level within the earth, presumably at a depth of some scores of kilometers, different segments of the crust, if of equal area, also have the same mass. The blocks thus balance one another.

GLOSSARY OF GEOLOGICAL TERMS   A-5

Joints (rock)   Rock fractures on which there has been no appreciable displacement parallel with the walls. They may sometimes be characterized, according to their relation with the rock bedding, as *strike joints* and *dip joints*.

Kaolin   A white, or slightly stained, clay, resulting from the decomposition of a highly felspathic rock and therefore also containing a variable proportion of other constituents derived from the parent rock.

Laccolith   A dome-shaped intrusion with both floor and roof concordant with the bedding planes of the invaded formation; the roof is arched upward as a result of the intrusion.

Laterite   A residual deposit, often concretionary, formed as a result of the decomposition of rocks by tropical weathering and groundwaters and consisting essentially of hydrated oxides of aluminum and iron.

Limestone   A general term for bedded rocks of exogenetic origin, consisting predominantly of calcium carbonate.

Lithology   The character of rocks as based on the megascopic observation of hand specimens. In its French usage, the term is synonymous with *petrography*.

Loam   A detrital deposit containing nearly equal proportions of sand, silt, and clay; these terms refer to the grain size of the particles.

Loess   A widespread deposit of aeolian silt, a buff-colored, porous, but coherent deposit traversed by a network of generations of grass roots.

Macroscopic   See under **Megascopic.**

Mafic   A mnemonic term for the ferromagnesian and other nonfelsic minerals actually present in an igneous rock.

Magma   A comprehensive term for the molten fluids generated within the earth from which igneous rocks are considered to have been derived by crystallization or other processes of consolidation.

Magnesian limestone   See under **Dolomite.**

Marble   A general term for any calcareous or other rock of similar hardness that can be polished for decorative purposes (petrologically restricted to granular crystalline limestones); it is derived from the metamorphism of limestone.

Marl   Freshwater lime deposit (lacustrine) usually with shells.

Megascopic   A general term, more appropriate than *macroscopic,* applied to observations made on minerals and rocks by means of the naked eye or pocket lens but not with a microscope.

Metamorphic   A term used for rocks derived from preexisting rocks by mineralogical, chemical, and structural alterations due to endogenic processes; the alteration is sufficiently complete throughout the body of the rock to produce a well-defined new type.

Mica schist   A schist composed essentially of micas and quartz, the foliation is mainly due to the parallel disposition of the mica flakes.

Mineral   A naturally occurring inorganic element or compound having an orderly internal structure and characteristic chemical and physical properties.

Mudstone   Consolidated mud (silt plus clay) lacking the fissility of shale.

Muskeg   A Canadian term of Cree Indian origin, meaning a moss-covered muck or peat bog. In the Far North all swamps are called muskegs.

Oölite   A rock made up of spheroidal or ellipsoidal grains formed by the deposition of successive coats of calcium carbonate around a nucleus.

Ophitic   A term applied to microscopic rock texture to designate a mass of longish, interlacing crystals, the spaces between which are filled with minerals of later crystallization.

Orogeny   The process of forming mountains, particularly by folding and thrusting.

Outcrops (rock)   Those places where the underlying bedrock comes to the surface of the ground and is exposed to view.

Outlier An isolated mass or detached remnant of younger rocks, or of rocks overthrust upon others, separated by erosion from the main mass to which they belong and now surrounded, areally, by older, or at least underlying, rocks; opposite of *inlier*.

Pebble A rock fragment between 4 and 64 mm in diameter, which has been rounded or otherwise abraded by the action of water, wind, or glacial ice.

Pegmatite A coarse-grained igneous dike rock, rich in quartz and feldspar.

Phanerocrystalline A term applied to igneous rocks in which all the crystals of the essential minerals can be distinguished individually by the naked eye; contrasted with *aphanitic*.

Phenocryst A term applied to isolated larger crystals visible to the unaided eye and lying in a finer mass of a rock of igneous origin.

Phreatic Relating to groundwater in the saturated zone.

Phyllite An argillaceous rock intermediate in metamorphic grade between slate and schist. The mica crystals impart a silky sheen to the surface of cleavage (or schistosity).

Plutonic A general term applied to major intrusions and to the rocks of which they are composed, suggestive of the depths at which they were formed in contradistinction to most minor intrusions and all volcanic rocks.

Porphyry Rocks containing conspicuous phenocrysts in a fine-grained or aphanitic groundmass. The resulting texture is described as *porphyritic*.

Pyroclasts A general term for fragmental deposits of volcanic *ejectamenta*, including volcanic conglomerates, agglomerates, tuffs, and ashes.

Quartzite A granulose metamorphic rock, representing a recrystallized sandstone, consisting predominantly of quartz. The name is also used for sandstones and grits cemented by silica that has grown in optical continuity around each fragment.

Regolith A general term for the superficial blanket of denudation products that is widely distributed over the more mature "solid" rocks. The term includes weathering residues, alluvium, and aeolian and glacial deposits.

Rhyolite A volcanic rock corresponding in chemical composition to granite.

Rock flour A general term for finely comminuted rock material corresponding in grade to mud but formed by the grinding action of glaciers and ice sheets and therefore composed largely of unweathered mineral particles.

Rudaceous A term applied to sedimentary rocks composed of coarsely grained detritus such as gravel, shingle, and pebbles.

Sand Fine, granular material derived from either the natural weathering or the artificial crushing of rocks, generally limited in diameter to between 2 and 0.05 mm.

Sandstone A cemented or otherwise compacted detrital sediment composed predominantly of quartz grains, the grades of the latter being those of sand. Varieties such as argillaceous, calcareous, and ferruginous are recognized according to the nature of the cementing material. Corresponding rocks of coarser grades are sometimes called *grits*.

Schist A general term for foliated metamorphic rocks, the structure of which is controlled by the prevalence of lamellar minerals such as micas—which have normally a flaky habit—or of stressed minerals crystallized in elongated forms rather than in the granular forms generally assumed.

Schistosity The property of a foliated rock whereby it can be divided into thin flakes or lenticles.

Sedimentary A general term for loose and cemented sediments of detrital origin, generally extended to include all exogenetic rocks.

Shale A laminated sediment, in which the constituent particles are predominantly of the clay type. The characteristic cleavage is that of bedding and such other secondary cleavage or fissibility as is approximately parallel to bedding.

Shingle Coarse beach gravel, relatively free from fine material and commonly of loose texture.

Siliceous Of or pertaining to silica; containing silica, or partaking of its nature. Containing abundant quartz.

Sill A tabular sheet of igneous rock injected along the bedding planes of sedimentary or volcanic formations.

Silt An unconsolidated clastic sediment, most of the particles of which are between 0.05 and 0.002 mm in diameter.

Slate A general term for compact aphanitic rocks, formed from shales, mudstones, or volcanic ashes and having the property of easy fissility along planes independent of the original bedding.

Slickensides Polished and grooved surfaces on rock faces that have been subjected to faulting.

Soapstone Rock consisting of the mineral talc or a closely related mineral species.

Stratum A layer that is separable along bedding planes from layers above and below; the separation arises from a break in deposition or a change in the character of the material deposited.

Strike (of rocks) The direction of the line of intersection of the plane of stratification with the horizontal plane.

Structure (rock) A term reserved for the larger features of rocks; e.g., a layered or laminated structure generally indicates sedimentary origin.

Syenite A term applied to phanerocrystalline rock composed essentially of alkali feldspars and one or more of the common mafic minerals; hornblende is especially characteristic.

Syncline A fold, generally concave upward, the core of which contains stratigraphically younger rocks.

Syngenetic A term now generally applied to mineral or ore deposits formed contemporaneously with the enclosing rocks, as contrasted with epigenetic deposits, which are of later origin than the enclosing rocks.

Talus A term used for the accumulation of fine, coarse, or mixed fragments and particles, fallen at or near the base of cliffs.

Texture (rock) The appearance, megascopic or microscopic, seen on a smooth surface of a homogeneous rock or mineral aggregate due to the degree of crystallization, the size of the crystals, and the shapes and interrelations of the crystals or other constituents.

Till The unstratified or little-stratified deposits of glaciers.

Tillite A term applied to consolidated till formed during glacial epochs anterior to that of the Pleistocene.

Trap An old Swedish name originally applied to igneous rocks that were neither coarsely crystalline, like granite, nor cellular and obviously volcanic, like pumice. The rocks so designated included basalts, dolerites, andesites, and altered varieties of some of these, such as epidiorite and diabase (types grouped as *greenstones*); and, finally, the mica traps, or lamprophyres. The word is often popularly used to describe local rock types and so is one to be generally avoided.

Tuff A rock formed of compacted pyroclastic fragments, some of which can generally be distinguished as such by the naked eye. The indurated equivalent of volcanic ash or dust (cf. *agglomerate*).

Unconformity A break in the geologic sequence where there is a surface of erosion or nondeposition separating two groups of strata.

Vadose A term applied to seepage waters occurring below the surface and above the water table; contrasted with *phreatic*.

Vein An irregular sinuous igneous injection or a tabular body of rock formed by deposition from solutions rich in water or other volatile substances.

Ventifact A wind-carved stone; a sandblasted pebble or larger fragment.

Water table The upper surface of a zone of saturation beneath the earth's crust, except where that surface is formed by an impermeable body.

Weathering The destructive alteration and decay of rocks by exogenetic processes acting at the surface and down to the depth to which atmospheric oxygen can penetrate.

A-8 APPENDIX A

There are now a considerable number of useful *dictionaries of geology* to which reference may be made to supplement this short glossary based on the reading and work of the authors. The varied uses of the words *soil, compaction,* and *consolidation* are discussed in the appropriate section of the text. The American Geological Institute has placed all geologists, and engineers, in its debt by publishing the 749-page *Glossary of Geology,* now in its second edition (1980). This may be rightly regarded as the final authority for all geological definitions.

appendix B

GEOLOGICAL SURVEYS OF THE WORLD

NORTH AMERICA

United States of America

U.S. Geological Survey, National Center
12201 Sunrise Valley Drive
Reston, Virginia 22092
(Major branches in Denver, Colorado and Menlo Park, California)

State Geological Surveys

Geological Survey of Alabama
Drawer O,
University of Alabama
Tuscaloosa, Alabama 35486

Alaska Division of Geological and Geophysical Surveys
3001 Porcupine Drive
Anchorage, Alaska 99501

Arizona Bureau of Geology
 & Mineral Technology
845 North Park Avenue
Tucson, Arizona 85719

Arkansas Geological Commission
Vardelle Parham Geology Center
3815 West Roosevelt Road
Little Rock, Arkansas 72204

California Division of Mines & Geology
Room 1341
1416 Ninth Street
Sacramento, California 95814

Colorado Geological Survey
Room 715
1313 Sherman Street
Denver, Colorado 80203

Connecticut Natural Resources Center
Room 553, State Office Building
165 Capitol Avenue
Hartford, Connecticut 06115

Delaware Geological Survey
University of Delaware
Newark, Delaware 19711

Florida Bureau of Geology
903 West Tennessee Street
Talahassee, Florida 32304.

Georgia Department of
  Natural Resources
Earth & Water Division
Room 400
19 Dr. Martin Luther King Jr Drive, SW
Atlanta, Georgia 30334

Hawaii Division of Water
  & Land Development
Box 373
Honolulu, Hawaii 96809

Idaho Bureau of Mines & Geology
Morrill Hall
Moscow, Idaho 83843

Illinois Geological Survey
Natural Resources Building
615 East Peabody Drive
Champaign, Illinois 61820

Indiana Geological Survey
611 North Walnut Grove
Bloomington, Indiana 47405

Iowa Geological Survey,
123 North Capitol Street
Iowa City, Iowa 52242

Kansas Geological Survey
1930 Avenue A, Campus West
University of Kansas
Lawrence, Kansas 66044

Kentucky Geological Survey
311 Breckinridge Hall
University of Kentucky
Lexington, Kentucky 40506

Louisiana Geological Survey
Box G, University Station
Baton Rouge, Louisiana 70803

Maine Geological Survey
Department of Conservation
State House Station #22
Augusta, Maine 04333

Maryland Geological Survey
Merryman Hall
Johns Hopkins University
Baltimore, Maryland 21218

Massachusetts Department of
  Environmental Quality Engineering
Division of Waterways
1 Winter Street
Boston, Massachusetts 02108

Michigan Geological Survey Division
Box 30028
Lansing, Michigan 48909

Minnesota Geological Survey
1633 Eustis Street
St. Paul, Minnesota 55108

Mississippi Geologic, Economic
  & Topographic Survey
Box 4915
Jackson, Mississippi 39216

Missouri Division of Geological
  & Land Survey
Box 250
Rolla, Missouri 65401

Montana Bureau of Mines & Geology
Montana College of Mineral Science
  & Technology
Butte, Montana 59701

Nebraska Conservation & Survey Division
University of Nebraska
Lincoln, Nebraska 68588

Nevada Bureau of Mines & Geology
University of Nevada
Reno, Nevada 89557

New Hampshire Department of
  Resources & Economic Development
James Hall
University of New Hampshire
Durham, New Hampshire 03824

New Jersey Bureau of Geology
  & Topography
CN-029
Trenton, New Jersey 08625

New Mexico Bureau of Mines
  & Mineral Resources
Socorro, New Mexico 87801

New York State Geological Survey
Room 3140
New York Cultural Education Center
Albany, New York 12230

North Carolina Department of Natural
  Resources & Community Development
Geological Survey Section
Box 27687
Raleigh, North Carolina 27611

North Dakota Geological Survey
University Station
Grand Forks, North Dakota 58202

Ohio Division of Geological Survey
Fountain Square, Building B
Columbus, Ohio 43224

Oklahoma Geological Survey
Room 163
830 Van Vleet Oval
Norman, Oklahoma 73019

Oregon Department of Geology
  & Mineral Industries
1005 State Office Building

GEOLOGICAL SURVEYS OF THE WORLD   B-3

1400 SW Fifth Avenue
Portland, Oregon 97201

Pennsylvania Bureau of Topographic
 & Geologic Survey
Department of Environmental Resources
Box 2357
Harrisburg, Pennsylvania 17120

Puerto Rico Servicio Geológico
Apartado 5887, Puerta de Tierra
San Juan, Puerto Rico 00906

South Carolina Geological Survey
State Development Board
Harbison Forest Road
Columbia, South Carolina 29210

South Dakota Geological Survey
Science Center
University of South Dakota
Vermillion, South Dakota 57069

Tennessee Division of Geology
G-5 State Office Building
Nashville, Tennessee 37219

Texas Bureau of Economic Geology
University of Texas
Box X, University Station
Austin, Texas 78712

Utah Geological & Mineral Survey
606 Black Hawk Way
Salt Lake City, Utah 84108

Vermont Geological Survey
Agency of Environmental Conservation
Montpelier, Vermont 05602

Virginia Division of Mineral Resources
Box 3667
Charlottesville, Virginia 22903

Washington Division of Geology
 & Earth Resources
Olympia, Washington 98504

West Virginia Geological
 & Economic Survey
Box 879
Morgantown, West Virginia 26505

Wisconsin Geological
 & Natural History Survey
1815 University Avenue
Madison, Wisconsin 53706

Wyoming Geological Survey
Box 3008, University Station
University of Wyoming
Laramie, Wyoming 82071

Canada

Geological Survey of Canada
Department of Energy, Mines
 and Resources
601 Booth Street
Ottawa, Ontario K1A OE8
(Major branch in Calgary)

Provincial Geological Surveys

Alberta Department of the Environment
Oxbridge Plaza
9820 - 106th Street
Edmonton, Alberta T5K 2J6

British Columbia Geological Division
Department of Mines
 and Petroleum Reserves
Legislative Buildings
Victoria, British Columbia V8V 1X4

Manitoba Department of Mines, Resources
 and Environmental Management
Legislative Building
Winnipeg, Manitoba R3C OV8

New Brunswick Department of
 Natural Resources
P.O. Box 6,000
Fredericton, New Brunswick E3B 5H1

Newfoundland Department of
 Mines and Energy
P.O. Box 4750
St. John's, Newfoundland A1C 5T7

Nova Scotia Department of Mines
1690 Hollis Street
P.O. Box 1087
Halifax, Nova Scotia B3J 2X1

Ontario Geological Survey
Ministry of Natural Resources
77 Grenville Street
Toronto, Ontario M5C 1B3

Prince Edward Island Department of
 the Environment
P.O. Box 2,000
Charlottetown
 Prince Edward Island C1A 7N8

APPENDIX B

Quebec Department of Natural Resources
1620 Boulevard de l'Entente
Quebec City, Quebec G1S 4N6

Saskatchewan Geological Survey
Department of Mines and Resources
1914 Hamilton Street
P.O. Box 5114
Regina, Saskatchewan S4P 3P5

Greenland

Grønlands Geologiske Undersøgelse
Østervoldgade 5-7
Tr. K1, DK-1350
København, Denmark

Mexico

Consejo de Recursos Naturales
  no Renovables
Niños Heroes 130
México 7, D.F.

CARIBBEAN ISLANDS

Barbados

Bellairs Research Institute
St. James

Cuba

Instituto de Geología
Academia de Ciencias de Cuba
Ave. Van-Troi no. 17203
Rancho Boyeros—Apartado Postal 10
La Habana

Dominican Republic

Dirección de Minería
Secretaría de Estado de Fomento
Santo Domingo

Guadeloupe and Dependencies

Arrondissement Minéralogique
  de la Guyane
B.P. 448
Pointe-à-Pitre
Guadeloupe

Haiti

Service Géologique
Department of Agriculture, Natural
  Resources and Rural Development
Damiens près Port-au-Prince

Jamaica

Geological Survey Department
Hope Gardens
Kingston 6

Martinique

Arrondissement Minéralogique
  de la Guyane
B.P. 458
Fort-de-France

Trinidad and Tobago

Ministry of Petroleum and Mines
P.O. Box 96
Port-of-Spain, Trinidad

CENTRAL AMERICA

Costa Rica

Dirección de Geología, Minas y Petróleo
Ministerio de Economía, Industria
  y Comercio
Aptdo. 10216
San José

El Salvador

Centro de Estudios
  y Investigaciones Geotécnicas
Ministerio de Obras Públicas
1-A Calle Poniente 925
San Salvador

Guatemala

División de Geología
Instituto Geográfico Nacional
Avenida las Américas 5-76, Zona 13
Guatemala City

Nicaragua

Servicio Geológico Nacional
Ministerio de Economía, Industria,
  y Comercio
Aptdo. Postal 1347
Managua, D.N.

Panama

Administración de Recursos Minerales
Aptdo. Postal 9658, Zona 4
Panama City

SOUTH AMERICA

Argentina

Dirección Nacional de Geología y Minería
Avenida Julio A. Roca 651
Buenos Aires

Bolivia

Servicio Geológico de Bolivia
Ministerio de Minas y Petróleo
Avenida 16 de Julio No. 1769
Casilla de Correo 2729
La Páz

Brazil

Divisão de Geología e Mineralogia
Departamento Nacional de
  Produção Mineral
Setor de Autarquia Norte
Anexo ao Edifício Petrobras
Brasília, D.F.

Chile

Instituto de Investigaciones Geológicas
Agustinas 785, Casilla 10465
Santiago

Colombia

Instituto Nacional de
  Investigaciones Geológico-Mineras
Carrera 30, No. 48-51
Aptdo. Nacional 2504
Bogotá, D.E.

Ecuador

Servicio Nacional de Geología y Minería
Ministerio de Industrias y Comercio
Casilla 23-A, Avenida 10 de Agosto 666
Quito

French Guiana

Bureau des Recherches Géologiques
  et Minières
B.P. 42
Cayenne

Guyana

Geological Surveys and
  Mines Department
P.O. Box 1028
Georgetown

Paraguay

Dirección de la Producción Mineral
Tucarí 271
Asunción

Peru

Servicio de Geología y Minería,
Paz Soldan 225
San Isidro, Aptdo. 889
Lima

Surinam

Geologisch Mijnbouwkundige Dienst
Klein Wasserstraat 1(2-6)
Paramaribo

Uruguay

Instituto Geológico del Uruguay
Calle J. Herrera y Obes 1239
Montevideo

Venezuela

Direccíon de Geología
Ministerio de Minas e Hidrocarburos
Torre Norte, Piso 19
Caracas

EUROPE

Albania
Ministry of Industry and Mining
Tirane

Austria
Geologische Bundesanstalt
Rasumofskygasse 23
A-1031 Vienna

Belgium
Service Géologique de Belgique
13 rue Jenner, Parc Leopold
Brussels 4

Bulgaria
Direktsiià za Geolozhki in
  Minni Prouchvaniià
Ministère des Mines et des
  Richesses du Sous-Sol
Sofia

Czechoslovakia
Ustav Geologický
(Geological Institute)
Ceskoslovenská Akademie ved
(Czechoslovak Academy of Sciences)
Lysolaje 6
Pota Suchodol 2
Praha 6

Denmark
Danmarks Geologiske Undersøgelse
31 Thoravej
DK 2400 København

Finland
Geologinen Tutkimuslaitos
02150 Otaniemi

France
Bureau des Recherches Géologiques
  et Minières (BRGM)
B.P. 6009
45108 Orléans—La Source 02

Federal Republic of Germany (West)
Geological Institut
TU Clausthal
3392 Clausthal-Zellerfeld
Bonn

German Democratic Republic (East)
Zentrales Geologisches Institut
104 Berlin Invalidenstrasse
Berlin

Great Britain
Institute of Geological Sciences
Exhibition Road, South Kensington
London, SW7 2DE

Greece
Institouton Geologias kai
  Ereynon Ypedaphous
(Institute of Geologic and
  Subsurface Research)
Amerikis No. 6
Athens

Hungary
Magyar Allami Földtani Intézet
(Hungarian State Geological Institute)
Népstadion-ut 14, Pf. 106
Budapest 1442

Iceland
Natural Heat Section
National Energy Authority
Reykjavík

Ireland
Geological Survey of Ireland
14 Hume Street
Dublin 2

Italy
Servizio Geologico d'Italia
Ministero dell'Industria del Commercio
  e dell'Artigianato
Direzione Generale della Miniere
Largo di Santa Susanna 13
Roma

Liechtenstein

Geological Survey of Liechtenstein
Vaduz

Luxembourg

Service Géologique de Luxembourg
Direction des Ponts et Chaussées
13, rue J.P. Koenig
Luxembourg

Malta

Department of Information
24 Merchants Street
Valletta

Netherlands

Rijks Geologische Dienst
Spaarne 17, P.O. Box 157
Haarlem

Northern Ireland

Geological Survey of Northern Ireland
20 College Gardens
Belfast

Norway

Geofysiske Kommisjon
Blindern
Oslo 3

Poland

Centrain Urzad Geologii
(Research Center for Geological Sciences)
Zwirki i Wigury 93
Warsaw

Portugal

Serviços Geológicos de Portugal
Rua da Academia das Ciências 19
Lisbon 2

Romania

Institutul de Geofizică Aplicată
Str. Izvor 78
Bucharest 5

Spain

Instituto Geológico y Minero de España
Rios Rosas 23
Madrid

Sweden

Sveriges Geologiske Undersökning
S10405 Stockholm

Switzerland

Schweizerische Geologische Kommission
Naturforschende Gesellschaft
4000 Bernoultistr. 32
Basel

Union of Soviet Socialist Republics

Geologic Institute
Akademiya Nauk SSSR
Pyzhevsky per. 7
Moscow G-17

Yugoslavia

Zavod za Geološki i
    Geofizička Istraživanja
(Institute for Geological and
    Geophysical Research)
Post-fah 227
Kamenicka 16
Belgrade

AFRICA

Algeria

Service Géologique de l'Algérie
Immeuble Mauretania—Agha
Boulevard Colonel Amirouche
Agha

Angola

Direcção des Serviços de Geología e Minas
Caixa Postal 1260-C
Luanda

Benin

Direction des Mines, de la Géologie
    et des Hydrocarbures

Ministère des Travaux Publics, des
  Transports et des Télécommunications,
B.P. 249
Cotonou

Botswana

Geological Survey and Mines Department
P.O. Box 94
Lobatsi

Burundi

Ministère des Affaires Economiques
  et Financières
Département de Géologie et Mines
B.P. 745
Bujumbura

Cameroon

Direction des Mines et de la Géologie
  du Cameroun
B.P. 70
Yaoundé

Central African Republic

Ministère des Travaux Publiques,
  des Transports, des Communications
  et des Mines
B.P. 737
Bangui

Chad

Direction des Mines et de la Géologie
B.P. 816
Fort-Lamy

Republic of Congo

Service des Mines et de la Géologie
B.P. 12
Brazzaville

Egypt

Geological Survey and Mining Authority
Ministry of Industry
Abbasia Post Office
Cairo

Ethiopia

Geological Survey
Ministry of Mines
P.O. Box 486
Addis Ababa

Gabon

Bureau de Recherches Géologiques
  et Minières
B.P. 175
Libreville

Gambia

Survey Department
Bathhurst

Ghana

Geological Survey of Ghana
P.O. Box M 80
Ministry Branch Post Office
Accra

Guinea

Service des Mines et de la Géologie
Conakry

Guinea-Bissau

Serviços de Geologia e Minas
Caixa Postal 399
Bissau

Ivory Coast

Service Géologique
Direction de la Géologie
  et de la Prospection Minière
Ministère de l'Economie et des Finances
B.P. 1368
Abidjan

Kenya

Geological Survey of Kenya
Mines and Geological Department
Ministry of Natural Resources

P.O. Box 3009
Nairobi

Lesotho

Geological Survey Department
Department of Mines and Geology
P.O. Box 750
Maseru

Liberia

Liberian Geological Survey
Ministry of Lands and Mines
P.O. Box 9024
Monrovia

Libya

Geological Research and
 Mining Department
Industrial Research Center
P.O. Box 3633
Tripoli

Malagasy Republic

Service Géologique
Ministère des Mines et de l'Energie
B.P. 280, Armandrianomby
Tananarive

Malawi

Geological Survey Department
Ministry of Natural Resources
P.O. Box 27, Liwonde Road
Zomba

Mali

Direction Nationale des Mines
 et de la Géologie
Koulouba, Bamako

Mauritania

Direction des Mines et de la Géologie
Ministère de l'Industrialisation
 et des Mines
B.P. 199
Nouakchott

Mauritius

Ministry of Agriculture and Natural
 Resources and Cooperative Development
Port Louis

Morocco

Division de la Géologie
Ministère du Commerce, de l'Industrie,
 des Mines, et de la Marine marchande
Rabat

Mozambique

Direcção dos Serviços de Geologia e Minas
Caixa Postal 217
Lourenço Marques

Namibia (South-West Africa)

Geological Survey
P.O. Box 2168
Windhoek

Niger

Service des Mines et de la Géologie
Ministère des Travaux Publics
 et des Mines
B.P. 257
Niamey

Nigeria

Geological Survey Division
Ministry of Mines and Power
P.M.B. 20007
Kaduna South

Réunion

Services des Travaux Publics
St. Denis

Rwanda

Service Géologique du Rwanda
Ministère du Commerce, des Mines
 et de l'Industrie
B.P. 15
Ruhengeria

Senegal

Bureau de Recherches Géologiques
  et Minières
B.P. 268
Dakar

Sierra Leone

Geological Survey Division
Ministry of Lands, Mines, and Labor
New England
Freetown

Somalia

Geological Survey Department
Ministry of Mining
P.O. Box 744
Mogadiscio

South Africa

Geological Survey of South Africa
Department of Mines
P.O. Box 112
Pretoria

Sudan

Geological Survey Department
Ministry of Mining and Industry
P.O. Box 410
Khartoum

Swaziland

Geological Survey and Mines Department
P. O. Box 9
Mbabane

Tanzania

Geological Survey
Mineral Resources Division
Ministry of Commerce and Industries
P.O. Box 903
Dodoma

Togo

Direction des Mines et de la Géologie
Ministère des Travaux Publics, Mines,
  Transports, et Télécommunications,
B.P. 356
Lomé

Tunisia

Service Géologique de Tunisie
95 Avenue Mohamed V
Tunis

Uganda

Geological Survey and Mines Department
P.O. Box 9
Entebbe

Upper Volta

Direction de la Géologie et des Mines
B.P. 601
Ouagadougou

Zaire

Service Géologique du Zaire
B.P. 898
44 Avenue des Huileries
Kinshasa

Zambia

Geological Survey of Zambia
Ministry of Mines
  and Mining Development
P.O. Box RW 135
Ridgeway, Lusaka

Zimbabwe

Geological Survey of Zimbabwe
Ministry of Mines and Lands
P.O. Box 8039, Causeway
Salisbury (Harare)

NEAR EAST-SOUTH ASIA

Afghanistan

Department of Mines and Geology
Ministry of Mines and Industries
Darulaman
Kabul

Bangladesh

Geological Survey
Pioneer Road
Segun Bagicha
Dacca

Cyprus

Geological Survey Department
Ministry of Agriculture
  and Natural Resources
Nicosia

India

Geological Survey of India
27 Jawaharlal Nehru Road
Calcutta 13

Iran

Geological Survey of Iran
P.O. Box 1964
Tehrãn

Iraq

Geological Survey Department
Directorate of Minerals
Ministry of Oil
P.O. Box 2330
Aliviya, Baghdad

Israel

Geological Survey of Israel
30 Malchei Israel Street
Jerusalem

Jordan

Natural Resources Activity
Box 7
Amman

Kuwait

Kuwait Institute for Scientific Research
Hilali Street
P.O. Box 12009
Al Kuwayt

Lebanon

Direction Générale des Travaux Publics
Ministère des Ressources Naturelles
  et Energie Hydraulique
Beirut

Nepal

Nepal Geological Survey
Lainchaur
Kathmandu

Oman

Ministry of Agriculture, Fisheries,
  Petroleum, and Minerals
P.O. Box 550/551
Muscat

Pakistan

Geological Survey of Pakistan
P.O. Box 15
Quetta

Saudi Arabia

Centre for Applied Geology
P.O. Box 1744
Jiddah

Sri Lanka (Ceylon)

Geological Survey Department
48 Sri Jinaratana Road
Colombo 2

Syria

Directorate of Geological Research
  and Mineral Resources
Ministry of Petroleum
Fardos Street
Damascus

Turkey

Mineral Research and Exploration
  Institute of Turkey
Posta Kutusu 116
Eskisehir Yolu
Ankara

Yemen

Central Planning Organization
P.O. Box 175
San'a', Y.A.R.

EAST ASIA-PACIFIC REGION

Australia

Bureau of Mineral Resources
B.M.R. Building
Constitution Avenue
Canberra, A.C.T.

Borneo

Geological Survey Department
Borneo Region, Malaysia
Kuching, Sawawak
Malaysia

Brunei

The Government Geologist
Brunei Town

Burma

Directorate of Geological Survey
  and Exploration
Ministry of Mines
Kanbe Road, Yankin P.O.
Rangoon

Peoples Republic of China

Institute of Geology
Department of Earth Sciences
Chinese Academy of Sciences
Peking

Republic of China (Taiwan)

Geological Survey of Taiwan
P.O. Box 1001
Taichung 400
Taipei

Fiji

Geological Survey Department
P.O. Box 2020, Mead Road
Government Buildings
Suva

Hong Kong

Mines Department, Branch Office
Canton Road Government Offices
Canton Road
Kowloon, Hong Kong

Indonesia

Direktorat Geologi
(Geological Survey of Indonesia)
Djalan Diponegoro 57
Bandung

Japan

Geological Survey of Japan, MITI
135 Hisamoto-cho
Kawasaki-shi, Kanagawa

Kampuchea (Cambodia)

Mines and Mineral Prospecting
Ministry of Industry
Phnom Penh

**Korea, Democratic Peoples
  Republic (North)**

Geology and Geography
  Research Institute
Academy of Sciences
Mammoon-dong
Central District
P'yongyang

Korea, Republic of (South)

Geological and Mineral Institute of Korea
219-3, Karibong-Dong, Yongdungpo-Ku
Seoul

Laos

Service Central des Mines du Laos
Ministère du Plan et de la Coopération
B.P. 46
Vientiane

Malaysia

Federal Geological Survey
Jalan Gurney
Kuala Lumpur 15-01

Mongolian People's Republic

Institute of Geology
Academy of Sciences
UI, Leniadom 2
Ulaanbaatar

New Caledonia

Service des Mines et de la Géologie
Rte. No. 1
B.P. 465
Nouméa

New Guinea

Geological Survey
Department of Lands, Surveys, and Mines
Konedobu, Papua

New Zealand

New Zealand Geological Survey
Department of Scientific
  and Industrial Research
P.O. Box 368
Lower Hutt, Wellington

Philippines

Bureau of Mines
P.O. Box 1595
Herran Street
Manila

Singapore

Survey Department
Ministry of Law
Singapore

Solomon Islands

Department of Geological Surveys
P.O. Box G24
Honiara, Guadalcanal

Thailand

Department of Mineral Resources
Ministry of National Development
Rama VI Road
Bangkok

Vanuatu

Geological Survey
British Residency
Vila

Socialist Republic of Vietnam

Geologic Section
State Committee of Sciences
Hanoi

Western Samoa

Lands and Survey Department
Apia

This short list has been abstracted, by permission, (with some amendments) from a much more complete listing contained in Circular 716 of the U.S. Geological Survey (*Worldwide Directory of National Earth-Science Agencies,* compiled by Anne Lucas Falk and Ralph L. Miller, 1975). This list has been restricted to one agency for each country, state, or province, even though in some cases several agencies give geological service; that which is listed will serve as an introduction to the services that are available if reference cannot be made to the list in the USGS Circular. The authors apologize for any inaccuracies that may have developed since the date of compiling this list. (As this volume is in press, the Circular has been reissued in expanded form as No. 834, 1981)

appendix C

GEOLOGICAL SOCIETIES OF THE WORLD

American Association of Petroleum Geologists

Box 979
Tulsa, Oklahoma 74101

American Association for Quaternary Research

Department of Geology
Arizona State University
Tempe, Arizona 85281

American Geological Institute

5205 Leesburg Pike
Falls Church, Virginia 22041

American Geophysical Union

2000 Florida Avenue N.W.
Washington, D.C. 20009

American Institute of Mining, Metallurgical & Petroleum Engineers

345 East 47th Street
New York, New York 10017

American Society of Civil Engineers

345 East 47th Street
New York, New York 10017

American Society of Photogrammetry

105 North Virginia Avenue
Falls Church, Virginia 22046

Association of American State Geologists

Florida Bureau of Geology
903 West Tennessee Street
Tallahassee, Florida 32304

Association of Engineering Geologists

8310 San Fernando Way
Dallas, Texas 75218

Association of Geoscientists for International Development

Asian Institute of Technology
Box 2754
Bangkok, Thailand

Bangladesh Geological Society

c/o Geological Survey of Bangladesh
Pioneer Road, Segunbagicha
Dacca 2, Bangladesh

British Quaternary Association

c/o W. G. Jardine
Department of Geology
University of Glasgow
Glasgow, Scotland

Canadian Geotechnical Society

c/o Engineering Institute of Canada
2050 Mansfield Street
Montreal, Quebec H3A IZ2.

APPENDIX C

Canadian Institute of Mining & Metallurgy

906–1117 Ste. Catherine Street West
Montreal, Quebec H3B 1J3

Canadian Society of Petroleum Geologists

612 Lougheed Building
Calgary, Alberta T2P 1M7

Clay Minerals Society

Box 19508
Houston, Texas 77024

Earthquake Engineering Research Institute

2620 Telegraph Avenue
Berkeley, California 94704

Ecuadorean Geological & Geophysical Society

Casilla 371A
Quito, Ecuador

European Association of Exploration Geophysicists

30 Carel van Bylandtlaan
The Hague, the Netherlands

Geochemical Society

Department of Geological Sciences
Virginia Polytechnic Institute
Blacksburg, Virginia 24061

Geological Association of Canada

Department of Earth Sciences
University of Waterloo
Waterloo, Ontario N2L 3G1

Geological & Mineral Institute of Korea

219-5, Karibong-Dong, Yongdungpo-ku
Seoul, Korea

Geological Society of Africa

Department of Geology
University of Khartoum
Khartoum, Sudan

Geological Society of America

3300 Penrose Place
Boulder, Colorado 80301

Geological Society of Australia

Perpetual Trustee Building
39 Hunter Street
Sydney, 2000, New South Wales
Australia

Geological Society of Denmark

Øster Voldgade 5-7
DK 1350 Copenhagen K
Denmark

Geological Society of Finland

Geologinen Tutkimuslaitos
F-02150 Espoo 15
Finland

Geological Society of Greece

34 Veranzeoon Street
Athens 102, Greece

Geological Society of Iceland

Science Institute
University of Iceland
Dunhaga 3
Reykjavík, Iceland

Geological Society of India

16/1, Ali Asar Road
Bangalore 560052, India

Geological Society of Iraq

Box 547
Baghdad, Iraq

Geological Society of London

Burlington House, Piccadilly
London, W1V OJU
United Kingdom

Geological Society of Malaysia

Department of Geology
University of Malaya
Pantai Valley
Kuala Lumpur, Malaysia

Geological Society of New Zealand

New Zealand Geological Survey
Box 30368
Lower Hutt, New Zealand

Geological Society of Norway

Sars gate 1
Oslo 5, Norway

Geological Society of the Philippines

Bureau of Mines Building
Pedro Gil Street
Manila, Philippines

Geological Society of Poland

Witsa Oleandry 2a
30-063 Cracow, Poland

Geological Society of Puerto Rico

Box 40575
Minillas Station
San Juan, Puerto Rico 00940

Geological Society of South Africa

Kelvin House
Holland Street
Johannesburg, South Africa

Geological Society of Sweden

Postfack
S-104 05 Stockholm, Sweden

Geological Society of Thailand

Economic Geology Division
Department of Mineral Resources
Rama VI Road
Bangkok 4, Thailand

Geological Society of Trinidad & Tobago

Box 771
Port of Spain
Trinidad, West Indies

Geothermal Energy Association

Box 752
Azusa, California 91702

Glaciological Society

Scott Polar Research Institute
Lensfield Road
Cambridge, U.K.

Guatemalan Geological Society

Instituto Centroamericano de
   Investigación y Technología Industrial
Avenida la Reforma 4-47
Zona 10
Guatemala City, Guatemala

Hungarian Mining & Metallurgical Society

1061 Budapest, V1
Anker köxl.l.em. 101-105
Hungary

Ikatan Ahli Geologi Indonesia

Jalan Diponegoro 57
Bandung, Indonesia

Institution of Mining & Metallurgy

44 Portland Place
London, W1N 4BR, U.K.

Instituto di Geologia Applicata

Via Eudossianna 18
00184 Rome, Italy

Instituto Geológico y Minero de España

Rios Rosas, 23
Madrid 3, Spain

International Association of Engineering Geology

Geologisches Landesamt NW
De Greiff Strasse 195
415 Krefeld, West Germany

International Association of Hydrogeologists

U.S. National Committee
2302 Fox Fire Court
Reston, Virginia 22091

International Quaternary Association (INQUA)

191, rue St. Jacques
Paris 5, France

International Union of Geological Sciences

Room 177
601 Booth Street
Ottawa, Ontario K1A OE8

International Union for Quaternary Research

Geografisch Instituut
Vrije Universiteit
A. Buyllaan 87
B-1050 Brussels, Belgium

Irish Geological Association

14 Hume Street
Dublin 2, Ireland

Israel Geological Society

Box 1239
Jerusalem, Israel

Koninklijk Nederlands Geologisch Mijnbouwkundig Genootschap (Royal Geological & Mining Society of the Netherlands)

Box 190
The Hague, the Netherlands

Mexican Geological Society

Instituto de Geología
Ciudad Universitaría
Apartado Postal 70-296
México, 20, D.F.
México

National Water Well Association

500 West Wilson Bridge Road
Worthington, Ohio 43085

Offshore Technology Conference

6200 North Central Expressway
Dallas, Texas 75206

Royal Society

6 Carlton House Terrace
London, SW1Y 5AG
U.K.

Royal Society of Canada

344 Wellington Street
Ottawa, Ontario K14 ON4

Royal Society of Edinburgh

22-24 George Street
Edinburgh, EH2 2PQ
Scotland

Royal Society of New Zealand

Box 12249
Wellington 1, New Zealand

Schweizerische Geologische Gesellschaft

(Swiss Geological Society)
Institut de Géologie
11, rue Emile-Argand
CH-2000 Neuchâtel 7, Switzerland

Seismological Society of America

2620 Telegraph Avenue
Berkeley, California 94701

Servicio Geologio de Bolivia

Ministerio de Minas y Petroleo
Avenida 16 de Julio no. 1769
La Paz, Bolivia

Sociedad Colombiana de Geologia

Facultad de Ciencias
Departamento de Geología
Apartado Postal 535
Bogotá, D.E., Colombia

Sociedad Geológica del Peru

Union Oil Company of Peru
R. Rivera Navarrete 791 6° Piso
San Isidro
Lima 1, Peru

Sociedade Brasileira de Geologia

Instituto de Geosciências
Cidade Universitária
Caixa Postal 20897
0100 Sâo Paulo, Brazil

Société Belge de Géologie

Rue Jenner, 13
B-1040 Brussels, Belgium

Société Géologique de Belgique

Université
Place du XX août, 7
B-4000 Liège, Belgium

Société Géologique de France

77 rue Claude Bernard
75005 Paris, France

Society of Economic Geologists

185 Estes Street
Lakewood, Colorado 80226

Soil Science Society of America

677 South Segoe Road
Madison, Wisconsin 53711

U.S. National Committee on Geology

c/o Linn Hoover
U.S. Geological Survey
Mail Stop 915
Reston, Virginia 22092

This short list has been taken, by permission, from a much longer list of approximately 500 societies in the earth sciences prepared by the American Geological Institute (listed above). Their list is regularly revised and published once a year in *Geotimes,* to which reference should be made for any society not listed herein.

appendix D

SOME USEFUL JOURNALS

This is a personal list of journals known to and used by the authors. There are doubtless other equally useful journals serving the field of geology and engineering, but it is hoped that this short list will prove of assistance at least as an introduction to current literature in the field. To this end, the following signs are provided to give some indication of the journals' special coverage:

- ● case histories are a regular feature
- ○ case histories are occasionally featured
- ■ current news items of relevant interest are a regular feature

GEOLOGICAL JOURNALS

- ○ **American Journal of Science,** monthly, from Kline Geological Laboratory, Yale University, New Haven, Conn., 06520.
- ● **Bulletin of the Association of Engineering Geologists,** quarterly, from F. T. Johnston, Exec. Dir., A.E.G., 8310 San Fernando Way, Dallas, Tex., 75218. Note that the first few issues of this excellent publication were called **Engineering Geology,** which causes confusion in referencing, see below.
- ○ **Bulletin of the Geological Society of America,** monthly, from G.S.A., 3300 Penrose Place, Boulder, Colo. 80301.
- ● **Bulletin of the International Association of Engineering Geology,** quarterly, from I.A.E.G., c/o Geologisches Landesamt N.W., DeGreiff Str. 195, Postfach 1080, D 415 Krefeld 1, F.R. Germany.
- ○ **Canadian Journal of Earth Sciences,** monthly, from the National Research Council of Canada, Ottawa, K1A OR6.
- ● **Case History Series,** of Engineering Geology Division of the Geological Society of America, 3300 Penrose Place, Boulder, Colo., 80301.
- ○ **Economic Geology,** monthly, from Economic Geology Publishing Co., Box 74, Michigan Technical University, Houghton, Mich., 49931.
- ● **Engineering Geology,** published quarterly by Elsevier Publishing Co., P.O. Box 211, Amsterdam, the Netherlands.

- ○ **Geological Magazine,** eight times a year, from Cambridge University Press, 32 East 57th Street, New York City, N.Y., 10022.
- ■ **Geotimes,** monthly, available from American Geological Institute, 5205 Leesburg Pike, Falls Church, Va., 22041.
- ○ **Journal of Geology,** eight times a year, University of Chicago Press, 5801 Ellis Avenue, Chicago, Ill., 60637.
- • **Quarterly Journal of Engineering Geology** (of the Geological Society of London), obtainable from Blackwell Scientific Publications, Ltd., Osney Mead, Oxford OX7 OEC, England.

GEOTECHNICAL JOURNALS

- • **Canadian Geotechnical Journal,** quarterly, from the National Research Council of Canada, Ottawa, K1A OR6, Canada.
- ○ **Geotechnique,** quarterly, from the British Geotechnical Society, c/o the Institution of Civil Engineers, Great George Street, London SWIP 3AA, England.
- ■ **Ground Engineering,** bimonthly, from Foundation Publications, Ltd., 33 Short Croft, Doddinghurst, Essex, England.
- ○ **International Journal of Rock Mechanics and Mining Science,** (including Geomechanics Abstract) monthly, from Pergamon Press, Head Hill Hall, Oxford OX3 OBW, England, or Maxwell House, Fairview Park, Elmsford, N.Y., 10523.
- ○ **Journal of the Geotechnical Engineering Division,** monthly, from the American Society of Civil Engineers, 345 East 47th Street, New York City, N.Y., 10017.
- ○ **Rock Mechanics,** quarterly, from Springer-Verlag, P.O.B. 367, A-1011 Wien, Austria.
- ■ **Tunnels and Tunnelling,** bimonthly, from Morgan-Grampian, Ltd., 30 Calderwood Street, London SE18 6QH, England.

CIVIL ENGINEERING JOURNALS

- ○ **Civil Engineering,** monthly, from the American Society of Civil Engineers, 345 East 47th Street, New York City, N.Y., 10017.
- ○ **Civil Engineering** (formerly *C.E. and Public Works Review*), monthly, from Calderwood Street, London, SE18 6QH, England.
- ○ **Engineering Journal,** bimonthly, from the Engineering Institute of Canada, 2050 Mansfield Street, Montreal, Que., 43A 122.
- ■ **Engineering News-Record,** weekly, from 1221 Avenue of the Americas, New York City, N.Y., 10020.
- ■ **New Civil Engineer,** weekly, and **New Civil Engineer International,** monthly, both from Institution of Civil Engineers, Great George Street, London SWIP 3AA, England.

OTHER SOURCES

There are many other sources of useful information in this field, published either regularly, irregularly, or only occasionally with other material. Typical of the second category are the renowned publications of the U.S. Geological Survey, the respective state geological surveys, and other national and local surveys (such as those of Canada and the Canadian provinces). Typical of the third are the publications of the Transportation Research Board (formerly the Highway Research Board), 2101 Constitution Avenue, Washington, D.C., which sometimes contain useful information on engineering geology.

GEOTECHNICAL ABSTRACTS

Geotechnical Abstracts (of the International Society of Soil Mechanics and Foundation Engineering) published by the German Society of S.M. and F.E. (Deutsche Gesellschaft für Erd-und Grundbau), 35a Kronprinzenstr., 43 Essen, W. Germany.

Soil Mechanics Information Service from Geodex International, Inc., P.O. Box 385, Glen Ellen, Calif., 95442.

GUIDE TO GEOLOGICAL LITERATURE

Sources of Information for the Literature of Geology by J. W. Mackay, published by the Geological Society of London, Burlington House, London W1V OJU, England, 2nd ed., 1974; an invaluable guide to the vast literature of geology, it is in effect a "bibliography of bibliographies of geology."

AND WHEN ALL ELSE FAILS...

Ulrich's International Periodicals Directory, 15th ed., 1973–1974, from R. R. Bowker Co. (a Xerox company), New York and London, will be found in most libraries; it will guide the inquirer to full details of publication of any periodical the title of which is known.

appendix E

ILLUSTRATION CREDITS

Acknowledgment of the source of illustrations is made in this way (rather than individually under each figure) so that we may make this public statement of our appreciation for the help we have received from all the individuals and organizations listed. Some of the figures appeared in the first and second editions of *Geology and Engineering;* others were used in *Cities and Geology.* More than half the illustrations, however, have been specially assembled for this Handbook. Every care has been taken to make the list which follows accurate, but if errors remain undetected we hereby tender our apologies to those concerned. No personal titles (of individuals) have been included to obviate any possibly invidious errors. We have overcome normal diffidence by including our own initials, when called for, so that the origin of all photographs may be known. To all who have kindly supplied photographs and drawings for the earlier books and now for this Handbook, we record our thanks for this valued assistance in thus adding corroborative detail to what might otherwise have been a bald and unconvincing narrative.

R.F.L. and P.F.K.

CHAPTER 1

1.1 Swiss National Tourist Office, Toronto.
1.2 Boston Museum of Natural Science; photo by Bradford Washburn.
1.3 Public Archives of Canada, No. PA 501,046.
1.4 The Royal Gorge Company, Canon City, Colorado.

CHAPTER 2

2.1 Snowy Mountains Hydro-Electric Authority, Cooma, N.S.W., Australia.
2.2 Snowy Mountains Hydro-Electric Authority, Cooma, N.S.W., Australia.
2.3 Snowy Mountains Hydro-Electric Authority, Cooma, N.S.W., Australia.
2.4 Baptie, Shaw, and Morton, Glasgow, Scotland, and the North of Scotland Hydro-Electric Board.
2.5 The Institution of Civil Engineers, London, from J. Paton, "The Glen Shira Hydro-Electric Project" (reference 2.2).
2.6 Copyright photograph, Gordon Mahan, Warren, Pennsylvania; U.S. Corps of Engineers, district engineer, Pittsburgh.
2.7 S. S. Philbrick, from "Kinzua Dam and the Glacial Foreground" (reference 2.3).
2.8 The Institution of Royal Engineers,

Chatham, England, from "The Work of the Royal Engineers ..." (reference 2.4).
2.9 Duisburg-Ruhrort Harbor Authority, Hafendirektor H. Bumm.
2.10 Duisburg-Ruhrort Harbor Authority and P. A. Haquebard, Geological Survey of Canada, Bedford, Nova Scotia.
2.11 Duisburg-Ruhrort Harbor Authority, Hafendirektor H. Bumm.
2.12 R.F.L.

CHAPTER 3

3.1 Geological Survey of Canada, Ottawa.
3.2 A. W. Joliffe, Queen's University, Kingston, Canada.
3.3 P.F.K.

CHAPTER 4

4.2 R.F.L.
4.3 U.S. Geological Survey, Washington.
4.4 U.S. Geological Survey, Washington, photo by H. E. Gregory.
4.5 U.S. Geological Survey, Washington, photo by J. Gilluly.
4.6 R.F.L.
4.8 R.F.L.
4.10 Geological Survey of Canada, Ottawa; photo no. 180345.
4.12 Geological Survey of Great Britain, Crown Copyright.

CHAPTER 5

5.1 Copyright photo *The Times* (London).
5.2 U.S. Geological Survey, Washington, D.C.; photo by F. E. Matthes.
5.3 R.F.L.
5.4 U.S. Geological Survey, Washington, D.C.; photo by Austin Post.
5.5 T. Vold, British Columbia Ministry of the Environment; and John Day, Ottawa.
5.6 R.F.L.
5.7 R.F.L.
5.8 R.F.L.
5.9 U. S. Geological Survey, Washington, D.C.; photo by Austin Post.
5.10 Geological Survey of Canada, Ottawa.
5.11 P.F.K.
5.12 Geological Survey of Canada, Ottawa; photo by A. M. Stalker, no. 147831.
5.13 After Chamberlin and Salisbury.
5.14 Based on a plate in W. B. Wright, *The Quaternary Ice Age*, 2d ed., Macmillan, London, 1937.
5.15 Geological Survey of Canada, Ottawa; photo no. 117969.
5.16 U.S. Geological Survey, Washington, D.C.; photo by W. C. Alden.
5.17 P.F.K.
5.18 U.S. Navy photograph, from U.S. Navy Alaskan Survey.
5.19 Photograph by Erwin Schneider, through Fritz Mueller and by permission of the Swiss Foundation for Alpine Research.
5.20 R.F.L.
5.22 Geological Survey of Canada, Ottawa; photo no. 202775, by J. S. Vincent.
5.23 Ontario Research Foundation; photo by L. J. Chapman.
5.24 Ontario Department of Mines; photo by A. P. Coleman.
5.25 R.F.L.
5.26 National Research Council of Canada, Division of Building Research; photo by I. C. MacFarlane.
5.27 National Research Council of Canada, Division of Building Research; photo by G. H. Johnston.

CHAPTER 6

6.1 National Research Council of Canada, Division of Building Research, Ottawa.
6.2 M. A. Hanna Co.; photo by G. D. Creelman.
6.3 Steep Rock Iron Mines, Ltd.; photo by N. Vincennes.
6.4 National Research Council of Canada, Division of Building Research, Ottawa.
6.5 Nairobi City Council, Kenya, and Howard Humphreys Ltd., Consultants, Kenya.
6.6 N. R. O'Brien, State University of New York at Potsdam.
6.7 National Research Council of Canada, Division of Building Research, Ottawa.
6.8 R.F.L.

ILLUSTRATION CREDITS   E-3

CHAPTER 7

7.1 American Society of Civil Engineers, from R. B. Peck, "Rock Foundations for Structures" (reference 7.2); and R. B. Peck.
7.2 M. Matula, Comenius University, Bratislava, Czechoslovakia; and the American Institute of Mining, Metallurgical, and Petroleum Engineers, New York.
7.3 Bureau of Topographic and Geologic Survey, Commenwealth of Pennsylvania, A. Socolow, director.
7.4 Department of Energy, Mines and Resources, Canada; Centre for Mineral and Energy Technology and D. G. Gladwin.
7.5 Shawinigan Consultants, Inc., and Hydro Quebec, Montreal.
7.6 Division of Geological Survey and Water Resources, Missouri; photo by E. E. Lutzen.
7.7 By permission from H. Xiang, chairman, Commission of Geotechnical Investigation of the Architectural Society of China, and the Institution of Mining and Metallurgy, London.
7.8 P. Lumb, University of Hong Kong.

CHAPTER 8

8.1 From Dr. O. E. Meinzer, Water Supply Paper no. 494, U.S. Geological Survey, Washington, D.C.
8.2 The superintendent of the Municipal Water Department, Riverside, California.
8.4 Metropolitan Water Board, London, England.
8.5 Ministry of Natural Resources of Ontario, Toronto (no. 7397).
8.7 R.F.L.

CHAPTER 9

9.1 Department of the Environment, Canada, by W. V. Morris.
9.2 R.F.L.
9.3 R.F.L.
9.4 National Research Council of Canada, Associate Committee on the National Building Code (from the 1980 Code Supplement), Ottawa.
9.5 National Research Council of Canada, Division of Building Research, Ottawa.

CHAPTER 10

10.1 Institution of Civil Engineers, London, from reference 10.1.
10.2 R.F.L.
10.3 British Aluminium Company, Ltd., Fort William, Scotland.
10.4 National Research Council of Canada, Division of Building Research, Ottawa.
10.5 American Society of Civil Engineers, from J. D. Parsons, "Glacial Like Formations ... ," (reference 10.7).
10.6 Geological Survey of Canada, Ottawa; photo by John S. Scott, no. 118541.
10.7 Institute of Geological Sciences, London, England; crown copyright.
10.8 Geological Survey of Canada, Ottawa; photo no. 92792.
10.9 N. W. Radforth, Parry Sound, Ontario.
10.10 National Air Photo Library of Canada, Ottawa, and J. D. Mollard, Regina (no. A11347).
10.11 R. A. Kreig and Associates, Anchorage.

CHAPTER 11

11.1 Deep Sea Drilling Project and C. J. Sisskind.
11.5 Chief of Engineers, U.S. Army.
11.6 George Wimpey & Co., Ltd.; photo by Skyphotos.
11.7 *Engineering News-Record,* New York.
11.8 British Columbia Hydro and Power Authority, Vancouver.
11.9 Q. Zaruba, Prague, Czechoslovakia.
11.10 R.F.L.
11.11 U.S. Corp of Engineers, Paul R. Fisher.
11.12 U.S. Bureau of Reclamation.
11.13 *Engineering News-Record,* New York.
11.14 Public Archives of Canada, Ottawa; no. C14413.
11.15 R.F.L.

11.16 Tennessee Valley Authority.
11.18 U.S. Bureau of Reclamation and L. R. Frei.
11.19 International Association for Engineering Geology, from J. Bazynski, "Studies of Landslide Areas . . . ," (reference 11.29), and J. Bazynski.
11.20 Ontario Hydro, Toronto, and J. L. Adams.

CHAPTER 12

12.1 International Nickel Company of Canada, Ltd.; photo by George Hunter.
12.2 U.S. Bureau of Public Roads, Washington.
12.3 National Building Research Institute, Pretoria, South Africa, and A. B. Williams.
12.4 From S. F. Kelly.
12.5 U.S. Bureau of Public Roads, Washington, D.C.
12.6 Geological Survey of South Africa, Pretoria.
12.7 B.C. Electric Co., Ltd.
12.8 Snowy Mountains Hydro-Electric Authority, Cooma, N.S.W., Australia.
12.9 International Geological Congress, Ottawa (from reference 12.15).

CHAPTER 13

13.1 Through Haley and Aldrich, A. W. Hatheway.
13.2 Boston Society of Civil Engineers (from reference 13.7).
13.3 U.S. Geological Survey, Washington, D.C.
13.4 National Research Council of Canada and Saskatchewan Research Council.
13.5 City Planning Department, Montreal.
13.6 R. E. Gray and The Pittsburgh Convention and Visitors Bureau.
13.7 Quido Zaruba, from *Geologic Features . . .*, (reference 13.15).
13.8 Port Authority of New York and New Jersey.
13.9 R.F.L.
13.10 Toronto Transportation Commission.
13.11 National Building Research Institute, Pretoria, South Africa, and A. B. Williams.
13.12 R.F.L.
13.13 J. Brierley, former city engineer of Exeter.
13.14 Japanese Building Research Institute, Tokyo, Japan.

CHAPTER 14

14.1 U.S. Geological Survey, Washington, D.C.
14.2 U.S. Geological Survey, Washington, D.C.
14.3 U.S. Geological Survey, Washington, D.C. with special acknowledgment to Robert E. Davies, Office of Scientific Information, for these three illustrations.
14.4 U.S. Geological Survey, Washington, D.C., after N. H. Darton.
14.5 From Z. Krelova, *Engineering-Geological Maps of the City of Prague*, (reference 14.7).

CHAPTER 15

15.1 R.F.L.
15.2 Southern California Edison Company.
15.3 Pacific Power and Light Company, Seattle.
15.4 Pacific Power and Light Company, Seattle.
15.5 Ontario Hydro, Toronto.
15.6 American Society of Civil Engineers, New York, from Proc. 83 (SM2) 1957 and J. W. Hilf.
15.7 Pacific Gas and Electric Company, San Francisco.
15.8 R.F.L.
15.9 J. N. Stothard, Cudworth, S. Yorkshire, England.
15.10 Geological Survey of Great Britain; crown copyright.
15.11 Rock of Ages Corporation, Barre, Vt.
15.12 University of Saskatchewan, Saskatoon; from J. J. Hamilton.

15.13 E. Winkler, University of Notre Dame, Notre Dame, Ind.
15.14 Carthage Marble Company, Inc.
15.15 Department of Geology, the Ohio State University, and R. Bates.

CHAPTER 16

16.1 Canada Lafarge Cement Company, Montreal.
16.2 Denver Hilton Hotel, Colo.
16.3 R.F.L.
16.4 R.F.L.
16.5 Consolidated Sand and Gravel Company, Toronto, through L. J. Chapman.
16.6 R.F.L.
16.7 National Research Council of Canada, Division of Building Research, Saskatoon; photo by J. J. Hamilton.
16.8 Cementation Piling and Foundations, Ltd., Croydon, Eng.
16.9 Cementation Piling and Foundations, Ltd., Croydon, Eng.
16.10 National Research Council of Canada, Division of Building Research, Ottawa; photo by E. G. Swenson.
16.11 R. Bates, The Ohio State University, Columbus, Ohio.

CHAPTER 17

17.1 The Structural Clay Products Institute, Washington, D.C.
17.2 Dinish Mohan, director, Central Building Research Institute, Roorkee, India.
17.3 Institute of Geological Sciences, London; crown copyright.
17.4 National Gypsum (Canada), Ltd., Montreal.
17.5 Central Electricity Generating Board, Ash Marketing Division, London, Eng.

CHAPTER 18

18.1 Public Record Office, London; by permission of British Rail.
18.2 North Yorkshire County Council; Col. G. A. Leech, county surveyor.
18.3 Institution of Mining and Metallurgy, London, from J. L. Francis and J. C. Allen, "Driving a Mines Drainage Tunnel..." (see reference 18.12).
18.4 U.S. Bureau of Mines, H. C. Stevens, Denver, Colo., and O. Tweto, USGS.
18.5 U.S. Bureau of Mines, H. C. Stevens, Denver, Colo. and O. Tweto, USGS.
18.6 Japanese National Railways, Tokyo.
18.7 Photo by G. J. Hucker through H. Merrill Roenke, Jr., both of Geneva, New York.
18.8 American Society of Civil Engineers, New York, from D. R. Warren, "Novel Construction Plan..." (see reference 18.21).
18.9 National Research Council of Canada, Ottawa, and Canadian Journals of Research.
18.11 The Moretrench Corporation, New York.
18.12 The Moretrench Corporation and Powell Brothers Construction Company, New York.
18.13 American Society of Civil Engineers, New York, from B. J. Prugh, "Excavation Tests Versatility..." (see reference 18.29).
18.14 R.F.L.
18.15 R.F.L.
18.16 Boston Public Library, Boston, Mass.
18.17 *Engineering News-Record,* New York.
18.18 Pennsylvania Bureau of Topographic and Geologic Survey, Harrisburg, A. A. Socolow, director.
18.19 New York City Transit Authority, Brooklyn, L. Rossum, deputy chief engineer.
18.20 Compagnie Nationale du Rhone, Lyons, France.

CHAPTER 19

19.1 Royal Ontario Museum, Toronto.
19.2 The St. Lawrence Seaway Development Corporation, Massena, N.Y., and John B. Adams III, chief engineer.

19.3 R.F.L.
19.4 The St. Lawrence Seaway Development Corporation, Massena, N.Y.
19.5 British Railways, Midland Region.
19.8 Ontario Hydro, Toronto.
19.9 Power Authority of the State of New York and Uhl, Hall, and Rich, consultants.
19.10 R.F.L., by permission of Steep Rock Iron Mines, Ltd., Toronto.
19.11 American Society of Civil Engineers, New York.
19.12 R.F.L.
19.13 Toronto Transit Commission.
19.14 Toronto Transit Commission.
19.16 Spencer, White and Prentis Inc., Rochelle Park, N.J.
19.17 Spencer, White and Prentis Inc., Rochelle Park, N.J.
19.18 Port Authority of New York and New Jersey.
19.19 J. F. Sadler, vice president, Potash Corporation of Saskatchewan, Saskatoon.
19.20 U.S. Bureau of Reclamation.
19.21 U.S. Bureau of Reclamation.
19.22 L. Casagrande, Casagrande Consultants, Arlington, Mass.
19.23 Ministry of Transportation and Communications, Ontario, Toronto.
19.24 The chief of engineers, U.S. Army.
19.25 Tennessee Valley Authority.
19.26 Skanska Cementjuteriet, Stockholm, Sweden; photo by Gosta Nordin.
19.27 National Astronomy and Ionosphere Center, Cornell University, Ithaca, N.Y.
19.28 North Carolina Department of Transportation, Raleigh.
19.29 Atlas Powder Company and Morrison-Knudsen Company, Inc.
19.30 J. B. Pritchard.

CHAPTER 20

20.1 The Institution of Civil Engineers, London, from Sir Francis Fox, "The Simplon Tunnel" (see reference 20.2).
20.2 California Department of Transportation, Sacramento.
20.3 From *Guidebook* no. 9 of the 16th International Geological Congress.
20.4 Bureau of Water Resource Development, City of New York.
20.5 From *Guidebook* no. 9 of the 16th International Geological Congress.
20.6 *Engineering News-Record*, New York.
20.7 *Engineering News-Record*, New York.
20.8 American Society of Civil Engineers, New York.
20.9 Japanese National Railways, Tokyo.
20.10 Ontario Hydro, Toronto.
20.11 Sir William Halcrow and Partners, London, Eng.
20.12 Sir William Halcrow and Partners, London, Eng.
20.13 New York Central Railroad Company.
20.14 Foundation Co. of Canada Ltd., Toronto.
20.15 Metropolitan Sanitary District of Greater Chicago; photo by Patricia Young.
20.16 Liverpool Geological Society, Liverpool, Eng.
20.17 Mott, Hay and Anderson, Westminster, Eng.
20.18 Mott, Hay and Anderson, Westminster, Eng.
20.19 U. S. Bureau of Reclamation; photo by E. S. Ensor.
20.20 Sir William Halcrow and Partners, Westminster, Eng.
20.21 Boston Society of Civil Engineers and M. P. Billings.
20.22 Metropolitan District Commission, Boston, Mass., and M. P. Billings.
20.23 From the report of the Channel Tunnel Committee, 1930; crown copyright.
20.24 Minneapolis-St. Paul Sanitary District.
20.25 Locher & Cie A.G.; Zürich; Switzerland.

CHAPTER 21

21.1 W. and D. Glowacki, Winchester, Mass.
21.2 *La Revue Vinicole,* Paris.

21.3 Artaud Freres, Nantes-Carquefou, France, and J. Tassie, Ottawa.
21.4 Missouri Division of Geological Survey and Water Resources; photo by J. D. Vineyard.
21.5 Missouri Division of Geological Survey and Water Resources; photo by J. D. Vineyard.
21.6 Missouri Division of Geological Survey and Water Resources; based on sketch in *Missouri Mineral News*.
21.7 Sir William Halcrow and Partners, Westminster, Eng.
21.8 Statens Vattenfallsverk, Valingby, Stockholm, Sweden.
21.9 Pacific Gas and Electric Company, San Francisco.
21.10 Aluminum Company of Canada, Ltd., Montreal.
21.11 Churchill Falls (Labrador) Corporation, Ltd., Montreal.
21.12 Societé d'Energie de la Baie James, Montreal, Quebec.
21.13 Skanska Cementgjuriet, Sweden.
21.14 General director, Fortifikationsfirvaätningen, Stockholm, Sweden.
21.15 Commanding officer, North American Air Defense Command, Colo.
21.16 Department of National Defence, Ottawa; Canadian Forces photo.
21.17 Genealogical department, Church of Jesus Christ of Latter-Day Saints, Salt Lake City, Utah.
21.18 J. Lankester, surveyor to the University of Oxford, Eng.
21.19 Queen's University, Kingston, Ontario; photo by W. Ryce.
21.20 Stockholms Vatten-och Avloppsverk, Stockholm, Sweden.

CHAPTER 22

22.1 W. A. Trow, Toronto.
22.2 Missouri Division of Geologic Survey and Water Resources; photo by J. D. Vineyard.
22.3 National Research Council of Canada, Division of Building Research, Ottawa.
22.4 Norwegian Geotechnical Institute.
22.5 American Society of Civil Engineers, New York.
22.7 National Museum of Canada.
22.8 Northwestern Mutual Life Insurance Company, Milwaukee.
22.9 James Drewitt and Son, Ltd., Bournmouth, Eng., and E. E. Green.
22.10 American Society of Civil Engineers, New York.
22.11 National Research Council of Canada, Division of Building Research, Ottawa.
22.12 A. W. Skempton, London, Eng.
22.14 A. B. A. Brink, Honeydew, South Africa.
22.15 Soil Mechanics, Ltd., London, Eng., and Demarara Sugar Terminals, Ltd., Guyana.
22.16 Takanaka Komnten Company, Ltd., and the Building Research Institute of Japan, Tokyo.
22.17 American Society of Civil Engineers, New York.
22.18 Société Anonym Conrad Zschokke, Geneva, Switzerland.
22.19 Stockholm Harbour Board and Skanska Cementgjuriet.
22.20 Turner Construction Company, New York.
22.21 Ripley, Klohn and Leonoff, Ltd., Vancouver, British Columbia.
22.22 Foundation Company of Canada Ltd., Toronto.
22.23 Ente Nazionale Italiano per il Turismo, Rome.
22.24 High Commission of India, Ottawa, and Lt. General Sir Harold Williams, Roorkee, India.
22.25 Departamento de Turismo, Mexico D.F.
22.26 National Research Council of Canada, Division of Building Research; photo by C. B. Crawford.
22.27 Ritchie and Partners, London, Eng.
22.28 County architect of Nottinghamshire, Eng.
22.29 American Society of Civil Engineers, New York.
22.30 J. Skopek, Prague, Czechoslovakia.
22.31 Royal Embassy of Sweden, Ottawa.
22.32 *The Oregonian*, Portland, and W. H. Stuart.
22.33 World Wide Photos.

22.34 Toyo Kogyo Company, Ltd., Hiroshima, Japan.
22.35 Société Anonym Conrad Zschokke, Geneva, Switzerland.
22.36 Société Anonym Conrad Zschokke, Geneva, Switzerland.
22.37 Public Utilities Board, Singapore.
22.38 The Institution of Civil Engineers, London, and Ove Arup and Partners, London, Eng.
22.39 Engineers Testing Laboratories, Inc., Phoenix, Ariz.

CHAPTER 23

23.1 Pennsylvania Electric Company, Johnstown.
23.2 Pennsylvania Power & Light Company.
23.3 Larderello S.p.A., Pisa, Italy.
23.4 Pacific Gas and Electric Company, San Francisco.
23.5 Director, Orkustofnun, Reykjavík, Iceland.
23.6 Electricité de France, Paris.
23.7 Press counsellor U.S.S.R. Embassy, Ottawa; Novosti photo.
23.8 F. W. Furkert, Public Works Department, New Zealand.
23.9 F. W. Furkert, Public Works Department, New Zealand.
23.10 Chief of engineers, U.S. Army.
23.11 United States Corps of Engineers, Portland District.
23.12 British Columbia Hydro and Power Authority, Vancouver.
23.13 National Research Council, Division of Building Research, Ottawa; photo by G. H. Johnston.
23.14 Institute of Geological Sciences; crown copyright; photo by D. R. Smith.
23.15 British Columbia Hydro and Power Authority, Vancouver.
23.16 Tennessee Valley Authority, Knoxville.
23.17 San Diego Gas and Electric Company.
23.18 From *Summary of Progress;* Geological Survey of Great Britain, 1930; crown copyright.
23.19 Bernard Johnson, Inc., Houston, Texas.
23.20 Central Office of Information, London, through press officer, British High Commission, Ottawa.
23.21 Public Service Electric and Gas Company, Newark, N.J.

CHAPTER 24

24.1 Public Archives of Canada, Ottawa, Dominion Bridge Collection (no. PA 108,929).
24.2 The Institution of Civil Engineers, London.
24.3 *Engineering News-Record,* New York.
24.4 California Department of Public Works, Sacramento.
24.5 American Society of Civil Engineers, New York.
24.6 New York State Thruway Authority.
24.7 From N. B. Hutcheon, Ottawa.
24.8 County engineer, Surrey, through British Ministry of Transport.
24.9 The Institution of Structural Engineers, London.
24.10 R.F.L.
24.11 Sir William Halcrow and Partners, London.
24.12 Chief engineer, Northwestern Railway of India.
24.13 British Steel Piling Company, Ltd., and British Railways.
24.14 City engineer and surveyor, Nottingham, Eng.
24.15 R.F.L.
24.16 Southern Pacific Transportation Company.
24.17 British Rail, London.
24.18 Fay, Spofford and Thorndyke.
24.19 Merritt, Chapman and Scott, Inc.
24.20 Scottish Tourist Board, Edinburgh.
24.21 R.F.L.
24.22 British Columbia Ministry of Transportation and Highways, Victoria.
24.23 British Columbia Ministry of Transportation and Highways, Victoria.
24.24 Michigan Department of Transportation.

ILLUSTRATION CREDITS   E-9

24.25 U.S. Steel Corporation, Pittsburgh.
24.26 Canadian Pacific Railway Company, Ltd., Montreal.
24.27 Canadian Pacific Railway Company, Ltd., Montreal.
24.28 R.F.L.
24.29 The Institution of Civil Engineers, London.
24.30 M. Matula, Bratislava.

CHAPTER 25

25.1 World Wide Photos.
25.2 Keystone Photos.
25.3 Department of Water and Power, Los Angeles; R. V. Phillips, chief engineer of water supply.
25.4 U.S. Bureau of Reclamation.
25.5 The Institution of Structural Engineers, London.
25.6 M. André Coyne, Paris, France.
25.7 U.S. Corps of Engineers, Mobile district engineer.
25.8 Water Service, Eastern Division, Department of the Environment for Northern Ireland, Belfast.
25.9 City water engineer, Liverpool, Eng.
25.10 Acres Consulting Services, Ltd., Niagara Falls, Ont.
25.11 Quido Zaruba, Prague, Czechoslovakia.
25.12 Société Forcés Motrices Bonne et Drac, Grenoble, France.
25.13 Chief engineer, Department of Public Works, Baltimore, Md.
25.14 Chief engineer, Department of Public Works, Baltimore, Md.
25.15 U.S. Bureau of Reclamation.
25.16 U.S. Bureau of Reclamation.
25.17 Electricité de France; photo by M. Barranger.
25.18 Metropolitan Water, Sewage and Drainage Board, Sydney, N.S.W., Australia.
25.19 Chief engineer and general manager, Water Department, City of Pasadena, Calif.
25.20 *Engineering News-Record*, New York.
25.21 U.S. Bureau of Reclamation.

25.22 Ambursen Dam Company, New York.
25.23 *Engineering News-Record*, New York.
25.24 Tennessee Valley Authority.
25.25 Tennessee Valley Authority.
25.26 *Civil Engineering*, ASCE, New York.
25.27 The chief civil engineer, Washington Water Power Company.
25.28 C. W. Isherwood, Edgar Morton & Partners, Altrincham, England; and K. T. Bass, Rofe, Kennard and Lapworth, Cardiff, Wales.
25.29 Chief engineer, Los Angeles County Flood Control District.
25.30 Ministry of Power and Natural Resources, Turkey; General Directorate of State Hydraulic Works.
25.31 R.F.L.
25.32 *Civil Engineering*, ASCE, New York.
25.33 *Engineering News-Record*, New York.
25.34 U.S. Corps of Engineers, district engineer, Sacramento.
25.35 *Engineering News-Record*, New York.
25.36 Department of Water and Power, Los Angeles.
25.37 G. T. Calder, director of southern division, Yorkshire Water Authority.
25.38 District engineer, U.S. Corp of Engineers, Nashville.
25.39 Director Gerente, Empressa Nacional Hidro-Electrica del Ribagorzana, Spain.

CHAPTER 26

26.1 From *Karst in China* (by permission).
26.4 John Blake, Ltd., Accrington, Eng.
26.5 Tennessee Valley Authority.
26.6 G. A. Kiersch, Ithaca, N.Y.
26.7 British Columbia Hydro and Power Authority, Vancouver.
26.8 Los Angeles County Flood Control District, Calif.

26.9 *Civil Engineering*, ASCE, New York; and E. O. Bengry, Santa Paula, California.
26.10 Ontario Hydro, Toronto.
26.11 Aluminum Company of Canada, Ltd., Montreal.
26.12 Ontario Hydro, Toronto.

CHAPTER 27

27.1 The governor, Panama Canal Zone.
27.2 Tennessee Valley Authority.
27.3 The Institution of Water Engineers, London.
27.4 Ontario Hydro, Toronto.
27.5 *Engineering News-Record*, New York.
27.6 New Brunswick Electric Power Commission, Fredericton; A. E. Strang, chief engineer.
27.7 Le Societé Nationale Electricité et Gaz d'Algerie and the editor, *Travaux*, Paris.
27.8 Wayss & Freytag A.G., Frankfurt.
27.9 Wayss & Freytag A.G., Frankfurt.
27.10 U.S. Corps of Engineers, district engineer, Mobile.
27.11 Pressure Grout Company (E.D. Graf) and H. J. Brunner Associates (E. G. Zacher), San Francisco.
27.12 Manitoba Hydro, Winnipeg; K. J. Fallis, executive engineer.
27.13 Manitoba Hydro, Winnipeg; K. J. Fallis, executive engineer.

CHAPTER 28

28.1 Scottish Tourist Board, Edinburgh.
28.2 Secretary and general manager, Des Moines Water Works, Des Moines, Iowa.
28.3 Binnie, Deacon and Gourlay, Consulting Engineers, Westminster, Eng.
28.4 *Engineering News-Record*, New York.
28.5 General manager and chief engineer, Metropolitan Water District of Southern California.
28.6 California Department of Water Resources.
28.7 Donald H. Kupfer, Louisiana State University.
28.8 Director of Mines, South Australia.
28.9 The Thames Conservancy; E. J. Brettell, chief engineer, Reading, Eng.
28.10 Chief engineer, Los Angeles County Flood Control District.
28.11 Mekorot Water Co., Ltd., and Y. Rabinowitch, city engineer, Tel Aviv, Israel.
28.12 Adolph Coors Company, Golden, Colo.
28.13 R.F.L.
28.14 Division of Water Power and Control, New York State Department of Conservation, Albany.
28.15 Spence Air Photos.
28.16 Greater Vancouver Visitors and Convention Bureau.
28.17 Binnie & Partners, Geoffrey Binnie, and Scott Wilson, Kirkpatrick & Partners, joint consulting engineers.
28.18 City engineer, Ogden, and U.S. Bureau of Reclamation.
28.19 Board of Water Supply, Honolulu, G. Yuen, manager and chief engineer.
28.20 Board of Water Supply, Honolulu, G. Yuen, manager and chief engineer.

CHAPTER 29

29.1 The Embassy of Canada, Athens, Greece.
29.2 R.F.L.
29.3 The governor, Panama Canal Zone.
29.4 The governor, Panama Canal Zone.
29.5 The governor, Panama Canal Zone.
29.6 Ontario Hydro, Toronto.
29.7 St. Lawrence Seaway Authority, Ottawa.
29.8 *Engineering News-Record*, New York.
29.9 Panama Canal Commission, R. D. Donaldson, chief of engineering division.

29.10 La Ministère des Travaux Publics et de la Reconstruction, Brussels, Belgium.
29.11 La Ministère des Travaux Publics et de la Reconstruction, Brussels, Belgium.

CHAPTER 30

30.1 U.S. Bureau of Public Roads, Washington, D.C.
30.2 Thurber Consultants, Victoria, British Columbia; photo by D. Maclean, Victoria.
30.3 Maryland State Roads Commission.
30.4 *Engineering News-Record,* New York.
30.5 U.S. Geological Survey; photo by E. W. Shaw.
30.6 Texas Highway Department and White's Mines, San Antonio, Texas.
30.7 R.F.L.
30.8 Freeman Fox & Partners, Westminster, Eng., and M. G. Lewis, resident engineer.
30.9 American Society of Civil Engineers, New York, and J. P. Nicoletti, and J. M. Keith.
30.10 W. M. Jollans, when he was engineer and manager, Water Works Department, Huddersfield, Eng.
30.11 West Riding County Council, Eng., S. M. Lovell, engineer, through J. N. Stothard.
30.12 California Division of Highways, Department of Public Works, Sacramento.
30.13 Pennsylvania Department of Transportation, Harrisburg.
30.14 H. W. Richardson.
30.15 Ente Nazionale Italiano per il Turismo, Rome.
30.16 United Nationas Photographic Service, New York.
30.17 Cleveland Museum of Natural History, Ohio.

CHAPTER 31

31.1 Schweizische Bundesbahnen, Zürich, Switzerland.
31.2 Thurber and Associates, Victoria, B.C.; photo by D. Maclean, Victoria.
31.3 *The Huddersfield Examiner,* Huddersfield, Eng.
31.4 Denver and Rio Grande Western Railroad.
31.5 Denver and Rio Grande Western Railroad.
31.6 The chief engineer, Southern Railway System.
31.7 U.S. Geological Survey; photo by R. F. Black.
31.8 Nederlandse Spoorwegen, Utrecht, Netherlands, through W. de Steur, chief engineer, Concrete Building.
31.9 F. L. Peckover, Vaudreuil, Quebec.
31.10 The Institution of Civil Engineers, London, Eng.
31.11 Western Australian Government Railways.
31.12 Canadian National Railways, Montreal.
31.13 Ferro-Carril de Antofagasta a Bolivia.
31.14 Berner Alpenbahn-Gesellschaft.
31.15 Furka-Oberalp Railway.
31.16 National Research Council of Canada, Division of Building Research, Ottawa; photo by P. Schaerer.
31.17 Derek Cross, Ayrshire, Scotland, and the editor, *The Dominion,* Wellington, N.Z.
31.18 New Zealand Government Railways.
31.19 The editor, *Calgary Herald,* and L. E. Jackson, Calgary, Alberta.
31.20 St. Lawrence Seaway Authority, Ottawa.
31.21 *Engineering News-Record,* New York.
31.22 R.F.L.
31.23 G. Plantema, director, Gemeentewerken, Rotterdam, Netherlands.
31.24 G. Plantema, director, Gemeentewerken, Rotterdam, Netherlands.

CHAPTER 32

32.1 Department of Transport, Ottawa, W. M. McLeish, administrator (Air).

32.2 Department of Transport, Ottawa, W. M. McLeish, administrator, (Air).
32.3 GRW Engineers Inc. (J. W. Dinning), Bowling Green, Kentucky.
32.4 Caterpillar Tractor Company, Inc., Peoria, Ill.
32.5 Atlas Copco (Canada) Inc., Montreal.

CHAPTER 33

33.1 Mississippi River Commission, U.S. Corps of Engineers.
33.2 Mississippi River Commission, U.S. Corps of Engineers.
33.3 American Association of Petroleum Geologists, Tulsa, Okla.
33.4 *Engineering News-Record*, New York.
33.5 Mississippi River Commission, U.S. Corps of Engineers.
33.6 Mississippi River Commission, U.S. Corps of Engineers.
33.7 The chief of engineers, U.S. Army.
33.8 The chief of engineers, U.S. Army.
33.9 National Research Council of Canada, Division of Building Research, Ottawa; photo by G. H. Johnston.
33.10 From *Water: A Primer* by L. B. Leopold, W. H. Freeman & Co., © 1974.
33.11 U.S. Soil Conservation Service, M. S. Wilder and E. A. Keller.
33.12 R.F.L.
33.13 Kennecott Minerals Company, from R. J. Anderson, Utah smelter plant superintendent.
33.14 New York Department of Transportation, Albany, from W. P. Moody, Soil Mechanics Bureau.
33.15 *Civil Engineering*, ASCE, New York; and divisional engineer, Portland, U.S. Corps of Engineers.
33.16 *Engineering News-Record*, New York.
33.17 Internationale Mosel-Gesellschaft mbH, Trier, W. Germany.
33.18 Internationale Mosel-Gesellschaft mbH, Trier, W. Germany.
33.19 District engineer, St. Paul, Minn., U.S. Corps of Engineers.
33.20 District engineer, Buffalo, N.Y., U.S. Corps of Engineers.

CHAPTER 34

34.1 New Brunswick Travel Bureau, Fredericton.
34.2 New Brunswick Travel Bureau, Fredericton.
34.4 Photo by J. K. St. Joseph, crown copyright reserved; by permission of British Air Ministry and Committee for Aerial Photography, Cambridge University.
34.5 Dover Harbour Board, W. Taylor Allen, general manager.
34.6 *Engineering News-Record*, New York.
34.7 Erie Mining Corporation.
34.8 Ente Provinciale per il Turismo di Napoli, Naples.
34.9 R.F.L.
34.10 F. Fox, *Sixty-Three Years of Engineering*, from J. Murray, London, (reference 34.25).
34.12 Toronto Harbour Commissioners, J. H. Jones, chief engineer.
34.13 American Society of Civil Engineers, New York.
34.14 U.S. Coast and Geodetic Survey, Washington.
34.15 The Institution of Civil Engineers, London, Eng.
34.16 Deputy administrative officer, Vizagapatam Port.
34.17 Milwaukee Public Museum and W. G. Murphy, Marquette University; photo by G. J. Gaenslen.
34.18 J. Duvivier, London, Eng.; photo by Photoflight, Ltd.
34.19 Divisional engineer, Jacksonville, Fla., U.S. Corps of Engineers.
34.20 Mersey Docks and Harbour Board, Liverpool, Eng.
34.21 Geological Survey of Canada, Ottawa.
34.22 T. Ritchie, Ottawa.
34.23 National Film Board of Canada, Ottawa.
34.24 *Bell System Journal*, Murray Hill, N.J.

34.25 Kemsley Picture Service, London, Eng.
34.26 The commandant, U.S. Coast Guard, Washington.
34.27 Phillips Petroleum Co., Norway, Sven Martinsen photographer; and O. Eide, Norwegian Geotechnical Institute, Oslo.
34.28 Richfield Oil Corporation, Inc.

CHAPTER 35

35.2 Photo by KLM Aerocarto n.v., Netherlands.
35.3 Photo by KLM Aerocarto n.v., Netherlands.
35.4 Ministerie Verkeer en Waterstaat, The Hague, Netherlands.
35.5 Hackensack Meadowlands Development Commission, G. Cascino, chief engineer.
35.6 Northwestern University, Evanston, Ill., and J. O. Osterberg.
35.7 The Port Authority of New York and New Jersey.
35.8 Geographical Survey Institute, Ministry of Construction, Tokyo, Japan, and K. Kamimura.
35.9 Toronto Harbour Commission, Ontario, Canada, and J. H. Jones, chief engineer.
35.10 The city engineer, Portsmouth, Eng.
35.11 Department of Transportation and Basic Services, Liverpool, Eng.

CHAPTER 36
36.1 R.F.L.
36.2 R.F.L.
36.3 National Research Council of Canada, Division of Building Research, Ottawa.
36.4 T. L. Pewe, Arizona State University, Tempe.
36.5 National Air Photo Library, Ottawa.
36.6 Geological Survey of Canada; photo by R. L. Christie.
36.7 R.F.L.
36.8 National Research Council of Canada, Division of Building Research, Ottawa; photo by G. H. Johnston.
36.9 R.F.L.
36.10 Geological Survey of Canada, Ottawa; photo by D. H. Yardley.
36.11 R.F.L.
36.12 National Research Council of Canada, Division of Building Research, Ottawa; photo by G. H. Johnston.
36.13 National Research Council of Canada, Division of Building Research, Ottawa; photo by R. J. E. Brown.
36.14 National Research Council of Canada, Division of Building Research, Ottawa; photo by R. J. E. Brown.
36.15 National Research Council of Canada, Division of Building Research, Ottawa; photo by G. H. Johnston.
36.16 J. R. Mackay, University of British Columbia, Vancouver.
36.17 B. Michel, Université Laval, Quebec City, Quebec.
36.18 Imperial Oil, Ltd., Toronto.

CHAPTER 37

37.1 Il direttore generale, Ente Nazionale Italiano per il Turismo, Rome.
37.2 The Commandant, U.S. Naval dockyard, U.S. Navy, Long Beach, Calif.
37.3 Il Direttore generale, Ente Nazionale Italiano per il Turismo, Rome.
37.4 C. B. Crawford, Ottawa.
37.5 S. Marayama, from his paper on land subsidence at Osaka in the proceedings of the Tokyo Symposium on Land Subsidence, Sept. 1969; by permission of International Association on Scientific Hydrology and UNESCO.
37.6 R.F.L.
37.7 R.F.L.
37.8 E. H. Scholes, Lancaster, Eng.
37.9 GAI Consultants, Inc., R. E. Gray, vice president, Pittsburgh.
37.10 The Institution of Civil Engineers, London, and F. W. L. Heathcote.
37.11 Wimpey Laboratories, Ltd., London, Eng. and D. G. Price.

CHAPTER 38

38.1 City of Springfield, Mo.; photo by W. S. Howe.

E-14 APPENDIX E

38.2 Pennsylvania Bureau of Topographic and Geologic Survey, A. A. Socolow, director.
38.3 Georgia Department of Transportation, Atlanta.
38.4 G. F. Sowers, Atlanta, Ga.
38.5 Canadian Salt Company, Ltd., Windsor, Ontario, W. S. Carpenter, chief engineer.
38.6 Cheshire County Council, Chester, Eng.
38.7 *Rand Daily Mail,* Johannesburg, and R. Procter-Sims.
38.8 R.F.L.
38.9 L. N. Stout and D. Hoffman, from *Introduction to Missouri's Geologic Environment,* Educational Series no. 3, Division of Geological Survey and Water Resources, Rolla, Mo., 1973, by permission.
38.10 From *Karst in China,* by permission.

CHAPTER 39

39.1 University of Toronto Press, and W. R. Schriever.
39.3 U.S. Geological Survey, Washington.
39.4 American Institute of Mining and Metallurgical Engineers, New York (reference 39.7, Alden).
39.5 City of Ottawa, and the National Capital Commission, Ottawa.
39.6 C. B. Crawford, Ottawa.
39.7 Department of National Defence, Ottawa; Canadian Forces photo.
39.8 R.F.L.
39.9 R.F.L.
39.10 Q. Zaruba, Prague, Czechoslovakia.
39.11 Q. Zaruba, Prague, Czechoslovakia.
39.12 Q. Zaruba, Prague, Czechoslovakia.
39.13 Department of Public Works, Canada, Ottawa.
39.14 R.F.L.
39.15 Municipal Engineering Publication, Ltd., London, Eng., from brochure of 8th annual conference of the Institute of Highway Superintendents, London, 1958.
39.16 Chief civil engineer, Southern Region, British Railways, London.
39.17 R.F.L.
39.18 Quinton, Code and Hill-Leeds and Barnard.
39.19 *Calgary Herald,* and L. Jackson, Geological Survey of Canada.

CHAPTER 40

40.1 Geological Survey of Canada, Ottawa.
40.2 From Cruden and Krahn, "A Re-examination . . . ," National Research Council of Canada, Ottawa (reference 40.2); and Canadian Journals of Research.
40.3 R.F.L.
40.4 Chief civil engineer, Western Region, British Railways, London.
40.5 Sir William Halcrow and Partners, London, Eng.
40.6 Aluminum Company of America, Pittsburgh.
40.7 Ron Ruels, Niagara Falls, Ontario.
40.8 U.S. Bureau of Reclamation; photo by Bethlehem Steel Company.
40.9 *Engineering News-Record,* New York.
40.10 Norwegian Geotechnical Institute, Oslo.
40.11 Department of Highways, Colorado, P. O. McCollough, district engineer, Aurora.
40.12 Geological Survey of Canada, Ottawa.
40.13 Ontario Hydro, Toronto.
40.14 A. B. Hawkins, University of Bristol, Eng.
40.15 Swiss Institute for Snow and Avalanche Research, Davos, M. de Quervain, director.
40.16 National Research Council of Canada, Division of Building Research, Vancouver; photo by P. Anhorn.
40.17 National Research Council of Canada, Division of Building Research, Vancouver; photo by P. A. Schaerer.
40.18 F. L. Peckover, Vaudreuil, Quebec.
40.19 Peckover and Carr, "The Treatment and Maintenance . . . ," National Research Council of Canada, Ottawa (reference 40.25); and Canadian Journals of Research.

40.20 Scottish Tourist Board, Edinburgh, Scotland.

CHAPTER 41

41.1 R.F.L.
41.2 From W. N. Benson, "Notes . . . ," the *New Zealand Journal of Science and Technology* (reference 41.2).
41.3 The secretary, Central Board of Irrigation, Simla, India (before 1947).
41.4 Aluminum Company of America, Pittsburgh.
41.5 U.S. Bureau of Reclamation.
41.6 U.S. Bureau of Reclamation; photo by E. E. Hertzog.
41.7 U.S. Soil Conservation Service.
41.8 R.F.L.

CHAPTER 42

42.1 R. J. Chandler, Imperial College of Science, London, England.
42.2 P. Lumb, University of Hong Kong.
42.3 From Brink *Engineering Geology of Southern Africa* (ref. 42.7) p. 31.
42.4 Central Building Research Institute, Roorkee, India, D. Mohan, director.
42.5 National Building Research Institute, South Africa; photo by A. A. B. Williams.
42.6 P. Lumb, University of Hong Kong.
42.7 A. Dreimanis, University of Western Ontario, London, Ont.
42.8 California Department of Water Resources, Sacramento.
42.9 R. M. Hardy, Edmonton, Alberta.
42.10 G. W. White, University of Illinois, Urbana.
42.11 M. A. Hanna Company; photo by G. D. Creelman.
42.12 National Research Council of Canada, Division of Building Research, Ottawa.

CHAPTER 43

43.1 Department of Regional Economic Expansion, Canada, Soil Mechanics and Materials Division, Saskatoon, N. Iverson, chief.
43.2 National Research Council of Canada, Division of Building Research, Ottawa.
43.3 New York State Geological Survey; photo by Donald H. Cadwell.
43.4 Acres Consulting Services, Ltd., Niagara Falls, Canada.
43.5 J. de Boer, Wesleyan University, Middletown, Conn.; and the Geological Society of America.
43.6 National Research Council of Canada, Division of Building Research, Ottawa; photo by P. Goudreau.
43.7 Q. Zaruba, Prague, Czechoslovakia.
43.8 A. V. Morgan, University of Waterloo, Ontario.
43.9 R.F.L.

CHAPTER 44

44.1 Scottish Tourist Board, Edinburgh.
44.2 U.S. Geological Survey, Washington, D.C.
44.3 Westinghouse Electric Corporation, M. B. Crowe, defense products manager, and C. F. Withington.
44.4 R.F.L.
44.5 U.S. Geological Survey, Washington, D.C.
44.6 Metropolitan Water District of Southern California.
44.7 New Brunswick Electric Power Commission, Fredericton; A. E. Strang, chief engineer.
44.8 U.S. Geological Survey, Washington, D.C.; photo by D. G. Herd.
44.9 Woodward Clyde Associates, L. S. Cluff, vice president, and H. R. Goode, Salt Lake City.
44.10 Jacob Feld, New York.
44.11 From Brink, *Engineering Geology of Southern Africa* (reference 42.7), p. 228; photo by Regal Studios.

CHAPTER 45

45.1 U.S. Geological Survey, EROS Data Center, Sioux Falls, S.D. (no. 80S3-137).
45.2 (a) Photo by Hördur Vilhjálmsson; (b) University of Iceland, Reykjavík; photo by Sigurdur Thorarinsson.
45.3 Departamento de Turismo, Mexico D.F.

45.4 H. J. Degenkolb, San Francisco; photo by El National, Caracas.
45.5 U.S. Geological Survey, Washington, D.C.
45.6 U.S. Geological Survey, Washington, D.C.
45.7 National Earthquake Information Center, National Oceanic and Atmospheric Administration, Washington, D.C.
45.8 From *Elementary Seismology,* by C. F. Richter, W. H. Freeman & Co., © 1958.
45.9 Japanese Society for Soil Mechanics and Foundation Engineering, from H. B. Seed, University of California at Berkeley.
45.10 U.S. Forest Service, Washington, D.C.
45.11 National Research Council of Canada, Associate Committee on the National Building Code, Ottawa.
45.12 G. A. Macdonald, University of Hawaii, Honolulu.

CHAPTER 46

46.1 Fox Photos, London, Eng.
46.2 U.S. Geological Survey, EROS Data Center, Sioux Falls, S.D. (no 8055-095).
46.3 *Providence Journal-Bulletin,* Providence, R.I.
46.4 City of Winnipeg, H. B. Young, director of civic properties.
46.5 U.S. Corps of Engineers, district engineer, Boston, Mass.; photo by C. E. Maguire, Inc., Waltham, Mass.
46.6 R.F.L.
46.7 National Research Council of Canada, Ottawa, and F. W. Render (reference 46.16), and Canadian Journals of Research.
46.8 Metropolitan Sanitary District of Greater Chicago; photo by Patricia Young.
46.9 Rendel, Palmer and Tritton, London, Eng., through P. A. Cox, senior partner.
46.10 P. H. Rahn, South Dakota School of Mines, Rapid City.
46.11 The Association of Engineering Geologists, Dallas, Texas, and P. H. Rahn (reference 46.22).
46.12 U.S. Corps of Engineers, Portland Division Office, and W. H. Stuart.
46.13 U.S. Corps of Engineers, district engineer, Pittsburgh, and S. B. Long.

CHAPTER 47

47.1 Metropolitan Washington Council of Governments (reference 47.12).
47.2 San Francisco Convention and Visitors Bureau.
47.3 Denver Convention and Visitors Bureau.
47.4 U.S. Geological Survey, Washington, D.C.; photo by J. S. Pomeroy.
47.5 St. Lawrence County Planning Board, Potsdam, N.Y.
47.6 Ontario Geological Survey, Toronto; O. L. White, chief of Engineering and Terrain Geology Section.

CHAPTER 48

48.1 U.S. Bureau of Reclamation, Washington, D.C.
48.2 Gibb, Coyne, Sogei (Kariba) (Pvt.), Ltd., and the Federal Power Board, Salisbury.
48.3 National Film Board, Ottawa.
48.4 Metropolitan Water District of Southern California, Los Angeles.
48.5 R.F.L.
48.6 J. Cherry, University of Waterloo, Ontario.
48.7 Central Electricity Generating Board, Ash Marketing Division, London, England.
48.8 U.S. Geological Survey, Washington, D.C.; photo by W. E. Davies.

CHAPTER 49

49.1 National Research Council of Canada, Division of Building Research, Ottawa; photo by W. J. Eden.
49.2 U.S. Corp of Engineers, district engineer, Norfolk, Va., and L. Jean Hunt.

49.3 Transport and Road Research Laboratory, Department of Transport, Crowthorne, Eng. and R. Murray.
49.4 R.F.L.
49.5 U.S. Soil Conservation Service, Washington, D.C.
49.6 Aggregate Producers Association of Ontario, Toronto, H. B. Cooke, executive director.
49.7 Flatiron Companies, Boulder, Colorado.
49.8 U.S. Geological Survey, Washington; photo by F. W. Osterwald.
49.9 Amax Coal Company, Indianapolis, Ind.
49.10 National Coal Board, London, Eng.
49.11 American Aggregates Corporation, Greenville, Ohio.
49.12 U.S. Soil Conservation Service, Washington, D.C.
49.13 U.S. Soil Conservation Service, Washington, D.C.
49.14 Conservation Branch, Department of Commerce and Development (now Ministry of Natural Resources), Toronto, Ontario.
49.15 Conservation Branch, Department of Commerce and Development (now Ministry of Natural Resources), Toronto, Ontario.
49.16 The Embassy of Japan, Ottawa.

CHAPTER 50

50.1 Geological Society of America, Boulder, Colo. (reference 50.5).
50.2 *Journal of Geology*, Chicago, Ill. (reference 50.6).
50.3 Connecticut Department of Transportation.
50.4 P.F.K.
50.5 Aerial Photos International Inc., Boston.
50.6 Connecticut Geological and Natural History Survey, J. W. Peoples, director; photo by J. Howard.
50.7 J. W. Peoples; photo by R. Arnold.
50.8 Boston Society of Civil Engineers (reference 50.23), and M. P. Billings.
50.9 Colorado Department of Highways, Denver, C. E. Shumate, chief engineer.
50.10 Geological Survey of Alabama, P. E. LaMoreaux, state geologist.
50.11 Armco Canada Ltd., Etobicoke, Ont.
50.12 R.F.L.
50.13 J. H. Williams, Division of Geology and Land Survey, Missouri; photo by G. St. Ivany.
50.14 J. H. Williams, Division of Geology and Land Survey, Missouri; photo by J. Vandike.

SUBJECT INDEX

This Index contains not only the subjects treated in the text but also the names of organizations associated with some of the works described. A few such organizations occur so frequently that it would be embarrassing, and probably less than useful, to list all the pages on which they are mentioned. The notable activity of the *U. S. Geological Survey* is reflected throughout the volume. Works of the *U. S. Corps of Engineers,* the *U. S. Bureau of Reclamation,* and the *Tennessee Valley Authority* all call for appropriate and frequent mention. It is hoped that this general acknowledgment of their work will ensure their necessary recognition by all users of this Handbook. The *American Society of Civil Engineers* and the *Institution of Civil Engineers* of London correspondingly call for repeated mention in view of the service they give in many of the major fields discussed herein as leading civil engineering professional societies of the English-speaking world. Some of the organizations which provided earlier photographs now have new names; for convenience the old names have sometimes been retained where this seemed desirable.

Page numbers in *italics* indicate illustrations.

Aberfan (U.K.) disaster, 48-14
Abutments, bridge, 24-11, 15, 18, 38
Accidents:
 from avalanches, 40-17
 Clifton Hall Tunnel (U.K.), 20-43
 gases in tunnels, 8-10, 20-39
 loss of crusher plant, 38-9
 methane in bridge piers, 48-13
 methane at Furnas (Brazil) hydroelectric project, 26-18
 Rapid City (S.Dak.) flood, 46-15
 from rockfalls, 26-18, 40-13
 over sinkholes, 38-2, 8
 Vaiont Dam (Italy) slide, 26-12
"Acts of God," 39-2, 40-19
Adobe, 5-11
Aerial photography:
 of Alaska pipeline, *10-18*

Aerial photography (*Cont.*):
 and ground truth, 10-17
 interpretation of, 10-15
 for locating fill, 15-16
 of Mackenzie River (N.W.T.) delta, 36-6
 of Mackenzie River (N.W.T.) pipeline, 10-17
 in northern Canada and U.S., 36-10
 for railroad location, 31-3
 start of, 10-14
Aggregate:
 alkali reaction (*see* Alkali-aggregate reaction)
 concrete, 16-6
 crushed rock (in U.K.), 2-10
 Rhine (West Germany) gravel, 2-14
Agriculture Department (U.S.), 33-15
Air pollution, 48-8
Airfields:
 drainage, 32-4

Airfields (*Cont.*):
 excavation for, 32-7
 and geology, 32-2
 grass cover, 32-9, 49-2
 growth of, 32-1
Airports:
 Bowling Green (Ky.), *32-5*
 Inuvik (N.W.T.), *36-16*
 Kanawha (W.Va.), 32-7
 La Guardia (N.Y.), 32-6
 Mirabel (Que.), *32-3*
 Newark (N.J.), 32-6
 St. Thomas (V.I.), *32-8*
 San Francisco (Calif.), 44-5
 Schipol (Netherlands), *31-9*
Alaska pipeline, 10-17
Alaskan developments, 36-1
Algeria:
 Algiers harbor, 34-11
 dams, 25-11, 27-11
Alkali-aggregate reaction:
 in concrete, 16-16, 43-7
 locations of: in California, 16-16
  in Nova Scotia, 16-17
  in Ontario, 16-17

S-1

Alkali-aggregate reaction (*Cont.*):
  tests for, 16-17
Alluvial areas, 5-13
American Society of Agricultural Engineers, 18-14
American Society for Testing and Materials, 15-22, 16-77, 30-2, 43-7
American Underground Space Association, 21-26
Andesite deposits, 4-11, 7-5
Anguar Island (Micronesia), groundwater contamination in, 8-20
Anhydrite deposits, 20-28, 43-6
Antarctica, 36-1
Anticline, 4-14, 15
Aqueducts, 8-10
Archeology:
  in Dover (U.K.), 30-27
  and roads, 50-13
  in Stockholm (Sweden), 22-44
  in Zurich (Switzerland), 13-14
Archives, underground, 21-22
Arctic, 36-1
Artesian basins:
  in Australia, 8-3, 4, 19, *8-18*
  in North and South Dakota, 8-17
Artesian water:
  causes of, 8-16
  under dams, 8-19
  deep wells to, 8-17
  in rivers, 8-17
Artesian wells, 28-32
Artois region (France), artesian water in, 8-16
As-constructed drawings, 10-21
Ash: fly, 17-9
  volcanic, 17-8
Asphalt for road paving, 30-13
Association of Engineering Geologists, 44-11
Atmosphere, 9-1
Atmospheric Environmental Service (Canada), 9-7

Atterberg soil tests, 6-5, 17
Auger boring, 11-7
Australia:
  bird problems, 49-15
  bridges in, 24-5
  calcined clay roads, 30-14
  Snowy Mountain scheme, *2-2*, *2-13*
  tunnels in, 20-21
  water supplies, 26-12, 28-17, 26
Austria, subways in, *31-28*
Avalanches:
  danger of, 40-17
  defense works, 40-19
  and railroads, 31-17, 21

Bacteria:
  and shales, 43-8
  and weathering, 5-5
Bank protection along Rhine River (West Germany), 15-14
Barachois, 34-22
Basalt, 4-11, 13
Beach Erosion Board (U.S.), 34-17, 26
Beach replenishment:
  Biloxi (Miss.), 34-22
  Erie (Pa.), 34-21, *34-22*
Beaufort Sea drillings, 36-2
Beaver dams, 33-15, 49-15
Bedding:
  cross, 4-9, 4-13
  false, 4-13
  sedimentary rocks. 4-12
Bedrock shattered by permafrost, 43-9
Beer, water for, 28-6
Belgium:
  Brussels subway, 31-28
  canals in, *29-14*
  shaft sinking, 19-25
  underground plants in, 21-13
Bell's bunds, 33-3
Bentonite deposits, 15-10, 20-29
  beneath bridges, 24-9
  slurry-trench, 19-19
Black cotton soil:
  African, 42-6, 7

Black cotton soil (*Cont.*):
  Indian, 42-7
Blasting, 19-34
  controlled, 19-37
  experiments on, 19-35
  at Ripple Rock (B.C.), 34-35
  for silt removal, 26-16
Block diagrams, 1-18
Bolivia, road building in, 30-24
Bolts, rock, 21-11
Boreholes:
  cameras for, 11-15
  inclined, 20-11
  interpretation of, 11-27
Boron in groundwater, 8-11
Boston Basin (Mass.), 50-7, 11
Boulder clay, 5-23
Boulder pavements, 5-25
Boulders, 5-24, 10-7, 22-8
  at bridge sites, 24-33
  and cofferdams, 24-28
  damage caused by, 40-7
  in excavations, 19-14
  locations of: in Hong Kong, 42-9
    in Singapore, 22-51
  in rivers, 24-4
  and till, 42-10
Bournes, 8-16
Braille trail (Mo.), *50-16*
Brazil, hydroelectric projects in, 26-18
Breakwaters:
  locations of: Forestville (Que.), 34-16
    Gulf of Ephesus (Greece), 34-15
    Haifa (Israel), 34-11
    Toronto (Ont.), 35-11
    Valparaiso (Chile), 34-17
  of scuttled ships, 34-24
  studies for, 34-16
Bricks:
  Staffordshire blue, 28-5
  from tunnel excavations, 31-4
  types of, 17-1, 2
Bridge foundations:
  adjustable, 24-21
  buoyant, 24-11
  construction, 24-25

Bridge foundations (*Cont.*):
  costs, 24-3
  historical aspects, 24-1
  inspection, 24-27, 37
  loads on, 24-9
  maintenance, 24-37
  prestressed, 24-34
  scouring at, 24-24
  seismic forces on, 24-22
  steel piles for, 24-25
Bridges:
  Anacostia River (D.C.), 24-39
  Bowling Green (Ky.), 24-29
  Broadway (Sask.), *24-13*
  Burford (U.K.), *24-14*
  Cape Cod Canal (Mass.), *24-29*
  Clifton (U.K.), *24-20*, 40-17
  Copper River (Alaska), 1-9
  Cornwall-Massena (Ont.-N.Y.), *24-4*
  Danube River (Czechoslovakia), *24-42*
  Decatur (Neb.), 33-9
  Deer Island (Me.), *24-30*
  Exeter medieval (U.K.), 13-18
  Firth of Forth (U.K.), *24-31*
  Fraser River (B.C.), *24-33*
  Georges River (Australia), 24-5
  Golden Gate (Calif.), 24-23
  Grey Street (Australia), 24-26
  Grosvenor (U.K.), 24-15, *24-16*
  Hargis (N.Mex.), 24-40
  Klosters (Switzerland), 24-37, *24-40*, 40-15
  Kocher Pass (West Germany), 27-13
  Kohala (Pakistan), *24-18*
  Lavertezzo Roman (Switzerland), 1-5
  Lethbridge (Alta.), 24-37, *24-38, 39*
  Lockwood (U.K.), 31-4
  London (U.K.), 24-1
  Mackinac (Mich.), *24-35*, 50-2
  Mid-Hudson (N.Y.), 24-26
  Miramichi River (N.B.), 11-20, *24-32*

Bridges (*Cont.*):
  Mortimer E. Cooley (Mich.), *24-10*, 24-11
  Pajaro River (Calif.), 24-22, *24-23*
  Peoria (Ill.), 24-30
  Port Mann (B.C.), *24-34*
  Quinnipiac (Conn.), 50-4
  Richmond-San Raphael (Calif.), 24-36
  Royal Albert (U.K.), 24-25, *24-26*
  Royal Gorge (Colo.), 1-17
  St. Norbert (Man.), 24-9
  San Francisco-Oakland (Calif.), 10-5, 11-8, *24-5*, 14
  17th of July (Iraq), *24-41*
  Steffenbach Gorge (Switzerland), 31-18
  Tangiwai (New Zealand), 31-21
  Tappan Zee (N.Y.), 11-9, 24-11, *24-12*
  Thousand Islands International (Ont.), *24-15*
  Vaudreuil (Que.), 24-13
  Volta River (Ghana), *24-17*
  Waterloo (U.K.), 25-12
  Wigan Canal (U.K.), *24-19*
Broken rock field trials, 15-16
Building foundations (*see* Foundations)
Building research:
  Canadian, 36-6
  Indian, 17-4
  South African, 12-11, 17
Building stones, 15-17
  costs of, 15-18
  decay of, 15-17
  durability of, 15-19
  from excavations, 31-4
  geographic types of: from Egypt, 15-20
  from Incas, 15-25
  from London, 15-17
  from Prague, 15-24
  in geological sequence, 15-24
  publications about, 15-24
Bulb of pressure, 22-11, 21
Bulldozers, 50-7

Bull's liver, 5-27, 10-11, 42-12
Bureau of Mines (U.S.), 18-10
Buried valleys, 5-22, 42-17
  in Chile, 20-7
  under Mersey River (U.K.), 20-34
  in St. Maurice Valley (Que.), 11-4
  tunnel boring operations, 20-30
  and water supply, 28-19
Burned clay, 17-2, 3

Cables, transatlantic, 34-36
Caissons:
  for bridge piers, 24-26
  tilting of, 24-26
  wartime, 35-5
California Water Plan, 28-14, *28-15*
Calyx drill holes, 11-18
Canadian Geotechnical Society, 13-6
Canalization of rivers, 33-18
  Mississippi River (U.S.), 33-22, *33-23*
  Mosel River (West Germany), 33-19, *33-21*
  Ohio River (U.S.), 33-18, 22, *33-20*
Canals:
  Albert (Belgium), 18-15, *29-14, 15*
  All-American (U.S.), 41-9
  Beauharnois (Que.), 19-7
  Brunswick, 50-2
  Caledonian (U.K.), 44-3
  Corinth (Greece), 29-2
  Cross-Florida (U.S.), 11-18
  Elbe-Seiten (West Germany), 50-7
  Erie (N.Y.), 50-2
  Grand Junction (U.K.), 18-4, 29-3
  Main-Danube (Europe), 29-15
  Manchester (U.K.), 41-5
  Massena Power (N.Y.), 19-7
  Motala River (Sweden), 19-34, *19-35*
  Panama, 29-5

Canals (Cont.):
  Rideau (Ont.), 16-3, 27-2, 29-4, 29-5
  St. Lawrence Seaway (U.S.-Canada), 29-9, *29-10*
  Welland (Ont.), 18-15, 19-7, 29-11
  Wiley-Dondero (N.Y.), 19-7
Canso Causeway (N.S.), 48-5
Carbon dioxide in groundwater, 8-10
Carbon monoxide in tunnels, 20-39
Casagrande "A" line, 6-12
Catchment areas:
  and geology, 28-4, 46-5
  ironclad, 28-26
Cathedrals:
  Baltimore (Md.), 22-28
  Chartres (France), 22-54
  St. Paul's (U.K.), 4-3, 18-26
  Strasbourg (France), 18-25
  Winchester (U.K.), 22-15
  York (U.K.), 22-54
Causeway in Nova Scotia, 48-5
Caves in France, 2-14, 16
Cement:
  low alkali, 16-17
  natural, 16-3
  Portland, 16-2
  Roman, 16-3
  special, 16-3
  sulfate resistant, 16-14
Cementation (*see* Grouting)
Chalk:
  cavities in, 38-3
  from Dover (U.K.), 34-11
  from Niobrara (Neb.), 15-13
  porosity of, 8-7
  pumping as slurry, 20-16, 49
  rockfalls from, 40-6
  soakaways, 18-7
  wine storage in, 21-4
Champlain Sea (Canada), 10-11
Chemical grouting (*see* Grouting)
Chert, 4-9
Chesapeake Bay (U.S.), 50-2
Cheshire (U.K.) meres, 38-2

Chicago (Ill.):
  drainage tunnels, *46-12*
  wells, 22-7
Chile:
  tunnels in, 20-6
  Valparaiso breakwater, 34-17
China:
  earthquakes, 45-6
  Liulang Cave project, 26-5
  loess deposits, 5-11
  training rivers, 33-2
Chlorite deposits in tunneling, 20-14
Churches, moving of, 22-42
Cirques, 5-18
Civil defense, use of underground shelters, 21-18
Claims for extras, 11-24, 19-5, 16
Clay, 5-27
  ancient, 5-31
  bloated, 17-8
  Bootlegger Cove (Alaska), 45-9
  boulder, 5-23
  for bricks, 17-2
  Cambrian (U.S.S.R.), 5-31
  for grouting, 27-3
  investigation of, 6-14
  ion exchange, 6-16
  Leda (Canada), 6-17
  London, 1-4, 23-24
  for military landings, 2-16
  Ordovician (U.S.S.R.), 5-31
  puddle, 15-11
  Sasamua (Kenya), 6-15
  sensitive, 6-10
  swelling, 42-15
  Triassic (U.K.), 5-31
  Upper Lias (U.K.), 42-2
  varved, 5-28, 6-11
  x-ray diffraction, 6-14
Climate:
  of cold regions, 36-2
  and floods, 46-6
  importance of, 9-1
  and road building, 30-6
  and rockfalls, 40-20
  and weathering, 42-4
Coal:
  atlas of workings, 22-39
  disposal of waste, 48-14
  dredging, 23-3

Coal (Cont.):
  machines for cutting, 19-33
Coal mining:
  in Czechoslovakia, 22-42
  in Norway, 36-2
  in United Kingdom, 31-4
  in West Germany, 2-17, 33-24
Coal tips, failures of, 48-14
Coast Guard (U.S.), 34-39
Coastal problems, 34-7
  landslides, 34-9
  reclamation, 34-8
  sand dunes, 34-9
  shore erosion, 34-9
  silting, 34-9
Coastal protection:
  Sheridan Park (Wis.), 34-26
  Sheringham (U.K.), 34-27
Cofferdams:
  protecting plants, 23-24
  St. Lawrence Seaway (U.S.-Canada), *29-10*
  tailor-made, 24-30
Cold storage plants, underground, 21-2
Compacted soil, foundations of, 22-30
  handmade, in China, 33-5
  for power stations, 22-22
Compressed air and sulfate attacks, 16-15
Concrete:
  alkali-aggregate reaction, 36-20, 43-7
  ready-mixed, 16-2
  Roman, 16-1
  and sulfate attacks, 16-15
  Tremie, 19-19
Concrete aggregates, 16-6
  impurities in, 16-7
  lightweight, 17-7
  limestone, 16-7
  selection of, 16-9
  unusual, 16-13
Concrete deterioration from sulfates, 8-11
Cone of depression, 8-21, 18-6, 28-17
Conglomerate, *4*-7, 4-9
  talpetate, 19-32
  torrential, 15-5
Contaminated water, 28-7

SUBJECT INDEX   S-5

Continental drift, 3-5, 45-1
Continental shelf, 34-6
Contracts, 1-11, 11-25
Coral:
  as aggregate, 16-13
  dredging of, 34-34
Corrasion by streams, 41-2
Crazing of concrete, *16-18*
Crushed rock:
  as aggregate, 16-5
  costs of, 16-5
  field experiments, 30-21
  hardness, 16-5
  as railroad ballast, 31-14
Crushers, rock, 16-5
Crushing plants, loss of, 18-22
Crust of earth, 3-5, 12-2
Crystallography, 3-3
Cuba, tunnels in, 20-16, *20-17*
Cut-off walls:
  for dams, 25-48, 52
  of puddle clay, 15-10
Cut-offs on Mississippi River (U.S.), *33-3*
Cycles, geological, 3-7
Czechoslovakia:
  bridges in, *24-42*
  church moving, 22-42
  dams in, 25-15
  karst fields, 38-11
  landslides in, *39-15*
  Prague, geology of, 13-10
  shattered rock, 43-10
  subsidence in, *37-14*
  test drilling, *11-13*
  weathered granite, *5-9*

Dams:
  early examples, 25-2
  earth fill placed against, *25-50*
  earthquake effects, 25-28
  effects of, 48-2
  failures of, 25-3
  foundations of: access to, 22-44
    anchored, 25-10
    bedrock soundness, 25-25
    caisson cutoff, 25-49

Dams, foundations of (*Cont.*):
  exploratory work, 25-18
  grouted, 27-5
  leakage through, 25-32
  oil under, 25-32
  poor rock, 25-40
  preliminary work, 25-14
  records of, 25-23
  rock weathering, 23-24
  "tied down," 25-26
  uplift under, 25-34
  inspections of, 25-8
  in karst fields, 26-4
  on landslide debris, 23-15
  leakage under, 25-51
  locations of: Abitibi Canyon (Ont.), 27-6
  Argal (U.K.), 25-13
  Assouan, First (Egypt), 23-24
  Assouan, High (Egypt), 21-23, 26-22, 27-17, 48-3
  Atiamuri (New Zealand), 25-46
  Auburn, proposed (Calif.), 25-17
  Austin (Tex.), 25-3
  Beechwood (N.B.), 27-10, 44-10
  Bersimis (Que.), 15-6
  Bhakra (India), 25-26
  Bonneville (Oreg.), 41-7
  Bort (France), 25-26
  Broomhead (U.K.), *25-51*
  Calderwood (Tenn.), 41-6
  Carillon (Que.), 7-11, 26-11
  Cheurfas (Algeria), 25-11
  Chickamauga (Tenn), 27-9
  Combie (Calif.), 17-9
  Conchas (N.Mex.), 8-19
  Conowingo (Pa.), 25-13
  Courtright (Calif.), 15-14
  Dalles (Oreg.), 11-16
  Denison (Okla.), 18-19
  Dez (Iran), 25-44
  Elephant Butte (N.Mex.), 26-15
  Englewood (Okla.), 12-14
  Escales (Spain), 25-53

Dams, locations of (*Cont.*):
  Estacada (Oreg.), 27-5
  Folsom (Calif.), 11-15, 25-47
  Fontana (N.C.), 16-6
  Fort Peck (Mont.), 19-32, 20-29
  Fort Randall (S.Dak.), *25-37*, 25-38
  Foum el Gherza (Algeria), 27-11
  Friant (Calif.), 16-13
  Gavins Point (S.Dak.), 15-13
  Gilbertsville (Tenn.), 11-11
  Glen Canyon (Ariz.), 40-9
  Grancarevo (Yugoslavia), 26-4
  Grand Coulee (Wash.), 11-17, 19-27, 27-14, 41-6
  Great Falls (Tenn.), 26-10, 27-4
  Hales Bar (Tenn.), 25-32, *25-33*
  Hanson (Wash.), 40-10
  Harlan County (Neb.), 25-26
  Hiwassee (N.C.), 19-38
  Hoover (Nev.-Ariz.), 25-19, 27-10, 40-9, 48-3
  Imperial (Ariz.-Calif.), 41-10
  Iriso (Venezuela), 43-7
  Islam (Pakistan), 41-5
  Jim Woodruff (Fla.), *27-15*, 27-16
  John Hollis Bankhead (Ala.), *25-12*, 25-13
  John Martin (Colo.), 49-7
  Karadj (Iran), 25-44
  Kariba (Zimbabwe), 48-3
  Keban (Turkey), 25-42
  Kenney (B.C.), 26-20
  Kielder (U.K.), 28-14
  Kinder (U.K.), 27-5
  Kingsley (Neb.), 17-3
  Kinzua (Pa.), 2-10
  Kippewa (Que.), 16-7
  Kukuan (Taiwan), 40-10

Dams, locations of (*Cont.*):
 Lake Waco (Tex.), 41-7
 Le Sautet (France), 25-19
 Madden (Panama), 25-36
 Malpasset (France), 25-4, *25-5*
 Mammoth Pool (Calif.), 15-4, *15-5*
 Merriman (N.Y.), 25-49
 Merwin (Wash.), 25-40
 Morris (Calif.), 25-28
 Mount Morris (N.Y.), 25-48
 Mulholland (Calif.), 25-50
 Neusa River (Colombia), 15-4
 Nickajack (Tenn.), 25-33
 Norris (Tenn.), 11-17, 27-8
 Noxon Rapids (Mont.), *25-39*, 25-40
 Oahe (S.Dak.), 20-29, 43-2
 Orlik (Czechoslovakia), 25-15, *25-16*
 Owyhee (Oreg.), 25-29
 Prettyboy (Md.), 11-18, 19-32, 25-19, *25-20*
 Quinze (Que.), 16-7
 Rodriguez (Mexico), 25-30, *25-31*
 Ryan (Mont.), 25-13
 St. Francis (Calif.), 25-3, *25-4*
 St. Pierrett-Cognet (France), 27-12
 San Andreas (Calif.), 25-28
 San Gabriel (Calif.), 25-41
 Santa Felicia (Calif.), 25-34, 26-17
 Sarrans (France), 12-13
 Sasamua (Kenya), 6-13
 Scammonden (U.K.), 30-21
 Scount Dyke (U.K.), 27-6
 Seminoe (Wyo.), 25-46
 Sennar (Sudan), 41-5
 Serre-Poncọn (France), 25-22
 Shand (Ont.), 15-3

Dams, locations of (*Cont.*):
 Silent Valley (U.K.), 5-24, *10-6*, *25-14*
 Spencer's Creek (Australia), 12-15
 Steenbras (South Africa), 25-11
 Tansa (India), 25-11
 Terzaghi (B.C.), 25-43
 Teton (Idaho), *25-7*
 Tumut Pond (Australia), 25-5
 Vaiont (Italy), 26-12, *26-13*
 Vyrnwy (U.K.), 25-14, *25-15*
 Warragamba (Australia), 25-26, *25-27*
 Watts Bar (Tenn.), *11-19*
 Wheeler (Ala.), 19-33
 White River (South Africa), 12-13
 Wilder (Conn.), 41-7
 Wimbleball (U.K.), *17-10*, 25-40, *25-41*
 Wolf Creek (Ky.), *25-52*
 Yale (Wash.), 15-5, *15-6, 7*
 maintenance of, 25-8
 on permeable strata, 25-36
 rockfill, 15-13, 25-41
 on rolled fill, 15-4
 in sea, 28-31
 subsurface, 28-9
 of till, 15-3
Debris collecting works, 41-13
Debris control, 33-16
Deep drilling, 11-2
Degree-days, 9-6
Deltas:
 glacial, 5-22
 Mackenzie River (N.W.T.), 33-11
 Mississippi River (La.), 6-17, *33-6*, 34-32
 Nile River (Egypt), 33-10
 Po River (Italy), 33-11
 Rhone River (France), 33-10
Denudation, 4-17
Deposits:
 aqueous, 5-12
 glacial, 5-12

Deposits (*Cont.*):
 interglacial, 5-16
 marine, 5-12
 morainal, 5-21
 outwash, 5-21
Desanding works, 41-11
Desilting works, 41-9
Deterioration of timber piles, 18-25
Dew ponds, 28-26
Dewatering, 18-21
Diatoms, detection of, 2-14
Dikes, 4-3
 on permafrost, 23-17
 of till, 15-7
Diorite deposits, 4-11
Dip of rock, 4-12
Disappearing river, Yugoslavia, 26-4
Diversion of rivers, 26-21, 48-4
Diving, 24-38, 34-7
Divining of water, 8-23
Drainage:
 acid mine, 18-26
 in clay, 19-14
 of dock sites, 18-15
 early examples, 18-3
 effects of, 18-22
 of glacial lakes, 18-23
 gravity, 18-9
 importance of, 18-31
 of landslides, 39-19, 21
 locations of: Casapalca Mines (Peru), 18-11
  Combegrove (U.K.), 8-3
  Dismal Swamp (N.C.), 18-8
  English fens (U.K.), 18-4
  Kilsby Tunnel (U.K.), 18-5
  Lake Copais (Greece), 18-3
  Lake Fucino (Italy), 20-2
  Lake Tequesquitengo (Mexico), 18-28
  Leadville Tunnel (Colo.), 18-10
  Miami oolite deposits, 18-20
  Prisley Bog (U.K.), 18-5
  South African mines (South Africa), 18-22

Drainage, locations of (*Cont.*):
    Welland Canal (Ont.), 18-15
    Zuider Zee (Netherlands), 18-4
  natural, 18-7
  for roads, 18-5
  in rock excavations, 19-32
  by shaft, 18-12
  by wellpoints, 18-18
Drains:
  soakway, 18-7
  tile, 18-13
Dredging:
  coal, 23-4
  disposal, 49-3
  equipment, 34-32
  locations of: Baltimore (Md.) harbor, 33-12
    Flood Rock (N.Y.), 34-33
    Hallsands (U.K.), 34-42
    Lake Okeechobee (Fla.), 34-33
    Pacific islands, 34-34
    St. Lawrence Seaway (U.S.-Canada), *34-34*
    Sunderland (U.K.), 34-34
  for reservoirs, 26-*16*
  of river silt, 41-10
  for roads, 30-8, 16
Drawings, 10-21
Docks:
  abandoned, 34-10
  Dover (U.K.), *34-12*
  San Diego (Calif.), 18-15
  Southampton (U.K.), 3-11, 18-15
  underground, in Sweden, *21-19*
Dolerite deposits, 4-11
Dolomite, cavities in, 38-2
Drift, 5-23
Drilling:
  deep, 11-1
  ocean, 11-2
  off-shore, 6-18
Dublin (Ireland), port project in, 35-13
Dust, volcanic, 4-3

Earth, physical characteristics of, 3-5, 12-2

Earthquakes:
  artificial, 12-5, 45-19
  damage from, 45-8
  and dams, 25-28
  effect on groundwater, 8-6
  intensity, 45-14
  minor effects, 45-11
  notable: Alaska (1964), 45-8
    Caracas (Venezuela, 1967), 45-7
    China (1556), 45-6
    Cornwall (Ont., 1944), 45-7
    Hebgen Lake (Mont., 1959), 45-14
    Quebec (1925), 45-7
    San Francisco (Calif., 1906), 45-7
  prediction of, 45-18
  recording of, 45-10
  and soil liquefaction, 45-11
  world distribution, 45-12
Edinburgh (U.K.):
  castle and rock, 40-23
  churchyards, 15-25
Egypt:
  dams in, 25-2, 24, 27-17, 48-3
  Qattara Depression, 48-5
  stonework in, 15-20
  tracing leakage, 26-9
  tunnels in, 20-2
  underground temples, 21-3
Electrical prospecting:
  and geophysical methods, 12-9
  and resistivity surveys, 12-11
Electroosmosis:
  at Bay City (Mich.), 19-28
  initiation of, 19-28
  Mahoning River Dam (Ohio), *19-29*
  for slope stability, 23-19
  for stabilizing silt, 23-20
El Salvador: bridge site drainage in, 18-20
  railroad excavation in, 19-32
Embankments, fill for, 19-30
Energy:
  conservation and underground space, 21-2, 26

Energy (*Cont.*):
  worldwide situation, 23-1
Engineering Foundation, 47-10
Engineering geology maps, 14-9, 10, 12, 14, 15
Environmental impact statements, 1-5, 49-12, 14
Eötvös torsion balance, 12-8
Erosion:
  bed retrogression, 41-7
  coastal: geological aspects, 34-18
    of Great Lakes (U.S.-Canada), 34-18
    at Lyme Regis (U.K.), 34-18, *34-19*
    worldwide, 34-17
  below dams, 41-5
  of Scarborough bluffs (U.K.), 16-11, 19-31
  by streamflow, 41-2
Erratic blocks, 5-15, 24
Escarpment, spoiling of, 49-5
Eskers, 5-22
Estuaries, tidal:
  Fraser River (B.C.), *34-31*, 34-32
  Garonne River (France), 34-30
  at Liverpool (U.K.), 34-29, *34-30*
  maintenance of, 34-28
  at New York (N.Y.), 34-29
Excavated material, disposal/use of, 35-9
Excavations, 19-2, 50-5
  for airfields, 32-6
  classifications of, 1-14
  costs of, 19-10
  locations of: Carlton Center (South Africa), *19-17*
    Los Angeles (Calif.), *19-15*
    New York (N.Y.), *19-18*
    Panama Canal (Panama), *19-14*
    Toronto (Ont.) subway, 19-2
  nuclear energy used for, 19-39
  in rock: in general, 19-31, 32

Excavations, in rock (Cont.):
   methods, 19-33
   shale, 19-33
   in sinkhole areas, 19-36
 in soil: in general, 19-11
   methods, 19-13
   supports for, 19-17
   unit prices, 19-5
   volume of, 48-2
Exeter (U.K.), medieval bridge at, *13-18*
Exfoliation of rock, 5-2, *5-4*

Fabrics, geotechnical, 49-5
Failures:
   from bed erosion, 41-7
   bridge foundation, 24-4, 9
   bridge piers, 24-39
   coal tips, 48-15
   cofferdams, 12-21
   of contractors, 19-5
   foundation, 22-7
   locations of: Aberfan (U.K.), 48-14
      Arapuni Plant (New Zealand), *23-11*
      Austin Dam (Tex.), 25-3
      Baldwin Hills Dam (Calif.), 25-6
      Buffalo Creek Dam (W.Va.), *48-15*
      Cedar Reservoir (Wash.), 26-7
      Gleno Dam (Italy), 25-39
      Hales Bar Dam (Tenn.), 25-33
      Hallsands (U.K.), 34-42
      Irish Canal (Ireland), 29-4
      Islam Weir (Pakistan), *41-5*
      Malpasset Dam (France), 25-4
      Puentes Dam (Spain), 25-2
      St. Francis Dam (Calif.), 25-2, *25-3*
      Sutton Bridge dock (U.K.), 24-10
      Teton Dam (Idaho), 25-7
      Transcona grain elevator (Man.), 22-30, *22-31*

Failures, locations of (Cont.):
      Wheeler Dam lock (Tenn.), 29-12
   pressure tunnels, 20-20
   and site investigations, 10-5
Fans, alluvial, 5-13
Faults:
   competent, 44-3
   and dam foundations, 25-28, 44-10
   dating, 44-12
   detection, 44-6
   displacements on, 44-2, 5
   effects of, 44-2
   evidence of, 44-13
   in excavations, 19-8
   importance of, 44-1
   locations of: Arizona, 44-13
      Beechwood Dam (N.B.), 25-40
      Bridge River Tunnel (B.C.), 12-12
      Delaware water supply tunnel (N.Y.), 20-12
      Folsom Dam (Calif.), *25-47*
      Great Glen (U.K.), 44-3
      Hayward (Calif.), 44-5
      Houston (Tex.), 44-13
      India, 44-6
      Mossend (U.K.), 4-16
      New Zealand, 44-6
      Ontario, 25-15
      Owyhee Dam (Oreg.), 25-30
      Red Canyon (Mont.), 44-8
      Red Point (N.B.), 44-7
      Rodriguez Dam (Mexico), *25-31*
      Rosamund (Calif.), 44-13
      San Andreas (Calif.), 12-20, *44-4, 5*
      Seminoe Dam (Wyo.), 25-46
      Sierra Madre (Mexico), 45-18
      South Africa, 44-15
      Wasatch (Utah), 44-13
   and nuclear plants, 44-11
   passive, 44-3
   surveys for, 44-7
   trenching for, 44-12

Faults, locations of (Cont.):
   and tunnels, 44-8
   types, 4-15, 44-2
Fauna, problems with, 49-13, 15
Federal Highway Act (U.S.), 50-12
Federal Housing Act (U.S.), 47-17
Federal Power Commission (U.S.), 25-13
Felsite deposits, 4-11
Fens, drainage of (U.K.), 18-4
Fetch, 34-2
Field testing of soil and rock, 11-19
Filled ground, 13-17
   for airfields, 32-1
   from disposal of excavated material, 19-18, 35-9
Filter sand, 15-12
Finland:
   road drainage, 18-9
   rock classifications, 15-15
Fire extinguished by grouting, 27-13
Flint deposits, 4-9
"Floating" foundations, 22-23, 49
Floating subway tunnels, 31-30
Flooding in subway tunnels, 18-28
Floodplains:
   control of, 46-15
   danger of, 46-14
   locations: Etobicoke Creek (Ont.), 46-15
      Prairie du Chien (Wis.), 46-15
      Rapid City (S.Dak.), 46-14
   use of, 46-15
Floods:
   and catchment areas, 46-4
   causes of, 46-6
   and climate, 46-6
   extreme, 28-5
   and ground conditions, 46-8
   locations: Denver (Colo., 1973), 46-8
      Mackenzie River (N.W.T.), 46-7

Floods, locations (*Cont.*):
  Mississippi River (U.S.), 33-4, *33-7*
  Nile River (Egypt), 33-10
  North America (1936), 46-6
  North Sea (1953), 35-4, 46-2
  Pakistan (1970), 46-1
  Pennsylvania, 46-17
  Putnam (Conn., 1955), *46-7*
  Red River (Man., 1950), 46-6, *46-7*
  Vanport (Oreg., 1948), 46-16
  prevention, by upstream engineering, 49-17
  protection from: Albuquerque (N.Mex.), 46-10
  Boston (Mass.), 46-9
  Chicago (Ill.), 46-11
  conservation areas, 46-9
  Denver (Colo.), 46-10
  Field (Ont.), 46-10
  London (U.K.), 46-13
  Pullman (Wash.), 46-10
  warning instruments, 31-25
  Winnipeg (Man.), 46-10
Floodways, effects of, 48-6
Flora, problems with, 49-2
Flow nets for dams, 25-38
Fly ash, 17-9
  disposal of, 23-5, 48-12
  use as fill, 48-12
Folding of sedimentary rocks, *4-14*, 4-15
Footwalls of faults, 4-16
Forensic geology, 42-17
Forests, submerged, 37-4
Fossil fuels, 23-2
Fossils, 2-14, 3-11, 30-27, 50-7, 13
Foundations:
  adjustable, 22-46
  building, 22-1
  caisson, 22-7, 25
  carried to rock, 22-6
  over coal workings, 22-39
  dam, 25-1
  on fill, 22-29
  flexible, 22-40
  "floating," 22-23

Foundations (*Cont.*):
  garage, 22-46
  geological factors, 22-2
  and groundwater, 22-13
  overloaded, 22-12
  on permafrost, 36-16
  piled, 22-20
  powerhouse, 23-1
  precast, 22-26
  preloading, 22-27
  on rock, 22-4
  seismic factors, 45-18
  settlement of, 22-30
  on sloping ground, 22-44
  for small buildings, 22-49
  on soil, 22-10
  three-point bearing, 22-39
  and trees, 22-18
France:
  dams in, 25-22, 27-12
  Donzère-Mondragon hydroelectric project, 18-30
  tidal power, 23-9
  tunnels in, 20-28
  underground stations, 21-12
  wine storage areas, 21-4
Frank "slide" (Alta.), 40-3
Freezing index, 9-9
Freezing process:
  early examples, 19-25
  in excavations, 19-25
  Milchbuck Tunnel (Switzerland), 20-50
  for railroad tunnels, 31-9
  in shaft sinking, 19-25
  to stop landslides, 19-27
  in tunnels, 20-36, 31-9

Gabbro deposits, 4-11
Gabions, 33-17
Garages:
  in Japan, 22-46
  in Los Angeles (Calif.), 19-15
  in Ottawa (Ont.), 13-12
  sunk in place, 22-27
  in Switzerland, 22-47
  underground, 21-24
Garbage:
  compressed blocks of, 35-14

Garbage (*Cont.*):
  dumps, 48-9
  General Accounting Office (U.S.), 22-50
  General Services Administration (U.S), 22-50
Geochemistry, 3-3
Geological maps, 3-3, 14-1
Geological Society of America, 3-5, 21-26, 23-13, 44-11
Geological Society of London, 3-3, 50-2
Geological surveys:
  Canada, 3-2, 13-9, 34-32
  France, 13-20
  India, 17-4
  United Kingdom, 1-9, 10-19, 14-8, 15-17, 19-8, 20-41, 23-17, 39-21
  United States, 2-16, 4-6, 7-8, 13-9, 14-10, 19-6, 30-5
Geology:
  beneath cities, 13-1
  cycles, 3-7
  education, 3-14
  extraterrestrial, 3-5
  field work, 3-13, 14-3
  forensic, 42-17
  history, 3-1
  outline of, 3-3
  physical, 3-4
  regional, 10-10
  and soil mechanics, 6-16
  structural, 3-4
  structures, 7-6, 8-12
  succession, 3-9
  surveying, 10-19
  terminology, 1-14
  time scales 3-10
Geomorphology, 3-4
Geophysics, 3-4, 12-1
Geothermal heating in Iceland, 23-7
Geothermal power:
  California, 23-6
  Italy, 23-5
  New Zealand, 23-6
Geysers, 23-7
Ghana, bridges in, 24-17
Ghyben-Hertzberg effect, 8-19, 28-33

S-10   SUBJECT INDEX

Gibraltar, water storage facilities in, 21-22
Glaciers:
  action of, 5-15
  boulders, 5-24
  continental, 5-17
  deltas of, 5-22
  deposits by, 5-19
  effects of, 5-22
  erosion caused by, 5-17
  extent of, 5-16
  moraines, 5-20
  mountains, 5-17
  notable: Hugh Miller (Alaska), 5-20
    Pattullo (B.C.), 5-12
    Taweche (Nepal), 5-21
    Yentna (Alaska), 5-5
  soils, 5-19
  striae, 5-18
  valley trains, 5-20
  valleys, 5-17
Glen Shira project (U.K.), 2-6
Glomar Challenger, 11-2
Gneiss, 4-7, 9
Gold deposits, 16-13, 18-23, 48-12
Gondwanaland, 45-1
Gorges:
  Avon (U.K.), 40-16
  Hoover Dam (Ariz.-Nev.), 25-21
  Lower Notch (Ont.), 25-15
  Merwin Dam (Wash.), 25-40
  Royal (Colo.), 31-5
  Yangtze River (China), 33-2
Grain elevators, foundation failures of, 22-30
Granite, 4-10, 5-9, 15-14
Granitic structures, 4-10
Grasses for airfields, 32-9
Gravel (*see* Sand and gravel)
Gravimeters, 12-8
Graving docks, 18-15
Great Lakes (U.S.-Canada), tilting of, 37-6
Greece:
  canals in, 29-2
  Lake Copias, drainage of, 18-3

Greece (*Cont.*):
  water power projects in, 26-8
Greenland, ice cover of, 5-17
Groins, permeable, 34-25, 26
"Ground truth," 10-17
Groundwater:
  annual variation, 8-5
  in Boston (Mass.), 18-25
  characteristics, 8-3
  in clay and shale, 8-8
  control of: under buildings, 22-16
    Mosel River, 33-21
  in crystalline rocks, 8-8
  depletion of, 9-3, 48-7
  discoveries, in Canada, 28-19
  distribution, 8-13
  effects of pumping, 8-21, 18-17
  and foundations, 22-13
  gases in, 8-9
  and geophysics, 12-19
  in Hanford facilities (Wash.), 48-16
  hardness, 8-10
  in Hershey Valley (Pa.), 18-22
  Historical aspects, 8-2
  hot, 8-4, 9
  importance of, 1-18
  impurities in, 8-12
  in limestone, 8-8
  lowering of, 18-25
  movement of, 8-5, 21
  for power stations, 23-2
  quality of, 8-8
  and radioactive waste, 48-16
  recharging, 22-14, 28-20, 48-7
  replenishment, 28-21
  rising, on Long Island (N.Y.), 18-28
  and rock mechanics, 7-6
  in sand and gravel, 8-7
  in sandstone, 8-8
  near sea, 8-19
  seawater contamination of, 8-20
  and site investigation, 10-9
  specific yield, 8-25
  subdivisions of, 8-4

Groundwater (*Cont.*):
  sulfate bearing, 16-14
  surveys, 8-22
  trace elements in, 8-11
  in tunnels, 20-33, 36
  use of, 8-1
  beneath Venice, 37-10
Grouting:
  with asphalt, 27-4
  and bridge design changes, 27-13
  with cement, 27-5
  chemical, 19-25
  for foundations, 27-17
  Joosten process, 27-16
  in New York (N.Y.), 22-8
  for tunnels, 20-32
  with clay, 27-2
  for cofferdams, 27-13
  high pressure, 27-13
  historical aspects of, 27-2
  Intrusion-Prepakt system, 27-13
  locations: Abitibi Canyon Dam (Ont.), 27-6
    Beechwood Dam (N.B.), 27-10
    Chickamauga Dam (Tenn.), 27-9
    Dartford Tunnel (U.K.), 27-17
    Dokan Dam (Iraq), 27-12
    Estacada Dam (Oreg.), 27-5
    Foum el Gherza Dam (Algeria), 27-11
    Grand Rapids hydroelectric project (Man.), 27-18
    Great Falls Dam (Tenn.), 26-11
    Hales Bar Dam (Tenn.), 23-33
    Hoover Dam (Ariz.-Nev.), 27-10
    Keban Dam (Turkey), 25-42
    Kinder Embankment (U.K.), 27-5
    Norris Dam (Tenn.), 27-8
    St. Pierret-Cognet Dam (France), 27-12
    Scout Dyke Dam (U.K.), 27-6

Grouting, locations (*Cont.*):
  Serre-Ponçon Dam
    (France), 25-23
  Severn Tunnel (U.K.),
    1-16, 20-32
  precementation, 27-6
  quantities used, 27-12
  special systems, 27-13
  in tunnels, 20-31
Guatemala, railroad excavations in, 19-32
Guyana, "floating" foundations in, 22-24
Gypsum:
  in beer, 28-6
  dams on, 43-7
  for plaster, 17-7
  under reservoirs, 26-3

Hackensack meadowlands (N.J.), development of, 35-8
Hade of fault, 4-15
Halloysite deposits, 6-15
Hallsands (U.K.), loss of, 34-42
Hanging valleys, 5-18
Hanging wall of fault, 4-16
Harbors:
  Duisberg (West Germany), 2-18
  Erie (Pa.), *34-22*
  Ischia (Italy), 34-14
  natural, 34-14
  Ostia (Italy), 34-11
  Palm Beach (Fla.), *34-28*
  St. Helier (Jersey), 34-34
  Taconite (Minn.). *34-12*
  Toronto (Ont.), *34-20*
  Vizagapatam (India), *34-23*
Hardness of water, 8-10
"Hardpan," 1-14, 22-7
Heavy media separation for aggregate, 16-10
Hogsback Cut (Colo.), *50-11*
Hong Kong:
  residual soils, 7-14, 42-4
  roads, 30-26
  water supply, 28-31
Hope "slide" (B.C.), *40-4*
Hospitals:
  in New York (N.Y.), 8-4, 19-18

Hospitals (*Cont.*):
  in Pittsburgh (Pa.), 22-8
  Portland (Oreg.), 22-5
  in St. Helier (Jersey), 16-15
  underground, 21-9
Hot springs, 45-2
Hotels:
  in Cuba, 27-13
  in Denver (Colo.), 16-7
  in London (U.K.), 22-54
  in Mexico, 18-29
  in Switzerland, 40-18
  in Victoria (B.C.), 22-37
Hurricanes, 2-10, 33-14, 34-39, 46-2, 8, 9, 15
Hydroelectric projects (*see* Water power projects)
Hydrological cycle, 9-2
Hythergraphs, 9-7

Ice:
  bridges, 36-19
  ground, 36-15
  Pleistocene cover, 37-5
  problems with, 36-18
  sea, 36-19
  wedges, 36-18
Iceland:
  geothermal heating in, 23-7
  magma eruptions in, 45-3
Ilmenite deposits, 16-13
Incas:
  roads of, 30-1
  stone work of, 15-24
India
  brick plants, 17-4
  dams, 25-11, 26
  harbors, 34-23
  irrigation projects, 41-10
  landslides, 39-6
  railroads, 31-11
Industrial minerals, 15-2, 25
Inspection, 1-16
  of bridge foundations, 24-27, 37
  of dams, 25-8
Interglacial deposits, 5-16, 19-2
International Association of Engineering Geology, 14-15, 38-10, 43-7

International Glaciological Society, 36-20
International Organization for Standards, 1-15
International Society of Rock Mechanics, 7-1
International Society of Soil Mechanics and Foundation Engineering, 6-2
International Union for Quaternary Research, 5-32
Intrusion-Prepakt method, 20-20
Iran, dams in, 25-44
Iraq:
  bridges in, 24-41
  dams in, 27-12
Ireland, port projects in, 35-13
Irrigation projects:
  India, 41-11
  Salt River (Ariz.), 18-13
Islands, artificial, 24-26, 34-41, 49-3
Isostasy, principle of, 12-8
Israel:
  Dead Sea power project, 48-5
  Haifa harbor, 34-11
Italy:
  geothermal power, 23-5
  harbors, 34-11
  highways, 30-25
  sea level changes, 37-4, 9
  underground stations, 21-10

Japan:
  building foundations in, 22-25, 47
  garages in, 22-46
  land conservation, 49-21
  tunnels in, 18-11, 20-18, 29
Jökulhlaup, 31-24, 46-3
Jordan, archeological excavations in, 19-40, 21-3
JOIDES program, 11-2
Jointing:
  columnar, *4-13*
  in granite, 4-10
  of igneous rocks, *4-13*

Joints:
  effect on porosity, 8-7
  slippage on, 44-14
  types of, 44-15

Kames, 5-22
Kansas City (Mo.), underground development in, 21-7, *21-8*
Kaolinite deposits, 6-15
Kaolinization of bedrock, 43-8
Karst fields, 38-9
  detection of sinkholes, 38-12
  geophysical surveys, 38-9, 13
  locations: in Czechoslovakia, 38-11
    in North America, 38-10
  radiometric surveys, 12-21
Karstic limestone:
  in cofferdams, 24-30
  at Keban Dam (Turkey), 25-43
Katavothras for drainage, 18-4
Kuwait, water supply in, 28-33

Lahars, 31-22
Lambert projection, 7-14
Land:
  covered-up, 48-1
  limitations of, 47-3
  restoration of, 49-2
  rising, 45-5
Land reclamation:
  in Bombay (India), 35-14
  in Dublin (Ireland), 35-13
  in Evanston (Ill.), 35-9
  in Hackensack meadowlands (N.J.), 35-7
  in Japan, 35-10, 13
  in Liverpool (U.K.), 35-13
  in Milwaukee (Wis.), 35-10
  in Portsmouth (U.K.), 35-12
  in Prescott (Ont.), 35-9
  in Puerto Rico, 35-7
  in Toronto (Ont.), 35-9

Land reclamation (*Cont.*):
  in West Miffling (Pa.), 35-7
  in Zuider Zee (Netherlands), 35-1
Land subsidence:
  backfilling for, 37-18
  designing for, 37-19
  locations: in Duisburg (West Germany), 2-18, 37-23
    off Florida coast, 37-4
    in Goose Creek (Tex.), 37-8
    in Houston (Tex.), 37-12
    in Japan, 37-11
    in Kutna Hora (Czechoslovakia), 37-14
    in London (U.K.), 37-7
    in Long Beach (Calif.), 37-8
    in Mexico City (Mexico), 37-11
    near New Madrid (Mo.), 3-8
    in Pittsburgh (Pa.), 37-17
    in Pozzuoli (Italy), 37-4
    in Venezuela, 3-8, 37-9
    in Venice (Italy), 37-9
  from mining, 37-14
  natural, 37-4
  from pumping, 37-6
  remedial measures for, 37-9
Landfills:
  leachate from, 48-10
  sanitary, 48-9
Landslides:
  causes of, 39-11
  classifications of, 39-2
  drainage for, 39-19
  dried-out, 39-22
  early studies of, 39-4
  guidelines for, 39-24
  in leda clay, 39-7
  locations of: Beaver County (Pa.), 47-16
    Bonneville (Oreg.), 23-14
    Chekamus (B.C.), 23-14
    Downie (B.C.), 26-13
    Farmer's Union (Colo.), 39-5
    Folkeston Warren (U.K.), 39-19

Landslides, locations of (*Cont.*):
  on French canals, 6-2, 39-3
  Gohna (India), 39-6
  Gros Ventre (Wyo.), 39-6
  Hanlova (Czechoslovakia), 39-15
  on Kiel Canal (West Germany), 6-2
  in New Zealand, 39-12
  Panama Canal, 6-2, 29-6
  Portland (Oreg.), 22-45
  Red Deer River (Alta.), 10-13
  St. Jean-Vianney (Que.), 39-9
  on Swedish railroads, 6-2
  Thompson River (B.C.), 39-6
  Vaiont Dam (Italy), 26-13
  Waikaremoana (New Zealand), 45-3
  natural, 39-5
  prevention of, 39-16
  remedial measures for, 39-16
  in sensitive clay, 39-8
  submarine, 34-36
Laterite deposits, 5-9, 42-6
Les Eboulemonts (Que.), landslides near, 39-8
Levees, natural, 5-13
Leakage:
  detection of, 26-8
  in reservoirs, 26-6, 7, 10
Lighthouses, offshore, 34-39
Lightships, 34-38
Lime, 17-6
Limestone, 4-9
  cavernous, 38-2, 43-5
  oölitic, 8-7
  pitting of, 15-19
  and reservoirs, 26-3
  sinkholes in, 38-4
  solution of, 38-1
Liquefaction of soil, 45-11
Littoral drift:
  bypassing, 34-25, *34-27*
  "in reverse," 23-5
  in India, 34-23
  at Rockaway Point (N.Y.), *34-23*

Littoral drift *(Cont.)*:
  in Toronto (Ont.), 16-11, 34-20
Load tests on bridge piers, 11-20
Locks, 29-11, 33-22
Loess, 5-10, 42-10
  and foundation, 42-11
  and roads, 30-10
  in San Joaquin Valley (Calif.), *42-11*
London Basin (U.K.), 8-14

Machinery foundations, 12-21
Mackenzie River Valley (N.W.T.), proposed pipeline in, 10-17
Magma, 45-3
Magnetic factors, 12-3
Magnetometers, 12-4, 21
Malenclaves, 48-10
Malta, underground space in 20-2
Manhattan (N.Y.), early geology of, *13-3*
Maps:
  end paper, 14-6
  engineering geology, 14-6, 14
  geological, 3-13, 10-19, 14-1
  trafficability, 2-16
  volcanic hazard, 45-5
Marble, 4-9, 15-23
Marine deposits, 5-14
Mass wasting, 39-1
Mediterranean Sea, salinity of, 48-4
Meres, 38-2
Methane:
  from landfills, 48-11
  problems with, 48-13
  in reservoirs, 26-18
  in tunnels, 20-38
Metric system, 1-15
Mexico:
  building settlements, 22-36
  dams in, 25-31
  land subsidence in, 37-11
Military engineering, 2-12
Millstone grit, 15-15

Mineralogy, 3-3
Minerals, 3-6
  clay, 5-7
  common, 4-3
  industrial, 15-2
  relative hardness of, 4-8
  weathering of, 5-6
Mines:
  Broken Hill (South Africa), 3-11
  Cripple Creek (Colorado), 20-2
  Steep Rock (Ont.), 19-14, 49-3
Mining:
  chemical grouting, 27-14
  drainage, 18-26
  hydraulic, waste, 17-9
  open pit rehabilitation, 49-11
  salt, 21-9, 36-6
  scars from old, 49-10
  strip rehabilitation, 49-12
  undersea, 34-35
  use of old, 21-6
Models, use of, 2-6, 8, 11-23, 24, 34-32
Mohole project, 11-2
Moon dust, 12-20
Montmorillonite deposits, 6-15, 42-3
Moraines:
  glacial, 5-20
  parallel, 5-23
  for railroad crossing, 31-2
  terminal, 5-22
Mud damage from floods, 46-3
Mud runs, 31-16, 39-10, 46-3
Multiple tills, 42-13
Muskeg, 5-29
  from air, 10-15
  in north, 36-13
  removal of, 36-14

N-value, Weinert's 42-4
National Academy of Sciences, 7-15, 29-6, 30-2, 36-6
National Building Code (Canada), 45-17
National Bureau of Standards, 15-22, 25-25

National Coal Board (U.K.), 37-23, 49-11
National Oceanic and Atmospheric Service (U.S.), 9-7
National Research Council of Canada, 45-17
Natural hazards, 47-9, 10
Netherlands:
  Delta project, 46-2
  Ijmuiden lock, 29-12
  Rotterdam subway, *31-28*
  Zuider Zee drainage, *35-1*
New Zealand:
  Arapuni plant, *23-10*
  dams in, 25-46
  earthquakes in, 45-3
  geothermal power, 23-6
  railroads in, 31-21
  slope instability in, 39-13
Niagara Falls (N.Y.-Ont.):
  dewatering, 33-23, 40-14
  erosion of, 41-4
Nickel deposits, 18-6
Nigeria, railroads in, 31-3
Nitrogen in tunnels, 20-39
Norfolk (U.K.), clay beach at, 2-16
North American Air Defense Command, *21-20, 21*
North Sea:
  floods, 35-4, 46-2
  soil studies, 42-17
Norway:
  building foundations in, 22-7
  Ekofisk offshore structures, 34-40
  rockfalls, 40-11
Nuclear energy:
  for excavations, 19-39
  fuel disposal, 48-17
Nuclear power plants:
  dismantling of, 48-16
  and faults, 44-11
Nuclear Regulatory Commission (U.S.), 19-39, 23-30

Obsidian, 4-11
Ocean, depths of, 3-5
Offshore structures:
  Ekofisk (Norway), 34-40

Offshore structures (Cont.):
  geotechnical studies for, 34-40
  for oil drilling, 34-39
Ogden (Utah), artesian wells in, 28-32
Oölite deposits, 18-21
Outwash:
  deposits, 5-21
  plains, 5-20
Overbreak in shale tunnels, 20-26
Overpumping, 37-7, 13
Overthrust belt (Wyo.), 6-7
Owens Valley (Calif.) aqueduct, 8-10

Pakistan:
  bridges in, 24-18
  dams in, 41-5
Paleosoils, 42-2
Palmdale "bulge" (Calif.), 37-3
Panama Canal (Panama), 6-2, 29-5
Pavement on clay, 42-17
Pearl Harbor (Hawaii), underground fuel storage at, 21-15
Peat, effect on water, 28-5
"Pebble marker," significance of, 19-17, 42-7
Pedology, 4-2, 5-7
Penetration tests, 24-32
Pennsylvania, rock characteristics of, 7-8
Permafrost, 5-29
  airstrips on, 36-16
  and Alaskan railroads, 31-13
  in Canada, 36-6
  beneath dikes, 23-17
  effects on rocks, 43-10
  explanation of, 5-30, 9-7, 36-4
  extent of, 36-6
  fossils in shafts, 20-39
  foundations on, 36-16
  patterned ground, 36-9
  problems with, 5-31
  terminology, 36-4
  terrain, in Northwest Territories, 5-30

Permafrost (Cont.):
  utilidors on, 36-17
Peru:
  jökulhlaup in, 46-3
  mine drainage in, 18-11
  railroads in, 31-7
Petrography, 17-11
Petrology, 3-4
Phenocrysts, 4-10, 4-11
Photogrammetry, 10-16
Photography:
  in boreholes, 11-16
  early processes, 25-15
Phreatophytes, 8-23
Physical geology, 3-4
Pile driving, effects of, 19-14
Piles:
  for foundations, 22-20
  steel sheet, 19-14
  timber, deterioration of, 18-25
Pingoes, 5-30
Pipelines:
  Alaskan, 36-10
  in Arctic, 36-21
  Mackenzie River Valley (N.W.T.), proposed, 36-10
Pisa, Leaning Tower of (Italy), 22-33
Pleistocene geology, 3-4
  canals in Germany, 50-7
  importance of, 5-32
  of Mississippi River delta (La.), 6-18
Plains, outwash, 5-20
Plants and groundwater, 8-23
Plaster from gypsum, 17-7
Plate tectonics, 3-4, 43-4
Pollen analyses, 50-4
Pollution:
  air, 48-8
  thermal, 48-6
  water, 33-24, 48-7
Pop-ups, quarry, 7-12
Porphyries, 4-11
Potholes:
  at Assouan Dam (Egypt), 41-3
  at Noxon Rapids Dam (Mont.), 25-39

Power stations, underground, 24-10
  in Canada, 21-13, 14, 15
  Kings River (Calif.), 21-12
  Marlengo (Italy), 41-11
  Snowy Mountain (Australia), 2-2
  in Sweden, 21-10
  in United Kingdom, 2-7, 21-10
Powerhouse foundations, 23-2
Pozzolan, 16-3
Prague (Czechoslovakia):
  building stones, 15-23
  urban geology of, 13-11
Precementation, 27-6
Preglacial drainage, 2-11
Preliminary studies, 25-17, 27
Preloading:
  for bridge piers, 24-32
  for building foundations, 22-27
  for power stations, 23-23
Pressure tunnels, problems with, 20-19
Prospecting:
  geobotanical, 10-20
  geochemical, 10-20
  geophysical, 12-3
Puddle clay, 15-10, 11
Puerto Rico:
  Arecibo telescope, 19-38
  market areas, 35-7
Pumice, 4-11
Pumped storage plants:
  in Belgium, 21-13
  in Switzerland, 21-12
  in United Kingdom, 23-17
  in West Germany, 21-12
Pumping:
  effect on subsidence, 37-11
  rock slurry, 20-49
Pumps, submersible, 18-14
Pyramids (Egypt), 15-20
Pyrite deposits:
  oxidation of, 16-16
  in shale, 43-7

Quarries, 19-37
  Haifa harbor (Israel), 34-11
  Hershey Valley (Pa.), 18-22

Quarries (*Cont.*):
  Hiwassee Dam (N.C.), 19-38
  Montclair (N.J.), *16-19*
  pop-ups, 7-12, 43-4
  Rock of Ages (Vt.), 15-20
  Salt Lake (Utah), 19-38
  San Gabriel Dam (Calif.), 25-40
  use of abandoned, 21-16
Quartz, 4-4
Quartzite, 4-9
Quaternary geology, 3-4, 5-32
Qattara Depression (Egypt), 48-5
Quicksand, 18-5, 19-17

Radiometers, microwave, 12-20
Radon from wells, 37-4
Railroads:
  ballast, 31-14
  over bogs, 30-9
  construction, 31-5
  over gorges, 31-5
  locating, 31-2
  maintenance of way, 31-10
  new, 31-1
  notable: in Australia, 31-14
    in Austria, 31-15
    in Canada, 1-13, 11-19, 18-23, 20-16, 23-16, 24-13, 32, 39, 30-7, 31-3, 8, 9, 20, 24, 39-11, 40-20, 42-15
    in Chile, 20-6
    in El Salvador, 19-32
    in France, 20-28
    in Guatemala, 19-32
    in India, 31-11
    in Netherlands, 31-7
    in New Zealand, 31-21, 39-13
    in Nigeria, 31-3
    in Peru, 31-7
    in Switzerland, 31-2, 17, 18, 40-15
    in U.S.S.R., 36-22
    in United Kingdom, 1-16, 13-13, 18-5, 19-8, 20-23, 32, 43, 27-13,

Railroads, notable (*Cont.*):
    31-4, 6, 11, 14, 34-8, 11, 39-19, 40-5, 6, 16
    in United States, 19-38, 20-28, 32, 26-7, 27-2, 31-5, 6, 8, 13, 14, 15, 24, 33-22, 39-12, 49-7
  protective fences for, 31-14
  relocation, 31-24
  track inspection, 31-14
  tunnels, 31-6
Rainfall:
  measurement, 9-4
  variation of, 9-3
Raised beaches, 34-6, 37-4
Regional geology, 10-10
Regoliths, 4-2
Rehabilitation:
  open pit mining, 49-10
  sand and gravel pits, 16-19, 49-8
Reservoirs:
  ancient landslides in, 26-13
  cleaning out, 26-15
  and coal mining, 26-12
  effects of, 48-3
  flooding of, 26-11
  geology of, 26-6
  leakage from, 26-2
  locations: Aguasabon (Ont.), *26-19*
    Baldwin Hills (Calif.), 25-6
    Boston (Mass.), 28-6
    Cat Creek (Nev.), 26-16
    Furnish Dam (Oreg.), 41-12
    Hondo (N. Mex.), 26-2
    Jerome (Idaho), 26-2
    Kensico (N.Y.), 20-12
    Madden (Panama), 27-3
    Malad (Idaho), 26-3
    Santa Felicia (Calif.), 26-17
    Silent Valley (U.K.), 28-6
  methane in, 26-18
  outlets from, 26-17
  peat in, 28-6
  perched, 26-19
  sealing of, 26-10

Reservoirs (*Cont.*):
  silting up, 26-13, 41-12
Residual soils, 5-7, 9
  in French Guiana, 6-13
  in Kenya, 6-13
  in South Africa, 42-5
  in Yukon, 36-2
Reversed river flow, 2-11
Revetment, stone, *15-14*
Rhabdomancy, 8-23
Rhine Graben (West Germany), 20-40
Rhyolite deposits, 4-11
Rippers for rock excavation, 15-15
Ripple Rock (B.C.), 11-17, 34-35
Riverbed retrogression, 41-7
Rivers:
  aggrading of, 5-13
  canalization of, 33-18
  disappearing, 24-14
  diversion of, 26-21
  training: bank protection, 33-17
    debris control, 33-16
    historical aspects, 33-2
Riverside (Calif.), hot groundwater in, 8-9
Roads:
  access, in tunnels, 25-44
  adjustable for subsidence, 37-21
  and archeology, 30-27
  over bogs, 30-9
  of calcined clay, 30-14
  construction of, 30-15
  damage, due to drainage lack, 30-9
  as dams, 30-20
  dredging for, 30-16
  and fossils, 30-27
  foundations for, 30-3
  guidebooks for, 30-3
  locations, 30-2, 4
    in Asia, 30-26
    in Bolivia, 30-24
    in Canada, 30-4, 24, 39-17, 21, 40-19
    in Finland, 18-9
    in Hong Kong, 30-26
    in Italy, 30-25
    in South Africa, 24-21
    in Spain, 30-24

Roads, locations (*Cont.*):
  in United Kingdom, 15-15, 18-7, 30-16, 21, 50-14
  in United States, 11-16, 12-16, 19-37, 20-32, 30-3, 5, 6, 10, 16, 17, 18, 22, 23, 25, 27, 34-22, 40-11, 50-11
  and loess, 30-10
  materials for, 30-12
  pumped fill for, 30-16
  ridge, 30-4
  rock asphalt for, 30-13
  seashells for, 30-14
  soil erosion from, 30-19
Roche Moutonné near Thessalon (Ont.), *5-19*
Rock:
  bolts, 40-9, 13
  broken, 15-12
  cores, 11-21
  cuts, 19-11
  discontinuities, 7-4
  drilling, close-lines, *19-33*
  engineering characteristics, 7-8
  extrusive, 4-3
  flour, 5-17, 6-10
  glaciers, 39-10
  hypabyssal, 4-3
  igneous, 3-6, 4-4
  intrusive, 4-3
  mechanical properties, 7-2
  mechanics: field measurements, 7-11
    field testing, 7-9
    in-situ tests, 7-10
    and joints, 7-13
    research, 7-15
    and structural geology, 7-7
  metamorphic, 3-7, 4-6
  movement of, 20-20, 25-27, 30-17, 33-19, 43-40
  permeability, 8-6
  porosity, 8-6
  for riprap, 15-14
  sedimentary, 3-6, 4-5
  strengths, 25-25
  stresses, in-situ, 7-12, 20-20, 25-27, 30-17, 43-40
  substance, 7-2
  testing, useless, 15-12

Rock (*Cont.*):
  uniformity, 7-4
  weathering, 7-4
  Young's modulus, 7-4
Rock-fill dams, 15-13
Rock quality designation, 7-9
Rockfalls and rockslides:
  concrete tunnels for, *40-10*
  and goats, 40-6
  locations: Big Creek plant (Calif.), 40-6
    Blackhawk (Calif.), 40-5
    Bristol (U.K.), 40-16
    Dover Cliffs (U.K.), 40-6
    Elm (Switzerland), 40-4
    Hope "slide" (B.C.), *40-4*
    Klosters (Switzerland), 24-38, 40-15
    in New Zealand, 39-14
    in New York, *44-14*
    Niagara Falls (N.Y.-Ont.). *40-13*
    in Norway, *40-11*
    Schoellkopf powerhouse (N.Y.), *40-8*
    Turtle Mountain (Alta.), *40-3*
    Vriog Cliffs (U.K.), *40-5*
  potential, 40-22
  protective works, 40-5, 8, 9, 17, 21
  and railroads, 31-14
  records of, 40-20
  on roads, 40-11
Romans:
  bathhouses, 50-13
  coastal works, 34-9
  drainage works, 18-3
  harbors, 34-11
  roads, 30-1
  tunnels, 20-2
  villas, 50-14
  water supply works, 28-2
Route location, 30-4
Royal Society, 12-4
Runoff, 28-3

St. Lawrence Seaway (U.S.-Canada), 15-7, 24-4, 29-9
  blasting experiments, 19-35
  excavations for, 19-3

St. Lawrence Seaway (U.S.-Canada) (*Cont.*):
  till stratification, 11-28
Sand dunes:
  engulfment by, *30-7*
  movement of, 49-6
  in Netherlands, 49-7
  in North Carolina, 40-8
  stabilization of, 49-7
Sand and gravel, 5-26, 27
  as aggregate, 16-4
  from air, 10-17
  for concrete, 16-10
  drains, 18-9
  for airfields, 32-5
  for roads, 30-9
  from dunes, 5-27
  for filters, 15-12
  in floodplains, 49-9
  impurities in, 16-5
  from lakes, 16-11
  loose, 42-12
  mineralogy of, 16-5
  mining from streams, 49-9
  pit rehabilitation, *49-8, 12*
  for powerhouse foundations, 23-27
  from rivers, 16-11
  from sea, 16-18
Sandstone, 4-9
  cross-bedded, 4-9
  pitting of, 15-19
Saprolite in tunnels, 20-39
Satellite photography, 10-18
Schist, 4-6, 9
Scouring:
  at bridge piers, 24-23
  of river beds, 24-41
Screes, 5-21
Sea:
  beach formation, 34-19
  changing levels, 34-6
  deep soundings, 34-5
  power of, 34-3, 42
  sediments from depths, 34-6
  submarine slides, 34-36
Sealing by ion exchange, 6-16
Seawater, intrusion of, 8-20, 15-10, 28-19
Sediment:
  in rivers, 41-4
  in sea, 34-6
Seiches, 34-5, 37-10

Seismology, 3-4
  forces, in design, 23-28, 24-22
  geophysical methods, 12-4, *12-6*
  prospecting, *12-7*
  regulations, 27-17, 45-16
  shocks, 34-35
  surveys, 23-17
  zones, 45-17
Serpentine, 4-9
Settlement:
  of bridge piers, 24-12, 19
  of buildings, 22-30
  over coal workings, 22-39
  prevention of, 22-36
Sewage:
  sludge disposal, 48-11
  underground disposal, 18-8
  and water supply, 28-23
Shafts and shaft sinking, 19-23
  in Catskill Mountains tunnel (N.Y.), 20-28
  geological factors, 50-8
  in Saskatchewan, *19-25*
  and shattered rock, 20-39
Shales, 4-8
  and bacteria, 43-8
  for bricks, 17-3
  in concrete, 16-10
  expanding, at dams, 25-50
  problems with, 21-26
  pyrite in, 43-7
  swelling, 42-3, *43-2*, 43-3
Shingle, 34-10, 42
Silt, 5-27
  dredging of, 41-10
  in harbors, 33-12
  in reservoirs, 41-12
  in rivers, 26-10, 15, 18
Sinkholes:
  and bridge collapse, *38-5*
  under crusher plants, *38-9*
  detection of, 38-13
  for drainage, *38-3*
  examples of, 38-1
    Cheshire meres (U.K.), 38-2, *38-6*
    "December Giant" (Ala.), 38-13
    in Florida, *38-6*
    in Pennsylvania, 18-22, 38-4, 8

Sinkholes (*Cont.*):
    in South Africa, *18-22*, 38-8
    in Windsor (Ont.), *38-6*
  filling of, 38-13
  from human activity, 38-4
  from pumping, 38-8
  from salt, 38-5
  used for telescopes, *19-36*
Site investigations, 10-3, 8
  adequacy of, 10-23
  costs of, 10-4
  detailed, 10-20
  and groundwater, 10-9
  and land subsidence, 37-22
  pattern of work, 10-7
Slag, 17-8, 10
Slate, 4-8
Slope of building site, 22-44
Slope protection, 39-3
  with fabrics, 49-5
  with wattles, 49-4
Sluicing:
  for excavation, 30-17
  silt from reservoirs, 26-16
"Slurry Mole," 20-29
Slurry-trench method, 19-19, 25-52
Soak holes, 38-3
Soil, 4-1
  aeolian, 5-10
  ancient, 42-2
  black cotton, 42-6
  characteristics, 5-26
  collapsible, 42-6
  compaction, 19-30
  conservation, 49-3, 19, 21
  for earth dams, 15-2, *15-9*
  erosion, 26-18, 30-19, 49-18
  in excavations, 1-14
  frozen, 36-14, 16
  glacial, 6-9
  lateritic, 42-6
  mechanics, 5-1
    analyses, 6-5
    and dike design, 15-8
    early, 6-2, 30-3
    experiments, 6-3
    and geology, 6-6
    testing, 6-4
    theories, 6-3
  organic, 5-28
  origins, 42-2
  preconsolidated, 6-11

Soil (*Cont.*):
  profiles, *5-8*, 5-26
  residual, 5-7, 6-12, 42-4
  sampling, in sea, 11-8
  sensitive, 42-13
  terminology, 5-26
  testing, 6-4
  transported, 5-10, 6-8
Soil Conservation Service (U.S.), 33-14, 49-20
South Africa:
  building foundations, 19-17, 22-22
  dams, 25-11
  roads, 24-21, 37-21
  soil weathering studies, 42-4
South Dakota, bentonite deposits in, 15-10
Spain:
  dams, 25-2, 54
  roads, 30-24
  underground streams, 21-5
  Urbasa Plain, 38-2
Specialists, use of, 1-7
Spits, recurved, 34-20
Spreading grounds for groundwater replenishment, 28-21
Springs, 8-15
  hot, 8-4
  intermittent, 8-16
Standard penetration test, 11-19
Steam power plants:
  Battersea (U.K.), 23-24
  Holtwood (Pa.), 23-4
  Johnsonville (Tenn.), 23-21
  in Louisiana, 23-23
  Sam Bertron (Tex.), *23-24*
  Shawnee (Tenn.), *23-20*
  South Bay (Calif.), *23-21*
  Warren (Pa.), *23-3*
Steel rail in pothole, 41-7
Stilling basins at dams, *41-6*
Stones, 15-21
Stratification of rock, 4-12
Stratigraphy, 3-4
Streams, 41-2, 8
Stress in rocks, 20-20, 25-27, 30-17, *43-4*
Striae, glacial, *5-18*
Strike of rock, 4-12
Structural geology, 3-4

Sturzstroms, 40-5
Subsurface investigations, 11-1
  contracts for, 11-25
  geological factors, 11-3, 29
  guidelines, 11-5
  interpretation, 11-26
  models, *11-23*
  reliance on, 11-28
  supervision of, 11-20
Subway systems:
  Amsterdam (Netherlands), 31-32
  Atlanta (Ga.), 11-11
  Brussels (Belgium), 31-28
  London (U.K.), 1-4, 31-26
  New York (N.Y.), 1-4, 18-28
  Paris (France), 31-28
  Rotterdam (Netherlands), *31-30*
  Toronto (Ont.), 13-16, 19-2, 19, 20-38, 31-28
  Vienna (Austria), *31-29*
Sudan, dams in, 41-5
Sulfates:
  attack on concrete, 16-14
  in groundwater, 8-11
Supports for excavations, 19-20, 21
Sweden:
  building foundations, 22-26
  civil defense measures, 21-18
  geophysical work, 12-17
  rock excavations, 19-34
  underground works, 21-11, *21-17*, 19
Switzerland:
  bridges, 40-15
  garages, 22-27, 47
  railroads, 31-2, 17
  tunnels, 20-4, 50
  underground works, 21-12
Syncline, 4-14, 15

Taiwan, dams in, 40-10
Taj Mahal (India), *22-34*
Talus, *5-11*
Tanks for excavations, 17-8
Television:
  in boreholes, 11-16

Television (*Cont.*):
  for detection of swallow holes, 30-16
Temperature: air, 9-6
  ground, 9-6, 21-2
  in tunnels, 20-27
  and weathering, 5-3
Terraces, river, *5-13*
Terrain analyses, 47-19
Test borings, 11-6, 7
  for Channel tunnel, 11-10
  geological use of, 50-2
  and geophysical surveys, 12-14
  inadequate, 19-16, 22-12, 51, 24-33
  television view of, 11-16
Test drillings, 11-13, 36-11
  inclined holes, 11-16
  inspection of, 11-17
  large diameter, 11-17
  records of, 11-21
  rock cores from, 11-21
Test embankments, 15-14
Test pits, 11-10, 19-5
Test shafts, 11-12
Test tunnels, 11-11, 12
Tests:
  for aggregate, 16-9
  on rock excavations, 15-15
Thermal pollution, 48-6
Throw of fault, 4-16
Tidal power projects:
  in Canada, 11-8, 23-8, 41-14
  in France, 23-9
  in U.S.S.R., 23-10
Tides, 35-2
  effect on groundwater, 8-6
  ranges, 24-4
Till:
  bouldery, 5-14
  compacted, 19-7
  composition of, 5-23
  for dams, 15-3
  excavations in, 1-14
  glacial, 5-20
  loose, 6-9, 42-13
  multiple, 19-6, 42-13, 50-5
  pile driving problems, 23-19
  in St. Lawrence Seaway (U.S.-Canada) project:
    dikes, 15-7

Till, in St. Lawrence Seaway (U.S.-Canada) project (*Cont.*):
  excavations, 19-5
  variations in, 11-28
Topsoil, 4-1
Trafficability maps, 2-16
Trace elements, in water, 8-11
Transportation Department (U.S.), 14-15, 19-38
Transported soils, 5-10
Trees and foundations, 22-18, *22-19*
Tsunami, 45-20
Tunnel boring machines, 20-29
Tunnels: alpine, 20-5
  break-ins, 20-4, 6
  in chalk, 20-47
  in clay, 20-47
  concrete, [from surface,] 20-23
  covered later, 20-2
  dip, 20-46
  drainage, 18-4, 25-46
  early examples, 20-2
  failures, 20-21, 42
  and faults, 44-8
  floating, for subways, 31-30
  gases in, 20-38
  geological factors, 20-3, 47
  linings, 20-19, 23
  locations of: Almendares (Cuba), 20-16, *20-17*
    Antwerp (Belgium), 19-25
    Beaucatcher (N.C.), 19-37
    Boston (Mass.), 1-10, 20-44, 50-11
    Box (U.K.), 20-23, 31-4
    Breakneck (N.Y.), *20-28*
    Bridge River (B.C.), *12-12*
    Broadway (Calif.), *20-7*
    Catskill Mountains (N.Y.), 10-5, *20-9*
    Channel, proposed (U.K.-France), 11-9, 20-47
    Chicago, drainage (Ill.), *20-31*

SUBJECT INDEX   S-19

Tunnels, location of (*Cont.*):
  Clifton Hall (U.K.), 20-42
  Cofton (U.K.), *19-8*
  Dartford (U.K.), 20-16, 27-17
  East River, gas (N.Y.), 20-8
  Fort Peck (Mont.), 20-29
  Halkyn, drainage (U.K.), 18-9
  Hoosac (Mass.), 20-23
  Hudson River (N.Y.), 20-11
  Kammon (Japan), 20-18
  Kilsby (U.K.), 18-5
  Las Raices (Chile), 20-6
  Leadville, drainage (Colo.), *18-11*
  Lochaber (U.K.), 20-25; 41
  London, subway (U.K.), 31-27
  Lorettoberg (West Germany), 20-40
  Lötschberg (Switzerland), 20-4
  Malden (Mass.), 50-11
  Mersey (U.K.), 1-8, 10-5, *20-35*
  Milwaukee (Wis.), 8-10
  Moffat (Colo.), 20-33
  Mont Blanc (France-Switzerland), 20-6
  Montreal, sewer (Que.), 20-30
  Oahe Dam (S. Dak.), 20-29
  Pennsylvania Turnpike (Pa.), 11-16
  Puketteraki (New Zealand), *39-13*
  Queens-Midtown (N.Y.), *20-13*
  Quintette (B.C.), *1-13*
  Rokko (Japan), 18-11
  Roman, 20-22
  St. Clair (Ont.-Mich.), 20-16
  St. Gotthard (Switzerland), 20-6
  San Jacinto (Calif.), *44-9*
  Seikan (Japan), 20-18

Tunnels, location of (*Cont.*):
  Severn (U.K.), 1-16, 20-32
  Simplon (Switzerland-Italy), 20-4, 23, 33
  Sir A. Beck project (Ont.), 19-10, 20-20
  Sydney (Australia), 20-21
  Tecolote (Calif.), 18-11, 20-36
  Totley (U.K.), 31-6
  Twin Cities, sewer (Minn.), 20-49
  Wapping, (U.K.), 20-14
  Wards Island (N.Y.), *20-15*
  Whatsham (B.C.), 20-21
  Wilson (Hawaii), 20-38
  Woodhead (U.K.), 20-41
  machines for, 20-29
  methods for, 20-27
  overbreak, 20-24
  pay lines, 20-24
  pressure in, 2-8, 20-18
  rapid, 31-32
  reinforcement, 30-19
  in sandstone, 20-49
  shapes of, 20-22
  spiral, at dams, 25-46
  temperatures in, 20-5
  underwater, 20-14
Turkey, dams in, 25-42

Underground space, 21-1
  for archives, 21-22
  for civil defense, 21-18
  early uses, 21-2
  and energy conservation, 21-2, 26
  for fuel storage, 21-14
  for garages, 21-24
  for libraries, 21-23
  problems with, 21-25
  for sewage plants, 21-25
  wartime uses, 21-6
  for wine storage, 21-5
Underpinning, 22-43
  in St. Helier Hospital (U.K.), 16-16
  in Singapore, 22-51
  in Sweden, 22-42
  in York (U.K.), 22-54

Underwater garages, *22-47*
"Undisturbed" soil samplings, 6-4, 11-8
Union of Soviet Socialist Republics:
  methane studies in, 26-18
  railroads in, 36-22
  Siberia, development of, 36-2
  tidal power plants in, 23-10
  underground hospitals in, 21-9
United States Steel Corporation, 24-36
Uplift pressures, 23-20, 25-35
Upstream engineering, 46-10
Urban geology, 13-2, 12
Utilidors, *36-17*

Valleys:
  buried, 5-22
  hanging, 5-18
"Vanishing rivers," 18-7
Vegetation and land restoration, 49-2
Venezuela:
  earthquakes in, 45-6
  gypsum studies in, 43-7
  land subsidence in, 37-9
Venice (Italy), groundwater beneath, 37-10
Vibrations, study of, 12-21
Vienna (Austria), subway in, 31-28
Volcanic ash, 17-8
Volcanoes, 45-3, 5

Warsaw (Poland), gravel dredging in, 16-11
Wash borings, 11-7
Washington Monument (D.C.), settlement of, 22-33
Waste:
  industrial, 48-11
  mining, 17-9
  problems with, 35-12, 48-9
  radioactive, 48-16
Water:
  artesian, 8-18
  brackish, in wells, 8-20
  connate (fossil), 8-5

Water (*Cont.*):
  contaminated, 28-7, 19
  divining, 8-23
  freshening, 35-4, 6
  hardness, 28-6
  impurities in, 28-5
  overpumping, 48-7
  phreatic, 8-3
  plutonic, 8-4
  pollution, 48-8, 13
  quality, 28-5
  quantities of, 9-2
  sea, 35-1
  softening, 8-10
  supplies: Denver (Colo.), 28-23
    Des Moines (Iowa), 28-8
    Gibraltar, 28-25
    Harrisonburg (Va.), 28-9
    Hong Kong, 28-31
    Honolulu (Hawaii), 28-33
    Juneau (Alaska), 28-26
    Kano (Nigeria), 28-8
    Kuwait, 28-33
    London (U.K.), 28-19, 27
    Long Island (N.Y.), 28-27
    Los Angeles (Calif.), 28-12, 29
    New York (N.Y.), 1-8, 28-12
    Newton-Longville (U.K.), 8-25
    Ogden (Utah), 28-31
    Richmond (U.K.), 8-15
    Riverside (Calif.), 8-9
    San Francisco (Calif.), 28-12
    Sydney (Australia), 28-30
    Toronto (Ont.), 28-8
    Vancouver (B.C.), 28-30
  supply, 28-1
    from groundwater, 28-15
    naturally filtered, 28-8
    from reservoirs, 28-10

Water, supply (*Cont.*):
    from rivers, 28-8
    from sewage effluent, 28-23
    table, 8-5, 28-17
    and weathering, 5-3
Water power projects:
  Aguasabon (Ont.), *26-19*
  Arapuni (New Zealand), *23-10*
  Bonneville (Oreg.), *23-13*
  Calderwood (Tenn.), *40-8*
  Cañon del Pato (Peru), 11-18
  Cephalonia (Greece), 26-8
  Cheakamus (B.C.), *23-14*
  Dead Sea, proposed (Israel), 48-5
  Donzère-Mondragon (France), *18-29*
  Dubrovnik (Yugoslavia), 26-4
  Fifteen Mile Falls (N.H.), 12-13
  Furnas (Brazil), 26-18
  Grand Rapids (Man.), 15-8, 27-18
  Kelsey (Man.), 23-16
  Kootenay (B.C.), 23-19
  Liulang Cave (China), *26-5*
  Lochaber (U.K.), *10-7*
  Marlengo (Italy), 41-11
  Niagara Falls (N.Y.), 19-11
  Qattara Depression, proposed (Egypt), 48-5
  Revelstoke (B.C.), 26-13
  Rhine River (West Germany), 1-11
  Schoellkopf (N.Y.), 40-7
  Sir A. Beck (Ont.), 19-11, 20-20
  Snowy Mountain (Australia), 2-2
Waves, 34-2
Weather services, 9-4, 7, 36-1
Weathering, 5-2
  and bacteria, 5-5

Weathering (*Cont.*):
  of dam foundations, 22-24
  extreme, 36-3
  of minerals, 5-6
  products of, 5-6, 42-3
  of rocks, 5-2, 6
  and temperature, 5-3
  and water, 5-3
  and wind, 5-3
Well point installations, 22-15
  at bridges, 18-20
  at dams, 18-19
  for drainage, 18-17
  at pumping stations, 18-18
  for sewerage, 18-19
Wells, 28-16
  deep: drainage, 18-15
    injection, 48-13
    pressure, 45-19
  dumb, 18-8
  soak, 28-17
  tray, 28-17
  variations in, 8-12
West Germany:
  Kocher River bridge, 27-13
  underground plants, 21-12
  water supply systems, 28-8
Wetlands, 46-9
Wind factors, 5-3, 9-7
Windmills for pumping, 35-2
Wine storage areas, underground, 21-4
Winnipeg (Man.), floodway, effects of, 48-6
World Trade Center (N.Y.), *13-13*, *35-10*

Yugoslavia:
  dams in, 26-4
  karst fields in, 38-10

Zimbabwe, hydroelectric projects in, 21-11, 48-3
Zuider Zee (Netherlands), drainage of, 18-4

NAME INDEX

This list will be some indication of the authors' appreciation of the valued information they have gleaned from the publications listed at the end of each chapter. The list contains the names of all those whose publications are there listed (except for those of this Handbook's two authors) but it does not include any names to be found in the Appendixes or in the lists of suggestions for further reading.

Adamson, W. S., 33-25
Agassiz, L., 3-2, 5-15
Agricola, G., 8-23
Ahearn, V. P., 49-22
Ahmed, A. A., 26-23, 48-17
Albin, P., 22-56
Albritton, B. C., 23-31
Alden, W. C., 39-24
Alexander, F. W., 24-44
Alexander the Great, 3-14
Alfors, J. T., 47-22
Allan, A., 20-51
Allan, J. C., 18-32
Allan, R. J., 39-25
Allen, A. S., 37-22
Allen, C. J., 31-33
Allen, H., 30-29
Allen, J. H., 23-31
Allen, P., 24-42
Alzura, P. R., 43-7, 12
Amberg, G. 40-23
Ami, H. M., 13-2, 22
Amrheim, J. E., 45-22
Andersen, P. F., 30-30
Anderson, A. P., 19-42
Anderson, D., 20-51
Arnold, H. M., 29-16
Ash, W. C., 34-44
Aspdin, J., 16-2
Aten, R. E., 28-38
Atkinson, J. D., 41-5, 14
Atwood, W. G., 39-5, 24
Auden, J. B., 44-6, 16

Bacon, F., 47-6
Baker, B., 6-2, 24-31, 25-24
Bailey, E. B., 20-41
Bailey, R. W., 49-23
Baird, D. M., 15-26
Bajournas, L., 34-43
Baldwin, M. W., 23-32
Balsley, J. B., 49-22
Banks, H. O., 28-36
Banks, J., 8-25
Baracos, A., 22-56
Bardes, J. E., 34-44
Barlow, P. H., 20-16
Barnes, A. A., 27-20
Barney, K. R., 45-23
Barrett, A., 50-17
Bartlett, J. V., 20-51, 31-33
Bass, K. T., 25-55
Bates, R. L., 16-21
Bauer, A. M., 49-22
Baum, A. L., 22-56
Bayley, S. H., 26-23
Bazynski, J., 11-30
Bean, D. C., 24-43
Bean, E. E., 30-28
Beard, F. D., 45-23
Beaumont, F., 20-29
Bechtold, L. C., 48-19
Beckman, P., 22-57
Behre, M. C., 27-21
Bell, A. L. L., 6-2, 19
Bell, J. H., 33-3
Bemelmans, L., 34-43

Bengry, E. O., 26-23
Bennett, F. J., 29-15, 30-29
Bennett, H. H., 49-22, 23
Benson, W. N., 39-12, 25, 41-14
Benton, P. H., 23-32
Berghinz, C., 37-21
Bergman, M., 21-28
Bergstrom, R. E., 14-16, 28-36
Berigny, C., 27-2, 20
Berkey, C. P., 20-8, 50, 23-31, 24-35, 25-21, 55
Bernstein, J., 45-23
Bernstein, L. B., 23-31
Berry, C. S., 23-32
Bertrand, M., 29-6
Bessey, G. E., 8-27
Bider, J. R., 49-22
Bierschenk, W. H., 48-19
Billings, M. P., 20-46, 52, 50-11, 18
Binnie, G. M., 27-21
Bishop, H. F., 46-18, 49-22
Bjerrum, L., 22-54, 40-23
Black, R. F., 31-33
Blake, N. M., 48-17
Blaney, H. F., 41-15
Blank, H. R., 50-17
Blaschke, T. O., 21-28
Blee, C. E., 19-42
Bligh, T. P., 21-2, 26
Blume, J. A., 34-45

NAME INDEX

Blumenthal, K. P., 35-15
Blyth, F. G. H., 38-13
Bolt, B. A., 47-22
Boniface, E. S., 28-37
Bonilla, M. G., 13-22, 44-15, 47-13, 22
Booth, W. S., 19-41
Borg, R. F., 11-26, 30
Borhek, R., 24-43
Boschen, H. C., 20-51, 27-21
Boswell, P. G. H., 1-8, 20-34, 51
Boughner, C. C., 46-18
Bouguer, P., 12-8
Boulton, G. O., 24-43
Bouwer, H., 48-18
Bowen, C. F. P., 43-12
Bowman, J., 28-36
Bowman, W. G., 21-27, 23-31, 25-55, 27-20
Boynton, R. M., 22-44
Bozozuk, M., 22-55
Bradley, O. R., 22-56
Bradley, W. H., 2-14
Brannfors, S., 19-41
Bray, J. W., 7-13
Brekke, T., 20-51
Brent-Jones, E., 49-22
Briedis, J., 44-16
Briggs, R. P., 47-22
Brink, A. B. A., 13-23, 22-55, 37-22, 38-8, 14, 42-4, 18, 49-22
Brobst, D. A., 15-26
Bromhead, E. N., 39-25
Brook, G. B., 16-20, 28-36
Brooke-Bradley, H. E., 24-43
Brooker, E. W., 10-11, 23
Brooks, A. H., 2-13, 20
Brown, C. B., 49-23
Brown, E. L., 22-55, 27-20
Brown, J. S., 50-16
Brown, P., 22-54
Brown, R. J. E., 5-33, 10-24, 36-22
Brown, R. L., 11-29
Brown, W. C., 10-24
Browne, W. R., 25-55, 26-23
Browning, B. C., 46-18
Bruce, A. W., 23-31
Brucker, E. E., 38-13
Bruckshaw, J. M., 11-29, 12-19, 22
Brune, G., 43-12

Brunel, I. K., 20-14, 24-25
Brunel, M. I., 20-14
Bryan, K., 5-30, 33, 36-22
Buckland, W., 20-23
Buckton, E. J., 24-43, 34-43
Bull, W. B., 49-22
Burckhardt, J., 21-4
Burdick, C. B., 48-18
Burn, K. N., 10-10, 23, 13-22, 17-11, 22-55, 48-18
Burnett, J. L., 47-22
Buro, M., 21-27
Burpee, L. H., 29-16
Burr, E., 34-44
Burton, V. R., 30-28
Buschbach, T. C., 46-18
Butcher, A. D., 41-5, 14
Buzzell, D. A., 17-11
By, J., 16-3, 29-5
Byers, A. F., 15-26

Cabanius, J., 25-55
Cady, D. R., 33-25
Calcano, C. E. F., 43-7, 12
Calembert, L., 38-10, 14
Caltrider, W. T., 26-23
Calvert, A. J., 38-14
Cameron, K. M., 16-20
Campbell, C. W., 22-54
Campbell, F. B., 32-9
Campbell, I., 47-22
Campbell, J. G., 27-21
Campbell, J. L., 20-51
Capps, S. R., 5-32
Carll, J. F., 2-11
Carlyle, T., 49-21
Carpmael, R., 1-19, 20-51, 40-23
Carson, R., 34-1, 43
Carter, J., 25-8
Casagrande, L., 19-28
Casey, T. L., 22-34
Casteret, N., 21-5, 27
Cawsey, D. C., 30-29
Cedergren, H., 30-9, 29
Cerutti, H. P., 19-40, 43-12
Chagnon, J. G., 39-25
Chaly, V., 16-20
Chandler, R. J., 42-2, 18
Chapman, L. J., 10-9, 23
Charlesworth, J. K., 5-32, 33
Chase, W. K., 32-9
Chatley, H., 33-24

Christians, G. W., 25-55
Christiansen, E. A., 13-6, 22, 28-19, 38-13, 47-23
Christman, H. E., 40-23
Church, M., 26-23
Clark, J. A. M., 31-33
Clark, K., 22-57
Clark, T. H., 13-9
Clarke, F. E., 49-22
Clarke, F. W., 5-32, 37-23
Cleaves, A. B., 19-40, 20-32, 30-28
Clevenger, W. A., 6-19
Cloos, H., 20-40, 51
Cluff, L. S., 44-15
Clyne, J. V., 20-51
Coates, D. R., 10-24, 33-25, 36-22
Cohen, P., 28-37
Collier, B. C., 26-23
Collin, A., 6-2, 19, 11-8, 29, 19-12, 41, 39-3, 24
Collins, J. A., 22-56
Collins, J. J., 37-23
Collins, L. E., 22-55
Colwell, R. N., 10-24
Conan Doyle, A., 42-18
Cooke-Yarborough, S. S., 35-15
Copp, H. D., 46-18
Cortes-Comerer, N., 50-18
Costes, A., 40-23
Cotton, C. A., 45-22
Cotton, J. C., 21-28
Couch, F. B., 25-57
Coulomb, C., 6-2
Counts, H. B., 37-24
Coupar, H., 50-2
Coveney, R. M., 21-28
Coyne, A., 19-21, 25-13, 54
Crandall, D. R., 45-5, 22
Crawford, C. B., 5-33, 6-20, 17-11, 22-55, 56, 39-25, 42-18, 48-18
Crawford, R. C., 33-25
Crichton, J. R., 23-32
Crocker, E. R., 20-51
Croft, A. R., 49-23
Cromwell, O., 18-4
Crook, R., 45-23
Crosby, I. B., 11-29, 12-22
Crosby, W. O., 5-23, 20-8
Cruden, D. M., 40-3, 23
Cullen, J., 45-5, 22

NAME INDEX   N-3

Cummings, A. E., 15-25
Cummins, G. A., 28-36
Cunney, C. A., 17-11

Daft, 12-9
Daiches, D., 28-36
Dallaire, G. E., 41-15, 48-18
Dalton, F. E., 46-18
Darton, N. H., 47-13, 22
Darwin, C., 3-2, 37-4
David, M. E., 2-20
David, T. E., 2-13
Davies, L. B., 26-23
Davies, W. E., 48-18
Davis, D. A., 8-27
Davis, E. J., 23-32
Davis, G. H., 37-24
Davis, H. S., 16-20
Davis, R. F., 10-23
Dawkins, B., 18-6, 32
Deacon, G. F., 25-14, 35, 55
Dean, A. C., 23-32
Dean, A. J., 18-31
Dean, A. W. H., 24-43
Dearman, W. R., 14-16
DeBruijn, C. M. A., 42-18
De Castro, 12-4
De Chardin, P. T., 50-15, 18
Deere, D. U., 7-3, 9, 13, 15
De France, H., 8-27
Degenkolb, H. J., 45-22
Deinhart, A. V., 41-14
De la Bleche, T., 15-17
De Lesseps, F., 29-5
DeLory, F. A., 39-4, 24
Denison, W., 27-2, 20
Dennis, T. H., 39-25
Denny, C. S., 30-29
De Saussure, H. B., 3-2
De Steur, W., 31-33
Devata, M., 20-51, 24-43
DiCervia, A. L. R., 25-57
Dickinson, C. H., 22-54
Dickinson, W. R., 44-15
Dietrich, 2-16
DiGiacinto, A., 19-41
DiGioia, A. M., 17-11
Dishaw, H. E., 10-16, 24
Disney, C. P., 24-43
Dixey, F., 12-19, 22
Dixey, R., 48-17
Dobrovolny, E., 45-22, 23
Doherty, R. E., 22-56

Dolan, R., 49-22
Dollerl, A., 31-34
Dolmage, V., 11-29, 12-11, 34-44
Doucet, G. J., 49-22
Douglas, J. N., 34-38
Dow, A. L., 11-29
Dowling, G. B., 37-23
Downing, R. A., 28-31
Dowrick, D. J., 22-57
Doyle, A. F., 46-18
Dreimanis, A., 6-17, 20
Drunkers, J. J., 35-15
Druppa, A., 38-14
Duane, D. B., 16-20
Dubost, R., 27-20
Dufour, H., 41-15
Duncan, M. A. G., 16-21
Durrell, L., 26-9, 22, 45-23
Dustin, T. G., 29-16
Duvivier, J., 34-44
Dymond, J., 12-22
Dzeguze, K., 38-13

Eads, J. B., 34-30
East, L. R., 28-37
Eaton, C., 50-2, 17
Eckart, N. A., 25-55
Eckel, E. B., 14-16, 15-26, 40-23, 45-22
Eckerman, J. P., 29-3, 16
Eden, W. J., 6-19, 39-25, 42-18, 43-12
Edholm, H., 21-15
Edison, T., 18-6, 19-24
Edlund, S., 21-28
Edmunds, F. H., 23-32
Edwards, A. T., 19-42
Ehlers, J., 50-7, 17
Eisbacher, G. H., 40-5, 23
Ekblaw, G. E., 30-28
Elliot, A. L., 45-23
Elliot, D. O., 33-24
Elliot, S. R., 35-15
Ellis, A. J., 8-27
Ellis, C. H., 31-32, 33
Ellis, S. H., 15-26
Ellson, G., 34-45
Elmendorf, C. H., 34-44
Emerson, F. E., 39-24
Emerson, R., 31-32
Emery, K. O., 15-26
Emrich, G. H., 48-18

Engle, W. W., 25-55
Enslin, J. F., 12-22
Ephros, A. Z., 25-55
Erasmus, D., 18-3
Escoffier, F. F., 34-44
Esmiol, E. E., 44-16
Ethelred, 24-1
Evans, O. M., 45-23
Evelyn, J., 4-1, 18
Everall, W. T., 24-43

Fairbridge, R. W., 34-43, 37-23
Fairchild, 19-6
Fairhurst, C., 21-28
Farey, J., 8-23
Farid, R. R., 23-32
Farquhar, O. C., 19-41
Farquharson, K. G., 47-22
Farvolden, R. N., 18-32
Fay, S., 37-23
Feld, J., 22-54, 44-16
Fereday, H. J., 24-43
Ferguson, H. A., 35-15
Ferguson, H. B., 33-3
Fiacka, J., 48-19
Finch, R. M., 24-43
Findlay, D. C., 11-29
Fintel, M., 45-22
Fischer, W. A., 10-24
Fisk, H. N., 6-18, 20, 33-5, 24, 50-4, 17
Fitch, H. R., 14-16
Fitzmaurice, M., 25-55, 41-14
Fleming, S., 11-19, 22-27, 24-32
Flemming, A. G., 16-20
Flett, J. S., 15-26
Flint, R. F., 5-33, 33-25, 38-13, 45-22
Flowers, B., 48-17, 19
Fluhr, T. W., 20-12, 50
Focht, J. A., 34-40, 44
Follenfant, H. G., 31-33
Follmer, I. R., 14-16
Fookes, P. G., 6-20, 43-13
Foose, R. M., 18-33, 38-14
Forney, F. H., 34-43
Fornwald, W. L., 11-29
Fortier, S., 41-15
Foster, H. L., 46-17
Fougerolles, M., 20-49
Fowler, J., 24-31

NAME INDEX

Fox, F., 8-26, 18-33, 20-50, 22-55, 26-9, 23, 31-33, 24-18, 43
Fox, J. R., 27-20
Foxworth, B. L., 28-37
Francis, J. C., 18-32
François, A., 27-13
Franke, O. L., 28-37
Freedman, J. L., 22-56
Frei, L. R., 25-55
Fritz, P., 48-17
Friz, T. O., 49-22
Frontinus, J., 28-2
Furkert, F. W., 23-12, 31, 45-23

Gardner, M. E., 14-16, 47-22
Gaskin, P. N., 48-18
Gay, T. W., 47-22
Geikie, A., 25-24
Geikie, J., 15-25, 26
Gerwick, B. C., 34-45
Gest, A. P., 16-20, 30-28
Geyer, A. R., 7-15, 38-14
Ghyben, W. B., 28-33
Gibbings, R., 21-27
Gibson, E. J., 21-27
Gilchrist, L., 12-21
Giles, W. H., 24-43
Gill, T. C., 18-32
Gilleland, J. E., 23-32
Gillette, D. H., 22-56
Gillott, G. A., 27-21
Gillott, J. E., 6-20, 16-21, 36-22, 43-12
Gimson, N. H., 27-21
Gish, 12-9
Glaeser, J. R., 20-51
Glossop, R., 27-2, 20, 48-18
Goddard, J. E., 46-18
Godfrey, K. A., 16-21, 42-18
Goethe, J. W. von, 29-3, 40-18
Goguel, J., 11-29
Golder, H. Q., 24-44
Goldstein, A., 24-43
Goodall, G. E., 25-55
Goodenough, X. H., 18-25
Goransson, T., 21-27
Gordon, G., 19-41
Gottschalk, L. C., 33-25
Gouch, J. W., 18-31
Gough, D. I., 12-17, 22

Gough, D. L., 48-17
Gough, H. B., 11-29
Gould, W. H., 34-43
Gourlaf, H. J. F., 28-36
Graff, E. D., 27-21
Grant, L. F., 28-13
Grant, U. S., 13-12, 33-4
Grantz, A., 44-15
Gray, R. E., 37-24
Greathead, J., 27-5
Green, E. E., 22-55
Gregory, J. W., 8-27, 30-1, 28
Grice, R. H., 15-25
Grieve, W. G., 20-51
Grimes, P., 28-37
Gronvold, K., 45-22
Gross, M. G., 48-18
Grothaus, W., 29-16
Gruner, H. E., 1-11, 19
Gustav Adolf VI, 22-44
Gwynne, C. S., 30-29
Gwyther, R. D., 34-44

Hack, J. T., 50-2, 17
Hadfield, C., 18-4, 31
Haefeli, R., 24-44, 40-23
Halcrow, W., 10-23, 16-20, 28-36
Halett, D. R., 47-23
Halevy, E., 26-23
Hall, J., 50-17
Halley, E., 8-2
Hallock, E. C., 15-25
Halperin, D., 31-34
Hamburger, P., 21-2, 26
Hamel, L., 21-27
Hamilton, R. M., 45-23
Hancock, N., 20-51
Hanes, F. E., 16-20
Hanna, B. E., 49-22
Hannant, R. B., 22-44
Hansen, T. C., 48-18
Hansen, W. R., 28-37, 46-18, 47-22
Hanshaw, B. B., 49-22
Happ, S. C., 25-55
Harding, H. J. B., 11-29
Harmon, B., 28-36
Harpaz, Y., 26-23
Harris, I. D., 14-16
Harrison, J. M., 3-13, 15
Harrison, W., 50-5, 17
Hart, S. S., 14-16

Harwood, F. L., 23-31
Haskins, G., 20-51
Hassall, E. R., 10-24
Hast, N., 7-12, 15
Haswell, C. K., 20-51, 43-12, 50-18
Hatheway, A. W., 44-16
Hatton, T. C., 8-27
Hau, H. J., 40-23
Hauser, E. O., 35-15
Hawkesley, T., 25-14, 35, 27-2
Hawkins, H. G., 50-17
Hayes, C. W., 29-8
Heathcote, F. W. L., 37-24
Hebda, R. J., 50-17
Hedstrom, H., 12-22
Heezen, B. C., 24-44
Heilprin, A., 13-2, 22
Heim, G., 46-18
Heine, W. N., 18-33
Helmholtz, H. von, 19-28
Henderson, L. H., 44-16
Henke, K. F., 43-12
Henkel, D. J., 40-24
Henry VIII, 24-11
Henson, F. I., 43-12
Herschel, C., 28-35
Herschel, J., 12-8
Herzberg, B., 28-34
Hewitt, D. F., 16-20
Heydenrych, R. A., 19-41
Higginbottom, I. E., 6-20, 43-13
Hill, H. E., 34-44
Hill, M., 22-54
Hill, R. A., 39-25
Hinds, J., 25-55, 28-36
Hitchings, D. C., 23-32
Hobson, J., 20-16
Hoeffer, H. J., 25-55
Hoek, E., 7-15
Hoffman, J. F., 8-27
Hogg, A. D., 20-51, 43-12
Hognstad, E., 34-45
Holmes, A., 42-18
Holtz, W. G., 42-18
Hondl, A., 31-34
Hook, G. S., 31-33
Hooke, R., 3-2
Hopper, H. R., 23-32
Horner, S. E., 30-28
Horton, R. E., 33-12
Howard, E. A., 34-44

NAME INDEX N-5

Howe, 29-8
Hromada, K., 37-24
Hubbert, M. K., 6-7, 19, 50-5, 17
Hucka, V., 7-15
Humery, R., 48-18
Hunt, G. S., 50-17
Hunt, H. W., 22-55
Hunt, I. A., 34-43
Hunt, L. H., 49-22
Hunter, W. H., 39-43
Hurst, W. D., 16-21
Huson, A., 40-23
Hutchinson, J. N., 39-25
Hutton, J., 3-2
Hvorslev, M. J., 11-1, 29
Hyde, J. E., 2-20

Idriss, I. M., 45-23
Imhoff, E. A., 49-22
Imrie, A., 11-29
Irvine, L. H. R., 30-14
Isaacs, B., 19-41
Isherwood, C. W., 25-56

Jack, R. L., 8-26, 28-36, 37
Jacks, G. V., 49-23
Jackson, L. E., 31-33
Jackson, M. W., 48-19
Jacoby, H. S., 10-23
Jahns, R. H., 48-19
James, A. N., 43-7, 12
Jennings, J. E., 13-23, 18-33, 38-14, 42-8, 18
Jessop, W., 18-4, 29-3
Jessup, W. E., 25-54
Johnson, A. W., 30-29
Johnson, F. G., 23-32
Johnson, J., 18-14
Johnson, W. A., 34-43, 44
Johnston, G. H., 23-32, 36-22
Johnston, H. A., 25-55, 57
Johnston, H. R., 11-29
Jones, D. E., 42-18
Jones, G., 45-23
Jones, J. H., 35-15
Jones, O. T., 12-22
Jorstad, F., 40-23
Joshua, 25-2
Justin, J. D., 25-3, 54

Kapp, M. S., 13-22, 19-41, 32-9, 35-15
Kassiff, G., 42-18
Kavanagh, T. C., 19-42
Kaye, C. A., 15-26, 43-12, 50-7, 17
Keaton, J. R., 44-16
Keays, R., 20-51
Keener, K. B., 41-14
Kehnemoyl, M., 23-32
Keith, A., 3-15
Keith, J. M., 30-29, 34-45
Kell, J., 27-21
Kellberg, J., 25-57
Kellberg, J. M., 14-16
Keller, E. A., 33-25
Kellerhals, R., 26-23
Kellog, F. H., 20-20
Kelly, S. F., 12-22
Kelvin, Lord, 1-18, 34-5
Kemp, J. F., 20-8
Kennedy, A., 21-27
Kennedy, J. M., 38-14
Kennedy, R. G., 41-10, 15
Kenney, T. C., 50-5, 17
Kerr, J. W. G., 40-20, .24
Kerr, L., 22-54
Kiersch, G. A., 26-22, 23
King, W. B. R., 2-13
Kingsley, C., 13-2, 22, 50-15, 18
Kinipple, W. R., 27-2
Kipling, R., 1-18
Kirkpatrick, C. R. S., 34-43
Knight, B. H., 17-11
Knight, P. G. K., 17-11
Knight, R. H., 17-11
Knightley, P., 37-23
Knill, J. L., 37-24, 40-24
Knopf, A., 33-25, 45-22
Kochis, F., 25-57
Kollert, R., 12-22
Korgin, C. E., 20-51
Kovari, K., 7-15
Kraft, L. M., 34-40, 44
Krahn, J., 40-3, 23
Kreig, R. A., 10-24, 36-22
Krelova, Z., 14-16
Krinitsky, E. L., 42-18
Kristinsson, G., 23-31
Kunze, W. E., 45-22
Kuznetson, A. M., 26-18, 23
Kyle, J. M., 32-9

Lacroix, Y., 25-56
Lacy, F. P., 25-55
Lado, G. E., 6-2, 39-25
LaFevers, J. L., 49-22
LaForge, L., 20-46
Lambe, T. W., 22-56
LaMoreaux, P. E., 50-18
Lancaster-Jones, P. F. F., 27-21
Land, D. D., 23-32
Lander, J. H., 23-32
Lane, D. A., 28-37
Lane, E. W., 41-15
Langbein, J. A., 44-16
Lankford, R. R., 50-17
Lanyon, R., 46-18
Laplace, P. S., 3-3
Lapworth, H., 8-5, 25-3, 54
La Rochelle, P., 39-25
Larsen, G., 45-22
Laurence, R. A., 14-16
Laurie, P., 21-27
Laverty, F. B., 28-37
Law, F. M., 28-37
Lea, F. M., 8-27, 16-15, 21
Leadbetter, B. C., 44-16
Le Clercq, A. G., 34-43
Lee, C. H., 6-20
Lee, C. S., 40-23
Lee, J. J., 31-33
Lehmann, J. G., 3-2
Leighton, F. B., 44-16
Lely, C., 35-3
Lemoine, P., 48-18
Leonardon, E. G., 12-22
Leopold, E. B., 50-17
Leopold, L. B., 8-26, 33-12, 25, 47-22, 49-22
Lerchen, F. H., 29-16
LeRoy, L. W., 50-18
Lessing, P., 49-22
Levin, P., 23-31
Lewis, A. F. G., 16-20
Linehan, D., 13-23
Livneh, M., 42-18
Lo, K. Y., 20-51
Lobdell, H. L., 35-15
Logan, W. E., 3-2
Lohman, E. H., 2-14
Long, N. S., 25-55
Longwell, C. R., 33-25, 38-13, 45-22
Low, G. J., 31-33
Lugeon, M., 12-13, 22

NAME INDEX

Lung, R., 39-25
Lupini, J. F., 39-25
Lupton, A. R. R., 43-7, 12
Lyell, C., 3-2, 37-4, 50-2, 17
Lynde, G. A., 34-43
Lynn, A. V., 11-30
Lyra, F. H., 26-23

McAdam, J. L., 30-1, 12, 28
MacBride, W. D., 18-33
McCahill, G., 21-28
McCallum, R. T., 1-19, 19-40
McCarthy, E., 50-7
McClelland, B., 6-18, 20, 50-4, 17
MacClintock, P., 19-6, 40
McClure, C. R., 44-16
McCue, J. H., 31-6
McCulloch, D., 29-5, 16
McDaniels, A. B., 28-36
MacDonald, D. F., 29-8
MacDonald, D. H., 22-56, 23-32, 43-12
MacDonald, G. A., 47-22
MacDonald, J. R., 7-15
McDunough, C., 25-54
MacFarlane, I. C., 5-32, 10-24, 36-22, 48-18
McGlade, W. G., 7-15
MacGregor, W. M., 26-23
McHaffie, M. G., 18-32
McHarg, I., 47-6, 22
McHenry, E., 34-44
McIldowie, G., 10-23, 25-55, 28-36
McKamey, J. B., 16-20
MacKenzie, C. J., 8-11, 24-43
MacKenzie, I. D., 7-15, 25-55, 27-20
McKeown, M. H., 45-23
Mackin, J. H., 26-7, 22
McLeod, N. W., 32-9
Maclure, W., 3-2
McMahon, C. A., 19-41
McMaster, W. M., 14-16
McNeill, M., 42-18
McTaggart, K. C., 40-23
Maigre, R., 25-55
Maillart, R., 40-15
Mair, C., 34-43
Malkin, A. B., 37-24
Mallet, R., 45-6, 22
Mandzel, S., 26-23

Mann, J. F., 46-17
Mann, W. K., 12-22
Mansur, C. I., 37-24
Marcinkevich, E. A., 25-55
Margison, A. D., 21-28
Marr, J. C., 18-32
Marr, N., 33-25
Marsh, G. P., 42-17, 19, 49-16, 23, 50-16, 18
Maskelyne, J., 12-8
Mason, A. C., 8-27, 46-17
Mason, E. E., 11-29, 34-44
Mason, P. L., 31-33
Mathews, W. H., 23-14, 40-23
Mathewson, C. C., 22-56
Matich, M. A. J., 30-30
Matthews, C. W., 15-25
Matula, M., 7-4, 15, 24-44
Mead, H. T., 23-32
Meinesz, V., 12-8
Meinzer, O. E., 8-23, 26, 27, 28-36, 37-8, 23, 48-18
Melhorn, W. N., 50-17
Melzer, K. J., 33-25
Mencl, V., 39-15, 25, 42-3, 18
Mensch, L., 25-54
Merrill, C. L., 36-22
Merrill, G. P., 4-1, 18, 5-2, 32
Metzger, H. P., 48-19
Meyers, J. F., 37-24
Millard, L. W., 24-42
Miller, R. E., 45-22
Milling, M. E., 42-19
Milton, J., 50-16
Mindess, S., 48-18
Minear, V. L., 27-20
Misiaszek, E. T., 50-5, 17
Mitchell, J. T., 22-56
Mitchelson, A. T., 28-37, 48-18
Mobley, C. F., 26-23
Mohan, D., 17-11
Mohr, C., 24-44, 40-24
Mohr, H. A., 22-54
Moisseiff, L. S., 24-43
Mollard, J. D., 10-16, 24, 36-22
Montgomery, Field Marshal, 2-16
Mooney, B. L., 48-19
Moore, D. W., 47-4
Moore, T. F., 18-32
Mooser, F., 37-24
Moran, B. J. J., 49-22

Morcos, S. A., 48-17
Moreland, T., 38-14
Moresco, R. L., 45-23
Morfeldt, C. O., 21-28
Morgan, A. E., 46-9
Morgan, H. D., 20-51, 31-33
Morgan, M. H., 24-43
Morgenstern, N. R., 40-3, 23
Morris, M. E., 24-44
Morris, S. B., 25-55
Morrison, F., 24-2
Morton, G. H., 20-34
Morum, S. W. F., 31-33
Moye, D. G., 2-19, 12-22, 25-55, 26-23, 43-12
Mudge, M. R., 15-25
Muller, L., 7-15, 16
Muller, S. W., 36-4, 22
Mumford, L., 47-5, 22
Murphy, P. J., 44-16
Murphy, V. J., 12-22
Murphy, W. J., 13-23
Murray, M. A., 21-27
Murray, R. C., 42-18

Nebuchadnezzar, 24-42
Neel, C. H., 37-23
Neill, F. C., 46-18
Nelson, J. C., 19-41
Newbery, J., 25-48, 57, 43-12
Newman, B., 20-51
Newton, I., 12-7
Nicol, T. B., 25-55, 43-12
Nicol, W., 3-2
Nicoletti, J. P., 30-29
Niederhoff, A. E., 40-23
Niini, H., 15-26
Nillson, T., 7-17
Nir, A., 26-23
Nixon, D., 21-27
Noble, C. B., 20-51
Noble, C. C., 33-25
Noetzli, F. A., 22-54
Nolan, T. R., 31-33
Northwood, T. D., 19-42
Norvitch, R. F., 13-23, 14-6, 47-23
Notardonato, F., 46-18
Nowson, W. J. R., 22-57
Nunan, J. P., 18-32
Nuzzo, W. L., 17-11
Nyzer, P. C., 34-44

NAME INDEX N-7

O'Dell, A. C., 31-32, 33
Odell, N. E., 31-35
Ogden, P. H., 22-56
Olaf, 24-1
Ollier, C., 42-4, 18
O'Neill, J. T., 28-37
Ongley, M., 39-25
O'Reilly, M. J., 13-22
Ostrowski, W. S. S., 19-41
O'Sullivan, P. M., 35-16

Page, B. N., 20-50
Paige, S., 24-35
Palissy, B., 8-2
Parillo, D. G., 35-15
Parizek, E. J., 21-28
Parizek, R. R., 46-18
Parker, C. G., 8-27
Parker, P. A. M., 28-37, 33-2
Parsons, J. D., 10-23, 22-55, 42-18
Parsons, M. F., 10-24
Passoneau, J. R., 13-22
Patin, D. R., 23-32
Paton, J., 2-19
Patterson, F. W., 27-21
Patton, F., 11-29
Patty, T. S., 30-13, 29
Payne, R. H., 37-23
Peach, B. N., 20-51
Pearce, C. E., 25-55
Peck, J. H., 44-16
Peck, R. B., 7-3, 15, 15-25, 22-55
Peckover, F. L., 11-29, 15-25, 24-44, 29-16, 31-33, 40-20, 24
Peel, C., 34-44
Penner, E., 6-19, 10-10, 23, 43-12
Perkins, R. L., 11-29
Perrault, P., 8-2
Perry, J. R., 16-20
Peterson, R., 43-12
Pettyjohn, W. A., 28-37
Philbrick, S. S., 2-10, 19, 22-54, 40-24
Philip, B., 50-18
Phillips, C., 34-44
Phillips, J., 18-31, 50-17
Picard, J., 34-7
Pigot, C. H., 7-15
Pihlainen, J. A., 36-22

Pike, C. W., 22-55
Pilkey, O. H., 24-43
Pillman, R. A., 23-32
Pinchback, R. T., 30-29
Pinkerton, I. L., 21-27
Piskin, K., 14-16
Pitt, M. E., 35-16
Plantema, G., 31-34
Playfair, J., 3-2
Poland, J. F., 37-24
Pollard, W. S., 47-4
Polo, M., 8-23
Pomeroy, J. S., 47-16, 22
Pratt, W. P., 15-26
Prelini, C., 25-54
Prentis, E. A., 22-54
Presant, E. W., 50-17
Prescott, W. H., 30-1, 29
Preston, R., 16-20, 28-36
Price, D. C., 37-24, 40-24
Prior, J. W., 12-22
Proctor, C. S., 10-23
Proctor, R. J., 39-25, 45-23
Proksch, E., 31-34
Prugh, B. J., 18-32, 33
Psutka, J. F., 11-29
Putnam, D. F., 10-9, 23

Quigley, L., 28-37
Quinn, A. de F. 34-43

Radbruch, D. H., 13-22, 47-13, 22
Radforth, N. W., 5-32
Rahm, D. A., 20-51
Rahn, P. H., 46-18
Rands, H. S., 27-20
Rankilor, P. T., 10-24
Rankine, J. M., 6-2
Ransome, F. L., 25-4, 54
Rao, K. L., 25-55
Rattenbury, O. B., 34-44
Ray, R. G., 10-24
Raymond, G. P., 31-33
Reade, T. M., 20-34, 29-4, 16
Reed, L. A., 30-29
Reed, V. S., 43-12
Reeves, F., 25-55
Reger, R. D., 10-24
Rehbock, 41-5
Reid, W. M., 44-16

Render, F. W., 45-22, 46-18, 48-17
Rennie, J., 18-26
Rettie, J. R., 27-21
Revelle, R., 11-29
Reynolds, O., 34-32
Rhoades, R., 11-29
Rhodes, A. D., 15-25
Rich, V., 48-17
Richardson, A. H., 33-25, 46-18
Richardson, E. C., 40-18
Richey, J. E., 24-44, 37-24
Richter, C. F., 45-10, 23
Rigour, J., 13-23
Ringers, J. A., 29-16
Ripley, D. M., 29-16
Roberts, C., 24-43
Robeson, F. A., 27-21
Robinson, T., 1-9
Roed, N. A., 47-23
Roger, R. D., 36-22
Rolt, L. T. C., 24-43
Rominger, J. F., 6-16, 20, 50-5, 17
Rooney, 12-9
Rose, D., 11-29
Ross, C. P., 25-55
Ross, J. C., 34-5
Rothe, J. P., 45-23
Rowe, P. W., 18-9, 32
Roy, T., 18-5, 32, 30-9, 29, 50-2
Rozycki, S. Z., 13-22
Rubey, W. W., 6-7, 19, 50-5, 17
Rubin, M., 50-17
Ruckman, D. W., 22-56
Ruffin, J. V., 34-45
Ruskin, J., 15-17
Russell, S. O., 47-22
Rutka, A., 30-30
Rutledge, P. C., 6-16, 20, 34-44, 50-5, 17
Ruttar, N. W., 47-22
Rysavy, P., 38-14

Saleem, Z. A., 48-18
Salsman, G., 37-23
Salvadori, F., 50-18
Sanborn, J. F., 20-8, 50
Sanders, J. E., 38-13
Sanderson, A. B., 24-44

NAME INDEX

Sanderson, I. T., 36-22
Sandstrom, G. E., 21-28
Sarrazin, J. P., 49-22
Saurin, B. F., 22-55
Savage, J. L., 24-43
Sawtell, H. E., 22-54
Sbar, M. L., 43-12
Scafer, C. T., 48-17
Schaerer, P. A., 40-24
Schatz, A., 15-26
Scheele, W. E., 30-30, 50-18
Schick, A. P., 41-13, 15
Schijf, J. B., 34-16, 43, 35-15
Schlocker, J., 13-22, 47-13, 22
Schlumberger, C., 12-9, 13, 22
Schmertmann, J. H., 18-32
Schmidt, F. E., 6-18
Schmidt, H. W., 3-5, 15
Schmidt, L. A., 26-23
Schmoll, H. R., 45-23
Schoklitsch, I., 20-25
Schriever, W. R., 13-22, 19-40, 41
Schulze, B. R., 30-29
Scott, A. C., 48-19
Scott, H. A., 11-29
Scott, J. D., 50-17
Scott, J. S., 10-11, 23, 13-23, 45-22, 50-17
Scott, K. M., 49-22
Scott, P. A., 20-51, 24-43
Scott, R. F., 47-22
Scott, W., 30-1
Scott, W. M., 24-42
Sebastyan, G. Y., 32-9
Seed, H. B., 45-23
Selby, K. C., 24-43
Seligman, G., 40-18
Sendlein, L. V. A., 49-22
Senour, C., 34-44
Shacketon, E., 2-13
Shadmon, A., 15-26
Shakespeare, William (playwright), 49-17
Shaler, N. S., 34-43
Shanley, W., 35-9
Shapley, H., 48-18
Sharpe, C. F. S., 39-1
Shaughnessy, T., 22-37
Shaw, H., 12-22
Sheets, M. M., 37-24
Shepard, F. P., 34-5, 43, 50-17

Sheppard, T., 3-15, 4-18, 8-26, 27, 18-31, 34-43
Sherard, J. L., 45-23
Sherlock, R. L., 48-17
Shreve, R. L., 40-23
Shumway, G., 37-23
Sikso, H. W., 15-26
Silar, J., 38-14
Simcic, E. P., 18-33
Simmons, M., 25-57
Simon, P., 6-20
Simpson, H. E., 14-16, 40-24
Simpson, T. A., 50-18
Sinclair, D. A., 31-34
Singer, C., 3-15
Singh, S. P., 30-29
Sissons, J. B., 22-55
Skempton, A. W., 18-4, 31, 22-56, 39-4, 24, 42-2, 18
Skermer, N. A., 47-22
Skopek, J., 22-56
Skouby, M. C., 37-24
Skrivanek, F., 38-14
Slemmons, D. R., 45-22
Slocum, H. S., 27-20
Small, J. B., 37-24
Smeaton, J., 34-37, 44
Smith, B., 28-36
Smith, C. H., 11-29
Smith, D. I., 23-17
Smith, D. W., 24-43
Smith, N. L., 30-29
Smith, R., 25-57
Smith, S., 18-23
Smith, W., 1-3, 19, 3-2, 4-1, 8-3, 25, 15-17, 18-4, 50-1
Smoots, V. A., 23-32
Smosna, R. A., 49-22
Snow, B. F., 18-33
Socolow, A. A., 38-14
Soderberg, A. D., 38-14
Sorsbie, R. F., 17-11
Soveri, U., 18-32
Sowers, G. B., 30-27, 50-12
Sowers, G. F., 11-30
Soyer, R., 48-18
Spangle, W., 47-23
Spencer, C. B., 22-56
Spring, F., 33-3, 34-25, 43
Springfield, V. T., 8-27
Squier, L. R., 22-54
Stamp, L. D., 1-19
Stanton, R. E., 39-24
Stanton, T. E., 16-21, 22-56
Stanton, W. F., 34-43

Starring, J. W., 30-29
Stearns, N. D., 8-27, 28-38
Steinbrugge, K. V., 44-15
Steno, N., 3-1
Stephan, H. J., 50-7, 17
Stephenson, G., 18-6, 30-9, 31-4
Stephenson, R., 1-3, 8-3, 18-5, 17
Stephenson, R. C., 30-28
Stermac, A. C., 11-30, 24-43
Stevens, A., 45-21
Stevens, J. C., 26-14, 23, 41-14
Stevens, J. F., 29-8
Stevenson, R. L., 35-2
Stevenson, T., 34-2, 43
Stewart, D. P., 19-6, 40
Stewart, J. W., 11-29, 34-44
Stickler, C. W., 20-51
Stohanet, E. E., 37-23
Stone, R. O., 46-17
Stothard, J. N., 15-26, 30-29
Stowe, L., 46-17
Strahan, C. M., 6-2, 30-2, 13
Strange, W. L., 26-10, 23
Strence, K. O., 25-55
Stuart, W. H., 25-35, 55
Suess, E., 13-2, 22
Supp, C. W. A., 30-28
Suskowski, Z., 13-22
Sutherland, H. B., 48-18
Sutherland, J. G., 22-56
Sutherland, R. A., 25-55
Swales, J. K., 25-57
Swenson, E. G., 8-27, 16-20, 21, 36-22
Sykes, L. R., 43-12

Taft, W. H., 29-8
Taggart, W. C., 46-18
Tanner, R. G., 43-12
Tavenas, F., 39-25
Taylor, 45-1
Taylor, G., 9-7, 11
Taylor, M., 27-20
Taylor, R. S., 13-22, 22-56
Taylor, T. U., 25-54
Tedrow, J. C. F., 42-18
Telford, T., 18-26, 30-1, 12
Terzaghi, K., 6-2, 8, 14, 19, 20, *6-18*, 22-55, 23-15, 31, 24-36, 25-5, 54, 55, 34-36, 44, 36-22, 50-5

Thomas, D. G., 31-33
Thomas, J. R., 31-33
Thompson, R. D. S., 28-36
Thomson, D. H., 28-4, 36, 46-5, 18
Thomson, F. G., 25-55
Thomson, J. J., 8-24, 27
Thorarinsson, S., 45-22
Thoresen, S. A., 19-41
Thorfinnson, S. T., 7-15
Thornley, J. H., 22-56
Thornthwaite, C. W., 9-3, 11, 30-29, 46-8
Thorvaldson, T., 8-11, 16-14
Tierney, F. L., 20-51
Tillman, R., 6-18
Todd, O. J., 33-24
Torpen, B. E., 33-25
Townsend, W. H., 24-44
Trimble, D. E., 14-16
Turnbull, W. J., 42-18
Tustin, T. G., 15-25
Tuttle, C. R., 12-22
Twain, M., 9-11
Twenhofel, W. H., 25-55

Underwood, L. B., 20-51
Upson, J. E., 50-17

Vallee, J., 6-20
Van Oosterom, H., 28-37
Varnes, D., 39-1, 24
Vermuyden, C., 18-4
Vetter, C. P., 41-15, 48-17
Vicks, G., 48-17
Viner-Bradley, N. E. V., 39-25
Vineyard, J. D., 38-4, 14
Vitruvius, 24-27, 43
Vivatrat, V., 22-56
Vogt, A., 8-23, 10-20
Volumard, P., 27-20
Von Moos, A., 13-22
Voorduin, W. L., 25-57

Waddell, J. A. L., 24-42
Wagner, W. R., 13-22, 47-22
Wahrhaftig, C., 31-33
Wales, W., 12-4, 22
Walker, D. A., 48-17
Walker, F., 20-51
Walker, T. A., 1-19
Wall, C. R., 46-18
Walsh, D., 34-7
Walton, I., 48-6
Walton, M. S., 13-23, 14-6, 47-23
Wantland, D., 12-22
Ward, W. H., 39-4, 24
Wardman, R. S., 13-22
Warren, D. R., 18-32
Waterhouse, L. L., 25-55, 26-23
Watkins, G. L., 48-14, 18
Watson, J. D., 22-56
Weaver, P., 37-24
Webb, W. E., 27-21
Weber, W. G., 30-29
Wegener, A. L., 45-1
Weimer, R. J., 50-18
Weinert, H. H., 30-8, 42-4, 18
Wells, J. D., 15-25
Wentworth, C. K., 8-27
Werner, A. G., 3-2
Wesley, R. H., 13-16, 23
Wheeler, M., 30-27, 50-14
Wheeler, W. H., 34-44
White, E. E., 21-27
White, G. W., 34-43, 42-18
White, L., 22-54
White, O. L., 7-15
White, R. E., 19-41
Whiteside, T., 20-51
Whitney, M., 6-2
Whittaker, T., 17-11
Whyte, R. O., 49-23
Widmer, K., 35-15
Wiggins, J. H., 47-22
Wilbur, E. A., 27-20
Wilkinson, D., 37-23
Willeumier, G. C., 24-44
Williams, 12-9
Williams, A. A. B., 42-18

Williams, H., 15-26, 22-56
Williams, H. M., 30-29
Williams, J. H., 38-4, 14
Williams, R. B. G., 43-13
Williamson, A. B., 23-32
Wilmot, C., 2-20
Wilshusen, J. P., 7-15
Wilson, A., 34-45
Wilson, A. E., 44-16
Wilson, J. T., 45-22
Wilson, K. C., 23-31
Winkler, E. M., 15-26
Winsor, L. M., 33-15, 25, 49-15, 22
Wiseman, G., 42-18
Wolman, M. G., 41-13, 15
Wood, A. M. M., 39-25
Woodring, W. P., 2-20
Woodruff, G. B., 24-43
Woods, K. B., 36-22
Woodward, H. B., 30-29, 42-18
Woodward, H. P., 31-33
Worcester, J. R., 13-4, 22
Worthen, W. E., 27-2
Wright, K. R., 46-18
Wulkowicz, G. M., 48-18
Wyllie, J., 8-27
Wyman, R. A., 16-20

Yardley, D. H., 36-22
Yokoyama, L., 45-22
Young, E. C., 48-17
Young, E. R., 30-29
Young, J. W., 28-37
Yuew, C. M. K., 20-51
Yundt, S. E., 16-20

Zacher, E. G., 27-21
Zaruba, Q., 13-11, 15, 15-23, 25-55, 37-24, 39-15, 25, 43-12
Zeevaert, L., 22-56
Zoega, J., 23-31
Zurcher, 29-6

About the Authors

Robert F. Legget has over 50 years' experience in field work, teaching, and research in civil engineering. After graduating from the University of Liverpool, he worked as a civil engineer in Scotland and Canada and taught at Queen's University (Kingston) and the University of Toronto. In 1951 he founded the Division of Building Research for the National Research Council of Canada, serving as its director until his retirement in 1969. Mr. Legget is a former president of the Geological Society of America.

Paul F. Karrow has traveled and worked in many parts of Canada and the United States. He was educated at Queen's University, Kingston, Ontario, and at the University of Illinois and has over twenty-five years of experience in geology and civil engineering, including extensive experience in aerial mapping with the Geological Survey of Canada and the Ontario Geological Survey. He is the author of numerous publications in geology and civil engineering.